Student Solutions Manual

for

Tussy and Gustafson's

Elementary and Intermediate Algebra

Third Edition

Kristy Hill
Hinds Community College

THOMSON
BROOKS/COLE

Australia • Canada • Mexico • Singapore • Spain • United Kingdom • United States

Cover image: Kevin Tolman

Printed in the United States of America
2 3 4 5 6 7 09 08 07

Printer: Globus

ISBN-13: 978-0-534-38690-0
ISBN-10: 0-534-38690-3

For more information about our products, contact us at:
Thomson Learning Academic Resource Center
1-800-423-0563

For permission to use material from this text or product, submit a request online at **http://www.thomsonrights.com.**
Any additional questions about permissions can be submitted by email to **thomsonrights@thomson.com.**

Thomson Higher Education
10 Davis Drive
Belmont, CA 94002-3098
USA

Asia (including India)
Thomson Learning
5 Shenton Way
#01-01 UIC Building
Singapore 068808

Australia/New Zealand
Thomson Learning Australia
102 Dodds Street
Southbank, Victoria 3006
Australia

Canada
Thomson Nelson
1120 Birchmount Road
Toronto, Ontario M1K 5G4
Canada

UK/Europe/Middle East/Africa
Thomson Learning
High Holborn House
50–51 Bedford Row
London WC1R 4LR
United Kingdom

Latin America
Thomson Learning
Seneca, 53
Colonia Polanco
11560 Mexico
D.F. Mexico

Spain (including Portugal)
Thomson Paraninfo
Calle Magallanes, 25
28015 Madrid, Spain

TABLE OF CONTENTS

Chapter 4 — Exponents and Polynomials

Chapter 5 — Factoring and Quadratic Equations

Chapter 6 — Rational Expressions and Equations

SECTION 1.1

VOCABULARY

1. The answer to an addition problem is called the **sum**. The answer to a subtraction problem is called the **difference**.

3. **Variables** are letters that stand for numbers.

5. An **equation** is a mathematical sentence that contains an = symbol.

7. The **horizontal** axis of a graph extends left and right and the **vertical** axis extends up and down.

CONCEPTS

9. equation

11. algebraic expression

13. algebraic expression

15. equation

17. a) multiplication and subtraction
    b) x

19. a) addition and subtraction
    b) m

21.

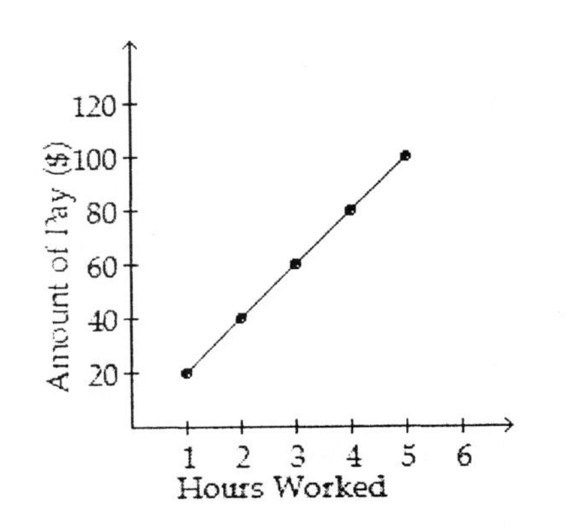

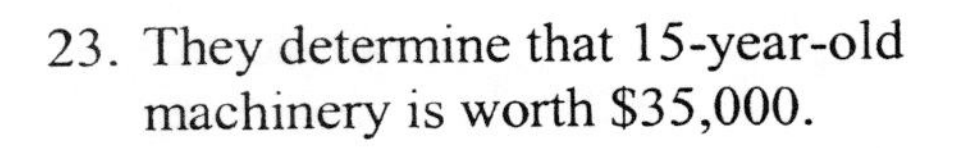

23. They determine that 15-year-old machinery is worth $35,000.

NOTATION

25. $5 \cdot 6$ or $5(6)$

27. $34 \cdot 75$ or $34(75)$

29. $4x$

31. $3rt$

33. lw

35. Prt

37. $2w$

39. xy

41. $\frac{32}{x}$

43. $\frac{90}{30}$

PRACTICE

45. The product of 8 and 2 is 16.

47. The difference of 11 and 9 is 2.

49. The product of 2 and x is 10.

51. The quotient of 66 and 11 is 6.

53. $p = 100 - d$

55. $7d = h$

57. $s = 3c$

59. $w = e + 1{,}200$

61. $p = r - 600$

63. $\frac{l}{4} = m$

65. $d = 360 + L$

Lunch time (minutes) L	School day (minutes) d
30	$d = 360 + \mathbf{30}$ $= \mathbf{390}$
40	$d = 360 + \mathbf{40}$ $= \mathbf{400}$
45	$d = 360 + \mathbf{45}$ $= \mathbf{405}$

67. $t = 1{,}500 - d$

Deductions d	Take-home pay t
200	$t = 1{,}500 - \mathbf{200}$ $= \mathbf{1{,}300}$
300	$t = 1{,}500 - \mathbf{300}$ $= \mathbf{1{,}200}$
400	$t = 1{,}500 - \mathbf{400}$ $= \mathbf{1{,}100}$

69. $d = \frac{e}{12}$

APPLICATIONS

71. $90° - \text{Angle } 1 = \text{Angle } 2$

Angle 1 (degrees)	Angle 2 (degrees)
0	**90**
30	**60**
45	**45**
60	**30**
90	**0**

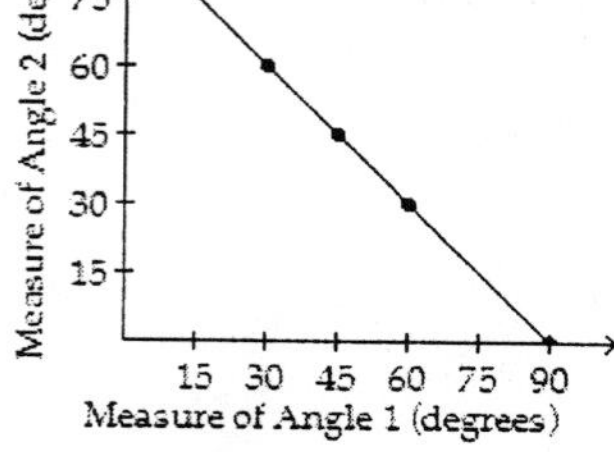

WRITING

73. Answers will vary.

75. Answers will vary.

CHALLENGE PROBLEMS

77. $s = t + 1$

s	t
10	**10 + 1 = 11**
18	19
34 – 1 = 33	34
47	48

SECTION 1.2

VOCABULARY

1. Numbers that have only 1 and themselves as factors, such as 23, 37, and 41, are called **prime** numbers.

3. The **numerator** of the fraction $\frac{3}{4}$ is 3, and the **denominator** is 4.

5. Two fractions that represent the same number, such as $\frac{1}{2}$ and $\frac{2}{4}$, are called **equivalent** fractions.

7. The **least or lowest** common denominator for a set of fractions is the smallest number each denominator will divide exactly.

CONCEPTS

9. $2 \cdot 2 \cdot 3 \cdot 5 = \mathbf{60}$

11. $\frac{4}{12} = \frac{1}{3}$

13. $\frac{\not{2} \bullet 2 \bullet \not{3}}{\not{2} \bullet \not{3} \bullet 5} = \frac{2}{5}$

15. a) To build up a fraction, we multiply it by $\boxed{1}$ in the form of $\frac{2}{2}$, $\frac{3}{3}$, or $\frac{4}{4}$, and so on.
    b) To simplify a fraction, we remove factors equal to $\boxed{1}$ in the form of $\frac{2}{2}$, $\frac{3}{3}$, or $\frac{4}{4}$, and so on.

17. a) 3 times
    b) 2 times

NOTATION

19. a) $\frac{5}{16}$

    b) $\frac{3}{3} = \boxed{1}$

    c) $\frac{5 \bullet 3}{16 \bullet 3} = \frac{15}{48}$

PRACTICE

21. 1, 2, 4, 5, 10, 20

23. 1, 2, 4, 7, 14, 28

25. $75 = 3 \cdot 5 \cdot 5$

27. $28 = 2 \cdot 2 \cdot 7$

29. $117 = 3 \cdot 3 \cdot 13$

31. $220 = 2 \cdot 2 \cdot 5 \cdot 11$

33. $\frac{1}{3} = \frac{1 \bullet 3}{3 \bullet 3}$
    $= \frac{3}{9}$

35. $\frac{4}{9} = \frac{4 \bullet 6}{9 \bullet 6}$
    $= \frac{24}{54}$

37. $\frac{7}{1} = \frac{7 \bullet 5}{1 \bullet 5}$
    $= \frac{35}{5}$

39. $\frac{6}{12} = \frac{\not{2} \bullet \not{3}}{\not{2} \bullet 2 \bullet \not{3}}$
    $= \frac{1}{2}$

41. $\dfrac{24}{18}=\dfrac{\cancel{2}\bullet 2\bullet 2\bullet \cancel{3}}{\cancel{2}\bullet \cancel{3}\bullet 3}$

$=\dfrac{4}{3}$

43. $\dfrac{15}{20}=\dfrac{\cancel{5}\bullet 3}{2\bullet 2\bullet \cancel{5}}$

$=\dfrac{3}{4}$

45. $\dfrac{72}{64}=\dfrac{\cancel{2}\bullet \cancel{2}\bullet \cancel{2}\bullet 3\bullet 3}{\cancel{2}\bullet \cancel{2}\bullet \cancel{2}\bullet 2\bullet 2\bullet 2}$

$=\dfrac{9}{8}$

47. $\dfrac{33}{56}=\dfrac{3\bullet 11}{2\bullet 2\bullet 2\bullet 7}$

$=\dfrac{33}{56}$

It is in lowest terms.

49. $\dfrac{36}{225}=\dfrac{2\bullet 2\bullet \cancel{3}\bullet \cancel{3}}{\cancel{3}\bullet \cancel{3}\bullet 5\bullet 5}$

$=\dfrac{4}{25}$

51. $\dfrac{1}{2}\bullet\dfrac{3}{5}=\dfrac{1\bullet 3}{2\bullet 5}$

$=\dfrac{3}{10}$

53. $\dfrac{4}{3}\left(\dfrac{6}{5}\right)=\dfrac{4(6)}{3(5)}$

$=\dfrac{24}{15}$

$=\dfrac{8}{5}$

55. $\dfrac{5}{12}\bullet\dfrac{18}{5}=\dfrac{5\bullet 18}{12\bullet 5}$

$=\dfrac{90}{60}$

$=\dfrac{3}{2}$

57. $\dfrac{21}{1}\left(\dfrac{10}{3}\right)=\dfrac{21\bullet 10}{1\bullet 3}$

$=\dfrac{210}{3}$

$=\dfrac{70}{1}$

$=70$

59. $7\dfrac{1}{2}\bullet 1\dfrac{2}{5}=\dfrac{15}{2}\bullet\dfrac{7}{5}$

$=\dfrac{15\bullet 7}{2\bullet 5}$

$=\dfrac{105}{10}$

$=\dfrac{21}{2}$

$=10\dfrac{1}{2}$

61. $6\bullet 2\dfrac{7}{24}=\dfrac{6}{1}\bullet\dfrac{55}{24}$

$=\dfrac{6\bullet 55}{1\bullet 24}$

$=\dfrac{330}{24}$

$=\dfrac{55}{4}$

$=13\dfrac{3}{4}$

63. $\dfrac{3}{5}\div\dfrac{2}{3}=\dfrac{3}{5}\bullet\dfrac{3}{2}$

$=\dfrac{9}{10}$

65. $\frac{3}{4} \div \frac{6}{5} = \frac{3}{4} \bullet \frac{5}{6}$
$= \frac{15}{24}$
$= \frac{5}{8}$

67. $\frac{21}{35} \div \frac{3}{14} = \frac{21}{35} \bullet \frac{14}{3}$
$= \frac{294}{105}$
$= \frac{14}{5}$

69. $6 \div \frac{3}{14} = \frac{6}{1} \bullet \frac{14}{3}$
$= \frac{84}{3}$
$= 28$

71. $3\frac{1}{3} \div 1\frac{5}{6} = \frac{10}{3} \div \frac{11}{6}$
$= \frac{10}{3} \bullet \frac{6}{11}$
$= \frac{60}{33}$
$= \frac{20}{11}$
$= 1\frac{9}{11}$

73. $8 \div 3\frac{1}{5} = \frac{8}{1} \div \frac{16}{5}$
$= \frac{8}{1} \bullet \frac{5}{16}$
$= \frac{40}{16}$
$= \frac{5}{2}$
$= 2\frac{1}{2}$

75. $\frac{3}{5} + \frac{3}{5} = \frac{3+3}{5}$
$= \frac{6}{5}$

77. $\frac{1}{6} + \frac{1}{24} = \frac{1}{6} \bullet \frac{4}{4} + \frac{1}{24}$
$= \frac{4}{24} + \frac{1}{24}$
$= \frac{4+1}{24}$
$= \frac{5}{24}$

79. $\frac{3}{5} + \frac{2}{3} = \frac{3}{5} \bullet \frac{3}{3} + \frac{2}{3} \bullet \frac{5}{5}$
$= \frac{9}{15} + \frac{10}{15}$
$= \frac{9+10}{15}$
$= \frac{19}{15}$

81. $\frac{5}{12} + \frac{1}{3} = \frac{5}{12} + \frac{1}{3} \bullet \frac{4}{4}$
$= \frac{5}{12} + \frac{4}{12}$
$= \frac{5+4}{12}$
$= \frac{9}{12}$
$= \frac{3}{4}$

83. $\frac{9}{4} - \frac{5}{6} = \frac{9}{4} \bullet \frac{3}{3} - \frac{5}{6} \bullet \frac{2}{2}$
$= \frac{27}{12} - \frac{10}{12}$
$= \frac{27-10}{12}$
$= \frac{17}{12}$

85. $\frac{7}{10}-\frac{1}{14}=\frac{7}{10}\bullet\frac{7}{7}-\frac{1}{14}\bullet\frac{5}{5}$
$=\frac{49}{70}-\frac{5}{70}$
$=\frac{49-5}{70}$
$=\frac{44}{70}$
$=\frac{22}{35}$

87. $\frac{5}{14}-\frac{4}{21}=\frac{5}{14}\bullet\frac{3}{3}-\frac{4}{21}\bullet\frac{2}{2}$
$=\frac{15}{42}-\frac{8}{42}$
$=\frac{15-8}{42}$
$=\frac{7}{42}$
$=\frac{1}{6}$

89. $3-\frac{3}{4}=\frac{3}{1}\bullet\frac{4}{4}-\frac{3}{4}$
$=\frac{12}{4}-\frac{3}{4}$
$=\frac{9}{4}$

91. $3\frac{3}{4}-2\frac{1}{2}=\frac{15}{4}-\frac{5}{2}$
$=\frac{15}{4}-\frac{5}{2}\bullet\frac{2}{2}$
$=\frac{15}{4}-\frac{10}{4}$
$=\frac{15-10}{4}$
$=\frac{5}{4}=1\frac{1}{4}$

93. $8\frac{2}{9}-7\frac{2}{3}=\frac{74}{9}-\frac{23}{3}$
$=\frac{74}{9}-\frac{23}{3}\bullet\frac{3}{3}$
$=\frac{74}{9}-\frac{69}{9}$
$=\frac{74-69}{9}$
$=\frac{5}{9}$

APPLICATIONS

95. BOTANY

a) Add the 2 widths of the growth rings.

$\frac{5}{32}+\frac{1}{16}=\frac{5}{32}+\frac{1}{16}\bullet\frac{2}{2}$
$=\frac{5}{32}+\frac{2}{32}$
$=\frac{7}{32}$

b) Subtract the 2 widths of the growth rings.

$\frac{5}{32}-\frac{1}{16}=\frac{5}{32}-\frac{1}{16}\bullet\frac{2}{2}$
$=\frac{5}{32}-\frac{2}{32}$
$=\frac{3}{32}$

97. FRAMES

Since there are 4 sides and each side is $10\frac{1}{8}$ inches long, multiply 4 times $10\frac{1}{8}$.

$$\begin{aligned}4\left(10\frac{1}{8}\right) &= \frac{4}{1} \bullet \frac{81}{8} \\ &= \frac{324}{8} \\ &= \frac{81}{2} \\ &= 40\frac{1}{2}\end{aligned}$$

$40\frac{1}{2}$ in of molding is needed.

WRITING

99. Answers will vary.

101. Answers will vary.

REVIEW

103. The difference of 7 and 5 is 2.

105. The quotient of 30 and 15 is 2.

107. $T = 15g$

Number of gears g	Number of teeth T
10	15(**10**) = **150**
12	15(**12**) = **180**

CHALLENGE PROBLEMS

109. Find a common denominator and compare numerators.

$$\frac{11}{12} \bullet \frac{3}{3} = \frac{33}{36}$$

$$\frac{8}{9} \bullet \frac{4}{4} = \frac{32}{36}$$

$$\frac{33}{36} > \frac{32}{36}$$

so $\frac{11}{12} > \frac{8}{9}$

SECTION 1.3

VOCABULARY

1. The set of **whole** numbers is $\{0, 1, 2, 3, 4, 5, \ldots\}$.

3. The set of **integers** is $\{\ldots -2, -1, 0, 1, 2, \ldots\}$.

5. Numbers less than zero are **negative**, and numbers greater than zero are **positive**.

7. A **rational** number is any number that can be expressed as a fraction with an integer numerator and a nonzero integer denominator.

9. An **irrational** number is a nonterminating, nonrepeating decimal.

11. Every point on the number line corresponds to exactly one **real** number.

CONCEPTS

13. opposites

15. $\frac{6}{1}, \frac{-9}{1}, \frac{-7}{8}, \frac{7}{2}, \frac{-3}{10}, \frac{283}{100}$

17. 13 and -3 are 8 units away from 5

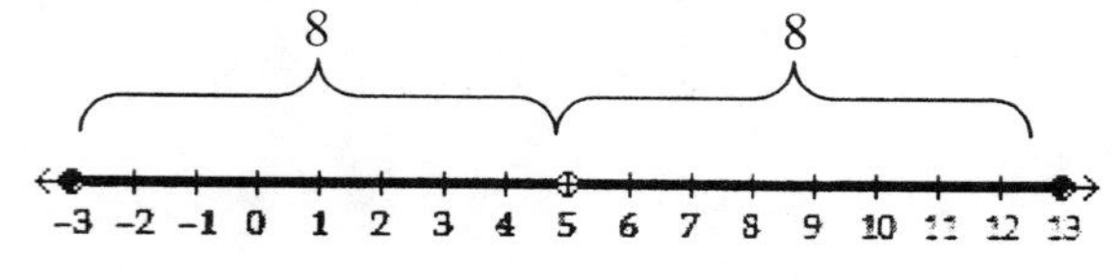

19. a) $a < b$
 b) $b > a$
 c) $b > 0$ and $a < 0$
 d) $|a| > |b|$

21. π inches

NOTATION

23. $\sqrt{2}$ is read "the **square root** of 2."

25. The symbol $\neq$ means **is not equal to**.

27. The symbol π is a letter from the **Greek** alphabet.

29. $$-\frac{4}{5} = \frac{-4}{5} = \frac{4}{-5}$$

PRACTICE

31. $$\frac{5}{8} = \begin{array}{r} 0.625 \\ 8\overline{)5.000} \end{array} = 0.625$$

33. $$\frac{1}{30} = \begin{array}{r} 0.0333 \\ 30\overline{)1.0000} \end{array} = 0.0\bar{3}$$

35. $$\frac{21}{50} = \begin{array}{r} 0.42 \\ 50\overline{)21.00} \end{array} = 0.42$$

37. $$\frac{5}{11} = \begin{array}{r} 0.454545 \\ 11\overline{)5.000000} \end{array} = 0.\overline{45}$$

39. $5 > 4$

41. $-2 > -3$

43. $|3.4| > \sqrt{2}$

45. $|-1.1| < 1.2$

47. $-\frac{5}{8} < -\frac{3}{8}$

49. $\left|-\frac{15}{2}\right| = 7.5$

51. $\frac{99}{100} = 0.99$

53. $0.333\ldots > 0.3$

55. $1 > \left|-\frac{15}{16}\right|$

57.
a) True
b) False
c) False
d) True

59.
a) $-5 > -6$
b) $-25 < 16$

61.

$-\frac{35}{8}$ $\quad -\pi \quad$ $-1\frac{1}{2}$ $\quad -0.33\ldots \quad \sqrt{2} \quad 3 \quad 4.25$

−5 −4 −3 −2 −1 0 1 2 3 4 5

APPLICATIONS

63. DRAFTING

natural: 9
whole: 9
integers: 9
rational: $9, \frac{15}{16}, 3\frac{1}{8}, 1.765$
irrational: $2\pi, 3\pi, \sqrt{89}$
real: all

65. TARGET PRACTICE

Shell 1 landed farther from the target.

$|-6| > |5|$

67. TRADE

a) 2000 had a deficit of -$81 billion.
2002 had a deficit of -$70 billion.
1999 had a deficit of -$74 billion.

b) The smallest deficit was in 1990 and was -$40 billion.

WRITING

69. Answers will vary.

71. Answers will vary.

REVIEW

73. $\frac{24}{54} = \frac{\cancel{2}\bullet 2\bullet 2\bullet \cancel{3}}{\cancel{2}\bullet \cancel{3}\bullet 3\bullet 3}$
$= \frac{4}{9}$

75. $\frac{3}{4}\left(\frac{8}{5}\right) = \frac{3\bullet 8}{4\bullet 5}$
$= \frac{24}{20}$
$= \frac{6}{5}$

77. $\frac{3}{10} + \frac{2}{15} = \frac{3}{10}\bullet\frac{3}{3} + \frac{2}{15}\bullet\frac{2}{2}$
$= \frac{9}{30} + \frac{4}{30}$
$= \frac{13}{30}$

CHALLENGE PROBLEMS

79. $\{0, 1, 2, 3, 4, 5, \ldots\}$

SECTION 1.4

VOCABULARY

1. Positive and negative numbers are called **signed** numbers.

3. Two numbers that are the same distance from 0 on a number line, but on opposite sides of it, are called **opposites**.

5. Since any number added to 0 remains the same (is identical), the number 0 is called the **identity** element for addition.

CONCEPTS

7. $-1 + (-3) = -4$

9. -5

11. -3

13. a) $a + (-a) = \boxed{0}$
    b) $a + 0 = \boxed{a}$
    c) $a + b = b + \boxed{a}$
    d) $(a + b) + c = a + \boxed{(b + c)}$

15. positive

17. a) $1 + (-5)$
    b) $-80.5 + 15$
    c) $(20 + 4)$

19. a) $5 + (-5) = 0$
    b) $-2.2 + 22 = 0$
    c) $0 + (-6) = -6$
    d) $-\frac{15}{16} + 0 = \frac{15}{16}$
    e) $-\frac{3}{4} + \frac{3}{4} = 0$
    f) $19 + (-19) = 0$

NOTATION

21. $x + y = y + x$

23. $8 + 9$

PRACTICE

25. $6 + (-8) = -2$

27. $-6 + 8 = 2$

29. $-4 + (-4) = -8$

31. $9 + (-1) = 8$

33. $-16 + 16 = 0$

35. $-65 + (-12) = -77$

37. $15 + (-11) = 4$

39. $300 + (-335) = -35$

41. $-10.5 + 2.3 = -8.2$

43. $-9.1 + (-11) = -20.1$

45. $0.7 + (-0.5) = 0.2$

47. $$-\frac{9}{16} + \frac{7}{16} = -\frac{2}{16} = -\frac{1}{8}$$

49. $$-\frac{1}{4} + \frac{2}{3} = -\frac{1}{4} \bullet \frac{3}{3} + \frac{2}{3} \bullet \frac{4}{4} = -\frac{3}{12} + \frac{8}{12} = \frac{5}{12}$$

51. $$-\frac{4}{5} + \left(-\frac{1}{10}\right) = -\frac{4}{5} \bullet \frac{2}{2} + \left(-\frac{1}{10}\right) = -\frac{8}{10} + \left(-\frac{1}{10}\right) = -\frac{9}{10}$$

53. $8 + (-5) + 13 = 3 + 13$
    $= 16$

55. $21 + (-27) + (-9) = -6 + (-9)$
$= -15$

57. $-27 + (-3) + (-13) + 22 = -30 + (-13) + 22$
$= -43 + 22$
$= -21$

59. $-20 + (-16 + 10) = -20 + (-6)$
$= -26$

61. $19 + (-20 + 1) = 19 + (-19)$
$= 0$

63. $(-7 + 8) + 2 + (-12 + 13) = 1 + 2 + 1$
$= 4$

65. $-7 + 5 + (-10) + 7 = -2 + (-10) + 7$
$= -12 + 7$
$= -5$

67. $-8 + 11 + (-11) + 8 + 1 = 3 + (-11) + 8 + 1$
$= -8 + 8 + 1$
$= 0 + 1$
$= 1$

69. $-2.1 + 6.5 + (-8.2) + 0.6 = 4.4 + (-8.2) + 0.6$
$= -3.8 + 0.6$
$= -3.2$

71. $-60 + 70 + (-10) + (-10) + 20$
$= 10 + (-10) + (-10) + 20$
$= 0 + (-10) + 20$
$= -10 + 20$
$= 10$

73. $-99 + (99 + 215) = (-99 + 99) + 215$
$= 0 + 215$
$= 215$

75. $(-112 + 56) + (-56) = -112 + [56 + (-56)]$
$= -112 + 0$
$= -112$

APPLICATIONS

77. MILITARY SCIENCE

$-1{,}500 + 2{,}400 + 1{,}250 = 900 + 1{,}250$
$= 2{,}150$

The net gain is 2, 150 m.

79. GOLF

Tiger Woods:
$-2 + (-6) + (-7) + (-3) = -8 + (-7) + (-3)$
$= -15 + (-3)$
$= -18$

Tom Kite:
$5 + (-3) + (-6) + (-2) = 2 + (-6) + (-2)$
$= -4 + (-2)$
$= -6$

Tommy Tolles:
$0 + 0 + 0 + (-5) = 0 + (-5)$
$= -5$

Tom Watson:
$3 + (-4) + (-3) + 0 = -1 + (-3) + 0$
$= -4 + 0$
$= -4$

81. CREDIT CARDS

a) Previous Balance: 3,660.66
New Purchases: 1,408.78

b) $-3{,}660.66 - 1{,}408.78 + 3{,}826.58$
$= -5{,}069.44 + 3{,}826.58$
$= -1{,}242.86$

83. MOVIE LOSSES

$-\$100 + \$11 = -\$89$ million

85. SAHARA DESERT

$-240 + 110 + 30 + (-55) + 100 + (-77)$
$= -130 + 30 + (-55) + 100 + (-77)$
$= -100 + (-55) + 100 + (-77)$
$= -155 + 100 + (-77)$
$= -55 + (-77)$
$= -132$

The net movement was 132 km southward.

87. PROFITS AND LOSSES

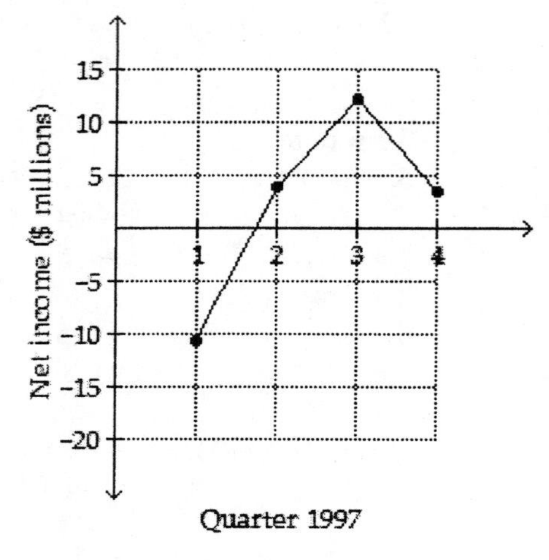

$$\begin{aligned}-10.7+4+12.2+(-3.4) \\ &= -6.7+12.2+(-3.4) \\ &= 5.5+(-3.4) \\ &= 2.1\end{aligned}$$

The net income for 2001 was $2.1 million.

REVIEW

89. True

91. The numbers –9 and 3 are 6 units away from –3.

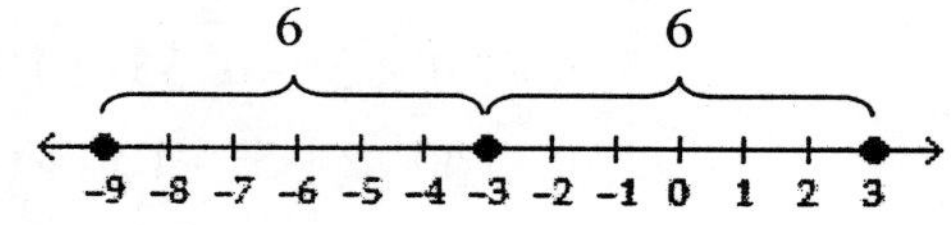

WRITING

93. Answers will vary.

CHALLENGE PROBLEMS

95. No. 1 + 1 = 2, and 2 is not a member of the set.

SECTION 1.5

VOCABULARY

1. Two numbers that are the same distance from 0 on a number line, but on opposite sides of it, are called **opposites**, or additive **inverses**.

3. The difference between the maximum and the minimum value of a collection of measurements is called the **range** of the values.

CONCEPTS

5. a) -12

   b) $\frac{1}{5}$

   c) -2.71

   d) 0

7. a) The opposite of the opposite of a number is that **number**.

   b) To subtract two numbers, add the first number to the **opposite** of the number to be subtracted.

9. $-1-9=-1\boxed{+}\boxed{(-9)}$

11. $15-(-8)=15+8$
    $=23$

    No. $15-(-8)\neq 7$

13. $-10+(-8)+(-23)+5+34$

15. a) $-|-500|=-500$

    b. $-(-y)=y$

PRACTICE

17. $4-7=4+(-7)$
    $=-3$

19. $2-15=2+(-15)$
    $=-13$

21. $8-(-3)=8+3$
    $=11$

23. $-12-9=-12+(-9)$
    $=-21$

25. $0-6=0+(-6)$
    $=-6$

27. $0-(-1)=0+1$
    $=1$

29. $10-(-2)=10+2$
    $=12$

31. $-1-(-3)=-1+3$
    $=2$

33. $20-(-20)=20+20$
    $=40$

35. $-3-(-3)=-3+3$
    $=0$

37. $-2-(-7)=-2+7$
    $=5$

39. $-4-5=-4+(-5)$
    $=-9$

41. $-44-44=-44+(-44)$
    $=-88$

43. $0-(-12)=0+12$
    $=12$

45. $-25-(-25)=-25+25$
    $=0$

47. $0-4=0+(-4)$
    $=-4$

49. $-19-(-17)=-19+17$
    $=-2$

51. $$-\frac{1}{8}-\frac{3}{8}=-\frac{1}{8}+\left(-\frac{3}{8}\right)$$
    $$=-\frac{4}{8}$$
    $$=-\frac{1}{2}$$

53. $-\frac{9}{16}-\left(-\frac{1}{4}\right)=-\frac{9}{16}+\left(\frac{1}{4}\right)$
$$=-\frac{9}{16}+\left(\frac{1}{4}\bullet\frac{4}{4}\right)$$
$$=-\frac{9}{16}+\frac{4}{16}$$
$$=-\frac{5}{16}$$

55. $\frac{1}{3}-\frac{3}{4}=\frac{1}{3}+\left(-\frac{3}{4}\right)$
$$=\frac{1}{3}\bullet\frac{4}{4}+\left(-\frac{3}{4}\bullet\frac{3}{3}\right)$$
$$=\frac{4}{12}+\left(-\frac{9}{12}\right)$$
$$=-\frac{5}{12}$$

57. $-0.9-0.2=-0.9+(-0.2)$
$=-1.1$

59. $6.3-9.8=6.3+(-9.8)$
$=-3.5$

61. $-1.5-0.8=-1.5+(-0.8)$
$=-2.3$

63. $2.8-(-1.8)=2.8+1.8$
$=4.6$

65. $17-(-5)=17+5$
$=22$

67. $-13-12=-13+(-12)$
$=-25$

69. $8-9-10\ =8+(-9)+(-10)$
$=-1+(-10)$
$=-11$

71. $-25-(-50)-75=-25+50+(-75)$
$=25+(-75)$
$=-50$

73. $-6+8-(-1)-10=-6+8+1+(-10)$
$=2+1+(-10)$
$=3+(-10)$
$=-7$

75. $61-(-62)+(-64)-60=61+62+(-64)+(-60)$
$=123+(-64)+(-60)$
$=59+(-60)$
$=-1$

77. $-6-7-(-3)+9=-6+(-7)+3+9$
$=-13+3+9$
$=-10+9$
$=-1$

79. $-20-(-30)-50+40=-20+30+(-50)+40$
$=10+(-50)+40$
$=-40+40$
$=0$

APPLICATIONS

81. TEMPERATURE RECORDS

$108-(-52)=108+52$
$=160°F$

83. LAND ELEVATIONS

$-282-(-1{,}312)=-282+1{,}312$
$=1{,}030$ ft
$=-23$

85. RACING

$-2.25-3.5=-5.75°$

87. HISTORY

$-347-81=-428$
He was born in 428 B.C.

89. GAUGES

$5-7-6=5+(-7)+(-6)$
$=-2+(-6)$
$=-8$
The arrow would move left and the new reading would be -8.

WRITING

91. Answers will vary.

93. Answers will vary.

REVIEW

95. $30 = 2 \cdot 3 \cdot 5$

97. $$\frac{3}{8} = \frac{3}{8} \bullet \frac{7}{7}$$
$$= \frac{21}{56}$$

99. true

CHALLENGE PROBLEMS

101. a) true; positive + positive = positive
b) true; negative - positive = negative
c) true; the opposite of a positive is a negative
d) false; the opposite of a negative is a positive

SECTION 1.6

VOCABULARY

1. The answer to a multiplication problem is called a **product**. The answer to a division problem is called a **quotient**.

3. The **commutative** property of multiplication states that changing the order when multiplying does not affect the answer.

5. Division of a nonzero number by zero is **undefined**.

CONCEPTS

7. $4(\boxed{-5})$

9. The product of two negative numbers is **positive**.

11. The product of **1** and any number is that number.

13. a) One of the numbers is 0.
    b) They are reciprocals or multiplicative inverses.

15. a) negative
    b) not possible to tell
    c) positive
    d) negative

17. a) $a \bullet b = b \bullet \boxed{a}$

    b) $(ab)(c) = \boxed{a(bc)}$

    c) $0 \bullet a = \boxed{0}$

    d) $1 \bullet a = \boxed{a}$

    e) $a\left(\frac{1}{a}\right) = \boxed{1}$ $(a \neq 0)$

19. a) associative property of multiplication
    b) multiplicative inverse
    c) commutative property of multiplication
    d) multiplication property of 1

NOTATION

21. a) $-4(-5) = 20$

    b) $\frac{16}{-8} = -2$

PRACTICE

23. $-2 \cdot 8 = -16$

25. $(-6)(-9) = 54$

27. $12(-5) = -60$

29. $-6 \cdot 4 = -24$

31. $-20(40) = -800$

33. $(-6)(-6) = 36$

35. $-0.6(-4) = 2.4$

37. $1.2(-0.4) = -0.48$

39. $-1.1(-0.9) = 0.99$

41. $7.2(-2.1) = -15.12$

43. $\frac{1}{2}\left(-\frac{3}{4}\right) = -\frac{3}{8}$

45.
$$\left(-\frac{7}{8}\right)\left(-\frac{2}{21}\right) = \frac{14}{168}$$
$$= \frac{1}{12}$$

47.
$$-\frac{16}{25} \bullet \frac{15}{64} = -\frac{240}{1600}$$
$$= -\frac{3}{20}$$

49.
$$-1\frac{1}{4}\left(-\frac{3}{4}\right) = -\frac{5}{4}\left(-\frac{3}{4}\right)$$
$$= \frac{15}{16}$$

51. -5.2 · 100 = -520

53. 0(-22) = 0

55. -3(-4)(0) = 0

57. 3(-4)(-5) = -12(-5)
= 60

59. (-4)(3)(-7) = -12(-7)
= 84

61. (-2)(-3)(-4)(-5) = 6(-4)(-5)
= -24(-5)
= 120

63.
$$\frac{1}{2}\left(-\frac{1}{3}\right)\left(-\frac{1}{4}\right)=-\frac{1}{6}\left(-\frac{1}{4}\right)$$
$$=\frac{1}{24}$$

65. -2(-3)(-4)(-5)(-6) = 6(-4)(-5)(-6)
= -24(-5)(-6)
= 120(-6)
= -720

67. -30 ÷ (-3) = 10

69. $\frac{-6}{-2}=3$

71. $\frac{4}{-2}=-2$

73. $\frac{80}{-20}=-4$

75. $\frac{17}{-17}=-1$

77. $\frac{-110}{-110}=1$

79. $\frac{-160}{-40}=-4$

81. $\frac{320}{-16}=-20$

83. $\frac{0.5}{-100}=-0.005$

85. $\frac{0}{150}=0$

87. $\frac{-17}{0}=$ **undefined**

89.
$$-\frac{1}{3}\div\frac{4}{5}=-\frac{1}{3}\bullet\frac{5}{4}$$
$$=-\frac{5}{12}$$

91.
$$-\frac{9}{16}\div\left(-\frac{3}{20}\right)=-\frac{9}{16}\bullet\left(-\frac{20}{3}\right)$$
$$=\frac{180}{48}$$
$$=\frac{15}{4}$$

93.
$$-3\frac{3}{8}\div\left(-2\frac{1}{4}\right)=-\frac{27}{8}\div\left(-\frac{9}{4}\right)$$
$$=-\frac{27}{8}\left(-\frac{4}{9}\right)$$
$$=\frac{108}{72}$$
$$=\frac{3}{2}$$
$$=1\frac{1}{2}$$

95. $\frac{-23.5}{5}=-4.7$

97. $\frac{-24.24}{-0.8}=30.3$

99.
$$-\frac{1}{2}(2\bullet 67)=\left(-\frac{1}{2}\bullet 2\right)(67)$$
$$=-1(67)$$
$$=-67$$

101. $-0.2(10 \cdot 3) = (-0.2 \cdot 10)3$
$= -2 \cdot 3$
$= -6$

APPLICATIONS

103. TEMPERATURE CHANGE

$12(-6) = -72°$

105. GAMBLING

First bet $40, loses $40
Second bet double $40, loses $80
Third bet doubles $80, loses $160

$(-40) + (-80) + (-160) = -240 + (-40)$
$= -280$

107. PLANETS

$$\frac{-386}{2} = -193\,°\text{ F}$$

109. ACCOUNTING

$1.9(-22.8) = -\$43.32$ million

111. THE QUEEN MARY

$$\frac{3{,}450{,}000 - 22{,}500{,}000}{31} = \frac{-19{,}050{,}000}{31} \approx -\$614{,}516$$

113. PHYSICS

a) normal: 5, -10

b) ×0.5: $5(0.5) = 2.5$
$-10(0.5) = -5$

c) ×1.5: $5(1.5) = 7.5$
$-10(1.5) = -15$

d) ×2: $5(2) = 10$
$-10(2) = -20$

WRITING

115. Answers will vary.

117. Answers will vary.

REVIEW

119. $-3 + (-4) + (-5) + 4 + 3$
$= -7 + (-5) + 4 + 3$
$= -12 + 4 + 3$
$= -8 + 3$
$= -5$

121. $$\frac{1}{2} + \frac{1}{4} + \frac{1}{3} = \frac{1}{2}\left(\frac{6}{6}\right) + \frac{1}{4}\left(\frac{3}{3}\right) + \frac{1}{3}\left(\frac{4}{4}\right)$$
$$= \frac{6}{12} + \frac{3}{12} + \frac{4}{12}$$
$$= \frac{6+3+4}{12}$$
$$= \frac{13}{12}$$
$$= 12\overline{)13.0000}\quad 1.0833$$
$$= 1.08\overline{3}$$

123. $$\frac{\cancel{2} \cdot 3 \cdot \cancel{5} \cdot \cancel{5}}{\cancel{2} \cdot \cancel{5} \cdot \cancel{5} \cdot 7} = \frac{3}{7}$$

CHALLENGE PROBLEMS

125. An odd number must be negative—one negative number, three negative numbers, or five negative numbers.

SECTION 1.7

VOCABULARY

1. In the exponential expression 3^2, 3 is the **base** and 2 is the **exponent**.

3. 7^5 is the fifth **power** of seven.

5. The rules for the **order** of operations guarantee that an evaluation of a numerical expression will result in a single answer.

CONCEPTS

7. a) $4 + 5 \cdot 6 = 9 \cdot 6$
   $= 54$
   $4 + 5 \cdot 6 = 4 + 30$
   $= 34$
   b) The correct answer is 34. Multiplication is to be done before addition.

9. The innermost grouping symbols are the parentheses. The outermost grouping symbols are the brackets.

11. a) subtraction, power, addition, multiplication
    b) power, multiplication, subtraction, addition

13. a) subtraction
    b) division

15. a) subtraction
    b) power
    c) power
    d) power

NOTATION

17. a) $3^1 = \boxed{3}$
    b) $x^1 = \boxed{x}$
    c) $9 = 9^{\boxed{1}}$
    d) $y = y^{\boxed{1}}$

19. a) -5
    b) 5

21. $-19 - 2[(1+2) \cdot 3] = -19 - 2[\boxed{3} \cdot 3]$
    $= -19 - 2[\boxed{9}]$
    $= -19 - \boxed{18}$
    $= -37$

PRACTICE

23. 3^4

25. 10^2k^3

27. $8\pi r^3$

29. $6x^2y^3$

31. $(-6)(-6) = 36$

33. $-(4)(4)(4)(4) = -256$

35. $(-5)(-5)(-5) = -125$

37. $-(-6)(-6)(-6)(-6) = -1{,}296$

39. $(-0.4)(-0.4) = 0.16$

41. $\left(-\frac{2}{5}\right)\left(-\frac{2}{5}\right)\left(-\frac{2}{5}\right) = -\frac{8}{125}$

43. $3 - 5 \cdot 4 = 3 - 20$
    $= -17$

45. $3 \cdot 8^2 = 3 \cdot 64$
    $= 192$

47. $8 \cdot 5^1 - 4 \div 2 = 8 \cdot 5 - 4 \div 2$
    $= 40 - 2$
    $= 38$

49. $100 - 8(10) + 60 = 100 - 80 + 60$
    $= 20 + 60$
    $= 80$

51. $-22 - (15 - 3) = -22 - 12$
    $= -34$

53. $-2(9) - 2(5) = -18 - 10$
    $= -28$

55. $5^2 + 13^2 = 25 + 169$
    $= 194$

57. $-4(6+5) = -4(11)$
$= -44$

59. $(9-3)(9-9) = (6)(0)$
$= 0$

61. $(-1-18)2 = (-19)2$
$= -38$

63. $-2(-1)^2 + 3(-1) - 3 = -2(1) + 3(-1) - 3$
$= -2 - 3 - 3$
$= -5 - 3$
$= -8$

65. $4^2 - (-2)^2 = 16 - 4$
$= 12$

67. $12 + 2\left(-\frac{9}{3}\right) - (-2) = 12 + 2(-3) + 2$
$= 12 - 6 + 2$
$= 6 + 2$
$= 8$

69. $\frac{-2-5}{-7-(-7)} = \frac{-7}{-7+7}$
$= \frac{-7}{0}$
$=$ undefined

71. $200 - (-6+5)^3 = 200 - (-1)^3$
$= 200 - (-1)$
$= 200 + 1$
$= 201$

73. $|5 \bullet 2^2 \bullet 4| - 30 = |5 \bullet 4 \bullet 4| - 30$
$= |20 \bullet 4| - 30$
$= |80| - 30$
$= 80 - 30$
$= 50$

75. $[6(5) - 5(5)]4 = [30 - 25]4$
$= [5]4$
$= 20$

77. $-6(130 - 4^3) = -6(130 - 64)$
$= -6(66)$
$= -396$

79. $(17 - 5 \cdot 2)^3 = (17 - 10)^3$
$= (7)^3$
$= 343$

81. $-5(-2)^3(3)^2 = -5(-8)(9)$
$= 40(9)$
$= 360$

83. $-2\left(\frac{15}{-5}\right) - \frac{6}{2} + 9 = -2(-3) - 3 + 9$
$= 6 - 3 + 9$
$= 3 + 9$
$= 12$

85. $\frac{5 \cdot 50 - 160}{-9} = \frac{250 - 160}{-9}$
$= \frac{90}{-9}$
$= -10$

87. $\frac{2(6-1)}{16-(-4)^2} = \frac{2(5)}{16-16}$
$= \frac{10}{0}$
$=$ undefined

89. $5(10+2) - 1 = 5(12) - 1$
$= 60 - 1$
$= 59$

91. $64 - 6[15+(-3)3] = 64 - 6[15 - 9]$
$= 64 - 6[6]$
$= 64 - 36$
$= 28$

93. $(12-2)^3 = (10)^3$
$= 1{,}000$

95. $(-3)^3\left(\frac{-4}{2}\right)(-1) = -27(-2)(-1)$
$= 54(-1)$
$= -54$

97. $$\frac{1}{2}\left(\frac{1}{8}\right)+\left(-\frac{1}{4}\right)^2=\frac{1}{16}+\frac{1}{16}=\frac{2}{16}=\frac{1}{8}$$

99. $$-2|4-8|=-2|-4|=-2(4)=-8$$

101. $$|7-8(4-7)|=|7-8(-3)|=|7-(-24)|=|7+24|=|31|=31$$

103. $$3+2[-1-4(5)]=3+2[-1-20]=3+2[-21]=3+(-42)=-39$$

105. $$-3[5^2-(7-3)^2]=-3[5^2-(4)^2]=-3[25-16]=-3[9]=-27$$

107. $$-(2\cdot 3-4)^3=-(6-4)^3=-(2)^3=-8$$

109. $$\frac{(3+5)^2+|-2|}{-2(5-8)}=\frac{(8)^2+2}{-2(-3)}=\frac{64+2}{6}=\frac{66}{6}=11$$

111. $$\frac{2[-4-2(3-1)}{3(-3)(-2)}=\frac{2[-4-2(2)]}{-9(-2)}=\frac{2[-4-4]}{18}=\frac{2[-8]}{18}=\frac{-16}{18}=-\frac{8}{9}$$

113. $$\frac{|6-4|+2|-4|}{26-2^4}=\frac{2+2(4)}{26-16}=\frac{2+8}{10}=\frac{10}{10}=1$$

115. $$\frac{(4^3-10)+(-4)}{5^2-(-4)(-5)}=\frac{(64-10)+(-4)}{25-20}=\frac{54+(-4)}{5}=\frac{50}{5}=10$$

117. $$\frac{72-(2-2\bullet 1)}{10^2-(90+2^2)}=\frac{72-(2-2)}{100-(90+4)}=\frac{72-0}{100-94}=\frac{72}{6}=12$$

119. $-\left(\frac{40-1^3-2^4}{3(2+5)+2}\right)=-\left(\frac{40-1-16}{3(7)+2}\right)$

$$=-\left(\frac{39-16}{21+2}\right)$$

$$=-\left(\frac{23}{23}\right)$$

$$=-1$$

APPLICATIONS

121. LIGHT

2 yards = 2^2 square units
3 yards = 3^2 square units
4 yards = 4^2 square units

123. AUTO INSURANCE

$$2,672+1,680+2,485=6,837$$

$$1,370+2,737+1,692=5,799$$

$$\frac{6 \text{ premiums}}{6}=\frac{6,837+5,799}{6}$$

$$=\frac{12,636}{6}$$

$$=2,106$$

The average is $2,106.

125. CASH AWARDS

a) $2,500+4(500)+35(150)+85(25)$

$=2,500+2,000+5,250+2,125$

$=11,875$

$11,875 will be awarded.

b) $\frac{11,875}{1+4+35+85}=\frac{11,875}{125}$

$=95$

The average cash prize is $95.

127. WRAPPING GIFTS

2(9) + 2(16) + 4(4) + 15
= 18 + 32 + 16 + 15
= 81
81 inches of ribbon is needed.

WRITING

129. Answers will vary.

131. Answers will vary.

REVIEW

133. a) ii
b) iii
c) iv
d) i

CHALLENGE PROBLEMS

135. $\left(3^2\right)^4=9^4$

$=6,561$

SECTION 1.8

VOCABULARY

1. Variables and/or numbers can be combined with the operations of arithmetic to create algebraic **expressions**.

3. A term, such as 27, that consists of a single number is called a **constant** term.

5. To **evaluate** an algebraic expression, we substitute the values for the variables and simplify.

7. $2x + 5$ is an example of an algebraic **expression**, whereas $2x + 5 = 7$ is an example of an **equation**.

CONCEPTS

9. $11x^2 - 6x - 9$
   a) 3 terms
   b) 11
   c) -6
   d) -9

11. a) term
    b) factor
    c) factor
    d) term
    e) factor
    f) term

13.
a)

Number of weeks	Number of days
1	$1 \cdot 7 = 7$
2	$2 \cdot 7 = 14$
3	$3 \cdot 7 = 21$
w	$w \cdot 7 = 7w$

b)

Number of seconds	Number of minutes
60	$60 \div 60 = 1$
120	$120 \div 60 = 2$
180	$180 \div 60 = 3$
s	$s \div 60 = \frac{s}{60}$

15. a) Let x = the weight of the car.
    $2x - 500$ = the weight of the van.
    b) Let $x = 2{,}000$.
    $2(\mathbf{2{,}000}) - 500 = 4{,}000 - 500$
    $= 3{,}500$
    The van weights 3,500 pounds.

17.

Type of coin	Number	· Value (in cents)	= Total value (in cents)
Nickel	6	5	30
Dime	d	10	$10d$
Half-dollar	$x + 5$	50	$50(x + 5)$

NOTATION

19. Let $a = 5$.
$$9a - a^2 = 9(\boxed{5}) - (5)^2 = 9(5) - \boxed{25} = \boxed{45} - 25 = 20$$

21. a) $8y$
    b) $2cd$
    c) $15sx$
    b) $-9a^3b^2$

PRACTICE

23. $l + 15$

25. $50x$

27. $\frac{w}{l}$

29. $P + p$

31. $k^2 - 2{,}005$

33. $J - 500$

35. $\frac{1{,}000}{n}$

37. $p + 90$

39. $35 + h + 300$

41. $p - 680$

43. $4d - 15$

45. $2(200 + t)$

47. $|a - 2|$

49. a) $5(60) = 300$ minutes
    b) $h(60) = 60h$

51. a) $3y$
    b) $\frac{f}{3}$

53. $29x$¢

55. $\frac{c}{6}$

57. $5b$

59. $\$5(x + 2)$

61.

x	$x^3 - 1$
0	$(\mathbf{0})^3 - 1 = 0 - 1$ $= 0$
-1	$(\mathbf{-1})^3 - 1 = -1 - 1$ $= -2$
-3	$(\mathbf{-3})^3 - 1 = -27 - 1$ $= -28$

63.

S	$\frac{5s + 36}{s}$
1	$\frac{5(\mathbf{1}) + 36}{\mathbf{1}} = \frac{5 + 36}{1}$ $= \frac{41}{1}$ $= 41$
6	$\frac{5(\mathbf{6}) + 36}{\mathbf{6}} = \frac{30 + 36}{6}$ $= \frac{66}{6}$ $= 11$
-12	$\frac{5(\mathbf{-12}) + 36}{\mathbf{-12}} = \frac{-60 + 36}{-12}$ $= \frac{-24}{-12}$ $= 2$

65.

Input x	Output $2x - \frac{x}{2}$
100	$2(\mathbf{100}) - \frac{\mathbf{100}}{2} = 200 - 50$ $= 150$
-300	$2(\mathbf{-300}) - \frac{\mathbf{-300}}{2} = -600 - (-150)$ $= -600 + 150$ $= -450$

67.

x	$(x + 1)(x + 5)$
-1	$(\mathbf{-1} + 1)(\mathbf{-1} + 5) = (0)(4)$ $= 0$
-5	$(\mathbf{-5} + 1)(\mathbf{-5} + 5) = (-4)(0)$ $= 0$
-6	$(\mathbf{-6} + 1)(\mathbf{-6} + 5) = (-5)(-1)$ $= 5$

69. $3y^2 - 6y - 4 = 3(\mathbf{-2})^2 - 6(\mathbf{-2}) - 4$
$= 3(4) - 6(-2) - 4$
$= 12 + 12 - 4$
$= 24 - 4$
$= 20$

71. $(3 + x)y = (3 + \mathbf{3}) \cdot \mathbf{-2}$
$= 6 \cdot -2$
$= -12$

73. $(x + y)^2 - |z + y| = (\mathbf{3} + \mathbf{-2})^2 - |\mathbf{-4} + \mathbf{-2}|$
$= (1)^2 - |-6|$
$= 1 - 6$
$= -5$

75. $(4x)^2 + 3y^2 = (4 \cdot \mathbf{3})^2 + 3(\mathbf{-2})^2$
$= (12)^2 + 3(4)$
$= 144 + 12$
$= 156$

77. $$-\frac{2x+y^3}{y+2z} = -\frac{2(\mathbf{3})+(\mathbf{-2})^3}{\mathbf{-2}+2(\mathbf{-4})}$$
$$= -\frac{6+(-8)}{-2+(-8)}$$
$$= -\frac{-2}{-10}$$
$$= -\frac{1}{5}$$

79. $b^2 - 4ac = (\mathbf{5})^2 - 4(\mathbf{-1})(\mathbf{-2})$
$= 25 - (-4)(-2)$
$= 25 - 8$
$= 17$

81. $a^2 + 2ab + b^2 = (\mathbf{-5})^2 + 2(\mathbf{-5})(\mathbf{-1}) + (\mathbf{-1})^2$
$= 25 + 2(-5)(-1) + 1$
$= 25 + 10 + 1$
$= 35 + 1$
$= 36$

83. $$\frac{n}{2}\left[2a+(n-1)d\right]$$
$$= \frac{10}{2}[2(-4)+(10-1)6]$$
$$= 5[2(-4)+(9)6]$$
$$= 5[-8+54]$$
$$= 5[46]$$
$$= 230$$

85. $$\frac{a^2+b^2}{2} = \frac{\mathbf{0}^2+(\mathbf{-10})^2}{2}$$
$$= \frac{0+100}{2}$$
$$= \frac{100}{2}$$
$$= 50$$

APPLICATIONS

87. ROCKETRY

t	$h = 64t - 16t^2$
1	$64(\mathbf{1}) - 16(\mathbf{1})^2 = 64(1) - 16(1)$ $= 64 - 16$ $= 48$
2	$64(\mathbf{2}) - 16(\mathbf{2})^2 = 64(2) - 16(4)$ $= 128 - 64$ $= 64$
3	$64(\mathbf{3}) - 16(\mathbf{3})^2 = 64(3) - 16(9)$ $= 192 - 144$ $= 48$
4	$64(\mathbf{4}) - 16(\mathbf{4})^2 = 64(4) - 16(16)$ $= 256 - 256$ $= 0$

89. ANTIFREEZE

For -34° F:
$$\frac{5(\mathbf{-34}-32)}{9} = \frac{5(-66)}{9}$$
$$= \frac{-330}{9}$$
$$\approx -37°C$$

For -84° F:
$$\frac{5(\mathbf{-84}-32)}{9} = \frac{5(-116)}{9}$$
$$= \frac{-580}{9}$$
$$\approx -64°C$$

91. TOOLS

$$A = \frac{1}{2}h(b + d)$$
$$= \frac{1}{2}\left(\frac{3}{4}\right)\left(1\frac{1}{4} + 2\frac{3}{8}\right)$$
$$= \frac{1}{2}\left(\frac{3}{4}\right)\left(\frac{5}{4} + \frac{19}{8}\right)$$
$$= \frac{1}{2}\left(\frac{3}{4}\right)\left(\frac{5}{4} \bullet \frac{2}{2} + \frac{19}{8}\right)$$
$$= \frac{1}{2}\left(\frac{3}{4}\right)\left(\frac{10}{8} + \frac{19}{8}\right)$$
$$= \frac{1}{2}\left(\frac{3}{4}\right)\left(\frac{29}{8}\right)$$
$$= \frac{3}{8}\left(\frac{29}{8}\right)$$
$$= \frac{87}{64}$$
$$= 1\frac{23}{64} \text{ in.}^2$$

The area is $1\frac{23}{64}$ in.^{2}

WRITING

93. Answers will vary.

94. Answers will vary.

REVIEW

97. $12 = 2 \cdot 2 \cdot 3$
$15 = 3 \cdot 5$
common denominator $= 2 \cdot 2 \cdot 3 \cdot 5 = 60$

99. $(-2 \cdot 3)^3 - 10 + 1 = (-6)^3 - 10 + 1$
$= -216 - 10 + 1$
$= -226 + 1$
$= -225$

CHALLENGE PROBLEMS

101. The answer would be 0, because
$(8 - 8) = 0$ is a factor of the expression.

CHAPTER 1 KEY CONCEPTS

1. f

2. j

3. h

4. a

5. b

6. e

7. c

8. i

9. d

10. g

11. $C = p + t$ (Answers may vary depending on the variables chosen.)

12. $b = 2t$ (Answers may vary depending on the variables chosen.)

13. $x + 4$ = amount of business ($ millions) in the year with the celebrity

14.

x	$3x^2 - 2x + 1$
0	$3(\mathbf{0})^2 - 2(\mathbf{0}) + 1 = 3(0) - 2(0) + 1$ $= 0 - 0 + 1$ $= 1$
4	$3(\mathbf{4})^2 - 2(\mathbf{4}) + 1 = 3(16) - 2(4) + 1$ $= 48 - 8 + 1$ $= 41$
6	$3(\mathbf{6})^2 - 2(\mathbf{6}) + 1 = 3(36) - 2(6) + 1$ $= 108 - 12 + 1$ $= 97$

CHAPTER 1 REVIEW

SECTION 1.1
Introducing the Language of Algebra

1. horizontal axis is 1 hour
   vertical axis is 100 cars

2. 100

3. 7 P.M.

4. The difference of 15 and 3 is 12.

5. The sum of 15 and 3 is 18.

6. The quotient of 15 and 3 is 5.

7. The product of 15 and 3 is 45.

8. $4 \cdot 9$; $4(9)$

9. $\frac{9}{3}$

10. $8b$

11. Prt

12. equation

13. algebraic expression

14.

Number of Brackets (b)	Number of Nails (n)
5	$\mathbf{5+5=10}$
10	$\mathbf{10+5=15}$
20	$\mathbf{20+5=25}$

SECTION 1.2
Fractions

15. a. 1(24), 2(12), 3(8), or 4(6)
    b. $2 \cdot 2 \cdot 6$ (answers will vary)
    c. 1, 2, 3, 4, 6, 8, 12, 24

16. $54 = 3^3 \cdot 2$

17. $147 = 7^2 \cdot 3$

18. $385 = 11 \cdot 7 \cdot 5$

19. $41 =$ prime

20. $\frac{12}{12} = 1$

21. $\frac{0}{10} = 0$

22. $\frac{20}{35} = \frac{2 \bullet 2 \bullet \cancel{5}}{\cancel{5} \bullet 7} = \frac{4}{7}$

23. $\frac{24}{18} = \frac{\cancel{2} \bullet 2 \bullet 2 \bullet \cancel{3}}{\cancel{2} \bullet \cancel{3} \bullet 3} = \frac{4}{3}$

24. $\frac{5}{8} \bullet \frac{8}{8} = \frac{40}{64}$

25. $\frac{12}{1} \bullet \frac{3}{3} = \frac{36}{3}$

26. $\frac{1}{8} \bullet \frac{7}{8} = \frac{1 \bullet 7}{8 \bullet 8} = \frac{7}{64}$

27. $\frac{16}{35} \bullet \frac{25}{48} = \frac{400}{1680} = \frac{5}{21}$

28. $\frac{1}{3} \div \frac{15}{16} = \frac{1}{3} \bullet \frac{16}{15} = \frac{16}{45}$

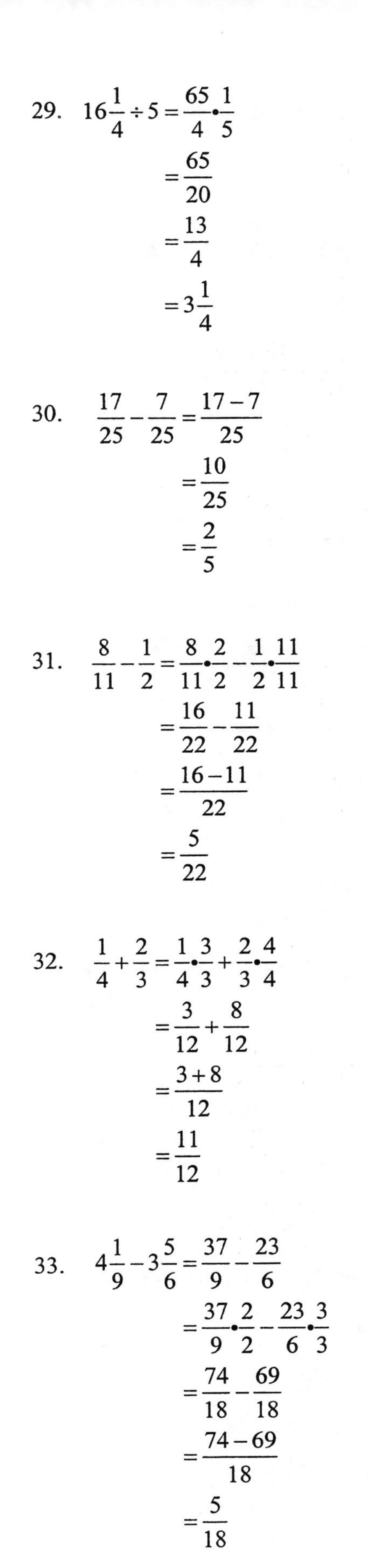

29. $16\frac{1}{4} \div 5 = \frac{65}{4} \cdot \frac{1}{5}$
$$= \frac{65}{20}$$
$$= \frac{13}{4}$$
$$= 3\frac{1}{4}$$

30. $\frac{17}{25} - \frac{7}{25} = \frac{17-7}{25}$
$$= \frac{10}{25}$$
$$= \frac{2}{5}$$

31. $\frac{8}{11} - \frac{1}{2} = \frac{8}{11} \cdot \frac{2}{2} - \frac{1}{2} \cdot \frac{11}{11}$
$$= \frac{16}{22} - \frac{11}{22}$$
$$= \frac{16-11}{22}$$
$$= \frac{5}{22}$$

32. $\frac{1}{4} + \frac{2}{3} = \frac{1}{4} \cdot \frac{3}{3} + \frac{2}{3} \cdot \frac{4}{4}$
$$= \frac{3}{12} + \frac{8}{12}$$
$$= \frac{3+8}{12}$$
$$= \frac{11}{12}$$

33. $4\frac{1}{9} - 3\frac{5}{6} = \frac{37}{9} - \frac{23}{6}$
$$= \frac{37}{9} \cdot \frac{2}{2} - \frac{23}{6} \cdot \frac{3}{3}$$
$$= \frac{74}{18} - \frac{69}{18}$$
$$= \frac{74-69}{18}$$
$$= \frac{5}{18}$$

34.
$$\frac{17}{24} - \frac{17}{32} = \frac{17}{24} \cdot \frac{4}{4} - \frac{17}{32} \cdot \frac{3}{3}$$
$$= \frac{68}{96} - \frac{51}{96}$$
$$= \frac{68-51}{96}$$
$$= \frac{17}{96} \text{ in.}$$

$\frac{17}{96}$ in. must be milled off.

SECTION 1.3
The Real Numbers

35. 0 is a whole number but is not a natural number.

36. –$65 billion

37. –206 feet

38. $0 \boxed{<} 5$

39. $-12 \boxed{>} -13$

40. $\frac{7}{10}$

41. $\frac{14}{3}$

42. $250\overline{)1.000}$ with quotient 0.004; $= 0.004$

43. $22\overline{)17.00000}$ with quotient 0.77272; $= 0.7\overline{72}$

44. $-\frac{17}{4}$ -2 $0.\overline{3}$ $\frac{7}{8}$ $\sqrt{2}$ π 3.75

–5 –4 –3 –2 –1 0 1 2 3 4 5

45. false

46. false

47. true

48. true

49. natural: 8
whole: 0, 8
integers: 0, –12, 8
rational: $-\frac{4}{5}$, 99.99, 0, –12, $4\frac{1}{2}$, 0.666..., 8
irrational: $\sqrt{2}$
real: all

50. $|-6| > |5|$

51. $-9 < |-10|$

SECTION 1.4
Adding Real Numbers

52. $-45 + (-37) = -82$

53. $25 + (-13) = 12$

54. $0 + (-7) = -7$

55. $-7 + 7 = 0$

56. $12 + (-8) + (-15) = 4 + (-15)$
$= -11$

57. $-9.9 + (-2.4) = -12.3$

58. $\frac{5}{16} + \left(-\frac{1}{2}\right) = \frac{5}{16} + \left(-\frac{1}{2} \cdot \frac{8}{8}\right)$
$= \frac{5}{16} + \left(-\frac{8}{16}\right)$
$= -\frac{3}{16}$

59. $35 + (-13) + (-17) + 6 = 22 + (-17) + 6$
$= 5 + 6$
$= 11$

60. commutative property of addition

61. associative property of addition

62. addition property of opposites

63. addition property of 0

64. $3(1) + 4(-1) = 3 + (-4)$
$= -1$

SECTION 1.5
Subtracting Real Numbers

65. –10

66. 3

67. $\frac{9}{16}$

68. –4

69. $45 - 64 = 45 + (-64)$
$= -19$

70. $-17 - 32 = -17 + (-32)$
$= -49$

71. $-27 - (-12) = -27 + 12$
$= -15$

72. $3.6 - (-2.1) = 3.6 + 2.1$
$= 5.7$

73. $0 - 10 = 0 + (-10)$
$= -10$

74. $-33 + 7 - 5 - (-2) = -33 + 7 + (-5) + 2$
$= -26 + (-5) + 2$
$= -31 + 2$
$= -29$

75. $29{,}028 - (-36{,}205) = 29{,}028 + 36{,}205$
$= 65{,}233$
The difference is 65,233 ft.

SECTION 1.6
Multiplying and Dividing Real Numbers

76. $-8 \cdot 7 = -56$

77. $(-9)(-6) = 54$

78. $2(-3)(-2) = -6(-2)$
$= 12$

79. $(-4)(-1)(-3)(-3) = 4(-3)(-3)$
$= -12(-3)$
$= 36$

80. $-1.2(-5.3) = 6.36$

81. $0.002(-1{,}000) = -2$

82. $-\frac{2}{3}\left(\frac{1}{5}\right) = -\frac{2 \bullet 1}{3 \bullet 5}$
$= -\frac{2}{15}$

83. $-6(-3)(0)(-1) = 18(0)(-1)$
$= 0(-1)$
$= 0$

84. associative property of multiplication

85. commutative property of multiplication

86. multiplication property of 1

87. inverse property of multiplication

88. 3

89. $-\frac{1}{3}$

90. $\frac{44}{-44} = -1$

91. $\frac{-100}{25} = -4$

92. $\frac{-81}{-27} = 3$

93. $-\frac{3}{5} \div \frac{1}{2} = -\frac{3}{5} \bullet \frac{2}{1}$
$= -\frac{6}{5}$

94. $\frac{-60}{0} = \text{undefined}$

95. $\frac{-4.5}{1} = -4.5$

96. The high reading is 2; the low reading is –3.

97. New high: $2(2) = 4$
New low: $2(-3) = -6$

SECTION 1.7
Exponents and Order of Operations

98. 8^5

99. a^4

100. $9\pi r^2$

101. x^3y^4

102. $9 \cdot 9 = 81$

103. $\left(-\frac{2}{3}\right)\left(-\frac{2}{3}\right)\left(-\frac{2}{3}\right) = -\frac{8}{27}$

104. $2(2)(2)(2)(2) = 32$

105. 50

106. 4 operations; power, multiplication, subtraction, addition

107. $2 + 5 \cdot 3 = 2 + 15$
$= 17$

108. $24 - 3(6)(4) = 24 - 18(4)$
$= 24 - 72$
$= -48$

109. $-(6 - 3)^2 = -(3)^2$
$= -9$

110. $4^3 + 2(-6 - 2 \cdot 2) = 64 + 2(-6 - 4)$
$= 64 + 2(-10)$
$= 64 - 20$
$= 44$

111. $10 - 5[-3 - 2(5 - 7^2)] - 5$
$= 10 - 5[-3 - 2(5 - 49)] - 5$
$= 10 - 5[-3 - 2(-44)] - 5$
$= 10 - 5[-3 + 88] - 5$
$= 10 - 5[85] - 5$
$= 10 - 425 - 5$
$= -415 - 5$
$= -420$

112. $\frac{-4(4+2)-4}{|-18-4(5)|} = \frac{-4(6)-4}{|-18-20|}$
$= \frac{-24-4}{|-38|}$
$= \frac{-28}{38}$
$= -\frac{14}{19}$

113. $(-3)^3\left(\frac{-8}{2}\right)+5 = -27(-4)+5$
$= 108+5$
$= 113$

114. $-9^2 + (-9)^2 = -81 + 81$
$= 0$

115. $20(5)+65(10)+25(20) = 1,250$
$5(50)+10(100) = 1,250$

$\frac{1,250+1,250}{20+65+25+5+10} = \frac{2,500}{125}$
$= 20$

The average donation is $20.

SECTION 1.8
Algebraic Expressions

116. 3

117. 1

118. 2, –5

119. 16, –5, 25

120. $\frac{1}{2}, 1$

121. 9.6, –1

122. $h + 25$

123. $x - 15$

124. $\frac{1}{2}t$

125. $(n + 4)$ in.

126. $(b - 4)$ in.

127. $10d$

128. $(x - 5)$ years

129.

Type of Coin	Number	Value (¢)	Total Value (¢)
Nickel	6	5	6(5) = 30
Dime	d	10	$10d$

130.

x	$20x - x^3$
0	$20(\mathbf{0}) - (\mathbf{0})^3 = 0 - 0$ $= 0$
1	$20(\mathbf{1}) - (\mathbf{1})^3 = 20 - 1$ $= 19$
–4	$20(\mathbf{-4}) - (\mathbf{-4})^3 = -80 - (-64)$ $= -80 + 64$ $= -16$

131. $(\mathbf{-10})^2 - 4(\mathbf{3})(\mathbf{5}) = 100 - 12(5)$
$= 100 - 60$
$= 40$

132. $\frac{19+17}{-(19)-(-18)} = \frac{36}{-19+18}$
$= \frac{36}{-1}$
$= -36$

CHAPTER 1 TEST

1. $24

2. 5 hours

3.

Area in square miles (a)	Number of fire stations (f) $f = \frac{a}{5}$
15	$\frac{\mathbf{15}}{5} = 3$
100	$\frac{\mathbf{100}}{5} = 20$
350	$\frac{\mathbf{350}}{5} = 70$

4. $180 = 5 \cdot 3 \cdot 3 \cdot 2 \cdot 2$
$= 2^2 \cdot 3^2 \cdot 5$

5. $\frac{42}{105} = \frac{2 \cdot \cancel{3} \cdot \cancel{7}}{\cancel{3} \cdot 5 \cdot \cancel{7}}$
$= \frac{2}{5}$

6. $\frac{15}{16} \div \frac{5}{8} = \frac{15}{16} \cdot \frac{8}{5}$
$= \frac{120}{80}$
$= \frac{3}{2}$
$= 1\frac{1}{2}$

7. $\frac{11}{12} - \frac{2}{9} = \frac{11}{12}\left(\frac{3}{3}\right) - \frac{2}{9}\left(\frac{4}{4}\right)$
$= \frac{33}{36} - \frac{8}{36}$
$= \frac{25}{36}$

8. $1\frac{2}{3} + 8\frac{2}{5} = \frac{5}{3} + \frac{42}{5}$
$= \frac{5}{3}\left(\frac{5}{5}\right) + \frac{42}{5}\left(\frac{3}{3}\right)$
$= \frac{25}{15} + \frac{126}{15}$
$= \frac{151}{15}$
$= 10\frac{1}{15}$

9. The oranges weigh 4.25 pounds.
The cost would be 4.25(0.84) = $3.57

10. $\frac{5}{6} = 6\overline{)5.0000}$ (quotient 0.8333) $= 0.8\overline{3}$

11. -3.75 -3 $-1\frac{1}{4}$ 0.5 $\sqrt{2}$ $\frac{7}{2}$

−5 −4 −3 −2 −1 0 1 2 3 4 5

12. a. true
b. false
c. true
d. true

13. The set of real numbers corresponds to all points on a number line. A real number is any number that is either a rational number or an irrational number.

14. a. $-2 > -3$
b. $-|-7| < 8$
c. $|-4| < -(-5)$
d. $|-\frac{7}{8}| > 0.5$

15.

$\frac{0.6 + (-0.3) + 1.7 + 1.5 + (-0.2) + 1.1 + (-0.2)}{7}$
$= \frac{4.2}{7}$
$= 0.6$

There was a gain of 0.6 of a rating point.

16. $(-6)+8+(-4)=2+(-4)$
$=-2$

17. $-\frac{1}{2}+\frac{7}{8}=-\frac{1}{2}\left(\frac{4}{4}\right)+\frac{7}{8}$
$=-\frac{4}{8}+\frac{7}{8}$
$=\frac{3}{8}$

18. $-10-(-4)=-10+4$
$=-6$

19. $(-2)(-3)(-5)=6(-5)$
$=-30$

20. $\frac{-22}{-11}=2$

21. $-6.1(0.4)=-2.44$

22. $\frac{0}{-3}=0$

23. $0-3=0+(-3)$
$=-3$

24. $3+(-3)=0$

25. $-30+50-10-(-40)=-30+50+(-10)+40$
$=20+(-10)+40$
$=10+40$
$=50$

26. $\left(-\frac{3}{5}\right)^3=\left(-\frac{3}{5}\right)\left(-\frac{3}{5}\right)\left(-\frac{3}{5}\right)$
$=-\frac{3\cdot3\cdot3}{5\cdot5\cdot5}$
$=-\frac{27}{125}$

27. $-12.5-(-26.5)=-12.5+26.5$
$=14$

28. associative property of addition

29. a. 9^5
b. $3x^2z^3$

30. $8+2\cdot3^4=8+2\cdot81$
$=8+162$
$=170$

31. $9^2-3[45-3(6+4)]=81-3[45-3(10)]$
$=81-3[45-30]$
$=81-3[15]$
$=81-45$
$=36$

32. $\frac{3(40-2^3)}{-2(6-4)^2}=\frac{3(40-8)}{-2(2)^2}$
$=\frac{3(32)}{-2(4)}$
$=\frac{96}{-8}$
$=-12$

33. $-10^2=-(10)(10)$
$=-100$

34. $3(x-y)-5(x+y)=3(\mathbf{2}-(\mathbf{-5}))-5(\mathbf{2}+(\mathbf{-5}))$
$=3(2+5)-5(2+(-5))$
$=3(7)-5(-3)$
$=21-(-15)$
$=21+15$
$=36$

35.

x	$2x-\frac{30}{x}$
5	$2(\mathbf{5})-\frac{30}{\mathbf{5}}=10-6$ $=4$
10	$2(\mathbf{10})-\frac{30}{\mathbf{10}}=20-3$ $=17$
−30	$2(\mathbf{-30})-\frac{30}{\mathbf{-30}}=-60+1$ $=-59$

36. $2w+7$

37. $x-2$ = number of songs on the CD

38. $25q$¢

39. An equation is a mathematical sentence that contains an = sign. An expression does not contain an = sign.

40. There are 3 terms. The coefficient of the second term is 5.

SECTION 2.1

VOCABULARY

1. An **equation** is a statement indicating that two expressions are equal.

3. To **check** the solution of an equation, we substitute the value for the variable in the original equation and see whether the result is a true statement.

5. Equations with the same solutions are called **equivalent** equations.

7. To solve an equation, we **isolate** the variable on one side of the equal symbol.

CONCEPTS

9. a) $x + 5 = 7$
   b) Subtract 5 from both sides.

11. a) $x + 6$
    b) It is neither since we do not know the value of x.
    c) $x = 5$ is not a solution because when 5 is substituted for x, the left side is 11.
    d) $x = 6$ is a solution because when 6 is substituted for x, the left side is 12.

13. **24**

15. ***n***

17. a) If $x = y$, then $x + c = y + \mathbf{c}$ and $x - c = y - \mathbf{c}$.
    Adding (or **subtracting**) the same number to (or from) **both** sides of an equation does not change the solution.
    b) If $x = y$, then $cx = \mathbf{c}y$ and $\frac{x}{c} = \frac{y}{c}$.
    Multiplying (or **dividing**) both sides of an **equation** by the same nonzero number does not change the solution.

19. a) Simplify: $x + 7 - 7$, x
    b) Simplify: $y - 2 + 2$, y
    c) Simplify: $\frac{5t}{5}$, t
    d) Simplify: $6 \cdot \frac{h}{6}$, h

NOTATION

21. Solve:

$$x + 15 = 45$$
$$x + 15 - \mathbf{15} = 45 - \mathbf{15}$$
$$\boxed{x = 30}$$

Check:

$$x + 15 = 45$$
$$30 + 15 \stackrel{?}{=} 45$$
$$45 = 45$$

30 is a solution.

23. a) The $\stackrel{?}{=}$ symbol means "possibly equal to".
    b) twenty–seven degrees, **27°**

PRACTICE

25. $\underline{\mathbf{6}} + 12 \stackrel{?}{=} 18$
    $18 \stackrel{?}{=} 18$ true

27. $2(\mathbf{-8}) + 3 \stackrel{?}{=} -15$
    $-16 + 3 \stackrel{?}{=} -15$
    $-13 \stackrel{?}{=} -15$ false

29. $0.5(\mathbf{5}) \stackrel{?}{=} 2.9$
    $2.5 \stackrel{?}{=} 2.9$ false

31. $33 - \frac{-6}{2} \stackrel{?}{=} 30$
    $33 + 3 \stackrel{?}{=} 30$
    $36 \stackrel{?}{=} 30$ false

33. $|\mathbf{20} - 8| \stackrel{?}{=} 10$
    $|12| \stackrel{?}{=} 10$
    $12 \stackrel{?}{=} 10$ false

35. $3(\mathbf{12}) - 2 \stackrel{?}{=} 4(\mathbf{12}) - 5$
    $36 - 2 \stackrel{?}{=} 48 - 5$
    $34 \stackrel{?}{=} 43$ false

37. $(\mathbf{-3})^2 - (\mathbf{-3}) - 6 \stackrel{?}{=} 0$
    $9 + 3 - 6 \stackrel{?}{=} 0$
    $12 - 6 \stackrel{?}{=} 0$
    $6 \stackrel{?}{=} 0$ false

39. $\dfrac{2}{1+1}+5 \stackrel{?}{=} \dfrac{12}{1+1}$

$\dfrac{2}{2}+5 \stackrel{?}{=} \dfrac{12}{2}$

$1+5 \stackrel{?}{=} 6$

$6 \stackrel{?}{=} 6$ true

41. $(-3-4)(-3+3) \stackrel{?}{=} 0$

$(-7)(0) \stackrel{?}{=} 0$

$0 \stackrel{?}{=} 0$ true

43. $x+7=10$

$x+7-7=10-7$

$\boxed{x=3}$

$3+7 \stackrel{?}{=} 10$ true

45. $a-5=66$

$a-5+5=66+5$

$\boxed{a=71}$

$71-5 \stackrel{?}{=} 66$ true

47. $0=n-9$

$0+9=n-9+9$

$\boxed{9=n}$

$0 \stackrel{?}{=} 9-9$ true

49. $9+p=9$

$9+p-9=9-9$

$\boxed{p=0}$

$9+0 \stackrel{?}{=} 9$ true

51. $x-16=-25$

$x-16+16=-25+16$

$\boxed{x=-9}$

$-9-16 \stackrel{?}{=} -25$ true

53. $a+3=0$

$a+3-3=0-3$

$\boxed{a=-3}$

$-3+3 \stackrel{?}{=} 0$ true

55. $f+3.5=1.2$

$f+3.5-3.5=1.2-3.5$

$\boxed{f=-2.3}$

$-2.3+3.5 \stackrel{?}{=} 1.2$ true

57. $-8+p=-44$

$-8+p+8=-44+8$

$\boxed{p=-36}$

$-8+(-36) \stackrel{?}{=} -44$ true

59. $8.9=-4.1+t$

$8.9+4.1=-4.1+4.1+t$

$\boxed{13=t}$

$8.9 \stackrel{?}{=} -4.1+13$ true

61. $d-\dfrac{1}{9}=\dfrac{7}{9}$

$d-\dfrac{1}{9}+\dfrac{1}{9}=\dfrac{7}{9}+\dfrac{1}{9}$

$\boxed{d=\dfrac{8}{9}}$

$\dfrac{8}{9}-\dfrac{1}{9} \stackrel{?}{=} \dfrac{7}{9}$ true

63. $s+\dfrac{4}{25}=\dfrac{11}{25}$

$s+\dfrac{4}{25}-\dfrac{4}{25}=\dfrac{11}{25}-\dfrac{4}{25}$

$\boxed{s=\dfrac{7}{25}}$

$\dfrac{7}{25}+\dfrac{4}{25} \stackrel{?}{=} \dfrac{11}{25}$ true

65. $4x=16$

$\dfrac{4x}{4}=\dfrac{16}{4}$

$\boxed{x=4}$

$4(4) \stackrel{?}{=} 16$ true

67.

$$369 = 9c$$
$$\frac{369}{9} = \frac{9c}{9}$$
$$\boxed{41 = c}$$

$369 \stackrel{?}{=} 9(41)$ true

69.

$$4f = 0$$
$$\frac{4f}{4} = \frac{0}{4}$$
$$\boxed{f = 0}$$

$4(0) \stackrel{?}{=} 0$ true

71.

$$23b = 23$$
$$\frac{23b}{23} = \frac{23}{23}$$
$$\boxed{b = 1}$$

$23(1) \stackrel{?}{=} 23$ true

73.

$$-8h = 48$$
$$\frac{-8h}{-8} = \frac{48}{-8}$$
$$\boxed{h = -6}$$

$-8(-6) \stackrel{?}{=} 48$ true

75.

$$100 = -5g$$
$$\frac{-100}{-5} = \frac{-5g}{-5}$$
$$\boxed{20 = g}$$

$-100 \stackrel{?}{=} -5(20)$ true

77.

$$-3.4y = -1.7$$
$$\frac{-3.4y}{-3.4} = \frac{-1.7}{-3.4}$$
$$\boxed{y = 0.5}$$

$-3.4(0.5) \stackrel{?}{=} -1.7$ true

79.

$$\frac{x}{15} = 3$$
$$\frac{15x}{15} = 15 \cdot 3$$
$$\boxed{x = 45}$$

$\frac{45}{15} \stackrel{?}{=} 3$ true

81.

$$0 = \frac{v}{11}$$
$$11 \cdot 0 = \frac{11v}{11}$$
$$\boxed{0 = v}$$

$0 \stackrel{?}{=} \frac{0}{11}$ true

83.

$$\frac{w}{-7} = 15$$
$$\frac{-7w}{-7} = -7 \cdot 15$$
$$\boxed{w = -105}$$

$\frac{-105}{-7} \stackrel{?}{=} 15$ true

85.

$$\frac{d}{-7} = -3$$
$$\frac{-7d}{-7} = -7(-3)$$
$$\boxed{d = 21}$$

$\frac{21}{-7} \stackrel{?}{=} -3$ true

87.

$$\frac{y}{0.6} = -4.4$$
$$\frac{0.6y}{0.6} = (0.6)(-4.4)$$
$$\boxed{y = -2.64}$$

$\frac{-2.64}{0.6} \stackrel{?}{=} -4.4$ true

89. $a + 456{,}932 = 1{,}708{,}921$

$$a + 456{,}932 - 456{,}932 = 1{,}708{,}921 - 456{,}932$$

$$\boxed{a = 1{,}251{,}989}$$

1,251,989 + 456,932 $\stackrel{?}{=}$ 1,708,921 true

91. $-1{,}563x = 43{,}764$

$$\frac{-1{,}563x}{-1{,}563} = \frac{43{,}764}{-1{,}563}$$

$$\boxed{x = -28}$$

(-1,563)(-28) $\stackrel{?}{=}$ 43,764 true

APPLICATIONS

93. SYNTHESIZERS

The two angles are supplementary.
The two angles total 180°.

Let x = measure of unknown angle

$$x + 115 = 180$$

$$x + 115 - 115 = 180 - 115$$

$$\boxed{x = 65}$$

65 + 115 $\stackrel{?}{=}$ 180 true

95. AVIATION

The two angles are complementary.
The two angles total 90°.

Let x = measure of unknown angle

$$x + 52 = 90$$

$$x + 52 - 52 = 90 - 52$$

$$\boxed{x = 38}$$

38 + 52 $\stackrel{?}{=}$ 90 true

WRITING

97. Answers will vary.

99. Answers will vary.

REVIEW

101. Evaluate: $-9 - 3x$ for $x = -3$

$$-9 - 3(-3) = -9 + 9$$

$$= 0$$

103. Translate: Subtract x from 45.

$$45 - x$$

CHALLENGE PROBLEMS

105. If $a + 80 = 50$, what is $a - 80$?

First solve: $a + 80 = 50$

$$a + 80 - 80 = 50 - 80$$

$$a = -30$$

Now substitute -30 in $a - 80$

$$-30 - 80 = -110$$

SECTION 2.2

VOCABULARY

1. A letter that is used to represent a number is called a **variable**.

3. To solve an applied problem, we let a **variable** represent the unknown quantity. Then we write an **equation** that models the situation. Finally, we **solve** the equation for the variable to find the unknown.

5. In the statement "10 is 50% of 20," 10 is called the **amount**, 50% is the **percent**, and 20 is the **base**.

7. $x + 371 + 479 = 1,240$

9. $x + 11,000 = 13,500$

11. $x + 5 + 8 + 16 = 31$

13. 

NOTATION

15. 12 is 40% of what number?
$12 = 40\% \cdot x$

17. a) $35\% = 0.35$
b) $3.5\% = 0.035$
c) $350\% = 3.5$
d) $\frac{1}{2}\% = 0.5\% = 0.005$

PRACTICE

19. What number is 48% of 650?
$$x = 48\% \cdot 650$$
$$x = 0.48 \cdot 650$$
$$\boxed{x = 312}$$

21. What percent of 300 is 78?
$$x \cdot 300 = 78$$
$$300x = 78$$
$$\frac{300x}{300} = \frac{78}{300}$$
$$x = 0.26$$
$$\boxed{x = 26\%}$$

23. 75 is 25% of what number?
$$75 = 25\% \cdot x$$
$$75 = 0.25x$$
$$\frac{75}{0.25} = \frac{0.25x}{0.25}$$
$$\boxed{300 = x}$$

25. What number is 92.4% of 50?
$$x = 92.4\% \cdot 50$$
$$x = 0.924 \cdot 50$$
$$\boxed{x = 46.2}$$

27. What percent of 16.8 is 0.42?
$$x \cdot 16.8 = 0.42$$
$$16.8x = 0.42$$
$$\frac{16.8x}{16.8} = \frac{0.42}{16.8}$$
$$x = 0.025$$
$$\boxed{x = 2.5\%}$$

29. 128.1 is 8.75% of what number?
$$128.1 = 8.75\% \cdot x$$
$$128.1 = 0.0875x$$
$$\frac{128.1}{0.0875} = \frac{0.0875x}{0.0875}$$
$$\boxed{1,464 = x}$$

APPLICATIONS

31. GRAVITY
$6x = 330$

33. MONARCHY
Given: George III reigned 59 years.
Find: Queen Victoria ruled how long?
Queen Victoria yrs – 4 = George III yrs
Let x = of years ruled by QV
$$x - 4 = 59$$
$$x - 4 + 4 = 59 + 4$$
$$\boxed{x = 63}$$

Queen Victoria ruled 63 years.

35. ATM RECEIPT

Given: Bal = \$287.00, WD = \$35.00

Find: Previous balance

PB – WD = Current Balance

Let x = previous balance

$$x-35=287$$

$$x-35+35=287+35$$

$$\boxed{x=322}$$

The previous balance was \$322.

37. TV NEWS

Given: aired 9-minute interview, over 3 days

Find: total length of interview

Total interview divided by 3 days = 9 min

Let x = total minutes of interview

$$\frac{x}{3}=9$$

$$\frac{3x}{3}=9 \cdot 3$$

$$\boxed{x=27}$$

The original interview was 27 minutes.

39. STATEHOOD

\# of states prior to 1800 = ?

\# of states between 1800-1850 = 15

\# of states between 1851-1900 = 14

\# of states between 1901-1950 = 3

\# of states between 1951-present = 2

total states = 50

Let x = # of states before 1800

$$x+15+14+3+2=50$$

$$x+34=50$$

$$x+34-34=50-34$$

$$\boxed{x=16}$$

16 states were part of the Union prior to 1800.

41. THEATER

\# of scenes 1st act = ?

\# of scenes 2nd act = 6

\# of scenes 3rd act = 5

\# of scenes 4th act = 5

\# of scenes 5th act = 3

total scenes = 24

Let x = # of scenes for 1st act

$$x+6+5+5+3=24$$

$$x+19=24$$

$$x+19-19=24-19$$

$$\boxed{x=5}$$

There are 5 scenes in the first act.

43. ORCHESTRAS

\# of woodwinds = 19

\# of brass = 23

\# of percussion = 2

\# of strings = ?

total musicians = 98

Let x = # of string musicians

$$19+23+2+x=98$$

$$x+44=98$$

$$x+44-44=98-44$$

$$\boxed{x=54}$$

There are 54 musicians in the string section.

45. BERMUDA TRIANGLE

Given: triangle

perimeter = 3,075 miles

1st side = 1,100 miles

2nd side = 1,000 miles

3rd side = ?

1st side + 2nd side + 3rd side = perimeter

Let x = length of 3rd side in miles

$$1{,}100 + 1{,}000 + x = 3{,}075$$

$$x + 2{,}100 = 3{,}075$$

$$x + 2{,}100-2{,}100 = 3{,}075-2{,}100$$

$$\boxed{x = 975}$$

It is 975 miles from Bermuda to Florida.

47. SPACE TRAVEL

Given: 364 foot tall rocket
1st stage = 138 feet
2nd stage = 98feet
3rd stage = 46 feet
escape tower = 28 feet
lunar module = ?

$1^{st} + 2^{nd} + 3^{rd}$ + escape + module = total

Let x = height of module in feet

$$138 + 98 + 46 + 28 + x = 364$$
$$x + 310 = 364$$
$$x + 310 - 310 = 364 - 310$$
$$\boxed{x = 54}$$

The lunar module was 54 ft tall.

49. STOP SIGNS

Given: Octagonal STOP sign = 8 sides
Total of measures of angles = 1,080°

Let x = measure of 1 of the 8 equal angles

$$8x = 1,080$$
$$\frac{8x}{8} = \frac{1,080}{8}$$
$$\boxed{x = 135}$$

Each angle is 135°.

PERCENT PROBLEMS

51. ANTISEPTICS

Given: base = 16 fl. oz.
percent = 3%
amount = ?

amount = percent • base

$$x = 3\% \cdot 16$$
$$x = 0.03 \cdot 16$$
$$\boxed{x = 0.48}$$

53. FEDERAL OUTLAYS

Given: base = \$1,900 billion
percent = 36%
amount = ?

amount = percent • base

$$x = 36\% \cdot 1,900$$
$$x = 0.36 \cdot 1,900$$
$$\boxed{x = 684}$$

The amount paid was \$684 billion.

55. COLLEGE ENTRANCE EXAMS

Given: base = 1,600 points
math amount = 550 points
verbal amount = 700 points
percent = ?

amount = percent • base

$$550 + 700 = x \cdot 1,600$$
$$1,250 = 1,600x$$
$$\frac{1,250}{1,600} = \frac{1,600x}{1,600}$$
$$0.78125 = x$$
$$\boxed{78.125\% = x}$$

57. DENTAL RECORDS

Given: base = total # of teeth = 32
filling teeth amount = 6
percent = ? (round to nearest %)

amount = percent • base

$$6 = x \cdot 32$$
$$6 = 32x$$
$$\frac{6}{32} = \frac{32x}{32}$$
$$0.1875 = x$$
$$18.75\% = x$$
$$\boxed{19\% = x}$$

59. CHILD CARE

Given: base = ? = maximum # of children
amount = 84 registered children
percent = 70% (round to nearest %)

$$\text{amount} = \text{percent} \bullet \text{base}$$
$$84 = 70\% \bullet x$$
$$84 = 0.7x$$
$$\frac{84}{0.7} = \frac{0.7x}{0.7}$$
$$\boxed{120 = x}$$

61. NUTRITION

Use the table

a) 5 g and 25%

b) Given: base = ? = total g of sat. fat
amount = 5 g
percent = 25% (round to nearest %)

$$\text{amount} = \text{percent} \bullet \text{base}$$
$$5 = 25\% \bullet x$$
$$5 = 0.25x$$
$$\frac{5}{0.25} = \frac{0.25x}{0.25}$$
$$\boxed{20 = x}$$

63. EXPORTS

Looking at the bar graph, one sees that there are decreases between the years of 1994 and 1995 and between the years of 2000 and 2001. By just looking, one can't really tell which is the larger percent decrease, so we have to calculate both of them and then make a decision.

Find the difference between the amounts exported for years 1994 and 1995.

$$51 - 46 = 5$$

Now find the percent of decrease.

$$\frac{5}{51} \approx 0.098039$$
$$\approx 0.098$$
$$= 9.8\%$$

Find the difference between the amounts exported for the years 2000 and 2001.

$$111 - 101 = 10$$

63. (cont.)

Now find the percent of decrease.

$$\frac{10}{111} \approx 0.090090$$
$$\approx 0.090$$
$$= 9.0\%$$

The greater percent of decrease is between the years of 1994 and 1995, and it is 9.8%.

65. INSURANCE COSTS

Find the difference between $1,050 and $925.

$$\$1{,}050 - \$925 = \$125$$

Now find the percent of decrease.

$$\frac{125}{1{,}050} \approx 0.11904$$
$$\approx 0.12$$
$$= 12\%$$

WRITING

67. Answers will vary.

69. Answers will vary.

REVIEW

71.

$$-\frac{16}{25} \div \left(-\frac{4}{15}\right) = -\frac{16}{25} \bullet \left(-\frac{15}{4}\right)$$
$$= \frac{12}{5}$$
$$= 2\frac{2}{5}$$

73.

$$x + 15 = -49$$
$$-34 + 15 \stackrel{?}{=} -49$$
$$-19 = -49$$
no

CHALLENGE PROBLEMS

75.

$$100\% - 99\frac{44}{100}\% = \frac{56}{100}\%$$
$$= 0.56\%$$
$$= 0.0056$$

SECTION 2.3

VOCABULARY

1. We can use the associative property of multiplication to **simplify** the expression $5(6x)$.

3. We simplify **expressions**, and we solve **equations**.

5. We call $-(c + 9)$ the **opposite** of a sum.

7. The **coefficient** of the term $-23y$ is -23.

CONCEPTS

9. Fill in the blanks to simplify each product.

   a) $5 \cdot 6t = (\boxed{5} \cdot \boxed{6})t$
   $= \boxed{30}t$

   b) $-8(2x)4 = (\boxed{-8} \cdot \boxed{2} \cdot \boxed{4})x$
   $= \boxed{-64}x$

11. They are not like terms.

13. Fill in the blanks.

   $-(x-10) = \boxed{-1}(x-10)$
   Distributing the multiplication by -1 changes the **sign** of each term within the parentheses.

15. Identify any like terms.

   a) $3a$, $2a$

   b) 10, 12

   c) none

   d) $9y^2$, $-8y^2$

17. Fill in the blanks.
   a) $4m + 6m = (\boxed{4+6})m$
   $= \boxed{10m}$

   b) $30n - 50n = (\boxed{30-50})n$
   $= \boxed{-20n}$

   c) $12 - 32d + 15 = -32d + \boxed{27}$

NOTATION

19. Translate to symbols.
   a) Six times the quantity of h minus four.
   $6(h - 4)$

   b) The opposite of the sum of z and sixteen.
   $-(z + 16)$

21.

student's	book's	equivalent?
$10x$	$10 + x$	no
$3 + y$	$y + 3$	yes
$5 - 8a$	$8a - 5$	no
$3x + 4$	$3(x + 4)$	no
$3 - 2x$	$-2x + 3$	yes
$h^2 + (-16)$	$h^2 - 16$	yes

PRACTICE

23. $63m$

25. $-35q$

27. $300t$

29. $11.2x$

31. g

33. $5x$

35. $6y$

37. s

39. $-20r$

41. $60c$

43. $-96m$

45. $5x + 15$

47. $36c - 42$

49. $24t + 16$

51. $0.4x - 1.6$

53. $5t + 5$

55. $-12x - 20$

57. $-78c + 18$

59. $-2w + 4$

61. $9x + 10$

63. $9r - 16$

65. $-x + 7$

67. $5.6y - 7$

69. $40d + 50$

71. $-12r - 60$

73. $x + y - 5$

75. $6x - 21y - 16z$

77. $20x$

79. 0

81. 0

83. r

85. $37y$

87. $-s^3$

89. 5

91. $-10r$

93. $3a$

95. $-3x$

97. x

99. $\frac{4}{5}t$

101. $0.4r$

103. $2z + 5(z-3) = 2z + 5z - 15$
$= 7z - 15$

105. $-2x + 5$

107. $10(2d-7) + 4 = 20d - 70 + 4$
$= 20d - 66$

109. $-(c+7) - 2(c-3) = -c - 7 - 2c + 6$
$= -3c - 1$

111. $2(s-7) - (s-2) = 2s - 14 - s + 2$
$= s - 12$

113. $6 - 4(-3c - 7) = 6 + 12c + 28$
$= 12c + 34$

115. $36\left(\frac{2}{9}x - \frac{3}{4}\right) + 36\left(\frac{1}{2}\right) = \frac{36}{1}\left(\frac{2}{9}x\right) - \frac{36}{1}\left(\frac{3}{4}\right) + \frac{36}{1}\left(\frac{1}{2}\right)$
$= 4(2x) - 9(3) + 18(1)$
$= 8x - 27 + 18$
$= 8x - 9$

APPLICATIONS

117. THE AMERICAN RED CROSS
The cross has 12 lengths of equal measurement, x. Its perimeter is $12x$.

119. PING-PONG
Width $= x$ ft
Length $= x + 4$ ft
$P = l + w + l + w$
$P = (x + 4) + x + (x + 4) + x$
$P = (4x + 8)$ft

WRITING

121. Answers will vary.

REVIEW

Evaluate each expression for $x = -3,\ y = -5,\ z = 0$.

123. $x^2z(y^3 - z) = (-3)(-3)(0)[(-5)(-5)(-5) - 0]$
$= 0$

125. $\dfrac{x-y^2}{2y-1+x}=\dfrac{-3-(-5)^2}{2(-5)-1+(-3)}$

$=\dfrac{-3-(25)}{-10-1-3}$

$=\dfrac{-28}{-14}$

$=2$

CHALLENGE PROBLEMS

127. 2 inches + 2 feet + 2 yards

One can convert each of the different units into a common unit. Changing feet and yards into inches would be the easier conversion.

$1 \text{ ft} = 12 \text{ inches}$

$2 \text{ ft} = 2(12 \text{ inches})$

$= 24 \text{ inches}$

$1 \text{ yd} = 36 \text{ inches}$

$2 \text{ yd} = 2(36 \text{ inches})$

$= 72 \text{ inches}$

$2 \text{ inches} + 2 \text{ feet} + 2 \text{ yards}$

$2 \text{ in} + 24 \text{ in} + 72 \text{ in} = \boxed{98 \text{ inches}}$

129. $-2[x+4(2x+1)] = -2[x+8x+4]$

$= -2[9x+4]$

$= -18x-8$

SECTION 2.4

VOCABULARY

1. An equation is a statement indicating that two expressions are **equal**.

3. After solving an equation, we can check our result by substituting that value for the variable in the **original** equation.

5. An equation that is true for all values of its variable is called an **identity**.

CONCEPTS

7. To solve the equation $2x - 7 = 21$, we first undo the **subtraction** of 7 by adding 7 to both sides. Then we undo the **multiplication** by 2 by dividing both sides by 2.

9. To solve $\frac{s}{3} + \frac{1}{4} = -\frac{1}{2}$, we can clear the equation of the fractions by **multiplying** both sides by 12.

11. One method of solving $-\frac{4}{5}x = 8$ is to multiply both sides of the equations by the reciprocal of $-\frac{4}{5}$. What is the reciprocal of $-\frac{4}{5}$? , $-\frac{5}{4}$

13. LCD = 30

15. Multiply by LCD 6.

NOTATION

17.
$$2x - 7 = 21$$
$$2x - 7 + 7 = 21 + 7$$
$$2x = 28$$
$$\frac{2x}{2} = \frac{28}{2}$$
$$\boxed{x = 14}$$

19. a) $-x = \boxed{-1}x$

b) $\frac{3x}{5} = \boxed{\frac{3}{5}}x$

PRACTICE

21.
$$2x + 5 = 17$$
$$2x + 5 - 5 = 17 - 5$$
$$2x = 12$$
$$\frac{2x}{2} = \frac{12}{2}$$
$$\boxed{x = 6}$$

23.
$$5q - 2 = 23$$
$$5q - 2 + 2 = 23 + 2$$
$$5q = 25$$
$$\frac{5q}{5} = \frac{25}{5}$$
$$\boxed{q = 5}$$

25.
$$-33 = 5t + 2$$
$$-33 - 2 = 5t + 2 - 2$$
$$-35 = 5t$$
$$\frac{-35}{5} = \frac{5t}{5}$$
$$\boxed{-7 = t}$$

27.
$$20 = -x$$
$$\frac{20}{-1} = \frac{-x}{-1}$$
$$\boxed{-20 = x}$$

29.
$$-g = -4$$
$$\frac{-g}{-1} = \frac{-4}{-1}$$
$$\boxed{g = 4}$$

31. $$\begin{aligned} 1.2-x&=-1.7\\ 1.2-x-1.2&=-1.7-1.2\\ -x&=-2.9\\ \frac{-x}{-1}&=\frac{-2.9}{-1}\\ &\boxed{x=2.9} \end{aligned}$$

33. $$\begin{aligned} -3p+7&=-3\\ -3p+7-7&=-3-7\\ -3p&=-10\\ \frac{-3p}{-3}&=\frac{-10}{-3}\\ &\boxed{p=\frac{10}{3}} \end{aligned}$$

35. $$\begin{aligned} 0-2y&=8\\ -2y&=8\\ \frac{-2y}{-2}&=\frac{8}{-2}\\ &\boxed{y=-4} \end{aligned}$$

37. $$\begin{aligned} -8-3c&=0\\ -8-3c+8&=0+8\\ -3c&=8\\ \frac{-3c}{-3}&=\frac{8}{-3}\\ &\boxed{c=-\frac{8}{3}} \end{aligned}$$

39. $$\begin{aligned} \frac{5}{6}k&=10\\ \frac{6}{5}\left(\frac{5}{6}k\right)&=\frac{6}{5}\left(\frac{10}{1}\right)\\ &\boxed{k=12} \end{aligned}$$

41. $$\begin{aligned} -\frac{7}{16}h&=21\\ -\frac{16}{7}\left(-\frac{7}{16}h\right)&=-\frac{16}{7}\left(\frac{21}{1}\right)\\ &\boxed{h=-48} \end{aligned}$$

43. $$\begin{aligned} -\frac{t}{3}+2&=6\\ -\frac{t}{3}+2-2&=6-2\\ -\frac{t}{3}&=4\\ \frac{-3}{1}\left(\frac{t}{3}\right)&=-3(4)\\ &\boxed{t=-12} \end{aligned}$$

45. $$\begin{aligned} 2(-3)+4y&=14\\ -6+4y&=14\\ -6+4y+6&=14+6\\ 4y&=20\\ \frac{4y}{4}&=\frac{20}{4}\\ &\boxed{y=5} \end{aligned}$$

47. $$\begin{aligned} 7(0)-4y&=17\\ 0-4y&=17\\ -4y&=17\\ \frac{-4y}{-4}&=\frac{17}{-4}\\ &\boxed{y=-\frac{17}{4}} \end{aligned}$$

49. $$\begin{aligned} 10.08&=4(0.5x+2.5)\\ 10.08&=2x+10\\ 10.08-10&=2x+10-10\\ 0.08&=2x\\ \frac{0.08}{2}&=\frac{2x}{2}\\ &\boxed{0.04=x} \end{aligned}$$

51.
$$-(4-m)=-10$$
$$-4+m=-10$$
$$-4+m+4=-10+4$$
$$\boxed{m=-6}$$

53.
$$15s+8-s=7+1$$
$$14s+8=8$$
$$14s+8-8=8-8$$
$$14s=0$$
$$\frac{14s}{14}=\frac{0}{14}$$
$$\boxed{s=0}$$

55.
$$-3(2y-2)-y=5$$
$$-6y+6-y=5$$
$$-7y+6=5$$
$$-7y+6-6=5-6$$
$$-7y=-1$$
$$\frac{-7y}{-7}=\frac{-1}{-7}$$
$$\boxed{y=\frac{1}{7}}$$

57.
$$3x-8-4x-7x=-2-8$$
$$-8x-8=-10$$
$$-8x-8+8=-10+8$$
$$-8x=-2$$
$$\frac{-8x}{-8}=\frac{-2}{-8}$$
$$\boxed{x=\frac{1}{4}}$$

59.
$$4(5b)+2(6b-1)=-34$$
$$20b+12b-2=-34$$
$$32b-2=-34$$
$$32b-2+2=-34+2$$
$$32b=-32$$
$$\frac{32b}{32}=\frac{-32}{32}$$
$$\boxed{b=-1}$$

61.
$$9(x+11)+5(13-x)=0$$
$$9x+99+65-5x=0$$
$$4x+164=0$$
$$4x+164-164=0-164$$
$$4x=-164$$
$$\frac{4x}{4}=\frac{-164}{4}$$
$$\boxed{x=-41}$$

63.
$$60r-50=15r-5$$
$$60r-50-15r=15r-5-15r$$
$$45r-50=-5$$
$$45r-50+50=-5+50$$
$$45r=45$$
$$\frac{45r}{45}=\frac{45}{45}$$
$$\boxed{r=1}$$

65.
$$8y-3=4y+15$$
$$8y-3-4y=4y+15-4y$$
$$4y-3=15$$
$$4y-3+3=15+3$$
$$4y=18$$
$$\frac{4y}{4}=\frac{18}{4}$$
$$\boxed{y=\frac{9}{2}}$$

67.
$$\begin{aligned} 5x+7.2&=4x\\ 5x+7.2-5x&=4x-5x\\ 7.2&=-x\\ \frac{7.2}{-1}&=\frac{-x}{-1}\\ \boxed{-7.2=x} \end{aligned}$$

69.
$$\begin{aligned} 8y+328&=4y\\ 8y+328-8y&=4y-8y\\ 328&=-4y\\ \frac{328}{-4}&=\frac{-4y}{-4}\\ \boxed{-82=y} \end{aligned}$$

71.
$$\begin{aligned} 15x&=x\\ 15x-x&=x-x\\ 14x&=0\\ \frac{14x}{14}&=\frac{0}{14}\\ \boxed{x=0} \end{aligned}$$

73.
$$\begin{aligned} 3(a+2)&=2(a-7)\\ 3a+6&=2a-14\\ 3a+6-2a&=2a-14-2a\\ a+6&=-14\\ a+6-6&=-14-6\\ \boxed{a=-20} \end{aligned}$$

75.
$$\begin{aligned} 2-3(x-5)&=4(x-1)\\ 2-3x+15&=4x-4\\ -3x+17&=4x-4\\ -3x+17+3x&=4x-4+3x\\ 17&=7x-4\\ 17+4&=7x-4+4\\ 21&=7x\\ \frac{21}{7}&=\frac{7x}{7}\\ \boxed{3=x} \end{aligned}$$

77.
$$\begin{aligned} \frac{x+5}{3}&=11\\ 3\left(\frac{x+5}{3}\right)&=3(11)\\ x+5&=33\\ x+5-5&=33-5\\ \boxed{x=28} \end{aligned}$$

79.
$$\begin{aligned} \frac{y}{6}+\frac{y}{4}&=-1\\ 12\left(\frac{y}{6}\right)+12\left(\frac{y}{4}\right)&=12(-1)\\ 2y+3y&=-12\\ 5y&=-12\\ \frac{5y}{5}&=\frac{-12}{5}\\ \boxed{y=\frac{-12}{5}} \end{aligned}$$

81.
$$\begin{aligned} -\frac{2}{9}&=\frac{5x}{6}-\frac{1}{3}\\ 18\left(-\frac{2}{9}\right)&=18\left(\frac{5x}{6}\right)-18\left(\frac{1}{3}\right)\\ -4&=15x-6\\ -4+6&=15x-6+6\\ 2&=15x\\ \frac{2}{15}&=\frac{15x}{15}\\ \boxed{\frac{2}{15}=x} \end{aligned}$$

83. $$\frac{2}{3}y+2=\frac{1}{5}+y$$
$$15\left(\frac{2}{3}y+2\right)=15\left(\frac{1}{5}+y\right)$$
$$10y+30=3+15y$$
$$10y+30-10y=3+15y-10y$$
$$30=3+5y$$
$$30-3=3+5y-3$$
$$27=5y$$
$$\frac{27}{5}=\frac{5y}{5}$$
$$\boxed{\frac{27}{5}=y}$$

85. $$-\frac{3}{4}n+2n=\frac{1}{2}n+\frac{13}{3}$$
$$12\left(-\frac{3}{4}n\right)+12(2n)=12\left(\frac{1}{2}n\right)+12\left(\frac{13}{3}\right)$$
$$-9n+24n=6n+52$$
$$15n=6n+52$$
$$15n-6n=6n+52-6n$$
$$9n=52$$
$$\frac{9n}{9}=\frac{52}{9}$$
$$\boxed{n=\frac{52}{9}}$$

87. $$\frac{10-5s}{3}=s$$
$$3\left(\frac{10-5s}{3}\right)=3(s)$$
$$10-5s=3s$$
$$10-5s+5s=3s+5s$$
$$10=8s$$
$$\frac{10}{8}=\frac{8x}{8}$$
$$\boxed{\frac{5}{4}=x}$$

89. $$\frac{5(1-x)}{6}=-x$$
$$6\left[\frac{5(1-x)}{6}\right]=6(-x)$$
$$5(1-x)=-6x$$
$$5-5x=-6x$$
$$5-5x+5x=-6x+5x$$
$$5=-x$$
$$\frac{5}{-1}=\frac{-x}{-1}$$
$$\boxed{-5=x}$$

91. $$\frac{3(d-8)}{4}=\frac{2(d+1)}{3}$$
$$\frac{3d-24}{4}=\frac{2d+2}{3}$$
$$12\left[\frac{3d-24}{4}\right]=12\left[\frac{2d+2}{3}\right]$$
$$3\left[3d-24\right]=4\left[2d+2\right]$$
$$9d-72=8d+8$$
$$9d-72+72=8d+8+72$$
$$9d=8d+80$$
$$9d-8d=8d+80-8d$$
$$\boxed{d=80}$$

93. $$\frac{1}{2}(x+3)+\frac{3}{4}(x-2)=x+1$$
$$\frac{1}{2}x+\frac{3}{2}+\frac{3}{4}x-\frac{6}{4}=x+1$$
$$\frac{4}{1}\left(\frac{1}{2}x\right)+\frac{4}{1}\left(\frac{3}{2}\right)+\frac{4}{1}\left(\frac{3}{4}x\right)-\frac{4}{1}\left(\frac{6}{4}\right)=4(x+1)$$
$$2x+6+3x-6=4x+4$$
$$5x=4x+4$$
$$5x-4x=4x+4-4x$$
$$\boxed{x=4}$$

95.

$$8x+3(2-x)=5(x+2)-4$$
$$8x+6-3x=5x+10-4$$
$$5x+6=5x+6$$
$$5x+6-5x=5x+6-5x$$
$$\boxed{6=6} \text{ true}$$

The terms involving x drop out and the result, 6 = 6, is true. This means **all real numbers** are a solution, and this equation is an **identity**.

97.

$$-3(s+2)=-2(s+4)-s$$
$$-3s-6=-2s-8-s$$
$$-3s-6=-3s-8$$
$$-3s-6+3s=-3s-8+3s$$
$$\boxed{-6=-8} \text{ false}$$

no solution

99.

$$2(3z+4)=2(3z-2)+13$$
$$6z+8=6z-4+13$$
$$6z+8=6z+9$$
$$6z+8-6z=6z+9-6z$$
$$\boxed{8=9} \text{ false}$$

no solution

101.

$$4(y-3)-y=3(y-4)$$
$$4y-12-y=3y-12$$
$$3y-12=3y-12$$
$$3y-12-3y=3y-12-3y$$
$$\boxed{-12=-12} \text{ true}$$

all real numbers

103.

$$\frac{h}{709}-23{,}898=-19{,}678$$
$$709\left(\frac{h}{709}\right)-709(23{,}898)=709(-19{,}678)$$
$$h-16{,}943{,}682=-13{,}951{,}702$$
$$\text{add } 16{,}943{,}682 \text{ to both sides}$$
$$+16{,}943{,}682=+16{,}943{,}682$$
$$\boxed{h=2{,}991{,}980}$$

WRITING

105. Answers will vary.

107. Answers will vary.

REVIEW

109. $-8-(-8)=-8+8$
$=0$

111. $\frac{1}{8}\bullet\frac{1}{8}=\frac{1}{64}$

113. $8x+8+8x-8=16x$

CHALLENGE PROBLEMS

115. $|x|=2$ or $x^2=4$ both have solutions of 2 and -2

SECTION 2.5

VOCABULARY

1. A **formula** is an equation that is used to state a known relationship between two or more variables.

3. The distance around a geometric figure is called its **perimeter**.

5. A segment drawn from the center of a circle to a point on the circle is called a **radius**.

7. The perimeter of a circle is called its **circumference**.

CONCEPTS

9. Use variables to write the formula relating the following:
   a) $d = rt$
   b) $r = c + m$
   c) $p = r - c$
   d) $I = Prt$
   e) $C = 2\pi r$

11. Complete the table.

	Rate	• time	= distance
Light	186,282 mi/sec	60 sec	**11,176,920 mi**
Sound	1,088 ft/sec	60 sec	**65,280 ft**

13. Tell which concept, perimeter, area, circumference, or volume should be used to find the following.
    a) volume
    b) circumference
    c) area
    d) perimeter

15. Write an expression for perimeter.
    a) $P = l + w + l + w$
    $P = (x+3)\text{cm} + 2\text{cm} + (x+3)\text{ cm} + 2\text{cm}$
    $\boxed{P = (2x + 10)\text{ cm}}$

    Write an expression for area.
    b) $A = lw$
    $A = 2\text{cm}\,(x + 3)\text{cm}$
    $\boxed{A = (2x + 6)\text{ cm}^2}$

NOTATION

17. Solve $Ax + By = C$ for y.

$$Ax + By = C$$
$$Ax + By - \boxed{Ax} = C - \boxed{Ax}$$
$$\boxed{By} = C - Ax$$
$$\frac{By}{\boxed{B}} = \frac{C - Ax}{\boxed{B}}$$
$$y = \frac{C - Ax}{B}$$

19.
   a) $\pi \doteq 3.14$
   b) 98π means $98 \cdot \pi$
   c) $V = \pi r^2 h$
   r represents the radius of the cylinder
   h represents the height of the cylinder

PRACTICE

21. SWIMMING
    d is 1,826 miles
    t is 743 hours
    r is ? mph

$$rt = d$$
$$r \cdot 743 = 1{,}826$$
$$\frac{r \cdot 743}{743} = \frac{1{,}826}{743}$$
$$\boxed{r = 2.45}$$

Rate is 2.5 mph.

23. HOLLYWOOD
    total receipts is $190 millions
    profits is $125 million
    cost is ? $

$$\text{cost} + \text{profit} = \text{total receipts}$$
$$\text{cost } + 125 = 190$$
$$\text{cost } + 125 - 125 = 190 - 125$$
$$\boxed{\text{cost } = \$65 \text{ million}}$$

25. ENTREPRENEURS

I is \$175
P is \$2,500
r is ?%
t is 2 years

$$P \bullet r \bullet t = I$$
$$2{,}500 \bullet r \bullet 2 = 175$$
$$\frac{2{,}500 \bullet r \bullet 2}{2{,}500 \bullet 2} = \frac{175}{2{,}500 \bullet 2}$$
$$r = 0.035$$
$$\boxed{r = 3.5\%}$$

Rate is 3.5%.

27. METALLURGY

C is $2{,}212°$

$$C = \frac{5}{9}(F - 32)$$
$$2{,}212 = \frac{5}{9}(F - 32)$$
$$\frac{9}{5}(2{,}212) = \frac{9}{5}\left[\frac{5}{9}(F - 32)\right]$$
$$\frac{19{,}908}{5} = F - 32$$
$$3981.6 + 32 = F - 32 + 32$$
$$4{,}013.6 = F$$
$$\boxed{4{,}014° = F}$$

The temperature is 4,014°F.

29. VALENTINE'S DAY

wholesale is \$12.95
retail is \$37.50
markup is \$?

wholesale + markup = retail
12.95 + markup = 37.50
12.95 + markup – 12.95 = 37.50 – 12.95
$\boxed{\text{markup} = \$24.55}$

31. YO-YOS

c is? inches
π is 3.14
r is 21 inches
Answers will vary due to approximation of π.

$$c = 2\pi r$$
$$c = 2 \bullet 3.14 \bullet 21$$
$$\boxed{c = 131.88}$$

The circumference is 132 in.

33. $E = IR$; for R

$$\frac{E}{I} = \frac{IR}{I}$$
$$\boxed{\frac{E}{I} = R}$$

35. $V = lwh$; for w

$$\frac{V}{lh} = \frac{lwh}{lh}$$
$$\boxed{\frac{V}{lh} = w}$$

37. $C = 2\pi r$; for r

$$\frac{C}{2\pi} = \frac{2\pi r}{2\pi}$$
$$\boxed{\frac{C}{2\pi} = r}$$

39. $A = \frac{Bh}{3}$; for h

$$3(A) = 3\left(\frac{Bh}{3}\right)$$
$$3A = Bh$$
$$\frac{3A}{B} = \frac{Bh}{B}$$
$$\boxed{\frac{3A}{B} = h}$$

41.

$$w=\frac{s}{f}\text{ ; for } f$$
$$f(w)=f\left(\frac{s}{f}\right)$$
$$fw=s$$
$$\frac{fw}{w}=\frac{s}{w}$$
$$\boxed{f=\frac{s}{w}}$$

43.

$$P=a+b+c\text{ , for } b$$
$$P=a+b+c$$
$$P-a-c=a+b+c-a-c$$
$$\boxed{P-a-c=b}$$

45.

$$T=2r+2t\text{ ; for } r$$
$$T=2r+2t$$
$$T-2t=2r+2t-2t$$
$$T-2t=2r$$
$$\frac{T-2t}{2}=\frac{2r}{2}$$
$$\boxed{\frac{T-2t}{2}=r}$$

47.

$$Ax+By=C\text{ ; for } x$$
$$Ax+By=C$$
$$Ax+By-By=C-By$$
$$Ax=C-By$$
$$\frac{Ax}{A}=\frac{C-By}{A}$$
$$\boxed{x=\frac{C-By}{A}}$$

49.

$$K=\frac{1}{2}mv^2\text{ ; for } m$$
$$K=\frac{1}{2}mv^2$$
$$2K=2\left(\frac{1}{2}mv^2\right)$$
$$2K=mv^2$$
$$\frac{2K}{v^2}=\frac{mv^2}{v^2}$$
$$\boxed{\frac{2K}{v^2}=m}$$

51.

$$A=\frac{a+b+c}{3}\text{ ; for } c$$
$$A=\frac{a+b+c}{3}$$
$$3A=3\left(\frac{a+b+c}{3}\right)$$
$$3A=a+b+c$$
$$3A-a-b=a+b+c-a-b$$
$$\boxed{3A-a-b=c}$$

53.

$$2E=\frac{T-t}{9}\text{ ; for } t$$
$$2E=\frac{T-t}{9}$$
$$9(2E)=9\left(\frac{T-t}{9}\right)$$
$$18E=T-t$$
$$18E+t=T-t+t$$
$$18E+t=T$$
$$18E+t-18E=T-18E$$
$$\boxed{t=T-18E}$$

55. $s = 4\pi r^2$; for r^2

$$s = 4\pi r^2$$

$$\frac{s}{4\pi} = \frac{4\pi r^2}{4\pi}$$

$$\boxed{\frac{s}{4\pi} = r^2}$$

57. $Kg = \frac{wv^2}{2}$; for v^2

$$Kg = \frac{wv^2}{2}$$

$$2(Kg) = 2\left(\frac{wv^2}{2}\right)$$

$$2Kg = wv^2$$

$$\frac{2Kg}{w} = \frac{wv^2}{w}$$

$$\boxed{\frac{2Kg}{w} = v^2}$$

59. $V = \frac{4}{3}\pi r^3$; for r^3

$$V = \frac{4}{3}\pi r^3$$

$$\frac{3}{4}(V) = \frac{3}{4}\left(\frac{4}{3}\pi r^3\right)$$

$$\frac{3V}{4} = \pi r^3$$

$$\frac{1}{\pi}\left(\frac{3V}{4}\right) = \frac{\pi r^3}{\pi}$$

$$\boxed{\frac{3V}{4\pi} = r^3}$$

61. $\frac{M}{2} - 9.9 = 2.1B$; for M

$$\frac{M}{2} - 9.9 = 2.1B$$

$$\frac{M}{2} - 9.9 + 9.9 = 2.1B + 9.9$$

$$\frac{M}{2} = 2.1B + 9.9$$

$$2\left(\frac{M}{2}\right) = 2(2.1B + 9.9)$$

$$\boxed{M = 4.2B + 19.8}$$

63. $S = 2\pi rh + 2\pi r^2$; for h

$$S = 2\pi rh + 2\pi r^2$$

$$S - 2\pi r^2 = 2\pi rh + 2\pi r^2 - 2\pi r^2$$

$$S - 2\pi r^2 = 2\pi rh$$

$$\frac{S - 2\pi r^2}{2\pi r} = \frac{2\pi rh}{2\pi r}$$

$$\boxed{\frac{S - 2\pi r^2}{2\pi r} = h}$$

Some students may attempt to cross out $2\pi r$. You can not divide out anything on the left side because the numerator contains two terms and the denominator contains a monomial factor.

65. $3x + y = 9$; for y

$$3x + y = 9$$

$$3x + y - 3x = 9 - 3x$$

$$\boxed{y = 9 - 3x}$$

67. $3y - 9 = x$; for y

$$3y - 9 = x$$

$$3y - 9 + 9 = x + 9$$

$$3y = x + 9$$

$$\frac{3y}{3} = \frac{x + 9}{3}$$

$$\boxed{y = \frac{1}{3}x + 3}$$

69. $4y + 16 = -3x$; for y

$$4y + 16 = -3x$$

$$4y + 16 - 16 = -3x - 16$$

$$4y = -3x - 16$$

$$\frac{4y}{4} = \frac{-3x - 16}{4}$$

$$\boxed{y = -\frac{3}{4}x - 4}$$

71. $A=\frac{1}{2}h(b+d)$; for b

$$A=\frac{1}{2}h(b+d)$$

$$2(A)=2\left(\frac{1}{2}h(b+d)\right)$$

$$2A=h(b+d)$$

$$\frac{2A}{h}=\frac{h(b+d)}{h}$$

$$\frac{2A}{h}=b+d$$

$$\frac{2A}{h}-d=b+d-d$$

$$\boxed{\frac{2A}{h}-d=b}$$

or

$$\boxed{b=\frac{2A-hd}{h}}$$

73. $\frac{7}{8}c+w=9$; for c

$$\frac{7}{8}c+w=9$$

$$\frac{7}{8}c+w-w=9-w$$

$$\frac{7}{8}c=9-w$$

$$\frac{8}{7}\left(\frac{7}{8}c\right)=\frac{8}{7}(9-w)$$

$$\boxed{c=\frac{72-8w}{7}}$$

APPLICATIONS

75. PROPERTIES OF WATER
In Fahrenheit, water freezes at 32° and boils at 212°.

In Celsius, water freezes at 0° and boils at 100°.

77. AVON PRODUCTS

Revenue	1,854.1	1,463.4
Cost of goods sold	679.5	506.5
Gross profit	**1,174.6**	**956.9**

79. CARPENTRY

$P=a+b+c$

$P=10+10+16$

$\boxed{P=36\text{ ft}}$

$A=½\,bh$

$A=½\cdot 16\cdot 6$

$\boxed{A=48\text{ ft}^2}$

81. ARCHERY

$C=2\pi r$

$C=2\cdot 3.14\cdot 8$

$C=50.24$

$\boxed{C=50.2\text{ in.}}$

Answers will vary due to the approximation of π.

$A=\pi r^2$

$A=3.14\cdot 8\cdot 8$

$A=200.96$

$\boxed{A=201.0\text{ in.}^2}$

83. LANDSCAPE

$P=a+b+c+d$

$P=10+12+10+24$

$\boxed{P=56\text{ in.}}$

$A=½\,h(B+b)$

$A=½\,8(24+12)$

$A=4(36)$

$\boxed{A=144\text{ in.}^2}$

85. MEMORIALS

Use the formula for the area of a triangle and multiply by 2.

$$A = 2\left(\frac{1}{2}bh\right)$$

$$A = bh$$

$$A = 245 \cdot 10$$

$$\boxed{A = 2{,}450 \text{ ft}^2}$$

87. RUBBER MEETS THE ROAD

$$P = 2l + 2w$$

$$P = 2 \cdot 6.375 + 2 \cdot 7.5$$

$$P = 12.75 + 15$$

$$\boxed{P = 27.75 \text{ in.}}$$

$$A = lw$$

$$A = 6.375 \cdot 7.5$$

$$\boxed{A = 47.8125 \text{ in.}^2}$$

89. FIREWOOD

$$A = lw$$

$$A = 8 \cdot 4$$

$$\boxed{A = 32 \text{ ft}^2}$$

$$V = lwh$$

$$V = 8 \cdot 4 \cdot 4$$

$$\boxed{V = 128 \text{ ft}^3}$$

91. IGLOOS

Use the formula for the volume of a sphere and multiply by ½.

$$V = \frac{1}{2}\left(\frac{4}{3}\pi r^3\right)$$

$$V = \frac{1}{2} \cdot \frac{4}{3} \cdot 3.14 \cdot 5.5 \cdot 5.5 \cdot 5.5$$

$$V = 348.28$$

$$\boxed{V = 348 \text{ ft}^3}$$

93. BARBECUING

$$A = \pi r^2$$

$$A = 3.14 \cdot 9 \cdot 9$$

$$A = 254.34$$

$$\boxed{A = 254 \text{ in.}^2}$$

Answers will vary due to the approximation of π.

95. PULLEYS

$L = 2D + 3.25(r + R)$; solve for D

$$L = 2D + 3.25(r + R)$$

$$L = 2D + 3.25r + 3.25R$$

$$L - 3.25r - 3.25R = 2D - 3.25r - 3.25R$$

$$L - 3.25r - 3.25R = 2D$$

$$\frac{L - 3.25r - 3.25R}{2} = \frac{2D}{2}$$

$$\boxed{\frac{L - 3.25r - 3.25R}{2} = D}$$

WRITING

97. Answers will vary.

98. Answers will vary.

REVIEW

99. Find 82% of 168.

$$0.82 \cdot 168$$

$$\boxed{137.76}$$

100. What percent of 200 is 30?

$$x \cdot 200 = 30$$

$$\frac{x \cdot 200}{200} = \frac{30}{200}$$

$$x = 0.15$$

$$\boxed{x = 15\%}$$

CHALLENGE PROBLEMS

101. $-7(\alpha-\beta)-(4\alpha-\theta)=\frac{\alpha}{2}$

solve for α

$$-7(\alpha-\beta)-(4\alpha-\theta)=\frac{\alpha}{2}$$

$$-7\alpha+7\beta-4\alpha+\theta=\frac{\alpha}{2}$$

$$-11\alpha+7\beta+\theta=\frac{\alpha}{2}$$

$$-11\alpha+7\beta+\theta+11\alpha=\frac{\alpha}{2}+11\alpha$$

$$7\beta+\theta=\frac{\alpha}{2}+11\alpha$$

$$7\beta+\theta=\frac{\alpha+22\alpha}{2}$$

$$7\beta+\theta=\frac{23\alpha}{2}$$

$$2(7\beta+\theta)=2\left(\frac{23\alpha}{2}\right)$$

$$14\beta+2\theta=23\alpha$$

$$\frac{14\beta+2\theta}{23}=\frac{23\alpha}{23}$$

$$\boxed{\frac{14\beta+2\theta}{23}=\alpha}$$

SECTION 2.6

VOCABULARY

1. The **perimeter** of a triangle or a rectangle is the distance around it.

3. The equal sides of an isosceles triangle meet to form the **vertex** angle. The angles opposite the equal sides are called **base** angles, and they have equal measures.

CONCEPTS

5. Pipe
   a) 17 ft = total length
      $x + 2$ = length of middle sized section
      $3x$ = length of longest section
   b) $x = 3$, short section
      $x + 2 = 3 + 2 = 5$ft, middle section
      $3x = 3(3) = 9$ ft, long section
      $3 + 5 + 9 = 17$ ft, total length of pipe

7. The sum of the measures of the angles of any triangle is **180°**.

9. Principal = $30,000
   rate = 14%
   time = 1 year

11. Complete the table.

	r •	t =	d
Husband	35	t	**35*t***
Wife	45	t	**45*t***

13. Complete the table.

a)

	Amount •	Strength =	Pure vinegar
Strong	x	0.06	**$0.06x$**
Weak	$10 - x$	0.03	**$0.03(10 - x)$**
Mixture	10	0.05	**0.05(10)**

b)

	Amount •	Strength =	Pure antifreeze
Strong	6	0.50	**0.50(6)**
Weak	x	0.25	**$0.25(x)$**
Mixture	$6 + x$	0.30	**$0.30(6 + x)$**

NOTATION

15. $100(0.08) = 8$
    To multiply a decimal by 100, move the decimal point two places to the right.

17. $2 \cdot 2w - 3 + 2w$ does not represent the perimeter of a rectangle because parentheses are needed $2 \cdot (2w - 3) + 2w$.

PRACTICE

19.
$$0.08x + 0.07(15{,}000 - x) = 1{,}110$$
$$0.08x + 1{,}050 - 0.07x = 1{,}110$$
$$0.01x + 1{,}050 = 1{,}110$$
subtract 1,050 from both sides
$$-1{,}050 = -1{,}050$$
$$0.01x = 60$$
$$\frac{0.01x}{0.01} = \frac{60}{0.01}$$
$$\boxed{x = 6{,}000}$$

APPLICATIONS

21. CARPENTRY

Short piece in feet	x
Long piece in feet	**$2x$**
Total feet	**12**

Let x = short piece in feet
$2x$ = long piece in feet
short + long = total
$$x + 2x = 12$$
$$3x = 12$$
$$\frac{3x}{3} = \frac{12}{3}$$
$$\boxed{x = 4}$$
$$2x = 2(4)$$
$$= 8$$
Short piece is 4 ft long.
Long piece is 8 ft long.

23. NATIONAL PARKS

Miles 1st day	x
Miles 2nd day	$x+6$
Miles 3rd day	$x+12$
Miles 4th day	$x+18$
Total miles	**444**

Let $x = 1^{st}$ day's travel in miles
$x+6 = 2^{nd}$ day's travel in miles
$x+12 = 3^{rd}$ day's travel in miles
$x+18 = 4^{th}$ day's travel in miles

$$1^{st} + 2^{nd} + 3^{rd} + 4^{th} = \text{total}$$
$$x+(x+6)+(x+12)+(x+18) = 444$$
$$4x+36 = 444$$
$$4x+36-36 = 444-36$$
$$4x = 408$$
$$\frac{4x}{4} = \frac{408}{4}$$
$$\boxed{x = 102}$$

1st day's travel is 102 miles
2nd day's travel is 108 miles
3rd day's travel is 114 miles
4th day's travel is 120 miles

25. TOURING

Weeks in Australia	x
Weeks in Japan	$x+4$
Weeks in Sweden	$x-2$
Total weeks	**38**

Let x = weeks in Australia
$x+4$ = weeks in Japan
x - 2 = weeks in Sweden

$$\text{Aus} + \text{Japan} + \text{Sweden} = 38$$
$$x+(x+4)+(x-2) = 38$$
$$3x+2 = 38$$
$$3x+2-2 = 38-2$$
$$3x = 36$$
$$\frac{3x}{3} = \frac{36}{3}$$
$$\boxed{x = 12}$$

12 weeks in Australia
16 weeks in Japan
10 weeks in Sweden

27. COUNTING CALORIES

Calories in ice cream	x
Calories in pie	$2x+100$
Total calories	**850**

Let x = calories in ice cream
$2x+100$ = calories in pie

$$\text{ice cream} + \text{pie} = \text{total}$$
$$x+(2x+100) = 850$$
$$3x+100 = 850$$
$$3x+100-100 = 850-100$$
$$3x = 750$$
$$\frac{3x}{3} = \frac{750}{3}$$
$$\boxed{x = 250}$$

250 calories in ice cream.
600 calories in pie.

29. ACCOUNTING

1st quarter income	x
2nd quarter income	$x+119$
3rd quarter income	$(x+119)-40$ $x+79$
4th quarter income	$7.7x$
Total income	**1,375**

Let $x = 1^{st}$ qt income in millions
$x+119 = 2^{nd}$ qt income in millions
$x+79 = 3^{rd}$ qt income in millions
$7.7x = 4^{th}$ qt income in millions

$$1^{st} + 2^{nd} + 3^{rd} + 4^{th} = \text{total}$$
$$x+(x+119)+(x+79)+7.7x = 1{,}375$$
$$10.7x+198 = 1{,}375$$
$$10.7x+198-198 = 1{,}375-198$$
$$10.7x = 1{,}177$$
$$\frac{10.7x}{10.7} = \frac{1{,}177}{10.7}$$
$$\boxed{x = 110}$$

1st qt income is $110 million
2nd qt income is $229 million
3rd qt income is $189 million
4th qt income is $847 million

31. ENGINEERING

3rd side in ft	x
1st equal side in ft	$x-4$
2nd equal side in ft	$x-4$
Perimeter in ft	**25**

Let $x = 3^{rd}$ side in ft
$x - 4$ = one equal side in ft
$x - 4$ = other equal side in ft

$$1^{st} + 2^{nd} + 3^{rd} = P$$
$$(x-4)+(x-4)+x = 25$$
$$3x - 8 = 25$$
$$3x - 8 + 8 = 25 + 8$$
$$3x = 33$$
$$\frac{3x}{3} = \frac{33}{3}$$
$$\boxed{x = 11}$$

3rd side is 11 ft.
One equal side is 7 ft.
Other equal side is 7 ft.

33. SWIMMING POOLS

width of pool in m	w
length of pool in m	$6w+30$
Perimeter in m	**1,110**

Let w = width in meters
$6w+30$ = length in meters

$$2l + 2w = P$$
$$2(6w+30)+2w = 1{,}110$$
$$12w + 60 + 2w = 1{,}110$$
$$14w + 60 = 1{,}110$$
$$14w + 60 - 60 = 1{,}110 - 60$$
$$14w = 1{,}050$$
$$\frac{14w}{14} = \frac{1{,}050}{14}$$
$$\boxed{w = 75}$$

Width is 75 meters.
Length is 480 meters.

35. TV TOWERS

Vertex angle (3rd angle)	x
1st base angle	$4x$
2nd base angle	$4x$
Total degrees	**180**

Let x = vertex angle
$4x$ = one base angle
$4x$ = other base angle

$$1^{st}\angle + 2^{nd}\angle + 3^{rd}\angle = \text{Total}$$
$$x + 4x + 4x = 180$$
$$9x = 180$$
$$\frac{9x}{9} = \frac{180}{9}$$
$$\boxed{x = 20}$$

The vertex angle is 20°.
One base angle is 80°.
Other base angle is 80°.

37. COMPLEMENTARY ANGLES

1st angle	$2x$
Complementary angle	$6x + 2$
Total degree	**90**

Let x = the number
$2x = 1^{st}$ angle
$6x + 2$ = complementary angle

$$1^{st}\angle + 2^{nd}\angle = \text{Total}$$
$$2x + (6x+2) = 90$$
$$8x + 2 = 90$$
$$8x + 2 - 2 = 90 - 2$$
$$8x = 88$$
$$\frac{8x}{8} = \frac{88}{8}$$
$$\boxed{x = 11}$$

The number is 11.
1st angle is 22°.
Complementary angle is 68°.

39. RENTALS

	#	• rent	= total
One bedroom	x	\$550	**550x**
Two bedroom	x	\$700	**700x**
Three bedroom	x	\$900	**900x**
All types			**\$36,550**

Let x = number of rentals

$$1 \text{ bed} + 2 \text{ bed} + 3 \text{ bed} = \text{total}$$
$$550x + 700x + 900x = 36,550$$
$$2,150x = 36,550$$
$$\frac{2,150x}{2,150} = \frac{36,550}{2,150}$$
$$\boxed{x = 17}$$

Number of rentals is 17.

41. SOFTWARE

	#	• cost	= total value
SS	x	\$150	**150x**
DB	x	\$195	**195x**
WP	$15 + 2x$	\$210	**210(2x + 15)**
All types			**\$72,000**

Let x = # of each sold for SS and DB
$2x + 15$ = # of WP sold

$$\text{SS val} + \text{BD val} + \text{WP val} = \text{total val}$$
$$150x + 195x + 210(2x + 15) = 72,000$$
$$150x + 195x + 420x + 3,150 = 72,000$$
$$765x + 3,150 = 72,000$$
$$765x + 3,150 - 3,150 = 72,000 - 3,150$$
$$765x = 68,850$$
$$\frac{765x}{765} = \frac{68,850}{765}$$
$$\boxed{x = 90}$$

90 SS were sold.

43. INVESTMENTS

	P •	r •	t =	I
7% int	x	0.07	1 yr	**0.07x**
8% int	x	0.08	1 yr	**0.08x**
10.5% int	x	0.105	1 yr	**0.105x**
T interest				**\$1,249.50**

Let x = the principal of each investment

$$7\% \text{ inv} + 8\% \text{ inv} + 10.5\% \text{ inv} = \text{Total inv}$$
$$0.07x + 0.08x + 0.105x = 1,249.50$$
$$0.255x = 1,249.50$$
$$\frac{0.255x}{0.255} = \frac{1,249.50}{0.255}$$
$$\boxed{x = 4,900}$$

\$4,900 was invested in each.

45. INVESTMENT PLANS

	P •	r •	t =	I
6.2% i	x	0.062	1 y	**0.062x**
12% i	65,000 - x	0.12	1 y	**0.12(65,000 – x)**
T i	65,000			**\$6,477.60**

Let x = amt invested at 6.2%
$65,000 - x$ = amt invested at 12%

$$6.2\% \text{ int} + 12\% \text{ int} = \text{Total int}$$
$$0.062x + 0.12(65,000 - x) = 6,477.60$$
$$0.062x + 7,800 - 0.12x = 6,477.60$$
$$-0.058x + 7,800 = 6,477.60$$
$$-0.058x + 7,800 - 7,800 = 6,477.60 - 7,800$$
$$-0.058x = -1,322.40$$
$$\frac{-0.058x}{-0.058} = \frac{-1,322.40}{-0.058}$$
$$\boxed{x = 22,800}$$

\$22,800 invested at 6.2%
\$42,200 invested at 12%

47. FINANCIAL PLANNING

	P •	r •	t =	I
11%	P	0.11	1	**0.11 *P***
13%	P	0.13	1	**0.13*P***
Total				**13% earns \$150 more**

Since the 13% investment earns \$150 more interest, add this amount to the 11% interest and then set them equal to each other.

Let P = amount invested in each

$$13\% \text{ int} = 11\% \text{ int} + 150$$
$$0.13P = 0.11P + 150$$
$$0.13P - 0.11P = 0.11P + 150 - 0.11P$$
$$0.02P = 150$$
$$\frac{0.02P}{0.02} = \frac{150}{0.02}$$
$$\boxed{P = 7{,}500}$$

\$7,500 invested in each

49. TORNADOS

	r •	t =	d
East van	20 mph	t	**20*t***
West van	25 mph	t	**25*t***
Total			**90 miles**

Let t = the time in hours each travels

East dist + West dist = total dist

$$20t + 25t = 90$$
$$45t = 90$$
$$\frac{45t}{45} = \frac{90}{45}$$
$$\boxed{t = 2}$$

2 hours each travels.

51. AIR TRAFFIC CONTROL

	r •	t =	d
1st plane	450 mph	t	**450*t***
2nd plane	500 mph	t	**500*t***
Total			**3,800 m**

Let t = the number of hours

1st dist + 2nd dist = total dist

$$450t + 500t = 3{,}800$$
$$950t = 3{,}800$$
$$\frac{950t}{950} = \frac{3{,}800}{950}$$
$$\boxed{t = 4}$$

4 hours each.

53. ROAD TRIPS

	r •	t =	d
slow part	40	t	**40*t***
fast part	50	$5 - t$	**50(5 – *t*)**
Total			**210 mi**

Let t = time of slow part in hours

slow dist + fast dist = total dist

$$40t + 50(5 - t) = 210$$
$$40t + 250 - 50t = 210$$
$$-10t + 250 = 210$$
$$-10t + 250 - 250 = 210 - 250$$
$$-10t = -40$$
$$\frac{-10t}{-10} = \frac{-40}{-10}$$
$$\boxed{t = 4}$$

It took 4 hours to travel the slow part.

55. PHOTOGRAPHIC CHEMICALS

	Amount	• Strength =	Pure acid
Weak	2	5%	**0.05(2)**
Strong	x	10%	**0.10(x)**
Mixture	$x+2$	7%	**0.07(x + 2)**

Let x = number of liters of 10% acetic acid

weak acid + strong acid = mixture

$$0.05(2)+0.10x=0.07(x+2)$$
$$0.1+0.10x=0.07x+0.14$$
$$0.1+0.10x-0.1=0.07x+0.14-0.1$$
$$0.10x=0.07x+0.04$$
$$0.10x-0.07x=0.07x+0.04-0.07x$$
$$0.03x=0.04$$
$$\frac{0.03x}{0.03}=\frac{0.04}{0.03}$$
$$\boxed{x=1\frac{1}{3}}$$

$1\frac{1}{3}$ liters of 10% acetic acid

57. ANTISEPTIC SOLUTION

	Amount	• Strength =	Pure BC
BC solution	30	10%	**0.10(30)**
Water	x	0%	**0.0(x)**
Mixture	$x+30$	8%	**0.08(x + 30)**

Let x = number of ounces of water

BC pure + water = Mixture pure

$$0.10(30)+0.0x=0.08(x+30)$$
$$3=0.08x+2.4$$
$$3-2.4=0.08x+2.4-2.4$$
$$0.6=0.08x$$
$$\frac{0.6}{0.08}=\frac{0.08x}{0.08}$$
$$\boxed{7.5=x}$$

7.5 ounces of water.

59. MIXING FUELS

	Amount	• Value =	Total value
$1.15 gas	x	1.15	**1.15(x)**
$0.85 gas	20	0.85	**20(0.85)**
Mixture	$x+20$	1.00	**1.00(x + 20)**

Let x = number of gallons of $1.15 gas

$1.15 val + $0.85 val = mixture val

$$1.15x+0.85(20)=1.00(x+20)$$
$$1.15x+17=x+20$$
$$1.15x+17-17=x+20-17$$
$$1.15x=x+3$$
$$1.15x-x=x+3-x$$
$$0.15x=3$$
$$\frac{0.15x}{0.15}=\frac{3}{0.15}$$
$$\boxed{x=20}$$

20 gallons of $1.15 gas

61. BLENDING GRASS SEED

	Amount	• Value =	Total value
Blue grass	100	$6	**$6(100)**
Rye grass	x	$3	**$3x**
Mixture	$x+100$	$5	**$5(x + 100)**

Let x = number of lbs of Ryegrass

Blue val + Rye val = Mixture val

$$6(100)+3x=5(x+100)$$
$$600+3x=5x+500$$
$$600+3x-500=5x+500-500$$
$$3x+100=5x$$
$$3x+100-3x=5x-3x$$
$$100=2x$$
$$\frac{100}{2}=\frac{2x}{2}$$
$$\boxed{50=x}$$

50 lbs of Ryegrass

63.MIXING CANDY

	Amount	• Value	= Total value
Lemon Drops	x	\$1.90	**\$1.90x**
Jelly Beans	$100-x$	\$1.20	**\$1.20(100–x)**
Mixture	100	\$1.48	**\$1.48(100)**

Let x = number of lbs. of lemon drops
$100-x$ = number of lbs. of jelly beans

LD val + JB val = Mixture val

$$1.90x+1.20(100-x)=1.48(100)$$
$$1.90x+120-1.20x=148$$
$$0.7x+120=148$$
$$0.7x+120-120=148-120$$
$$0.7x=28$$
$$\frac{0.7x}{0.7}=\frac{28}{0.7}$$
$$\boxed{x=40}$$

40 lbs. of lemon drops
60 lbs. of jelly beans

CHALLENGE PROBLEMS

75. EVAPORATION

	Amount	• Strength =	Pure BC
Salt solution	300	2%	**0.02(300)**
Water	x	0%	**0.0(x)**
Mixture	300-x	3%	**0.03(300-x)**

Let x = ml of water to be boiled away

Old pure - water = New pure

$$300(0.02)+0.0x=0.03(300-x)$$
$$6=9-0.03x$$
$$6-9=9-0.03x-9$$
$$-3=-0.03x$$
$$\frac{-3}{-0.03}=\frac{-0.03x}{-0.03}$$
$$\boxed{100=x}$$

100 ml of water to be boiled away.

WRITING

65. Answers will vary.

67. Answers will vary.

REVIEW

69. $-25(2x-5)=-50x+125$

71. $-(-3x-3)=3x+3$

73. $(4y-4)(4)=16y-16$

SECTION 2.7

VOCABULARY

1. An **inequality** is a statement that contains one of the following symbols: >, ≥, <, or ≤

3. To **solve** an inequality means to find all the values of the variable that make the inequality true.

5. The solution set of $x > 2$ can be expressed in **interval** notation as (2, ∞).

CONCEPTS

7.
   a) $35 \geq 34$, **true**
   b) $-16 \leq -17$, **false**
   c) $\frac{3}{4} \leq 0.75$, **true**
   d) $-0.6 \geq -0.5$, **false**

9.
   a) $17 \geq -2$, **$-2 \leq 17$**
   b) $32 < x$, **$x > 32$**

11. Fill in the blanks.
   a) Adding the same number to, or subtracting the **same** number from, both sides of an inequality does not change the inequality.

   b) Multiplying or dividing both sides of an inequality by the same **positive** number does not change the solutions.

13. Solve $x + 2 > 10$ and give the solution sets:

$$x+2>10$$
$$x+2-2>10-2$$
$$x>8$$

   a) ←——(——→ 8

   b) (8, ∞)

   c) All real numbers greater than 8.

15. To solve compound inequalities, the properties of inequalities are applied to all **three** parts of the inequality.

NOTATION

17.
   a) The symbol < means "**is less than**,"

$$x+2>10$$
$$x+2-2>10-2$$
$$x>8$$

   and the symbol > means " **is greater than**."

   b) The symbol ≥ means "**is greater than** or equal to," and symbol ≤ means "is less than **or equal to**."

19. Give an example of each symbol: bracket, parenthesis, infinity, negative infinity.

   [, (, ∞, -∞

21. Complete the solution to solve each inequality.

$$4x-5 \geq 7$$
$$4x-5+\boxed{5} \geq 7+\boxed{5}$$
$$4x \geq \boxed{12}$$
$$\frac{4x}{\boxed{4}} \geq \frac{12}{\boxed{4}}$$
$$x \geq 3$$

   Solution set: $[\boxed{3}, \infty)$

PRACTICE

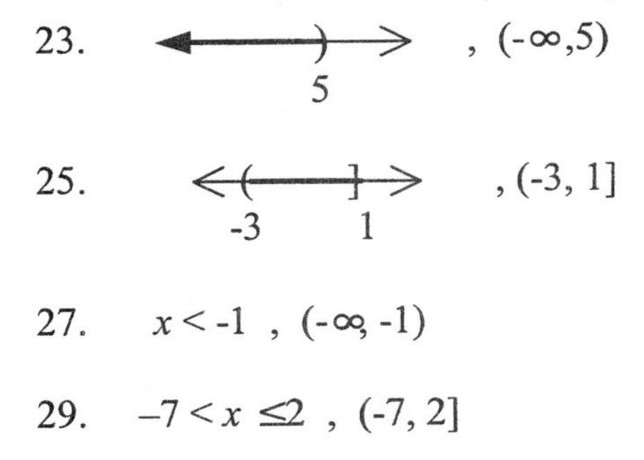

23. ←——)——→ 5 , (-∞,5)

25. ←(——]——→ -3 1 , (-3, 1]

27. $x < -1$, (-∞, -1)

29. $-7 < x \leq 2$, (-7, 2]

31.

$$x + 2 > 5$$
$$x + 2 - 2 > 5 - 2$$
$$x > 3$$
$$(3, \infty)$$

3

33.

$$3 + x < 2$$
$$3 + x - 3 < 2 - 3$$
$$x < -1$$
$$(-\infty, -1)$$

-1

35.

$$g - 30 \geq -20$$
$$g - 30 + 30 \geq -20 + 30$$
$$x \geq 10$$
$$[10, \infty)$$

10

37.

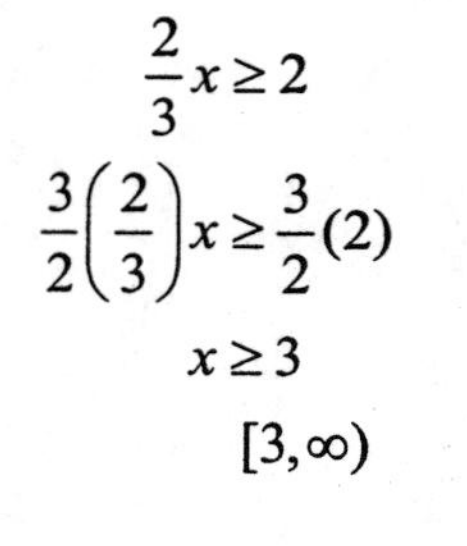

$$\frac{2}{3}x \geq 2$$
$$\frac{3}{2}\left(\frac{2}{3}\right)x \geq \frac{3}{2}(2)$$
$$x \geq 3$$
$$[3, \infty)$$

3

39.

$$\frac{y}{4} + 1 \leq -9$$
$$\frac{y}{4} + 1 - 1 \leq -9 - 1$$
$$\frac{y}{4} \leq -10$$
$$\frac{4}{1}\left(\frac{y}{4}\right) \leq 4(-10)$$
$$y \leq -40$$
$$(-\infty, -40]$$

-40

41.

$$7x - 1 > 5$$
$$7x - 1 + 1 > 5 + 1$$
$$7x > 6$$
$$\frac{7x}{7} > \frac{6}{7}$$
$$x > \frac{6}{7}$$
$$\left(\frac{6}{7}, \infty\right)$$

$\frac{6}{7}$

43.

$$0.5 \geq 2x - 0.3$$
$$0.5 + 0.3 \geq 2x - 0.3 + 0.3$$
$$0.8 \geq 2x$$
$$\frac{0.8}{2} \geq \frac{2x}{2}$$
$$0.4 \geq x$$
$$x \leq 0.4$$
$$(-\infty, 0.4]$$

0.4

45.

$$-30y \le -600$$
$$\frac{-30y}{-30} \ge \frac{-600}{-30}$$
$$y \ge 20$$
$$[20, \infty)$$

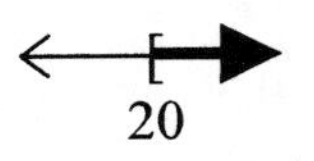

47.

$$-\frac{7}{8}x \le 21$$
$$-\frac{8}{7}\left(-\frac{7}{8}x\right) \ge -\frac{8}{7}(21)$$
$$x \ge -24$$
$$[-24, \infty)$$

-24

49.

$$-1 \le -\frac{1}{2}n$$
$$-\frac{2}{1}(-1) \ge -\frac{2}{1}\left(-\frac{1}{2}n\right)$$
$$2 \ge n$$
$$n \le 2$$
$$(-\infty, 2]$$

2

51.

$$\frac{m}{-42} - 1 > -1$$
$$\frac{m}{-42} - 1 + 1 > -1 + 1$$
$$\frac{m}{-42} > 0$$
$$\frac{-42}{1}\left(\frac{m}{-42}\right) < \frac{-42}{1}(0)$$
$$m < 0$$
$$(-\infty, 0)$$

0

53.

$$-x - 3 \le 7$$
$$-x - 3 + 3 \le 7 + 3$$
$$-x \le 10$$
$$\frac{-x}{-1} \ge \frac{10}{-1}$$
$$x \ge -10$$
$$[-10, \infty)$$

-10

55.

$$-3x - 7 > -1$$
$$-3x - 7 + 7 > -1 + 7$$
$$-3x > 6$$
$$\frac{-3x}{-3} < \frac{6}{-3}$$
$$x < -2$$
$$(-\infty, -2)$$

-2

57.

$$-4x + 6 > 17$$
$$-4x + 6 - 6 > 17 - 6$$
$$-4x > 11$$
$$\frac{-4x}{-4} < \frac{11}{-4}$$
$$x < -\frac{11}{4}$$
$$\left(-\infty, -\frac{11}{4}\right)$$

$-\frac{11}{4}$

59.

$$2x+9\le x+8$$
$$2x+9-9\le x+8-9$$
$$2x\le x-1$$
$$2x-x\le x-1-x$$
$$x\le -1$$
$$(-\infty,-1]$$

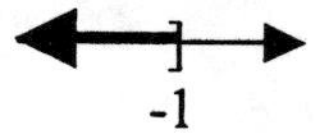

-1

61.

$$9x+13\ge 8x$$
$$9x+13-13\ge 8x-13$$
$$9x\ge 8x-13$$
$$9x-8x\ge 8x-13-8x$$
$$x\ge -13$$
$$[-13,\infty)$$

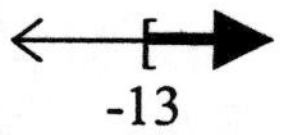

-13

63.

$$8x+4>-(3x-4)$$
$$8x+4>-3x+4$$
$$8x+4-4>-3x+4-4$$
$$8x>-3x$$
$$8x+3x>-3x+3x$$
$$11x>0$$
$$\frac{11x}{11}>\frac{0}{11}$$
$$x>0$$
$$(0,\infty)$$

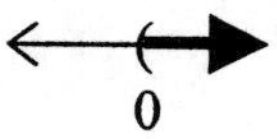

0

65.

$$0.4x+0.4\le 0.1x+0.85$$
$$0.4x+0.4-0.4\le 0.1x+0.85-0.4$$
$$0.4x\le 0.1x+0.45$$
$$0.4x-0.1\le 0.1x+0.45-0.1x$$
$$0.3x\le 0.45$$
$$\frac{0.3x}{0.3}\le\frac{0.45}{0.3}$$
$$x\le 1.5$$
$$(-\infty,1.5]$$

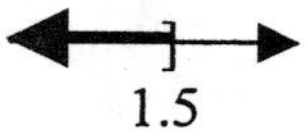

1.5

67.

$$7<\frac{5}{3}a-3$$
$$7+3<\frac{5}{3}a-3+3$$
$$10<\frac{5}{3}a$$
$$\frac{3}{5}(10)<\frac{3}{5}\left(\frac{5}{3}a\right)$$
$$6<a$$
$$a>6$$
$$(6,\infty)$$

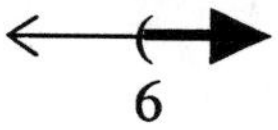

6

69.

$$7-x\le 3x-2$$
$$7-x+2\le 3x-2+2$$
$$9-x\le 3x$$
$$9-x+x\le 3x+x$$
$$9\le 4x$$
$$\frac{9}{4}\le\frac{4x}{4}$$
$$\frac{9}{4}\le x$$
$$x\ge\frac{9}{4}$$
$$\left[\frac{9}{4},\infty\right)$$

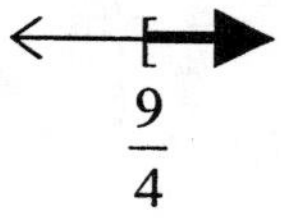

$\frac{9}{4}$

71.

$$8(5-x) \le 10(8-x)$$
$$40-8x \le 80-10x$$
$$40-8x-40 \le 80-10x-40$$
$$-8x \le 40-10x$$
$$-8x+10x \le 40-10x+10x$$
$$2x \le 40$$
$$\frac{2x}{2} \le \frac{40}{2}$$
$$x \le 20$$
$$(-\infty, 20]$$

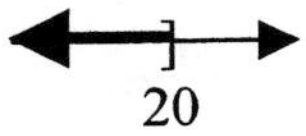

20

73.

$$\frac{1}{2}+\frac{n}{5} > \frac{3}{4}$$
$$20\left(\frac{1}{2}\right)+20\left(\frac{n}{5}\right) > 20\left(\frac{3}{4}\right)$$
$$10+4n > 15$$
$$10+4n-10 > 15-10$$
$$4n > 5$$
$$\frac{4n}{4} > \frac{5}{4}$$
$$n > \frac{5}{4}$$
$$\left(\frac{5}{4}, \infty\right)$$

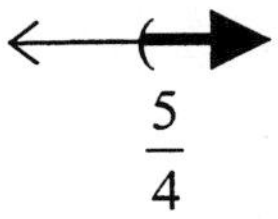

$\frac{5}{4}$

75.

$$-\frac{2}{3} \ge \frac{2y}{3}-\frac{3}{4}$$
$$12\left(-\frac{2}{3}\right) \ge 12\left(\frac{2y}{3}\right)-12\left(\frac{3}{4}\right)$$
$$-8 \ge 8y-9$$
$$-8+9 \ge 8y-9+9$$
$$1 \ge 8y$$
$$\frac{1}{8} \ge \frac{8y}{8}$$
$$\frac{1}{8} \ge y$$
$$y \le \frac{1}{8}$$
$$\left(-\infty, \frac{1}{8}\right]$$

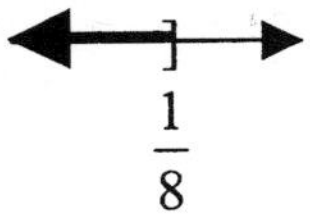

$\frac{1}{8}$

77.

$$\frac{6x+1}{4} \le x+1$$
$$4\left(\frac{6x+1}{4}\right) \le 4(x+1)$$
$$6x+1 \le 4x+4$$
$$6x+1-1 \le 4x+4-1$$
$$6x \le 4x+3$$
$$6x-4x \le 4x+3-4x$$
$$2x \le 3$$
$$\frac{2x}{2} \le \frac{3}{2}$$
$$x \le \frac{3}{2}$$
$$\left(-\infty, \frac{3}{2}\right]$$

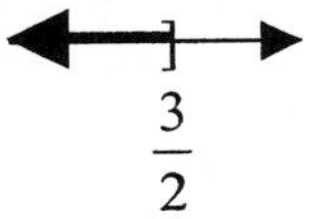

$\frac{3}{2}$

79.

$$\frac{5}{2}(7x-15)+x \ge \frac{13}{2}x-\frac{3}{2}$$
$$2\left(\frac{5}{2}(7x-15)\right)+2(x) \ge 2\left(\frac{13}{2}x\right)-2\left(\frac{3}{2}\right)$$
$$5(7x-15)+2x \ge 13x-3$$
$$35x-75+2x \ge 13x-3$$
$$37x-75 \ge 13x-3$$
$$37x-75+75 \ge 13x-3+75$$
$$37x \ge 13x+72$$
$$37x-13x \ge 13x+72-13x$$
$$24x \ge 72$$
$$\frac{24x}{24} \ge \frac{72}{24}$$
$$x \ge 3$$
$$[3,\infty)$$

3

81.

$$2 < x-5 < 5$$
$$2+5 < x-5+5 < 5+5$$
$$7 < x < 10$$
$$(7,10)$$

7 10

83.

$$0 \le x+10 \le 10$$
$$0-10 \le x+10-10 \le 10-10$$
$$-10 \le x \le 0$$
$$[-10,0]$$

-10 0

85.

$$-3 \le \frac{c}{2} \le 5$$
$$2(-3) \le 2\left(\frac{c}{2}\right) \le 2(5)$$
$$-6 \le c \le 10$$
$$[-6,10]$$

-6 10

87.

$$3 \le 2x-1 < 5$$
$$3+1 \le 2x-1+1 < 5+1$$
$$4 \le 2x < 6$$
$$\frac{4}{2} \le \frac{2x}{2} < \frac{6}{2}$$
$$2 \le x < 3$$
$$[2,3)$$

2 3

89.

$$4 < -2x < 10$$
$$-\frac{1}{2}(4) > -\frac{1}{2}(-2x) > -\frac{1}{2}(10)$$
$$-2 > x > -5$$
$$-5 < x < -2$$
$$(-5,-2)$$

-5 -2

91.

$$0 < 10-5x \le 15$$
$$0-10 < 10-5x-10 \le 15-10$$
$$-10 < -5x \le 5$$
$$\frac{-10}{-5} > \frac{-5x}{-5} \ge \frac{5}{-5}$$
$$2 > x \ge -1$$
$$-1 \le x < 2$$
$$[-1,2)$$

-1 2

93.

$$9(0.05-0.3x)+0.162 \le 0.081+15x$$
$$0.45-2.7x+0.162 \le 0.081+15x$$
$$-2.7x+0.612-0.612 \le 0.081+15x-0.612$$
$$-2.7x \le 15x-0.531$$
$$-2.7x-15x \le 15x-0.531-15x$$
$$-17.7x \le -0.531$$
$$\frac{-17.7x}{-17.7} \ge \frac{-0.531}{-17.7}$$
$$x \ge 0.03$$
$$[0.03,\infty)$$

0.03

APPLICATIONS

95. GRADES

Let x = % score needed on last exam

$$\frac{68+75+79+x}{4} \ge 80$$

$$\frac{222+x}{4} \ge 80$$

$$4\left(\frac{222+x}{4}\right) \ge 4(80)$$

$$222+x \ge 320$$

$$222+x-222 \ge 320-222$$

$$x \ge 98$$

97. FLEET AVERAGES

Let x = mpg for third model car

$$\frac{17+19+x}{3} \ge 21$$

$$\frac{36+x}{3} \ge 21$$

$$3\left(\frac{36+x}{3}\right) \ge 3(21)$$

$$36+x \ge 63$$

$$36+x-36 \ge 63-36$$

$$x \ge 27$$

99. GEOMETRY

Let s = length of one equal side

$$0 < 3s \le 57$$

$$\frac{0}{3} < \frac{3s}{3} \le \frac{57}{3}$$

$$0 \text{ ft} < s \le 19 \text{ ft}$$

101. COUNTER SPACE

Let x = the acceptable value

$$2x+2(x+5) \ge 150$$

$$2x+2x+10 \ge 150$$

$$4x+10 \ge 150$$

$$4x+10-10 \ge 150-10$$

$$4x \ge 140$$

$$\frac{4x}{4} \ge \frac{140}{4}$$

$$x \ge 35 \text{ ft}$$

103. SAFETY CODES

a) Ramps or inclines

$0^\circ < a \le 18^\circ$

b) Stairs

$18^\circ \le a \le 50^\circ$

c) Preferred range for stairs

$30^\circ \le a \le 37^\circ$

d) Ladders with cleats

$75^\circ \le a < 90^\circ$

105. WEIGHT CHART

a) 10 months old, "heavy"

$26 \text{ lb} \le w \le 31 \text{ lb}$

b) 5 months old, "light"

$12 \text{ lb} \le w \le 14 \text{ lb}$

c) 8 months old, "average"

$18.5 \text{ lb} \le w \le 20.5 \text{ lb}$

d) 3 months old, moderately light"

$11 \text{ lb} \le w \le 13 \text{ lb}$

WRITING

107. Answers will vary.

REVIEW

109.

$$-5^3 = -(5)(5)(5)$$

$$= -125$$

111. Complete the table

x	x^2-3
2	2^2-3
	$4-3$
	1
0	0^2-3
	$0-3$
	-3
3	3^2-3
	$9-3$
	6

CHALLENGE PROBLEMS

113.

$$3-x<5<7-x$$
$$3-x+x<5+x<7-x+x$$
$$3<5+x<7$$
$$3-5<5+x-5<7-5$$
$$-2<x<2$$
$$(-2,2)$$

-2 2

CHAPTER 2 KEY CONCEPTS

1.

a. $-3x+2+5x-10=2x-8$

b. $-3x+2+5x-10=4$

$$2x-8=4$$
$$2x-8+8=4+8$$
$$2x=12$$
$$\frac{2x}{2}=\frac{12}{2}$$
$$\boxed{x=6}$$

2.

a. $4(y+2)-3(y+1)=4y+8-3y-3$

$$=y+5$$

b. $4(y+2)=3(y+1)$

$$4y+8=3y+3$$
$$4y+8-8=3y+3-8$$
$$4y=3y-5$$
$$4y-3y=3y-5-3y$$
$$\boxed{y=-5}$$

3.

a. $\frac{1}{3}a+\frac{1}{3}a=\frac{2}{3}a$

b. $\frac{1}{3}a+\frac{1}{3}=\frac{1}{2}$

$$6\left(\frac{1}{3}a+\frac{1}{3}\right)=6\left(\frac{1}{2}\right)$$
$$2a+2=3$$
$$2a+2-2=3-2$$
$$2a=1$$
$$\frac{2a}{2}=\frac{1}{2}$$
$$\boxed{a=\frac{1}{2}}$$

4.

a. $-(2x+10)=-2x-10$

b. $-2x\geq-10$

$$\frac{-2x}{-2}\leq\frac{-10}{-2}$$
$$\boxed{x\leq5}$$

Reverse the symbol's direction.

5.

a. $\frac{2}{3}(x-2)-\frac{1}{6}(4x-8)$

$$=\frac{2}{3}x-\frac{4}{3}-\frac{4}{6}x+\frac{8}{6}$$
$$=\frac{4}{6}x-\frac{4}{6}x-\frac{8}{6}+\frac{8}{6}$$
$$=0$$

b. $\frac{2}{3}(x-2)-\frac{1}{6}(4x-8)=0$

$$\frac{2}{3}x-\frac{4}{3}-\frac{4}{6}x+\frac{8}{6}=0$$
$$\frac{4}{6}x-\frac{4}{6}x-\frac{8}{6}+\frac{8}{6}=0$$
$$\boxed{0=0}$$

all real numbers

6. The mistake is on the third line. The student made an equation out of the answer, which is $x-6$, by writing "0 =" on the left. Then the student solved the equation.

CHAPTER 2 REVIEW

SECTION 2.1
Solving Equations

1.

$$x - 34 = 50, \quad 84$$
$$84 - 34 \stackrel{?}{=} 50$$
$$50 = 50$$
yes

2.

$$5y + 2 = 12 \ , \ 3$$
$$5(3) + 2 \stackrel{?}{=} 12$$
$$15 + 2 \stackrel{?}{=} 12$$
no

3.

$$\frac{x}{5} = 6 \ , \ -30$$
$$\frac{-30}{5} \stackrel{?}{=} 6$$
$$-6 = 6$$
no

4.

$$a^2 - a - 1 = 0 \ , \ 2$$
$$(2)^2 - 2 - 1 \stackrel{?}{=} 0$$
$$4 - 2 - 1 \stackrel{?}{=} 0$$
$$1 \neq 0$$
no

5.

$$5b - 2 = 3b - 8, \ -3$$
$$5(-3) - 2 \stackrel{?}{=} 3(-3) - 8$$
$$-15 - 2 \stackrel{?}{=} -9 - 8$$
$$-17 = -17$$
yes

6.

$$\frac{2}{y+1} = \frac{12}{y+1} - 5 \ , \ 1$$
$$\frac{2}{1+1} \stackrel{?}{=} \frac{12}{1+1} - 5$$
$$\frac{2}{2} \stackrel{?}{=} \frac{12}{2} - 5$$
$$1 \stackrel{?}{=} 6 - 5$$
$$1 = 1$$
yes

7. To solve $x + 8 = 10$ means to find all the values of the **variable** that make the equation a **true** statement.

8.

$$x - 9 = 12$$
$$x - 9 + 9 = 12 + 9$$
$$\boxed{x = 21}$$
check
$$21 - 9 \stackrel{?}{=} 12$$
$$12 = 12$$

9.

$$y + 15 = -32$$
$$y + 15 - 15 = -32 - 15$$
$$\boxed{y = -47}$$
check
$$-47 + 15 \stackrel{?}{=} -32$$
$$-32 = -32$$

10.

$$a + 3.7 = 16.9$$
$$a + 3.7 - 3.7 = 16.9 - 3.7$$
$$\boxed{a = 13.2}$$
check
$$13.2 + 3.7 \stackrel{?}{=} 16.9$$
$$16.9 = 16.9$$

11.
$$100 = -7 + r$$
$$100 + 7 = -7 + r + 7$$
$$\boxed{107 = r}$$
check
$$100 \stackrel{?}{=} -7 + 107$$
$$100 = 100$$

12.
$$120 = 15c$$
$$\frac{120}{15} = \frac{15c}{15}$$
$$\boxed{8 = c}$$
check
$$120 \stackrel{?}{=} 15(8)$$
$$120 = 120$$

13.
$$t - \frac{1}{2} = \frac{1}{2}$$
$$t - \frac{1}{2} + \frac{1}{2} = \frac{1}{2} + \frac{1}{2}$$
$$\boxed{t = 1}$$
check
$$1 - \frac{1}{2} \stackrel{?}{=} \frac{1}{2}$$
$$\frac{1}{2} = \frac{1}{2}$$

14.
$$\frac{t}{8} = -12$$
$$8\left(\frac{t}{8}\right) = 8(-12)$$
$$\boxed{t = -96}$$
check
$$\frac{-96}{8} \stackrel{?}{=} -12$$
$$-12 = -12$$

15.
$$3 = \frac{q}{2.6}$$
$$2.6(3) = 2.6\left(\frac{q}{2.6}\right)$$
$$\boxed{7.8 = q}$$
check
$$3 \stackrel{?}{=} \frac{7.8}{2.6}$$
$$3 = 3$$

16.
$$6b = 0$$
$$\frac{6b}{6} = \frac{0}{6}$$
$$\boxed{b = 0}$$
check
$$6(0) \stackrel{?}{=} 0$$
$$0 = 0$$

17.
$$\frac{x}{14} = 0$$
$$14\left(\frac{x}{14}\right) = 14(0)$$
$$\boxed{x = 0}$$
check
$$\frac{0}{14} \stackrel{?}{=} 0$$
$$0 = 0$$

18.
$$x + 20 = 180$$
$$x + 20 - 20 = 180 - 20$$
$$\boxed{x = 160^\circ}$$
check
$$160 + 20 \stackrel{?}{=} 180$$
$$180 = 180$$

SECTION 2.2
Problem Solving

19.

goalkeepers	3
defenders	6
midfielders	6
forwards	x
Total	20

Let x = number of forwards

$$3+6+6+x=20$$
$$x+15=20$$
$$x+15-15=20-15$$
$$\boxed{x=5}$$

check

$$3+6+6+5\stackrel{?}{=}20$$
$$20=20$$

20.

To Philadelphia	296 mi
To Washington D.C.	133 mi
To Boston	x mi
Total	858 mi

Let x = miles back to Boston

$$296+133+x=858$$
$$x+429=858$$
$$x+429-429=858-429$$
$$\boxed{x=429}$$

check

$$296+133+429\stackrel{?}{=}858$$
$$858=858$$

21.

Let x = measure of one equal angle

$$6x=360$$
$$\frac{6x}{6}=\frac{360}{6}$$
$$\boxed{x=60}$$

check

$$6(60)\stackrel{?}{=}360$$
$$360=360$$

22.

Amount = Percent • Base

$A=23.5\% \cdot \$231$ billions

$A=0.235 \cdot \$231$ billion

$A=\$54.285$ billion

$\boxed{A=\$54 \text{ billion}}$

23.

Amount = Percent • Base

$A=3.5\% \cdot \$764$

$A=0.035 \cdot \$764$

$\boxed{A=\$26.74}$

24.

Amount = Percent • Base

$$4.81=2.5\% \cdot b$$
$$4.81=0.025 \cdot b$$
$$\frac{4.81}{0.025}=\frac{0.025b}{0.025}$$
$$\boxed{192.4=b}$$

25.

Amount = Percent • Base

$A=30\% \cdot \$1{,}890$

$A=0.3 \cdot \$1{,}890$

$\boxed{A=\$567}$

No, their present monthly payment of \$625 is more than 30% of their income.

26.

Find the difference between \$6 and \$100.

$$100-6=94$$

Now find the percent of increase.

$$\frac{94}{6}\approx 15.666$$
$$\approx 15.67$$
$$\approx 1{,}567\%$$

The increase is 1,567%.

SECTION 2.3
Simplifying Algebraic Expressions

27. $-4(7w) = -28w$

28. $-3r(-5) = 15r$

29. $3(-2x)(-4) = -6x(-4)$
$= 24x$

30. $0.4(5.2f) = 2.08f$

31. $15(\frac{3}{5}a) = 9a$

32. $\frac{7}{2} \bullet \frac{2}{7}r = r$

33. $5(x+3) = 5x+15$

34. $-2(2x+3-y) = -4x-6+2y$

35. $-(a-4) = -a+4$

36. $\frac{3}{4}(4c-8) = 3c-6$

37. $40\left(\frac{x}{2}+\frac{4}{5}\right) = 20x+32$

38. $2(-3c-7)(2.1) = (-6c-14)(2.1)$
$= -12.6c-29.4$

39. $8p+5p-4p = 13p-4p$
$= 9p$

40. $-5m+2-2m-2 = -5m-2m+2-2$
$= -7m$

41. $n+n+n+n = 4n$

42. $5(p-2)-2(3p+4) = 5p-10-6p-8$
$= 5p-6p-10-8$
$= -p-18$

43. $55.7k-55.6k = 0.1k$

44. $8a^3+4a^3-20a^3 = 12a^3-20a^3$
$= -8a^3$

45. $\frac{3}{5}w-\left(-\frac{2}{5}w\right) = \frac{3}{5}w+\frac{2}{5}w$
$= \frac{5}{5}w$
$= w$

46. $36\left(\frac{1}{9}h-\frac{3}{4}\right)+36\left(\frac{1}{3}\right) = 4h-27+12$
$= 4h-15$

47. $(x+7)$ ft $+(2x-3)$ ft $+x$ ft
$= [(x+2x+x)+(7-3)]$ ft
$= (4x+4)$ ft

SECTION 2.4
More about Solving Equations

48.
$$5x+4 = 14$$
$$5x+4-4 = 14-4$$
$$5x = 10$$
$$\frac{5x}{5} = \frac{10}{5}$$
$$\boxed{x = 2}$$

check

$$5(2)+4 \stackrel{?}{=} 14$$
$$10+4 \stackrel{?}{=} 14$$
$$14 = 14$$

49.
$$98.6-t = 129.2$$
$$98.6-t+t = 129.2+t$$
$$98.6 = 129.2+t$$
$$98.6-129.2 = 129.2+t-129.2$$
$$\boxed{-30.6 = t}$$

check

$$98.6-(-30.6) \stackrel{?}{=} 129.2$$
$$98.6+30.6 \stackrel{?}{=} 129.2$$
$$129.2 = 129.2$$

50.

$$\frac{n}{5}-2=4$$
$$\frac{n}{5}-2+2=4+2$$
$$\frac{n}{5}=6$$
$$5\left(\frac{n}{5}\right)=5(6)$$
$$\boxed{n=30}$$

check

$$\frac{30}{5}-2\stackrel{?}{=}4$$
$$6-2\stackrel{?}{=}4$$
$$4=4$$

51.

$$\frac{b-5}{4}=-6$$
$$4\left(\frac{b-5}{4}\right)=4(-6)$$
$$b-5=-24$$
$$b-5+5=-24+5$$
$$\boxed{b=-19}$$

check

$$\frac{-19-5}{4}\stackrel{?}{=}-6$$
$$\frac{-24}{4}\stackrel{?}{=}-6$$
$$-6=-6$$

52.

$$5(2x-4)-5x=0$$
$$10x-20-5x=0$$
$$5x-20=0$$
$$5x-20+20=0+20$$
$$5x=20$$
$$\frac{5x}{5}=\frac{20}{5}$$
$$\boxed{x=4}$$

check

$$5(2\bullet 4-4)-5\bullet 4\stackrel{?}{=}0$$
$$5(8-4)-20\stackrel{?}{=}0$$
$$5(4)-20\stackrel{?}{=}0$$
$$20-20\stackrel{?}{=}0$$
$$0=0$$

53.

$$-2(x-5)=5(-3x+4)+3$$
$$-2x+10=-15x+20+3$$
$$-2x+10=-15x+23$$
$$-2x+10-10=-15x+23-10$$
$$-2x=-15x+13$$
$$-2x+15x=-15x+13+15x$$
$$13x=13$$
$$\frac{13x}{13}=\frac{13}{13}$$
$$\boxed{x=1}$$

check

$$-2(1-5)\stackrel{?}{=}5(-3\bullet 1+4)+3$$
$$-2(-4)\stackrel{?}{=}5(1)+3$$
$$8=8$$

54.

$$\frac{3}{4}=\frac{1}{2}+\frac{d}{5}$$
$$20\left(\frac{3}{4}\right)=20\left(\frac{1}{2}\right)+20\left(\frac{d}{5}\right)$$
$$15=10+4d$$
$$15-10=10+4d-10$$
$$5=4d$$
$$\frac{5}{4}=\frac{4d}{4}$$
$$\boxed{\frac{5}{4}=d}$$

check

$$\frac{3}{4}\stackrel{?}{=}\frac{1}{2}+\frac{\frac{5}{4}}{5}$$
$$\frac{3}{4}\stackrel{?}{=}\frac{1}{2}+\left(\frac{5}{4}\bullet\frac{1}{5}\right)$$
$$\frac{3}{4}\stackrel{?}{=}\frac{1}{2}+\frac{1}{4}$$
$$\frac{3}{4}\stackrel{?}{=}\frac{2}{4}+\frac{1}{4}$$
$$\frac{3}{4}=\frac{3}{4}$$

55.

$$-\frac{2}{3}f=4$$
$$-\frac{3}{2}\left(-\frac{2}{3}f\right)=-\frac{3}{2}(4)$$
$$\boxed{f=-6}$$

check

$$-\frac{2}{3}(-6)\overset{?}{=}4$$
$$4=4$$

56.

$$\frac{3(2-c)}{2}=\frac{-2(2c+3)}{5}$$
$$10\left(\frac{3(2-c)}{2}\right)=10\left(\frac{-2(2c+3)}{5}\right)$$
$$5[3(2-c)]=2[-2(2c+3)]$$
$$15(2-c)=-4(2c+3)$$
$$30-15c=-8c-12$$
$$30-15c+15c=-8c-12+15c$$
$$30=7c-12$$
$$30+12=7c-12+12$$
$$42=7c$$
$$\frac{42}{7}=\frac{7c}{7}$$
$$\boxed{6=c}$$

check

$$\frac{3(2-6)}{2}\overset{?}{=}\frac{-2(2\cdot 6+3)}{5}$$
$$\frac{3(-4)}{2}\overset{?}{=}\frac{-2(15)}{5}$$
$$\frac{-12}{2}\overset{?}{=}\frac{-30}{5}$$
$$-6=-6$$

57.

$$\frac{b}{3}+\frac{11}{9}+3b=-\frac{5}{6}b$$
$$18\left(\frac{b}{3}\right)+18\left(\frac{11}{9}\right)+18(3b)=18\left(-\frac{5}{6}b\right)$$
$$6b+22+54b=-15b$$
$$60b+22=-15b$$
$$60b+22-22=-15b-22$$
$$60b=-15b-22$$
$$60b+15b=-15b-22+15b$$
$$75b=-22$$
$$\frac{75b}{75}=\frac{-22}{75}$$
$$\boxed{b=\frac{-22}{75}}$$

check

$$\frac{-\frac{22}{75}}{3}+\frac{11}{9}+3\left(-\frac{22}{75}\right)\overset{?}{=}-\frac{5}{6}\left(-\frac{22}{75}\right)$$
$$-\frac{22}{75}\left(\frac{1}{3}\right)+\frac{11}{9}-\frac{22}{25}\overset{?}{=}\frac{11}{45}$$
$$-\frac{22}{225}+\frac{11}{9}-\frac{22}{25}\overset{?}{=}\frac{11}{45}$$
$$225\left(-\frac{22}{225}+\frac{11}{9}-\frac{22}{25}\right)\overset{?}{=}225\left(\frac{11}{45}\right)$$
$$-22+275-198\overset{?}{=}55$$
$$55=55$$

58.

$$3(a+8)=6(a+4)-3a$$
$$3a+24=6a+24-3a$$
$$3a+24=3a+24$$
$$3a+24a=3a+24-3a$$
$$\boxed{24=24}$$

identity; all real numbers

59.

$$2(y+10)+y=3(y+8)$$
$$2y+20+y=3y+24$$
$$3y+20=3y+24$$
$$3y+20-3y=3y+24-3y$$
$$\boxed{20\neq 24}$$

contradiction; no solution

SECTION 2.5
Formulas

60.

$$\text{Wholesale} = \$219$$
$$\text{Markup} = ?$$
$$\text{Retail} = \$395$$
$$\text{WS} + \text{MUp} = \text{Retail}$$
$$219 + \text{MUp} = 395$$
$$219 + \text{MUp} - 219 = 395 - 219$$
$$\boxed{\text{MUp} = 176}$$

The markup is \$176.

61.

$$\text{Sales} = \$13{,}500$$
$$\text{Profit} = \$1{,}700$$
$$\text{Expenses} = ?$$
$$\text{Exp} + \text{Prof} = \text{Sales}$$
$$\text{Exp} + 1{,}700 = 13{,}500$$
$$\text{Exp} + 1{,}700 - 1{,}700 = 13{,}500 - 1{,}700$$
$$\boxed{\text{Exp} = 11{,}800}$$

The expenses are \$11,800.

62.

$d = 500$ miles

$r = 166.499$ mph

$t = ?$ hrs

$$rt = d$$
$$166.499t = 500$$
$$\frac{166.499t}{166.499} = \frac{500}{166.499}$$
$$t = 3.003$$
$$\boxed{t = 3.00}$$

The amount of time is about 3 hr.

63.

$$C^{o} = 1{,}056$$
$$\frac{5}{9}(F - 32) = C$$
$$\frac{5}{9}(F - 32) = 1{,}065$$
$$9\left(\frac{5}{9}(F - 32)\right) = 9(1{,}065)$$
$$5F - 160 = 9{,}585$$
$$5F - 160 + 160 = 9{,}585 + 160$$
$$5F = 9{,}745$$
$$\frac{5F}{5} = \frac{9{,}745}{5}$$
$$\boxed{F = 1{,}949}$$

The temperature is 1,949°F.

64.

$l = 60$ inches

$w = 24$ inches

$P = ?$ inches

$P = 2l + 2w$

$P = 2 \bullet 60 + 2 \bullet 24$

$P = 120 + 48$

$\boxed{P = 168}$

Perimeter is 168 inches.

65.

$l = 60$ inches

$w = 24$ inches

$A = ?$ sq in

$A = lw$

$A = 60 \bullet 24$

$\boxed{A = 1{,}440}$

Area is 1,440 in.^{2}

66.

$b = 17$ meters

$h = 9$ meters

$A = ?$ sq m

$A = \frac{1}{2}bh$

$A = 0.5 \bullet 17 \bullet 9$

$\boxed{A = 76.5}$

Area is 76.5 m^2.

67.

$b = 11$ inches

$d = 13$ inches

$h = 12$ inches

$A = ?$ sq in.

$A = \frac{1}{2}h(b+d)$

$A = 0.5 \cdot 12 \cdot (11+13)$

$A = 6(24)$

$\boxed{A = 144}$

Area is 144 in.^{2}

68.

$r = 8$ cm

$C = ?$ cm

$C = 2\pi r$

$C = 2 \cdot 3.14 \cdot 8$

$\boxed{C = 50.24}$

Circumference is 50.24 cm.

Answers will vary due to the approximation of π.

69.

$r = 8$ cm

$A = ?$ sq cm

$A = \pi r^2$

$A = 3.14 \cdot 8 \cdot 8$

$\boxed{A = 200.96}$

Area is 200.96 cm^2.

Answers will vary due to the approximation of π.

70.

$l = 60$ inches

$w = 24$ inches

$h = 3$ inches

$V = ?$ cu in.

$V = lwh$

$V = (60)(24)(3)$

$\boxed{V = 4{,}320}$

Volume is 4,320 in.^{3}

71.

$h = 12$ feet

$r = 0.5$ foot

$V = ?$ cu ft

$V = \pi r^2 h$

$V = 3.14 \cdot 0.5 \cdot 0.5 \cdot 12$

$V = 9.42$

$\boxed{V = 9.4}$

Volume is 9.4 ft^3.

Answers will vary due to the approximation of π.

72.

$b = 6$ feet

$h = 10$ feet

$V = ?$ cu ft

$V = \frac{1}{3}Bh$

$V = \frac{1}{3} \cdot 6 \cdot 6 \cdot 10$

$\boxed{V = 120}$

Volume is 120 ft^3.

73.

$d = 9$ inches

$r = 4.5$ inches

$V = ?$ cu in.

$V = \frac{4}{3}\pi r^3$

$V = \frac{4}{3} \cdot 3.14 \cdot 4.5 \cdot 4.5 \cdot 4.5$

$\boxed{V = 381.51}$

Volume is 381.51 in.^{3}

Answers will vary due to the approximation of π.

74.

$A = 2\pi rh$, for h

$A = 2\pi rh$

$\frac{A}{2\pi r} = \frac{2\pi rh}{2\pi r}$

$\boxed{\frac{A}{2\pi r} = h}$

75. $A - BC = \dfrac{G-K}{3}$, for G

$$A - BC = \frac{G-K}{3}$$
$$3(A - BC) = 3\left(\frac{G-K}{3}\right)$$
$$3(A - BC) = G - K$$
$$3A - 3BC + K = G - K + K$$
$$\boxed{3A - 3BC + K = G}$$

76. $a^2 + b^2 = c^2$, for b^2

$$a^2 + b^2 = c^2$$
$$a^2 + b^2 - a^2 = c^2 - a^2$$
$$\boxed{b^2 = c^2 - a^2}$$

77. $4y - 16 = 3x$, for y

$$4y - 16 = 3x$$
$$4y - 16 + 16 = 3x + 16$$
$$4y = 3x + 16$$
$$\frac{4y}{4} = \frac{3x+16}{4}$$
$$\boxed{y = \frac{3}{4}x + 4}$$

SECTION 2.6
More about Problem Solving

78.

one piece = 15 ft
other piece = x ft
last piece = $3x - 2$ ft
total = 45 feet

Let x = length of other piece in ft
$3x - 2$ = length of last piece in ft

$$x + (3x - 2) + 15 = 45$$
$$4x + 13 = 45$$
$$4x + 13 - 13 = 45 - 13$$
$$4x = 32$$
$$\frac{4x}{4} = \frac{32}{4}$$
$$\boxed{x = 8}$$

Shorter piece is 8 ft.

79.

	#	• cost	= value
Movie stars	x	\$250	\$250 x
TV celebrities	$x + 8$	\$75	\$75(x + 8)
total			\$1,900

Let x = # of movie star autographs
$x + 8$ = # of TV stars autographs

$$250x + 75(x + 8) = 1{,}900$$
$$250x + 75x + 600 = 1{,}900$$
$$325x + 600 = 1{,}900$$
$$325x + 600 - 600 = 1{,}900 - 600$$
$$325x = 1{,}300$$
$$\frac{325x}{325} = \frac{1{,}300}{325}$$
$$\boxed{x = 4}$$
$$\boxed{\begin{aligned} x + 8 &= 4 + 8 \\ &= 12 \end{aligned}}$$

4 movie star autographs.
12 TV stars autographs.

80.

l = 5 inches more than width
w = ? inches
$P = 109\frac{1}{2}$ = 109.5 inches

Let w = width of painting
$w + 5$ = length of painting

$$2x + 2(x + 5) = 109.5$$
$$2x + 2x + 10 = 109.5$$
$$4x + 10 = 109.5$$
$$4x + 10 - 10 = 109.5 - 10$$
$$4x = 99.5$$
$$\frac{4x}{4} = \frac{99.5}{4}$$
$$\boxed{x = 24.875}$$
$$\boxed{\begin{aligned} x + 5 &= 24.875 + 5 \\ &= 29.875 \end{aligned}}$$

Width is 24.875 inches.
Length is 29.875 inches.

81. base angle = ?°

vertex angle = 27°

total degrees = 180°

Let x = one base angle

$$x + x + 27 = 180$$
$$2x + 27 = 180$$
$$2x + 27 - 27 = 180 - 27$$
$$2x = 153$$
$$\frac{2x}{2} = \frac{153}{2}$$
$$\boxed{x = 76.5}$$

Both base angles are 76.5°.

82. $45x

83.

	P	• r	• t =	I
7%	P	0.07	1	**0.07*P***
9%	27,000 - P	0.09	1	**0.09(27,000 – *P*)**
Total interest				**$2,110**

Let P = amount invested at 7%

27,000 - P = amount invested at 9%

7% int + 9% int = Total int

$$0.07P + 0.09(27,000 - P) = 2,110$$
$$0.07P + 2,430 - 0.09P = 2,110$$
$$-0.02P + 2,430 = 2,110$$
$$-0.02P + 2,430 - 2,430 = 2,110 - 2,430$$
$$-0.02P = -320$$
$$\frac{-0.02P}{-0.02} = \frac{-320}{-0.02}$$
$$\boxed{P = 16,000}$$
$$\boxed{\begin{aligned} 27,000 - P &= 27,000 - 16,000 \\ &= 11,000 \end{aligned}}$$

$16,000 earned at 7%.

$11,000 earned at 9%.

84.

	r	• t =	d
Walker	3 mph	t	**3*t***
Bike rider	12 mph	t	**12*t***
Total			**5 miles**

Let t = the # of minutes they meet

1st dist + 2nd dist = total dist

$$3t + 12t = 5$$
$$15t = 5$$
$$\frac{15t}{15} = \frac{5}{15}$$
$$t = \frac{1}{3} \text{, convert}$$
$$\boxed{\begin{aligned} t &= \frac{1}{3} \bullet 60 \text{ minutes} \\ &= 20 \end{aligned}}$$

They will meet in 20 minutes.

85.

	Amount	• Value	= Total value
Candy	x	$0.90	**$0.90*x***
Gum Drops	$20 - x$	$1.50	**$1.50(20–*x*)**
Mixture	20	$1.20	**$1.20(20)**

Let x = number of lbs. of candy

20 - x = number of lbs. of gumdrops

Candy val + GD val = Mixture val

$$0.90x + 1.50(20 - x) = 1.20(20)$$
$$0.90x + 30 - 1.50x = 24$$
$$-0.60x + 30 = 24$$
$$-0.60x + 30 - 30 = 24 - 30$$
$$-0.60x = -6$$
$$\frac{-0.60x}{-0.60} = \frac{-6}{-0.60}$$
$$\boxed{x = 10}$$
$$\boxed{\begin{aligned} 20 - x &= 20 - 10 \\ &= 10 \end{aligned}}$$

10 of lbs. of each needed.

86.

	Amount	• Strength =	Pure salt
Acid	x	12%	**$0.12x$**

$0.12x$ gallon is pure acetic acid.

SECTION 2.7
Inequalities

87.

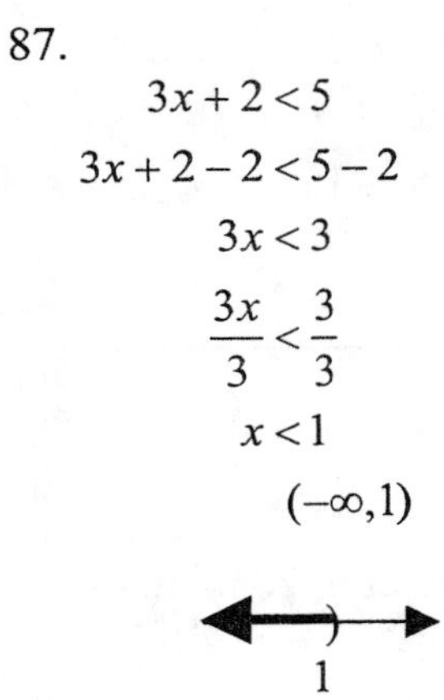

$$3x + 2 < 5$$
$$3x + 2 - 2 < 5 - 2$$
$$3x < 3$$
$$\frac{3x}{3} < \frac{3}{3}$$
$$x < 1$$
$$(-\infty, 1)$$

88.

$$-\frac{3}{4}x \geq -9$$
$$-\frac{4}{3}\left(-\frac{3}{4}x\right) \leq -\frac{4}{3}(-9)$$
$$x \leq 12$$
$$(-\infty, 12]$$

89.

$$\frac{3}{4} < \frac{d}{5} + \frac{1}{2}$$
$$20\left(\frac{3}{4}\right) < 20\left(\frac{d}{5}\right) + 20\left(\frac{1}{2}\right)$$
$$15 < 4d + 10$$
$$15 - 10 < 4d + 10 - 10$$
$$5 < 4d$$
$$\frac{5}{4} < \frac{4d}{4}$$
$$\frac{5}{4} < d$$
$$d > \frac{5}{4}$$
$$\left(\frac{5}{4}, \infty\right)$$

90.

$$5(3 - x) \leq 3(x - 3)$$
$$15 - 5x \leq 3x - 9$$
$$15 - 5x - 15 \leq 3x - 9 - 15$$
$$-5x \leq 3x - 24$$
$$-5x - 3x \leq 3x - 24 - 3x$$
$$-8x \leq -24$$
$$\frac{-8x}{-8} \geq \frac{-24}{-8}$$
$$x \geq 3$$
$$[3, \infty)$$

91.

$$8 < x + 2 < 13$$
$$8 - 2 < x + 2 - 2 < 13 - 2$$
$$6 < x < 11$$
$$(6, 11)$$

92.
$$0 \le 3-2x < 10$$
$$0-3 \le 3-2x-3 < 10-3$$
$$-3 \le -2x < 7$$
$$\frac{-3}{-2} \ge \frac{-2x}{-2} > \frac{7}{-2}$$
$$\frac{3}{2} \ge x > -\frac{7}{2}$$
$$-\frac{7}{2} < x \le \frac{3}{2}$$
$$\left(-\frac{7}{2}, \frac{3}{2}\right]$$

$-\frac{7}{2}$ $\frac{3}{2}$

93. $2.40 \text{ g} \le w \le 2.53 \text{ g}$

94. Let l = the maximum length

$w = 18$ inches

P = not greater than 132 inches
$$2l + 2w \le P$$
$$2l + 2(18) \le 132$$
$$2l + 36 \le 132$$
$$2l + 36 - 36 \le 132 - 36$$
$$2l \le 96$$
$$\frac{2l}{2} \le \frac{96}{2}$$
$$\boxed{l \le 48}$$
Length is ≤ 48 inches.

CHAPTER 2 TEST

1. Is 3 a solution of $x+3=4x-6$?

$$x+3=4x-6$$
$$3+3\overset{?}{=}4(3)-6$$
$$9\overset{?}{=}12-6$$
$$9\neq 6 \text{, no}$$

2. Let x = the angle

Angle + 60 = 180

$$x+60=180$$
$$x+60-60=180-60$$
$$\boxed{x=120}$$

The angle is 120°.

3. Let x = # of multiple births in 1997

$7x$ = # of multiple births in 2000

7,322 = total # of births of in 2000

$$7x = 7{,}322$$
$$\frac{7x}{7}=\frac{7{,}322}{7}$$
$$\boxed{x=1{,}046}$$

1,046 births in 1971.

4. Let x = selling price of house

$15\%x$ = down payment

\$11,400 = down payment

$$0.15x = 11{,}400$$
$$\frac{0.15x}{0.15}=\frac{11{,}400}{0.15}$$
$$\boxed{x=76{,}000}$$

Selling price of house is \$76,000.

5. Bulls win-lost record 72-10.

Find total games played. 82

Find games won. 72

Find winning percentage. $\frac{72}{82}$

$$\frac{72}{82}=0.8780$$
$$=0.878$$

Dolphins win-lost record 14-0.

Find total games played. 14

Find games won. 14

Find winning percentage. $\frac{14}{14}$

$$\frac{14}{14}=1.000$$

6. Starting temperature. 98.6°F

Ending temperature. 101.6°F

Find the difference. 3°F

Find percent increase.

$$\frac{3}{98.6}=0.0304$$
$$=0.03$$
$$=3\%$$

7.
$$P=2l+2w$$
$$l=(x+3)\text{ feet}$$
$$w=x\text{ feet}$$

$$P=2(x+3)+2x$$
$$P=2x+6+2x$$
$$\boxed{P=(4x+6)\text{ feet}}$$

8. $2(x+7)=2x+2(7)$

The distributive property.

9. $5(-4x) = -20x$

10. $-8(-7t)(4) = -224t$

11. $\frac{4}{5}(15a+5) - 16a = 12a + 4 - 16a$
$$= -4a + 4$$

12. $-1.1d^2 - 3.8d^2 - d^2 = -5.9d^2$

13.
$$5h + 8 - 3h + h = 8$$
$$3h + 8 = 8$$
$$3h + 8 - 8 = 8 - 8$$
$$3h = 0$$
$$\frac{3h}{3} = \frac{0}{3}$$
$$\boxed{h = 0}$$

14.
$$\frac{4}{5}t = -4$$
$$\frac{5}{4}\left(\frac{4}{5}t\right) = \frac{5}{4}(-4)$$
$$\boxed{t = -5}$$

15.
$$\frac{11(b-1)}{5} = 3b - 2$$
$$\frac{11b - 11}{5} = 3b - 2$$
$$5\left(\frac{11b-11}{5}\right) = 5(3b-2)$$
$$11b - 11 = 15b - 10$$
$$11b - 11 + 11 = 15b - 10 + 11$$
$$11b = 15b + 1$$
$$11b - 15b = 15b + 1 - 15b$$
$$-4b = 1$$
$$\frac{-4b}{-4} = \frac{1}{-4}$$
$$\boxed{b = -\frac{1}{4}}$$

16.
$$0.8x + 1.4 = 2.9 + 0.2x$$
$$0.8x + 1.4 - 1.4 = 2.9 + 0.2x - 1.4$$
$$0.8x = 1.5 + 0.2x$$
$$0.8x - 0.2x = 1.5 + 0.2x - 0.2x$$
$$0.6x = 1.5$$
$$\frac{0.6x}{0.6} = \frac{1.5}{0.6}$$
$$\boxed{x = 2.5}$$

17.
$$\frac{m}{2} - \frac{1}{3} = \frac{1}{4}$$
$$12\left(\frac{m}{2} - \frac{1}{3}\right) = 12\left(\frac{1}{4}\right)$$
$$6m - 4 = 3$$
$$6m - 4 + 4 = 3 + 4$$
$$6m = 7$$
$$\frac{6m}{6} = \frac{7}{6}$$
$$\boxed{m = \frac{7}{6}}$$

18.
$$23 - 5(x+10) = -12$$
$$23 - 5x - 50 = -12$$
$$-5x - 27 = -12$$
$$-5x - 27 + 27 = -12 + 27$$
$$-5x = 15$$
$$\frac{-5x}{-5} = \frac{15}{-5}$$
$$\boxed{x = -3}$$

19.
$$A = P + Prt, \text{ for } r$$
$$A - P = P + Prt - P$$
$$A - P = Prt$$
$$\frac{A-P}{Pt} = \frac{Prt}{Pt}$$
$$\boxed{\frac{A-P}{Pt} = r}$$

20.

Revenue = \$445

Cost = \$295

Profit = ?

Profit = Revenue − Cost

$P = 445 - 295$

$\boxed{P = 150}$

Profit is \$150.

21.

$F = 14°$

$C = \frac{5}{9}(F - 32)$

$C = \frac{5}{9}(14 - 32)$

$C = \frac{5}{9}(-18)$

$\boxed{C = -10}$

22.

$d = 10$ in.

$r = 5$ in.

$\text{tank} = \frac{3}{4}$ full

$V = \frac{4}{3}\pi r^3$

$V = \frac{3}{4}\left[\frac{4}{3}(3.14)(5 \text{ in.})^3\right]$

$V = (3.14)(125) \text{ in.}^3$

$V = 392.5$

$\boxed{V = 393}$

Volume is 393 in^3.

Answers may vary due to the approximation of π.

23.

	r •	t =	d
slow car	55	t	**55*t***
fast car	65	t	**65*t***
Total miles			**72**

Let t = time in hr for each car

slow dis + fast dis = total dis

$55t + 65t = 72$

$120t = 72$

$\frac{120t}{120} = \frac{72}{120}$

$\boxed{t = \frac{3}{5} \text{ hr}}$

24.

	Amount •	Strength =	Pure salt
Weak	x	2%	**0.02*x***
Strong	30	10%	**0.10(30)**
Mixture	$x + 30$	8%	**0.08(*x* + 30)**

Let x = # of liters of 2% brine

weak salt + strong salt = mixture

$0.02x + 0.10(30) = 0.08(x + 30)$

$0.02x + 3 = 0.08x + 2.4$

$0.02x + 3 - 2.4 = 0.08x + 2.4 - 2.4$

$0.02x + 0.6 = 0.08x$

$-0.02x + 0.6 - 0.02x = 0.08x - 0.02x$

$0.6 = 0.06x$

$\frac{0.6}{0.06} = \frac{0.06x}{0.06}$

$\boxed{10 = x}$

10 liters of 2% brine needed.

25.

Vertex angle (3rd angle)	**44**
1st base angle	x
2nd base angle	x
Total degrees	**180**

Let x = measure of both base angles

vertex $\angle$ + base $\angle$ + base $\angle$ = total

$$44+x+x=180$$
$$2x+44=180$$
$$2x+44-44=180-44$$
$$2x=136$$
$$\frac{2x}{2}=\frac{136}{2}$$
$$\boxed{x=68}$$

Each base angle is 68°.

26.

	P	r	t	I
6.2%	P	0.08	1	**0.08*P***
12%	13,750 - P	0.09	1	**0.09(13,750 – *P*)**
Total				**$1,185**

Let P = amount invested at 8%

8% int + 9% int = Total int

$$0.08P+0.09(13,750-P)=1,185$$
$$0.08P+1,237.50-0.09P=1,185$$
$$-0.01P+1,237.50=1,185$$
$$-0.01P+1,237.50-1,237.50=1,185-1,237.50$$
$$-0.01P=-52.50$$
$$\frac{-0.01P}{-0.01}=\frac{-52.5}{-0.01}$$
$$\boxed{P=5,250}$$

$5,250 invested at 8%

27.
$$-8x-20\le 4$$
$$-8x-20+20\le 4+20$$
$$-8x\le 24$$
$$\frac{-8x}{-8}\ge\frac{24}{-8}$$
$$x\ge -3$$
$$[-3,\infty)$$

–3

28.
$$-4\le 2(x+1)<10$$
$$\frac{-4}{2}\le\frac{2(x+1)}{2}<\frac{10}{2}$$
$$-2\le x+1<5$$
$$-2-1\le x+1-1<5-1$$
$$-3\le x<4$$
$$[-3,4)$$

–3 4

29. $1.497-0.001\le w\le 1.497+0.001$

$$\boxed{1.496\text{ in.}\le w\le 1.498\text{ in.}}$$

30.
$$2(y-7)-3y=-(y-3)-17$$
$$2y-14-3y=-y+3-17$$
$$-y-14=-y-14$$
$$-y-14+y=-y-14+y$$
$$\boxed{-14=-14},\ \text{true}$$

When the left side is equal to the right side, this means that any real number is a solution.

CHAPTER 2 CUMULATIVE REVIEW

1. a. $4m-3+2m$, expression
   b. $4m=3+2m$, equation

2.

Weight (lb)	Cooking times (hr)
15	$\frac{15}{5}=3$
20	$\frac{20}{5}=4$
25	$\frac{25}{5}=5$

3. Prime Factorization of 200.
   $2 \bullet 2 \bullet 2 \bullet 5 \bullet 5 = 2^3 \bullet 5^2$

4. $\frac{24}{36}=\frac{2}{3}$

5. $\frac{11}{21}\left(-\frac{14}{33}\right)=-\frac{2}{9}$

6. $\frac{3}{4}\div\frac{1}{8}=\frac{3}{4}\bullet\frac{8}{1}=6$
   6 - $\frac{1}{8}$ cups of flour.

7. $\frac{4}{5}+\frac{2}{3}=\frac{12}{15}+\frac{10}{15}$
   $=\frac{22}{15}$
   $=1\frac{7}{15}$

8. $42\frac{1}{8}-29\frac{2}{3}=42\frac{3}{24}-29\frac{16}{24}$
   $=41\frac{27}{24}-29\frac{16}{24}$
   $=12\frac{11}{24}$

9. $\frac{15}{16}=0.9375$

10. $0.45(100)=45$

11. a. $|-65|=65$
    b. $-|-12|=-12$

12. $x \bullet 5=5x$
    The commutative property of multiplication.

13. 3
    natural number, whole number, integer, rational number, real number

14. -1.95
    rational number, real number

15. $\frac{17}{20}$
    rational number, real number

16. π
    irrational number, real number

17. a. $4 \bullet 4 \bullet 4 = 4^3$
    b. $\pi \bullet r \bullet r \bullet h = \pi r^2 h$

18. a. $-6+(-12)+8=-18+8$
       $=-10$
    b. $-15-(-1)=-15+1$
       $=-14$
    c. $2(-32)=-64$
    d. $\frac{0}{35}=0$
    e. $\frac{-11}{11}=-1$

19. a. The sum of the width w and 12.
       $w + 12$
    b. Four less than a number n.
       $n - 4$

20. total days $= 4+8+3+2+0+5+4+6$
$=32$
number of people $= 8$
average number of sick days $= \frac{32}{8}$
$=4$

21.

x	$x^2 - 3$
-2	$(-2)^2 - 3$ $4 - 3$ 1
0	$(0)^2 - 3$ $0 - 3$ -3
3	$(3)^2 - 3$ $9 - 3$ 6

22. $l = \frac{2,000}{d^2}$
Answers may vary depending on the variable chosen.

23. a. $A = lw$
$A = (2 \text{ ft})(3 \text{ ft})$
$A = 32 \text{ ft}^2$

b. $A = \pi r^2$
$A = (3.14)(0.625 \text{ ft})(0.625 \text{ ft})$
$A = 1.22$
$A = 1.2 \text{ ft}^2$
Answers will vary due to the approximation of π.

c. Percent of area of red dot to area of the whole flag.
$\frac{1.2}{6} = 0.2$
$= 20\%$

24. 45 is 15% of what number?
$45 = 0.15 \cdot x$
$\frac{45}{0.15} = \frac{0.15x}{0.15}$
$300 = x$

25. Let $x = -5, y = 3$, and $z = 0$.
$(3x - 2y)z = [(3 \cdot -5) - (2 \cdot 3)]0$
$= 0$

26. Let $x = -5, y = 3$, and $z = 0$.
$\frac{x - 3y + |z|}{2 - x} = \frac{-5 - 3(3) + |0|}{2 - (-5)}$
$= \frac{-5 - 9 + 0}{2 + 5}$
$= \frac{-14}{7}$
$= -2$

27. Let $x = -5, y = 3$, and $z = 0$.
$x^2 - y^2 + z^2 = (-5)^2 - (3)^2 + 0^2$
$= 25 - 9$
$= 16$

28. Let $x = -5, y = 3$, and $z = 0$.
$\frac{x}{y} + \frac{y+2}{3-z} = \frac{-5}{3} + \frac{3+2}{3-0}$
$= \frac{-5}{3} + \frac{5}{3}$
$= 0$

29. $-8(4d) = -32d$

30. $5(2x - 3y + 1) = 10x - 15y + 5$

31. $2x + 3x - x = 4x$

32. $3a^2 + 6a^2 - 17a^2 = -8a^2$

33. $\frac{2}{3}(15t - 30) + t - 30 = 10t - 20 + t - 30$
$= 11t - 50$

34. $5(t - 4) + 3t = 5t - 20 + 3t$
$= 8t - 20$

35. $(x + 3)\text{ft}$

36. $(x + 3)\text{ft} + (x - 3)\text{ft} + x \text{ ft} = 3x \text{ ft}$

37.
$$3x-4=23$$
$$3x-4+4=23+4$$
$$3x=27$$
$$\frac{3x}{3}=\frac{27}{3}$$
$$x=9$$

38.
$$\frac{x}{5}+3=7$$
$$\frac{x}{5}+3-3=7-3$$
$$\frac{x}{5}=4$$
$$5\left(\frac{x}{5}\right)=5(4)$$
$$x=20$$

39.
$$-5p+0.7=3.7$$
$$-5p+0.7-0.7=3.7-0.7$$
$$-5p=3$$
$$\frac{-5p}{-5}=\frac{3}{-5}$$
$$p=-\frac{3}{5}$$
$$p=-0.6$$

40.
$$\frac{y-4}{5}=3-y$$
$$5\left(\frac{y-4}{5}\right)=5(3-y)$$
$$y-4=15-5y$$
$$y-4+4=15-5y+4$$
$$y=-5y+19$$
$$y+5y=-5y+19+5y$$
$$6y=19$$
$$\frac{6y}{6}=\frac{19}{6}$$
$$y=\frac{19}{6}$$

41.
$$-\frac{4}{5}x=16$$
$$5\left(-\frac{4}{5}x\right)=5(16)$$
$$-4x=80$$
$$\frac{-4x}{-4}=\frac{80}{-4}$$
$$x=-20$$

42.
$$\frac{1}{2}+\frac{x}{5}=\frac{3}{4}$$
$$20\left(\frac{1}{2}\right)+20\left(\frac{x}{5}\right)=20\left(\frac{3}{4}\right)$$
$$10+4x=15$$
$$10+4x-10=15-10$$
$$4x=5$$
$$\frac{4x}{4}=\frac{5}{4}$$
$$x=\frac{5}{4}$$

43.
$$-9(n+2)-2(n-3)=10$$
$$-9n-18-2n+6=10$$
$$-11n-12=10$$
$$-11n-12+12=10+12$$
$$-11n=22$$
$$\frac{-11n}{-11}=\frac{22}{-11}$$
$$n=-2$$

44.
$$\frac{2}{3}(r-2)=\frac{1}{6}(4r-1)+1$$
$$6\left(\frac{2}{3}(r-2)\right)=6\left(\frac{1}{6}(4r-1)\right)+6(1)$$
$$4(r-2)=(4r-1)+6$$
$$4r-8=4r-1+6$$
$$4r-8=4r+5$$
$$4r-8-4r=4r+5-4r$$
$$-8=5$$

no solution, contradiction

45. width = 5 meters

length = 13 meters

$$A = lw$$
$$A = 5(13)$$
$$A = 65 \text{ m}^2$$

46.

diameter = 12 cm

radius = 6 cm

height = 10 cm

$$V = \frac{1}{3}\pi r^2 h$$
$$V = \frac{1}{3}(3.14)(6 \text{ cm})(6 \text{ cm})(10 \text{ cm})$$
$$V = 376.80 \text{ cm}^3$$

Answers will vary due to the approximation of π.

47.

$$V = \frac{1}{3}\pi r^2 h$$
$$3(V) = 3\left(\frac{1}{3}\pi r^2 h\right)$$
$$3V = \pi r^2 h$$
$$\frac{3V}{\pi h} = \frac{\pi r^2 h}{\pi h}$$
$$\frac{3V}{\pi h} = r^2$$

48. distance = 3 ft.

force = 12.5 lb

work = ?

$$W = Fd$$
$$W = (12.5)(3)$$
$$W = 37.5 \text{ft-lb}$$

49. distance = 3 ft.

weight (Force) = ?

work = 28.35 ft-lb

$$W = Fd$$
$$28.35 = 3F$$
$$\frac{28.35}{3} = \frac{3F}{3}$$
$$9.45 = F$$

1 – gallon of paint weighs 9.45 lb.

50. both base angles = x

vertex angle = 70°

base ∠ + base ∠ + vertex ∠ = total

$$x + x + 70 = 180$$
$$2x + 70 = 180$$
$$2x + 70 - 70 = 180 - 70$$
$$2x = 110$$
$$\frac{2x}{2} = \frac{110}{2}$$
$$x = 55$$

The base angles are 55°.

51.

	P	• r	• t =	I
8%	P	0.08	1	**0.08*P***
9%	10,000 - P	0.09	1	**0.09(10,000 – *P*)**
T				**$860**

Let P = amount invested at 8%

$10{,}000 - P$ = amount invested at 9%

8% int + 9% int = Total int

$$0.08P + 0.09(10{,}000 - P) = 860$$
$$0.08P + 900 - 0.09P = 860$$
$$-0.01P + 900 = 860$$
$$-0.01P + 900 - 900 = 860 - 900$$
$$-0.01P = -40$$
$$\frac{-0.01P}{-0.01} = \frac{-40}{-0.01}$$
$$P = 4{,}000$$

$4,000 invested at 8%.

52.

	Amount	• Strength =	Pure gold
40%	x	40%	**0.40x**
10%	10	10%	**0.10(10)**
25%	$x + 10$	25%	**0.25(x + 10)**

Let x = oz of 40% gold

40% gold + 10% gold = 25% gold

$$0.40x + 0.10(10) = 0.25(x+10)$$
$$0.40x + 1 = 0.25x + 2.5$$
$$0.40x + 1 - 1 = 0.25x + 2.5 - 1$$
$$0.40x = 0.25x + 1.5$$
$$0.40x - 0.25x = 0.25x + 1.5 - 0.25x$$
$$0.15x = 1.5$$
$$\frac{0.15x}{0.15} = \frac{1.5}{0.15}$$
$$x = 10$$

10 oz of 40% gold needed.

53.

$$x - 4 > -6$$
$$x - 4 + 4 > -6 + 4$$
$$x > -2$$
$$(-2, \infty)$$

54.

$$-6x \geq -12$$
$$\frac{-6x}{-6} \leq \frac{-12}{-6}$$
$$x \leq 2$$
$$(-\infty, 2]$$

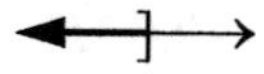

55.

$$8x + 4 \geq 5x + 1$$
$$8x + 4 - 4 \geq 5x + 1 - 4$$
$$8x \geq 5x - 3$$
$$8x - 5x \geq 5x - 3 - 5x$$
$$3x \geq -3$$
$$\frac{3x}{3} \geq \frac{-3}{3}$$
$$x \geq -1$$
$$[-1, \infty)$$

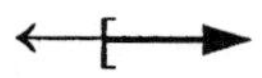

56.

$$-1 \leq 2x + 1 < 5$$
$$-1 - 1 \leq 2x + 1 - 1 < 5 - 1$$
$$-2 \leq 2x < 4$$
$$\frac{-2}{2} \leq \frac{2x}{2} < \frac{4}{2}$$
$$-1 \leq x < 2$$
$$[-1, 2)$$

SECTION 3.1

VOCABULARY

1. (-1, -5) is called an **ordered** pair.

3. A rectangular coordinate system is formed by two perpendicular number lines called the **x-axis** and the **y-axis**. The point where the axes cross is called the **origin**.

5. The point with coordinates (4, 2) can be graphed on a **rectangular** coordinate system.

CONCEPTS

7. a) To plot the point with coordinates (-5, 4), we start at the **origin** and move 5 units to the **left** and then move 4 units **up**.
   b) To plot the point with coordinates $\left(6,-\frac{3}{2}\right)$, we start at the **origin** and move 6 units to the **right** and then move $\frac{3}{2}$ units **down**.

9. a) Quadrants **I and II**.
   b) Quadrants **II and III**.
   c) Quadrant **II**.
   d) Quadrant **IV**.

11. 60 beats/min

13. 140 beats/min

15. 5 min and 50 min after starting

17. No difference

19. Europe sold about 17 million and the United States sold about 11 million so the difference is **about 6 million.**

NOTATION

21. (3, 5) is an ordered pair
    3(5) indicates multiplication
    5(3 +5) an expression containing grouping symbols

23. Yes, $\left(2.5,-\frac{7}{2}\right),\left(2\frac{1}{2},-3.5\right),\left(2.5,-3\frac{1}{2}\right)$ all name the same point because all of the x-values are equal to each other and all the y-values are equal to each other.

25. The 4 of the ordered pair (4, 5) is associated with the **horizontal** axis.

PRACTICE

27.

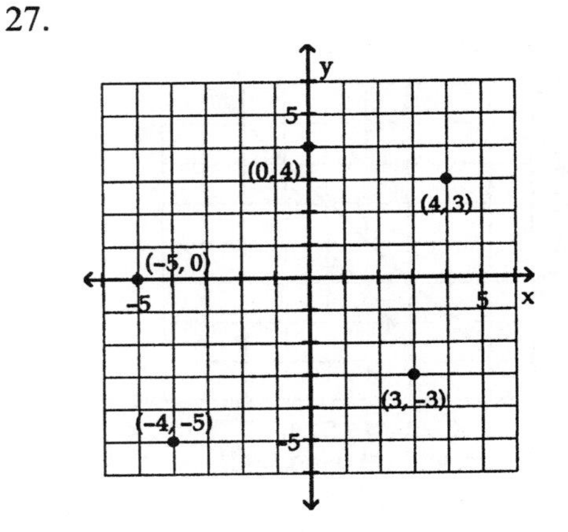

29.

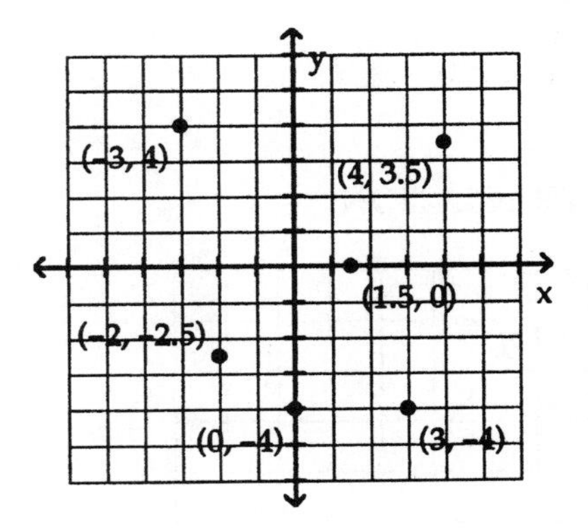

APPLICATIONS

31. CONSTRUCTION
    rivets: (-6, 0), (-2, 0), (2, 0), (6, 0)
    welds: (-4, 3), (0, 3), (4, 3)
    anchors: (-6, -3), (6, -3)

33. BATTLESHIP
    (E, 4), (F, 3), (G, 2)

35. MAPS
Rockford (5, B), Mount Carroll (1, C),
Harvard (7, A), intersection (5, E)

37. GEOMETRY
Fourth vertex is (2, 4)
Area = lw
Area = (3)(4)
Area = 12 sq. units

39. THE MILITARY

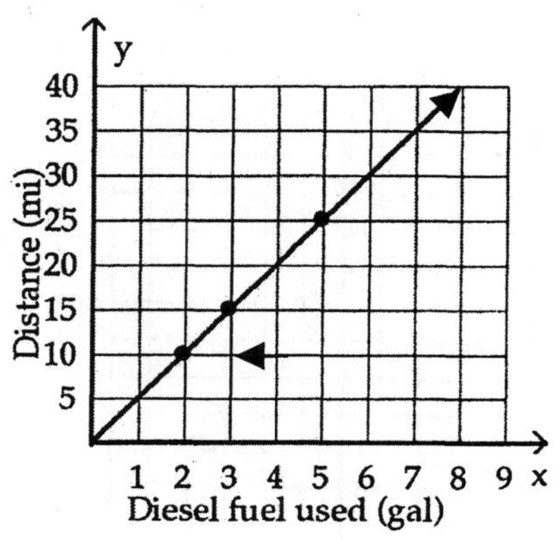

a) 35 mi
b) 4 gal
c) 32.5 mi

41. DEPRECIATION

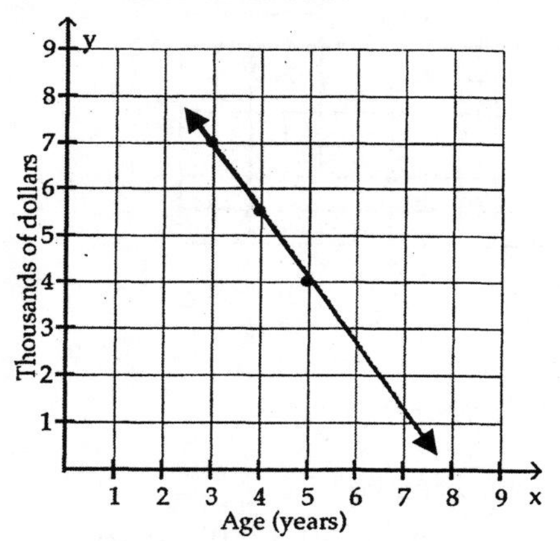

a) A 3-year old car is worth **$7,000**
b) Value of 7-year old car is **$1,000**
c) $2,500 worth after **6 yr**.

43. CENSUS
(1950, 152 million)
(1960, 179 million)
(1970, 202 million)
(1980, 227 million)
(1990, 252 million)
(2000, 277 million)

WRITING

45. Answers will vary.

47. Answers will vary.

REVIEW

49. Solve $AC = \frac{2}{3}h - T$ for h.

$$AC = \frac{2}{3}h - T$$

$$AC + \boldsymbol{T} = \frac{2}{3}h - T + \boldsymbol{T}$$

$$AC + T = \frac{2}{3}h$$

$$\left(\frac{3}{2}\right)(AC + T) = \left(\frac{3}{2}\right)\left(\frac{2}{3}h\right)$$

$$\boxed{\frac{3(AC+T)}{2} = h}$$

51. Evaluate: $\frac{-4(4+2)-2^3}{|-12-20|} = \frac{-4(6)-8}{|-32|}$

$$= \frac{-24-8}{|-32|}$$

$$= \frac{-32}{32}$$

$$\boxed{=-1}$$

CHALLENGE PROBLEMS

53. Sum of its coordinates is negative and product of its coordinates is positive. Such coordinates lie in **Quadrant III**.
Example: (-2, -5)
-2 + -5 = -10
(-2)(-5) = 10

SECTION 3.2

VOCABULARY

1. We say $y = 2x + 5$ is an equation in **two** variables, x and y.

3. Solutions of equations in two variables are often listed in a **table** of solutions.

5. The equation $y = 3x + 8$ is said to be **linear** because its graph is a line.

7. A linear equation in two variables has **infinitely** many solutions.

CONCEPTS

9. Consider the equation $y = -2x + 6$
   a) There **two** variables, $\boldsymbol{x}$ and $\boldsymbol{y}$.
   b) Yes, (4, -2) satisfies the equation.
   c) No, (-3, 0) is not a solution.
   d) Infinitely many.

11. Every point on the graph represents an ordered-pair **solution** of y = -2x –3 and every ordered-pair solution is a **point** on the graph.

13.
   a) At least one point is in error. The 3 points should lie on a straight line. One of the points is not correct. Check the computations.
   b) The line is too short. Arrowheads are not drawn.

15. Solve each equation for y.

a)
$$5y = 10x + 5$$
$$\frac{5y}{5} = \frac{10x}{5} + \frac{5}{5}$$
$$\boxed{y = 2x + 1}$$

b)
$$3y = -5x - 6$$
$$\frac{3y}{3} = \frac{-5x}{3} - \frac{6}{3}$$
$$\boxed{y = -\frac{5}{3}x - 2}$$

c)
$$-7y = -x + 21$$
$$\frac{-7y}{-7} = \frac{-x}{-7} + \frac{21}{-7}$$
$$\boxed{y = \frac{1}{7}x - 3}$$

NOTATION

17. Verify (-2,6) is a solution of $y = -x + 4$

$$y = -x + 4$$
$$\mathbf{6 \stackrel{?}{=} -(-2) + 4}$$
$$6 \stackrel{?}{=} 2 + 4$$
$$\mathbf{6 = 6}$$
(-2, 6) is a solution

19. Complete the labeling of the table of solutions and graph of $c = -a + 4$.

a	c	(a, c)
-1	5	(-1, 5)
0	4	(0, 4)
2	2	(2, 2)

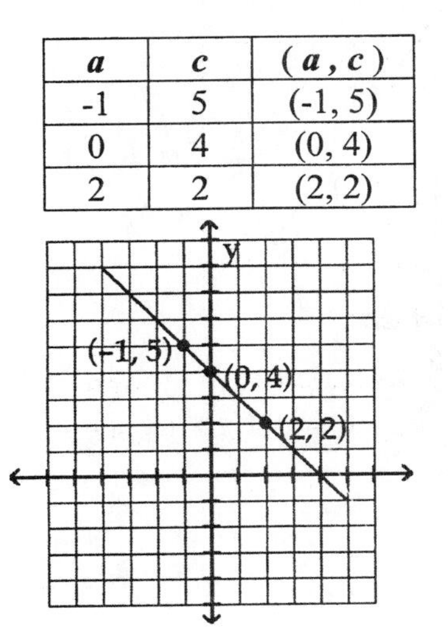

PRACTICE

21. $y = 5x - 4$; (1,1)
$1 \stackrel{?}{=} 5(1) - 4$
$1 \stackrel{?}{=} 5 - 4$
$1 = 1$
yes

23. $y = \left(-\frac{3}{4}\right)x + 8$; (-8, 12)
$12 \stackrel{?}{=} \left(-\frac{3}{4}\right)(-8) + 8$
$12 \stackrel{?}{=} 6 + 8$
$12 \stackrel{?}{=} 14$
no

25. $7x - 2y = 3$; (2, 6)
$7(2) - 2(6) \stackrel{?}{=} 3$
$14 - 12 \stackrel{?}{=} 3$
$2 \stackrel{?}{=} 3$
no

27. $x + 12y = -12$; (0, -1)
$0 + 12(-1) \stackrel{?}{=} -12$
$0 - 12 \stackrel{?}{=} -12$
$-12 \stackrel{?}{=} -12$
yes

29. $y = -5x - 4$; (-3, ?)
$y = -5(-3) - 4$
$y = 15 - 4$
$\mathbf{y = 11}$
(-3, **11**)

31. $4x - 5y = -4$; (?, 4)
$4x - 5(4) = -4$
$4x - 20 = -4$
$4x - 20 + 20 = -4 + 20$
$4x = 16$
$\mathbf{x = 4}$
(**4**, 4)

33. $y = 2x - 4$

x	y	(x, y)
8	$y = 2x - 4$ $y = 2(8) - 4$ $y = 16 - 4$ $\mathbf{y = 12}$	(8, **12**)
$y = 2x - 4$ $8 = 2x - 4$ $8+4 = 2x - 4+4$ $\mathbf{6 = x}$	8	(**6**, 8)

35. $3x - y = -2$

x	y	(x, y)
-5	**-13**	(-5, **-13**)
-1	-1	(**-1**, -1)

37. $y = 2x - 3$

x	y	(x, y)
-2	$y = 2x - 3$ $y = 2(-2) - 3$ $y = -4 - 3$ $y = -7$	(-2, -7)
0	$y = 2x - 3$ $y = 2(0) - 3$ $y = 0 - 3$ $y = -3$	(0, -3)
2	$y = 2x - 3$ $y = 2(2) - 3$ $y = 4 - 3$ $y = 1$	(2, 1)

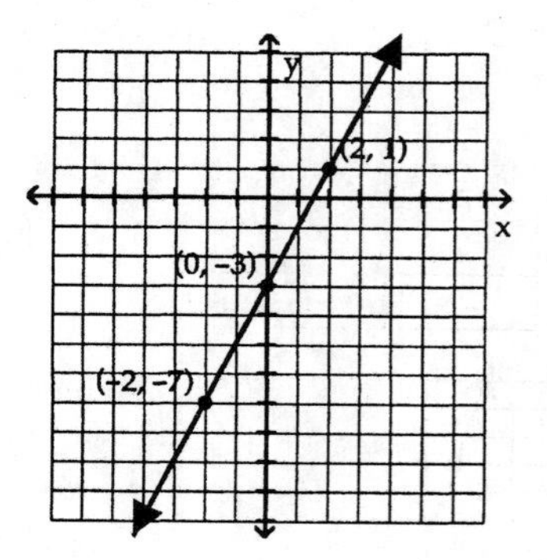

39. $y = 5x - 4$

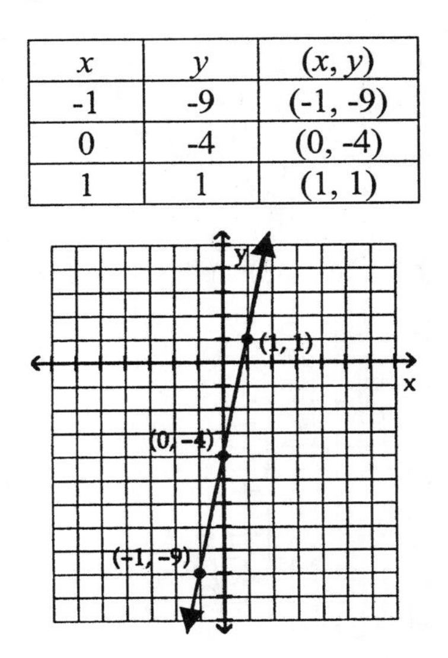

x	y	(x, y)
-1	-9	(-1, -9)
0	-4	(0, -4)
1	1	(1, 1)

41. $y = x$

x	y	(x, y)
-3	-3	(-3, -3)
0	0	(0, 0)
4	4	(4, 4)

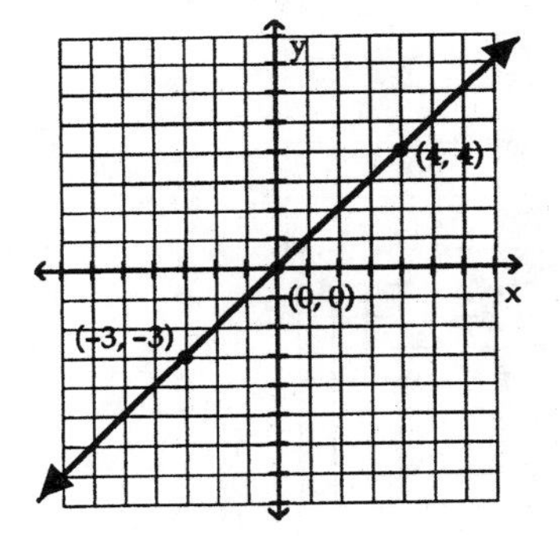

43. $y = -3x + 2$

x	y	(x, y)
-1	5	(-1, 5)
0	2	(0, 2)
1	-1	(1, -1)

45. $y = -x - 1$

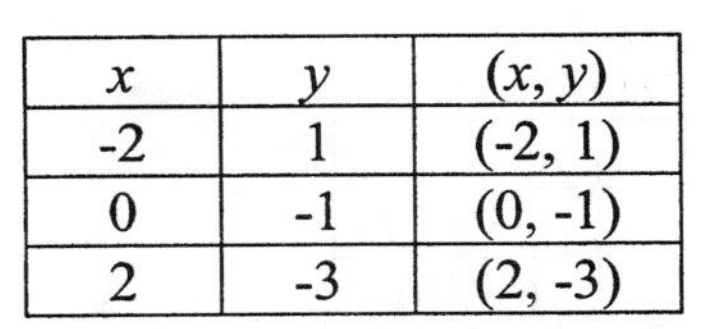

x	y	(x, y)
-2	1	(-2, 1)
0	-1	(0, -1)
2	-3	(2, -3)

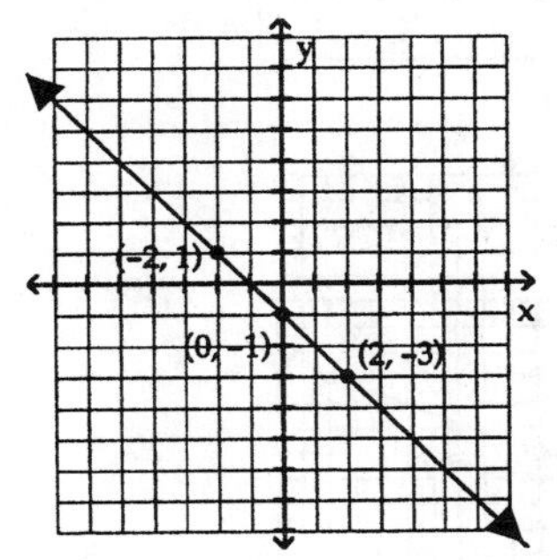

47. $y = \dfrac{x}{3}$

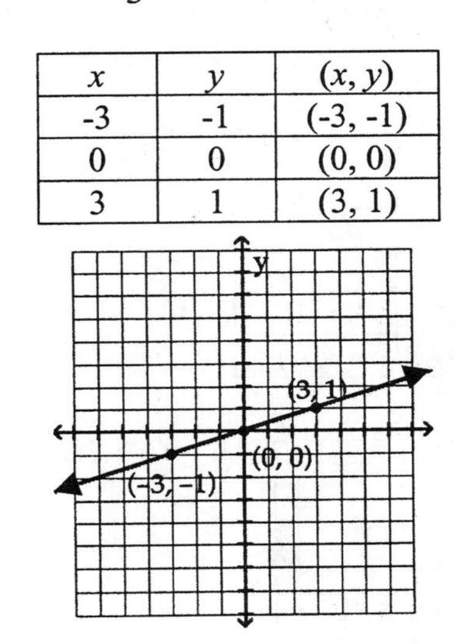

x	y	(x, y)
-3	-1	(-3, -1)
0	0	(0, 0)
3	1	(3, 1)

49. $y = -\dfrac{1}{2}x$

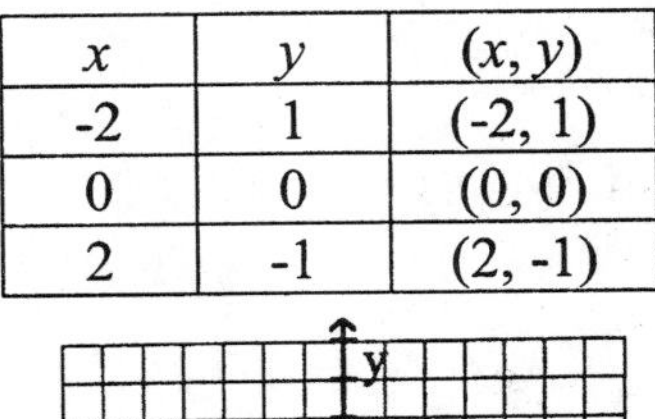

x	y	(x, y)
-2	1	(-2, 1)
0	0	(0, 0)
2	-1	(2, -1)

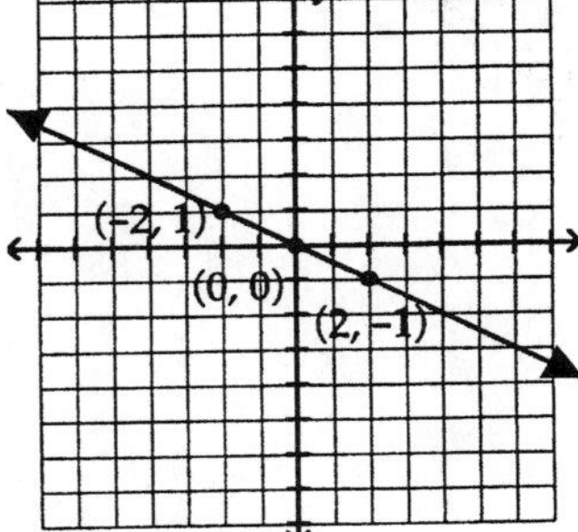

51. $y = \frac{3}{8}x - 6$

x	y	(x, y)
-8	-9	(-8, -9)
0	-6	(0, -6)
8	-3	(8, -3)

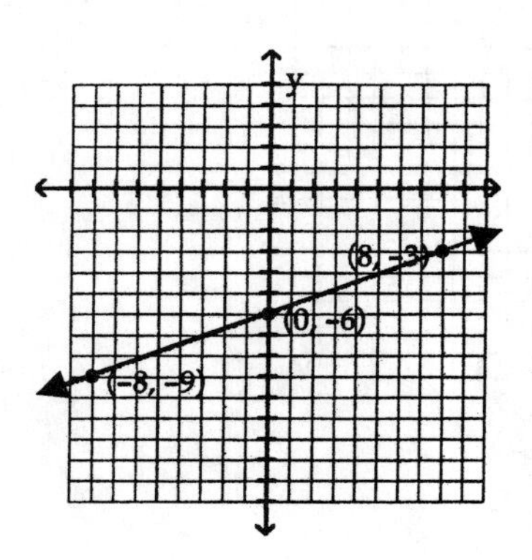

53. $y = \frac{2}{3}x - 2$

x	y	(x, y)
-3	-4	(-3, -4)
0	-2	(0, -2)
3	0	(3, 0)

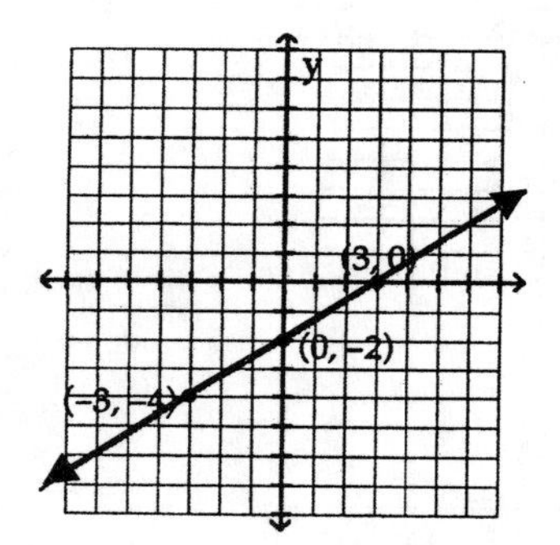

55. $7y = -2x$; divide both terms by 7

$y = -\frac{2}{7}x$

x	y	(x, y)
-7	2	(-7, 2)
0	0	(0, 0)
7	-2	(7, -2)

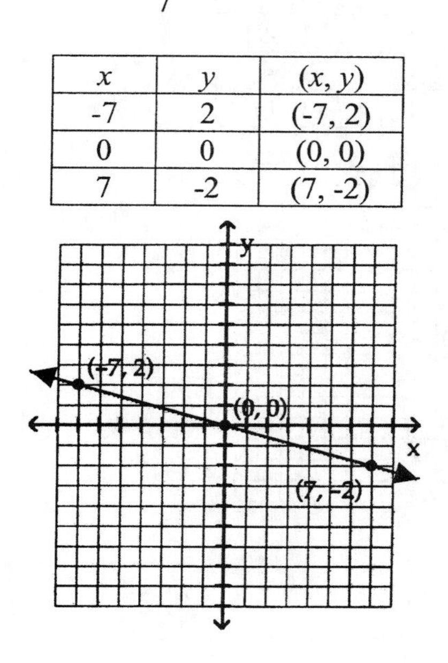

57. $3y = 12x + 15$; divide all 3 terms by 3

$y = 4x + 5$

x	y	(x, y)
-1	1	(-1, 1)
0	5	(0, 5)
1	9	(1, 9)

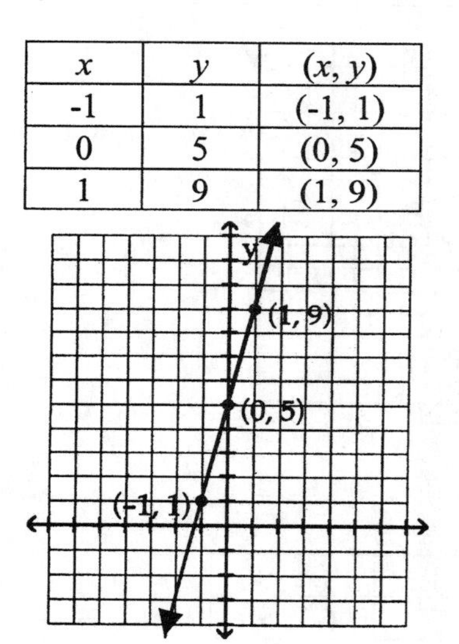

59. $5y = x + 20$; divide all 3 terms by 5

$$y = \frac{x}{5} + 4$$

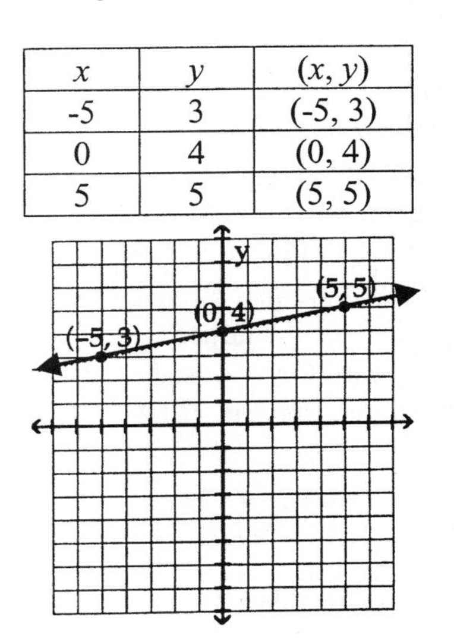

x	y	(x, y)
-5	3	(-5, 3)
0	4	(0, 4)
5	5	(5, 5)

61. $y + 1 = 7x$; subtract 1 from both sides
$y = 7x - 1$

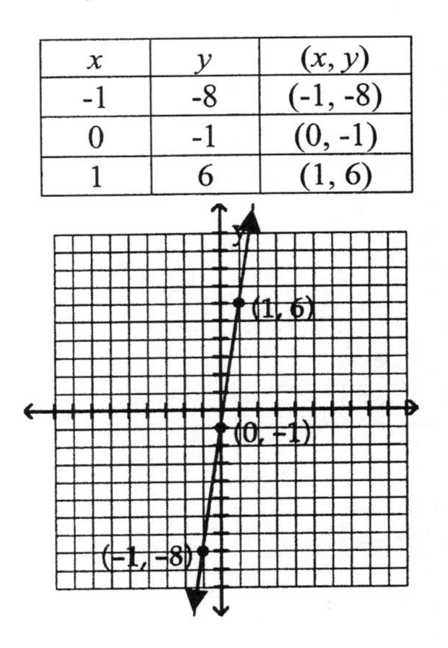

x	y	(x, y)
-1	-8	(-1, -8)
0	-1	(0, -1)
1	6	(1, 6)

APPLICATIONS

63. BILLIARDS

$y = 2x - 4$

x	y	(x, y)
1	**-2**	**(1, -2)**
2	**0**	**(2, 0)**
4	**4**	**(4, 4)**

$y = -2x + 12$

x	y	(x, y)
4	**4**	**(4, 4)**
6	**0**	**(6, 0)**
8	**-4**	**(8, -4)**

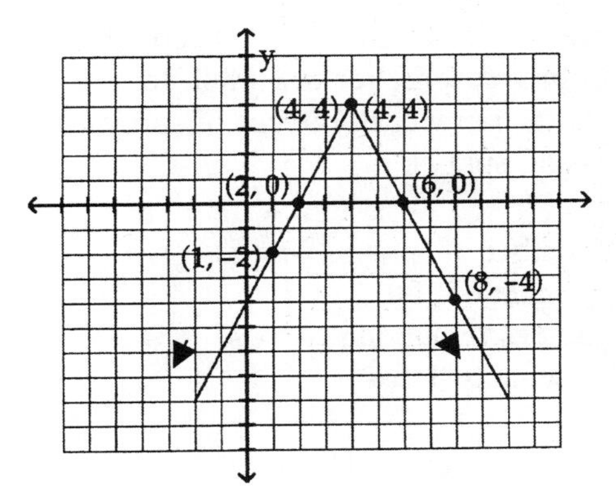

Did you notice that the angle the ball makes as it approaches the rail is the same angle when the ball rebounds off the rail and heads towards the hole?

65. HOUSEKEEPING

The problem wants us to estimate (not calculate) the amount of polish left in the bottle after **650** sprays. The numbers selected for **"n"** were used to calculate the values for **"A"** but you will notice that 650 was not a selected value to be used in the calculations.

Two different scales are used to keep the graph within reason. See the graph for the answer to the number of ounces of spray left after 650 sprays.

n = number of sprays (x-axis)
scale of 100
A = ounces remaining (y-axis)
scale of 1

$$A = -0.02n + 16$$

n	A	(n, A)
100	$-0.02(100) + 16$ $-2 + 16$ 14	(100, 14)
200	12	(200, 12)
300	$-0.02(300) + 16$ $-6 + 16$ 10	(300, 10)
500	6	(500, 6)

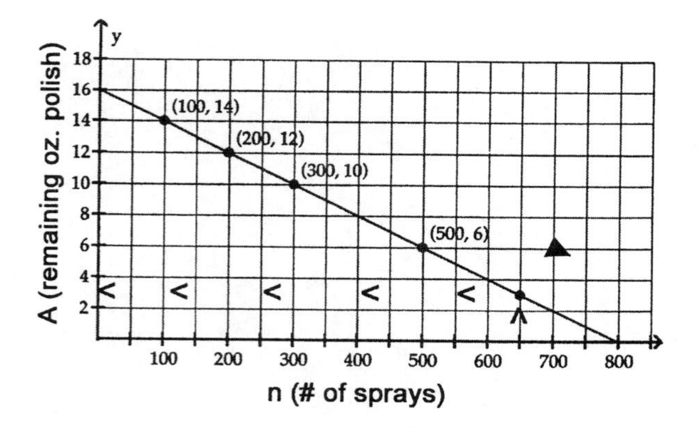

67. NFL TICKETS

This problem wants us to use a value for **"t"** that requires a little thought. The problem states the years (1990 – 2002) are to be considered for use with the formula, but it doesn't want us to use those large numbers. So we assign the following numbers to the stated years: 0 = 1990, 1 = 1991, 2 = 1992, and so on. Since we are asked to predict the price for the year 2010, we have to find the appropriate number for the year 2010. Based upon the previous pattern, we find **20 = 2010.**

t = number of years (x-axis)
p = price of ticket (y-axis)

$$p = \frac{9}{4}t + 23$$

t	p	(t, p)
0	\$23	(0, \$23)
4	$\left(\frac{9}{4}\right)(4) + 23$ $9 + 23$ \$32	(4, \$32)
8	$\left(\frac{9}{4}\right)(8) + 23$ $18 + 23$ \$41	(8, \$41)
12	$\left(\frac{9}{4}\right)(12) + 23$ $27 + 23$ \$50	(12, \$50)
16	\$59	(16, \$59)

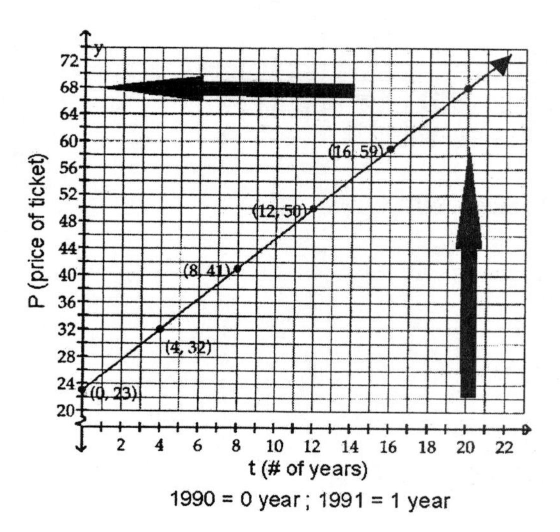

69. RAFFLES

You are to predict the number of raffle tickets to be sold for $6.

p = price of raffle ticket (x-axis)
n = # of tickets sold (y-axis)

$n = -20p + 300$

p	n	(p, n)
\$1	-20(1) + 300 -20 + 300 280	(\$1, 280)
\$3	-20(3)+ 300 -60 + 300 240	(\$3, 240)
\$5	-20(5) + 300 -100 + 300 200	(\$5, 200)

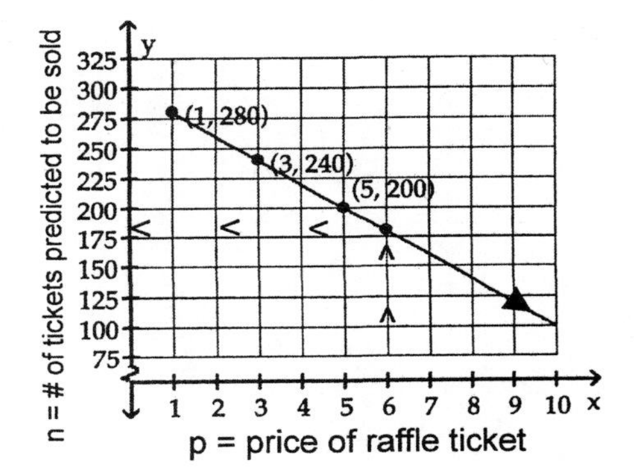

The predicted number of tickets is about 180.

WRITING

71. Answers will vary.

73. Answers will vary.

75. Answers will vary.

REVIEW

77. Simplify:
$-(-5 - 4c) = 5 + 4c$

79. $-2^2 + 2^2 = -(2)(2) + (2)(2)$
$= -4 + 4$
$= 0$

81. $1 + 2[-3 - 4(2 - 8^2)] = 1 + 2[-3 - 4(2 - 64)]$
$= 1 + 2[-3 - 4(-62)]$
$= 1 + 2[-3 + 248]$
$= 1 + 2[245]$
$= 1 + 490$
$= 491$

CHALLENGE PROBLEMS

83.

x	$y = x^2 + 1$	(x, y)
-3	$(-3)^2 + 1 = 9 + 1$ $= 10$	(-3, 10)
-2	$(-2)^2 + 1 = 4 + 1$ $= 5$	(-2, 5)
-1	$(-1)^2 + 1 = 1 + 1$ $= 2$	(-1, 2)
0	$0^2 + 1 = 0 + 1$ $= 1$	(0, 1)
1	$1^2 + 1 = 1 + 1$ $= 2$	(1, 2)
2	$2^2 + 1 = 4 + 1$ $= 5$	(2, 5)
3	$3^2 + 1 = 9 + 1$ $= 10$	(3, 10)

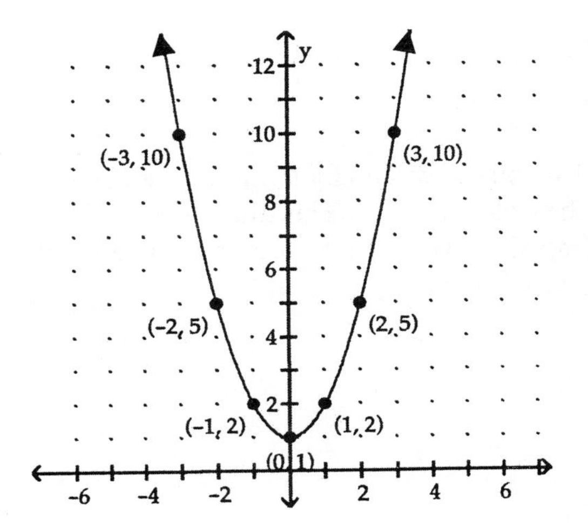

What did you notice about these values? This type of graph is called a **Parabola** and it has lines of symmetry.

85.

x	$y = \|x\| - 2$	(x, y)
-3	$\|-3\| - 2$ $3 - 2$ 1	(-3, 1)
-2	$\|-2\| - 2$ $2 - 2$ 0	(-2, 0)
-1	$\|-1\| - 2$ $1 - 2$ -1	(-1, -1)
0	$\|0\| - 2$ $0 - 2$ -2	(0, -2)
1	$\|1\| - 2$ $1 - 2$ -1	(1, -1)
2	$\|2\| - 2$ $2 - 2$ 0	(2, 0)
3	$\|3\| - 2$ $3 - 2$ 1	(3, 1)

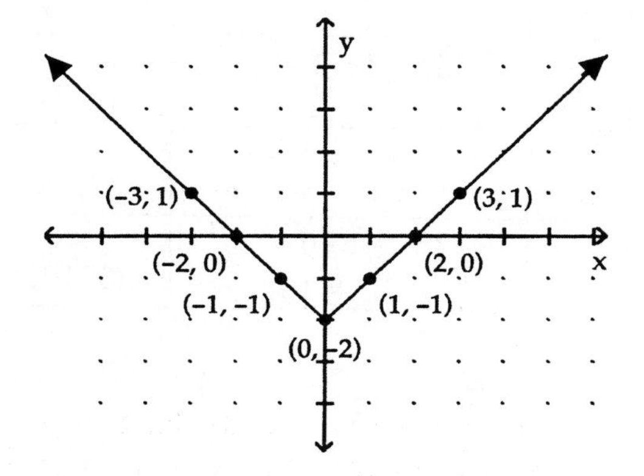

Do you notice anything interesting about the ordered pairs? This kind of graph is called an **Absolute Value** graph, and it has line of symmetry.

SECTION 3.3

VOCABULARY

1. We say $5x + 3y = 10$ is an equation in **two** variables, x and y.

3. The equation $2x - 3y = 7$ is written in **general or standard** form.

5. The y-intercept of a line is the point where the line **intersects or crosses** the y-axis.

CONCEPTS

7. The x-intercept is (4, 0) and the y-intercept is (0, 3).

9. The x-intercept is (-5, 0) and the y-intercept is (0, -4).

11. The y-intercept is (0, 2). There is no x-intercept because the line does not cross the x-axis.

13. The x-intercept is $\left(-2\frac{1}{2}, 0\right)$ and the y-intercept is $\left(0, \frac{2}{3}\right)$.

15. a) The term $3x$ would be equal to 0.
 b) Let $x = 0$ for $3x + 2y = 6$:
 $3(\mathbf{0}) + 2y = 6$
 $\mathbf{0} + 2y = 6$
 $2y = 6$
 $y = 3$
 So the y-intercept is (0, 3)

17. A line may have at most **two** intercepts. A line must have at least **one** intercept.

19. a) ii
 b) iv
 c) vi
 d) i
 e) iii
 f) v

NOTATION

21. a) The point (0, 6) lies on the y-axis.
 b) Yes, the point (0, 0) lies on both the x-axis and y-axis.

23. The equation of the x-axis is $y = 0$. The equation of the y-axis is $x = 0$.

PRACTICE

25. $4x + 5y = 20$

y-intercept:	x-intercept:
If $x = 0$,	If $y = 0$
$4(\mathbf{0}) + 5y = 20$	$4x + 5(\mathbf{0}) = 20$
$5y = 20$	$4x = 20$
$y = 4$	$x = 5$

The y-intercept is (0, 4), and the x-intercept is (5, 0).

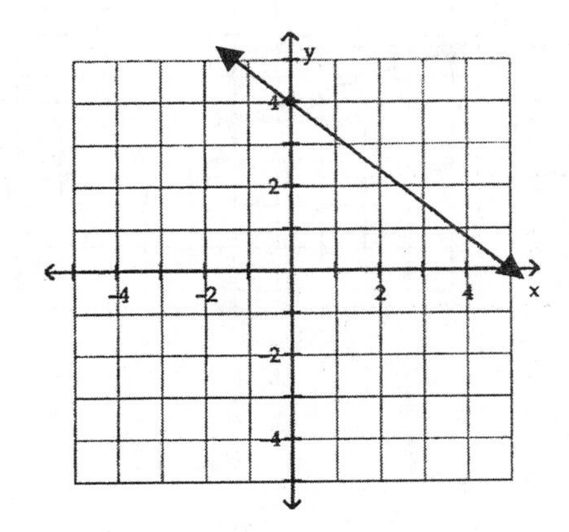

27. $x - y = -3$

y-intercept:	x-intercept:
If $x = 0$,	If $y = 0$
$(\mathbf{0}) - y = -3$	$x - (\mathbf{0}) = -3$
$-y = -3$	$x = -3$
$y = 3$	

The y-intercept is (0, 3), and the x-intercept is (-3, 0).

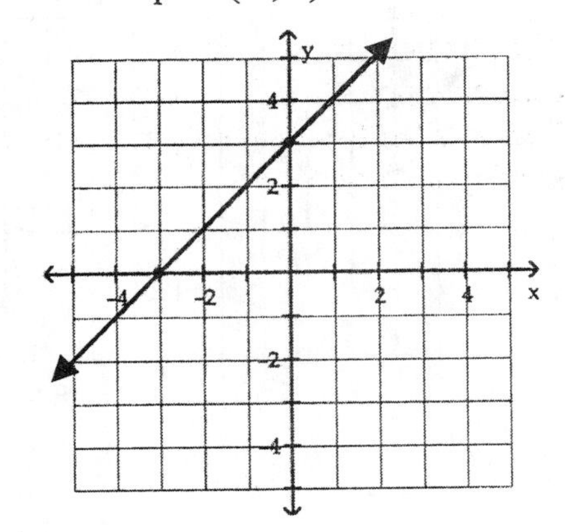

29. $5x + 15y = -15$

y-intercept:	x-intercept:
If $x = 0$,	If $y = 0$
$5(\mathbf{0}) + 15y = -15$	$5x + 15(\mathbf{0}) = -15$
$15y = -15$	$5x = -15$
$y = -1$	$x = -3$

The y-intercept is (0, -1), and the x-intercept is (-3, 0).

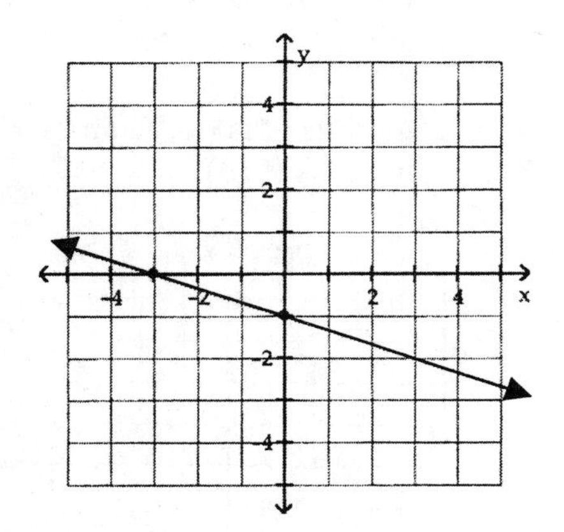

33. $4x - 3y = 12$

y-intercept:	x-intercept:
If $x = 0$,	If $y = 0$
$4(\mathbf{0}) - 3y = 12$	$4x - 3(\mathbf{0}) = 12$
$-3y = 12$	$4x = 12$
$y = -4$	$x = 3$

The y-intercept is (0, -4), and the x-intercept is (3, 0).

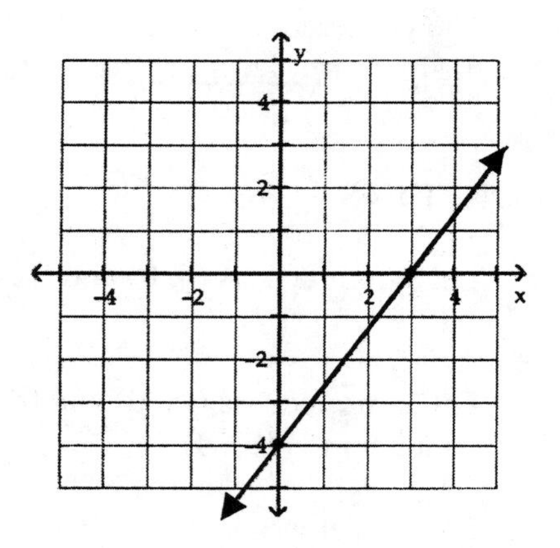

31. $x + 2y = -2$

y-intercept:	x-intercept:
If $x = 0$,	If $y = 0$
$(\mathbf{0}) + 2y = -2$	$x + 2(\mathbf{0}) = -2$
$2y = -2$	$x = -2$
$y = -1$	

The y-intercept is (0, -1), and the x-intercept is (-2, 0).

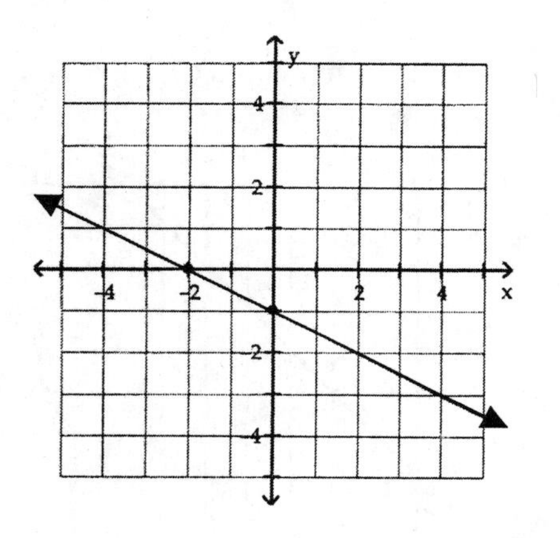

35. $3x + y = -3$

y-intercept:	x-intercept:
If $x = 0$,	If $y = 0$
$3(\mathbf{0}) + y = -3$	$3x + (\mathbf{0}) = -3$
$y = -3$	$3x = -3$
	$x = -1$

The y-intercept is (0, -3), and the x-intercept is (-1, 0).

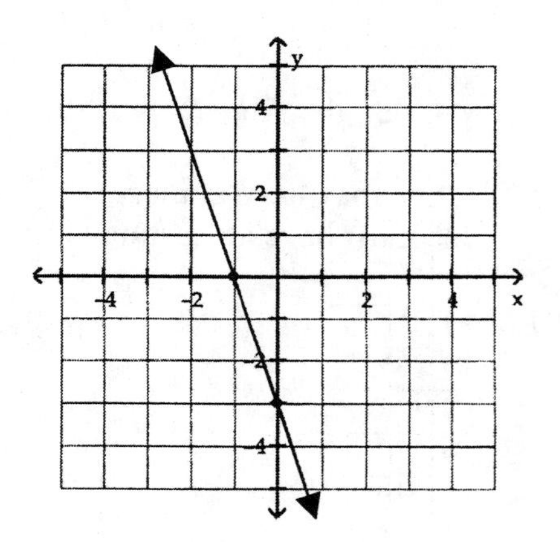

37. $9x - 4y = -9$

y-intercept:	x-intercept:
If $x = 0$,	If $y = 0$
$9(\mathbf{0}) - 4y = -9$	$9x - 4(\mathbf{0}) = -9$
$-4y = -9$	$9x = -9$
$y = \frac{9}{4}$	$x = -1$

The y-intercept is $(0, \frac{9}{4})$, and the x-intercept is $(-1, 0)$.

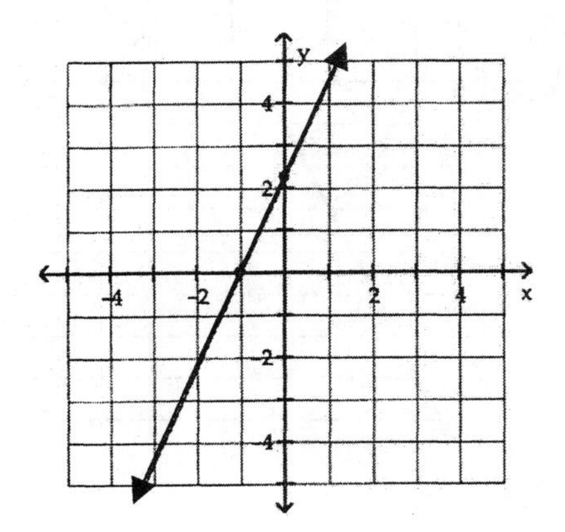

41. $4x - 2y = 6$

y-intercept:	x-intercept:
If $x = 0$,	If $y = 0$
$4(\mathbf{0}) - 2y = 6$	$4x - 2(\mathbf{0}) = 6$
$-2y = 6$	$4x = 6$
$y = -3$	$x = \frac{6}{4} = \frac{3}{2}$

The y-intercept is $(0, -3)$, and the x-intercept is $(\frac{3}{2}, 0)$.

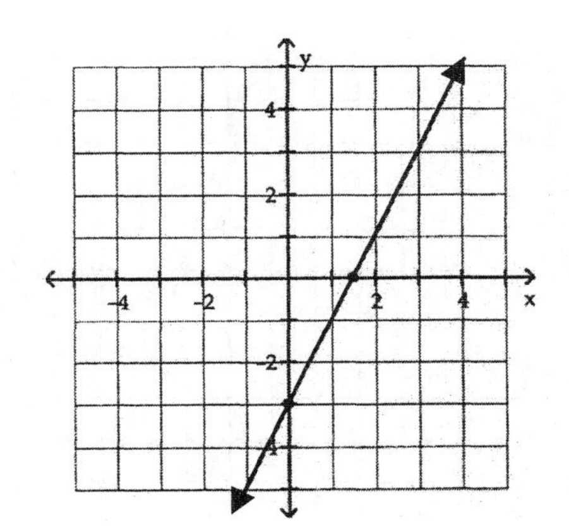

39. $8 = 3x + 4y$

y-intercept:	x-intercept:
If $x = 0$,	If $y = 0$
$8 = 3(\mathbf{0}) + 4y$	$8 = 3x + 4(\mathbf{0})$
$8 = 4y$	$8 = 3x$
$2 = y$	$\frac{8}{3} = x$

The y-intercept is $(0, 2)$, and the x-intercept is $(\frac{8}{3}, 0)$.

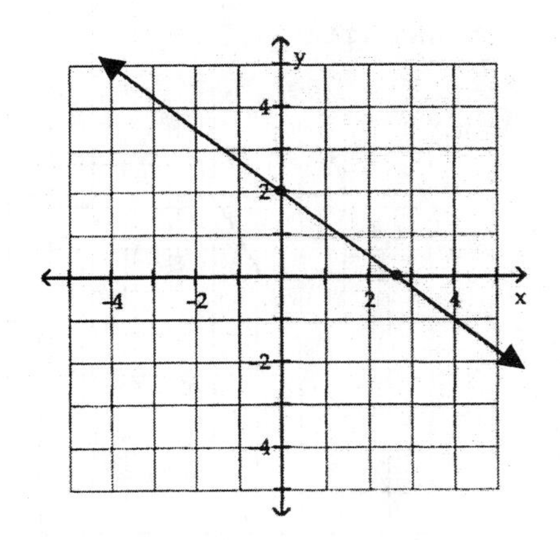

43. $3x - 4y = 11$

y-intercept:	x-intercept:
If $x = 0$,	If $y = 0$
$3(\mathbf{0}) - 4y = 11$	$3x - 4(\mathbf{0}) = 11$
$-4y = 11$	$3x = 11$
$y = -\frac{11}{4}$	$x = \frac{11}{3}$

The y-intercept is $(0, -\frac{11}{4})$, and the x-intercept is $(\frac{11}{3}, 0)$.

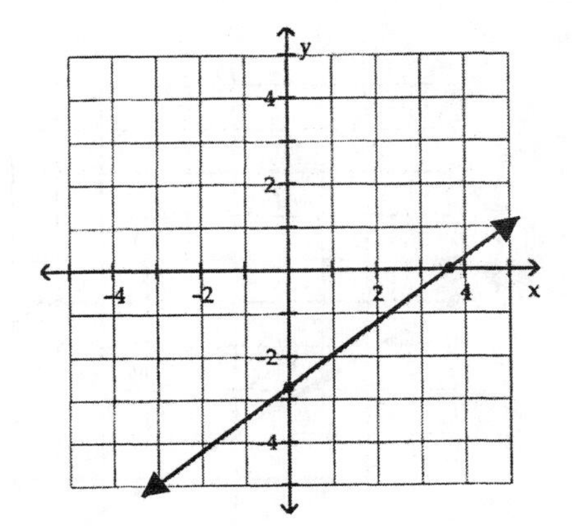

45. $9x + 3y = 10$

y-intercept:	x-intercept:
If $x = 0$,	If $y = 0$
$9(\mathbf{0}) + 3y = 10$	$9x + 3(\mathbf{0}) = 10$
$3y = 10$	$9x = 10$
$y = \frac{10}{3}$	$x = \frac{10}{9}$

The y-intercept is $(0, \frac{10}{3})$, and the x-intercept is $(\frac{10}{9}, 0)$.

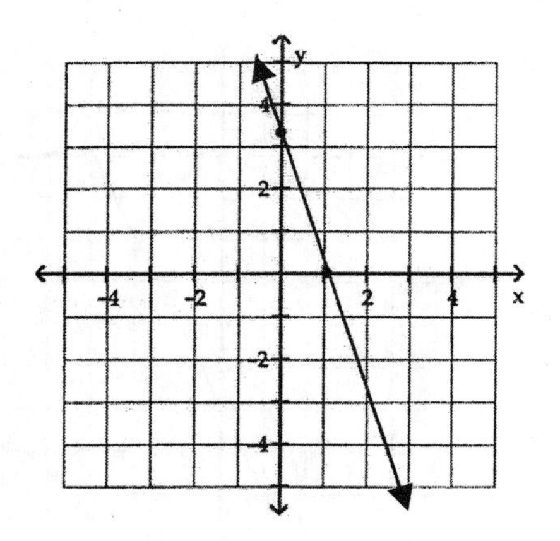

49. $-4x = 8 - 2y$

y-intercept:	x-intercept:
If $x = 0$,	If $y = 0$
$-4(\mathbf{0}) = 8 - 2y$	$-4x = 8 - 2(\mathbf{0})$
$0 = 8 - 2y$	$-4x = 8$
$2y = 8$	$x = -2$
$y = 4$	

The y-intercept is $(0, 4)$, and the x-intercept is $(-2, 0)$.

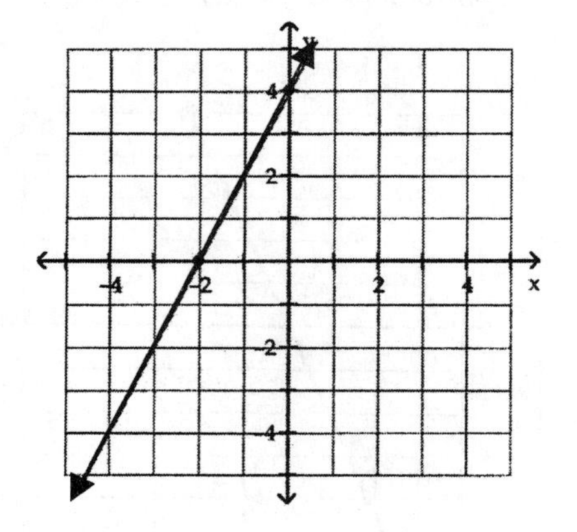

47. $3x = -15 - 5y$

y-intercept:	x-intercept:
If $x = 0$,	If $y = 0$
$3(\mathbf{0}) = -15 - 5y$	$3x = -15 - 5(\mathbf{0})$
$0 = -15 - 5y$	$3x = -15$
$5y = -15$	$x = -5$
$y = -3$	

The y-intercept is $(0, -3)$, and the x-intercept is $(-5, 0)$.

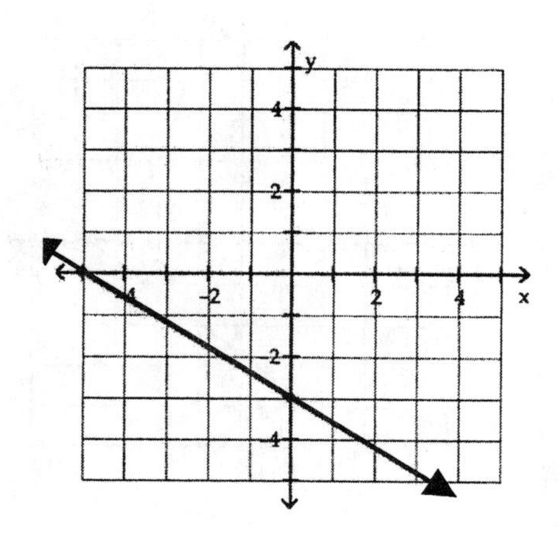

51. $7x = 4y - 12$

y-intercept:	x-intercept:
If $x = 0$,	If $y = 0$
$7(\mathbf{0}) = 4y - 12$	$7x = 4(\mathbf{0}) - 12$
$0 = 4y - 12$	$7x = -12$
$-4y = -12$	$x = -\frac{12}{7}$
$y = 3$	

The y-intercept is $(0, 3)$, and the x-intercept is $(-\frac{12}{7}, 0)$.

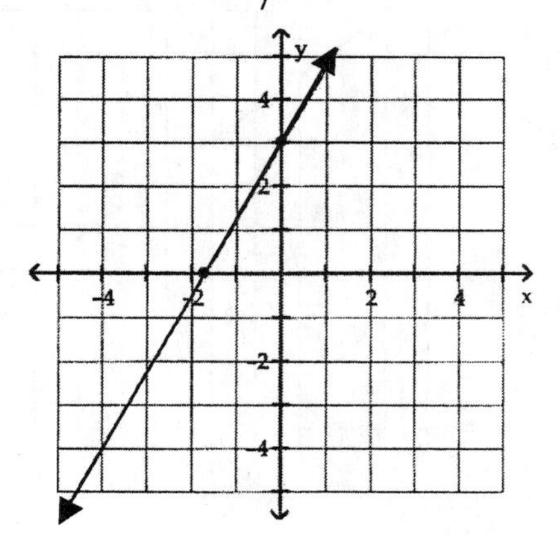

53. $y - 3x = -\frac{4}{3}$

y-intercept:	x-intercept:
If $x = 0$,	If $y = 0$
$y - 3(\mathbf{0}) = -\frac{4}{3}$	$\mathbf{0} - 3x = -\frac{4}{3}$
$y = -\frac{4}{3}$	$-3x = -\frac{4}{3}$
	$x = -\frac{4}{3} \cdot -\frac{1}{3} = \frac{4}{9}$

The y-intercept is $(0, -\frac{4}{3})$, and the x-intercept is $(\frac{4}{9}, 0)$.

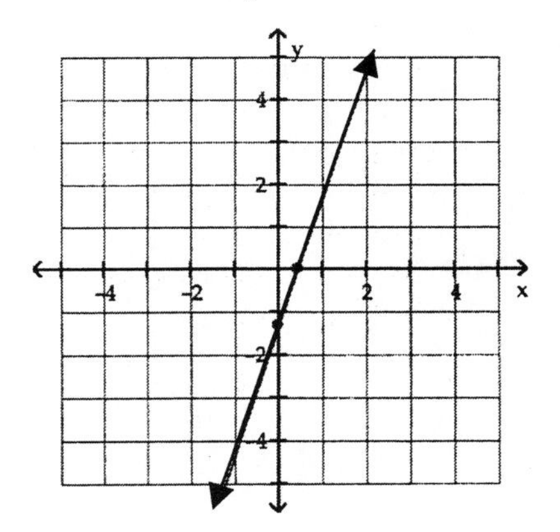

55. $y = 4$ is a horizontal line through (0, 4).

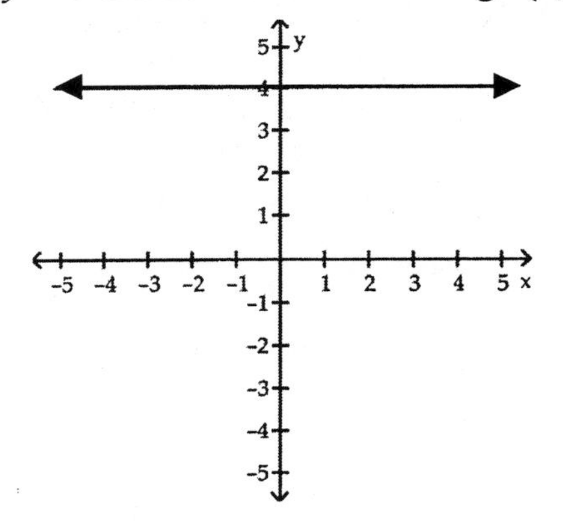

57. $x = -2$ is a vertical line through (-2, 0).

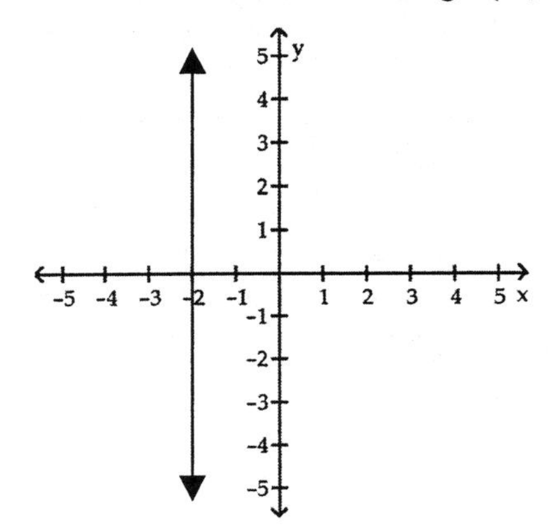

59. $y = -\frac{1}{2}$ is a horizontal line through $(0, -\frac{1}{2})$.

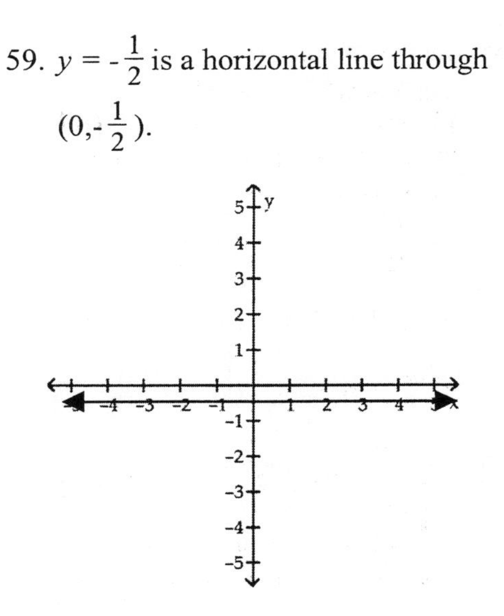

61. $x = \frac{4}{3}$ is a vertical line through $(\frac{4}{3}, 0)$

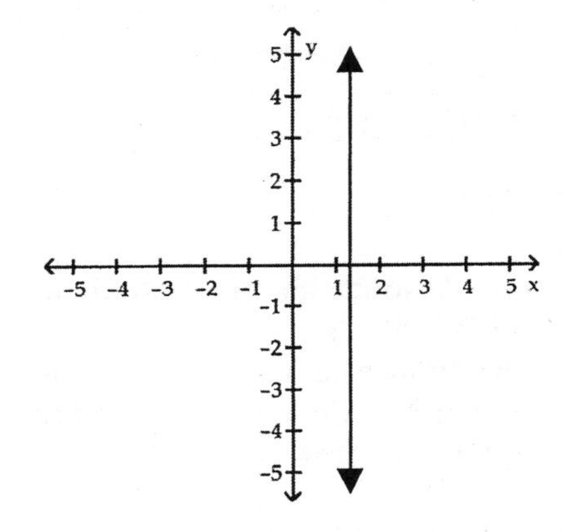

63. $y - 2 = 0$
$y - 2 + 2 = 0 + 2$
$y = 2$ is a horizontal line through (0, 2).

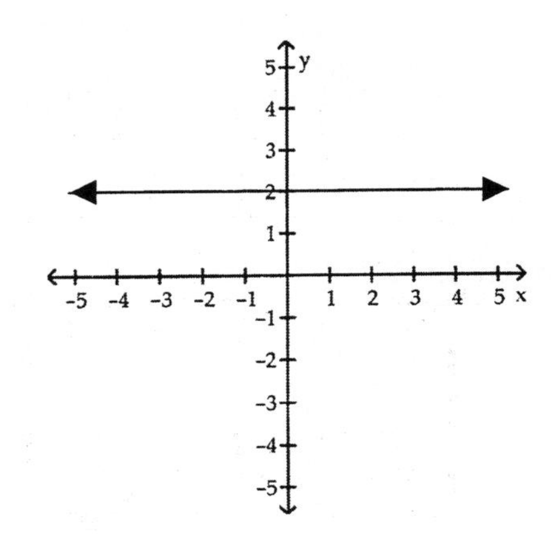

65. $-2x + 3 = 11$

$-2x + 3 - 3 = 11 - 3$

$-2x = 8$

$\frac{-2x}{-2} = \frac{8}{-2}$

$x = -4$ is a vertical line through (-4, 0)

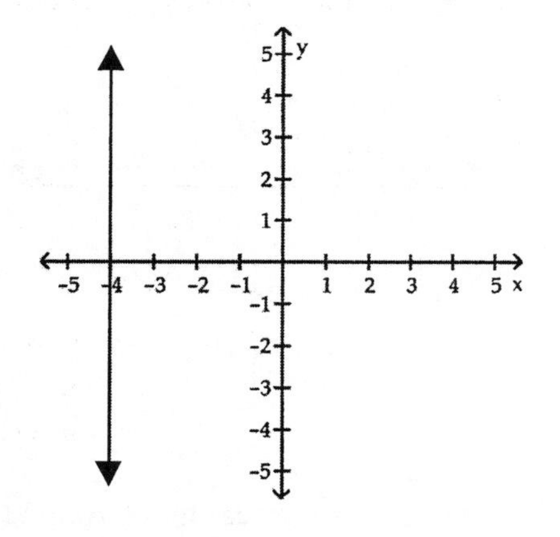

APPLICATIONS

67. CHEMISTRY

a) Absolute zero is approximately **-270° C**.

b) When the temperature is absolute zero, the volume of the gas is **zero milliliters**.

69. LANDSCAPING

a) The y-intercept is (0, 200) which indicates that if only shrubs are purchased, he can buy 200.

b) The x-intercept is (100, 0) which indicates that if only trees are purchased, he can buy 100.

WRITING

71. Answers will vary.

73. Answers will vary.

REVIEW

75. $\frac{\cancel{3}\cdot\cancel{5}\cdot\cancel{5}}{\cancel{3}\cdot\cancel{5}\cdot\cancel{5}\cdot 5} = \frac{1}{5}$

77. $2x - 6$

CHALLENGE PROBLEMS

79. Line $y = b$ will intersect line $x = a$ at the ordered pair (a, b).

SECTION 3.4

VOCABULARY

1. A **ratio** is the quotient of two numbers.

3. The **slope** of a line is defined to the ratio of the change in y to the change in x.

5. The rate of **change** of a linear relationship can be found by finding the slope of the graph of the line and attaching the proper units.

CONCEPTS

7. a) A line with *positive* slope **rises** from left to right.

   b). A line with *negative* slope **falls** from left to right.

9. A rise of 2 with a run of 15 would be written as a ratio of $\frac{2}{15}$.

11. a) The rise is –3.
    b) The run is 8.
    c) The slope would be the ratio of the rise over the run or $-\frac{3}{8}$.

13. a) Count the lines starting at point **A** and move up that vertical line. Stop when you reach the horizontal line that contains point **B**. Now count the lines until you reach point B. The rise is positive 3 and the run is positive 6. The slope is $\frac{3}{6}$, reduced to $\frac{1}{2}$.

    b) Count the lines starting at point **B** and move up that vertical line. Stop when you reach the horizontal line that contains point **C**. Now count the lines until you reach point C. The rise is positive 1 and the run is positive 2. The slope is $\frac{1}{2}$.

13.c) Count the lines starting at point **A** and move up that vertical line. Stop when you reach the horizontal line that contains point **C**. Now count the lines until you reach point C. The rise is positive 4 and the run is positive 8. The slope is $\frac{4}{8}$, reduced to $\frac{1}{2}$.

   d) Using any two points on the line will determine the same slope because all the points belong to the same line.

15. a) The fraction can be written simply as 0.
    b) When the slope contains a 0 in the denominator, we write it as *undefined*.

17. To convert a ratio to a percent, change the fraction to a decimal and then change the decimal to a percent.
    a) $\frac{2}{5} = 0.4 = 40\%$
    b) $\frac{3}{20} = 0.15 = 15\%$

NOTATION

19. The formula for finding the slope is given by:

$$m = \frac{y_2 - y_1}{x_2 - x_1}$$

21. y^2 means to raise y to the second power. y_2 means y sub two.

PRACTICE

23. $m = \frac{2}{3}$

25. $m = \frac{4}{3}$

27. $m = -2$

29. $m = 0$

31. $m = -\frac{1}{5}$

33. $(2,4)$ and $(1,3)$

(x_1, y_1) and (x_2, y_2)

$$m = \frac{y_2 - y_1}{x_2 - x_1}$$
$$= \frac{3-4}{1-2}$$
$$= \frac{-1}{-1}$$
$$= 1$$

35. $m = -3$

37. $m = \frac{5}{4}$

39. $(-3,5)$ and $(-5,6)$

(x_1, y_1) and (x_2, y_2)

$$m = \frac{y_2 - y_1}{x_2 - x_1}$$
$$= \frac{6-5}{-5-(-3)}$$
$$= \frac{1}{-5+3}$$
$$= \frac{1}{-2}$$

41. $m = \frac{3}{5}$

43. $(5,7)$ and $(-4,7)$

(x_1, y_1) and (x_2, y_2)

$$m = \frac{y_2 - y_1}{x_2 - x_1}$$
$$= \frac{7-7}{-4-5}$$
$$= \frac{0}{-9}$$
$$= 0$$

45. $(8,-4)$ and $(8,-3)$

(x_1, y_1) and (x_2, y_2)

$$m = \frac{y_2 - y_1}{x_2 - x_1}$$
$$= \frac{-3-(-4)}{8-8}$$
$$= \frac{1}{0}$$

$m =$ undefined

47. $m = -\frac{2}{3}$

49. $(-2.5, 1.75)$ and $(-0.5, -7.75)$

(x_1, y_1) and (x_2, y_2)

$$m = \frac{y_2 - y_1}{x_2 - x_1}$$
$$= \frac{-7.75 - 1.75}{-0.5-(-2.5)}$$
$$= \frac{-9.5}{2}$$
$$= -4.75$$

51. $m = \frac{3}{4}$

53. $m = 0$

55. $y = -2$, This means that all the points of the line have y-coordinates of -2 and many different x-coordinates. Using the slope formula, means that the y values will always produce a **0** for an answer. And this means that one is dividing into **0**, which is a valid operation. **Slope is 0.**

57. $m =$ undefined

59. $m = 0$

61. $m =$ undefined

APPLICATIONS

63. DRAINAGE

First determine the rise of the slope by looking at the ruler. **3 inches**
The run is determined from the length of the 2-by-4. **10 feet**
Since the two unit measurements are different, one must convert one to the other. Since converting **feet into inches** stays a whole number, this may keep the math simpler. 10 feet times 12 inches = **120 inches**.
Compute the ratio.

$$m = \frac{\text{rise}}{\text{run}}$$
$$= \frac{3}{120}$$
$$= \frac{1}{40}$$

Determine that the slope rises from left to right thus indicating a **positive** slope.

$$\boxed{m = \frac{1}{40}}$$

65. GRADE OF A ROAD

$$m = \frac{\text{change in } y}{\text{change in } x}$$
$$= \frac{264}{5280}$$
$$= \frac{1}{20}$$

Change the fraction to a percent by dividing: $\frac{1}{20} = 0.05 = 5\%$

67. ENGINEERING

a) The change in y is 2 and the change in x is 16.

$$m = \frac{\text{change in } y}{\text{change in } x}$$
$$= \frac{2}{16}$$
$$= \frac{1}{8}$$

b) For each ramp, the change in y is 1 and the change in x is 12.

$$m = \frac{\text{change in } y}{\text{change in } x}$$
$$= \frac{1}{12}$$

c) Design #1 would be less expensive, but it would also be steeper. Design #2 is not as steep but would be more expensive.

69. IRRIGATION

Find an order pair for each of the end-points: (0, 8000) and (8, 1000). Then apply the following formula:

$$m = \frac{y_2 - y_1}{x_2 - x_1}$$
$$= \frac{1000 - 8000}{8 - 0}$$
$$= \frac{-7000}{8}$$
$$= -875$$

The rate of change would be -875 gallons per hour.

71. MILK PRODUCTION

Find an order pair for each of the end-points: (1993, 15,700) and (2002, 18,400). Then apply the following formula:

$$
\begin{aligned}
m &= \frac{y_2 - y_1}{x_2 - x_1} \\
&= \frac{18,400 - 15,700}{2002 - 1993} \\
&= \frac{2,700}{9} \\
&= 300
\end{aligned}
$$

So, the rate of change would be 300 pounds per year.

WRITING

73. Answers will vary.

75. Answers will vary.

REVIEW

77. HALLOWEEN CANDY

Let x = the pounds of licorice.
Then $60 - x$ = the pounds of gumdrops.
The value of the licorice would be \$1.90x and the value of the gumdrops would be \$2.20$(60 - x)$. The value of the mixture would be \$2(60).

$$
\begin{aligned}
1.90x + 2.20(60 - x) &= 2(60) \\
1.90x + 2.20(60) + 2.20(-x) &= 2(60) \\
1.90x + 132 - 2.20x &= 120 \\
1.90x - 2.20x + 132 &= 120 \\
-0.3x + 132 &= 120 \\
-0.3x + 132 - 132 &= 120 - 132 \\
-0.3x &= -12 \\
\boxed{x = 40}
\end{aligned}
$$

Thus, he would need 40 lbs of licorice and 60 – (40) = 20 lbs of gumdrops.

CHALLENGE PROBLEMS

79. Find the slope from A to B:

$$
\begin{aligned}
m &= \frac{y_2 - y_1}{x_2 - x_1} \\
&= \frac{0 - (-10)}{20 - (-50)} \\
&= \frac{10}{70} \\
&= \frac{1}{7}
\end{aligned}
$$

Find the slope from A to C:

$$
\begin{aligned}
m &= \frac{y_2 - y_1}{x_2 - x_1} \\
&= \frac{2 - (-10)}{34 - (-50)} \\
&= \frac{12}{84} \\
&= \frac{1}{7}
\end{aligned}
$$

Find the slope from C to B:

$$
\begin{aligned}
m &= \frac{y_2 - y_1}{x_2 - x_1} \\
&= \frac{0 - 2}{20 - 34} \\
&= \frac{-2}{-14} \\
&= \frac{1}{7}
\end{aligned}
$$

Since the slope is the same for each, all three points would lie on the same line.

SECTION 3.5

VOCABULARY

1. The equation $y = mx + b$ is called the **slope-intercept** form of the equation of a line.

3. **Parallel** lines do not intersect.

5. The numbers $\frac{5}{6}$ and $-\frac{6}{5}$ are called negative **reciprocals**. Their product is -1.

CONCEPTS

7. a) No. The equation needs to be solved y for the equation to be in slope-intercept form.
   b) No. The equation needs to be solved y for the equation to be in slope-intercept form.
   c) Yes
   d) No. The equation needs to be solved y for the equation to be in slope-intercept form.
   e) Yes
   f) Yes, the right side of the equation just doesn't have a visible y-intercept. The equation could be written as $y = 2x + \mathbf{0}$.

9. a)

$$5y = 10x + 20$$
$$\frac{5y}{5} = \frac{10x}{5} + \frac{20}{5}$$
$$\boxed{y} = \boxed{2x} + \boxed{4}$$

   b)

$$-2y = 6x - 12$$
$$\frac{-2y}{-2} = \frac{6x}{-2} - \frac{12}{-2}$$
$$\boxed{y} = \boxed{-3x}\ \boxed{+}\ 6$$

11. Select the point that the line goes through and also lies on the y-axis. The y-intercept is **"0"**. Start from this point (0, 0), count **down** (negative) **5** places and then count **right** (positive) **4** places stopping at the second point on the line. Slope is $-\frac{5}{4}$. Now put these values into the standard equation of $y = mx + b$.

$$y = -\frac{5}{4}x + 0$$
$$y = -\frac{5}{4}x$$

13. a) Two different lines with the same slope are **parallel**.
    b) If the slopes of two lines are negative reciprocals, the lines are **perpendicular**.
    c) The product of the slopes of perpendicular lines is **–1**.

15. a) Line 1's slope is 2. Line 2 is perpendicular to Line 1; therefore the slope of Line 2 is $-\frac{1}{2}$.
    b) Line 2's slope is $-\frac{1}{2}$. Line 3 is perpendicular to Line 1; therefore the slope of Line 3 is 2. Line 3 is also parallel to Line 1 thus making the two slopes the same.
    c) Line 3's slope is 2. Line 4 is perpendicular to Line 3; therefore the slope of Line 4 is $-\frac{1}{2}$. Line 4 is also parallel to Line 2 thus making the two slopes the same.
    d) Lines 1 and 2 have the same y-intercept because they both intersect the y-axis at the same place.

NOTATION

17.
$$2x+5y=15$$
$$2x+5y-\boxed{2x}=\boxed{-2x}+15$$
$$\boxed{5y}=-2x+15$$
$$\frac{5y}{\boxed{5}}=\frac{-2x}{\boxed{5}}+\frac{15}{\boxed{5}}$$
$$y=-\frac{2}{5}x+\boxed{3}$$

The slope is $\boxed{-\frac{2}{5}}$ and the y-intercept is $\boxed{(0,3)}$.

19. a) $\frac{8x}{2}=4x$

b) $\frac{8x}{6}=\frac{4x}{3}$

c) $\frac{-8x}{-8}=-x$

d) $\frac{-16}{8}=-2$

21. **This is the right angle symbol.** It states that two lines are perpendicular to each other.

PRACTICE

23. $y = 4x + 2$
slope = 4 ; *y*-intercept = (0, 2)

25. $y = -5x - 8$
slope = -5 ; *y*-intercept = (0, -8)

27. $4x - 2 = y$
slope = 4 ; *y*-intercept = (0, -2)
It doesn't matter if the "y" is on the other side of the equal marks, the coefficient (4) of the "x" is still the slope and the constant (-2) is still the y-intercept.

29. $y=\frac{x}{4}-\frac{1}{2}$

slope = $\frac{1}{4}$; *y*-intercept = $\left(0,\ -\frac{1}{2}\right)$

Same reason as for #27.

31. $y=\frac{1}{2}x+6$

slope = $\frac{1}{2}$; *y*-intercept = (0, 6)

Same reason as for #27.

33. $y = 6 - x$
slope = -1 ; *y*-intercept = (0, 6)
It doesn't matter if the "x" comes after the constant, the coefficient of the "x" (-1) is still the slope and the constant (6) is still the y-intercept.

35. $x + y = 8$
Isolate the "y" and then find the slope and y-intercept.
$$y = -x + 8$$
slope = -1 ; *y*-intercept = (0, 8)

37. $6y = x - 6$
Isolate the "y" and then find the slope and y-intercept.
$$y=\frac{1}{6}x-1$$
slope = $\frac{1}{6}$; *y*-intercept = (0, -1)

39. $7y = -14x + 49$
Solve for "y" and then find the slope and y-intercept.
$$y = -2x + 7$$
slope = -2 ; *y*-intercept = (0, 7)

41. $-4y = 6x - 4$
Solve for "y" and then find the slope and y-intercept.
$$y=-\frac{3}{2}x+1$$
slope = $-\frac{3}{2}$; *y*-intercept = (0, 1)

43. $2x + 3y = 6$
$$3y = -2x + 6$$
$$y=-\frac{2}{3}x+2$$
slope = $-\frac{2}{3}$; *y*-intercept = (0, 2)

45. $3x - 5y = 15$

$-5y = -3x + 15$

$y = \frac{3}{5}x - 3$

slope = $\frac{3}{5}$ **; *y*-intercept = (0, -3)**

47. $-6x + 6y = -11$

$6y = 6x - 11$

$y = x - \frac{11}{6}$

slope = 1 ; *y*-intercept = $\left(0, -\frac{11}{6}\right)$

49. $y = x$

You can put a constant of "0" after the x and make the equation look like the previous equations.

$y = x + 0$

slope = 1 ; *y*-intercept = (0, 0)

51. $y = -5x$

$y = -5x + 0$

slope = -5 ; *y*-intercept = (0,0)

53. $y = -2$

You may put "0x" in the equation so it will look like the previous one.

$y = 0x - 2$

slope = 0 ; *y*-intercept = (0, -2)

This is a special line, which is horizontal and passes through the y-axis at -2.

55.

$-5y - 2 = 0$

$-5y = 2$

$y = -\frac{2}{5}$

$y = 0x - \frac{2}{5}$

slope = 0 ; *y*-intercept = $\left(0, -\frac{2}{5}\right)$

57. $y = 5x - 3$

Plot the y-intercept (0, -3) first. Use this as the starting point for placing the other two points by using the slope (m = 5).

Slope is rise over run. So $m = 5 = \frac{5}{1}$.

Start at (0, -3), go **up** the vertical axis 5 places and then go **right** 1 place, **stop.**

$m = \frac{5}{1} = \frac{-5}{-1}$ (Change both values to their opposites.) Start at (0, -3), go **down** the vertical axis 5 places and then go **left** 1 place, **stop**. This gives you three points in a straight line.

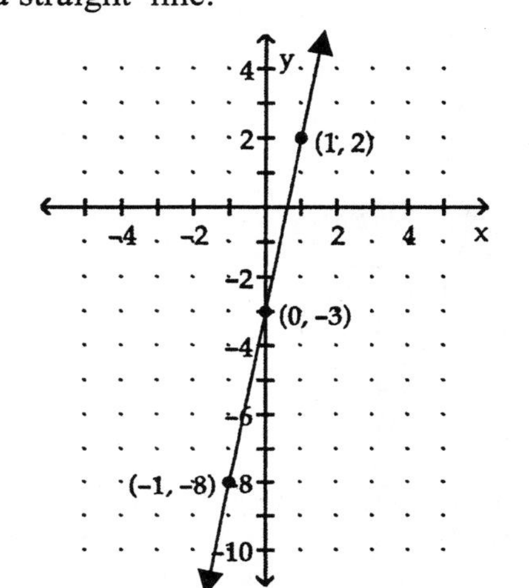

59. $y = \frac{1}{4}x - 2$

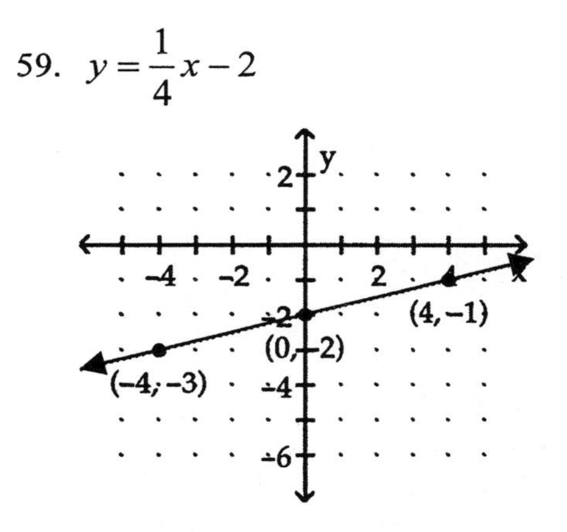

61. $y = -3x + 6$

63. $y = -\frac{8}{3}x + 5$

65. $m = 3$, y-intercept = (0,3)

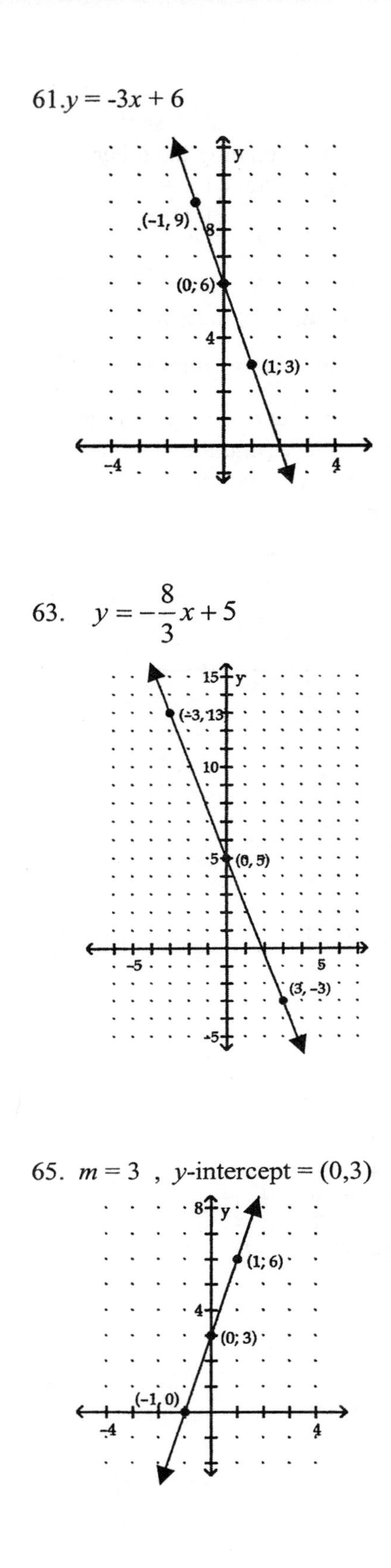

67. $m = -\frac{1}{2}$, y-intercept = (0,2)

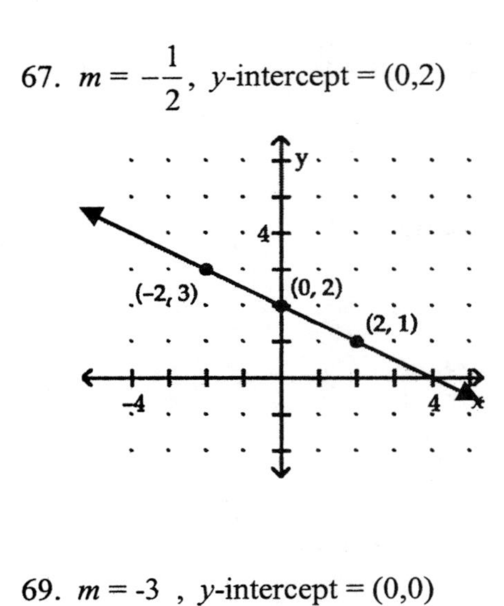

69. $m = -3$, y-intercept = (0,0)

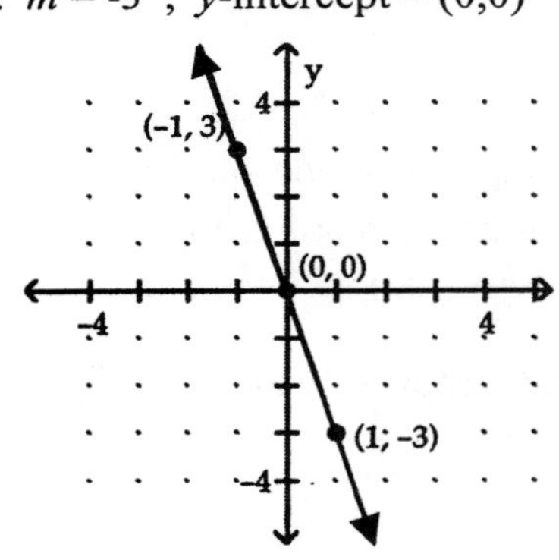

71. $m = -4$, y-intercept = (0,-4)
Remember to solve for "y" first.

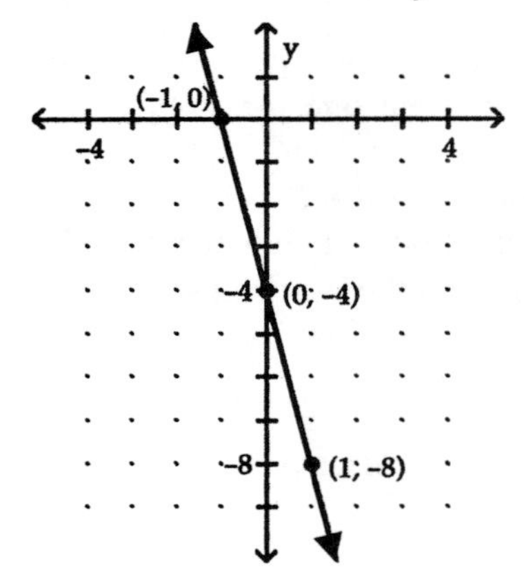

73. $m = -\frac{3}{4}$, y-intercept = (0,4)

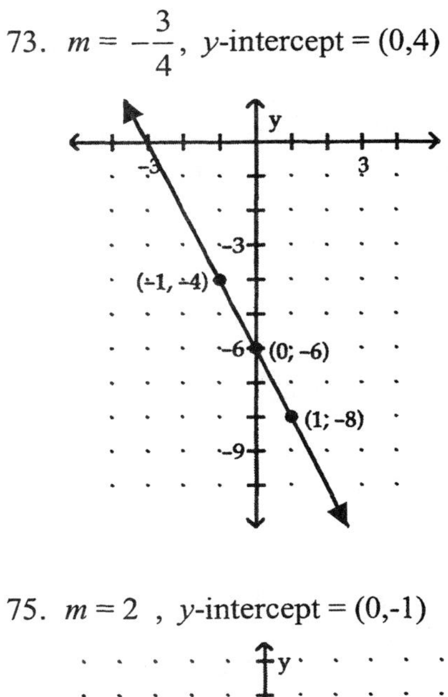

75. $m = 2$, y-intercept = (0,-1)

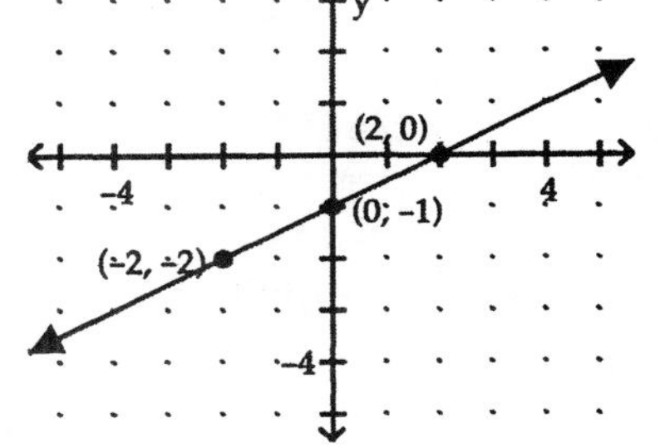

77. To determine if a pair of equations is parallel, perpendicular, or neither, just find each slope and then compare. If the slopes are equal, then the lines are parallel. If the slopes are negative reciprocals of each other or if their product is –1, then the lines are perpendicular.
$y = 6x + 8$, $m = 6$
$y = 6x$, $m = 6$

Slopes are equal; lines are parallel.

79. $y = x$, $m = 1$
$y = -x$, $m = -1$

Product is –1; lines are perpendicular.

81. $y = -2x - 9$, $m = -2$
$y = 2x - 9$, $m = 2$

Slopes are not equal nor is their product –1; therefore they are neither.

83. $x - y = 12$, $m = 1$
$-2x + 2y = -23$, $m = 1$
Parallel.

85. $x = 9$, m = undefined, vertical line
$y = 8$, $m = 0$, horizontal line

Perpendicular.

APPLICATIONS

87. PRODUCTION COSTS:
a) basic fee = \$5,000 (constant)
extra cost = \$2,000/hr (not constant)
Total Prod Cost = extra cost + basic cost
$\mathbf{c = \$2{,}000h + \$5{,}000}$

b) The cost for 8 hours of filming is found by replacing "h" with 8 and then calculating the results.
$\mathbf{c = \$2{,}000h + \$5{,}000}$
$c = \$2{,}000(8) + \$5{,}000$
$c = \$16{,}000 + \$5{,}000$
$\mathbf{c = \$21{,}000}$

89. CHEMISTRY
Step 1: -10° F (constant)
Step 2: raise temperature 5° F every minute (not constant)
Find final temperature F for t minutes

$F = (5° \text{ F})(t \text{ minutes}) + (-10)$
$\mathbf{F = 5t - 10}$

91. EMPLOYMENT SERVICE
Step 1: \$500 fee (constant)
Step 2: \$20/month (not constant)
Find actual cost "c" after m months

$c = (\$20)(m \text{ months}) + \$500$
$\mathbf{c = 20m + 500}$

93. SEWING COSTS

a) Step 1: basic fee = \$20 (constant)
Step 2: \$5/letter (not constant)
Find total "c" with x letters

c = (\$5/letter)(x letters) + \$20
$\mathbf{c = 5x + 20}$

b) Solid line is the graph for part "**b**".

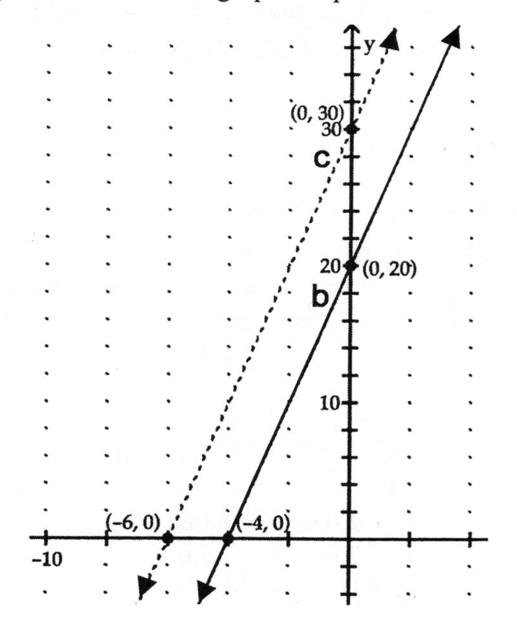

c) Dotted line of the graph is part "**c**".

95. PROFESSIONAL HOCKEY
Step 1: average price = \$33.50 (constant)
Step 2: increase of \$2.00/year (not constant)
Find average cost "c" for t years from 1995-2002

c = (\$2/year)(t years) + \$33.50
$\mathbf{c = 2t + 33.50}$

WRITING

97. Answers will vary.

REVIEW

99. CABLE TV
Step 1: 186 foot cable cut into 4 pieces
Step 2: Each piece is 3 ft longer than the previous one.

Let x = 1st piece in feet
$1^{ST} + 3$ = 2nd piece in feet
$2^{nd} + 3$ = 3rd piece in feet
$3^{rd} + 3$ = 4th piece in feet

$$x + (1^{st} + 3) + (2^{nd} + 3) + (3^{rd} + 3) = 186$$

$$x + (x + 3) + [(x + 3) + 3] + \{[(x + 3) + 3] + 3\} = 186$$

$$x + (x + 3) + (x + 6) + (x + 9) = 186$$

$$4x + 18 = 186$$

$$4x = 168$$

$x = 42$ ft
$x + 3 = 42 + 3 = 45$ ft
$x + 6 = 42 + 6 = 48$ ft
$x + 9 = 42 + 9 = 51$ ft

1st piece is 42 feet.
2nd piece is 45 feet.
3rd piece is 48 feet.
4th piece is 51 feet.

CHALLENGE PROBLEMS

101. If one would draw several lines such that they went through quadrants I, II, and IV, one would see that the **slope** of all the lines would be **negative** and that all the ***y*-intercepts** would be **positive**.
$m < 0; b > 0$

SECTION 3.6

VOCABULARY

1. $y - y_1 = m(x - x_1)$ is called the **point-slope** form of the equation of a line.

CONCEPTS

3. a) $y - 4 = 2(x - 5)$; **point–slope form**
   b) $y = 2x + 15$; **slope-intercept form**

5. a) $x - (-6) = x + 6$
   b) $y - (-9) = y + 9$

7. Find the slope given (2, -3) and (-4, 12)

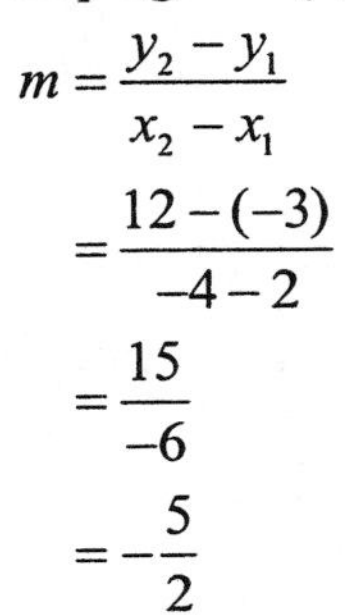

$$m = \frac{y_2 - y_1}{x_2 - x_1} = \frac{12-(-3)}{-4-2} = \frac{15}{-6} = -\frac{5}{2}$$

9. a) Does $y = 4x + 7$ have the required slope?
   Yes, it does. According to $y = mx + b$, $m = 4$.
   b) Does $y = 4x + 7$ pass through the point (-1, 3)? You can verify this by substituting the given points into the given equation. $3 \stackrel{?}{=} 4(-1) + 7$
   $3 \stackrel{?}{=} -4 + 7$
   $3 \stackrel{?}{=} 3$, yes

11. a) Bottom left point (-2, -3)
    b) $m = \dfrac{5}{6}$
    c) Second point (4, 2)

13. a) No, because it is only one point.
    b) No, because along with the slope one point needs to be known.
    c) Yes, because one point and one slope is all that is needed to write the equation of a line.
    d) Yes, because more than one point is given and the slope can be determined from the two points.

15. Choose the two points that lie on the line.
    (67, 170) and (79, 220)

NOTATION

17. Point-slope form: $y - y_1 = m(x - x_1)$

19. The original equation was in **point-slope** form. After solving for y, we obtain an equation in **slope-intercept** form.

PRACTICE

21. $y - 1 = 3(x - 2)$

23. $y + 1 = -\dfrac{4}{5}(x + 5)$

25.

$$y-1=\frac{1}{5}(x-10)$$
$$y-1=\frac{1}{5}x-2$$
$$y-1+1=\frac{1}{5}x-2+1$$
$$y=\frac{1}{5}x-1$$

27.

$$y-8=-5(x-(-9))$$
$$y-8=-5(x+9)$$
$$y-8=-5x-45$$
$$y-8+8=-5x-45+8$$
$$y=-5x-37$$

29.

$$y-(-4)=-\frac{4}{3}(x-6)$$
$$y+4=-\frac{4}{3}x+8$$
$$y+4-4=-\frac{4}{3}x+8-4$$
$$y=-\frac{4}{3}x+4$$

31.

$$y-(-6)=-\frac{11}{6}(x-2)$$
$$y+6=-\frac{11}{6}x+\frac{11}{3}$$
$$y+6-6=-\frac{11}{6}x+\frac{11}{3}-6$$
$$y=-\frac{11}{6}x+\frac{11}{3}-\frac{18}{3}$$
$$y=-\frac{11}{6}x-\frac{7}{3}$$

33.

$$y-0=-\frac{2}{3}(x-3)$$
$$y=-\frac{2}{3}x+2$$

35.

$$y-4=8(x-0)$$
$$y-4=8x$$
$$y-4+4=8x+4$$
$$y=8x+4$$

37.

$$y-0=-3(x-0)$$
$$y=-3x$$

39. Use the slope formula to find the slope.

$$m=\frac{y_2-y_1}{x_2-x_1}$$
$$=\frac{7-1}{1-(-2)}$$
$$=\frac{6}{3}$$
$$=2$$

Use the point (1, 7) with the slope-point formula:

$$y-y_1=m(x-x_1)$$
$$y-7=2(x-1)$$
$$y-7=2x-2$$
$$y-7+7=2x-2+7$$
$$y=2x+5$$

41. Use the slope formula to find the slope.

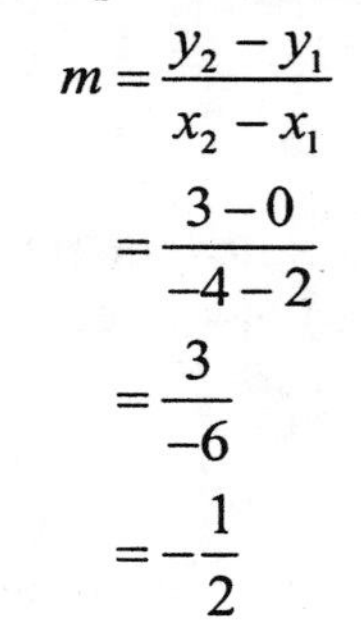

$$m=\frac{y_2-y_1}{x_2-x_1}$$
$$=\frac{3-0}{-4-2}$$
$$=\frac{3}{-6}$$
$$=-\frac{1}{2}$$

Use the point (2, 0) with the slope-point formula:

$$y-y_1=m(x-x_1)$$
$$y-0=-\frac{1}{2}(x-2)$$
$$y=-\frac{1}{2}x+1$$

43. Use the slope formula to find the slope.

$$m=\frac{y_2-y_1}{x_2-x_1}$$
$$=\frac{5-5}{7-5}$$
$$=\frac{0}{2}$$
$$=0$$

Use the point (5, 5) with the slope-point formula:

$$y-y_1=m(x-x_1)$$
$$y-5=0(x-5)$$
$$y-5=0$$
$$y-5+5=0+5$$
$$y=5$$

45. Use the slope formula to find the slope.

$$m = \frac{y_2 - y_1}{x_2 - x_1} = \frac{1-0}{5-(-5)} = \frac{1}{10}$$

Use the point (5, 1) with the slope-point formula:

$$y - y_1 = m(x - x_1)$$
$$y - 1 = \frac{1}{10}(x-5)$$
$$y - 1 = \frac{1}{10}x - \frac{1}{2}$$
$$y - 1 + 1 = \frac{1}{10}x - \frac{1}{2} + 1$$
$$y = \frac{1}{10}x - \frac{1}{2} + \frac{2}{2}$$
$$y = \frac{1}{10}x + \frac{1}{2}$$

47. Use the slope formula to find the slope.

$$m = \frac{y_2 - y_1}{x_2 - x_1} = \frac{17-2}{-8-(-8)} = \frac{15}{0} = \text{undefined}$$

A horizontal line has an undefined slope.

The equation is $x = -8$

49. Use the slope formula to find the slope.

$$m = \frac{y_2 - y_1}{x_2 - x_1} = \frac{-4-0}{-11-5} = \frac{-4}{-16} = \frac{1}{4}$$

Use the point (5, 0) with the slope-point formula:

$$y - y_1 = m(x - x_1)$$
$$y - 0 = \frac{1}{4}(x-5)$$
$$y = \frac{1}{4}x - \frac{5}{4}$$

51. $x = 4$
The equation of a **vertical line** has the form $x = a$.

53. $y = 5$
The equation of a **horizontal line** has the form $y = b$.

55.

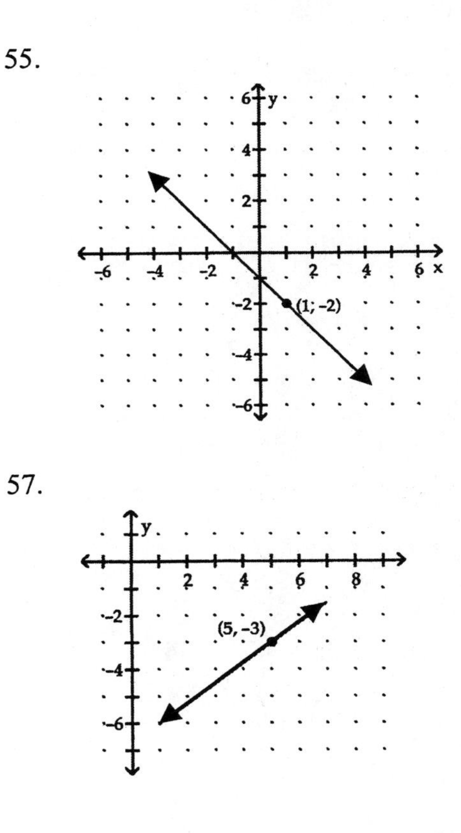

57.

59.

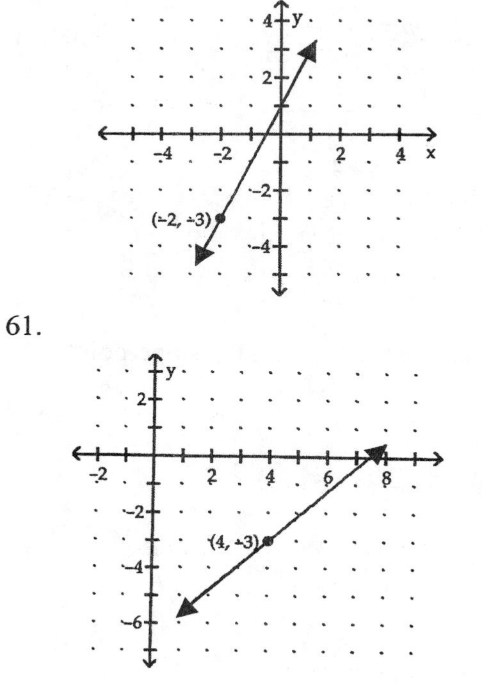

61.

APPLICATIONS

63. ANATOMY

In this problem, we are given two different letters *r and h* to use other than *x* and *y*. One has to determine which letter correlates with which letter.

$r = x$ and $h = y$.

Find the slope which is given as 3.9 inches for each 1-inch for the radius.

$$m = \frac{\text{rise}}{\text{run}} = \frac{3.9}{1} = 3.9$$

Given: 64-inch tall woman is h_1.
Given: 9-inch-long radius bone is r_1.
Now use the point-slope formula.

$$h - h_1 = m(r - r_1)$$
$$h - 64 = 3.9(r - 9)$$
$$h - 64 = 3.9r - 35.1$$

$$h - 64 + 64 = 3.9r - 35.1 + 64$$

$$\boxed{h = 3.9r + 28.9}$$

65. POLE VAULTING

Part 1: Given (5, 2) and (10, 0)
Find the slope:

$$m = \frac{y_2 - y_1}{x_2 - x_1} = \frac{2-0}{5-10} = \frac{2}{-5} = -\frac{2}{5}$$

Now pick one point (10, 0) and use

$$y - y_1 = m(x - x_1)$$
$$y - 0 = -\frac{2}{5}(x - 10)$$
$$\boxed{y = -\frac{2}{5}x + 4}$$

Part 2: Given (9, 7) and (10, 0)
Find the slope:

$$m = \frac{y_2 - y_1}{x_2 - x_1} = \frac{7-0}{9-10} = \frac{7}{-1} = -7$$

Now pick one point (10, 0) and use

$$y - y_1 = m(x - x_1)$$
$$y - 0 = -7(x - 10)$$
$$\boxed{y = -7x + 70}$$

Part 3: Given (10, 7) and (10, 0)
Find the slope:

$$m = \frac{y_2 - y_1}{x_2 - x_1} = \frac{7-0}{10-10} = \frac{7}{0}$$

Slope is undefined.

Undefined slope means the line is vertical and $x = a$ is the equation. $\boxed{x = 10}$

67. TOXIC CLEANUP

a) In this problem, we are given two different letters m and y *(yard)* to use for x and y. One has to determine which letter correlates with which letter, $m = x$ and $y = y$.

Find the slope which is given as two ordered pairs (3, 800) and (5, 720). The reason for the numbers 3 and 5 is because after "3 months" and then "2 months later".

$$m = \frac{y_2 - y_1}{m_2 - m_1}$$
$$= \frac{800 - 720}{3 - 5}$$
$$= \frac{80}{-2}$$
$$= -40$$

Now pick one point (3, 800) and use

$$y - y_1 = m(m - m_1)$$
$$y - 800 = -40(m - 3)$$
$$y - 800 = -40m + 120$$
$$y - 800 + 800 = -40m + 120 + 800$$
$$\boxed{y = -40m + 920}$$

b) To predict the number of cubic yards of waste on the site one year after the cleanup project began, let
$m = 1$ year $= 12$ months.

$$y = -40(12) + 920$$
$$= -480 + 920$$
$$= 440 \text{ yd}^3$$

69. TRAMPOLINES

In this problem, we are given two different letters r and l to use for x and y. One has to determine which letter correlates with which letter, $r = x$ and $l = y$.

Find the slope which is given as two ordered pairs (3, 19) and (7, 44).

$$m = \frac{l_2 - l_1}{r_2 - r_1}$$
$$= \frac{44 - 19}{7 - 3}$$
$$= \frac{25}{4}$$

Now pick one point (3, 19) and use
$l - l_1 = m(r - r_1)$

$$l - 19 = \frac{25}{4}(r - 3)$$
$$l - 19 = \frac{25}{4}r - \frac{75}{4}$$
$$l - 19 + 19 = \frac{25}{4}r - \frac{75}{4} + 19$$
$$l = \frac{25}{4}r - \frac{75}{4} + \frac{76}{4}$$
$$\boxed{l = \frac{25}{4}r - \frac{1}{4}}$$

71. GOT MILK

a) Two points must be selected (yr, gal). (11, 26.5) and (19, 24.5) were selected because they both lie on the line. Now you must find the slope of the two selected ordered pairs.

$$m = \frac{y_2 - y_1}{x_2 - x_1}$$
$$= \frac{26.5 - 24.5}{11 - 19}$$
$$= \frac{2}{-8}$$
$$= -\frac{1}{4}$$

Now pick one point (11, 26.5) and use

$$y - y_1 = m(x - x_1)$$
$$y - 26.5 = -\frac{1}{4}(x - 11)$$
$$y - 26.5 = -\frac{1}{4}x + \frac{11}{4}$$
$$y - 26.5 + 26.5 = -\frac{1}{4}x + 2.75 + 26.5$$
$$y = -\frac{1}{4}x + 29.25$$
$$\boxed{y = -0.25 + 29.25}$$

or

$$\boxed{y = -\frac{1}{4}x + \frac{117}{4}}$$

b) Use $x = 40$ (the year 2015 correlates with 40) to calculate the number of gallons of milk to be consumed.

$$y = -0.25x + 29.25$$
$$y = -0.25 \cdot 40 + 29.25$$
$$y = -10 + 29.25$$
$$\boxed{y = 19.25 \text{ gal}}$$

WRITING

73. Answers will vary.

75. Answers will vary.

REVIEW

77. FRAMING PICTURES

Given: length is 5 inches more than 2(width)
perimeter is 112 inches

Implies: width = w

Formula: $l + w + l + w = P$

Let w = width in inches
$2w + 5$ = length in inches

$$l + w + l + w = P$$
$$(2w + 5) + w + (2w + 5) + w = 112$$
$$6w + 10 = 112$$
$$6w + 10 - 10 = 112 - 10$$
$$6w = 102$$
$$\frac{6}{6}w = \frac{102}{6}$$
$$\boxed{w = 17}$$

$$l = 2w + 5$$
$$l = 2 \cdot 17 + 5$$
$$l = 34 + 5$$
$$\boxed{l = 39}$$

Dimensions: 17 in. by 39 in.

CHALLENGE PROBLEMS

79. Given (2, 5)
Find equation of line parallel to $y = 4x - 7$.

First, find the slope from $y = 4x - 7$, $m = 4$.

The two lines are parallel and have the same slope, so use $m = 4$ and (2, 5) to write the equation of the line.

$$y - y_1 = m(x - x_1)$$
$$y - 5 = 4(x - 2)$$
$$y - 5 = 4x - 8$$
$$y - 5 + 5 = 4x - 8 + 5$$
$$\boxed{y = 4x - 3}$$

SECTION 3.7

VOCABULARY

1. An **inequality** is a statement that contains one of the symbols $<$, $\leq$, $>$, or $\geq$

3. A **solution** of a linear inequality is an ordered pair of numbers that makes the inequality true.

5. In the graph, the line $2x - y = 4$ is the **boundary**.

7. When graphing a linear inequality, we determine which half-plane to shade by substituting the coordinates of a test **point** into the inequality.

CONCEPTS

9. a) False
   b) False
   c) True
   d) False

11. a)

$$x + 4y < -1$$
$$(\mathbf{3}) + 4(\mathbf{1}) \overset{?}{<} -1$$
$$3 + 4 \overset{?}{<} -1$$
$$7 \overset{?}{<} -1$$

No, (3, 1) is not a solution

b)

$$x + 4y < -1$$
$$(\mathbf{-2}) + 4(\mathbf{0}) \overset{?}{<} -1$$
$$-2 + 0 \overset{?}{<} -1$$
$$-2 < -1$$

Yes, (-2, 0) is a solution.

c)

$$x + 4y < -1$$
$$(\mathbf{-0.5}) + 4(\mathbf{0.2}) \overset{?}{<} -1$$
$$-0.5 + 0.8 \overset{?}{<} -1$$
$$0.3 \overset{?}{<} -1$$

No, (-0.5, 0.2) is not a solution.

11. d)

$$x + 4y < -1$$
$$(\mathbf{-2}) + 4\left(\frac{\mathbf{1}}{\mathbf{4}}\right) \overset{?}{<} -1$$
$$-2 + 1 \overset{?}{<} -1$$
$$-1 \overset{?}{<} -1$$

No, $\left(-2, \frac{1}{4}\right)$ is not a solution.

13. a) **No,** $y > -x$ would not include the boundary line.
    b) **Yes,** $5x - 3y \leq -2$ would include the boundary line.

15. If the test point is false, you shade the half-plane opposite that in which the test point lies.

17. a) **Yes,** the graph is shaded over (2, 1).
    b) **No,** the graph is not shaded over (-2, -4).
    c) **No,** the graph is not shaded over (4, -2).
    d) **Yes,** the graph is shaded over (-3, 4).

19. a) The graph of $y = 5$ is a horizontal line.
    b) The graph of $x = -6$ is a vertical line.

NOTATION

21. a) $<$ is the symbol for less than
    b) $>$ is the symbol for greater than
    c) $\leq$ is the symbol for less than or equal to
    d) $\geq$ is the symbol for greater than or equal to

23. The symbols $\leq$ and $\geq$ are associated with a solid boundary line.

PRACTICE

25. $x - y \geq -2$

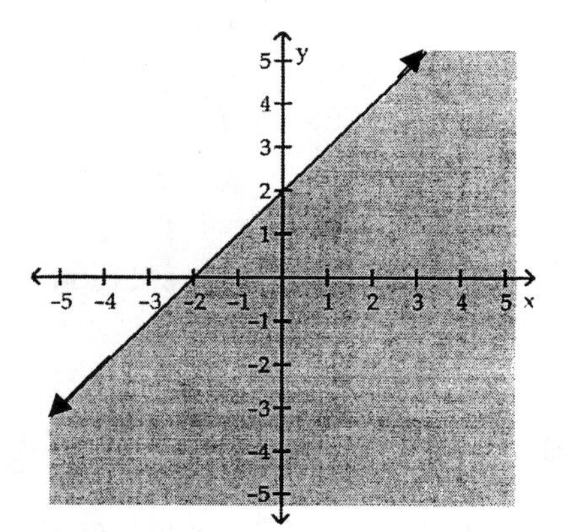

27. $y > 2x - 4$

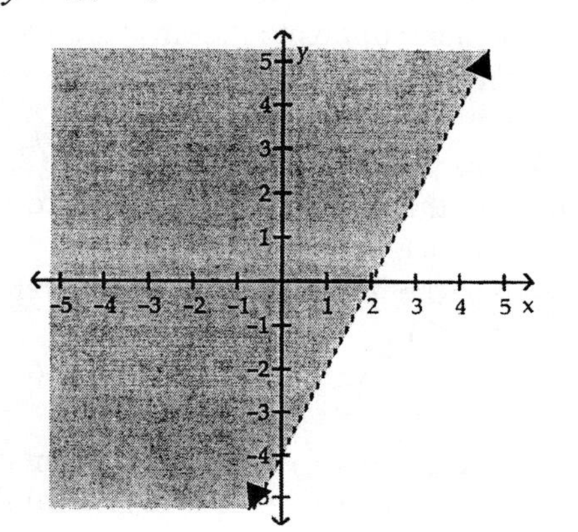

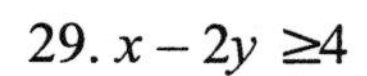

29. $x - 2y \geq 4$

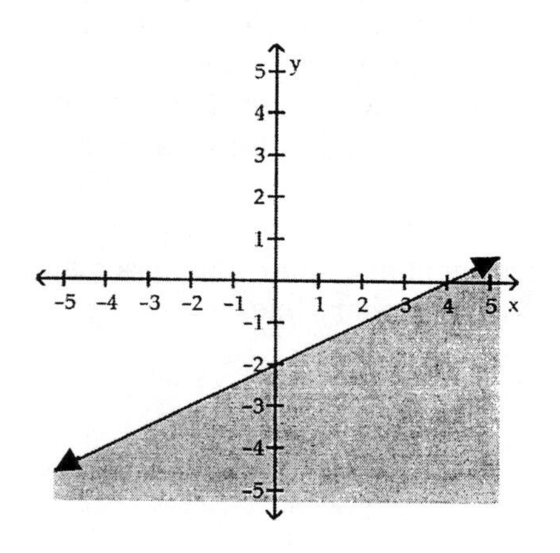

31. $y \leq 4x$

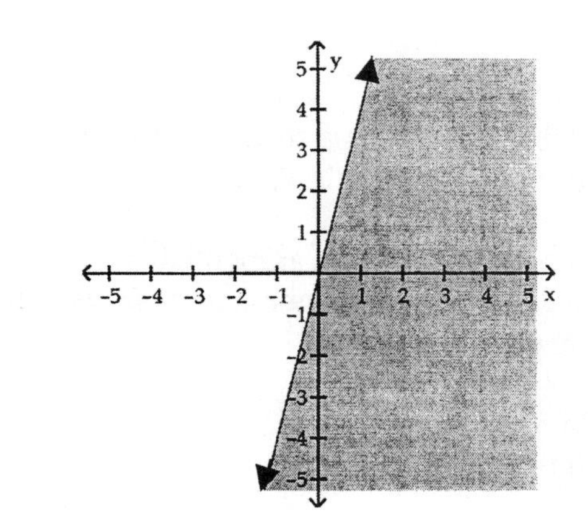

33. $x + y \geq 3$

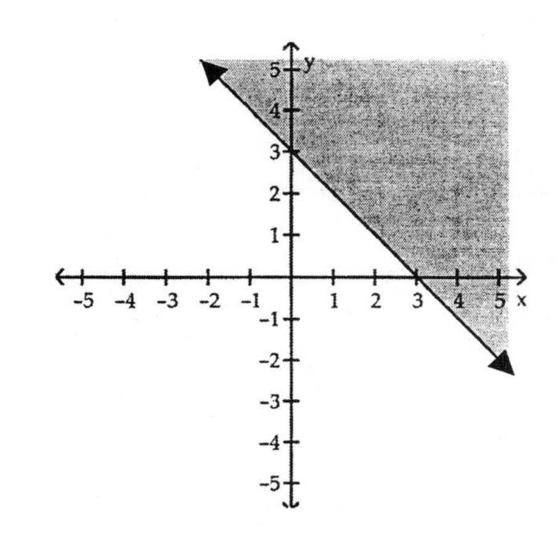

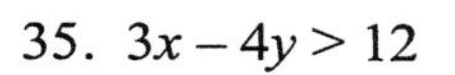

35. $3x - 4y > 12$

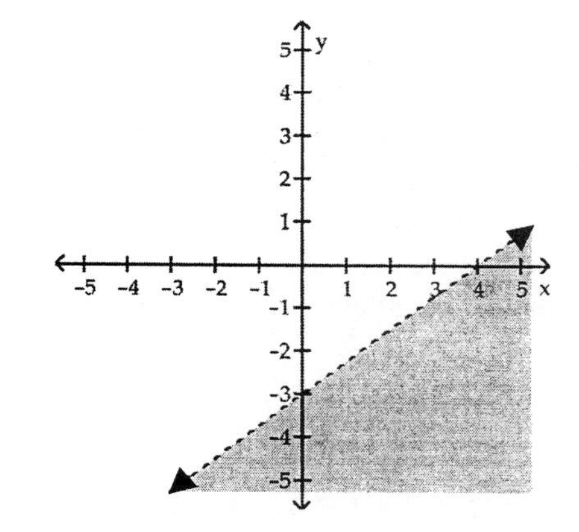

37. $2x + 3y \leq -12$

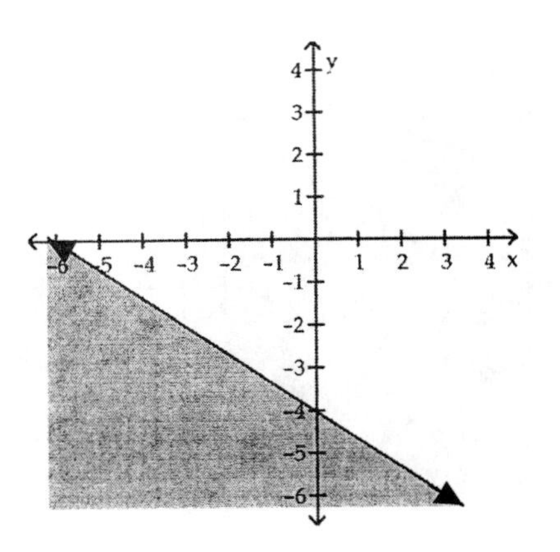

39. $y < 2x - 1$

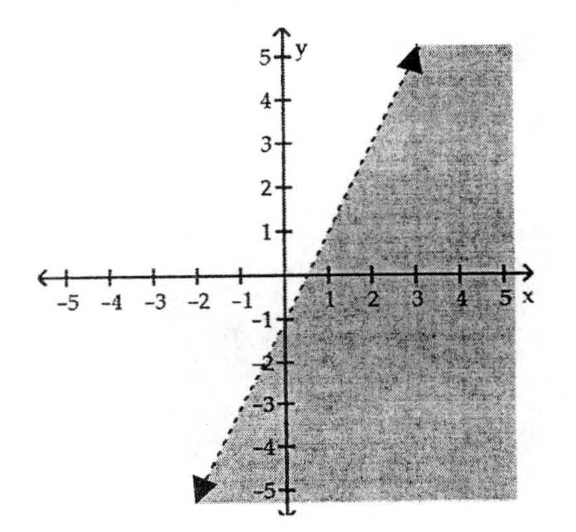

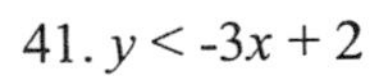

41. $y < -3x + 2$

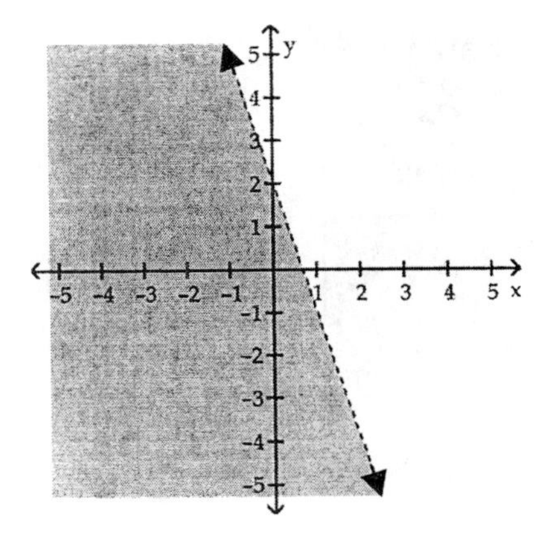

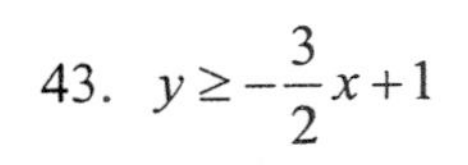

43. $y \geq -\frac{3}{2}x + 1$

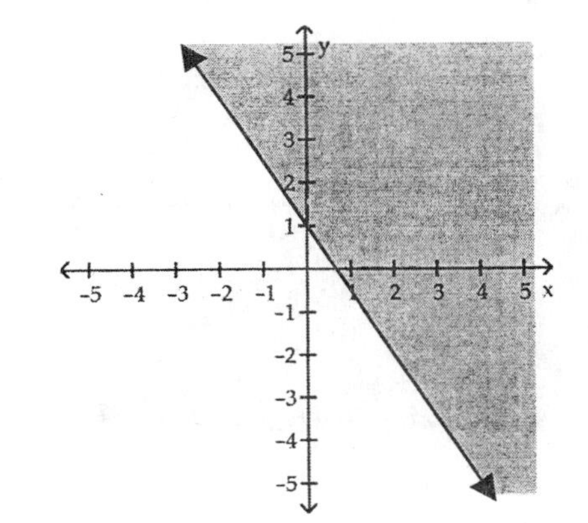

45. $x - 2y \geq 4$

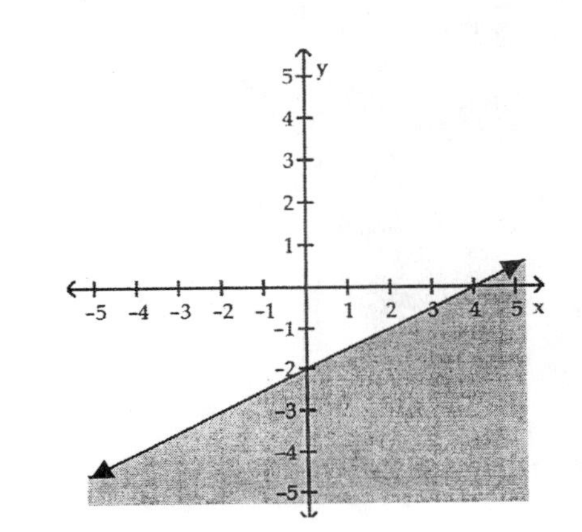

47. $2y - x < 8$

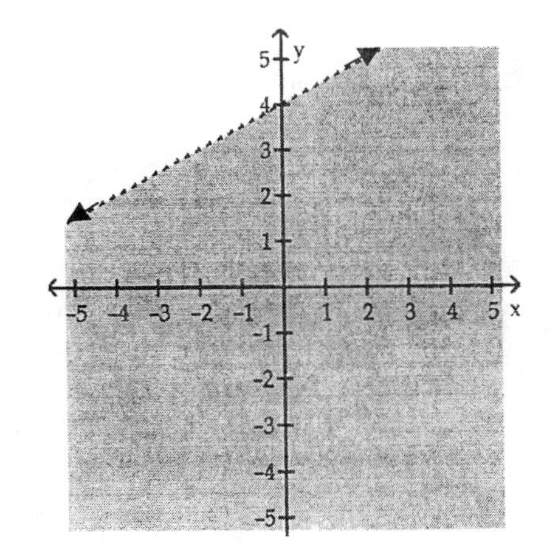

49. $y + x < 0$

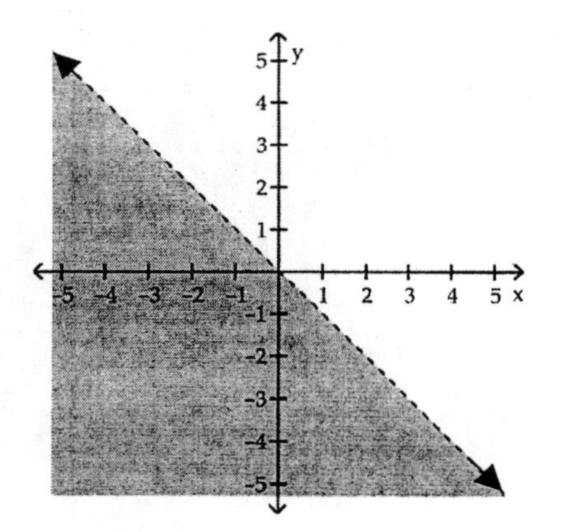

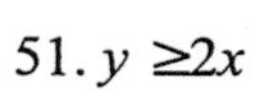
51. $y \geq 2x$

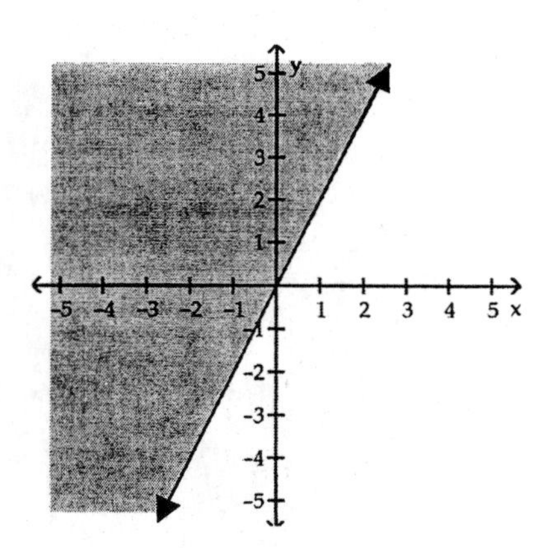

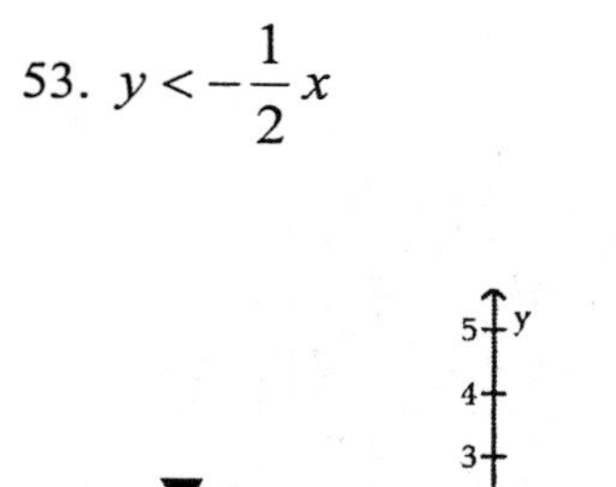
53. $y < -\frac{1}{2}x$

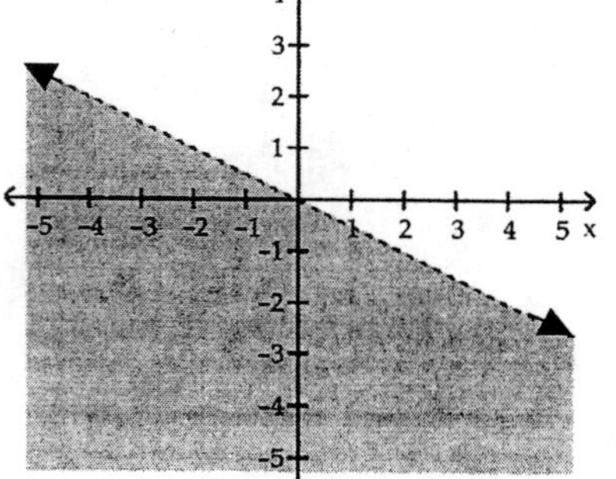

55. $x < 2$

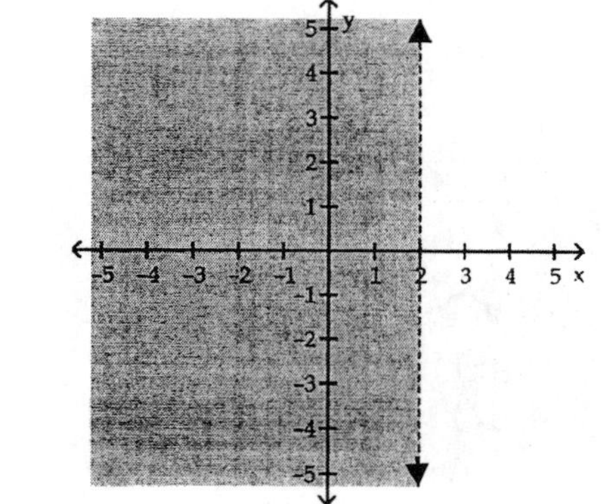

57. $y \leq 1$

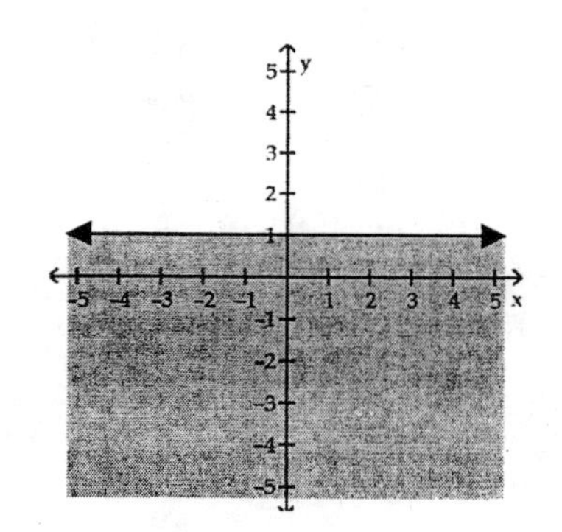

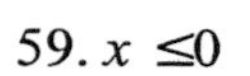
59. $x \leq 0$

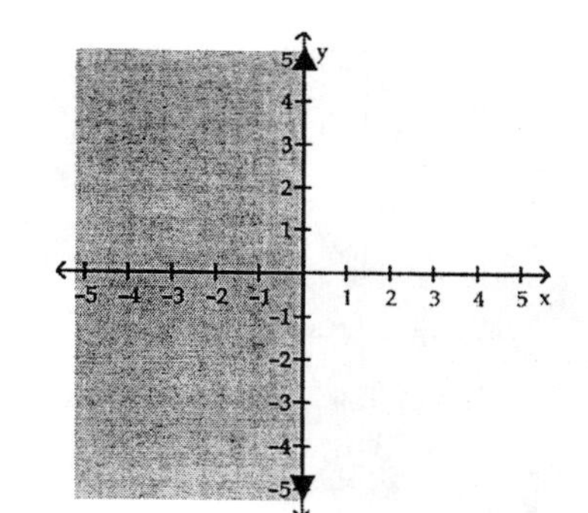

61. $7x - 2y < 21$

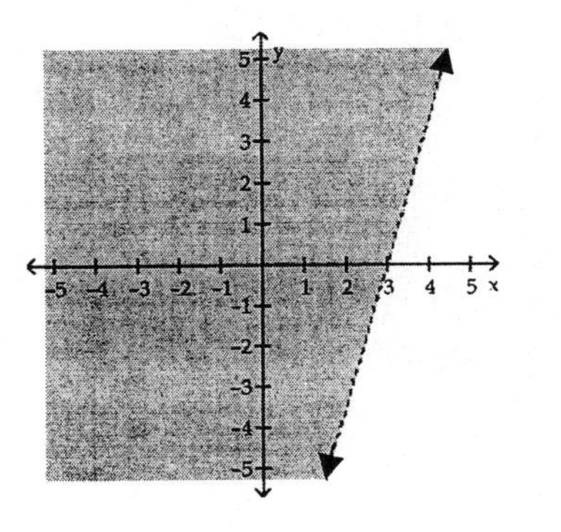

63. $2x - 3y \geq 4$

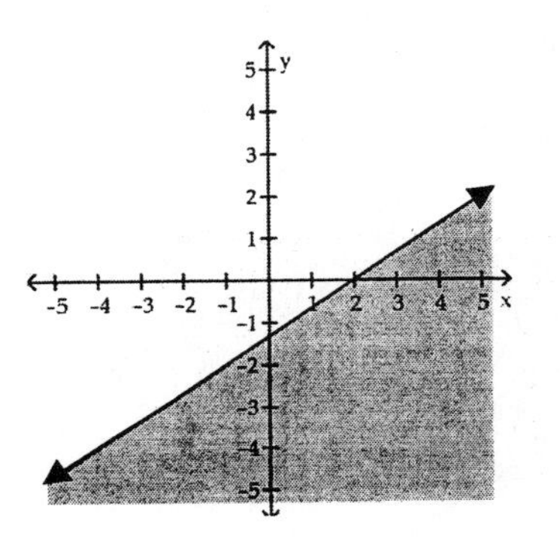

65. $5x + 3y < 0$

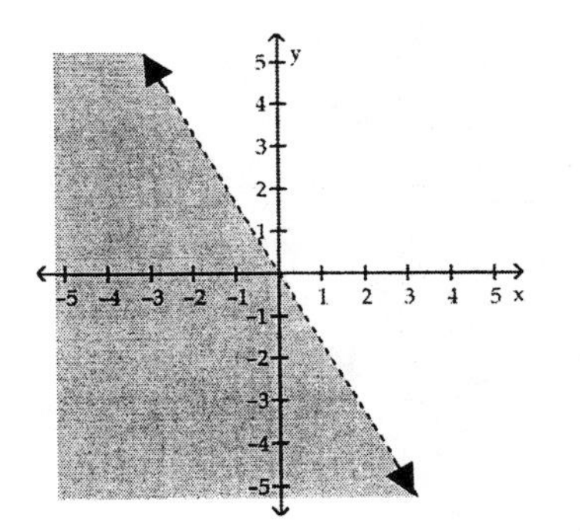

APPLICATIONS

67. ii—On all of the shaded areas, the sum exceeds 6.

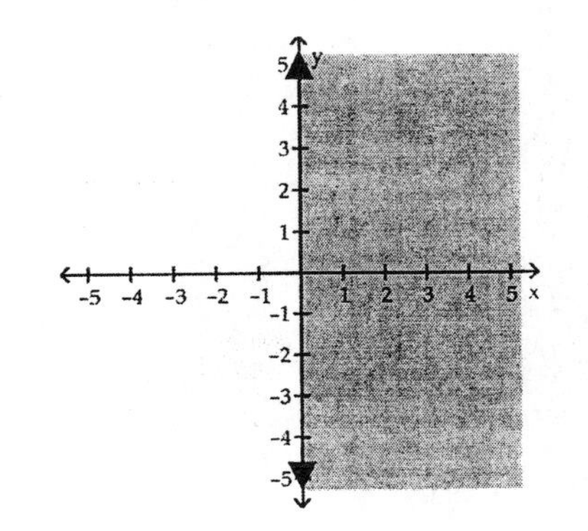

69. $3x + 4y \leq 120$

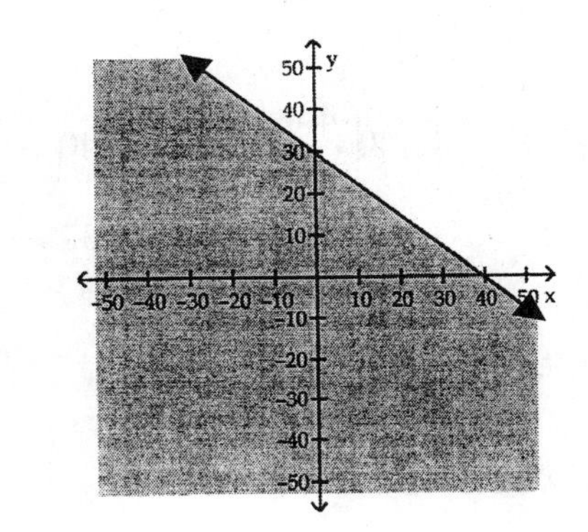

71. $100x + 88y \geq 4{,}400$

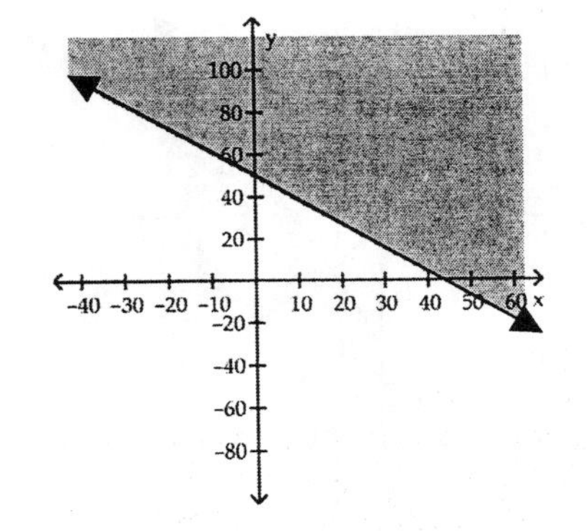

WRITING

73. Answers will vary.

75. Answers will vary.

REVIEW

77.

$$A = P + Prt$$
$$A - P = Prt$$
$$\frac{A - P}{Pr} = \frac{Prt}{Pr}$$
$$\frac{A - P}{Pr} = t$$

79.

$$40\left(\frac{3}{8}x - \frac{1}{4}\right) + 40\left(\frac{4}{5}\right) = 40\left(\frac{3}{8}x\right) + 40\left(-\frac{1}{4}\right) + 40\left(\frac{4}{5}\right)$$
$$= 15x - 10 + 32$$
$$= 15x + 22$$

CHALLENGE PROBLEMS

81. $3x - 2y \geq 6$

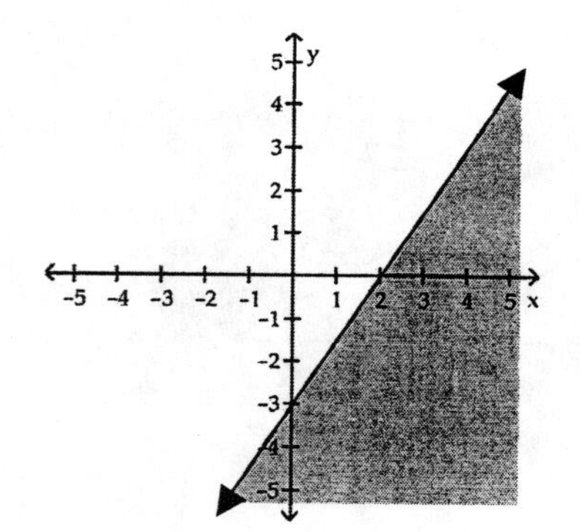

CHAPTER 3 KEY CONCEPTS

1. Use point–slope form of a line:
   $y - y_1 = m(x - x_1)$
   $y - (\mathbf{-4}) = \mathbf{-3}(x - \mathbf{0})$
   $y + 4 = -3x + 0$
   $y + 4 - 4 = -3x + 0 - 4$
   $y = -3x - 4$

2. Find the slope of the line containing the two points:
   $$m = \frac{y_2 - y_1}{x_2 - x_1} = \frac{0-2}{-5-5} = \frac{-2}{-10} = \frac{1}{5}$$

   Use point–slope form of a line:
   $y - y_1 = m(x - x_1)$

   $$y - \mathbf{2} = \frac{\mathbf{1}}{\mathbf{5}}(x - \mathbf{5})$$
   $$y - 2 = \frac{1}{5}x - 1$$
   $$y - 2 + 2 = \frac{1}{5}x - 1 + 2$$
   $$y = \frac{1}{5}x + 1$$

3. Let $c = 160$.
   $$T = \frac{1}{4}c + 40$$
   $$T = \frac{1}{4}(\mathbf{160}) + 40$$
   $$T = 40 + 40$$
   $$T = 80$$

   So T = 80° F

4. $y = 209x + 2{,}660$

   To find the expenditure in 2020, let $x = 2020 - 1990 = 30$.
   $y = 209(\mathbf{30}) + 2{,}660$
   $y = 6{,}270 + 2{,}660$
   $y = 8{,}930$
   The expenditure in 2020 would be $8,930.

5. $2x - 4y = 8$
   Let $x = 0$ and solve for y.
   $2(\mathbf{0}) - 4y = 8$
   $0 - 4y = 8$
   $-4y = 8$
   $y = -2$

   Let $y = 0$ and solve for x.
   $2x - 4(\mathbf{0}) = 8$
   $2x - 0 = 8$
   $2x = 8$
   $x = 4$
   Let $x = -2$ and solve for y.
   $2(\mathbf{-2}) - 4y = 8$
   $-4 - 4y = 8$
   $-4 - 4y + 4 = 8 + 4$
   $-4y = 12$
   $y = -3$

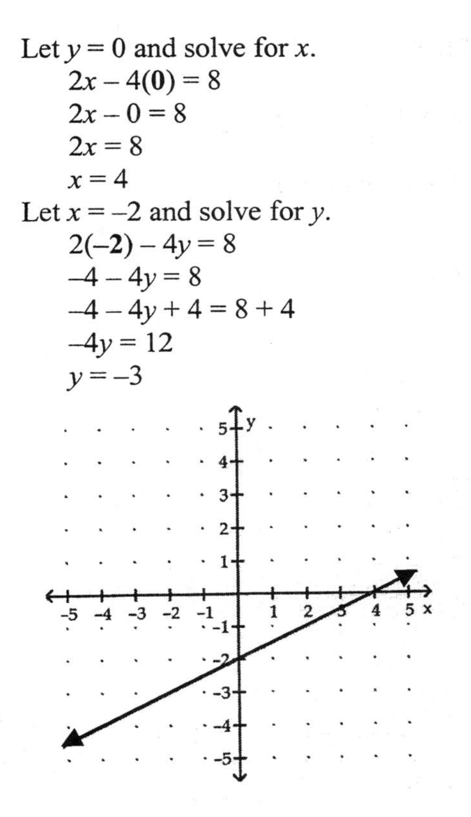

6. a. The y–intercept indicates that the cost of the press is $40,000 when new.
   b. The slope of the line is –5,000. This indicates that the value of the press decreased $5,000 each year.

7. a. Answers may vary. Possible answers are (–4, 2), (–3, 0), (–2, –2), (–1, –4) or (0, –6)
   b. Pick two points and use the formula for slope:
   $$m = \frac{y_2 - y_1}{x_2 - x_1} = \frac{0-2}{-3-(-4)} = \frac{-2}{1} = -2$$
   c. Use the point–slope form of a line with $m = -2$ and any point on the line such as (–3, 0).
   $y - y_1 = m(x - x_1)$
   $y - \mathbf{0} = \mathbf{-2}(x - (\mathbf{-3}))$
   $y = -2(x + 3)$
   $y = -2x - 6$

8. Parallel lines have the same slope. A line parallel to the graphed line would be a vertical line going through the point (1, –1). The equation of a vertical line through that point would be $\boldsymbol{x = 1}$.

CHAPTER 3 REVIEW

SECTION 3.1
Graphing Using the Rectangular Coordinate System

1.

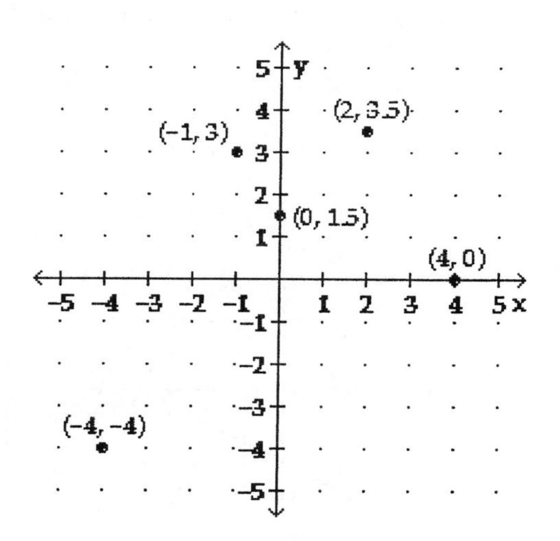

2. HAWAIIAN ISLANDS

   The coordinates of Oahu are approximately (158, 21.5).

3. Since both coordinates are negative, the point (-3, -4) lies in Quadrant III.

4. The coordinates of the origin are (0, 0).

5. GEOMETRY

   The coordinates of the fourth vertex would be (1, 4). Each side of the square is 6 units long.

   A = (length)(width)
   A = (6)(6)
   A = 36 square units

6. COLLEGE ENROLLMENTS

   a. The maximum enrollment was 2,500 students and occurred in Week 2.
   b. Two weeks before the semester began, the number of students enrolled was 1,000.
   c. The enrollment was 2,250 during week 1 and week 5.

SECTION 3.2
Graphing Linear Equations

7. Check to see if (-3, -2) is a solution of $y = 2x + 4$ by substituting –3 for x and –2 for y.

   $\mathbf{-2} \stackrel{?}{=} 2(\mathbf{-3}) + 4$
   $-2 \stackrel{?}{=} -6 + 4$
   $-2 \stackrel{?}{=} -2$
   $-2 = -2$

   So, (-3, -2) is a solution of $y = 2x + 4$.

8. For $x = -2$, substitute –2 in for x and solve for y.

   $3x + 2y = -18$
   $3(\mathbf{-2}) + 2y = -18$
   $-6 + 2y = -18$
   $-6 + 6 + 2y = -18 + 6$
   $2y = -12$
   $y = -6$

   The ordered pair would be (-2, -6).

   If $y = 3$, then substitute in 3 for y and solve the equation for x.

   $3x + 2(\mathbf{3}) = -18$
   $3x + 6 = -18$
   $3x + 6 - 6 = -18 - 6$
   $3x = -24$
   $x = -8$

   The ordered pair would be (-8, 3).

9. The equations ii and v are not linear equations because the x term in both equations have term that contains an exponent higher than 1.

10. $y = 4x - 2$

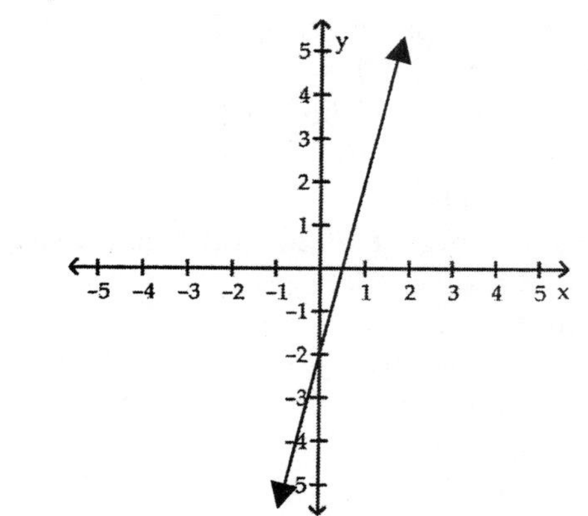

11. Solve the equation for y.

$5y = -5x + 15$

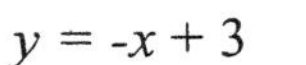

$\frac{5y}{5} = \frac{-5x}{5} + \frac{15}{5}$

$y = -x + 3$

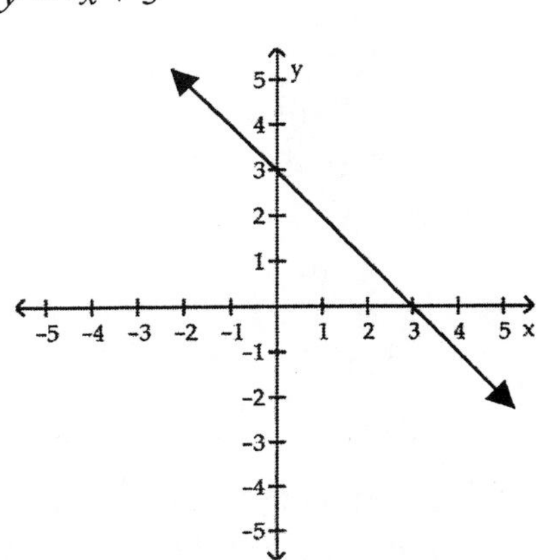

12. a) Since point A lies on the line, substituting the coordinates into the equation would result in a **true** statement.

b) Since point B does not lie on the line, substituting the coordinates into the equation would result in a **false** statement.

13. BIRTHDAY PARTIES

$c = 8n + 50$

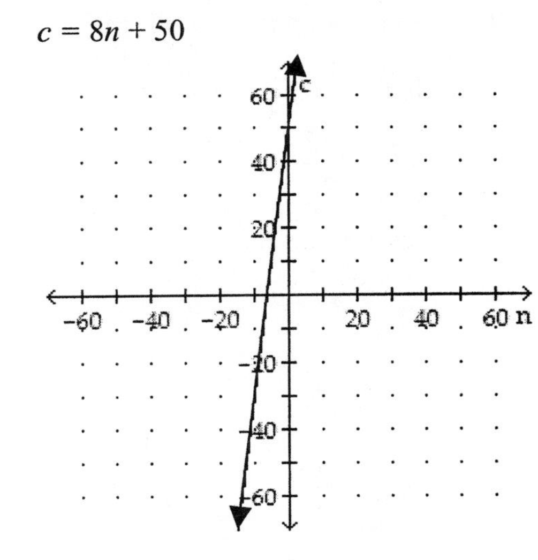

According to the graph, the cost for a party with 18 children attending would be approximately $195.

SECTION 3.3
More about Graphing Linear Equations

14. The x-intercept is the point the line crosses the x-axis, so the x-intercept is (-3, 0). The y-intercept is the point the line crosses the y-axis, so the x-intercept is (0, 2.5).

15. To find the x-intercept, let $y = 0$.

$-4x + 2y = 8$
$-4x + 2(\mathbf{0}) = 8$
$-4x + 0 = 8$
$-4x = 8$
$x = -2$

So, the x-intercept is (-2, 0).

To find the y-intercept, let $x = 0$.

$-4x + 2y = 8$
$-4(\mathbf{0}) + 2y = 8$
$0 + 2y = 8$
$2y = 8$
$y = 4$

So, the y-intercept is (0, 4).

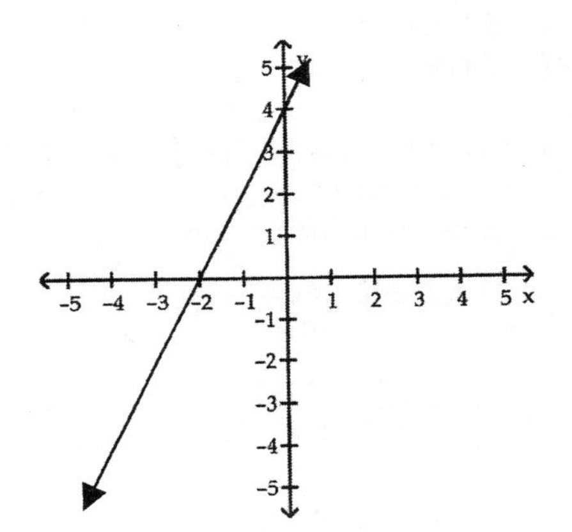

16. The graph $y = 4$ is a horizontal line through (0, 4).

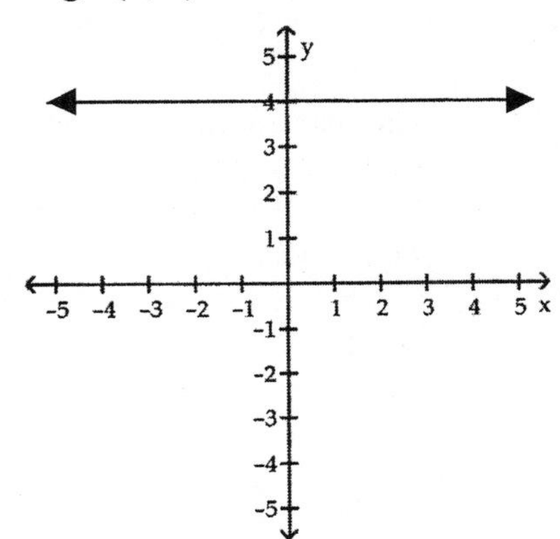

17. The graph $x = -1$ is a vertical line through (-1, 0).

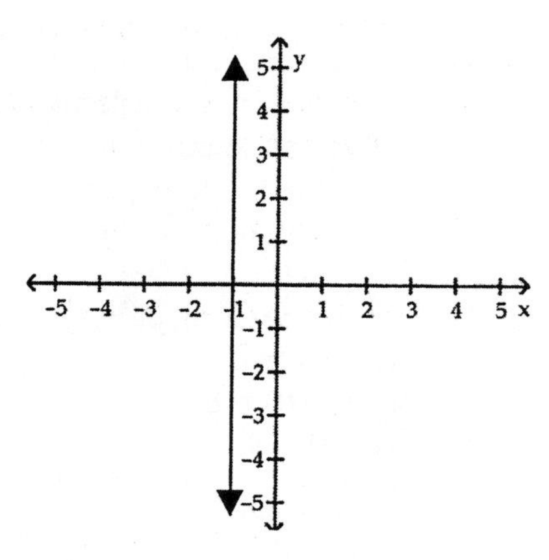

18. DEPRECIATION

The y-intercept is (0, 25,000). This indicates that when the equipment was new, it was originally valued at $25,000. The x-intercept is (10, 0). This indicates that in 10 years, the equipment will have no value.

SECTION 3.4
The Slope of a Line

19. Use (-4, -4) and (4, -2) for the points (x_1, y_1) and (x_2, y_2) and substitute the values into the equation:

$$\begin{aligned} m &= \frac{y_2 - y_1}{x_2 - x_1} \\ &= \frac{-2-(-4)}{4-(-4)} \\ &= \frac{-2+4}{4+4} \\ &= \frac{2}{8} \\ &= \frac{1}{4} \end{aligned}$$

20. The slope of a horizontal line is always **zero**. You can pick any two points such as (1, -3) and (4, -3) to substitute into the equation:

$$\begin{aligned} m &= \frac{y_2 - y_1}{x_2 - x_1} \\ &= \frac{-3-(-3)}{1-4} \\ &= \frac{-3+3}{1-4} \\ &= \frac{0}{-3} \\ &= 0 \end{aligned}$$

21. Use (2, -3) and (4, -17) for the points (x_1, y_1) and (x_2, y_2) and substitute the values into the equation:

$$\begin{aligned} m &= \frac{y_2 - y_1}{x_2 - x_1} \\ &= \frac{-17-(-3)}{4-2} \\ &= \frac{-17+3}{4-2} \\ &= \frac{-14}{2} \\ &= -7 \end{aligned}$$

22. Use (1, -4) and (3, -7) for the points (x_1, y_1) and (x_2, y_2) and substitute the values into the equation:

$$\begin{aligned} m &= \frac{y_2 - y_1}{x_2 - x_1} \\ &= \frac{-7-(-4)}{3-1} \\ &= \frac{-7+4}{3-1} \\ &= \frac{-3}{2} \\ &= -\frac{3}{2} \end{aligned}$$

23. CARPENTRY

$$\begin{aligned} m &= \frac{\text{change in } y}{\text{change in } x} \\ &= \frac{6}{16} \\ &= \frac{3}{8} \end{aligned}$$

24. HANDICAP ACCESSIBILITY

$$m = \frac{\text{change in } y}{\text{change in } x}$$
$$= \frac{2}{24}$$
$$= \frac{1}{12}$$
$$\approx 0.083$$
$$\approx 8.3\%$$

25. TOURISM

a. The y values would be the number of visitors and the x values would be the years.

$$m = \frac{\text{change in } y}{\text{change in } x}$$
$$= \frac{44.8 - 47.3}{1994 - 1992}$$
$$= \frac{-2.5}{2}$$
$$= -1.25$$

The rate of change was -1.25 million people per year.

b. The y values would be the number of visitors and the x values would be the years.

$$m = \frac{\text{change in } y}{\text{change in } x}$$
$$= \frac{34.1 - 26.0}{1988 - 1986}$$
$$= \frac{8.1}{2}$$
$$= 4.05$$

The rate of change was 4.05 million people per year.

SECTION 3.5
Slope-Intercept Form

26. Use the formula $y = mx + b$.
In $y = \frac{3}{4}x - 2$, $m = \frac{3}{4}$ and $b = -2$. So, the slope is $\frac{3}{4}$ and the y-intercept is $(0, -2)$.

27. Use the formula $y = mx + b$.
In $y = -4x$, $m = -4$ and $b = 0$. So, the slope is -4 and the y-intercept is $(0, 0)$.

28. Use the formula $y = mx + b$. Rewrite $\frac{x}{8}$ as $\frac{1}{8}x$. Then the equation is $y = \frac{1}{8}x + 10$, so $m = \frac{1}{8}$ and $b = 10$. So, the slope is $\frac{1}{8}$ and the y-intercept is $(0, 10)$.

29. Solve the equation for y.

$$7x + 5y = -21$$
$$7x - 7x + 5y = -7x - 21$$
$$5y = -7x - 21$$
$$\frac{5y}{5} = \frac{-7x}{5} - \frac{21}{5}$$
$$y = -\frac{7}{5}x - \frac{21}{5}$$

$m = -\frac{7}{5}$ and $b = -\frac{21}{5}$

The slope is $-\frac{7}{5}$, and the y-intercept is $(0, -\frac{21}{5})$.

30. Let $m = -6$ and $b = 4$ and substitute the values into the equation:

$$y = mx + b$$
$$y = -6x + 4$$

31. Solve the equation for y.

$$9x - 3y = 15$$
$$9x - 9x - 3y = -9x + 15$$
$$-3y = -9x + 15$$
$$\frac{-3y}{-3} = \frac{-9x}{-3} + \frac{15}{-3}$$
$$y = 3x - 5$$

Since $m = 3$ and $b = -5$, the slope is 3 and the y-intercept is $(0, -5)$.

32. COPIERS

a. 300 is the slope (rate of change) and 75,000 is the y-intercept. The equation would be

$$c = 300w + 75{,}000.$$

b. Let $w = 52$ and substitute it into the equation $c = 300w + 75{,}000$.

$c = 300(\mathbf{52}) + 75{,}000$

$c = 15{,}600 + 75{,}000$

$c = 90{,}600$

In 1 year (52 weeks), the number of copies made would be 90,600.

33. a. Both lines have a slope of $-\frac{2}{3}$. Since the slope is identical for both lines, the lines must be **parallel**.

b. Solve the first equation for y to find the slope.

$x + 5y = -10$

$x + 5y - x = -x - 10$

$5y = -x - 10$

$\frac{5y}{5} = \frac{-x}{5} - \frac{10}{5}$

$y = -\frac{1}{5}x - 2$

The slope of the first equation is $-\frac{1}{5}$ and the slope of the second equation is 5. Since $-\frac{1}{5}$ and 5 are negative reciprocals, the lines must be **perpendicular**.

SECTION 3.6
Point-Slope Form

34. $m = 3$ and $(x_1, y_1) = (1, 5)$

Use point-slope form of the equation:

$y - y_1 = m(x - x_1)$

$y - \mathbf{5} = \mathbf{3}(x - \mathbf{1})$

$y - 5 = 3(x) + 3(-1)$

$y - 5 = 3x - 3$

$y - 5 + 5 = 3x - 3 + 5$

$y = 3x + 2$

35. $m = -\frac{1}{2}$ and $(x_1, y_1) = (-4, -1)$

Use point-slope form of the equation:

$y - y_1 = m(x - x_1)$

$y - (\mathbf{-1}) = -\frac{1}{2}(x - (\mathbf{-4}))$

$y + 1 = -\frac{1}{2}(x + 4)$

$y + 1 = -\frac{1}{2}(x) - \frac{1}{2}(4)$

$y + 1 = -\frac{1}{2}x - 2$

$y + 1 - 1 = -\frac{1}{2}x - 2 - 1$

$y = -\frac{1}{2}x - 3$

36. Find the slope of the line containing the two points (3, 7) and (-6, 1):

$m = \frac{y_2 - y_1}{x_2 - x_1}$

$= \frac{1-7}{-6-3}$

$= \frac{-6}{-9}$

$= \frac{2}{3}$

Use the slope of ⅔ and one of the points to substitute into the point-slope form of a line.

Let $m = \frac{2}{3}$ and $(x_1, y_1) = (3, 7)$.

$y - y_1 = m(x - x_1)$

$y - 7 = \frac{2}{3}(x - 3)$

$y - 7 = \frac{2}{3}(x) + \frac{2}{3}(-3)$

$y - 7 = \frac{2}{3}x - 2$

$y - 7 + 7 = \frac{2}{3}x - 2 + 7$

$y = \frac{2}{3}x + 5$

37. The slope of a horizontal line is always zero. Use the point-slope formula with $m = 0$ and $(x_1, y_1) = (6, -8)$.

$y - y_1 = m(x - x_1)$

$y - (-8) = 0(x - 6)$

$y + 8 = 0(x) + 0(-6)$

$y + 8 = 0x - 0$

$y + 8 - 8 = 0x - 0 - 8$

$y = 0x - 8$

$y = -8$

38. CAR REGISTRATION

Let the x-values be the number of years old and the y-values be the amount of the registration fee. As an ordered pair, it would be (2, 380) and (4, 310). First find the rate of change, or slope.

$$m = \frac{y_2 - y_1}{x_2 - x_1}$$

$$= \frac{310 - 380}{4 - 2}$$

$$= \frac{-70}{2}$$

$$= -35$$

Use $m = -35$ and the point (2, 380) to substitute the values into the point-slope formula.

$y - y_1 = m(x - x_1)$

$y - \mathbf{380} = -35(x - \mathbf{2})$

$y - 380 = -35(x) + -35(-2)$

$y - 380 = -35x + 70$

$y - 380 + 380 = -35x + 70 + 380$

$y = -35x + 450$

SECTION 3.7
Graphing Linear Inequalities

39. a. $2x - y \leq -4$

$2(\mathbf{0}) - \mathbf{5} \stackrel{?}{=} -4$

$0 - 5 \stackrel{?}{=} -4$

$-5 \leq -4$

So yes, (0, 5) is a solution.

b. $2x - y \leq -4$

$2(\mathbf{2}) - \mathbf{8} \stackrel{?}{=} -4$

$4 - 8 \stackrel{?}{=} -4$

$-4 \leq -4$

So yes, (2, 8) is a solution.

c. $2x - y \leq -4$

$2(\mathbf{-3}) - (\mathbf{-2}) \stackrel{?}{=} -4$

$-6 + 2 \stackrel{?}{=} -4$

$-4 \leq -4$

So yes, (-3, -2) is a solution.

d. $2x - y \leq -4$

$2(-½) - (\mathbf{-5}) \stackrel{?}{=} -4$

$-1 + 5 \stackrel{?}{=} -4$

$4 \geq -4$

No, (-½, -5) is not a solution.

40. Solve the equation for y:

$x - y < 5$

$-y < -x + 5$

$\frac{-y}{-1} < \frac{-x}{-1} + \frac{5}{-1}$

$y > x - 5$

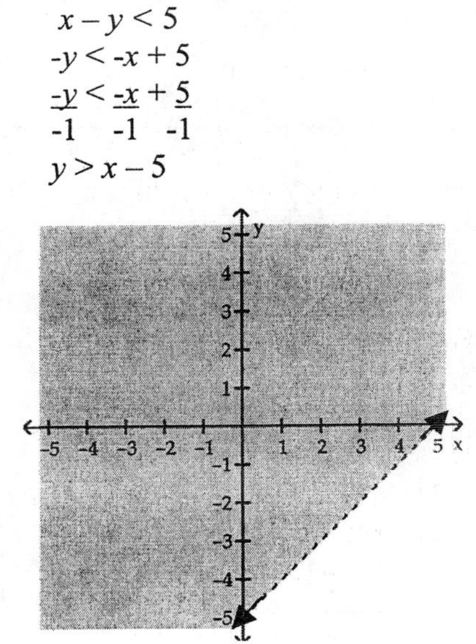

41. Solve the equation for y.

$$2x - 3y \geq 6$$

$$-3y \geq -2x + 6$$

$$\frac{-3y}{-3} \geq \frac{-2x}{-3} + \frac{6}{-3}$$

$$y \leq \frac{2}{3}x - 2$$

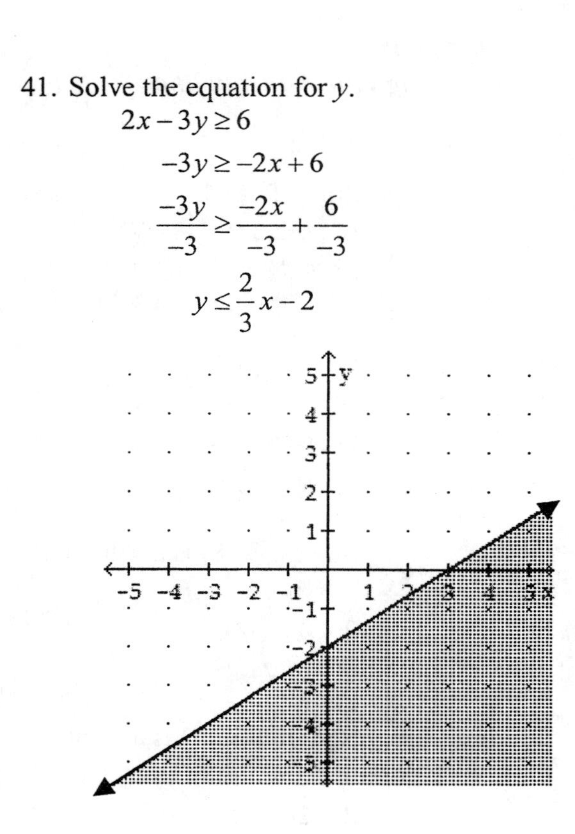

42. $y \leq -2x$

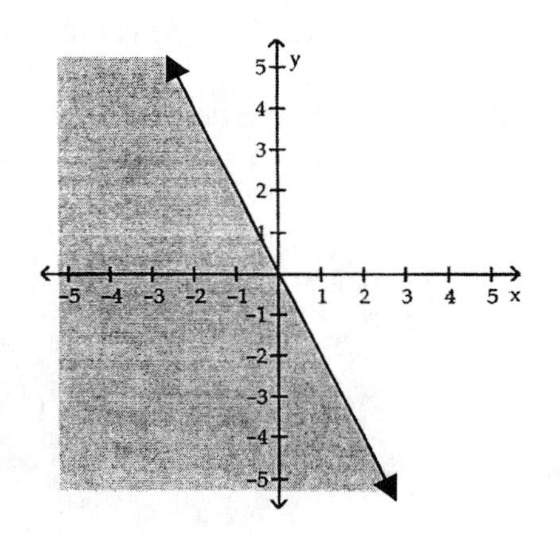

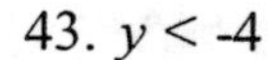

43. $y < -4$

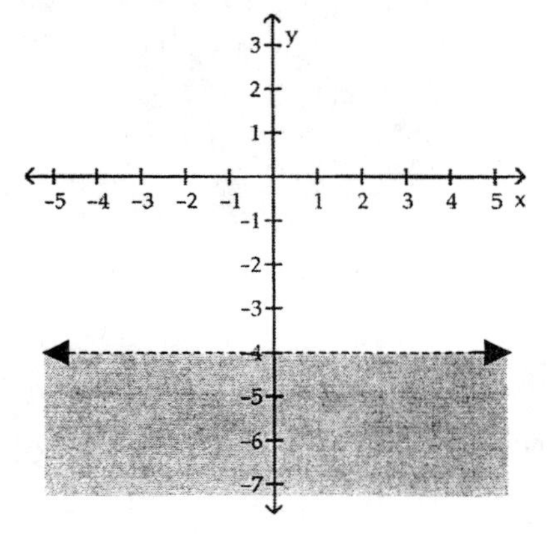

44. a. true
 b. false
 c. false

45. WORK SCHEDULES

Work Schedules: Answers may vary. Possible answers are (2, 4), (5, 3), and (6, 2).

46. An equation contains an equal sign (=). An inequality contains an inequality symbol such as <, >, ≤, or ≥

CHAPTER 3 TEST

1. 10

2. 60

3. There were 30 dogs in the kennel **1 day before** the holiday began and on the **3rd day** of the holiday.

4. The y–intercept shows there were **50** dogs in the kennel when the holiday began.

5. 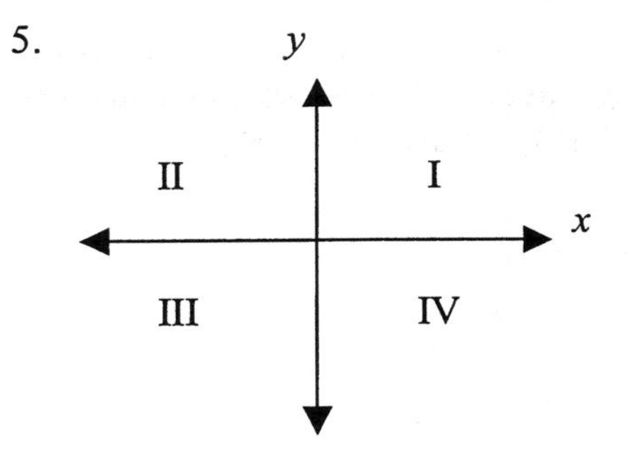

6. Substitute in 2 for x and solve for y:
$x + 4y = 6$
$\mathbf{2} + 4y = 6$
$2 - 2 + 4y = 6 - 2$
$4y = 4$
$y = 1$
So the ordered pair would be (2, 1).

Substitute in 3 for y and solve for x:
$x + 4y = 6$
$x + 4(\mathbf{3}) = 6$
$x + 12 = 6$
$x + 12 - 12 = 6 - 12$
$x = -6$
So the ordered pair would be (–6, 3).

7. Substitute $x = -3$ and $y = -4$ into the equation:
$3x - 4y = 7$
$3(\mathbf{-3}) - 4(\mathbf{-4}) \stackrel{?}{=} 7$
$-9 + 16 \stackrel{?}{=} 7$
$7 = 7$
Yes, (–3, –4) is a solution.

8. a. The result would be **false** because the point does not lie on the line.
b. The result would be **true** because the point does lie on the line.

9. $y = \frac{1}{3}x$

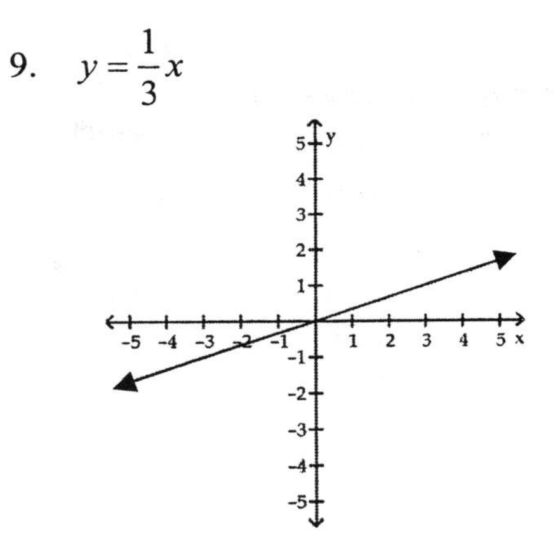

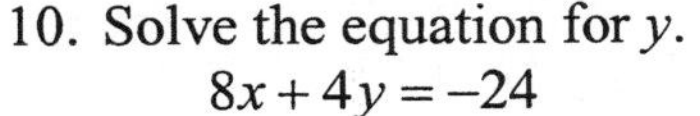

10. Solve the equation for y.
$8x + 4y = -24$
$8x + 4y - 8x = -24 - 8x$
$4y = -8x - 24$
$\frac{4y}{4} = \frac{-8x}{4} - \frac{24}{4}$
$y = -2x - 6$

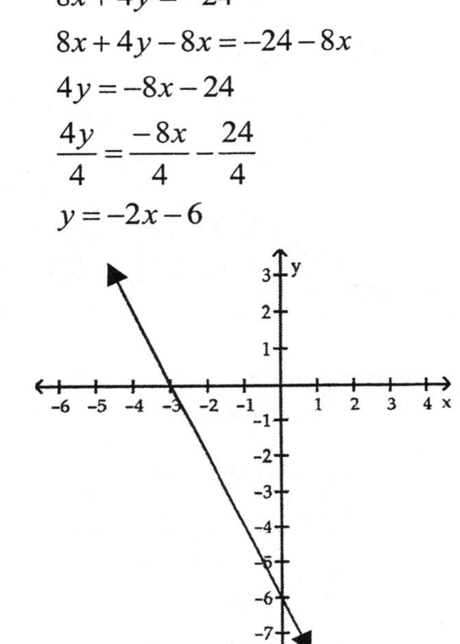

11. To find the x–intercept, let $y = 0$.
$2x - 3y = 6$
$2x - 3(\mathbf{0}) = 6$
$2x - 0 = 6$
$2x = 6$
$\frac{2x}{2} = \frac{6}{2}$
$x = 3$
The x–intercept would be (3, 0).

To find the y–intercept, let $x = 0$.
$2x - 3y = 6$
$2(\mathbf{0}) - 3y = 6$
$0 - 3y = 6$
$-3y = 6$
$\frac{-3y}{-3} = \frac{6}{-3}$
$y = -2$
The y–intercept would be (0, –2).

12. Pick two points such as (4, 4) and (–3, –4) and use the formula for slope.

$$m=\frac{y_2-y_1}{x_2-x_1}$$
$$=\frac{-4-4}{-3-4}$$
$$=\frac{-8}{-7}$$
$$=\frac{8}{7}$$

13. For (–1, 3) and (3, –1):

$$m=\frac{y_2-y_1}{x_2-x_1}$$
$$=\frac{-1-3}{3-(-1)}$$
$$=\frac{-4}{4}$$
$$=-1$$

14. The slope of a vertical line is **undefined.**

15. Perpendicular lines have slopes that are negative reciprocals of each other.

Therefore, if the slope of one line is $-\frac{7}{8}$, the line perpendicular would have a slope of $-\frac{8}{7}$.

16. For both lines, the slope is 2. Therefore, the lines would be **parallel.**

17.

$$m=\frac{\text{change in } y}{\text{change in } x}$$
$$=\frac{y_2-y_1}{x_2-x_1}$$
$$=\frac{250-100}{2-12}$$
$$=\frac{150}{-10}$$
$$=-15$$

So the rate of change would be –15 feet per mile.

18.

$$m=\frac{\text{change in } y}{\text{change in } x}$$
$$=\frac{y_2-y_1}{x_2-x_1}$$
$$=\frac{100-300}{12-20}$$
$$=\frac{-200}{-8}$$
$$=25$$

So the rate of change would be 25 feet per mile.

19. $x=-4$

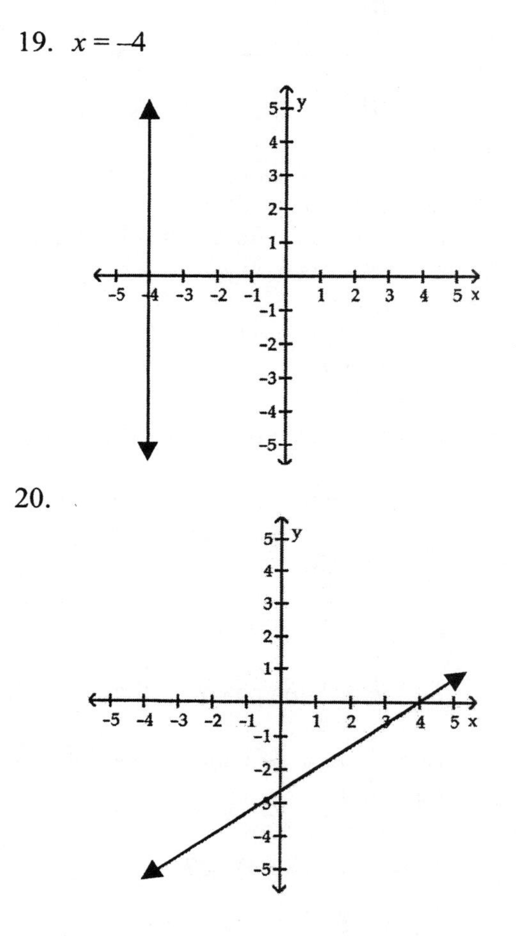

20.

21. Solve the equation for y and use slope–intercept form: $y = mx + b$.

$x + 2y = 8$
$x - x + 2y = -x + 8$
$2y = -x + 8$
$\frac{2y}{2} = \frac{-x}{2} + \frac{8}{2}$
$y = -\frac{1}{2}x + 4$

So, $m = -\frac{1}{2}$ and the y–intercept is (0, 4).

22. $m = 7$ and $(x_1, y_1) = (-2, 5)$

Use point–slope form of the equation:

$y - y_1 = m(x - x_1)$
$y - \mathbf{5} = \mathbf{7}(x - (\mathbf{-2}))$
$y - 5 = 7(x + 2)$
$y - 5 = 7(x) + 7(2)$
$y - 5 = 7x + 14$
$y - 5 + 5 = 7x + 14 + 5$
$y = 7x + 19$

23. DEPRECIATION

$y = -1500x + 15{,}000$

24. Substitute (6, 1) into the inequality.

$2x - 4y \geq 8$
$2(\mathbf{6}) - 4(\mathbf{1}) \overset{?}{\geq} 8$
$12 - 4 \overset{?}{\geq} 8$
$8 \geq 8$

Yes, (6, 1) is a solution.

25. WATER HEATERS

Find the slope of the line using the two points (140, 13) and (180, 5):

$m = \dfrac{y_2 - y_1}{x_2 - x_1}$
$= \dfrac{5-13}{180-140}$
$= \dfrac{-8}{40}$
$= -\dfrac{1}{5}$

Use the slope of $-\frac{1}{5}$ and one of the points to substitute into the point–slope form of a line.

Let $m = -\frac{1}{5}$ and $(x_1, y_1) = (180, 5)$.

$y - y_1 = m(x - x_1)$
$y - \mathbf{5} = -\dfrac{1}{5}(x - \mathbf{180})$
$y - 5 = -\dfrac{1}{5}(x) + -\dfrac{1}{5}(-180)$
$y - 5 = -\dfrac{1}{5}x + 36$
$y - 5 + 5 = -\dfrac{1}{5}x + 36 + 5$
$y = -\dfrac{1}{5}x + 41$

26. Substitute (6, 1) into the inequality.

$2x - 4y \geq 8$
$2(\mathbf{6}) - 4(\mathbf{1}) \overset{?}{\geq} 8$
$12 - 4 \overset{?}{\geq} 8$
$8 \geq 8$

Yes, (6, 1) is a solution

27. Solve the inequality for y.

$x - y > -2$
$-y > -x - 2$
$\dfrac{-y}{-1} > \dfrac{-x}{-1} - \dfrac{2}{-1}$
$y < x + 2$

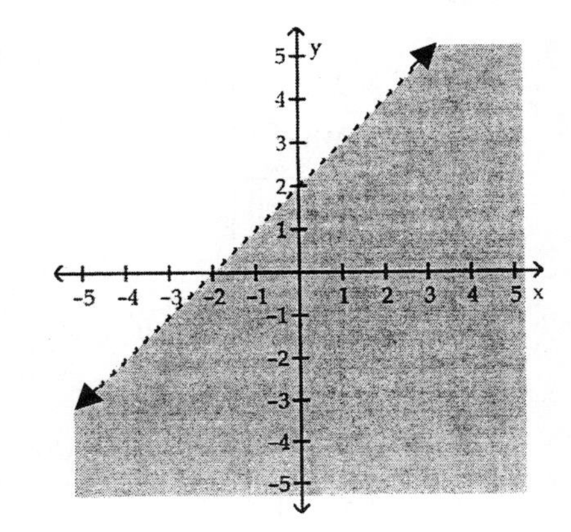

SECTION 4.1

VOCABULARY

1. Expressions such as x^4, 10^3, and $(5t)^2$ are called **exponential** expressions.

3. The expression x^4 represents a repeated multiplication where x is to be written as a **factor** four times.

5. $(h^3)^7$ is a **power** of an exponential expression.

CONCEPTS

7. a) $(3x)^4$ means $\boxed{3x} \cdot \boxed{3x} \cdot \boxed{3x} \cdot \boxed{3x}$.

   b) $(-5y)(-5y)(-5y)$ can be written as $\boxed{(-5y)^3}$.

9. a) Answers may vary. An example is $(3x^2)^6$.

   b) Answers may vary. An example is

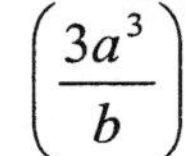

$\left(\frac{3a^3}{b}\right)^2$.

11. a) $2x^2$
    b) 0
    c) x^4
    d) a^7

13. a) doesn't simplify
    b) doesn't simplify
    c) x^5
    d) x

15. a) $(-4)^2 = -4 \cdot -4 = 16$
    b) $-4^2 = -(4 \cdot 4) = -16$

NOTATION

17. $\left(x^4x^2\right)^3 = \left(\boxed{x^6}\right)^3$
    $= x^{\boxed{18}}$

19. base 4, exponent 3

21. base x, exponent 5

23. base $-3x$, exponent 2

25. base y, exponent 6

27. base $y + 9$, exponent 4

29. base $-3ab$, exponent 7

31. $x \cdot x \cdot x \cdot x \cdot x$

33. $\left(\frac{t}{2}\right)\left(\frac{t}{2}\right)\left(\frac{t}{2}\right)$

35. $(x - 5)(x - 5)$

37. $(4t)^4$

39. $-4t^3$

41. $(x - y)^3$

PRACTICE

43. $12^3 \cdot 12^4 = 12^{3+4}$
    $= 12^7$

45. $2(2^3)(2^2) = 2^{1+3+2}$
    $= 2^6$

47. $a^3 \cdot a^3 = a^{3+3}$
    $= a^6$

49. $x^4x^3 = x^{4+3}$
    $= x^7$

51. $a^3aa^5 = a^{3+1+5}$
    $= a^9$

53. $(-7)^2(-7)^3 = (-7)^{2+3}$
    $= (-7)^5$

55. $(8t)^{20}(8t)^{40} = (8t)^{20+40}$
    $= (8t)^{60}$

57. $(n - 1)^2(n - 1) = (n - 1)^{2+1}$
    $= (n - 1)^3$

59. $y^3(y^2y^4) = y^3(y^{2+4})$
$= y^3(y^6)$
$= y^{3+6}$
$= y^9$

61. $\frac{8^{12}}{8^4} = 8^{12-4}$
$= 8^8$

63. $\frac{x^{15}}{x^3} = x^{15-3}$
$= x^{12}$

65. $\frac{c^{10}}{c^9} = c^{10-9}$
$= c^1$
$= c$

67. $\frac{(k-2)^{15}}{(k-2)} = (k-2)^{15-1}$
$= (k-2)^{14}$

69. $(3^2)^4 = 3^{2\bullet 4}$
$= 3^8$

71. $(y^5)^3 = y^{5\bullet 3}$
$= y^{15}$

73. $(m^{50})^{10} = m^{50\bullet 10}$
$= m^{500}$

75. $(a^2b^3)(a^3b^3) = a^{2+3}b^{3+3}$
$= a^5b^6$

77. $(cd^4)(cd) = c^{1+1}d^{4+1}$
$= c^2d^5$

79. $xy^2 \bullet x^2y = x^{1+2}y^{2+1}$
$= x^3y^3$

81. $\frac{y^3y^4}{yy^2} = \frac{y^{3+4}}{y^{1+2}}$
$= \frac{y^7}{y^3}$
$= y^{7-3}$
$= y^4$

83. $\frac{c^3d^7}{cd} = c^{3-1}d^{7-1}$
$= c^2d^6$

85. $(x^2x^3)^5 = (x^{2+3})^5$
$= (x^5)^5$
$= x^{5\bullet 5}$
$= x^{25}$

87. $(3zz^2z^3)^5 = (3z^6)^5$
$= 3^5z^{6\bullet 5}$
$= 243z^{30}$

89. $(3n^8)^2 = 3^2n^{8\bullet 2}$
$= 9n^{16}$

91. $(uv)^4 = u^{1\bullet 4}v^{1\bullet 4}$
$= u^4v^4$

93. $(a^3b^2)^3 = a^{3\bullet 3}b^{2\bullet 3}$
$= a^9b^6$

95. $(-2r^2s^3)^3 = (-2)^3r^{2\bullet 3}s^{3\bullet 3}$
$= -8r^6s^9$

97. $\left(\dfrac{a}{b}\right)^3 = \dfrac{a^3}{b^3}$

99. $\left(\dfrac{x^2}{y^3}\right)^5 = \dfrac{x^{2\cdot5}}{y^{3\cdot5}}$

$= \dfrac{x^{10}}{y^{15}}$

101. $\left(\dfrac{-2a}{b}\right)^5 = \dfrac{(-2)^5 a^{1\cdot5}}{b^{1\cdot5}}$

$= \dfrac{-32a^5}{b^5}$

103. $\dfrac{(6k)^7}{(6k)^4} = (6k)^{7-4}$

$= (6k)^3$

$= 6^3 k^3$

$= 216k^3$

105. $\dfrac{(a^2b)^{15}}{(a^2b)^9} = (a^2b)^{15-9}$

$= (a^2b)^6$

$= a^{2\cdot6} b^{1\cdot6}$

$= a^{12} b^6$

107. $\dfrac{a^2a^3a^4}{(a^4)^2} = \dfrac{a^{2+3+4}}{a^{4\cdot2}}$

$= \dfrac{a^9}{a^8}$

$= a^{9-8}$

$= a$

109. $\dfrac{(ab^2)^3}{(ab)^2} = \dfrac{a^{1\cdot3}b^{2\cdot3}}{a^{1\cdot2}b^{1\cdot2}}$

$= \dfrac{a^3b^6}{a^2b^2}$

$= a^{3-2}b^{6-2}$

$= ab^4$

111. $\dfrac{(r^4s^3)^4}{(rs^3)^3} = \dfrac{r^{4\cdot4}s^{3\cdot4}}{r^{1\cdot3}s^{3\cdot3}}$

$= \dfrac{r^{16}s^{12}}{r^3s^9}$

$= r^{16-3}s^{12-9}$

$= r^{13}s^3$

113. $\left(\dfrac{y^3y}{2yy^2}\right)^3 = \left(\dfrac{y^{3+1}}{2y^{1+2}}\right)^3$

$= \left(\dfrac{y^4}{2y^3}\right)^3$

$= \left(\dfrac{y^{4-3}}{2}\right)^3$

$= \left(\dfrac{y}{2}\right)^3$

$= \dfrac{y^{1\cdot3}}{2^3}$

$= \dfrac{y^3}{8}$

115. $$\left(\frac{3t^3t^4t^5}{4t^2t^6}\right)^3 = \left(\frac{3t^{12}}{4t^8}\right)^3$$
$$= \left(\frac{3t^{12-8}}{4}\right)^3$$
$$= \left(\frac{3t^4}{4}\right)^3$$
$$= \frac{3^3t^{4\bullet 3}}{4^3}$$
$$= \frac{27t^{12}}{64}$$

APPLICATIONS

117. Area of a square:
$A = s^2$
$= (a^5)^2$
$= a^{5\bullet 2}$
$= a^{10}$ mi^2

119. Volume of a prism:
$V = blh$
$= x^4 \cdot x^3 \cdot x^2$
$= x^{4+3+2}$
$= x^9$ ft^3

121. ART HISTORY

a) $(5x)(5x) = 25x^2$ ft^2
b) $\pi(3a)^2 = 9a^2\pi$ ft^2

123. BOUNCING BALLS

$$32\left(\frac{1}{2}\right) = 16 \text{ ft}$$
$$32\left(\frac{1}{2}\right)^2 = 32\left(\frac{1}{4}\right)$$
$$= 8 \text{ ft}$$
$$32\left(\frac{1}{2}\right)^3 = 32\left(\frac{1}{8}\right)$$
$$= 4 \text{ ft}$$
$$32\left(\frac{1}{2}\right)^4 = 32\left(\frac{1}{16}\right)$$
$$= 2 \text{ ft}$$

WRITING

125. Answers will vary.

127. Answers will vary.

REVIEW

129. c

131. d

CHALLENGE PROBLEMS

133. a) $x^{2m}x^{3m} = x^{2m+3m}$
$= x^{5m}$

b) $\left(y^{5c}\right)^4 = y^{5c\bullet 4}$
$= y^{20c}$

c) $\dfrac{m^{8x}}{m^{4x}} = m^{8x-4x}$
$= m^{4x}$

d) $\left(2a^{6y}\right)^4 = 2^4a^{6y\bullet 4}$
$= 16a^{24y}$

SECTION 4.2

VOCABULARY

1. In the expression 8^{-3}, 8 is the **base**, and -3 is the **exponent**.

3. Another way to write 2^{-3} is to write its **reciprocal** and to change the sign of the exponent:

$$2^{-3} = \frac{1}{2^{\boxed{3}}}$$

CONCEPTS

5. a) $\frac{6^4}{6^4} = 6^{\boxed{4-4}} = 6^{\boxed{0}}$ $\qquad \frac{6^4}{6^4} = \frac{\boxed{6}\bullet\boxed{6}\bullet\boxed{6}\bullet\boxed{6}}{6\bullet6\bullet6\bullet6} = \boxed{1}$

b) So we define 6^0 to be $\boxed{1}$, and in general, if x is any nonzero real number, then $x^0 = \boxed{1}$.

7.

x	3^x
2	$3^2 = 9$
1	$3^1 = 3$
0	$3^0 = 1$
-1	$3^{-1} = \frac{1}{3}$
-2	$3^{-2} = \frac{1}{3^2} = \frac{1}{9}$

9.

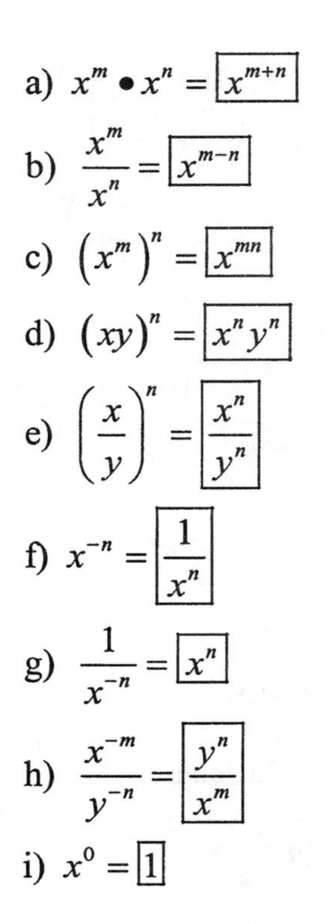

11. Each base occurs only **once**.
There are **no** parentheses
There are no negative or zero **exponents**

NOTATION

13. $(y^5y^3)^{-5} = \left(\boxed{y^8}\right)^{-5}$
$= y^{\boxed{-40}}$
$= \frac{1}{y^{\boxed{40}}}$

15.

Expression	Base	Exponent
4^{-2}	**4**	**-2**
$6x^{-5}$	$\boldsymbol{x}$	**-5**
$\left(\frac{3}{y}\right)^{-8}$	$\frac{\mathbf{3}}{\mathbf{y}}$	**-8**
-7^{-1}	**7**	**-1**
$(-2)^{-3}$	**-2**	**-3**
$10a^0b$	$\boldsymbol{a}$	**0**

PRACTICE

17. $7^0 = 1$

19. $\left(\frac{1}{4}\right)^0 = 1$

21. $2x^0 = 2 \cdot 1$
$= 2$

23. $(-x)^0 = 1$

25. $\left(\dfrac{a^2b^3}{ab^4}\right)^0 = 1$

27. $\dfrac{5}{2x^0} = \dfrac{5}{2(1)}$
$= \dfrac{5}{2}$

29. $-15x^0y = -15(1)(y)$
$= -15y$

31. $12^{-2} = \dfrac{1}{12^2}$
$= \dfrac{1}{144}$

33. $(-4)^{-1} = \dfrac{1}{(-4)^1}$
$= -\dfrac{1}{4}$

35. $44g^{-6} = 44 \bullet \dfrac{1}{g^6}$
$= \dfrac{44}{g^6}$

37. $(-10)^{-3} = \dfrac{1}{(-10)^3}$
$= -\dfrac{1}{1{,}000}$

39. $-4^{-3} = -\dfrac{1}{4^3}$
$= -\dfrac{1}{64}$

41. $-(-4)^{-3} = -\dfrac{1}{(-4)^3}$
$= -\dfrac{1}{-64}$
$= \dfrac{1}{64}$

43. $x^{-2} = \dfrac{1}{x^2}$

45. $-b^{-5} = -\dfrac{1}{b^5}$

47. $\left(\dfrac{1}{6}\right)^{-2} = \left(\dfrac{6}{1}\right)^2$
$= \dfrac{36}{1}$
$= 36$

49. $\left(-\dfrac{1}{2}\right)^{-3} = \left(-\dfrac{2}{1}\right)^3$
$= -\dfrac{8}{1}$
$= -8$

51. $\left(\dfrac{7}{8}\right)^{-1} = \left(\dfrac{8}{7}\right)^1$
$= \dfrac{8}{7}$

53. $\dfrac{1}{5^{-3}} = 5^3$
$= 125$

55. $\dfrac{2^{-4}}{3^{-1}} = \dfrac{3^1}{2^4}$
$= \dfrac{3}{16}$

57. $\dfrac{a^{-5}}{b^{-2}} = \dfrac{b^2}{a^5}$

59. $\dfrac{r^{-50}}{r^{-70}} = r^{-50-(-70)}$

$= r^{-50+70}$

$= r^{20}$

61. $-\dfrac{1}{p^{-10}} = -p^{10}$

63. $\dfrac{h^{-5}}{h^{2}} = h^{-5-2}$

$= h^{-7}$

$= \dfrac{1}{h^{7}}$

65. $\dfrac{8}{s^{-1}} = 8s$

67. $\dfrac{h}{h^{-6}} = h^{1-(-6)}$

$= h^{1+6}$

$= h^{7}$

69. $(2y)^{-4} = \dfrac{1}{(2y)^{4}}$

$= \dfrac{1}{16y^{4}}$

71. $(ab^{2})^{-3} = \dfrac{1}{(ab^{2})^{3}}$

$= \dfrac{1}{a^{1\bullet 3}b^{2\bullet 3}}$

$= \dfrac{1}{a^{3}b^{6}}$

73. $2^{5} \bullet 2^{-2} = 2^{5+(-2)}$

$= 2^{3}$

$= 8$

75. $\left(\dfrac{y^{4}}{3}\right)^{-2} = \left(\dfrac{3}{y^{4}}\right)^{2}$

$= \dfrac{3^{2}}{y^{4\bullet 2}}$

$= \dfrac{9}{y^{8}}$

77. $\dfrac{y^{4}}{y^{5}} = y^{4-5}$

$= y^{-1}$

$= \dfrac{1}{y}$

79. $\dfrac{(r^{2})^{3}}{(r^{3})^{4}} = \dfrac{r^{6}}{r^{12}}$

$= r^{6-12}$

$= r^{-6}$

$= \dfrac{1}{r^{6}}$

81. $\dfrac{4s^{-5}}{t^{-2}} = \dfrac{4t^{2}}{s^{5}}$

83. $(5d^{-2})^{3} = 5^{3}d^{-2\bullet 3}$

$= 125d^{-6}$

$= \dfrac{125}{d^{6}}$

85. $\dfrac{-2a^{-4}}{a^{-8}} = -2a^{-4-(-8)}$

$= -2a^{-4+8}$

$= -2a^{4}$

87. $$x^{-3}x^{-3}x^{-3} = x^{-3+(-3)+(-3)} = x^{-9} = \frac{1}{x^9}$$

89. $$\frac{t\left(t^{-2}\right)^{-2}}{t^{-5}} = \frac{t(t^4)}{t^{-5}} = \frac{t^{1+4}}{t^{-5}} = \frac{t^5}{t^{-5}} = t^{5-(-5)} = t^{10}$$

91. $$\frac{y^4y^3}{y^4y^{-2}} = \frac{y^{4+3}}{y^{4+(-2)}} = \frac{y^7}{y^2} = y^{7-2} = y^5$$

93. $$\frac{2a^4a^{-2}}{a^2a^0} = \frac{2a^{4+(-2)}}{a^{2+0}} = \frac{2a^2}{a^2} = 2a^0 = 2(1) = 2$$

95. $$\left(ab^2\right)^{-2} = \frac{1}{\left(ab^2\right)^2} = \frac{1}{a^{1\bullet 2}b^{2\bullet 2}} = \frac{1}{a^2b^4}$$

97. $$\left(x^2y\right)^{-3} = \frac{1}{\left(x^2y\right)^3} = \frac{1}{x^{2\bullet 3}y^{1\bullet 3}} = \frac{1}{x^6y^3}$$

99. $$\left(x^{-4}x^3\right)^3 = \left(x^{-4+3}\right)^3 = \left(x^{-1}\right)^3 = x^{-3} = \frac{1}{x^3}$$

101. $$\left(-2x^3y^{-2}\right)^{-5} = \frac{1}{\left(-2x^3y^{-2}\right)^5} = \frac{1}{-32x^{15}y^{-10}} = -\frac{y^{10}}{32x^{15}}$$

103. $$\left(\frac{a^3}{a^{-4}}\right)^2 = \left(a^{3-(-4)}\right)^2 = \left(a^7\right)^2 = a^{14}$$

105. $$\left(\frac{4x^2}{3x^{-5}}\right)^4 = \left(\frac{4x^{2-(-5)}}{3}\right)^4 = \left(\frac{4x^7}{3}\right)^4 = \frac{4^4x^{7\bullet 4}}{3^4} = \frac{256x^{28}}{81}$$

107. $\left(\dfrac{y^3z^{-2}}{3y^{-4}z^3}\right)^2 = \left(\dfrac{y^{3-(-4)}z^{-2-3}}{3}\right)^2$

$= \left(\dfrac{y^7z^{-5}}{3}\right)^2$

$= \left(\dfrac{y^7}{3z^5}\right)^2$

$= \dfrac{y^{7\bullet 2}}{3^2 z^{5\bullet 2}}$

$= \dfrac{y^{14}}{9z^{10}}$

109. $\dfrac{2^{-2}g^{-2}h^{-3}}{9^{-1}h^{-3}} = \dfrac{9^1h^3}{2^2g^2h^3}$

$= \dfrac{9h^{3-3}}{4g^2}$

$= \dfrac{9}{4g^2}$

111. $5^0 + (-7)^0 = 1 + 1$

$= 2$

113. $2^{-2} + 4^{-1} = \dfrac{1}{2^2} + \dfrac{1}{4^1}$

$= \dfrac{1}{4} + \dfrac{1}{4}$

$= \dfrac{2}{4}$

$= \dfrac{1}{2}$

115. $9^0 - 9^{-1} = 1 - \dfrac{1}{9}$

$= \dfrac{9}{9} - \dfrac{1}{9}$

$= \dfrac{8}{9}$

APPLICATIONS

117. THE DECIMAL NUMERATION SYSTEM

$10^2, 10^1, 10^0, 10^{-1}, 10^{-2}, 10^{-3}, 10^{-4}$

119. RETIREMENT

$P = A(1+i)^{-n}$

$P = \mathbf{100,000}(1+\mathbf{0.08})^{-40}$

$P = 100,000(1.08)^{-40}$

$P = 100,000\left(\dfrac{1}{1.08}\right)^{40}$

$P \approx \$4,603.09$

121. BIOLOGY

If $n = 0$, it gives the initial number of bacteria b.

WRITING

123. Answers will vary.

REVIEW

125. IQ TESTS

$\mathbf{135} = \dfrac{x}{\mathbf{10}} \bullet 100$

$135 = 10x$

$13.5 = x$

The mental age is 13.5 years

127. $y = mx + b$

$y = \dfrac{3}{4}x - 5$

CHALLENGE PROBLEMS

129. a) $r^{5m}r^{-6m} = r^{5m+(-6m)}$

$$= r^{-m}$$

$$= \frac{1}{r^m}$$

b) $\frac{x^{3n}}{x^{6n}} = \frac{x^{3n-(6n)}}{1}$

$$= \frac{x^{-3n}}{1}$$

$$= \frac{1}{x^{3n}}$$

SECTION 4.3

VOCABULARY

1. 4.84×10^9 and 1.05×10^{-2} are written in **scientific** notation.

3. Scientific **notation** provides a compact way of writing very large or very small numbers.

5. 10^{34}, 10^{50}, and 10^{-14} are **powers** of 10.

CONCEPTS

7. When we multiply a decimal by 10^5, the decimal point moves 5 places to the **right**. When we multiply a decimal by 10^{-7}, the decimal point moves 7 places to the **left**.

9. a) 10^{-7}
   b) 10^9

11. a) When a real number great than 1 is written in scientific notation, the exponent on 10 is a **positive** number.

    b) When a real number between 0 and 1 is written in scientific notation, the exponent on 10 is a **negative** number.

13. a) $7{,}700 = \boxed{7.7} \times 10^3$
    b) $500{,}000 = \boxed{5.0} \times 10^5$
    c) $114{,}000{,}000 = 1.14 \times 10^{\boxed{8}}$

15. a) $(5.1 \times 1.5)(10^9 \times 10^{22})$
    b) $\dfrac{8.8}{2.2} \times \dfrac{10^{30}}{10^{19}}$

NOTATION

17. A positive number is written in scientific notation when it is written in the form $N \times 10^n$, where **1** $\leq N <$ **10** and n is an **integer**.

PRACTICE

19. 230; Since the exponent is 2, the decimal moves 2 places to the right.

21. 812,000; Since the exponent is 5, the decimal moves 5 places to the right.

23. 0.00115; Since the exponent is -3, the decimal moves 3 places to the left.

25. 0.000976; Since the exponent is -4, the decimal moves 4 places to the left.

27. 6,001,000; Since the exponent is 6, the decimal moves 6 places to the right.

29. 2.718; Since the exponent is 0, the decimal does not move.

31. 0.06789; Since the exponent is -2, the decimal moves 2 places to the left.

33. 0.00002; Since the exponent is -5, the decimal moves 5 places to the left.

35. 9,000,000,000; Since the exponent is 9, the decimal moves 9 places to the right.

37. 2.3×10^4; Move the decimal 4 places to the left.

39. 1.7×10^6; Move the decimal 6 places to the left.

41. 6.2×10^{-2}; Move the decimal 2 places to the right.

43 5.1×10^{-6}; Move the decimal 6 places to the right.

45. 5.0×10^9; Move the decimal 9 places to the left.

47. 3.0×10^{-7}; Move the decimal 7 places to the right.

49. 9.09×10^8; Move the decimal 8 places to the left.

51. 3.45×10^{-2}; Move the decimal 2 places to the right.

53. 9.0×10^{0}; The decimal does not move.

55. 1.718×10^{18}; Move the decimal 18 places to the left.

57. 1.23×10^{-14}; Move the decimal 14 places to the right.

59. $(3.4 \times 10^{2})(2.1 \times 10^{3})$
$= (3.4 \times 2.1)(10^{2} \times 10^{3})$
$= 7.14 \times 10^{5}$
$= 714{,}000$

61. $(8.4 \times 10^{-13})(4.8 \times 10^{9})$
$= (8.4 \times 4.8)(10^{-13} \times 10^{9})$
$= 40.32 \times 10^{-4}$
$= 0.004032$

63. $\frac{9.3\times10^{2}}{3.1\times10^{-2}} = \frac{9.3}{3.1}\times\frac{10^{2}}{10^{-2}}$
$= 3.0\times10^{2-(-2)}$
$= 3.0\times10^{4}$
$= 30{,}000$

65. $\frac{1.29\times10^{-6}}{3\times10^{-4}} = \frac{1.29}{3}\times\frac{10^{-6}}{10^{-4}}$
$= 0.43\times10^{-6-(-4)}$
$= 0.43\times10^{-2}$
$= 0.0043$

67. $(6 \times 10^{-9})(5.5 \times 10^{6})$
$= (5.6 \times 5.5)(10^{-9} \times 10^{6})$
$= 30.8 \times 10^{-3}$
$= 0.0308$

69. $\frac{9.6\times10^{4}}{(1.2\times10^{4})(4\times10^{-5})} = \frac{9.6\times10^{4}}{(1.2\times4)(10^{4}\times10^{-5})}$
$= \frac{9.6\times10^{4}}{4.8\times10^{4+(-5)}}$
$= \frac{9.6\times10^{4}}{4.8\times10^{-1}}$
$= \frac{9.6}{4.8}\times\frac{10^{4}}{10^{-1}}$
$= 2.0\times10^{4-(-1)}$
$= 2.0\times10^{5}$
$= 200{,}000$

71. $(4.564\times10^{2})^{6} = (4.564)^{6}\times(10^{2})^{6}$
$= 9038.030748\times10^{12}$
$= 9.038030748\times10^{15}$

73. $(2.25\times10^{2})^{-5} = (2.25)^{-5}\times(10^{2})^{-5}$
$= 0.0173415299\times10^{-10}$
$= 1.73415299\times10^{-12}$

APPLICATIONS

75. ASTRONOMY

2.57×10^{13} miles; move the decimal to the left 13 places

77. GEOGRAPHY

63,800,000 mi^{2}; move the decimal to the right 7 places

79. LENGTH OF A METER

6.22×10^{-3} miles; move the decimal to the right 3 places

81. WAVELENGTHS

gamma ray, x-ray, ultraviolet, visible light, infrared, microwave, radio wave

83. PROTONS

$(1.7 \times 10^{-24})(1 \times 10^{6})$
$= (1.7 \times 1)(10^{-24} \times 10^{6})$
$= 1.7 \times 10^{-24+6}$
$= 1.7 \times 10^{-18}$ g

85. LIGHT YEARS

$(5.87 \times 10^{12}$ miles$)(5.28 \times 10^{3}$ feet$)$
$= (5.87 \times 5.28)(10^{12} \times 10^{3})$
$= 30.9936 \times 10^{12+3}$
$= 30.9936 \times 10^{15}$
$= 3.09936 \times 10^{16}$ feet

87. INTEREST

$I = Prt$
$= (5.2 \times 10^{12})(0.04)(1)$
$= (5.2 \times 10^{12})(4 \times 10^{-2})(1)$
$= (5.2 \times 4 \times 1)(10^{12} \times 10^{-2})$
$= 20.88 \times 10^{12 + (-2)}$
$= 20.88 \times 10^{10}$
$= 2.088 \times 10^{11}$ dollars

89. THE MILITARY

a) 1.7×10^6; 1,700,000
b) 1986 ; 2.05×10^6 ; 2,050,000

WRITING

91. Answers will vary.

93. Answers will vary.

REVIEW

95. $-5y^{55} = -5(-1)^{55}$
$= -5(-1)$
$= 5$

97. $c = 30t + 45$

CHALLENGE PROBLEMS

99. a) -2.5×10^{-4}

b) $$\frac{1}{2.5\times10^{-4}} = \frac{1\times10^{0}}{2.5\times10^{-4}}$$
$$= \frac{1}{2.5}\times\frac{10^{0}}{10^{-4}}$$
$$= 0.4\times10^{0-(-4)}$$
$$= 0.4\times10^{4}$$
$$= 4.0\times10^{3}$$

SECTION 4.4

VOCABULARY

1. A **polynomial** is a term or a sum of terms in which all variable have whole-number exponents.

3. The degree of a polynomial is the same as the degree of its **term** with the highest degree.

5. The **degree** of the monomial $3x^7$ is 7.

7. $x^3 - 6x^2 + 9x - 2$ is a polynomial in ***x*** and is written in **decreasing or descending** powers of x.

9. Because the graph of $y = x^3$ is not a straight line, we call $y = x^3$ a **nonlinear** equation.

CONCEPTS

11. a) Yes
    b) No
    c) No
    d) Yes
    e) Yes
    f) Yes

13. $y = x^2 + 3$

    $\mathbf{2} \stackrel{?}{=} (\mathbf{-1})^2 + 3$

    $2 \stackrel{?}{=} 1 + 3$

    $2 \stackrel{?}{=} 4$

    No, (-1, 2) is not a solution.

15. Not enough ordered pairs were found—the correct graph is a parabola.

NOTATION

17. $$\begin{aligned} -2x^2 + 3x - 1 &= -2(\boxed{-2})^2 + 3(\boxed{-2}) - 1 \\ &= -2(\boxed{4}) + 3(\boxed{-2}) - 1 \\ &= \boxed{-8} + (-6) - 1 \\ &= \boxed{-14} - 1 \\ &= \boxed{-15} \end{aligned}$$

19. $y = x^2 - 1$ and $y = x^3 - 1$

PRACTICE

21. two terms - binomial

23. three terms - trinomial

25. one term - monomial

27. two terms - binomial

29. three terms - trinomial

31. four terms - none of these

33. three terms - trinomial

35. 4th

37. 2nd

39. 1st

41. 4th (2+2 and 3+1)

43. 12th

45. 0th

47. $-9x + 1$

    a) $$\begin{aligned} -9(\mathbf{5}) + 1 &= -45 + 1 \\ &= -44 \end{aligned}$$

    b) $$\begin{aligned} -9(\mathbf{-4}) + 1 &= 36 + 1 \\ &= 37 \end{aligned}$$

49. $3x^2 - 2x + 8$

    a) $$\begin{aligned} 3(\mathbf{1})^2 - 2(\mathbf{1}) + 8 &= 3(1) - 2(1) + 8 \\ &= 3 - 2 + 8 \\ &= 1 + 8 \\ &= 9 \end{aligned}$$

    b) $$\begin{aligned} 3(\mathbf{0})^2 - 2(\mathbf{0}) + 8 &= 3(0) - 2(0) + 8 \\ &= 0 - 0 + 8 \\ &= 8 \end{aligned}$$

51. $-x^2 - 6$

    a) $$\begin{aligned} -(\mathbf{-4})^2 - 6 &= -(16) - 6 \\ &= -22 \end{aligned}$$

    b) $$\begin{aligned} -(\mathbf{20})^2 - 6 &= -(400) - 6 \\ &= -406 \end{aligned}$$

53. $x^3 + 3x^2 + 2x + 4$

a) $(\mathbf{2})^3 + 3(\mathbf{2})^2 + 2(\mathbf{2}) + 4$
$= 8 + 3(4) + 2(2) + 4$
$= 8 + 12 + 4 + 4$
$= 28$

b) $(\mathbf{-2})^3 + 3(\mathbf{-2})^2 + 2(\mathbf{-2}) + 4$
$= -8 + 3(4) + 2(-2) + 4$
$= -8 + 12 - 4 + 4$
$= 4$

55. $x^4 - x^3 + x^2 + 2x - 1$

a) $(\mathbf{1})^4 - (\mathbf{1})^3 + (\mathbf{1})^2 + 2(\mathbf{1}) - 1$
$= 1 - 1 + 1 + 2(1) - 1$
$= 1 - 1 + 1 + 2 - 1$
$= 2$

b) $(\mathbf{-1})^4 - (\mathbf{-1})^3 + (\mathbf{-1})^2 + 2(\mathbf{-1}) - 1$
$= 1 - (-1) + 1 + 2(-1) - 1$
$= 1 + 1 + 1 - 2 - 1$
$= 0$

57.

x	$y = x^2$
-3	$(\mathbf{-3})^2 = 9$
-2	$(\mathbf{-2})^2 = 4$
-1	$(\mathbf{-1})^2 = 1$
0	$(\mathbf{0})^2 = 0$
1	$(\mathbf{1})^2 = 1$
2	$(\mathbf{4})^2 = 4$
3	$(\mathbf{3})^2 = 9$

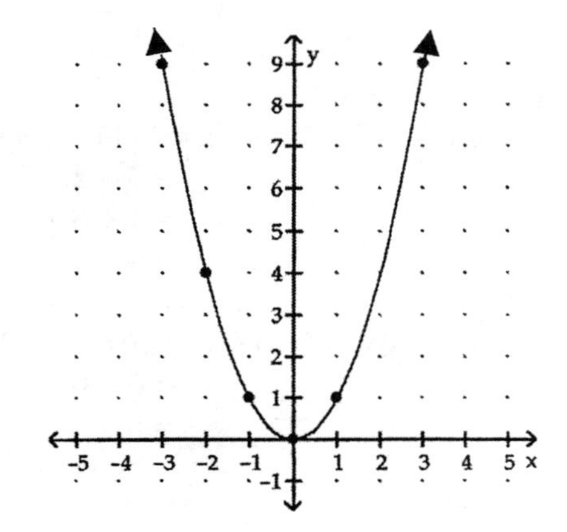

59.

x	$y = x^2 + 1$
-2	$(\mathbf{-2})^2 + 1 = 4 + 1$ $= 5$
-1	$(\mathbf{-1})^2 + 1 = 1 + 1$ $= 2$
0	$(\mathbf{0})^2 + 1 = 0 + 1$ $= 1$
1	$(\mathbf{1})^2 + 1 = 1 + 1$ $= 2$
2	$(\mathbf{2})^2 + 1 = 4 + 1$ $= 5$

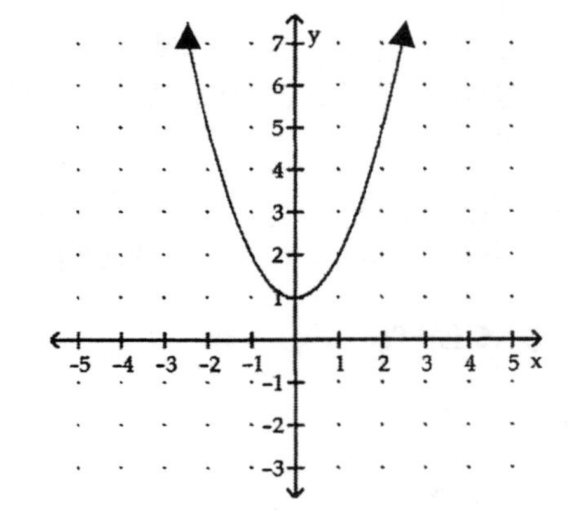

61.

x	$y = -x^2 - 2$
-2	$-(\mathbf{-2})^2 - 2 = -4 - 2$ $= -6$
-1	$-(\mathbf{-1})^2 - 2 = -1 - 2$ $= -3$
0	$-(\mathbf{0})^2 - 2 = -0 - 2$ $= -2$
1	$-(\mathbf{1})^2 - 2 = -1 - 2$ $= -3$
2	$-(\mathbf{2})^2 - 2 = -4 - 2$ $= -6$

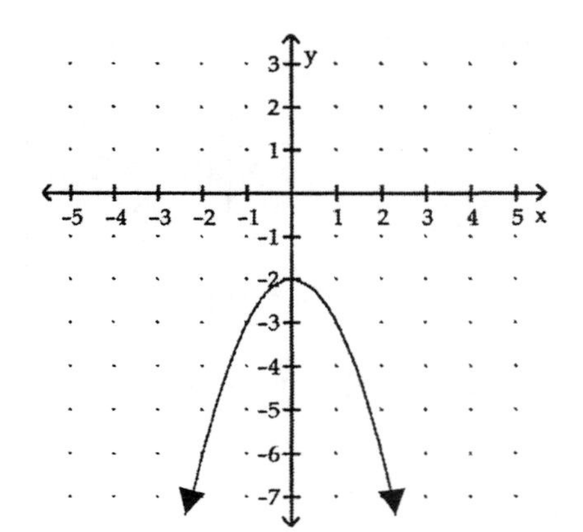

63.

x	$y = 2x^2 - 3$
-2	$2(\mathbf{-2})^2 - 3 = 2(4) - 3$ $= 8 - 3$ $= 5$
-1	$2(\mathbf{-1})^2 - 3 = 2(1) - 3$ $= 2 - 3$ $= -1$
0	$2(\mathbf{0})^2 - 3 = 2(0) - 3$ $= 0 - 3$ $= -3$
1	$2(\mathbf{1})^2 - 3 = 2(1) - 3$ $= 2 - 3$ $= -1$
2	$2(\mathbf{2})^2 - 3 = 2(4) - 3$ $= 8 - 3$ $= 5$

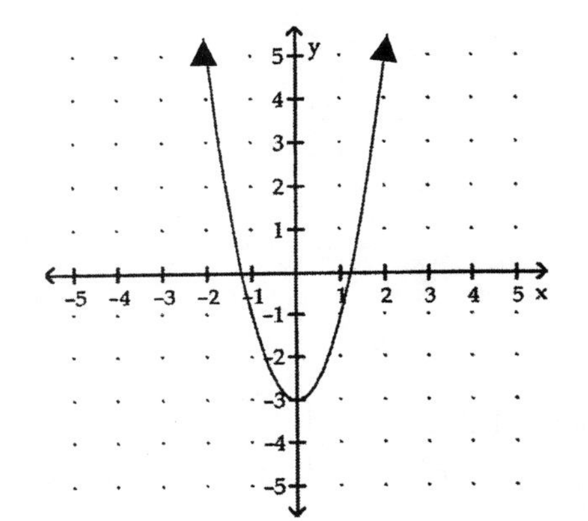

65.

x	$y = x^3 + 2$
-2	$(\mathbf{-2})^3 + 2 = -8 + 2$ $= -6$
-1	$(\mathbf{-1})^3 + 2 = -1 + 2$ $= 1$
0	$(\mathbf{0})^3 + 2 = 0 + 2$ $= 2$
1	$(\mathbf{1})^3 + 2 = 1 + 2$ $= 3$
2	$(\mathbf{2})^3 + 2 = 8 + 2$ $= 10$

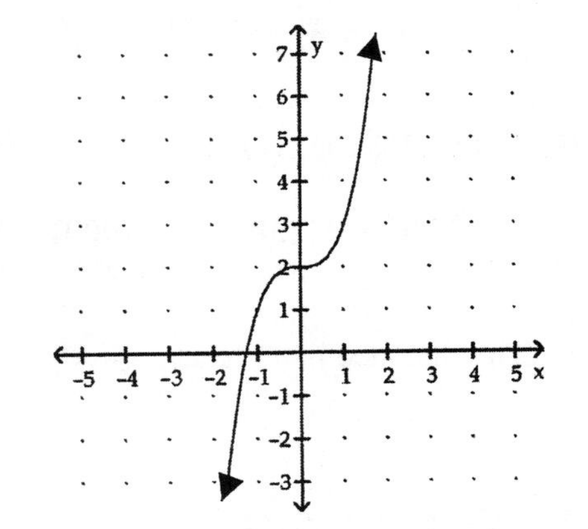

67.

x	$y = -x^3 - 1$
-2	$-(\mathbf{-2})^3 - 1 = -(-8) - 1$ $= 7$
-1	$-(\mathbf{-1})^3 - 1 = -(-1) - 1$ $= 0$
0	$-(\mathbf{0})^3 - 1 = -0 - 1$ $= -1$
1	$-(\mathbf{1})^3 - 1 = -1 - 1$ $= -2$
2	$-(\mathbf{2})^3 - 1 = -8 - 1$ $= -9$

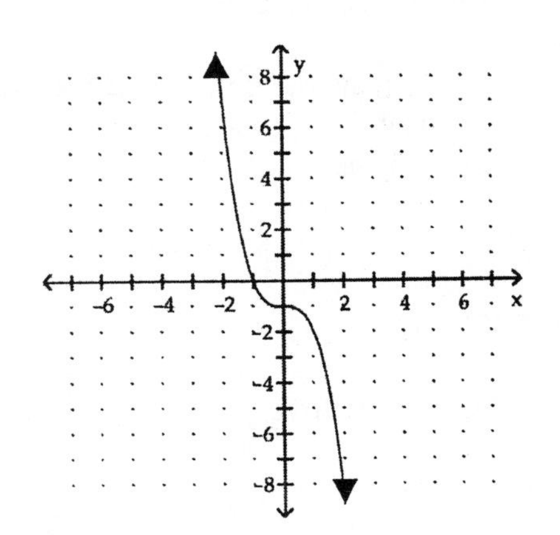

APPLICATIONS

69. SUPERMARKETS

Use the formula from Example 3 and let $c = 6$.

$$\frac{1}{3}c^3 + \frac{1}{2}c^2 + \frac{1}{6}c = \frac{1}{3}(6)^3 + \frac{1}{2}(6)^2 + \frac{1}{6}(6)$$
$$= \frac{1}{3}(216) + \frac{1}{2}(36) + \frac{1}{6}(6)$$
$$= 72 + 18 + 1$$
$$= 91$$

91 cantaloupes are used.

71. STOPPING DISTANCE

$$0.04(\mathbf{30})^2 + 0.9(\mathbf{30}) = 0.04(900) + 27$$
$$= 36 + 27$$
$$= 63$$

The distance is 63 feet.

73. SCIENCE HISTORY

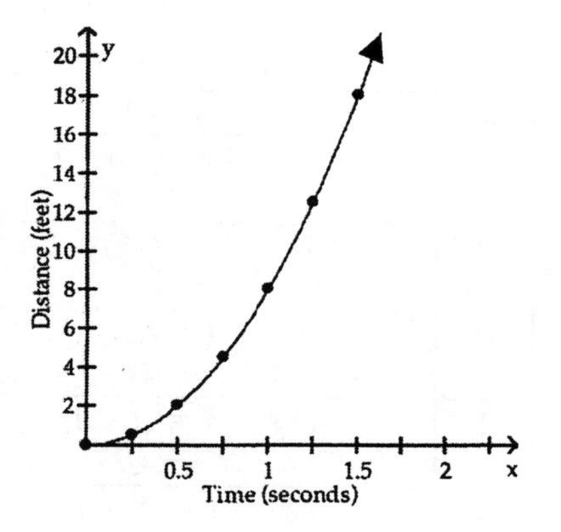

75. MANUFACTURING

a) It costs 8¢ to make a 2-inch bolt.
b) 12¢
c) a 4-inch bolt

WRITING

77. Answers will vary.

REVIEW

79.
$$-4(3y+2) \le 28$$
$$-4(3y) - 4(2) \le 28$$
$$-12y - 8 \le 28$$
$$-12y - 8 + 8 \le 28 + 8$$
$$-12y \le 36$$
$$y \ge -3$$
$$[-3, \infty)$$

-8 -7 -6 -5 -4 -3 -2 -1 0 1 2

81.
$$\left(x^2x^4\right)^3 = \left(x^{2+4}\right)^3$$
$$= \left(x^6\right)^3$$
$$= x^{6\bullet 3}$$
$$= x^{18}$$

83.
$$\left(\frac{y^2y^5}{y^4}\right)^3 = \left(\frac{y^{2+5}}{y^4}\right)^3$$
$$= \left(\frac{y^7}{y^4}\right)^3$$
$$= \left(y^{7-4}\right)^3$$
$$= \left(y^3\right)^3$$
$$= y^{3\bullet 3}$$
$$= y^9$$

CHALLENGE PROBLEMS

85. Answers will vary.
Example: $x^2 + x - 1$

SECTION 4.5

VOCABULARY

1. The expression $(b^3 - b^2 - 9b + 1) +$ $(b^3 - b^2 - 9b + 1)$ is the sum of two **polynomials**.

3. **Like** terms have the same variables with the same exponents.

5. The polynomial $2t^4 + 3t^3 - 4t^2 + 5t - 6$ is written in **descending** powers of t.

CONCEPTS

7. To add polynomials, **combine** like terms.

9. $$\begin{array}{r} 8x^2 - 7x - 1 \\ + \boxed{-4x^2} - 6x \boxed{+} 9 \\ \hline \end{array}$$

11. a) $2x^2 + 3x^2 = (2 + 3)x^2 = 5x^2$
    b) $15m^3 - m^3 = (15 - 1)m^3 = 14m^3$
    c) $8a^3b - ab$; unlike terms, so you cannot combine them
    d) $6cd + 4c^2d$; unlike terms, so you cannot combine them

13. a) $-(5x^2 - 8x + 23) = -5x^2 + 8x - 23$
    b) $-(-5y^4 + 3y^2 - 7) = 5y^4 - 3y^2 + 7$

NOTATION

15. $(6x^2 + 2x + 3) - (4x^2 - 7x + 1)$
$$= 6x^2 + 2x + 3 \boxed{-} 4x^2 \boxed{+} 7x \boxed{-} 1$$
$$= \boxed{2x^2} + 9x + \boxed{2}$$

17. True

PRACTICE

19. $8t^2 + 4t^2 = (8 + 4)t^2$
$= 12t^2$

21. $-32u^3 - 16u^3 = (-32 - 16)u^3$
$= -48u^3$

23. $1.8x - 1.9x = (1.8 - 1.9)x$
$= -0.1x$

25. $\frac{1}{2}st + \frac{3}{2}st = \left(\frac{1}{2} + \frac{3}{2}\right)st$
$= \frac{4}{2}st$
$= 2st$

27. $3r - 4r + 7r = (3 - 4 + 7)r$
$= 6r$

29. $-4ab + 4ab - ab = (-4 + 4 - 1)ab$
$= -ab$

31. $10x^2 - 8x + 9x - 9x^2$
$= (10 - 9)x^2 + (-8 + 9)x$
$= 1x^2 + 1x$
$= x^2 + x$

33. $6x^3 + 8x^4 + 7x^3 + (-8x^4)$
$= (8 - 8)x^4 + (6 + 7)x^3$
$= 0x^4 + 13x^3$
$= 13x^3$

35. $4x^2y + 5 - 6x^3y - 3x^2y + 2x^3y$
$= (-6 + 2)x^3y + (4 - 3)x^2y + 5$
$= -4x^3y + x^2y + 5$

37. $\frac{2}{3}d^2 - \frac{1}{4}c^2 + \frac{5}{6}c^2 - \frac{1}{2}cd + \frac{1}{3}d^2$
$= \left(-\frac{1}{4} + \frac{5}{6}\right)c^2 - \frac{1}{2}cd + \left(\frac{2}{3} + \frac{1}{3}\right)d^2$
$= \frac{14}{24}c^2 - \frac{1}{2}cd + \frac{3}{3}d^2$
$= \frac{7}{12}c^2 - \frac{1}{2}cd + d^2$

39. $(3x + 7) + (4x - 3) = (3x + 4x) + (7 - 3)$
$= 7x + 4$

41. $(9a^2 + 3a) - (2a - 4a^2)$
$= 9a^2 + 3a - 2a + 4a^2$
$= (9a^2 + 4a^2) + (3a - 2a)$
$= 13a^2 + a$

43. $(2x + 3y) + (5x - 10y)$
$= (2x + 5x) + (3y - 10y)$
$= 7x - 7y$

45. $(-8x - 3y) - (-11x + y)$
$= -8x - 3y + 11x - y$
$= (-8x + 11x) + (-3y - y)$
$= 3x - 4y$

47. $(3x^2 - 3x - 2) + (3x^2 + 4x - 3)$
$= (3x^2 + 3x^2) + (-3x + 4x) + (-2 - 3)$
$= 6x^2 + x - 5$

49. $(2b^2 + 3b - 5) - (2b^2 - 4b - 9)$
$= 2b^2 + 3b - 5 - 2b^2 + 4b + 9$
$= (2b^2 - 2b^2) + (3b + 4b) + (-5 + 9)$
$= 0b^2 + 7b + 4$
$= 7b + 4$

51. $(2x^2-3x+1) - (4x^2-3x+2) + (2x^2+3x+2)$
$= 2x^2 - 3x + 1 - 4x^2 + 3x - 2 + 2x^2 + 3x + 2$
$= (2x^2-4x^2+2x^2) + (-3x+3x+3x) + (1-2+2)$
$= 3x + 1$

53. $(-4h^3 + 5h^2 + 15) - (h^3 - 15)$
$= -4h^3 + 5h^2 + 15) - h^3 + 15$
$= (-4h^3 - h^3) + 5h^2 + (15 + 15)$
$= -5h^3 + 5h^2 + 30$

55. $(1.04x^2+2.07x-5.01)+(1.33x-2.98x^2+5.02)$
$= (1.04-2.98)x^2 + (2.07+1.33)x + (-5.01+5.02)$
$= 1.94x^2 + 3.4x + 0.01$

57. $\left(\frac{7}{8}r^4 + \frac{5}{3}r^2 - \frac{9}{4}\right) - \left(-\frac{3}{8}r^4 - \frac{2}{3}r^2 - \frac{1}{4}\right)$

$= \frac{7}{8}r^4 + \frac{5}{3}r^2 - \frac{9}{4} + \frac{3}{8}r^4 + \frac{2}{3}r^2 + \frac{1}{4}$

$= \left(\frac{7}{8}r^4 + \frac{3}{8}r^4\right) + \left(\frac{5}{3}r^2 + \frac{2}{3}r^2\right) + \left(-\frac{9}{4} + \frac{1}{4}\right)$

$= \frac{10}{8}r^4 + \frac{7}{3}r^2 - \frac{8}{4}$

$= \frac{5}{4}r^4 + \frac{7}{3}r^2 - 2$

59.
$$\begin{array}{rr} & 3x^2 + 4x + 5 \\ + & 2x^2 - 3x + 6 \\ \hline & 5x^2 + x + 11 \end{array}$$

61.
$$\begin{array}{rr} & 3x^2 + 4x - 5 \\ - & -2x^2 - 2x + 3 \\ \hline \end{array}$$

$$\begin{array}{rr} & 3x^2 + 4x - 5 \\ + & 2x^2 + 2x - 3 \\ \hline & 5x^2 + 6x - 8 \end{array}$$

63.
$$\begin{array}{rr} & 4x^3 + 4x^2 - 3x + 10 \\ - & 5x^3 - 2x^2 - 4x - 4 \\ \hline \end{array}$$

$$\begin{array}{rr} & 4x^3 + 4x^2 - 3x + 10 \\ + & -5x^3 + 2x^2 + 4x + 4 \\ \hline & -x^3 + 6x^2 + x + 14 \end{array}$$

65.
$$\begin{array}{rr} & -3x^3 + 4x^2 - 4x + 9 \\ + & 2x^3 \qquad + 9x - 3 \\ \hline & -x^3 + 4x^2 + 5x + 6 \end{array}$$

67.
$$\begin{array}{rr} & 3x^3 + 4x^2 + 7x + 12 \\ - & -4x^3 + 6x^2 + 9x - 3 \\ \hline \end{array}$$

$$\begin{array}{rr} & 3x^3 + 4x^2 + 7x + 12 \\ + & 4x^3 - 6x^2 - 9x + 3 \\ \hline & 7x^3 - 2x^2 - 2x + 15 \end{array}$$

69. $2(x + 3) + 4(x - 2)$
$= 2x + 6 + 4x - 8$
$= (2x + 4x) + (6 - 8)$
$= 6x - 2$

71. $-2(x^2 + 7x - 1) - 3(x^2 - 2x + 2)$
$= -2x^2 - 14x + 2 - 3x^2 + 6x - 6$
$= (-2x^2 - 3x^2) + (-14x + 6x) + (2 - 6)$
$= -5x^2 - 8x - 4$

73. $2(2y^2-2y+2) - 4(3y^2-4y-1) + 4(y^3-y^2-y)$
$= 4y^2 - 4y + 4 - 12y^2 + 16y + 4 + 4y^3 - 4y^2 - 4y$
$= 4y^3 + (4-12-4)y^2 + (-4+16-4)y + (4+4)$
$= 4y^3 - 12y^2 + 8y + 8$

75. $(5s^2 - s + 9) - (s^2 + 4s + 2)$
$= 5s^2 - s + 9 - s^2 - 4s - 2$
$= (5s^2 - s^2) + (-s - 4s) + (9 - 2)$
$= 4s^2 - 5s + 7$

77. $(2y^5 - y^4) - (-y^5 + 5y^4 - 1.2)$
$= 2y^5 - y^4 + y^5 - 5y^4 + 1.2$
$= (2y^5 + y^5) + (-y^4 - 5y^4) + 1.2$
$= 3y^5 - 6y^4 + 1.2$

79. $(3t^3 + t^2) + (-t^3 + 6t - 3) - (t^3 - 2t^2 + 2)$
$= 3t^3 + t^2 - t^3 + 6t - 3 - t^3 + 2t^2 - 2$
$= (3t^3 - t^3 - t^3) + (t^2 + 2t^2) + 6t + (-3-2)$
$= t^3 + 3t^2 + 6t - 5$

81. $(-2x^2-7x+1)+(-4x^2+8x-1)+(3x^2+4x-7)$
$= -2x^2 - 7x + 1 - 4x^2 + 8x - 1 + 3x^2 + 4x - 7$
$= (-2x^2-4x^2+3x^2) + (-7x+8x+4x) + (1-1-7)$
$= -3x^2 + 5x - 7$

APPLICATIONS

83. GREEK ARCHITECTURE

a) $(x^2 - 3x + 2) - (5x - 10)$
$= x^2 - 3x + 2 - 5x + 10$
$= x^2 + (-3x - 5x) + (2 + 10)$
$= (x^2 - 8x + 12)$ feet

b) $(x^2 - 3x + 2) + (5x - 10)$
$= x^2 + (-3x + 5x) + (2 - 10)$
$= (x^2 + 2x - 8)$ feet

85. JETS

$(9x - 15) + (2x + 3)$
$= (9x + 2x) + (-15 + 3)$
$= (11x - 12)$ ft

87. READING BLUEPRINTS

a) length:
$(x^2-x+6)+(4x+3) = x^2 + 3x + 9$
width:
$(x^2-6x+3)+(3x+1) = x^2 - 3x + 4$

length – width =
$(x^2+3x+9) - (x^2-3x+4)$
$= x^2 + 3x + 9 - x^2 + 3x + 4$
$= (x^2 - x^2)+(3x + 3x)+(9 + 4)$
$= (6x + 13)$ ft

b) Use the length and widths from part a:
$2(x^2 + 3x + 9) + 2(x^2 - 3x + 4)$
$= 2x^2 + 6x + 18 + 2x^2 - 6x + 8$
$= (2x^2 + 2x^2)+(6x - 6x)+(18 + 8)$
$= (4x^2 + 26)$ ft

89. NAVAL OPERATIONS

a) $(-16t^2+150t+40)-(-16t^2+128t+20)$
$= -16t^2+150t + 40 + 16t^2 - 128t - 20$
$= (-16t^2+16t^2)+(150t-128t)+(40-20)$
$= (22t + 20)$ ft

b) Let $t = 4$ and substitute it into the answer from part a.
$22(4) + 20 = 88 + 20$
$= 108$ ft

WRITING

91. Answers will vary.

93. Answers will vary.

95. Answers will vary.

REVIEW

97. The sum of the measure of the angles of a triangle is 180°.

99. $2x + 3y = 9$
$2x + 3y - 2x = 9 - 2x$
$3y = -2x + 9$
$y = -\frac{2}{3}x + 3$

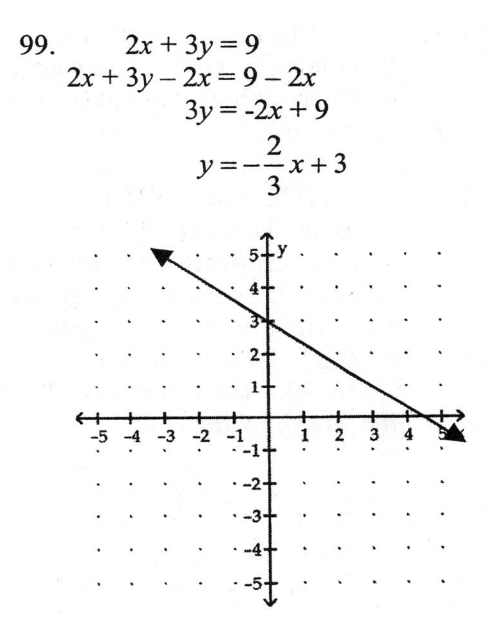

CHALLENGE PROBLEMS

101. $(6x^2 - 7x - 8) - (2x^2 - x + 3)$
$= 6x^2 - 7x - 8 - 2x^2 + x - 3$
$= (6x^2 - 2x^2) + (-7x + x) + (-8 - 3)$
$= 4x^2 - 6x - 11$

SECTION 4.6

VOCABULARY

1. The expression $(2x^3)(3x^4)$ is the product of two **monomials**.

3. The expression $(2a - 4)(3a + 5)$ is the product of two **binomials**.

5. To simplify $y^2 + 3y + 2y + 6$, we combine the **like** terms $3y$ and $2y$ to get $y^2 + 5y + 6$.

CONCEPTS

7. a) To multiply two monomials, multiply the numerical factors (the **coefficients**) and then multiply the **variable** factors.
   b) To multiply two polynomials, multiply **each** term of one polynomial by **each** term of the other polynomial, and then combine like terms.
   c) When multiplying three polynomials, we begin by multiplying **any** two of them, and then we multiply that result by the **third** polynomial.
   d) To find the area of a rectangle, multiply the **length** by the width.

9. $(2x + 5)(3x + 4) = \boxed{6x^2} + \boxed{8x} + \boxed{15x} + \boxed{20}$

11. a) $(x - 4) + (x + 8) = 2x + 4$

    b) $(x - 4) - (x + 8) = x - 4 - x - 8$
    $= -12$

    c) $(x - 4)(x + 8) = x^2 + 8x - 4x - 32$
    $= x^2 + 4x - 32$

13. $7x(3x^2 - 2x + 5) = \boxed{7x}(3x^2) - \boxed{7x}(2x) + \boxed{7x}(5)$
    $= \boxed{21x^3} - 14x^2 + 35x$

PRACTICE

15. $15m(m) = 15m^2$

17. $3x^2(4x^3) = (3 \cdot 4)(x^2 \cdot x^3)$
    $= 12x^5$

19. $(3b^2)(-2b)(4b^3) = (3 \cdot -2 \cdot 4)(b^2 \cdot b \cdot b^3)$
    $= -24b^6$

21. $(2x^2y^3)(3x^3y^2) = (2 \cdot 3)(x^2y^3 \cdot x^3y^2)$
    $= 6x^5y^5$

23. $\left(8a^5\right)\left(-\frac{1}{4}a^6\right) = \left(8 \bullet -\frac{1}{4}\right)\left(a^5a^6\right)$
    $= -2a^{11}$

25. $(1.2c^3)(5c^3) = (1.2 \cdot 5)(c^3c^3)$
    $= 6c^6$

27. $\left(\frac{1}{2}a\right)\left(\frac{1}{4}a^4\right)\left(a^5\right) = \left(\frac{1}{2} \bullet \frac{1}{8}\right)\left(aa^4a^5\right)$
    $= \frac{1}{16}a^{10}$

29. $(x^2y^5)(x^2z^5)(-3z^3) = -3(x^2x^2y^5z^5z^3)$
    $= -3x^4y^5z^8$

31. $3(x + 4) = 3(x) + 3(4)$
    $= 3x + 12$

33. $-4(t^2 + 7) = -4(t^2) - 4(7)$
    $= -4t^2 - 28$

35. $3x(x - 2) = 3x(x) + 3x(-2)$
    $= 3x^2 - 6x$

37. $-2x^2(3x^2 - x) = -2x^2(3x^2) - 2x^2(-x)$
    $= -6x^4 + 2x^3$

39. $(x^2 - 12x)(6x^{12}) = x^2(6x^{12}) - 12x(6x^{12})$
    $= 6x^{14} - 72x^{13}$

41. $\frac{5}{8}t^2\left(t^6 + 8t^2\right) = \frac{5}{8}t^2\left(t^6\right) + \frac{5}{8}t^2\left(8t^2\right)$
    $= \frac{5}{8}t^8 + 5t^4$

43. $0.3p^5(0.4p^4 - 6p^2) = 0.3p^5(0.4p^4) + 0.3p^5(-6p^2)$
    $= 0.12p^9 - 1.8p^7$

45. $-4x^2z(3x^2 - z) = -4x^2z(3x^2) - 4x^2z(-z)$
    $= -12x^4z + 4x^2z^2$

47. $2x^2(3x^2 + 4x - 7)$
    $= 2x^2(3x^2) + 2x^2(4x) + 2x^2(-7)$
    $= 6x^4 + 8x^3 - 14x^2$

49. $3a(4a^2 + 3a - 4)$
    $= 3a(4a^2) + 3a(3a) + 3a(-4)$
    $= 12a^3 + 9a^2 - 12a$

51. $(-2a^2)(-3a^3)(3a-2) = (6a^5)(3a-2)$
$= 6a^5(3a) + 6a^5(-2)$
$= 18a^6 - 12a^5$

53. $(y-3)(y+5) = y(y) + y(5) - 3(y) - 3(5)$
$= y^2 + 5y - 3y - 15$
$= y^2 + 2y - 15$

55. $(t+4)(2t-3) = t(2t) + t(-3) + 4(2t) + 4(-3)$
$= 2t^2 - 3t + 8t - 12$
$= 2t^2 + 5t - 12$

57. $(2y-5)(3y+7)$
$= 2y(3y) + 2y(7) - 5(3y) - 5(7)$
$= 6y^2 + 14y - 15y - 35$
$= 6y^2 - y - 35$

59. $(2x+3)(2x-5)$
$= 2x(2x) + 2x(-5) + 3(2x) + 3(-5)$
$= 4x^2 - 10x + 6x - 15$
$= 4x^2 - 4x - 15$

61. $(a+b)(a+b)$
$= a(a) + a(b) + b(a) + b(b)$
$= a^2 + ab + ab + b^2$
$= a^2 + 2ab + b^2$

63. $(3a-2b)(4a+b)$
$= 3a(4a) + 3a(b) - 2b(4a) - 2b(b)$
$= 12a^2 + 3ab - 8ab - 2b^2$
$= 12a^2 - 5ab - 2b^2$

65. $(t^2-3)(t^2+4)$
$= t^2(t^2) + t^2(4) - 3(t^2) - 3(4)$
$= t^4 + 4t^2 - 3t^2 - 12$
$= t^4 + t^2 - 12$

67. $(-3t+2s)(2t-3s)$
$= -3t(2t) - 3t(-3s) + 2s(2t) + 2s(-3s)$
$= -6t^2 + 9st + 4st - 6s^2$
$= -6t^2 + 13st - 6s^2$

69. $\left(4a - \frac{5}{9}r\right)\left(2a + \frac{3}{4}r\right)$

$= 4a(2a) + 4a\left(\frac{3}{4}r\right) - \frac{5}{9}r(2a) - \frac{5}{9}r\left(\frac{3}{4}r\right)$

$= 8a^2 + 3ar - \frac{10}{9}ar - \frac{5}{12}r^2$

$= 8a^2 + \frac{17}{9}ar - \frac{5}{12}r^2$

71. $4(2x+1)(x-2)$
$= 4(2x \cdot x + 2x \cdot -2 + 1 \cdot x + 1 \cdot -2)$
$= 4(2x^2 - 4x + x - 2)$
$= 4(2x^2 - 3x - 2)$
$= 4(2x^2) + 4(-3x) + 4(-2)$
$= 8x^2 - 12x - 8$

73. $3a(a+b)(a-b)$
$= 3a(a \cdot a + a \cdot -b + b \cdot a + b \cdot -b)$
$= 3a(a^2 - ab + ab - b^2)$
$= 3a(a^2 - b^2)$
$= 3a^3 - 3ab^2$

75. $(x+2)(x^2-2x+3)$
$= x(x^2-2x+3) + 2(x^2-2x+3)$
$= x \cdot x^2 + x \cdot -2x + x \cdot 3 + 2 \cdot x^2 + 2 \cdot -2x + 2 \cdot 3$
$= x^3 - 2x^2 + 3x + 2x^2 - 4x + 6$
$= x^3 + (-2x^2 + 2x^2) + (3x - 4x) + 6$
$= x^3 - x + 6$

77. $(4t+3)(t^2+2t+3)$
$= 4t(t^2+2t+3) + 3(t^2+2t+3)$
$= 4t^3 + 8t^2 + 12t + 3t^2 + 6t + 9$
$= 4t^3 + (8t^2 + 3t^2) + (12t + 6t) + 9$
$= 4^3 + 11t^2 + 18t + 9$

79. $(x^2+6x+7)(2x-5)$
$= x^2(2x-5) + 6x(2x-5) + 7(2x-5)$
$= 2x^3 - 5x^2 + 12x^2 - 30x + 14x - 35$
$= 2x^3 + (-5x^2 + 12x^2) + (-30x + 14x) - 35$
$= 2x^3 + 7x^2 - 16x - 35$

81. $(-3x+y)(x^2-8xy+16y^2)$
$= -3x(x^2-8xy+16y^2) + y(x^2-8xy+16y^2)$
$= -3x^3 + 24x^2y - 48xy^2 + x^2y - 8xy^2 + 16y^3$
$= -3x^3 + (24x^2y + x^2y) + (-48xy^2 - 8xy^2) + 16y^3$
$= -3x^3 + 25x^2y - 56xy^2 + 16y^3$

83. $(r^2-r+3)(r^2-4r-5)$
$= r^2(r^2-4r-5) - r(r^2-4r-5) + 3(r^2-4r-5)$
$= r^4 - 4r^3 - 5r^2 - r^3 + 4r^2 + 5r + 3r^2 - 12r - 15$
$= r^4 - 4r^3 - r^3 - 5r^2 + 4r^2 + 3r^2 + 5r - 12r - 15$
$= r^4 - 5r^3 + 2r^2 - 7r - 15$

85.
$$\begin{array}{r} x^2 - 2x + 1 \\ x + 2 \\ \hline 2x^2 - 4x + 2 \\ x^3 - 2x^2 + x \\ \hline x^3 + 0x^2 - 3x + 2 \\ x^3 - 3x + 2 \end{array}$$

87.

$$\begin{array}{r} 4x^2 + 3x - 4 \\ 3x + 2 \\ \hline 8x^2 + 6x - 8 \\ 12x^3 + 9x^2 - 12x \quad \\ \hline 12x^3 + 17x^2 - 6x - 8 \end{array}$$

89.

$$\begin{array}{r} 2a^2 + 3a + 1 \\ 3a^2 - 2a + 4 \\ \hline 8a^2 + 12a + 4 \\ -4a^3 - 6a^2 - 2a \quad \\ 6a^4 + 9a^3 + 3a^2 \quad \quad \\ \hline 6a^4 + 5a^3 + 5a^2 + 10a + 4 \end{array}$$

APPLICATIONS

91. STAMPS

$$\begin{aligned} &(2x + 1)(3x - 1) \\ &= 2x \cdot 3x + 2x \cdot -1 + 1 \cdot 3x + 1 \cdot -1 \\ &= 6x^2 - 2x + 3x - 1 \\ &= (6x^2 + x - 1) \text{ cm}^2 \end{aligned}$$

93. $A = s^2$

$$\begin{aligned} &(3x + 1)(3x + 1) \\ &= 3x \cdot 3x + 3x \cdot 1 + 1 \cdot 3x + 1 \cdot 1 \\ &= 9x^2 + 3x + 3x + 1 \\ &= (9x^2 + 6x + 1) \text{ ft}^2 \end{aligned}$$

95. $A = ½\, bh$

$$\begin{aligned} \frac{1}{2}(4x - 2)(2x - 2) &= \frac{1}{2}(8x^2 - 8x - 4x + 4) \\ &= \frac{1}{2}(8x^2 - 12x + 4) \\ &= (4x^2 - 6x + 2)\text{cm}^2 \end{aligned}$$

97. SUNGLASSES

$$\begin{aligned} &2 \cdot 3.14(x + 1)(x - 1) \\ &= 2 \cdot 3.14(x^2 - x + x - 1) \\ &= 2 \cdot 3.14(x^2 - 1) \\ &= 6.28(x^2 - 1) \\ &= (6.28x^2 - 6.28) \text{ in}^2 \end{aligned}$$

99. TOYS

$$\begin{aligned} (5x + 4)(7x + 3) &= 35x^2 + 15x + 28x + 12 \\ &= (35x^2 + 43x + 12) \text{ cm}^2 \end{aligned}$$

101. LUGGAGE

$$\begin{aligned} x(x - 3)(2x + 2) &= x(2x^2 + 2x - 6x - 6) \\ &= x(2x^2 - 4x - 6) \\ &= (2x^3 - 4x^2 - 6x) \text{ in}^3 \end{aligned}$$

WRITING

103. Answers will vary.

105. Answers will vary.

REVIEW

107. Use any two points on the line such as (-2, 0) and (0, 2):

$$\begin{aligned} m &= \frac{y_2 - y_1}{x_2 - x_1} \\ &= \frac{2 - 0}{0 - (-2)} \\ &= \frac{2}{2} \\ &= 1 \end{aligned}$$

109. Use any two points on the line such as (-1, 0) and (2, -2):

$$\begin{aligned} m &= \frac{y_2 - y_1}{x_2 - x_1} \\ &= \frac{-2 - 0}{2 - (-1)} \\ &= -\frac{2}{3} \end{aligned}$$

111. Line 1 intersects the y-axis at (0, 2).

CHALLENGE PROBLEMS

113. a) $(x - 1)(x + 1) = x^2 - 1$

b)

$$\begin{aligned} &(x - 1)(x^2 + x + 1) \\ &= x(x^2 + x + 1) - 1(x^2 + x + 1) \\ &= x^3 + x^2 + x - x^2 - x - 1 \\ &= x^3 - 1 \end{aligned}$$

c)

$$\begin{aligned} &(x - 1)(x^3 + x^2 + x + 1) \\ &= x(x^3+x^2+x+1) - 1(x^3+x^2+x+1) \\ &= x^4 + x^3 + x^2 + x - x^3 - x^2 - x - 1 \\ &= x^4 - 1 \end{aligned}$$

SECTION 4.7

VOCABULARY

1. The first **term** of $3x + 6$ is $3x$ and the second **term** is 6.

3. $(b + 1)(b - 1)$ is the product of the **sum** and difference of two terms.

5. An expression of the form $x^2 - y^2$ is called a **difference** of two squares.

CONCEPTS

7. a) $(x^m)^n = \boxed{x^{m \cdot n}}$
   b) $(xy)^n = \boxed{x^n y^n}$

9. a) $(a + 5)(a + 5) = (a + 5)^{\boxed{2}}$
   b) $(n - 12)(n - 12) = (n - 12)^{\boxed{2}}$

11. $(x + y)(x - y) = x^2 - y^2$

The square of the **second** term

The **square** of the first term

NOTATION

13. $(x + 4)^2 = \boxed{x}^2 + 2(x)(\boxed{4}) + \boxed{4}^2$
$= x^2 + \boxed{8x} + 16$

15. $(s + 5)(s - 5) = \boxed{s}^2 \boxed{-} \boxed{5}^2$
$= s^2 - 25$

PRACTICE

17. $(x + 1)^2 = x^2 + 2(x)(1) + 1^2$
$= x^2 + 2x + 1$

19. $(r + 2)^2 = r^2 + 2(r)(2) + 2^2$
$= r^2 + 4r + 4$

21. $(m - 6)^2 = m^2 - 2(m)(6) + (-6)^2$
$= m^2 - 12m + 36$

23. $(f - 8)^2 = f^2 - 2(f)(8) + (-8)^2$
$= f^2 - 16f + 64$

25. $(d + 7)(d - 7) = d^2 - 7^2$
$= d^2 - 49$

27. $(n + 6)(n - 6) = n^2 - 6^2$
$= n^2 - 36$

29. $(4x + 5)^2 = (4x)^2 + 2(4x)(5) + (5)^2$
$= 16x^2 + 40x + 25$

31. $(7m - 2)^2 = (7m)^2 - 2(7m)(2) + (-2)^2$
$= 49m^2 - 28m + 4$

33. $(y^2 + 9)^2 = (y^2)^2 + 2(y^2)(9) + (9)^2$
$= y^4 + 18y^2 + 81$

35. $(2v^3 - 8)^2 = (2v^3)^2 - 2(2v^3)(8) + (-8)^2$
$= 4v^6 - 32v^3 + 64$

37. $(4f + 0.4)(4f - 0.4) = (4f)^2 - (0.4)^2$
$= 16f^2 - 0.16$

39. $(3n + 1)(3n - 1) = (3n)^2 - (1)^2$
$= 9n^2 - 1$

41. $(1 - 3y)^2 = (1)^2 - 2(1)(3y) + (-3y)^2$
$= 1 - 6y + 9y^2$

43. $(x - 2y)^2 = (x)^2 - 2(x)(2y) + (-2y)^2$
$= x^2 - 4xy + 4y^2$

45. $(2a - 3b)^2 = (2a)^2 - 2(2a)(3b) + (-3b)^2$
$= 4a^2 - 12ab + 9b^2$

47. $\left(s + \frac{3}{4}\right)^2 = s^2 + 2(s)\left(\frac{3}{4}\right) + \left(\frac{3}{4}\right)^2$
$= s^2 + \frac{6}{4}s + \frac{9}{16}$
$= s^2 + \frac{3}{2}s + \frac{9}{16}$

49. $(a + b)^2 = a^2 + 2(a)(b) + b^2$
$= a^2 + 2ab + b^2$

51. $(r - s)^2 = r^2 - 2(r)(s) + (-s)^2$
$= r^2 - 2rs + s^2$

53. $\left(6b + \frac{1}{2}\right)\left(6b - \frac{1}{2}\right) = (6b)^2 - \left(\frac{1}{2}\right)^2$
$= 36b^2 - \frac{1}{4}$

55. $(r + 10s)^2 = (r)^2 + 2(r)(10s) + (10s)^2$
$= r^2 + 20rs + 100s^2$

57. $(6-2d^3)^2 = (6)^2 - 2(6)(2d^3) + (-2d^3)^2$
$= 36 - 24d^3 + 4d^6$

59. $-(8x+3)^2 = -[(8x)^2 + 2(8x)(3) + (3)^2]$
$= -[64x^2 + 48x + 9]$
$= -64x^2 - 48x - 9$

61. $-(5-6g)(5+6g) = -[(5)^2 - (6g)^2]$
$= -[25 - 36g^2]$
$= -25 + 36g^2$

63. $3x(2x+3)^2 = 3x[(2x)^2 + 2(2x)(3) + (3)^2]$
$= 3x[4x^2 + 12x + 9]$
$= 12x^3 + 36x^2 + 27x$

65. $-5d(4d-1)^2 = -5d[(4d)^2 - 2(4d)(1) + (-1)^2]$
$= -5d[16d^2 - 8d + 1]$
$= -80d^3 + 40d^2 - 5d$

67. $4d(d^2+g^3)(d^2-g^3) = 4d[(d^2)^2 - (g^3)^2]$
$= 4d[d^4 - g^6]$
$= 4d^5 - 4dg^6$

69. $(x+4)^3 = (x+4)(x+4)^2$
$= (x+4)[x^2 + 2(x)(4) + 4^2]$
$= (x+4)[x^2 + 8x + 16]$
$= x(x^2+8x+16) + 4(x^2+8x+16)$
$= x^3 + 8x^2 + 16x + 4x^2 + 32x + 64$
$= x^3 + 12x^2 + 48x + 64$

71. $(n-6)^3 = (n-6)(n-6)^2$
$= (n-6)[n^2 - 2(n)(6) + (-6)^2]$
$= (n-6)[n^2 - 12n + 36]$
$= n(n^2 - 12n + 36) - 6(n^2 - 12n + 36)$
$= n^3 - 12n^2 + 36n - 6n^2 + 72n - 216$
$= n^3 - 18n^2 + 108n - 216$

73. $(2g-3)^3 = (2g-3)(2g-3)^2$
$= (2g-3)[(2g)^2 - 2(2g)(3) + (-3)^2]$
$= (2g-3)[4g^2 - 12g + 9]$
$= 2g(4g^2-12g+9) - 3(4g^2-12g+9)$
$= 8g^3 - 24g^2 + 18g - 12g^2 + 36g - 27$
$= 8g^3 - 36g^2 + 54g - 27$

75. $(n-2)^4$
$= (n-2)^2(n-2)^2$
$= [(n)^2-2(n)(2)+(-2)^2][(n)^2-2(n)(2)+(-2)^2]$
$= [n^2 - 4n + 4][n^2 - 4n + 4]$
$= n^2(n^2-4n+4) - 4n(n^2-4n+4) + 4(n^2-4n+4)$
$= n^4 - 4n^3 + 4n^2 - 4n^3 + 16n^2 - 16n + 4n^2 - 16n + 16$
$= n^4 - 8n^3 + 24n^2 - 32n + 16$

77. $2t(t+2) + (t-1)(t+9)$
$= 2t^2 + 4t + t^2 + 9t - 1t - 9$
$= 3t^2 + 12t - 9$

79. $(x+y)(x-y) + x(x+y)$
$= (x^2 - y^2) + x^2 + xy$
$= 2x^2 + xy - y^2$

81. $(3x-2)^2 + (2x+1)^2$
$= [(3x)^2 - 2(3x)(2) + (2)^2] + [(2x)^2 + 2(2x)(1) + 1^2]$
$= 9x^2 - 12x + 4 + 4x^2 + 4x + 1$
$= 13x^2 - 8x + 5$

83. $(m+10)^2 - (m-8)^2$
$= [(m)^2 + 2(m)(10) + (10)^2] - [(m)^2 - 2(m)(8) + 8^2]$
$= [m^2 + 20m + 100] - [m^2 - 16m + 64]$
$= m^2 + 20m + 100 - m^2 + 16m - 64$
$= 36m + 36$

APPLICATIONS

85. DINNER PLATES

$$\frac{\pi}{4}(D-d)(D+d) = \frac{\pi}{4}(D^2 - d^2)$$
$$= \frac{\pi}{4}D^2 - \frac{\pi}{4}d^2$$

87. PLAYPENS

$(x+6)^2 = (x^2 + 2(x)(6) + 6^2)$
$= (x^2 + 12x + 36)$ in.^{2}

89. PAINTING

Find the length: $(4x+2)+(8x+4) = 12x + 6$

Multiply the length times the width:

$(4x+1)(12x+6)$
$= 48x^2 + 24x + 12x + 6$
$= 48x^2 + 36x + 6$

Find the area of one window: $x \cdot x = x^2$

Area of 12 windows $= 12x^2$

Subtract the area of the 12 windows from the total area:

$48x^2 + 36x + 6 - 12x^2$
$= (36x^2 + 36x + 6)$ ft^2

WRITING

91. Answers will vary.

93. Answers will vary.

REVIEW

95. $189 = 3 \cdot 3 \cdot 3 \cdot 7$
$= 3^3 \cdot 7$

97. $\frac{30}{36} = \frac{\cancel{2} \bullet \cancel{3} \bullet 5}{\cancel{2} \bullet 2 \bullet \cancel{3} \bullet 3} = \frac{5}{6}$

99. $\frac{7}{8} \bullet \frac{3}{5} = \frac{7 \bullet 3}{8 \bullet 5}$
$= \frac{21}{40}$

CHALLENGE PROBLEMS

101. a) $(x + 1)(x - 1) = x^2 - 1$
b) $(x + 1)(x + 1) = x^2 + 2x + 1$
c) $(x + a)(x + b) = x^2 + ax + bx + ab$

SECTION 4.8

VOCABULARY

1. The **numerator** of $\frac{15x^2-25x}{5x}$ is $15x^2 - 25x$ and the **denominator** is $5x$.

3. The expression $\frac{6x^3y-4x^2y^2+8xy^3-2y^4}{2x^4}$ is a **polynomial** divided by a monomial.

5. The powers of x in $2x^4 + 3x^3 + 4x^2 - 7x - 8$ are written in **descending** order.

7. The expression $5x^2 + 6$ is missing an x-term. We can insert a **placeholder** $0x$ term and write it as $5x^2 + 0x + 6$.

CONCEPTS

9. a) $\frac{x^m}{x^n} = x^{m-n}$

   b) $x^{-n} = \frac{1}{x^n}$

11. a) $7x^3 + 5x^2 - 3x - 9$

    b) $6x^4 - x^3 + 2x^2 + 9x$

13. $-9x - (-7x) = -9x + 7x$
    $= -2x$

15. It is correct.
    $(3x + 2)(x + 2) = 3x^2 + 6x + 2x + 4$
    $= 3x^2 + 8x + 4$

NOTATION

17. $\frac{28x^5 - x^3 + 7x^2}{7x^2} = \frac{28x^5}{\boxed{7x^2}} - \frac{\boxed{x^3}}{7x^2} + \frac{7x^2}{\boxed{7x^2}}$

$$= 4x^{5-2} - \frac{x^{3-2}}{\boxed{7}} + x^{\boxed{2-2}}$$

$$= \boxed{4x^3} - \frac{x}{7} + \boxed{1}$$

19. a) $5x^4 + 0x^3 + 2x^2 + 0x - 1$

    b) $-3x^5 + 0x^4 - 2x^3 + 0x^2 + 4x - 6$

21. $\frac{8}{6} = \frac{4}{3}$

23. $\frac{x^5}{x^2} = x^{5-2}$
    $= x^3$

25. $\frac{45m^{10}}{9m^5} = 5m^{10-5}$
    $= 5m^5$

27. $\frac{12h^8}{9h^6} = \frac{4h^{8-6}}{3}$
    $= \frac{4h^2}{3}$

29. $\frac{-3d^4}{15d^8} = \frac{-1}{5}d^{4-8}$
    $= -\frac{1}{5}d^{-4}$
    $= -\frac{1}{5d^4}$

31. $\frac{r^3s^2}{rs^3} = r^{3-1}s^{2-3}$
    $= r^2s^{-1}$
    $= \frac{r^2}{s}$

33. $\frac{8x^3y^2}{4xy^3} = 2x^{3-1}y^{2-3}$
    $= 2x^2y^{-1}$
    $= \frac{2x^2}{y}$

35. $\dfrac{-16r^3y^2}{-4r^2y^4} = 4r^{3-2}y^{2-4}$

$= 4r^1y^{-2}$

$= \dfrac{4r}{y^2}$

37. $\dfrac{-65rs^2t}{15r^2s^3t} = -\dfrac{13}{3}r^{1-2}s^{2-3}t^{1-1}$

$= -\dfrac{13}{3}r^{-1}s^{-1}t^0$

$= -\dfrac{13}{3rs}$

39. $\dfrac{6x+9}{3} = \dfrac{6x}{3} + \dfrac{9}{3}$

$= 2x+3$

41. $\dfrac{8x^9 - 32x^6}{4x^4} = \dfrac{8x^9}{4x^4} - \dfrac{32x^6}{4x^4}$

$= 2x^5 - 8x^2$

43. $\dfrac{6h^{12} + 48h^9}{24h^{10}} = \dfrac{6h^{12}}{24h^{10}} + \dfrac{48h^9}{24h^{10}}$

$= \dfrac{1}{4}h^{12-10} + 2h^{9-10}$

$= \dfrac{1}{4}h^2 + 2h^{-1}$

$= \dfrac{h^2}{4} + \dfrac{2}{h}$

45. $\dfrac{-18w^6 - 9}{9w^4} = \dfrac{-18w^6}{9w^4} - \dfrac{9}{9w^4}$

$= -2w^{6-4} - 1w^{-4}$

$= -2w^2 - \dfrac{1}{w^4}$

47. $\dfrac{9s^8 - 18s^5 + 12s^4}{3s^3} = \dfrac{9s^8}{3s^3} - \dfrac{18s^5}{3s^3} + \dfrac{12s^4}{3s^3}$

$= 3s^{8-3} - 6s^{5-3} + 4s^{4-3}$

$= 3s^5 - 6s^2 + 4s$

49. $\dfrac{7c^5 + 21c^4 - 14c^3 - 35c}{7c^2}$

$= \dfrac{7c^5}{7c^2} + \dfrac{21c^4}{7c^2} - \dfrac{14c^3}{7c^2} - \dfrac{35c}{7c^2}$

$= c^{5-2} + 3c^{4-2} - 2c^{3-2} - 5c^{1-2}$

$= c^3 + 3c^2 - 2c^1 - 5c^{-1}$

$= c^3 + 3c^2 - 2c - \dfrac{5}{c}$

51. $\dfrac{5x - 10y}{25xy} = \dfrac{5x}{25xy} - \dfrac{10y}{25xy}$

$= \dfrac{1}{5y} - \dfrac{2}{5x}$

53. $\dfrac{15a^3b^2 - 10a^2b^3}{5a^2b^2} = \dfrac{15a^3b^2}{5a^2b^2} - \dfrac{10a^2b^3}{5a^2b^2}$

$= 3a^{3-2}b^{2-2} - 2a^{2-2}b^{3-2}$

$= 3a^1b^0 - 2a^0b^1$

$= 3a - 2b$

55. $\dfrac{12x^3y^2 - 8x^2y - 4x}{4xy}$

$= \dfrac{12x^3y^2}{4xy} - \dfrac{8x^2y}{4xy} - \dfrac{4x}{4xy}$

$= 3x^{3-1}y^{2-1} - 2x^{2-1}y^{1-1} - 4x^{1-1}y^{-1}$

$= 3x^2y - 2x - \dfrac{1}{y}$

57. $\dfrac{-25x^2y + 30xy^2 - 5xy}{-5xy}$

$= \dfrac{-25x^2y}{-5xy} + \dfrac{30xy^2}{-5xy} + \dfrac{-5xy}{-5xy}$

$= 5x^{2-1}y^{1-1} - 6x^{1-1}y^{2-1} + 1x^{1-1}y^{1-1}$

$= 5x^1y^0 - 6x^0y^1 + 1x^0y^0$

$= 5x - 6y + 1$

59. $x+6$

$$\begin{array}{r} x+6 \\ x+2\overline{)x^2+8x+12} \\ \underline{x^2+2x} \\ 6x+12 \\ \underline{6x+12} \\ 0 \end{array}$$

61. $y+12$

$$\begin{array}{r} y+12 \\ y+1\overline{)y^2+13y+12} \\ \underline{y^2+1y} \\ 12y+12 \\ \underline{12y+12} \\ 0 \end{array}$$

63. $3a-2$

$$\begin{array}{r} 3a-2 \\ 2a+3\overline{)6a^2+5a-6} \\ \underline{6a^2+9a} \\ -4a-6 \\ \underline{-4a-6} \\ 0 \end{array}$$

65. $b+3$

$$\begin{array}{r} b+3 \\ 3b+2\overline{)3b^2+11b+6} \\ \underline{3b^2+2b} \\ 9b+6 \\ \underline{9b+6} \\ 0 \end{array}$$

67. $2x+1$

Write the dividend in descending order.

$$\begin{array}{r} 2x+1 \\ 5x+3\overline{)10x^2+11x+3} \\ \underline{10x^2+6x} \\ 5x+3 \\ \underline{5x+3} \\ 0 \end{array}$$

69. $x-7$

Write the dividend and the divisor in descending order.

$$\begin{array}{r} x-7 \\ 2x+4\overline{)2x^2-10x-28} \\ \underline{2x^2+4x} \\ -14x-28 \\ \underline{-14x-28} \\ 0 \end{array}$$

71. $3x+2$

Write the dividend in descending order.

$$\begin{array}{r} 3x+2 \\ 2x-1\overline{)6x^2+x-2} \\ \underline{6x^2-3x} \\ 4x-2 \\ \underline{4x-2} \\ 0 \end{array}$$

73. $x^2 + 2x - 1$

$$
\begin{array}{r}
x^2 + 2x - 1 \\
2x + 3 \overline{\smash{)}\, 2x^3 + 7x^2 + 4x - 3} \\
\underline{2x^3 + 3x^2} \qquad\qquad \\
4x^2 + 4x \quad\; \\
\underline{4x^2 + 6x} \quad\; \\
-2x - 3 \\
\underline{-2x - 3} \\
0
\end{array}
$$

75. $2x^2 + 2x + 1$

$$
\begin{array}{r}
2x^2 + 2x + 1 \\
3x + 2 \overline{\smash{)}\, 6x^3 + 10x^2 + 7x + 2} \\
\underline{6x^3 + 4x^2} \qquad\qquad \\
6x^2 + 7x \quad\; \\
\underline{6x^2 + 4x} \quad\; \\
3x + 2 \\
\underline{3x + 2} \\
0
\end{array}
$$

77. $x^2 + x + 1$

$$
\begin{array}{r}
x^2 + x + 1 \\
2x + 1 \overline{\smash{)}\, 2x^3 + 3x^2 + 3x + 1} \\
\underline{2x^3 + x^2} \qquad\qquad \\
2x^2 + 3x \quad\; \\
\underline{2x^2 + x} \quad\; \\
2x + 1 \\
\underline{2x + 1} \\
0
\end{array}
$$

79. $x + 1$

Insert a $0x$ term as a placeholder.

$$
\begin{array}{r}
x + 1 \\
x - 1 \overline{\smash{)}\, x^2 + 0x - 1} \\
\underline{x^2 - x} \quad\; \\
x - 1 \\
\underline{x - 1} \\
0
\end{array}
$$

81. $2x - 3$

Insert a $0x$ term as a placeholder.

$$
\begin{array}{r}
2x - 3 \\
2x + 3 \overline{\smash{)}\, 4x^2 + 0x - 9} \\
\underline{4x^2 + 6x} \quad\; \\
-6x - 9 \\
\underline{-6x - 9} \\
0
\end{array}
$$

83. $x^2 - x + 1$

Insert a $0x^2$ term and a $0x$ term as placeholders.

$$
\begin{array}{r}
x^2 - x + 1 \\
x + 1 \overline{\smash{)}\, x^3 + 0x^2 + 0x + 1} \\
\underline{x^3 + x^2} \qquad\qquad \\
-x^2 + 0x \quad\; \\
\underline{-x^2 - x} \quad\; \\
x + 1 \\
\underline{x + 1} \\
0
\end{array}
$$

85. $a^2 - 3a + 10 + \frac{-30}{a+3}$

Insert a $0a^2$ term and a 0 constant term as placeholders.

$$
\begin{array}{r}
a^2 - 3a + 10 + \frac{-30}{a+3} \\
a + 3 \overline{\smash{)}\, a^3 + 0a^2 + a + 0} \\
\underline{a^3 + 3a^2} \qquad\qquad \\
-3a^2 + a \quad\; \\
\underline{-3a^2 - 9a} \quad\; \\
10a + 0 \\
\underline{10a + 30} \\
-30
\end{array}
$$

87. $x+1+\frac{-1}{2x+3}$

$$\begin{array}{r} x+1+\frac{-1}{2x+3} \\ 2x+3\overline{)2x^2+5x+2} \\ \underline{2x^2+3x} \\ 2x+2 \\ \underline{2x+3} \\ -1 \end{array}$$

89. $2x+2+\frac{-3}{2x+1}$

$$\begin{array}{r} 2x+2+\frac{-3}{2x+1} \\ 2x+1\overline{)4x^2+6x-1} \\ \underline{4x^2+2x} \\ 4x-1 \\ \underline{4x+2} \\ -3 \end{array}$$

91. $x^2+2x-1+\frac{6}{2x+3}$

$$\begin{array}{r} x^2+2x-1+\frac{6}{2x+3} \\ 2x+3\overline{)2x^3+7x^2+4x+3} \\ \underline{2x^3+3x^2} \\ 4x^2+4x \\ \underline{4x^2+6x} \\ -2x+3 \\ \underline{-2x-3} \\ 6 \end{array}$$

93. $2x^2+x+1+\frac{2}{3x-1}$

$$\begin{array}{r} 2x^2+x+1+\frac{2}{3x-1} \\ 3x-1\overline{)6x^3+x^2+2x+1} \\ \underline{6x^3-2x^2} \\ 3x^2+2x \\ \underline{3x^2-x} \\ 3x+1 \\ \underline{3x-1} \\ 2 \end{array}$$

APPLICATIONS

95. POOL

$$\frac{6x^2-3x+9}{3}=\frac{6x^2}{3}-\frac{3x}{3}+\frac{9}{3}$$
$$=\left(2x^2-x+3\right)\text{in.}$$

97. a) $d=rt$

$$\frac{d}{r}=\frac{rt}{r}$$
$$\frac{d}{r}=t$$
$$t=\frac{d}{r}$$

b) $t=\frac{d}{r}$

$$=\frac{6x^3}{2x}$$
$$=3x^{3-1}$$
$$=3x^2$$

c) $t=\frac{d}{r}$

$$=\frac{x^2+x-12}{x-3}$$

$$=\begin{array}{r} x+4 \\ x-3\overline{)x^2+x-12} \\ \underline{x^2-3x} \\ 4x-12 \\ \underline{4x-12} \\ 0 \end{array}$$

99. AIR CONDITIONING

Multiply $3x \cdot 4x = 12x^2$. Then divide $12x^2$ into the volume of $(36x^3 - 24x^2)$ ft^3.

$$\frac{36x^3-24x^2}{12x^2}=\frac{36x^3}{12x^2}-\frac{24x^2}{12x^2}$$
$$=(3x-2)\text{ft}$$

101. COMMUNICATION

$4x^2+3x+7$

$$\begin{array}{r} 4x^2+3x+7 \\ 2x-3\overline{)8x^3-6x^2+5x-21} \\ \underline{8x^3-12x^2} \qquad\qquad \\ 6x^2+5x \qquad \\ \underline{6x^2-9x} \qquad \\ 14x-21 \\ \underline{14x-21} \\ 0 \end{array}$$

WRITING

103. Answers will vary.

105. Answers will vary.

REVIEW

107.
$$\begin{aligned} y-y_1&=m(x-x_1)\\ y-(-6)&=-\frac{11}{6}(x-2)\\ y+6&=-\frac{11}{6}x-\frac{11}{6}(-2)\\ y+6&=-\frac{11}{6}x+\frac{11}{3}\\ y+6-6&=-\frac{11}{6}x+\frac{11}{3}-6\\ y&=-\frac{11}{6}x+\frac{11}{3}-\frac{18}{3}\\ y&=-\frac{11}{6}x-\frac{7}{3} \end{aligned}$$

109.
$$\begin{aligned} -10(18-4^2)^3&=-10(18-16)^3\\ &=-10(2)^3\\ &=-10(8)\\ &=-80 \end{aligned}$$

CHALLENGE PROBLEMS

111.
$$\begin{aligned} &\frac{6x^{6m}y^{6n}+15x^{4m}y^{7n}-24x^{2m}y^{8n}}{3x^{2m}y^n}\\ &=\frac{6x^{6m}y^{6n}}{3x^{2m}y^n}+\frac{15x^{4m}y^{7n}}{3x^{2m}y^n}-\frac{24x^{2m}y^{8n}}{3x^{2m}y^n}\\ &=2x^{6m-2m}y^{6n-n}+5x^{4m-2m}y^{7n-n}\\ &\qquad -8x^{2m-2m}y^{8n-n}\\ &=2x^{4m}y^{5n}+5x^{2m}y^{6n}-8y^{7n} \end{aligned}$$

CHAPTER 4 KEY CONCEPTS

1. a. This is a polynomial in **x**. It is written in **descending** powers of x.
   b. It has 4 terms.
   c. The degrees of each term are 3, 2, 1, and 0.
   d. The degree of the polynomial is 3.
   e. The coefficients are 1, -2, 6, and -8.

2. a. binomial
   b. none of these
   c. trinomial
   d. monomial

3. $4x^3 + 3x^3 = (4+3)x^3$
   $= 7x^3$

4. $7m^{10} + (-6m^{10}) = (7-6)m^{10}$
   $= 1m^{10}$
   $= m^{10}$

5. $7a^2b - 9a^2b = (7-9)a^2b$
   $= -2a^2b$

6. $(6y^5)(-7y^8) = (6 \cdot -7)y^{5+8}$
   $= -42y^{13}$

7. $\dfrac{16c^4d^5}{8c^2d^6} = 2c^{4-2}d^{5-6}$
   $= 2c^2d^{-1}$
   $= \dfrac{2c^2}{d}$

8. $(5f^3)^2 = 5^2 f^{3\cdot 2}$
   $= 25f^6$

9. To add polynomials, **combine** like terms.

10. To subtract two polynomials, change the **signs** of the terms of the polynomial being subtracted, and combine like terms.

11. To multiply two polynomials, multiply **each** term of one polynomial by **each** term of the other polynomial and combine like terms.

12. To divide a polynomial by a monomial, divide each **term** of the polynomial by the monomial.

13. To divide two polynomials, use the **long** division method.

14. $(8x^3 + 4x^2 - 8x + 1) + (6x^3 - 5x^2 - 2x + 3)$
    $= (8x^3 + 6x^3) + (4x^2 - 5x^2) + (-8x - 2x) + (1+3)$
    $= 14x^3 - x^2 - 10x + 4$

15. $(20s^3t + s^2t^2 - 6st^3) - (8s^3t - 9s^2t^2 + 12st^3)$
    $= 20s^3t + s^2t^2 - 6st^3 - 8s^3t + 9s^2t^2 - 12st^3$
    $= (20s^3t - 8s^3t) + (s^2t^2 + 9s^2t^2) + (-6st^3 - 12st^3)$
    $= 12s^3t + 10s^2t^2 - 18st^3$

16. $(2x+3)(x-8) = 2x(x) + 2x(-8) + 3(x) + 3(-8)$
    $= 2x^2 - 16x + 3x - 24$
    $= 2x^2 - 13x - 24$

17. $(2x^2+3)^2 = (2x^2)^2 + 2(2x^2)(3) + 3^2$
    $= 4x^4 + 12x^2 + 9$

18. $(4h^5 + 8t)(4h^5 + 8t) = (4h^5)^2 - (8t)^2$
    $= 16h^{10} - 64t^2$

19. $(y^2+y-6)(y+3) = y^2(y+3) + y(y+3) - 6(y+3)$
    $= y^3 + 3y^2 + y^2 + 3y - 6y - 18$
    $= y^3 + (3y^2+y^2) + (3y-6y) - 18$
    $= y^3 + 4y^2 - 3y - 18$

20. $\dfrac{9x^6 + 27x^7 - 18x^5}{3x^2} = \dfrac{9x^6}{3x^2} + \dfrac{27x^7}{3x^2} - \dfrac{18x^5}{3x^2}$
    $= 3x^{6-2} + 9x^{7-2} - 6x^{5-2}$
    $= 3x^4 + 9x^5 - 6x^3$

21. $x^2 + 2x + 3$

$$
\begin{array}{r}
x^2 + 2x + 3 \\
x+1 \overline{)\, x^3 + 3x^2 + 5x + 3} \\
\underline{x^3 + x^2} \qquad\qquad \\
2x^2 + 5x \qquad \\
\underline{2x^2 + 2x} \qquad \\
3x + 3 \\
\underline{3x + 3} \\
0
\end{array}
$$

CHAPTER 4 REVIEW

SECTION 4.1
Rules for Exponents

1. a. $-3x^4 = -3 \cdot x \cdot x \cdot x \cdot x$

   b. $\left(\frac{1}{2}pq\right)^3 = \left(\frac{1}{2}pq\right)\left(\frac{1}{2}pq\right)\left(\frac{1}{2}pq\right)$

2. a. base x, exponent 6
   b. base $2x$, exponent 6

3. $5^3 = 5 \cdot 5 \cdot 5$
   $= 125$

4. $(-8)^2 = (-8)(-8)$
   $= 64$

5. $-8^2 = -(8)(8)$
   $= -64$

6. $(5-3)^2 = (2)^2$
   $= 2(2)$
   $= 4$

7. $7^4 \cdot 7^8 = 7^{4+8}$
   $= 7^{12}$

8. $mmnn = m^2n^2$

9. $(y^7)^3 = y^{7 \cdot 3}$
   $= y^{21}$

10. $(3x)^4 = 3^4 \cdot x^4$
    $= 81x^4$

11. $b^3b^4b^5 = b^{3+4+5}$
    $= b^{12}$

12. $-z^2(z^3y^2) = -y^2z^{2+3}$
    $= -y^2z^5$

13. $(-16s)^2s = (-16)^2s^2s$
    $= 256s^{2+1}$
    $= 256s^3$

14. $(2x^2y)^2 = 2^2 \cdot x^{2\cdot 2} \cdot y^{1\cdot 2}$
    $= 4x^4y^2$

15. $(x^2x^3)^3 = (x^{2+3})^3$
    $= (x^5)^3$
    $= x^{5\cdot 3}$
    $= x^{15}$

16. $\left(\frac{x^2y}{xy^2}\right)^2 = \left(x^{2-1}y^{1-2}\right)^2$
    $= \left(xy^{-1}\right)^2$
    $= \left(x^{1\cdot 2}y^{-1\cdot 2}\right)$
    $= x^2y^{-2}$
    $= \frac{x^2}{y^2}$

17. $\frac{(m-25)^{16}}{(m-25)^4} = (m-25)^{16-4}$
    $= (m-25)^{12}$

18. $\frac{\left(5y^2z^3\right)^3}{\left(yz\right)^5} = \frac{5^3y^{2\cdot 3}z^{3\cdot 3}}{y^{1\cdot 5}z^{1\cdot 5}}$
    $= \frac{125y^6z^9}{y^5z^5}$
    $= 125y^{6-5}z^{9-5}$
    $= 125yz^4$

19. $V = (4x^4)^3$
    $= 4^3 \cdot x^{4\cdot 3}$
    $= 64x^{12}$

20. $A = (y^2)^2$
    $= y^{2\cdot 2}$
    $= y^4$

SECTION 4.2
Zero and Negative Exponents

21. $x^0 = 1$

22. $(3x^2y^2)^0 = 1$

23. $(3x^0)^2 = 3^2 \cdot x^{0\cdot 2}$
    $= 9 \cdot 1$
    $= 9$

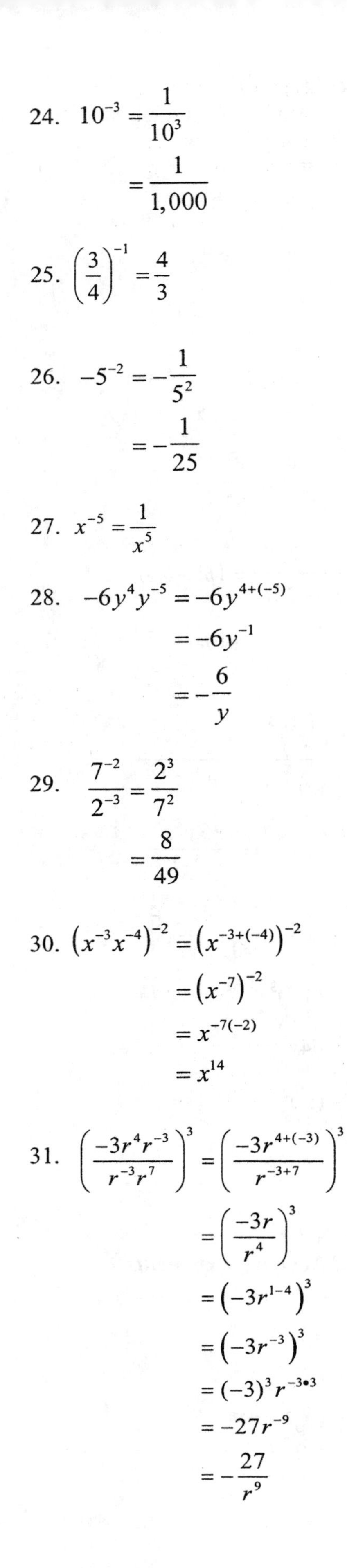

24. $10^{-3} = \frac{1}{10^3}$
$= \frac{1}{1,000}$

25. $\left(\frac{3}{4}\right)^{-1} = \frac{4}{3}$

26. $-5^{-2} = -\frac{1}{5^2}$
$= -\frac{1}{25}$

27. $x^{-5} = \frac{1}{x^5}$

28. $-6y^4y^{-5} = -6y^{4+(-5)}$
$= -6y^{-1}$
$= -\frac{6}{y}$

29. $\frac{7^{-2}}{2^{-3}} = \frac{2^3}{7^2}$
$= \frac{8}{49}$

30. $\left(x^{-3}x^{-4}\right)^{-2} = \left(x^{-3+(-4)}\right)^{-2}$
$= \left(x^{-7}\right)^{-2}$
$= x^{-7(-2)}$
$= x^{14}$

31. $\left(\frac{-3r^4r^{-3}}{r^{-3}r^7}\right)^3 = \left(\frac{-3r^{4+(-3)}}{r^{-3+7}}\right)^3$
$= \left(\frac{-3r}{r^4}\right)^3$
$= \left(-3r^{1-4}\right)^3$
$= \left(-3r^{-3}\right)^3$
$= (-3)^3 r^{-3\bullet 3}$
$= -27r^{-9}$
$= -\frac{27}{r^9}$

32. $\left(\frac{4z^4}{z^3}\right)^{-2} = \left(4z^{4-3}\right)^{-2}$
$= (4z)^{-2}$
$= \frac{1}{(4z)^2}$
$= \frac{1}{4^2z^2}$
$= \frac{1}{16z^2}$

SECTION 4.3
Scientific Notation

33. $728 = 7.28 \times 10^2$

34. $9,370,000,000,000,000 = 9.37 \times 10^{15}$

35. $0.0136 = 1.36 \times 10^{-2}$

36. $0.00942 = 9.42 \times 10^{-3}$

37. $0.018 \times 10^{-2} = 1.8 \times 10^{-4}$

38. $753 \times 10^3 = 7.53 \times 10^{2+3}$
$= 7.53 \times 10^5$

39. $7.26 \times 10^5 = 726,000$; move the decimal 5 places to the right

40. $3.91 \times 10^{-8} = 0.0000000391$; move the decimal 8 places to the left

41. $2.68 \times 10^0 = 2.68$; the decimal does not move

42. $5.76 \times 10^1 = 57.6$; move the decimal 1 place to the right

43. $\dfrac{(0.00012)(0.00004)}{0.00000016} = \dfrac{(1.2\times10^{-4})(4\times10^{-5})}{1.6\times10^{-7}}$
$= \dfrac{(1.2\bullet4)(10^{-4+(-5)})}{1.6\times10^{-7}}$
$= \dfrac{4.8\times10^{-9}}{1.6\times10^{-7}}$
$= \dfrac{4.8}{1.6}\times\dfrac{10^{-9}}{10^{-7}}$
$= 3\times10^{-9-(-7)}$
$= 3\times10^{-2}$
$= 0.03$

44. $\dfrac{(4{,}800)(20{,}000)}{600{,}000} = \dfrac{(4.8\times10^{3})(2\times10^{4})}{6\times10^{5}}$
$= \dfrac{(4.8\bullet2)(10^{3+4})}{6\times10^{5}}$
$= \dfrac{9.6\times10^{7}}{6\times10^{5}}$
$= \dfrac{9.6}{6}\times\dfrac{10^{7}}{10^{5}}$
$= 1.6\times10^{7-5}$
$= 1.6\times10^{2}$
$= 160$

45. 6.31 billion = 6,310,000,000
= 6.31×10^9

46. $\dfrac{1.0\times10^{-8}}{1.0\times10^{-13}} = 1.0\times10^{-8-(-13)}$
$= 1.0\times10^{5}$
$= 100{,}000$ nuclei

SECTION 4.4
Polynomials

47. a. 4 terms
b. $3x^3$
c. 3, –1, 1, 10
d. 10

48. a. 7th degree; monomial
b. 3rd degree; monomial
c. 2nd degree; binomial
d. 5th degree; trinomial
e. 6th degree; binomial
f. 4th degree; none of these

49. $-x^5 - 3x^4 + 3$

Let $x = 0$.
$-(\mathbf{0})^5 - 3(\mathbf{0})^4 + 3 = -0 - 3(0) + 3$
$= 0 - 0 + 3$
$= 3$

Let $x = -2$.
$-(\mathbf{-2})^5 - 3(\mathbf{-2})^4 + 3 = -(-32) - 3(16) + 3$
$= 32 - 48 + 3$
$= -13$

50. DIVING

$0.1875x^2 - 0.0078125x^3$
$= 0.1875(\mathbf{8})^2 - 0.0078125(\mathbf{8})^3$
$= 0.1875(\mathbf{8})^2 - 0.0078125(\mathbf{8})^3$
$= 0.1875(64) - 0.0078125(512)$
$= 12 - 4$
$= 8$
The amount of deflection is 8 in.

51.

x	$y = x^2$
–3	$(\mathbf{-3})^2 = 9$
–2	$(\mathbf{-2})^2 = 4$
–1	$(\mathbf{-1})^2 = 1$
0	$(\mathbf{0})^2 = 0$
1	$(\mathbf{1})^2 = 1$
2	$(\mathbf{2})^2 = 4$
3	$(\mathbf{3})^2 = 9$

52.

x	$y = x^3 + 1$
–3	$(-3)^3 + 1 = -27 + 1 = -26$
–2	$(-2)^3 + 1 = -8 + 1 = -7$
–1	$(-1)^3 + 1 = -1 + 1 = 0$
0	$(0)^3 + 1 = 0 + 1 = 1$
1	$(1)^3 + 1 = 1 + 1 = 2$
2	$(2)^3 + 1 = 8 + 1 = 9$
3	$(3)^3 + 1 = 27 + 1 = 28$

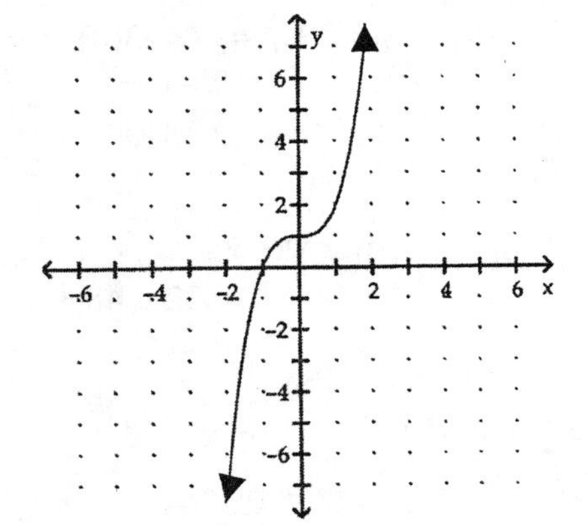

SECTION 4.5
Adding and Subtracting Polynomials

53. $6y^3 + 8y^4 + 7y^3 + (-8y^4)$
$= (8y^4 - 8y^4) + (6y^3 + 7y^3)$
$= 0y^4 + 13y^3$
$= 13y^3$

54. $4a^2b + 5 - 6a^3b - 3a^2b + 2a^3b$
$= (-6a^3b + 2a^3b) + (4a^2b - 3a^2b) + 5$
$= -4a^3b + a^2b + 5$

55. $\frac{5}{6}x^2 + \frac{1}{3}y^2 - \frac{1}{4}x^2 - \frac{3}{4}xy + \frac{2}{3}y^2$
$= \left(\frac{5}{6}x^2 - \frac{1}{4}x^2\right) - \frac{3}{4}xy + \left(\frac{1}{3}y^2 + \frac{2}{3}y^2\right)$
$= \left(\frac{5}{6} \bullet \frac{2}{2}x^2 - \frac{1}{4} \bullet \frac{3}{3}x^2\right) - \frac{3}{4}xy + \left(\frac{1}{3}y^2 + \frac{2}{3}y^2\right)$
$= \left(\frac{10}{12}x^2 - \frac{3}{12}x^2\right) - \frac{3}{4}xy + \left(\frac{1}{3}y^2 + \frac{2}{3}y^2\right)$
$= \frac{7}{12}x^2 - \frac{3}{4}xy + \frac{3}{3}y^2$
$= \frac{7}{12}x^2 - \frac{3}{4}xy + y^2$

56. $-(c^5 + 5c^4 - 12) = -c^5 - 5c^4 + 12$

57. $(2r^6 + 14r^3) + (23r^6 - 5r^3 + 5r)$
$= (2r^6 + 23r^6) + (14r^3 - 5r^3) + 5r$
$= 25r^6 + 9r^3 + 5r$

58. $(7a^2 + 2a - 5) - (3a^2 - 2a + 1)$
$= 7a^2 + 2a - 5 - 3a^2 + 2a - 1$
$= (7a^2 - 3a^2) + (2a + 2a) (-5 - 1)$
$= 4a^2 + 4a - 6$

59. $(3r^3s + r^2s^2 - 3rs^3 - 3s^4) + (r^3s - 8r^2s^2 - 4rs^3 + s^4)$
$= (3r^3s + r^3s) + (r^2s^2 - 8r^2s^2) + (-3rs^3 - 4rs^3) +$
$(-3s^4 + s^4)$
$= 4r^3s - 7r^2s^2 - 7rs^3 - 2s^4$

60. $3(9x^2 + 3x + 7) - 2(11x^2 - 5x + 9)$
$= 3(9x^2) + 3(3x) + 3(7) - 2(11x^2) - 2(5x) - 2(9)$
$= 27x^2 + 9x + 21 - 22x^2 + 10x - 18$
$= (27x^2 - 22x^2) + (9x + 10x) + (21 - 18)$
$= 5x^2 + 19x + 3$

61. $(2z^2 + 3z - 7) + (-4z^3 - 2z - 3) - (-3z^3 - 4z + 7)$
$= 2z^2 + 3z - 7 - 4z^3 - 2z - 3 + 3z^3 + 4z - 7$
$= (-4z^3 + 3z^3) + 2z^2 + (3z - 2z + 4z) + (-7-3-7)$
$= -z^3 + 2z^2 + 5z - 17$

62.
$$\begin{array}{r} 3x^2 + 5x + 2 \\ + \quad \underline{x^2 - 3x + 6} \\ 4x^2 + 2x + 8 \end{array}$$

63.
$$\begin{array}{r} 20x^3 \qquad\quad + 12x \\ - \ \underline{12x^3 - 7x^2 - \ 7x} \end{array} \longrightarrow \begin{array}{r} 20x^3 \qquad\quad + 12x \\ \underline{+ \ 12x^3 + 7x^2 + \ 7x} \\ 32x^3 + 7x^2 + 19x \end{array}$$

64. GARDENING

$(2x^2 + x + 1) - (x^2 - 2)$
$= 2x^2 + x + 1 - x^2 + 2$
$= (2x^2 - x^2) + x + (1 + 2)$
$= (x^2 + x + 3)$ in.

SECTION 4.6
Multiplying Polynomials

65. $(2x^2)(5x) = (2 \cdot 5)(x^{2+1})$
$= 10x^3$

66. $(-6x^4z^3)(x^6z^2) = -6x^{4+6}z^{3+2}$
$= -6x^{10}z^5$

67. $5b^3 \cdot 6b^2 \cdot 4b^6 = (5 \cdot 6 \cdot 4)b^{3+2+6}$
$= 120b^{11}$

68. $\frac{2}{3}h^5\left(3h^9+12h^6\right)=\frac{2}{3}h^5\left(3h^9\right)+\frac{2}{3}h^5\left(12h^6\right)$
$=\left(\frac{2}{3}\bullet\frac{3}{1}\right)h^{5+9}+\left(\frac{2}{3}\bullet\frac{12}{1}\right)h^{5+6}$
$=2h^{14}+8h^{11}$

69. $x^2y(y^2-xy)=x^2y(y^2)-x^2y(xy)$
$=x^2y^{1+2}-x^{2+1}y^{1+1}$
$=x^2y^3-x^3y^2$

70. $3n^2(3n^2-5n+2)=3n^2(3n^2)+3n^2(-5n)+3n^2(2)$
$=9n^4-15n^3+6n^2$

71. $2x(3x^4)(x+2)=(2\cdot3\cdot x^{1+4})(x+2)$
$=6x^5(x+2)$
$=6x^5(x)+6x^5(2)$
$=6x^6+12x^5$

72. $-a^2b^2(-a^4b^2+a^3b^3-ab^4+7a)$
$=-a^2b^2(-a^4b^2)-a^2b^2(a^3b^3)-a^2b^2(-ab^4)-a^2b^2(7a)$
$=a^6b^4-a^5b^5+a^3b^6-7a^3b^2$

73. $(x+3)(x+2)=x\cdot x+x\cdot2+3\cdot x+3\cdot2$
$=x^2+2x+3x+6$
$=x^2+5x+6$

74. $(2x+1)(x-1)=2x\cdot x+2x\cdot(-1)+1\cdot x+1\cdot(-1)$
$=2x^2-2x+x-1$
$=2x^2-x-1$

75. $(3a-3)(2a+2)=3a\cdot2a+3a\cdot2-3\cdot2a-3\cdot2$
$=6a^2+6a-6a-6$
$=6a^2-6$

76. $6(a-1)(a+1)=6(a\cdot a+a\cdot1-1\cdot a-1\cdot1)$
$=6(a^2+a-a-1)$
$=6(a^2-1)$
$=6a^2-6$

77. $(a-b)(2a+b)=a\cdot2a+a\cdot b-b\cdot2a-b\cdot b$
$=2a^2+ab-2ab-b^2$
$=2a^2-ab-b^2$

78. $(3n^4-5n^2)(2n^4-n^2)$
$=3n^4\cdot2n^4+3n^4\cdot(-n^2)-5n^2\cdot2n^4-5n^2\cdot(-n^2)$
$=6n^8-3n^6-10n^6+5n^4$
$=6n^8-13n^6+5n^4$

79. $(2a-3)(4a^2+6a+9)$
$=2a(4a^2+6a+9)-3(4a^2+6a+9)$
$=2a(4a^2)+2a(6a)+2a(9)-3(4a^2)-3(6a)-3(9)$
$=8a^3+12a^2+18a-12a^2-18a-27$
$=8a^3+(12a^2-12a^2)+(18a-18a)-27$
$=8a^3-27$

80. $(8x^2+x-2)(7x^2+x-1)$
$=8x^2(7x^2+x-1)+x(7x^2+x-1)-2(7x^2+x-1)$
$=56x^4+8x^3-8x^2+7x^3+x^2-x-14x^2-2x+2$
$=56x^4+(8x^3+7x^3)+(-8x^2+x^2-14x^2)+(-x-2x)+2$
$=56x^4+15x^3-21x^2-3x+2$

81.
$$\begin{array}{r} 4x^2-2x+1 \\ \underline{2x+1} \\ 4x^2-2x+1 \\ \underline{8x^3-4x^2+2x} \\ 8x^3+1 \end{array}$$

82. APPLIANCE

perimeter:
$(x+6)+(2x-1)=(x+2x)+(6-1)$
$=(3x+5)$ in

area:
$(x+6)(2x-1)=x(2x)+x(-1)+6(2x)+6(-1)$
$=2x^2-x+12x-6$
$=(2x^2+11x-6)$ in^2

volume:
$3x(x+6)(2x-1)=3x[x(2x)+x(-1)+6(2x)+6(-1)$
$=3x[2x^2-x+12x-6]$
$=3x(2x^2+11x-6)$
$=(6x^3+33x^2-18x)$ in^3

SECTION 4.7
Special Products

83. $(x+3)(x+3)=x(x)+x(3)+3(x)+3(3)$
$=x^2+3x+3x+9$
$=x^2+6x+9$

84. $(2x-0.9)(2x+0.9)=(2x)^2-(0.9)^2$
$=4x^2-0.81$

85. $(a-3)^2=a^2-2(a)(3)+(-3)^2$
$=a^2-6a+9$

86. $(x+4)^2=x^2+2(x)(4)+4^2$
$=x^2+8x+16$

87. $(-2y+1)^2 = (-2y)^2 + 2(-2y)(1) + 1^2$
$= 4y^2 - 4y + 1$

88. $(y^2+1)(y^2-1) = (y^2)^2 - 1^2$
$= y^4 - 1$

89. $(6r^2+10s)^2 = (6r^2)^2 + 2(6r^2)(10s) + (10s)^2$
$= 36r^4 + 120r^2s + 100s^2$

90. $-(8a-3)^2 = -[(8a)^2 - 2(8a)(3) + (-3)^2]$
$= -[64a^2 - 48a + 9]$
$= -64a^2 + 48a - 9$

91. $80s(r^2+s^2)(r^2-s^2) = 80s[(r^2)^2 - (s^2)^2]$
$= 80s(r^4 - s^4)$
$= 80r^4s - 80s^5$

92. $4b(3b-4)^2 = 4b[(3b)^2 - 2(3b)(4) + (-4)^2]$
$= 4b[9b^2 - 24b + 16]$
$= 36b^3 - 96b^2 + 64b$

93. $\left(t - \frac{3}{4}\right)^2 = t^2 - 2(t)\left(\frac{3}{4}\right) + \left(-\frac{3}{4}\right)^2$
$= t^2 - \frac{3}{2}t + \frac{9}{16}$

94. $(m+2)^3 = (m+2)^2(m+2)$
$= (m^2 + 2\cdot m\cdot 2 + 2^2)(m+2)$
$= (m^2 + 4m + 4)(m+2)$
$= m^2(m+2) + 4m(m+2) + 4(m+2)$
$= m^3 + 2m^2 + 4m^2 + 8m + 4m + 8$
$= m^3 + 6m^2 + 12m + 8$

95. $(5c-1)^2 - (c+6)(c-6)$
$= [(5c)^2 - 2(5c)(1) + (-1)^2] - [c^2 - 6^2]$
$= (25c^2 - 10c + 1) - (c^2 - 36)$
$= 25c^2 - 10c + 1 - c^2 + 36$
$= 24c^2 - 10c + 37$

96. GRAPHIC ARTS

To find the length and width of the picture, subtract the border on each end from the total length and width.

The length of the picture would be
$(x+3) - 2(\frac{1}{2}) = x + 3 - 1$
$= x + 2.$
The width of the picture would be
$(x-3) - 2(\frac{1}{2}) = x - 3 - 1$
$= x - 4.$
The area of the picture would be
$(x+2)(x-4) = x(x) + x(-4) + 2(x) + 2(-4)$
$= x^2 - 4x + 2x - 8$
$= (x^2 - 2x - 8)$ in.^{2}

SECTION 4.8
Division of Polynomials

97. $\frac{16n^8}{8n^5} = 2n^{8-5}$
$= 2n^3$

98. $\frac{-14x^2y}{21xy^3} = -\frac{2}{3}x^{2-1}y^{1-3}$
$= -\frac{2}{3}x^1y^{-2}$
$= -\frac{2x}{3y^2}$

99. $\frac{a^{15} - 24a^8}{6a^{12}} = \frac{a^{15}}{6a^{12}} - \frac{24a^8}{6a^{12}}$
$= \frac{1}{6}a^{15-12} - 4a^{8-12}$
$= \frac{1}{6}a^3 - 4a^{-4}$
$= \frac{a^3}{6} - \frac{4}{a^4}$

100. $\frac{15a^2b + 20ab^2 - 25ab}{5ab}$
$= \frac{15a^2b}{5ab} + \frac{20ab^2}{5ab} - \frac{25ab}{5ab}$
$= 3a + 4b - 5$

101.

$$
\begin{array}{r}
x-5 \\
x-1\overline{)x^2-6x+5} \\
\underline{x^2-x} \\
-5x+5 \\
\underline{-5x+5} \\
0
\end{array}
$$

102. Make sure the dividend is in descending order

$$
\begin{array}{r}
2x+1 \\
x+3\overline{)2x^2+7x+3} \\
\underline{2x^2+6x} \\
x+3 \\
\underline{x+3} \\
0
\end{array}
$$

103. Insert $0x$ for the missing x term.

$$
\begin{array}{r}
3x^2+2x+1+\frac{2}{2x-1} \\
2x-1\overline{)6x^3+x^2+0x+1} \\
\underline{6x^3-3x^2} \\
4x^2+0x \\
\underline{4x^2-2x} \\
2x+1 \\
\underline{2x-1} \\
2
\end{array}
$$

104. Make sure the dividend is in descending order and insert $0x^2$ for the missing x^2 term.

$$
\begin{array}{r}
3x^2-x-4 \\
3x+1\overline{)9x^3+0x^2-13x-4} \\
\underline{9x^3+3x^2} \\
-3x^2-13x \\
\underline{-3x^2-x} \\
-12x-4 \\
\underline{-12x+4} \\
0
\end{array}
$$

105.

$$
\begin{aligned}
(y+3)(3y+2) &= y\cdot 3y+y\cdot(2)+3\cdot 3y+3\cdot(2) \\
&= 3y^2+2y+9y+6 \\
&= 3y^2+11y+6
\end{aligned}
$$

106. SAVING BONDS

$$
\begin{aligned}
\frac{50x+250}{50} &= \frac{50x}{50}+\frac{250}{50} \\
&= x+5
\end{aligned}
$$

The number would be $(x + 5)$.

CHAPTER 4 TEST

1. $2xxxyyyy = 2x^3y^4$

2. $-6^2 = -(6)(6)$
$= -36$

3. $y^2(yy^3) = y^2(y^{1+3})$
$= y^2(y^4)$
$= y^{2+4}$
$= y^6$

4. $(2x^3)^5(x^2)^3 = 2^5 \cdot x^{3\cdot 5} \cdot x^{2\cdot 3}$
$= 32x^{15} \cdot x^6$
$= 32x^{15+6}$
$= 32x^{21}$

5. $3x^0 = 3 \cdot 1$
$= 3$

6. $2y^{-5}y^2 = 2y^{-5+2}$
$= 2y^{-3}$
$= \dfrac{2}{y^3}$

7. $5^{-3} = \dfrac{1}{5^3}$
$= \dfrac{1}{125}$

8. $\dfrac{(x+1)^{15}}{(x+1)^6} = (x+1)^{15-6}$
$= (x+1)^9$

9. $\dfrac{y^2}{yy^{-2}} = \dfrac{y^2}{y^{1+(-2)}}$
$= \dfrac{y^2}{y^{-1}}$
$= y^{2-(-1)}$
$= y^3$

10. $\left(\dfrac{a^2b^{-1}}{4a^3b^{-2}}\right)^{-3} = \left(\dfrac{4a^3b^{-2}}{a^2b^{-1}}\right)^3$
$= \left(4a^{3-2}b^{-2-(-1)}\right)^3$
$= \left(4ab^{-1}\right)^3$
$= \left(\dfrac{4a}{b}\right)^3$
$= \dfrac{4^3a^3}{b^3}$
$= \dfrac{64a^3}{b^3}$

11. $\left(10y^4\right)^3 = 10^3y^{4\cdot 3}$
$= 1{,}000y^{12}$ in.^{3}

12. ELECTRICITY

$6{,}250{,}000{,}000{,}000{,}000{,}000 = 6.25 \times 10^{18}$

13. $9.3 \times 10^{-5} = 0.000093$

14. $(2.3\times10^{18})(4.0\times10^{-15}) = (2.3\cdot 4.0)\times10^{18+(-15)}$
$= 9.2 \times 10^3$
$= 9{,}200$

15.

Term	Coefficient	Degree
x^4	1	4
$2x^2$	2	2
−12	−12	0
Degree of the polynomial		4

16. binomial

17. The degree of $3x^2y^3$ is $2+3 = 5$.
The degree of $2x^3y$ is $3 + 1 = 4$.
The degree of $-5x^2y$ is $2 + 1 = 3$.

So, the degree of the polynomial is 5.

18. $y = x^2 + 2$

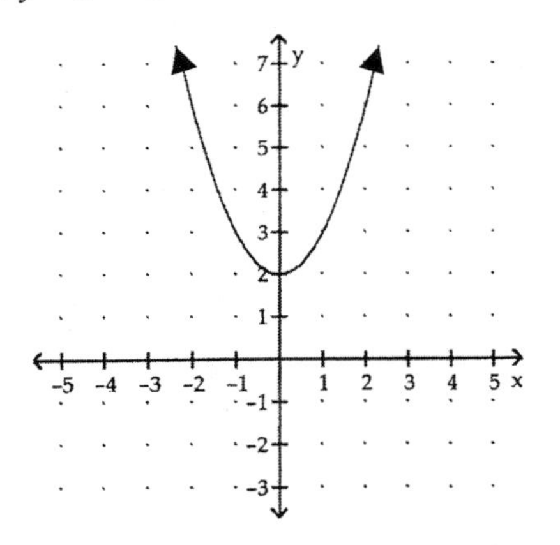

19. FREE FALL

Let $t = 18$.

$$-16(\mathbf{18})^2 + 5 = -16(324) + 5{,}184$$
$$= -5{,}184 + 5{,}184$$
$$= 0 \text{ ft}$$

The rock hits the canyon floor 18 seconds after being dropped.

20. $4a^2b + 5 - 6a^3b - 3a^2b + 2a^3b$
$$= (-6a^3b + 2a^3b) + (4a^2b - 3a^2b) + 5$$
$$= -4a^3b + a^2b + 5$$

21. $(3a^2 - 4a - 6) + (2a^2 - a + 9)$
$$= (3a^2 + 2a^2) + (-4a - a) + (-6 + 9)$$
$$= 5a^2 - 5a + 3$$

22. $(7b^3c - 5bc) - (b^3c - 3bc + 12)$
$$= 7b^3c - 5bc - b^3c + 3bc - 12$$
$$= (7b^3c - b^3c) + (-5bc + 3bc) - 12$$
$$= 6b^3c - 2bc - 12$$

23.
$$\begin{array}{r} -5y^3 + 4y^2 - 11y + 3 \\ -\ \underline{-2y^3 - 14y^2 + 17y - 32} \end{array}$$

$$\begin{array}{r} -5y^3 + 4y^2 - 11y + 3 \\ +\ \underline{2y^3 + 14y^2 - 17y + 32} \\ -3y^3 + 18y^2 - 28y + 35 \end{array}$$

24. $-6(x - y) + 2(x + y)$
$$= -6x + 6y + 2x + 2y$$
$$= (-6x + 2x) + (6y + 2y)$$
$$= -4x + 8y$$

25. $(-2x^3)(2x^2y) = -4x^{3+2}y$
$$= -4x^5y$$

26. $3y^2(y^2 - 2y + 3)$
$$= 3y^2(y^2) + 3y^2(-2y) + 3y^2(3)$$
$$= 3y^4 - 6y^3 + 9y^2$$

27. $(2x - 5)(3x + 4)$
$$= 2x(3x) + 2x(4) - 5(3x) - 5(4)$$
$$= 6x^2 + 8x - 15x - 20$$
$$= 6x^2 - 7x - 20$$

28. $(2x - 3)(x^2 - 2x + 4)$
$$= 2x(x^2 - 2x + 4) - 3(x^2 - 2x + 4)$$
$$= 2x^3 - 4x^2 + 8x - 3x^2 + 6x - 12$$
$$= 2x^3 + (-4x^2 - 3x^2) + (8x + 6x) - 12$$
$$= 2x^3 - 7x^2 + 14x - 12$$

29. $(1 + 10c)(1 - 10c) = 1^2 - (10c)^2$
$$= 1 - 100c^2$$

30. $(7b^3 - 3)^2 = (7b^3)^2 - 2(7b^3)(3) + 3^2$
$$= 49b^6 - 42b^3 + 9$$

31. $(x + y)(x - y) + x(x + y)$
$$= (x^2 - y^2) + (x^2 + xy)$$
$$= 2x^2 + xy - y^2$$

32.
$$\frac{6a^2 - 12b^2}{24ab} = \frac{6a^2}{24ab} - \frac{12b^2}{24ab}$$
$$= \frac{1}{4}a^{2-1}b^{-1} - \frac{1}{2}a^{-1}b^{2-1}$$
$$= \frac{1}{4}a^1b^{-1} - \frac{1}{2}a^{-1}b^1$$
$$= \frac{a}{4b} - \frac{b}{2a}$$

33.
$$\begin{array}{r} x - 2 \\ 2x + 3 \overline{\smash{\big)}\, 2x^2 - x - 6} \\ \underline{2x^2 + 3x} \\ -4x - 6 \\ \underline{-4x - 6} \\ 0 \end{array}$$

34. $2x-1\overline{)6x^3+x^2+0x+1}$ with quotient $3x^2+2x+1+\frac{2}{2x-1}$

$$
\begin{array}{r}
6x^3-3x^2 \\ \hline
3x^2+0x \\
3x^2-2x \\ \hline
2x+1 \\
2x-1 \\ \hline
2
\end{array}
$$

35. $x-1\overline{)x^2-6x+5}$ with quotient $x-5$

$$
\begin{array}{r}
x^2-x \\ \hline
-5x+5 \\
-5x+5 \\ \hline
0
\end{array}
$$

Width of rectangle is $(x-5)$ ft.

CHAPTER 4 CUMULATIVE REVIEW

1. a. 1993
   b. 1986
   c. 1996–1997

2. $\frac{3}{4} \div \frac{6}{5} = \frac{3}{4} \bullet \frac{5}{6}$
$= \frac{3(5)}{4(6)}$
$= \frac{15}{24}$
$= \frac{5}{8}$

3. $\frac{7}{10} - \frac{1}{14} = \frac{7}{10}\left(\frac{7}{7}\right) - \frac{1}{14}\left(\frac{5}{5}\right)$
$= \frac{49}{70} - \frac{5}{70}$
$= \frac{44}{70}$
$= \frac{22}{35}$

4. π is an irrational number.

5. RACING

$250 - x$

6. CLINICAL TRIALS

6% of 300 = 0.06(300) = 18

5% of 320 = 0.05(320) = 16

7. $100 = 2 \cdot 2 \cdot 5 \cdot 5$
$= 2^2 \cdot 5^2$

8.

$-2\frac{1}{4}$ -1.75 0.5 $\sqrt{2}$ $\frac{7}{2}$

-4 -3 -2 -1 0 1 2 3 4

9. $\frac{2}{3} = 3\overline{)2.000}$ with quotient 0.666, $= 0.\overline{6}$

10. associative property of multiplication

11. $10d$ cents

12. $13r - 12r = 1r$
$= r$

13. $27\left(\frac{2}{3}x\right) = \frac{27}{1}\left(\frac{2}{3}\right)x$
$= \frac{54}{3}x$
$= 18x$

14. $4(d-3) - (d-1) = 4d - 12 - d + 1$
$= (4d - d) + (-12 + 1)$
$= 3d - 11$

15. $(13c - 3)(-6) = 13c(-6) - 3(-6)$
$= -78c + 18$

16. $-3^2 + |4^2 - 5^2| = -9 + |16 - 25|$
$= -9 + |-9|$
$= -9 + 9$
$= 0$

17. $(4-5)^{20} = (-1)^{20}$
$= 1$

18. $\frac{-3-(-7)}{2^2 - 3} = \frac{-3+7}{4-3}$
$= \frac{4}{1}$
$= 4$

19. $12 - 2[1 - (-8 + 2)] = 12 - 2[1 - (-6)]$
$= 12 - 2[1 + 6]$
$= 12 - 2[7]$
$= 12 - 14$
$= -2$

20. $3(x-5) + 2 = 2x$
$3x - 15 + 2 = 2x$
$3x - 13 = 2x$
$3x - 13 - 3x = 2x - 3x$
$-13 = -1x$
$\frac{-13}{-1} = \frac{-1x}{-1}$
$\boxed{13 = x}$

21.
$$\frac{x-5}{3}-5=7$$
$$\frac{x-5}{3}-5+5=7+5$$
$$\frac{x-5}{3}=12$$
$$3\left(\frac{x-5}{3}\right)=3(12)$$
$$x-5=36$$
$$x-5+5=36+5$$
$$\boxed{x=41}$$

22.
$$\frac{2}{5}x+1=\frac{1}{3}+x$$
$$15\left(\frac{2}{5}x+1\right)=15\left(\frac{1}{3}+x\right)$$
$$6x+15=5+15x$$
$$6x+15-15=5+15x-15$$
$$6x=15x-10$$
$$6x-15x=15x-10-15x$$
$$-9x=-10$$
$$\frac{-9x}{-9}=\frac{-10}{-9}$$
$$\boxed{x=\frac{10}{9}}$$

23.
$$-\frac{5}{8}h=15$$
$$8\left(-\frac{5}{8}h\right)=8(15)$$
$$-5h=120$$
$$\frac{-5h}{-5}=\frac{120}{-5}$$
$$\boxed{h=-24}$$

24.
$$8(4+x)>10(6+x)$$
$$32+8x>60+10x$$
$$32+8x-10x>60+10x-10x$$
$$32-2x>60$$
$$32-2x-32>60-32$$
$$-2x>28$$
$$\frac{-2x}{-2}>\frac{28}{-2}$$
$$x<-14$$
$$(-\infty,-14)$$

14

25.
$$A=\frac{1}{2}h(b+B)$$
$$2(A)=2\left[\frac{1}{2}h(b+B)\right]$$
$$2A=h(b+B)$$
$$\frac{2a}{(b+B)}=\frac{h(b+B)}{(b+B)}$$
$$\frac{2A}{b+B}=h$$
$$\boxed{h=\frac{2A}{b+B}}$$

26. CANDY

Let x = the number of pounds of red licorice.
$30-x$ = the number of pounds of lemon gumdrops.

$\$1.90x + \$2.20(30-x) = \$2(30)$
$1.90x + 66 - 2.2x = 60$
$-0.3x + 66 = 60$
$-0.3x + 66 - 66 = 60 - 66$
$-0.3x = -6$
$\frac{-0.3x}{-0.3} = \frac{-6}{-0.3}$
$x = 20$
$30 - x = 30 - 20$
$= 10$

20 lb of \$1.90 candy
10 lb of \$2.20 candy

27. Graph: $4x - 3y = 12$

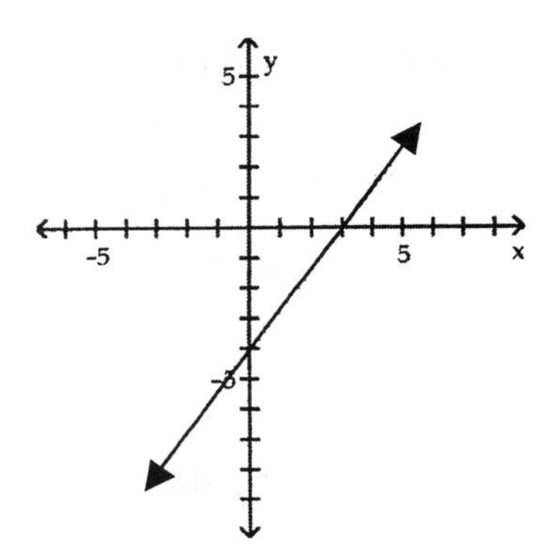

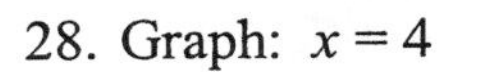

28. Graph: $x = 4$

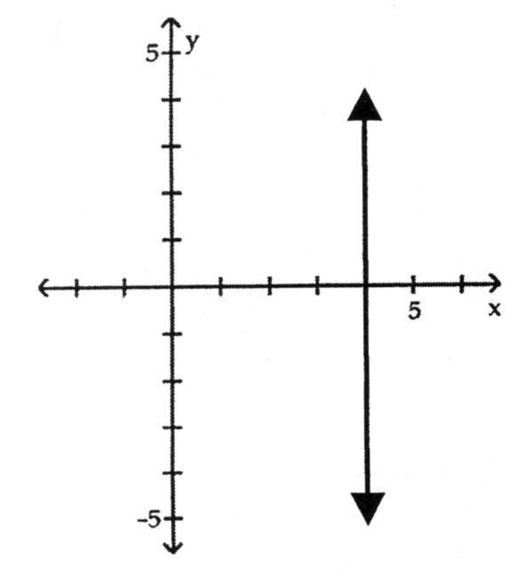

29.
$$m = \frac{y_2 - y_1}{x_2 - x_1} = \frac{8-4}{6-(-2)} = \frac{4}{8} = \frac{1}{2}$$

30. The slope of a horizontal line is always **0**.

31. Solve the equation for y.

$$2x - 3y = 12$$
$$2x - 3y - 2x = 12 - 2x$$
$$-3y = -2x + 12$$
$$\frac{-3y}{-3} = \frac{-2x}{-3} + \frac{12}{-3}$$
$$y = \frac{2}{3}x - 4$$
$$m = \frac{2}{3}$$

32. $m = \frac{2}{3}$ and $b = 5$

$$y = mx + b$$
$$y = \frac{2}{3}x + 5$$

33. Find the slope:

$$m = \frac{y_2 - y_1}{x_2 - x_1} = \frac{10-4}{6-(-2)} = \frac{6}{8} = \frac{3}{4}$$

Use the point–slope formula:

$$y - y_1 = m(x - x_1)$$
$$y - 10 = \frac{3}{4}(x - 6)$$
$$y - 10 = \frac{3}{4}(x) - \frac{3}{4}\left(\frac{6}{1}\right)$$
$$y - 10 = \frac{3}{4}x - \frac{9}{2}$$
$$y - 10 + 10 = \frac{3}{4}x - \frac{9}{2} + 10$$
$$y = \frac{3}{4}x - \frac{11}{2}$$
$$4(y) = 4\left(\frac{3}{4}x - \frac{11}{2}\right)$$
$$4y = 3x - 22$$
$$-3x + 4y = 3x - 22 - 3x$$
$$-3x + 4y = -22$$
$$\frac{-3x}{-1} + \frac{4y}{-1} = \frac{-22}{-1}$$
$$3x - 4y = 22$$

34. $y = 4$
A horizontal line has the equation $y = b$.

35. Find the slope of both lines.

$$y = -\frac{3}{4}x + \frac{15}{4}$$

so the slope is $-\frac{3}{4}$.

$$4x - 3y = 25$$
$$4x - 3y - 4x = 25 - 4x$$
$$-3y = -4x + 25$$
$$\frac{-3y}{-3} = \frac{-4x}{-3} + \frac{25}{-3}$$
$$y = \frac{4}{3}x - \frac{25}{3}$$

The slope is $\frac{4}{3}$.

Since $-\frac{3}{4}$ and $\frac{4}{3}$ are negative reciprocals, the two lines are **perpendicular**.

36. $(17x^4-3x^2-65x-12)-(23x^4+14x^2+3x-23)$
$= 17x^4 -3x^2 -65x -12 -23x^4 -14x^2 -3x +23$
$= 17x^4 -23x^4 -3x^2 -14x^2 -65x -3x -12 +23$
$= -6x^4 - 17x^2 - 68x + 11$

37. $(-3x^2y^4)^2 = (-3)^2x^{2\bullet 2}y^{4\bullet 2}$
$= 9x^4y^8$

38. $$(2y)^{-4} = \frac{1}{(2y)^4} = \frac{1}{2^4y^4} = \frac{1}{16y^4}$$

39. $(x^3x^4)^2 = (x^{3+4})^2$
$= (x^7)^2$
$= x^{7\cdot 2}$
$= x^{14}$

40. $ab^3c^4 \cdot ab^4c^2 = a^{1+1}b^{3+4}c^{4+2}$
$= a^2b^7c^6$

41. $$\frac{a^4b^0}{a^{-3}} = a^{4-(-3)} \bullet b^0 = a^7 \bullet 1 = a^7$$

42. $$\left(\frac{4t^3t^4t^5}{3t^2t^6}\right)^3 = \left(\frac{4t^{3+4+5}}{3t^{2+6}}\right)^3 = \left(\frac{4t^{12}}{3t^8}\right)^3 = \left(\frac{4}{3}t^{12-8}\right)^3 = \left(\frac{4t^4}{3}\right)^3 = \frac{4^3t^{4\bullet 3}}{3^3} = \frac{64t^{12}}{27}$$

43. $(4c^2 + 3c - 2) + (3c^2 + 4c + 2)$
$= (4c^2 + 3c^2) + (3c + 4c) + (-2 + 2)$
$= 7c^2 + 7c$

44. $3x(2x + 3)^2 = 3x[(2x)^2 + 2(2x)(3) + 3^2]$
$= 3x(4x^2 + 12x + 9)$
$= 12x^3 + 36x^2 + 27x$

45. $(2t+3s)(3t-s) = 2t(3t)+2t(-s) +3s(3t)+3s(-s)$
$= 6t^2 - 2st + 9st - 3s^2$
$= 6t^2 + 7st - 3s^2$

46. Write the dividend in descending order.

$$\begin{array}{r} 2x+1 \\ 5x+3\overline{)10x^2+11x+3} \\ \underline{10x^2+6x} \\ 5x+3 \\ \underline{5x+3} \\ 0 \end{array}$$

47. Graph: x^2

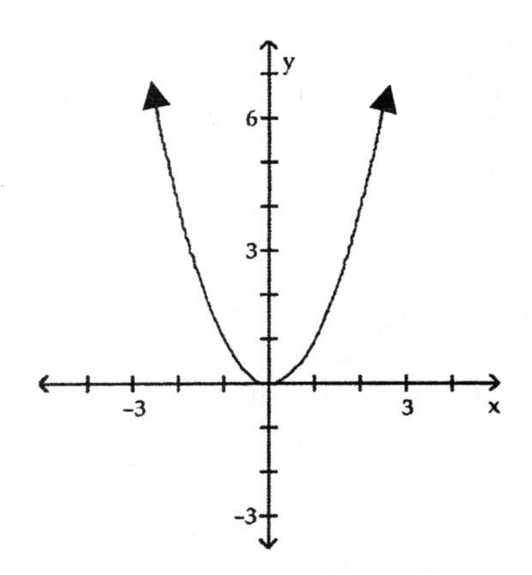

48. $615{,}000 = 6.15 \times 10^5$; move the decimal 5 places to the left

49. $0.0000013 = 1.3 \times 10^{-6}$; move the decimal 6 places to the right

50. MUSICAL INSTRUMENTS

Let $x = 2$ (half of the distance of 4 feet)

$$\begin{aligned} &0.01875(\mathbf{2})^4 - 0.15(\mathbf{2})^3 + 1.2(\mathbf{2}) \\ &\quad = 0.01875(16) - 0.15(8) + 1.2(2) \\ &\quad = 0.3 - 1.2 + 2.4 \\ &\quad = 1.5 \end{aligned}$$

The amount of deflection is 1.5 in.

SECTION 5.1

VOCABULARY

1. The letters GCF stand for **greatest common factor**.

3. When we write 24 as $2 \cdot 2 \cdot 2 \cdot 3$, we say that 24 has been written as a product of **primes**.

5. To factor $m^3 + 3m^2 + 4m + 12$ by **grouping**, we begin by writing $m^2(m + 3) + 4(m + 3)$.

CONCEPTS

7. a. $15 = 3 \cdot 7$
   b. $8 = 2 \cdot 2 \cdot 2$
   c. $36 = 2 \cdot 2 \cdot 3 \cdot 3$
   d. $98 = 2 \cdot 7 \cdot 7$

9. $12a^2b^2 = 2 \cdot 2 \cdot 3 \cdot a \cdot a \cdot b \cdot b$
   $15a^3b = 3 \cdot 5 \cdot a \cdot a \cdot a \cdot b$
   $75a^4b^2 = 3 \cdot 5 \cdot 5 \cdot a \cdot a \cdot a \cdot a \cdot b \cdot b$
   $\text{GCF} = 3 \cdot a \cdot a \cdot b = 3a^2b$

11. a. the distributive property

    b. $2x^2 - 6x = 2x \cdot x - 2x \cdot 3$
    $= \boxed{2x}(x - 3)$

13. a. The GCF is $6a^2$, not $6a$.
    b. The 0 in the first line should be 1.

15. No. The common factor $(c - d)$ still needs to be factored out.

NOTATION

17. $40m^4 - 8m^3 + 32m =$ **$8m$**$(5m^3 - m^2 +$ **4**$)$

PRACTICE

19. $12 = 2 \cdot 2 \cdot 3$
    $= 2^2 \cdot 3$

21. $40 = 2 \cdot 2 \cdot 2 \cdot 5$
    $= 2^3 \cdot 5$

23. $98 = 2 \cdot 7 \cdot 7$
    $= 2 \cdot 7^2$

25. $225 = 3 \cdot 3 \cdot 5 \cdot 5$
    $= 3^2 \cdot 5^2$

27. $18 = 2 \cdot 3 \cdot 3$
    $24 = 2 \cdot 2 \cdot 2 \cdot 3$
    $\text{GCF} = 2 \cdot 3$
    $= 6$

29. $m^4 = m \cdot m \cdot m \cdot m$
    $m^3 = m \cdot m \cdot m$
    $\text{GCF} = m \cdot m \cdot m$
    $= m^3$

31. $20c^2 = 2 \cdot 2 \cdot 5 \cdot c \cdot c$
    $12c = 2 \cdot 2 \cdot 3 \cdot c$
    $\text{GCF} = 2 \cdot 2 \cdot c$
    $= 4c$

33. $6m^4n = 2 \cdot 3 \cdot m \cdot m \cdot m \cdot m \cdot n$
    $12m^3n^2 = 2 \cdot 2 \cdot 3 \cdot m \cdot m \cdot m \cdot n \cdot n$
    $9m^3n^3 = 3 \cdot 3 \cdot m \cdot m \cdot m \cdot n \cdot n \cdot n$
    $\text{GCF} = 3 \cdot m \cdot m \cdot m \cdot n$
    $= 3m^3n$

35. $6x = 3(2x)$

37. $24y^2 = 8y \cdot 3y$

39. $30t^3 = 15t(2t^2)$

41. $32x^3z^4 = 4xz^2(8x^2z^2)$

43. $3x + 6 = 3 \cdot x + 3 \cdot 2$
    $= 3(x + 2)$

45. $18m - 9 = 9 \cdot 2m - 9 \cdot 1$
    $= 9(2m - 1)$

47. $18x + 24 = 6 \cdot 3x + 6 \cdot 4$
    $= 6(3x + 4)$

49. $d^2 - 7d = d \cdot d - d \cdot 7$
    $= d(d - 7)$

51. $14c^3 + 63 = 7 \cdot 2c^3 + 7 \cdot 9$
    $= 7(2c^3 + 9)$

53. $12x^2 - 6x - 24 = 6 \cdot 2x^2 - 6 \cdot x - 6 \cdot 4$
    $= 6(2x^2 - x - 4)$

55. $t^3 + 2t^2 = t^2 \cdot t + t^2 \cdot 2$
    $= t^2(t + 2)$

57. $ab + ac - ad = a \cdot b + a \cdot c - a \cdot d$
$= a(b + c - d)$

59. $a^3 - a^2 = a^2 \cdot a - a^2 \cdot 1$
$= a^2(a - 1)$

61. $24x^2y^3 + 8xy^2 = 8xy^2 \cdot 3xy + 8xy^2 \cdot 1$
$= 8xy^2(3xy + 1)$

63. $12uvw^3 - 18uv^2w^2 = 6uvw^2 \cdot 2w - 6uvw^2 \cdot 3v$
$= 6uvw^2(2w - 3v)$

65. $12r^2 - 3rs + 9r^2s^2 = 3r \cdot 4r - 3r \cdot s + 3r \cdot 3rs^2$
$= 3r(4r - s + 3rs^2)$

67. $\pi R^2 - \pi ab = \pi \cdot R^2 - \pi \cdot ab$
$= \pi(R^2 - ab)$

69. $3(x + 2) - x(x + 2) = (x + 2)(3 - x)$

71. $h^2(14 + r) + 14 + r = h^2(14 + r) + 1(14 + r)$
$= (14 + r)(h^2 + 1)$

73. $-a - b = -(a + b)$

75. $-2x + 5y = -(2x - 5y)$

77. $-3r + 2s - 3 = -(3r - 2s + 3)$

79. $-3x^2 - 6x = -3x \cdot x - 3x \cdot 2$
$= -3x(x + 2)$

81. $-4a^2b^3 + 12a^3b^2 = -4a^2b^2 \cdot b - 4a^2b^2 \cdot -3a$
$= -4a^2b^2(b - 3a)$

83. $-4a^2b^2c^2 + 14a^2b^2c - 10ab^2c^2$
$= -2ab^2c \cdot 2ac - 2ab^2c \cdot -7a - 2ab^2c \cdot 5c$
$= -2ab^2c(2ac - 7a + 5c)$

85. $2x + 2y + ax + ay = (2x + 2y) + (ax + ay)$
$= 2(x + y) + a(x + y)$
$= (x + y)(2 + a)$

87. $9p - 9q + mp - mq = (9p - 9q) + (mp - mq)$
$= 9(p - q) + m(p - q)$
$= (p - q)(9 + m)$

89. $pm - pn + qm - qn = (pm - pn) + (qm - qn)$
$= p(m - n) + q(m - n)$
$= (m - n)(p + q)$

91. $2xy - 3y^2 + 2x - 3y$
$= (2xy - 3y^2) + (2x - 3y)$
$= y(2x - 3y) + 1(2x - 3y)$
$= (2x - 3y)(y + 1)$

93. $ax + bx - a - b = (ax + bx) - (a + b)$
$= x(a + b) - 1(a + b)$
$= (a + b)(x - 1)$

95. $ax^3 + bx^3 + 2ax^2y + 2bx^2y$
$= x^2[ax + bx + 2ay + 2by]$
$= x^2[(ax + bx) + (2ay + 2by)]$
$= x^2[x(a + b) + 2y(a + b)]$
$= x^2(a + b)(x + 2y)$

97. $2x^3z - 4x^2z + 32xz - 64z$
$= 2z[x^3 - 2x^2 + 16x - 32]$
$= 2z[(x^3 - 2x^2) + (16x - 32)]$
$= 2z[x^2(x - 2) + 16(x - 2)]$
$= 2z(x - 2)(x^2 + 16)$

99. $x(y + 9) - 11(y + 9) = (y + 9)(x - 11)$

101. $9mp + 3mq - 3np - nq$
$= (9mp + 3mq) - (3np + nq)$
$= 3m(3p + q) - n(3p + q)$
$= (3p + q)(3m - n)$

103. $25x^5y^7z^3 - 45x^3y^2z^6$
$= 5x^3y^2z^3 \cdot 5x^2y^5 - 5x^3y^2z^3 \cdot 9z^3$
$= 5x^3y^2z^3(5x^2y^5 - 9z^3)$

105. $24m - 12n + 16$
$= 4 \cdot 6m - 4 \cdot 3n + 4 \cdot 4$
$= 4(6m - 3n + 4)$

107. $-60P^2 - 80P = -20P \cdot 3P - 20P \cdot 4$
$= -20P(3P + 4)$

APPLICATIONS

109. REARVIEW MIRRORS

a. Windshield:
$A = lw$
$= (2x)(3x^2)$
$= 6x^3$ cm^2

b. Driver's side door:
$A = lw$
$= (3x)(4x)$
$= 12x^2$ cm^2

Passenger side door:
$A = lw$
$= (3x)(4x)$
$= 12x^2$ cm^2

Area of both mirrors is $24x^2$ cm^2

c. Total area:
$6x^3 + 24x^2 = 6x^2 \cdot x + 6x^2 \cdot 4$
$= 6x^2(x + 4)$ cm^2

111. AIRCRAFT CARRIERS

$x^3 + 4x^2 + 5x + 20$

$= (x^3 + 4x^2) + (5x + 20)$

$= x^2(x + 4) + 5(x + 4)$

$= (x + 4)(x^2 + 5)$

The length and width are $(x^2 + 5)$ ft. and $(x + 4)$ ft.

WRITING

113. Answers will vary.

115. Answers will vary.

REVIEW

117. INSURANCE COST

Find the difference in the two premiums:

$1,050 – $925 = $125

$125 is what percent of $1,050

$$125 = x \cdot 1{,}050$$

$$125 = 1{,}050x$$

$$\frac{125}{1{,}050} = \frac{1{,}050x}{1{,}050}$$

$$0.119 \approx x$$

$$\boxed{12\% \approx x}$$

CHALLENGE PROBLEMS

119. $6x^{4m}y^n + 21x^{3m}y^{2n} - 15x^{2m}y^{3n}$

$= 3x^{2m}y^n \cdot 2x^{2m} + 3x^{2m}y^n \cdot 7x^m y^n - 3x^{2m}y^n \cdot 5y^{2n}$

$= 3x^{2m}y^n(2x^{2m} + 7x^m y^n - 5y^{2n})$

SECTION 5.2

VOCABULARY

1. A polynomial that has three terms is called a **trinomial**. A polynomial that has two terms is called a **binomial**.

3. Since $10 = (-5)(-2)$, we say that -5 and -2 are **factors** of 10.

5. The **lead** coefficient of the trinomial $x^2 - 3x + 2$ is 1, the **coefficient** of the middle term is -3, and the last term is 2.

CONCEPTS

7. Two factorizations of 4 that involve only positive numbers are **$4 \cdot 1$** and **$2 \cdot 2$**. Two factorizations of 4 that involve only negative numbers are **-4(-1)** and **-2(-2).**

9. To factor $x^2 + x - 56$, we must find two integers whose **product** is -56 and whose **sum** is 1.

11. $x^2 + 5x + 3$ cannot be factored because we cannot find two integers whose product is **3** and whose sum is **5**.

13. a) the Last step: $5 \cdot 4 = 20$
    b) the Outer step: $x \cdot 4 = 4x$
    the Inner step: $5 \cdot x = 5x$

15. a) 1
    b) 15, 8
    c) 5 and 3

17. a) They are both positive or both negative.
    b) One will be positive, the other negative.

19. $x(x + 1) + 5(x + 1) = (x + 1)(x + 5)$

21.

Negative factors of 8	Sum of factors of 8
$-1(-8) = 8$	**$-1 + (-8) = -9$**
$-2(-4) = 8$	$-2 + (-4) = -6$

NOTATION

23. $(x + 3)(x - 2)$

PRACTICE

25. $x^2 + 3x + 2 = (x + 2)(x + 1)$

27. $t^2 - 9t + 14 = (t - 7)(t - 2)$

29. $a^2 + 6a - 16 = (a + 8)(a - 2)$

31. The factors of 11 whose sum is 12 are 11 and 1.
    $z^2 + 12z + 11 = (z + 11)(z + 1)$

33. The factors of 6 whose sum is -5 are -2 and -3.
    $m^2 - 5m + 6 = (m - 3)(m - 2)$

35. The factors of -5 whose sum is -4 are -5 and 1.
    $a^2 - 4a - 5 = (a - 5)(a + 1)$

37. The factors of -24 whose sum is 5 are 8 and -3.
    $x^2 + 5x - 24 = (x + 8)(x - 3)$

39. The factors of -39 whose sum is -10 are -13 and 3.
    $a^2 - 10a - 39 = (a - 13)(a + 3)$

41. Prime. There are no factors of 15 whose sum is 10.

43. Prime. There are no factors of 4 whose sum is -2.

45. The factors of 18 whose sum is -9 are -3 and -6.
    $r^2 - 9r + 18 = (r - 3)(r - 6)$

47. The factors of 4 whose sum is 4 are 2 and 2.
    $x^2 + 4xy + 4y^2 = (x + 2y)(x + 2y)$

49. The factors of -12 whose sum is -4 are -6 and 2.
    $a^2 - 4ab - 12b^2 = (a - 6b)(a + 2b)$

51. Prime. There are no factors of 4 whose sum is 2.

53. $-x^2 - 7x - 10 = -1(x^2 + 7x + 10)$
$= -(x + 5)(x + 2)$

55. $-t^2 - 15t + 34 = -1(t^2 + 15t - 34)$
$= -(t + 17)(t - 2)$

57. $-a^2 - 4ab - 3b^2 = -(a^2 + 4ab + 3b^2)$
$= -(a + 3b)(a + b)$

59. $-x^2 + 6xy + 7y^2 = -1(x^2 - 6xy - 7y^2)$
$= -(x - 7y)(x + y)$

61. $4 - 5x + x^2 = x^2 - 5x + 4$
$= (x - 4)(x - 1)$

63. $10y + 9 + y^2 = y^2 + 10y + 9$
$= (y + 9)(y + 1)$

65. $r^2 - 2 - r = r^2 - r - 2$
$= (r - 2)(r + 1)$

67. $4rx + r^2 + 3x^2 = r^2 + 4rx + 3x^2$
$= (r + 3x)(r + x)$

69. $2x^2 + 10x + 12 = 2(x^2 + 5x + 6)$
$= 2(x + 3)(x + 2)$

71. $-5a^2 + 25a - 30 = -5(a^2 - 5a + 6)$
$= -5(a - 3)(a - 2)$

73. $z^3 - 29z^2 + 100z = z(z^2 - 29z + 100)$
$= z(z - 4)(z - 25)$

75. $12xy + 4x^2y - 72y$
$= 4x^2y + 12xy - 72y$
$= 4y(x^2 + 3x - 18)$
$= 4y(x + 6)(x - 3)$

77. $-r^2 + 14r - 40 = -1(r^2 - 14r + 40)$
$= -(r - 10)(r - 4)$

79. $-13yz + y^2 - 14z^2 = y^2 - 13yz - 14z^2$
$= (y - 14z)(y + z)$

81. $s^2 + 11s - 26 = (s + 13)(s - 2)$

83. $a^2 + 10ab + 9b^2 = (a + 9b)(a + b)$

85. $-x^2 + 21x + 22 = -1(x^2 - 21x - 22)$
$= -(x - 22)(x + 1)$

87. $d^3 - 11d^2 - 26d = d(d^2 - 11d - 26)$
$= d(d - 13)(d + 2)$

APPLICATIONS

89. PETS
$x^3 + 12x^2 + 27x = x(x^2 + 12x + 27)$
$= x(x + 3)(x + 9)$

The length is $(x + 9)$ in.
The width is x in.
The height is $(x + 3)$ in.

WRITING

91. Answers will vary.

93. Answers will vary.

95. Answers will vary.

REVIEW

97. $\frac{x^{12}x^{-7}}{x^3x^4} = \frac{x^{12+(-7)}}{x^{3+4}}$
$= \frac{x^5}{x^7}$
$= x^{5-7}$
$= x^{-2}$
$= \frac{1}{x^2}$

99. $\frac{a^4a^{-2}}{a^2a^0} = \frac{a^{4+(-2)}}{a^{2+0}}$
$= \frac{a^2}{a^2}$
$= a^{2-2}$
$= a^0$
$= 1$

CHALLENGE PROBLEMS

101. $x^2 - \frac{6}{5}x + \frac{9}{25} = \left(x - \frac{3}{5}\right)\left(x - \frac{3}{5}\right)$

103. $x^{2m} - 12x^m - 45 = (x^m + 3)(x^m - 15)$

105. $c = 5; 5 + 1 = 6$
$c = 8; 2 + 4 = 6$
$c = 9; 3 + 3 = 6$

SECTION 5.3

VOCABULARY

1. The trinomial $3x^2 - x - 12$ has a **lead** coefficient of 3. The middle term is $-x$ and the last **term** is -12.

3. The numbers 6 and -2 are two integers whose **product** is -12 and whose **sum** is 4.

5. To factor a trinomial by grouping, we begin by finding ac, the key number.

CONCEPTS

7. a) $1 \cdot 9 = 9$ or $3 \cdot 3 = 9$
   b) $-2(-8) = 16$, $-1(-16) = 16$, or $-4(-4) = 16$
   c) $-2(5) = -10$, $2(-5) = -10$, $-1(10) = -10$, or $1(-10) = -10$

9. a) $4y \cdot 2 = 8y$
   b) $-8 \cdot 3y = -24y$
   c) $8y + (-24y) = -16y$

11. $10x$ and x, or $5x$ and $2x$

13. a) When factoring a trinomial, we write it in **descending** powers of the variable. Then we factor out any **GCF** (including -1 if that is necessary to make the lead coefficient **positive**).
    b) $3s^2$
    c) $-(2d^2 - 19d + 8)$

15. Since the last term of the trinomial is **positive** and the middle term is **negative**, the integers must be **negative** factors of 6.

17. Since the last term of the trinomial is **negative** , the signs of the integers will be **different**.

19. $3x(5x - 2) + 2(5x - 2) = (5x - 2)(3x + 2)$

21.

Negative factors of 12	Sum of factors of 12
-1(-12) = 12	**-1 + (-12) = -13**
-2(-6) = 12	**-2 + (-6) = -8**
-3(-4) = 12	**-3 + (-4) = -7**

NOTATION

23. a) Answers will vary. $x^2 + 6x + 1$
    b) Answers will vary. $3x^2 + 6x + 1$

25. a) $a = 12, b = 20, c = -9$
    b) key number $= ac = 12(-9) = -108$

PRACTICE

27. $3a^2 + 13a + 4 = (3a + 1)(a + 4)$

29. $4z^2 - 13z + 3 = (z - 3)(4z - 1)$

31. $2m^2 + 5m - 12 = (2m - 3)(m + 4)$

33. $12t^2 + 17t + 6 = 12t^2 + \underline{\mathbf{9}}t + \underline{\mathbf{8}}t + 6$
    $= \underline{\mathbf{3t}}(4t + 3) + \underline{\mathbf{2}}(4t + 3)$
    $= (\underline{\mathbf{4t + 3}})(3t + 2)$

35. $3a^2 + 13a + 4 = 3a^2 + 1a + 12a + 4$
    $= a(3a + 1) + 4(3a + 1)$
    $= (3a + 1)(a + 4)$

37. $5x^2 + 11x + 2 = 5x^2 + 10x + 1x + 2$
    $= 5x(x + 2) + 1(x + 2)$
    $= (x + 2)(5x + 1)$

39. $4x^2 + 8x + 3 = 4x^2 + 2x + 6x + 3$
    $= 2x(2x + 1) + 3(2x + 1)$
    $= (2x + 1)(2x + 3)$

41. $6x^2 + 25x + 21 = 6x^2 + 18x + 7x + 21$
    $= 6x(x + 3) + 7(x + 3)$
    $= (x + 3)(6x + 7)$

43. $2x^2 - 3x + 1 = 2x^2 - 2x - x + 1$
    $= 2x(x - 1) - 1(x - 1)$
    $= (x - 1)(2x - 1)$

45. $4t^2 - 4t + 1 = 4t^2 - 2t - 2t + 1$
    $= 2t(2t - 1) - 1(2t - 1)$
    $= (2t - 1)(2t - 1)$

47. $15t^2 - 34t + 8 = 15t^2 - 30t - 4t + 8$
    $= 15t(t - 2) - 4(t - 2)$
    $= (t - 2)(15t - 4)$

49. $2x^2 - 3x - 2 = 2x^2 - 4x + x - 2$
    $= 2x(x - 2) + 1(x - 2)$
    $= (x - 2)(2x + 1)$
    $= (a - 2)(3a + 2)$

51. $12y^2 - y - 1 = 12y^2 - 4y + 3y - 1$
$= 4y(3y - 1) + 1(3y - 1)$
$= (3y - 1)(4y + 1)$

53. $10y^2 - 3y - 1 = 10y^2 - 5y + 2y - 1$
$= 5y(2y - 1) + 1(2y - 1)$
$= (2y - 1)(5y + 1)$

55. $12y^2 - 5y - 2 = 12y^2 - 8y + 3y - 2$
$= 4y(3y - 2) + 1(3y - 2)$
$= (3y - 2)(4y + 1)$

57. $2m^2 + 5m - 10$
prime

59. $-13x + 3x^2 - 10 = 3x^2 - 13x - 10$
$= 3x^2 - 15x + 2x - 10$
$= 3x(x - 5) + 2(x - 5)$
$= (x - 5)(3x + 2)$

61. $6r^2 + rs - 2s^2 = 6r^2 + 4rs - 3rs - 2s^2$
$= 2r(3r + 2s) - s(3r + 2s)$
$= (3r + 2s)(2r - s)$

63. $8n^2 + 91n + 33 = 8n^2 + 88n + 3n + 33$
$= 8n(n + 11) + 3(n + 11)$
$= (n + 11)(8n + 3)$

65. $4a^2 - 15ab + 9b^2$
$= 4a^2 - 3ab - 12ab + 9b^2$
$= a(4a - 3b) - 3b(4a - 3b)$
$= (4a - 3b)(a - 3b)$

67. $130r^2 + 20r - 110$
$= 10(13r^2 + 2r - 11)$
$= 10(13r^2 + 13r - 11r - 11)$
$= 10[13r(r + 1) - 11(r + 1)]$
$= 10(13r - 11)(r + 1)$

69. $-5t^2 - 13t - 6 = -(5t^2 + 13t + 6)$
$= -(5t^2 + 10t + 3t + 6)$
$= -[5t(t + 2) + 3(t + 2)]$
$= -(t + 2)(5t + 3)$

71. $36y^2 - 88 + 32 = 4(9y^2 - 22y + 8)$
$= 4[9y^2 - 18y - 4y + 8]$
$= 4[9y(y - 2) - 4(y - 2)]$
$= 4(y - 2)(9y - 4)$
$= 5(2a - 1)(7a - 6)$

73. $4x^2 + 8xy + 3y^2$
$= 4x^2 + 2xy + 6xy + 3y^2$
$= 2x(2x + y) + 3y(2x + y)$
$= (2x + y)(2x + 3y)$

75. $12y^2 + 12 - 25y = 12y^2 - 25y + 12$
$= 12y^2 - 16y - 9y + 12$
$= 4y(3y - 4) - 3(3y - 4)$
$= (3y - 4)(4y - 3)$

77. $18x^2 + 31x - 10 = 18x^2 + 36x - 5x - 10$
$= 18x(x + 2) - 5(x + 2)$
$= (x + 2)(18x - 5)$

79. $-y^3 - 13y^2 - 12y = -y(y^2 + 13y + 12)$
$= -y(y + 1)(y + 12)$

81. $3x^2 + 6 + x = 3x^2 + x + 6$
prime

83. $30r^5 + 63r^4 - 30r^3$
$= 3r^3(10r^2 + 21r - 10)$
$= 3r^3(10r^2 + 25r - 4r - 10)$
$= 3r^3[5r(2r + 5) - 2(2r + 5)]$
$= 3r^3(2r + 5)(5r - 2)$

85. $2a^2 + 3b^2 + 5ab = 2a^2 + 5ab + 3b^2$
$= 2a^2 + 2ab + 3ab + 3b^2$
$= 2a(a + b) + 3b(a + b)$
$= (a + b)(2a + 3b)$

87. $pq + 6p^2 - q^2 = 6p^2 + pq - q^2$
$= 6p^2 + 3pq - 2pq - q^2$
$= 3p(2p + q) - q(2p + q)$
$= (2p + q)(3p - q)$

89. $6x^3 - 15x^2 - 9x = 3x(2x^2 - 5x - 3)$
$= 3x(2x^2 - 6x + x - 3)$
$= 3x[2x(x - 3) + 1(x - 3)]$
$= 3x(x - 3)(2x + 1)$

91. $15 + 8a^2 - 26a = 8a^2 - 26a + 15$
$= 8a^2 - 20a - 6a + 15$
$= 4a(2a - 5) - 3(2a - 5)$
$= (2a - 5)(4a - 3)$

93. $16m^3n + 20m^2n^2 + 6mn^3$
$= 2mn(8m^2 + 10mn + 3n^2)$
$= 2mn(8m^2 + 4mn + 6mn + 3n^2)$
$= 2mn[4m(2m + n) + 3n(2m + n)]$
$= 2mn(2m + n)(4m + 3n)$

APPLICATIONS

95. FURNITURE

$$\begin{aligned} 4x^2 + 20x - 11 &= 4x^2 - 2x + 22x - 11 \\ &= 2x(2x - 1) + 11(2x - 1) \\ &= (2x - 1)(2x + 11) \end{aligned}$$

The length is $(2x + 11)$ in.
The width is $(2x - 1)$ in.

To find the difference, subtract:

$$\begin{aligned} (2x + 11) - (2x - 1) &= 2x + 11 - 2x + 1 \\ &= 12 \text{ in.} \end{aligned}$$

The difference is 12 in.

WRITING

97. Answers will vary.

99. Answers will vary.

101. Answers will vary.

REVIEW

103. $\begin{aligned} -7^2 &= -(7)(7) \\ &= -49 \end{aligned}$

105. $7^0 = 1$

107. $\begin{aligned} \frac{1}{7^{-2}} &= 7^2 \\ &= 49 \end{aligned}$

CHALLENGE PROBLEMS

109. $6a^{10} + 5a^5 - 21$

$$\begin{aligned} &= 6a^{10} + 14a^5 - 9a^5 - 21 \\ &= 2a^5(3a^5 + 7) - 3(3a^5 + 7) \\ &= (3a^5 + 7)(2a^5 - 3) \end{aligned}$$

111. $8x^2(c^2+c-2) - 2x(c^2+c-2) - 1(c^2+c-2)$

$$\begin{aligned} &= (c^2 + c - 2)(8x^2 - 2x - 1) \\ &= (c + 2)(c - 1)(4x + 1)(2x - 1) \end{aligned}$$

SECTION 5.4

VOCABULARY

1. $x^2 + 6x + 9$ is a **perfect** square trinomial because it is the square of the binomial $(x + 3)$.

CONCEPTS

3. a) The first term is the square of **5x**.
   b) The last term is the square of **3**.
   c) The middle term is twice the product of **5x** and **3**.

5. a) $x^2 + 2xy + y^2 = (\underline{x} + \underline{y})^2$
   b) $x^2 - 2xy + y^2 = (x - y)^2$
   c) $x^2 - y^2 = (x + y)(x - y)$

7. 1, 4, 9, 16, 25, 36, 47, 64, 81, 100

9. a) $(x + 6)(x - 6) = x^2 - 6x + 6x - 36$
   $= x^2 - 36$

   $(x + 6)(x + 6) = x^2 + 6x + 6x + 36$
   $= x^2 + 12x + 36$

   $(x - 6)(x - 6) = x^2 - 6x - 6x + 36$
   $= x^2 - 12x + 36$

   b) No

11. $(3x + 4y)(3x - 4y)$
    $= 9x^2 - 12xy + 12xy - 16y^2$
    $= 9x^2 - 16y^2$

NOTATION

13. Answers will vary.
    a) $x^2 - 9$
    b) $(x - 9)^2$
    c) $x^2 + 2xy + y^2$
    d) $x^2 + 25$

15. a) base: $(x + 3)$
    exponent: 2

    b) $(x - 8)(x - 8)$

PRACTICE

17. $a^2 - 6a + 9 = (a - \underline{3})^2$

19. $4x^2 + 4x + 1 = (2x + 1)^2$

21. $x^2 + 6x + 9 = (x + 3)^2$

23. $b^2 + 2b + 1 = (b + 1)^2$

25. $c^2 - 12c + 36 = (c - 6)^2$

27. $y^2 - 8y + 16 = (y - 4)^2$

29. $t^2 + 20t + 100 = (t + 10)^2$

31. $2u^2 - 36u + 162 = 2(u^2 - 18u + 81)$
    $= 2(u - 9)^2$

33. $36x^3 + 12x^2 + x = x(36x^2 + 12x + 1)$
    $= x(6x + 1)^2$

35. $9 + 4x^2 + 12x = 4x^2 + 12x + 9$
    $= (2x + 3)^2$

37. $a^2 + 2ab + b^2 = (a + b)^2$

39. $25m^2 + 70mn + 49n^2 = (5m + 7n)^2$

41. $9x^2y^2 + 30xy + 25 = (3xy + 5)^2$

43. $t^2 - \frac{2}{3}t + \frac{1}{9} = \left(t - \frac{1}{3}\right)^2$

45. $s^2 - 1.2s + 0.36 = (s - 0.6)^2$

47. $y^2 - 49 = (y + \underline{7})(y - \underline{7})$

49. $t^2 - w^2 = (\underline{t} + \underline{w})(t - w)$

51. $x^2 - 16 = (x - 4)(x + 4)$

53. $4y^2 - 1 = (2y + 1)(2y - 1)$

55. $9x^2 - y^2 = (3x + y)(3x - y)$

57. $16a^2 - 25b^2 = (4a + 5b)(4a - 5b)$

59. $36 - y^2 = (6 + y)(6 - y)$

61. $a^2 + b^2$
    prime

63. $a^4 - 144b^2 = (a^2 + 12b)(a^2 - 12b)$

65. $t^2z^2 - 64 = (tz + 8)(tz - 8)$

67. $y^2 - 63$
    prime

69. $8x^2 - 32y^2 = 8(x^2 - 4y^2)$
$= 8(x + 2y)(x - 2y)$

71. $7a^2 - 7 = 7(a^2 - 1)$
$= 7(a + 1)(a - 1)$

73. $-25 + v^2 = v^2 - 25$
$= (v + 5)(v - 5)$

75. $6x^4 - 6x^2y^2 = 6x^2(x^2 - y^2)$
$= 6x^2(x + y)(x - y)$

77. $x^4 - 81 = (x^2 + 9)(x^2 - 9)$
$= (x^2 + 9)(x + 3)(x - 3)$

79. $a^4 - 16 = (a^2 + 4)(a^2 - 4)$
$= (a^2 + 4)(a + 2)(a - 2)$

81. $c^2 - \frac{1}{16} = \left(c + \frac{1}{4}\right)\left(c - \frac{1}{4}\right)$

APPLICATIONS

83. GENETICS
$p^2 + 2pq + q^2 = 1$
$(p + q)^2 = 1$

85. PHYSICS
$0.5gt_1^2 - 0.5gt_2^2 = 0.5g(t_1^2 - t_2^2)$
$= 0.5g(t_1 + t_2)(t_1 - t_2)$
The distance is $0.5g(t_1 + t_2)(t_1 - t_2)$.

WRITING

87. Answers will vary.

89. Answers will vary.

91. Answers will vary.

REVIEW

93. $\frac{5x^2 + 10y^2 - 15xy}{5xy} = \frac{5x^2}{5xy} + \frac{10y^2}{5xy} - \frac{15xy}{5xy}$
$= \frac{x}{y} + \frac{2y}{x} - 3$

95. Write the dividend in descending order of exponents.

$$\begin{array}{r} 3a + 2 \\ 2a - 1 \overline{)\, 6a^2 + a - 2} \\ \underline{6a^2 - 3a} \\ 4a - 2 \\ \underline{4a - 2} \\ 0 \end{array}$$

CHALLENGE PROBLEMS

97. $5(4x + 3)(4x - 3) = 5(16x^2 - 9)$
$= 80x^2 - 45$

$c = 45$

99. $81x^6 + 36x^3y^2 + 4y^4 = (9x^3 + 2y^2)^2$

101. $(x + 5)^2 - y^2 = (x + 5 + y)(x + 5 - y)$

103. $(x+5)^2 - y^2 = [(x+5) + y][(x+5) - y]$
$= (x + 5 + y)(x + 5 - y)$

SECTION 5.5

VOCABULARY

1. $x^3 + 27$ is the **sum** of two cubes.

3. The factorization of $x^3 + 8$ is $(x + 2)(x^2 - 2x + 4)$. The first factor is a **binomial** and the second is a trinomial.

CONCEPTS

5. a) $F^3 + L^3 = (\boxed{F} + \boxed{L})(F^2 - FL + L^2)$

   b) $F^3 - L^3 = (F\boxed{-}L)(\boxed{F^2} + FL + \boxed{L^2})$

   a) $125 = (\boxed{5})^3$

   b) $27m^3 = (\boxed{3m})^3$

   c) $8x^3 - 27 = (\boxed{2x})^3 - (\boxed{3})^3$

   d) $x^3 + 64y^3 = (\boxed{x})^3 + (\boxed{4y})^3$

7. This is $\boxed{m}$ cubed.

   This is $\boxed{4}$ cubed.

9. 1, 8, 27, 64, 125, 216

11. It is factored completely.

NOTATION

13. Answers will vary.
   a) $x^3 + 8$
   b) $(x + 8)^3$

PRACTICE

15. $a^3 + 8 = (a)^3 + (2)^3$
   $= (a + 2)[(a)^2 - a(2) + (2)^2]$
   $= (a + 2)(a^2 - \boxed{2a} + 4)$

17. $b^3 + 27 = (b)^3 + (3)^3$
   $= (b + 3)[(b)^2 - b(3) + (3)^2]$
   $= (\boxed{b + 3})(b^2 - 3b + 9)$

19. $y^3 + 1 = (y)^3 + (1)^3$
   $= (y + 1)[(y)^2 - y(1) + (1)^2]$
   $= (y + 1)(y^2 - y + 1)$

21. $a^3 - 27 = (a)^3 - (3)^3$
   $= (a - 3)[(a)^2 + a(3) + (3)^2]$
   $= (a - 3)(a^2 + 3a + 9)$

23. $8 + x^3 = (2)^3 + (x)^3$
   $= (2 + x)[(2)^2 - 2(x) + (x)^2]$
   $= (2 + x)(4 - 2x + x^2)$

25. $s^3 - t^3 = (s)^3 - (t)^3$
   $= (s - t)[(s)^2 + s(t) + (t)^2]$
   $= (s - t)(s^2 + st + t^2)$

27. $a^3 + 8b^3 = (a)^3 + (2b)^3$
   $= (a + 2b)[(a)^2 - a(2b) + (2b)^2]$
   $= (a + 2b)(a^2 - 2ab + 4b^2)$

29. $64x^3 - 27 = (4x)^3 - (3)^3$
   $= (4x - 3)[(4x)^2 + 4x(3) + (3)^2]$
   $= (4x - 3)(16a^2 + 12x + 9)$

31. $a^6 - b^3 = (a^2)^3 - (b)^3$
   $= (a^2 - b)[(a^2)^2 + a^2(b) + (b)^2]$
   $= (a^2 - b)(a^4 + a^2b + b^2)$

33. $2x^3 + 54 = 2(x^3 + 27)$
   $= 2[(x)^3 + (3)^3]$
   $= 2(x + 3)[(x)^2 - x(3) + (3)^2]$
   $= 2(x + 3)(x^2 - 3x + 9)$

35. $-x^3 + 216 = -(x^3 - 216)$
   $= -(x)^3 - (6)^3$
   $= -(x - 6)[(x)^2 + x(6) + (6)^2]$
   $= -(x - 6)(x^2 + 6x + 36)$

37. $64m^3x - 8n^3x$
   $= 8x(8m^3 - n^3)$
   $= 8x[(2m)^3 - (n)^3]$
   $= 8x(2m - n)[(2m)^2 + 2m(n) + (n)^2]$
   $= 8x(2m - n)(4m^2 + 2mn + n^2)$

APPLICATIONS

39. MAILING BREAKABLES

Volume of larger cube:

$V = 10^3$
$= 1{,}000 \text{ in.}^3$

Volume of smaller cube:

$V = x^3 \text{ in.}^3$

Volume of empty space:

$V = (1{,}000 - x^3) \text{ in.}^3$
$= (10)^3 - (x)^3$
$= (10 - x)(10^2 + 10(x) + x^2)$
$= (10 - x)(100 + 10x + x^2)$

WRITING

41. Answers will vary.

REVIEW

43. $\frac{7}{9} = 0.\overline{7}$, repeating decimal

45. $\{\ldots, -4, -3, -2, -1, 0, 1, 2, 3, 4, \ldots\}$

47. $2x^2 + 5x - 3$, $x = -3$

$2x^2 + 5x - 3 = 2(\mathbf{-3})^2 + 5(\mathbf{-3}) - 3$
$= 2(9) + 5(-3) - 3$
$= 18 - 15 - 3$
$= 0$

CHALLENGE PROBLEMS

49. a) $(x^3)^2 - 1^2$
$= (x^3 + 1)(x^3 - 1)$
$= (x + 1)(x^2 - x + 1)(x + 1)(x^2 - x + 1)$

b) $(x^2)^3 - 1^3$
$= (x^2 - 1)(x^4 + x^2 + 1)$
$= (x + 1)(x - 1)(x^4 + x^2 + 1)$

c) They appear to be different because the trinomial factor $(x^4 + x^2 + 1)$ of part "b" has not been factored completely.
$(x^4 + x^2 + 1) = (x^2 - x + 1)(x^2 + x + 1)$

51. $x^{3m} - y^{3n} = (x^m)^3 - (y^n)^3$
$= (x^m - y^n)(x^{2m} + x^m y^n + y^{2n})$

SECTION 5.6

VOCABULARY

1. A polynomial is factored **completely** when no factor can be factored further.

3. A polynomial that does not factor using integers is called a **prime** polynomial.

5. $4x^2 - 12x + 9$ is a perfect square **trinomial** because it is the square of $(2x - 3)$.

7. When factoring a polynomial, always factor out the **GCF** first.

CONCEPTS

9. factor out the GCF

11. perfect square trinomial

13. sum of two cubes

15. trinomial factoring

NOTATION

17. $8m^3$

PRACTICE

19. $2ab^2 + 8ab - 24a = 2a(b^2 + 4a - 12)$
$= 2a(b + 6)(b - 2)$

21. $-8p^3q^7 - 4p^2q^3 = -4p^2q^3(2pq^4 + 1)$

23. $20m^2 + 100m + 125 = 5(4m^2 + 20m + 25)$
$= 5(2m + 5)^2$

25. $x^2 + 7x + 1$
prime

27. $-2x^5 + 128x^2 = -2x^2(x^3 - 64)$
$= -2x^2[(x)^3 - (4)^3]$
$= -2x^2(x - 4)(x^2 - 4x + 16)$

29. $a^2c + a^2d^2 + bc + bd^2$
$= (a^2c + a^2d^2) + (bc + bd^2)$
$= a^2(c + d^2) + b(c + d^2)$
$= (c + d^2)(a^2 + b)$
$= 2t^2(3t - 5)(t + 4)$

31. $-9x^2y^2 + 6xy - 1 = -(9x^2y^2 - 6xy + 1)$
$= -(3xy - 1)^2$

33. $5x^3y^3z^4 + 25x^2y^3z^2 - 35x^3y^2z^5$
$= 5x^2y^2z^2(xyz^2 + 5y - 7xz^3)$

35. $2c^2 + 5cd - 3d^2 = (2c + d)(c - 3d)$

37. $8a^2x^3y - 2b^2xy = 2xy(4a^2x^2 - b^2)$
$= 2xy(2ax + b)(2ax - b)$

39. $a^2(x - a) - b^2(x - a) = (x - a)(a^2 - b^2)$
$= (x - a)(a + b)(a - b)$

41. $a^2b^2 - 144 = (ab + 12)(ab - 12)$

43. $2ac + 4ad + bc + 2bd$
$= (2ac + 4ad) + (bc + 2bd)$
$= 2a(c + 2d) + b(c + 2d)$
$= (c + 2d)(2a + b)$

45. $v^2 - 14v + 49 = (v - 7)^2$

47. $39 + 10n - n^2 = -n^2 + 10n + 39$
$= -(n^2 - 10n - 39)$
$= -(n + 3)(n - 13)$

49. $16y^8 - 81z^4$
$= (4y^4 + 9z^2)(4y^4 - 9z^2)$
$= (4y^4 + 9z^2)(2y^2 + 3z)(2y^2 - 3z)$

51. $6x^2 - 14x + 8 = 2(3x^2 - 7x + 4)$
$= 2(3x - 4)(x - 1)$

53. $4x^2y^2 + 4xy^2 + y^2 = y^2(4x^2 + 4x + 1)$
$= y^2(2x + 1)^2$

55. $4m^5n + 500m^2n^4$
$= 4m^2n(m^3 + 125n^3)$
$= 4m^2n[(m)^3 + (5n)^3]$
$= 4m^2n(m + 5n)(m^2 - 5mn + 25n^2)$

57. $a^2x^2 + b^2y^2 + b^2x^2 + a^2y^2$
$= (a^2x^2 + a^2y^2) + (b^2x^2 + b^2x^2)$
$= a^2(x^2 + y^2) + b^2(x^2 + y^2)$
$= (x^2 + y^2)(a^2 + b^2)$

59. $4x^2 + 9y^2$
prime

61. $16a^5 - 54a^2b^3$
$= 2a^2(8a^3 - 27b^3)$
$= 2a^2[(2a)^3 - (3b)^3]$
$= 2a^2(2a - 3b)(4a^2 + 6ab + 9b^2)$

63. $27x - 27y - 27z = 27(x - y - z)$

65. $xy - ty + xs - ts = (xy - ty) + (xs - ts)$
$= y(x - t) + s(x - t)$
$= (x - t)(y + s)$

67. $35x^8 - 2x^7 - x^6 = x^6(35x^2 - 2x - 1)$
$= x^6(7x + 1)(5x - 1)$

69. $5(x - 2) + 10y(x - 2) = (x - 2)(5 + 10y)$
$= (x - 2)(5)(1 + 2y)$
$= 5(x - 2)(1 + 2y)$

71. $49p^2 + 28pq + 4q^2 = (7p + 2q)^2$

73. $4t^2 + 36 = 4(t^2 + 9)$

75. $p^3 - 2p^2 + 3p - 6 = (p^3 - 2p^2) + (3p - 6)$
$= p^2(p - 2) + 3(p - 2)$
$= (p - 2)(p^2 + 3)$

WRITING

77. Answers will vary.

79. Answers will vary.

REVIEW

81.

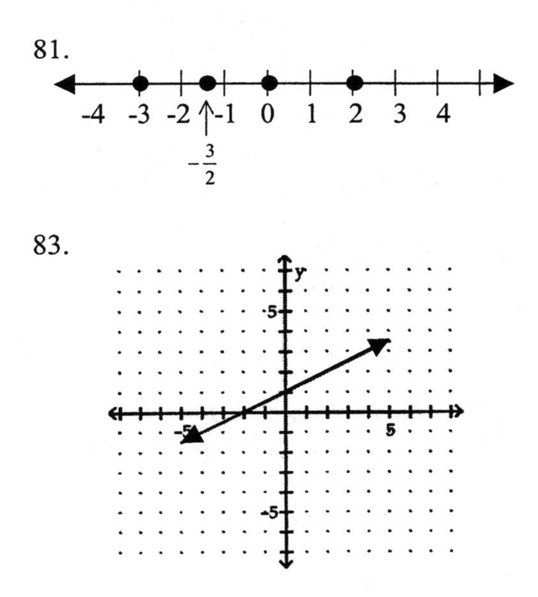

CHALLENGE PROBLEMS

85. $x^4 - 2x^2 - 8 = (x^2 + 2)(x^2 - 4)$
$= (x^2 + 2)(x + 2)(x - 2)$

87. $24 - x^3 + 8x^2 - 3x = 24 - 3x + 8x^2 - x^3$
$= (24 - 3x) + (8x^2 - x^3)$
$= 3(8 - x) + x^2(8 - x)$
$= (8 - x)(3 + x^2)$

89. $x^9 + y^6 = [(x^3)^3 + (y^2)^3]$
$= (x^3 + y^2)(x^6 - x^3y^2 + y^4)$

SECTION 5.7

VOCABULARY

1. Any equation that can be written in the form $ax^2 + bx + c = 0$ ($a \neq 0$) is called a **quadratic** equation.

3. Integers that are 1 unit apart, such as 8 and 9, are called **consecutive** integers.

5. The longest side of a right triangle is the **hypotenuse**. The remaining two sides are the **legs** of the triangle.

CONCEPTS

7. a. yes
   b. no
   c. yes
   d. no

9.
$$5x+4=0$$
$$5x+4-4=0-4$$
$$5x=-4$$
$$\frac{5x}{5}=\frac{-4}{5}$$
$$\boxed{x=-\frac{4}{5}}$$

11. a. $x^2+7x=-6$
    Add 6 to both sides.
    b. $x(x+7)=3$
    Distribute the multiplication by x and subtract 3 from both sides.

13. a. $x^2+6x-16=(x-2)(x+8)$

b.
$$x^2+6x-16=0$$
$$(x-2)(x+8)=0$$
$$x-2=0$$
$$x-2+2=0+2$$
$$\boxed{x=2}$$
$$x+8=0$$
$$x+8-8=0-8$$
$$\boxed{x=-8}$$

15. The value of h is 0.

17. If the length of the hypotenuse of a right triangle is c and the lengths of the other two legs are a and b, then $\boxed{a^2+b^2}=c^2$.

NOTATION

19.
$$7y^2+14y=0$$
$$\boxed{7y}(y+2)=0$$
$$7y=0 \quad \text{or} \quad \boxed{y+2}=0$$
$$y=\boxed{0} \qquad y=\boxed{-2}$$

PRACTICE

21.
$$(x-2)(x+3)=0$$
$$x-2=0 \quad \text{or} \quad x+3=0$$
$$x=2 \qquad x=-3$$

23.
$$(2s-5)(s+6)=0$$
$$2s-5=0 \quad \text{or} \quad s+6=0$$
$$2s=5 \qquad s=-6$$
$$s=\frac{5}{2}$$

25.
$$2(t-7)(t+8)=0$$
$$t-7=0 \quad \text{or} \quad t+8=0$$
$$t=7 \qquad t=-8$$

27.
$$(x-1)(x+2)(x-3)=0$$
$$x-1=0 \qquad x+2=0 \qquad x-3=0$$
$$x=1 \qquad x=-2 \qquad x=3$$

29.
$$x(x-3)=0$$
$$x=0 \quad \text{or} \quad x-3=0$$
$$x=3$$

31.

$$6x(2x-5)=0$$
$$6x=0 \quad \text{or} \quad 2x-5=0$$
$$x=0 \qquad 2x=5$$
$$x=\frac{5}{2}$$

33.

$$w^2-7w=0$$
$$w(w-7)=0$$
$$w=0 \quad \text{or} \quad w-7=0$$
$$w=7$$

35.

$$3x^2+8x=0$$
$$x(3x+8)=0$$
$$x=0 \quad \text{or} \quad 3x+8=0$$
$$3x=-8$$
$$x=-\frac{8}{3}$$

37.

$$8s^2-16s=0$$
$$8s(s-2)=0$$
$$8s=0 \quad \text{or} \quad s-2=0$$
$$s=0 \qquad s=2$$

39.

$$x^2-25=0$$
$$(x+5)(x-5)=0$$
$$x+5=0 \quad \text{or} \quad x-5=0$$
$$x=-5 \qquad x=5$$

41.

$$4x^2-1=0$$
$$(2x+1)(2x-1)=0$$
$$2x+1=0 \quad \text{or} \quad 2x-1=0$$
$$2x=-1 \qquad 2x=1$$
$$x=-\frac{1}{2} \qquad x=\frac{1}{2}$$

43.

$$9y^2-4=0$$
$$(3y+2)(3y-2)=0$$
$$3y+2=0 \quad \text{or} \quad 3y-2=0$$
$$3y=-2 \qquad 3y=2$$
$$y=-\frac{2}{3} \qquad y=\frac{2}{3}$$

45.

$$x^2=100$$
$$x^2-100=0$$
$$(x+10)(x-10)=0$$
$$x+10=0 \quad \text{or} \quad x-10=0$$
$$x=-10 \qquad x=10$$

47.

$$4x^2=81$$
$$4x^2-81=0$$
$$(2x+9)(2x-9)=0$$
$$2x+9=0 \quad \text{or} \quad 2x-9=0$$
$$2x=-9 \qquad 2x=9$$
$$x=-\frac{9}{2} \qquad x=\frac{9}{2}$$

49.

$$x^2-13x+12=0$$
$$(x-12)(x-1)=0$$
$$x-12=0 \quad \text{or} \quad x-1=0$$
$$x=12 \qquad x=1$$

51.

$$x^2-4x-21=0$$
$$(x-7)(x+3)=0$$
$$x-7=0 \quad \text{or} \quad x+3=0$$
$$x=7 \qquad x=-3$$

53.

$$x^2-9x+8=0$$
$$(x-8)(x-1)=0$$
$$x-8=0 \quad \text{or} \quad x-1=0$$
$$x=8 \qquad x=1$$

55.

$$a^2 + 8a = -15$$
$$a^2 + 8a + 15 = 0$$
$$(a+5)(a+3) = 0$$
$$a+5=0 \quad \text{or} \quad a+3=0$$
$$a=-5 \qquad\qquad a=-3$$

57.

$$4y+4=-y^2$$
$$y^2+4y+4=0$$
$$(y+2)(y+2)=0$$
$$y+2=0 \quad \text{or} \quad y+2=0$$
$$y=-2 \qquad\qquad y=-2$$

59.

$$0=x^2-16x+64$$
$$x^2-16x+64=0$$
$$(x-8)(x-8)=0$$
$$x-8=0 \quad \text{or} \quad x-8=0$$
$$x=8 \qquad\qquad x=8$$

61.

$$2x^2-5x+2=0$$
$$(2x-1)(x-2)=0$$
$$2x-1=0 \quad \text{or}$$
$$2x=1 \qquad\qquad x-2=0$$
$$x=\frac{1}{2} \qquad\qquad x=2$$

63.

$$5x^2-6x+1=0$$
$$(5x-1)(x-1)=0$$
$$5x-1=0 \quad \text{or} \quad x-1=0$$
$$5x=1 \qquad\qquad x=1$$
$$x=\frac{1}{5}$$

65.

$$4r^2+4r=-1$$
$$4r^2+4r+1=0$$
$$(2r+1)(2r+1)=0$$
$$2r+1=0 \quad \text{or} \quad 2r+1=0$$
$$2r=-1 \qquad\qquad 2r=-1$$
$$r=-\frac{1}{2} \qquad\qquad r=-\frac{1}{2}$$

67.

$$12b^2+26b+12=0$$
$$2(6b^2+13b+6)=0$$
$$2(2b+3)(3b+2)=0$$
$$2b+3=0 \quad \text{or} \quad 3b+2=0$$
$$2b=-3 \qquad\qquad 3b=-2$$
$$b=-\frac{3}{2} \qquad\qquad b=-\frac{2}{3}$$

69.

$$-15x^2+2=-7x$$
$$-15x^2+7x+2=0$$
$$-(15x^2-7x-2)=0$$
$$-(5x+1)(3x-2)=0$$
$$5x+1=0 \quad \text{or} \quad 3x-2=0$$
$$5x=-1 \qquad\qquad 3x=2$$
$$x=-\frac{1}{5} \qquad\qquad x=\frac{2}{3}$$

71.

$$x(2x-3)=20$$
$$2x^2-3x=20$$
$$2x^2-3x-20=0$$
$$(2x+5)(x-4)=0$$
$$2x+5=0 \quad \text{or} \quad x-4=0$$
$$2x=-5 \qquad\qquad x=4$$
$$x=-\frac{5}{2}$$

73.

$$(n+8)(n-3)=-30$$
$$n^2+8n-3n-24=-30$$
$$n^2+5n-24=-30$$
$$n^2+5n+6=0$$
$$(n+2)(n-3)=0$$
$$n+2=0 \quad \text{or} \quad n+3=0$$
$$n=-2 \qquad n=-3$$

75.

$$3b^2-30b=6b-60$$
$$3b^2-30b-6b+60=0$$
$$3b^2-36b+60=0$$
$$3(b^2-12b+20)=0$$
$$3(b-2)(b-10)=0$$
$$b-2=0 \quad \text{or} \quad b-10=0$$
$$b=2 \qquad b=10$$

77.

$$(d+1)(8d+1)=18d$$
$$8d^2+d+8d+1=18d$$
$$8d^2+9d+1=18d$$
$$8d^2+9d+1-18d=0$$
$$8d^2-9d+1=0$$
$$(8d-1)(d-1)=0$$
$$8d-1=0 \quad \text{or} \quad d-1=0$$
$$8d=1 \qquad d=1$$
$$d=\frac{1}{8}$$

79.

$$(x-2)(x^2-8x+7)=0$$
$$(x-2)(x-7)(x-1)=0$$
$$x-2=0, \quad x-7=0 \quad \text{or} \quad x-1=0$$
$$x=2 \qquad x=7 \qquad x=1$$

81.

$$x^3+3x^2+2x=0$$
$$x(x^2+3x+2)=0$$
$$x(x+1)(x+2)=0 \quad x+1=0, \quad \text{or} \quad x+2=0$$
$$x=0, \qquad x=-1 \qquad x=-2$$

83.

$$k^3-27k-6k^2=0$$
$$k^3-6k^2-27k=0$$
$$k(k^2-6k-27)=0$$
$$k(k-9)(k+3)=0 \quad k-9=0, \quad \text{or} \quad k+3=0$$
$$k=0, \qquad k=9 \qquad k=-3$$

85.

$$2x^3=2x(x+2)$$
$$2x^3=2x^2+4x$$
$$2x^3-2x^2-4x=0$$
$$2x(x^2-x-2)=0$$
$$2x(x+1)(x-2)=0$$
$$2x=0, \quad x+1=0, \quad \text{or} \quad x-2=0$$
$$x=0 \qquad x=-1 \qquad x=2$$

APPLICATIONS

87. OFFICIATING

height is 0 feet (hits the ground)

$$h=-16t^2+22t+3$$
$$0=-16t^2+22t+3$$
$$16t^2-22t-3=0$$
$$(8t+1)(2t-3)=0$$
$$8t+1=0 \quad \text{or} \quad 2t-3=0$$
$$8t=-1 \qquad 2t=3$$
$$t=-\frac{1}{8} \qquad \boxed{t=\frac{3}{2}}$$

Time is 1.5 sec.

89. EXHIBITION DIVING

height is 0 feet (enters the water)

$$h = -16t^2 + 64$$
$$0 = -16t^2 + 64$$
$$16t^2 - 64 = 0$$
$$16(t^2 - 4) = 0$$
$$16(t+2)(t-2) = 0$$
$$t + 2 = 0 \qquad t - 2 = 0$$
$$t = -2 \qquad \boxed{t = 2}$$

Time is 2 sec.

91. CHOREOGRAPHY

# of dancers is 36.

# of rows is unknown.

$$d = \frac{1}{2}r(r+1)$$
$$36 = \frac{1}{2}r(r+1)$$
$$2 \cdot 36 = 2 \cdot \frac{1}{2}r(r+1)$$
$$72 = r(r+1)$$
$$72 = r^2 + r$$
$$0 = r^2 + r - 72$$
$$0 = (r+9)(r-8)$$
$$r + 9 = 0 \qquad r - 8 = 0$$
$$r = -9 \qquad \boxed{r = 8}$$

# of rows is 8.

93. CUSTOMER SERVICE

x is the current number.

$x + 1$ is the next number.

Their product is 156.

$$x(x+1) = 156$$
$$x^2 + x = 156$$
$$x^2 + x - 156 = 0$$
$$(x+13)(x-12) = 0$$
$$x + 13 = 0 \qquad x - 12 = 0$$
$$x = -13 \qquad \boxed{x = 12}$$

Current # is 12.

95. PLOTTING POINTS

x is odd x-cor in QI (all are positive).

$x + 2$ is odd y-cor in QI.

Their product is 143.

$$x(x+2) = 143$$
$$x^2 + 2x = 143$$
$$x^2 + 2x - 143 = 0$$
$$(x+13)(x-11) = 0$$
$$x + 13 = 0 \qquad x - 11 = 0$$
$$x = -13 \qquad \boxed{x = 11}$$

Coordinate is (11,13).

97. INSULATION

Length is $(2w+1)$ m.

Width is w m.

Area is 36 sq. m.

$A = lw$

$$w(2w+1) = 36$$
$$2w^2 + w = 36$$
$$2w^2 + w - 36 = 0$$
$$(2w+9)(w-4) = 0$$
$$2w + 9 = 0 \qquad w - 4 = 0$$
$$2w = -9 \qquad \boxed{w = 4}$$
$$w = -\frac{9}{2}$$

Dimensions are 4 m by 9 m.

99. DESIGNING A TENT

Height is x ft.

Length(base) is $(2x+2)$ ft

Area is 30 sq. ft.

$$A=\frac{1}{2}bh$$

$$\frac{1}{2}x(2x+2)=30$$

$$2\cdot\frac{1}{2}x(2x+2)=2\cdot 30$$

$$x(2x+2)=60$$

$$2x^2+2x=60$$

$$2x^2+2x-60=0$$

$$2(x^2+x-30)=0$$

$$2(x+6)(x-5)=0$$

$$x+6=0 \qquad x-5=0$$

$$x=-6 \qquad \boxed{x=5}$$

Height is 5 ft.

Base is 12 ft.

101. WIND DAMAGE

Short leg is x ft.

Long leg is $(x+2)$ ft.

Hypotenuse is $(x+4)$ ft.

$$a^2+b^2=c^2$$

Let x = short leg (trunk) in ft

$x+2$ = long leg in ft

$x+4$ = hypotenuse (top) in ft.

$$x^2+(x+2)^2=(x+4)^2$$

$$x^2+x^2+4x+4=x^2+8x+16$$

$$2x^2+4x+4=x^2+8x+16$$

Subtract $(x^2+8x+16)$ from both sides.

$$x^2-4x-12=0$$

$$(x+2)(x-6)=0$$

$$x+2=0 \qquad x-6=0$$

$$x=-2 \qquad \boxed{x=6}$$

Trunk is 6 ft.

Tree upright is 16 ft.

103. CAR REPAIRS

Back is unknown in ft.

Base is 2 ft longer than back.

Ramp is 1 ft longer than back.

$$a^2+b^2=c^2$$

Let x = back (short leg) in ft

$x+1$ = ramp (long leg) in ft

$x+2$ = base (hypotenuse) in ft

$$x^2+(x+1)^2=(x+2)^2$$

$$x^2+x^2+2x+1=x^2+4x+4$$

$$x^2-2x-3=0$$

$$(x+1)(x-3)=0$$

$$x+1=0 \qquad x-3=0$$

$$x=-1 \qquad \boxed{x=3}$$

Back is 3 feet.

Ramp is 4 feet.

Base is 5 feet.

WRITING

105. Answers will vary.

107. Answers will vary.

REVIEW

109. EXERCISE

$15 \text{ min} \le t < 30 \text{ min}$

CHALLENGE PROBLEMS

111.

$$x^4-625=0$$

$$(x^2+25)(x^2-25)=0$$

$$x^2+25=0 \qquad x^2-25=0$$

$$x^2=-25 \qquad (x+5)(x-5)=0$$

not a real # $\qquad x+5=0 \quad x-5=0$

$$\boxed{x=-5} \qquad \boxed{x=5}$$

CHAPTER 5 KEY CONCEPTS

1. You would start with $3x + 27$.
   The answer would be $3(x + 9)$.

2. You would start with $x^2 + 12x + 27$.
   The answer would be $(x + 3)(x + 9)$.

3.

$$\begin{aligned} -3a^2 + 21a - 36 &= -3(a^2 - 7a + 12) \\ &= -3(a-3)(a-4) \end{aligned}$$

4.

$$x^2 - 121y^2 = (x + 11y)(x - 11y)$$

5.

$$\begin{aligned} rt + 2r + st + 2s &= (rt + 2r) + (st + 2s) \\ &= r(t+2) + s(t+2) \\ &= (r+s)(t+2) \end{aligned}$$

6.

$$\begin{aligned} v^3 - 8 &= (v)^3 - (2)^3 \\ &= (v-2)(v^2 + 2v + 4) \end{aligned}$$

7.

$$6t^2 - 19t + 15 = (3t - 5)(2t - 3)$$

8.

$$\begin{aligned} 25y^2 - 20y + 4 &= (5y-2)(5y-2) \\ &= (5y-2)^2 \end{aligned}$$

9.

$$\begin{aligned} 2r^3 - 50r &= 2r(r^2 - 25) \\ &= 2r(r+5)(r-5) \end{aligned}$$

10.

$$\begin{aligned} 46w - 6 + 16w^2 &= 16w^2 + 46w - 6 \\ &= 2(8w^2 + 23w - 3) \\ &= 2(8w-1)(w+3) \end{aligned}$$

CHAPTER 5 REVIEW

SECTION 5.1
The Greatest Common Factor; Factoring by Grouping

1. $5 \cdot 7$

2. $2^5 \cdot 3$

3. 7

4. $18a^3$

5. $3x+9y=3(x+3y)$

6. $5ax^2+15a=5a(x^2+3)$

7. $7s^5+14s^3=7s^3(s^2+2)$

8. $\pi ab-\pi ac=\pi a(b-c)$

9. $2x^3+4x^2-8x=2x(x^2+2x-4)$

10. $x^2y^2z+xy^3z+xy^2z=xy^2z(x+yz-1)$

11. $-5ab^2+10a^2b-15ab=-5ab(b-2a+3)$

12. $4(x-2)-x(x-2)=(x-2)(4-x)$

13. $-a-7=-(a+7)$

14. $-4t^2+3t-1=-(4t^2-3t+1)$

15.
$$\begin{aligned}2c+2d+ac+ad&=(2c+2d)+(ac+ad)\\&=2(c+d)+a(c+d)\\&=(c+d)(2+a)\end{aligned}$$

16.
$$\begin{aligned}3xy+9x-2y-6&=(3xy+9x)+(-2y-6)\\&=3x(y+3)-2(y+3)\\&=(y+3)(3x-2)\end{aligned}$$

17.
$$\begin{aligned}2a^3-a+2a^2-1&=(2a^3-a)+(2a^2-1)\\&=a(2a^2-1)+1(2a^2-1)\\&=(2a^2-1)(a+1)\end{aligned}$$

18.
$$\begin{aligned}&4m^2n+12m^2-8mn-24m\\&=4m(mn+3m-2n-6)\\&=4m[(mn+3m)-(2n+6)]\\&=4m[m(n+3)-2(n+3)]\\&=4m(n+3)(m-2)\end{aligned}$$

SECTION 5.2
Factoring Trinomials of the Form x^2+bx+c

19. 1

20.

Factors of 6	Sum of the Factors of 6
1(6)	$1+6=7$
2(3)	$2+3=5$
−1(−6)	$-1+(-6)=-7$
−2(−3)	$-2+(-3)=-5$

21.
$x^2+2x-24=(x+6)(x-4)$

22.
$x^2-4x-12=(x-6)(x+2)$

23.
$x^2-7x+10=(x-5)(x-2)$

24.
$t^2+10t+15$ is prime

25.
$$\begin{aligned}-y^2+9y-20&=-(y^2-9y+20)\\&=-(y-5)(y-4)\end{aligned}$$

26.
$$\begin{aligned}10y+9+y^2&=y^2+10y+9\\&=(y+9)(y+1)\end{aligned}$$

27.
$c^2+3cd-10d^2=(c+5d)(c-2d)$

28.
$$\begin{aligned}-mn+m^2+2n^2&=m^2-mn+2n^2\\&=(m-2n)(m-n)\end{aligned}$$

29. Multiply to see if $(x-4)(x+5) = x^2 + x - 20$.

30. There are no two integers whose product is 11 and whose sum is 7.

31.

$$5a^5 + 45a^4 - 50a^3 = 5a^3\left(a^2 + 9a - 10\right) = 5a^3(a+10)(a-1)$$

32.

$$-4x^2y - 4x^3 + 24xy^2 = -4x^3 - 4x^2y + 24xy^2 = -4x\left(x^2 + xy - 6y^2\right) = -4x(x+3y)(x-2y)$$

SECTION 5.3

Factoring Trinomials of the Form $ax^2 + bx + c$

33.

$$2x^2 - 5x - 3 = (2x+1)(x-3)$$

34.

$$10y^2 + 21y - 10 = (2y+5)(5y-2)$$

35.

$$-3x^2 + 14x + 5 = -\left(3x^2 - 14x - 5\right) = -(3x+1)(x-5)$$

36.

$$-9p^2 - 6p + 6p^3 = 6p^3 - 9p^2 - 6p = 3p\left(2p^2 - 3p - 2\right) = 3p(2p+1)(p-2)$$

37.

$$4b^2 - 17bc + 4c^2 = (4b-c)(b-4c)$$

38.

$7y^2 + 7y - 18$ is prime

39.

$12x^2 - x - 1 = (4x+1)(3x-1)$

The dimensions are $(4x+1)$ in. by $(3x-1)$ in.

40. The signs of the second terms must be negative.

SECTION 5.4

Factoring Perfect Square Trinomials and the Difference of Two Squares

41.

$$x^2 + 10x + 25 = (x+5)^2$$

42.

$$9y^2 + 16 - 24y = 9y^2 - 24y + 16 = (3y-4)^2$$

43.

$$-z^2 + 2z - 1 = -\left(z^2 - 2z + 1\right) = -(z-1)^2$$

44.

$$25a^2 + 20ab + 4b^2 = (5a+2b)^2$$

45.

$$x^2 - 9 = (x+3)(x-3)$$

46.

$$49t^2 - 25y^2 = (7t-5y)(7t+5y)$$

47.

$$x^2y^2 - 400 = (xy-20)(xy+20)$$

48.

$$8at^2 - 32a = 8a\left(t^2 - 4\right) = 8a(t+2)(t-2)$$

49.

$$c^4 - 256 = \left(c^2 + 16\right)\left(c^2 - 16\right) = \left(c^2 + 16\right)(c-4)(c+4)$$

50.

$h^2 + 36$ is prime

SECTION 5.5
Factoring the Sum and Difference of Two Cubes

51.

$$h^3+1=(h)^3+(1)^3$$
$$=(h+1)(h^2-h+1)$$

52.

$$125p^3+q^3=(5p)^3+(q)^3$$
$$=(5p+q)(25p^2-5pq+q^2)$$

53.

$$x^3-27=(x)^3-(3)^3$$
$$=(x-3)(x^2+3x+9)$$

54.

$$16x^5-54x^2y^3=2x^2(8x^3-27y^3)$$
$$=2x^2\left[(2x)^3-(3y)^3\right]$$
$$=2x^2(2x-3y)(4x^2+6xy+9y^2)$$

59.

$$12w^2-36w+27=3(4w^2-12w+9)$$
$$=3(2w-3)^2$$

60.

$121p^2+36q^2$ is prime

61.

$$2t^3+10=2(t^3+5)$$

62.

$$400-m^2=(20-m)(20+m)$$

63.

$$x^2+64y^2+16xy=x^2+16xy+64y^2$$
$$=(x+8y)^2$$

64.

$$18c^3d^2-12c^3d-24c^2d=6c^2d(3cd-2c-4)$$

SECTION 5.6
A Factoring Strategy

55.

$$14y^3+6y^4-40y^2=6y^4+14y^3-40y^2$$
$$=2y^2(3y^2+7y-20)$$
$$=2y^2(3y-5)(y+4)$$

56.

$$s^2t+s^2u^2+tv+u^2v=s^2(t+u^2)+v(t+u^2)$$
$$=(t+u^2)(s^2+v)$$

57.

$$j^4-16=(j^2+4)(j^2-4)$$
$$=(j^2+4)(j+2)(j-2)$$

58.

$$-3j^3-24k^3=-3(j^3+8k^3)$$
$$=-3\left[(j)^3+(2k)^3\right]$$
$$=-3(j+2k)(j^2-2jk+4k^2)$$

SECTION 5.7
Solving Quadratic Equations by Factoring

65.

$$x^2+2x=0$$
$$x(x+2)=0$$
$$x=0 \text{ or } x+2=0$$
$$x=-2$$

66.

$$x(x-6)=0$$
$$x=0 \text{ or } x-6=0$$
$$x=6$$

67.

$$x^2-9=0$$
$$(x-3)(x+3)=0$$
$$x-3=0 \text{ or } x+3=0$$
$$x=3 \qquad x=-3$$

68.

$$a^2 - 7a + 12 = 0$$
$$(a-4)(a-3) = 0$$
$$a - 4 = 0 \quad \text{or} \quad a - 3 = 0$$
$$a = 4 \qquad\qquad a = 3$$

69.

$$2t^2 + 8t + 8 = 0$$
$$2(t^2 + 4t + 4) = 0$$
$$2(t+2)(t+2) = 0$$
$$t + 2 = 0 \quad \text{or} \quad t + 2 = 0$$
$$t = -2 \qquad\qquad t = -2$$

70.

$$2x - x^2 = -24$$
$$0 = x^2 - 2x - 24$$
$$x^2 - 2x - 24 = 0$$
$$(x-6)(x+4) = 0$$
$$x - 6 = 0 \quad \text{or} \quad x + 4 = 0$$
$$x = 6 \qquad\qquad x = -4$$

71.

$$5a^2 - 6a + 1 = 0$$
$$(5a-1)(a-1) = 0$$
$$5a - 1 = 0 \quad \text{or} \quad a - 1 = 0$$
$$5a = 1 \qquad\qquad a = 1$$
$$a = \frac{1}{5}$$

72.

$$2p^3 = 2p(p+2)$$
$$2p^3 = 2p^2 + 4p$$
$$2p^3 - 2p^2 - 4p = 0$$
$$2p(p^2 - p - 2) = 0$$
$$2p(p-2)(p+1) = 0$$
$$2p = 0, \quad p - 2 = 0, \quad \text{or} \quad p + 1 = 0$$
$$p = 0 \qquad p = 2 \qquad p = -1$$

73.

Height is unknown.

Length (base) is 3 ft more than twice ht.

Area is 45 sq m.

$$A = \frac{1}{2}bh$$

Let x = height in ft

$2x + 3$ = length (base) in ft

$$\frac{1}{2}x(2x+3) = 45$$
$$2 \bullet \frac{1}{2}x(2x+3) = 2 \bullet 45$$
$$x(2x+3) = 90$$
$$2x^2 + 3x = 90$$
$$2x^2 + 3x - 90 = 0$$
$$(2x+15)(x-6) = 0$$
$$2x + 15 = 0 \qquad\qquad x - 6 = 0$$
$$2x = -15 \qquad\qquad \boxed{x = 6}$$
$$x = -\frac{15}{2}$$

Base is 15 m.

74.

x is the # of times for Hepburn (less).

$x + 1$ is the # of times for Streep (more).

Their product is 156.

Let x = Hepburn's #

$x + 1$ = Streep's #

$$x(x+1) = 156$$
$$x^2 + x = 156$$
$$x^2 + x - 156 = 0$$
$$(x+13)(x-12) = 0$$
$$x + 13 = 0 \qquad\qquad x + 12 = 0$$
$$x = -13 \qquad\qquad \boxed{x = 12}$$

Hepburn was nominated 12 times.

Streep was nominated 13 times.

75.

Pole is x meters tall.

Ground is $(x+7)$ meters long.

Rope is $(x+8)$ meters long.

$$a^2+b^2=c^2$$

Let $x=$ pole (short leg) in m

$x+7=$ ground (long leg) in m

$x+8=$ rope (hypotenuse) in m

$$x^2+(x+7)^2=(x+8)^2$$

$$x^2+x^2+14x+49=x^2+16x+64$$

$$x^2-2x-15=0$$

$$(x+3)(x-5)=0$$

$$x+3=0 \qquad x-5=0$$

$$x=-3 \qquad \boxed{x=5}$$

Pole is 5 meters tall.

76.

The height is 0 feet when it hits the ground.

$$h=-16t^2+1,600$$

$$0=-16t^2+1,600$$

$$16t^2-1,600=0$$

$$16t(t^2-100)=0$$

$$16t(t-10)(t+10)=0$$

$$t=0 \qquad t-10=0 \qquad t+10=0$$

$$\boxed{t=0} \qquad \boxed{t=10} \qquad \boxed{t=-10}$$

Time is 10 sec.

CHAPTER 5 TEST

1. $196 = 2 \cdot 2 \cdot 7 \cdot 7$
   $= 2^2 \cdot 7^2$

2. $111 = 3 \cdot 37$

3. $45x^4y^6 = 3 \cdot 3 \cdot 5 \cdot x \cdot x \cdot x \cdot x \cdot y \cdot y \cdot y \cdot y \cdot y \cdot y$
   $30x^3y^8 = 2 \cdot 3 \cdot 5 \cdot x \cdot x \cdot x \cdot y \cdot y \cdot y \cdot y \cdot y \cdot y \cdot y \cdot y$
   $\text{GCF} = 3 \cdot 5 \cdot x \cdot x \cdot x \cdot y \cdot y \cdot y \cdot y \cdot y \cdot y$
   $= 15x^3y^6$

4.
$$4x + 16 = 4(x+4)$$

5.
$$30a^2b^3 - 20a^3b^2 + 5abc = 5abc\left(6ab^2 - 4a^2b + 1\right)$$

6.
$$q^2 - 81 = (q+9)(q-9)$$

7.
$x^2 + 9$ is prime

8.
$$16x^4 - 81 = \left(4x^2 + 9\right)\left(4x^2 - 9\right)$$
$$= \left(4x^2 + 9\right)(2x+3)(2x-3)$$

9.
$$x^2 + 4x + 3 = (x+3)(x+1)$$

10.
$$-x^2 + 9x + 22 = -\left(x^2 - 9x - 22\right)$$
$$= -(x+2)(x-11)$$

11.
$$9a - 9b + ax - bx = 9(a-b) + x(a-b)$$
$$= (a-b)(9+x)$$

12.
$$2a^2 + 5a - 12 = (2a-3)(a+4)$$

13.
$$18x^2 - 60xy + 50y^2 = 2\left(9x^2 - 30xy + 25y^2\right)$$
$$= 2(3x - 5y)^2$$

14.
$$x^3 + 8 = (x)^3 + (2)^3$$
$$= (x+2)\left(x^2 - 2x + 4\right)$$

15.
$$20m^8 - 15m^6 = 5m^6\left(4m^2 - 3\right)$$

16.
$$3a^3 - 81 = 3\left(a^3 - 27\right)$$
$$= 3\left[(a)^3 - (3)^3\right]$$
$$= 3(a-3)\left(a^2 + 3a + 9\right)$$

17.
$$4r^2 - \pi r^2 = r^2(4 - \pi)$$

18.
$$25x^2 - 40x + 16 = (5x-4)^2$$
Each side is $5x - 4$.

19.
$$x^2 - 3x - 54 = (x-9)(x+6)$$

Check:
$$(x-9)(x+6) = x^2 + 6x - 9x - 54$$
$$= x^2 - 3x - 54$$

20.
$$(x+3)(x-2) = 0$$
$x + 3 = 0$ or $x - 2 = 0$
$x = -3$ $\qquad x = 2$

21.
$$x^2 - 25 = 0$$
$$(x-5)(x+5) = 0$$
$x - 5 = 0$ or $x + 5 = 0$
$x = 5$ $\qquad x = -5$

22.
$$36x^2 - 6x = 0$$
$$6x(6x-1) = 0$$
$6x = 0$ or $6x - 1 = 0$
$x = 0$ $\qquad x = \frac{1}{6}$

23.

$$x(x+6)=-9$$
$$x^2+6x=-9$$
$$x^2+6x+9=0$$
$$(x+3)(x+3)=0$$
$$x+3=0 \quad \text{or} \quad x+3=0$$
$$x=-3 \qquad x=-3$$

24.

$$6x^2+x-1=0$$
$$(2x+1)(3x-1)=0$$
$$2x+1=0 \quad \text{or} \quad 3x-1=0$$
$$2x=-1 \qquad 3x=1$$
$$x=-\frac{1}{2} \qquad x=\frac{1}{3}$$

25.

$$(a-2)(a-5)=28$$
$$a^2-5a-2a+10=28$$
$$a^2-7a+10=28$$
$$a^2-7a-18=0$$
$$(a-9)(a+2)=0$$
$$a-9=0 \quad \text{or} \quad a+2=0$$
$$a=9 \qquad a=-2$$

26.

$$x^3+7x^2=-6x$$
$$x^3+7x^2+6x=0$$
$$x(x^2+7x+6)=0$$
$$x(x+1)(x+6)=0$$
$$x=0, \quad x+1=0, \quad \text{or} \quad x+6=0$$
$$x=-1 \qquad x=-6$$

27.

Length is $(w+3)$ feet.

Width is w feet.

Area is 54 sq. ft.

$A=lw$

Let $w=$ width in ft

$w+3=$ length in ft

$$w(w+3)=54$$
$$w^2+3w=54$$
$$w^2+3w-54=0$$
$$(w+9)(w-6)=0$$
$$w+9=0 \qquad w-6=0$$
$$w=-9 \qquad \boxed{w=6}$$

Dimensions are 6 ft by 9 ft.

28. A quadratic equation is an equation that can be written in the form $ax^2+bx+c=0$. An example of a quadratic equation is $x^2+2x+1=0$. (Answers will vary.)

29.

Short leg is $(x-4)$.

Long leg is $(x-2)$.

Hypotenuse is x.

$a^2+b^2=c^2$

Let $x-4=$ short leg

$x-2=$ long leg

$x=$ hypotenuse

$$(x-4)^2+(x-2)^2=x^2$$
$$x^2-8x+16+x^2-4x+4=x^2$$
$$2x^2-12x+20=x^2$$
$$2x^2-12x+20-x^2=x^2-x^2$$
$$x^2-12x+20=0$$
$$(x-2)(x-10)=0$$
$$x-2=0 \qquad x-10=0$$
$$x=2 \qquad \boxed{x=10}$$

Hypotenuse is 10.

30. At least one of the numbers is 0.

SECTION 6.1

VOCABULARY

1. A quotient of two polynomials, such as $\frac{x^2+x}{x^2-3x}$, is called a **rational** expression.

3. Because of the division by 0, the expression $\frac{8}{0}$ is **undefined**.

5. To **simplify** a rational expression, we remove factors common to the numerator and denominator.

CONCEPTS

7. To find what value of x makes each rational expression undefined, set each denominator equal to zero and solve for x.

   a) $x = 0$
   b) $x - 6 = 0$
   $x - 6 + 6 = 0 + 6$
   $x = 6$
   c) $x + 6 = 0$
   $x + 6 - 6 = 0 - 6$
   $x = -6$

9. a) $x + 1$
   b)To simplify the rational expression, we replace $\frac{x+1}{x+1}$ with the equivalent fraction $\frac{1}{1}$. This removes the factor $\frac{x+1}{x+1}$, which is equal to 1.

11. a) The factors $x - 2$ and $2 - x$ are opposites.
    b) To simplify the rational expression, we replace $\frac{x-2}{2-x}$ with the equivalent fraction $\frac{-1}{1}$. This removes the factor $\frac{x-2}{2-x}$, which is equal to -1.

13. a) $\frac{\cancel{(x+2)}(x-2)}{(x+1)\cancel{(x+2)}} = \frac{x-2}{x+1}$

    b) $\frac{y(y-2)}{9(2-y)} = \frac{-y\cancel{(2-y)}}{9\cancel{(2-y)}}$
    $= -\frac{y}{9}$

    c) $\frac{\cancel{(2m+7)}(m-5)}{\cancel{(2m+7)}} = \frac{m-5}{1}$
    $= m - 5$

    d) $\frac{\cancel{x} \bullet \cancel{x}}{\cancel{x} \bullet \cancel{x}(x-30)} = \frac{1}{x-30}$

NOTATION

15. Yes, all three answers are equivalent.

PRACTICE

17. $\frac{x-2}{x-5} = \frac{6-2}{6-5}$
    $= \frac{4}{1}$
    $= 4$

19. $\frac{x^2-4x-12}{x^2+x-2} = \frac{6^2-4(6)-12}{6^2+6-2}$
    $= \frac{36-24-12}{36+6-2}$
    $= \frac{0}{40}$
    $= 0$

21. $\frac{y+5}{3y-2} = \frac{-3+5}{3(-3)-2}$
    $= \frac{2}{-9-2}$
    $= \frac{2}{-11}$
    $= -\frac{2}{11}$

23. $$\frac{y^3}{3y+1} = \frac{(-3)^3}{3(-3)^2+1}$$
$$= \frac{-27}{3(9)+1}$$
$$= \frac{-27}{27+1}$$
$$= -\frac{27}{28}$$

25. $$8x = 0$$
$$\frac{8x}{8} = \frac{0}{8}$$
$$\boxed{x = 0}$$

27. $$x - 2 = 0$$
$$x - 2 + 2 = 0 + 2$$
$$\boxed{x = 2}$$

29. No matter what real number is substituted for x, $x^2 + 6$ will not be 0. Thus, **no real numbers** will make the rational expression undefined.

31. $$2x - 1 = 0$$
$$2x - 1 + 1 = 0 + 1$$
$$2x = 1$$
$$\frac{2x}{2} = \frac{1}{2}$$
$$\boxed{x = \frac{1}{2}}$$

33. $$x^2 - 36 = 0$$
$$x^2 - 36 + 36 = 0 + 36$$
$$x^2 = 36$$
$$\boxed{x = 6 \text{ and } x = -6}$$

Both 6^2 and $(-6)^2 = 36$

35. $$x^2 + x - 2 = 0$$
$$(x + 2)(x - 1) = 0$$
$$x + 2 = 0 \quad \text{and} \quad x - 1 = 0$$
$$x + 2 - 2 = 0 - 2 \quad x - 1 + 1 = 0 + 1$$
$$\boxed{x = -2 \quad \text{and} \quad x = 1}$$

37. $$\frac{45}{9a} = \frac{5 \cdot \cancel{3} \cdot \cancel{3}}{\cancel{3} \cdot \cancel{3} \cdot a}$$
$$= \frac{5}{a}$$

39. $$\frac{6x^2}{4x^2} = \frac{\cancel{2} \cdot 3 \cdot \cancel{x} \cdot \cancel{x}}{\cancel{2} \cdot 2 \cdot \cancel{x} \cdot \cancel{x}}$$
$$= \frac{3}{2}$$

41. $$\frac{2x^2}{x+2} = \text{does not simplify}$$

43. $$\frac{15x^2y}{5xy^2} = \frac{\cancel{5} \cdot 3 \cdot \cancel{x} \cdot x \cdot \cancel{y}}{\cancel{5} \cdot \cancel{x} \cdot \cancel{y} \cdot y}$$
$$= \frac{3x}{y}$$

45. $$\frac{6x+3}{3y} = \frac{\cancel{3}(2x+1)}{\cancel{3}y}$$
$$= \frac{2x+1}{y}$$

47. $$\frac{x+3}{3x+9} = \frac{\cancel{x+3}}{3(\cancel{x+3})}$$
$$= \frac{1}{3}$$

49. $$\frac{a^3 - a^2}{a^4 - a^3} = \frac{a^2(a-1)}{a^3(a-1)}$$
$$= \frac{\cancel{a} \cdot \cancel{a}(\cancel{a-1})}{\cancel{a} \cdot \cancel{a} \cdot a(\cancel{a-1})}$$
$$= \frac{1}{a}$$

51. $$\frac{4x+16}{5x+20} = \frac{4(\cancel{x+4})}{5(\cancel{x+4})}$$
$$= \frac{4}{5}$$

53. $$\frac{4c+4d}{d+c}=\frac{4(c+d)}{d+c}=\frac{4\cancel{(c+d)}}{\cancel{c+d}}=\frac{4}{1}=4$$

55. $$\frac{x-7}{7-x}=\frac{\cancel{x-7}}{-1\cancel{(x-7)}}=\frac{1}{-1}=-1$$

57. $$\frac{6x-30}{5-x}=\frac{6(x-5)}{5-x}=\frac{6\cancel{(x-5)}}{-1\cancel{(x-5)}}=\frac{6}{-1}=-6$$

59. $$\frac{x^2-4}{x^2-x-2}=\frac{\cancel{(x-2)}(x+2)}{\cancel{(x-2)}(x+1)}=\frac{x+2}{x+1}$$

61. $$\frac{x^2+3x+2}{x^2+x-2}=\frac{\cancel{(x+2)}(x+1)}{\cancel{(x+2)}(x-1)}=\frac{x+1}{x-1}$$

63. $$\frac{2x^2-8x}{x^2-6x+8}=\frac{2x\cancel{(x-4)}}{(x-2)\cancel{(x-4)}}=\frac{2x}{x-2}$$

65. $$\frac{2-a}{a^2-a-2}=\frac{2-a}{(a-2)(a+1)}=\frac{-1\cancel{(a-2)}}{\cancel{(a-2)}(a+1)}=-\frac{1}{a+1}$$

67. $$\frac{b^2+2b+1}{(b+1)^3}=\frac{\cancel{(b+1)}\,\cancel{(b+1)}}{\cancel{(b+1)}\,\cancel{(b+1)}\,(b+1)}=\frac{1}{b+1}$$

69. $$\frac{6x^2-13x+6}{3x^2+x-2}=\frac{(2x-3)\cancel{(3x-2)}}{\cancel{(3x-2)}(x+1)}=\frac{2x-3}{x+1}$$

71. $$\frac{10(c-3)+10}{3(c-3)+3}=\frac{10c-30+10}{3c-9+3}=\frac{10c-20}{3c-6}=\frac{10\cancel{(c-2)}}{3\cancel{(c-2)}}=\frac{10}{3}$$

73. $$\frac{6a+3(a+2)+12}{a+2}=\frac{6a+3a+6+12}{a+2}=\frac{9a+18}{a+2}=\frac{9\cancel{(a+2)}}{\cancel{a+2}}=\frac{9}{1}=9$$

75. $$\frac{3x^2-27}{2x^2-5x-3}=\frac{3(x^2-9)}{(2x+1)(x-3)}=\frac{3(x+3)\cancel{(x-3)}}{(2x+1)\cancel{(x-3)}}=\frac{3(x+3)}{2x+1}$$

77.

$$\frac{-x^2-4x+77}{x^2-4x-21}=\frac{-(x^2+4x-77)}{(x-7)(x+3)}$$
$$=-\frac{(x+11)\cancel{(x-7)}}{\cancel{(x-7)}(x+3)}$$
$$=-\frac{x+11}{x+3}$$

79.

$$\frac{x(x-8)+16}{16-x^2}=\frac{x^2-8x+16}{(4-x)(4+x)}$$
$$=\frac{(x-4)(x-4)}{(4+x)(4-x)}$$
$$=-\frac{(x-4)\cancel{(x-4)}}{(4+x)\cancel{(x-4)}}$$
$$=-\frac{x-4}{x+4}$$
$$\text{or } \frac{4-x}{4+x}$$

81.

$$\frac{m^2-2mn+n^2}{2m^2-2n^2}=\frac{m^2-2mn+n^2}{2(m^2-n^2)}$$
$$=\frac{(m-n)\cancel{(m-n)}}{2(m+n)\cancel{(m-n)}}$$
$$=\frac{m-n}{2(m+n)}$$

83.

$$\frac{16a^2-1}{4a+4}=\frac{(4a-1)(4a+1)}{4(a+1)}$$
$$=\text{does not simplify}$$

85.

$$\frac{8u^2-2u-15}{4u^4+5u^3}=\frac{(2u-3)\cancel{(4u+5)}}{u^3\cancel{(4u+5)}}$$
$$=\frac{2u-3}{u^3}$$

87.

$$\frac{y-xy}{xy-x}=\frac{y(1-x)}{x(y-1)}$$
$$=\text{does not simplify}$$

89.

$$\frac{6a-6b+6c}{9a-9b+9c}=\frac{6\cancel{(a-b+c)}}{9\cancel{(a-b+c)}}$$
$$=\frac{6}{9}$$
$$=\frac{2}{3}$$

91.

$$\frac{15x-3x^2}{25y-5xy}=\frac{3x\cancel{(5-x)}}{5y\cancel{(5-x)}}$$
$$=\frac{3x}{5y}$$

APPLICATIONS

93. ROOFING

$$\frac{rise}{run}=\frac{x^2+4x+4}{x^2-4}$$
$$=\frac{\cancel{(x+2)}(x+2)}{\cancel{(x+2)}(x-2)}$$
$$=\frac{x+2}{x-2}$$

95. ORGAN PIPES

Let $L=6$

$$n=\frac{512}{\mathbf{6}}$$
$$=\frac{\cancel{2}\cdot 2\cdot 2\cdot 2\cdot 2\cdot 2\cdot 2\cdot 2\cdot 2}{\cancel{2}\cdot 3}$$
$$=\frac{256}{3}$$
$$=85\frac{1}{3}$$

97. MEDICAL SCHOOL

For 1:00, let $c = 1$.

$$c = \frac{4(1)}{(1)^2 + 1} = \frac{4}{1+1} = \frac{4}{2} = 2 \text{ mg per liter}$$

For 2:00, let $c = 2$.

$$c = \frac{4(2)}{(2)^2 + 1} = \frac{8}{4+1} = \frac{8}{5} = 1.6 \text{ mg per liter}$$

For 3:00, let $c = 3$.

$$c = \frac{4(3)}{(3)^2 + 1} = \frac{12}{9+1} = \frac{12}{10} = 1.2 \text{ mg per liter}$$

WRITING

99. Answers will vary.

101. Answers will vary.

REVIEW

103. $(a + b) + c = a + (b + c)$

105. At least one of them is zero.

CHALLENGE PROBLEMS

107. $$\frac{(x^2+2x+1)(x^2-2x+1)}{(x^2-1)^2} = \frac{[(x+1)(x+1)][(x-1)(x-1)]}{(x^2-1)(x^2-1)} = \frac{\cancel{(x+1)}\,\cancel{(x+1)}\,\cancel{(x-1)}\,\cancel{(x-1)}}{\cancel{(x+1)}\,\cancel{(x-1)}\,\cancel{(x+1)}\,\cancel{(x-1)}} = 1$$

109. $$\frac{x^3-27}{x^3-9x} = \frac{(x-3)(x^2+3x+9)}{x(x^2-9)} = \frac{\cancel{(x-3)}(x^2+3x+9)}{x\,\cancel{(x-3)}(x+3)} = \frac{x^2+3x+9}{x(x+3)}$$

SECTION 6.2

VOCABULARY

1. A quotient of **opposites** is -1. For example, $\frac{x-8}{8-x} = -1$.

3. To find the reciprocal of a rational expression, we **invert** its numerator and denominator.

CONCEPTS

5. To multiply rational expressions, multiply their **numerators** and multiply their **denominators**. In symbols,

$$\frac{A}{B} \bullet \frac{C}{D} = \boxed{\frac{AC}{BD}}$$

7. $\frac{\cancel{(x+7)} \bullet 2 \bullet \cancel{5}}{\cancel{5}(x+1)\cancel{(x+7)}(x-9)} = \frac{2}{(x+1)(x-9)}$

9. $6n = \frac{6n}{1}$

11. $\frac{1}{18x}$

13. $\frac{x^2+2x+1}{x^2+1}$

15. feet

NOTATION

17. a) m^2
    b) m^3

PRACTICE

19. $\frac{3}{y} \cdot \frac{y}{2} = \frac{3y}{2y}$

$$= \frac{3}{2}$$

21.

$$\frac{35n}{12} \cdot \frac{16}{7n^2} = \frac{35n \cdot 16}{12 \cdot 7n^2}$$
$$= \frac{5 \cdot \cancel{7} \cdot \cancel{n} \cdot \cancel{2} \cdot \cancel{2} \cdot 2 \cdot 2}{3 \cdot \cancel{2} \cdot \cancel{2} \cdot \cancel{7} \cdot \cancel{n} \cdot n}$$
$$= \frac{20}{3n}$$

23.

$$\frac{2x^2y}{3xy} \cdot \frac{3xy^2}{2} = \frac{2x^2y \cdot 3xy^2}{3xy \cdot 2}$$
$$= \frac{\cancel{2} \cdot \cancel{x} \cdot x \cdot \cancel{y} \cdot \cancel{3} \cdot x \cdot y \cdot y}{\cancel{3} \cdot \cancel{x} \cdot \cancel{y} \cdot \cancel{2}}$$
$$= x^2y^2$$

25.

$$\frac{10r^2st^3}{6rs^2} \cdot \frac{3r^3t}{2rst} = \frac{10r^2st^3 \cdot 3r^3t}{6rs^2 \cdot 2rst}$$
$$= \frac{\cancel{2} \cdot 5 \cdot \cancel{r} \cdot \cancel{r} \cdot \cancel{s} \cdot t \cdot t \cdot t \cdot \cancel{3} \cdot r \cdot r \cdot r \cdot \cancel{t}}{\cancel{2} \cdot \cancel{3} \cdot \cancel{r} \cdot \cancel{s} \cdot s \cdot 2 \cdot \cancel{r} \cdot s \cdot \cancel{t}}$$
$$= \frac{5r^3t^3}{2s^2}$$

27.

$$\frac{z+7}{7} \cdot \frac{z+2}{z} = \frac{(z+7)(z+2)}{7z}$$

29.

$$\frac{x+5}{5} \cdot \frac{x}{x+5} = \frac{\cancel{(x+5)}x}{5\cancel{(x+5)}}$$
$$= \frac{x}{5}$$

31.

$$\frac{x-2}{2} \cdot \frac{2x}{2-x} = \frac{(x-2) \cdot 2x}{2(2-x)}$$
$$= -\frac{\cancel{(x-2)} \cdot \cancel{2}x}{\cancel{2}\cancel{(x-2)}}$$
$$= -x$$

33.

$$\frac{5}{m} \cdot m = \frac{5}{m} \cdot \frac{m}{1}$$
$$= \frac{5\cancel{m}}{\cancel{m}}$$
$$= 5$$

35.

$$15x\left(\frac{x+1}{15x}\right)=\frac{15x}{1}\left(\frac{x+1}{15x}\right)$$
$$=\frac{\cancel{15x}(x+1)}{\cancel{15x}}$$
$$=x+1$$

37.

$$12y\left(\frac{y+8}{6y}\right)=\frac{12y}{1}\left(\frac{y+8}{6y}\right)$$
$$=\frac{2\cdot\cancel{2}\cdot\cancel{3y}(y+8)}{\cancel{2}\cdot\cancel{3y}}$$
$$=2(y+8)$$
$$=2y+16$$

39.

$$x+8\left(\frac{x+5}{x+8}\right)=\frac{x+8}{1}\left(\frac{x+5}{x+8}\right)$$
$$=\frac{\cancel{(x+8)}(x+5)}{\cancel{(x+8)}}$$
$$=x+5$$

41.

$$10(h+9)\frac{h-3}{h+9}=\frac{10\cancel{(h+9)}(h-3)}{\cancel{h+9}}$$
$$=10(h-3)$$
$$=10h-30$$

43.

$$\frac{(x+1)^2}{x+1}\cdot\frac{x+2}{x+1}=\frac{\cancel{(x+1)}\cancel{(x+1)}(x+2)}{\cancel{(x+1)}\cancel{(x+1)}}$$
$$=x+2$$

45.

$$\frac{2x+6}{x+3}\cdot\frac{3}{4x}=\frac{2(x+3)}{x+3}\cdot\frac{3}{4x}$$
$$=\frac{\cancel{2}\cancel{(x+3)}\cdot 3}{\cancel{(x+3)}\cdot\cancel{2}\cdot 2x}$$
$$=\frac{3}{2x}$$

47.

$$\frac{x^2-x}{x}\cdot\frac{3x-6}{3-3x}=\frac{x(x-1)}{x}\cdot\frac{3(x-2)}{3(1-x)}$$
$$=\frac{\cancel{3}\cdot\cancel{x}\cancel{(x-1)}(x-2)}{-\cancel{3}\cdot\cancel{x}\cancel{(x-1)}}$$
$$=\frac{x-2}{-1}$$
$$=-(x-2)$$

49.

$$\frac{x^2+x-6}{5x}\cdot\frac{5x-10}{x+3}=\frac{(x+3)(x-2)}{5x}\cdot\frac{5(x-2)}{x+3}$$
$$=\frac{\cancel{5}\cancel{(x+3)}(x-2)(x-2)}{\cancel{5}x\cancel{(x+3)}}$$
$$=\frac{(x-2)^2}{x}$$

51.

$$\frac{m^2-2m-3}{2m+4}\cdot\frac{m^2-4}{m^2+3m+2}$$
$$=\frac{(m-3)(m+1)}{2(m+2)}\cdot\frac{(m-2)(m+2)}{(m+2)(m+1)}$$
$$=\frac{(m-3)\cancel{(m+1)}(m-2)\cancel{(m+2)}}{2\cancel{(m+2)}(m+2)\cancel{(m+1)}}$$
$$=\frac{(m-2)(m-3)}{2(m+2)}$$

53.

$$\frac{2x^2+17x+21}{x^2+2x-35}\cdot\frac{x^2-25}{2x^2-7x-15}$$
$$=\frac{(2x+3)(x+7)}{(x-5)(x+7)}\cdot\frac{(x-5)(x+5)}{(2x+3)(x-5)}$$
$$=\frac{\cancel{(2x+3)}\cancel{(x+7)}\cancel{(x-5)}(x+5)}{\cancel{(x-5)}\cancel{(x+7)}\cancel{(2x+3)}(x-5)}$$
$$=\frac{x+5}{x-5}$$

55.

$$\frac{4x^2-12xy+9y^2}{x^3y^2}\cdot\frac{x^2y}{9y^2-4x^2}$$
$$=\frac{(2x-3y)(2x-3y)}{x\cdot x\cdot x\cdot y\cdot y}\cdot\frac{x\cdot x\cdot y}{(3y+2x)(3y-2x)}$$
$$=\frac{x\cdot x\cdot y(2x-3y)(2x-3y)}{x\cdot x\cdot x\cdot y\cdot y(3y+2x)(3y-2x)}$$
$$=-\frac{\cancel{x}\cdot\cancel{x}\cdot\cancel{y}(2x-3y)\cancel{(2x-3y)}}{\cancel{x}\cdot\cancel{x}\cdot x\cdot\cancel{y}\cdot y(3y+2x)\cancel{(2x-3y)}}$$
$$=-\frac{2x-3y}{xy(3y+2x)}$$

57.

$$\frac{3x^2+5x+2}{x^2-9}\cdot\frac{x-3}{x^2-4}\cdot\frac{x^2+5x+6}{6x+4}$$
$$=\frac{(3x+2)(x+1)}{(x+3)(x-3)}\cdot\frac{x-3}{(x-2)(x+2)}\cdot\frac{(x+2)(x+3)}{2(3x+2)}$$
$$=\frac{\cancel{(3x+2)}(x+1)\cancel{(x-3)}\cancel{(x+2)}\cancel{(x+3)}}{2\cancel{(x+3)}\cancel{(x-3)}(x-2)\cancel{(x+2)}\cancel{(3x+2)}}$$
$$=\frac{x+1}{2(x-2)}$$

59.

$$\frac{2}{y}\div\frac{4}{3}=\frac{2}{y}\cdot\frac{3}{4}$$
$$=\frac{6}{4y}$$
$$=\frac{\cancel{2}\cdot 3}{\cancel{2}\cdot 2\cdot y}$$
$$=\frac{3}{2y}$$

61.

$$\frac{3x}{y}\div\frac{2x}{4}=\frac{3x}{y}\cdot\frac{4}{2x}$$
$$=\frac{12x}{2xy}$$
$$=\frac{\cancel{2}\cdot 2\cdot 3\cdot\cancel{x}}{\cancel{2}\cdot\cancel{x}\cdot y}$$
$$=\frac{6}{y}$$

63.

$$\frac{x^2y}{3xy}\div\frac{xy^2}{6y}=\frac{x^2y}{3xy}\cdot\frac{6y}{xy^2}$$
$$=\frac{6x^2y^2}{3x^2y^3}$$
$$=\frac{2\cdot\cancel{3}\cdot\cancel{x}\cdot\cancel{x}\cdot\cancel{y}\cdot\cancel{y}}{\cancel{3}\cdot\cancel{x}\cdot\cancel{x}\cdot y\cdot\cancel{y}\cdot\cancel{y}}$$
$$=\frac{2}{y}$$

65.

$$24n^2\div\frac{18n^3}{n-1}=\frac{24n^2}{1}\cdot\frac{n-1}{18n^3}$$
$$=\frac{24n^2(n-1)}{18n^3}$$
$$=\frac{\cancel{2}\cdot 2\cdot 2\cdot\cancel{3}\cdot\cancel{n}\cdot\cancel{n}(n-1)}{\cancel{2}\cdot\cancel{3}\cdot 3\cdot\cancel{n}\cdot\cancel{n}\cdot n}$$
$$=\frac{4(n-1)}{3n}$$

67.

$$\frac{x+2}{3x}\div\frac{x+2}{2}=\frac{x+2}{3x}\cdot\frac{2}{x+2}$$
$$=\frac{2\cancel{(x+2)}}{3x\cancel{(x+2)}}$$
$$=\frac{2}{3x}$$

69.

$$\frac{(z-2)^2}{3z^2}\div\frac{z-2}{6z}=\frac{(z-2)^2}{3z^2}\cdot\frac{6z}{z-2}$$
$$=\frac{6z(z-2)^2}{3z^2(z-2)}$$
$$=\frac{2\cdot\cancel{3}\cancel{z}\cancel{(z-2)}(z-2)}{\cancel{3}\cancel{z}\cdot z\cancel{(z-2)}}$$
$$=\frac{2(z-2)}{z}$$

71.

$$\frac{9a-18}{28}\div\frac{9a^3}{35}=\frac{9a-18}{28}\cdot\frac{35}{9a^3}$$
$$=\frac{\cancel{3}\cdot\cancel{3}(a-2)\cdot 5\cdot\cancel{7}}{2\cdot 2\cdot\cancel{7}\cdot\cancel{3}\cdot\cancel{3}\cdot a\cdot a\cdot a}$$
$$=\frac{5(a-2)}{4a^3}$$

73.

$$\frac{x^2-4}{3x+6} \div \frac{2-x}{x+2} = \frac{x^2-4}{3x+6} \cdot \frac{x+2}{2-x}$$
$$= \frac{(x-2)(x+2)(x+2)}{3(x+2)(2-x)}$$
$$= -\frac{(x-2)(x+2)(x+2)}{3(x+2)(x-2)}$$
$$= -\frac{x+2}{3}$$

75.

$$\frac{x^2-1}{3x-3} \div (x+1) = \frac{x^2-1}{3x-3} \cdot \frac{1}{x+1}$$
$$= \frac{(x-1)(x+1)}{3(x-1)(x+1)}$$
$$= \frac{1}{3}$$

77.

$$\frac{x^2-2x-35}{3x^2+27x} \div \frac{x^2+7x+10}{6x^2+12x}$$
$$= \frac{x^2-2x-35}{3x^2+27x} \cdot \frac{6x^2+12x}{x^2+7x+10}$$
$$= \frac{(x-7)(x+5)2\cdot 3x(x+2)}{3x(x+9)(x+2)(x+5)}$$
$$= \frac{2(x-7)}{x+9}$$

79.

$$\frac{36c^2-49d^2}{3d^3} \div \frac{12c+14d}{d^4}$$
$$= \frac{36c^2-49d^2}{3d^3} \cdot \frac{d^4}{12c+14d}$$
$$= \frac{(6c+7d)(6c-7d)\cdot d\cdot d\cdot d\cdot d}{3\cdot d\cdot d\cdot d\cdot 2(6c+7d)}$$
$$= \frac{d(6c-7d)}{6}$$

81.

$$\frac{2d^2+8d-42}{3-d} \div \frac{2d^2+14d}{d^2+5d}$$
$$= \frac{2d^2+8d-42}{3-d} \cdot \frac{d^2+5d}{2d^2+14d}$$
$$= \frac{2(d^2+4d-21)}{3-d} \cdot \frac{d(d+5)}{2d(d+7)}$$
$$= \frac{2(d+7)(d-3)d(d+5)}{(3-d)\cdot 2d(d+7)}$$
$$= -\frac{2(d+7)(d-3)d(d+5)}{(d-3)\cdot 2\cdot d(d+7)}$$
$$= -(d+5)$$

83.

$$\frac{2r-3s}{12} \div \left(4r^2-12rs+9s^2\right)$$
$$= \frac{2r-3s}{12} \cdot \frac{1}{\left(4r^2-12rs+9s^2\right)}$$
$$= \frac{(2r-3s)}{2\cdot 2\cdot 3(2r-3s)(2r-3s)}$$
$$= \frac{1}{12(2r-3s)}$$

85. $\frac{150 \text{ yd}}{1} \cdot \frac{3 \text{ ft}}{1 \text{ yd}} = 450 \text{ ft}$

87. $\frac{30 \text{ meters}}{1 \text{ sec}} \cdot \frac{60 \text{ sec}}{1 \text{ min}} = 1{,}800$ meters per min.

APPLICATIONS

89. INTERNATIONAL ALPHABET

Multiply the number of squares (6) times the area of one square:

$$6\left(\frac{2x+1}{2}\right)\left(\frac{2x+1}{2}\right)=\frac{2\cdot 3(2x+1)(2x+1)}{2\cdot 2}$$
$$=\frac{3(2x+1)(2x+1)}{2}$$
$$=\frac{3\left((2x)^2+2(2x)(1)+1^2\right)}{2}$$
$$=\frac{3\left(4x^2+4x+1\right)}{2}$$
$$=\frac{12x^2+12x+3}{2}\text{ in.}^2$$

91. TALKING

There are 365 days in 1 year.

$$\frac{12{,}000}{1\text{ day}}\cdot\frac{365\text{ days}}{1\text{ year}}=4{,}380{,}000\text{ words per year}$$

93. NATURAL LIGHT

9 square feet = 1 square yard

$$\frac{72\text{ ft}^2}{1}\cdot\frac{1\text{ yd}^2}{9\text{ ft}^2}=\frac{72}{9}$$
$$=8\text{ yd}^2$$

95. BEARS

60 minutes = 1 hour

$$\frac{30\text{ miles}}{1\text{ hour}}\cdot\frac{1\text{ hour}}{60\text{ minutes}}=\frac{30}{60}$$
$$=\frac{1}{2}\text{ mile per minute}$$

97. TV TRIVIA

1 square mile = 640 acres

$$\frac{160\text{ acres}}{1}\cdot\frac{1\text{ mi}^2}{640\text{ acres}}=\frac{160}{640}$$
$$=\frac{1}{4}\text{ mi}^2$$

WRITING

99. Answers will vary.

101. Answers will vary.

103. Answers will vary.

REVIEW

105. HARDWARE

Let the width = x.
Then the length = $2x-2$.
The brace, wall, and shelf form a right triangle, so use the Pythagorean Theorem to solve the equation.

$$8^2+x^2=(2x-2)^2$$
$$64+x^2=(2x)^2-2(2x)(2)+(-2)^2$$

$$64+x^2=4x^2-8x+4$$
$$64+x^2-64-x^2=4x^2-8x+4-64-x^2$$
$$0=3x^2-8x-60$$
$$0=(3x+10)(x-6)$$

$3x+10=0$ and $x-6=0$

$3x+10-10=0-10$ $\quad x-6+6=0+6$

$3x=-10$ $\quad x=6$

$x=-\frac{10}{3}$

Since the width cannot be negative, the only valid answer for x is 6.

width = x
= 6 in.

length = $2x-2$
= $2(6)-2$
= $12-2$
= 10 in.

CHALLENGE PROBLEMS

107.

$$\frac{c^3-2c^2+5c-10}{c^2-c-2}\cdot\frac{c^3+c^2-5c-5}{c^4-25}$$

$$=\frac{c^2(c-2)+5(c-2)}{(c-2)(c+1)}\cdot\frac{c^2(c+1)-5(c+1)}{(c^2-5)(c^2+5)}$$

$$=\frac{\cancel{(c^2+5)}\,\cancel{(c-2)}\,\cancel{(c^2-5)}\,\cancel{(c+1)}}{\cancel{(c-2)}\,\cancel{(c+1)}\,\cancel{(c^2-5)}\,\cancel{(c^2+5)}}$$

$$=1$$

109.

$$\frac{y^2}{x+1}\cdot\frac{x^2+2x+1}{x^2-1}\div\frac{3y}{xy-y}$$

$$=\frac{y^2}{x+1}\cdot\frac{x^2+2x+1}{x^2-1}\cdot\frac{xy-y}{3y}$$

$$=\frac{y\cdot y\cdot\cancel{(x+1)}\,\cancel{(x+1)}\cdot\cancel{y}\,\cancel{(x-1)}}{\cancel{(x+1)}\,\cancel{(x-1)}\,\cancel{(x+1)}3\cancel{y}}$$

$$=\frac{y^2}{3}$$

SECTION 6.3

VOCABULARY

1. The rational expressions $\frac{7}{6n}$ and $\frac{n+1}{6n}$ have a **common** denominator of $6n$.

3. The **least common denominator** of $\frac{x-8}{x+6}$ and $\frac{6-5x}{x}$ is $x(x+6)$.

5. To simplify $5y+8-y+4$, we combine **like** terms.

CONCEPTS

7. To add or subtract rational expressions that have the same denominator, add or subtract the **numerators**, and write the sum or difference over the common **denominator**. In symbols, if $\frac{A}{D}$ and $\frac{B}{D}$ are rational expressions,

$$\frac{A}{D}+\frac{B}{D}=\boxed{\frac{A+B}{D}} \quad \text{or} \quad \frac{A}{D}-\frac{B}{D}=\boxed{\frac{A-B}{D}}$$

9. a) $-(6x+9)=-6x-9$
   b) $x^2+3x-1+x^2-5x$
   $=(x^2+x^2)+(3x-5x)-1$
   $=2x^2-2x-1$
   c) $7x-1-(5x-6)$
   $=7x-1-5x+6$
   $=(7x-5x)+(-1+6)$
   $=2x+5$
   d) $4x^2-2x-(x^2+x)$
   $=4x^2-2x-x^2-x$
   $=(4x^2-x^2)+(-2x-x)$
   $=3x^2-3x$

11.
$$\frac{x^2-14x+49}{x^2-49}=\frac{\cancel{(x-7)}(x-7)}{\cancel{(x-7)}(x+7)}$$
$$=\frac{x-7}{x+7}$$

13. a) $40x^2=2\cdot2\cdot2\cdot5\cdot x\cdot x$
    b) $2x^2-6x=2x(x-3)$
    c) $n^2-64=(n-8)(n+8)$

15. a) $\frac{5}{21a}$
    b) $\frac{a+4}{a+4}$

NOTATION

17. yes

PRACTICE

19.
$$\frac{9}{x}+\frac{2}{x}=\frac{9+2}{x}$$
$$=\frac{11}{x}$$

21.
$$\frac{x}{18}+\frac{5}{18}=\frac{x+5}{18}$$

23.
$$\frac{m-3}{m^3}-\frac{5}{m^3}=\frac{m-3-5}{m^3}$$
$$=\frac{m-8}{m^3}$$

25.
$$\frac{2x}{y}-\frac{x}{y}=\frac{2x-x}{y}$$
$$=\frac{x}{y}$$

27.
$$\frac{13t}{99}-\frac{35t}{99}=\frac{13t-35t}{99}$$
$$=\frac{-22t}{99}$$
$$=-\frac{\cancel{11}\cdot2\cdot t}{\cancel{11}\cdot9}$$
$$=-\frac{2t}{9}$$

29.

$$\frac{x}{9}+\frac{2x}{9}=\frac{x+2x}{9}$$
$$=\frac{3x}{9}$$
$$=\frac{\cancel{3}\cdot x}{\cancel{3}\cdot 3}$$
$$=\frac{x}{3}$$

31.

$$\frac{50}{r^3-25}+\frac{r}{r^3-25}=\frac{r+50}{r^3-25}$$

33.

$$\frac{6a}{a+2}-\frac{4a}{a+2}=\frac{6a-4a}{a+2}$$
$$=\frac{2a}{a+2}$$

35.

$$\frac{7}{t+5}-\frac{9}{t+5}=\frac{7-9}{t+5}$$
$$=\frac{-2}{t+5}$$
$$=-\frac{2}{t+5}$$

37.

$$\frac{3x-5}{x-2}+\frac{6x-13}{x-2}=\frac{3x-5+6x-13}{x-2}$$
$$=\frac{9x-18}{x-2}$$
$$=\frac{9\cancel{(x-2)}}{\cancel{x-2}}$$
$$=9$$

39.

$$\frac{6x-5}{3xy}-\frac{3x-5}{3xy}=\frac{6x-5-(3x-5)}{3xy}$$
$$=\frac{6x-5-3x+5}{3xy}$$
$$=\frac{\cancel{3x}}{\cancel{3x}y}$$
$$=\frac{1}{y}$$

41.

$$\frac{3y-2}{2y+6}-\frac{2y-5}{2y+6}=\frac{3y-2-(2y-5)}{2y+6}$$
$$=\frac{3y-2-2y+5}{2y+6}$$
$$=\frac{\cancel{y+3}}{2\cancel{(y+3)}}$$
$$=\frac{1}{2}$$

43.

$$\frac{x+3}{2y}+\frac{x+5}{2y}=\frac{x+3+x+5}{2y}$$
$$=\frac{2x+8}{2y}$$
$$=\frac{\cancel{2}(x+4)}{\cancel{2}y}$$
$$=\frac{x+4}{y}$$

45.

$$\frac{6x^2-11x}{3x+2}-\frac{10}{3x+2}=\frac{6x^2-11x-10}{3x+2}$$
$$=\frac{\cancel{(3x+2)}(2x-5)}{\cancel{3x+2}}$$
$$=2x-5$$

47.

$$\frac{2-p}{p^2-p}-\frac{-p+2}{p^2-p}=\frac{2-p-(-p+2)}{p^2-p}$$
$$=\frac{2-p+p-2}{p^2-p}$$
$$=\frac{0}{p(p-1)}$$
$$=0$$

49.

$$\frac{3x^2}{x+1}+\frac{x-2}{x+1}=\frac{3x^2+x-2}{x+1}$$
$$=\frac{(3x-2)\cancel{(x+1)}}{\cancel{x+1}}$$
$$=3x-2$$

51.

$$\frac{11w+1}{3w(w-9)}-\frac{11w}{3w(w-9)}=\frac{11w+1-11w}{3w(w-9)}$$
$$=\frac{1}{3w(w-9)}$$

53.
$$\frac{a}{a^2+5a+6}+\frac{3}{a^2+5a+6}=\frac{a+3}{a^2+5a+6}$$
$$=\frac{\cancel{a+3}}{(\cancel{a+3})(a+2)}$$
$$=\frac{1}{a+2}$$

55.
$$\frac{2c}{c^2-d^2}-\frac{2d}{c^2-d^2}=\frac{2c-2d}{c^2-d^2}$$
$$=\frac{2(\cancel{c-d})}{(\cancel{c-d})(c+d)}$$
$$=\frac{2}{c+d}$$

57.
$$\frac{11n-1}{(n+4)(n-2)}-\frac{4n}{(n+4)(n-2)}=\frac{11n-1-4n}{(n+4)(n-2)}$$
$$=\frac{7n-1}{(n+4)(n-2)}$$

59.
$$\frac{1}{t^2-2t+1}-\frac{6-t}{t^2-2t+1}=\frac{1+(6-t)}{t^2-2t+1}$$
$$=\frac{1-6+t}{t^2-2t+1}$$
$$=\frac{t-5}{(t-1)(t-1)}$$
$$=\frac{t-5}{t^2-2t+1}$$

61. $2x = 2\cdot x$
$6x = 2\cdot 3\cdot x$
LCD $= 2\cdot 3\cdot x = \mathbf{6x}$

63. $3a^2b = 3\cdot a\cdot a\cdot b$
$a^2b^3 = a\cdot a\cdot b\cdot b\cdot b$
LCD $= 3\cdot a\cdot a\cdot b\cdot b\cdot b = \mathbf{3a^2b^3}$

65. c cannot be factored
$c+2$ cannot be factored
LCD $= \mathbf{c(c+2)}$

67. $15a^3 = 3\cdot 5\cdot a\cdot a\cdot a$
$10a = 2\cdot 5\cdot a$
LCD $= 2\cdot 3\cdot 5\cdot a\cdot a\cdot a = \mathbf{30a^3}$

69. $4b+8 = 4(b+2) = 2\cdot 2(b+2)$
$6 = 2\cdot 3$
LCD $= 2\cdot 2\cdot 3(b+2) = \mathbf{12(b+2)}$

71. $x^2-1 = (x-1)(x+1)$
$x+1$ cannot be factored
LCD $= \mathbf{(x-1)(x+1)}$

73. $3x-1$ cannot be factored
$3x+1$ cannot be factored
LCD $= \mathbf{(3x+1)(3x-1)}$

75. $x^2-4x-5 = (x+1)(x-5)$
$x^2-25 = (x-5)(x+5)$
LCD $= \mathbf{(x+1)(x-5)(x+5)}$

77. $2n^2+13n+20 = (2n+5)(n+4)$
$n^2+8n+16 = (n+4)(n+4) = (n+4)^2$
LCD $= \mathbf{(2n+5)(n+4)^2}$

79. $\frac{25}{4}\cdot\frac{5x}{5x}=\frac{125x}{20x}$

81. $\frac{8}{x}\cdot\frac{xy}{xy}=\frac{8xy}{x^2y}$

83.
$$\frac{3x}{x+1}\cdot\frac{x+1}{x+1}=\frac{3x(x+1)}{(x+1)^2}$$
$$=\frac{3x^2+3x}{(x+1)^2}$$

85.
$$\frac{2y}{x}\cdot\frac{x+3}{x+3}=\frac{2y(x+3)}{x(x+3)}$$
$$=\frac{2xy+6y}{x(x+3)}$$

87.
$$\frac{10}{b-1}\cdot\frac{3}{3}=\frac{10\cdot 3}{3(b-1)}$$
$$=\frac{30}{3(b-1)}$$

89.

$$\frac{t+5}{4t+8}=\frac{t+5}{4(t+2)}$$
$$=\frac{t+5}{4(t+2)}\cdot\frac{5}{5}$$
$$=\frac{5(t+5)}{4\cdot 5(t+2)}$$
$$=\frac{5t+25}{20(t+2)}$$

91.

$$\frac{y+3}{y^2-5y+6}=\frac{y+3}{(y-2)(y-3)}$$
$$=\frac{y+3}{(y-2)(y-3)}\cdot\frac{4y}{4y}$$
$$=\frac{4y(y+3)}{4y(y-2)(y-3)}$$
$$=\frac{4y^2+12y}{4y(y-2)(y-3)}$$

93.

$$\frac{12-h}{h^2-81}=\frac{12-h}{(h-9)(h+9)}$$
$$=\frac{12-h}{(h-9)(h+9)}\cdot\frac{3}{3}$$
$$=\frac{3(12-h)}{3(h-9)(h+9)}$$
$$=\frac{36-3h}{3(h-9)(h+9)}$$

95.

$$\frac{6t}{t^2+4t+3}=\frac{6t}{(t+1)(t+3)}$$
$$=\frac{6t}{(t+1)(t+3)}\cdot\frac{t-2}{t-2}$$
$$=\frac{6t(t-2)}{(t+1)(t+3)(t-2)}$$
$$=\frac{6t^2-12t}{(t+1)(t-2)(t+3)}$$

APPLICATIONS

97. DOING LAUNDRY

Find the LCD of 30 and 45.
$30 = 2\cdot 3\cdot 5$
$45 = 3\cdot 3\cdot 5$
LCD $= 2\cdot 3\cdot 3\cdot 5 =$ **90**
In 90 minutes, the washer and dryer will end simultaneously.

WRITING

99. Answers will vary.

101. Answers will vary.

103. Answers will vary.

REVIEW

105. $I = Prt$

107. $A=\frac{1}{2}bh$

109. $d = rt$

CHALLENGE PROBLEMS

111. $\frac{3xy}{x-y}-\frac{x(3y-x)}{x-y}-\frac{x(x-y)}{x-y}$

$$=\frac{3xy-x(3y-x)-x(x-y)}{x-y}$$
$$=\frac{3xy-3xy+x^2-x^2+xy}{x-y}$$
$$=\frac{xy}{x-y}$$

SECTION 6.4

VOCABULARY

1. The rational expressions $\frac{x}{x-7}$ and $\frac{1}{x-7}$ have like denominators. The rational expressions $\frac{x+5}{x-7}$ and $\frac{4x}{x+7}$ have **unlike** denominators.

CONCEPTS

3. a) $20x^2 = 2 \cdot 2 \cdot 5 \cdot x \cdot x$
   b) $x^2 + 4x - 12 = (x+6)(x-2)$

5. LCD = $(x+6)(x+3)$

7. To build $\frac{x}{x+2}$ so that it has a denominator of $5(x+2)$, we multiply it by 1 in the form of $\frac{5}{5}$.

9. a) $\frac{5n}{5n}$
   b) $\frac{3}{3}$

11.
$$\frac{x-3}{x-4} \cdot \frac{8}{8} = \frac{8(x-3)}{8(x-4)} = \frac{8x-24}{8(x-4)}$$

13. $$\frac{7x}{3x^2(x-5)} + \frac{10}{3x^2(x-5)} = \frac{7x+10}{3x^2(x-5)}$$

15. $-1(x-10) = -x + 10 = 10 - x$

17. You must multiply **−1** by $(y-4)$ to obtain $4-y$: $-1(y-4) = -y + 4 = 4 - y$

19. $x = \frac{x}{1}$

NOTATION

21.

Yes. $\frac{m^2+2m}{(m-1)(m-4)} = \frac{m^2+2m}{(m-4)(m-1)}$

No. $\frac{-5x^2-7}{4x(x+3)} = -\frac{5x^2+7}{4x(x+3)} \neq -\frac{5x^2-7}{4x(x+3)}$

Yes. $\frac{-2x}{x-y} = \frac{-2x}{-(-x+y)} = \frac{2x}{-x+y} = \frac{2x}{y-x}$

PRACTICE

23.
$$\frac{x}{3} + \frac{2x}{7} = \frac{x}{3}\left(\frac{7}{7}\right) + \frac{2x}{7}\left(\frac{3}{3}\right) = \frac{7x}{21} + \frac{6x}{21} = \frac{7x+6x}{21} = \frac{13x}{21}$$

25.
$$\frac{2y}{9} + \frac{y}{3} = \frac{2y}{9} + \frac{y}{3}\left(\frac{3}{3}\right) = \frac{2y}{9} + \frac{3y}{9} = \frac{2y+3y}{9} = \frac{5y}{9}$$

27.
$$\frac{11}{5x} - \frac{5}{6x} = \frac{11}{5x} - \frac{5}{2 \cdot 3x} = \frac{11}{5x}\left(\frac{2\cdot 3}{2\cdot 3}\right) - \frac{5}{2\cdot 3x}\left(\frac{5}{5}\right) = \frac{66}{30x} - \frac{25}{30x} = \frac{66-25}{30x} = \frac{41}{30x}$$

29.

$$\frac{7}{m^2}-\frac{2}{m}=\frac{7}{m\cdot m}-\frac{2}{m}$$
$$=\frac{7}{m\cdot m}-\frac{2}{m}\left(\frac{m}{m}\right)$$
$$=\frac{7}{m^2}-\frac{2m}{m^2}$$
$$=\frac{7-2m}{m^2}$$

31.

$$\frac{4x}{3}+\frac{2x}{y}=\frac{4x}{3}\left(\frac{y}{y}\right)+\frac{2x}{y}\left(\frac{3}{3}\right)$$
$$=\frac{4xy}{3y}+\frac{6x}{3y}$$
$$=\frac{4xy+6x}{3y}$$

33.

$$\frac{1}{6c^4}+\frac{8}{9c^2}=\frac{1}{2\cdot 3c\cdot c\cdot c\cdot c}+\frac{8}{3\cdot 3c\cdot c}$$
$$=\frac{1}{2\cdot 3c\cdot c\cdot c\cdot c}\left(\frac{3}{3}\right)+\frac{8}{3\cdot 3c\cdot c}\left(\frac{2c^2}{2c^2}\right)$$
$$=\frac{3}{18c^4}+\frac{16c^2}{18c^4}$$
$$=\frac{3+16c^2}{18c^4}$$
$$=\frac{16c^2+3}{18c^4}$$

35.

$$\frac{y}{8}+\frac{y-3}{16}=\frac{y}{2\cdot 2\cdot 2}+\frac{y-3}{2\cdot 2\cdot 2\cdot 2}$$
$$=\frac{y}{2\cdot 2\cdot 2}\left(\frac{2}{2}\right)+\frac{y-3}{2\cdot 2\cdot 2\cdot 2}$$
$$=\frac{2y}{16}+\frac{y-3}{16}$$
$$=\frac{2y+y-3}{16}$$
$$=\frac{3y-3}{16}$$

37.

$$\frac{n}{5}-\frac{n-2}{15}=\frac{n}{5}-\frac{n-2}{3\cdot 5}$$
$$=\frac{n}{5}\left(\frac{3}{3}\right)-\frac{n-2}{3\cdot 5}$$
$$=\frac{3n}{15}-\frac{n-2}{15}$$
$$=\frac{3n-(n-2)}{15}$$
$$=\frac{3n-n+2}{15}$$
$$=\frac{2n+2}{15}$$

39.

$$\frac{2-b}{6}-\frac{b-7}{21}=\frac{2-b}{2\cdot 3}-\frac{b-7}{3\cdot 7}$$
$$=\frac{2-b}{2\cdot 3}\left(\frac{7}{7}\right)-\frac{b-7}{3\cdot 7}\left(\frac{2}{2}\right)$$
$$=\frac{7(2-b)-2(b-7)}{42}$$
$$=\frac{14-7b-2b+14}{42}$$
$$=\frac{-9b+28}{42}$$

41.

$$\frac{x-1}{9x}-\frac{x-2}{x^3}=\frac{x-1}{3\cdot 3\cdot x}-\frac{x-2}{x\cdot x\cdot x}$$
$$=\frac{x-1}{3\cdot 3\cdot x}\left(\frac{x\cdot x}{x\cdot x}\right)-\frac{x-2}{x\cdot x\cdot x}\left(\frac{3\cdot 3}{3\cdot 3}\right)$$
$$=\frac{x^2(x-1)}{9x^3}-\frac{9(x-2)}{9x^3}$$
$$=\frac{x^2(x-1)-9(x-2)}{9x^3}$$
$$=\frac{x^3-x^2-9x+18}{9x^3}$$

43.

$$\frac{y+2}{5y^2}+\frac{y+4}{15y}=\frac{y+2}{5\cdot y\cdot y}+\frac{y+4}{5\cdot 3\cdot y}$$
$$=\frac{y+2}{5\cdot y\cdot y}\left(\frac{3}{3}\right)+\frac{y+4}{5\cdot 3\cdot y}\left(\frac{y}{y}\right)$$
$$=\frac{3(y+2)}{15y^2}+\frac{y(y+4)}{15y^2}$$
$$=\frac{3(y+2)+y(y+4)}{15y^2}$$
$$=\frac{3y+6+y^2+4y}{15y^2}$$
$$=\frac{y^2+7y+6}{15y^2}$$

45.

$$\frac{x+5}{xy}-\frac{x-1}{x^2y}=\frac{x+5}{x\cdot y}-\frac{x-1}{x\cdot x\cdot y}$$
$$=\frac{x+5}{x\cdot y}\left(\frac{x}{x}\right)-\frac{x-1}{x\cdot x\cdot y}$$
$$=\frac{x(x+5)}{x^2y}-\frac{x-1}{x^2y}$$
$$=\frac{x(x+5)-(x-1)}{x^2y}$$
$$=\frac{x^2+5x-x+1}{x^2y}$$
$$=\frac{x^2+4x+1}{x^2y}$$

47.

$$\frac{x-3}{6x}+\frac{x+4}{8x}=\frac{x-3}{2\cdot 3\cdot x}+\frac{x+4}{2\cdot 2\cdot 2\cdot x}$$
$$=\frac{x-3}{2\cdot 3\cdot x}\left(\frac{2\cdot 2}{2\cdot 2}\right)+\frac{x+4}{2\cdot 2\cdot 2\cdot x}\left(\frac{3}{3}\right)$$
$$=\frac{4(x-3)}{24x}+\frac{3(x+4)}{24x}$$
$$=\frac{4(x-3)+3(x+4)}{24x}$$
$$=\frac{4x-12+3x+12}{24x}$$
$$=\frac{7x}{24x}$$
$$=\frac{7}{24}$$

49.

$$\frac{a+2}{b}+\frac{b-2}{a}=\frac{a+2}{b}\left(\frac{a}{a}\right)+\frac{b-2}{a}\left(\frac{b}{b}\right)$$
$$=\frac{a(a+2)+b(b-2)}{ab}$$
$$=\frac{a^2+2a+b^2-2b}{ab}$$

51.

$$\frac{x}{x+1}+\frac{x-1}{x}=\frac{x}{x+1}\left(\frac{x}{x}\right)+\frac{x-1}{x}\left(\frac{x+1}{x+1}\right)$$
$$=\frac{x(x)}{x(x+1)}+\frac{(x-1)(x+1)}{x(x+1)}$$
$$=\frac{x^2}{x(x+1)}+\frac{x^2-1}{x(x+1)}$$
$$=\frac{x^2+x^2-1}{x(x+1)}$$
$$=\frac{2x^2-1}{x(x+1)}$$

53.

$$\frac{1}{5x}+\frac{7x}{x+5}=\frac{1}{5x}\left(\frac{x+5}{x+5}\right)+\frac{7x}{x+5}\left(\frac{5x}{5x}\right)$$
$$=\frac{(x+5)}{5x(x+5)}+\frac{7x(5x)}{5x(x+5)}$$
$$=\frac{x+5}{5x(x+5)}+\frac{35x^2}{5x(x+5)}$$
$$=\frac{x+5+35x^2}{5x(x+5)}$$
$$=\frac{35x^2+x+5}{5x(x+5)}$$

55.

$$\frac{9}{t+3}+\frac{8}{t+2}=\frac{9}{t+3}\left(\frac{t+2}{t+2}\right)+\frac{8}{t+2}\left(\frac{t+3}{t+3}\right)$$
$$=\frac{9(t+2)}{(t+3)(t+2)}+\frac{8(t+3)}{(t+3)(t+2)}$$
$$=\frac{9t+18}{(t+3)(t+2)}+\frac{8t+24}{(t+3)(t+2)}$$
$$=\frac{9t+18+8t+24}{(t+3)(t+2)}$$
$$=\frac{17t+42}{(t+3)(t+2)}$$

57.

$$\frac{3x}{2x-1}-\frac{2x}{2x+3}=\frac{3x}{2x-1}\left(\frac{2x+3}{2x+3}\right)-\frac{2x}{2x+3}\left(\frac{2x-1}{2x-1}\right)$$
$$=\frac{3x(2x+3)}{(2x-1)(2x+3)}-\frac{2x(2x-1)}{(2x-1)(2x+3)}$$
$$=\frac{6x^2+9x}{(2x-1)(2x+3)}-\frac{4x^2-2x}{(2x-1)(2x+3)}$$
$$=\frac{6x^2+9x-(4x^2-2x)}{(2x-1)(2x+3)}$$
$$=\frac{6x^2+9x-4x^2+2x}{(2x-1)(2x+3)}$$
$$=\frac{2x^2+11x}{(2x-1)(2x+3)}$$

59.

$$\frac{4}{a+2}-\frac{7}{(a+2)^2}=\frac{4}{a+2}\left(\frac{a+2}{a+2}\right)-\frac{7}{(a+2)^2}$$
$$=\frac{4(a+2)}{(a+2)^2}-\frac{7}{(a+2)^2}$$
$$=\frac{4a+8}{(a+2)^2}-\frac{7}{(a+2)^2}$$
$$=\frac{4a+8-7}{(a+2)^2}$$
$$=\frac{4a+1}{(a+2)^2}$$

61.

$$\frac{3m}{m-2}+\frac{m-3}{m+5}=\frac{3m}{m-2}\left(\frac{m+5}{m+5}\right)+\frac{m-3}{m+5}\left(\frac{m-2}{m-2}\right)$$
$$=\frac{3m(m+5)}{(m-2)(m+5)}+\frac{(m-3)(m-2)}{(m-2)(m+5)}$$
$$=\frac{3m^2+15m}{(m-2)(m+5)}+\frac{m^2-2m-3m+6}{(m-2)(m+5)}$$
$$=\frac{3m^2+15m+m^2-2m-3m+6}{(m-2)(m+5)}$$
$$=\frac{4m^2+10m+6}{(m-2)(m+5)}$$

63.

$$\frac{s+7}{s+3}-\frac{s-3}{s+7}=\frac{s+7}{s+3}\left(\frac{s+7}{s+7}\right)-\frac{s-3}{s+7}\left(\frac{s+3}{s+3}\right)$$
$$=\frac{(s+7)(s+7)}{(s+3)(s+7)}-\frac{(s-3)(s+3)}{(s+3)(s+7)}$$
$$=\frac{s^2+7s+7s+49}{(s+3)(s+7)}-\frac{s^2-9}{(s+3)(s+7)}$$
$$=\frac{s^2+7s+7s+49-(s^2-9)}{(s+3)(s+7)}$$
$$=\frac{s^2+7s+7s+49-s^2+9}{(s+3)(s+7)}$$
$$=\frac{14s+58}{(s+3)(s+7)}$$

65.

$$\frac{7}{(a+1)(a+3)}+\frac{5}{(a+3)^2}=\frac{7}{(a+1)(a+3)}\left(\frac{a+3}{a+3}\right)+\frac{5}{(a+3)^2}\left(\frac{a+1}{a+1}\right)$$
$$=\frac{7(a+3)}{(a+1)(a+3)^2}+\frac{5(a+1)}{(a+1)(a+3)^2}$$
$$=\frac{7a+21}{(a+1)(a+3)^2}+\frac{5a+5}{(a+1)(a+3)^2}$$
$$=\frac{7a+21+5a+5}{(a+1)(a+3)^2}$$
$$=\frac{12a+26}{(a+1)(a+3)^2}$$

67.

$$\frac{2c-3}{(2c+1)(c-3)}-\frac{3}{c-3}=\frac{2c-3}{(2c+1)(c-3)}-\frac{3}{c-3}\left(\frac{2c+1}{2c+1}\right)$$
$$=\frac{2c-3}{(2c+1)(c-3)}-\frac{6c+3}{(2c+1)(c-3)}$$
$$=\frac{2c-3-(6c+3)}{(2c+1)(c-3)}$$
$$=\frac{2c-3-6c-3}{(2c+1)(c-3)}$$
$$=\frac{-4c-6}{(2c+1)(c-3)}$$
$$=-\frac{4c+6}{(2c+1)(c-3)}$$

69.

$$\frac{b}{b+1}-\frac{b+1}{2b+2}=\frac{b}{b+1}-\frac{b+1}{2(b+1)}$$
$$=\frac{b}{b+1}\left(\frac{2}{2}\right)-\frac{b+1}{2(b+1)}$$
$$=\frac{2b}{2(b+1)}-\frac{b+1}{2(b+1)}$$
$$=\frac{2b-(b+1)}{2(b+1)}$$
$$=\frac{2b-b-1}{2(b+1)}$$
$$=\frac{b-1}{2(b+1)}$$

71.

$$\frac{2}{a^2+4a+3}+\frac{1}{a+3}=\frac{2}{(a+1)(a+3)}+\frac{1}{a+3}$$
$$=\frac{2}{(a+1)(a+3)}+\frac{1}{a+3}\left(\frac{a+1}{a+1}\right)$$
$$=\frac{2}{(a+1)(a+3)}+\frac{a+1}{a+3}$$
$$=\frac{2+a+1}{(a+1)(a+3)}$$
$$=\frac{(a+3)}{(a+1)(a+3)}$$
$$=\frac{1}{a+1}$$

73.

$$\frac{7s}{s^2+s-12}-\frac{4}{s+4}=\frac{7s}{(s+4)(s-3)}-\frac{4}{s+4}$$
$$=\frac{7s}{(s+4)(s-3)}-\frac{4}{s+4}\left(\frac{s-3}{s-3}\right)$$
$$=\frac{7s}{(s+4)(s-3)}-\frac{4s-12}{(s+4)(s-3)}$$
$$=\frac{7s-(4s-12)}{(s+4)(s-3)}$$
$$=\frac{7s-4s+12}{(s+4)(s-3)}$$
$$=\frac{3s+12}{(s+4)(s-3)}$$
$$=\frac{3(s+4)}{(s+4)(s-3)}$$
$$=\frac{3}{s-3}$$

75.

$$\frac{x}{x-2}+\frac{4+2x}{x^2-4}=\frac{x}{x-2}+\frac{4+2x}{(x+2)(x-2)}$$
$$=\frac{x}{x-2}\left(\frac{x+2}{x+2}\right)+\frac{4+2x}{(x+2)(x-2)}$$
$$=\frac{x^2+2x}{(x+2)(x-2)}+\frac{4+2x}{(x+2)(x-2)}$$
$$=\frac{x^2+2x+4+2x}{(x+2)(x-2)}$$
$$=\frac{x^2+4x+4}{(x+2)(x-2)}$$
$$=\frac{(x+2)(x+2)}{(x+2)(x-2)}$$
$$=\frac{x+2}{x-2}$$

77.

$$\frac{x+1}{x+2}-\frac{x^2+1}{x^2-x-6}=\frac{x+1}{x+2}-\frac{x^2+1}{(x+2)(x-3)}$$
$$=\frac{x+1}{x+2}\left(\frac{x-3}{x-3}\right)-\frac{x^2+1}{(x+2)(x-3)}$$
$$=\frac{x^2-3x+x-3}{(x+2)(x-3)}-\frac{x^2+1}{(x+2)(x-3)}$$
$$=\frac{x^2-2x-3-(x^2+1)}{(x+2)(x-3)}$$
$$=\frac{x^2-2x-3-x^2-1}{(x+2)(x-3)}$$
$$=\frac{-2x-4}{(x+2)(x-3)}$$
$$=\frac{-2(x+2)}{(x+2)(x-3)}$$
$$=-\frac{2}{x-3}$$

79.

$$\frac{2}{3h-6}+\frac{3}{4h+8}=\frac{2}{3(h-2)}+\frac{3}{4(h+2)}$$
$$=\frac{2}{3(h-2)}\left(\frac{4(h+2)}{4(h+2)}\right)+\frac{3}{4(h+2)}\left(\frac{3(h-2)}{3(h-2)}\right)$$
$$=\frac{8(h+2)}{12(h-2)(h+2)}+\frac{9(h-2)}{12(h-2)(h+2)}$$
$$=\frac{8h+16}{12(h-2)(h+2)}+\frac{9h-18}{12(h-2)(h+2)}$$
$$=\frac{8h+16+9h-18}{12(h-2)(h+2)}$$
$$=\frac{17h-2}{12(h-2)(h+2)}$$

81.

$$\frac{8}{y^2-16}-\frac{7}{y^2-y-12}=\frac{8}{(y+4)(y-4)}-\frac{7}{(y-4)(y+3)}$$
$$=\frac{8}{(y+4)(y-4)}\left(\frac{y+3}{y+3}\right)-\frac{7}{(y-4)(y+3)}\left(\frac{y+4}{y+4}\right)$$
$$=\frac{8y+24}{(y+4)(y-4)(y+3)}-\frac{7y+28}{(y+4)(y-4)(y+3)}$$
$$=\frac{8y+24-(7y+28)}{(y+4)(y-4)(y+3)}$$
$$=\frac{8y+24-7y-28}{(y+4)(y-4)(y+3)}$$
$$=\frac{y-4}{(y+4)(y-4)(y+3)}$$
$$=\frac{1}{(y+4)(y+3)}$$

83.

$$\frac{4}{s^2+5s+4}+\frac{s}{s^2+2s+1}$$
$$=\frac{4}{(s+1)(s+4)}+\frac{s}{(s+1)^2}$$
$$=\frac{4}{(s+1)(s+4)}\left(\frac{s+1}{s+1}\right)+\frac{s}{(s+1)(s+1)}\left(\frac{s+4}{s+4}\right)$$
$$=\frac{4s+4}{(s+4)(s+1)^2}+\frac{s^2+4s}{(s+4)(s+1)^2}$$
$$=\frac{4s+4+s^2+4s}{(s+4)(s+1)^2}$$
$$=\frac{s^2+8s+4}{(s+4)(s+1)^2}$$

85.

$$\frac{5}{x^2-9x+8}-\frac{3}{x^2-6x-16}=\frac{5}{(x-8)(x-1)}-\frac{3}{(x-8)(x+2)}$$
$$=\frac{5}{(x-8)(x-1)}\left(\frac{x+2}{x+2}\right)-\frac{3}{(x-8)(x+2)}\left(\frac{x-1}{x-1}\right)$$
$$=\frac{5x+10}{(x-8)(x-1)(x+2)}-\frac{3x-3}{(x-8)(x-1)(x+2)}$$
$$=\frac{5x+10-(3x-3)}{(x-8)(x-1)(x+2)}$$
$$=\frac{5x+10-3x+3}{(x-8)(x-1)(x+2)}$$
$$=\frac{2x+13}{(x-8)(x-1)(x+2)}$$

87.

$$\frac{x+1}{2x+4}-\frac{x^2}{2x^2-8}=\frac{x+1}{2(x+2)}-\frac{x^2}{2(x^2-4)}$$
$$=\frac{x+1}{2(x+2)}-\frac{x^2}{2(x+2)(x-2)}$$
$$=\frac{x+1}{2(x+2)}\left(\frac{x-2}{x-2}\right)-\frac{x^2}{2(x+2)(x-2)}$$
$$=\frac{x^2-2x+x-2}{2(x+2)(x-2)}-\frac{x^2}{2(x+2)(x-2)}$$
$$=\frac{x^2-2x+x-2-x^2}{2(x+2)(x-2)}$$
$$=\frac{-x-2}{2(x+2)(x-2)}$$
$$=-\frac{x+2}{2(x+2)(x-2)}$$
$$=-\frac{1}{2(x-2)}$$

89.

$$\frac{8}{x}+6=\frac{8}{x}+\frac{6}{1}\left(\frac{x}{x}\right)$$
$$=\frac{8}{x}+\frac{6x}{x}$$
$$=\frac{8+6x}{x}$$
$$=\frac{6x+8}{x}$$

91.

$$b-\frac{3}{a^2}=\frac{b}{1}\left(\frac{a^2}{a^2}\right)-\frac{3}{a^2}$$
$$=\frac{a^2b}{a^2}-\frac{3}{a^2}$$
$$=\frac{a^2b-3}{a^2}$$

93.

$$\frac{9}{x-4}+x=\frac{9}{x-4}+\frac{x}{1}\left(\frac{x-4}{x-4}\right)$$
$$=\frac{9}{x-4}+\frac{x^2-4x}{x-4}$$
$$=\frac{9+x^2-4x}{x-4}$$
$$=\frac{x^2-4x+9}{x-4}$$

95.

$$\frac{x+2}{x+1}-5=\frac{x+2}{x+1}-\frac{5}{1}\left(\frac{x+1}{x+1}\right)$$
$$=\frac{x+2}{x+1}-\frac{5x+5}{x+1}$$
$$=\frac{x+2-(5x+5)}{x+1}$$
$$=\frac{x+2-5x-5}{x+1}$$
$$=\frac{-4x-3}{x+1}$$
$$=-\frac{4x+3}{x+1}$$

97.

$$\frac{5}{a-4}+\frac{7}{4-a}=\frac{5}{a-4}+\frac{7}{4-a}\left(\frac{-1}{-1}\right)$$
$$=\frac{5}{a-4}+\frac{-7}{-4+a}$$
$$=\frac{5}{a-4}+\frac{-7}{a-4}$$
$$=\frac{5-7}{a-4}$$
$$=\frac{-2}{a-4}$$
$$=-\frac{2}{a-4}$$

99.

$$\frac{r+2}{r^2-4}+\frac{4}{4-r^2}=\frac{r+2}{r^2-4}+\frac{4}{4-r^2}\left(\frac{-1}{-1}\right)$$
$$=\frac{r+2}{r^2-4}+\frac{-4}{-4+r^2}$$
$$=\frac{r+2}{r^2-4}+\frac{-4}{r^2-4}$$
$$=\frac{r+2-4}{r^2-4}$$
$$=\frac{r-2}{(r-2)(r+2)}$$
$$=\frac{1}{r+2}$$

101.

$$\frac{y+3}{y-1}-\frac{y+4}{1-y}=\frac{y+3}{y-1}-\frac{y+4}{1-y}\left(\frac{-1}{-1}\right)$$
$$=\frac{y+3}{y-1}-\frac{-y-4}{-1+y}$$
$$=\frac{y+3}{y-1}-\frac{-y-4}{y-1}$$
$$=\frac{y+3-(-y-4)}{y-1}$$
$$=\frac{y+3+y+4}{y-1}$$
$$=\frac{2y+7}{y-1}$$

APPLICATIONS

103.

$$\frac{3}{2x^2}+\frac{10}{3x}=\frac{3}{2x^2}\left(\frac{3}{3}\right)+\frac{10}{3x}\left(\frac{2x}{2x}\right)$$
$$=\frac{9}{6x^2}+\frac{20x}{6x^2}$$
$$=\frac{20x+9}{6x^2}\text{ cm}$$

WRITING

105. Answers will vary.

107. Answers will vary.

REVIEW

109. Use $y = mx + b$ for $y = 8x + 2$.
Then, $m = 8$ and $b = (0, 2)$.

111. $y = -2$
$y = 0x - 2$
$m = 0$

CHALLENGE PROBLEMS

113.

$$\frac{a}{a-1}-\frac{2}{a+2}+\frac{3(a-2)}{a^2+a-2}=\frac{a}{a-1}-\frac{2}{a+2}+\frac{3(a-2)}{(a+2)(a-1)}$$
$$=\frac{a}{a-1}\left(\frac{a+2}{a+2}\right)-\frac{2}{a+2}\left(\frac{a-1}{a-1}\right)+\frac{3(a-2)}{(a+2)(a-1)}$$
$$=\frac{a^2+2a}{(a+2)(a-1)}-\frac{2a-2}{(a+2)(a-1)}+\frac{3a-6}{(a+2)(a-1)}$$
$$=\frac{a^2+2a-(2a-2)+(3a-6)}{(a+2)(a-1)}$$
$$=\frac{a^2+2a-2a+2+3a-6}{(a+2)(a-1)}$$
$$=\frac{a^2+3a-4}{(a+2)(a-1)}$$
$$=\frac{(a+4)(a-1)}{(a+2)(a-1)}$$
$$=\frac{a+4}{a+2}$$

115.

$$\frac{1}{a+1}+\frac{a^2-7a+10}{2a^2-2a-4}\cdot\frac{2a^2-50}{a^2+10a+25}$$
$$=\frac{1}{a+1}+\left(\frac{(a-2)(a-5)}{2(a^2-a-2)}\cdot\frac{2(a^2-25)}{(a+5)(a+5)}\right)$$
$$=\frac{1}{a+1}+\left(\frac{(a-2)(a-5)}{2(a-2)(a+1)}\cdot\frac{2(a-5)(a+5)}{(a+5)(a+5)}\right)$$
$$=\frac{1}{a+1}+\left(\frac{(a-2)(a-5)}{\not{2}(a-2)(a+1)}\cdot\frac{\not{2}(a-5)(a+5)}{(a+5)(a+5)}\right)$$
$$=\frac{1}{a+1}+\frac{(a-5)(a-5)}{(a+1)(a+5)}$$
$$=\frac{1}{a+1}\left(\frac{a+5}{a+5}\right)+\frac{(a-5)(a-5)}{(a+1)(a+5)}$$
$$=\frac{a+5}{(a+1)(a+5)}+\frac{a^2-5a-5a+25}{(a+1)(a+5)}$$
$$=\frac{a+5+a^2-10a+25}{(a+1)(a+5)}$$
$$=\frac{a^2-9a+30}{(a+1)(a+5)}$$

SECTION 6.5

VOCABULARY

1. The expression $\frac{\frac{2}{3}-\frac{1}{x}}{\frac{x-3}{4}}$ is called a **complex** rational expression or, more simply, a **complex** fraction.

3. To find the **reciprocal** of $\frac{x+8}{x+7}$, we invert it.

CONCEPTS

5. To simplify a complex fraction, write its numerator and denominator as **single** rational expressions. Then perform the division by multiplying the numerator of the complex fraction by the **reciprocal** of the denominator of the complex fraction. **Simplify** the result, if possible.

7. a) $\frac{x-3}{4}$
   b) yes
   c) $\frac{1}{12}-\frac{x}{6}$
   d) no

9. a) y, 3, and 6
   b) $6y$
   c) $\frac{6y}{6y}$

11. $\frac{x}{12}(24)=\frac{24x}{12}$
    $=2x$

13. $\frac{2}{3}(6y^2)=\frac{12y^2}{3}$
    $=4y^2$

NOTATION

15. $\frac{4x^2}{15}\div\frac{16x}{25}=\frac{\frac{4x^2}{15}}{\frac{16x}{25}}$

PRACTICE

17.
$$\frac{\frac{x}{2}}{\frac{6}{5}}=\frac{x}{2}\div\frac{6}{5}$$
$$=\frac{x}{2}\bullet\frac{5}{6}$$
$$=\frac{5x}{12}$$

19.
$$\frac{\frac{2}{3}}{\frac{3}{4}}=\frac{2}{3}\div\frac{3}{4}$$
$$=\frac{2}{3}\bullet\frac{4}{3}$$
$$=\frac{8}{9}$$

21.
$$\frac{\frac{x}{y}}{\frac{1}{x}}=\frac{x}{y}\div\frac{1}{x}$$
$$=\frac{x}{y}\bullet\frac{x}{1}$$
$$=\frac{x^2}{y}$$

23.
$$\frac{\frac{n}{8}}{\frac{1}{n^2}}=\frac{n}{8}\div\frac{1}{n^2}$$
$$=\frac{n}{8}\bullet\frac{n^2}{1}$$
$$=\frac{n^3}{8}$$

25. Multiply by the LCD of x.

$$\frac{\frac{1}{x}-3}{\frac{5}{x}+2}=\frac{\frac{1}{x}-3}{\frac{5}{x}+2}\bullet\frac{x}{x}$$

$$=\frac{x\bullet\frac{1}{x}-x\bullet 3}{x\bullet\frac{5}{x}+x\bullet 2}$$

$$=\frac{1-3x}{5+2x}$$

27. Multiply by the LCD of 3.

$$\frac{\frac{2}{3}+1}{\frac{1}{3}+1}=\frac{\frac{2}{3}+1}{\frac{1}{3}+1}\bullet\frac{3}{3}$$

$$=\frac{3\left(\frac{2}{3}+1\right)}{3\left(\frac{1}{3}+1\right)}$$

$$=\frac{3\bullet\frac{2}{3}+3\bullet 1}{3\bullet\frac{1}{3}+3\bullet 1}$$

$$=\frac{2+3}{1+3}$$

$$=\frac{5}{4}$$

29. Simplify using division.

$$\frac{\frac{4a}{11}}{\frac{6a}{55}}=\frac{4a}{11}\div\frac{6a}{55}$$

$$=\frac{4a}{11}\bullet\frac{55}{6a}$$

$$=\frac{220a}{66a}$$

$$=\frac{10}{3}$$

31. Simplify using division.

$$\frac{40x^2}{\frac{20x}{9}}=\frac{40x^2}{1}\div\frac{20x}{9}$$

$$=\frac{40x^2}{1}\bullet\frac{9}{20x}$$

$$=\frac{360x^2}{20x}$$

$$=18x$$

33. Multiply by the LCD of $2y$.

$$\frac{\frac{1}{y}-\frac{5}{2}}{\frac{3}{y}}=\frac{\frac{1}{y}-\frac{5}{2}}{\frac{3}{y}}\bullet\frac{2y}{2y}$$

$$=\frac{\left(\frac{1}{y}-\frac{5}{2}\right)2y}{\left(\frac{3}{y}\right)2y}$$

$$=\frac{\frac{1}{y}(2y)-\frac{5}{2}(2y)}{\frac{3}{y}(2y)}$$

$$=\frac{2-5y}{6}$$

35. Multiply by the LCD of 12.

$$\frac{\frac{d+2}{2}}{\frac{d}{3}-\frac{d}{4}}=\frac{\frac{d+2}{2}}{\frac{d}{3}-\frac{d}{4}}\bullet\frac{12}{12}$$

$$=\frac{\left(\frac{d+2}{2}\right)12}{\left(\frac{d}{3}\right)12-\left(\frac{d}{4}\right)12}$$

$$=\frac{6(d+2)}{4d-3d}$$

$$=\frac{6d+12}{d}$$

37. Multiply by the LCD of 16.

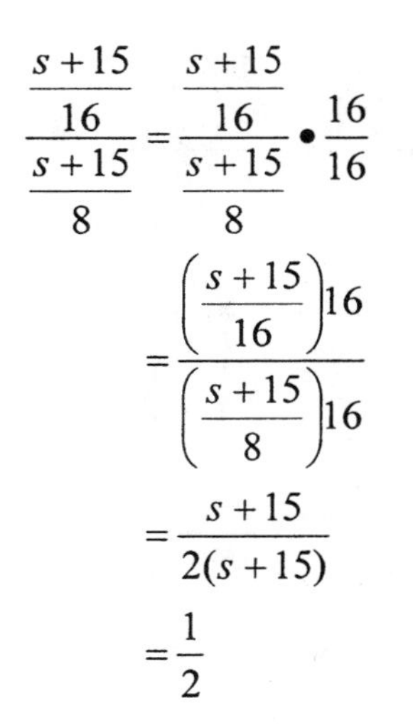

$$\frac{\frac{s+15}{16}}{\frac{s+15}{8}}=\frac{\frac{s+15}{16}}{\frac{s+15}{8}}\bullet\frac{16}{16}$$
$$=\frac{\left(\frac{s+15}{16}\right)16}{\left(\frac{s+15}{8}\right)16}$$
$$=\frac{s+15}{2(s+15)}$$
$$=\frac{1}{2}$$

39. Multiply by the LCD of 4.

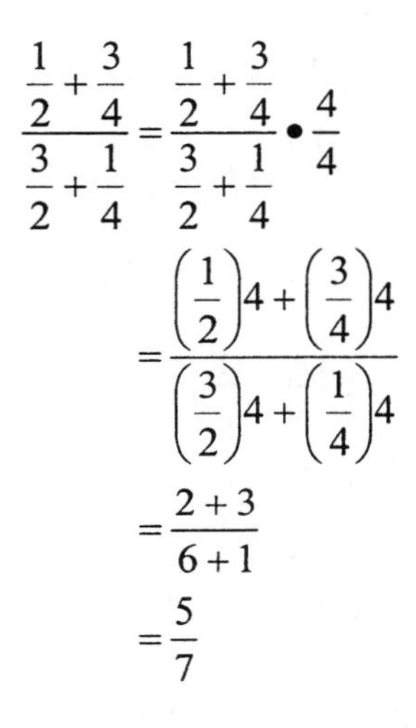

$$\frac{\frac{1}{2}+\frac{3}{4}}{\frac{3}{2}+\frac{1}{4}}=\frac{\frac{1}{2}+\frac{3}{4}}{\frac{3}{2}+\frac{1}{4}}\bullet\frac{4}{4}$$
$$=\frac{\left(\frac{1}{2}\right)4+\left(\frac{3}{4}\right)4}{\left(\frac{3}{2}\right)4+\left(\frac{1}{4}\right)4}$$
$$=\frac{2+3}{6+1}$$
$$=\frac{5}{7}$$

41. Multiply by the LCD of 30.

$$\frac{\frac{x^4}{30}}{\frac{7x}{15}}=\frac{\frac{x^4}{30}}{\frac{7x}{15}}\bullet\frac{30}{30}$$
$$=\frac{\left(\frac{x^4}{30}\right)30}{\left(\frac{7x}{15}\right)30}$$
$$=\frac{x^4}{14x}$$
$$=\frac{x^3}{14}$$

43. Multiply by the LCD of s^3.

$$\frac{\frac{2}{s}-\frac{2}{s^2}}{\frac{4}{s^3}+\frac{4}{s^2}}=\frac{\frac{2}{s}-\frac{2}{s^2}}{\frac{4}{s^3}+\frac{4}{s^2}}\bullet\frac{s^3}{s^3}$$
$$=\frac{\left(\frac{2}{s}\right)s^3-\left(\frac{2}{s^2}\right)s^3}{\left(\frac{4}{s^3}\right)s^3+\left(\frac{4}{s^2}\right)s^3}$$
$$=\frac{2s^2-2s}{4+4s}$$
$$=\frac{2s(s-1)}{2\cdot 2(1+s)}$$
$$=\frac{s(s-1)}{2(1+s)}$$
$$=\frac{s^2-s}{2+2s}$$

45. Simplify using division.

$$\frac{-\frac{3}{x^3}}{\frac{6}{x^5}}=-\frac{3}{x^3}\div\frac{6}{x^5}$$
$$=-\frac{3}{x^3}\bullet\frac{x^5}{6}$$
$$=-\frac{3x^5}{6x^3}$$
$$=-\frac{x^2}{2}$$

47. Multiply by the LCD of x.

$$\frac{\frac{2}{x}+2}{\frac{4}{x}+2}=\frac{\frac{2}{x}+2}{\frac{4}{x}+2}\bullet\frac{x}{x}$$
$$=\frac{\frac{2}{x}(x)+2(x)}{\frac{4}{x}(x)+2(x)}$$
$$=\frac{2+2x}{4+2x}$$
$$=\frac{2(1+x)}{2(2+x)}$$
$$=\frac{1+x}{2+x}$$

49. Simplify using division.

$$\frac{\frac{2x-8}{15}}{\frac{3x-12}{35x}}=\frac{2x-8}{15}\div\frac{3x-12}{35x}$$
$$=\frac{2x-8}{15}\bullet\frac{35x}{3x-12}$$
$$=\frac{2(x-4)\cdot 5\cdot 7\cdot x}{3\cdot 5\cdot 3(x-4)}$$
$$=\frac{14x}{9}$$

51. Simplify using division.

$$\frac{\frac{t-6}{16}}{\frac{12-2t}{t}}=\frac{t-6}{16}\div\frac{12-2t}{t}$$
$$=\frac{t-6}{16}\bullet\frac{t}{12-2t}$$
$$=\frac{(t-6)\cdot t}{2\cdot 2\cdot 2\cdot 2\cdot 2(6-t)}$$
$$=-\frac{(t-6)\cdot t}{2\cdot 2\cdot 2\cdot 2\cdot 2(t-6)}$$
$$=-\frac{t}{32}$$

53. Multiply by the LCD of $4c^2$.

$$\frac{\frac{2}{c^2}}{\frac{1}{c}+\frac{5}{4}}=\frac{\frac{2}{c^2}}{\frac{1}{c}+\frac{5}{4}}\bullet\frac{4c^2}{4c^2}$$
$$=\frac{\frac{2}{c^2}\left(4c^2\right)}{\frac{1}{c}\left(4c^2\right)+\frac{5}{4}\left(4c^2\right)}$$
$$=\frac{8}{4c+5c^2}$$

55. Multiply by the LCD of a^2b^2.

$$\frac{\frac{1}{a^2b}-\frac{5}{ab}}{\frac{3}{ab}-\frac{7}{ab^2}}=\frac{\frac{1}{a^2b}-\frac{5}{ab}}{\frac{3}{ab}-\frac{7}{ab^2}}\bullet\frac{a^2b^2}{a^2b^2}$$
$$=\frac{\frac{1}{a^2b}\left(a^2b^2\right)-\frac{5}{ab}\left(a^2b^2\right)}{\frac{3}{ab}\left(a^2b^2\right)-\frac{7}{ab^2}\left(a^2b^2\right)}$$
$$=\frac{b-5ab}{3ab-7a}$$

57. Multiply by the LCD of x.

$$\frac{\frac{3y}{x}-y}{y-\frac{y}{x}}=\frac{\frac{3y}{x}-y}{y-\frac{y}{x}}\bullet\frac{x}{x}$$
$$=\frac{\frac{3y}{x}(x)-y(x)}{y(x)-\frac{y}{x}(x)}$$
$$=\frac{3y-xy}{xy-y}$$
$$=\frac{y(3-x)}{y(x-1)}$$
$$=\frac{3-x}{x-1}$$

59. Simplify using division.

$$\frac{\frac{b^2-81}{18a^2}}{\frac{4b-36}{9a}}=\frac{b^2-81}{18a^2}\div\frac{4b-36}{9a}$$
$$=\frac{b^2-81}{18a^2}\bullet\frac{9a}{4b-36}$$
$$=\frac{(b-9)(b+9)\bullet 3\cdot 3\cdot a}{2\cdot 3\cdot 3\cdot a\cdot a\bullet 2\cdot 2(b-9)}$$
$$=\frac{b+9}{8a}$$

61. Multiply by the LCD of $8h$.

$$\frac{4-\frac{1}{8h}}{12+\frac{3}{4h}}=\frac{4-\frac{1}{8h}}{12+\frac{3}{4h}}\bullet\frac{8h}{8h}$$

$$=\frac{4(8h)-\frac{1}{8h}(8h)}{12(8h)+\frac{3}{4h}(8h)}$$

$$=\frac{32h-1}{96h+6}$$

63. Multiply by a LCD of xy.

$$\frac{1}{\frac{1}{x}+\frac{1}{y}}=\frac{1}{\frac{1}{x}+\frac{1}{y}}\bullet\frac{xy}{xy}$$

$$=\frac{1(xy)}{\frac{1}{x}(xy)+\frac{1}{y}(xy)}$$

$$=\frac{xy}{y+x}$$

65. Multiply by the LCD of $(x+1)$.

$$\frac{\frac{1}{x+1}}{1+\frac{1}{x+1}}=\frac{\frac{1}{x+1}}{1+\frac{1}{x+1}}\bullet\frac{x+1}{x+1}$$

$$=\frac{\frac{1}{x+1}(x+1)}{1(x+1)+\frac{1}{x+1}(x+1)}$$

$$=\frac{1}{x+1+1}$$

$$=\frac{1}{x+2}$$

67. Multiply by the LCD of $(x+2)$.

$$\frac{\frac{x}{x+2}}{\frac{x}{x+2}+x}=\frac{\frac{x}{x+2}}{\frac{x}{x+2}+x}\bullet\frac{x+2}{x+2}$$

$$=\frac{\frac{x}{x+2}(x+2)}{\frac{x}{x+2}(x+2)+x(x+2)}$$

$$=\frac{x}{x+x^2+2x}$$

$$=\frac{x}{x^2+3x}$$

$$=\frac{x}{x(x+3)}$$

$$=\frac{1}{x+3}$$

69. Simplify using division.

$$\frac{\frac{5t^2}{9x^2}}{\frac{3t}{x^2t}}=\frac{5t^2}{9x^2}\div\frac{3t}{x^2t}$$

$$=\frac{5t^2}{9x^2}\bullet\frac{x^2t}{3t}$$

$$=\frac{5\cdot t\cdot t\cdot x\cdot x\cdot t}{3\cdot 3\cdot x\cdot x\cdot 3\cdot t}$$

$$=\frac{5t^2}{27}$$

71. Simplify using division.

$$\frac{\frac{m^2-4}{3m+3}}{\frac{2m+4}{m+1}}=\frac{m^2-4}{3m+3}\div\frac{2m+4}{m+1}$$

$$=\frac{m^2-4}{3m+3}\bullet\frac{m+1}{2m+4}$$

$$=\frac{(m+2)(m-2)(m+1)}{3(m+1)2(m+2)}$$

$$=\frac{m-2}{6}$$

73. Multiply by a LCD of xy.

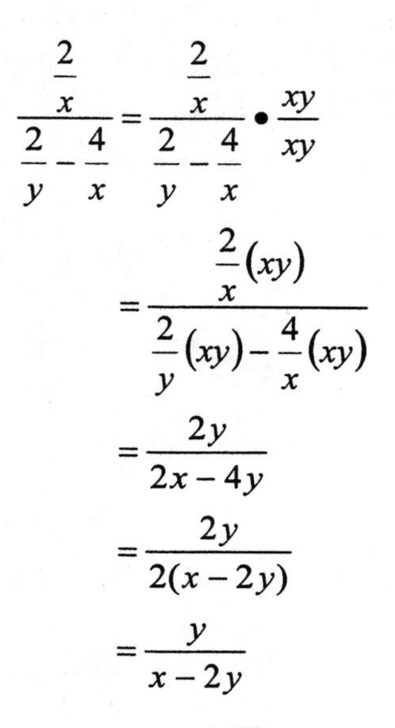

$$\frac{\frac{2}{x}}{\frac{2}{y}-\frac{4}{x}}=\frac{\frac{2}{x}}{\frac{2}{y}-\frac{4}{x}}\bullet\frac{xy}{xy}$$

$$=\frac{\frac{2}{x}(xy)}{\frac{2}{y}(xy)-\frac{4}{x}(xy)}$$

$$=\frac{2y}{2x-4y}$$

$$=\frac{2y}{2(x-2y)}$$

$$=\frac{y}{x-2y}$$

75. Multiply by the LCD of mn.

$$\frac{\frac{m}{n}+\frac{n}{m}}{\frac{m}{n}-\frac{n}{m}}=\frac{\frac{m}{n}+\frac{n}{m}}{\frac{m}{n}-\frac{n}{m}}\bullet\frac{mn}{mn}$$

$$=\frac{\frac{m}{n}(mn)+\frac{n}{m}(mn)}{\frac{m}{n}(mn)-\frac{n}{m}(mn)}$$

$$=\frac{m^2+n^2}{m^2-n^2}$$

77. Multiply by the LCD of $(x-1)$.

$$\frac{3+\frac{3}{x-1}}{3-\frac{3}{x-1}}=\frac{3+\frac{3}{x-1}}{3-\frac{3}{x-1}}\bullet\frac{x-1}{x-1}$$

$$=\frac{3(x-1)+\frac{3}{x-1}(x-1)}{3(x-1)-\frac{3}{x-1}(x-1)}$$

$$=\frac{3x-3+3}{3x-3-3}$$

$$=\frac{3x}{3x-6}$$

$$=\frac{3x}{3(x-2)}$$

$$=\frac{x}{x-2}$$

APPLICATIONS

79. GARDENING TOOLS

$$\frac{\frac{x}{2}}{\frac{7x}{3}}=\frac{x}{2}\div\frac{7x}{3}$$

$$=\frac{x}{2}\bullet\frac{3}{7x}$$

$$=\frac{3x}{14x}$$

$$=\frac{3}{14}\text{ in.}$$

81. ELECTRONICS

Multiply by the LCD of R_1R_2.

$$\frac{1}{\frac{1}{R_1}+\frac{1}{R_2}}=\frac{1}{\frac{1}{R_1}+\frac{1}{R_2}}\bullet\frac{R_1R_2}{R_1R_2}$$

$$=\frac{1(R_1R_2)}{\frac{1}{R_1}(R_1R_2)+\frac{1}{R_2}(R_1R_2)}$$

$$=\frac{R_1R_2}{R_2+R_1}$$

WRITING

83. Answers will vary.

85. Answers will vary.

REVIEW

87. $(8x)^0=1$

89. $$\left(\frac{3r}{4r^3}\right)^4=\left(\frac{3}{4}r^{1-3}\right)^4$$

$$=\left(\frac{3}{4}r^{-2}\right)^4$$

$$=\left(\frac{3}{4r^2}\right)^4$$

$$=\frac{3^4}{4^4r^{2\bullet4}}$$

$$=\frac{81}{256r^8}$$

91.

$$\left(\frac{6r^{-2}}{2r^{3}}\right)^{-2}=\left(3r^{-2-3}\right)^{-2}$$
$$=\left(3r^{-5}\right)^{-2}$$
$$=3^{-2}r^{-5\bullet-2}$$
$$=\frac{1}{3^2}\bullet r^{10}$$
$$=\frac{1}{9}\bullet r^{10}$$
$$=\frac{r^{10}}{9}$$

CHALLENGE PROBLEMS

93. Multiply by the LCD of $(h+1)(h+2)$.

$$\frac{\frac{h}{h^2+3h+2}}{\frac{4}{h+2}-\frac{4}{h+1}}=\frac{\frac{h}{(h+1)(h+2)}}{\frac{4}{h+2}-\frac{4}{h+1}}$$
$$=\frac{\frac{h}{(h+1)(h+2)}}{\frac{4}{h+2}-\frac{4}{h+1}}\bullet\frac{(h+1)(h+2)}{(h+1)(h+2)}$$
$$=\frac{\frac{h}{(h+1)(h+2)}\bullet(h+1)(h+2)}{\frac{4}{h+2}\bullet(h+1)(h+2)-\frac{4}{h+1}\bullet(h+1)(h+2)}$$
$$=\frac{h}{4(h+1)-4(h+2)}$$
$$=\frac{h}{4h+4-4h-8}$$
$$=\frac{h}{-4}$$
$$=-\frac{h}{4}$$

95. Multiply the fraction by the LCD of $(a+1)$. Then add the two rational expressions by finding a common denominator.

$$a+\frac{a}{1+\frac{a}{a+1}}=a+\left(\frac{a}{1+\frac{a}{a+1}}\bullet\frac{a+1}{a+1}\right)$$
$$=a+\left(\frac{a(a+1)}{1(a+1)+\frac{a}{a+1}(a+1)}\right)$$
$$=a+\left(\frac{a^2+a}{a+1+a}\right)$$
$$=a+\frac{a^2+a}{2a+1}$$
$$=\frac{a}{1}\left(\frac{2a+1}{2a+1}\right)+\frac{a^2+a}{2a+1}$$
$$=\frac{2a^2+a}{2a+1}+\frac{a^2+a}{2a+1}$$
$$=\frac{2a^2+a+a^2+a}{2a+1}$$
$$=\frac{3a^2+2a}{2a+1}$$

SECTION 6.6

VOCABULARY

1. Equations that contain one or more rational expressions, such as $\frac{x}{x+2}=4+\frac{10}{x+2}$, are called **rational** equations.

3. To **clear** a rational equation of fractions, multiply both sides by the LCD of all rational expressions in the equation.

5. $x^2-x+2=0$ is a **quadratic** equation.

CONCEPTS

7. a) Yes.

$$\frac{1}{\mathbf{5}-1}\overset{?}{=}1-\frac{3}{\mathbf{5}-1}$$
$$\frac{1}{4}\overset{?}{=}1-\frac{3}{4}$$
$$\frac{1}{4}\overset{?}{=}\frac{4}{4}-\frac{3}{4}$$
$$\frac{1}{4}=\frac{1}{4}$$

b) No, 5 is an extraneous solution because it makes the denominator 0.

$$\frac{\mathbf{5}}{\mathbf{5}-5}\overset{?}{=}3+\frac{5}{\mathbf{5}-5}$$
$$\frac{5}{0}\overset{?}{=}3+\frac{5}{0}$$

9. a) 3 and 0
   b) 3 and 0
   c) 3 and 0

11. y

13. $(x+8)(x-8)$

15. $(x+2)(x-2)$

17. $12=1+10x$

NOTATION

19.
$$\frac{2}{a}+\frac{1}{2}=\frac{7}{2a}$$
$$\boxed{2a}\left(\frac{2}{a}+\frac{1}{2}\right)=\boxed{2a}\left(\frac{7}{2a}\right)$$
$$\boxed{2a}\left(\frac{2}{a}\right)+\boxed{2a}\left(\frac{1}{2}\right)=\boxed{2a}\left(\frac{7}{2a}\right)$$
$$\boxed{4}+a=\boxed{7}$$
$$4+a-4=7-4$$
$$\boxed{a=3}$$

21. multiplication: $fpq=f\cdot p\cdot q$

PRACTICE

23. Multiply both sides of the equation by 6.

$$\frac{2}{3}=\frac{1}{2}+\frac{x}{6}$$
$$6\left(\frac{2}{3}\right)=6\left(\frac{1}{2}+\frac{x}{6}\right)$$
$$6\left(\frac{2}{3}\right)=6\left(\frac{1}{2}\right)+6\left(\frac{x}{6}\right)$$
$$4=3+x$$
$$4-3=3+x-3$$
$$1=x$$
$$\boxed{x=1}$$

Check :

$$\frac{2}{3}\overset{?}{=}\frac{1}{2}+\frac{\mathbf{1}}{6}$$
$$\frac{2}{3}\overset{?}{=}\frac{3}{6}+\frac{1}{6}$$
$$\frac{2}{3}\overset{?}{=}\frac{4}{6}$$
$$\frac{2}{3}=\frac{2}{3}$$

25. Multiply both sides of the equation by 18.

$$\frac{x}{18}=\frac{1}{3}-\frac{x}{2}$$

$$18\left(\frac{x}{18}\right)=18\left(\frac{1}{3}-\frac{x}{2}\right)$$

$$18\left(\frac{x}{18}\right)=18\left(\frac{1}{3}\right)-18\left(\frac{x}{2}\right)$$

$$x=6-9x$$

$$x+9x=6-9x+9x$$

$$10x=6$$

$$\frac{10x}{10}=\frac{6}{10}$$

$$\boxed{x=\frac{3}{5}}$$

Check:

$$\frac{\frac{3}{5}}{18}\overset{?}{=}\frac{1}{3}-\frac{\frac{3}{5}}{2}$$

$$\frac{3}{5}\div 18\overset{?}{=}\frac{1}{3}-\frac{3}{5}\div 2$$

$$\frac{3}{5}\bullet\frac{1}{18}\overset{?}{=}\frac{1}{3}-\frac{3}{5}\bullet\frac{1}{2}$$

$$\frac{1}{30}\overset{?}{=}\frac{1}{3}-\frac{3}{10}$$

$$\frac{1}{30}\overset{?}{=}\frac{10}{30}-\frac{9}{30}$$

$$\frac{1}{30}=\frac{1}{30}$$

27. Multiply both sides of the equation by 14.

$$\frac{a-1}{7}-\frac{a-2}{14}=\frac{1}{2}$$

$$14\left(\frac{a-1}{7}-\frac{a-2}{14}\right)=14\left(\frac{1}{2}\right)$$

$$14\left(\frac{a-1}{7}\right)-14\left(\frac{a-2}{14}\right)=14\left(\frac{1}{2}\right)$$

$$2(a-1)-(a-2)=7$$

$$2a-2-a+2=7$$

$$(2a-a)+(-2+2)=7$$

$$\boxed{a=7}$$

Check:

$$\frac{7-1}{7}-\frac{7-2}{14}\overset{?}{=}\frac{1}{2}$$

$$\frac{6}{7}-\frac{5}{14}\overset{?}{=}\frac{1}{2}$$

$$\frac{12}{14}-\frac{5}{14}\overset{?}{=}\frac{1}{2}$$

$$\frac{7}{14}\overset{?}{=}\frac{1}{2}$$

$$\frac{1}{2}=\frac{1}{2}$$

29. Multiply both sides of the equation by x.

$$\frac{3}{x}+2=3$$

$$x\left(\frac{3}{x}+2\right)=x(3)$$

$$x\left(\frac{3}{x}\right)+x(2)=x(3)$$

$$3+2x=3x$$

$$3+2x-2x=3x-2x$$

$$3=x$$

$$\boxed{x=3}$$

Check:

$$\frac{3}{3}+2\overset{?}{=}3$$

$$1+2\overset{?}{=}3$$

$$3=3$$

31. Multiply both sides of the equation by $(x-5)$.

$$\frac{x}{x-5}-\frac{5}{x-5}=3$$
$$(x-5)\left(\frac{x}{x-5}-\frac{5}{x-5}\right)=(x-5)(3)$$
$$(x-5)\frac{x}{x-5}-(x-5)\frac{5}{x-5}=3(x-5)$$
$$x-5=3x-15$$
$$x-5+5=3x-15+5$$
$$x=3x-10$$
$$x-3x=3x-10-3x$$
$$-2x=-10$$
$$\frac{-2x}{-2}=\frac{-10}{-2}$$
$$\boxed{x=5}$$

$\boxed{\text{No solution; 5 is extraneous}}$

Check:

$$\frac{5}{5-5}-\frac{5}{5-5}\stackrel{?}{=}3$$
$$\frac{5}{0}-\frac{5}{0}\stackrel{?}{=}3$$

If $x=5$, the denominators of both fractions are 0 and the expressions are undefined; so 5 is extraneous and there are no solutions.

33. Multiply both sides of the equation by $4a$.

$$\frac{a}{4}-\frac{4}{a}=0$$
$$4a\left(\frac{a}{4}-\frac{4}{a}\right)=4a(0)$$
$$4a\left(\frac{a}{4}\right)-4a\left(\frac{4}{a}\right)=4a(0)$$
$$a^2-16=0$$
$$(a-4)(a+4)=0$$
$$a-4=0 \quad \text{and} \quad a+4=0$$
$$a-4+4=0+4 \quad a+4-4=0-4$$
$$\boxed{a=4 \qquad a=-4}$$

Check:

$$\frac{4}{4}-\frac{4}{4}\stackrel{?}{=}0 \qquad \frac{-4}{4}-\frac{4}{-4}\stackrel{?}{=}0$$
$$1-1\stackrel{?}{=}0 \qquad -1-(-1)\stackrel{?}{=}0$$
$$0=0 \qquad 0=0$$

35. Multiply both sides of the equation by $(y+1)$.

$$\frac{2}{y+1}+5=\frac{12}{y+1}$$
$$(y+1)\left(\frac{2}{y+1}+5\right)=(y+1)\left(\frac{12}{y+1}\right)$$
$$(y+1)\bullet\frac{2}{y+1}+(y+1)\bullet 5=(y+1)\bullet\frac{12}{y+1}$$
$$2+5(y+1)=12$$
$$2+5y+5=12$$
$$5y+7=12$$
$$5y+7-7=12-7$$
$$5y=5$$
$$\frac{5y}{5}=\frac{5}{5}$$
$$\boxed{y=1}$$

Check:

$$\frac{2}{1+1}+5\stackrel{?}{=}\frac{12}{1+1}$$
$$\frac{2}{2}+5\stackrel{?}{=}\frac{12}{2}$$
$$1+5\stackrel{?}{=}6$$
$$6=6$$

37. Multiply both sides of the equation by b.

$$-\frac{1}{b}=\frac{5}{b}-9-\frac{6}{b}$$
$$b\left(-\frac{1}{b}\right)=b\left(\frac{5}{b}-9-\frac{6}{b}\right)$$
$$b\left(-\frac{1}{b}\right)=b\left(\frac{5}{b}\right)-b(9)-b\left(\frac{6}{b}\right)$$
$$-1=5-9b-6$$
$$-1=-1-9b$$
$$-1+1=-1-9b+1$$
$$0=9b$$
$$\frac{0}{9}=\frac{9b}{9}$$
$$\boxed{0=b}$$

$\boxed{\text{No solution; } 0 \text{ is extraneous}}$

Check:
$$-\frac{1}{0}\overset{?}{=}\frac{5}{0}-9-\frac{6}{0}$$

39. Multiply both sides of the equation by $(t+2)(t+7)$.

$$\frac{1}{t+2}=\frac{t-1}{t+7}$$
$$(t+2)(t+7)\frac{1}{t+2}=(t+2)(t+7)\frac{t-1}{t+7}$$
$$1(t+7)=(t+2)(t-1)$$
$$t+7=t^2-t+2t-2$$
$$t+7=t^2+t-2$$
$$t+7-t-7=t^2+t-2-t-7$$
$$0=t^2-9$$
$$0=(t-3)(t+3)$$
$$t-3=0 \quad \text{and} \quad t+3=0$$
$$\boxed{t=3 \quad \text{and} \quad t=-3}$$

Check:
$$\frac{1}{3+2}\overset{?}{=}\frac{3-1}{3+7} \qquad \frac{1}{-3+2}\overset{?}{=}\frac{-3-1}{-3+7}$$
$$\frac{1}{5}\overset{?}{=}\frac{2}{10} \qquad \frac{1}{-1}\overset{?}{=}\frac{-4}{-4}$$
$$\frac{1}{5}=\frac{1}{5} \qquad -1=-1$$

41. Multiply both sides of the equation by $12n$.

$$\frac{5}{n}+\frac{5}{12}=0$$
$$12n\left(\frac{5}{n}+\frac{5}{12}\right)=12n(0)$$
$$12n\left(\frac{5}{n}\right)+12n\left(\frac{5}{12}\right)=12n(0)$$
$$60+5n=0$$
$$60+5n-60=0-60$$
$$5n=-60$$
$$\frac{5n}{5}=\frac{-60}{5}$$
$$\boxed{n=-12}$$

Check:
$$\frac{5}{-12}+\frac{5}{12}\overset{?}{=}0$$
$$0=0$$

43. Multiply both sides of the equation by $40y$.

$$\frac{1}{8}+\frac{2}{y}=\frac{1}{y}+\frac{1}{10}$$
$$40y\left(\frac{1}{8}+\frac{2}{y}\right)=40y\left(\frac{1}{y}+\frac{1}{10}\right)$$
$$40y\left(\frac{1}{8}\right)+40y\left(\frac{2}{y}\right)=40y\left(\frac{1}{y}\right)+40y\left(\frac{1}{10}\right)$$
$$5y+80=40+4y$$
$$5y+80-80=40+4y-80$$
$$5y=-40+4y$$
$$5y-4y=-40+4y-4y$$
$$\boxed{y=-40}$$

Check:
$$\frac{1}{8}+\frac{2}{-40}\overset{?}{=}\frac{1}{-40}+\frac{1}{10}$$
$$\frac{5}{40}-\frac{2}{40}\overset{?}{=}-\frac{1}{40}+\frac{4}{40}$$
$$\frac{3}{40}=\frac{3}{40}$$

45. Multiply both sides of the equation by $24b$.

$$\frac{1}{8}+\frac{2}{b}-\frac{1}{12}=0$$
$$24b\left(\frac{1}{8}+\frac{2}{b}-\frac{1}{12}\right)=24b(0)$$
$$24b\left(\frac{1}{8}\right)+24b\left(\frac{2}{b}\right)-24b\left(\frac{1}{12}\right)=24b(0)$$
$$3b+48-2b=0$$
$$b+48=0$$
$$b+48-48=0-48$$
$$\boxed{b=-48}$$

Check:

$$\frac{1}{8}+\frac{2}{-48}-\frac{1}{12}\overset{?}{=}0$$
$$\frac{6}{48}-\frac{2}{48}-\frac{4}{48}\overset{?}{=}0$$
$$0=0$$

47. Multiply both sides of the equation by $5x(x-5)$.

$$\frac{4}{5x}=\frac{8}{x-5}$$
$$5x(x-5)\bullet\frac{4}{5x}=5x(x-5)\bullet\frac{8}{x-5}$$
$$4(x-5)=5x(8)$$
$$4x-20=40x$$
$$4x-20-4x=40x-4x$$
$$-20=36x$$
$$\frac{-20}{36}=x$$
$$-\frac{5}{9}=x$$
$$\boxed{x=-\frac{5}{9}}$$

Check:

$$\frac{4}{5(-\frac{5}{9})}\overset{?}{=}\frac{8}{-\frac{5}{9}-5}$$
$$\frac{4}{-\frac{25}{9}}\overset{?}{=}\frac{8}{-\frac{50}{9}}$$
$$4\div-\tfrac{25}{9}\overset{?}{=}8\div-\tfrac{50}{9}$$
$$-\frac{36}{25}=-\frac{36}{25}$$

49. Multiply both sides of the equation by $12a$.

$$\frac{1}{4}-\frac{5}{6}=\frac{1}{a}$$
$$12a\left(\frac{1}{4}-\frac{5}{6}\right)=12a\left(\frac{1}{a}\right)$$
$$12a\bullet\frac{1}{4}-12a\bullet\frac{5}{6}=12a\bullet\frac{1}{a}$$
$$3a-10a=12$$
$$-7a=12$$
$$\boxed{a=-\frac{12}{7}}$$

Check:

$$\frac{1}{4}-\frac{5}{6}\overset{?}{=}\frac{1}{-\frac{12}{7}}$$
$$\frac{3}{12}-\frac{10}{12}\overset{?}{=}1\div-\tfrac{12}{7}$$
$$-\frac{7}{12}=-\frac{7}{12}$$

51. Multiply both sides of the equation by $4h$.

$$\frac{3}{4h}+\frac{2}{h}=1$$
$$4h\left(\frac{3}{4h}+\frac{2}{h}\right)=4h(1)$$
$$4h\bullet\frac{3}{4h}+4h\bullet\frac{2}{h}=4h\bullet1$$
$$3+8=4h$$
$$11=4h$$
$$\boxed{\frac{11}{4}=h}$$

Check:

$$\frac{3}{4(\frac{11}{4})}+\frac{2}{\frac{11}{4}}\overset{?}{=}1$$
$$\frac{3}{11}+\frac{8}{11}\overset{?}{=}1$$
$$\frac{11}{11}=1$$

53. Multiply both sides of the equation by $2r$.

$$\frac{3r}{2}-\frac{3}{r}=\frac{3r}{2}+3$$

$$2r\left(\frac{3r}{2}-\frac{3}{r}\right)=2r\left(\frac{3r}{2}+3\right)$$

$$2r\bullet\frac{3r}{2}-2r\bullet\frac{3}{r}=2r\bullet\frac{3r}{2}+2r\bullet 3$$

$$3r^2-6=3r^2+6r$$

$$3r^2-6-3r^2=3r^2+6r-3r^2$$

$$-6=6r$$

$$\frac{-6}{6}=\frac{6r}{6}$$

$$-1=r$$

$$\boxed{r=-1}$$

Check:

$$\frac{3(-1)}{2}-\frac{3}{-1}\overset{?}{=}\frac{3(-1)}{2}+3$$

$$\frac{-3}{2}+\frac{3}{1}\overset{?}{=}\frac{-3}{2}+3$$

$$\frac{-3}{2}+\frac{6}{2}\overset{?}{=}\frac{-3}{2}+\frac{6}{2}$$

$$\frac{3}{2}=\frac{3}{2}$$

55. Multiply both sides of the equation by $3(x-3)$.

$$\frac{1}{3}+\frac{2}{x-3}=1$$

$$3(x-3)\left(\frac{1}{3}+\frac{2}{x-3}\right)=3(x-3)(1)$$

$$3(x-3)\bullet\frac{1}{3}+3(x-3)\bullet\frac{2}{x-3}=3(x-3)\bullet 1$$

$$x-3+3\bullet 2=3(x-3)$$

$$x-3+6=3x-9$$

$$x+3=3x-9$$

$$x+3-3=3x-9-3$$

$$x=3x-12$$

$$x-3x=3x-12-3x$$

$$-2x=-12$$

$$\frac{-2x}{-2}=\frac{-12}{-2}$$

$$\boxed{x=6}$$

Check:

$$\frac{1}{3}+\frac{2}{6-3}\overset{?}{=}1$$

$$\frac{1}{3}+\frac{2}{3}\overset{?}{=}1$$

$$\frac{3}{3}=1$$

57. Multiply both sides of the equation by $(z-3)(z+1)$.

$$\frac{z-4}{z-3}=\frac{z+2}{x+1}$$

$$(z-3)(z+1)\left(\frac{z-4}{z-3}\right)=(z-3)(z+1)\left(\frac{z+2}{x+1}\right)$$

$$(z+1)(z-4)=(z-3)(z+2)$$

$$z^2-4z+z-4=z^2+2z-3z-6$$

$$z^2-3z-4=z^2-z-6$$

$$z^2-3z-4-z^2=z^2-z-6-z^2$$

$$-3z-4=-z-6$$

$$-3z-4+z=-z-6+z$$

$$-2z-4=-6$$

$$-2z-4+4=-6+4$$

$$-2z=-2$$

$$z=1$$

Check:

$$\frac{1-4}{1-3}\overset{?}{=}\frac{1+2}{1+1}$$

$$\frac{-3}{-2}\overset{?}{=}\frac{3}{2}$$

$$\frac{3}{2}=\frac{3}{2}$$

59. Multiply both sides of the equation by $(v+2)(v-1)$.

$$\frac{v}{v+2}+\frac{1}{v-1}=1$$
$$(v+2)(v-1)\left(\frac{v}{v+2}+\frac{1}{v-1}\right)=(v+2)(v-1)(1)$$
$$v(v-1)+1(v+2)=(v+2)(v-1)$$
$$v^2-v+v+2=v^2-v+2v-2$$
$$v^2+2=v^2+v-2$$
$$v^2+2-v^2=v^2+v-2-v^2$$
$$2=v-2$$
$$2+2=v-2+2$$
$$\boxed{4=v}$$

Check:

$$\frac{4}{4+2}+\frac{1}{4-1}\stackrel{?}{=}1$$
$$\frac{4}{6}+\frac{1}{3}\stackrel{?}{=}1$$
$$\frac{4}{6}+\frac{2}{6}\stackrel{?}{=}1$$
$$\frac{6}{6}=1$$

61. Multiply both sides of the equation by $(a+2)$

$$\frac{a^2}{a+2}-\frac{4}{a+2}=a$$
$$(a+2)\left(\frac{a^2}{a+2}-\frac{4}{a+2}\right)=(a+2)(a)$$
$$(a+2)\left(\frac{a^2}{a+2}\right)-(a+2)\left(\frac{4}{a+2}\right)=(a+2)(a)$$
$$a^2-4=a^2+2a$$
$$a^2-4-a^2=a^2+2a-a^2$$
$$-4=2a$$
$$\frac{-4}{2}=\frac{2a}{2}$$
$$\boxed{-2=a}$$

$\boxed{\text{No solution; -2 is extraneous}}$

Check:

$$\frac{(-2)^2}{-2+2}-\frac{4}{-2+2}\stackrel{?}{=}-2$$
$$\frac{4}{0}-\frac{4}{0}\stackrel{?}{=}-2$$

63. Multiply both sides of the equation by $(x+4)$.

$$\frac{5}{x+4}+\frac{1}{x+4}=x-1$$
$$(x+4)\left(\frac{5}{x+4}+\frac{1}{x+4}\right)=(x+4)(x-1)$$
$$(x+4)\left(\frac{5}{x+4}\right)+(x+4)\left(\frac{1}{x+4}\right)=(x+4)(x-1)$$
$$5+1=x^2-x+4x-4$$
$$6=x^2+3x-4$$
$$6-6=x^2+3x-4-6$$
$$0=x^2+3x-10$$
$$0=(x+5)(x-2)$$
$$0=x+5 \quad\text{and}\quad 0=x-2$$
$$0-5=x+5-5 \quad 0+2=x-2+2$$
$$\boxed{-5=x \quad\text{and}\quad x=2}$$

Check:

$$\frac{5}{-5+4}+\frac{1}{-5+4}\stackrel{?}{=}-5-1 \qquad \frac{5}{2+4}+\frac{1}{2+4}\stackrel{?}{=}2-1$$
$$\frac{5}{-1}+\frac{1}{-1}\stackrel{?}{=}-6 \qquad \frac{5}{6}+\frac{1}{6}\stackrel{?}{=}1$$
$$-5-1\stackrel{?}{=}-6 \qquad \frac{6}{6}=1$$
$$-6=-6$$

65. Multiply both sides of the equation by $2(x+1)$.

$$\frac{3}{x+1}-\frac{x-2}{2}=\frac{x-2}{x+1}$$

$$2(x+1)\left(\frac{3}{x+1}-\frac{x-2}{2}\right)=2(x+1)\left(\frac{x-2}{x+1}\right)$$

$$2(3)-(x+1)(x-2)=2(x-2)$$

$$6-\left(x^2-2x+x-2\right)=2x-4$$

$$6-x^2+2x-x+2=2x-4$$

$$-x^2+x+8=2x-4$$

$$-x^2+x+8+x^2-x-8=2x-4+x^2-x-8$$

$$0=x^2+x-12$$

$$0=(x-3)(x+4)$$

$$0=x-3 \text{ and } 0=x+4$$

$$\boxed{x=3 \quad \text{and} \quad x=-4}$$

Check:

$$\frac{3}{3+1}-\frac{3-2}{2}\overset{?}{=}\frac{3-2}{3+1} \qquad \frac{3}{-4+1}-\frac{-4-2}{2}\overset{?}{=}\frac{-4-2}{-4+1}$$

$$\frac{3}{4}-\frac{1}{2}\overset{?}{=}\frac{1}{4} \qquad \frac{3}{-3}-\frac{-6}{2}\overset{?}{=}\frac{-6}{-3}$$

$$\frac{1}{4}=\frac{1}{4} \qquad -1+3\overset{?}{=}2$$

$$2=2$$

67. Multiply both sides of the equation by $(b+3)(b-5)$.

$$\frac{b+2}{b+3}+1=\frac{-7}{b-5}$$

$$(b+3)(b-5)\left(\frac{b+2}{b+3}+1\right)=(b+3)(b-5)\left(\frac{-7}{b-5}\right)$$

$$(b-5)(b+2)+(b+3)(b-5)=-7(b+3)$$

$$b^2+2b-5b-10+b^2-5b+3b-15=-7b-21$$

$$2b^2-5b-25=-7b-21$$

$$2b^2-5b-25+7b+21=-7b-21+7b+21$$

$$2b^2+2b-4=0$$

$$2(b^2+b-2)=0$$

$$2(b+2)(b-1)=0$$

$$b+2=0 \text{ and } b-1=0$$

$$\boxed{b=-2 \text{ and } b=1}$$

Check:

$$\frac{-2+2}{-2+3}+1\overset{?}{=}\frac{-7}{-2-5} \qquad \frac{1+2}{1+3}+1\overset{?}{=}\frac{-7}{1-5}$$

$$\frac{0}{1}+1\overset{?}{=}\frac{-7}{-7} \qquad \frac{3}{4}+1\overset{?}{=}\frac{-7}{-4}$$

$$0+1\overset{?}{=}1 \qquad \frac{3}{4}+\frac{4}{4}\overset{?}{=}\frac{7}{4}$$

$$1=1 \qquad \frac{7}{4}=\frac{7}{4}$$

69. Multiply both sides of the equation by $u(u-1)$.

$$\frac{u}{u-1}+\frac{1}{u}=\frac{u^2+1}{u^2-u}$$

$$\frac{u}{u-1}+\frac{1}{u}=\frac{u^2+1}{u(u-1)}$$

$$u(u-1)\left(\frac{u}{u-1}+\frac{1}{u}\right)=u(u-1)\left(\frac{u^2+1}{u(u-1)}\right)$$

$$u(u)+1(u-1)=u^2+1$$

$$u^2+u-1=u^2+1$$

$$u^2+u-1-u^2=u^2+1-u^2$$

$$u-1=1$$

$$u-1+1=1+1$$

$$\boxed{u=2}$$

Check:

$$\frac{2}{2-1}+\frac{1}{2}\overset{?}{=}\frac{2^2+1}{2^2-2}$$

$$\frac{2}{1}+\frac{1}{2}\overset{?}{=}\frac{4+1}{4-2}$$

$$\frac{4}{2}+\frac{1}{2}\overset{?}{=}\frac{5}{2}$$

$$\frac{5}{2}=\frac{5}{2}$$

71. Multiply both sides of the equation by $(n-3)(n+3)$.

$$\frac{n}{n^2-9}+\frac{n+8}{n+3}=\frac{n-8}{n-3}$$

$$\frac{n}{(n-3)(n+3)}+\frac{n+8}{n+3}=\frac{n-8}{n-3}$$

$$(n+3)(n-3)\left(\frac{n}{(n+3)(n-3)}+\frac{n+8}{n+3}\right)=(n+3)(n-3)\left(\frac{n-8}{n-3}\right)$$

$$n+(n-3)(n+8)=(n+3)(n-8)$$

$$n+n^2+8n-3n-24=n^2-8n+3n-24$$

$$n^2+6n-24=n^2-5n-24$$

$$n^2+6n-24-n^2=n^2-5n-24-n^2$$

$$6n-24=-5n-24$$

$$6n-24+5n=-5n-24+5n$$

$$11n-24=-24$$

$$11n-24+24=-24+24$$

$$11n=0$$

$$\frac{11n}{11}=\frac{0}{11}$$

$$\boxed{n=0}$$

Check:

$$\frac{0}{0^2-9}+\frac{0+8}{0+3}\stackrel{?}{=}\frac{0-8}{0-3}$$

$$\frac{0}{-9}+\frac{8}{3}\stackrel{?}{=}\frac{-8}{-3}$$

$$\frac{8}{3}=\frac{8}{3}$$

73. Multiply both sides of the equation by $x(x-1)$.

$$\frac{x}{x-1}-\frac{12}{x^2-x}=\frac{-1}{x-1}$$

$$\frac{x}{x-1}-\frac{12}{x(x-1)}=\frac{-1}{x-1}$$

$$x(x-1)\left(\frac{x}{x-1}-\frac{12}{x(x-1)}\right)=x(x-1)\left(\frac{-1}{x-1}\right)$$

$$x(x)-12=x(-1)$$

$$x^2-12=-x$$

$$x^2-12+x=-x+x$$

$$x^2+x-12=0$$

$$(x+4)(x-3)=0$$

$$x+4=0 \quad \text{and} \quad x-3=0$$

$$x+4-4=0-4 \quad x-3+3=0+3$$

$$\boxed{x=-4 \quad \text{and} \quad x=3}$$

Check:

$$\frac{-4}{-4-1}-\frac{12}{(-4)^2-(-4)}\stackrel{?}{=}\frac{-1}{-4-1} \qquad \frac{3}{3-1}-\frac{12}{(3)^2-3}\stackrel{?}{=}\frac{-1}{3-1}$$

$$\frac{-4}{-5}-\frac{12}{16+4}\stackrel{?}{=}\frac{-1}{-5} \qquad \frac{3}{2}-\frac{12}{9-3}\stackrel{?}{=}\frac{-1}{2}$$

$$\frac{4}{5}-\frac{12}{20}\stackrel{?}{=}\frac{1}{5} \qquad \frac{3}{2}-\frac{12}{6}\stackrel{?}{=}-\frac{1}{2}$$

$$\frac{16}{20}-\frac{12}{20}\stackrel{?}{=}\frac{1}{5} \qquad \frac{9}{6}-\frac{12}{6}\stackrel{?}{=}-\frac{1}{2}$$

$$\frac{4}{20}=\frac{1}{5} \qquad -\frac{3}{6}=-\frac{1}{2}$$

75. Multiply both sides of the equation by $b(b+3)$.

$$1-\frac{3}{b}=\frac{-8b}{b^2+3b}$$

$$1-\frac{3}{b}=\frac{-8b}{b(b+3)}$$

$$b(b+3)\left(1-\frac{3}{b}\right)=b(b+3)\left(\frac{-8b}{b(b+3)}\right)$$

$$b(b+3)-3(b+3)=-8b$$

$$b^2+3b-3b-9=-8b$$

$$b^2-9=-8b$$

$$b^2-9+8b=-8b+8b$$

$$b^2+8b-9=0$$

$$(b+9)(b-1)=0$$

$$b+9=0 \quad \text{and} \quad b-1=0$$

$$b+9-9=0-9 \quad \text{and} \quad b-1+1=0+1$$

$$\boxed{b=-9 \quad \text{and} \quad b=1}$$

Check:

$$1-\frac{3}{-9}\stackrel{?}{=}\frac{-8(-9)}{(-9)^2+3(-9)} \qquad 1-\frac{3}{1}\stackrel{?}{=}\frac{-8(1)}{(1)^2+3(1)}$$

$$1+\frac{1}{3}\stackrel{?}{=}\frac{72}{81-27} \qquad 1-3\stackrel{?}{=}\frac{-8}{1+3}$$

$$\frac{4}{3}\stackrel{?}{=}\frac{72}{54} \qquad -2\stackrel{?}{=}\frac{-8}{4}$$

$$\frac{4}{3}=\frac{4}{3} \qquad -2=-2$$

77. Multiply both sides of the equation by $(x-1)(x+9)$.

$$\frac{3}{x-1}-\frac{1}{x+9}=\frac{18}{x^2+8x-9}$$
$$\frac{3}{x-1}-\frac{1}{x+9}=\frac{18}{(x-1)(x+9)}$$
$$(x-1)(x+9)\left(\frac{3}{x-1}-\frac{1}{x+9}\right)=(x-1)(x+9)\left(\frac{18}{(x-1)(x+9)}\right)$$
$$3(x+9)-(x-1)=18$$
$$3x+27-x+1=18$$
$$2x+28=18$$
$$2x+28-28=18-28$$
$$2x=-10$$
$$\frac{2x}{2}=\frac{-10}{2}$$
$$\boxed{x=-5}$$

Check:

$$\frac{3}{-5-1}-\frac{1}{-5+9}\stackrel{?}{=}\frac{18}{(-5)^2+8(-5)-9}$$
$$\frac{3}{-6}-\frac{1}{4}\stackrel{?}{=}\frac{18}{25-40-9}$$
$$-\frac{6}{12}-\frac{3}{12}\stackrel{?}{=}\frac{18}{-24}$$
$$-\frac{9}{12}\stackrel{?}{=}-\frac{18}{24}$$
$$-\frac{3}{4}=-\frac{3}{4}$$

79. Muliply both sides of the equation by n.

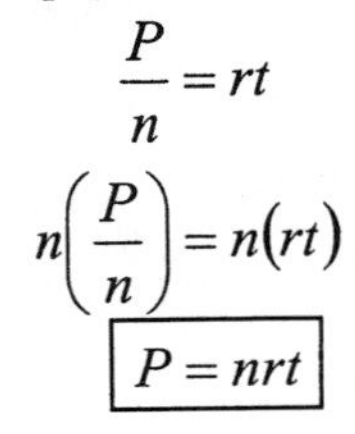

$$\frac{P}{n}=rt$$
$$n\left(\frac{P}{n}\right)=n(rt)$$
$$\boxed{P=nrt}$$

81. Multiply both sides of the equation by bd.

$$\frac{a}{b}=\frac{c}{d}$$
$$bd\left(\frac{a}{b}\right)=bd\left(\frac{c}{d}\right)$$
$$da=bc$$
$$\frac{da}{a}=\frac{bc}{a}$$
$$\boxed{d=\frac{bc}{a}}$$

83. Multiply both sides of the equation by $(b+d)$.

$$h=\frac{2A}{b+d}$$
$$(b+d)h=(b+d)\left(\frac{2A}{b+d}\right)$$
$$h(b+d)=2A$$
$$\frac{h(b+d)}{2}=\frac{2A}{2}$$
$$\boxed{\frac{h(b+d)}{2}=A}$$

85. Multiply both sides of the equation by ab.

$$\frac{1}{a}+\frac{1}{b}=1$$
$$(ab)\frac{1}{a}+(ab)\frac{1}{b}=(ab)1$$
$$b+a=ab$$
$$b+a-a=ab-a$$
$$b=ab-a$$
$$b=a(b-1)$$
$$\frac{b}{b-1}=\frac{a(b-1)}{b-1}$$
$$\boxed{\frac{b}{b-1}=a}$$

87. Multiply both sides of the equation by $(R + r)$.

$$I = \frac{E}{R+r}$$
$$(R+r)I = (R+r)\frac{E}{R+r}$$
$$I(R+r) = E$$
$$IR + Ir = E$$
$$IR + Ir - IR = E - IR$$
$$Ir = E - IR$$
$$\frac{Ir}{I} = \frac{E - IR}{I}$$
$$\boxed{r = \frac{E - IR}{I}}$$

89. Multiply both sides of the equation by rst.

$$\frac{1}{r} + \frac{1}{s} = \frac{1}{t}$$
$$(rst)\frac{1}{r} + (rst)\frac{1}{s} = (rst)\frac{1}{t}$$
$$st + rt = rs$$
$$st + rt - rt = rs - rt$$
$$st = rs - rt$$
$$st = r(s - t)$$
$$\frac{st}{s-t} = \frac{r(s-t)}{s-t}$$
$$\boxed{\frac{st}{s-t} = r}$$

91. Multiply both sides of the equation by $6d$.

$$F = \frac{L^2}{6d} + \frac{d}{2}$$
$$(6d)F = (6d)\frac{L^2}{6d} + (6d)\frac{d}{2}$$
$$6dF = L^2 + 3d^2$$
$$6dF - 3d^2 = L^2 + 3d^2 - 3d^2$$
$$\boxed{6dF - 3d^2 = L^2}$$

APPLICATIONS

93. MEDICINE

Multiply both sides of the equation by $(R + B)$.

$$H = \frac{RB}{R+B}$$
$$(R+B)H = (R+B)\frac{RB}{R+B}$$
$$HR + HB = RB$$
$$HR + HB - HR = RB - HR$$
$$HB = RB - HR$$
$$HB = R(B - H)$$
$$\boxed{\frac{HB}{B-H} = R}$$

95. ELECTRONICS

Multiply both sides of the equation by (rr_1r_2).

$$\frac{1}{r} = \frac{1}{r_1} + \frac{1}{r_2}$$
$$(rr_1r_2)\frac{1}{r} = (rr_1r_2)\frac{1}{r_1} + (rr_1r_2)\frac{1}{r_2}$$
$$r_1r_2 = rr_2 + rr_1$$
$$r_1r_2 = r(r_2 + r_1)$$
$$\frac{r_1r_2}{r_2 + r_1} = \frac{r(r_2 + r_1)}{r_2 + r_1}$$
$$\boxed{\frac{r_1r_2}{r_2 + r_1} = r}$$

WRITING

97. Answers will vary.

99. Answers will vary.

REVIEW

101. Factor out the GCF.
$x^2 + 4x = x(x + 4)$

103. Factor using FOIL
$2x^2 + x - 3 = (2x + 3)(x - 1)$

105. Factor Difference of Two Squares
$x^4 - 81 = (x^2 + 9)(x^2 - 9)$
$= (x^2 + 9)(x + 3)(x - 3)$

CHALLENGE PROBLEMS

107. Rewrite the equation without negative exponents. Then multiply both sides of the equation by x^2.

$$x^{-2} + 2x^{-1} + 1 = 0$$

$$\frac{1}{x^2} + \frac{2}{x} + 1 = 0$$

$$(x^2)\frac{1}{x^2} + (x^2)\frac{2}{x} + (x^2)\cdot 1 = (x^2)0$$

$$1 + 2x + x^2 = 0$$

$$x^2 + 2x + 1 = 0$$

$$(x + 1)(x + 1) = 0$$

$$x + 1 = 0$$

$$x + 1 - 1 = 0 - 1$$

$$\boxed{x = -1}$$

SECTION 6.7

VOCABULARY

1. In the formula $d = rt$, the variable d stands for the **distance** traveled, r is the **rate**, and t is the **time**.

3. In the formula $I = Prt$, the variable I stands for the amount of **interest** earned, P is the **principal**, r stands for the annual interest **rate**, and t is the **time**.

CONCEPTS

5. iii. $\frac{5+x}{8+x} = \frac{2}{3}$

7. Divide both sides by r.

$$d = rt$$
$$\frac{d}{r} = \frac{rt}{r}$$
$$\frac{d}{r} = t$$
$$t = \frac{d}{r}$$

9. a) $9\% = 0.09$
   b) $0.035 = 3.5\%$

11. $\frac{1}{3}$

13. $\frac{x}{4}$

15. City savings: $\frac{50}{r} \cdot r \cdot 1 = 50$

Credit union: $\left(\frac{75}{r-0.02}\right)(r-0.02)(1) = 75$

NOTATION

17. $\frac{55}{9} = 6\frac{2}{9}$ days

PRACTICE

19. NUMBER PROBLEM

Let the number be x.

$$\frac{2+x}{5+x} = \frac{2}{3}$$
$$3(5+x)\frac{2+x}{5+x} = 3(5+x)\frac{2}{3}$$
$$3(2+x) = 2(5+x)$$
$$6 + 3x = 10 + 2x$$
$$6 + 3x - 2x = 10 + 2x - 2x$$
$$6 + x = 10$$
$$6 + x - 6 = 10 - 6$$
$$\boxed{x = 4}$$

21. NUMBER PROBLEM

Let the number be x.

$$\frac{3 \bullet 2}{4+x} = 1$$
$$(4+x)\frac{6}{4+x} = 1(4+x)$$
$$6 = 4 + x$$
$$6 - 4 = 4 + x - 4$$
$$\boxed{2 = x}$$

23. NUMBER PROBLEM

Let the number be x.

$$\frac{3+x}{4+2x} = \frac{4}{7}$$
$$7(4+2x)\frac{3+x}{4+2x} = 7(4+2x)\frac{4}{7}$$
$$7(3+x) = 4(4+2x)$$
$$21 + 7x = 16 + 8x$$
$$21 + 7x - 7x = 16 + 8x - 7x$$
$$\boxed{21} = 16 + x$$
$$21 - 16 = 16 + x - 16$$
$$\boxed{5 = x}$$

25. NUMBER PROBLEM

Let the number be x.

$$x+\frac{1}{x}=\frac{13}{6}$$

$$(6x)\left(x+\frac{1}{x}\right)=(6x)\left(\frac{13}{6}\right)$$

$$(6x)(x)+(6x)\left(\frac{1}{x}\right)=(6x)\left(\frac{13}{6}\right)$$

$$6x^2+6=13x$$

$$6x^2+6-13x=13x-13x$$

$$6x^2-13x+6=0$$

$$(2x-3)(3x-2)=0$$

$$2x-3=0 \text{ and } 3x-2=0$$

$$2x=3 \text{ and } 3x=2$$

$$\boxed{x=\frac{3}{2} \text{ and } x=\frac{2}{3}}$$

APPLICATIONS

27. COOKING

$$\frac{1+x}{4+x}=\frac{3}{4}$$

$$4(4+x)\left(\frac{1+x}{4+x}\right)=4(4+x)\left(\frac{3}{4}\right)$$

$$4(1+x)=3(4+x)$$

$$4+4x=12+3x$$

$$4+4x-3x=12+3x-3x$$

$$4+x=12$$

$$4+x-4=12-4$$

$$\boxed{x=8}$$

29. FILLING A POOL

Let x be the number of hours it takes both pipes to fill the pool working together.

Amount of work 1st pipe in 1 hour = $\frac{1}{5}$

Amount of work 2nd pipe in 1 hour = $\frac{1}{4}$

Amount of work together in 1 hour = $\frac{1}{x}$

1st pipe + 2nd pipe = work together

$$\frac{1}{5}+\frac{1}{4}=\frac{1}{x}$$

$$20x\left(\frac{1}{5}+\frac{1}{4}\right)=20x\left(\frac{1}{x}\right)$$

$$4x+5x=20$$

$$9x=20$$

$$x=\frac{20}{9}$$

$$\boxed{x=2\frac{2}{9}}$$

It will take both $2\frac{2}{9}$ hours.

31. HOLIDAY DECORATING

Let x be the number of hours it takes both crews to put up the decorations working together.

1st Crew's work in 1 hour = $\frac{1}{8}$

2nd Crew's work in 1 hour = $\frac{1}{10}$

Amount of work together in 1 hour = $\frac{1}{x}$

1st crew + 2nd crew = work together

$$\frac{1}{8}+\frac{1}{10}=\frac{1}{x}$$

$$40x\left(\frac{1}{8}+\frac{1}{10}\right)=40x\left(\frac{1}{x}\right)$$

$$5x+4x=40$$

$$9x=40$$

$$x=\frac{40}{9}$$

$$\boxed{x=4\frac{4}{9}}$$

It will take both $4\frac{4}{9}$ hours.

33. FILLING A POOL

Let x be the number of hours it takes both pipes to fill the pool working together. Since the drain is EMPTYING the pool, subtract the work of the drain from the work of the pipe.

Amount of work pipe in 1 hour = $\frac{1}{4}$

Amount of work drain in 1 hour = $\frac{1}{8}$

Amount of work together in 1 hour = $\frac{1}{x}$

pipe's work - drain's work = work together

$$\frac{1}{4}-\frac{1}{8}=\frac{1}{x}$$

$$8x\left(\frac{1}{4}-\frac{1}{8}\right)=8x\left(\frac{1}{x}\right)$$

$$2x-x=8$$

$$\boxed{x=8}$$

It will take the pipe 8 hours.

35. GRADING PAPERS

Let x be the number of minutes it takes the teacher and the aide to grade the papers working together.

Teacher's work in 1 minute = $\frac{1}{30}$

Aide's work in 1 minute = $\frac{1}{2(30)}=\frac{1}{60}$

Amount of work together in 1 minute = $\frac{1}{x}$

$$\frac{1}{30}+\frac{1}{60}=\frac{1}{x}$$

$$60x\left(\frac{1}{30}+\frac{1}{60}\right)=60x\left(\frac{1}{x}\right)$$

$$2x+x=60$$

$$3x=60$$

$$\frac{3x}{3}=\frac{60}{3}$$

$$\boxed{x=20}$$

It will take both 20 minutes.

37. PRINTERS

Let x be the number of hours it takes both printers to print the schedules working together.

Slow printer's work in 1 hour = $\frac{1}{6}$

Old printer's work in 1 hour = $\frac{1}{4}$

$\frac{3}{4}$ of work together in 1 hour = $\frac{3}{4}\bullet\frac{1}{x}=\frac{3}{4x}$

$$\frac{1}{6}+\frac{1}{4}=\frac{3}{4x}$$

$$12x\left(\frac{1}{6}+\frac{1}{4}\right)=12x\left(\frac{3}{4x}\right)$$

$$2x+3x=9$$

$$5x=9$$

$$x=\frac{9}{5}$$

$$x=1\frac{4}{5}$$

$$\boxed{x=1.8}$$

It will take both 1.8 hours.

39. PHYSICAL FITNESS

Use $t=\dfrac{d}{r}$ to find the time for both riding and walking. Let the rate walking = r. Then the rate riding would be $(r + 10)$ since it is 10 mph faster.

Riding: $t=\dfrac{d}{r}$

$=\dfrac{28}{r+10}$

Walking: $t=\dfrac{d}{r}$

$=\dfrac{8}{r}$

Since the time is the same for both riding and walking, set both times equal to each other and solve the equation for r.

$$\frac{28}{r+10}=\frac{8}{r}$$
$$r(r+10)\left(\frac{28}{r+10}\right)=r(r+10)\left(\frac{8}{r}\right)$$
$$28r=8(r+10)$$
$$28r=8r+80$$
$$28r-8r=8r+80-8r$$
$$20r=80$$
$$\frac{20r}{20}=\frac{80}{20}$$
$$\boxed{r=4}$$

Her rate walking is 4 mph.

41. PACKING FRUIT

Use $t=\dfrac{d}{r}$ to find the time for both conveyor belts. Let the rate of the 1st belt be r, then the rate of the 2nd belt (1 ft per second slower) would be $r-1$.

1st belt: $t=\dfrac{d}{r}$

$=\dfrac{300}{r}$

2nd belt: $t=\dfrac{d}{r}$

$=\dfrac{10}{r-1}$

Set both times equal to each other and solve the equation for r.

$$\frac{300}{r}=\frac{100}{r-1}$$
$$r(r-1)\left(\frac{300}{r}\right)=r(r-1)\left(\frac{100}{r-1}\right)$$
$$300(r-1)=100r$$
$$300r-300=100r$$
$$300r-300-300r=100r-300r$$
$$-300=-200r$$
$$\frac{-300}{-200}=\frac{-200r}{-200}$$
$$\frac{3}{2}=r$$
$$r=\frac{3}{2}$$
$$\boxed{\begin{aligned} r&=1.5\\ r-1&=1.5-1\\ &=0.5\end{aligned}}$$

The rate of the 1st conveyor belt is 1.5 mph or $1\frac{1}{2}$ mph, and the rate of the 2nd conveyor belt would be $1.5 - 1 = 0.5$ mph or $\frac{1}{2}$ mph.

43. WIND SPEED

Use $t=\frac{d}{r}$ to find the time for both planes.

Let x be the rate of the wind. The rate downwind would be $255+x$ and the rate of the plane upwind would be $255-x$.

Downwind: $t=\frac{d}{r}$

$$=\frac{300}{255+x}$$

Upwind: $t=\frac{d}{r}$

$$=\frac{210}{255-x}$$

Since the flying time is the same for both planes, set both times equal to each other and solve the equation for x.

$$\frac{300}{255+x}=\frac{210}{255-x}$$

$$(255+x)(255-x)\left(\frac{300}{255+x}\right)=(255+x)(255-x)\left(\frac{210}{255-x}\right)$$

$$300(255-x)=210(255+x)$$

$$76{,}500-300x=53{,}550+210x$$

$$76{,}500-300x-210x=53{,}550+210x-210x$$

$$76{,}500-510x=53{,}550$$

$$76{,}500-510x-76{,}500=53{,}550-76{,}500$$

$$-510x=-22{,}950$$

$$\frac{-510x}{-510}=\frac{-22{,}950}{-510}$$

$$\boxed{x=45}$$

The rate of the wind is 45 mph.

45. COMPARING INVESTMENTS

Use $P=\frac{I}{rt}$ to find the rate of interest paid by each investment. Let x be the rate of the interest paid by the tax-free bonds. Then the rate of the interest paid by the credit union would be $x-0.02$ (2% less than). Let $t=1$ year.

Bonds: $P=\frac{I}{rt}$

$$=\frac{300}{x(1)}$$

$$=\frac{300}{x}$$

Credit Union: $P=\frac{I}{rt}$

$$=\frac{200}{(x-0.02)(1)}$$

$$=\frac{200}{x-0.02}$$

Since the principal is the same for both investments, set both rational expressions equal to each other and solve the equation for x.

$$\frac{300}{x}=\frac{200}{x-0.02}$$

$$x(x-0.02)\left(\frac{300}{x}\right)=x(x-0.02)\left(\frac{200}{x-0.02}\right)$$

$$300(x-0.02)=200x$$

$$300x-6=200x$$

$$300x-6-300x=200x-300x$$

$$-6=-100x$$

$$\frac{-6}{-100}=x$$

$$0.06=x$$

$$\boxed{x=6\%}$$

rate of interest of bonds = 6%

rate of interest of credit union = 6–2

= 4%

47. COMPARING INVESTMENTS

Use $P = \frac{I}{rt}$ to find the rate of interest paid by each investment. Let x be the rate of the interest paid by the 1st CD. Then the rate of the interest paid by the 2nd CD would be $x + 0.01$ (differ by 1%). Let $t = 1$ year.

1st CD:
$$P = \frac{I}{rt} = \frac{175}{x(1)} = \frac{175}{x}$$

2nd CD:
$$P = \frac{I}{rt} = \frac{200}{(x+0.01)(1)} = \frac{200}{x+0.01}$$

Since the principal is the same for both accounts, set both rational expressions equal to each other and solve the equation for x.

$$\frac{175}{x} = \frac{200}{x+0.01}$$
$$x(x+0.01)\left(\frac{175}{x}\right) = x(x+0.01)\left(\frac{200}{x+0.01}\right)$$
$$175(x+0.01) = 200x$$
$$175x + 1.75 = 200x$$
$$175x + 1.75 - 175x = 200x - 175x$$
$$1.75 = 25x$$
$$\frac{1.75}{25} = x$$
$$0.07 = x$$
$$\boxed{7\% = x}$$

The rate of interest of the 1st CD is 7%.
The rate of interest of 2nd CD is 7 +1 = 8%.

WRITING

49. Answers will vary.

REVIEW

51. repeating; $\frac{7}{9} = 9\overline{)7.000}$ (quotient 0.777) $= 0.\overline{7}$

53. {..., -4, -3, -2, -1, 0, 1, 2, 3, 4, ...}

55. $2(\mathbf{-3})^2 + 5(\mathbf{-3}) - 3 = 2(9) + 5(-3) - 3$
$= 18 - 15 - 3$
$= 3 - 3$
$= 0$

CHALLENGE PROBLEMS

57. RIVER TOURS

Use $t = \frac{d}{r}$ to find the time for both trips.

Let x equal the rate of the boat in still water. The rate of the boat downstream would be $x + 5$ and the rate of the boat upstream would be $x - 5$. $d = 60$.

Downstream: $t = \frac{d}{r}$

$= \frac{60}{x+5}$

Upstream: $t = \frac{d}{r}$

$= \frac{60}{x-5}$

Since the total time is 5, find the sum of the times and set it equal to 5.

$$\frac{60}{x+5}+\frac{60}{x-5}=5$$
$$(x+5)(x-5)\left(\frac{60}{x+5}+\frac{60}{x-5}\right)=(x+5)(x-5)(5)$$
$$60(x-5)+60(x+5)=5(x^2-25)$$
$$60x-300+60x+300=5x^2-125$$
$$120x=5x^2-125$$
$$120x-120x=5x^2-125-120x$$
$$0=5x^2-120x-125$$
$$\frac{0}{5}=\frac{5x^2}{5}-\frac{120x}{5}-\frac{125}{5}$$
$$0=x^2-24x-25$$
$$0=(x-25)(x+1)$$
$$x-25=0 \text{ and } x+1=0$$
$$x=25 \text{ and } x=-1$$

The rate cannot be negative, so $\boxed{x = 25}$

The still-water rate of the boat is 25 mph.

59. SALES

Let x = the number of radios bought.

Cost of each radio $= \frac{1,200}{x}$

$x - 6$ = the number of radios sold

Cost of each radio + \$10 $= \frac{1,200}{x-6}$

Cost of each radio+\$10–\$10 $= \frac{1,200}{x-6} - \$10$

Cost of each radio $= \frac{1,200}{x-6} - 10$

Set the two expressions equal to each other and solve for x.

$$\frac{1,200}{x}=\frac{1,200}{x-6}-10$$
$$x(x-6)\left(\frac{1,200}{x}\right)=x(x-6)\left(\frac{1,200}{x-6}-10\right)$$
$$1,200(x-6)=1,200x-10x(x-6)$$
$$1,200x-7,200=1,200x-10x^2+60x$$
$$1,200x-7,200-1,200x=1,200x-10x^2+60x-1,200x$$
$$-7,200=-10x^2+60x$$
$$-7,200+10x^2-60x=-10x^2+60x+10x^2-60x$$
$$10x^2-60x-7,200=0$$
$$10(x^2-6x-720)=0$$
$$10(x-30)(x+24)=0$$
$$x-30=0 \text{ and } x+24=0$$
$$\boxed{x=30} \text{ and } x=-24$$

Since x cannot be negative, $x = 30$.
She bought 30 radios.

SECTION 6.8

VOCABULARY

1. A **ratio** is the quotient of two numbers or the quotient of two quantities with the same units. A **rate** is a quotient of two quantities that have different units.

3. The **terms** of the proportion $\frac{2}{x}=\frac{16}{40}$ are 2, x, 16, and 40.

5. The product of the extremes and the product of the means of a proportion are also known as **cross** products.

7. Examples of **unit** prices are $1.65 per gallon, 17¢ per day, and $50 per foot.

CONCEPTS

9. WEST AFRICA

$$\frac{\text{red stripes}}{\text{white stripes}}=\frac{6}{5}$$

11. a) $6(10)=60$
    $5(12)=60$
    b) $15(x)=15x$
    $2(45)=90$

13. SNACK FOODS

$$\frac{\boxed{25}}{\boxed{2}}=\frac{\boxed{1{,}000}}{\boxed{x}}$$

15. GROCERY SHOPPING

The 24 twelve ounce bottles would be the better buy.

12 - eight oz bottles =

$$\frac{\text{price}}{\text{number of ounces}}=\frac{1.79}{12(8)}$$
$$=\frac{1.79}{96}$$
$$=1.9\text{ cents per oz}$$

24 - twelve oz bottles =

$$\frac{\text{price}}{\text{number of ounces}}=\frac{4.49}{24(12)}$$
$$=\frac{4.49}{288}$$
$$=1.6\text{ cents per oz}$$

17. $$\frac{AB}{DE}=\frac{\boxed{BC}}{EF}$$
$$\frac{BC}{\boxed{EF}}=\frac{CA}{FD}$$
$$\frac{CA}{FD}=\frac{AB}{\boxed{DE}}$$

NOTATION

19. $$\frac{12}{18}=\frac{x}{24}$$
$$12\cdot 24=18\cdot\boxed{x}$$
$$\boxed{288}=18x$$
$$\frac{288}{\boxed{18}}=\frac{18x}{\boxed{18}}$$
$$\boxed{16}=x$$

21. a) $\frac{12}{15}=\frac{4}{5}$
    b) $\frac{9\text{ crates}}{7\text{ crates}}=\frac{9}{7}$

23. We read $\Delta XYZ \sim \Delta MNO$ as: triangle XYZ is **similar** to triangle MNO.

PRACTICE

25. $\frac{4}{15}$

27. $\frac{30}{24} = \frac{5}{4}$

29. 60 minutes = 1 hour

$$\frac{90 \text{ minutes}}{3 \text{ hours}} = \frac{90 \text{ minutes}}{3(60) \text{ minutes}} = \frac{90}{180} = \frac{1}{2}$$

31. 4 quarts = 1 gallon

$$\frac{13 \text{ quarts}}{2 \text{ gallons}} = \frac{13 \text{ quarts}}{2(4) \text{ quarts}} = \frac{13}{8}$$

33. no; $9(70) \stackrel{?}{=} 7(81)$
$630 \neq 567$

35. yes; $7(6) \stackrel{?}{=} 3(14)$
$42 = 42$

37. no; $9(80) \stackrel{?}{=} 19(38)$
$720 \neq 722$

39.
$$\begin{aligned} 2(6) &= 3 \cdot x \\ 12 &= 3x \\ \frac{12}{3} &= \frac{3x}{3} \\ \boxed{4 = x} \end{aligned}$$

41.
$$\begin{aligned} 5 \cdot c &= 10(3) \\ 5c &= 30 \\ \frac{5c}{5} &= \frac{30}{5} \\ \boxed{c = 6} \end{aligned}$$

43.
$$\begin{aligned} 6(4) &= 8 \cdot x \\ 24 &= 8x \\ \frac{24}{8} &= \frac{8x}{8} \\ \boxed{3 = x} \end{aligned}$$

45.
$$\begin{aligned} (x+1) \cdot 15 &= 5 \cdot 3 \\ 15(x+1) &= 15 \\ 15x + 15 &= 15 \\ 15x + 15 - 15 &= 15 - 15 \\ 15x &= 0 \\ \frac{15x}{15} &= \frac{0}{15} \\ \boxed{x = 0} \end{aligned}$$

47.
$$\begin{aligned} (x+7) \cdot 4 &= -4 \cdot 1 \\ 4(x+7) &= -4 \\ 4x + 28 &= -4 \\ 4x + 28 - 28 &= -4 - 28 \\ 4x &= -32 \\ \frac{4x}{4} &= \frac{-32}{4} \\ \boxed{x = -8} \end{aligned}$$

49.
$$\begin{aligned} (5-x) \cdot 34 &= 17 \cdot 13 \\ 34(5-x) &= 221 \\ 170 - 34x &= 221 \\ 170 - 34x - 170 &= 221 - 170 \\ -34x &= 51 \\ \frac{-34x}{-34} &= \frac{51}{-34} \\ \boxed{x = -\frac{3}{2}} \end{aligned}$$

51.
$$\begin{aligned} (2x-1) \cdot 54 &= 18 \cdot 9 \\ 54(2x-1) &= 162 \\ 108x - 54 &= 162 \\ 108x - 54 + 54 &= 162 + 54 \\ 108x &= 216 \\ \frac{108x}{108} &= \frac{216}{108} \\ \boxed{x = 2} \end{aligned}$$

53.
$$\begin{aligned} (x-1) \cdot 3 &= 9 \cdot 2x \\ 3(x-1) &= 18x \\ 3x - 3 &= 18x \\ 3x - 3 - 3x &= 18x - 3x \\ -3 &= 15x \\ \frac{-3}{15} &= \frac{15x}{15} \\ \boxed{-\frac{1}{5} = x} \end{aligned}$$

55.
$$\begin{aligned} 8x \cdot 4 &= 3 \cdot (11x + 9) \\ 32x &= 3(11x + 9) \\ 32x &= 33x + 27 \\ 32x - 33x &= 33x + 27 - 33x \\ -x &= 27 \\ \frac{-x}{-1} &= \frac{27}{-1} \\ \end{aligned}$$
$$\boxed{x = -27}$$

57.
$$\begin{aligned} 2 \cdot 6 &= 3x \cdot x \\ 12 &= 3x^2 \\ 12 - 12 &= 3x^2 - 12 \\ 0 &= 3x^2 - 12 \\ 0 &= 3(x^2 - 4) \\ 0 &= 3(x - 2)(x + 2) \end{aligned}$$
$$x - 2 = 0 \quad \text{and} \quad x + 2 = 0$$
$$x - 2 + 2 = 0 + 2 \qquad x + 2 - 2 = 0 - 2$$
$$x = 2 \qquad x = -2$$
$$\boxed{x = 2, -2}$$

59.
$$\begin{aligned} (b - 5) \cdot b &= 3 \cdot 2 \\ b(b - 5) &= 6 \\ b^2 - 5b &= 6 \\ b^2 - 5b - 6 &= 6 - 6 \\ b^2 - 5b - 6 &= 0 \\ (b - 6)(b + 1) &= 0 \end{aligned}$$
$$b - 6 = 0 \quad \text{and} \quad b + 1 = 0$$
$$b - 6 + 6 = 0 + 6 \qquad b + 1 - 1 = 0 - 1$$
$$b = 6 \qquad b = -1$$
$$\boxed{b = 6, -1}$$

61.
$$\begin{aligned} (x - 1) \cdot 3x &= (x + 1) \cdot 2 \\ 3x(x - 1) &= 2(x + 1) \\ 3x^2 - 3x &= 2x + 2 \\ 3x^2 - 3x - 2x - 2 &= 2x + 2 - 2x - 2 \\ 3x^2 - 5x - 2 &= 0 \\ (3x + 1)(x - 2) &= 0 \end{aligned}$$
$$3x + 1 = 0 \quad \text{and} \quad x - 2 = 0$$
$$3x + 1 - 1 = 0 - 1 \qquad x - 2 + 2 = 0 + 2$$
$$3x = -1 \qquad x = 2$$
$$\frac{3x}{3} = \frac{-1}{3}$$
$$x = -\frac{1}{3}$$
$$\boxed{x = -\frac{1}{3}, 2}$$

63.
$$\begin{aligned} \frac{4}{12} &= \frac{5}{x} \\ 4(x) &= 12(5) \\ 4x &= 60 \\ \frac{4x}{4} &= \frac{60}{4} \end{aligned}$$
$$\boxed{x = 15}$$

65.
$$\begin{aligned} \frac{x}{12} &= \frac{6}{9} \\ x(9) &= 12(6) \\ 9x &= 72 \\ \frac{9x}{9} &= \frac{72}{9} \end{aligned}$$
$$\boxed{x = 8}$$

APPLICATIONS

67. GEAR RATIOS

a) $\frac{18}{12} = \frac{3}{2}$, 3:2

b) $\frac{12}{18} = \frac{2}{3}$, 2:3

69. SHOPPING FOR CLOTHES

$$\begin{aligned} \frac{2 \text{ shirts}}{\$25} &= \frac{5 \text{ shirts}}{x} \\ 2(x) &= 25(5) \\ 2x &= 125 \\ \frac{2x}{2} &= \frac{125}{2} \end{aligned}$$
$$\boxed{x = \$62.50}$$

71. COOKING

$$\frac{\text{bottles}}{\text{gallons}}$$
$$\begin{aligned} \frac{4}{2} &= \frac{x}{10} \\ 4(10) &= 2(x) \\ 40 &= 2x \\ \frac{40}{2} &= \frac{2x}{2} \end{aligned}$$
$$\boxed{20 = x}$$

73. CPR

$$\frac{\text{compressions}}{\text{breaths}}$$

$$\frac{5}{2}=\frac{210}{x}$$
$$5(x)=2(210)$$
$$5x=420$$
$$\frac{5x}{5}=\frac{420}{5}$$
$$\boxed{x=84}$$

75. NUTRITION

calories:

$$\frac{10\text{-oz}}{16\text{-oz}}=\frac{355}{x}$$
$$10(x)=16(355)$$
$$10x=5{,}680$$
$$\frac{10x}{10}=\frac{5{,}680}{10}$$
$$\boxed{x=568}$$

fat:

$$\frac{10\text{-oz}}{16\text{-oz}}=\frac{8}{x}$$
$$10(x)=16(8)$$
$$10x=128$$
$$\frac{10x}{10}=\frac{128}{10}$$
$$x=12.8$$
$$\boxed{x\approx 13}$$

protein:

$$\frac{10\text{-oz}}{16\text{-oz}}=\frac{9}{x}$$
$$10(x)=16(9)$$
$$10x=144$$
$$\frac{10x}{10}=\frac{144}{10}$$
$$x=14.4$$
$$\boxed{x\approx 14}$$

77. QUALITY CONTROL

$$\frac{17}{500}=\frac{x}{15{,}000}$$
$$17(15{,}000)=500(x)$$
$$255{,}000=500x$$
$$\frac{255{,}000}{500}=\frac{500x}{500}$$
$$\boxed{510=x}$$

79. MIXING FUELS

$$\frac{50}{1}=\frac{6(128)}{x}$$
$$\frac{50}{1}=\frac{768}{x}$$
$$50x=768$$
$$\frac{50x}{50}=\frac{768}{50}$$
$$\boxed{x=15.36}$$

The instructions are not exactly right, but close.

81. CROP DAMAGE

$$\frac{35}{3}=\frac{x}{12}$$
$$35(12)=3(x)$$
$$420=3x$$
$$\frac{420}{3}=\frac{3x}{3}$$
$$\boxed{140=x}$$

83. MODEL RAILROADS

Change 9 inches to $\frac{3}{4}$ foot.

$$\frac{87}{1}=\frac{x}{9\text{ inches}}$$
$$\frac{87}{1}=\frac{x}{\frac{9}{12}\text{ foot}}$$
$$\frac{87}{1}=\frac{x}{\frac{3}{4}}$$
$$87\left(\frac{3}{4}\right)=1(x)$$
$$\frac{261}{4}=x$$
$$64\frac{1}{4}=x$$
$$\boxed{\begin{array}{l} x=65\frac{1}{4}\text{ feet} \\ x=65\text{ feet, 3 inches}\end{array}}$$

85. BLUEPRINTS

Change 1 foot to 12 inches.

$$\frac{\frac{1}{4}}{12} = \frac{2\frac{1}{2}}{x}$$

$$\frac{1}{4}x = 12\left(2\frac{1}{2}\right)$$

$$\frac{1}{4}x = 12\left(\frac{5}{2}\right)$$

$$\frac{1}{4}x = 30$$

$$4 \bullet \frac{1}{4}x = 4 \bullet 30$$

$$x = 120 \text{ inches}$$

$$x = \frac{120}{12}$$

$$\boxed{x = 10 \text{ feet}}$$

87. $\frac{\$25}{45 \text{ minutes}} \approx \$0.56$ per minute

$\frac{\$35}{60 \text{ minutes}} \approx \$0.58$ per minute

45 minutes for $25 is the better buy.

89. $\frac{\$9.99}{100 \text{ cards}} \approx \$0.10$ per card

$\frac{\$12.99}{150 \text{ cards}} \approx \$0.09$ per card

150 cards for $12.99 is the better buy.

91. $\frac{\$1.50}{6 \text{ drinks}} = \$0.25$ per drink

$\frac{\$6.25}{24 \text{ drinks}} \approx \$0.26$ per drink

6-pack for $1.50 is the better buy.

93. HEIGHT OF A TREE

$$\frac{6}{h} = \frac{4}{26}$$

$$6(26) = h(4)$$

$$156 = 4h$$

$$\frac{156}{4} = \frac{4h}{4}$$

$$\boxed{39 \text{ ft} = h}$$

95. SURVEYING

$$\frac{20}{32} = \frac{w}{75}$$

$$32w = 20(75)$$

$$32w = 1{,}500$$

$$\frac{32w}{32} = \frac{1{,}500}{32}$$

$$w = 46.875$$

$$\boxed{w = 46\tfrac{7}{8} \text{ ft.}}$$

WRITING

97. Answers will vary.

99. Answers will vary.

REVIEW

101. $\frac{9}{10} = 10\overline{)9.0} = 0.9 = 0.9 = 90\%$ (quotient 0.9)

103. $0.30(1{,}600) = 480$

CHALLENGE PROBLEMS

105. $\frac{b}{a} = \frac{d}{c}, \quad \frac{c}{d} = \frac{a}{b}, \quad \frac{c}{a} = \frac{d}{b}$

CHAPTER 6: KEY CONCEPTS

1. a.

$$\frac{2x^2 - 8x}{x^2 - 6x + 8} = \frac{2x(x-4)}{(x-2)(x-4)}$$
$$= \frac{2x}{x-2}$$

   b. $x - 4$

2. a.

$$\frac{x^2 + 2x + 1}{x} \bullet \frac{x^2 - x}{x^2 - 1} = \frac{(x+1)(x+1) \cdot x(x-1)}{x \cdot (x+1)(x-1)}$$
$$= x + 1$$

   b. $x, x - 1, x + 1$

3. a.

$$\frac{x}{x+1} + \frac{x-1}{x} = \frac{x}{x+1} \bullet \frac{x}{x} + \frac{x-1}{x} \bullet \frac{x+1}{x+1}$$
$$= \frac{x^2 + (x-1)(x+1)}{x(x+1)}$$
$$= \frac{x^2 + x^2 - 1}{x(x+1)}$$
$$= \frac{2x^2 - 1}{x(x+1)}$$

   b. $\frac{x}{x}, \frac{x+1}{x+1}$

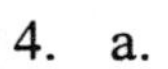

4. a.

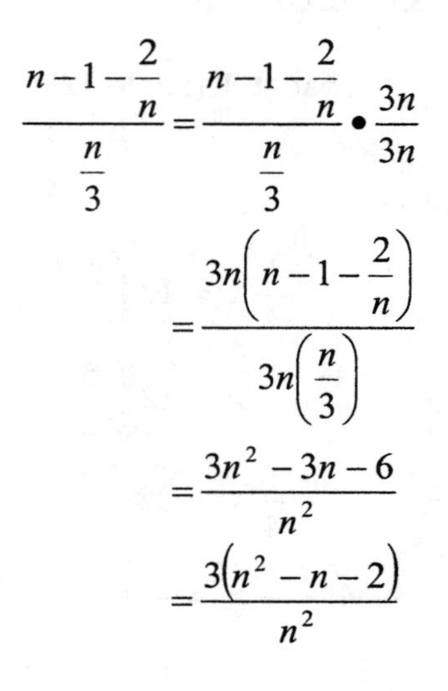

$$\frac{n - 1 - \frac{2}{n}}{\frac{n}{3}} = \frac{n - 1 - \frac{2}{n}}{\frac{n}{3}} \bullet \frac{3n}{3n}$$
$$= \frac{3n\left(n - 1 - \frac{2}{n}\right)}{3n\left(\frac{n}{3}\right)}$$
$$= \frac{3n^2 - 3n - 6}{n^2}$$
$$= \frac{3(n^2 - n - 2)}{n^2}$$

   b. $3n$

5. a.

$$\frac{11}{b} + \frac{13}{b} = 12$$
$$b\left(\frac{11}{b} + \frac{13}{b}\right) = b(12)$$
$$b\left(\frac{11}{b}\right) + b\left(\frac{13}{b}\right) = b(12)$$
$$11 + 13 = 12b$$
$$24 = 12b$$
$$\frac{24}{12} = \frac{12b}{12}$$
$$\boxed{2 = b}$$

   b. b

6. a.

$$\frac{3}{s+2}-\frac{5}{s^2+s-2}=\frac{1}{s-1}$$
$$\frac{3}{s+2}-\frac{5}{(s+2)(s-1)}=\frac{1}{s-1}$$
$$(s+2)(s-1)\left(\frac{3}{s+2}-\frac{5}{(s+2)(s-1)}\right)=(s+2)(s-1)\left(\frac{1}{s-1}\right)$$
$$3(s-1)-5=s+2$$
$$3s-3-5=s+2$$
$$3s-8=s+2$$
$$3s-8-s=s+2-s$$
$$2s-8=2$$
$$2s-8+8=2+8$$
$$2s=10$$
$$\frac{2s}{2}=\frac{10}{2}$$
$$\boxed{s=5}$$

b. $(s+2)(s-1)$

7. a.

$$y+\frac{3}{4}=\frac{3y-50}{4y-24}$$
$$y+\frac{3}{4}=\frac{3y-50}{4(y-6)}$$
$$4(y-6)\left(y+\frac{3}{4}\right)=4(y-6)\left(\frac{3y-50}{4(y-6)}\right)$$
$$4y(y-6)+3(y-6)=3y-50$$
$$4y^2-24y+3y-18=3y-50$$
$$4y^2-21y-18=3y-50$$
$$4y^2-21y-18-3y+50=3y-50-3y+50$$
$$4y^2-24y+32=0$$
$$4(y^2-6y+8)=0$$
$$4(y-2)(y-4)=0$$
$$y-2=0 \quad \text{and} \quad y-4=0$$
$$y-2+2=0+2 \quad y-4+4=0+4$$
$$\boxed{y=2 \quad \text{and} \quad y=4}$$

b. $4(y-6)$

8. a.

$$\frac{1}{a}-\frac{1}{b}=1$$
$$ab\left(\frac{1}{a}-\frac{1}{b}\right)=ab\bullet 1$$
$$b-a=ab$$
$$b-a-b=ab-b$$
$$-a=b(a-1)$$
$$-\frac{a}{a-1}=\frac{b(a-1)}{a-1}$$
$$\frac{a}{1-a}=b$$
$$b=\frac{a}{1-a}$$

b. ab

CHAPTER 6 REVIEW

SECTION 6.1
Simplifying Rational Expressions

1. To find the values of x for which the rational expression is undefined, set the denominator equal to zero and solve for x.
$x^2 - 16 = 0$
$(x-4)(x+4) = 0$
$x - 4 = 0$ and $x + 4 = 0$
$\boxed{x = 4 \text{ and } x = -4}$

2.
$$\frac{(-2)^2 - 1}{-2-5} = \frac{4-1}{-2-5} = \frac{3}{-7} = -\frac{3}{7}$$

3.
$$\frac{3x^2}{6x^3} = \frac{1}{2}x^{2-3} = \frac{1}{2}x^{-1} = \frac{1}{2x}$$

4.
$$\frac{5xy^2}{2x^2y^2} = \frac{5}{2}x^{1-2}y^{2-2} = \frac{5}{2}x^{-1}y^0 = \frac{5}{2x}$$

5.
$$\frac{x^2}{x^2+x} = \frac{\cancel{x} \bullet x}{\cancel{x}(x+1)} = \frac{x}{x+1}$$

6.
$$\frac{a^2-4}{a+2} = \frac{\cancel{(a+2)}(a-2)}{\cancel{a+2}} = a - 2$$

7.
$$\frac{3p-2}{2-3p} = -\frac{\cancel{3p-2}}{\cancel{3p-2}} = -1$$

8.
$$\frac{8-x}{x^2-5x-24} = \frac{8-x}{(x-8)(x+3)} = -\frac{\cancel{x-8}}{\cancel{(x-8)}(x+3)} = -\frac{1}{x+3}$$

9.
$$\frac{2x^2-16x}{2x^2-18x+16} = \frac{2x(x-8)}{2(x^2-9x+8)} = \frac{\cancel{2}x\cancel{(x-8)}}{\cancel{2}\cancel{(x-8)}(x-1)} = \frac{x}{x-1}$$

10. $\dfrac{x^2+x-2}{x^2-x-2} = \dfrac{(x+2)(x-1)}{(x-2)(x+1)}$

It does not simplify

11.
$$\frac{4(t+3)+8}{3(t+3)+6} = \frac{4t+12+8}{3t+9+6} = \frac{4t+20}{3t+15} = \frac{4\cancel{(t+5)}}{3\cancel{(t+5)}} = \frac{4}{3}$$

12. x is not a common factor of the numerator and the denominator; x is a term of the numerator.

13. DOSAGES

Let $A = 11$ and $D = 300$.

$$C = \frac{\mathbf{300(11+1)}}{24} = \frac{300(12)}{24} = \frac{3{,}600}{24} = 150 \text{ mg}$$

SECTION 6.2
Multiplying and Dividing Rational Expressions

14.

$$\frac{3xy}{2x} \bullet \frac{4x}{2y^2} = \frac{3\cdot\cancel{x}\cdot\cancel{y}\cdot\cancel{2}\cdot\cancel{2}\cdot x}{\cancel{2}\cdot\cancel{x}\cdot\cancel{2}\cdot\cancel{y}\cdot y} = \frac{3x}{y}$$

15.

$$56x\left(\frac{12}{7x}\right) = \frac{56x}{1}\left(\frac{12}{7x}\right) = \frac{2\cdot 2\cdot 2\cdot\cancel{7}\cdot\cancel{x}\cdot 2\cdot 2\cdot 3}{\cancel{7}\cdot\cancel{x}} = 96$$

16.

$$\frac{x^2-1}{x^2+2x} \bullet \frac{x}{x+1} = \frac{\cancel{(x+1)}(x-1)\cancel{x}}{\cancel{x}(x+2)\cancel{(x+1)}} = \frac{x-1}{x+2}$$

17.

$$\frac{x^2+x}{3x-15} \bullet \frac{6x-30}{x^2+2x+1} = \frac{x\cancel{(x+1)}\cdot 2\cdot\cancel{3}\cancel{(x-5)}}{\cancel{3}\cancel{(x-5)}\cancel{(x+1)}(x+1)} = \frac{2x}{x+1}$$

18.

$$\frac{3x^2}{5x^2y} \div \frac{6x}{15xy^2} = \frac{3x^2}{5x^2y} \bullet \frac{15xy^2}{6x} = \frac{\cancel{3}\cdot\cancel{x}\cdot\cancel{x}\cdot 3\cdot\cancel{5}\cdot\cancel{x}\cdot\cancel{y}\cdot y}{\cancel{5}\cdot\cancel{x}\cdot\cancel{x}\cdot\cancel{y}\cdot 2\cdot\cancel{3}\cdot\cancel{x}} = \frac{3y}{2}$$

19.

$$\frac{x^2+5x}{x^2+4x-5} \div x^2 = \frac{x^2+5x}{x^2+4x-5} \bullet \frac{1}{x^2} = \frac{\cancel{x}\cancel{(x+5)}}{\cancel{(x+5)}(x-1)\cdot\cancel{x}\cdot x} = \frac{1}{x(x-1)}$$

20.

$$\frac{x^2-x-6}{1-2x} \div \frac{x^2-2x-3}{2x^2+x-1} = \frac{x^2-x-6}{1-2x} \bullet \frac{2x^2+x-1}{x^2-2x-3}$$
$$= \frac{(x-3)(x+2)(2x-1)(x+1)}{(1-2x)(x-3)(x+1)}$$
$$= \frac{\cancel{(x-3)}(x+2)\cancel{(2x-1)}\cancel{(x+1)}}{\cancel{(2x-1)}\cancel{(x-3)}\cancel{(x+1)}}$$
$$= -(x+2)$$
$$= -x-2$$

21. a) Yes, 1 ft = 12 in.
    b) No, 60 min $\neq$1 day
    c) Yes, 2,000 lbs = 1 ton
    d) Yes, 1 gal = 4 qt.

22. TRAFFIC SIGNS

$$\frac{20 \text{ miles}}{\text{hour}} \bullet \frac{1 \text{ hour}}{60 \text{ minues}} = \frac{20 \text{ miles}}{60 \text{ minutes}} = \frac{1}{3} \text{ mile per minute}$$

SECTION 6.3
Addition and Subtraction with Like Denominator; Least Common Denominators

23. $\frac{13c}{15d} - \frac{8}{15d} = \frac{13c-8}{15d}$

24.

$$\frac{x}{x+y}+\frac{y}{x+y}=\frac{\cancel{x+y}}{\cancel{x+y}}$$
$$=1$$

25.

$$\frac{3x}{x-7}-\frac{x-2}{x-7}=\frac{3x-(x-2)}{x-7}$$
$$=\frac{3x-x+2}{x-7}$$
$$=\frac{2x+2}{x-7}$$

26.

$$\frac{a}{a^2-2a-8}+\frac{2}{a^2-2a-8}=\frac{a+2}{a^2-2a-8}$$
$$=\frac{\cancel{a+2}}{(a-4)(\cancel{a+2})}$$
$$=\frac{1}{a-4}$$

27. x = does not simplify
$9=3\cdot 3$
LCD $=3\cdot 3\cdot x=\mathbf{9x}$

28. $2x^3=2\cdot x\cdot x\cdot x$
$8x=2\cdot 2\cdot 2\cdot x$
LCD $=2\cdot 2\cdot 2\cdot x\cdot x\cdot x=\mathbf{8x^3}$

29. m = does not simplify
$m-8$ = does not simplify
LCD $=m(m-8)$

30. $5x+1$ = does not simplify
$5x-1$ = does not simplify
LCD $=(5x+1)(5x-1)$

31. $a^2-25=(a-5)(a+5)$
$a-5$ = does not simplify
LCD $=(a-5)(a+5)$

32. $t^2+10t+25=(t+5)(t+5)$
$2t^2+17t+35=(t+5)(2t+7)$
LCD $=(t+5)(t+5)(2t+7)$
$=(2t+7)\,(t+5)^2$

33.

$$\frac{9}{a}=\frac{9}{a}\bullet\frac{7}{7}$$
$$=\frac{63}{7a}$$

34.

$$\frac{2y+1}{x-9}=\frac{2y+1}{x-9}\left(\frac{x}{x}\right)$$
$$=\frac{x(2y+1)}{x(x-9)}$$
$$=\frac{2xy+x}{x(x-9)}$$

35.

$$\frac{b+7}{3b-15}=\frac{b+7}{3(b-5)}$$
$$=\frac{b+7}{3(b-5)}\left(\frac{2}{2}\right)$$
$$=\frac{2b+14}{6(b-5)}$$

36.

$$\frac{9r}{r^2+6r+5}=\frac{9r}{(r+1)(r+5)}\left(\frac{r-4}{r-4}\right)$$
$$=\frac{9r^2-36r}{(r+1)(r+5)(r-4)}$$

SECTION 6.4
Addition and Subtraction with Unlike Denominator

37.

$$\frac{1}{7}-\frac{1}{a}=\frac{1}{7}\left(\frac{a}{a}\right)-\frac{1}{a}\left(\frac{7}{7}\right)$$
$$=\frac{a}{7a}-\frac{7}{7a}$$
$$=\frac{a-7}{7a}$$

38.

$$\frac{x}{x-1}+\frac{1}{x}=\frac{x}{x-1}\left(\frac{x}{x}\right)+\frac{1}{x}\left(\frac{x-1}{x-1}\right)$$
$$=\frac{x^2}{x(x-1)}+\frac{x-1}{x(x-1)}$$
$$=\frac{x^2+x-1}{x(x-1)}$$

39.

$$\frac{2t+2}{t^2+2t+1}-\frac{1}{t+1}=\frac{2t+2}{(t+1)(t+1)}-\frac{1}{t+1}$$
$$=\frac{2t+2}{(t+1)(t+1)}-\frac{1}{t+1}\left(\frac{t+1}{t+1}\right)$$
$$=\frac{2t+2}{(t+1)(t+1)}-\frac{t+1}{(t+1)(t+1)}$$
$$=\frac{2t+2-(t+1)}{(t+1)(t+1)}$$
$$=\frac{2t+2-t-1}{(t+1)(t+1)}$$
$$=\frac{\cancel{t+1}}{(\cancel{t+1})(t+1)}$$
$$=\frac{1}{t+1}$$

40.

$$\frac{x+2}{2x}-\frac{2-x}{x^2}=\frac{x+2}{2x}\left(\frac{x}{x}\right)-\frac{2-x}{x^2}\left(\frac{2}{2}\right)$$
$$=\frac{x^2+2x}{2x^2}-\frac{4-2x}{2x^2}$$
$$=\frac{x^2+2x-(4-2x)}{2x^2}$$
$$=\frac{x^2+2x-4+2x}{2x^2}$$
$$=\frac{x^2+4x-4}{2x^2}$$

41.

$$\frac{6}{b-1}-\frac{b}{1-b}=\frac{6}{b-1}-\frac{-b}{b-1}$$
$$=\frac{6+b}{b-1}$$
$$=\frac{b+6}{b-1}$$

42.

$$\frac{8}{c}+6=\frac{8}{c}+\frac{6}{1}$$
$$=\frac{8}{c}+\frac{6}{1}\left(\frac{c}{c}\right)$$
$$=\frac{8}{c}+\frac{6c}{c}$$
$$=\frac{8+6c}{c}$$
$$=\frac{6c+8}{c}$$

43.

$$\frac{n+7}{n+3}-\frac{n-3}{n+7}=\frac{n+7}{n+3}\left(\frac{n+7}{n+7}\right)-\frac{n-3}{n+7}\left(\frac{n+3}{n+3}\right)$$
$$=\frac{n^2+14n+49}{(n+3)(n+7)}-\frac{n^2-9}{(n+3)(n+7)}$$
$$=\frac{n^2+14n+49-(n^2-9)}{(n+3)(n+7)}$$
$$=\frac{n^2+14n+49-n^2+9}{(n+3)(n+7)}$$
$$=\frac{14n+58}{(n+3)(n+7)}$$

44.

$$\frac{4}{t+2}-\frac{7}{(t+2)^2}=\frac{4}{t+2}\left(\frac{t+2}{t+2}\right)-\frac{7}{(t+2)^2}$$
$$=\frac{4t+8}{(t+2)^2}-\frac{7}{(t+2)^2}$$
$$=\frac{4t+8-7}{(t+2)^2}$$
$$=\frac{4t+1}{(t+2)^2}$$

45.

$$\frac{6}{a^2-9}-\frac{5}{a^2-a-6}=\frac{6}{(a-3)(a+3)}-\frac{5}{(a-3)(a+2)}$$

$$=\frac{6}{(a-3)(a+3)}\left(\frac{a+2}{a+2}\right)-\frac{5}{(a-3)(a+2)}\left(\frac{a+3}{a+3}\right)$$

$$=\frac{6a+12}{(a-3)(a+3)(a+2)}-\frac{5a+15}{(a-3)(a+3)(a+2)}$$

$$=\frac{6a+12-(5a+15)}{(a-3)(a+3)(a+2)}$$

$$=\frac{6a+12-5a-15}{(a-3)(a+3)(a+2)}$$

$$=\frac{\cancel{a-3}}{\cancel{(a-3)}(a+3)(a+2)}$$

$$=\frac{1}{(a+3)(a+2)}$$

46.

$$\frac{2}{3y-6}+\frac{3}{4y+8}=\frac{2}{3(y-2)}+\frac{3}{4(y+2)}$$

$$=\frac{2}{3(y-2)}\left(\frac{4(y+2)}{4(y+2)}\right)+\frac{3}{4(y+2)}\left(\frac{3(y-2)}{3(y-2)}\right)$$

$$=\frac{8(y+2)}{12(y-2)(y+2)}+\frac{9(y-2)}{12(y-2)(y+2)}$$

$$=\frac{8y+16+9y-18}{12(y-2)(y+2)}$$

$$=\frac{17y-2}{12(y-2)(y+2)}$$

47. Yes. Factor a -1 out of the numerator and the answers would be the same.

48. VIDEO CAMERA

Perimeter $= 2l + 2w$

$$2\left(\frac{3}{x-1}\right)+2\left(\frac{4}{x+6}\right)=\frac{6}{x-1}+\frac{8}{x+6}$$

$$=\frac{6}{x-1}\left(\frac{x+6}{x+6}\right)+\frac{8}{x+6}\left(\frac{x-1}{x-1}\right)$$

$$=\frac{6x+36+8x-8}{(x+6)(x-1)}$$

$$=\frac{14x+28}{(x+6)(x-1)}\text{ units}$$

Area $= l \cdot w$

$$\frac{3}{x-1}\bullet\frac{4}{x+6}=\frac{12}{(x-1)(x+6)}\text{ sq units}$$

SECTION 6.5
Simplifying Complex Fractions

49.

$$\frac{\frac{n^4}{30}}{\frac{7n}{15}}=\frac{n^4}{30}\div\frac{7n}{15}$$

$$=\frac{n^4}{30}\bullet\frac{15}{7n}$$

$$=\frac{15n^4}{210n}$$

$$=\frac{n^3}{14}$$

50.

$$\frac{\frac{r^2-81}{18s^2}}{\frac{4r-36}{9s}}=\frac{\frac{r^2-81}{18s^2}}{\frac{4r-36}{9s}}\bullet\frac{18s^2}{18s^2}$$

$$=\frac{18s^2\left(\frac{r^2-81}{18s^2}\right)}{18s^2\left(\frac{4r-36}{9s}\right)}$$

$$=\frac{r^2-81}{2s(4r-36)}$$

$$=\frac{r^2-81}{8rs-72}$$

$$=\frac{(r-9)(r+9)}{8s(r-9)}$$

$$=\frac{r+9}{8s}$$

51.

$$\frac{\frac{1}{y}+1}{\frac{1}{y}-1}=\frac{\frac{1}{y}+1}{\frac{1}{y}-1}\bullet\frac{y}{y}$$

$$=\frac{y\left(\frac{1}{y}+1\right)}{y\left(\frac{1}{y}-1\right)}$$

$$=\frac{1+y}{1-y}$$

52.

$$\frac{\frac{7}{a^2}}{\frac{1}{a}+\frac{10}{3}}=\frac{\frac{7}{a^2}}{\frac{1}{a}+\frac{10}{3}}\bullet\frac{3a^2}{3a^2}$$

$$=\frac{3a^2\left(\frac{7}{a^2}\right)}{3a^2\left(\frac{1}{a}\right)+3a^2\left(\frac{10}{3}\right)}$$

$$=\frac{21}{3a+10a^2}$$

53.

$$\frac{\frac{2}{x-1}+\frac{x-1}{x+1}}{\frac{1}{x^2-1}}=\frac{\frac{2}{x-1}+\frac{x-1}{x+1}}{\frac{1}{(x-1)(x+1)}}$$

$$=\frac{\frac{2}{x-1}+\frac{x-1}{x+1}}{\frac{1}{(x-1)(x+1)}}\bullet\frac{(x-1)(x+1)}{(x-1)(x+1)}$$

$$=\frac{(x-1)(x+1)\left(\frac{2}{x-1}\right)+(x-1)(x+1)\left(\frac{x-1}{x+1}\right)}{(x-1)(x+1)\left(\frac{1}{(x-1)(x+1)}\right)}$$

$$=\frac{2(x+1)+(x-1)(x-1)}{1}$$

$$=2x+2+x^2-x-x+1$$

$$=x^2+3$$

54.

$$\frac{\frac{1}{x^2y}-\frac{5}{xy}}{\frac{3}{xy}-\frac{7}{xy^2}}=\frac{\frac{1}{x^2y}-\frac{5}{xy}}{\frac{3}{xy}-\frac{7}{xy^2}}\bullet\frac{x^2y^2}{x^2y^2}$$

$$=\frac{x^2y^2\left(\frac{1}{x^2y}\right)-x^2y^2\left(\frac{5}{xy}\right)}{x^2y^2\left(\frac{3}{xy}\right)-x^2y^2\left(\frac{7}{xy^2}\right)}$$

$$=\frac{y-5xy}{3xy-7x}$$

SECTION 6.6
Solving Rational Equations

55.

$$\frac{3}{x} = \frac{2}{x-1}$$
$$x(x-1)\left(\frac{3}{x}\right) = x(x-1)\left(\frac{2}{x-1}\right)$$
$$3(x-1) = 2x$$
$$3x - 3 = 2x$$
$$3x - 3 - 3x = 2x - 3x$$
$$-3 = -x$$
$$\frac{-3}{-1} = \frac{-x}{-1}$$
$$\boxed{3 = x}$$

Check:

$$\frac{3}{3} \stackrel{?}{=} \frac{2}{3-1}$$
$$\frac{3}{3} \stackrel{?}{=} \frac{2}{2}$$
$$1 = 1$$

56.

$$\frac{a}{a-5} = 3 + \frac{5}{a-5}$$
$$(a-5)\left(\frac{a}{a-5}\right) = (a-5)\left(3 + \frac{5}{a-5}\right)$$
$$a = 3(a-5) + 5$$
$$a = 3a - 15 + 5$$
$$a = 3a - 10$$
$$a - 3a = 3a - 10 - 3a$$
$$-2a = -10$$
$$\frac{-2a}{-2} = \frac{-10}{-2}$$
$$a = 5$$

No solution; 5 is extraneous since it makes the denominator zero

Check:

$$\frac{5}{5-5} \stackrel{?}{=} 3 + \frac{5}{5-5}$$
$$\frac{5}{0} = 3 + \frac{5}{0}$$

57.

$$\frac{2}{3t} + \frac{1}{t} = \frac{5}{9}$$
$$9t\left(\frac{2}{3t} + \frac{1}{t}\right) = 9t\left(\frac{5}{9}\right)$$
$$3(2) + 9(1) = t(5)$$
$$6 + 9 = 5t$$
$$15 = 5t$$
$$\frac{15}{5} = \frac{5t}{5}$$
$$\boxed{3 = t}$$

Check:

$$\frac{2}{3(3)} + \frac{1}{3} \stackrel{?}{=} \frac{5}{9}$$
$$\frac{2}{9} + \frac{1}{3} \stackrel{?}{=} \frac{5}{9}$$
$$\frac{2}{9} + \frac{3}{9} \stackrel{?}{=} \frac{5}{9}$$
$$\frac{5}{9} = \frac{5}{9}$$

58.

$$a = \frac{3a-50}{4a-24} - \frac{3}{4}$$
$$a = \frac{3a-50}{4(a-6)} - \frac{3}{4}$$
$$4(a-6)(a) = 4(a-6)\left(\frac{3a-50}{4(a-6)} - \frac{3}{4}\right)$$
$$4a(a-6) = 3a - 50 - 3(a-6)$$
$$4a^2 - 24a = 3a - 50 - 3a + 18$$
$$4a^2 - 24a = -32$$
$$4a^2 - 24a + 32 = -32 + 32$$
$$4a^2 - 24a + 32 = 0$$
$$4(a^2 - 6a + 8) = 0$$
$$4(a-2)(a-4) = 0$$
$$a - 2 = 0 \quad \text{and} \quad a - 4 = 0$$
$$a - 2 + 2 = 0 + 2 \quad a - 4 + 4 = 0 + 4$$
$$\boxed{a = 2 \quad \text{and} \quad a = 4}$$

Check:

$$2 \stackrel{?}{=} \frac{3(2)-50}{4(2)-24} - \frac{3}{4}$$
$$2 \stackrel{?}{=} \frac{6-50}{8-24} - \frac{3}{4}$$
$$2 \stackrel{?}{=} \frac{-44}{-16} - \frac{3}{4}$$
$$2 \stackrel{?}{=} \frac{11}{4} - \frac{3}{4}$$
$$2 \stackrel{?}{=} \frac{8}{4}$$
$$2 = 2$$

$$4 \stackrel{?}{=} \frac{3(4)-50}{4(4)-24} - \frac{3}{4}$$
$$4 \stackrel{?}{=} \frac{12-50}{16-24} - \frac{3}{4}$$
$$4 \stackrel{?}{=} \frac{-38}{-8} - \frac{3}{4}$$
$$4 \stackrel{?}{=} \frac{19}{4} - \frac{3}{4}$$
$$4 \stackrel{?}{=} \frac{16}{4}$$
$$4 = 4$$

59.

$$\frac{4}{x+2} - \frac{3}{x+3} = \frac{6}{x^2+5x+6}$$
$$\frac{4}{x+2} - \frac{3}{x+3} = \frac{6}{(x+2)(x+3)}$$
$$(x+2)(x+3)\left(\frac{4}{x+2} - \frac{3}{x+3}\right) = (x+2)(x+3)\left(\frac{6}{(x+2)(x+3)}\right)$$
$$4(x+3) - 3(x+2) = 6$$
$$4x + 12 - 3x - 6 = 6$$
$$x + 6 = 6$$
$$x + 6 - 6 = 6 - 6$$
$$\boxed{x = 0}$$

Check:

$$\frac{4}{0+2} - \frac{3}{0+3} \stackrel{?}{=} \frac{6}{0^2 + 5(0) + 6}$$
$$\frac{4}{2} - \frac{3}{3} \stackrel{?}{=} \frac{6}{6}$$
$$2 - 1 \stackrel{?}{=} 1$$
$$1 = 1$$

60.

$$\frac{3}{x+1} - \frac{x-2}{2} = \frac{x-2}{x+1}$$
$$2(x+1)\left(\frac{3}{x+1} - \frac{x-2}{2}\right) = 2(x+1)\left(\frac{x-2}{x+1}\right)$$
$$2(3) - (x+1)(x-2) = 2(x-2)$$
$$6 - (x^2 - 2x + x - 2) = 2x - 4$$
$$6 - x^2 + 2x - x + 2 = 2x - 4$$
$$-x^2 + x + 8 = 2x - 4$$
$$-x^2 + x + 8 + x^2 - x - 8 = 2x - 4 + x^2 - x - 8$$
$$0 = x^2 + x - 12$$
$$0 = (x+4)(x-3)$$
$$x + 4 = 0 \quad \text{and} \quad x - 3 = 0$$
$$x + 4 - 4 = 0 - 4 \quad x - 3 + 3 = 0 + 3$$
$$\boxed{x = -4 \quad \text{and} \quad x = 3}$$

Check:

$$\frac{3}{-4+1} - \frac{-4-2}{2} \stackrel{?}{=} \frac{-4-2}{-4+1}$$
$$\frac{3}{-3} - \frac{-6}{2} \stackrel{?}{=} \frac{-6}{-3}$$
$$-1 + 3 \stackrel{?}{=} 2$$
$$2 = 2$$

$$\frac{3}{3+1} - \frac{3-2}{2} \stackrel{?}{=} \frac{3-2}{3+1}$$
$$\frac{3}{4} - \frac{1}{2} \stackrel{?}{=} \frac{1}{4}$$
$$\frac{3}{4} - \frac{2}{4} \stackrel{?}{=} \frac{1}{4}$$
$$\frac{1}{4} = \frac{1}{4}$$

61. ENGINEERING

$$E = 1 - \frac{T_2}{T_1}$$
$$T_1(E) = T_1\left(1 - \frac{T_2}{T_1}\right)$$
$$T_1E = T_1 - T_2$$
$$T_1E - T_1 = T_1 - T_2 - T_1$$
$$T_1E - T_1 = -T_2$$
$$T_1(E-1) = -T_2$$
$$\frac{T_1(E-1)}{E-1} = \frac{-T_2}{E-1}$$
$$T_1 = \frac{-T_2}{E-1}$$
$$\boxed{T_1 = \frac{T_2}{1-E}}$$

62.

$$\frac{1}{x} = \frac{1}{y} + \frac{1}{z}$$
$$xyz\left(\frac{1}{x}\right) = xyz\left(\frac{1}{y} + \frac{1}{z}\right)$$
$$yz = xz + xy$$
$$yz - xy = xz + xy - xy$$
$$yz - xy = xz$$
$$y(z-x) = xz$$
$$\frac{y(z-x)}{z-x} = \frac{xz}{z-x}$$
$$\boxed{y = \frac{xz}{z-x}}$$

SECTION 6.7
Problem Solving Using Rational Equations

63. NUMBER PROBLEMS

$$\frac{4+2x}{5-x} = 5$$
$$(5-x)\left(\frac{4+2x}{5-x}\right) = (5-x)(5)$$
$$4 + 2x = 25 - 5x$$
$$4 + 2x + 5x = 25 - 5x + 5x$$
$$4 + 7x = 25$$
$$4 + 7x - 4 = 25 - 4$$
$$7x = 21$$
$$\frac{7x}{7} = \frac{21}{7}$$
$$\boxed{x = 3}$$

The number is 3.

64. EXERCISE

Use the formula $t = \frac{d}{r}$.

Biking: $t = \frac{30}{x+10}$

Jogging: $t = \frac{10}{x}$

Since the times are the same, set the times equal and solve the equation for x.

$$\frac{30}{x+10} = \frac{10}{x}$$
$$x(x+10)\left(\frac{30}{x+10}\right) = x(x+10)\left(\frac{10}{x}\right)$$
$$30x = 10(x+10)$$
$$30x = 10x + 100$$
$$30x - 10x = 10x + 100 - 10x$$
$$20x = 100$$
$$\frac{20x}{20} = \frac{100}{20}$$
$$\boxed{x = 5}$$

She can jog 5 mph.

65. HOUSE CLEANING

$$r = \frac{W}{t}$$
$$= \frac{1}{4} \text{ of the job per hour}$$

66. HOUSE PAINTING

rate of homeowner $= \frac{1}{14}$

rate of professional $= \frac{1}{10}$

rate together $= \frac{1}{x}$

$$\frac{1}{14} + \frac{1}{10} = \frac{1}{x}$$
$$70x\left(\frac{1}{14} + \frac{1}{10}\right) = 70x\left(\frac{1}{x}\right)$$
$$5x + 7x = 70$$
$$12x = 70$$
$$\frac{12x}{12} = \frac{70}{12}$$
$$x = 5\frac{10}{12}$$
$$\boxed{x = 5\frac{5}{6} \text{ days}}$$

67. INVESTMENTS

Use the formula $P = \frac{I}{rt}$.

Savings: $P = \frac{100}{x(1)}$
$$= \frac{100}{x}$$

Credit Union: $P = \frac{120}{(x+0.01)1}$
$$= \frac{120}{x+0.01}$$

Since the principal is the same for both investments, set them equal and solve the equation for x.

$$\frac{100}{x} = \frac{120}{x+0.01}$$
$$x(x+0.01)\left(\frac{100}{x}\right) = x(x+0.01)\left(\frac{120}{x+0.01}\right)$$
$$100(x+0.01) = 120x$$
$$100x + 1 = 120x$$
$$100x + 1 - 100x = 120x - 100x$$
$$1 = 20x$$
$$\frac{1}{20} = \frac{20x}{20}$$
$$\frac{1}{20} = x$$
$$\boxed{x = 0.05 = 5\%}$$

68. WIND SPEED

Use the formula $t = \frac{d}{r}$.

Downwind: $t = \frac{400}{360 + x}$

Upwind: $t = \frac{320}{360 - x}$

Since the times are the same, set them equal and solve the equation for x.

$$\frac{400}{360+x} = \frac{320}{360-x}$$
$$(360+x)(360-x)\left(\frac{400}{360+x}\right) = (360+x)(360-x)\left(\frac{320}{360-x}\right)$$
$$400(360-x) = 320(360+x)$$
$$144{,}000 - 400x = 115{,}200 + 320x$$
$$144{,}000 - 400x + 400x = 115{,}200 + 320x + 400x$$
$$144{,}000 = 115{,}200 + 720x$$
$$144{,}000 - 115{,}200 = 115{,}200 + 720x - 115{,}200$$
$$28{,}800 = 720x$$
$$\frac{28{,}800}{720} = \frac{720x}{720}$$
$$\boxed{40 = x}$$

The rate of the wind is 40 mph.

SECTION 6.8
Proportions and Similar Triangles

69. No.

$4(34) \stackrel{?}{=} 7(20)$
$156 \neq 140$

70. Yes.

$5(42) \stackrel{?}{=} 7(30)$
$210 = 210$

71. $3(9) = x(6)$
$$27 = 6x$$
$$\frac{27}{6} = \frac{6x}{6}$$
$$\boxed{\frac{9}{2} = x}$$

72.
$$x(5) = 3(x)$$
$$5x = 3x$$
$$5x - 3x = 3x - 3x$$
$$2x = 0$$
$$\frac{2x}{2} = \frac{0}{2}$$
$$\boxed{x = 0}$$

73.
$$(x - 2) \cdot 7 = 5(x)$$
$$7x - 14 = 5x$$
$$7x - 14 - 7x = 5x - 7x$$
$$-14 = -2x$$
$$\frac{-14}{-2} = \frac{-2x}{-2}$$
$$\boxed{7 = x}$$

74.
$$2x(x - 1) = (x + 4) \cdot 3$$
$$2x^2 - 2x = 3x + 12$$
$$2x^2 - 2x - 3x - 12 = 3x + 12 - 3x - 12$$
$$2x^2 - 5x - 12 = 0$$
$$(2x + 3)(x - 4) = 0$$
$$2x + 3 = 0 \quad \text{and} \quad x - 4 = 0$$
$$2x = -3 \quad \text{and} \quad x = 4$$
$$x = -\frac{3}{2}$$
$$\boxed{x = -\frac{3}{2}, 4}$$

75. DENTISTRY

$$\frac{3}{4} = \frac{x}{340}$$
$$3(340) = 4(x)$$
$$1{,}020 = 4x$$
$$\frac{1{,}020}{4} = \frac{4x}{4}$$
$$\boxed{255 = x}$$

225 will develop gum disease.

76.
$$\frac{x}{12} = \frac{6}{3.6}$$
$$x(3.6) = 12(6)$$
$$3.6x = 72$$
$$\frac{3.6x}{3.6} = \frac{72}{3.6}$$
$$\boxed{x = 20}$$

The pole is 20 ft. tall.

77. PORCELAIN FIGURINE

$$\frac{1}{12} = \frac{5.5}{x}$$
$$1(x) = 12(5.5)$$
$$\boxed{x = 66}$$

The flutist is 66 in. or 5 ft. 6 in. tall.

78. COMPARISON SHOPPING

$$\frac{60}{150} = \$0.40 \text{ per compact disc}$$
$$\frac{98}{250} = \$0.392 \text{ per compact disc}$$

250 for $98 is the better buy.

CHAPTER 6 TEST

1. $5x = 0$
   $\frac{5x}{5} = \frac{0}{5}$
   $\boxed{x = 0}$

2. $x^2 + x - 6 = 0$
   $(x + 3)(x - 2) = 0$
   $x + 3 = 0$ and $x - 2 = 0$
   $x + 3 - 3 = 0 - 3$ $\quad$ $x - 2 + 2 = 0 + 2$
   $\boxed{x = -3 \text{ and } x = 2}$

3. MEMORY

   Let $d = 7$ days in one week.
   $$n = \frac{35 + 5(7)}{7} = \frac{35 + 35}{7} = \frac{70}{7} = 10$$
   A person can remember 10 words.

4. U.S. SCHOOL CONSTRUCTION
   $$\frac{109{,}512 \text{ ft}^2}{1} \bullet \frac{1 \text{ yd}^2}{9 \text{ ft}^2} = \frac{109{,}512}{9} = 12{,}168 \text{ yd}^2$$

5.
   $$\frac{48x^2y}{54xy^2} = \frac{\not{2} \cdot 2 \cdot 2 \cdot 2 \cdot \not{3} \cdot \not{x} \cdot x \cdot \not{y}}{\not{2} \cdot 3 \cdot 3 \cdot \not{3} \cdot \not{x} \cdot y \cdot \not{y}} = \frac{8x}{9y}$$

6.
   $$\frac{7m - 49}{7 - m} = \frac{7(m - 7)}{-(-7 + m)} = -\frac{7(\cancel{m - 7})}{\cancel{m - 7}} = -7$$

7.
   $$\frac{2x^2 - x - 3}{4x^2 - 9} = \frac{(\cancel{2x - 3})(x + 1)}{(\cancel{2x - 3})(2x + 3)} = \frac{x + 1}{2x + 3}$$

8.
   $$\frac{3(x + 2) - 3}{6x + 5 - (3x + 2)} = \frac{3x + 6 - 3}{6x + 5 - 3x - 2} = \frac{\cancel{3x + 3}}{\cancel{3x + 3}} = 1$$

9. $3c^2d = 3 \cdot c \cdot c \cdot d$
   $c^2d^3 = c \cdot c \cdot d \cdot d \cdot d$
   $\text{LCD} = 3 \cdot c \cdot c \cdot d \cdot d \cdot d = 3c^2d^3$

10. $n^2 - 4n - 5 = (n - 5)(n + 1)$
    $n^2 - 25 = (n - 5)(n + 5)$
    $\text{LCD} = (n + 1)(n + 5)(n - 5)$

11.
    $$\frac{12x^2y}{15xy} \bullet \frac{25y^2}{16x} = \frac{\not{2} \cdot \not{2} \cdot \not{3} \cdot \not{x} \cdot \not{x} \cdot \not{y} \cdot \not{5} \cdot 5 \cdot y \cdot y}{\not{3} \cdot \not{5} \cdot \not{x} \cdot \not{y} \cdot \not{2} \cdot \not{2} \cdot 2 \cdot 2 \cdot \not{x}} = \frac{5y^2}{4}$$

12.
    $$\frac{x^2 + 3x + 2}{3x + 9} \bullet \frac{x + 3}{x^2 - 4} = \frac{(x + 1)(\cancel{x + 2})(\cancel{x + 3})}{3(\cancel{x + 3})(\cancel{x + 2})(x - 2)} = \frac{x + 1}{3(x - 2)}$$

13.
    $$\frac{x - x^2}{3x^2 + 6x} \div \frac{3x - 3}{3x^3 + 6x^2} = \frac{x - x^2}{3x^2 + 6x} \bullet \frac{3x^3 + 6x^2}{3x - 3}$$
    $$= \frac{x(1 - x) \cdot 3x^2(x + 2)}{3x(x + 2) \cdot 3(x - 1)}$$
    $$= -\frac{x(\cancel{x - 1}) \cdot \cancel{3x} \cdot x(\cancel{x + 2})}{\cancel{3x}(\cancel{x + 2}) \cdot 3(\cancel{x - 1})}$$
    $$= -\frac{x^2}{3}$$

14.
    $$\frac{a^2 - 16}{a - 4} \div (6a + 24) = \frac{a^2 - 16}{a - 4} \bullet \frac{1}{6a + 24} = \frac{(\cancel{a - 4})(\cancel{a + 4})}{(\cancel{a - 4}) \cdot 6(\cancel{a + 4})} = \frac{1}{6}$$

15.

$$\frac{3y+7}{2y+3}-\frac{3y-6}{2y+3}=\frac{3y+7-(3y-6)}{2y+3}$$
$$=\frac{3y+7-3y+6}{2y+3}$$
$$=\frac{13}{2y+3}$$

16.

$$\frac{2n}{5m}-\frac{n}{2}=\frac{2n}{5m}\left(\frac{2}{2}\right)-\frac{n}{2}\left(\frac{5m}{5m}\right)$$
$$=\frac{4n}{10m}-\frac{5mn}{10m}$$
$$=\frac{4n-5mn}{10m}$$

17.

$$\frac{x+1}{x}+\frac{x-1}{x+1}=\frac{x+1}{x}\left(\frac{x+1}{x+1}\right)+\frac{x-1}{x+1}\left(\frac{x}{x}\right)$$
$$=\frac{x^2+x+x+1}{x(x+1)}+\frac{x^2-x}{x(x+1)}$$
$$=\frac{x^2+x+x+1+x^2-x}{x(x+1)}$$
$$=\frac{2x^2+x+1}{x(x+1)}$$

18.

$$\frac{a+3}{a-1}-\frac{a+4}{1-a}=\frac{a+3}{a-1}+\frac{a+4}{a-1}$$
$$=\frac{a+3+a+4}{a-1}$$
$$=\frac{2a+7}{a-1}$$

19.

$$\frac{9}{c-4}+c=\frac{9}{c-4}+\frac{c}{1}\left(\frac{c-4}{c-4}\right)$$
$$=\frac{9}{c-4}+\frac{c^2-4c}{c-4}$$
$$=\frac{c^2-4c+9}{c-4}$$

20.

$$\frac{6}{t^2-9}-\frac{5}{t^2-t-6}=\frac{6}{(t-3)(t+3)}-\frac{5}{(t-3)(t+2)}$$
$$=\frac{6}{(t-3)(t+3)}\left(\frac{t+2}{t+2}\right)-\frac{5}{(t-3)(t+2)}\left(\frac{t+3}{t+3}\right)$$
$$=\frac{6t+12}{(t-3)(t+3)(t+2)}-\frac{5t+15}{(t-3)(t+3)(t+2)}$$
$$=\frac{6t+12-(5t+15)}{(t-3)(t+3)(t+2)}$$
$$=\frac{6t+12-5t-15}{(t-3)(t+3)(t+2)}$$
$$=\frac{\cancel{t-3}}{\cancel{(t-3)}(t+3)(t+2)}$$
$$=\frac{1}{(t+3)(t+2)}$$

21.

$$\frac{\frac{3m-9}{8m}}{\frac{5m-15}{32}}=\frac{\frac{3m-9}{8m}}{\frac{5m-15}{32}}\bullet\frac{32m}{32m}$$
$$=\frac{32m\left(\frac{3m-9}{8m}\right)}{32m\left(\frac{5m-15}{32}\right)}$$
$$=\frac{4(3m-9)}{m(5m-15)}$$
$$=\frac{12m-36}{5m^2-15m}$$
$$=\frac{12\cancel{(m-3)}}{5m\cancel{(m-3)}}$$
$$=\frac{12}{5m}$$

22.

$$\frac{\frac{3}{as^2}+\frac{6}{a^2s}}{\frac{6}{a}-\frac{9}{s^2}}=\frac{\frac{3}{as^2}+\frac{6}{a^2s}}{\frac{6}{a}-\frac{9}{s^2}}\bullet\frac{a^2s^2}{a^2s^2}$$

$$=\frac{a^2s^2\left(\frac{3}{as^2}\right)+a^2s^2\left(\frac{6}{a^2s}\right)}{a^2s^2\left(\frac{6}{a}\right)-a^2s^2\left(\frac{9}{s^2}\right)}$$

$$=\frac{3a+6s}{6as^2-9a^2}$$

$$=\frac{\cancel{3}(a+2s)}{\cancel{3}(2as^2-3a^2)}$$

$$=\frac{a+2s}{2as^2-3a^2}$$

23.

$$\frac{1}{3}+\frac{4}{3y}=\frac{5}{y}$$

$$3y\left(\frac{1}{3}+\frac{4}{3y}\right)=3y\left(\frac{5}{y}\right)$$

$$y+4=15$$

$$y+4-4=15-4$$

$$\boxed{y=11}$$

Check:

$$\frac{1}{3}+\frac{4}{3(11)}\overset{?}{=}\frac{5}{11}$$

$$\frac{1}{3}+\frac{4}{33}\overset{?}{=}\frac{5}{11}$$

$$\frac{11}{33}+\frac{4}{33}\overset{?}{=}\frac{5}{11}$$

$$\frac{15}{33}\overset{?}{=}\frac{5}{11}$$

$$\frac{5}{11}=\frac{5}{11}$$

24.

$$\frac{9n}{n-6}=3+\frac{54}{n-6}$$

$$(n-6)\left(\frac{9n}{n-6}\right)=(n-6)\left(3+\frac{54}{n-6}\right)$$

$$9n=3(n-6)+54$$

$$9n=3n-18+54$$

$$9n=3n+36$$

$$9n-3n=3n+36-3n$$

$$6n=36$$

$$\frac{6n}{6}=\frac{36}{6}$$

$$n=6$$

$\boxed{\text{No Solution; 6 is extraneous}}$

Check:

$$\frac{9(6)}{6-6}\overset{?}{=}3+\frac{54}{6-6}$$

$$\frac{54}{0}\overset{?}{=}3+\frac{54}{0}$$

25.

$$\frac{7}{q^2-q-2}+\frac{1}{q+1}=\frac{3}{q-2}$$

$$\frac{7}{(q-2)(q+1)}+\frac{1}{q+1}=\frac{3}{q-2}$$

$$(q-2)(q+1)\left(\frac{7}{(q-2)(q+1)}+\frac{1}{q+1}\right)=(q-2)(q+1)\left(\frac{3}{q-2}\right)$$

$$(q-2)(q+1)\frac{7}{(q-2)(q+1)}+(q-2)(q+1)\frac{1}{q+1}=(q-2)(q+1)\frac{3}{q-2}$$

$$7+q-2=3(q+1)$$

$$q+5=3q+3$$

$$q+5-q=3q+3-q$$

$$5=2q+3$$

$$5-3=2q+3-3$$

$$2=2q$$

$$\frac{2}{2}=\frac{2q}{2}$$

$$\boxed{1=q}$$

Check:

$$\frac{7}{1^2-1-2}+\frac{1}{1+1}\overset{?}{=}\frac{3}{1-2}$$

$$\frac{7}{-2}+\frac{1}{2}\overset{?}{=}\frac{3}{-1}$$

$$-\frac{7}{2}+\frac{1}{2}\overset{?}{=}-3$$

$$-\frac{6}{2}\overset{?}{=}-3$$

$$-3=-3$$

26.

$$\frac{2}{3}=\frac{2c-12}{3c-9}-c$$

$$\frac{2}{3}=\frac{2c-12}{3(c-3)}-\frac{c}{1}$$

$$3(c-3)\frac{2}{3}=3(c-3)\frac{2c-12}{3(c-3)}-3(c-3)\frac{c}{1}$$

$$2(c-3)=2c-12-3c(c-3)$$

$$2c-6=2c-12-3c^2+9c$$

$$2c-6=-3c^2+11c-12$$

$$2c-6+3c^2-11c+12=-3c^2+11c-12+3c^2-11c+12$$

$$3c^2-9c+6=0$$

$$3(c^2-3c+2)=0$$

$$3(c-2)(c-1)=0$$

$$c-2=0 \quad \text{and} \quad c-1=0$$

$$c-2+2=0+2 \quad c-1+1=0+1$$

$$\boxed{c=2 \qquad c=1}$$

Check:

$$\frac{2}{3}\stackrel{?}{=}\frac{2(1)-12}{3(1)-9}-1 \qquad \frac{2}{3}\stackrel{?}{=}\frac{2(2)-12}{3(2)-9}-2$$

$$\frac{2}{3}\stackrel{?}{=}\frac{-10}{-6}-1 \qquad \frac{2}{3}\stackrel{?}{=}\frac{-8}{-3}-2$$

$$\frac{2}{3}\stackrel{?}{=}\frac{10}{6}-\frac{6}{6} \qquad \frac{2}{3}\stackrel{?}{=}\frac{8}{3}-\frac{6}{3}$$

$$\frac{2}{3}\stackrel{?}{=}\frac{4}{6} \qquad \frac{2}{3}=\frac{2}{3}$$

$$\frac{2}{3}=\frac{2}{3}$$

27.

$$\frac{y}{y-1}=\frac{y-2}{y}$$

$$y(y-1)\left(\frac{y}{y-1}\right)=y(y-1)\left(\frac{y-2}{y}\right)$$

$$y(y)=(y-1)(y-2)$$

$$y^2=y^2-2y-y+2$$

$$y^2=y^2-3y+2$$

$$y^2-y^2=y^2-3y+2-y^2$$

$$0=-3y+2$$

$$0+3y=-3y+2+3y$$

$$3y=2$$

$$\boxed{y=\frac{2}{3}}$$

Check:

$$\frac{\frac{2}{3}}{\frac{2}{3}-1}\stackrel{?}{=}\frac{\frac{2}{3}-2}{\frac{2}{3}}$$

$$\frac{\frac{2}{3}}{\frac{2}{3}-\frac{3}{3}}\stackrel{?}{=}\frac{\frac{2}{3}-\frac{6}{3}}{\frac{2}{3}}$$

$$\frac{\frac{2}{3}}{-\frac{1}{3}}\stackrel{?}{=}\frac{-\frac{4}{3}}{\frac{2}{3}}$$

$$-2=-2$$

28.

$$H = \frac{RB}{R+B}$$
$$(R+B)(H) = (R+B)\left(\frac{RB}{R+B}\right)$$
$$HR + HB = RB$$
$$HR + HB - HB = RB - HB$$
$$HR = RB - HB$$
$$HR = B(R-H)$$
$$\frac{HR}{R-H} = \frac{B\cancel{(R-H)}}{\cancel{R-H}}$$
$$\boxed{\frac{HR}{R-H} = B}$$

29. HEALTH RISK

Yes.

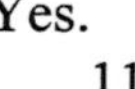

$$\frac{114}{120} = \frac{\cancel{2}\cdot\cancel{3}\cdot 19}{\cancel{2}\cdot 2\cdot 2\cdot\cancel{3}\cdot 5}$$
$$= \frac{19}{20}$$

30. CURRENCY EXCHANGE RATE

$$\frac{5}{3} = \frac{1{,}500}{x}$$
$$5(x) = 3(1{,}500)$$
$$5x = 4{,}500$$
$$\frac{5x}{5} = \frac{4{,}500}{5}$$
$$\boxed{x = 900}$$

The traveler received 900 British pounds.

31. TV TOWERS

$$\frac{x}{114} = \frac{6}{4}$$
$$x(4) = 114(6)$$
$$4x = 684$$
$$\frac{4x}{4} = \frac{684}{4}$$
$$\boxed{x = 171}$$

The tower is 171 ft. tall.

32. COMPARISON SHOPPING

$$\frac{\$3.89}{80} = \$0.048625 \text{ per sheet}$$
$$\frac{\$6.19}{120} = \$0.051583 \text{ per sheet}$$

80 sheets for \$3.89 is the better buy.

33. CLEANING HIGHWAYS

amount of work of 1st worker $= \frac{1}{7}$

amount of work of 2nd worker $= \frac{1}{9}$

amount of work together $= \frac{1}{x}$

$$\frac{1}{7} + \frac{1}{9} = \frac{1}{x}$$
$$63x\left(\frac{1}{7} + \frac{1}{9}\right) = 63x\left(\frac{1}{x}\right)$$
$$9x + 7x = 63$$
$$16x = 63$$
$$\frac{16x}{16} = \frac{63}{16}$$
$$\boxed{x = 3\frac{15}{16}}$$

It will take both of them $3\frac{15}{16}$ hrs.

34. PHYSICAL FITNESS

Use the formula $t = \frac{d}{r}$.

roller blades: $t = \frac{d}{r}$

$$= \frac{5}{x+6}$$

jogging: $t = \frac{d}{r}$

$$= \frac{2}{x}$$

Since the traveling time is the same, set the times equal and solve for x.

$$\frac{5}{x+6} = \frac{2}{x}$$

$$x(x+6)\frac{5}{x+6} = x(x+6)\frac{2}{x}$$

$$5x = 2(x+6)$$

$$5x = 2x + 12$$

$$5x - 2x = 2x + 12 - 2x$$

$$3x = 12$$

$$\frac{3x}{3} = \frac{12}{3}$$

$$\boxed{x = 4}$$

He jogs 4 mph.

35. 5 is not a common factor of the numerator and denominator.

36. We multiply both sides of the equation by the LCD of the rational expressions appearing in the equation. The resulting equation is easier to solve.

CHAPTER 6 CUMULATIVE REVIEW

1. $9^2 - 3[45 - 3(6+4)] = 81 - 3[45 - 3(10)]$
$= 81 - 3[45 - 30]$
$= 81 - 3[15]$
$= 81 - 45$
$= 36$

2. GRAND KING SIZE BED

$$\frac{7{,}840 - 6{,}240}{6{,}240} = \frac{1{,}600}{6{,}240}$$
$$\approx 0.26$$
$$\approx 26\%$$

3. $|2-4| \boxed{<} -(-6)$
$|-2| < 6$
$2 < 6$

4.
$$\frac{80+73+61+73+98}{5} = \frac{385}{5}$$
$$= 77$$

5.
$$F = \frac{9}{5}C + 32$$
$$F = \frac{9}{5}(\mathbf{40}) + 32$$
$$= 72 + 32$$
$$= 104^\circ \text{ C}$$

6.
$$V = \frac{1}{3}Bh$$
$$= \frac{1}{3}(\mathbf{6^2})(\mathbf{20})$$
$$= \frac{1}{3}(36)(20)$$
$$= 240 \text{ ft}^3$$

7. a. false
b. false
c. true
d. true

8. $2 - 3(x-5) = 4(x-1)$
$2 - 3x + 15 = 4x - 4$
$17 - 3x = 4x - 4$
$17 - 3x - 4x = 4x - 4 - 4x$
$17 - 7x = -4$
$17 - 7x - 17 = -4 - 17$
$-7x = -21$
$\frac{-7x}{-7} = \frac{-21}{-7}$
$\boxed{x = 3}$

9. $8(c+7) - 2(c-3) = 8c + 56 - 2c + 6$
$= 6c + 62$

10.
$$A - c = 2B + r$$
$$A - c - r = 2B + r - r$$
$$A - c - r = 2B$$
$$\frac{A-c-r}{2} = \frac{2B}{2}$$
$$\boxed{\frac{A-c-r}{2} = B}$$

11. $7x + 2 \geq 4x - 1$
$7x + 2 - 4x \geq 4x - 1 - 4x$
$3x + 2 \geq -1$
$3x + 2 - 2 \geq -1 - 2$
$3x \geq -3$
$\frac{3x}{3} \geq \frac{-3}{3}$
$x \geq -1$

$[-1, \infty)$

[(at) -1

12.
$$\frac{4}{5}d = -4$$
$$\frac{5}{4} \bullet \frac{4}{5}d = -4 \bullet \frac{5}{4}$$
$$\boxed{d = -5}$$

13. BLENDING TEA

Let x = pounds of \$3.20 tea and
$(20 - x)$ = pounds of \$2 tea.

$$3.20x + 2(20 - x) = 2.72(20)$$
$$3.2x + 40 - 2x = 54.4$$
$$10(3.2x + 40 - 2x) = 10(54.4)$$
$$32x + 400 - 20x = 544$$
$$12x + 400 = 544$$
$$12x + 400 - 400 = 544 - 400$$
$$12x = 144$$
$$\frac{12x}{12} = \frac{144}{12}$$
$$\boxed{\begin{array}{c} x = 12 \\ 20 - x = 20 - 12 = 8 \end{array}}$$

12 lbs of \$3.20 tea and 8 lbs of \$2 tea

14. SPEED OF A PLANE

$d = rt$
Let x = rate of plane 1 and
$x + 200$ = rate of plane 2.
The time for both planes is 5 hours.
d for plane 1 = $5x$
d for plane 2 = $5(x + 200)$
Plane 1's dis + plane 2's dis = 6,000 miles.

$$5x + 5(x + 200) = 6{,}000$$
$$5x + 5x + 1{,}000 = 6{,}000$$
$$10x + 1{,}000 = 6{,}000$$
$$10x + 1{,}000 - 1{,}000 = 6{,}000 - 1{,}000$$
$$10x = 5{,}000$$
$$\frac{10x}{10} = \frac{5{,}000}{10}$$
$$\boxed{x = 500}$$

15. $y = 2x - 3$

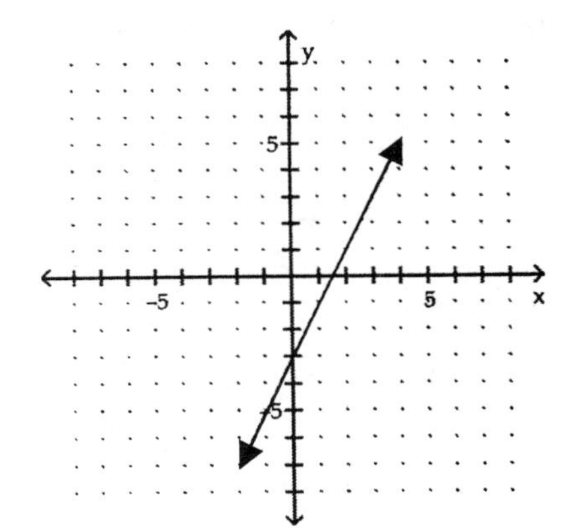

16.

$$3x - 2y \le 6$$
$$3x - 2y - 3x \le 6 - 3x$$
$$-2y \le -3x + 6$$
$$\frac{-2y}{-2} \le \frac{-3x}{-2} + \frac{6}{-2}$$
$$y \ge \frac{3}{2}x - 3$$

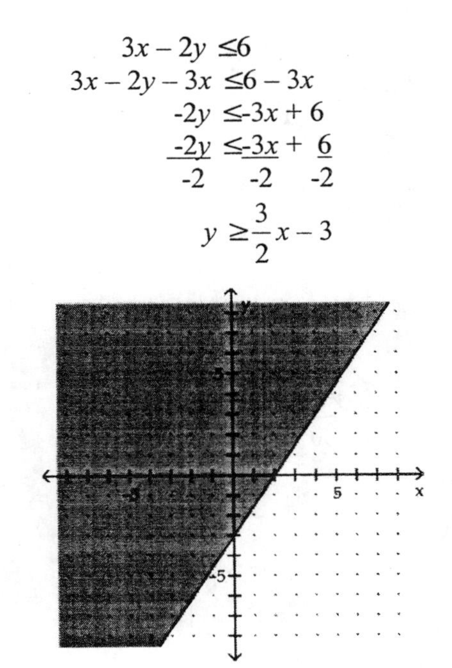

17. Perpendicular lines have slopes that are opposite reciprocals. The slope of the given line is $-\frac{7}{8}$, so the slope of the line perpendicular to it is $\mathbf{\frac{8}{7}}$.

18.

$$m = \frac{y_2 - y_1}{x_2 - x_1}$$
$$= \frac{-1 - 3}{3 - (-1)}$$
$$= \frac{-4}{4}$$
$$= -1$$

19.

$$y - y_1 = m(x - x_1)$$
$$y - 5 = 3(x - 1)$$
$$y - 5 = 3x - 3$$
$$y - 5 + 5 = 3x - 3 + 5$$
$$y = 3x + 2$$

20. CUTTING STEEL

$$m = \frac{y_2 - y_1}{x_2 - x_1}$$
$$= \frac{0.4 - 0}{50 - 0}$$
$$= \frac{0.4}{50}$$
$$= 0.008 \text{ mm/m}$$

21. $x^4x^3 = x^{4+3}$
$= x^7$

22. $(x^2x^3)^5 = (x^{2+3})^5$
$= (x^5)^5$
$= x^{5 \bullet 5}$
$= x^{25}$

23.
$$\left(\frac{y^3y}{2yy^2}\right)^3 = \left(\frac{y^4}{2y^3}\right)^3$$
$$= \left(\frac{1}{2}y^{4-3}\right)^3$$
$$= \left(\frac{y}{2}\right)^3$$
$$= \frac{y^3}{8}$$

24.
$$\left(\frac{-2a}{b}\right)^5 = \frac{(-2)^5a^5}{b^5}$$
$$= -\frac{32a^5}{b^5}$$

25.
$$\left(a^{-2}b^3\right)^{-4} = a^{-2\bullet-4}b^{3\bullet-4}$$
$$= a^8b^{-12}$$
$$= \frac{a^8}{b^{12}}$$

26.
$$\frac{9b^0b^3}{3b^{-3}b^4} = \frac{9b^{0+3}}{3b^{-3+4}}$$
$$= \frac{9b^3}{3b^1}$$
$$= 3b^{3-1}$$
$$= 3b^2$$

27. $290{,}000 = 2.9 \times 10^5$

28. 3

29. CONCENTRIC CIRCLES

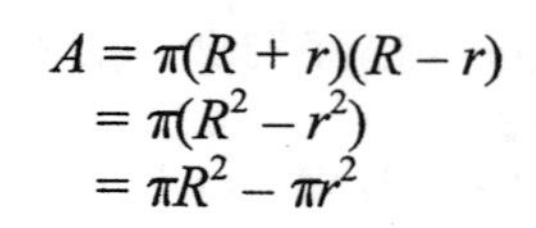

30. $y = -x^3$

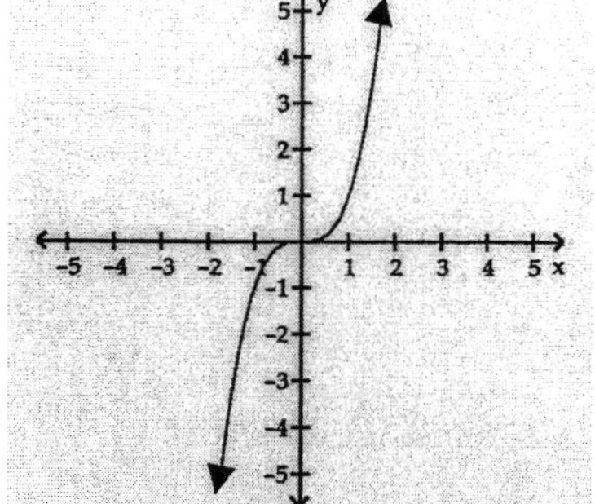

31. $(3x^2 - 3x - 2) + (3x^2 + 4x - 3)$
$= (3x^2 + 3x^2) + (-3x + 4x) + (-2 - 3)$
$= 6x^2 + x - 5$

32. $(2x^2y^3)(3x^3y^2) = (2 \cdot 3)x^{2+3}y^{3+2}$
$= 6x^5y^5$

33. $(2y - 5)(3y + 7) = 6y^2 + 14y - 15y - 35$
$= 6y^2 - y - 35$

34. $-4x^2z(3x^2 - z) = -12x^4z + 4x^2z^2$

35.
$$\frac{6x+9}{3} = \frac{6x}{3} + \frac{9}{3}$$
$$= 2x + 3$$

36. $x^2 + 2x - 1$

$$\begin{array}{r} x^2 + 2x - 1 \\ 2x + 3\overline{)2x^3 + 7x^2 + 4x - 3} \\ \underline{2x^3 + 3x^2} \qquad\quad \\ 4x^2 + 4x \qquad \\ \underline{4x^2 + 6x} \qquad \\ -2x - 3 \\ \underline{-2x - 3} \\ 0 \end{array}$$

37. $k^3t - 3k^2t = k^2t(k - 3)$

38. $2ab + 2ac + 3b + 3c = (2ab+2ac)+(3b+3c)$
$= 2a(b + c) + 3(b + c)$
$= (b + c)(2a + 3)$

39. $2a^2 - 200b^2 = 2(a^2 - 100b^2)$
$= 2(a + 10b)(a - 10b)$

40. $b^3 + 125 = (b + 5)(b^2 - b \cdot 5 + 5^2)$
$= (b + 5)(b^2 - 5b + 25)$

41. $u^2 - 18u + 81 = (u - 9)(u - 9)$
$= (u - 9)^2$

42. $6x^2 - 13x - 63 = (2x - 9)(3x + 7)$

43. $-r^2 + r + 2 = -(r^2 - r - 2)$
$= -(r - 2)(r + 1)$

44. $u^2 + 10u + 15$
prime

45. $5x^2 + x = 0$
$x(5x + 1) = 0$
$x = 0$ or $5x + 1 = 0$
$5x = -1$
$x = -\frac{1}{5}$

$$\boxed{x = 0, -\frac{1}{5}}$$

46. $6x^2 - 5x = -1$
$6x^2 - 5x + 1 = -1 + 1$
$6x^2 - 5x + 1 = 0$
$(2x - 1)(3x - 1) = 0$
$2x - 1 = 0$ or $3x - 1 = 0$
$2x = 1$ $\quad$ $3x = 1$
$x = \frac{1}{2}$ $\quad$ $x = \frac{1}{3}$

$$\boxed{x = \frac{1}{2}, \frac{1}{3}}$$

47. COOKING

$w(w + 6) = 160$
$w^2 + 6w = 160$
$w^2 + 6w - 160 = 160 - 160$
$w^2 + 6w - 160 = 0$
$(w + 16)(w - 10) = 0$
$w + 16 = 0$ or $w - 10 = 0$
$w = -16$ $\quad$ $w = 10$

The width cannot be negative.

$$\boxed{\begin{array}{l} w = 10 \text{ in.} \\ l = w + 6 = 10 + 6 = 16 \text{ in.} \end{array}}$$

48. $x^2 - 25 = 0$
$(x - 5)(x + 5) = 0$
$x - 5 = 0$ or $x + 5 = 0$
$x = 5$ $\quad$ $x = -5$

$$\boxed{x = 5, -5}$$

49.

$$\frac{x^2 - 16}{x - 4} \div \frac{3x + 12}{x} = \frac{x^2 - 16}{x - 4} \bullet \frac{x}{3x + 12}$$
$$= \frac{\cancel{(x-4)}\cancel{(x+4)}}{\cancel{x-4}} \bullet \frac{x}{3\cancel{(x+4)}}$$
$$= \frac{x}{3}$$

50.

$$\frac{4}{x - 3} + \frac{5}{3 - x} = \frac{4}{x - 3} + \frac{-5}{x - 3}$$
$$= \frac{4 + (-5)}{x - 3}$$
$$= \frac{-1}{x - 3}$$
$$= -\frac{1}{x - 3}$$

51.

$$\frac{m}{m^2+5m+6}-\frac{2}{m^2+3m+2}=\frac{m}{(m+2)(m+3)}-\frac{2}{(m+1)(m+2)}$$

$$=\frac{m}{(m+2)(m+3)}\left(\frac{m+1}{m+1}\right)-\frac{2}{(m+1)(m+2)}\left(\frac{m+3}{m+3}\right)$$

$$=\frac{m^2+m}{(m+1)(m+2)(m+3)}-\frac{2m+6}{(m+1)(m+2)(m+3)}$$

$$=\frac{m^2+m-(2m+6)}{(m+1)(m+2)(m+3)}$$

$$=\frac{m^2+m-2m-6}{(m+1)(m+2)(m+3)}$$

$$=\frac{m^2-m-6}{(m+1)(m+2)(m+3)}$$

$$=\frac{(m-3)\cancel{(m+2)}}{(m+1)\cancel{(m+2)}(m+3)}$$

$$=\frac{m-3}{(m+1)(m+3)}$$

52.

$$\frac{2-\frac{2}{x+1}}{2+\frac{2}{x}}=\frac{2-\frac{2}{x+1}}{2+\frac{2}{x}}\bullet\frac{x(x+1)}{x(x+1)}$$

$$=\frac{2x(x+1)-\frac{2}{x+1}x(x+1)}{2x(x+1)+\frac{2}{x}x(x+1)}$$

$$=\frac{2x^2+2x-2x}{2x^2+2x+2x+2}$$

$$=\frac{2x^2}{2x^2+4x+2}$$

$$=\frac{\cancel{2}x^2}{\cancel{2}(x^2+2x+1)}$$

$$=\frac{x^2}{(x+1)^2}$$

53.

$$\frac{7}{5x}-\frac{1}{2}=\frac{5}{6x}+\frac{1}{3}$$

$$30x\left(\frac{7}{5x}-\frac{1}{2}\right)=30x\left(\frac{5}{6x}+\frac{1}{3}\right)$$

$$42-15x=25+10x$$

$$42-15x-10x=25+10x-10x$$

$$42-25x=25$$

$$42-25x-42=25-42$$

$$-25x=-17$$

$$\frac{-25x}{-25}=\frac{-17}{-25}$$

$$\boxed{x=\frac{17}{25}}$$

54.

$$\frac{u}{u-1}+\frac{1}{u}=\frac{u^2+1}{u^2-u}$$

$$\frac{u}{u-1}+\frac{1}{u}=\frac{u^2+1}{u(u-1)}$$

$$u(u-1)\left(\frac{u}{u-1}+\frac{1}{u}\right)=u(u-1)\left(\frac{u^2+1}{u(u-1)}\right)$$

$$u^2+u-1=u^2+1$$

$$u^2+u-1-u^2=u^2+1-u^2$$

$$u-1=1$$

$$u-1+1=1+1$$

$$\boxed{u=2}$$

55. DRAINING A TANK

$$\frac{1}{24}+\frac{1}{36}=\frac{1}{x}$$

$$72x\left(\frac{1}{24}+\frac{1}{36}\right)=72x\left(\frac{1}{x}\right)$$

$$3x+2x=72$$

$$5x=72$$

$$\frac{5x}{5}=\frac{72}{5}$$

$$\boxed{x=14\frac{2}{5}}$$

It will take both pipes $14\frac{2}{5}$ hours.

56. HEIGHT OF A TREE

$$\frac{3}{2.5} = \frac{x}{29}$$
$$3(29) = 2.5x$$
$$87 = 2.5x$$
$$\frac{87}{2.5} = \frac{2.5x}{2.5}$$
$$\boxed{34.8 = x}$$

Height of tree is 34.8 ft.

SECTION 7.1

VOCABULARY

1. A pair of equations $\begin{cases} x-y=-1 \\ 2x-y=1 \end{cases}$ is called a **system** of equations.

3. The x-coordinate of the ordered pairs (-4, 7) is –4 and a **y-coordinate** is 7.

5. The point of **intersection** of the lines graphed in part (a) of the following figure is (1, 2).

7. A system of equations that has at least one solution is called a **consistent** system. A system with no solution is called an **inconsistent** system.

CONCEPTS

9. The solution is the pair of coordinates that are the same in both tables. **(-4, -1).**

11. true

13. false

15. $$\begin{cases} y=5x-1 \\ 30x-6=6y \end{cases}$$

$$30x-6=6y$$
$$\frac{30x}{6}-\frac{6}{6}=\frac{6y}{6}$$
$$5x-1=y$$

17. No solution; independent

NOTATION

19. is possibly equal to

PRACTICE

21. solution $(1,1)$

$$x+y=2$$
$$1+1\overset{?}{=}2$$
$$2=2$$
yes
$$2x-y=1$$
$$2(1)-1\overset{?}{=}1$$
$$2-1\overset{?}{=}1$$
$$1=1$$
yes

23. solution $(3,-2)$

$$2x+y=4$$
$$2(3)+(-2)\overset{?}{=}4$$
$$6-2\overset{?}{=}4$$
$$4=4$$
yes
$$y=1-x$$
$$-2\overset{?}{=}1-3$$
$$-2=-2$$
yes

25. solution $(-2,-4)$

$$4x+5y=-23$$
$$4(-2)+5(-4)\overset{?}{=}-23$$
$$-8-20\overset{?}{=}-23$$
$$-28\neq-23$$
no

There is no need to check the second equation.

27.

solution $\left(\frac{1}{2},3\right)$

$$2x+y=4$$

$$2\left(\frac{1}{2}\right)+3\stackrel{?}{=}4$$

$$1+3\stackrel{?}{=}4$$

$$4=4$$

yes

$$4x-11=3y$$

$$4\left(\frac{1}{2}\right)-11\stackrel{?}{=}3(3)$$

$$2-11\stackrel{?}{=}9$$

$$-9\neq 9$$

no

29.

solution (2.5, 3.5)

$$4x-3=2y$$

$$4(2.5)-3\stackrel{?}{=}2(3.5)$$

$$10-3\stackrel{?}{=}7$$

$$7=7$$

yes

$$4y+1=6x$$

$$4(3.5)+1\stackrel{?}{=}6(2.5)$$

$$14+1\stackrel{?}{=}15$$

$$15=15$$

yes

31. (3, 2)

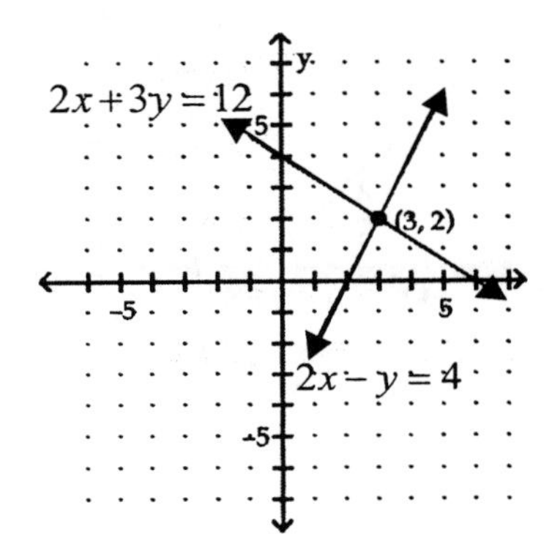

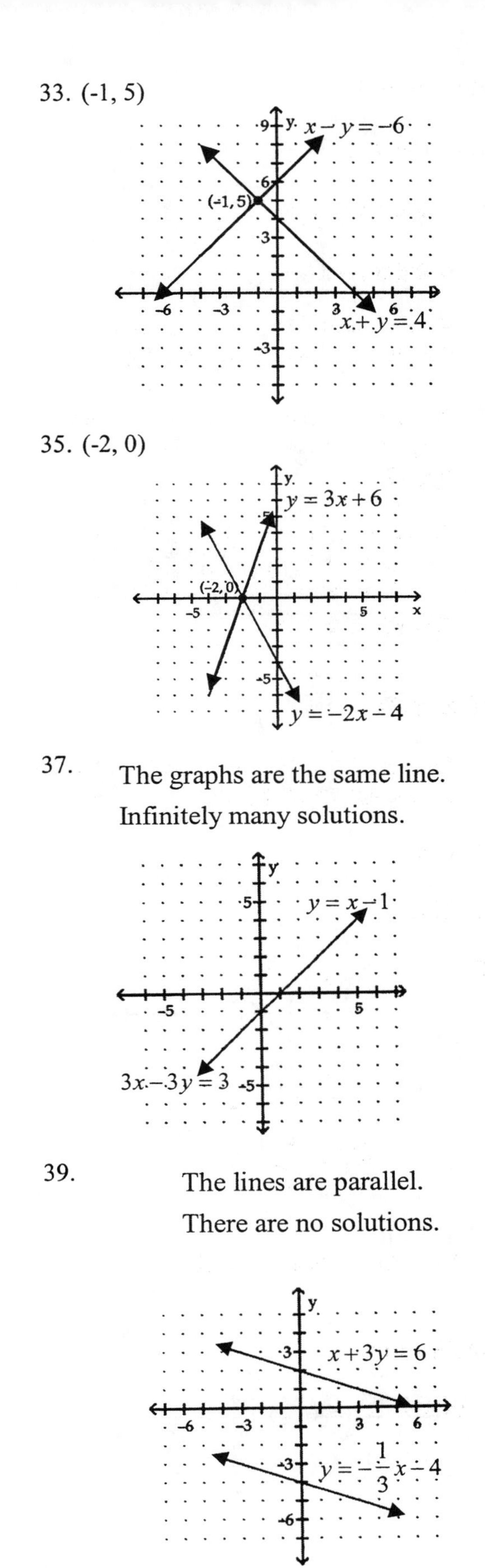

33. (-1, 5)

35. (-2, 0)

37. The graphs are the same line.
Infinitely many solutions.

39. The lines are parallel.
There are no solutions.

41. (4, -6)

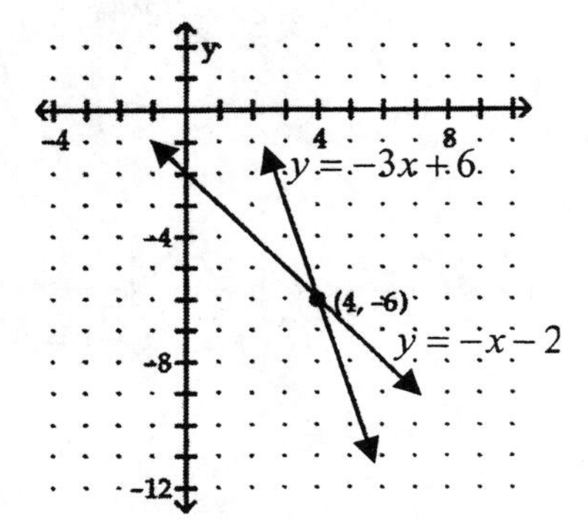

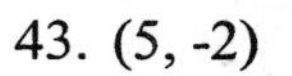

43. (5, -2)

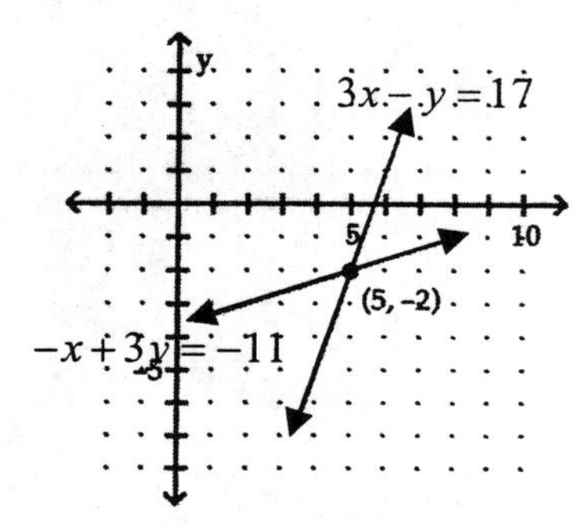

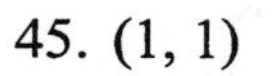

45. (1, 1)

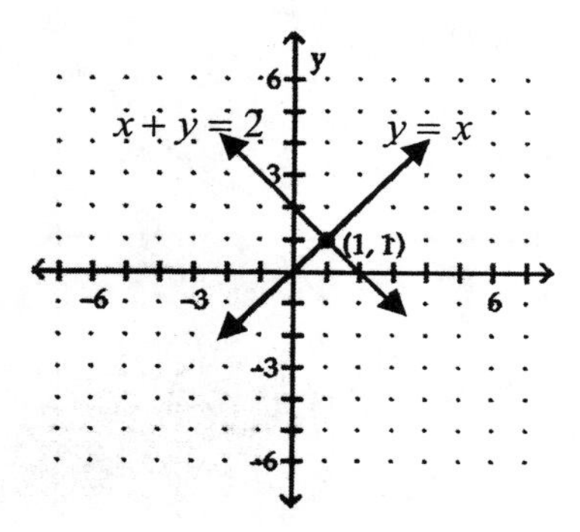

47. (-4, 0)

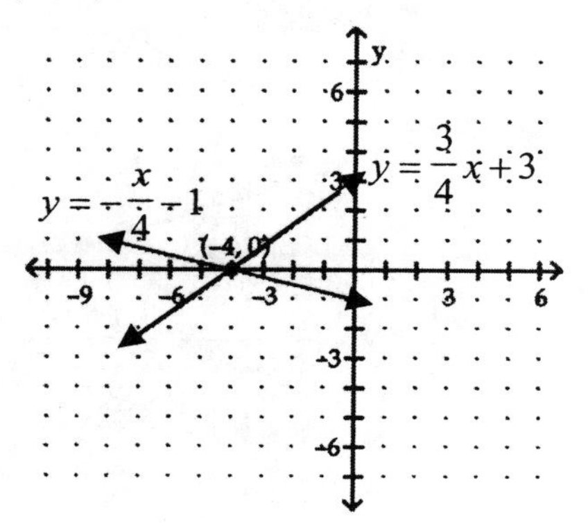

49. The lines are parallel.
There are no solutions.

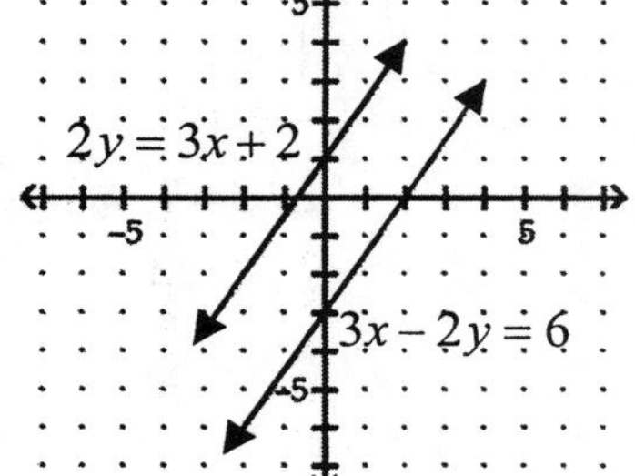

51. The graphs are the same line.
Infinitely many solutions.

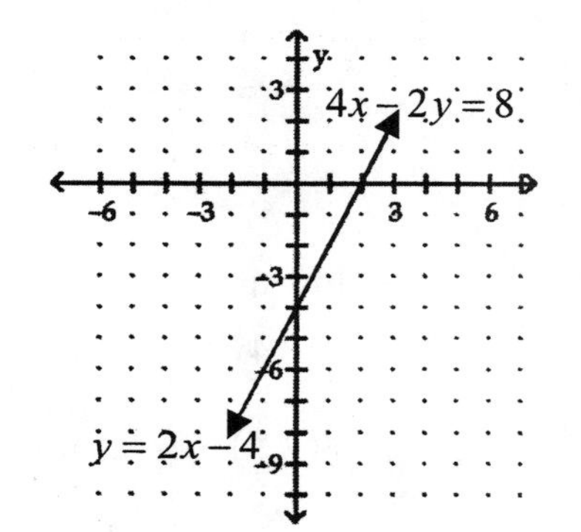

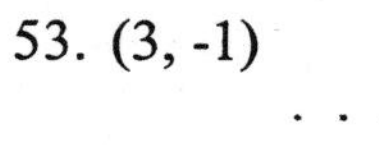

53. (3, -1)

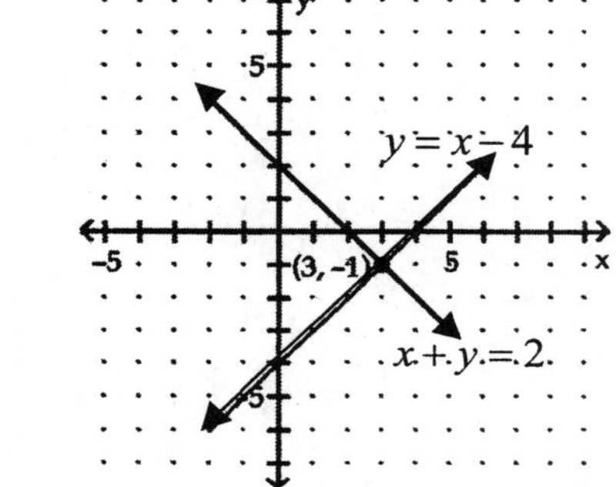

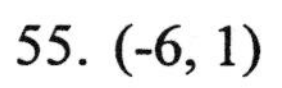

55. (-6, 1)

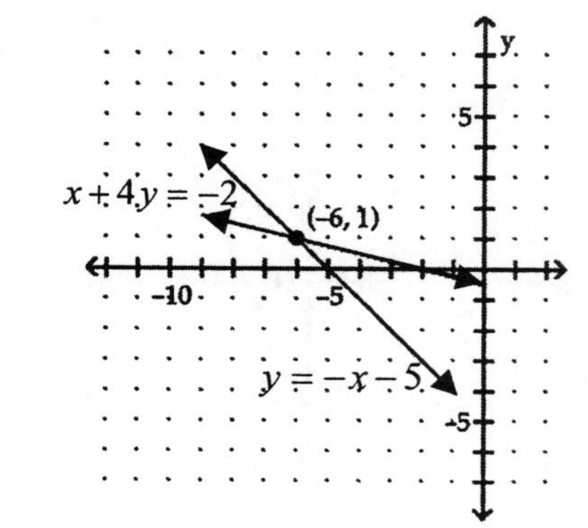

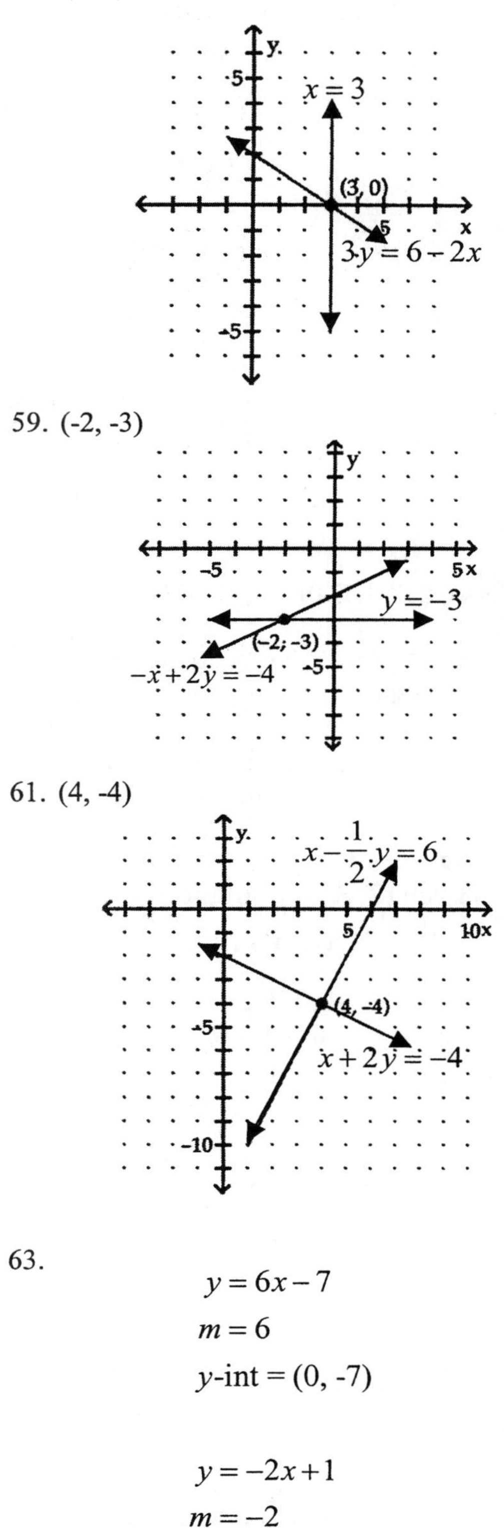

63.

$y = 6x - 7$
$m = 6$
y-int = (0, -7)

$y = -2x + 1$
$m = -2$
y-int = (0, 1)
1 solution

65.

$3x - y = -3$
$y = 3x + 3$
$m = 3$
y-int = (0, 3)

$y - 3x = 3$
$y = 3x + 3$
$m = 3$
y-int = (0, 3)
same line
infinitely many solutions

67.

$y = -x + 6$
$m = -1$
y-int = (0, 6)

$x + y = 8$
$y = -x + 8$
$m = -1$
y-int = (0, 8)
parallel lines
no solution

69.

$6x + y = 0$
$y = -6x + 0$
$m = -6$
y-int = (0, 0)

$2x + 2y = 0$
$y = -x + 0$
$m = -1$
y-int = (0, 0)
one solution

71. (1, 3)

73. no solution

APPLICATIONS

75. TRANSPLANTS

1994, 4,100

77. LATITUDE AND LONGITUDE

a) Houston, New Orleans, St. Augustine
b) St. Louis, Memphis, New Orleans
c) New Orleans

79. DAILY TRACKING POLL
a) the incumbent, 7%
b) November 2
c) the challenger, 3

81. TV COVERAGE
10 miles

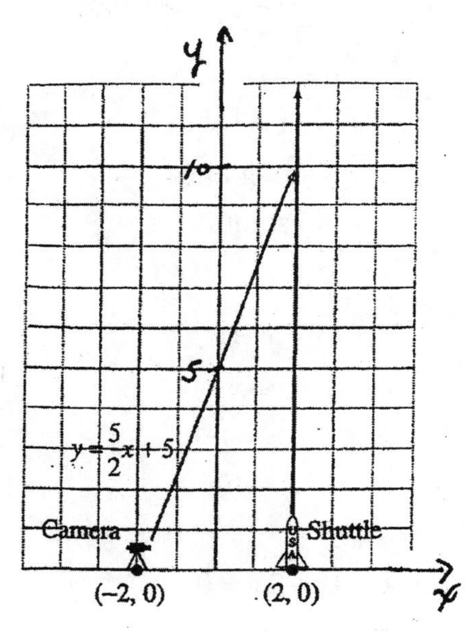

WRITING

83. Answers will vary.

85. Answers will vary.

REVIEW

87. $$\frac{x+3}{x^2}+\frac{x+5}{2x}=\frac{2}{2}\left(\frac{x+3}{x^2}\right)+\frac{x}{x}\left(\frac{x+5}{2x}\right)$$
$$=\frac{2x+6}{2x^2}+\frac{x^2+5x}{2x^2}$$
$$=\frac{x^2+7x+6}{2x^2}$$

89. $$\frac{z^2+4z-5}{5z-5}\bullet\frac{5z}{z+5}=\frac{(z+5)(z-1)}{5(z-1)}\bullet\frac{5z}{z+5}$$
$$=z$$

CHALLENGE PROBLEMS

91. No. Suppose the equations are dependent. Then they have the same graph, and would intersect at an infinite number of points. Therefore, the system would have at least one solution, and be consistent.

93. Answers will vary.
Given (2, 3) is a solution.

$x + y = 5$

$x - y = 10$

There are many more systems than these two.

SECTION 7.2

VOCABULARY

1. $\begin{cases} y = x + 3 \\ 3x - y = -1 \end{cases}$ is called a **system** of equations.

3. When checking a proposed solution of a system of equations, always use the **original** equations.
5. In mathematics, "to **substitute**" means to replace an expression with one that has the same value.

7. With the substitution method, the basic objective is to use an appropriate substitution to obtain one equation in **one** variable.

9. When the graphs of the equations of a system are identical lines, the equations are called dependent and the system has an **infinitely** many solutions.

CONCEPTS

11. a) $y = -3x$
    b) $x = 2y + 1$

13. a) $x = 2 + 2y$
    b) $y = -11 + 2x$

15.
$$x + 3x - 4 = 8$$
$$x + 3(x - 4) = 8$$

17. a) LCD = 15

b)
$$\frac{x}{5} + \frac{2y}{3} = 1$$
$$15\left(\frac{x}{5}\right) + 15\left(\frac{2y}{3}\right) = 15(1)$$
$$3x + 10y = 15$$

19. a) no
    b) ii

NOTATION

21. $\begin{cases} y = 3x \\ x - y = 4 \end{cases}$

$x - y = 4$ The 2nd equation

$x - (\boxed{3x}) = 4$

$-2x = \boxed{4}$

$x = \boxed{-2}$

$y = 3x$ The 1st equation

$y = 3(\boxed{-2})$

$y = \boxed{-6}$

The solution is $(\boxed{-2}, \boxed{-6})$.

PRACTICE

23. $\begin{cases} y = 2x \\ x + y = 6 \end{cases}$

$x + y = 6$ The 2nd equation

$x + (2x) = 6$

$3x = 6$

$\boxed{x = 2}$

$y = 2x$ The 1st equation

$y = 2(2)$

$\boxed{y = 4}$

The solution is (2, 4).

25. $\begin{cases} y = 2x - 6 \\ 2x + y = 6 \end{cases}$

$2x + y = 6$ The 2nd equation

$2x + (2x - 6) = 6$

$4x - 6 = 6$

$4x - 6 + 6 = 6 + 6$

$4x = 12$

$\frac{4x}{4} = \frac{12}{4}$

$\boxed{x = 3}$

$y = 2x - 6$ The 1st equation

$y = 2(3) - 6$

$\boxed{y = 0}$

The solution is (3, 0).

27. $\begin{cases} y = 2x+5 \\ x+2y = -5 \end{cases}$

$$x+2y=-5 \quad \text{The 2}^{\text{nd}} \text{ equation}$$
$$x+2(2x+5)=-5$$
$$x+4x+10=-5$$
$$5x+10=-5$$
$$5x+10-10=-5-10$$
$$5x=-15$$
$$\frac{5x}{5}=\frac{-15}{5}$$
$$\boxed{x=-3}$$
$$y=2x+5 \quad \text{The 1}^{\text{st}} \text{ equation}$$
$$y=2(-3)+5$$
$$\boxed{y=\ -1}$$

The solution is (-3, -1).

29. $\begin{cases} 3x+y=-4 \\ x=y \end{cases}$

$$x=y \quad \text{The 2}^{\text{nd}} \text{ equation}$$
$$3x+x=-4$$
$$4x=-4$$
$$\frac{4x}{4}=\frac{-4}{4}$$
$$\boxed{x=-1}$$
$$3x+y=-4 \quad \text{The 1}^{\text{st}} \text{ equation}$$
$$3(-1)+y=-4$$
$$-3+y+3=-4+3$$
$$\boxed{y=-1}$$

The solution is (-1, -1).

31. $\begin{cases} 2a+4b=-24 \\ a=20-2b \end{cases}$

$$2a+4b=-24 \quad \text{The 1}^{\text{st}} \text{ equation}$$
$$2(20-2b)+4b=-24$$
$$40-4b+4b=-24$$
$$40=-24$$

false

No solution.

33. $\begin{cases} 2a-3b=-13 \\ -b=-2a-7 \end{cases}$

solve for b in 2[nd]equation

$\begin{cases} 2a-3b=-13 \\ b=2a+7 \end{cases}$

$$2a-3b=-13 \quad \text{The 1}^{\text{st}} \text{ equation}$$
$$2a-3(2a+7)=-13$$
$$2a-6a-21=-13$$
$$-4a-21=-13$$
$$-4a-21+21=-13+21$$
$$-4a=8$$
$$\frac{-4a}{-4}=\frac{8}{-4}$$
$$\boxed{a=-2}$$
$$b=2a+7 \quad \text{The 2}^{\text{nd}} \text{ equation}$$
$$b=2(-2)+7$$
$$b=-4+7$$
$$\boxed{b=3}$$

The solution is (-2, 3).

35. $\begin{cases} -y=11-3x \\ 2x+5y=-4 \end{cases}$

solve for y in 1[st]equation

$\begin{cases} y=3x-11 \\ 2x+5y=-4 \end{cases}$

$$2x+5y=-4 \quad \text{The 2}^{\text{nd}} \text{ equation}$$
$$2x+5(3x-11)=-4$$
$$2x+15x-55=-4$$
$$17x-55=-4$$
$$17x-55+55=-4+55$$
$$17x=51$$
$$\frac{17x}{17}=\frac{51}{17}$$
$$\boxed{x=3}$$
$$y=3x-11 \quad \text{The 1}^{\text{st}} \text{ equation}$$
$$y=3(3)-11$$
$$y=9-11$$
$$\boxed{y=-2}$$

The solution is (3, -2).

37. $\begin{cases} 2x+3=-4y \\ x-6=-8y \end{cases}$

solve for x in 2^{nd}equation

$\begin{cases} 2x+3=-4y \\ x=6-8y \end{cases}$

$2x+3=-4y$ The 1st equation

$2(6-8y)+3=-4y$

$12-16y+3=-4y$

$-16y+15=-4y$

$-16y+15+16y=-4y+16y$

$15=12y$

$\frac{15}{12}=\frac{12y}{12}$

$\boxed{\frac{5}{4}=y}$

$x=6-8y$ The 2nd equation

$x=6-8\left(\frac{5}{4}\right)$

$x=6-10$

$\boxed{x=-4}$

The solution is $\left(-4,\frac{5}{4}\right)$.

39. $\begin{cases} r+3s=9 \\ 3r+2s=13 \end{cases}$

solve for r in 1^{st}equation

$\begin{cases} r=9-3s \\ 3r+2s=13 \end{cases}$

$3r+2s=13$ The 2nd equation

$3(9-3s)+2s=13$

$27-9s+2s=13$

$27-7s=13$

$27-7s-27=13-27$

$-7s=-14$

$\frac{-7s}{-7}=\frac{-14}{-7}$

$\boxed{s=2}$

$r=9-3s$ The 1st equation

$r=9-3(2)$

$r=9-6$

$\boxed{r=3}$

The solution is (3, 2).

41. $\begin{cases} 6x-3y=5 \\ 2y+x=0 \end{cases}$

solve for x in 2nd equation

$\begin{cases} 6x-3y=5 \\ x=-2y \end{cases}$

$6x-3y=5$ The 1st equation

$6(-2y)-3y=5$

$-12y-3y=5$

$-15y=5$

$\frac{-15y}{-15}=\frac{5}{-15}$

$\boxed{y=-\frac{1}{3}}$

$x=-2y$ The 2nd equation

$x=-2\left(-\frac{1}{3}\right)$

$\boxed{x=\frac{2}{3}}$

The solution is $(\frac{2}{3}, -\frac{1}{3})$.

43. $\begin{cases} y-3x=-5 \\ 21x=7y+35 \end{cases}$

solve for y in 1st equation

$\begin{cases} y=-5+3x \\ 21x=7y+35 \end{cases}$

$21x=7y+35$ The 2nd equation

$21x=7(-5+3x)+35$

$21x=-35+21x+35$

$21x=21x$

$21x-21x=21x-21x$

$\boxed{0=0}$

true

Infinitely many solutions.

45. $\begin{cases} y = 3x+6 \\ y = -2x-4 \end{cases}$

$$y = 3x+6 \quad \text{The 1}^{\text{st}} \text{ equation}$$
$$-2x-4 = 3x+6$$
$$-2x-4+4 = 3x+6+4$$
$$-2x = 3x+10$$
$$-2x-3x = 3x+10-3x$$
$$-5x = 10$$
$$\frac{-5x}{-5} = \frac{10}{-5}$$
$$\boxed{x = -2}$$
$$y = -2x-4 \quad \text{The 2}^{\text{nd}} \text{ equation}$$
$$y = -2(-2)-4$$
$$y = 4-4$$
$$\boxed{y = 0}$$

The solution is (-2, 0).

47. $\begin{cases} x = \frac{1}{3}y-1 \\ x = y+1 \end{cases}$

$$x = \frac{1}{3}y-1 \quad \text{The 1}^{\text{st}} \text{ equation}$$
$$y+1 = \frac{1}{3}y-1$$
$$3(y)+3(1) = 3\left(\frac{1}{3}y\right)-3(1)$$
$$3y+3 = y-3$$
$$3y+3-3 = y-3-3$$
$$3y = y-6$$
$$3y-y = y-6-y$$
$$2y = -6$$
$$\frac{2y}{2} = \frac{-6}{2}$$
$$\boxed{y = -3}$$
$$x = y+1 \quad \text{The 2}^{\text{nd}} \text{ equation}$$
$$x = -3+1$$
$$\boxed{x = -2}$$

The solution is (-2, -3).

49. $\begin{cases} 4x+5y = 2 \\ 3x-y = 11 \end{cases}$

solve for y in 2ndequation

$\begin{cases} 4x+5y = 2 \\ 3x-11 = y \end{cases}$

$$4x+5y = 2 \quad \text{The 1}^{\text{st}} \text{ equation}$$
$$4x+5(3x-11) = 2$$
$$4x+15x-55 = 2$$
$$19x-55 = 2$$
$$19x-55+55 = 2+55$$
$$19x = 57$$
$$\frac{19x}{19} = \frac{57}{19}$$
$$\boxed{x = 3}$$
$$3x-11 = y \quad \text{The 2}^{\text{nd}} \text{ equation}$$
$$3(3)-11 = y$$
$$9-11 = y$$
$$\boxed{-2 = y}$$

The solution is (3, -2).

51. $\begin{cases} 3x+4y = -7 \\ 2y-x = -1 \end{cases}$

solve for x in 2ndequation

$\begin{cases} 3x+4y = -7 \\ 2y+1 = x \end{cases}$

$$3x+4y = -7 \quad \text{The 1}^{\text{st}} \text{ equation}$$
$$3(2y+1)+4y = -7$$
$$6y+3+4y = -7$$
$$10y+3 = -7$$
$$10y+3-3 = -7-3$$
$$10y = -10$$
$$\frac{10y}{10} = \frac{-10}{10}$$
$$\boxed{y = -1}$$
$$2y+1 = x \quad \text{The 2}^{\text{nd}} \text{ equation}$$
$$2(-1)+1 = x$$
$$-2+1 = x$$
$$\boxed{-1 = x}$$

The solution is (-1, -1).

53.
$$\begin{cases} 6 - y = 4x \\ 2y = -8x - 20 \end{cases}$$

solve for y in 1st equation

$$\begin{cases} 6 - 4x = y \\ 2y = -8x - 20 \end{cases}$$

$2y = -8x - 20$ The 2nd equation

$$2(6 - 4x) = -8x - 20$$
$$12 - 8x = -8x - 20$$
$$12 - 8x - 12 = -8x - 20 - 12$$
$$-8x = -8x - 32$$
$$-8x + 8x = -8x - 32 + 8x$$
$$\boxed{0 = -32}$$

false

No solution.

55.
$$\begin{cases} b = \frac{2}{3}a \\ 8a - 3b = 3 \end{cases}$$

$8a - 3b = 3$ The 2nd equation

$$8a - 3\left(\frac{2}{3}a\right) = 3$$
$$8a - 2a = 3$$
$$6a = 3$$
$$\frac{6a}{6} = \frac{3}{6}$$
$$\boxed{a = \frac{1}{2}}$$

$b = \frac{2}{3}a$ The 1st equation

$$b = \frac{2}{3}\left(\frac{1}{2}\right)$$
$$\boxed{b = \frac{1}{3}}$$

The solution is $\left(\frac{1}{2}, \frac{1}{3}\right)$.

57.
$$\begin{cases} 2x + 5y = -2 \\ -\frac{x}{2} = y \end{cases}$$

re-arrange 2nd equation

$$\begin{cases} 2x + 5y = -2 \\ x = -2y \end{cases}$$

$2x + 5y = -2$ The 1st equation

$$2(-2y) + 5y = -2$$
$$-4y + 5y = -2$$
$$\boxed{y = -2}$$

$x = -2y$ The 2nd equation

$$x = -2(-2)$$
$$\boxed{x = 4}$$

The solution is (4, -2).

59. $\begin{cases} \frac{x}{2}+\frac{y}{2}=-1 \\ \frac{x}{3}-\frac{y}{2}=-4 \end{cases}$

clear both equations of fractions

$\begin{cases} 2\left(\frac{x}{2}\right)+2\left(\frac{y}{2}\right)=2(-1) \\ 6\left(\frac{x}{3}\right)-6\left(\frac{y}{2}\right)=6(-4) \end{cases}$

$\begin{cases} x+y=-2 \\ 2x-3y=-24 \end{cases}$

solve for x in 1st equation

$\begin{cases} x=-2-y \\ 2x-3y=-24 \end{cases}$

$2x-3y=-24$ The 2nd equation

$2(-2-y)-3y=-24$

$-4-2y-3y=-24$

$-5y-4=-24$

$-5y-4+4=-24+4$

$-5y=-20$

$\frac{-5y}{-5}=\frac{-20}{-5}$

$\boxed{y=4}$

$x=-2-y$ The 1st equation

$x=-2-4$

$x=-6$

$\boxed{x=-6}$

The solution is (-6, 4).

61. $\begin{cases} 5x=\frac{1}{2}y-1 \\ \frac{1}{4}y=10x-1 \end{cases}$

clear both equations of fractions

$\begin{cases} 2(5x)=2\left(\frac{1}{2}y\right)-2(1) \\ 4\left(\frac{1}{4}y\right)=4(10x)-4(1) \end{cases}$

$\begin{cases} 10x=y-2 \\ y=40x-4 \end{cases}$

$10x=y-2$ The 1st equation

$10x=(40x-4)-2$

$10x=40x-6$

$10x-40x=40x-6-40x$

$-30x=-6$

$\frac{-30x}{-30}=\frac{-6}{-30}$

$\boxed{x=\frac{1}{5}}$

$10x=y-2$ The 1st equation

$10\left(\frac{1}{5}\right)=y-2$

$2=y-2$

$2+2=y-2+2$

$\boxed{4=y}$

The solution is $\left(\frac{1}{5}, 4\right)$.

63. $\begin{cases} x - \dfrac{4y}{5} = 4 \\ \dfrac{y}{3} = \dfrac{x}{2} - \dfrac{5}{2} \end{cases}$

clear both equations of fractions

$$\begin{cases} 5(x) - 5\left(\dfrac{4y}{5}\right) = 5(4) \\ 6\left(\dfrac{y}{3}\right) = 6\left(\dfrac{x}{2}\right) - 6\left(\dfrac{5}{2}\right) \end{cases}$$

$$\begin{cases} 5x - 4y = 20 \\ 2y = 3x - 15 \end{cases}$$

solve for y in 2nd equation

$$\begin{cases} 5x - 4y = 20 \\ y = \dfrac{3}{2}x - \dfrac{15}{2} \end{cases}$$

$5x - 4y = 20$ The 1st equation

$$5x - 4\left(\frac{3}{2}x - \frac{15}{2}\right) = 20$$

$$5x - 6x + 30 = 20$$

$$-x + 30 = 20$$

$$-x + 30 - 30 = 20 - 30$$

$$-x = -10$$

$$\frac{-x}{-1} = \frac{-10}{-1}$$

$$\boxed{x = 10}$$

$y = \dfrac{3}{2}x - \dfrac{15}{2}$ The 2nd equation

$$y = \frac{3}{2}(10) - \frac{15}{2}$$

$$y = \frac{30}{2} - \frac{15}{2}$$

$$\boxed{y = \frac{15}{2}}$$

The solution is $\left(10, \dfrac{15}{2}\right)$.

65. $\begin{cases} y + x = 2x + 2 \\ 6x - 4y = 21 - y \end{cases}$

put both equations in general form

$$\begin{cases} y - x = 2 \\ 6x - 3y = 21 \end{cases}$$

solve for y in 1st equation

$$\begin{cases} y = 2 + x \\ 6x - 3y = 21 \end{cases}$$

$6x - 3y = 21$ The 2nd equation

$$6x - 3(2 + x) = 21$$

$$6x - 6 - 3x = 21$$

$$3x - 6 = 21$$

$$3x - 6 + 6 = 21 + 6$$

$$3x = 27$$

$$\frac{3x}{3} = \frac{27}{3}$$

$$\boxed{x = 9}$$

$y = 2 + x$ The 1st equation

$$y = 2 + 9$$

$$\boxed{y = 11}$$

The solution is $(9, 11)$.

67. $\begin{cases} x = -3y + 6 \\ 2x + 4y = 6 + x + y \end{cases}$

put 2nd equation in general form

$$\begin{cases} x = -3y + 6 \\ x + 3y = 6 \end{cases}$$

$x + 3y = 6$ The 2nd equation

$$-3y + 6 + 3y = 6$$

$$\boxed{6 = 6}$$

true

Infinitely many solutions.

69.

$$\begin{cases} 4x+5y+1=-12+2x \\ x-3y+2=-3-x \end{cases}$$

put both equations in general form

$$\begin{cases} 2x+5y=-13 \\ 2x-3y=-5 \end{cases}$$

solve for $2x$ in 1st equation

$$\begin{cases} 2x=-13-5y \\ 2x-3y=-5 \end{cases}$$

$2x-3y=-5$ The 2nd equation

$(-13-5y)-3y=-5$

$-8y-13=-5$

$-8y-13+13=-5+13$

$-8y=8$

$\frac{-8y}{-8}=\frac{8}{-8}$

$\boxed{y=-1}$

$2x=-13-5y$ The 1st equation

$2x=-13-5(-1)$

$2x=-13+5$

$2x=-8$

$\frac{2x}{2}=\frac{-8}{2}$

$\boxed{x=-4}$

The solution is (-4, -1).

71.

$$\begin{cases} 3(x-1)+3=8+2y \\ 2(x+1)=8+y \end{cases}$$

distribute both equations

$$\begin{cases} 3x-3+3=8+2y \\ 2x+2=8+y \end{cases}$$

put 1st equation in general form

solve for y in 2nd equation

$$\begin{cases} 3x-2y=8 \\ 2x-6=y \end{cases}$$

$3x-2y=8$ The 1st equation

$3x-2(2x-6)=8$

$3x-4x+12=8$

continued

71. continued

$-x+12=8$

$-x+12-12=8-12$

$-x=-4$

$\frac{-x}{-1}=\frac{-4}{-1}$

$\boxed{x=4}$

$2x-6=y$ The 2nd equation

$2(4)-6=y$

$8-6=y$

$\boxed{2=y}$

The solution is (4, 2).

APPLICATIONS

73. DINING
He can pick melon, because it's the same price as the hash browns.

WRITING

75. Answers will vary.

77. Answers will vary.

REVIEW

79. $3x-8=1$; check 3

$3(3)-8 \stackrel{?}{=} 1$

$9-8 \stackrel{?}{=} 1$

$1=1$ true

yes

81. $3(x+8)+5x=2(12+4x)$; check 3

$3(3+8)+5(3) \stackrel{?}{=} 2[12+4(3)]$

$3(11)+15 \stackrel{?}{=} 2(12+12)$

$33+15 \stackrel{?}{=} 2(24)$

$48=48$ true

yes

83. $x^3+7x^2=x^2-9x$; check 3

$$(3)^3+7(3)^2 \stackrel{?}{=} (3)^2-9(3)$$
$$27+7(9) \stackrel{?}{=} 9-27$$
$$27+63 \stackrel{?}{=} -18$$
$$90=-18 \text{ false}$$

no

CHALLENGE PROBLEMS

85. $$\begin{cases} \dfrac{6x-1}{3}-\dfrac{5}{3}=\dfrac{3y+1}{2} \\ \dfrac{1+5y}{4}+\dfrac{x+3}{4}=\dfrac{17}{2} \end{cases}$$

clear both equations of fractions

$$\begin{cases} 6\left(\dfrac{6x-1}{3}\right)-6\left(\dfrac{5}{3}\right)=6\left(\dfrac{3y+1}{2}\right) \\ 4\left(\dfrac{1+5y}{4}\right)+4\left(\dfrac{x+3}{4}\right)=4\left(\dfrac{17}{2}\right) \end{cases}$$

$$\begin{cases} 12x-2-10=9y+3 \\ 1+5y+x+3=34 \end{cases}$$

put 1st equation in general form

solve for x in 2nd equation

$$\begin{cases} 12x-9y=15 \\ x=30-5y \end{cases}$$

$12x-9y=15$ The 1st equation

$$12(30-5y)-9y=15$$
$$360-60y-9y=15$$
$$360-69y=15$$
$$360-69y-360=15-360$$
$$-69y=345$$
$$\frac{-69y}{-69}=\frac{345}{-69}$$
$$\boxed{y=5}$$

$x=30-5y$ The 2nd equation

$$x=30-5(5)$$
$$x=30-25$$
$$\boxed{x=5}$$

The solution is $(5, 5)$.

87. $$\begin{cases} \dfrac{1}{2}x=y+3 \\ x-2y=6 \end{cases}$$

solve 2nd equation for x

$$x=2y+6$$

To find one solution, let $y=-3$. Substitute that value into 2nd equation for y.

let $y=-3$

$$x=2y+6$$
$$x=2(-3)+6$$
$$x=-6+6$$
$$x=0$$

$\boxed{\text{1st solution is (0, -3)}}$

solve 2nd equation for x

$$x=2y+6$$

To find one solution let $y=0$. Substitute that value into 2nd equation for y.

let $y=0$

$$x=2y+6$$
$$x=2(0)+6$$
$$x=0+6$$
$$x=6$$

$\boxed{\text{2nd solution is (6, 0)}}$

solve 2nd equation for x

$$x=2y+6$$

To find one solution let $y=-2$. Substitute that value into 2nd equation for y.

let $y=-2$

$$x=2y+6$$
$$x=2(-2)+6$$
$$x=-4+6$$
$$x=2$$

$\boxed{\text{3rd solution is (2, -2)}}$

You can check each of these solutions by substituting them into the 1st equation.

SECTION 7.3

VOCABULARY

1. The coefficient of $3x$ and $-3x$ are **opposites**.

3. When the following equations are added, the variable y will be **eliminated**.

5. The elimination method for solving a system is based on the **addition** property of equality: *When equal quantities are added to both sides of an equation, the results are equal.*

CONCEPTS

7. $7y$ and $-7y$

9. a) $4x + y = 2$

   multiply both sides by 3

   $3(4x + y) = 3(2)$

   simplify

   $\boxed{12x + 3y = 6}$

   b) $x - 3y = 1$

   multiply both sides by -2

   $-2(x - 3y) = -2(1)$

   simplify

   $\boxed{-2x + 6y = -2}$

11. $\begin{cases} 4x + 3y = 11 \\ 3x - 2y = 4 \end{cases}$

$$x = 2$$

pick either equation to work with to solve for y

$$x = 2$$
$$4x + 3y = 11$$
$$4(2) + 3y = 11$$
$$8 + 3y = 11$$
$$8 + 3y - 8 = 11 - 8$$
$$3y = 3$$
$$\frac{3y}{3} = \frac{3}{3}$$
$$\boxed{y = 1}$$

13. $\begin{cases} 2x + 3y = -1 \\ 3x + 5y = -2 \end{cases}$

$$2x + 3y = -1$$

Is (1, -1) is solution?

$$2x + 3y = -1$$
$$2(1) + 3(-1) \stackrel{?}{=} -1$$
$$2 - 3 \stackrel{?}{=} -1$$
$$-1 = -1$$

true

$$3x + 5y = -2$$

Is (1, -1) is solution?

$$3x + 5y = -2$$
$$3(1) + 5(-1) \stackrel{?}{=} -2$$
$$3 - 5 \stackrel{?}{=} -2$$
$$-2 = -2$$

true

NOTATION

15. Solve $\begin{cases} x + y = 5 \\ x - y = -3 \end{cases}$

$$x + y = 5$$
$$x - y = -3$$
$$\boxed{2x} = 2$$
$$\frac{2x}{2} = \frac{2}{2}$$
$$x = \boxed{1}$$

$x + y = 5$ The 1st equation

$$\boxed{1} + y = 5$$
$$y = 4$$

The solution is $\boxed{(1, 4)}$.

17. $\begin{cases} x+y=5 \\ x-y=1 \end{cases}$

eliminate the y

$$\begin{array}{l} x+y=5 \\ \underline{x-y=1} \\ 2x \quad =6 \end{array}$$

$$\frac{2x}{2}=\frac{6}{2}$$

$$\boxed{x=3}$$

$x+y=5$ 1st equation

$$3+y=5$$

$$3+y-3=5-3$$

$$\boxed{y=2}$$

solution is (3, 2)

19. $\begin{cases} x+y=1 \\ x-y=5 \end{cases}$

eliminate the y

$$\begin{array}{l} x+y=1 \\ \underline{x-y=5} \\ 2x \quad =6 \end{array}$$

$$\frac{2x}{2}=\frac{6}{2}$$

$$\boxed{x=3}$$

$x+y=1$ 1st equation

$$3+y=1$$

$$3+y-3=1-3$$

$$\boxed{y=-2}$$

solution is (3, -2)

21. $\begin{cases} x+y=-5 \\ -x+y=-1 \end{cases}$

eliminate the x

$$\begin{array}{l} x+y=-5 \\ \underline{-x+y=-1} \\ \quad 2y=-6 \end{array}$$

$$\frac{2y}{2}=\frac{-6}{2}$$

$$\boxed{y=-3}$$

$x+y=-5$ 1st equation

$$x+(-3)=-5$$

$$x-3+3=-5+3$$

$$\boxed{x=-2}$$

solution is (-2, -3)

23. $\begin{cases} 4x+3y=24 \\ 4x-3y=-24 \end{cases}$

eliminate the y

$$\begin{array}{l} 4x+3y=24 \\ \underline{4x-3y=-24} \\ 8x \quad =0 \end{array}$$

$$\frac{8x}{8}=\frac{0}{8}$$

$$\boxed{x=0}$$

$4x+3y=24$ 1st equation

$$4(0)+3y=24$$

$$3y=24$$

$$\frac{3y}{3}=\frac{24}{3}$$

$$\boxed{y=8}$$

solution is (0, 8)

25.

$$\begin{cases} 2s+t=-2 \\ -2s-3t=-6 \end{cases}$$

eliminate the s

$$2s+\ t=-2$$
$$\underline{-2s-3t=-6}$$
$$-2t=-8$$
$$\frac{-2t}{-2}=\frac{-8}{-2}$$
$$\boxed{t=4}$$

$2s+t=-2$ 1st equation

$$2s+4=-2$$
$$2s+4-4=-2-4$$
$$2s=-6$$
$$\frac{2s}{2}=\frac{-6}{2}$$
$$\boxed{s=-3}$$

solution is (-3, 4)

27.

$$\begin{cases} 5x-4y=8 \\ -5x-4y=8 \end{cases}$$

eliminate the x

$$5x-4y=8$$
$$\underline{-5x-4y=8}$$
$$-8y=16$$
$$\frac{-8y}{-8}=\frac{16}{-8}$$
$$\boxed{y=-2}$$

$5x-4y=8$ 1st equation

$$5x-4(-2)=8$$
$$5x+8=8$$
$$5x+8-8=8-8$$
$$5x=0$$
$$\frac{5x}{5}=\frac{0}{5}$$
$$\boxed{x=0}$$

solution is (0, -2)

29.

$$\begin{cases} 4x-7y=-19 \\ -4x+7y=19 \end{cases}$$

eliminate the x or y

$$4x-7y=-19$$
$$\underline{-4x+7y=19}$$
$$0=0$$

true

infinitely many solutions

31.

$$\begin{cases} x+3y=-9 \\ x+8y=-4 \end{cases}$$

multiply 1st equation by -1

$$-x-3y=9$$
$$\underline{x+8y=-4}$$
$$5y=5$$
$$\frac{5y}{5}=\frac{5}{5}$$
$$\boxed{y=1}$$

$x+3y=-9$ 1st equation

$$x+3(1)=-9$$
$$x+3-3=-9-3$$
$$\boxed{x=-12}$$

solution is (-12, 1)

33.

$$\begin{cases} 5c+d=-15 \\ 6c+d=-20 \end{cases}$$

multiply 1st equation by -1

$$-5c-d=15$$
$$\underline{6c+d=-20}$$
$$\boxed{c=-5}$$

$5c+d=-15$ 1st equation

$$5(-5)+d=-15$$
$$-25+d=-15$$
$$-25+d+25=-15+25$$
$$\boxed{d=10}$$

solution is (-5, 10)

35.

$$\begin{cases} 7x - y = 10 \\ 8x - y = 13 \end{cases}$$

multiply 1st equation by -1

$$-7x + y = -10$$
$$8x - y = 13$$
$$\boxed{x = 3}$$

$7x - y = 10$ 1st equation

$$7(3) - y = 10$$
$$21 - y = 10$$
$$21 - y + y = 10 + y$$
$$21 - 10 = 10 + y - 10$$
$$\boxed{11 = y}$$

solution is (3, 11)

37.

$$\begin{cases} 3x - 5y = -29 \\ 3x - 5y = 15 \end{cases}$$

multiply 1st equation by -1

$$-3x + 5y = 29$$
$$3x - 5y = 15$$
$$\boxed{0 = 44}$$

false

no solution

39.

$$\begin{cases} 6x - 3y = -7 \\ 9x + y = 6 \end{cases}$$

multiply 2nd equation by 3

$$6x - 3y = -7$$
$$27x + 3y = 18$$
$$33x = 11$$
$$\frac{33x}{33} = \frac{11}{33}$$
$$\boxed{x = \frac{1}{3}}$$

$6x - 3y = -7$ 1st equation

$$6\left(\frac{1}{3}\right) - 3y = -7$$
$$2 - 3y = -7$$
$$2 - 3y - 2 = -7 - 2$$
$$-3y = -9$$
$$\frac{-3y}{-3} = \frac{-9}{-3}$$
$$\boxed{y = 3}$$

solution is $\left(\frac{1}{3}, 3\right)$

41.

$$\begin{cases} 9x+4y=31 \\ 6x-y=-5 \end{cases}$$

multiply 2nd equation by 4

$$9x+4y=31$$
$$\underline{24x-4y=-20}$$
$$33x=11$$
$$\frac{33x}{33}=\frac{11}{33}$$
$$\boxed{x=\frac{1}{3}}$$

$6x-y=-5$ 2nd equation

$$6\left(\frac{1}{3}\right)-y=-5$$
$$2-y=-5$$
$$2-y+y=-5+y$$
$$2=-5+y$$
$$2+5=-5+y+5$$
$$\boxed{7=y}$$

solution is $\left(\frac{1}{3}, 7\right)$

43.

$$\begin{cases} 8x+8y=-16 \\ 3x+y=-4 \end{cases}$$

divide 1st equation by -8

$$-x-y=2$$
$$\underline{3x+y=-4}$$
$$2x=-2$$
$$\frac{2x}{2}=\frac{-2}{2}$$
$$\boxed{x=-1}$$

$3x+y=-4$ 2nd equation

$$3(-1)+y=-4$$
$$-3+y=-4$$
$$-3+y+3=-4+3$$
$$\boxed{y=-1}$$

solution is (-1, -1)

45.

$$\begin{cases} 7x-50y=-43 \\ x+3y=4 \end{cases}$$

multiply 2nd equation by -7

$$7x-50y=-43$$
$$\underline{-7x-21y=-28}$$
$$-71y=-71$$
$$\frac{-71y}{-71}=\frac{-71}{-71}$$
$$\boxed{y=1}$$

$x+3y=4$ 2nd equation

$$x+3(1)=4$$
$$x+3-3=4-3$$
$$\boxed{x=1}$$

solution is (1, 1)

47.

$$\begin{cases} 8x-4y=18 \\ 3x-2y=8 \end{cases}$$

multiply 2nd equation by -2

$$\begin{array}{r} 8x-4y=18 \\ -6x+4y=-16 \\ \hline 2x=2 \end{array}$$

$$\frac{2x}{2}=\frac{2}{2}$$

$$\boxed{x=1}$$

$3x-2y=8$ $\quad 2^{nd}$ equation

$$3(1)-2y=8$$

$$3-2y-3=8-3$$

$$-2y=5$$

$$\frac{-2y}{-2}=\frac{5}{-2}$$

$$\boxed{y=-\frac{5}{2}}$$

solution is $\left(1,\ -\frac{5}{2}\right)$

49.

$$\begin{cases} 4x+3y=7 \\ 3x-2y=-16 \end{cases}$$

multiply 1st equation by 2

multiply 2nd equation by 3

$$\begin{array}{r} 8x+6y=14 \\ 9x-6y=-48 \\ \hline 17x=-34 \end{array}$$

$$\frac{17x}{17}=\frac{-34}{17}$$

$$\boxed{x=-2}$$

$4x+3y=7$ $\quad 1^{st}$ equation

$$4(-2)+3y=7$$

$$-8+3y+8=7+8$$

$$3y=15$$

$$\frac{3y}{3}=\frac{15}{3}$$

$$\boxed{y=5}$$

solution is (-2, 5)

51.

$$\begin{cases} 3x+4y=12 \\ 4x+5y=17 \end{cases}$$

multiply 1st equation by 5

multiply 2nd equation by -4

$$\begin{array}{r} 15x+20y=60 \\ -16x-20y=-68 \\ \hline -x=-8 \end{array}$$

$$\frac{-x}{-1}=\frac{-8}{-1}$$

$$\boxed{x=8}$$

$3x+4y=12$ $\quad 1^{st}$ equation

$$3(8)+4y=12$$

$$24+4y-24=12-24$$

$$4y=-12$$

$$\frac{4y}{4}=\frac{-12}{4}$$

$$\boxed{y=-3}$$

solution is (8, -3)

53.

$$\begin{cases} -3x+6y=-9 \\ -5x+4y=-15 \end{cases}$$

multiply 1st equation by 2

multiply 2nd equation by -3

$$\begin{array}{r} -6x+12y=-18 \\ 15x-12y=45 \\ \hline 9x=27 \end{array}$$

$$\frac{9x}{9}=\frac{27}{9}$$

$$\boxed{x=3}$$

$-3x+6y=-9$ $\quad 1^{st}$ equation

$$-3(3)+6y=-9$$

$$-9+6y+9=-9+9$$

$$6y=0$$

$$\frac{6y}{6}=\frac{0}{6}$$

$$\boxed{y=0}$$

solution is (3, 0)

55.

$$\begin{cases} 4a+7b=2 \\ 9a-3b=1 \end{cases}$$

multiply 1st equation by 3

multiply 2nd equation by 7

$$12a+21b=6$$
$$\underline{63a-21b=7}$$
$$75a=13$$
$$\frac{75a}{75}=\frac{13}{75}$$
$$\boxed{a=\frac{13}{75}}$$

$4a+7b=2$ 1st equation

$$4\left(\frac{13}{75}\right)+7b=2$$
$$\frac{52}{75}+7b=2$$
$$\frac{52}{75}+7b-\frac{52}{75}=2-\frac{52}{75}$$
$$7b=\frac{150}{75}-\frac{52}{75}$$
$$7b=\frac{98}{75}$$
$$7b\left(\frac{1}{7}\right)=\frac{98}{75}\left(\frac{1}{7}\right)$$
$$\boxed{b=\frac{14}{75}}$$

solution is $\left(\frac{13}{75}, \frac{14}{75}\right)$

57.

$$\begin{cases} 9x=10y \\ 3x-2y=12 \end{cases}$$

put 1st equation in general form

$$\begin{cases} 9x-10y=0 \\ 3x-2y=12 \end{cases}$$

multiply 2nd equation by -3

$$9x-10y=0$$
$$\underline{-9x+6y=-36}$$
$$-4y=-36$$
$$\frac{-4y}{-4}=\frac{-36}{-4}$$
$$\boxed{y=9}$$

$9x=10y$ 1st equation

$$9x=10(9)$$
$$9x=90$$
$$\frac{9x}{9}=\frac{90}{9}$$
$$\boxed{x=10}$$

solution is (10, 9)

59.

$$\begin{cases} 2x+5y+13=0 \\ 2x+5=3y \end{cases}$$

put both equations in general form

$$\begin{cases} 2x+5y=-13 \\ 2x-3y=-5 \end{cases}$$

multiply 2nd equation by -1

$$2x+5y=-13$$
$$\underline{-2x+3y=5}$$
$$8y=-8$$
$$\frac{8y}{8}=\frac{-8}{8}$$
$$\boxed{y=-1}$$

$2x+5=3y$ 2nd equation

$$2x+5=3(-1)$$
$$2x+5-5=-3-5$$
$$2x=-8$$
$$\frac{2x}{2}=\frac{-8}{2}$$
$$\boxed{x=-4}$$

solution is (-4, -1)

61.

$$\begin{cases} 0 = 4x - 3y \\ 5x = 4y - 2 \end{cases}$$

put 2nd equation in general form

$$\begin{cases} 4x - 3y = 0 \\ 5x - 4y = -2 \end{cases}$$

multiply 1st equation by 4

multiply 2nd equation by -3

$$16x - 12y = 0$$
$$\underline{-15x + 12y = 6}$$
$$\boxed{x = 6}$$

$4x - 3y = 0$ 1st equation

$$4(6) - 3y = 0$$
$$24 - 3y - 24 = 0 - 24$$
$$-3y = -24$$
$$\frac{-3y}{-3} = \frac{-24}{-3}$$
$$\boxed{y = 8}$$

solution is (6, 8)

63.

$$\begin{cases} 3x - 16 = 5y \\ -3x + 5y - 33 = 0 \end{cases}$$

put both equations in general form

$$\begin{cases} 3x - 5y = 16 \\ -3x + 5y = 33 \end{cases}$$

$$3x - 5y = 16$$
$$\underline{-3x + 5y = 33}$$
$$\boxed{0 = 49}$$

false

no solution

65.

$$\begin{cases} \frac{3}{5}s + \frac{4}{5}t = 1 \\ -\frac{1}{4}s + \frac{3}{8}t = 1 \end{cases}$$

clear both equations of fractions

$$\begin{cases} 5\left(\frac{3}{5}s\right) + 5\left(\frac{4}{5}t\right) = 5(1) \\ 8\left(-\frac{1}{4}s\right) + 8\left(\frac{3}{8}t\right) = 8(1) \end{cases}$$

$$\begin{cases} 3s + 4t = 5 \\ -2s + 3t = 8 \end{cases}$$

multiply 1st equation by 2

multiply 2nd equation by 3

$$6s + 8t = 10$$
$$\underline{-6s + 9t = 24}$$
$$17t = 34$$
$$\frac{17t}{17} = \frac{34}{17}$$
$$\boxed{t = 2}$$

$3s + 4t = 5$ 1st equation

$$3s + 4(2) = 5$$
$$3s + 8 - 8 = 5 - 8$$
$$3s = -3$$
$$\frac{3s}{3} = \frac{-3}{3}$$
$$\boxed{s = -1}$$

solution is (-1, 2)

67.

$$\begin{cases}\frac{1}{2}s-\frac{1}{4}t=1\\ \frac{1}{3}s+t=3\end{cases}$$

clear both equations of fractions

$$\begin{cases}4\left(\frac{1}{2}s\right)-4\left(\frac{1}{4}t\right)=4(1)\\ 3\left(\frac{1}{3}s\right)+3(t)=3(3)\end{cases}$$

$$\begin{cases}2s-t=4\\ s+3t=9\end{cases}$$

multiply 2nd equation by -2

$$2s-t=4$$
$$\underline{-2s-6t=-18}$$
$$-7t=-14$$
$$\frac{-7t}{-7}=\frac{-14}{-7}$$
$$\boxed{t=2}$$

$\frac{1}{3}s+t=3$ 2nd equation

$$\frac{1}{3}s+(2)=3$$
$$\frac{1}{3}s+2-2=3-2$$
$$\frac{1}{3}s=1$$
$$\frac{3}{1}\left(\frac{1}{3}s\right)=3(1)$$
$$\boxed{s=3}$$

solution is (3, 2)

69.

$$\begin{cases}-\frac{m}{4}-\frac{n}{3}=\frac{1}{12}\\ \frac{m}{2}-\frac{5n}{4}=\frac{7}{4}\end{cases}$$

clear both equations of fractions

$$\begin{cases}12\left(-\frac{m}{4}\right)-12\left(\frac{n}{3}\right)=12\left(\frac{1}{12}\right)\\ 4\left(\frac{m}{2}\right)-4\left(\frac{5n}{4}\right)=4\left(\frac{7}{4}\right)\end{cases}$$

$$\begin{cases}-3m-4n=1\\ 2m-5n=7\end{cases}$$

multiply 1st equation by 2

multiply 2nd equation by 3

$$-6m-8n=2$$
$$\underline{6m-15n=21}$$
$$-23n=23$$
$$\frac{-23n}{-23}=\frac{23}{-23}$$
$$\boxed{n=-1}$$

$2m-5n=7$ 2nd equation

$$2m-5(-1)=7$$
$$2m+5-5=7-5$$
$$2m=2$$
$$\frac{2m}{2}=\frac{2}{2}$$
$$\boxed{m=1}$$

solution is (1, -1)

71.

$$\begin{cases} \dfrac{x}{2} - 3y = 7 \\ -x + 6y = -14 \end{cases}$$

clear 1st equation of fractions

$$\begin{cases} 2\left(\dfrac{x}{2}\right) - 2(3y) = 2(7) \\ -x + 6y = -14 \end{cases}$$

$$\begin{cases} x - 6y = 14 \\ -x + 6y = -14 \end{cases}$$

$$\begin{array}{r} x - 6y = 14 \\ -x + 6y = -14 \\ \hline \boxed{0 = 0} \end{array}$$

true

infinitely many solutions

73.

$$\begin{cases} 2x + y = 10 \\ 0.1x + 0.2y = 1.0 \end{cases}$$

multiply 2nd equation by 10

$$\begin{cases} 2x + y = 10 \\ 1x + 2y = 10 \end{cases}$$

multiply 2nd equation by -2

$$\begin{array}{r} 2x + y = 10 \\ -2x - 4y = -20 \\ \hline -3y = -10 \end{array}$$

$$\frac{-3y}{-3} = \frac{-10}{-3}$$

$$\boxed{y = \frac{10}{3}}$$

$x + 2y = 10$ 2nd equation

$$x + 2\left(\frac{10}{3}\right) = 10$$

$$x + \frac{20}{3} = 10$$

$$x + \frac{20}{3} - \frac{20}{3} = 10 - \frac{20}{3}$$

$$x = \frac{30}{3} - \frac{20}{3}$$

$$\boxed{x = \frac{10}{3}}$$

solution is $\left(\dfrac{10}{3}, \dfrac{10}{3}\right)$

75.

$$\begin{cases} 2x - y = 16 \\ 0.03x + 0.02y = 0.03 \end{cases}$$

multiply 2nd equation by 100

$$\begin{cases} 2x - y = 16 \\ 3x + 2y = 3 \end{cases}$$

multiply 1st equation by 2

$$\begin{array}{r} 4x - 2y = 32 \\ 3x + 2y = 3 \\ \hline 7x = 35 \end{array}$$

$$\frac{7x}{7} = \frac{35}{7}$$

$$\boxed{x = 5}$$

$2x - y = 16$ 1st equation

$$2(5) - y = 16$$

$$10 - y - 10 = 16 - 10$$

$$-y = 6$$

$$\frac{-y}{-1} = \frac{6}{-1}$$

$$\boxed{y = -6}$$

solution is (5, -6)

APPLICATIONS

77.

$$\begin{cases} 9x + 11y = 352 \\ 5x - 11y = -198 \end{cases}$$

$$\begin{array}{r} 9x + 11y = 352 \\ 5x - 11y = -198 \\ \hline 14x = 154 \end{array}$$

$$\frac{14x}{14} = \frac{154}{14}$$

$$\boxed{x = 11}$$

$x = 0$ implies 1980

$x = 11$ implies $1980 + 11$

The year when the percents are equal is 1991.

WRITING

79. Answers will vary.

81. Answers will vary.

REVIEW

83. 10 is less than x

$$\boxed{x-10}$$

85. $x-x$

$$\boxed{0}$$

87. $A=\frac{1}{2}bh$

$b=4$ feet

$h=3.75$ feet

$A=??$

$A=0.5(4 \text{ ft})(3.75 \text{ ft})$

$A=7.5 \text{ ft}^2$

CHALLENGE PROBLEMS

89.
$$\begin{cases}\frac{x-3}{2}+\frac{y+5}{3}=\frac{11}{6}\\ \frac{x+3}{3}-\frac{5}{12}=\frac{y+3}{4}\end{cases}$$

clear both equations of fractions

$$\begin{cases}6\left(\frac{x-3}{2}\right)+6\left(\frac{y+5}{3}\right)=6\left(\frac{11}{6}\right)\\ 12\left(\frac{x+3}{3}\right)-12\left(\frac{5}{12}\right)=12\left(\frac{y+3}{4}\right)\end{cases}$$

$$\begin{cases}3(x-3)+2(y+5)=11\\ 4(x+3)-1(5)=3(y+3)\end{cases}$$

$$\begin{cases}3x-9+2y+10=11\\ 4x+12-5=3y+9\end{cases}$$

$$\begin{cases}3x+2y+1=11\\ 4x+7=3y+9\end{cases}$$

put both equations in general form

$$\begin{cases}3x+2y=10\\ 4x-3y=2\end{cases}$$

multiply 1st equation by 3

multiply 2nd equation by 2

$$9x+6y=30$$
$$\underline{8x-6y=4}$$
$$17x=34$$
$$\frac{17x}{17}=\frac{34}{17}$$
$$\boxed{x=2}$$

$3x+2y=10$ 1st equation

$$3(2)+2y=10$$
$$6+2y-6=10-6$$
$$2y=4$$
$$\frac{2y}{2}=\frac{4}{2}$$
$$\boxed{y=2}$$

solution is (2, 2)

SECTION 7.4

VOCABULARY

1. A **variable** is a letter that stands for a number.

3. $\begin{cases} a+b=20 \\ a=2b+4 \end{cases}$ is a **system** of equations.

5. Two angles are said to be **complementary** if the sum of their measures is 90°.

CONCEPTS

7.

Let $x =$ length of shorter piece

$y =$ length of longer piece

20 in. = total length

longer = 1 in. less than twice shorter

$\boxed{\begin{array}{l} x+y=20\text{ , 1}^{\text{st}}\text{ equation} \\ y=2x-1\text{ , 2}^{\text{nd}}\text{ equation} \end{array}}$

9.

Let $x =$ larger $\angle$

$y =$ smaller $\angle$

sum of supplementary $\angle = 180^0$

smaller $\angle = 25°$ less than larger $\angle$

$\boxed{\begin{array}{l} x+y=180\text{ , 1}^{\text{st}}\text{ equation} \\ y=x-25\text{ , 2}^{\text{nd}}\text{ equation} \end{array}}$

11.

let $x =$ cost of a chicken taco

$y =$ cost of a beef taco

$\boxed{5x+2y=10}$

13.

a) Downstream: $x+c$

b) Upstream: $x-c$

15.

a) x ml of 30% acid solution

y ml of 40% acid solution

$(x+y)$ ml

b) 33% acid solution

PRACTICE

17.

Let $x =$ measure of one $\angle$

$y =$ measure of other $\angle$

sum of complementary $\angle = 90^0$

one $\angle$ is 10° more than three times other $\angle$

$$\begin{cases} x+y=90 \\ x=10+3y \end{cases}$$

$$\begin{cases} x+y=90 \\ x-3y=10 \end{cases}$$

$$\begin{array}{r} x+y=90 \\ -x+3y=-10 \\ \hline 4y=80 \end{array}$$

$$\frac{4y}{4}=\frac{80}{4}$$

$$\boxed{y=20}$$

$$x=10+3y$$

$$x=10+3(20)$$

$$\boxed{x=70}$$

One $\angle$ is 20°.

Other $\angle$ is 70°.

19.

Let $x =$ large $\angle$

$y =$ small $\angle$

sum of supplementary $\angle = 180^0$

large $\angle$ − small $\angle = 80°$

$$\begin{cases} x+y=180 \\ x-y=80 \end{cases}$$

$$\begin{array}{r} x+y=180 \\ x-y=80 \\ \hline 2x=260 \end{array}$$

$$\frac{2x}{2}=\frac{260}{2}$$

$$\boxed{x=130}$$

$$x+y=180$$

$$130+y=180$$

$$130+y-130=180-130$$

$$\boxed{y=50}$$

Large $\angle$ is 130°.

Small $\angle$ is 50°.

APPLICATIONS

21. TREE TRIMMING

Let x = length of upper arm

y = length of lower arm

total length of both arms = 51 ft.

upper arm = 7 ft. shorter than lower arm

$$\begin{cases} x+y=51 \\ x=y-7 \end{cases}$$

$$\begin{cases} x+y=51 \\ x-y=-7 \end{cases}$$

$$\begin{array}{r} x+y=51 \\ x-y=-7 \\ \hline 2x=44 \end{array}$$

$$\frac{2x}{2}=\frac{44}{2}$$

$$\boxed{x=22}$$

$$x+y=51$$

$$22+y=51$$

$$22+y-22=51-22$$

$$\boxed{y=29}$$

Upper arm is 22 ft.

Lower arm is 29 ft.

23. EXECUTIVE BRANCH

Let x = president's salary (large)

y = vice-president salary (small)

total of both salaries = \$592,600

president makes \$207,400 more than VP

$$\begin{cases} x+y=592,600 \\ x=y+207,400 \end{cases}$$

$$\begin{cases} x+y=592,600 \\ x-y=207,400 \end{cases}$$

$$\begin{array}{r} x+y=592,600 \\ x-y=207,400 \\ \hline 2x=800,000 \end{array}$$

$$\frac{2x}{2}=\frac{800,000}{2}$$

$$\boxed{x=400,000}$$

$$x+y=592,600$$

$$400,000+y=592,600$$

subtract 400,000 from both sides

-400,000 = -400,000

$$\boxed{y=192,600}$$

President's salary is \$400,000.

Vice-President salary is \$192,600.

25. MARINE CORPS

Let $x = \angle 1$

$y = \angle 2$

sum of supplementary $\angle = 180^0$

$\angle 2 = 15°$ less than twice $\angle 1$

$$\begin{cases} x+y=180 \\ y=2x-15 \end{cases}$$

$$\begin{cases} x+y=180 \\ -2x+y=-15 \end{cases}$$

$$\begin{array}{r} x+y=180 \\ 2x-y=15 \\ \hline 3x=195 \end{array}$$

$$\frac{3x}{3}=\frac{195}{3}$$

$$\boxed{x=65}$$

$$y=2x-15$$

$$y=2(65)-15$$

$$y=130-15$$

$$\boxed{y=115}$$

$\angle 1$ is 65°.

$\angle 2$ is 115°.

27. THEATER SCREEN

Let l = length in ft.

w = width in ft.

width = 26 ft. less than length

perimeter = 332 ft.

perimeter = $2l + 2w$

$$\begin{cases} 2l+2w=332 \\ w=l-26 \end{cases}$$

$$2l+2(l-26)=332$$

$$2l+2l-52=332$$

$$4l-52+52=332+52$$

$$4l=384$$

$$\frac{4l}{4}=\frac{384}{4}$$

$$\boxed{l=96}$$

$$w=l-26$$

$$w=96-26$$

$$\boxed{w=70}$$

Length is 96 ft.

Width is 70 ft.

29. GEOMETRY

Let l = length in meters (large)
w = width in meters (small)
perimeter of path is 50 meters
$2l + 2w =$ perimeter
width is two-thirds its length

$$\begin{cases} 2l + 2w = 50 \\ w = \frac{2}{3}l \end{cases}$$

$$2l + 2\left(\frac{2}{3}l\right) = 50$$

$$3(2l) + 3\left(\frac{4}{3}l\right) = 3(50)$$

$$6l + 4l = 150$$

$$10l = 150$$

$$\frac{10l}{10} = \frac{150}{10}$$

$$\boxed{l = 15}$$

$$2l + 2w = 50$$

$$2(15) + 2w = 50$$

$$30 + 2w - 30 = 50 - 30$$

$$2w = 20$$

$$\frac{2w}{2} = \frac{20}{2}$$

$$\boxed{w = 10}$$

Length is 15 m.
Width is 10 m.

31. BUYING PAINTING SUPPLIES

Let x = price of one gallon of paint
y = price of one brush
cost of 8 gal and 3 brushes is \$270
cost of 6 gal and 2 brushes is \$200

$$\begin{cases} 8x + 3y = 270 \\ 6x + 2y = 200 \end{cases}$$

$$\begin{cases} -2(8x + 3y) = -2(270) \\ 3(6x + 2y) = 3(200) \end{cases}$$

$$-16x - 6y = -540$$

$$18x + 6y = 600$$

$$2x = 60$$

$$\frac{2x}{2} = \frac{60}{2}$$

$$\boxed{x = 30}$$

$$6x + 2y = 200$$

$$6(30) + 2y = 200$$

$$180 + 2y - 180 = 200 - 180$$

$$2y = 20$$

$$\frac{2y}{2} = \frac{20}{2}$$

$$\boxed{y = 10}$$

A gallon of paint costs \$30.
One brush costs \$10.

33. SELLING ICE CREAM

Let x = # of cones sold

y = # of sundaes sold

a cone cost \$0.90 , a sundae cost \$1.65

# of cones and # of sundaes = 148

total receipts = \$180.45

$$\begin{cases} x+y=148 \\ 0.90x+1.65y=180.45 \end{cases}$$

$$\begin{cases} x=148-y \\ 0.90x+1.65y=180.45 \end{cases}$$

$$0.90x+1.65y=180.45$$

$$0.90(148-y)+1.65y=180.45$$

$$133.20-0.90y+1.65y=180.45$$

$$133.20+0.75y-133.20=180.45-133.20$$

$$0.75y=47.25$$

$$\frac{0.75y}{0.75}=\frac{47.25}{0.75}$$

$$\boxed{y=63}$$

$$x+y=148$$

$$x+63=148$$

$$x+63-63=148-63$$

$$\boxed{x=85}$$

85 cones were sold.

63 sundaes were sold.

35. MAKING TIRES

a) cost of 1st mold $=\$1,000+\$15t$

cost of 2nd mold $=\$3,000+\$10t$

Let t = break point for tires

$$\begin{cases} c=1,000+15t \\ c=3,000+10t \end{cases}$$

cost of 1st mold = cost of 2nd mold

$$1,000+15t=3,000+10t$$

$$1,000+15t-1,000=3,000+10t-1,000$$

$$15t=2,000+10t$$

$$15t-10t=2,000+10t-10t$$

$$5t=2,000$$

$$\frac{5t}{5}=\frac{2,000}{5}$$

$$\boxed{t=400}$$

If 400 tires are made, the cost will be the same on either machine.

b)

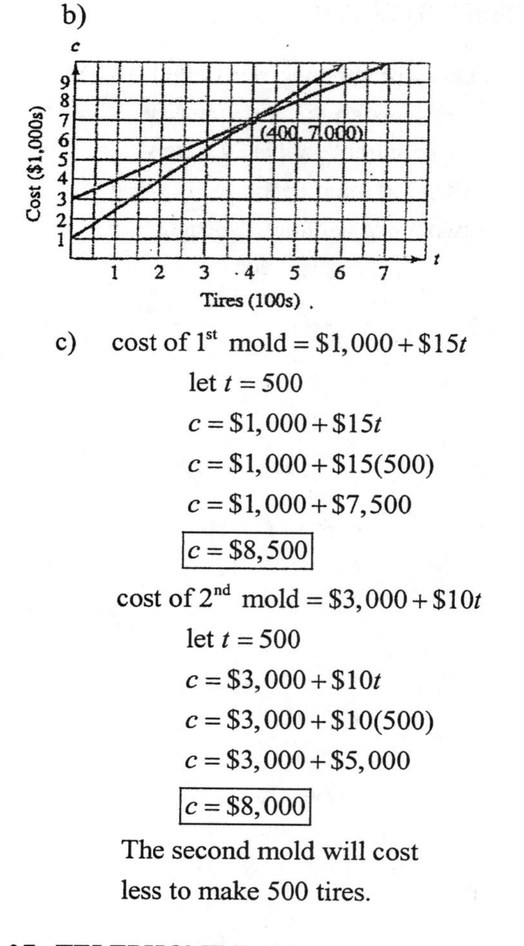

c) cost of 1st mold $=\$1,000+\$15t$

let $t=500$

$$c=\$1,000+\$15t$$

$$c=\$1,000+\$15(500)$$

$$c=\$1,000+\$7,500$$

$$\boxed{c=\$8,500}$$

cost of 2nd mold $=\$3,000+\$10t$

let $t=500$

$$c=\$3,000+\$10t$$

$$c=\$3,000+\$10(500)$$

$$c=\$3,000+\$5,000$$

$$\boxed{c=\$8,000}$$

The second mold will cost less to make 500 tires.

37. TELEPHONE RATES

Plan $1=\$10+\$0.08m$

Plan $2=\$15+\$0.04m$

Let m = break point in minutes

$$\begin{cases} c=10+0.08m \\ c=15+0.04m \end{cases}$$

$$10+0.08m=15+0.04m$$

$$10+0.08m-10=15+0.04m-10$$

$$0.08m=5+0.04m$$

$$0.08m-0.04m=5+0.04m-0.04m$$

$$0.04m=5$$

$$\frac{0.04m}{0.04}=\frac{5}{0.04}$$

$$\boxed{m=125}$$

It will take 125 minutes for each cost to be the same for either plan.

39. GULF STREAM

Let s = speed of boat in still water (mph)

c = speed of current (mph)

rate • time = distance

north: 300 mi. in 10 hr. with current

south: 300 mi. in 15 hr. against current

$$\begin{cases} 10(s+c)=300 \\ 15(s-c)=300 \end{cases}$$

$$\begin{cases} \frac{10(s+c)}{10}=\frac{300}{10} \\ \frac{15(s-c)}{15}=\frac{300}{15} \end{cases}$$

$$\begin{cases} s+c=30 \\ s-c=20 \end{cases}$$

$$\begin{array}{r} s+c=30 \\ s-c=20 \\ \hline 2s=50 \end{array}$$

$$\frac{2s}{2}=\frac{50}{2}$$

$$\boxed{s=25}$$

$$s+c=30$$

$$25+c=30$$

$$25+c-25=30-25$$

$$\boxed{c=5}$$

Speed of boat is 25 mph.

Speed of current is 5 mph.

41. AVIATION

Let s = speed of plane in still air (mph)

c = speed of wind (mph)

rate • time = distance

with wind: 800 mi. in 4 hr.

against wind: 800 mi. in 5 hr.

$$\begin{cases} 4(s+c)=800 \\ 5(s-c)=800 \end{cases}$$

$$\begin{cases} \frac{4(s+c)}{4}=\frac{800}{4} \\ \frac{5(s-c)}{5}=\frac{800}{5} \end{cases}$$

$$\begin{cases} s+c=200 \\ s-c=160 \end{cases}$$

$$\begin{array}{r} s+c=200 \\ s-c=160 \\ \hline 2s=360 \end{array}$$

$$\frac{2s}{2}=\frac{360}{2}$$

$$\boxed{s=180}$$

$$s+c=200$$

$$180+c=200$$

$$180+c-180=200-180$$

$$\boxed{c=20}$$

Speed of plane is 180 mph.

Speed of jet stream is 20 mph.

43.STUDENT LOANS

Let x = amt loaned to nursing student

y = amt loaned to business student

principal • rate • time = interest

total gift: $5,000

interest rate for nurse: 5%

interest rate for business: 7%

collected interest first year: $310

$$\begin{cases} x+y=5,000 \\ 0.05x+0.07y=310 \end{cases}$$

$$\begin{cases} x=5,000-y \\ 0.05x+0.07y=310 \end{cases}$$

$$0.05x+0.07y=310$$
$$0.05(5,000-y)+0.07y=310$$
$$250-0.05y+0.07y=310$$
$$250+0.02y-250=310-250$$
$$0.02y=60$$
$$\frac{0.02y}{0.02}=\frac{60}{0.02}$$
$$\boxed{y=3,000}$$
$$x+y=5,000$$
$$x+3,000=5,000$$
$$x+3,000-3,000=5,000-3,000$$
$$\boxed{x=2,000}$$

Nursing student loan is $2,000.

Business student loan is $3,000.

45.MARINE BIOLOGY

Let x = # of liters of 6% water

y = # of liters of 2% water

percent • amt of salt water = pure salt

total water: 16 liters

new percent of salt water: 3%

$$\begin{cases} x+y=16 \\ 0.06x+0.02y=0.03(16) \end{cases}$$

$$\begin{cases} x=16-y \\ 0.06x+0.02y=0.48 \end{cases}$$

$$0.06x+0.02y=0.48$$
$$0.06(16-y)+0.02y=0.48$$
$$0.96-0.06y+0.02y=0.48$$
$$0.96-0.04y-0.96=0.48-0.96$$
$$-0.04y=-0.48$$
$$\frac{-0.04y}{-0.04}=\frac{-0.48}{-0.04}$$
$$\boxed{y=12}$$
$$x+y=16$$
$$x+12=16$$
$$x+12-12=16-12$$
$$\boxed{x=4}$$

4 liters of 6% salt water needed.

12 liters of 2% salt water needed.

47. COFFEE SALES

Let x = # of lb of Colombian coffee

y = # of lb of Brazilian coffee

cost/lb • amt = value

Colombian coffee: \$8.75/lb

Brazilian coffee: \$3.75/lb

blend: \$6.35/lb

total lb.: 100 lb

$$\begin{cases} x+y=100 \\ 8.75x+3.75y=6.35(100) \end{cases}$$

$$\begin{cases} x=100-y \\ 8.75x+3.75y=635 \end{cases}$$

$$8.75x+3.75y=635$$
$$8.75(100-y)+3.75y=635$$
$$875-8.75y+3.75y=635$$
$$875-5y-875=635-875$$
$$-5y=-240$$
$$\frac{-5y}{-5}=\frac{-240}{-5}$$
$$\boxed{y=48}$$
$$x+y=100$$
$$x+48=100$$
$$x+48-48=100-48$$
$$\boxed{x=52}$$

52 lb of Colombian (\$8.75/lb) coffee needed.

48 lb of Brazilian (\$3.75/lb) coffee needed.

WRITING

49. Answers will vary.

REVIEW

51. $x<4$

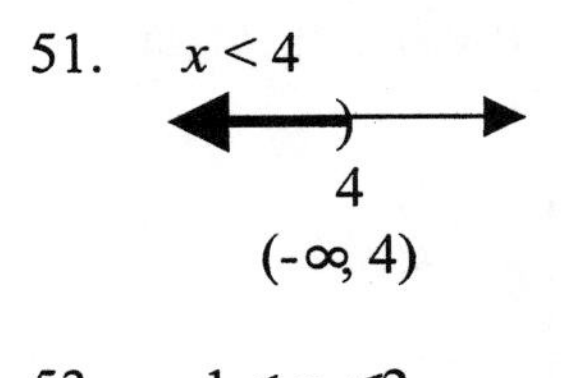

$(-\infty, 4)$

53. $-1<x\leq 2$

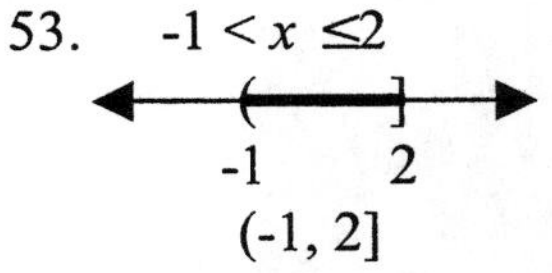

$(-1, 2]$

CHALLENGE PROBLEMS

55.

$$\begin{cases} 3 \text{ nails} + 1 \text{ bolt} = 3 \text{ nuts} \\ 1 \text{ bolt} + 1 \text{ nut} = 5 \text{ nails} \end{cases}$$

$$\begin{cases} 3 \text{ nails} + 1 \text{ bolt} = 3 \text{ nuts} \\ 1 \text{ bolt} = 5 \text{ nails} - 1 \text{ nut} \end{cases}$$

$$3 \text{ nails} + 1 \text{ bolt} = 3 \text{ nuts}$$
$$3 \text{ nails} + (5 \text{ nails} - 1 \text{ nut}) = 3 \text{ nuts}$$
$$8 \text{ nails} - 1 \text{ nut} = 3 \text{ nuts}$$
$$8 \text{ nails} - 1 \text{ nut} + 1 \text{ nut} = 3 \text{ nuts} + 1 \text{ nut}$$
$$8 \text{ nails} = 4 \text{ nuts}$$
$$\frac{8 \text{ nails}}{4} = \frac{4 \text{ nuts}}{4}$$
$$\boxed{2 \text{ nails} = 1 \text{ nut}}$$

The weight of 1 nut is the same as 2 nails.

SECTION 7.5

VOCABULARY

1. $\begin{cases} x+y>2 \\ x+y<4 \end{cases}$ is a system of linear **inequalities**.

3. To **solve** a system of linear inequalities means to find all of its solutions.

5. To find the solutions of a system of two linear inequalities graphically, look for the **intersection**, or overlap, of the two shaded regions.

CONCEPTS

7. a) $3x-y=5$
   b) dashed

9. slope: $4=\frac{4}{1}$; y-intercept: (0, -3)

11.
   a) Does (0, 0) make $2x+y>4$ true?

$$2x+y>4$$
$$2(0)+(0)\overset{?}{>}4$$
$$\boxed{2>4} \text{ false}$$

   b) above

13.
   a) (4, -2) ; yes
   b) (1, 3) ; no
   c) the origin ; no

15.
   a) $x+y=2$, ii
   b) $x+y\geq 2$, iii
   c) $\begin{cases} x+y=2 \\ x-y=2 \end{cases}$, iv
   d) $\begin{cases} x+y=2 \\ x-y=2 \end{cases}$, i

NOTATION

17. ABC

PRACTICE

Use the next three detailed exercises as guides for this set of PRACTICE exercises.

19. $\begin{cases} x+2y\leq 3 \\ 2x-y\geq 1 \end{cases}$

Step 1, graph the boundary of 1st inequality, as $x+2y=3$. Shade the region.

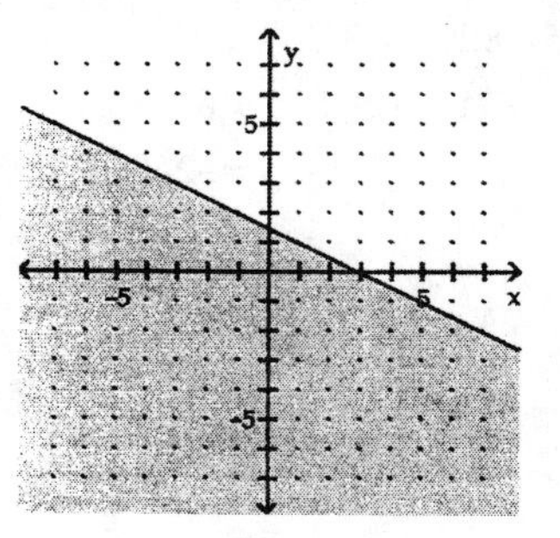

Step 2, graph the boundary of 2nd inequality, as $2x-y=1$. Shade the region.

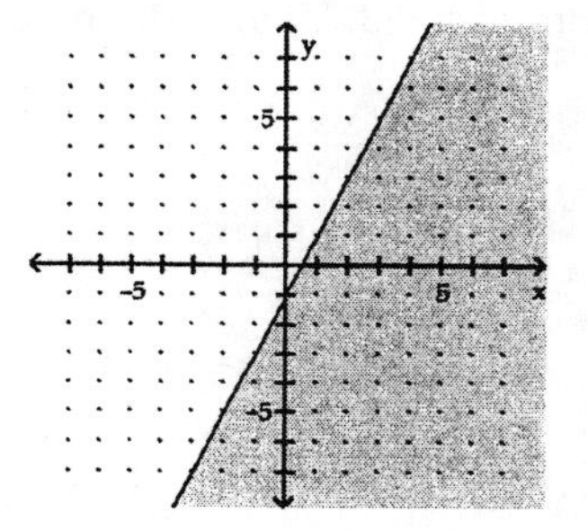

Step 3, identify the intersection of the two regions.

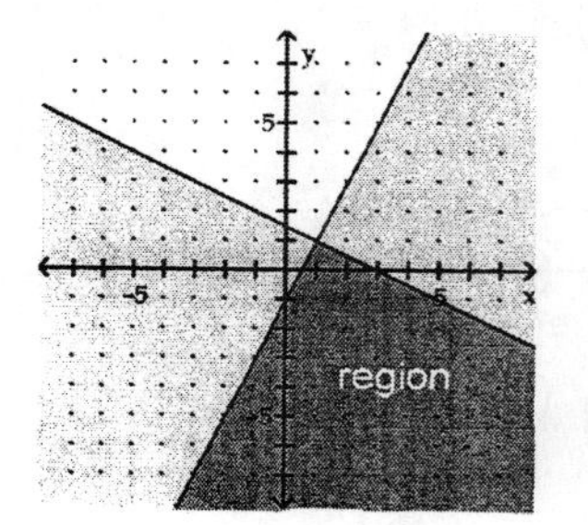

21. $\begin{cases} x+y<-1 \\ x-y>-1 \end{cases}$

Step 1, graph the boundary of 1st inequality, as $x+y=-1$. Shade the region.

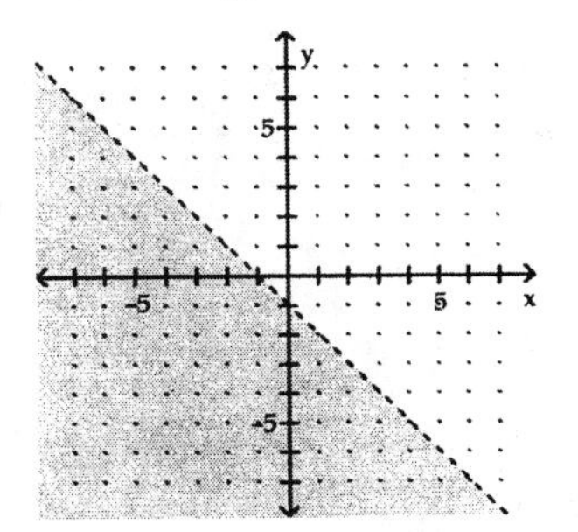

Step 2, graph the boundary of 2nd inequality, as $x-y=-1$. Shade the region.

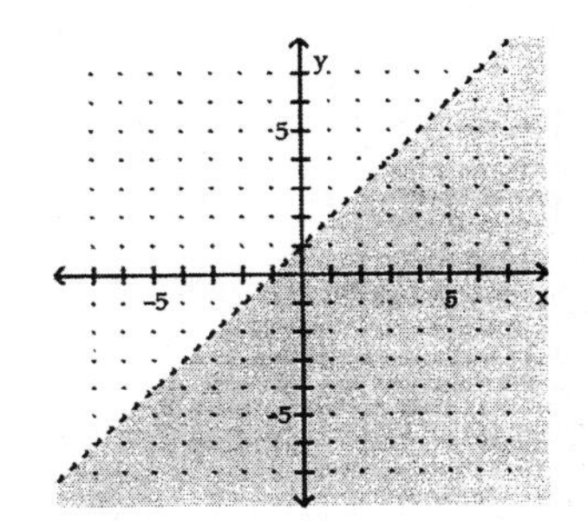

Step 3, identify the intersection of the two regions.

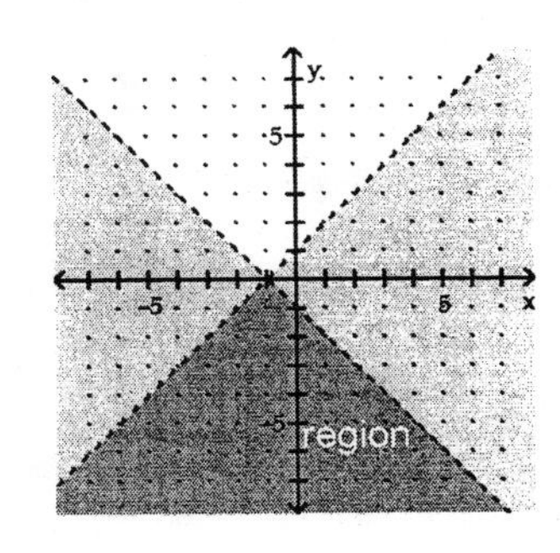

23. $\begin{cases} 2x-3y \le 0 \\ y \ge x-1 \end{cases}$

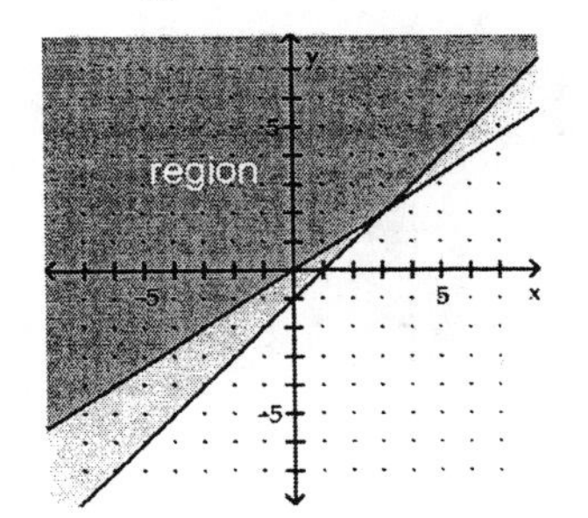

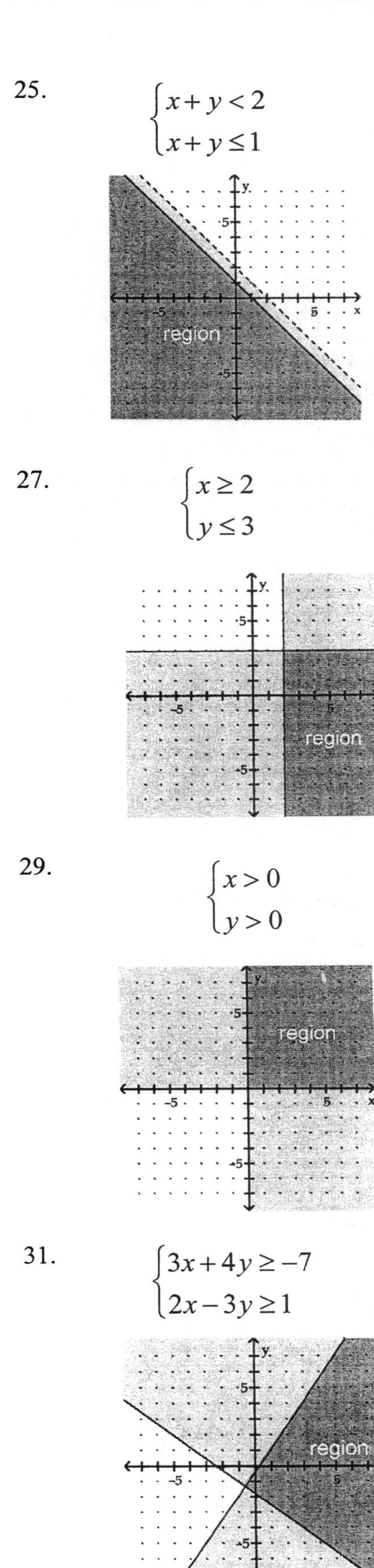

25. $\begin{cases} x+y<2 \\ x+y \le 1 \end{cases}$

27. $\begin{cases} x \ge 2 \\ y \le 3 \end{cases}$

29. $\begin{cases} x>0 \\ y>0 \end{cases}$

31. $\begin{cases} 3x+4y \ge -7 \\ 2x-3y \ge 1 \end{cases}$

33.

$$\begin{cases} 2x + y < 7 \\ y > 2 - 2x \end{cases}$$

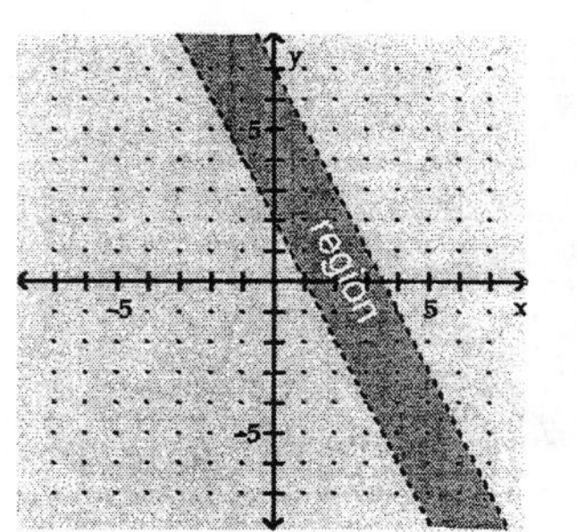

35.

$$\begin{cases} 2x - 4y > -6 \\ 3x + y \geq 5 \end{cases}$$

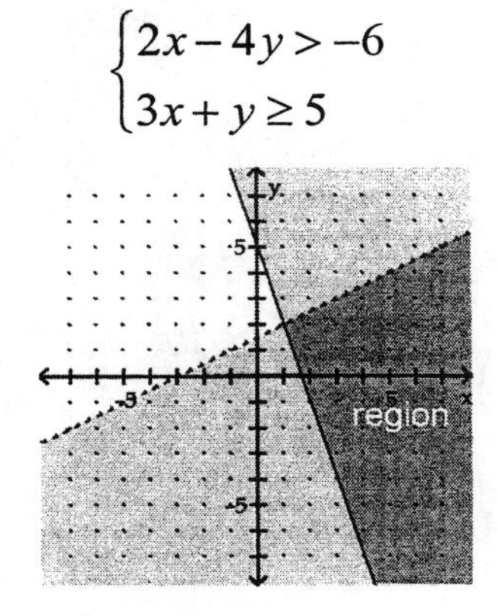

37.

$$\begin{cases} 3x - y + 4 \leq 0 \\ 3y > -2x - 10 \end{cases}$$

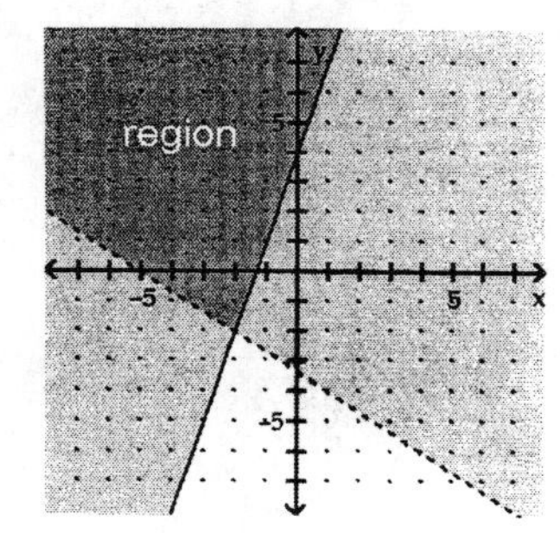

39.

$$\begin{cases} y \geq x \\ y \leq \frac{1}{3}x + 1 \end{cases}$$

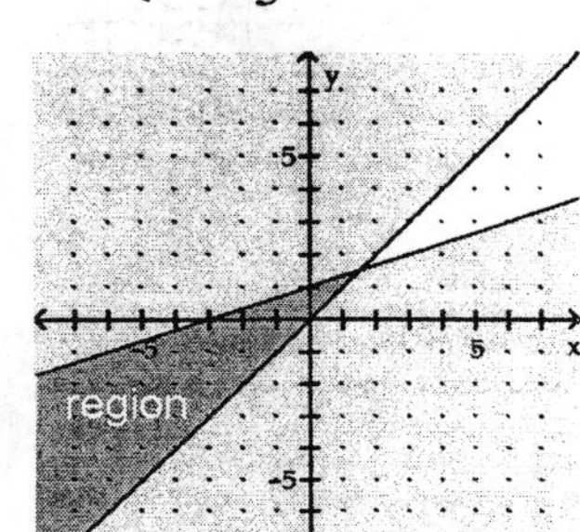

41.

$$\begin{cases} x + y > 0 \\ y - x < -2 \end{cases}$$

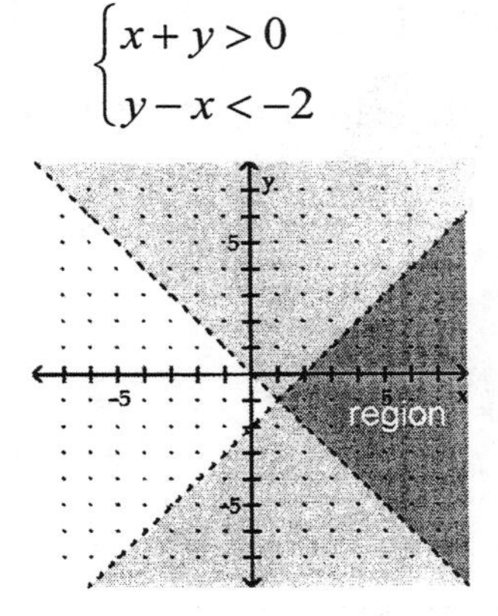

43.

$$\begin{cases} x \geq 0 \\ y \geq 0 \\ x + y \leq 3 \end{cases}$$

With this system, one extra boundary is set up before identifying the intersection of the system.

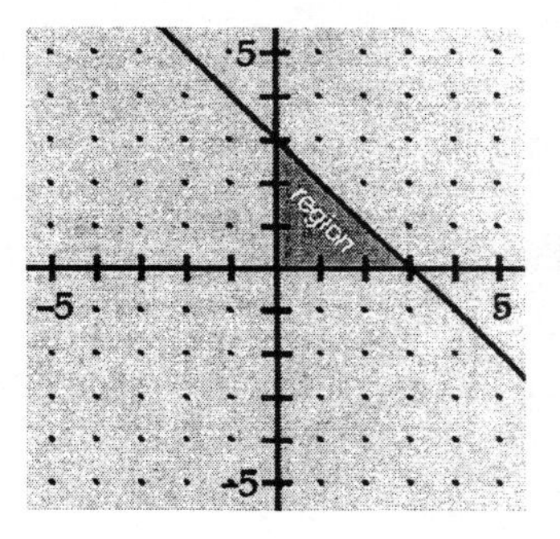

45.

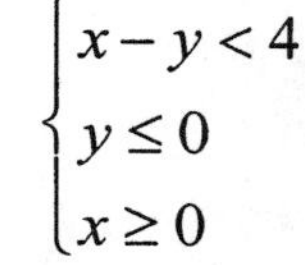

$$\begin{cases} x - y < 4 \\ y \leq 0 \\ x \geq 0 \end{cases}$$

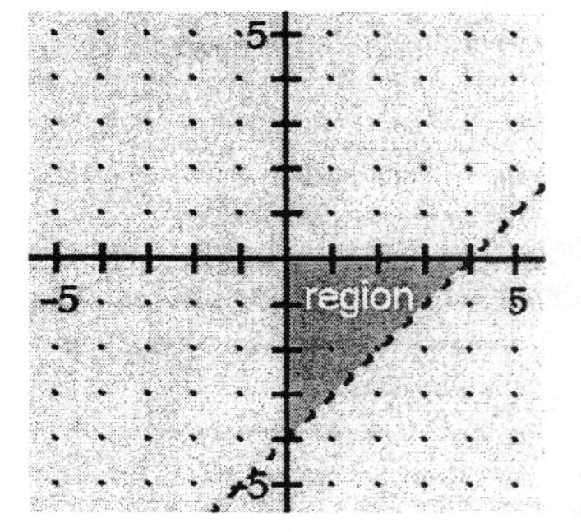

APPLICATIONS

47. BIRDS OF PREY

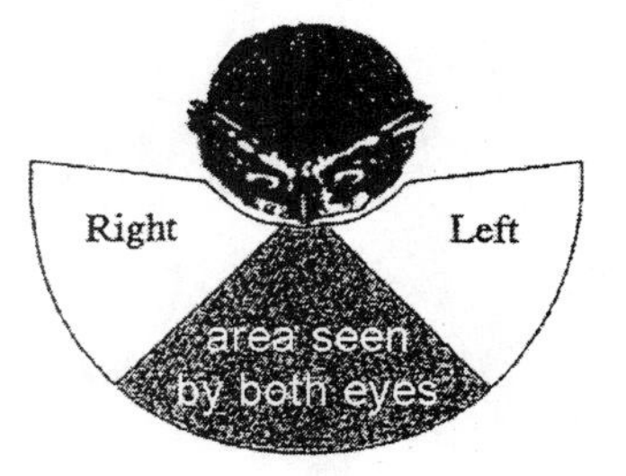

49. BUYING COMPACT DISCS

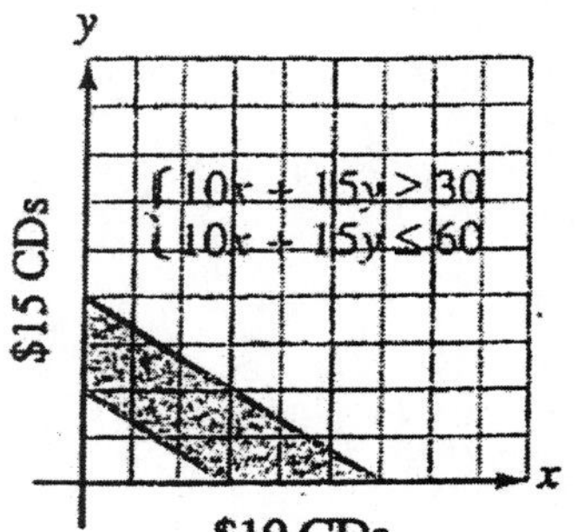

1 $10 CD and 2 $15 CDs
1 $10 CD and 3 $15 CDs
2 $10 CDs and 1 $15 CD
2 $10 CDs and 2 $15 CDs
3 $10 CDs and 1 $15 CD
3 $10 CDs and 2 $15 CDs
4 $10 CDs and 1 $15 CD

51. BUYING FURNITURE

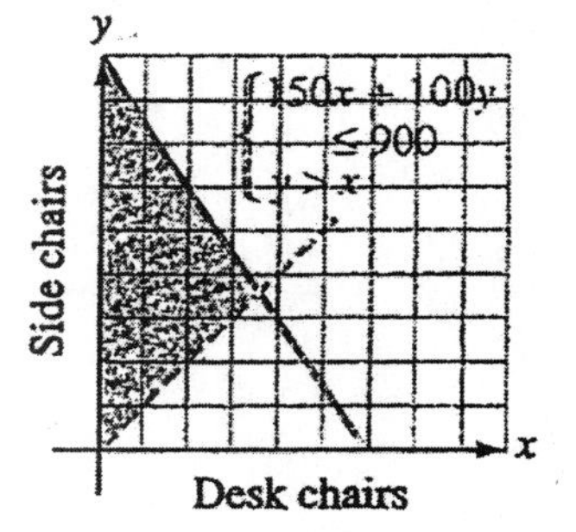

1 desk chair and 2 side chairs
1 desk chair and 3 side chairs
1 desk chair and 4 side chairs
1 desk chair and 5 side chairs
1 desk chair and 6 side chairs
1 desk chair and 7 side chairs
2 desk chair and 3 side chairs
2 desk chair and 4 side chairs
2 desk chair and 5 side chairs
2 desk chair and 6 side chairs
3 desk chair and 4 side chairs

53. PESTICIDES

$$\begin{cases} y \geq -2x + 1 \\ y \geq \frac{1}{4}x - 4 \end{cases}$$

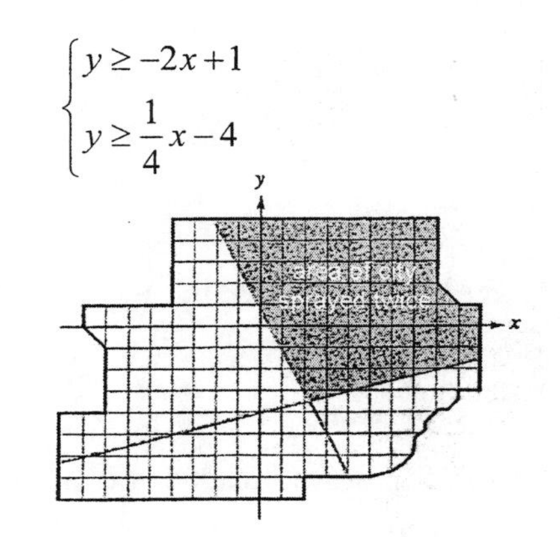

WRITING

55. Answers will vary.

57. Answers will vary.

REVIEW

59. $2.3 \times 10^2 = 230$

61. $9.76 \times 10^{-4} = 0.000976$

63. $290{,}000 = 2.9 \times 10^5$

65. $0.0000051 = 5.1 \times 10^{-6}$

CHALLENGE PROBLEMS

67.

$$\begin{cases} \dfrac{x}{3} - \dfrac{y}{2} < -3 \\ \dfrac{x}{3} + \dfrac{y}{2} > -1 \end{cases}$$

Clear both of fractions

$$\begin{cases} 2x - 3y < -18 \\ 2x + 3y > -6 \end{cases}$$

Graph the boundaries of both.

Identify the intersection.

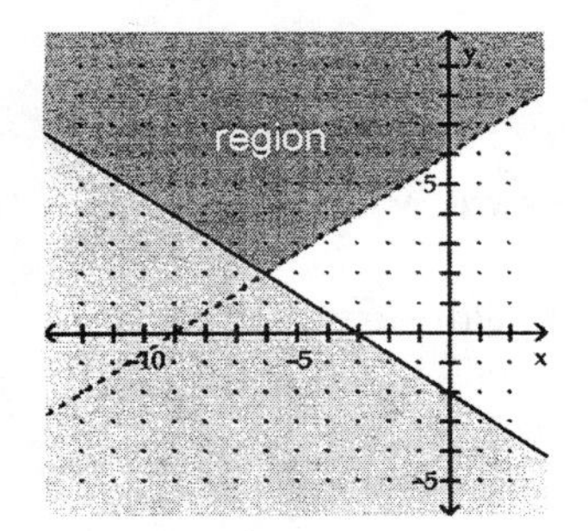

69.

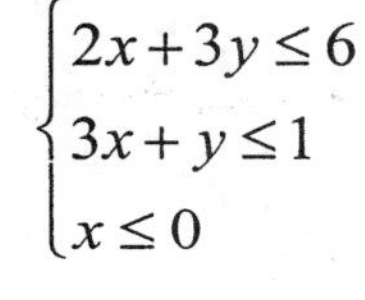

$$\begin{cases} 2x + 3y \le 6 \\ 3x + y \le 1 \\ x \le 0 \end{cases}$$

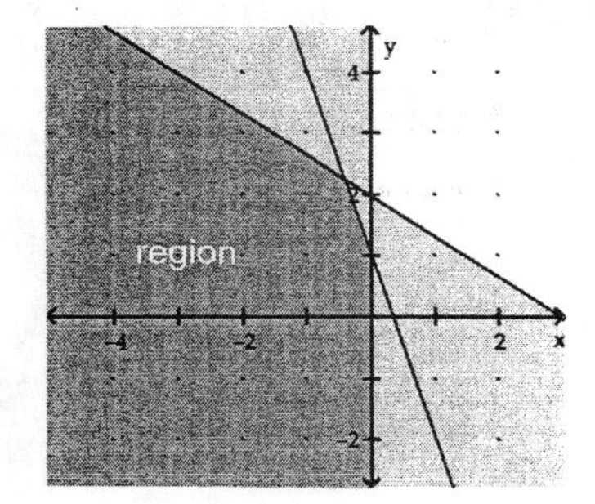

CHAPTER 7 KEY CONCEPTS

Solving Systems of Equations by Graphing

1. FOOD SERVICES

Caterer	Setup Fee	Cost per Meal	Equation
Sunshine	$1,000	$4	$y = 4x + 1{,}000$
Lucy's	$500	$5	$y = 5x + 500$

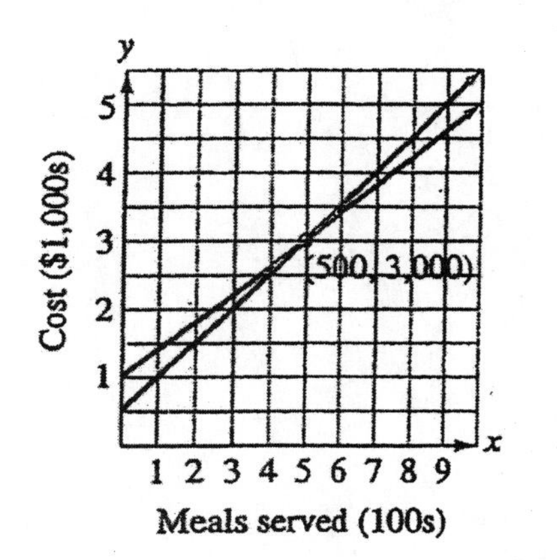

Point of intersection: (500, 3,000)

500 meals, $3,000

Solving Systems of Equations by Substitution

2.

$$\begin{cases} y = 2x - 9 \\ x + 3y = 8 \end{cases}$$

$$x + 3(2x - 9) = 8$$

$$x + 6x - 27 = 8$$

$$7x - 27 + 27 = 8 + 27$$

$$7x = 35$$

$$\frac{7x}{7} = \frac{35}{7}$$

$$\boxed{x = 5}$$

$$y = 2x - 9$$

$$y = 2(5) - 9$$

$$y = 10 - 9$$

$$\boxed{y = 1}$$

solution is (5, 1)

3.

$$\begin{cases} 3x + 4y = -7 \\ 2y - x = -1 \end{cases}$$

$$\begin{cases} 3x + 4y = -7 \\ 2y + 1 = x \end{cases}$$

$$3x + 4y = -7$$

$$3(2y + 1) + 4y = -7$$

$$6y + 3 + 4y = -7$$

$$10y + 3 - 3 = -7 - 3$$

$$10y = -10$$

$$\frac{10y}{10} = \frac{-10}{10}$$

$$\boxed{y = -1}$$

$$x = 2y + 1$$

$$x = 2(-1) + 1$$

$$x = -2 + 1$$

$$\boxed{x = -1}$$

solution is (-1, -1)

Solving Systems of Equations by Elimination

4.

$$\begin{cases} x + y = 1 \\ x - y = 5 \end{cases}$$

eliminate the y

$$\begin{array}{r} x + y = 1 \\ x - y = 5 \\ \hline 2x = 6 \end{array}$$

$$\frac{2x}{2} = \frac{6}{2}$$

$$\boxed{x = 3}$$

$$x + y = 1$$

$$3 + y = 1$$

$$3 + y - 3 = 1 - 3$$

$$\boxed{y = -2}$$

solution is (3, -2)

5.

$$\begin{cases} 2x-3y=-18 \\ 3x+2y=-1 \end{cases}$$

multiply 1st equation by 2

multiply 2nd equation by 3

$$\begin{cases} 4x-6y=-36 \\ 9x+6y=-3 \end{cases}$$

eliminate the y

$$\begin{array}{r} 4x-6y=-36 \\ 9x+6y=-3 \\ \hline 13x=-39 \end{array}$$

$$\frac{13x}{13}=\frac{-39}{13}$$

$$\boxed{x=-3}$$

$$3x+2y=-1$$

$$3(-3)+2y=-1$$

$$-9+2y+9=-1+9$$

$$2y=8$$

$$\frac{2y}{2}=\frac{8}{2}$$

$$\boxed{y=4}$$

solution is (-3, 4)

CHAPTER 7 REVIEW

SECTION 7.1
Solving Systems of Equations by Graphing

1. $(2, \text{-}3)\ \begin{cases} 3x-2y=12 \\ 2x+3y=-5 \end{cases}$

substitute (2, -3) into

$3x-2y=12$

$3(2)-2(\text{-}3)\overset{?}{=}12$

$6+6\overset{?}{=}12$

$\boxed{12=12}$ yes

substitute (2, -3) into

$2x+3y=\text{-}5$

$2(2)+3(\text{-}3)\overset{?}{=}\text{-}5$

$4-9\overset{?}{=}\text{-}5$

$\boxed{\text{-}5=\text{-}5}$ yes

(2, -3) is a solution.

2. $\left(\frac{7}{2}, \frac{\text{-}2}{3}\right)\ \begin{cases} 3y=2x-9 \\ 2x+3y=5 \end{cases}$

substitute $\left(\frac{7}{2}, \frac{\text{-}2}{3}\right)$ into

$3y=2x-9$

$3\left(\frac{\text{-}2}{3}\right)\overset{?}{=}2\left(\frac{7}{2}\right)-9$

$-2\overset{?}{=}7-9$

$\boxed{-2=-2}$ yes

substitute $\left(\frac{7}{2}, \frac{\text{-}2}{3}\right)$ into

$2x+3y=5$

$2\left(\frac{7}{2}\right)+3\left(\frac{\text{-}2}{3}\right)\overset{?}{=}5$

$7-2\overset{?}{=}5$

$\boxed{5=5}$ yes

$\left(\frac{7}{2}, \frac{\text{-}2}{3}\right)$ is a solution.

3. COLLEGE ENROLLMENT

(1978, 5.6)

In 1978, the same number of men as women were enrolled in college, about 5.6 million of each.

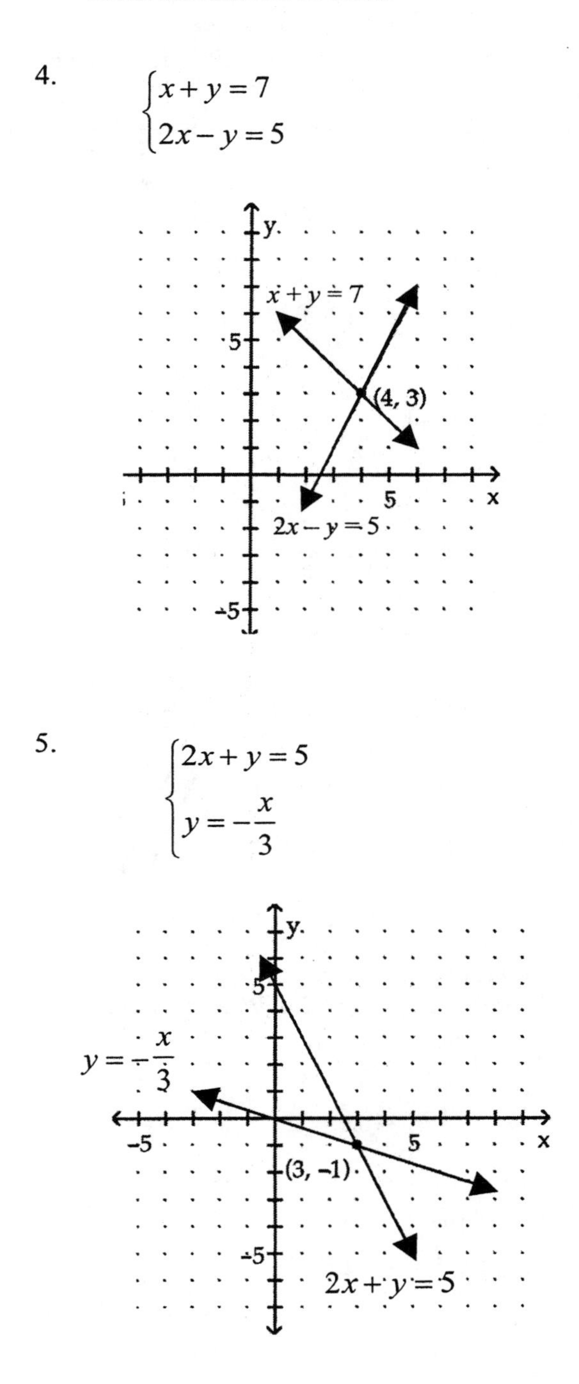

6. $\begin{cases} 3x+6y=6 \\ x+2y-2=0 \end{cases}$

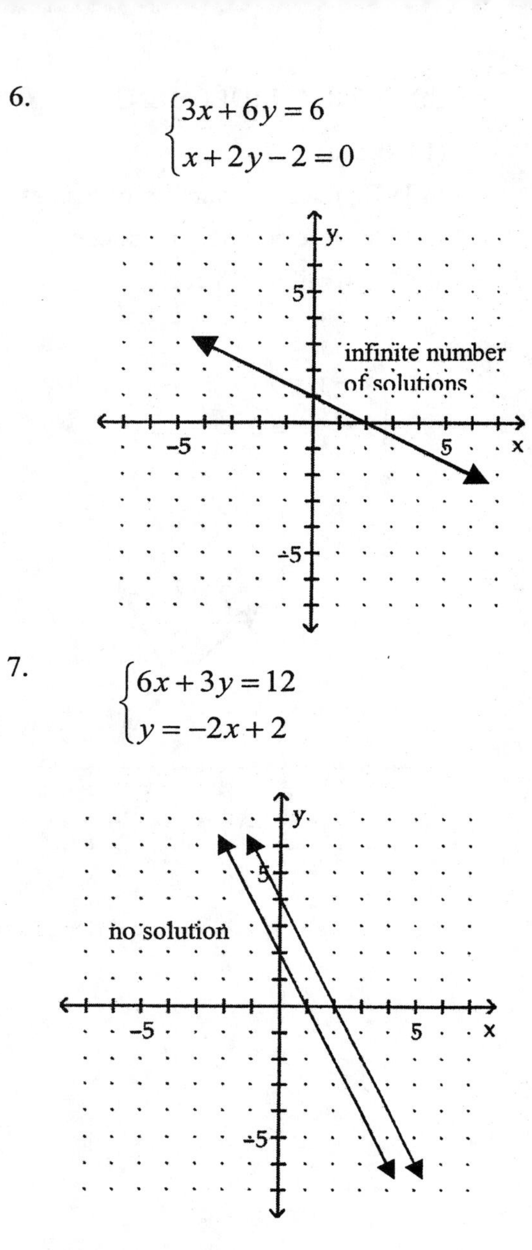

7. $\begin{cases} 6x+3y=12 \\ y=-2x+2 \end{cases}$

SECTION 7.2
Solving Systems of Equations by Substitution

8. $\begin{cases} y=15-3x \\ 7y+3x=15 \end{cases}$

$$7y+3x=15$$
$$7(15-3x)+3x=15$$
$$105-21x+3x=15$$
$$105-18x-105=15-105$$
$$-18x=-90$$
$$\frac{-18x}{-18}=\frac{-90}{-18}$$
$$\boxed{x=5}$$

continued

8. continued

$$y=15-3x$$
$$y=15-3(5)$$
$$y=15-15$$
$$\boxed{y=0}$$

solution is (5, 0)

9. $\begin{cases} x=y \\ 5x-4y=3 \end{cases}$

$$5x-4y=3$$
$$5(y)-4y=3$$
$$\boxed{y=3}$$
$$x=y$$
$$\boxed{x=3}$$

solution is (3, 3)

10. $\begin{cases} 6x+2y=8-y+x \\ 3x=2-y \end{cases}$

$\begin{cases} 5x+3y=8 \\ y=2-3x \end{cases}$

$$5x+3y=8$$
$$5x+3(2-3x)=8$$
$$5x+6-9x=8$$
$$-4x+6-6=8-6$$
$$-4x=2$$
$$\frac{-4x}{-4}=\frac{2}{-4}$$
$$\boxed{x=-\frac{1}{2}}$$
$$y=2-3x$$
$$y=2-3\left(-\frac{1}{2}\right)$$
$$y=2+\frac{3}{2}$$
$$\boxed{y=\frac{7}{2}}$$

solution is $\left(-\frac{1}{2},\frac{7}{2}\right)$

11.

$$\begin{cases} r = 3s + 7 \\ r = 2s + 5 \end{cases}$$

$$3s + 7 = 2s + 5$$
$$3s + 7 - 7 = 2s + 5 - 7$$
$$3s = 2s - 2$$
$$3s - 2s = 2s - 2 - 2s$$
$$\boxed{s = -2}$$
$$r = 2(-2) + 5$$
$$\boxed{r = 1}$$

solution is (1, -2)

12.

$$\begin{cases} 9x + 3y - 5 = 0 \\ 3x + y = \frac{5}{3} \end{cases}$$

$$\begin{cases} 9x + 3y - 5 = 0 \\ y = \frac{5}{3} - 3x \end{cases}$$

$$9x + 3y - 5 = 0$$
$$9x + 3\left(\frac{5}{3} - 3x\right) - 5 = 0$$
$$9x + 5 - 9x - 5 = 0$$
$$\boxed{0 = 0}$$

infinitely many solutions

13.

$$\begin{cases} \frac{x}{2} + \frac{y}{2} = 11 \\ \frac{5x}{16} - \frac{3y}{16} = \frac{15}{8} \end{cases}$$

clear both equations of fractions

$$\begin{cases} 2\left(\frac{x}{2}\right) + 2\left(\frac{y}{2}\right) = 2(11) \\ 16\left(\frac{5x}{16}\right) - 16\left(\frac{3y}{16}\right) = 16\left(\frac{15}{8}\right) \end{cases}$$

$$\begin{cases} x + y = 22 \\ 5x - 3y = 30 \end{cases}$$

$$\begin{cases} x = 22 - y \\ 5x - 3y = 30 \end{cases}$$

continued

13. continued

$$5x - 3y = 30$$
$$5(22 - y) - 3y = 30$$
$$110 - 5y - 3y = 30$$
$$-8y + 110 - 110 = 30 - 110$$
$$-8y = -80$$
$$\frac{-8y}{-8} = \frac{-80}{-8}$$
$$\boxed{y = 10}$$
$$x = 22 - y$$
$$x = 22 - 10$$
$$\boxed{x = 12}$$

solution is (12, 10)

14. a. no solutions
b. two parallel lines
c. inconsistent system

SECTION 7.3
Solving Systems of Equations by Elimination

15.

$$\begin{cases} 2x + y = 1 \\ 5x - y = 20 \end{cases}$$

eliminate the y

$$2x + y = 1$$
$$\underline{5x - y = 20}$$
$$7x = 21$$
$$\frac{7x}{7} = \frac{21}{7}$$
$$\boxed{x = 3}$$

$2x + y = 1$ 1st equation

$$2(3) + y = 1$$
$$6 + y - 6 = 1 - 6$$
$$\boxed{y = -5}$$

solution is (3, -5)

16. $\begin{cases} x+8y=7 \\ x-4y=1 \end{cases}$

multiply 2nd equation by -1

eliminate the x

$$\begin{array}{r} x+8y=7 \\ -x+4y=-1 \\ \hline 12y=6 \end{array}$$

$$\frac{12y}{12}=\frac{6}{12}$$

$$\boxed{y=\frac{1}{2}}$$

$x+8y=7$ 1st equation

$$x+8\left(\frac{1}{2}\right)=7$$

$$x+4-4=7-4$$

$$\boxed{x=3}$$

solution is $\left(3,\ \frac{1}{2}\right)$

17. $\begin{cases} 5a+b=2 \\ 3a+2b=11 \end{cases}$

multiply 1st equation by -2

to eliminate the b

$$\begin{array}{r} -10a-2b=-4 \\ 3a+2b=11 \\ \hline -7a=7 \end{array}$$

$$\frac{-7a}{-7}=\frac{7}{-7}$$

$$\boxed{a=-1}$$

$5a+b=2$ 1st equation

$$5(-1)+b=2$$

$$-5+b+5=2+5$$

$$\boxed{b=7}$$

solution is (-1, 7)

18. $\begin{cases} 11x+3y=27 \\ 8x+4y=36 \end{cases}$

multiply 1st equation by 4

multiply 2nd equation by -3

eliminate the y

$$\begin{array}{r} 44x+12y=108 \\ -24x-12y=-108 \\ \hline 20x=0 \end{array}$$

$$\frac{20x}{20}=\frac{0}{20}$$

$$\boxed{x=0}$$

$11x+3y=27$ 1st equation

$$11(0)+3y=27$$

$$\frac{3y}{3}=\frac{27}{3}$$

$$\boxed{y=9}$$

solution is (0, 9)

19. $\begin{cases} 9x+3y=15 \\ 3x=5-y \end{cases}$

divide 1st equation by 3

put 2nd equation in standard form

$\begin{cases} 3x+y=5 \\ 3x+y=5 \end{cases}$

both equations are exactly the same

infinitely many solutions

20.

$$\begin{cases} -\frac{a}{4}-\frac{b}{3}=\frac{1}{12} \\ \frac{a}{2}-\frac{5b}{4}=\frac{7}{4} \end{cases}$$

remove fractions from both

multiply 1st equation by 12

multiply 2nd equation by 4

$$\begin{cases} -3a-4b=1 \\ 2a-5b=7 \end{cases}$$

multiply 1st equation by 2

multiply 2nd equation by 3

to eliminate the a

$$\begin{aligned} -6a-8b&=2 \\ 6a-15b&=21 \\ \hline -23b&=23 \end{aligned}$$

$$\frac{-23b}{-23}=\frac{23}{-23}$$

$$\boxed{b=-1}$$

$2a-5b=7$ 2nd equation

$$2a-5(-1)=7$$

$$2a+5-5=7-5$$

$$2a=2$$

$$\frac{2a}{2}=\frac{2}{2}$$

$$\boxed{a=1}$$

solution is (1,-1)

21.

$$\begin{cases} 0.02x+0.05y=0 \\ 0.3x-0.2y=-1.9 \end{cases}$$

remove the decimals

multiply 1st equation by 100

multiply 2nd equation by 10

$$\begin{cases} 2x+5y=0 \\ 3x-2y=-19 \end{cases}$$

multiply 1st equation by 2

multiply 2nd equation by 5

eliminate the y

$$\begin{aligned} 4x+10y&=0 \\ 15x-5y&=-95 \\ \hline 19x&=-95 \end{aligned}$$

$$\frac{19x}{19}=\frac{-95}{19}$$

$$\boxed{x=-5}$$

$2x+5y=0$ 1st equation

$$2(-5)+5y=0$$

$$-10+5y+10=0+10$$

$$5y=10$$

$$\frac{5y}{5}=\frac{10}{5}$$

$$\boxed{y=2}$$

solution is (-5, 2)

22.

$$\begin{cases} -\frac{1}{4}x=1-\frac{2}{3}y \\ 6x-18y=5-2y \end{cases}$$

remove the fractions from 1st equation by multiplying by 12

put both equations in standard form

$$\begin{cases} -3x+8y=12 \\ 6x-16y=5 \end{cases}$$

multiply 1st equation by 2

eliminate the y

$$\begin{aligned} -6x+16y&=24 \\ 6x-16y&=5 \\ \hline 0&=29 \end{aligned}$$

$0=29$ false

no solution

23. Elimination; no varibles have a coefficient of 1 or -1

24. Substitution; equation 1 is solved for x

SECTION 7.4

Problem Solving Using Systems of Equations

25. ELEVATION

Let x = Las Vegas' elevation (higher)

y = Baltimore's elevation (lower)

total of both elevations = 2,100 ft.

LV ele = 20 times greater than Balt ele

$$\begin{cases} x+y=2{,}100 \\ x=20y \end{cases}$$

$$x+y=2{,}100$$
$$20y+y=2{,}100$$
$$21y=2{,}100$$
$$\frac{21y}{21}=\frac{2{,}100}{21}$$
$$\boxed{y=100}$$
$$x+y=2{,}100$$
$$x+100=2{,}100$$
$$x+100-100=2{,}100-100$$
$$\boxed{x=2{,}000}$$

Las Vegas' elevation is 2,000 ft.

Balimore's elevation is 100 ft.

26. PAINTING EQUIPMENT

Let x = length of base part of ladder (longer)

y = length of extension (shorter)

fully extended = 35 ft.

extension = 7 ft. shorter than base

$$\begin{cases} x+y=35 \\ y=x-7 \end{cases}$$

$$x+y=35$$
$$x+x-7=35$$
$$2x-7+7=35+7$$
$$2x=42$$
$$\frac{2x}{2}=\frac{42}{2}$$
$$\boxed{x=21}$$
$$x+y=35$$
$$21+y=35$$
$$21+y-21=35-21$$
$$\boxed{y=14}$$

Base part of ladder is 21 ft.

Extension part of ladder is 14 ft.

27. CRASH INVESTIGATION

Let l = length in yards (large)

w = width in yards (small)

perimeter of scene is 420 yards

$2l+2w$ = perimeter

if width is $\frac{3}{4}$ of the length

find the area to be searched

$$\begin{cases} 2l+2w=420 \\ w=\frac{3}{4}l \end{cases}$$

$$2l+2(0.75l)=420$$
$$2l+1.50l=420$$
$$\frac{3.5l}{3.5}=\frac{420}{3.5}$$
$$\boxed{l=120}$$
$$w=0.75l$$
$$w=0.75(120)$$
$$\boxed{w=90}$$

area to be searched = (120 yds)(90 yds)

= 10,800 yd^2

28. CELEBRITY ENDORSEMENT

a. Athlete' s equation:

$y = 5x + 30{,}000$

Actor's equation:

$y = 10x + 20{,}000$

b. earned by Athlete: $y = 5x + 30{,}000$

earned by Actor: $y = 10x + 20{,}000$

Let x = break point for juice machines

$$\begin{cases} y = 5x + 30{,}000 \\ y = 10x + 20{,}000 \end{cases}$$

earned by Athlete = earned by Actor

$$5x + 30{,}000 = 10x + 20{,}000$$

subtract 20,000 from both sides

$$20{,}000 = 20{,}000$$

$$5x + 10{,}000 = 10x$$

$$5x + 10{,}000 - 5x = 10x - 5x$$

$$10{,}000 = 5x$$

$$\frac{10{,}000}{5} = \frac{5x}{5}$$

$$\boxed{2{,}000 = x}$$

When 2,000 juice machines are sold, both would earn the same.

c.

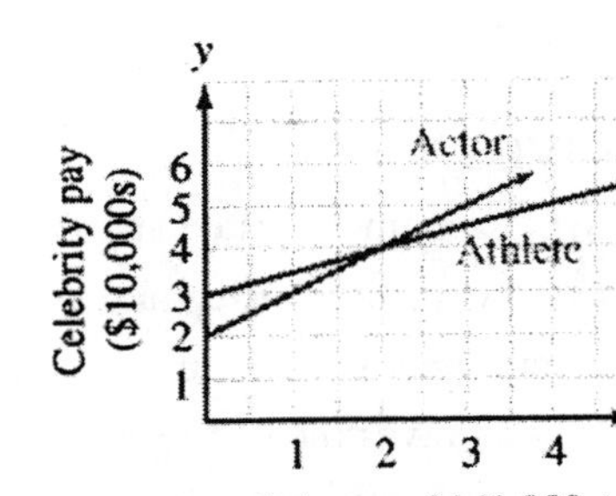

29.CANDY OUTLET STORE

Let x = # of lb of gummy worms

y = # of lb of gummy bears

cost/lb • amt = value

gummy worms: \$3.00/lb

gummy bears: \$1.50/lb

mixture: \$2.10/lb

total pounds: 30

$$\begin{cases} x + y = 30 \\ 3x + 1.5y = 2.1(30) \end{cases}$$

$$\begin{cases} x = 30 - y \\ 3x + 1.5y = 63 \end{cases}$$

$$3x + 1.5y = 63$$

$$3(30 - y) + 1.5y = 63$$

$$90 - 3y + 1.5y = 63$$

$$90 - 1.5y - 90 = 63 - 90$$

$$-1.5y = -27$$

$$\frac{-1.5y}{-1.5} = \frac{-27}{-1.5}$$

$$\boxed{y = 18}$$

$$x + y = 30$$

$$x + 18 = 30$$

$$x + 18 - 18 = 30 - 18$$

$$\boxed{x = 12}$$

12 lb of g. worms (\$3/lb) is needed.

18 lb of g. bears (\$6/lb) is needed.

30. BOATING

Let s = speed of boat in still water (mph)

c = speed of current (mph)

rate • time = distance

with current: 56 mi. in 4 hr.

against current: 56 mi. in $4+3$ (hr. longer)

$$\begin{cases} 4(s+c)=56 \\ 7(s-c)=56 \end{cases}$$

$$\begin{cases} \dfrac{4(s+c)}{4}=\dfrac{56}{4} \\ \dfrac{7(s-c)}{7}=\dfrac{56}{7} \end{cases}$$

$$\begin{cases} s+c=14 \\ s-c=8 \end{cases}$$

$$\begin{aligned} s+c&=14 \\ s-c&=8 \\ \hline 2s&=22 \\ \frac{2s}{2}&=\frac{22}{2} \\ \boxed{s=11} \\ s+c&=14 \\ 11+c&=14 \\ 11+c-11&=14-11 \\ \boxed{c=3} \end{aligned}$$

Speed of boat is 11 mph.

Speed of current is 3 mph.

31. SHOPPING

Let x = price of one bottle of cleaner

y = price of one bottle of soaking

2 cleaners and 3 soaking is \$63.40

3 cleaners and 2 soaking is \$69.60

$$\begin{cases} 2x+3y=63.40 \\ 3x+2y=69.60 \end{cases}$$

$$\begin{aligned} -2(2x+3y)&=-2(63.40) \\ 3(3x+2y)&=3(69.60) \\ \hline -4x-6y&=-126.80 \\ 9x+6y&=208.80 \\ \hline 5x&=82 \\ \frac{5x}{5}&=\frac{82}{5} \\ \boxed{x=16.40} \\ 2x+3y&=63.40 \\ 2(16.4)+3y&=63.40 \\ 32.8+3y-32.8&=63.40-32.8 \\ 3y&=30.60 \\ \frac{3y}{3}&=\frac{30.60}{3} \\ \boxed{y=10.20} \end{aligned}$$

A bottle of cleaner cost \$16.40.

A bottle of soaking cost \$10.20.

32. INVESTING

Let x = amt invested at 10%

y = amt invested at 6%

principal • rate • time = interest

total investment: \$3,000

annual interest for both: \$270

$$\begin{cases} x+y=3{,}000 \\ 0.10x+0.06y=270 \end{cases}$$

$$\begin{cases} x=3{,}000-y \\ 0.10x+0.06y=270 \end{cases}$$

$$\begin{aligned} 0.10x+0.06y&=270 \\ 0.10(3{,}000-y)+0.06y&=270 \\ 300-0.10y+0.06y&=270 \\ 300-0.04y-300&=270-300 \\ -0.04y&=-30 \\ \frac{-0.04y}{-0.04}&=\frac{-30}{-0.04} \\ \boxed{y=750} \end{aligned}$$

\$750 invested at 6%.

33. ANTIFREEZE

Let x = # of gal of 40% antifreeze

y = # of gal of 70% antifreeze

percent • amt of antifreeze = pure antifreeze

total gallons: 20

final mixture percent: 50%

$$\begin{cases} x+y=20 \\ 0.4x+0.7y=0.5(20) \end{cases}$$

$$\begin{cases} x=20-y \\ 0.4x+0.7y=10 \end{cases}$$

$$0.4x+0.7y=10$$

$$0.4(20-y)+0.7y=10$$

$$8-0.4y+0.7y=10$$

$$8+0.3y-8=10-8$$

$$0.3y=2$$

$$\frac{0.3y}{0.3}=\frac{2}{0.3}$$

$$\boxed{y=6.666 \text{ or } y=6\frac{2}{3}}$$

$$x+y=20$$

$$x+6\frac{2}{3}=20$$

$$x+6\frac{2}{3}-6\frac{2}{3}=20-6\frac{2}{3}$$

$$\boxed{x=13\frac{1}{3}}$$

$13\frac{1}{3}$ gal of 40% antifreeze is needed.

$6\frac{2}{3}$ gal of 70% antifreeze is needed.

SECTION 7.5

Solving Systems of Inequalities

34. $\begin{cases} 5x+3y<15 \\ 3x-y>3 \end{cases}$

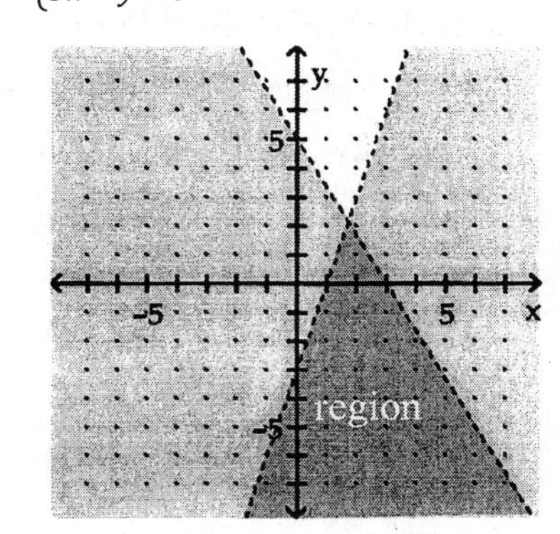

35. $\begin{cases} 3y \le x \\ y>3x \end{cases}$

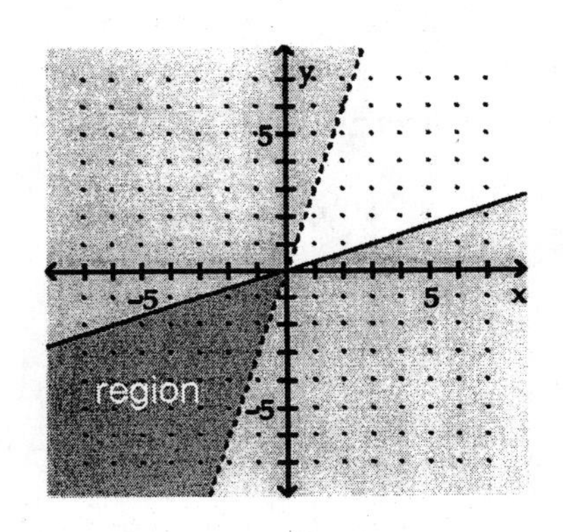

36. $\begin{cases} 10x+20y \ge 40 \\ 10x+20y \le 60 \end{cases}$

Let x = # of T-shirts

y = # of pants

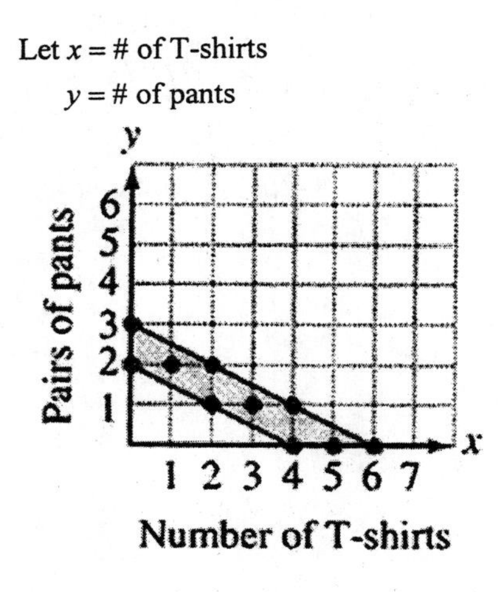

CHAPTER 7 TEST

1. $(5, 3)\ \begin{cases} 3x + 2y = 21 \\ x + y = 8 \end{cases}$

substitute (5, 3) into

$3x + 2y = 21$

$3(5) + 2(3) \stackrel{?}{=} 21$

$15 + 6 \stackrel{?}{=} 21$

$\boxed{21 = 21}$ yes

substitute (5, 3) into

$x + y = 8$

$5 + 3 \stackrel{?}{=} 8$

$\boxed{8 = 8}$ yes

(5, 3) is a solution.

2. $(-2, -1)\ \begin{cases} 4x + y = -9 \\ 2x - 3y = -7 \end{cases}$

substitute (-2, -1) into

$4x + y = -9$

$4(-2) + (-1) \stackrel{?}{=} -9$

$-8 - 1 \stackrel{?}{=} -9$

$\boxed{-9 = -9}$ yes

substitute (-2, -1) into

$2x - 3y = -7$

$2(-2) - 3(-1) \stackrel{?}{=} -7$

$-4 + 3 \stackrel{?}{=} -7$

$\boxed{-1 \neq -7}$ no

(-2, -1) is not a solution.

3.

a. A **solution** of a system of linear equations is an ordered pair that satisfies each equation.

b. A system of equations that has at least one solution is called a **consistent** system.

c. A system of equations that has no solution is called an **inconsistent** system.

d. Equations with different graphs are called **independent** equations.

e. A system of **dependent** equations has an infinite number of solutions

4. $\begin{cases} y = 2x - 1 \\ x - 2y = -4 \end{cases}$

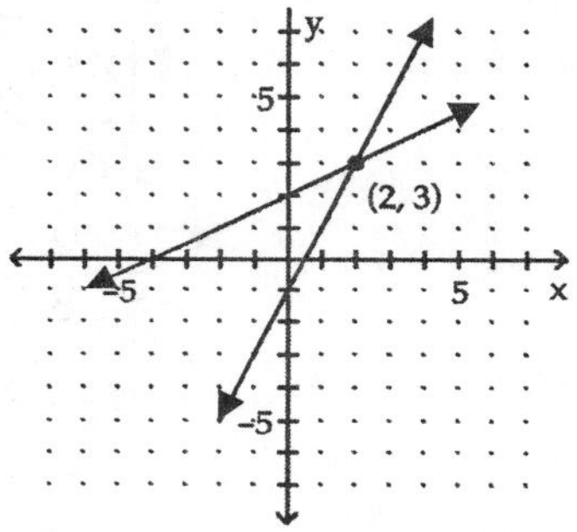

5. $\begin{cases} x + y = 5 \\ y = -x \end{cases}$

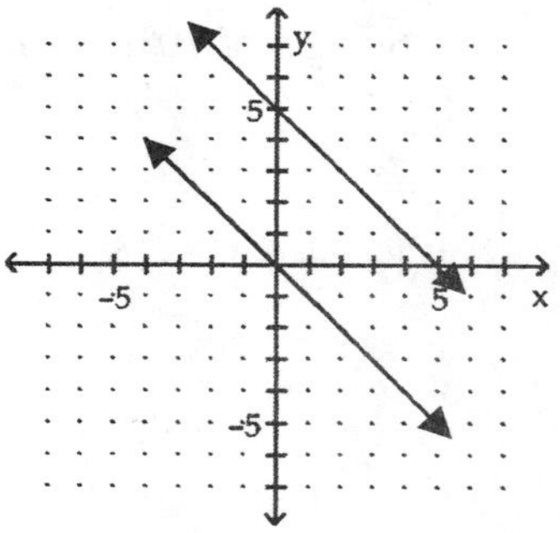

$\boxed{\text{no solution}}$

6. (30, 3,000); if 30 items are sold, the salesperson gets paid the same by both plans, \$3,000.

7. Plan 1

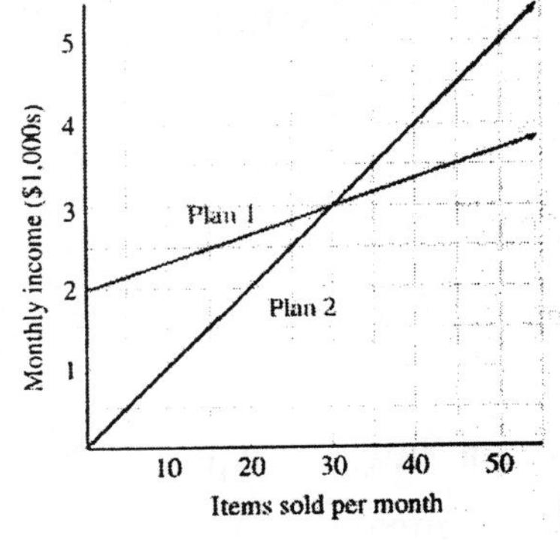

8.
$$\begin{cases} y = x - 1 \\ 2x + y = -7 \end{cases}$$

$$2x + y = -7$$
$$2x + (x - 1) = -7$$
$$3x - 1 = -7$$
$$3x - 1 + 1 = -7 + 1$$
$$3x = -6$$
$$\frac{3x}{3} = \frac{-6}{3}$$
$$\boxed{x = -2}$$
$$y = x - 1$$
$$y = -2 - 1$$
$$\boxed{y = -3}$$

solution is (-2, -3)

9.
$$\begin{cases} 3x + 6y = -15 \\ x + 2y = -5 \end{cases}$$
$$\begin{cases} 3x + 6y = -15 \\ x = -5 - 2y \end{cases}$$

$$3x + 6y = -15$$
$$(-5 - 2y) + 6y = -15$$
$$-15 - 6y + 6y = -15$$
$$-15 = -15 \text{ true}$$

infinitely many solutions

10.
$$\begin{cases} 3a + 4b = -7 \\ 2b - a = -1 \end{cases}$$
$$\begin{cases} 3a + 4b = -7 \\ 2b + 1 = a \end{cases}$$

$$3a + 4b = -7$$
$$3(2b + 1) + 4b = -7$$
$$6b + 3 + 4b = -7$$
$$10b + 3 - 3 = -7 - 3$$
$$10b = -10$$
$$\frac{10b}{10} = \frac{-10}{10}$$
$$\boxed{b = -1}$$
$$a = 2b + 1$$
$$a = 2(-1) + 1$$
$$a = -2 + 1$$
$$\boxed{a = -1}$$

solution is (-1, -1)

11.
$$\begin{cases} 3x - y = 2 \\ 2x + y = 8 \end{cases}$$

eliminate the y

$$3x - y = 2$$
$$\underline{2x + y = 8}$$
$$5x = 10$$
$$\frac{5x}{5} = \frac{10}{5}$$
$$\boxed{x = 2}$$
$$2x + y = 8 \quad 2^{nd} \text{ equation}$$
$$2(2) + y = 8$$
$$4 + y - 4 = 8 - 4$$
$$\boxed{y = 4}$$

solution is (2, 4)

12. $\begin{cases} 4x + 3y = -3 \\ -3x = -4y + 21 \end{cases}$

put 2nd equation in standard form

$\begin{cases} 4x + 3y = -3 \\ -3x + 4y = 21 \end{cases}$

multiply 1st equation by 3

multiply 2nd equation by 4

$\begin{cases} 12x + 9y = -9 \\ -12x + 16y = 84 \end{cases}$

eliminate the x

$$\begin{array}{r} 12x + 9y = -9 \\ -12x + 16y = 84 \\ \hline 25y = 75 \end{array}$$

$$\frac{25y}{25} = \frac{75}{25}$$

$$\boxed{y = 3}$$

$4x + 3y = -3$ 1st equation

$$4x + 3(3) = -3$$

$$4x + 9 - 9 = -3 - 9$$

$$4x = -12$$

$$\frac{4x}{4} = \frac{-12}{4}$$

$$\boxed{x = -3}$$

solution is (-3, 3)

13. $\begin{cases} 3x - 5y - 16 = 0 \\ \frac{x}{2} - \frac{5}{6}y = \frac{1}{3} \end{cases}$

put 1st equation in standard form

remove fractions from 2nd equation

$\begin{cases} 3x - 5y - 16 + 16 = 0 + 16 \\ 6\left(\frac{x}{2}\right) - 6\left(\frac{5}{6}y\right) = 6\left(\frac{1}{3}\right) \end{cases}$

$\begin{cases} 3x - 5y = 16 \\ 3x - 5y = 2 \end{cases}$

multiply 2nd equation by -1

eliminate the x

$$\begin{array}{r} 3x - 5y = 16 \\ -3x + 5y = -2 \\ \hline 0 = 14 \text{ false} \end{array}$$

no solution

14. Elimination method; the terms involving y can be eliminated easily.

15. CHILD CARE

Let x = first part of commute

y = second part of commute

total commute to work: 22 miles

1st part is 6 miles less than 2nd part

$$\begin{cases} x + y = 22 \\ x = y - 6 \end{cases}$$

put 2nd equation in standard form

$$\begin{cases} x + y = 22 \\ x - y = -6 \end{cases}$$

eliminate the y

$$\begin{array}{r} x + y = 22 \\ x - y = -6 \\ \hline 2x = 16 \end{array}$$

$$\frac{2x}{2} = \frac{16}{2}$$

$$\boxed{x = 8}$$

$$x + y = 22$$

$$8 + y = 22$$

$$8 + y - 8 = 22 - 8$$

$$\boxed{y = 14}$$

1st part is 8 miles.

2nd part is 14 miles.

16. VACATIONING

Let x = number of adult tickets

y = number of child tickets

an adult ticket cost $21

a child ticket cost $14

# of adult and # of children = 7

total cost $119

$$\begin{cases} x + y = 7 \\ 21x + 14y = 119 \end{cases}$$

multiply 1st equation by -21

$$\begin{cases} -21x - 21y = -147 \\ 21x + 14y = 119 \end{cases}$$

eliminate the x

$$\begin{array}{r} -21x - 21y = -147 \\ 21x + 14y = 119 \\ \hline -7y = -28 \end{array}$$

$$\frac{-7y}{-7} = \frac{-28}{-7}$$

$$\boxed{y = 4}$$

$$x + y = 7$$

$$x + 4 = 7$$

$$x + 4 - 4 = 7 - 4$$

$$\boxed{x = 3}$$

3 adult tickets were bought.

4 child tickets were bought.

17. FINANCIAL PLANNING

Let $x =$ amt put into 8% account

$y =$ amt put into 9% acount

principal • rate • time = interest

total investment: \$10,000

first year combined interest: \$840

$$\begin{cases} x+y=10{,}000 \\ 0.08x+0.09y=840 \end{cases}$$

$$\begin{cases} x=10{,}000-y \\ 0.08x+0.09y=840 \end{cases}$$

$$0.08x+0.09y=840$$

$$0.08(10{,}000-y)+0.09y=840$$

$$800-0.08y+0.09y=840$$

$$800+0.01y-800=840-800$$

$$0.01y=40$$

$$\frac{0.01y}{0.01}=\frac{40}{0.01}$$

$$\boxed{y=4{,}000}$$

$$x+y=10{,}000$$

$$x+4{,}000=10{,}000$$

$$x+4{,}000-4{,}000=10{,}000-4{,}000$$

$$\boxed{x=6{,}000}$$

\$6,000 is put into 1st account.

\$4,000 is put into 2nd account.

18. KAYAKING

Let $x =$ speed of kayak in miles/hour

$c =$ speed of current in miles/hour

$(x - c)$ mph

19. TETHER BALL

Let $x =$ top $\angle$

$y =$ bottom $\angle$

sum of 2 complementary $\angle$'s $= 90$

$$x+y=90$$

20.

$$\begin{cases} 2x+3y\le 6 \\ x>2 \end{cases}$$

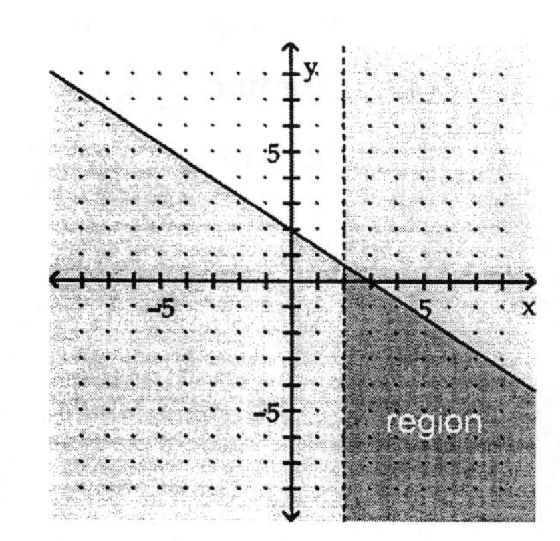

21. CLOTHES SHOPPING

$$\begin{cases} 20x+20y\ge 80 \\ 20x+40y\le 120 \end{cases}$$

Let $x =$ # of \$20 shirts

$y =$ # of \$40 pants

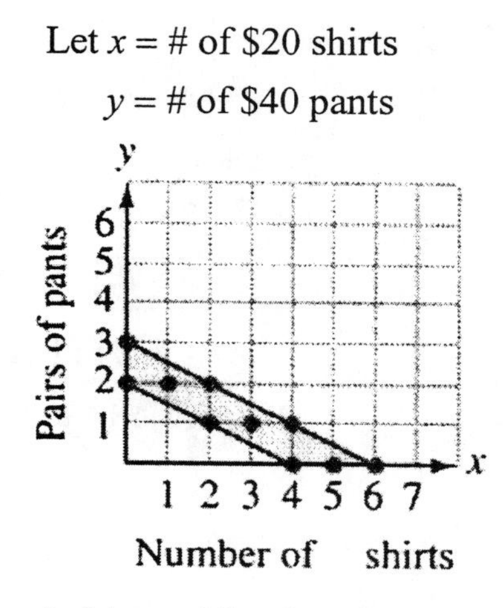

1 shirt and 2 pairs of pants

2 shirts and 2 pairs of pants

3 shirts and 1 pair of pants

There are other answers.

SECTION 8.1

VOCABULARY

1. An **equation** is a statement indicating that two expressions are equal.

3. An equation that is true for all values of its variable is called an **identity**.

5. $<, >, \leq$, and $\geq$ are **inequality** symbols.

7. $x \geq 3$ and $x < 4$ is a **compound** inequality and $-6 < x+1 \leq 1$ is a **double** linear inequality.

9. The **union** of two sets is the set of elements that are in one set, or the other, or both.

CONCEPTS

11. If any quantity is **added** to (or **subtracted** from) both sides of an equation, a new equation is formed that is equivalent to the original one.

13. If both sides of an inequality are multiplied by a **positive** number, a new inequality is formed that has the same direction as the original one.

15. The solution set of a compound inequality containing the word *and* includes all numbers that make **both** inequalities true.

17. The double inequality $4 < 3x+5 \leq 15$ is equivalent to $4 < 3x + 5$ **and** $3x+5 \leq 15$.

19. When solving a compound inequality containing the word *or*, the solution set is the **union** of the solution sets of the inequalities.

21. a. No.

$$\frac{-3}{3}+1 \overset{?}{\geq} 0 \quad \text{and} \quad 2(-3)-3 \overset{?}{<} -10$$

$$-1+1 \overset{?}{\geq} 0 \qquad -6-3 \overset{?}{<} -10$$

$$0 \geq 0 \qquad -9 \not< -10$$

b. Yes.

$$2(-3) \overset{?}{\leq} 0 \quad \text{or} \quad -3(-3) \overset{?}{<} -5$$

$$-6 \leq 0 \qquad 9 \not< -5$$

23. $[-2,1)$

NOTATION

25. We read $\cup$ as **union** and $\cap$ as **intersection**.

PRACTICE

27.

$$2x+1=13$$
$$2x+1-1=13-1$$
$$2x=12$$
$$x=6$$

29.

$$3x+1=3$$
$$3x+1-1=3-1$$
$$3x=2$$
$$x=\frac{2}{3}$$

31.

$$3(x+1)=15$$
$$3x+3=15$$
$$3x+3-3=15-3$$
$$3x=12$$
$$x=4$$

33.

$$\begin{aligned} 2r-5&=1-r\\ 2r-5+r&=1-r+r\\ 3r-5&=1\\ 3r-5+5&=1+5\\ 3r&=6\\ r&=2 \end{aligned}$$

35.

$$\begin{aligned} 3(2y-4)-6&=3y\\ 6y-12-6&=3y\\ 6y-18&=3y\\ 6y-18-6y&=3y-6y\\ -18&=-3y\\ 6&=y\\ y&=6 \end{aligned}$$

37.

$$\begin{aligned} 5(5-a)&=37-2a\\ 25-5a&=37-2a\\ 25-5a+2a&=37-2a+2a\\ 25-3a&=37\\ 25-3a-25&=37-25\\ -3a&=12\\ a&=-4 \end{aligned}$$

39.

$$\begin{aligned} 2(a-5)-(3a+1)&=0\\ 2a-10-3a-1&=0\\ -a-11&=0\\ -a-11+11&=0+11\\ -a&=11\\ a&=-11 \end{aligned}$$

41.

$$\begin{aligned} 2x-6&=-2x+4(x-2)\\ 2x-6&=-2x+4x-8\\ 2x-6&=2x-8\\ 2x-6-2x&=2x-8-2x\\ -6&\neq-8 \end{aligned}$$

$\varnothing$; contradiction

43.

$$\begin{aligned} 9(x+2)&=-6(4-x)+18\\ 9x+18&=-24+6x+18\\ 9x+18&=-6+6x\\ 9x+18-6x&=-6+6x-6x\\ 3x+18&=-6\\ 3x+18-18&=-6-18\\ 3x&=-24\\ x&=-8 \end{aligned}$$

45.

$$\begin{aligned} 12+3(x-4)-21&=5[5-4(4-x)]\\ 12+3x-12-21&=5[5-16+4x]\\ 3x-21&=5[-11+4x]\\ 3x-21&=-55+20x\\ 3x-21-20x&=-55+20x-20x\\ -17x-21&=-55\\ -17x-21+21&=-55+21\\ -17x&=-34\\ x&=2 \end{aligned}$$

47.

$$\begin{aligned} \frac{1}{2}x-4&=-1+2x\\ 2\left(\frac{1}{2}x-4\right)&=2(-1+2x)\\ x-8&=-2+4x\\ x-8-4x&=-2+4x-4x\\ -3x-8&=-2\\ -3x-8+8&=-2+8\\ -3x&=6\\ x&=-2 \end{aligned}$$

49.

$$\begin{aligned} 2y+1&=5(0.2y+1)-(4-y)\\ 2y+1&=y+5-4+y\\ 2y+1&=2y+1\\ 2y+1-2y&=2y+1-2y\\ 1&=1 \end{aligned}$$

$\mathbb{R}$; identity

51.

$$\frac{x}{2}-\frac{x}{3}=4$$
$$6\left(\frac{x}{2}-\frac{x}{3}\right)=6(4)$$
$$3x-2x=24$$
$$x=24$$

53.

$$\frac{x}{6}+1=\frac{x}{3}$$
$$6\left(\frac{x}{6}+1\right)=6\left(\frac{x}{3}\right)$$
$$x+6=2x$$
$$x+6-x=2x-x$$
$$6=x$$
$$x=6$$

55.

$$\frac{3+p}{3}-4p=1-\frac{p+7}{2}$$
$$6\left(\frac{3+p}{3}-4p\right)=6\left(1-\frac{p+7}{2}\right)$$
$$2(3+p)-24p=6-3(p+7)$$
$$6+2p-24p=6-3p-21$$
$$6-22p=-15-3p$$
$$6-22p+3p=-15-3p+3p$$
$$6-19p=-15$$
$$6-19p-6=-15-6$$
$$-19p=-21$$
$$p=\frac{21}{19}$$

57.

$$3(x-4)+6=-2(x+4)+5x$$
$$3x-12+6=-2x-8+5x$$
$$3x-6=3x-8$$
$$3x-6+6=3x-8+6$$
$$3x=3x-2$$
$$3x-3x=3x-2-3x$$
$$0\neq-2$$

$\varnothing$; contradiction

59.

$$\frac{4}{5}(x+5)=\frac{7}{8}(3x+23)-7$$
$$\frac{4}{5}x+4=\frac{21}{8}x+\frac{161}{8}-7$$
$$40\left(\frac{4}{5}x+4\right)=40\left(\frac{21}{8}x+\frac{161}{8}-7\right)$$
$$32x+160=105x+805-280$$
$$32x+160=105x+525$$
$$32x+160-105x=105x+525-105x$$
$$-73x+160=525$$
$$-73x+160-160=525-160$$
$$-73x=365$$
$$x=-5$$

61.

$$V=\frac{1}{3}Bh$$
$$3(V)=3\left(\frac{1}{3}Bh\right)$$
$$3V=Bh$$
$$\frac{3V}{h}=\frac{Bh}{h}$$
$$\frac{3V}{h}=B$$
$$B=\frac{3V}{h}$$

63.

$$p=2\ell+2w$$
$$p-2\ell=2\ell+2w-2\ell$$
$$p-2\ell=2w$$
$$\frac{p-2\ell}{2}=\frac{2w}{2}$$
$$\frac{p-2\ell}{2}=w$$
$$w=\frac{p-2\ell}{2}$$

65.

$$z = \frac{x-\mu}{\sigma}$$
$$\sigma(z) = \sigma\left(\frac{x-\mu}{\sigma}\right)$$
$$\sigma z = x-\mu$$
$$\sigma z + \mu = x - \mu + \mu$$
$$\sigma z + \mu = x$$
$$x = \sigma z + \mu$$

67.

$$z = \frac{x-\mu}{\sigma}$$
$$\sigma(z) = \sigma\left(\frac{x-\mu}{\sigma}\right)$$
$$\sigma z = x-\mu$$
$$\sigma z - x = x - \mu - x$$
$$\sigma z - x = -\mu$$
$$\frac{\sigma z - x}{-1} = \frac{-\mu}{-1}$$
$$-\sigma z + x = \mu$$
$$x - \sigma z = \mu$$
$$\mu = x - \sigma z$$

69.

$$S = \frac{n(a+l)}{2}$$
$$2(S) = 2\left(\frac{n(a+l)}{2}\right)$$
$$2S = n(a+l)$$
$$2S = na + nl$$
$$2S - na = na + nl - na$$
$$2S - na = nl$$
$$\frac{2S-na}{n} = \frac{nl}{n}$$
$$\frac{2S-na}{n} = l$$
$$l = \frac{2S-na}{n}$$

or

$$l = \frac{2S}{n} - a$$

71.

$$5x - 3 > 7$$
$$5x - 3 + 3 > 7 + 3$$
$$5x > 10$$
$$x > 2$$
$$(2,\infty)$$

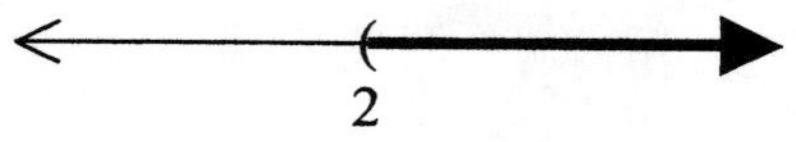

73.

$$-3x - 1 \le 5$$
$$-3x - 1 + 1 \le 5 + 1$$
$$-3x \le 6$$
$$x \ge -2$$
$$[-2,\infty)$$

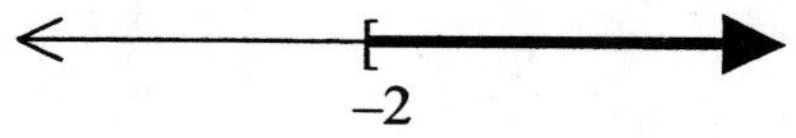

75.

$$8 - 9y \ge -y$$
$$8 - 9y + 9y \ge -y + 9y$$
$$8 \ge 8y$$
$$1 \ge y$$
$$y \le 1$$
$$(-\infty,1]$$

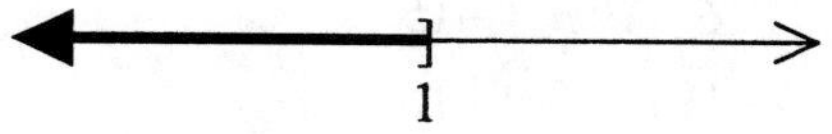

77.

$$7 < \frac{5}{3}a - 3$$
$$3(7) < 3\left(\frac{5}{3}a - 3\right)$$
$$21 < 5a - 9$$
$$21 + 9 < 5a - 9 + 9$$
$$30 < 5a$$
$$6 < a$$
$$a > 6$$
$$(6,\infty)$$

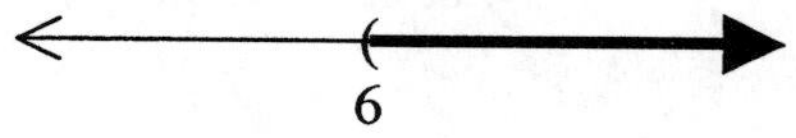

79.

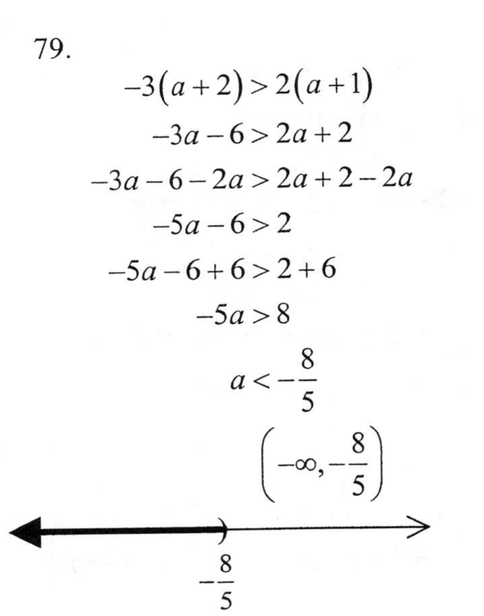

$$-3(a+2) > 2(a+1)$$
$$-3a-6 > 2a+2$$
$$-3a-6-2a > 2a+2-2a$$
$$-5a-6 > 2$$
$$-5a-6+6 > 2+6$$
$$-5a > 8$$
$$a < -\frac{8}{5}$$
$$\left(-\infty, -\frac{8}{5}\right)$$

81.

$$\frac{1}{2}y+2 \ge \frac{1}{3}y-4$$
$$6\left(\frac{1}{2}y+2\right) \ge 6\left(\frac{1}{3}y-4\right)$$
$$3y+12 \ge 2y-24$$
$$3y+12-2y \ge 2y-24-2y$$
$$y+12 \ge -24$$
$$y+12-12 \ge -24-12$$
$$y \ge -36$$
$$[-36, \infty)$$

−36

83.

$$\frac{x-7}{2} - \frac{x-1}{5} \le -\frac{x}{4}$$
$$20\left(\frac{x-7}{2} - \frac{x-1}{5}\right) \le 20\left(-\frac{x}{4}\right)$$
$$10(x-7) - 4(x-1) \le -5x$$
$$10x - 70 - 4x + 4 \le -5x$$
$$6x - 66 \le -5x$$
$$6x - 66 - 6x \le -5x - 6x$$
$$-66 \le -11x$$
$$6 \ge x$$
$$x \le 6$$
$$(-\infty, 6]$$

6

85.

$x > -2$ and $x \le 5$

$$(-2, 5]$$

−2 5

87.

$x+3 < 3x-1$ and $4x-3 \le 3x$

$$x+3-3x < 3x-1-3x \qquad 4x-3-4x \le 3x-4x$$
$$-2x+3 < -1 \qquad -3 \le -x$$
$$-2x+3-3 < -1-3 \qquad 3 \ge x$$
$$-2x < -4 \qquad x \le 3$$
$$x > 2$$
$$(2, 3]$$

2 3

89.

$x-1 \le 2(x+2)$ and $x \le 2x-5$

$$x-1 \le 2x+4 \qquad x-2x \le 2x-5-2x$$
$$x-1-2x \le 2x+4-2x \qquad -x \le -5$$
$$-x-1 \le 4 \qquad x \ge 5$$
$$-x-1+1 \le 4+1$$
$$-x \le 5$$
$$x \ge -5$$
$$[5, \infty)$$

5

91.

$$4 \le x+3 \le 7$$
$$4-3 \le x+3-3 \le 7-3$$
$$1 \le x \le 4$$
$$[1, 4]$$

1 4

93.

$$-6 \le \frac{1}{3}a + 1 < 0$$
$$3(-6) \le 3\left(\frac{1}{3}a + 1\right) < 3(0)$$
$$-18 \le a + 3 < 0$$
$$-18 - 3 \le a + 3 - 3 < 0 - 3$$
$$-21 \le a < -3$$
$$[-21,\ -3)$$

−21 −3

95.

$$-2 < -b + 3 < 5$$
$$-2 - 3 < -b + 3 - 3 < 5 - 3$$
$$-5 < -b < 2$$
$$\frac{-5}{-1} > \frac{-b}{-1} > \frac{2}{-1}$$
$$5 > b > -2$$
$$-2 < b < 5$$
$$(-2, 5)$$

−2 5

97.

$$0 \le \frac{4-x}{3} \le 2$$
$$3(0) \le 3\left(\frac{4-x}{3}\right) \le 3(2)$$
$$0 \le 4 - x \le 6$$
$$0 - 4 \le 4 - x - 4 \le 6 - 4$$
$$-4 \le -x \le 2$$
$$\frac{-4}{-1} \ge \frac{-x}{-1} \ge \frac{2}{-1}$$
$$4 \ge x \ge -2$$
$$-2 \le x \le 4$$
$$[-2,\ 4]$$

−2 4

99

$$x \le -2 \quad \text{or} \quad x > 6$$
$$(-\infty, -2] \cup (6, \infty)$$

−2 6

101.

$$x - 3 < -4 \quad \text{or} \quad x - 2 > 0$$
$$x - 3 + 3 < -4 + 3 \qquad x - 2 + 2 > 0 + 2$$
$$x < -1 \qquad x > 2$$
$$(-\infty, -1) \cup (2, \infty)$$

−1 2

103.

$$3x + 2 < 8 \quad \text{or} \quad 2x - 3 > 11$$
$$3x + 2 - 2 < 8 - 2 \qquad 2x - 3 + 3 > 11 + 3$$
$$3x < 6 \qquad 2x > 14$$
$$x < 2 \qquad x > 7$$
$$(-\infty, 2) \cup (7, \infty)$$

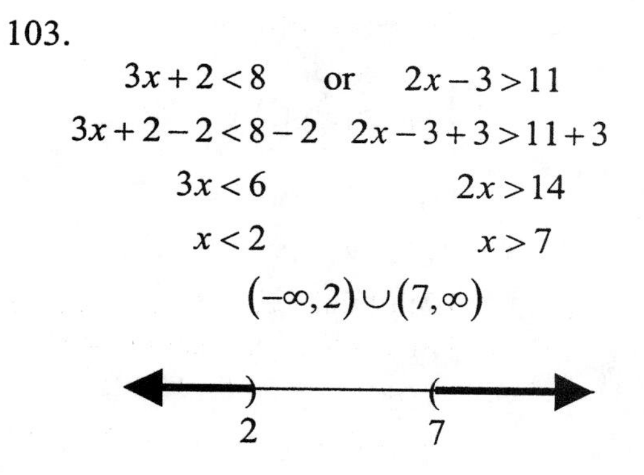

APPLICATIONS

105. SPRING TOURS

Let x = the # of students they supervise.

$$1{,}810 - 15.5x = 1{,}500$$
$$1{,}810 - 15.5x - 1{,}810 = 1{,}500 - 1{,}810$$
$$-15.5x = -310$$
$$x = 20$$

He or she must supervise 20 students to get the reduced cost of \$1,500.

107. FENCING PENS
There are 5 sides of the pen that are x ft and 2 sides that are $(x + 5)$ ft. He has 150 feet of fencing.

$$5x+2(x+5)=150$$
$$5x+2x+10=150$$
$$7x+10=150$$
$$7x+10-10=150-10$$
$$7x=140$$
$$x=20$$

The width of the pen is $x = 20$.
The length of the pen is
$x + (x + 5) = 2x + 5 = 2(20) + 5 = 45$.
The dimensions are 20 ft by 45 ft.

109. PROFITS
Let p = profit.

$$p+27<42$$
$$p+27-27<42-27$$
$$p<\$15$$

111. BUYING COMPACT DISCS
Let x = # of compact discs.

$$175+8.50x\le 330$$
$$175+8.50x-175\le 330-175$$
$$8.50x\le 155$$
$$x\le 18.2$$

The student can buy up to 18 compact discs.

113. BABY FURNITURE

a. The inequality for the perimeter of the playpen is $128\le 4s\le 192$.

b. $128\le 4s\le 192$

$$\frac{128}{4}\le\frac{4s}{4}\le\frac{192}{4}$$
$$32\le s\le 48$$

WRITING

115. Answers will vary.

REVIEW

117.

$$\left(\frac{t^3t^5t^{-6}}{t^2t^{-4}}\right)^{-3}=\left(\frac{t^{3+5+(-6)}}{t^{2+(-4)}}\right)^{-3}$$
$$=\left(\frac{t^2}{t^{-2}}\right)^{-3}$$
$$=\left(t^{2-(-2)}\right)^{-3}$$
$$=\left(t^4\right)^{-3}$$
$$=\frac{1}{\left(t^4\right)^3}$$
$$=\frac{1}{t^{4\cdot 3}}$$
$$=\frac{1}{t^{12}}$$

CHALLENGE PROBLEMS

119.

$$2(-2)\le 3x-1 \quad\text{and}\quad 3x-1\le -1-3$$
$$-4\le 3x-1 \qquad 3x-1\le -4$$
$$-3\le 3x \qquad 3x\le -3$$
$$-1\le x \qquad x\le -1$$
$$x\ge -1$$

$$[-1,-1]$$

−1

121.

$$x-1\le 2(x+2) \quad\text{and}\quad x\le 2x-5$$
$$x-1\le 2x+4 \qquad -x\le -5$$
$$x\le 2x+5 \qquad x\ge 5$$
$$-x\le 5$$
$$x\ge -5$$

$$[5,\infty)$$

5

SECTION 8.2

VOCABULARY

1. The **absolute value** of a number is its distance from 0 on a number line.

3. $|2x-1|>10$ is an absolute value **inequality**.

5. To **isolate** the absolute value in $|3-x|-4=5$, we add 4 to both sides.

7. When we say that the absolute value equation and a compound equation are equivalent, we mean that they have the same **solution(s)**.

CONCEPTS

9. The absolute value of a real number is greater than or equal to 0, but never **negative**.

11. To solve $|x|>5$, we must find the coordinates of all points on a number line that are **more than** 5 units from the origin.

13. To solve $|x|=5$, we must find the coordinates of all points on a number line that are **5** units from the origin.

15. a. -2 and 2
    b. $-1.99, -1, 0, 1,$ and 1.99
    c. $-4, -3, -2.01, 2.01, 3,$ and 4

17. a. $x-7=8$ or $x-7=-8$
    b. $x+10=x-3$ or $x+10=-(x-3)$

19. a. $x=8$ or $x=-8$
    b. $x\geq 8$ or $x\leq -8$
    c. $-8\leq x\leq 8$
    d. $5x-1=x+3$ or $5x-1=-(x+3)$

NOTATION

21. a. ii
    b. iii
    c. i

23. $(-\infty,-1)\cup(3,\infty)$

PRACTICE

25.
$$|x|=23$$
$$x=23 \quad\text{or}\quad x=-23$$

27.
$$|5x|=20$$
$$5x=20 \quad\text{or}\quad 5x=-20$$
$$x=4 \qquad\qquad x=-4$$

29.
$$|x-3.1|=6$$
$$x-3.1=6 \quad\text{or}\quad x-3.1=-6$$
$$x=9.1 \qquad\qquad x=-2.9$$

31.
$$|3x+2|=16$$
$$3x+2=16 \quad\text{or}\quad 3x+2=-16$$
$$3x=14 \qquad\qquad 3x=-18$$
$$x=\frac{14}{3} \qquad\qquad x=-6$$

33.
$$|x|-3=9$$
$$|x|=12$$
$$x=12 \quad\text{or}\quad x=-12$$

35.
$$\left|\frac{7}{2}x+3\right|=-5$$

Since the absolute value of an expression can never be negative, there is no solution.

37.
$$|3-4x|+1=6$$
$$|3-4x|=5$$
$$3-4x=5 \quad\text{or}\quad 3-4x=-5$$
$$-4x=2 \qquad\qquad -4x=-8$$
$$x=-\frac{1}{2} \qquad\qquad x=2$$

39.

$$2|3x+24|=0$$
$$\frac{2|3x+24|}{2}=\frac{0}{2}$$
$$|3x+24|=0$$
$$3x+24=0$$
$$3x=-24$$
$$x=-8$$

41.

$$\left|\frac{3x+48}{3}\right|=12$$
$$\frac{3x+48}{3}=12 \quad \text{or} \quad \frac{3x+48}{3}=-12$$
$$3\left(\frac{3x+48}{3}\right)=3(12) \quad 3\left(\frac{3x+48}{3}\right)=3(-12)$$
$$3x+48=36 \qquad 3x+48=-36$$
$$3x=-12 \qquad 3x=-84$$
$$x=-4 \qquad x=-28$$

43.

$$-7=2-|0.3x-3|$$
$$-9=-|0.3x-3|$$
$$\frac{-9}{-1}=\frac{-|0.3x-3|}{-1}$$
$$9=|0.3x-3|$$
$$0.3x-3=9 \quad \text{or} \quad 0.3x-3=-9$$
$$0.3x=12 \qquad 0.3x=-6$$
$$x=40 \qquad x=-20$$

45.

$$\frac{6}{5}=\left|\frac{3x}{5}+\frac{x}{2}\right|$$
$$\frac{3x}{5}+\frac{x}{2}=\frac{6}{5} \quad \text{or} \quad \frac{3x}{5}+\frac{x}{2}=-\frac{6}{5}$$
$$10\left(\frac{3x}{5}+\frac{x}{2}\right)=10\left(\frac{6}{5}\right) \quad 10\left(\frac{3x}{5}+\frac{x}{2}\right)=10\left(-\frac{6}{5}\right)$$
$$6x+5x=12 \qquad 6x+5x=-12$$
$$11x=12 \qquad 11x=-12$$
$$x=\frac{12}{11} \qquad x=-\frac{12}{11}$$

47.

$$|2x+1|=|3(x+1)|$$
$$|2x+1|=|3x+3|$$
$$2x+1=3x+3 \quad \text{or} \quad 2x+1=-(3x+3)$$
$$2x+1=3x+3 \qquad 2x+1=-3x-3$$
$$-x+1=3 \qquad 5x+1=-3$$
$$-x=2 \qquad 5x=-4$$
$$x=-2 \qquad x=-\frac{4}{5}$$

49.

$$|2-x|=|3x+2|$$
$$2-x=3x+2 \quad \text{or} \quad 2-x=-(3x+2)$$
$$2-x=3x+2 \qquad 2-x=-3x-2$$
$$2-4x=2 \qquad 2+2x=-2$$
$$-4x=0 \qquad 2x=-4$$
$$x=0 \qquad x=-2$$

51.

$$\left|\frac{x}{2}+2\right|=\left|\frac{x}{2}-2\right|$$
$$\frac{x}{2}+2=\frac{x}{2}-2 \quad \text{or} \quad \frac{x}{2}+2=-\left(\frac{x}{2}-2\right)$$
$$\frac{x}{2}+2=\frac{x}{2}-2 \qquad \frac{x}{2}+2=-\frac{x}{2}+2$$
$$2\left(\frac{x}{2}+2\right)=2\left(\frac{x}{2}-2\right) \quad 2\left(\frac{x}{2}+2\right)=2\left(-\frac{x}{2}+2\right)$$
$$x+4=x-4 \qquad x+4=-x+4$$
$$4\neq-4 \qquad 2x=0$$
$$\varnothing \qquad x=0$$

53.

$$\left|x+\frac{1}{3}\right|=|x-3|$$

$$x+\frac{1}{3}=x-3 \quad \text{or} \quad x+\frac{1}{3}=-(x-3)$$

$$x+\frac{1}{3}=x-3 \qquad x+\frac{1}{3}=-x+3$$

$$3\left(x+\frac{1}{3}\right)=3(x-3) \qquad 3\left(x+\frac{1}{3}\right)=3(-x+3)$$

$$3x+1=3x-9 \qquad 3x+1=-3x+9$$

$$3x=3x-10 \qquad 3x=-3x+8$$

$$0x=-10 \qquad 6x=8$$

$$0\neq-10 \qquad x=\frac{4}{3}$$

$$\varnothing$$

55.

$$|x|<4$$

$$-4<x<4$$

$$(-4,4)$$

$-4 \quad 4$

57.

$$|x+9|\le 12$$

$$-12\le x+9\le 12$$

$$-12-9\le x+9-9\le 12-9$$

$$-21\le x\le 3$$

$$[-21,\ 3]$$

$-21 \quad 3$

59.

$$|3x-2|<10$$

$$-10<3x-2<10$$

$$-10+2<3x-2+2<10+2$$

$$-8<3x<12$$

$$-\frac{8}{3}<x<4$$

$$\left(-\frac{8}{3},\ 4\right)$$

$-\frac{8}{3} \quad 4$

61.

$$|3x+2|\le -3$$

Since the absolute value can never be less than or equal to -3, there is **no solution.**

63.

$$|x|>3$$

$$x>3 \quad \text{or} \quad x<-3$$

$$(-\infty,-3)\cup(3,\infty)$$

$-3 \quad 3$

65.

$$|x-12|>24$$

$$x-12>24 \quad \text{or} \quad x-12<-24$$

$$x>36 \qquad x<-12$$

$$(-\infty,-12)\cup(36,\infty)$$

$-12 \quad 36$

67.

$$|3x+2|>14$$

$$3x+2>14 \quad \text{or} \quad 3x+2<-14$$

$$3x>12 \qquad 3x<-16$$

$$x>4 \qquad x<-\frac{16}{3}$$

$$\left(-\infty,-\frac{16}{3}\right)\cup(4,\infty)$$

$-\frac{16}{3} \quad 4$

69.

$$|4x+3|>-5$$

$$(-\infty,\infty)$$

The absolute value will always be greater than -5, so the answer is all real numbers.

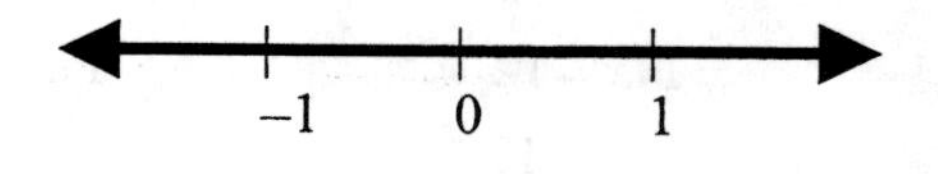

71.

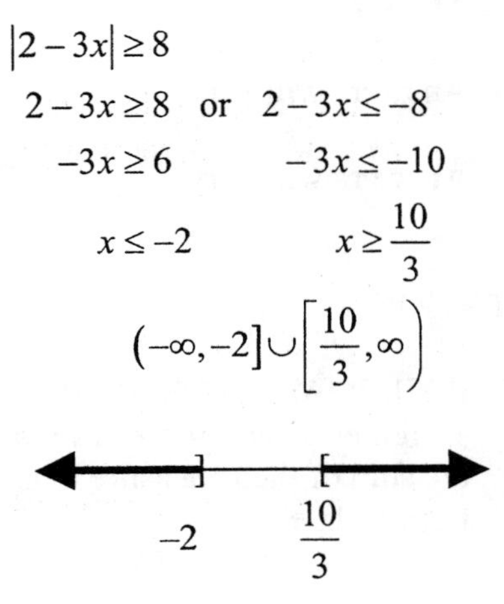

$$|2-3x| \geq 8$$

$$2-3x \geq 8 \quad \text{or} \quad 2-3x \leq -8$$

$$-3x \geq 6 \qquad -3x \leq -10$$

$$x \leq -2 \qquad x \geq \frac{10}{3}$$

$$(-\infty,-2] \cup \left[\frac{10}{3},\infty\right)$$

-2 $\quad \frac{10}{3}$

73.

$$-|2x-3| < -7$$

$$\frac{-|2x-3|}{-1} > \frac{-7}{-1}$$

$$|2x-3| > 7$$

$$2x-3 > 7 \quad \text{or} \quad 2x-3 < -7$$

$$2x > 10 \qquad 2x < -4$$

$$x > 5 \qquad x < -2$$

$$(-\infty,-2) \cup (5,\infty)$$

-2 $\quad 5$

75.

$$\left|\frac{x-2}{3}\right| \leq 4$$

$$-4 \leq \frac{x-2}{3} \leq 4$$

$$3(-4) \leq 3\left(\frac{x-2}{3}\right) \leq 3(4)$$

$$-12 \leq x-2 \leq 12$$

$$-12+2 \leq x-2+2 \leq 12+2$$

$$-10 \leq x \leq 14$$

$$[-10,\ 14]$$

-10 $\quad 14$

77.

$$|3x+1|+2 < 6$$

$$|3x+1| < 4$$

$$-4 < 3x+1 < 4$$

$$-4-1 < 3x+1-1 < 4-1$$

$$-5 < 3x < 3$$

$$-\frac{5}{3} < x < 1$$

$$\left(-\frac{5}{3},\ 1\right)$$

$-\frac{5}{3}$ $\quad 1$

79.

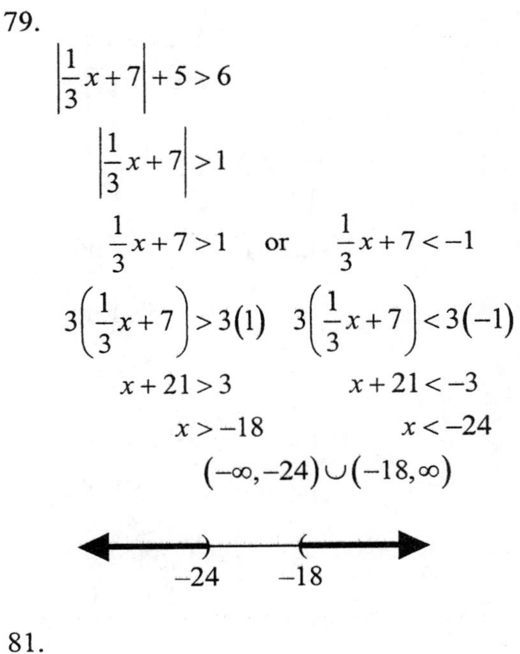

$$\left|\frac{1}{3}x+7\right|+5 > 6$$

$$\left|\frac{1}{3}x+7\right| > 1$$

$$\frac{1}{3}x+7 > 1 \quad \text{or} \quad \frac{1}{3}x+7 < -1$$

$$3\left(\frac{1}{3}x+7\right) > 3(1) \qquad 3\left(\frac{1}{3}x+7\right) < 3(-1)$$

$$x+21 > 3 \qquad x+21 < -3$$

$$x > -18 \qquad x < -24$$

$$(-\infty,-24) \cup (-18,\infty)$$

-24 $\quad -18$

81.

$$|0.5x+1|+2 \leq 0$$

$$|0.5x+1| \leq -2$$

Since the absolute value can never be less than or equal to –2, there is **no solution.**

APPLICATIONS

83. TEMPERATURE RANGES

$$|t-78|\le 8$$
$$-8\le t-78\le 8$$
$$-8+78\le t-78+78\le 8+78$$
$$70^\circ\le t\le 86^\circ$$

85. AUTO MECHANICS

a. $\left|c-0.6^\circ\right|\le 0.5^\circ$

b.

$$|c-0.6|\le 0.5$$
$$-0.5\le c-0.6\le 0.5$$
$$-0.5+0.6\le c-0.6+0.6\le 0.5+0.6$$
$$0.1^\circ\le c\le 1.1^\circ$$
$$\left[0.1^\circ,\ 1.1^\circ\right]$$

87. ERROR ANALYSIS

a. Trial 1: $p=22.91\%$

$$|\mathbf{22.91}-25.46|\overset{?}{\le}1.00$$
$$|-2.55|\overset{?}{\le}1.00$$
$$2.55\not\le 1.00$$

no

Trial 2: $p=26.45\%$

$$|\mathbf{26.45}-25.46|\overset{?}{\le}1.00$$
$$|0.99|\overset{?}{\le}1.00$$
$$0.99\le 1.00$$

yes

Trial 3: $p=26.49\%$

$$|\mathbf{26.49}-25.46|\overset{?}{\le}1.00$$
$$|1.03|\overset{?}{\le}1.00$$
$$1.03\not\le 1.00$$

no

Trial 4: $p=24.76\%$

$$|\mathbf{24.76}-25.46|\overset{?}{\le}1.00$$
$$|-0.70|\overset{?}{\le}1.00$$
$$0.70\le 1.00$$

yes

b. It is less than or equal to 1%.

WRITING

89. Answers will vary.

91. Answers will vary.

REVIEW

93. RAILROAD CROSSINGS

Together, x° and y° form a straight line, so the sum of their measure must equal 180°. Let $y = 2x + 30$.

$$x+y=180$$
$$x+(2x+30)=180$$
$$3x+30=180$$
$$3x=150$$
$$x=50$$

$$x=50^\circ$$
$$y=2(50)+30=130^\circ$$

CHALLENGE PROBLEMS

95. $k<0$

97. x and y must have different signs.

SECTION 8.3

VOCABULARY

1. When we write $2x + 4$ as $2(x + 2)$, we say that we have **factored** $2x + 4$.

3. The abbreviation GCF stands for **greatest common factor**.

5. To factor means to factor **completely**. Each factor of a completely factored expression will be **prime**.

7. A polynomial with three terms, such as $3x^2 - 2x + 4$, is called a **trinomial**.

9. The **lead** coefficient of the trinomial $x^2 - 3x + 2$ is 1, the **coefficient** of the middle term is -3, and the last **term** is 2.

CONCEPTS

11. $\text{GCF} = 2 \cdot 3 \cdot x \cdot y \cdot y = 6xy^2$

13.

Factors of 8	Sum of the factors of 8
1(8) = 8	1 + 8 = 9
2(4) = 8	2 + 4 = 6
−1(−8) = 8	−1 + (−8) = −9
−2(−4) = 8	−2 + (−4) = −6

15.

Negative factors of 12	Sum of factors of 12
−1(−12) = 12	−1 + (−12) = −13
−2(−6) = 12	−2 + (−6) = −8
−3(−4) = 12	−3 + (−4) = −7

NOTATION

17.
$$15c^3d^4 - 25c^2d^4 + 5c^3d^6 = \boxed{5c^2d^4}\left(3c - 5 + cd^2\right)$$

19.
$$6m^2 + 7m - 3 = \left(\boxed{3m} - 1\right)\left(2m + \boxed{3}\right)$$

PRACTICE

21.
$$2x^2 - 6x = 2x(x-3)$$

23.
$$15x^2y - 10x^2y^2 = 5x^2y(3-2y)$$

25.
$$27z^3 + 12z^2 + 3z = 3z\left(9z^2 + 4z + 1\right)$$

27.
$$11m^3n^2 - 12x^2y \quad \text{is prime}$$

29.
$$-5xy + y - 4 = -(5xy - y + 4)$$

31.
$$24s^3 - 12s^2t + 6st^2 = 6s\left(4s^2 - 2st + t^2\right)$$

33.
$$(x+y)u + (x+y)v = (x+y)(u+v)$$

35.
$$-18a^2b - 12ab^2 = -6ab(3a + 2b)$$

37.
$$\frac{3}{5}ax^4 + \frac{1}{5}bx^2 - \frac{4}{5}ax^3 = \frac{1}{5}x^2\left(3ax^2 + b - 4ax\right)$$

39.
$$5(a-b) - t(a-b) = (a-b)(5-t)$$

41.
$$3(m+n+p) + x(m+n+p) = (m+n+p)(3+x)$$

43.
$$-63u^3v^6z^9 + 28u^2v^7z^2 - 21u^3v^3z^4$$
$$= -7u^2v^3z^2\left(9uv^3z^7 - 4v^4 + 3uz^2\right)$$

45.
$$4\left(x^2+1\right)^2 + 2\left(x^2+1\right)^3 = 2\left(x^2+1\right)^2\left(2 + x^2 + 1\right)$$
$$= 2\left(x^2+1\right)^2\left(x^2+3\right)$$

47.

$$r_1r_2 = rr_2 + rr_1$$
$$r_1r_2 - rr_1 = rr_2 + rr_1 - rr_1$$
$$r_1(r_2 - r) = rr_2$$
$$\frac{r_1(r_2 - r)}{r_2 - r} = \frac{rr_2}{r_2 - r}$$
$$r_1 = \frac{rr_2}{r_2 - r}$$

49.

$$S(1-r) = a - \ell r$$
$$S - Sr = a - \ell r$$
$$S - Sr + \ell r = a - \ell r + \ell r$$
$$S - Sr + \ell r = a$$
$$S - Sr + \ell r - S = a - S$$
$$\ell r - Sr = a - S$$
$$r(\ell - S) = a - S$$
$$\frac{r(\ell - S)}{\ell - S} = \frac{a - S}{\ell - S}$$
$$r = \frac{a - S}{\ell - S} \cdot \frac{-1}{-1}$$
$$r = \frac{S - a}{S - \ell}$$

51.

$$b^2x^2 + a^2y^2 = a^2b^2$$
$$b^2x^2 + a^2y^2 - a^2y^2 = a^2b^2 - a^2y^2$$
$$b^2x^2 = a^2b^2 - a^2y^2$$
$$b^2x^2 = a^2(b^2 - y^2)$$
$$\frac{b^2x^2}{b^2 - y^2} = \frac{a^2(b^2 - y^2)}{b^2 - y^2}$$
$$\frac{b^2x^2}{b^2 - y^2} = a^2$$
$$a^2 = \frac{b^2x^2}{b^2 - y^2}$$

53.

$$ax + bx + ay + by = x(a+b) + y(a+b)$$
$$= (a+b)(x+y)$$

55.

$$x^2 + yx + x + y = x(x+y) + 1(x+y)$$
$$= (x+y)(x+1)$$

57.

$$a^2 - 4b + ab - 4a = a^2 + ab - 4a - 4b$$
$$= a(a+b) - 4(a+b)$$
$$= (a+b)(a-4)$$

59.

$$x^2 + 4y - xy - 4x = x^2 - xy + 4y - 4x$$
$$= x(x-y) + 4(y-x)$$
$$= x(x-y) - 4(x-y)$$
$$= (x-y)(x-4)$$

61.

$$a^2x + bx - a^2 - b = x(a^2 + b) - 1(a^2 + b)$$
$$= (a^2 + b)(x-1)$$

63.

$$x^2 + xy + xz + xy + y^2 + zy$$
$$= (x^2 + xy + xz) + (xy + y^2 + zy)$$
$$= x(x+y+z) + y(x+y+z)$$
$$= (x+y+z)(x+y)$$

65.

$$1 - m + mn - n = 1(1-m) + n(m-1)$$
$$= 1(1-m) - n(1-m)$$
$$= (1-m)(1-n)$$

67.

$$2ax^2 - 4 + a - 8x^2 = 2ax^2 - 8x^2 + a - 4$$
$$= 2x^2(a-4) + 1(a-4)$$
$$= (a-4)(2x^2 + 1)$$

69.

$$mpx + mqx + npx + nqx$$
$$= x[mp + mq + np + nq]$$
$$= x[m(p+q) + n(p+q)]$$
$$= x(p+q)(m+n)$$

71.

$$x^2y + xy^2 + 2xyz + xy^2 + y^3 + 2y^2z$$
$$= y[x^2 + xy + 2xz + xy + y^2 + 2yz]$$
$$= y[(x^2 + xy + 2xz) + (xy + y^2 + 2yz)]$$
$$= y[x(x + y + 2z) + y(x + y + 2z)]$$
$$= y(x + y + 2z)(x + y)$$

73. $x^2 - 5x + 6 = (x - 2)(x - 3)$

75. $x^2 - 7x + 10 = (x - 5)(x - 2)$

77. $b^2 + 8b + 18$ is prime

79. $-x + x^2 - 30 = x^2 - x - 30$
$= (x - 6)(x + 5)$

81. $a^2 - 18a + 81 = (a - 9)^2$

83. $x^2 - 4xy - 21y^2 = (x - 7y)(x + 3y)$

85. $s^2 - 10st + 16t^2 = (s - 2t)(s - 8t)$

87. $3x^2 + 12x - 63 = 3(x^2 + 4x - 21)$
$= 3(x + 7)(x - 3)$

89. $32 - a^2 + 4a = -a^2 + 4a + 32$
$= -(a^2 - 4a - 32)$
$= -(a - 8)(a + 4)$

91. $-3a^2x^2 + 15a^2x - 18a^2 = -3a^2(x^2 - 5x + 6)$
$= -3a^2(x - 2)(x - 3)$

93. $y^4 - 13y^2 + 30 = (y^2 - 10)(y^2 - 3)$

95. $b^4x^2 - 12b^2x^2 + 35x^2 = x^2(b^4 - 12b^2 + 35)$
$= x^2(b^2 - 5)(b^2 - 7)$

97. $6y^2 + 7y + 2 = (3y + 2)(2y + 1)$

99. $8a^2 + 6a - 9 = (4a - 3)(2a + 3)$

101. $6x^2 - 5xy - 4y^2 = (3x - 4y)(2x + y)$

103. $5x^2 + 4x + 1$ is prime

105. $6z^2 + 17z + 12 = (2z + 3)(3z + 4)$

107. $4y^2 + 4y + 1 = (2y + 1)^2$

109. $-3a^2 + ab + 2b^2 = -(3a^2 - ab - 2b^2)$
$= -(3a + 2b)(a - b)$

111. $20a^2 + 60ab + 45b^2$
$= 5(4a^2 + 12ab + 9b^2)$
$= 5(2a + 3b)(2a + 3b)$
$= 5(2a + 3b)^2$

113. $64h^6 + 24h^5 - 4h^4$
$= 4h^4(16h^2 + 6h - 1)$
$= 4h^4(8h - 1)(2h + 1)$

115. $6a^2(m + n) + 13a(m + n) - 15(m + n)$
$= (m + n)(6a^2 + 13a - 15)$
$= (m + n)(6a - 5)(a + 3)$

117. $(x + a)^2 + 2(x + a) + 1$

Let $y = (x + a)$
$y^2 + 2y + 1 = (y + 1)(y + 1)$
$= (y + 1)^2$
Substitute $(x + a)$ for y.
$= (x + a + 1)^2$

119. $(a + b)^2 - 2(a + b) - 24$

Let $x = (a + b)$
$x^2 - 2x - 24 = (x - 6)(x + 4)$
Substitute $(a + b)$ for x.
$= (a + b - 6)(a + b + 4)$

121. $14(q - r)^2 - 17(q - r) - 6$

Let $x = (q - r)$.
$14x^2 - 17x - 6 = (2x - 3)(7x + 2)$
Substitute $(q - r)$ for x.
$= [2(q - r) - 3][7(q - r) + 2]$
$= (2q - 2r - 3)(7q - 7r + 2)$

APPLICATIONS

123. CRAYONS

$$V = \pi r^2 h_1 + \frac{1}{3}\pi r^2 h_2$$
$$= \pi r^2\left(h_1 + \frac{1}{3}h_2\right)$$

125. ICE

Area of square $= s^2$

$$s^2 = 6x^2 + 36x + 54$$
$$= 6\left(x^2 + 6x + 9\right)$$
$$= 6(x+3)(x+3)$$
$$= 6(x+3)^2$$

The length of an edge of the block is $x + 3$.

WRITING

127. Answers will vary.

REVIEW

129. INVESTMENTS

Let x = the amount invested in each account.

Account	Principal	· Rate	· Time	= Interest
7%	x	0.07	1	$0.07x$
8%	x	0.08	1	$0.08x$
10.5%	x	0.105	1	$0.105x$
Combined Interest		$1,249.50		

7% int. + 8% int. + 10.5% int.= Combined int.

$$0.07x + 0.08x + 0.105x = 1{,}249.50$$
$$0.255x = 1{,}249.50$$
$$\frac{0.255x}{0.255} = \frac{1{,}249.50}{0.255}$$
$$x = \$4{,}900$$

$4,900 is invested in each account.

CHALLENGE PROBLEMS

131.

$$x^{n+2} + x^{n+3} = x^2\left(x^{n+2-2} + x^{n+3-2}\right)$$
$$= x^2\left(x^n + x^{n+1}\right)$$

133.

$$x^{2n} + 2x^n + 1 = \left(x^n + 1\right)\left(x^n + 1\right)$$
$$= \left(x^n + 1\right)^2$$

135.

$$x^{4n} + 2x^{2n}y^{2n} + y^{4n} = \left(x^{2n} + y^{2n}\right)\left(x^{2n} + y^{2n}\right)$$
$$= \left(x^{2n} + y^{2n}\right)^2$$

SECTION 8.4

VOCABULARY

1. When the polynomial $4x^2 - 25$ is written as $(2x)^2 - (5)^2$, we see that it is the difference of two **squares**.

CONCEPTS

3. 1, 4, 9, 16, 25, 36, 49, 64, 81, 100

5. a. $(x + 5)(x + 5) = x^2 + 10x + 25$
   b. $(x + 5)(x - 5) = x^2 - 25$

7. By the commutative property of multiplication, $(9t - 4)(9t + 4) = (9t + 4)(9t - 4)$.

9. $p^2 - q^2 = (p+q)\boxed{(p-q)}$

11. $p^2q + pq^2 = \boxed{pq}(p+q)$

13. $p^3 - q^3 = (p-q)\boxed{(p^2+pq+q^2)}$

NOTATION

15. Answers will vary.
   a. $x^2 - 4$
   b. $(x - 4)^2$
   c. $x^2 + 4$
   d. $x^3 + 8$
   e. $(x + 8)^3$

PRACTICE

17.
$$\begin{aligned} x^2 - 4 &= (x)^2 - (2)^2 \\ &= (x+2)(x-2) \end{aligned}$$

19.
$$\begin{aligned} 9y^2 - 64 &= (3y)^2 - (8)^2 \\ &= (3y+8)(3y-8) \end{aligned}$$

21. $x^2 + 25$
prime

23.
$$\begin{aligned} 400 - c^2 &= (20)^2 - (c)^2 \\ &= (20-c)(20+c) \end{aligned}$$

25.
$$\begin{aligned} 625a^2 - 169b^4 &= (25a)^2 - (13b^2)^2 \\ &= (25a - 13b^2)(25a + 13b^2) \end{aligned}$$

27.
$$\begin{aligned} 81a^4 - 49b^2 &= (9a^2)^2 - (7b)^2 \\ &= (9a^2 - 7b)(9a^2 + 7b) \end{aligned}$$

29.
$$\begin{aligned} 36x^4y^2 - 49z^4 &= (6x^2y)^2 - (7z^2)^2 \\ &= (6x^2y - 7z^2)(6x^2y + 7z^2) \end{aligned}$$

31.
$$(x+y)^2 - z^2 = (x+y+z)(x+y-z)$$

33.
$$(a-b)^2 - c^2 = (a-b+c)(a-b-c)$$

35.
$$\begin{aligned} x^4 - y^4 &= (x^2)^2 - (y^2)^2 \\ &= (x^2+y^2)(x^2-y^2) \\ &= (x^2+y^2)(x+y)(x-y) \end{aligned}$$

37.
$$\begin{aligned} 256x^4y^4 - z^8 &= (16x^2y^2)^2 - (z^4)^2 \\ &= (16x^2y^2 + z^4)(16x^2y^2 - z^4) \\ &= (16x^2y^2 + z^4)(4xy + z^2)(4xy - z^2) \end{aligned}$$

39.
$$\begin{aligned} \frac{1}{36} - y^4 &= \left(\frac{1}{6}\right)^2 - (y^2)^2 \\ &= \left(\frac{1}{6} + y^2\right)\left(\frac{1}{6} - y^2\right) \end{aligned}$$

41.
$$2x^2 - 288 = 2(x^2 - 144) = 2(x+12)(x-12)$$

43.
$$2x^3 - 32x = 2x(x^2 - 16) = 2x(x+4)(x-4)$$

45.
$$5x^3 - 125x = 5x(x^2 - 25) = 5x(x+5)(x-5)$$

47.
$$r^2s^2t^2 - t^2x^4y^2 = t^2(r^2s^2 - x^4y^2) = t^2(rs + x^2y)(rs - x^2y)$$

49.
$$a^2 - b^2 + a + b = (a^2 - b^2) + (a+b) = (a+b)(a-b) + 1(a+b) = (a+b)(a-b+1)$$

51.
$$a^2 - b^2 + 2a - 2b = (a^2 - b^2) + (2a - 2b) = (a+b)(a-b) + 2(a-b) = (a-b)(a+b+2)$$

53.
$$2x + y + 4x^2 - y^2 = (2x+y) + (4x^2 - y^2) = 1(2x+y) + (2x+y)(2x-y) = (2x+y)(1+2x-y)$$

55.
$$x^3 - xy^2 - 4x^2 + 4y^2 = (x^3 - xy^2) + (-4x^2 + 4y^2) = x(x^2 - y^2) - 4(x^2 - y^2) = (x^2 - y^2)(x-4) = (x+y)(x-y)(x-4)$$

57.
$$x^2 + 4x + 4 - y^2 = (x^2 + 4x + 4) - y^2 = (x+2)^2 - y^2 = (x+2-y)(x+2+y)$$

59.
$$x^2 + 2x + 1 - 9z^2 = (x^2 + 2x + 1) - 9z^2 = (x+1)^2 - 9z^2 = (x+1-3z)(x+1+3z)$$

61.
$$c^2 - 4a^2 + 4ab - b^2 = c^2 - (4a^2 - 4ab + b^2) = c^2 - (2a-b)^2 = (c+2a-b)(c-(2a-b)) = (c+2a-b)(c-2a+b)$$

63. $r^3 + s^3 = (r+s)(r^2 - rs + s^2)$

65.
$$x^3 - 8y^3 = (x)^3 - (2y)^3 = (x-2y)(x^2 + 2xy + 4y^2)$$

67.
$$64a^3 - 125b^6 = (4a)^3 - (5b^2)^3 = (4a - 5b^2)(16a^2 + 20ab^2 + 25b^4)$$

69.
$$125x^3y^6 + 216z^9 = (5xy^2)^3 + (6z^3)^3 = (5xy^2 + 6z^3)(25x^2y^4 - 30xy^2z^3 + 36z^6)$$

71.
$$x^6 + y^6 = (x^2)^3 + (y^2)^3 = (x^2 + y^2)(x^4 - x^2y^2 + y^4)$$

73.
$$5x^3 + 625 = 5(x^3 + 125) = 5(x+5)(x^2 - 5x + 25)$$

75.
$$4x^5 - 256x^2 = 4x^2(x^3 - 64) = 4x^2(x-4)(x^2 + 4x + 16)$$

77.

$$128u^2v^3 - 2t^3u^2 = 2u^2\left(64v^3 - t^3\right)$$
$$= 2u^2\left(4v - t\right)\left(16v^2 + 4tv + t^2\right)$$

79.

$$\left(a+b\right)x^3 + 27\left(a+b\right)$$
$$= \left(a+b\right)\left(x^3 + 27\right)$$
$$= \left(a+b\right)\left(x+3\right)\left(x^2 - 3x + 9\right)$$

81.

$$x^9 - y^{12}z^{15} = \left(x^3\right)^3 - \left(y^4z^5\right)^3$$
$$= \left(x^3 - y^4z^5\right)\left(x^6 + x^3y^4z^5 + y^8z^{10}\right)$$

83.

$$\left(a+b\right)^3 + 27$$
$$= \left(a+b\right)^3 + \left(3\right)^3$$
$$= \left(\left(a+b\right)+3\right)\left(\left(a+b\right)^2 - 3\left(a+b\right) + 9\right)$$
$$= \left(a+b+3\right)\left(a^2 + 2ab + b^2 - 3a - 3b + 9\right)$$

85.

$$y^3\left(y^2 - 1\right) - 27\left(y^2 - 1\right)$$
$$= \left(y^2 - 1\right)\left(y^3 - 27\right)$$
$$= \left(y+1\right)\left(y-1\right)\left(y-3\right)\left(y^2 + 3y + 9\right)$$

87.

$$x^6 - 1$$
$$= \left(x^3 + 1\right)\left(x^3 - 1\right)$$
$$= \left(x+1\right)\left(x^2 - x + 1\right)\left(x-1\right)\left(x^2 + x + 1\right)$$

89.

$$x^{12} - y^6$$
$$= \left(x^6 + y^3\right)\left(x^6 - y^3\right)$$
$$= \left(x^2 + y\right)\left(x^4 - x^2y + y^2\right)\left(x^2 - y\right)\left(x^4 + x^2y + y^2\right)$$

91.

$$a^4 - 13a^2 + 36 = \left(a^2 - 4\right)\left(a^2 - 9\right)$$
$$= \left(a+2\right)\left(a-2\right)\left(a+3\right)\left(a-3\right)$$

APPLICATIONS

93. CANDY

$$V = \frac{4}{3}\pi r_1^3 - \frac{4}{3}\pi r_2^3$$
$$= \frac{4}{3}\pi\left(r_1^3 - r_2^3\right)$$
$$= \frac{4}{3}\pi\left(r_1 - r_2\right)\left(r_1^2 + r_1r_2 + r_2^2\right)$$

WRITING

95. Answers will vary.

REVIEW

97. Determine the cost per minute.

45 minutes for \$25:

$$\frac{25}{45} \approx \$0.55 \text{ per minute}$$

1 hour = 60 minutes for \$35:

$$\frac{35}{60} \approx \$0.58 \text{ per minute}$$

The better buy is 45 minutes for \$25.

CHALLENGE PROBLEMS

99

$$4x^{2n} - 9y^{2n} = \left(2x^n\right)^2 - \left(3y^n\right)^2$$
$$= \left(2x^n - 3y^n\right)\left(2x^n + 3y^n\right)$$

101.

$$a^{3b} - c^{3b} = \left(a^b\right)^3 - \left(c^b\right)^3$$
$$= \left(a^b - c^b\right)\left(a^{2b} + a^bc^b + c^{2b}\right)$$

103.

$$27x^{3n} + y^{3n} = \left(3x^n\right)^3 + \left(y^n\right)^3$$
$$= \left(3x^n + y^n\right)\left(9x^{2n} - 3x^ny^n + y^{2n}\right)$$

105.

$$x^{32}-y^{32}$$
$$=\left(x^{16}+y^{16}\right)\left(x^{16}-y^{16}\right)$$
$$=\left(x^{16}+y^{16}\right)\left(x^{8}+y^{8}\right)\left(x^{8}-y^{8}\right)$$
$$=\left(x^{16}+y^{16}\right)\left(x^{8}+y^{8}\right)\left(x^{4}+y^{4}\right)\left(x^{4}-y^{4}\right)$$
$$=\left(x^{16}+y^{16}\right)\left(x^{8}+y^{8}\right)\left(x^{4}+y^{4}\right)\left(x^{2}+y^{2}\right)\left(x^{2}-y^{2}\right)$$
$$=\left(x^{16}+y^{16}\right)\left(x^{8}+y^{8}\right)\left(x^{4}+y^{4}\right)\left(x^{2}+y^{2}\right)(x+y)(x-y)$$

SECTION 8.5

VOCABULARY

1. A quotient of two polynomials, such as $\frac{x^2+x}{x^2-3x}$, is called a **rational** expression.

3. The quotient of **opposites** is –1. For example, $\frac{x-8}{8-x}=-1$.

5. The **reciprocal** of $\frac{a+3}{a+7}$ is $\frac{a+7}{a+3}$.

7. To **build** a rational expression, we multiply it by a form of 1. For example, $\frac{2}{n^2}\bullet\frac{8}{8}=\frac{16}{8n^2}$.

CONCEPTS

9. To multiply rational expressions, multiply their **numerators** and multiply their **denominators**. To divide two rational expressions, multiply the first by the **reciprocal** of the second.

$$\frac{A}{B}\bullet\frac{C}{D}=\boxed{\frac{AC}{BD}} \qquad \frac{A}{B}\div\frac{C}{D}=\boxed{\frac{A}{B}\bullet\frac{D}{C}}$$

11. To find the least common denominator of several rational expressions, **factor** each denominator completely. The LCD is a product that uses each different factor the **greatest** number of times it appears in any one factorization.

13. a. ii
    b. adding or subtracting rational expressions
    c. simplifying a rational expression

15. a. twice
    b. once

17. a. c
    b. $(p+2)(p+3)$

NOTATION

19. $$\frac{x^2+3x}{x-1}-\frac{2x-1}{x-1}=\frac{x^2+3x-\left(\boxed{2x-1}\right)}{x-1}$$

PRACTICE

21. $$\frac{24x^3y^4}{18x^4y^3}=\frac{\cancel{2}\bullet2\bullet2\bullet\cancel{3}\bullet\cancel{x}\bullet\cancel{x}\bullet\cancel{x}\bullet\cancel{y}\bullet\cancel{y}\bullet\cancel{y}\bullet y}{\cancel{2}\bullet\cancel{3}\bullet3\bullet\cancel{x}\bullet\cancel{x}\bullet\cancel{x}\bullet x\bullet\cancel{y}\bullet\cancel{y}\bullet\cancel{y}}=\frac{4y}{3x}$$

23. $$\frac{9y^2(y-z)}{21y(y-z)^2}=\frac{\cancel{3}\bullet3\bullet\cancel{y}\bullet y\bullet\cancel{(y-z)}}{\cancel{3}\bullet7\bullet\cancel{y}\bullet\cancel{(y-z)}(y-z)}=\frac{3y}{7(y-z)}$$

25. $$\frac{(a-b)(b-c)(c-d)}{(c-d)(b-c)(a-b)}=\frac{\cancel{(a-b)}\cancel{(b-c)}\cancel{(c-d)}}{\cancel{(c-d)}\cancel{(b-c)}\cancel{(a-b)}}=1$$

27. $$\frac{3m-6n}{3n-6m}=\frac{\cancel{3}(m-2n)}{\cancel{3}(n-2m)}=\frac{m-2n}{n-2m}$$

29. $$\frac{x^2+2x+1}{x^2+4x+3}=\frac{(x+1)\cancel{(x+1)}}{(x+3)\cancel{(x+1)}}=\frac{x+1}{x+3}$$

31. $$\frac{6x^2-7x-5}{2x^2+5x+2}=\frac{(3x-5)\cancel{(2x+1)}}{\cancel{(2x+1)}(x+2)}=\frac{3x-5}{x+2}$$

33.

$$\frac{ax+by+ay+bx}{a^2-b^2}=\frac{(ax+bx)+(ay+by)}{(a+b)(a-b)}$$
$$=\frac{x(a+b)+y(a+b)}{(a+b)(a-b)}$$
$$=\frac{(x+y)\cancel{(a+b)}}{\cancel{(a+b)}(a-b)}$$
$$=\frac{x+y}{a-b}$$

35.

$$\frac{12-3x^2}{x^2-x-2}=\frac{3(4-x^2)}{(x-2)(x+1)}$$
$$=\frac{3(2-x)(2+x)}{(x-2)(x+1)}$$
$$=\frac{-3\cancel{(x-2)}(x+2)}{\cancel{(x-2)}(x+1)}$$
$$=\frac{-3(x+2)}{(x+1)}$$
$$=-\frac{3(x+2)}{(x+1)}$$

37.

$$\frac{a^3+27}{4a^2-36}=\frac{(a+3)(a^2-3a+9)}{4(a^2-9)}$$
$$=\frac{\cancel{(a+3)}(a^2-3a+9)}{4(a-3)\cancel{(a+3)}}$$
$$=\frac{a^2-3a+9}{4(a-3)}$$

39.

$$\frac{2x^2+2x-12}{x^3+3x^2-4x-12}=\frac{2(x^2+x-6)}{x^2(x+3)-4(x+3)}$$
$$=\frac{2(x+3)(x-2)}{(x+3)(x^2-4)}$$
$$=\frac{2\cancel{(x+3)}\cancel{(x-2)}}{\cancel{(x+3)}\cancel{(x-2)}(x+2)}$$
$$=\frac{2}{(x+2)}$$

41.

$$\frac{p^3+p^2q-2pq^2}{pq^2+p^2q-2p^3}=\frac{p(p^2+pq-2q^2)}{p(q^2+pq-2p^2)}$$
$$=\frac{p(p+2q)(p-q)}{p(q-p)(q+2p)}$$
$$=\frac{-1\cancel{p}(p+2q)\cancel{(q-p)}}{\cancel{p}\cancel{(q-p)}(q+2p)}$$
$$=-\frac{p+2q}{q+2p}\quad\text{or}\quad\frac{-p-2q}{q+2p}$$

43.

$$\frac{(2x^2+3xy+y^2)(3a+b)}{(x+y)(2xy+2bx+y^2+by)}$$
$$=\frac{(2x+y)(x+y)(3a+b)}{(x+y)[2x(y+b)+y(y+b)]}$$
$$=\frac{\cancel{(2x+y)}\cancel{(x+y)}(3a+b)}{\cancel{(x+y)}(y+b)\cancel{(2x+y)}}$$
$$=\frac{3a+b}{y+b}$$

45.

$$\frac{10a^2}{3b^4}\bullet\frac{12b^3}{5a^2}=\frac{2\bullet\cancel{5}\bullet\cancel{a}\bullet\cancel{a}\bullet2\bullet2\bullet\cancel{3}\bullet\cancel{b}\bullet\cancel{b}\bullet\cancel{b}}{\cancel{3}\bullet\cancel{b}\bullet\cancel{b}\bullet\cancel{b}\bullet b\bullet\cancel{5}\bullet\cancel{a}\bullet\cancel{a}}$$
$$=\frac{8}{b}$$

47.

$$\frac{m^2n}{4}\div\frac{mn^3}{6}=\frac{m^2n}{4}\bullet\frac{6}{mn^3}$$
$$=\frac{\cancel{m}\bullet m\bullet\cancel{n}\bullet\cancel{2}\bullet3}{\cancel{2}\bullet2\bullet\cancel{m}\bullet\cancel{n}\bullet n\bullet n}$$
$$=\frac{3m}{2n^2}$$

49.

$$12y\left(\frac{y+8}{6y}\right)=\frac{2\bullet\cancel{6y}(y+8)}{\cancel{6y}}$$
$$=2(y+8)$$
$$=2y+16$$

51.

$$\frac{x^2-16}{x^2-25} \div \frac{x+4}{x-5} = \frac{(x^2-16)(x-5)}{(x^2-25)(x+4)}$$

$$= \frac{\cancel{(x+4)}(x-4)\cancel{(x-5)}}{(x+5)\cancel{(x-5)}\cancel{(x+4)}}$$

$$= \frac{x-4}{x+5}$$

53.

$$\frac{x^2+2x+1}{9x} \cdot \frac{2x^2-2x}{2x^2-2} = \frac{(x^2+2x+1)(2x^2-2x)}{9x(2x^2-2)}$$

$$= \frac{(x+1)(x+1)2x(x-1)}{9x(2)(x^2-1)}$$

$$= \frac{\cancel{(x+1)}(x+1)\cancel{(2)}\cancel{(x)}\cancel{(x-1)}}{9\cancel{x}\cancel{(2)}\cancel{(x-1)}\cancel{(x+1)}}$$

$$= \frac{x+1}{9}$$

55.

$$\frac{2x^2-x-3}{x^2-1} \cdot \frac{x^2+x-2}{2x^2+x-6}$$

$$= \frac{(2x^2-x-3)(x^2+x-2)}{(x^2-1)(2x^2+x-6)}$$

$$= \frac{\cancel{(2x-3)}\cancel{(x+1)}\cancel{(x+2)}\cancel{(x-1)}}{\cancel{(x+1)}\cancel{(x-1)}\cancel{(2x-3)}\cancel{(x+2)}}$$

$$= 1$$

57.

$$(2x^2-15x+25) \div \frac{2x^2-3x-5}{x+1}$$

$$= \frac{2x^2-15x+25}{1} \cdot \frac{x+1}{2x^2-3x-5}$$

$$= \frac{\cancel{(2x-5)}(x-5)}{1} \cdot \frac{\cancel{(x+1)}}{\cancel{(2x-5)}\cancel{(x+1)}}$$

$$= x-5$$

59.

$$\frac{3n^2+5n-2}{12n^2-13n+3} \div \frac{n^2+3n+2}{4n^2+5n-6}$$

$$= \frac{3n^2+5n-2}{12n^2-13n+3} \cdot \frac{4n^2+5n-6}{n^2+3n+2}$$

$$= \frac{(3n^2+5n-2)(4n^2+5n-6)}{(12n^2-13n+3)(n^2+3n+2)}$$

$$= \frac{\cancel{(3n-1)}\cancel{(n+2)}\cancel{(4n-3)}(n+2)}{\cancel{(4n-3)}\cancel{(3n-1)}\cancel{(n+2)}(n+1)}$$

$$= \frac{n+2}{n+1}$$

61.

$$\frac{2x^2+5xy+3y^2}{3x^2-5xy+2y^2} \div \frac{2x^2+xy-3y^2}{-3x^2+5xy-2y^2}$$

$$= \frac{2x^2+5xy+3y^2}{3x^2-5xy+2y^2} \cdot \frac{-(3x^2-5xy+2y^2)}{2x^2+xy-3y^2}$$

$$= \frac{(2x^2+5xy+3y^2)(-1)(3x^2-5xy+2y^2)}{(3x^2-5xy+2y^2)(2x^2+xy-3y^2)}$$

$$= \frac{-\cancel{(2x+3y)}(x+y)\cancel{(3x-2y)}\cancel{(x-y)}}{\cancel{(3x-2y)}(x-y)\cancel{(2x+3y)}\cancel{(x-y)}}$$

$$= -\frac{x+y}{x-y}$$

63.

$$\frac{p^3-q^3}{q^2-p^2} \cdot \frac{q^2+pq}{p^3+p^2q+pq^2}$$

$$= \frac{(p^3-q^3)(q^2+pq)}{(q^2-p^2)(p^3+p^2q+pq^2)}$$

$$= \frac{(p-q)(p^2+pq+q^2)(q)(q+p)}{(q+p)(q-p)(p)(p^2+pq+q^2)}$$

$$= -\frac{\cancel{(q-p)}\cancel{(p^2+pq+q^2)}(q)\cancel{(q+p)}}{\cancel{(q+p)}\cancel{(q-p)}(p)\cancel{(p^2+pq+q^2)}}$$

$$= -\frac{q}{p}$$

65.

$$\frac{y^3-x^3}{2x^2+2xy+x+y}\bullet\frac{2x^2-5x-3}{yx-3y-x^2+3x}$$

$$=\frac{(y-x)(y^2+xy+x^2)}{2x(x+y)+1(x+y)}\bullet\frac{(2x+1)(x-3)}{y(x-3)-x(x-3)}$$

$$=\frac{\cancel{(y-x)}(y^2+xy+x^2)}{(x+y)\cancel{(2x+1)}}\bullet\frac{\cancel{(2x+1)}\cancel{(x-3)}}{\cancel{(x-3)}\cancel{(y-x)}}$$

$$=\frac{y^2+xy+x^2}{x+y}$$

67.

$$\left(x^2+x-2cx-2c\right)\bullet\frac{x^2+3x+2}{x^2-4c^2}$$

$$=\frac{\left(x^2+x-2cx-2c\right)\left(x^2+3x+2\right)}{x^2-4c^2}$$

$$=\frac{\left(x(x+1)-2c(x+1)\right)(x+2)(x+1)}{(x+2c)(x-2c)}$$

$$=\frac{(x+1)\cancel{(x-2c)}(x+2)(x+1)}{(x+2c)\cancel{(x-2c)}}$$

$$=\frac{(x+1)^2(x+2)}{(x+2c)}$$

69.

$$\frac{15c^2d^3}{8x}\div\frac{25cd^4x}{16}\bullet\frac{5x^2}{4d}$$

$$=\frac{15c^2d^3}{8x}\bullet\frac{16}{25cd^4x}\bullet\frac{5x^2}{4d}$$

$$=\frac{3\cdot\cancel{5}\cdot\cancel{c}\cdot c\cdot\cancel{d}\cdot\cancel{d}\cdot\cancel{d}\cdot\cancel{2}\cdot\cancel{2}\cdot\cancel{2}\cdot\cancel{2}\cdot\cancel{5}\cdot\cancel{x}\cdot\cancel{x}}{\cancel{2}\cdot\cancel{2}\cdot\cancel{2}\cdot\cancel{x}\cdot\cancel{5}\cdot\cancel{5}\cdot\cancel{5}\cdot\cancel{c}\cdot\cancel{d}\cdot\cancel{d}\cdot\cancel{d}\cdot d\cdot\cancel{x}\cdot\cancel{2}\cdot 2\cdot d}$$

$$=\frac{3c}{2d^2}$$

71.

$$\frac{4x^2-10x+6}{x^4-3x^3}\div\frac{2x-3}{2x^3}\bullet\frac{x-3}{2-2x}$$

$$=\frac{4x^2-10x+6}{x^4-3x^3}\bullet\frac{2x^3}{2x-3}\bullet\frac{x-3}{2-2x}$$

$$=\frac{2(2x-3)(x-1)}{x\bullet x\bullet x(x-3)}\bullet\frac{2\bullet x\bullet x\bullet x}{(2x-3)}\bullet\frac{(x-3)}{2(1-x)}$$

$$=\frac{\cancel{2}\cancel{(2x-3)}\cancel{(x-1)}}{\cancel{x}\bullet\cancel{x}\bullet\cancel{x}\cancel{(x-3)}}\bullet\frac{2\bullet\cancel{x}\bullet\cancel{x}\bullet\cancel{x}}{\cancel{(2x-3)}}\bullet\frac{\cancel{(x-3)}}{-\cancel{2}\cancel{(x-1)}}$$

$$=-2$$

73.

$$\frac{2x^2+x-1}{x^2-1}\div\left(\frac{x^2+2x-35}{x^2-6x+5}\div\frac{x^2-9x+14}{2x^2-5x+2}\right)$$

$$=\frac{2x^2+x-1}{x^2-1}\div\left(\frac{x^2+2x-35}{x^2-6x+5}\bullet\frac{2x^2-5x+2}{x^2-9x+14}\right)$$

$$=\frac{2x^2+x-1}{x^2-1}\div\left(\frac{(x+7)\cancel{(x-5)}}{(x-1)\cancel{(x-5)}}\bullet\frac{(2x-1)\cancel{(x-2)}}{(x-7)\cancel{(x-2)}}\right)$$

$$=\frac{2x^2+x-1}{x^2-1}\div\frac{(x+7)(2x-1)}{(x-1)(x-7)}$$

$$=\frac{2x^2+x-1}{x^2-1}\bullet\frac{(x-1)(x-7)}{(x+7)(2x-1)}$$

$$=\frac{\cancel{(2x-1)}\cancel{(x+1)}}{\cancel{(x+1)}\cancel{(x-1)}}\bullet\frac{\cancel{(x-1)}(x-7)}{(x+7)\cancel{(2x-1)}}$$

$$=\frac{x-7}{x+7}$$

75.

$$\frac{3}{a+b}-\frac{a}{a+b}=\frac{3-a}{a+b}$$

77.

$$\frac{3x}{2x+2}+\frac{x+4}{2x+2}=\frac{3x+x+4}{2x+2}$$

$$=\frac{4x+4}{2x+2}$$

$$=\frac{\overset{2}{\cancel{4}}\cancel{(x+1)}}{\cancel{2}\cancel{(x+1)}}$$

$$=2$$

79.

$$\frac{5x}{x+1}+\frac{3}{x+1}-\frac{2x}{x+1}=\frac{5x+3-2x}{x+1}$$

$$=\frac{3x+3}{x+1}$$

$$=\frac{3\cancel{(x+1)}}{\cancel{x+1}}$$

$$=3$$

81.

$$\frac{3y-2}{2y+6}-\frac{2y-5}{2y+6}=\frac{3y-2-(2y-5)}{2y+6}$$
$$=\frac{3y-2-2y+5}{2y+6}$$
$$=\frac{y+3}{2y+6}$$
$$=\frac{\cancel{y+3}}{2\cancel{(y+3)}}$$
$$=\frac{1}{2}$$

83.

$$\frac{2x+1}{x^4-81}+\frac{2-x}{x^4-81}=\frac{2x+1+2-x}{x^4-81}$$
$$=\frac{x+3}{x^4-81}$$
$$=\frac{x+3}{(x^2+9)(x^2-9)}$$
$$=\frac{\cancel{x+3}}{(x^2+9)\cancel{(x+3)}(x-3)}$$
$$=\frac{1}{(x^2+9)(x-3)}$$

85.

$$\frac{3}{4x}+\frac{2}{3x}=\frac{3}{4x}\left(\frac{3}{3}\right)+\frac{2}{3x}\left(\frac{4}{4}\right)$$
$$=\frac{9}{12x}+\frac{8}{12x}$$
$$=\frac{9+8}{12x}$$
$$=\frac{17}{12x}$$

87.

$$\frac{3}{x+2}+\frac{5}{x-4}=\frac{3}{x+2}\left(\frac{x-4}{x-4}\right)+\frac{5}{x-4}\left(\frac{x+2}{x+2}\right)$$
$$=\frac{3(x-4)}{(x+2)(x-4)}+\frac{5(x+2)}{(x+2)(x-4)}$$
$$=\frac{3x-12}{(x+2)(x-4)}+\frac{5x+10}{(x+2)(x-4)}$$
$$=\frac{3x-12+5x+10}{(x+2)(x-4)}$$
$$=\frac{8x-2}{(x+2)(x-4)}$$
$$=\frac{2(4x-1)}{(x+2)(x-4)}$$

89.

$$\frac{x+2}{x+5}-\frac{x-3}{x+7}$$
$$=\frac{x+2}{x+5}\left(\frac{x+7}{x+7}\right)-\frac{x-3}{x+7}\left(\frac{x+5}{x+5}\right)$$
$$=\frac{(x+2)(x+7)}{(x+5)(x+7)}-\frac{(x-3)(x+5)}{(x+5)(x+7)}$$
$$=\frac{x^2+7x+2x+14}{(x+5)(x+7)}-\frac{x^2+5x-3x-15}{(x+5)(x+7)}$$
$$=\frac{x^2+9x+14}{(x+5)(x+7)}-\frac{x^2+2x-15}{(x+5)(x+7)}$$
$$=\frac{x^2+9x+14-(x^2+2x-15)}{(x+5)(x+7)}$$
$$=\frac{x^2+9x+14-x^2-2x+15}{(x+5)(x+7)}$$
$$=\frac{7x+29}{(x+5)(x+7)}$$

91.

$$\frac{x+8}{x-3}-\frac{x-14}{3-x}=\frac{x+8}{x-3}+\frac{x-14}{x-3}$$
$$=\frac{x+8+x-14}{x-3}$$
$$=\frac{2x-6}{x-3}$$
$$=\frac{2\cancel{(x-3)}}{\cancel{x-3}}$$
$$=2$$

93.

$$\frac{a^2+ab}{a^3-b^3}-\frac{b^2}{b^3-a^3}=\frac{a^2+ab}{a^3-b^3}+\frac{b^2}{a^3-b^3}$$
$$=\frac{a^2+ab+b^2}{a^3-b^3}$$
$$=\frac{\cancel{a^2+ab+b^2}}{(a-b)\left(\cancel{a^2+ab+b^2}\right)}$$
$$=\frac{1}{a-b}$$

95.

$$\frac{x+5}{xy}-\frac{x-1}{x^2y}=\frac{x+5}{xy}\left(\frac{x}{x}\right)-\frac{x-1}{x^2y}$$
$$=\frac{x^2+5x}{x^2y}-\frac{x-1}{x^2y}$$
$$=\frac{x^2+5x-(x-1)}{x^2y}$$
$$=\frac{x^2+5x-x+1}{x^2y}$$
$$=\frac{x^2+4x+1}{x^2y}$$

97.

$$\frac{x}{x^2+5x+6}+\frac{x}{x^2-4}$$
$$=\frac{x}{(x+2)(x+3)}+\frac{x}{(x+2)(x-2)}$$
$$=\frac{x}{(x+2)(x+3)}\left(\frac{x-2}{x-2}\right)+\frac{x}{(x+2)(x-2)}\left(\frac{x+3}{x+3}\right)$$
$$=\frac{x^2-2x}{(x+2)(x-2)(x+3)}+\frac{x^2+3x}{(x+2)(x-2)(x+3)}$$
$$=\frac{x^2-2x+x^2+3x}{(x+2)(x-2)(x+3)}$$
$$=\frac{2x^2+x}{(x+2)(x-2)(x+3)}$$
$$=\frac{x(2x+1)}{(x+2)(x-2)(x+3)}$$

99.

$$\frac{4}{x^2-2x-3}-\frac{x}{3x^2-7x-6}$$
$$=\frac{4}{(x-3)(x+1)}-\frac{x}{(3x+2)(x-3)}$$
$$=\frac{4}{(x-3)(x+1)}\left(\frac{3x+2}{3x+2}\right)-\frac{x}{(3x+2)(x-3)}\left(\frac{x+1}{x+1}\right)$$
$$=\frac{12x+8}{(x-3)(x+1)(3x+2)}-\frac{x^2+x}{(3x+2)(x-3)(x+1)}$$
$$=\frac{12x+8-\left(x^2+x\right)}{(x-3)(x+1)(3x+2)}$$
$$=\frac{12x+8-x^2-x}{(x-3)(x+1)(3x+2)}$$
$$=\frac{-x^2+11x+8}{(x-3)(x+1)(3x+2)}$$

101.

$$\frac{x}{x^2-4}-\frac{x}{x+2}+\frac{2}{x}$$
$$=\frac{x}{(x+2)(x-2)}-\frac{x}{x+2}+\frac{2}{x}$$
$$=\frac{x}{(x+2)(x-2)}\left(\frac{x}{x}\right)-\frac{x}{x+2}\left(\frac{x(x-2)}{x(x-2)}\right)+\frac{2}{x}\left(\frac{(x+2)(x-2)}{(x+2)(x-2)}\right)$$
$$=\frac{x^2}{x(x+2)(x-2)}-\frac{x^2(x-2)}{x(x+2)(x-2)}+\frac{2\left(x^2-4\right)}{x(x+2)(x-2)}$$
$$=\frac{x^2}{x(x+2)(x-2)}-\frac{x^3-2x^2}{x(x+2)(x-2)}+\frac{2x^2-8}{x(x+2)(x-2)}$$
$$=\frac{x^2-\left(x^3-2x^2\right)+2x^2-8}{x(x+2)(x-2)}$$
$$=\frac{x^2-x^3+2x^2+2x^2-8}{x(x+2)(x-2)}$$
$$=\frac{-x^3+5x^2-8}{x(x+2)(x-2)}$$

103.

$$1+x-\frac{x}{x-5}=1\left(\frac{x-5}{x-5}\right)+x\left(\frac{x-5}{x-5}\right)-\frac{x}{x-5}$$
$$=\frac{x-5}{x-5}+\frac{x^2-5x}{x-5}-\frac{x}{x-5}$$
$$=\frac{x-5+x^2-5x-x}{x-5}$$
$$=\frac{x^2-5x-5}{x-5}$$

105.

$$\frac{2}{x-1}-\frac{2x}{x^2-1}-\frac{x}{x^2+2x+1}$$
$$=\frac{2}{x-1}-\frac{2x}{(x+1)(x-1)}-\frac{x}{(x+1)(x+1)}$$
$$=\frac{2}{x-1}\left(\frac{(x+1)(x+1)}{(x+1)(x+1)}\right)-\frac{2x}{(x+1)(x-1)}\left(\frac{x+1}{x+1}\right)-\frac{x}{(x+1)(x+1)}\left(\frac{x-1}{x-1}\right)$$
$$=\frac{2(x^2+2x+1)}{(x-1)(x+1)^2}-\frac{2x(x+1)}{(x-1)(x+1)^2}-\frac{x(x-1)}{(x-1)(x+1)^2}$$
$$=\frac{2x^2+4x+2}{(x-1)(x+1)^2}-\frac{2x^2+2x}{(x-1)(x+1)^2}-\frac{x^2-x}{(x-1)(x+1)^2}$$
$$=\frac{2x^2+4x+2-(2x^2+2x)-(x^2-x)}{(x-1)(x+1)^2}$$
$$=\frac{2x^2+4x+2-2x^2-2x-x^2+x}{(x-1)(x+1)^2}$$
$$=\frac{-x^2+3x+2}{(x-1)(x+1)^2}$$

107.

$$\frac{\frac{4x}{y}}{\frac{6xz}{y^2}}=\frac{4x}{y}\div\frac{6xz}{y^2}$$
$$=\frac{4x}{y}\cdot\frac{y^2}{6xz}$$
$$=\frac{\cancel{2}\cdot 2\cdot\cancel{x}\cdot\cancel{y}\cdot y}{\cancel{y}\cdot 3\cdot\cancel{2}\cdot\cancel{x}\cdot z}$$
$$=\frac{2y}{3z}$$

109.

$$\frac{\frac{x-y}{xy}}{\frac{y-x}{x}}=\frac{x-y}{xy}\div\frac{y-x}{x}$$
$$=\frac{x-y}{xy}\cdot\frac{x}{y-x}$$
$$=\frac{x(x-y)}{xy(y-x)}$$
$$=-\frac{\cancel{x}\,\cancel{(x-y)}}{\cancel{x}\,y\,\cancel{(x-y)}}$$
$$=-\frac{1}{y}$$

111.

$$\frac{\frac{ac-ad-c+d}{a^3-1}}{\frac{c^2-2cd+d^2}{a^2+a+1}}=\frac{ac-ad-c+d}{a^3-1}\div\frac{c^2-2cd+d^2}{a^2+a+1}$$
$$=\frac{ac-ad-c+d}{a^3-1}\cdot\frac{a^2+a+1}{c^2-2cd+d^2}$$
$$=\frac{a(c-d)-1(c-d)}{(a-1)(a^2+a+1)}\cdot\frac{(a^2+a+1)}{(c-d)(c-d)}$$
$$=\frac{\cancel{(a-1)}\,\cancel{(c-d)}\,\cancel{(a^2+a+1)}}{\cancel{(a-1)}\,\cancel{(a^2+a+1)}\,\cancel{(c-d)}\,(c-d)}$$
$$=\frac{1}{c-d}$$

113.

$$\frac{\frac{1}{a}+\frac{1}{b}}{\frac{1}{a}}=\frac{\frac{1}{a}+\frac{1}{b}}{\frac{1}{a}}\cdot\frac{ab}{ab}$$
$$=\frac{\frac{ab}{a}+\frac{ab}{b}}{\frac{ab}{a}}$$
$$=\frac{b+a}{b}$$

115.

$$\frac{\frac{y}{x}-\frac{x}{y}}{\frac{1}{x}+\frac{1}{y}}=\frac{\frac{y}{x}-\frac{x}{y}}{\frac{1}{x}+\frac{1}{y}}\bullet\frac{xy}{xy}$$

$$=\frac{\frac{xy^2}{x}-\frac{x^2y}{y}}{\frac{xy}{x}+\frac{xy}{y}}$$

$$=\frac{y^2-x^2}{y+x}$$

$$=\frac{(y-x)\cancel{(y+x)}}{\cancel{y+x}}$$

$$=y-x$$

117.

$$\frac{1+\frac{6}{x}+\frac{8}{x^2}}{1+\frac{1}{x}-\frac{12}{x^2}}=\frac{1+\frac{6}{x}+\frac{8}{x^2}}{1+\frac{1}{x}-\frac{12}{x^2}}\bullet\frac{x^2}{x^2}$$

$$=\frac{x^2+\frac{6x^2}{x}+\frac{8x^2}{x^2}}{x^2+\frac{x^2}{x}-\frac{12x^2}{x^2}}$$

$$=\frac{x^2+6x+8}{x^2+x-12}$$

$$=\frac{(x+2)\cancel{(x+4)}}{(x-3)\cancel{(x+4)}}$$

$$=\frac{x+2}{x-3}$$

119.

$$\frac{\frac{1}{a+1}+1}{\frac{3}{a-1}+1}=\frac{\frac{1}{a+1}+1}{\frac{3}{a-1}+1}\bullet\frac{(a+1)(a-1)}{(a+1)(a-1)}$$

$$=\frac{\frac{(a+1)(a-1)}{a+1}+(a+1)(a-1)}{\frac{3(a+1)(a-1)}{a-1}+(a+1)(a-1)}$$

$$=\frac{a-1+a^2-1}{3(a+1)+a^2-1}$$

$$=\frac{a^2+a-2}{3a+3+a^2-1}$$

$$=\frac{a^2+a-2}{a^2+3a+2}$$

$$=\frac{\cancel{(a+2)}(a-1)}{\cancel{(a+2)}(a+1)}$$

$$=\frac{a-1}{a+1}$$

121.

$$\frac{x^{-1}+y^{-1}}{x^{-1}-y^{-1}}=\frac{\frac{1}{x}+\frac{1}{y}}{\frac{1}{x}-\frac{1}{y}}$$

$$=\frac{\frac{1}{x}+\frac{1}{y}}{\frac{1}{x}-\frac{1}{y}}\bullet\frac{xy}{xy}$$

$$=\frac{\frac{xy}{x}+\frac{xy}{y}}{\frac{xy}{x}-\frac{xy}{y}}$$

$$=\frac{y+x}{y-x}$$

123.

$$\frac{\frac{2}{x+3}-\frac{1}{x-3}}{\frac{3}{x^2-9}}$$

$$=\frac{\frac{2}{x+3}-\frac{1}{x-3}}{\frac{3}{(x+3)(x-3)}}$$

$$=\frac{\frac{2}{x+3}-\frac{1}{x-3}}{\frac{3}{(x+3)(x-3)}}\bullet\frac{(x+3)(x-3)}{(x+3)(x-3)}$$

$$=\frac{\frac{2(x+3)(x-3)}{x+3}-\frac{1(x+3)(x-3)}{x-3}}{\frac{3(x+3)(x-3)}{(x+3)(x-3)}}$$

$$=\frac{2(x-3)-(x+3)}{3}$$

$$=\frac{2x-6-x-3}{3}$$

$$=\frac{x-9}{3}$$

125.

$$\frac{\frac{h}{h^2+3h+2}}{\frac{4}{h+2}-\frac{4}{h+1}}=\frac{\frac{h}{(h+2)(h+1)}}{\frac{4}{h+2}-\frac{4}{h+1}}$$

$$=\frac{\frac{h}{(h+2)(h+1)}}{\frac{4}{h+2}-\frac{4}{h+1}}\bullet\frac{(h+2)(h+1)}{(h+2)(h+1)}$$

$$=\frac{\frac{h(h+2)(h+1)}{(h+2)(h+1)}}{\frac{4(h+2)(h+1)}{h+2}-\frac{4(h+2)(h+1)}{h+1}}$$

$$=\frac{h}{4(h+1)-4(h+2)}$$

$$=\frac{h}{4h+4-4h-8}$$

$$=\frac{h}{-4}$$

$$=-\frac{h}{4}$$

127.

$$\frac{1+\frac{a}{b}}{1-\frac{a}{1-\frac{a}{b}}}=\frac{\frac{b}{b}+\frac{a}{b}}{1-\frac{a}{\frac{b}{b}-\frac{a}{b}}}$$

$$=\frac{\frac{b+a}{b}}{1-\frac{a}{\frac{b-a}{b}}}$$

$$=\frac{\frac{b+a}{b}}{1-\frac{ab}{b-a}}$$

$$=\frac{\frac{b+a}{b}}{\frac{b-a}{b-a}-\frac{ab}{b-a}}$$

$$=\frac{\frac{b+a}{b}}{\frac{b-a-ab}{b-a}}$$

$$=\frac{b+a}{b}\div\frac{b-a-ab}{b-a}$$

$$=\frac{b+a}{b}\bullet\frac{b-a}{b-a-ab}$$

$$=\frac{(b+a)(b-a)}{b(b-a-ab)}$$

APPLICATIONS

129. PHYSICS

Trial 1:

$$D = rt$$
$$= \left(\frac{k_1^2 + 3k_1 + 2}{k_1 - 3}\right)\left(\frac{k_1^2 - 3k_1}{k_1 + 1}\right)$$
$$= \frac{\cancel{(k_1+1)}(k_1+2)(k_1)\cancel{(k_1-3)}}{\cancel{(k_1-3)}\cancel{(k_1+1)}}$$
$$= k_1(k_1+2)$$

Trial 2:

$$t = D \div r$$
$$= (k_2^2 + 11k_2 + 30) \div \left(\frac{k_2^2 + 6k_2 + 5}{k_2 + 1}\right)$$
$$= \frac{k_2^2 + 11k_2 + 30}{1} \bullet \frac{k_2 + 1}{k_2^2 + 6k_2 + 5}$$
$$= \frac{\cancel{(k_2+5)}(k_2+6)\cancel{(k_2+1)}}{\cancel{(k_2+5)}\cancel{(k_2+1)}}$$
$$= k_2 + 6$$

131. THE AMAZON

a.

	Rate	Time	Distance
downriver	$r+5$	$\frac{3}{r+5}$	3
upriver	$r-5$	$\frac{3}{r-5}$	3

b.

$$\frac{3}{r-5} - \frac{3}{r+5} = \frac{3}{r-5}\left(\frac{r+5}{r+5}\right) - \frac{3}{r+5}\left(\frac{r-5}{r-5}\right)$$
$$= \frac{3r+15}{(r+5)(r-5)} - \frac{3r-15}{(r+5)(r-5)}$$
$$= \frac{3r+15-(3r-15)}{(r+5)(r-5)}$$
$$= \frac{3r+15-3r+15}{(r+5)(r-5)}$$
$$= \frac{30}{(r+5)(r-5)}$$

WRITING

133. Answers will vary.

135. Answers will vary.

REVIEW

137. Let the integers be x and y.

$$\begin{cases} x + y = 38 \\ x - y = 12 \end{cases}$$

$$x + y = 38$$
$$x - y = 12$$
$$2x = 50$$
$$x = 25$$

$$x + y = 38$$
$$25 + y = 38$$
$$y = 13$$

The two integers are 25 and 13.

CHALLENGE PROBLEMS

139.

$$\frac{a^6 - 64}{(a^2 + 2a + 4)(a^2 - 2a + 4)}$$
$$= \frac{(a^3 + 8)(a^3 - 8)}{(a^2 + 2a + 4)(a^2 - 2a + 4)}$$
$$= \frac{(a+2)\cancel{(a^2 - 2a + 4)}(a-2)\cancel{(a^2 + 2a + 4)}}{\cancel{(a^2 + 2a + 4)}\cancel{(a^2 - 2a + 4)}}$$
$$= (a+2)(a-2)$$

141.

$$\left[\left(x^{-1}+1\right)^{-1}+1\right]^{-1}=\frac{1}{\left[\left(x^{-1}+1\right)^{-1}+1\right]}$$

$$=\frac{1}{\frac{1}{\left(x^{-1}+1\right)}+1}$$

$$=\frac{1}{\frac{1}{\frac{1}{x}+1}+1}$$

$$=\frac{1}{\frac{1}{\frac{1}{x}+\frac{1}{1}\bullet\frac{x}{x}}+1}$$

$$=\frac{1}{\frac{1}{\frac{1}{x}+\frac{x}{x}}+1}$$

$$=\frac{1}{\frac{1}{\frac{x+1}{x}}+1}$$

$$=\frac{1}{\frac{x}{x+1}+1}$$

$$=\frac{1}{\frac{x}{x+1}+\frac{1}{1}\bullet\frac{x+1}{x+1}}$$

$$=\frac{1}{\frac{x}{x+1}+\frac{x+1}{x+1}}$$

$$=\frac{1}{\frac{2x+1}{x+1}}$$

$$=\frac{x+1}{2x+1}$$

SECTION 8.6

VOCABULARY

1. $y = 3x - 1$ is an equation in **two** variables, x and y.

3. Solutions of equations in two variables are often listed in a **table** of solutions.

5. The equation $y = 3x + 8$ is said to be **linear** because its graph is a line.

7. The point where a graph intersects the y–axis is called the **y–intercept** and the point where it intersects the x–axis is called the **x–intercept**.

CONCEPTS

9. The graph of any equation of the form $x = a$ is a **vertical** line.

11. The formula to compute slope is $m = \dfrac{y_2 - y_1}{x_2 - x_1}$.

13. The graph of the equation $y = 0$ is the **x–axis**. The graph of the equation $x = 0$ is the **y–axis**.

15. **Parallel** lines have the same slope.

17. The point–slope form of the equation of a line is $y - y_1 = m(x - x_1)$.

19. The general form of the equation of a line is $Ax + By = C$.

21. a. The x–intercept is the point the line crosses the x–axis: (–3, 0).
The y–intercept is the point the line crosses the y–axis: (0, 4).
b. False

NOTATION

23. The symbol x_1 is read as x **sub 1**.

PRACTICE

25. $y = x - 2$

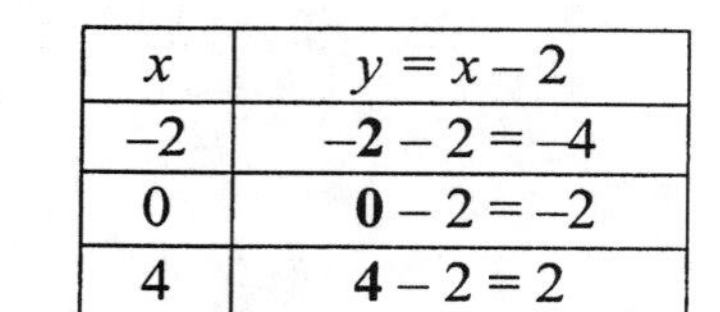

x	$y = x - 2$
–2	**–2 – 2 = –4**
0	**0 – 2 = –2**
4	**4 – 2 = 2**

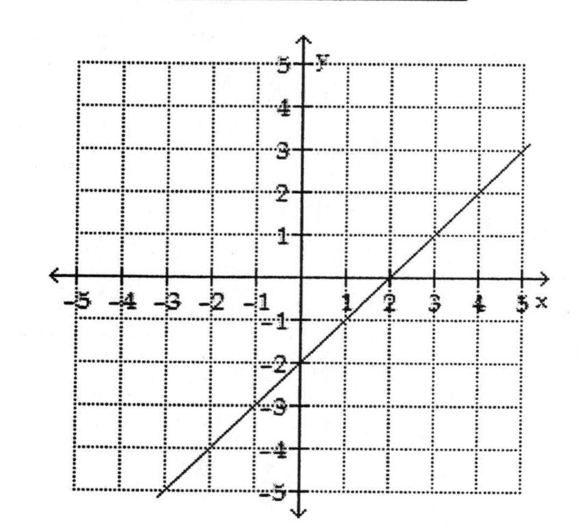

27. $y = x$

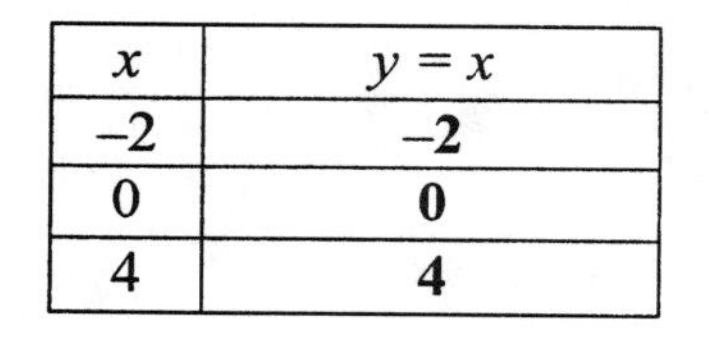

x	$y = x$
–2	**–2**
0	**0**
4	**4**

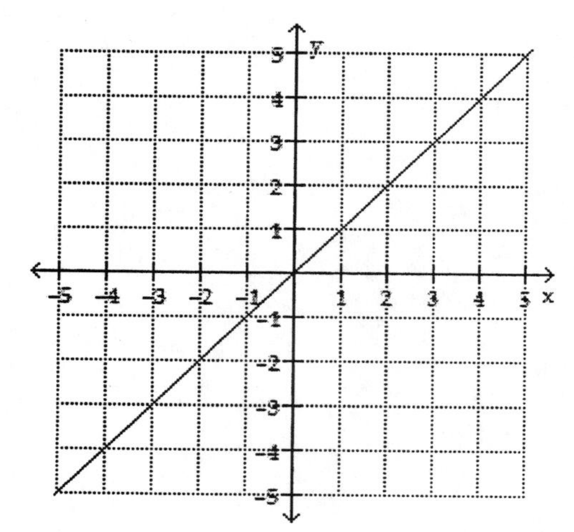

29. $y = 2x - 3$

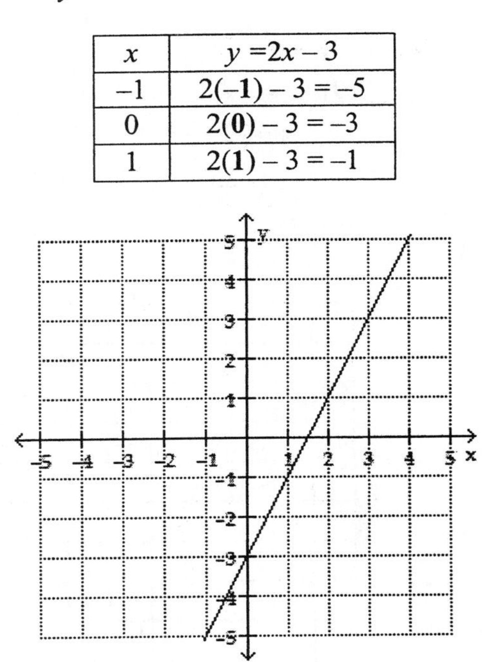

x	$y = 2x - 3$
-1	$2(\mathbf{-1}) - 3 = -5$
0	$2(\mathbf{0}) - 3 = -3$
1	$2(\mathbf{1}) - 3 = -1$

31. $y = -\frac{1}{3}x - 1$

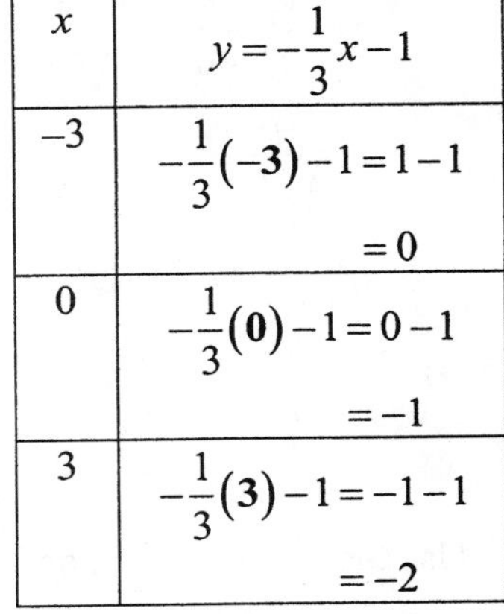

x	$y = -\frac{1}{3}x - 1$
-3	$-\frac{1}{3}(\mathbf{-3}) - 1 = 1 - 1$ $= 0$
0	$-\frac{1}{3}(\mathbf{0}) - 1 = 0 - 1$ $= -1$
3	$-\frac{1}{3}(\mathbf{3}) - 1 = -1 - 1$ $= -2$

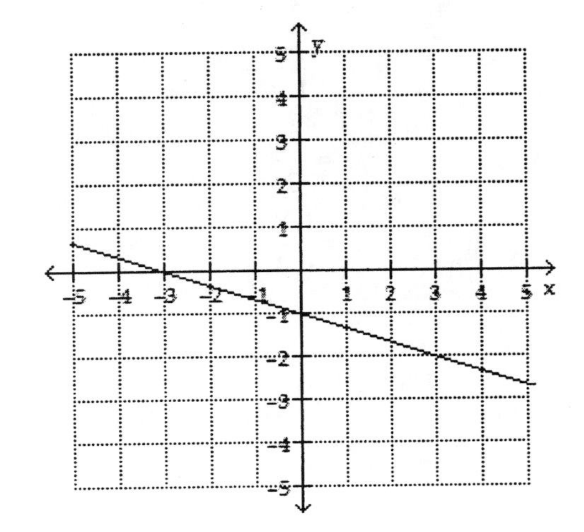

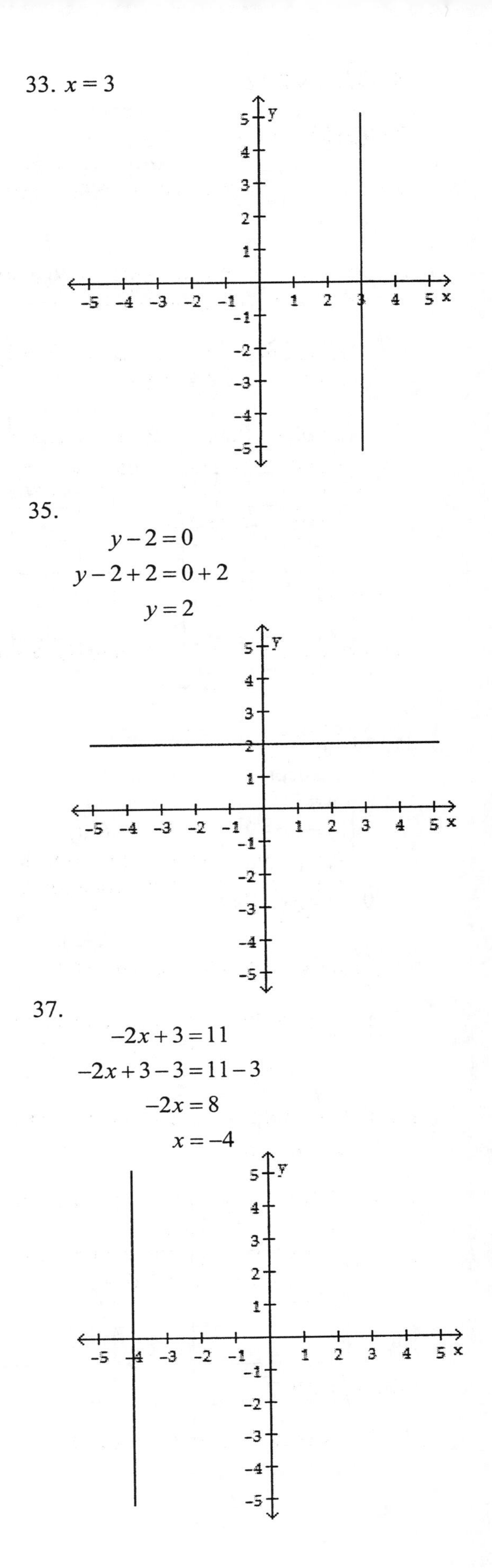

33. $x = 3$

35.

$$y - 2 = 0$$
$$y - 2 + 2 = 0 + 2$$
$$y = 2$$

37.

$$-2x + 3 = 11$$
$$-2x + 3 - 3 = 11 - 3$$
$$-2x = 8$$
$$x = -4$$

39. $3x + 4y = 12$

x–intercept:	y–intercept:
Let $y = 0$.	Let $x = 0$.
$3x + 4(0) = 12$	$3(0) + 4y = 12$
$3x = 12$	$4y = 12$
$x = 4$	$y = 3$
$(4,0)$	$(0,3)$

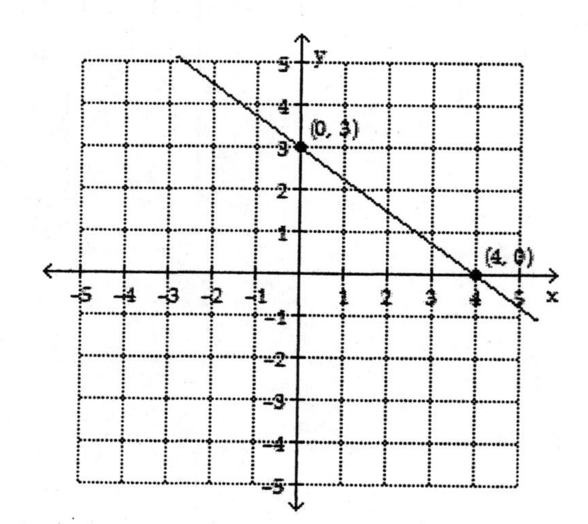

41. $3y = 6x - 9$

x–intercept:	y–intercept:
Let $y = 0$.	Let $x = 0$.
$3(0) = 6x - 9$	$3y = 6(0) - 9$
$0 = 6x - 9$	$3y = -9$
$0 + 9 = 6x - 9 + 9$	$y = -3$
$9 = 6x$	$(0,-3)$
$\frac{9}{6} = x$	
$\frac{3}{2} = x$	
$\left(\frac{3}{2}, 0\right)$	

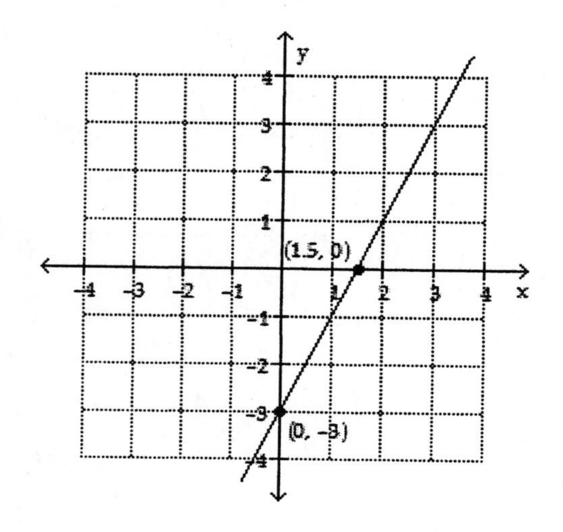

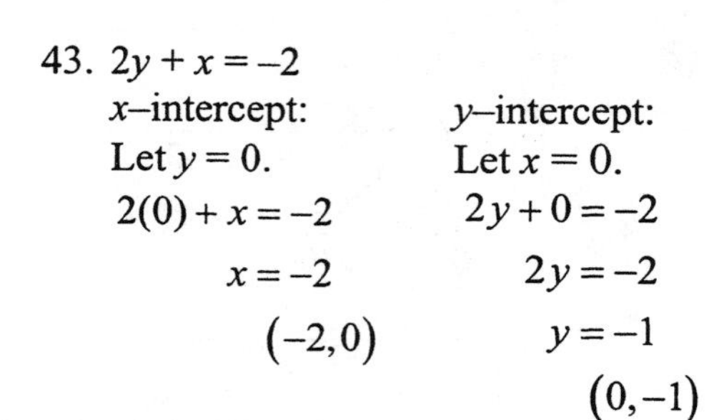

43. $2y + x = -2$

x–intercept:	y–intercept:
Let $y = 0$.	Let $x = 0$.
$2(0) + x = -2$	$2y + 0 = -2$
$x = -2$	$2y = -2$
$(-2,0)$	$y = -1$
	$(0,-1)$

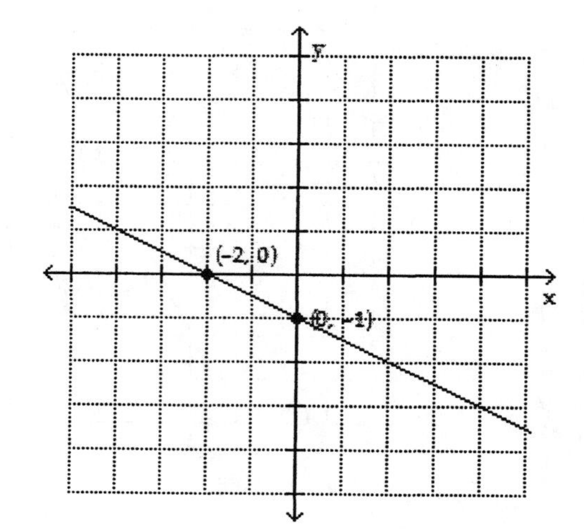

45. Use the points (5, 0) and (–2, –6).

$$m = \frac{y_2 - y_1}{x_2 - x_1} = \frac{-6 - 0}{-2 - 5} = \frac{-6}{-7} = \frac{6}{7}$$

47. Use the points (–2, 1) and (1, –7).

$$m = \frac{y_2 - y_1}{x_2 - x_1} = \frac{-7 - 1}{1 - (-2)} = \frac{-8}{3} = -\frac{8}{3}$$

49. (0, 0), (3, 9)

$$m = \frac{y_2 - y_1}{x_2 - x_1} = \frac{9-0}{3-0} = \frac{9}{3} = 3$$

51. (–1, 8), (6, 1)

$$m = \frac{y_2 - y_1}{x_2 - x_1} = \frac{1-8}{6-(-1)} = \frac{-7}{7} = -1$$

53. (3, –1), (–6, 2)

$$m = \frac{y_2 - y_1}{x_2 - x_1} = \frac{2-(-1)}{-6-3} = \frac{3}{-9} = -\frac{1}{3}$$

55. (–7, 5), (–7, 2)

$$m = \frac{y_2 - y_1}{x_2 - x_1} = \frac{2-5}{-7-(-7)} = \frac{-3}{0} = \text{undefined}$$

57. Write the equation in the form $y = mx + b$.

$$3x + 2y = 12$$
$$2y = -3x + 12$$
$$y = -\frac{3}{2}x + 6$$
$$m = -\frac{3}{2}$$

59. Write the equation in the form $y = mx + b$.

$$3x = 4y - 2$$
$$3x + 2 = 4y$$
$$4y = 3x + 2$$
$$y = \frac{3}{4}x + \frac{1}{2}$$
$$m = \frac{3}{4}$$

61.

$$3x - y = 4 \qquad 3x - y = 7$$
$$y = 3x - 4 \qquad y = 3x - 7$$
$$m = 3 \qquad m = 3$$

parallel lines

63.

$$x + y = 2 \qquad y = x + 5$$
$$y = -x + 2 \qquad m = 1$$
$$m = -1$$

perpendicular lines

65.

$$9x - 12y = 17 \qquad 3x - 4y = 17$$
$$-12y = -9x + 17 \qquad -4y = -3x + 17$$
$$y = \frac{-9}{-12}x + \frac{17}{-12} \qquad y = \frac{-3}{-4}x + \frac{17}{-4}$$
$$y = \frac{3}{4}x - \frac{17}{12} \qquad y = \frac{3}{4}x - \frac{17}{4}$$
$$m = \frac{3}{4} \qquad m = \frac{3}{4}$$

parallel lines

67.

$$y - y_1 = m(x - x_1)$$
$$y - 7 = 5(x - 0)$$
$$y - 7 = 5x - 0$$
$$-7 = 5x - y$$
$$5x - y = -7$$

69. Find m.

$$m = \frac{y_2 - y_1}{x_2 - x_1}$$
$$= \frac{4-0}{4-0}$$
$$= \frac{4}{4}$$
$$= 1$$

Use the point (0, 0) and $m = 1$.

$$y - y_1 = m(x - x_1)$$
$$y - 0 = 1(x - 0)$$
$$y = 1x - 0$$
$$0 = x - y$$
$$x - y = 0$$

71. Find m.

$$m = \frac{y_2 - y_1}{x_2 - x_1}$$
$$= \frac{-3-4}{0-3}$$
$$= \frac{-7}{-3}$$
$$= \frac{7}{3}$$

Use the point (0, –3) and $m = \frac{7}{3}$.

$$y - y_1 = m(x - x_1)$$
$$y - (-3) = \frac{7}{3}(x - 0)$$
$$y + 3 = \frac{7}{3}x - 0$$
$$3 = \frac{7}{3}x - y$$
$$\frac{7}{3}x - y = 3$$
$$3\left(\frac{7}{3}x - y\right) = 3(3)$$
$$7x - 3y = 9$$

73. $y = mx + b$
$y = 3x + 17$

75. Use $m = -7, x = 7, y = 5$ to find b.

$$y = mx + b$$
$$5 = -7(7) + b$$
$$5 = -49 + b$$
$$5 + 49 = -49 + b + 49$$
$$54 = b$$

$m = -7$ and $b = 54$
$y = -7x + 54$

77. Find m.

$$m = \frac{y_2 - y_1}{x_2 - x_1}$$
$$= \frac{10-8}{2-6}$$
$$= \frac{2}{-4}$$
$$= -\frac{1}{2}$$

Use $m = -\frac{1}{2}, x = 2, y = 10$ to find b.

$$y = mx + b$$
$$10 = -\frac{1}{2}(2) + b$$
$$10 = -1 + b$$
$$10 + 1 = -1 + b + 1$$
$$11 = b$$

$m = -\frac{1}{2}$ and $b = 11$

$y = -\frac{1}{2}x + 11$

79. The slopes of parallel lines are the same. Since the slope of $y = 4x - 7$ is 4, the slope of the line parallel would also be 4. Use $m = 4$ and (0, 0) to find the equation of the line.

$$y - y_1 = m(x - x_1)$$
$$y - 0 = 4(x - 0)$$
$$y = 4x - 0$$
$$y = 4x$$

81. Find the slope of the given line.

$$4x - y = 7$$
$$4x - y - 4x = 7 - 4x$$
$$-y = -4x + 7$$
$$\frac{-y}{-1} = \frac{-4x}{-1} + \frac{7}{-1}$$
$$y = 4x - 7$$
$$m = 4$$

Use $m = 4$ and (2, 5) to find the equation of the line.

$$y - y_1 = m(x - x_1)$$
$$y - 5 = 4(x - 2)$$
$$y - 5 = 4x - 8$$
$$y - 5 + 5 = 4x - 8 + 5$$
$$y = 4x - 3$$

83. The slopes of perpendicular lines are opposite reciprocals. Since the slope of $y = 4x - 7$ is 4, the slope of the line perpendicular would be $-\frac{1}{4}$.

Use $m = -\frac{1}{4}$ and (0, 0) to find the equation of the line.

$$y - y_1 = m(x - x_1)$$
$$y - 0 = -\frac{1}{4}(x - 0)$$
$$y = -\frac{1}{4}x - 0$$
$$y = -\frac{1}{4}x$$

85. Find the slope of the given line.

$$4x - y = 7$$
$$4x - y - 4x = 7 - 4x$$
$$-y = -4x + 7$$
$$\frac{-y}{-1} = \frac{-4x}{-1} + \frac{7}{-1}$$
$$y = 4x - 7$$
$$m = 4$$

opposite reciprocal $= -\frac{1}{4}$

Use $m = -\frac{1}{4}$ and (2, 5) to find the equation of the line.

$$y - y_1 = m(x - x_1)$$
$$y - 5 = -\frac{1}{4}(x - 2)$$
$$y - 5 = -\frac{1}{4}x + \frac{1}{2}$$
$$y - 5 + 5 = -\frac{1}{4}x + \frac{1}{2} + 5$$
$$y = -\frac{1}{4}x + \frac{11}{2}$$

87. $P(0, 0)$, $Q(6, 8)$

$$\text{midpoint} = \left(\frac{x_1 + x_2}{2}, \frac{y_1 + y_2}{2}\right)$$
$$= \left(\frac{0+6}{2}, \frac{0+8}{2}\right)$$
$$= \left(\frac{6}{2}, \frac{8}{2}\right)$$
$$= (3,4)$$

89. $P(6, 8)$, $Q(12, 16)$

$$\text{midpoint} = \left(\frac{x_1 + x_2}{2}, \frac{y_1 + y_2}{2}\right)$$
$$= \left(\frac{6+12}{2}, \frac{8+16}{2}\right)$$
$$= \left(\frac{18}{2}, \frac{24}{2}\right)$$
$$= (9,12)$$

91. $P(-2,-8)$, $Q(3,4)$

$$\text{midpoint} = \left(\frac{x_1+x_2}{2}, \frac{y_1+y_2}{2}\right)$$
$$= \left(\frac{-2+3}{2}, \frac{-8+4}{2}\right)$$
$$= \left(\frac{1}{2}, \frac{-4}{2}\right)$$
$$= \left(\frac{1}{2}, -2\right)$$

93. Let the coordinates of Q be (x_2, y_2).

$$\text{midpoint} = \left(\frac{x_1+x_2}{2}, \frac{y_1+y_2}{2}\right)$$
$$(-2,3) = \left(\frac{x_1+x_2}{2}, \frac{y_1+y_2}{2}\right) \text{ so}$$
$$-2 = \frac{x_1+x_2}{2} \quad \text{and} \quad 3 = \frac{y_1+y_2}{2}$$

x–coordinate:
$$-2 = \frac{-8+x_2}{2}$$
$$-4 = -8 + x_2$$
$$4 = x_2$$

y–coordinate:
$$3 = \frac{5+y_2}{2}$$
$$6 = 5 + y_2$$
$$1 = y_2$$

The coordinates of Q are (4, 1).

APPLICATIONS

95. HIGHWAY GRADES

Change 2.5 miles to feet by multiplying it by 5,280.

$$m = \frac{528 \text{ ft}}{2.5 \text{ mi}}$$
$$= \frac{528}{2.5(5,280)}$$
$$= \frac{528}{13,200}$$
$$= \frac{1}{25}$$
$$= 4\%$$

97. RATES OF GROWTH

The rate of growth is the slope of the line.

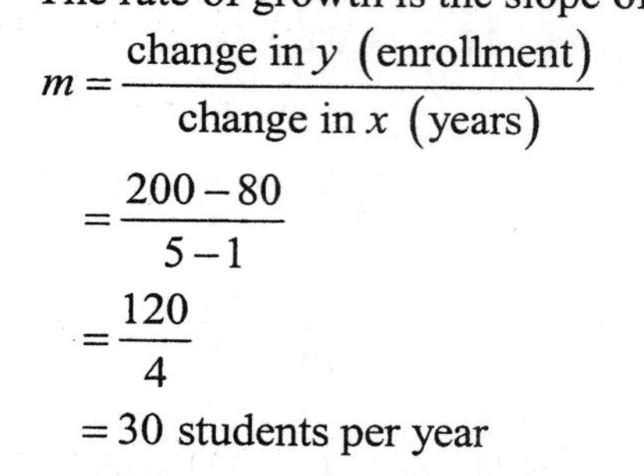

$$m = \frac{\text{change in } y \text{ (enrollment)}}{\text{change in } x \text{ (years)}}$$
$$= \frac{200-80}{5-1}$$
$$= \frac{120}{4}$$
$$= 30 \text{ students per year}$$

99. SALVAGE VALUES

Write ordered pairs in the form (age, value). Use (0, 19,984) for the purchase value and (8, 1,600) for the present value.

Find the rate of change (m).

$$m = \frac{y_2 - y_1}{x_2 - x_1}$$
$$= \frac{1,600 - 19,984}{8-0}$$
$$= \frac{-18,384}{8}$$
$$= -2,298$$

Use the point (0, 19,984) and $m = -2{,}298$.

$$y - y_1 = m(x - x_1)$$
$$y - 19,984 = -2,298(x-0)$$
$$y - 19,984 = -2,298x + 0$$
$$y = -2,298x + 19,984$$

101. CRIMINOLOGY

a. Use (77,000, 575) and $m = \frac{1}{100}$

$$B - B_1 = m(p - p_1)$$
$$B - 575 = \frac{1}{100}(p - 77{,}000)$$
$$B - 575 = \frac{1}{100}p - 770$$
$$B - 575 + 575 = \frac{1}{100}p - 770 + 575$$
$$B = \frac{1}{100}p - 195$$

b. Find B if $p = 110{,}000$

$$B = \frac{1}{100}p - 195$$
$$= \frac{1}{100}(110{,}000) - 195$$
$$= 1{,}100 - 195$$
$$= 905$$

WRITING

103. Answers will vary.

105. Answers will vary.

REVIEW

107. Let x = the amount invested at each %.

$$0.06x + 0.07x + 0.08x = 2{,}037$$
$$0.21x = 2{,}037$$
$$\frac{0.21x}{0.21} = \frac{2{,}037}{0.21}$$
$$x = \$9{,}700 \text{ in each account}$$
$$3(\$9{,}700) = \$29{,}100 \text{ in all}$$

CHALLENGE PROBLEMS

109. Find the slope of the line through (1, 3) and (–2, 7).

$$m = \frac{y_2 - y_1}{x_2 - x_1}$$
$$= \frac{7 - 3}{-2 - 1}$$
$$= \frac{4}{-3}$$
$$= -\frac{4}{3}$$

The slope of the line perpendicular must be the opposite reciprocal, or $\frac{3}{4}$. Use the ordered pairs (4, b) and (8, –1).

$$m = \frac{y_2 - y_1}{x_2 - x_1}$$
$$\frac{3}{4} = \frac{-1 - b}{8 - 4}$$
$$\frac{3}{4} = \frac{-1 - b}{4}$$
$$3(4) = 4(-1 - b)$$
$$12 = -4 - 4b$$
$$12 + 4 = -4 - 4b + 4$$
$$16 = -4b$$
$$-4 = b$$

SECTION 8.7

VOCABULARY

1. A **function** is a rule (or correspondence) that assigns to each value of one variable (called the independent variable) exactly **one** value of another variable (called the dependent variable).

3. We can think of a function as a machine that takes some **input** x and turns it into some **output** $f(x)$.

5. If $f(2) = -1$, we call -1 a function **value**.

CONCEPTS

7. YEAR COST PER LB

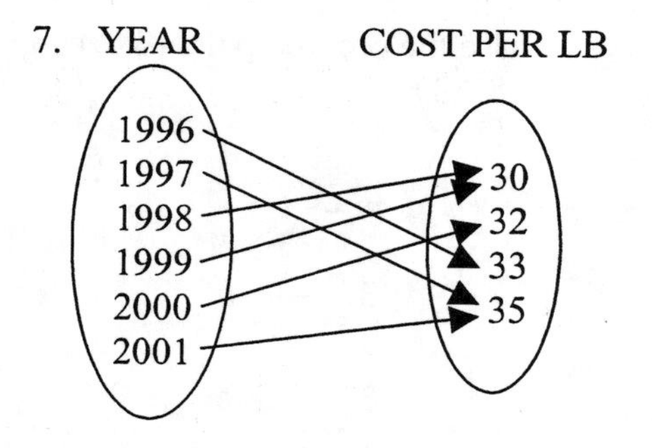

9. a. If $y = 5x + 1$, find the value of y when $x = 8$.
   b. If $f(x) = 5x + 1$, find $\boxed{f(8)}$.

11. $f(x) = 2x^2 - 1$

x	$f(x) = 2x^2 - 1$	
-3	$2(-3)^2 - 1 = 2(9) - 1$ $= 18 - 1$ $= 17$	$\rightarrow (-3, 17)$
0	$2(0)^2 - 1 = 2(0) - 1$ $= 0 - 1$ $= -1$	$\rightarrow (0, -1)$
2	$2(2)^2 - 1 = 2(4) - 1$ $= 8 - 1$ $= 7$	$\rightarrow (2, 7)$

13. a. $(-2, 4)$ and $(-2, -4)$
    b. No; the x–value -2 is assigned more than one y–value (4 and -4).

NOTATION

15. We read $f(x) = 5x - 6$ as "f $\boxed{\text{of}}$ x is $5x$ minus 6."

17. Since $y = \boxed{f(x)}$, the equations $y = 3x + 2$ and $f(x) = 3x + 2$ are equivalent.

19. When graphing the function $f(x) = -x + 5$, the vertical axis of the rectangular coordinate system can be labeled $\boxed{f(x)}$ or $\boxed{y}$.

PRACTICE

21. yes

23. no; (4, 2), (4, 4), (4, 6)

25. yes

27. no; (–1, 0) and (–1, 2)

29. no; (3, 4) and (3, –4) or (4, 3) and (4, –3)

31. yes

33. yes

35. no

37. no

39. $f(x) = 3x$

$$f(3) = 3(3) = 9 \qquad f(-1) = 3(-1) = -3$$

41. $f(x) = 2x - 3$

$$f(3) = 2(3) - 3 = 6 - 3 = 3 \qquad f(-1) = 2(-1) - 3 = -2 - 3 = -5$$

43. $f(x) = 7 + 5x$

$$f(3) = 7 + 5(3) = 7 + 15 = 22 \qquad f(-1) = 7 + 5(-1) = 7 - 5 = 2$$

45. $f(x)=9-2x$

$f(3)=9-2(3)$
$=9-6$
$=3$

$f(-1)=9-2(-1)$
$=9+2$
$=11$

47. $g(x)=x^2$

$g(2)=\mathbf{2}^2$
$=4$

$g(3)=\mathbf{3}^2$
$=9$

49. $g(x)=x^3-1$

$g(2)=\mathbf{2}^3-1$
$=8-1$
$=7$

$g(3)=\mathbf{3}^3-1$
$=27-1$
$=26$

51. $g(x)=(x+1)^2$

$g(2)=(\mathbf{2}+1)^2$
$=3^2$
$=9$

$g(3)=(\mathbf{3}+1)^2$
$=4^2$
$=16$

53. $g(x)=2x^2-x$

$g(2)=2(\mathbf{2})^2-\mathbf{2}$
$=2(4)-2$
$=8-2$
$=6$

$g(3)=2(\mathbf{3})^2-\mathbf{3}$
$=2(9)-3$
$=18-3$
$=15$

55. $h(x)=|x|+2$

$h(2)=|\mathbf{2}|+2$
$=2+2$
$=4$

$h(-2)=|\mathbf{-2}|+2$
$=2+2$
$=4$

57. $h(x)=x^2-2$

$h(2)=(\mathbf{2})^2-2$
$=4-2$
$=2$

$h(-2)=(\mathbf{-2})^2-2$
$=4-2$
$=2$

59. $h(x)=\dfrac{1}{x+3}$

$h(2)=\dfrac{1}{\mathbf{2}+3}$
$=\dfrac{1}{5}$

$h(-2)=\dfrac{1}{\mathbf{-2}+3}$
$=\dfrac{1}{1}$
$=1$

61. $h(x)=\dfrac{x}{x-3}$

$h(2)=\dfrac{\mathbf{2}}{\mathbf{2}-3}$
$=\dfrac{2}{-1}$
$=-2$

$h(-2)=\dfrac{\mathbf{-2}}{\mathbf{-2}-3}$
$=\dfrac{-2}{-5}$
$=\dfrac{2}{5}$

63. $f(t)=|t-2|$

t	$f(x)=\lvert t-2\rvert$
-1.7	$\lvert\mathbf{-1.7}-2\rvert=\lvert-3.7\rvert$ $=3.7$
0.9	$\lvert\mathbf{0.9}-2\rvert=\lvert-1.1\rvert$ $=1.1$
5.4	$\lvert\mathbf{5.4}-2\rvert=\lvert 3.4\rvert$ $=3.4$

65. $g(x)=x^3$

Input	Output
$-\frac{3}{4}$	$\left(-\frac{3}{4}\right)^3=-\frac{3}{4}\left(-\frac{3}{4}\right)\left(-\frac{3}{4}\right)$ $=-\frac{27}{64}$
$\frac{1}{6}$	$\left(\frac{1}{6}\right)^3=\frac{1}{6}\left(\frac{1}{6}\right)\left(\frac{1}{6}\right)$ $=\frac{1}{216}$
$\frac{5}{2}$	$\left(\frac{5}{2}\right)^3=\frac{5}{2}\left(\frac{5}{2}\right)\left(\frac{5}{2}\right)$ $=\frac{125}{8}$

67. $g(x)=2x$

$g(w)=2(w)$
$=2w$

$g(w+1)=2(w+1)$
$=2w+2$

69. $g(x)=3x-5$

$$g(w)=3(w)-5 \qquad g(w+1)=3(w+1)-5$$
$$=3w-5 \qquad =3w+3-5$$
$$=3w-2$$

71.

$$f(x)=-2x+5$$
$$f(x)=5$$
$$-2x+5=5$$
$$-2x+5-5=5-5$$
$$-2x=0$$
$$x=0$$

73.

$$f(x)=\frac{3}{2}x-2$$
$$f(x)=-\frac{1}{2}$$
$$\frac{3}{2}x-2=-\frac{1}{2}$$
$$2\left(\frac{3}{2}x-2\right)=2\left(-\frac{1}{2}\right)$$
$$3x-4=-1$$
$$3x-4+4=-1+4$$
$$3x=3$$
$$x=1$$

75. *D:* {−2, 4, 6}
R: {3, 5, 7}

77. *D:* the set of all real numbers
R: the set of all real numbers

79. *D:* the set of all real numbers
R: the set of all nonnegative real numbers (whenever you square a number, the result will always be positive)

81. *D:* the set of all real numbers
R: the set of all nonnegative real numbers (whenever you take the absolute value of a number, the result is always positive)

83. *D:* the set of all real numbers expect 4 (4 makes the denominator 0 and undefined)
R: the set of all real numbers except 0

85. not a function

87. function

89. function

91. function

93. $f(x)=2x-1$

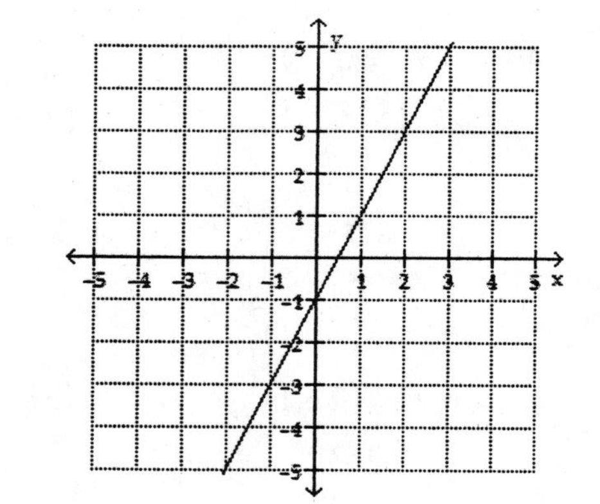

95. $f(x)=\frac{2}{3}x-2$

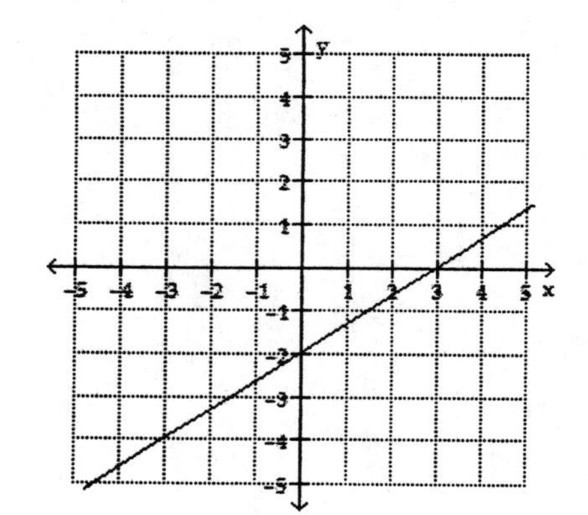

APPLICATIONS

97. DECONGESTANTS

$$C(F)=\frac{5}{9}(F-32)$$

$$C(68)=\frac{5}{9}(68-32)$$
$$=\frac{5}{9}(36)$$
$$=20$$

$$C(77)=\frac{5}{9}(77-32)$$
$$=\frac{5}{9}(45)$$
$$=25$$

between 20° C and 25° C

99. CONCESSIONAIRES

a. $f(b)=1.75b-50$

b.

$$f(110)=1.75(110)-50$$
$$=192.5-50$$
$$=\$142.50$$

101. EARTH'S ATMOSPHERE

a. (200, 25), (200, 90), (200, 105)

b. It doesn't pass the vertical line test.

103. TAXES

a.

$$T(a)=700+0.15(a-7,000)$$
$$T(25,000)=700+0.15(25,000-7,000)$$
$$=700+0.15(18,000)$$
$$=700+2,700$$
$$=\$3,400$$

The tax on an income of $25,000 is $3,400.

b. $T(a)=3,910+0.25(a-28,400)$

WRITING

105. Answers will vary.

REVIEW

107. $-3\frac{3}{4}=-\frac{15}{4}$

109. $0.333...=\frac{1}{3}$

CHALLENGE PROBLEMS

111. Yes.

$$f(x)+g(x)=2x+1+x^2$$

$$g(x)+f(x)=x^2+2x+1$$

$$f(x)+g(x)=g(x)+f(x)$$

SECTION 8.8

VOCABULARY

1. Functions whose graphs are not straight lines are called **nonlinear** functions.

3. The graph of $f(x) = x^2$ is a cup-like shape called a **parabola**.

5. The function $f(x) = x^3$ is called the **cubing** function.

CONCEPTS

7. a. –4
   b. 0
   c. 2
   d. –1

9. a. 4
   b. 3
   c. 0 and 2
   d. 1

11. a.

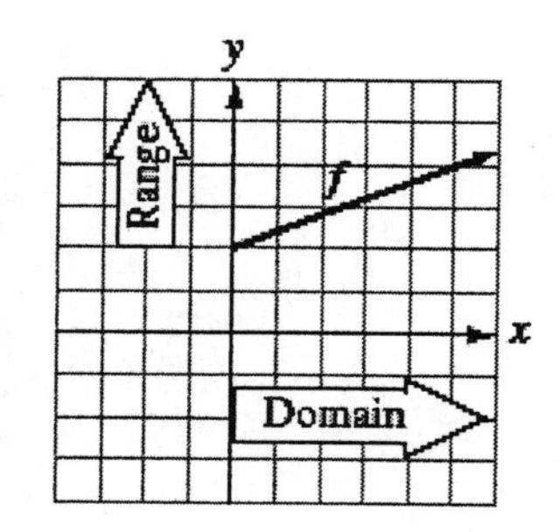

b. D: all nonnegative real numbers
   R: all real numbers greater than or equal to 2

13. h is the x-value of the vertex: 4, 0, and –2

15.

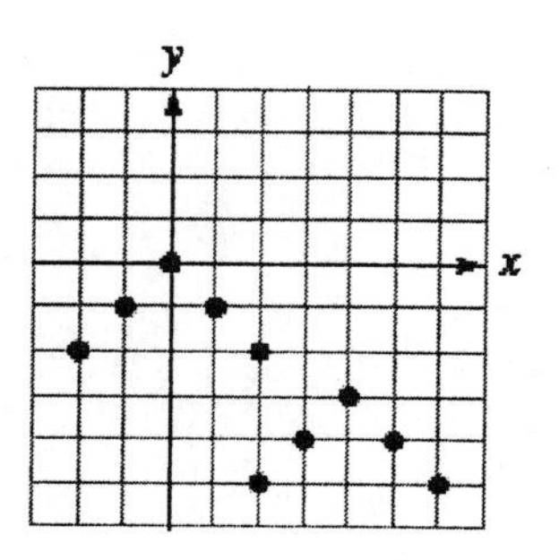

NOTATION

17. The graph of $f(x) = (x + 4)^3$ is the same as the graph of $f(x) = x^3$ except that it is shifted **4** units to the **left**.

19. The graph of $f(x) = x^2 + 5$ is the same as the graph of $f(x) = x^2$ except that it is shifted **5** units **up**.

PRACTICE

21. $f(x) = x^2 - 3$

x	y
2	$2^2 - 3 = 1$
1	$1^2 - 3 = -2$
0	$0^2 - 3 = -3$
–1	$(-1)^2 - 3 = -2$
–2	$(-2)^2 - 3 = 1$

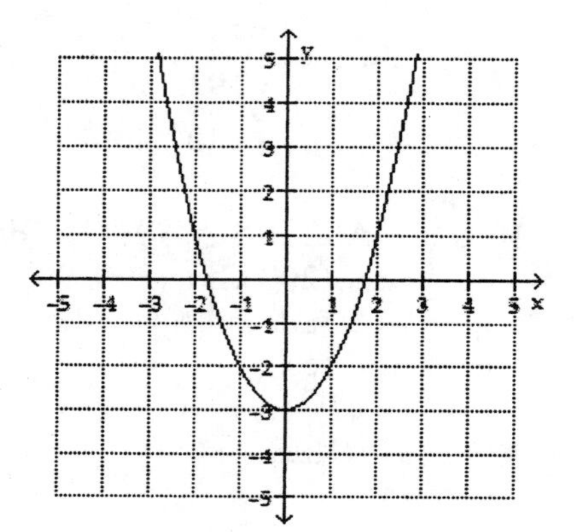

D: the set of real numbers
R: the set of real numbers greater than or equal to –3

23. $f(x) = (x-1)^3$

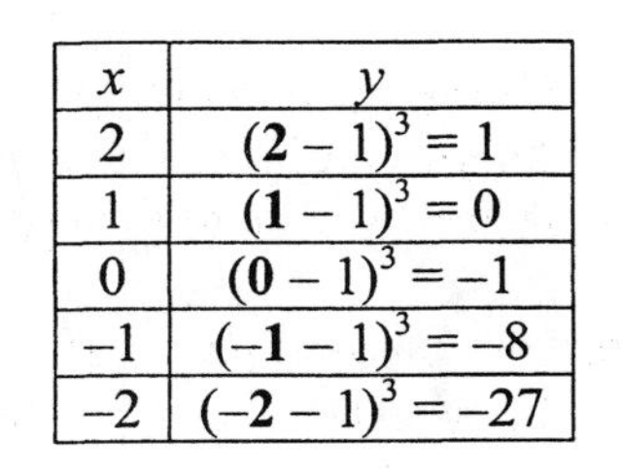

x	y
2	$(\mathbf{2}-1)^3 = 1$
1	$(\mathbf{1}-1)^3 = 0$
0	$(\mathbf{0}-1)^3 = -1$
−1	$(\mathbf{-1}-1)^3 = -8$
−2	$(\mathbf{-2}-1)^3 = -27$

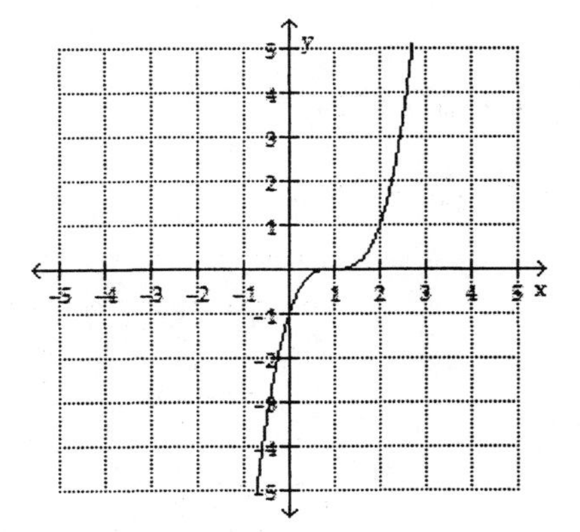

D: the set of real numbers
R: the set of real numbers

25. $f(x) = |x| - 2$

x	y
2	$|\mathbf{2}| - 2 = 0$
1	$|\mathbf{1}| - 2 = -1$
0	$|\mathbf{0}| - 2 = -2$
−1	$|\mathbf{-1}| - 2 = -1$
−2	$|\mathbf{-2}| - 2 = 0$

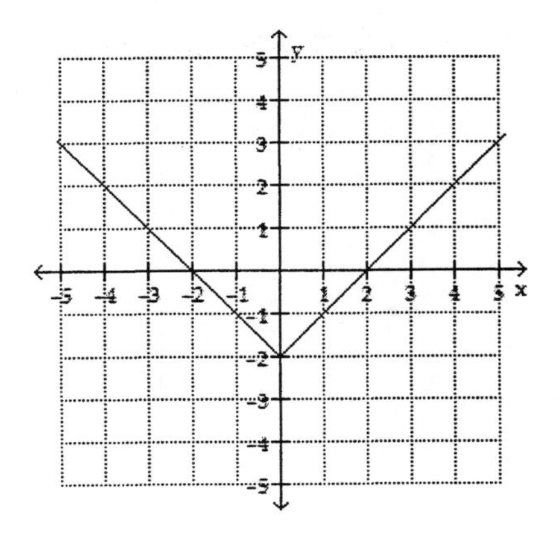

D: the set of real numbers
R: the set of real numbers greater than or equal to −2

27. $f(x) = |x-1|$

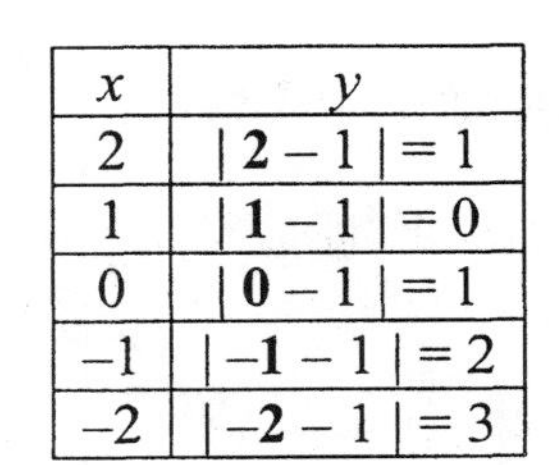

x	y
2	$|\mathbf{2}-1| = 1$
1	$|\mathbf{1}-1| = 0$
0	$|\mathbf{0}-1| = 1$
−1	$|\mathbf{-1}-1| = 2$
−2	$|\mathbf{-2}-1| = 3$

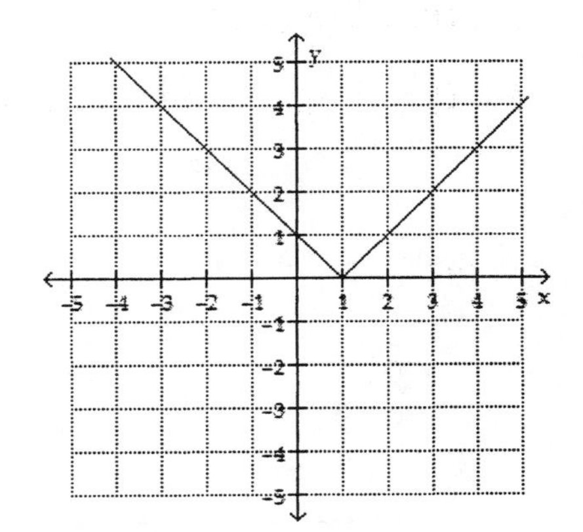

D: the set of real numbers
R: the set of real numbers greater than or equal to 0

29. $f(x) = -3x$

x	y
2	$-3(\mathbf{2}) = -6$
1	$-3(\mathbf{1}) = -3$
0	$-3(\mathbf{0}) = 0$
−1	$-3(\mathbf{-1}) = 3$
−2	$-3(\mathbf{-2}) = 6$

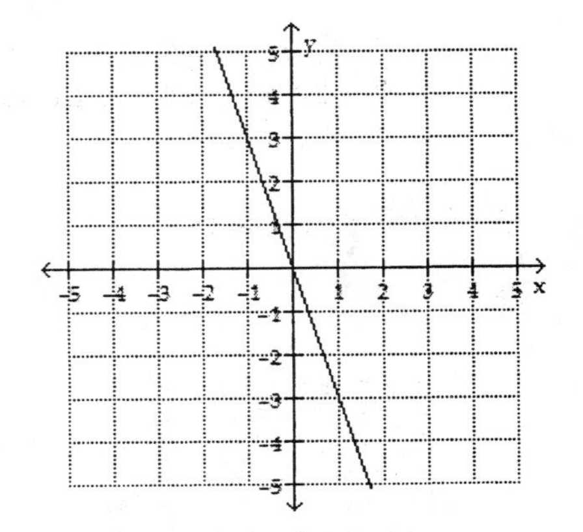

D: the set of real numbers
R: the set of real numbers

31. $f(x) = x^2 + 8$

33. $f(x) = | x + 5 |$

35. $f(x) = (x - 6)^2$

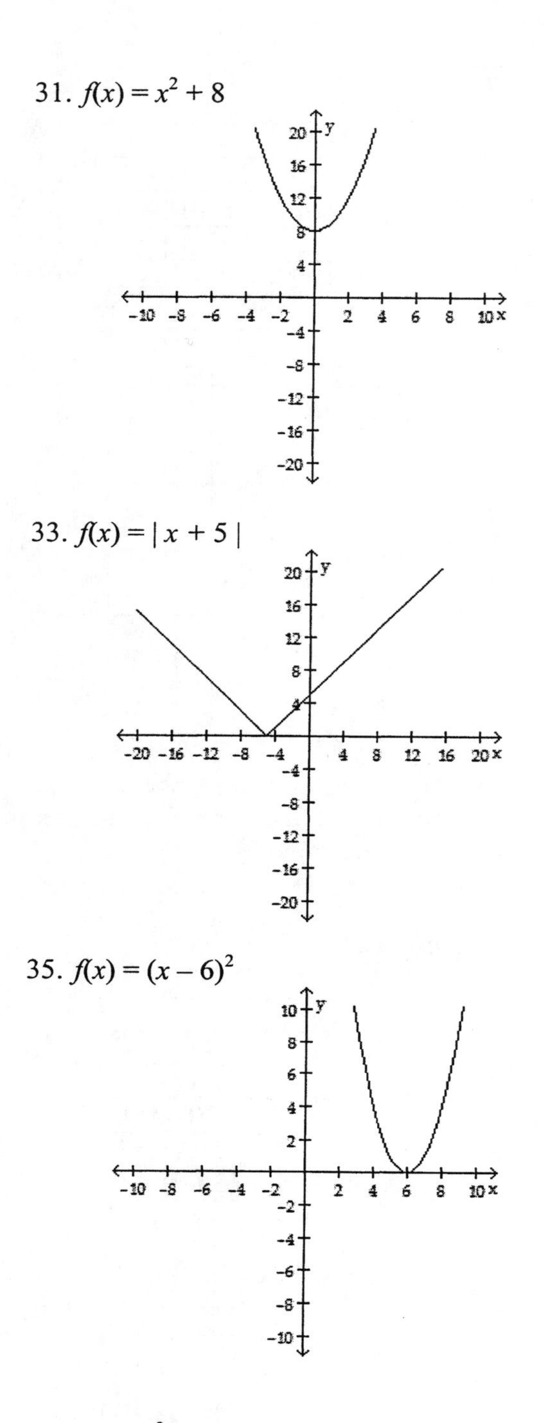

37. $f(x) = x^3 + 8$

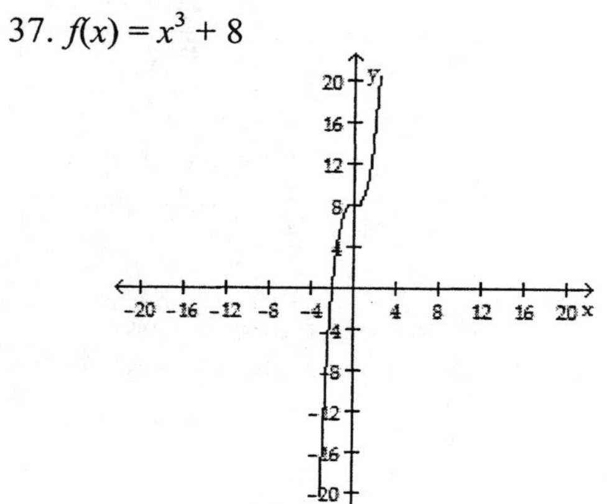

39. $f(x) = x^2 - 5$

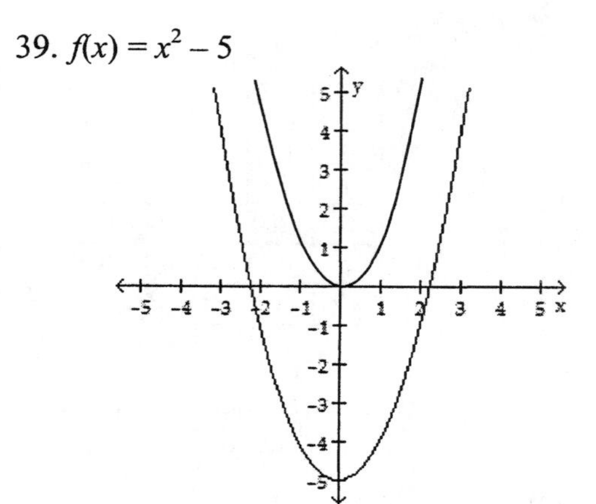

41. $f(x) = (x - 1)^3$

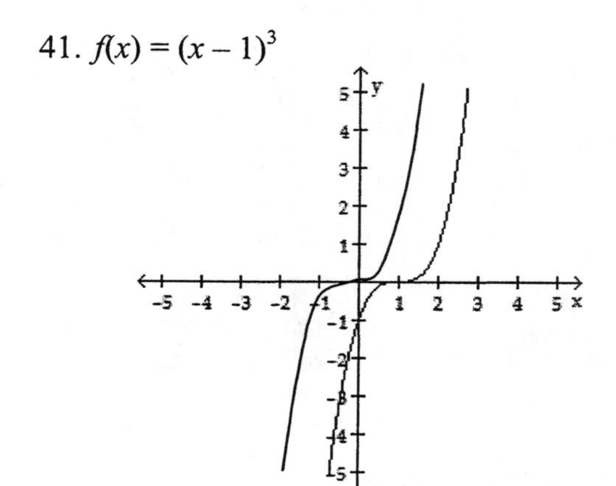

43. $f(x) = | x - 2 | - 1$

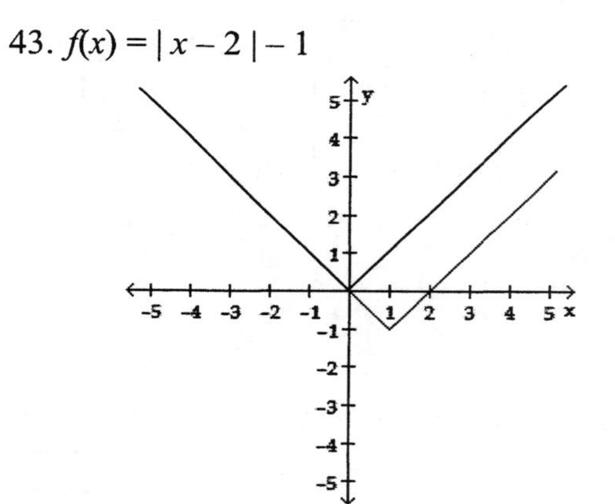

45. $f(x) = (x + 1)^3 - 2$

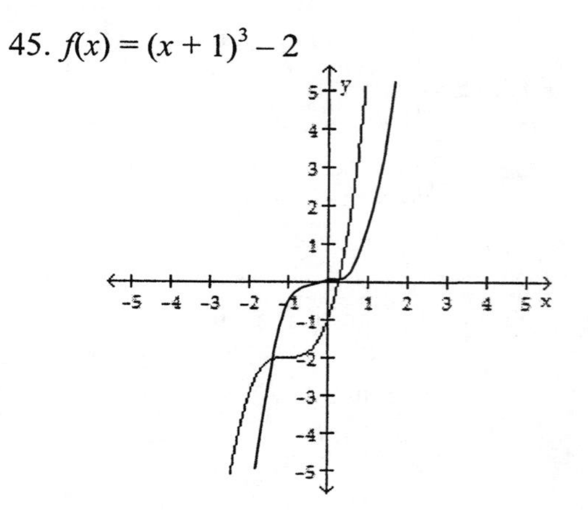

47. $f(x) = -x^3$

49. $f(x) = -x^2$

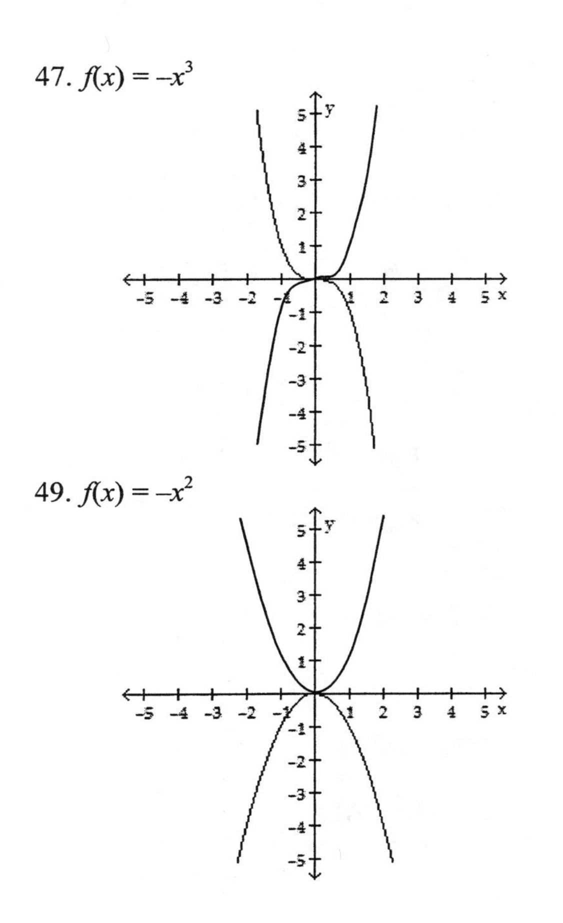

APPLICATIONS

51. OPTICS
$f(x) = |x|$

53. CENTER OF GRAVITY
a parabola

WRITING

55. Answers will vary.

57. Answers will vary.

REVIEW

59.

$$T - W = ma$$
$$T - W - T = ma - T$$
$$-W = ma - T$$
$$\frac{-W}{-1} = \frac{ma - T}{-1}$$
$$W = -ma + T$$
$$W = T - ma$$

61.

$$s = \frac{1}{2}gt^2 + vt$$
$$s - vt = \frac{1}{2}gt^2 + vt - vt$$
$$s - vt = \frac{1}{2}gt^2$$
$$2(s - vt) = 2\left(\frac{1}{2}gt^2\right)$$
$$2(s - vt) = gt^2$$
$$\frac{2(s - vt)}{t^2} = \frac{gt^2}{t^2}$$
$$\frac{2(s - vt)}{t^2} = g$$

63. Find 20% of \$4.5 million.
$0.20 \cdot 4.5 = 0.9$ million

\$4.5 million + \$0.9 million = \$5.4 million

CHALLENGE PROBLEMS

65. $f(x) = \begin{cases} |x| & \text{for } x \geq 0 \\ x^3 & \text{for } x < 0 \end{cases}$

x	y
–3	$(-3)^3 = -27$
–2	$(-2)^3 = -8$
–1	$(-1)^3 = -1$
0	$\|0\| = 0$
1	$\|1\| = 1$
2	$\|2\| = 2$
3	$\|3\| = 3$

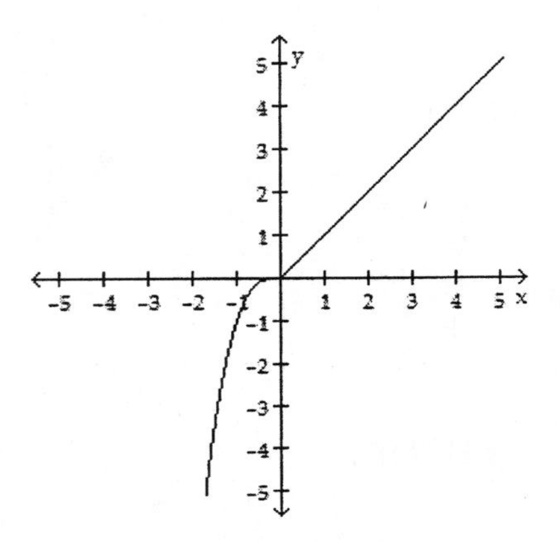

SECTION 8.9

VOCABULARY

1. The equation $y = kx$ defines **direct** variation.

3. In $y = kx$, the **constant** of variation is k.

5. The equation $y = kxz$ represents **joint** variation.

CONCEPTS

7.

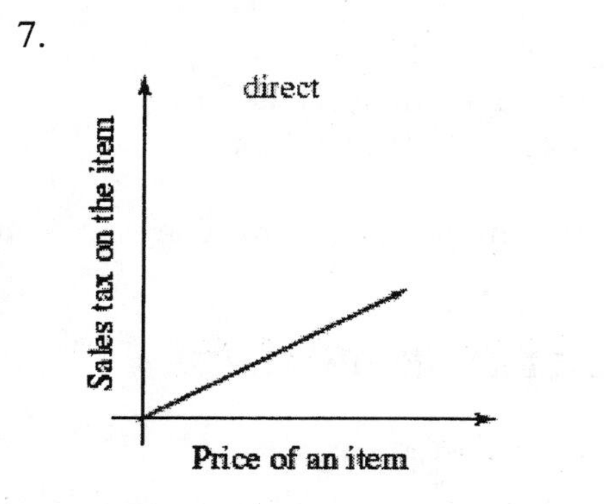

9.

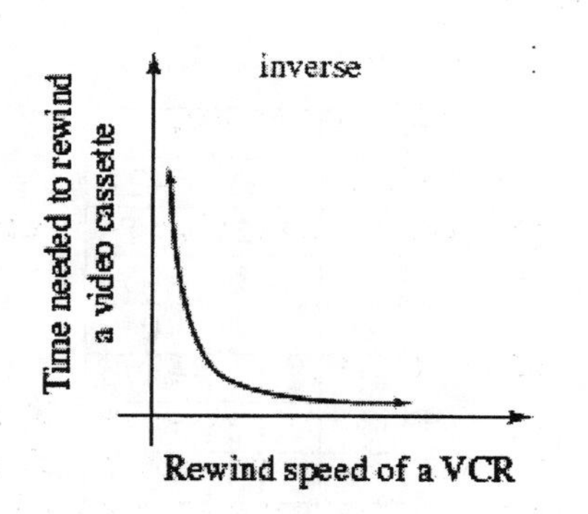

NOTATION

11. a. yes
    b. no
    c. no
    d. yes

PRACTICE

13. $A = kp^2$

15. $v = \frac{k}{r^3}$

17. $B = kmn$

19. $P = \frac{ka^2}{j^3}$

21. L varies jointly with m and n.

23. E varies jointly with a and the square of b.

25. X varies directly with x^2 and inversely with y^2.

27. R varies directly with L and inversely with d^2.

APPLICATIONS

29. FREE FALL
Let s = distance and t = time.
Find k if s = 1,024 and t = 8.

$$s = kt^2$$
$$1{,}024 = k(8)^2$$
$$1{,}024 = 64k$$
$$16 = k$$

Find s if k = 16 and t = 10.

$$s = 16(10)^2$$
$$s = 16(100)$$
$$s = 1{,}600$$

It will fall 1,600 ft.

31. FARMING
Let a = acres planted and b = number of bushels.
Find k if a = 8 and b = 144.

$$b = ka$$
$$144 = k(8)$$
$$144 = 8k$$
$$18 = k$$

Find a if k = 18 and b = 1,152.

$$b = ka$$
$$1{,}152 = (18)(a)$$
$$1{,}152 = 18k$$
$$64 = k$$

The farmer needs 64 acres.

33. FARMING

Let d = days feed will last and a = number of animals.

Find k if $d = 10$ and $a = 25$.

$$d = \frac{k}{a}$$
$$10 = \frac{k}{25}$$
$$25(10) = 25\left(\frac{k}{25}\right)$$
$$250 = k$$

Find d if $k = 250$ and $a = 10$.

$$d = \frac{k}{a}$$
$$d = \frac{250}{10}$$
$$d = 25$$

The feed will last 25 days.

35. GAS PRESSURE

Let v = volume of gas and p = pressure.

Find k when $v = 20$ and $p = 6$.

$$v = \frac{k}{p}$$
$$20 = \frac{k}{6}$$
$$6(20) = 6\left(\frac{k}{6}\right)$$
$$120 = k$$

Find v when $k = 120$ and $p = 10$.

$$v = \frac{k}{p}$$
$$v = \frac{120}{10}$$
$$v = 12 \text{ in.}^3$$

The pressure would be 12 in.^3

37. ORGAN PIPES

Let v = frequency of vibration and l = length of the pipe.

Find k if $v = 256$ and $l = 2$

$$v = \frac{k}{l}$$
$$256 = \frac{k}{2}$$
$$2(256) = 2\left(\frac{k}{2}\right)$$
$$512 = k$$

Find v if $k = 512$ and $l = 6$.

$$v = \frac{k}{l}$$
$$v = \frac{512}{6}$$
$$v = \frac{256}{3}$$
$$v = 85\frac{1}{3}$$

The pipe will vibrate $85\frac{1}{3}$ times.

39. GEOMETRY

Let A = area, l = length, w = width, and h = height.

$$V = klwh$$
$$V = k(2l)(3w)(2h)$$
$$V = 12klwh$$

The volume is multiplied by a factor of 12.

41. STORING OIL

Let g = gallons of oil, h = height of tank, r = radius of the tank, and $k = 23.5$.

Since the diameter is 15, the radius would be half that distance, or 7.5 ft.

$$g = khr^2$$
$$g = (23.5)(20)(7.5)^2$$
$$g = (23.5)(20)(56.25)$$
$$g = 26,437.5$$

26,437.5 gallons of oil can be stored in the tank.

43. ELECTRONICS
Let v = voltage, r = resistance, and c = current.

$$v = rc$$
$$6 = r \bullet 2$$
$$6 = 2r$$
$$3 = r$$
$$r = 3 \text{ ohms}$$

45. STRUCTURAL ENGINEERING
Let F = force of deflection, w = width, d = depth, and k = constant. Find k using the given information that $F = 1.1$, $w = 4$, and $d = 4$.

$$F = \frac{k}{wd^3}$$
$$1.1 = \frac{k}{4(4)^3}$$
$$1.1 = \frac{k}{4(64)}$$
$$1.1 = \frac{k}{256}$$
$$256(1.1) = 256\left(\frac{k}{256}\right)$$
$$281.6 = k$$

Let $k = 281.6$, $w = 2$, and $d = 8$ to find the force of deflection.

$$F = \frac{k}{wd^3}$$
$$= \frac{281.6}{2(8)^3}$$
$$= \frac{281.6}{2(512)}$$
$$= \frac{281.6}{1{,}024}$$
$$= 0.275 \text{ in.}$$

The deflection is 0.275 in.

47. GAS PRESSURE
Let P = pressure, T = temperature, k = constant, and V = volume.
Find k when $T = 273$, $P = 1$, and $V = 1$.

$$P = \frac{kT}{V}$$
$$1 = \frac{k(273)}{1}$$
$$1 = 273k$$
$$\frac{1}{273} = k$$

Find T when $P = 1$ and $V = 2$.

$$P = \frac{kT}{V}$$
$$1 = \frac{\left(\frac{1}{273}\right)T}{2}$$
$$2(1) = 2\left(\frac{\left(\frac{1}{273}\right)T}{2}\right)$$
$$2 = \frac{1}{273}T$$
$$273(2) = 273\left(\frac{1}{273}T\right)$$
$$546 = T$$
$$T = 546 \text{ Kelvin}$$

The temperature is 546 Kelvin.

WRITING

49. Answers will vary.

51. Answers will vary.

REVIEW

53. 3.5×10^4

55. 0.0025

CHALLENGE PROBLEMS

57. Answers will vary.

CHAPTER 8 KEY CONCEPTS

1. u
2. j
3. w
4. p
5. q
6. r
7. v
8. f
9. s
10. k
11. m
12. o
13. z
14. c
15. b
16. n
17. d
18. i
19. x
20. g
21. a
22. y
23. t
24. h
25. l
26. e

CHAPTER 8 REVIEW

SECTION 8.1
Review of Equations and Inequalities

1.

$$5x+12=0$$
$$5x+12-12=0-12$$
$$5x=-12$$
$$\frac{5x}{5}=\frac{-12}{5}$$
$$x=-\frac{12}{5}$$

2.

$$-3x-7+x=6x+20-5x$$
$$-2x-7=x+20$$
$$-2x-7-x=x+20-x$$
$$-3x-7=20$$
$$-3x-7+7=20+7$$
$$-3x=27$$
$$\frac{-3x}{-3}=\frac{27}{-3}$$
$$x=-9$$

3.

$$4(y-1)=28$$
$$4y-4=28$$
$$4y-4+4=28+4$$
$$4y=32$$
$$\frac{4y}{4}=\frac{32}{4}$$
$$y=8$$

4.

$$2-13(x-1)=4-6x$$
$$2-13x+13=4-6x$$
$$15-13x=4-6x$$
$$15-13x+6x=4-6x+6x$$
$$15-7x=4$$
$$15-7x-15=4-15$$
$$-7x=-11$$
$$\frac{-7x}{-7}=\frac{-11}{-7}$$
$$x=\frac{11}{7}$$

5.

$$\frac{8}{3}(x-5)=\frac{2}{5}(x-4)$$
$$15\cdot\frac{8}{3}(x-5)=15\cdot\frac{2}{5}(x-4)$$
$$40(x-5)=6(x-4)$$
$$40x-200=6x-24$$
$$40x-200-6x=6x-24-6x$$
$$34x-200=-24$$
$$34x-200+200=-24+200$$
$$34x=176$$
$$\frac{34x}{34}=\frac{176}{34}$$
$$x=\frac{88}{17}$$

6.

$$\frac{3y}{4}-14=-\frac{y}{3}-1$$
$$12\left(\frac{3y}{4}-14\right)=12\left(-\frac{y}{3}-1\right)$$
$$9y-168=-4y-12$$
$$9y-168+4y=-4y-12+4y$$
$$13y-168=-12$$
$$13y-168+168=-12+168$$
$$13y=156$$
$$\frac{13y}{13}=\frac{156}{13}$$
$$y=12$$

7.

$$2x+4=2(x+3)-2$$
$$2x+4=2x+6-2$$
$$2x+4=2x+4$$
$$4=4$$

$\mathbb{R}$; identity

8.

$$(3x-2)-x=2(x-4)$$
$$2x-2=2x-8$$
$$-2\neq-8$$

$\varnothing$; contradiction

9.

$$-\frac{5}{4}p = 10$$
$$-\frac{4}{5}\left(-\frac{5}{4}p\right) = -\frac{4}{5}(10)$$
$$p = -8$$

10.

$$\frac{4t+1}{3} - \frac{t+5}{6} = \frac{t-3}{6}$$
$$6\left(\frac{4t+1}{3} - \frac{t+5}{6}\right) = 6\left(\frac{t-3}{6}\right)$$
$$2(4t+1) - (t+5) = t-3$$
$$8t + 2 - t - 5 = t - 3$$
$$7t - 3 = t - 3$$
$$6t - 3 = -3$$
$$6t = 0$$
$$t = 0$$

11.

$$V = \pi r^2 h$$
$$\frac{V}{\pi r^2} = \frac{\pi r^2 h}{\pi r^2}$$
$$\frac{V}{\pi r^2} = h$$

12.

$$v = \frac{1}{6}ab(x+y)$$
$$6 \cdot v = 6 \cdot \frac{1}{6}ab(x+y)$$
$$6v = ab(x+y)$$
$$\frac{6v}{ab} = \frac{ab(x+y)}{ab}$$
$$\frac{6v}{ab} = x + y$$
$$\frac{6v}{ab} - y = x + y - y$$
$$\frac{6v}{ab} - y = x$$
$$x = \frac{6v}{ab} - y \quad \text{or}$$
$$x = \frac{6v - aby}{ab}$$

13.

$$0.3x - 0.4 \geq 1.2 - 0.1x$$
$$10(0.3x - 0.4) \geq 10(1.2 - 0.1x)$$
$$3x - 4 \geq 12 - x$$
$$3x - 4 + x \geq 12 - x + x$$
$$4x - 4 \geq 12$$
$$4x - 4 + 4 \geq 12 + 4$$
$$4x \geq 16$$
$$x \geq 4; \quad [4, \infty)$$

4

14.

$$\frac{7}{4}(x+3) < \frac{3}{8}(x-3)$$
$$8 \cdot \frac{7}{4}(x+3) < 8 \cdot \frac{3}{8}(x-3)$$
$$14(x+3) < 3(x-3)$$
$$14x + 42 < 3x - 9$$
$$14x + 42 - 3x < 3x - 9 - 3x$$
$$11x + 42 < -9$$
$$11x + 42 - 42 < -9 - 42$$
$$11x < -51$$
$$x < -\frac{51}{11}; \quad \left(-\infty, -\frac{51}{11}\right)$$

$-\frac{51}{11}$

15.

$$-16 < -\frac{4}{5}x$$
$$-\frac{5}{4}(-16) > -\frac{5}{4}\left(-\frac{4}{5}x\right)$$
$$20 > x$$
$$x < 20; \quad (-\infty, 20)$$

20

16.

$$3 < 3x + 4 < 10$$
$$3 - 4 < 3x + 4 - 4 < 10 - 4$$
$$-1 < 3x < 6$$
$$\frac{-1}{3} < \frac{3x}{3} < \frac{6}{3}$$
$$-\frac{1}{3} < x < 2$$
$$\left(-\frac{1}{3}, 2\right)$$

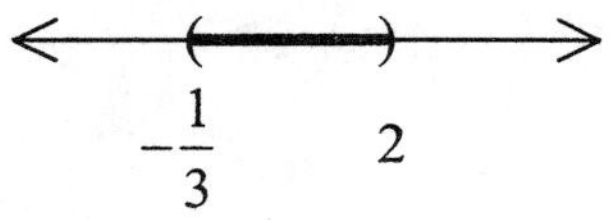

17.

$$-2x > 8 \quad \text{and} \quad x + 4 \geq -6$$
$$\frac{-2x}{-2} < \frac{8}{-2} \qquad x + 4 - 4 \geq -6 - 4$$
$$x < -4 \qquad x \geq -10$$
$$-10 \leq x < -4$$
$$[-10, -4)$$

18.

$$x + 1 < -4 \quad \text{or} \quad x - 4 > 0$$
$$x + 1 - 1 < -4 - 1 \quad x - 4 + 4 > 0 + 4$$
$$x < -5 \qquad x > 4$$
$$(-\infty - 5) \cup (4, \infty)$$

19. Let x = length of one piece.
Then $3x$ = length of the other piece.

$$x + 3x = 20$$
$$4x = 20$$
$$x = 5$$

He should cut the board 5 ft from one end.

20. Let width = x. Then length = $x + 4$.

$$P = 2l + 2w$$
$$28 = 2(x + 4) + 2(x)$$
$$28 = 2x + 8 + 2x$$
$$28 = 4x + 8$$
$$20 = 4x$$
$$5 = x$$

The width = 5 meters and the length = 5 + 4 = 9 meters.
Find the area if width = 5 and length = 9.

$$A = lw$$
$$= 5(9)$$
$$= 45 \text{ m}^2$$

The area is 45 m^2.

SECTION 8.2
Solving Absolute Value Equations and Inequalities.

21.

$$|4x| = 8$$
$$4x = 8 \quad \text{or} \quad 4x = -8$$
$$x = 2 \qquad x = -2$$

22.

$$2|3x + 1| - 1 = 19$$
$$2|3x + 1| - 1 + 1 = 19 + 1$$
$$2|3x + 1| = 20$$
$$\frac{2|3x + 1|}{2} = \frac{20}{2}$$
$$|3x + 1| = 10$$
$$3x + 1 = 10 \quad \text{or} \quad 3x + 1 = -10$$
$$3x + 1 - 1 = 10 - 1 \quad 3x + 1 - 1 = -10 - 1$$
$$3x = 9 \qquad 3x = -11$$
$$x = 3 \qquad x = -\frac{11}{3}$$

23.

$$\left|\frac{3}{2}x-4\right|-10=-1$$

$$\left|\frac{3}{2}x-4\right|-10+10=-1+10$$

$$\left|\frac{3}{2}x-4\right|=9$$

$$\frac{3}{2}x-4=9 \quad \text{or} \quad \frac{3}{2}x-4=-9$$

$$2\left(\frac{3}{2}x-4\right)=2(9) \qquad 2\left(\frac{3}{2}x-4\right)=2(-9)$$

$$3x-8=18 \qquad 3x-8=-18$$

$$3x-8+8=18+8 \qquad 3x-8+8=-18+8$$

$$3x=26 \qquad 3x=-10$$

$$x=\frac{26}{3} \qquad x=-\frac{10}{3}$$

24.

$$\left|\frac{2-x}{3}\right|=-4$$

No solution. Since an absolute value can never be negative, there are no real numbers x that can make the equation true.

25.

$$|3x+2|=|2x-3|$$

$$3x+2=2x-3 \quad \text{or} \quad 3x+2=-(2x-3)$$

$$3x+2=2x-3 \qquad 3x+2=-2x+3$$

$$3x+2-2x=2x-3-2x \qquad 3x+2+2x=-2x+3+2x$$

$$x+2=-3 \qquad 5x+2=3$$

$$x+2-2=-3-2 \qquad 5x+2-2=3-2$$

$$x=-5 \qquad 5x=1$$

$$x=-5 \qquad x=\frac{1}{5}$$

26.

$$\left|\frac{2(1-x)+1}{2}\right|=\left|\frac{3x-2}{3}\right|$$

$$\left|\frac{2-2x+1}{2}\right|=\left|\frac{3x-2}{3}\right|$$

$$\left|\frac{-2x+3}{2}\right|=\left|\frac{3x-2}{3}\right|$$

$$\frac{-2x+3}{2}=\frac{3x-2}{3} \quad \text{or} \quad \frac{-2x+3}{2}=-\left(\frac{3x-2}{3}\right)$$

$$6\left(\frac{-2x+3}{2}\right)=6\left(\frac{3x-2}{3}\right) \qquad 6\left(\frac{-2x+3}{2}\right)=-6\left(\frac{3x-2}{3}\right)$$

$$3(-2x+3)=2(3x-2) \qquad 3(-2x+3)=-2(3x-2)$$

$$-6x+9=6x-4 \qquad -6x+9=-6x-4$$

$$-6x+9-6x=6x-4-6x \qquad -6x+9+6x=-6x-4+6x$$

$$-12x+9=-4 \qquad 9\neq-4$$

$$-12x+9-9=-4-9$$

$$-12x=-13$$

$$x=\frac{13}{12}$$

$$\boxed{x=\frac{13}{12}}$$

27.

$$|x|\le 3$$

$$-3\le x\le 3$$

$$[-3,3]$$

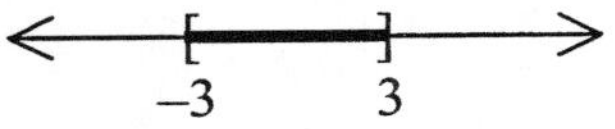

28.

$$|2x+7|<3$$

$$-3<2x+7<3$$

$$-3-7<2x+7-7<3-7$$

$$-10<2x<-4$$

$$-5<x<-2$$

$$(-5,-2)$$

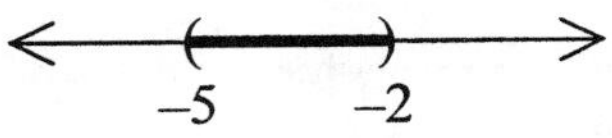

29.

$$2|5-3x| \le 28$$
$$\frac{2|5-3x|}{2} \le \frac{28}{2}$$
$$|5-3x| \le 14$$
$$-14 \le 5-3x \le 14$$
$$-14-5 \le 5-3x-5 \le 14-5$$
$$-19 \le -3x \le 9$$
$$\frac{-19}{-3} \ge \frac{-3x}{-3} \ge \frac{9}{-3}$$
$$\frac{19}{3} \ge x \ge -3$$
$$-3 \le x \le \frac{19}{3}$$
$$\left[-3, \frac{19}{3}\right]$$

-3 $\frac{19}{3}$

30.

$$\left|\frac{2}{3}x+14\right|+6<6$$
$$\left|\frac{2}{3}x+14\right|+6-6<6-6$$
$$\left|\frac{2}{3}x+14\right|<0$$

No solution. Since an absolute value can never be negative (less than zero), there are no real numbers x that can make the equation true.

31.

$$|x|>1$$
$$x>1 \text{ or } x<-1$$
$$(-\infty,-1)\cup(1,\infty)$$

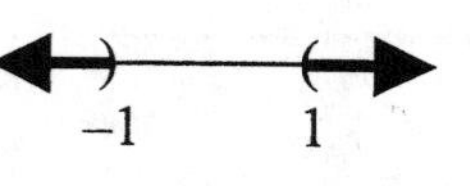

32.

$$|3x-8|-4 \ge 0$$
$$|3x-8|-4+4 \ge 0+4$$
$$|3x-8| \ge 4$$
$$3x-8 \le -4 \quad \text{or} \quad 3x-8 \ge 4$$
$$3x-8+8 \le -4+8 \quad 3x-8+8 \ge 4+8$$
$$3x \le 4 \qquad 3x \ge 12$$
$$x \le \frac{4}{3} \qquad x \ge 4$$
$$\left(-\infty, \frac{4}{3}\right] \cup [4,\infty)$$

$\frac{4}{3}$ 4

SECTION 8.3
Review of Factoring

33.

$$z^2-11z+30=(z-5)(z-6)$$

34.

$$\begin{aligned} x^4+4y+4x^2+x^2y &= x^4+x^2y+4x^2+4y \\ &= x^2(x^2+y)+4(x^2+y) \\ &= (x^2+y)(x^2+4) \end{aligned}$$

35.

$$4a^2-5a+1=(4a-1)(a-1)$$

36.

$$\begin{aligned} &27x^3y^3z^3+81x^4y^5z^2-90x^2y^3z^7 \\ &= 9x^2y^3z^2(3xz+9x^2y^2-10z^5) \end{aligned}$$

37.

$$\begin{aligned} &y^2+3y+2+2x+xy \\ &= (y^2+3y+2)+(2x+xy) \\ &= (y+1)(y+2)+x(2+y) \\ &= (y+2)(y+1+x) \end{aligned}$$

38.

$$-x^2-3x+28=-(x^2+3x-28)$$
$$=-(x+7)(x-4)$$

39.

$$15x^2-57xy-12y^2=3(5x^2-19xy-4y^2)$$
$$=3(5x+y)(x-4y)$$

40.

$w^8-w^4-90=(w^4-10)(w^4+9)$

41.

$$r^2y-ar-ry+a+r-1$$
$$=(r^2y-ar+r)+(-ry+a-1)$$
$$=r(ry-a+1)-(ry-a+1)$$
$$=(ry-a+1)(r-1)$$

42.

$$49a^6+84a^3b^2+36b^4=(7a^3+6b^2)^2$$

43.

$3b^2+2b+1$
prime

44.

$$2a^4+4a^3-6a^2=2a^2(a^2+2a-3)$$
$$=2a^2(a+3)(a-1)$$

45.

$(s+t)^2-2(s+t)+1$
Let $x=(s+t)$
$x^2-2x+1=(x-1)^2$
Substitute $(s+t)$ for x in the answer.
$(x-1)^2=(s+t-1)^2$

46.

$$m_1m_2=mm_2+mm_1$$
$$m_1m_2-mm_1=mm_2+mm_1-mm_1$$
$$m_1m_2-mm_1=mm_2$$
$$m_1(m_2-m)=mm_2$$
$$\frac{m_1(m_2-m)}{m_2-m}=\frac{mm_2}{m_2-m}$$
$$m_1=\frac{mm_2}{m_2-m}$$

SECTION 8.4
The Difference of Two Squares; the Sum and Difference of Two Cubes

47. $z^2-16=(z+4)(z-4)$

48. $x^2y^4-64z^6=(xy^2+8z^3)(xy^2-8z^3)$

49. $a^2b^2+c^2$
prime

50. $c^2-(a+b)^2=(c+a+b)(c-a-b)$

51.

$$32a^4c-162b^4c$$
$$=2c(16a^4-81b^4)$$
$$=2c(4a^2+9b^2)(4a^2-9b^2)$$
$$=2c(4a^2+9b^2)(2a+3b)(2a-3b)$$

52.

$$k^2+2k+1-9m^2=(k^2+2k+1)-9m^2$$
$$=(k+1)^2-9m^2$$
$$=(k+1+3m)(k+1-3m)$$

53.

$$m^2-n^2-m-n=(m^2-n^2)-(m+n)$$
$$=(m+n)(m-n)-1(m+n)$$
$$=(m+n)(m-n-1)$$

54.

$$t^3+64=(t+4)(t^2-4t+16)$$

55.

$$8a^3-125b^9$$
$$=(2a-5b^3)\left((2a)^2+(2a)(5b^3)+(5b^3)^2\right)$$
$$=(2a-5b^3)(4a^2+10ab^3+25b^6)$$

56.

$$V=\frac{\pi}{2}r_1^2h-\frac{\pi}{2}r_2^2h$$
$$=\frac{\pi}{2}h(r_1^2-r_2^2)$$
$$=\frac{\pi}{2}h(r_1+r_2)(r_1-r_2)$$

SECTION 8.5
Review of Rational Expressions

57.

$$\frac{248x^2y}{576xy^2}=\frac{\cancel{2}\cdot\cancel{2}\cdot\cancel{2}\cdot 31\cdot\cancel{x}\cdot x\cdot\cancel{y}}{\cancel{2}\cdot\cancel{2}\cdot\cancel{2}\cdot 2\cdot 2\cdot 2\cdot 3\cdot 3\cdot\cancel{x}\cdot\cancel{y}\cdot y}$$

$$=\frac{31x}{72y}$$

58.

$$\frac{2m-2n}{n-m}=\frac{2(m-n)}{n-m}$$

$$=-\frac{2\cancel{(m-n)}}{\cancel{m-n}}$$

$$=-2$$

59.

$$\frac{3x^3y^4}{c^2d}\cdot\frac{c^3d^2}{21x^5y^4}=\frac{\cancel{3}\cancel{x}\cancel{x}\cancel{x}\cancel{y}\cancel{y}\cancel{y}\cancel{y}\cancel{c}\cancel{c}c\cancel{d}d}{\cancel{3}\cdot 7\cancel{c}\cancel{c}\cancel{d}\cancel{x}\cancel{x}\cancel{x}xx\cancel{y}\cancel{y}\cancel{y}\cancel{y}}$$

$$=\frac{cd}{7x^2}$$

60.

$$\frac{2a^2-5a-3}{a^2-9}\div\frac{2a^2+5a+2}{2a^2+5a-3}$$

$$=\frac{2a^2-5a-3}{a^2-9}\cdot\frac{2a^2+5a-3}{2a^2+5a+2}$$

$$=\frac{\cancel{(2a+1)}\cancel{(a-3)}}{\cancel{(a-3)}\cancel{(a+3)}}\cdot\frac{(2a-1)\cancel{(a+3)}}{\cancel{(2a+1)}(a+2)}$$

$$=\frac{2a-1}{a+2}$$

61.

$$\frac{m^2+3m+9}{m^2+mp+mr+pr}\div\frac{m^3-27}{am+ar+bm+br}$$

$$=\frac{m^2+3m+9}{m^2+mp+mr+pr}\cdot\frac{am+ar+bm+br}{m^3-27}$$

$$=\frac{m^2+3m+9}{m(m+p)+r(m+p)}\cdot\frac{a(m+r)+b(m+r)}{(m-3)(m^2+3m+9)}$$

$$=\frac{\cancel{m^2+3m+9}}{\cancel{(m+r)}(m+p)}\cdot\frac{(a+b)\cancel{(m+r)}}{(m-3)\cancel{(m^2+3m+9)}}$$

$$=\frac{a+b}{(m+p)(m-3)}$$

62.

$$\frac{x^3+3x^2+2x}{2x^2-2x-12}\cdot\frac{3x^2-3x}{x^3-3x^2-4x}\div\frac{x^2+3x+2}{2x^2-4x-16}$$

$$=\frac{x^3+3x^2+2x}{2x^2-2x-12}\cdot\frac{3x^2-3x}{x^3-3x^2-4x}\cdot\frac{2x^2-4x-16}{x^2+3x+2}$$

$$=\frac{x(x^2+3x+2)}{2(x^2-x-6)}\cdot\frac{3x(x-1)}{x(x^2-3x-4)}\cdot\frac{2(x^2-2x-8)}{x^2+3x+2}$$

$$=\frac{\cancel{x}\cancel{(x+1)}\cancel{(x+2)}}{\cancel{2}(x-3)\cancel{(x+2)}}\cdot\frac{3x(x-1)}{\cancel{x}\cancel{(x-4)}\cancel{(x+1)}}\cdot\frac{\cancel{2}\cancel{(x-4)}\cancel{(x+2)}}{(x+1)\cancel{(x+2)}}$$

$$=\frac{3x(x-1)}{(x-3)(x+1)}$$

63.

$$\frac{d^2}{c^3-d^3}+\frac{c^2+cd}{c^3-d^3}=\frac{d^2+c^2+cd}{c^3-d^3}$$

$$=\frac{\cancel{c^2+cd+d^2}}{(c-d)\cancel{(c^2+cd+d^2)}}$$

$$=\frac{1}{c-d}$$

64.

$$\frac{4}{t-3}+\frac{6}{3-t}=\frac{4}{t-3}-\frac{6}{t-3}$$
$$=\frac{4-6}{t-3}$$
$$=\frac{-2}{t-3}$$
$$=-\frac{2}{t-3}$$

65.

$$\frac{5x}{14z^2}+\frac{y^2}{16z}=\frac{5x}{14z^2}\left(\frac{8}{8}\right)+\frac{y^2}{16z}\left(\frac{7z}{7z}\right)$$
$$=\frac{40x}{112z^2}+\frac{7y^2z}{112z^2}$$
$$=\frac{40x+7y^2z}{112z^2}$$

66.

$$\frac{4}{3xy-6y}-\frac{4}{10-5x}=\frac{4}{3y(x-2)}-\frac{4}{5(2-x)}$$
$$=\frac{4}{3y(x-2)}+\frac{4}{5(x-2)}$$
$$=\frac{4}{3y(x-2)}\left(\frac{5}{5}\right)+\frac{4}{5(x-2)}\left(\frac{3y}{3y}\right)$$
$$=\frac{20}{15y(x-2)}+\frac{12y}{15y(x-2)}$$
$$=\frac{20+12y}{15y(x-2)}$$

67.

$$\frac{y+7}{y+3}-\frac{y-3}{y+7}=\frac{y+7}{y+3}\left(\frac{y+7}{y+7}\right)-\frac{y-3}{y+7}\left(\frac{y+3}{y+3}\right)$$
$$=\frac{y^2+14y+49}{(y+3)(y+7)}-\frac{y^2-9}{(y+3)(y+7)}$$
$$=\frac{y^2+14y+49-(y^2-9)}{(y+3)(y+7)}$$
$$=\frac{y^2+14y+49-y^2+9}{(y+3)(y+7)}$$
$$=\frac{14y+58}{(y+3)(y+7)}$$

68.

$$\frac{2x}{x+1}+\frac{3x}{x+2}+\frac{4x}{x^2+3x+2}$$
$$=\frac{2x}{x+1}+\frac{3x}{x+2}+\frac{4x}{(x+1)(x+2)}$$
$$=\frac{2x}{x+1}\left(\frac{x+2}{x+2}\right)+\frac{3x}{x+2}\left(\frac{x+1}{x+1}\right)+\frac{4x}{(x+1)(x+2)}$$
$$=\frac{2x^2+4x}{(x+1)(x+2)}+\frac{3x^2+3x}{(x+1)(x+2)}+\frac{4x}{(x+1)(x+2)}$$
$$=\frac{2x^2+4x+3x^2+3x+4x}{(x+1)(x+2)}$$
$$=\frac{5x^2+11x}{(x+1)(x+2)}$$

69.

$$\frac{\frac{4a^3b^2}{9c}}{\frac{14a^3b}{9c^4}}=\frac{\frac{4a^3b^2}{9c}}{\frac{14a^3b}{9c^4}}\bullet\frac{9c^4}{9c^4}$$
$$=\frac{(9c^4)\left(\frac{4a^3b^2}{9c}\right)}{(9c^4)\left(\frac{14a^3b}{9c^4}\right)}$$
$$=\frac{c^3(4a^3b^2)}{14a^3b}$$
$$=\frac{4a^3b^2c^3}{14a^3b}$$
$$=\frac{2bc^3}{7}$$

70.

$$\frac{\frac{p^2-9}{6pt}}{\frac{p^2+5p+6}{3pt}}=\frac{\frac{p^2-9}{6pt}}{\frac{p^2+5p+6}{3pt}}\bullet\frac{6pt}{6pt}$$

$$=\frac{6pt\left(\frac{p^2-9}{6pt}\right)}{6pt\left(\frac{p^2+5p+6}{3pt}\right)}$$

$$=\frac{p^2-9}{2\left(p^2+5p+6\right)}$$

$$=\frac{\cancel{(p+3)}(p-3)}{2(p+2)\cancel{(p+3)}}$$

$$=\frac{p-3}{2(p+2)}$$

71.

$$\frac{1-\frac{1}{x}-\frac{2}{x^2}}{1+\frac{4}{x}+\frac{3}{x^2}}=\frac{1-\frac{1}{x}-\frac{2}{x^2}}{1+\frac{4}{x}+\frac{3}{x^2}}\bullet\frac{x^2}{x^2}$$

$$=\frac{x^2\left(1-\frac{1}{x}-\frac{2}{x^2}\right)}{x^2\left(1+\frac{4}{x}+\frac{3}{x^2}\right)}$$

$$=\frac{x^2-x-2}{x^2+4x+3}$$

$$=\frac{(x-2)\cancel{(x+1)}}{\cancel{(x+1)}(x+3)}$$

$$=\frac{x-2}{x+3}$$

72.

$$\frac{\frac{2}{b^2-1}-\frac{3}{ab-a}}{\frac{3}{ab-a}-\frac{2}{b^2-1}}$$

$$=\frac{\frac{2}{(b+1)(b-1)}-\frac{3}{a(b-1)}}{\frac{3}{a(b-1)}-\frac{2}{(b+1)(b-1)}}$$

$$=\frac{\frac{2}{(b+1)(b-1)}-\frac{3}{a(b-1)}}{\frac{3}{a(b-1)}-\frac{2}{(b+1)(b-1)}}\bullet\frac{a(b+1)(b-1)}{a(b+1)(b-1)}$$

$$=\frac{2a-3(b+1)}{3(b+1)-2a}$$

$$=\frac{2a-3b-3}{3b+3-2a}$$

$$=\frac{2a-3b-3}{-2a+3b+3}$$

$$=-\frac{2a-3b-3}{2a-3b-3}$$

$$=-1$$

SECTION 8.6
Review of Linear Equations in Two Variables

73. $y=-\frac{1}{3}x-1$

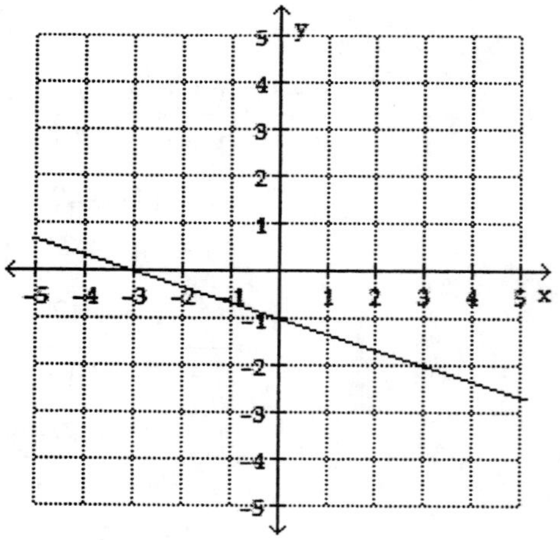

74. $x = -2$

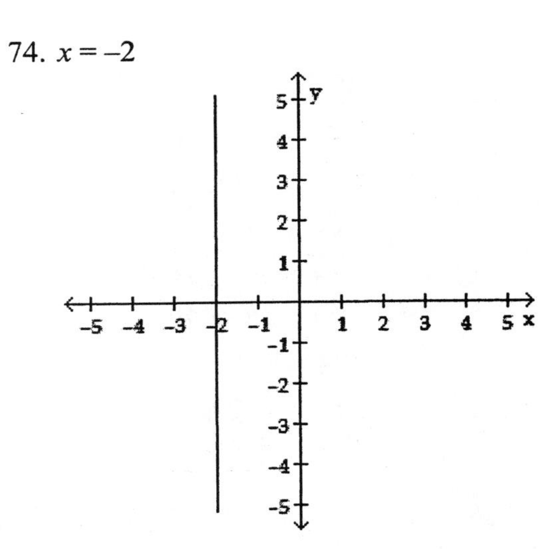

75. $2x + y = 4$

x–intercept:
Let $y = 0$.
$$2x + (0) = 4$$
$$2x = 4$$
$$x = 2$$
$$(2,0)$$

y–intercept:
Let $x = 0$.
$$2(0) + y = 4$$
$$y = 4$$
$$(0,4)$$

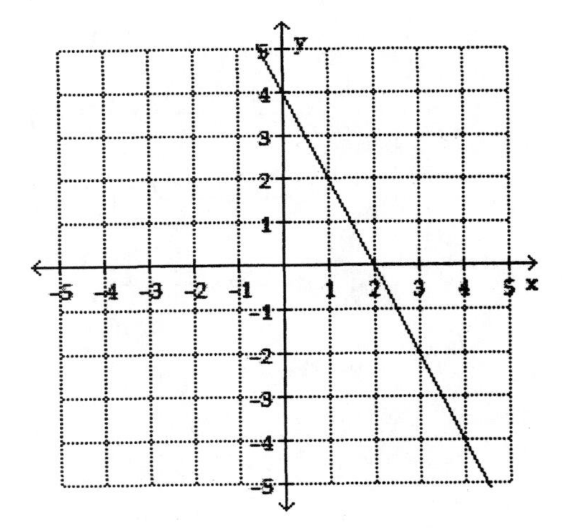

76. $3x - 4y - 8 = 0$

x–intercept:
Let $y = 0$.
$$3x - 4(0) - 8 = 0$$
$$3x - 8 = 0$$
$$3x = 8$$
$$x = 2.\overline{7}$$
$$(2.\overline{7},0)$$

y–intercept:
Let $x = 0$.
$$3(0) - 4y - 8 = 0$$
$$-4y - 8 = 0$$
$$-4y = 8$$
$$y = -2$$
$$(0,-2)$$

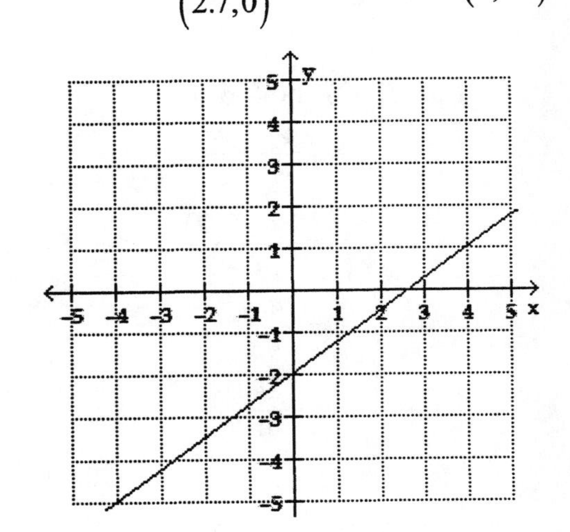

77. Write the equation in the form $y = mx + b$.
$$2x - 3y = 18$$
$$-3y = -2x + 18$$
$$y = \frac{2}{3}x - 6$$
$$m = \frac{2}{3}$$

78. For l_1, use (–3, –2) and (2, 2).
$$m = \frac{y_2 - y_1}{x_2 - x_1}$$
$$= \frac{2-(-2)}{2-(-3)}$$
$$= \frac{4}{5}$$

For l_2, use (0, 4) and (5, –4).
$$m = \frac{y_2 - y_1}{x_2 - x_1}$$
$$= \frac{-4-4}{5-0}$$
$$= -\frac{8}{5}$$

79. (2, 5) and (5, 8)

$$m = \frac{y_2 - y_1}{x_2 - x_1} = \frac{8-5}{5-2} = \frac{3}{3} = 1$$

80. (3, –2) and (–6, 12)

$$m = \frac{y_2 - y_1}{x_2 - x_1} = \frac{12-(-2)}{-6-3} = \frac{14}{-9} = -\frac{14}{9}$$

81. (–2, 4) and (8, 4)

$$m = \frac{y_2 - y_1}{x_2 - x_1} = \frac{4-4}{8-(-2)} = \frac{0}{10} = 0$$

82. (–5, –4) and (–5, 8)

$$m = \frac{y_2 - y_1}{x_2 - x_1} = \frac{8-(-4)}{-5-(-5)} = \frac{12}{0} = \text{undefined}$$

83. perpendicular; $4\left(-\frac{1}{4}\right) = -1$

84. parallel; $0.5 = \frac{1}{2}$

85. $m = 3$, $(x_1, y_1) = (-8, 5)$

$$\begin{aligned} y - y_1 &= m(x - x_1) \\ y - 5 &= 3(x - (-8)) \\ y - 5 &= 3(x + 8) \\ y - 5 &= 3x + 24 \\ y - 5 + 5 &= 3x + 24 + 5 \\ y &= 3x + 29 \\ -3x + y &= 29 \\ 3x - y &= -29 \end{aligned}$$

86. Find the slope of the line through (–2, 4) and (6, –9).

$$m = \frac{y_2 - y_1}{x_2 - x_1} = \frac{-9-4}{6-(-2)} = \frac{-13}{8}$$

$m = -\frac{13}{8}$, $(x_1, y_1) = (-2, 4)$

$$\begin{aligned} y - y_1 &= m(x - x_1) \\ y - 4 &= -\frac{13}{8}(x - (-2)) \\ y - 4 &= -\frac{13}{8}(x + 2) \\ y - 4 &= -\frac{13}{8}x - \frac{13}{4} \\ y - 4 + 4 &= -\frac{13}{8}x - \frac{13}{4} + 4 \\ y &= -\frac{13}{8}x + \frac{3}{4} \\ 8(y) &= 8\left(-\frac{13}{8}x + \frac{3}{4}\right) \\ 8y &= -13x + 6 \\ 13x + 8y &= 6 \end{aligned}$$

87. To find the slope of $3x - 2y = 7$, solve the equation for y.

$$3x - 2y = 7$$
$$-2y = -3x + 7$$
$$y = \frac{3}{2}x - \frac{7}{2}$$
$$m = \frac{3}{2}$$

Parallel lines have the same slope.

$m = \frac{3}{2}$, $(x_1, y_1) = (-3, -5)$

$$y - y_1 = m(x - x_1)$$
$$y - (-5) = \frac{3}{2}(x - (-3))$$
$$y + 5 = \frac{3}{2}(x + 3)$$
$$y + 5 = \frac{3}{2}x + \frac{9}{2}$$
$$y + 5 - 5 = \frac{3}{2}x + \frac{9}{2} - 5$$
$$y = \frac{3}{2}x - \frac{1}{2}$$

88. To find the slope of $3x - 2y = 7$, solve the equation for y.

$$3x - 2y = 7$$
$$-2y = -3x + 7$$
$$y = \frac{3}{2}x - \frac{7}{2}$$
$$m = \frac{3}{2}$$

Perpendicular lines have slopes that are opposite reciprocals. The opposite reciprocal of $\frac{3}{2}$ is $-\frac{2}{3}$.

$m = -\frac{2}{3}$, $(x_1, y_1) = (-3, -5)$

$$y - y_1 = m(x - x_1)$$
$$y - (-5) = -\frac{2}{3}(x - (-3))$$
$$y + 5 = -\frac{2}{3}(x + 3)$$
$$y + 5 = -\frac{2}{3}x - 2$$
$$y + 5 - 5 = -\frac{2}{3}x - 2 - 5$$
$$y = -\frac{2}{3}x - 7$$

89. Let the ordered pairs be (0, 8,700) and (5, 100). Find the slope (rate of change).

$$m = \frac{y_2 - y_1}{x_2 - x_1}$$
$$= \frac{100 - 8,700}{5 - 0}$$
$$= \frac{-8,600}{5}$$
$$= -1,720$$

Use (0, 8,700) and $m = -1,720$.

$$y - y_1 = m(x - x_1)$$
$$y - 8,700 = -1,720(x - 0)$$
$$y - 8,700 = -1,720x - 0$$
$$y - 8,700 + 8,700 = -1,720x - 0 + 8,700$$
$$y = -1,720x + 8,700$$

90.

$$\left(\frac{x_1 + x_2}{2}, \frac{y_1 + y_2}{2}\right) = \left(\frac{-3 + 6}{2}, \frac{5 + 11}{2}\right)$$
$$= \left(\frac{3}{2}, \frac{16}{2}\right)$$
$$= \left(\frac{3}{2}, 8\right)$$

SECTION 8.7
An Introduction to Functions

91. yes

92. no

93. yes

94. no

95. $f(x) = 3x + 2$

$$f(-3) = 3(-3) + 2$$
$$= -9 + 2$$
$$= -7$$

96. $g(x)=\dfrac{x^2-4x+4}{2}$

$$g(8)=\frac{(\mathbf{8})^2-4(\mathbf{8})+4}{2}$$
$$=\frac{64-32+4}{2}$$
$$=\frac{36}{2}$$
$$=18$$

97. $g(x)=\dfrac{x^2-4x+4}{2}$

$$g(8)=\frac{(\mathbf{-2})^2-4(\mathbf{-2})+4}{2}$$
$$=\frac{4+8+4}{2}$$
$$=\frac{16}{2}$$
$$=8$$

98. $f(x)=3x+2$

$$f(t)=3(\boldsymbol{t})+2$$
$$=3t+2$$

99.

$$-5x+7=-8$$
$$-5x+7-7=-8-7$$
$$-5x=-15$$
$$x=3$$

100.

$$\frac{3}{4}x-1=0$$
$$4\left(\frac{3}{4}x-1\right)=4(0)$$
$$3x-4=0$$
$$3x-4+4=0+4$$
$$3x=4$$
$$x=\frac{4}{3}$$

101. D: the set of real numbers
R: the set of real numbers

102. D: the set of real numbers except 2 (2 makes the denominator 0 and undefined)
R: the set of real numbers except 0 (the numerator cannot be 0)

103. It is a function; it passes the vertical line test.

104. It is not a function; it does not pass the vertical line test.

105. (0, 21.2) and (20, 48.6)

$$m=\frac{y_2-y_1}{x_2-x_1}$$
$$=\frac{48.6-21.2}{20-0}$$
$$=\frac{27.4}{20}$$
$$=1.37\% \text{ per year}$$

The y–intercept is (0, 21.2).

Use the slope–intercept form of an equation.

$$f(t)=mt+b$$
$$f(t)=1.37t+21.2$$

Find $f(24)$ since $2004-1980=24$.

$$f(24)=1.37(24)+21.2$$
$$=32.88+21.2$$
$$=54.08$$
$$\approx 54\%$$

106. $f(x)=\frac{2}{3}x-2$

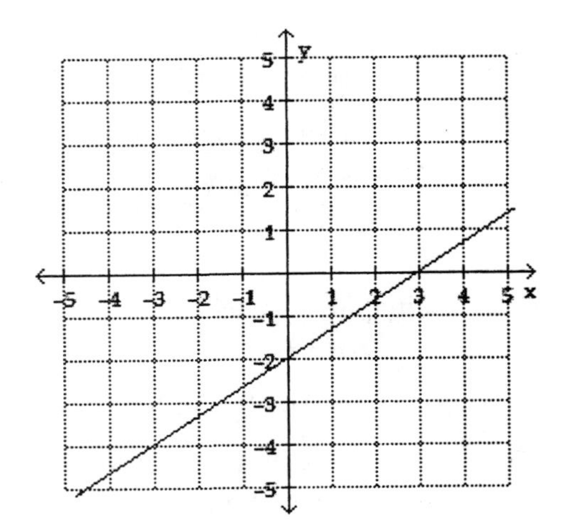

109. $f(x)=x^2-3$

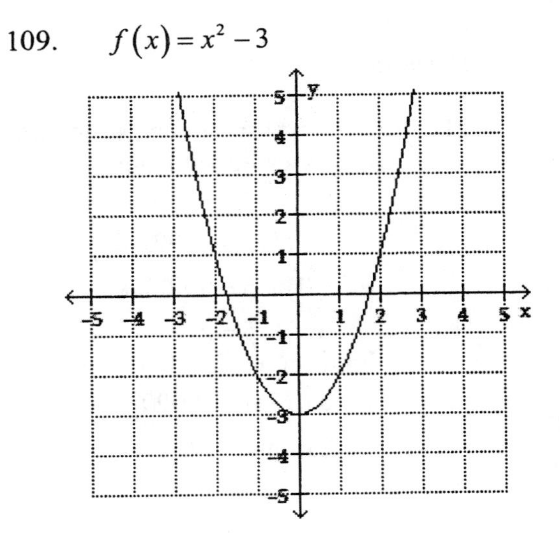

110. $f(x)=(x-2)^3+1$

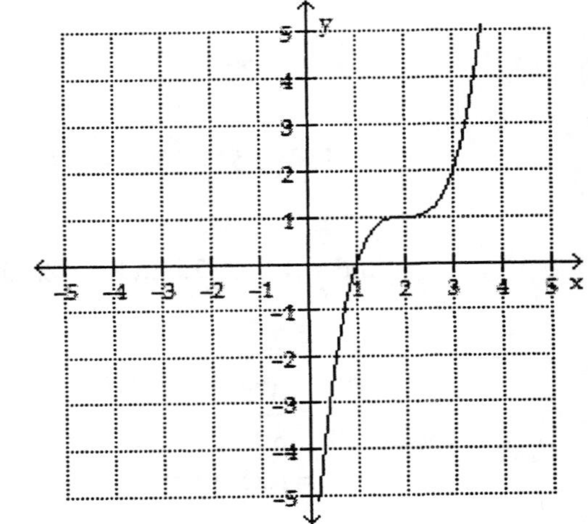

SECTION 8.8
Graphs of Functions

107. a. -4
 b. 3
 c. $x=0$

108.
$f(x)=|x+2|$
$g(x)=|x|-3$
$h(x)=-|x|$

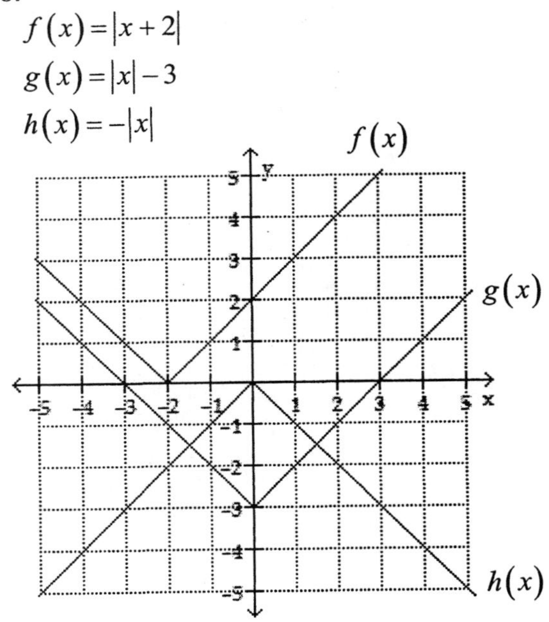

111. D: the set of real numbers
 R: the set of real numbers

112. D: the set of real numbers
 R: the set of real numbers greater than or equal to 1

SECTION 8.9
Variation

113. Let t = property tax, k = constant, and a = assessed valuation.

$$t = ka$$
$$1{,}575 = k(90{,}000)$$
$$1{,}575 = 90{,}000k$$
$$0.0175 = k$$

Find t if $a = 312{,}000$ and $k = 0.0175$.

$$t = 0.0175(312{,}000)$$
$$t = \$5{,}460$$

114. Let c = current, k = constant, and r = resistance.

$$c = \frac{k}{r}$$
$$2.5 = \frac{k}{150}$$
$$150(2.5) = 150\left(\frac{k}{150}\right)$$
$$375 = k$$

Find c if $k = 375$ and $r = 2(150) = 300$.

$$c = \frac{k}{r}$$
$$= \frac{375}{300}$$
$$= 1.25 \text{ amps}$$

115.

$$y = kxz$$
$$3 = k(24)(4)$$
$$3 = 96k$$
$$\frac{3}{96} = k$$
$$\frac{1}{32} = k$$
$$k = \frac{1}{32}$$

116. f = force of wind, k = constant, A = area of sign, and v = velocity of wind.

$$f = kAv^2$$
$$19.8 = k(3 \cdot 1.5)(10)^2$$
$$19.8 = k(4.5)(100)$$
$$19.8 = 450k$$
$$0.044 = k$$

Find f if $k = 0.044$ and $v = 80$.

$$f = kAv^2$$
$$f = 0.044(1.5 \cdot 3)(80)^2$$
$$f = 0.044(4.5)(6{,}400)$$
$$f = 1{,}267.2 \text{ lb}$$

117. inverse variation

118.

$$x_1 = \frac{kt^3}{x_2}$$
$$1.6 = \frac{k(8)^3}{64}$$
$$1.6 = \frac{512k}{64}$$
$$1.6 = 8k$$
$$0.2 = k$$

CHAPTER 8 TEST

1.

$$t + 18 = 5t - 3 + t$$
$$t + 18 = 6t - 3$$
$$-5t + 18 = -3$$
$$-5t = -21$$
$$t = \frac{-21}{-5}$$
$$t = \frac{21}{5}$$

2.

$$\frac{2}{3}(2s + 2) = \frac{1}{6}(5s + 29) - 4$$
$$6\left[\frac{2}{3}(2s + 2)\right] = 6\left[\frac{1}{6}(5s + 29) - 4\right]$$
$$4(2s + 2) = 1(5s + 29) - 24$$
$$8s + 8 = 5s + 29 - 24$$
$$8s + 8 = 5s + 5$$
$$3s + 8 = 5$$
$$3s = -3$$
$$s = -1$$

3.

$$6 - (x - 3) - 5x = 3[1 - 2(x + 2)]$$
$$6 - x + 3 - 5x = 3[1 - 2x - 4]$$
$$-6x + 9 = 3[-2x - 3]$$
$$-6x + 9 = -6x - 9$$
$$-6x + 9 + 6x = -6x - 9 + 6x$$
$$9 \neq -9$$

$\varnothing$ (no solution); contradiction

4.

$$n = \frac{360}{180 - a}$$
$$(180 - a)n = (180 - a)\left(\frac{360}{180 - a}\right)$$
$$180n - an = 360$$
$$-an = 360 - 180n$$
$$a = \frac{360 - 180n}{-n}$$
$$a = \frac{-360 + 180n}{n}$$
$$a = \frac{180n - 360}{n}$$

5.

width	x
length	$x + 5$
perimeter	26

$$P = 2L + 2W$$
$$26 = 2(x + 5) + 2(x)$$
$$26 = 2x + 10 + 2x$$
$$26 = 4x + 10$$
$$26 - 10 = 4x + 10 - 10$$
$$16 = 4x$$
$$\frac{16}{4} = \frac{4x}{4}$$
$$4 = x$$
$$9 = x + 5$$

The length is 9 cm and the width is 4 cm.

6. Let x = grade on Exam 5.

$$\frac{70 + 79 + 85 + 88 + x}{5} > 80$$
$$\frac{322 + x}{5} > 80$$
$$5\left(\frac{322 + x}{5}\right) > 5(80)$$
$$322 + x > 400$$
$$322 + x - 322 > 400 - 322$$
$$x > 78$$

She must make higher than 78.

7.

$$-2(2x + 3) \geq 14$$
$$-4x - 6 \geq 14$$
$$-4x - 6 + 6 \geq 14 + 6$$
$$-4x \geq 20$$
$$x \leq -5; \quad (-\infty, -5]$$

−5

8.

$$-2 < \frac{x-4}{3} < 4$$
$$3(-2) < 3\left(\frac{x-4}{3}\right) < 3(4)$$
$$-6 < x-4 < 12$$
$$-6+4 < x-4+4 < 12+4$$
$$-2 < x < 16$$
$$(-2, 16)$$

−2 16

9.

$$3x \geq -2x+5 \quad \text{and} \quad 7 \geq 4x-2$$
$$3x+2x \geq -2x+5+2x \quad 7+2 \geq 4x-2+2$$
$$5x \geq 5 \quad 9 \geq 4x$$
$$x \geq 1 \quad \frac{9}{4} \geq x$$
$$\left[1, \frac{9}{4}\right]$$

1 $\frac{9}{4}$

10.

$$3x < -9 \quad \text{or} \quad -\frac{x}{4} < -2$$
$$\frac{3x}{3} < \frac{-9}{3} \quad -4\left(-\frac{x}{4}\right) > -4(-2)$$
$$x < -3 \quad x > 8$$
$$(-\infty, -3) \cup (8, \infty)$$

−3 8

11.

$$|2x-4| > 22$$
$$2x-4 < -22 \quad \text{or} \quad 2x-4 > 22$$
$$2x-4+4 < -22+4 \quad 2x-4+4 > 22+4$$
$$2x < -18 \quad 2x > 26$$
$$x < -9 \quad x > 13$$
$$(-\infty, -9) \cup (13, \infty)$$

−9 13

12.

$$2|3(x-2)| \leq 4$$
$$2|3x-6| \leq 4$$
$$\frac{2|3x-6|}{2} \leq \frac{4}{2}$$
$$|3x-6| \leq 2$$
$$-2 \leq 3x-6 \leq 2$$
$$-2+6 \leq 3x-6+6 \leq 2+6$$
$$4 \leq 3x \leq 8$$
$$\frac{4}{3} \leq x \leq \frac{8}{3}$$
$$\left[\frac{4}{3}, \frac{8}{3}\right]$$

$\frac{4}{3}$ $\frac{8}{3}$

13.

$$|2x+3| - 19 = 0$$
$$|2x+3| = 19$$
$$2x+3 = 19 \quad \text{or} \quad 2x+3 = -19$$
$$2x = 16 \quad 2x = -22$$
$$x = 8 \quad x = -11$$

14.

$$|3x+4| = |x+12|$$
$$3x+4 = x+12 \quad \text{or} \quad 3x+4 = -(x+12)$$
$$3x+4 = x+12 \quad 3x+4 = -x-12$$
$$3x+4-4 = x+12-4 \quad 3x+4-4 = -x-12-4$$
$$3x = x+8 \quad 3x = -x-16$$
$$3x-x = x+8-x \quad 3x+x = -x-16+x$$
$$2x = 8 \quad 4x = -16$$
$$x = 4 \quad x = -4$$

15.

$$12a^3b^2c - 3a^2b^2c^2 + 6abc^3$$
$$= 3abc\left(4a^2b - abc + 2c^2\right)$$

16.

$$\begin{aligned} 4y^4 - 64 &= 4(y^4 - 16) \\ &= 4(y^2 + 4)(y^2 - 4) \\ &= 4(y^2 + 4)(y + 2)(y - 2) \end{aligned}$$

17.

$$\begin{aligned} b^3 + 125 &= (b)^3 + (5)^3 \\ &= (b + 5)(b^2 - 5b + 25) \end{aligned}$$

18.

$$\begin{aligned} 6u^2 + 9u - 6 &= 3(2u^2 + 3u - 2) \\ &= 3(2u - 1)(u + 2) \end{aligned}$$

19.

$$\begin{aligned} ax - xy + ay - y^2 &= x(a - y) + y(a - y) \\ &= (a - y)(x + y) \end{aligned}$$

20.

$$25m^8 - 60m^4n + 36n^2 = (5m^4 - 6n)^2$$

21.

$144b^2 + 25$
prime

22.

$$\begin{aligned} x^2 + 6x + 9 - y^2 &= (x^2 + 6x + 9) - y^2 \\ &= (x + 3)^2 - y^2 \\ &= (x + 3 - y)(x + 3 + y) \end{aligned}$$

23.

$$\begin{aligned} &64a^3 - 125b^6 \\ &= (4a - 5b^2)\left[(4a)^2 + (4a)(5b^2) + (5b^2)^2\right] \\ &= (4a - 5b^2)(16a^2 + 20ab^2 + 25b^4) \end{aligned}$$

24.

$$\begin{aligned} &(x - y)^2 + 3(x - y) - 10 \\ &= [(x - y) + 5][(x - y) - 2] \\ &= (x - y + 5)(x - y - 2) \end{aligned}$$

25.

$$\begin{aligned} \frac{3y - 6z}{2z - y} &= \frac{3(y - 2z)}{2z - y} \\ &= \frac{3\cancel{(y - 2z)}}{-1\cancel{(y - 2z)}} \\ &= -3 \end{aligned}$$

26.

$$\begin{aligned} \frac{2x^2 + 7xy + 3y^2}{4xy + 12y^2} &= \frac{(2x + y)\cancel{(x + 3y)}}{4y\cancel{(x + 3y)}} \\ &= \frac{2x + y}{4y} \end{aligned}$$

27.

$$\begin{aligned} \frac{x^3 + y^3}{4} \div \frac{x^2 - xy + y^2}{2x + 2y} &= \frac{x^3 + y^3}{4} \bullet \frac{2x + 2y}{x^2 - xy + y^2} \\ &= \frac{(x + y)\cancel{(x^2 - xy + y^2)}}{\cancel{2} \bullet 2} \bullet \frac{\cancel{2}(x + y)}{\cancel{x^2 - xy + y^2}} \\ &= \frac{(x + y)^2}{2} \end{aligned}$$

28.

$$\begin{aligned} &\frac{xu + 2u + 3x + 6}{u^2 - 9} \bullet \frac{13u - 39}{x^2 + 3x + 2} \\ &= \frac{u(x + 2) + 3(x + 2)}{(u + 3)(u - 3)} \bullet \frac{13(u - 3)}{(x + 1)(x + 2)} \\ &= \frac{\cancel{(u + 3)}\cancel{(x + 2)}}{\cancel{(u + 3)}\cancel{(u - 3)}} \bullet \frac{13\cancel{(u - 3)}}{(x + 1)\cancel{(x + 2)}} \\ &= \frac{13}{x + 1} \end{aligned}$$

29.

$$\begin{aligned} \frac{-3t + 4}{t^2 + t - 20} + \frac{6 + 5t}{t^2 + t - 20} &= \frac{-3t + 4 + 6 + 5t}{t^2 + t - 20} \\ &= \frac{2t + 10}{t^2 + t - 20} \\ &= \frac{2\cancel{(t + 5)}}{\cancel{(t + 5)}(t - 4)} \\ &= \frac{2}{t - 4} \end{aligned}$$

30.

$$\frac{a+3}{a^2-a-2}-\frac{a-4}{a^2-2a-3}$$

$$=\frac{a+3}{(a-2)(a+1)}-\frac{a-4}{(a-3)(a+1)}$$

$$=\frac{a+3}{(a-2)(a+1)}\left(\frac{a-3}{a-3}\right)-\frac{a-4}{(a-3)(a+1)}\left(\frac{a-2}{a-2}\right)$$

$$=\frac{a^2-9}{(a-2)(a+1)(a-3)}-\frac{a^2-2a-4a+8}{(a-2)(a+1)(a-3)}$$

$$=\frac{a^2-9}{(a-2)(a+1)(a-3)}-\frac{a^2-6a+8}{(a-2)(a+1)(a-3)}$$

$$=\frac{a^2-9-(a^2-6a+8)}{(a-2)(a+1)(a-3)}$$

$$=\frac{a^2-9-a^2+6a-8}{(a-2)(a+1)(a-3)}$$

$$=\frac{6a-17}{(a-2)(a+1)(a-3)}$$

31.

$$\frac{\frac{2u^2w^3}{v^2}}{\frac{4uw^4}{uv}}=\frac{\frac{2u^2w^3}{v^2}}{\frac{4uw^4}{uv}}\bullet\frac{uv^2}{uv^2}$$

$$=\frac{\left(\frac{uv^2}{1}\right)\left(\frac{2u^2w^3}{v^2}\right)}{\left(\frac{uv^2}{1}\right)\left(\frac{4uw^4}{uv}\right)}$$

$$=\frac{2u^3w^3}{4uvw^4}$$

$$=\frac{u^2}{2vw}$$

32.

$$\frac{\frac{4}{3k}+\frac{k}{k+1}}{\frac{k}{k+1}-\frac{3}{k}}=\frac{\frac{4}{3k}+\frac{k}{k+1}}{\frac{k}{k+1}-\frac{3}{k}}\bullet\frac{3k(k+1)}{3k(k+1)}$$

$$=\frac{3k(k+1)\left(\frac{4}{3k}\right)+3k(k+1)\left(\frac{k}{k+1}\right)}{3k(k+1)\left(\frac{k}{k+1}\right)-3k(k+1)\left(\frac{3}{k}\right)}$$

$$=\frac{4(k+1)+3k(k)}{3k(k)-3(k+1)(3)}$$

$$=\frac{4k+4+3k^2}{3k^2-9(k+1)}$$

$$=\frac{3k^2+4k+4}{3k^2-9k-9}$$

33. The slope of $y=-\frac{3}{2}x-7$ is $-\frac{3}{2}$. The slope of the line parallel to that would also be $-\frac{3}{2}$. Use the point-slope form of an equation with (–2, 1) and $m=-\frac{3}{2}$.

$$y-y_1=m(x-x_1)$$

$$y-1=-\frac{3}{2}(x-(-2))$$

$$y-1=-\frac{3}{2}(x+2)$$

$$y-1=-\frac{3}{2}x-3$$

$$y-1+1=-\frac{3}{2}x-3+1$$

$$y=-\frac{3}{2}x-2$$

$$2(y)=2\left(-\frac{3}{2}x-2\right)$$

$$2y=-3x-4$$

$$3x+2y=-4$$

34. Solve the equation for y.

$$2x+9=-6y$$
$$6y=-2x-9$$
$$\frac{6y}{6}=\frac{-2x}{6}-\frac{9}{6}$$
$$y=-\frac{1}{3}x-\frac{3}{2}$$
$$m=-\frac{1}{3};\ b=\left(0,-\frac{3}{2}\right)$$

35. (–3, 5) and (4, –6)

$$m=\frac{y_2-y_1}{x_2-x_1}$$
$$=\frac{-6-5}{4-(-3)}$$
$$=\frac{-11}{7}$$
$$=-\frac{11}{7}$$

36. a. Find the rate of change (m) using the ordered pairs (0, 4,000) and (6, 400).

$$m=\frac{y_2-y_1}{x_2-x_1}$$
$$=\frac{400-4,000}{6-0}$$
$$=\frac{-3,600}{6}$$
$$=-600$$

Use $m=-600$ and (0, 4,000)

$$y-y_1=m(x-x_1)$$
$$y-4,000=-600(x-0)$$
$$y-4,000=-600x$$
$$y-4,000+4,000=-600x+4,000$$
$$y=-600x+4,000$$

b. (0, 4,000); It gives the value of the copier when new, \$4,000.

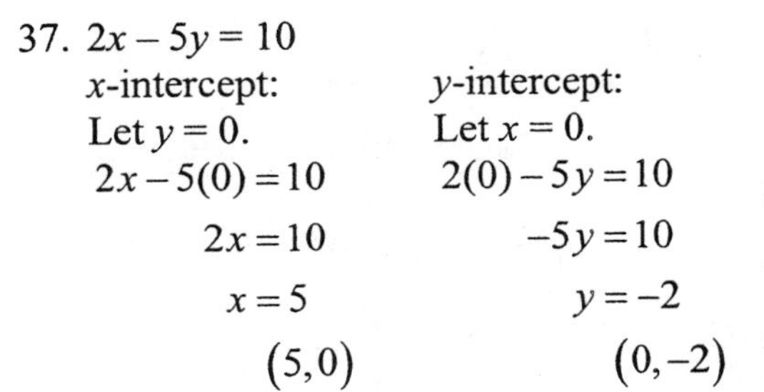

37. $2x-5y=10$

x-intercept:
Let $y=0$.
$$2x-5(0)=10$$
$$2x=10$$
$$x=5$$
$$(5,0)$$

y-intercept:
Let $x=0$.
$$2(0)-5y=10$$
$$-5y=10$$
$$y=-2$$
$$(0,-2)$$

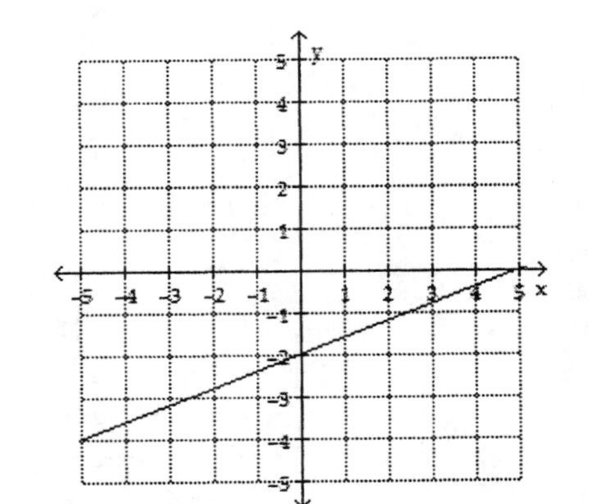

38. $y=-2$

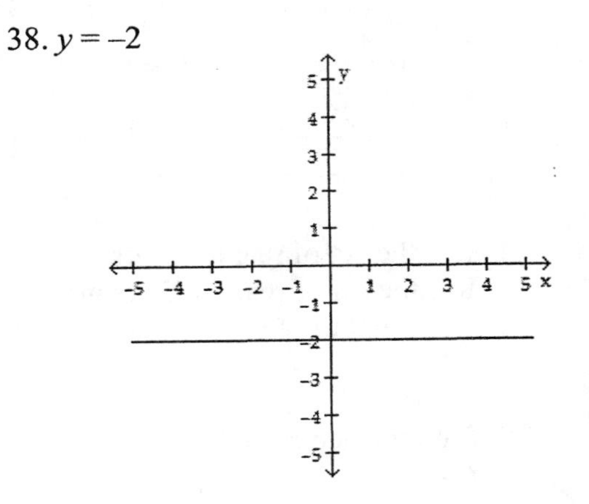

39. Yes. There is only one y-value assigned to each x-value.

40. No. It does not pass the vertical line test.

41. $g(x)=x^2-2x-1$

$$g(0)=0^2-2(0)-1$$
$$=0-0-1$$
$$=-1$$

42. Let $f(x)=4$.

$$-\frac{4}{5}x-12=4$$
$$5\left(-\frac{4}{5}x-12\right)=5(4)$$
$$-4x-60=20$$
$$-4x-60+60=20+60$$
$$-4x=80$$
$$x=-20$$

43. $g(x)=-|x+2|$

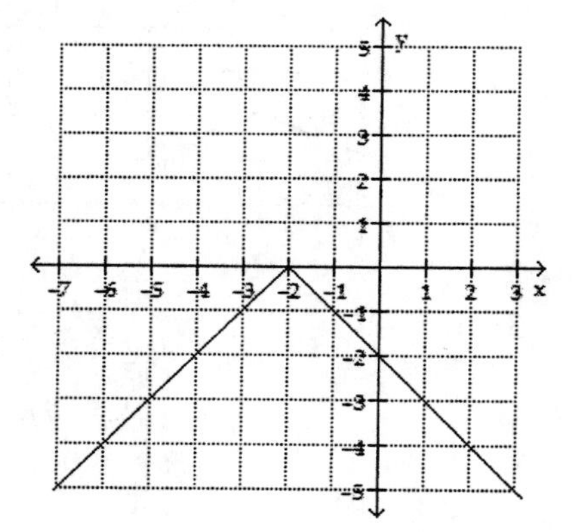

44. D: the set of real numbers
R: the set of real numbers greater than or equal to –5

45. Find y when $x=-2$.
$f(-2)=-2$

46. Find x when $y=3$.
$x=2$

47. Find k when $y=4$ and $x=30$.

$$y=kx$$
$$4=k(30)$$
$$4=30k$$
$$\frac{4}{30}=k$$
$$\frac{2}{15}=k$$
$$k=\frac{2}{15}$$

Find y when $x=9$ and $k=\frac{2}{15}$.

$$y=kx$$
$$y=\left(\frac{2}{15}\right)(9)$$
$$y=\frac{18}{15}$$
$$y=\frac{6}{5}$$

48. Let L = loudness and d = distance.

$$L=\frac{k}{d^2}$$
$$100=\frac{k}{30^2}$$
$$100=\frac{k}{900}$$
$$90{,}000=k$$

Find L if $k=90{,}000$ and $d=60$.

$$L=\frac{k}{d^2}$$
$$=\frac{90{,}000}{60^2}$$
$$=\frac{90{,}000}{3{,}600}$$
$$=25 \text{ decibels}$$

CHAPTER 8 CUMULATIVE REVIEW

1. According to the circle graph, 32% of the candy sales were for Halloween.

   Find 32% of $6,300,000,000

   $0.32(6{,}300{,}000{,}000) = 2{,}016{,}000{,}000$

   The candy sales for Halloween were $2,016,000,000.

2. $\{\ldots, -3, -2, -1, 0, 1, 2, 3, \ldots\}$

3.
$$\begin{aligned} 3-4[-10-4(-5)] &= 3-4[-10+20] \\ &= 3-4[10] \\ &= 3-40 \\ &= -37 \end{aligned}$$

4.
$$\begin{aligned} \frac{|-45|-2(-5)+1^5}{2\cdot 9-2^4} &= \frac{45-2(-5)+1}{2\cdot 9-16} \\ &= \frac{45+10+1}{18-16} \\ &= \frac{56}{2} \\ &= 28 \end{aligned}$$

5. The diameter is 6 inches, so the radius (half the diameter) is 3 inches. Since there are 12 inches in one foot, change 3 inches to feet by dividing by 12:

$$r = \frac{3}{12} = \frac{1}{4}$$

   Find the volume with $h = 6$ ft. and $r = \frac{1}{4}$ ft.

$$\begin{aligned} V &= \pi r^2 h \\ &= \pi\left(\frac{1}{4}\right)^2(6) \\ &= \pi\left(\frac{1}{16}\right)(6) \\ &= \frac{6\pi}{16} \\ &= \frac{3\pi}{8} \\ &\approx 1.2 \text{ ft}^3 \end{aligned}$$

6. rational numbers: terminating and repeating decimals
   Irrational numbers: nonterminating, nonrepeating decimals

7.
$$\begin{aligned} |x| - xy &= |2| - 2(-4) \\ &= 2+8 \\ &= 10 \end{aligned}$$

8.
$$\begin{aligned} \frac{x^2-y^2}{3x+y} &= \frac{(2)^2-(-4)^2}{3(2)+(-4)} \\ &= \frac{4-16}{6-4} \\ &= \frac{-12}{2} \\ &= -6 \end{aligned}$$

9.
$$\begin{aligned} 3p^2-6(5p^2+p)+p^2 &= 3p^2-30p^2-6p+p^2 \\ &= -26p^2-6p \end{aligned}$$

10.
$$\begin{aligned} -(a+2)-(a-b) &= -a-2-a+b \\ &= -2a+b-2 \end{aligned}$$

11. The sum of the angles of a triangle is 180° and the right angle is 90°.

$$\begin{aligned} x+2x+90 &= 180 \\ 3x+90 &= 180 \\ 3x &= 90 \\ x &= 30 \end{aligned}$$

12. Let x = the amount of money in the 12% fund.

	Principal ·	rate ·	time =	Interest
mutual fund	x	0.12	1	$0.12x$
bonds	$45{,}000-x$	0.065	1	$0.065(45{,}000-x)$

$$\begin{aligned} 0.12x+0.065(45{,}000-x)&=4{,}300\\ 0.12x+2{,}925-0.065x&=4{,}300\\ 0.055x&=1{,}375\\ x&=25{,}000 \end{aligned}$$

$$\begin{aligned} 45{,}000-x&=45{,}000-25{,}000\\ &=20{,}000 \end{aligned}$$

\$25,000 was invested in mutual funds and \$20,000 was invested in bonds.

13.

Change $11\frac{3}{4}$ inches to feet by dividing by 12.

$11\frac{3}{4}\div 12=0.97916\overline{6}$

Area = length · width
$\approx 205\cdot 0.979$
$\approx 201\text{ ft}^2$

14. a. $P=2l+2w$
b. $A=lw$
c. $A=\pi r^2$
d. $d=rt$

15.

$$\begin{aligned}(5x-8y)-(-2x+5y)&=5x-8y+2x-5y\\&=7x-13y\end{aligned}$$

16.

$$2x^2(3x^2+4x-7)=6x^4+8x^3-14x^2$$

17.

$$\begin{aligned}(6p-5q)^2&=(6p-5q)(6p-5q)\\&=36p^2-30pq-30pq+25q^2\\&=36p^2-60pq+25q^2\end{aligned}$$

18.

$$\begin{aligned}(x+3)(2x-3)&=2x^2-3x+6x-9\\&=2x^2+3x-9\end{aligned}$$

19.

$$\begin{aligned}\frac{2x-32}{16x}&=\frac{2x}{16}-\frac{32}{16}\\&=\frac{1}{8}-\frac{2}{x}\end{aligned}$$

20.

$$\begin{array}{r} 3x+1 \\ 3x+1\overline{)9x^2+6x+1} \\ \underline{9x^2+3x} \\ 3x+1 \\ \underline{3x+1} \\ 0 \end{array}$$

21.

$$\begin{aligned}&(-3x+y)(x^2-8xy+16y^2)\\&=-3x(x^2-8xy+16y^2)+y(x^2-8xy+16y^2)\\&=-3x^3+24x^2y-48xy^2+x^2y-8xy^2+16y^3\\&=-3x^3+25x^2y-56xy^2+16y^3\end{aligned}$$

22.

$$\begin{aligned}(x+y)(x-y)+x(x+y)&=x^2-y^2+x^2+xy\\&=2x^2+xy-y^2\end{aligned}$$

23.

$$\begin{aligned}(x^5)^2(x^7)^3&=x^{5\cdot 2}x^{7\cdot 3}\\&=x^{10}x^{21}\\&=x^{10+21}\\&=x^{31}\end{aligned}$$

24.

$$\begin{aligned}\frac{16(aa^2)^3}{2a^2a^3}&=\frac{16(a^3)^3}{2a^{2+3}}\\&=\frac{16a^{3\cdot 3}}{2a^5}\\&=\frac{16a^9}{2a^5}\\&=8a^{9-5}\\&=8a^4\end{aligned}$$

25.

$$\frac{2^{-4}}{3^{-1}} = \frac{3^1}{2^4}$$
$$= \frac{3}{16}$$

26.

$$(2x)^0 = 1$$

27.

$$\left(-3u^{-2}v^3\right)^{-3} = \left(-\frac{3v^3}{u^2}\right)^{-3}$$
$$= \left(-\frac{u^2}{3v^3}\right)^3$$
$$= -\frac{u^{2\cdot 3}}{3^3 v^{3\cdot 3}}$$
$$= -\frac{u^6}{27v^9}$$

28.

$$\left(\frac{12y^3z^{-2}}{3y^{-4}z^3}\right)^2 = \left(4y^{3-(-4)}z^{-2-3}\right)^2$$
$$= \left(4y^7z^{-5}\right)^2$$
$$= \left(\frac{4y^7}{z^5}\right)^2$$
$$= \frac{4^2 y^{7\cdot 2}}{z^{5\cdot 2}}$$
$$= \frac{16y^{14}}{z^{10}}$$

29.

$$\frac{3}{x+1} - \frac{x-2}{2} = \frac{x-2}{x+1}$$
$$2(x+1)\left(\frac{3}{x+1} - \frac{x-2}{2}\right) = 2(x+1)\left(\frac{x-2}{x+1}\right)$$
$$2(3) - (x+1)(x-2) = 2(x-2)$$
$$6 - \left(x^2 - 2x + x - 2\right) = 2x - 4$$
$$6 - x^2 + 2x - x + 2 = 2x - 4$$
$$-x^2 + x + 8 = 2x - 4$$
$$0 = x^2 + x - 12$$
$$0 = (x+4)(x-3)$$
$$x + 4 = 0 \quad \text{or} \quad x - 3 = 0$$
$$x = -4 \qquad\qquad x = 3$$

30. Find a.

$$\frac{15}{20} = \frac{12}{a}$$
$$15a = 12(20)$$
$$15a = 240$$
$$a = 16$$

Find b.

$$\frac{15}{20} = \frac{6}{b}$$
$$15b = 6(20)$$
$$15b = 120$$
$$b = 8$$

31.

$$8s^2 - 16s = 0$$
$$8s(s-2) = 0$$
$$8s = 0 \quad \text{or} \quad s - 2 = 0$$
$$s = 0 \qquad\qquad s = 2$$

32.

$$x^2 + 2x - 15 = 0$$
$$(x+5)(x-3) = 0$$
$$x + 5 = 0 \quad \text{or} \quad x - 3 = 0$$
$$x = -5 \qquad\qquad x = 3$$

33. $\begin{cases} x+4y=-2 \\ y=-x-5 \end{cases}$

Graph both equations on the same coordinate plane. The lines intersect and the point (–6, 1), so the solution to the system of equations is (–6, 1).

34. $\begin{cases} 2x-3y<0 \\ y>x-1 \end{cases}$

Graph both inequalities on the same coordinate plane. The region of intersection is the solution to the system of inequalities.

35. $\begin{cases} x-2y=2 \\ 2x+3y=11 \end{cases}$

Solve the first equation for x.

$\begin{cases} x=2y+2 \\ 2x+3y=11 \end{cases}$

Substitute $x = 2y + 2$ into the second equation and solve for y.

$$2(2y+2)+3y=11$$
$$4y+4+3y=11$$
$$7y+4=11$$
$$7y=7$$
$$y=1$$

Substitute $y = 1$ into the first equation and solve for x.

$$x=2y+2$$
$$x=2(1)+2$$
$$x=2+2$$
$$x=4$$

The solution is (4, 1).

36. Let x = servings of egg noodles and y = servings of rice pilaf. Write a system of equations using the servings of protein and fat in egg noodles and rice pilaf.

$\begin{cases} 5x+4y=22 \\ 3x+5y=21 \end{cases}$

Multiply the first equation by 3 and the second equation by –5. Use the elimination method to solve for y.

$$\begin{array}{r} 15x+12y=66 \\ -15x-25y=-105 \\ \hline -13y=-39 \end{array}$$
$$y=3$$

Substitute $y = 3$ into the first equation and solve for x.

$$5x+4y=22$$
$$5x+4(3)=22$$
$$5x+12=22$$
$$5x=10$$
$$x=2$$

2 servings of egg noodles and 3 servings of rice pilaf must be eaten.

SECTION 9.1

VOCABULARY

1. $5x^2$ is the **square** root of $25x^4$ because $(5x^2)^2 = 25x^4$.

3. In the expression $\sqrt[3]{27x^6}$, the **index** is 3 and $27x^6$ is the **radicand**.

5. When n is an odd number, $\sqrt[n]{x}$ represents an **odd** root. When n is an **even** number, $\sqrt[n]{x}$ represents an even root.

CONCEPTS

7. b is a square root of a if $b^2 = \boxed{a}$.

9. The number 25 has **two** square roots. The principal square root of 25 is $\boxed{5}$.

11. $\sqrt[3]{x} = y$ if $y^3 = \boxed{x}$.

13. The graph of $f(x) = \sqrt{x} + 3$ is the graph of $f(x) = \sqrt{x}$ translated $\boxed{3}$ units **up**.

15. a. $f(11) = 3$
    b. $f(2) = 0$
    c. $x = 6$
    d. D: $[2, \infty)$ and R: $[0, \infty)$

17. $f(x) = \sqrt{x}$

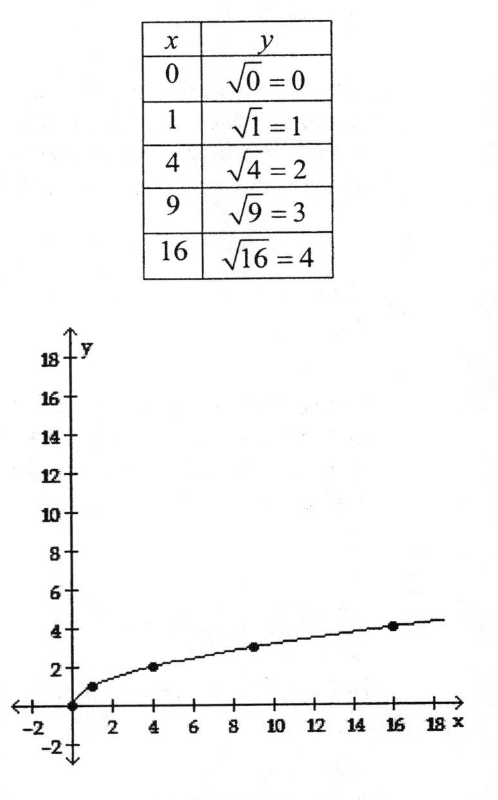

x	y
0	$\sqrt{0} = 0$
1	$\sqrt{1} = 1$
4	$\sqrt{4} = 2$
9	$\sqrt{9} = 3$
16	$\sqrt{16} = 4$

D: $[0, \infty)$ and R: $[0, \infty)$

NOTATION

19. $\sqrt{x^2} = |x|$

21. $f(x) = \sqrt{x - 5}$

PRACTICE

23.
$$\sqrt{121} = 11$$

25.
$$-\sqrt{64} = -1(8)$$
$$= -8$$

27.
$$\sqrt{\frac{1}{9}} = \frac{\sqrt{1}}{\sqrt{9}}$$
$$= \frac{1}{3}$$

29.

$$\sqrt{0.25} = 0.5$$

31.

$\sqrt{-25}$ is not real

33.

$$\sqrt{(-4)^2} = \sqrt{16}$$
$$= 4$$

35.

$$\sqrt{12} = 3.4641$$

37.

$$\sqrt{679.25} = 26.0624$$

39.

$$\sqrt{4x^2} = \sqrt{(2x)^2}$$
$$= |2x|$$
$$= 2|x|$$

41.

$$\sqrt{(t+5)^2} = |t+5|$$

43.

$$\sqrt{(-5b)^2} = |-5b|$$
$$= 5|b|$$

45.

$$\sqrt{a^2 + 6a + 9} = \sqrt{(a+3)^2}$$
$$= |a+3|$$

47.

$$f(x) = \sqrt{x-4}$$
$$f(4) = \sqrt{4-4}$$
$$= \sqrt{0}$$
$$= 0$$

49.

$$f(x) = \sqrt{x-4}$$
$$f(20) = \sqrt{20-4}$$
$$= \sqrt{16}$$
$$= 4$$

51.

$$g(x) = \sqrt[3]{x-4}$$
$$g(12) = \sqrt[3]{12-4}$$
$$= \sqrt[3]{8}$$
$$= 2$$

53.

$$g(x) = \sqrt[3]{x-4}$$
$$g(-996) = \sqrt[3]{-996-4}$$
$$= \sqrt[3]{-1{,}000}$$
$$= -10$$

55.

$$f(x) = \sqrt{x^2 + 1}$$
$$f(4) = \sqrt{4^2 + 1}$$
$$= \sqrt{16+1}$$
$$= \sqrt{17}$$
$$= 4.1231$$

57.

$$g(x) = \sqrt[3]{x^2 + 1}$$
$$g(6) = \sqrt[3]{6^2 + 1}$$
$$= \sqrt[3]{36+1}$$
$$= \sqrt[3]{37}$$
$$= 3.3322$$

59.

x	y
0	$-\sqrt{0}=0$
1	$-\sqrt{1}=-1$
4	$-\sqrt{4}=-2$
9	$-\sqrt{9}=-3$
16	$-\sqrt{16}=-4$

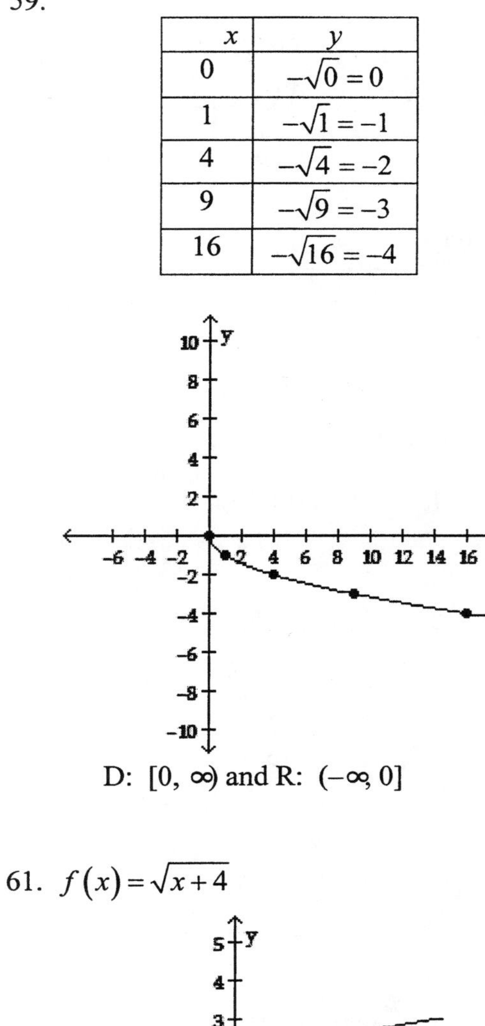

D: $[0, \infty)$ and R: $(-\infty, 0]$

61. $f(x)=\sqrt{x+4}$

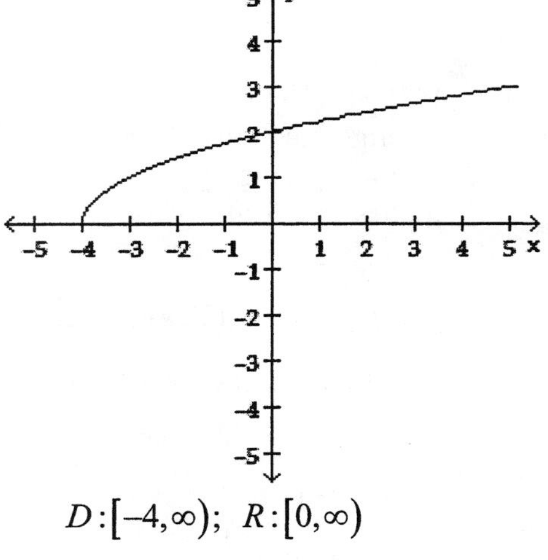

$D:[-4,\infty);\ \ R:[0,\infty)$

63. $f(x)=\sqrt[3]{x-3}$

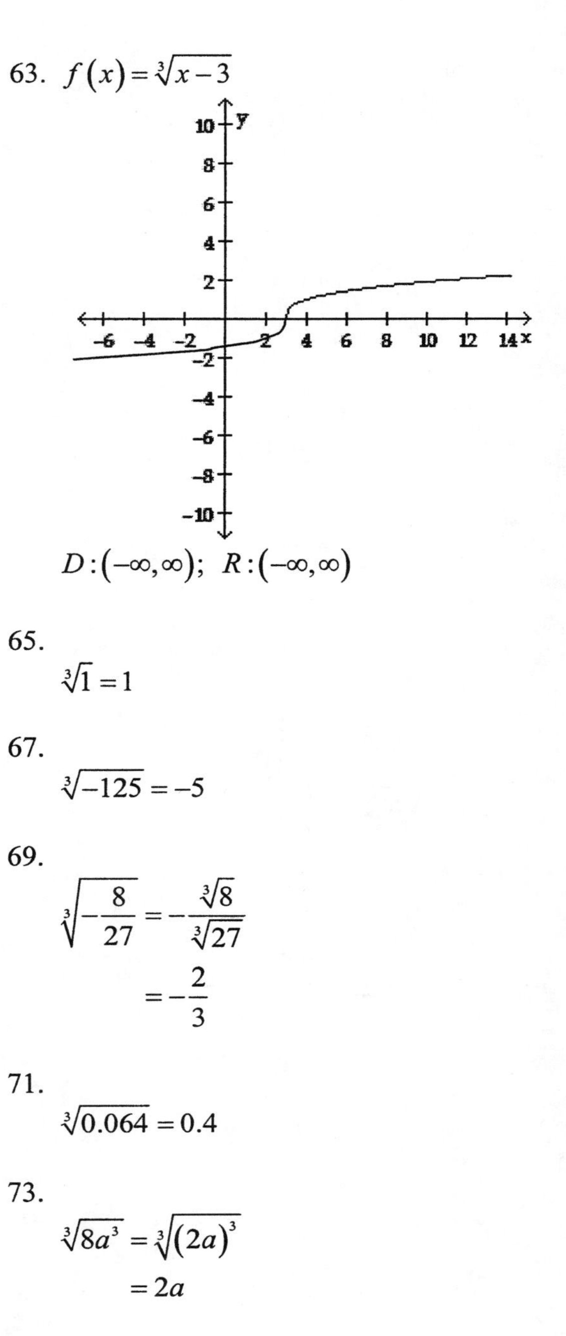

$D:(-\infty,\infty);\ \ R:(-\infty,\infty)$

65.

$$\sqrt[3]{1}=1$$

67.

$$\sqrt[3]{-125}=-5$$

69.

$$\sqrt[3]{-\frac{8}{27}}=-\frac{\sqrt[3]{8}}{\sqrt[3]{27}}$$
$$=-\frac{2}{3}$$

71.

$$\sqrt[3]{0.064}=0.4$$

73.

$$\sqrt[3]{8a^3}=\sqrt[3]{(2a)^3}$$
$$=2a$$

75.

$$\sqrt[3]{-1{,}000p^3q^3}=\sqrt[3]{(-10pq)^3}$$
$$=-10pq$$

77.

$$\sqrt[3]{-\frac{1}{8}m^6n^3}=\sqrt[3]{\left(-\frac{1}{2}m^2n\right)^3}$$
$$=-\frac{1}{2}m^2n$$

79.

$$\sqrt[3]{-0.064s^9t^6} = \sqrt[3]{\left(-0.4s^3t^2\right)^3}$$
$$= -0.4s^3t^2$$

81.

$$\sqrt[4]{81} = \sqrt[4]{(3)^4}$$
$$= 3$$

83.

$$-\sqrt[5]{243} = -\sqrt[5]{(3)^5}$$
$$= -1(3)$$
$$= -3$$

85.

$\sqrt[4]{-256}$ is not a real number

87.

$$\sqrt[4]{\frac{16}{625}} = \sqrt[4]{\frac{(2)^4}{(5)^4}}$$
$$= \frac{2}{5}$$

89.

$$-\sqrt[5]{-\frac{1}{32}} = -\sqrt[5]{-\frac{(1)^5}{(2)^5}}$$
$$= -1\left(-\frac{1}{2}\right)$$
$$= \frac{1}{2}$$

91.

$$\sqrt[5]{32a^5} = \sqrt[5]{(2a)^5}$$
$$= 2a$$

93.

$$\sqrt[4]{16a^4} = \sqrt[4]{(2a)^4}$$
$$= 2a$$

95.

$$\sqrt[4]{k^{12}} = \sqrt[4]{\left(k^3\right)^4}$$
$$= k^3$$

97.

$$\sqrt[4]{\frac{1}{16}m^4} = \sqrt[4]{\left(\frac{1}{2}m\right)^4}$$
$$= \frac{1}{2}m$$

99.

$$\sqrt[25]{(x+2)^{25}} = x+2$$

APPLICATIONS

101. EMBROIDERY

$$r = \sqrt{\frac{A}{\pi}}$$
$$= \sqrt{\frac{38.5}{\pi}}$$
$$\approx \sqrt{12.3}$$
$$\approx 3.5$$

$$d = 2r$$
$$= 2(3.5)$$
$$= 7.0 \text{ in.}$$

103. PULSE RATES
Change his height from feet to inches by multiplying by 12 since there are 12 inches in 1 foot.

$$7 \text{ ft } 8\frac{1}{2} \text{ in.} = 7(12) + 8\frac{1}{2} \text{ inches}$$
$$= 84 + 8.5 \text{ inches}$$
$$= 92.5 \text{ inches}$$

$$p(t) = \frac{590}{\sqrt{t}}$$
$$p(92.5) = \frac{590}{\sqrt{92.5}}$$
$$= \frac{590}{9.6}$$
$$= 61.3 \text{ beats/min.}$$

105. BIOLOGY

The volume required for each rat is 125 ft^3, so to find the volume for 5 rats, multiply 125 times 5.

$$\begin{aligned} d(V) &= \sqrt[3]{12\left(\frac{V}{\pi}\right)} \\ &= \sqrt[3]{12\left(\frac{5 \cdot 125}{\pi}\right)} \\ &= \sqrt[3]{12\left(\frac{625}{\pi}\right)} \\ &= \sqrt[3]{12(198.9)} \\ &= \sqrt[3]{2,386.8} \\ &= 13.4 \text{ ft} \end{aligned}$$

107. COLLECTIBLES

$$\begin{aligned} r &= \sqrt[n]{\frac{A}{P}} - 1 \\ &= \sqrt[5]{\frac{950}{800}} - 1 \\ &= \sqrt[5]{1.1875} - 1 \\ &= 1.035 - 1 \\ &= 0.035 \\ &= 3.5\% \end{aligned}$$

WRITING

109. Answers will vary.

111. Answers will vary.

REVIEW

113.

$$\begin{aligned} &\frac{x^2 - x - 6}{x^2 - 2x - 3} \cdot \frac{x^2 - 1}{x^2 + x - 2} \\ &= \frac{(x-3)(x+2)}{(x-3)(x+1)} \cdot \frac{(x+1)(x-1)}{(x+2)(x-1)} \\ &= 1 \end{aligned}$$

115.

$$\begin{aligned} \frac{3}{m+1} + \frac{3m}{m-1} &= \frac{3}{m+1}\left(\frac{m-1}{m-1}\right) + \frac{3m}{m-1}\left(\frac{m+1}{m+1}\right) \\ &= \frac{3(m-1)}{(m+1)(m-1)} + \frac{3m(m+1)}{(m+1)(m-1)} \\ &= \frac{3m-3}{(m+1)(m-1)} + \frac{3m^2+3m}{(m+1)(m-1)} \\ &= \frac{3m-3+3m^2+3m}{(m+1)(m-1)} \\ &= \frac{3m^2+6m-3}{(m+1)(m-1)} \\ &= \frac{3(m^2+2m-1)}{(m+1)(m-1)} \end{aligned}$$

CHALLENGE PROBLEMS

117. $f(x) = -\sqrt{x-2} + 3$

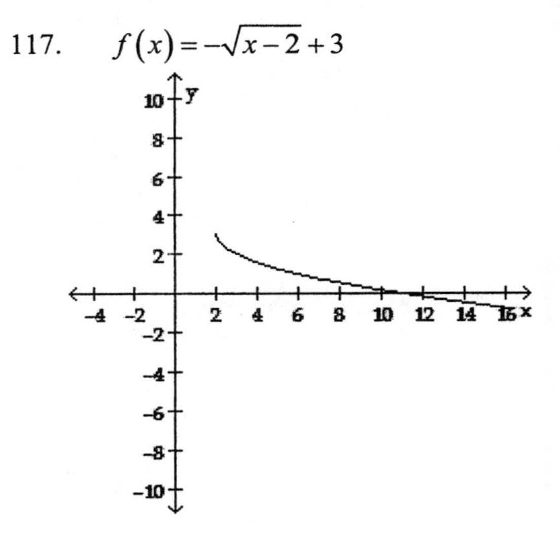

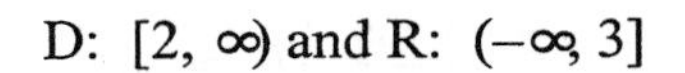

SECTION 9.2

VOCABULARY

1. The expressions $4^{1/2}$ and $(-8)^{-2/3}$ have **rational (or fractional)** exponents.

3. In the radical expression $\sqrt[3]{4{,}096x^{12}}$, 3 is the **index**, and $4{,}096x^{12}$ is the **radicand**.

CONCEPTS

5.

Radical form	Exponential form
$\sqrt[5]{25}$	$25^{1/5}$
$\left(\sqrt[3]{-27}\right)^2$	$(-27)^{2/3}$
$\left(\sqrt[4]{16}\right)^{-3}$	$16^{-3/4}$
$\left(\sqrt{81}\right)^3$	$81^{3/2}$
$-\sqrt{\dfrac{9}{64}}$	$-\left(\dfrac{9}{64}\right)^{1/2}$

7. Simplify each number.

$$8^{2/3} = 4$$
$$(-125)^{1/3} = -5$$
$$-16^{-1/4} = -\frac{1}{2}$$
$$4^{3/2} = 8$$
$$-\left(\frac{9}{100}\right)^{-1/2} = -\frac{10}{3}$$

$(-125)^{1/3}$ $-\left(\frac{9}{100}\right)^{-1/2}$ $-16^{-1/4}$ $8^{2/3}$ $4^{3/2}$

-6 -5 -4 -3 -2 -1 0 1 2 3 4 5 6 7 8

9. $x^{1/n} = \boxed{\sqrt[n]{x}}$

11. $x^{-m/n} = \boxed{\dfrac{1}{x^{m/n}}}$

NOTATION

13.

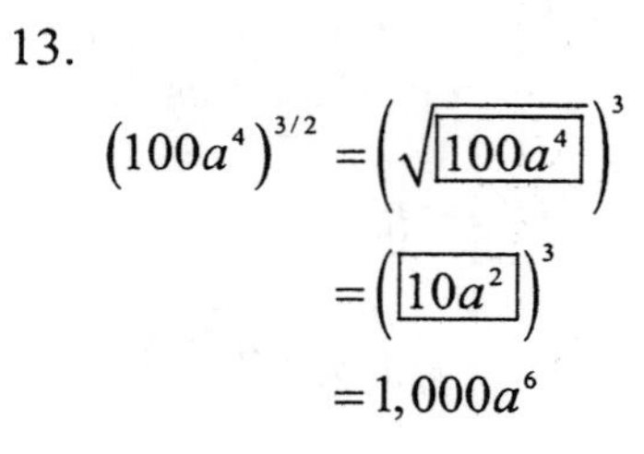

$$(100a^4)^{3/2} = \left(\sqrt{\boxed{100a^4}}\right)^3$$
$$= \left(\boxed{10a^2}\right)^3$$
$$= 1{,}000a^6$$

PRACTICE

15. $x^{1/3} = \sqrt[3]{x}$

17. $(3x)^{1/4} = \sqrt[4]{3x}$

19. $(6x^3y)^{1/4} = \sqrt[4]{6x^3y}$

21. $(x^2+y^2)^{1/2} = \sqrt{x^2+y^2}$

23. $\sqrt{m} = m^{1/2}$

25. $\sqrt[4]{3a} = (3a)^{1/4}$

27. $\sqrt[6]{8abc} = (8abc)^{1/6}$

29. $\sqrt[3]{a^2-b^2} = (a^2-b^2)^{1/3}$

31.
$$4^{1/2} = \sqrt{4}$$
$$= 2$$

33.
$$125^{1/3} = \sqrt[3]{125}$$
$$= 5$$

35.
$$81^{1/4} = \sqrt[4]{81}$$
$$= 3$$

37.
$$32^{1/5} = \sqrt[5]{32}$$
$$= 2$$

39.

$$\left(\frac{1}{4}\right)^{1/2}=\sqrt{\frac{1}{4}}=\frac{\sqrt{1}}{\sqrt{4}}=\frac{1}{2}$$

41.

$$-16^{1/4}=-\sqrt[4]{16}=-2$$

43.

$$(-64)^{1/2}=\sqrt{-64}$$

is not a real number

45.

$$(-216)^{1/3}=\sqrt[3]{-216}=-6$$

47.

$$\left(x^2\right)^{1/2}=\sqrt{x^2}=|x|$$

49.

$$\left(m^4\right)^{1/2}=\sqrt{m^4}=m^2$$

51.

$$\left(n^9\right)^{1/3}=\sqrt[3]{n^9}=n^3$$

53.

$$\left(25y^2\right)^{1/2}=\sqrt{25y^2}=|5y|=5|y|$$

55.

$$\left(16x^4\right)^{1/4}=\sqrt[4]{16x^4}=|2x|=2|x|$$

57.

$$\left(-64x^8\right)^{1/4}=\sqrt[4]{-64x^8}$$

is not a real number

59.

$$\left[(x+1)^4\right]^{1/4}=\sqrt[4]{(x+1)^4}=|x+1|$$

61.

$$36^{3/2}=\left(\sqrt{36}\right)^3=(6)^3=216$$

63.

$$81^{3/4}=\left(\sqrt[4]{81}\right)^3=(3)^3=27$$

65.

$$144^{3/2}=\left(\sqrt{144}\right)^3=(12)^3=1{,}728$$

67.

$$\left(\frac{1}{8}\right)^{2/3}=\left(\sqrt[3]{\frac{1}{8}}\right)^2=\left(\frac{1}{2}\right)^2=\frac{1}{4}$$

69.

$$\left(25x^4\right)^{3/2}=\left(\sqrt{25x^4}\right)^3=\left(5x^2\right)^3=125x^6$$

71.

$$\left(\frac{8x^3}{27}\right)^{2/3} = \left(\sqrt[3]{\frac{8x^3}{27}}\right)^2 = \left(\frac{2x}{3}\right)^2 = \frac{4x^2}{9}$$

73.

$$4^{-1/2} = \frac{1}{4^{1/2}} = \frac{1}{\sqrt{4}} = \frac{1}{2}$$

75.

$$25^{-5/2} = \frac{1}{25^{5/2}} = \frac{1}{\left(\sqrt{25}\right)^5} = \frac{1}{5^5} = \frac{1}{3{,}125}$$

77.

$$\left(16x^2\right)^{-3/2} = \frac{1}{\left(16x^2\right)^{3/2}} = \frac{1}{\left(\sqrt{16x^2}\right)^3} = \frac{1}{(4x)^3} = \frac{1}{64x^3}$$

79.

$$\left(-27y^3\right)^{-2/3} = \frac{1}{\left(-27y^3\right)^{2/3}} = \frac{1}{\left(\sqrt[3]{-27y^3}\right)^2} = \frac{1}{(-3y)^2} = \frac{1}{9y^2}$$

81.

$$\left(\frac{27}{8}\right)^{-4/3} = \left(\frac{8}{27}\right)^{4/3} = \left(\sqrt[3]{\frac{8}{27}}\right)^4 = \left(\frac{2}{3}\right)^4 = \frac{16}{81}$$

83.

$$\left(-\frac{8x^3}{27}\right)^{-1/3} = \left(-\frac{27}{8x^3}\right)^{1/3} = \sqrt[3]{-\frac{27}{8x^3}} = -\frac{3}{2x}$$

85.

$$9^{3/7}9^{2/7} = 9^{3/7+2/7} = 9^{5/7}$$

87.

$$6^{-2/3}6^{-4/3} = 6^{-6/3} = 6^{-2} = \frac{1}{6^2} = \frac{1}{36}$$

89.

$$\frac{3^{4/3}3^{1/3}}{3^{2/3}} = \frac{3^{4/3+1/3}}{3^{2/3}} = \frac{3^{5/3}}{3^{2/3}} = 3^{5/3-2/3} = 3^{3/3} = 3^1 = 3$$

91.

$$a^{2/3}a^{1/3} = a^{2/3+1/3} = a^{3/3} = a^1 = a$$

93.

$$\left(a^{2/3}\right)^{1/3} = a^{(2/3)(1/3)} = a^{2/9}$$

95.

$$\left(a^{1/2}b^{1/3}\right)^{3/2} = a^{(1/2)(3/2)}b^{(1/3)(3/2)} = a^{3/4}b^{1/2}$$

97.

$$\left(27x^{-3}\right)^{-1/3} = \left(\frac{27}{x^3}\right)^{-1/3} = \left(\frac{x^3}{27}\right)^{1/3} = \sqrt[3]{\frac{x^3}{27}} = \frac{x}{3} = \frac{1}{3}x$$

99.

$$y^{1/3}\left(y^{2/3} + y^{5/3}\right) = y^{1/3+2/3} + y^{1/3+5/3} = y^{3/3} + y^{6/3} = y + y^2$$

101.

$$x^{3/5}\left(x^{7/5} - x^{2/5} + 1\right) = x^{3/5+7/5} - x^{3/5+2/5} + x^{3/5} = x^{10/5} - x^{5/5} + x^{3/5} = x^2 - x + x^{3/5}$$

103.

$$\sqrt[6]{p^3} = p^{3/6} = p^{1/2} = \sqrt{p}$$

105.

$$\sqrt[4]{25b^2} = \left(5^2b^2\right)^{1/4} = 5^{2(1/4)}b^{2(1/4)} = 5^{2/4}b^{2/4} = 5^{1/2}b^{1/2} = \sqrt{5b}$$

107.

$$\sqrt[10]{x^2y^2} = \left(x^2y^2\right)^{1/10} = x^{2(1/10)}y^{2(1/10)} = x^{2/10}y^{2/10} = x^{1/5}y^{1/5} = \sqrt[5]{xy}$$

109.

$$\sqrt[9]{\sqrt{c}} = \left(c^{1/2}\right)^{1/9} = c^{(1/2)(1/9)} = c^{1/18} = \sqrt[18]{c}$$

111.

$$\sqrt[5]{\sqrt[3]{7m}} = \left[(7m)^{1/3}\right]^{1/5} = (7m)^{(1/3)(1/5)} = (7m)^{1/15} = \sqrt[15]{7m}$$

113.

$$\sqrt[3]{15} = 2.47$$

115.

$$\sqrt[5]{1.045} = 1.01$$

APPLICATIONS

117. BALLISTIC PENDULUMS

Let $M = 6.0$, $m = 0.0625$, $g = 32$, and $h = 0.9$.

$$\begin{aligned} v &= \frac{m+M}{m}(2gh)^{1/2} \\ &= \frac{0.0625+6.0}{0.0625}(2 \cdot 32 \cdot 0.9)^{1/2} \\ &= \frac{6.0625}{0.0625}(57.6)^{1/2} \\ &= 97\left(\sqrt{57.6}\right) \\ &= 97(7.589) \\ &= 736 \text{ ft/sec} \end{aligned}$$

119. RELATIVITY

Let $c = 186{,}000$, $v = 160{,}000$, and $m_0 = 1$.

$$\begin{aligned} m &= m_0\left(1-\frac{v^2}{c^2}\right)^{-1/2} \\ &= 1\left(1-\frac{160{,}000^2}{186{,}000^2}\right)^{-1/2} \\ &= 1\left(1-\frac{2.56\times10^{10}}{3.4596\times10^{10}}\right)^{-1/2} \\ &= 1(1-0.74)^{-1/2} \\ &= 1(0.26)^{-1/2} \\ &= 1\left(\frac{1}{0.26}\right)^{1/2} \\ &= 1\left(\sqrt{\frac{1}{0.26}}\right) \\ &= 1(1.96) \\ &= 1.96 \text{ units} \end{aligned}$$

121. CUBICLES

The area of the base of one cube is $A(V) = V^{2/3}$. To find the area of all 18 cubes, multiply that value by 18.

$$\begin{aligned} A &= 18V^{2/3} \\ &= 18(4{,}096)^{2/3} \\ &= 18\left(\sqrt[3]{4{,}096}\right)^2 \\ &= 18(16)^2 \\ &= 18(256) \\ &= 4{,}608 \text{ in.}^2 \end{aligned}$$

$$\begin{aligned} A &= \frac{4{,}608}{12^2} \\ &= \frac{4{,}608}{144} \\ &= 32 \text{ ft}^2 \end{aligned}$$

WRITING

123. Answers will vary.

REVIEW

125. COMMUTING TIME

$$\begin{aligned} t &= \frac{k}{r} \\ 3 &= \frac{k}{50} \\ 50(3) &= 50\left(\frac{k}{50}\right) \\ 150 &= k \end{aligned}$$

$$\begin{aligned} t &= \frac{150}{60} \\ t &= 2.5 \text{ hours} \end{aligned}$$

CHALLENGE PROBLEMS

127. Yes.

$$\begin{aligned} 16^{2/4} &= \left(\sqrt[4]{16}\right)^2 \\ &= 2^2 \\ &= 4 \end{aligned} \qquad \begin{aligned} 16^{1/2} &= \sqrt{16} \\ &= 4 \end{aligned}$$

SECTION 9.3

VOCABULARY

1. Radical expressions such as $\sqrt[3]{4}$ and $6\sqrt[3]{4}$ with the same index and the same radicand are called **like** radicals.

3. The largest perfect square **factor** of 27 is 9.

CONCEPTS

5. $\sqrt[n]{ab}=\sqrt[n]{a}\sqrt[n]{b}$
   In words, the *n*th root of the **product** of two numbers is equal to the product of their *n*th **roots**.

7. a. $\sqrt{4\bullet5}$
   b. $\sqrt{4}\sqrt{5}$
   c. $\sqrt{4\bullet5}=\sqrt{4}\sqrt{5}$

9. a. Answers will vary. Possible answers are $\sqrt{5}$ and $\sqrt[3]{5}$. No, the expressions cannot be added.
   b. Answers will vary. Possible answers are $\sqrt{5}$ and $\sqrt{6}$. No, the expressions cannot be added.

NOTATION

11.
$$\begin{aligned}\sqrt[3]{32k^4}&=\sqrt[3]{\boxed{8k^3}\bullet4k}\\&=\sqrt[3]{\boxed{8k^3}}\sqrt[3]{4k}\\&=2k\sqrt[3]{\boxed{4k}}\end{aligned}$$

PRACTICE

13.
$$\begin{aligned}\sqrt{20}&=\sqrt{4\bullet5}\\&=\sqrt{4}\sqrt{5}\\&=2\sqrt{5}\end{aligned}$$

15.
$$\begin{aligned}\sqrt{200}&=\sqrt{100\bullet2}\\&=\sqrt{100}\sqrt{2}\\&=10\sqrt{2}\end{aligned}$$

17.
$$\begin{aligned}\sqrt[3]{80}&=\sqrt[3]{8\bullet10}\\&=\sqrt[3]{8}\sqrt[3]{10}\\&=2\sqrt[3]{10}\end{aligned}$$

19.
$$\begin{aligned}\sqrt[3]{-81}&=\sqrt[3]{-27\bullet3}\\&=\sqrt[3]{-27}\sqrt[3]{3}\\&=-3\sqrt[3]{3}\end{aligned}$$

21.
$$\begin{aligned}\sqrt[4]{32}&=\sqrt[4]{16\bullet2}\\&=\sqrt[4]{16}\sqrt[4]{2}\\&=2\sqrt[4]{2}\end{aligned}$$

23.
$$\begin{aligned}-\sqrt[5]{96}&=-\sqrt[5]{32\bullet3}\\&=-\sqrt[5]{32}\sqrt[5]{3}\\&=-2\sqrt[5]{3}\end{aligned}$$

25.
$$\begin{aligned}\sqrt[6]{320}&=\sqrt[6]{64\bullet5}\\&=\sqrt[6]{64}\sqrt[6]{5}\\&=2\sqrt[6]{5}\end{aligned}$$

27.
$$\begin{aligned}\sqrt{\frac{7}{9}}&=\frac{\sqrt{7}}{\sqrt{9}}\\&=\frac{\sqrt{7}}{3}\end{aligned}$$

29.

$$\sqrt[3]{\frac{7}{64}} = \frac{\sqrt[3]{7}}{\sqrt[3]{64}}$$
$$= \frac{\sqrt[3]{7}}{4}$$

31.

$$\sqrt[4]{\frac{3}{10,000}} = \frac{\sqrt[4]{3}}{\sqrt[4]{10,000}}$$
$$= \frac{\sqrt[4]{3}}{10}$$

33.

$$\sqrt[5]{\frac{3}{32}} = \frac{\sqrt[5]{3}}{\sqrt[5]{32}}$$
$$= \frac{\sqrt[5]{3}}{2}$$

35.

$$\frac{\sqrt{500}}{\sqrt{5}} = \sqrt{\frac{500}{5}}$$
$$= \sqrt{100}$$
$$= 10$$

37.

$$\frac{\sqrt[3]{48}}{\sqrt[3]{6}} = \sqrt[3]{\frac{48}{6}}$$
$$= \sqrt[3]{8}$$
$$= 2$$

39.

$$\sqrt{50x^2} = \sqrt{25x^2}\sqrt{2}$$
$$= 5x\sqrt{2}$$

41.

$$\sqrt{32b} = \sqrt{16}\sqrt{2b}$$
$$= 4\sqrt{2b}$$

43.

$$-\sqrt{112a^3} = -\sqrt{16a^2}\sqrt{7a}$$
$$= -4a\sqrt{7a}$$

45.

$$\sqrt{175a^2b^3} = \sqrt{25a^2b^2}\sqrt{7b}$$
$$= 5ab\sqrt{7b}$$

47.

$$-\sqrt{300xy} = -\sqrt{100}\sqrt{3xy}$$
$$= -10\sqrt{3xy}$$

49.

$$\sqrt[3]{-54x^6} = \sqrt[3]{-27x^6}\sqrt[3]{2}$$
$$= -3x^2\sqrt[3]{2}$$

51.

$$\sqrt[3]{16x^{12}y^3} = \sqrt[3]{8x^{12}y^3}\sqrt[3]{2}$$
$$= 2x^4y\sqrt[3]{2}$$

53.

$$\sqrt[4]{32x^{12}y^4} = \sqrt[4]{16x^{12}y^4}\sqrt[4]{2}$$
$$= 2x^3y\sqrt[4]{2}$$

55.

$$\sqrt[5]{a^7} = \sqrt[5]{a^5}\sqrt[5]{a^2}$$
$$= a\sqrt[5]{a^2}$$

57.

$$\sqrt[6]{m^{11}} = \sqrt[6]{m^6}\sqrt[6]{m^5}$$
$$= m\sqrt[6]{m^5}$$

59.

$$\sqrt[5]{32t^{11}} = \sqrt[5]{32t^{10}}\sqrt[5]{t}$$
$$= 2t^2\sqrt[5]{t}$$

61.
$$\frac{\sqrt[3]{189a^4}}{\sqrt[3]{7a}} = \sqrt[3]{\frac{189a^4}{7a}}$$
$$= \sqrt[3]{27a^3}$$
$$= 3a$$

63.
$$\frac{\sqrt{98x^3}}{\sqrt{2x}} = \sqrt{\frac{98x^3}{2x}}$$
$$= \sqrt{49x^2}$$
$$= 7x$$

65.
$$\sqrt{\frac{z^2}{16x^2}} = \frac{\sqrt{z^2}}{\sqrt{16x^2}}$$
$$= \frac{z}{4x}$$

67.
$$\sqrt[4]{\frac{5x}{16z^4}} = \frac{\sqrt[4]{5x}}{\sqrt[4]{16z^4}}$$
$$= \frac{\sqrt[4]{5x}}{2z}$$

69.
$$4\sqrt{2x} + 6\sqrt{2x} = 10\sqrt{2x}$$

71.
$$8\sqrt[5]{7a^2} - 7\sqrt[5]{7a^2} = \sqrt[5]{7a^2}$$

73.
$$\sqrt{2} - \sqrt{8} = \sqrt{2} - \sqrt{4}\sqrt{2}$$
$$= \sqrt{2} - 2\sqrt{2}$$
$$= -\sqrt{2}$$

75.
$$\sqrt{98} - \sqrt{50} = \sqrt{49}\sqrt{2} - \sqrt{25}\sqrt{2}$$
$$= 7\sqrt{2} - 5\sqrt{2}$$
$$= 2\sqrt{2}$$

77.
$$3\sqrt{24} + \sqrt{54} = 3\sqrt{4}\sqrt{6} + \sqrt{9}\sqrt{6}$$
$$= 3\left(2\sqrt{6}\right) + 3\sqrt{6}$$
$$= 6\sqrt{6} + 3\sqrt{6}$$
$$= 9\sqrt{6}$$

79.
$$\sqrt[3]{24x} + \sqrt[3]{3x} = \sqrt[3]{8}\sqrt[3]{3x} + \sqrt[3]{3x}$$
$$= 2\sqrt[3]{3x} + \sqrt[3]{3x}$$
$$= 3\sqrt[3]{3x}$$

81.
$$\sqrt[3]{32} - \sqrt[3]{108} = \sqrt[3]{8}\sqrt[3]{4} - \sqrt[3]{27}\sqrt[3]{4}$$
$$= 2\sqrt[3]{4} - 3\sqrt[3]{4}$$
$$= -\sqrt[3]{4}$$

83.
$$2\sqrt[3]{125} - 5\sqrt[3]{64} = 2(5) - 5(4)$$
$$= 10 - 20$$
$$= -10$$

85.
$$14\sqrt[4]{32} - 15\sqrt[4]{162} = 14\sqrt[4]{16}\sqrt[4]{2} - 15\sqrt[4]{81}\sqrt[4]{2}$$
$$= 14\left(2\sqrt[4]{2}\right) - 15\left(3\sqrt[4]{2}\right)$$
$$= 28\sqrt[4]{2} - 45\sqrt[4]{2}$$
$$= -17\sqrt[4]{2}$$

87.
$$3\sqrt[4]{512} + 2\sqrt[4]{32} = 3\sqrt[4]{256}\sqrt[4]{2} + 2\sqrt[4]{16}\sqrt[4]{2}$$
$$= 3\left(4\sqrt[4]{2}\right) + 2\left(2\sqrt[4]{2}\right)$$
$$= 12\sqrt[4]{2} + 4\sqrt[4]{2}$$
$$= 16\sqrt[4]{2}$$

89.

$$\sqrt{98}-\sqrt{50}-\sqrt{72}=\sqrt{49}\sqrt{2}-\sqrt{25}\sqrt{2}-\sqrt{36}\sqrt{2}$$
$$=7\sqrt{2}-5\sqrt{2}-6\sqrt{2}$$
$$=-4\sqrt{2}$$

91.

$$\sqrt{18t}+\sqrt{300t}-\sqrt{243t}$$
$$=\sqrt{9}\sqrt{2t}+\sqrt{100}\sqrt{3t}-\sqrt{81}\sqrt{3t}$$
$$=3\sqrt{2t}+10\sqrt{3t}-9\sqrt{3t}$$
$$=3\sqrt{2t}+\sqrt{3t}$$

93.

$$2\sqrt[3]{16}-\sqrt[3]{54}-3\sqrt[3]{128}$$
$$=2\sqrt[3]{8}\sqrt[3]{2}-\sqrt[3]{27}\sqrt[3]{2}-3\sqrt[3]{64}\sqrt[3]{2}$$
$$=2\left(2\sqrt[3]{2}\right)-3\sqrt[3]{2}-3\left(4\sqrt[3]{2}\right)$$
$$=4\sqrt[3]{2}-3\sqrt[3]{2}-12\sqrt[3]{2}$$
$$=-11\sqrt[3]{2}$$

95.

$$\sqrt{25y^2z}-\sqrt{16y^2z}=\sqrt{25y^2}\sqrt{z}-\sqrt{16y^2}\sqrt{z}$$
$$=5y\sqrt{z}-4y\sqrt{z}$$
$$=y\sqrt{z}$$

97.

$$\sqrt{36xy^2}+\sqrt{49xy^2}=\sqrt{36y^2}\sqrt{x}+\sqrt{49y^2}\sqrt{x}$$
$$=6y\sqrt{x}+7y\sqrt{x}$$
$$=13y\sqrt{x}$$

99.

$$2\sqrt[3]{64a}+2\sqrt[3]{8a}=2\sqrt[3]{64}\sqrt[3]{a}+2\sqrt[3]{8}\sqrt[3]{a}$$
$$=2\left(4\sqrt[3]{a}\right)+2\left(2\sqrt[3]{a}\right)$$
$$=8\sqrt[3]{a}+4\sqrt[3]{a}$$
$$=12\sqrt[3]{a}$$

101.

$$\sqrt{y^5}-\sqrt{9y^5}-\sqrt{25y^5}$$
$$=\sqrt{y^4}\sqrt{y}-\sqrt{9y^4}\sqrt{y}-\sqrt{25y^4}\sqrt{y}$$
$$=y^2\sqrt{y}-3y^2\sqrt{y}-5y^2\sqrt{y}$$
$$=-7y^2\sqrt{y}$$

103.

$$\sqrt[5]{x^6y^2}+\sqrt[5]{32x^6y^2}+\sqrt[5]{x^6y^2}$$
$$=\sqrt[5]{x^5}\sqrt[5]{xy^2}+\sqrt[5]{32x^5}\sqrt[5]{xy^2}+\sqrt[5]{x^5}\sqrt[5]{xy^2}$$
$$=x\sqrt[5]{xy^2}+2x\sqrt[5]{xy^2}+x\sqrt[5]{xy^2}$$
$$=4x\sqrt[5]{xy^2}$$

APPLICATIONS

105. UMBRELLAS

Let $r=4$ and $h=2$.

$$S=\pi r\sqrt{r^2+h^2}$$
$$=\pi(4)\sqrt{4^2+2^2}$$
$$=\pi(4)\sqrt{16+4}$$
$$=4\pi\sqrt{20}$$
$$=4\pi\sqrt{4}\sqrt{5}$$
$$=4\pi\left(2\sqrt{5}\right)$$
$$=8\pi\sqrt{5}\text{ ft}^2$$

$$S\approx 56.2\text{ ft}^2$$

107. BLOW DRYERS

Let $P = 1{,}200$ and $R = 16$.

$$I=\sqrt{\frac{P}{R}}$$
$$=\sqrt{\frac{1{,}200}{16}}$$
$$=\sqrt{75}$$
$$=\sqrt{25}\sqrt{3}$$
$$=5\sqrt{3}\text{ amps}$$

$$I\approx 8.7\text{ amps}$$

109. DUCTWORK
Add up all of the sides.

$$
\begin{aligned}
&4\sqrt{80}+2\sqrt{20}+2\sqrt{45}+2\sqrt{75}\\
&\quad=4\sqrt{16}\sqrt{5}+2\sqrt{4}\sqrt{5}+2\sqrt{9}\sqrt{5}+2\sqrt{25}\sqrt{3}\\
&\quad=4\left(4\sqrt{5}\right)+2\left(2\sqrt{5}\right)+2\left(3\sqrt{5}\right)+2\left(5\sqrt{3}\right)\\
&\quad=16\sqrt{5}+4\sqrt{5}+6\sqrt{5}+10\sqrt{3}\\
&\quad=\left(26\sqrt{5}+10\sqrt{3}\right)\text{ in.}\\
&\quad\approx 75.5\text{ in.}
\end{aligned}
$$

WRITING

111. Answers will vary.

113. Answers will vary.

REVIEW

115.

$$
\begin{aligned}
3x^2y^3\left(-5x^3y^{-4}\right)&=-15x^5y^{-1}\\
&=-\frac{15x^5}{y}
\end{aligned}
$$

117.

$$
\begin{array}{r}
3p+4-\dfrac{5}{2p-5}\\
2p-5\overline{)\,6p^2-7p-25}\\
\underline{6p^2-15p}\\
8p-25\\
\underline{8p-20}\\
-5
\end{array}
$$

CHALLENGE PROBLEMS

119. If $a=0$, then b can be any real number. If $b=0$, then a can be any real number.

VOCABULARY

1. To multiply $(\sqrt{3}+\sqrt{2})(\sqrt{3}-2\sqrt{2})$, we can use the **FOIL** method.

3. The denominator of the fraction $\frac{4}{\sqrt{5}}$ is an **irrational** number.

5. To obtain a **perfect** cube radicand in the denominator of $\frac{\sqrt[3]{7}}{\sqrt[3]{5n}}$, we multiply the fraction by $\frac{\sqrt[3]{25n^2}}{\sqrt[3]{25n^2}}$.

CONCEPTS

7. a. $4\sqrt{6}+2\sqrt{6}=6\sqrt{6}$

   b.
$$\begin{aligned}4\sqrt{6}(2\sqrt{6})&=8\sqrt{36}\\&=8(6)\\&=48\end{aligned}$$

   c. cannot be simplified

   d. $3\sqrt{2}(-2\sqrt{3})=-6\sqrt{6}$

9. Any number multiplied by 1 is the same number. $\frac{\sqrt{7}}{\sqrt{7}}=1$

11. A radical appears in the denominator.

NOTATION

13.
$$\begin{aligned}5\sqrt{8}\cdot7\sqrt{6}&=5(7)\sqrt{8}\boxed{\sqrt{6}}\\&=35\sqrt{\boxed{48}}\\&=35\sqrt{\boxed{16}\cdot3}\\&=35(\boxed{4})\sqrt{3}\\&=140\sqrt{3}\end{aligned}$$

PRACTICE

15.
$$\begin{aligned}\sqrt{11}\sqrt{11}&=\sqrt{121}\\&=11\end{aligned}$$

17.
$$\begin{aligned}(\sqrt{7})^2&=\sqrt{7}\sqrt{7}\\&=\sqrt{49}\\&=7\end{aligned}$$

19.
$$\begin{aligned}\sqrt{2}\sqrt{8}&=\sqrt{16}\\&=4\end{aligned}$$

21.
$$\begin{aligned}\sqrt{5}\sqrt{10}&=\sqrt{50}\\&=\sqrt{25}\sqrt{2}\\&=5\sqrt{2}\end{aligned}$$

23.
$$\begin{aligned}2\sqrt{3}\sqrt{6}&=2\sqrt{18}\\&=2\sqrt{9}\sqrt{2}\\&=2(3\sqrt{2})\\&=6\sqrt{2}\end{aligned}$$

25.
$$\begin{aligned}\sqrt[3]{5}\sqrt[3]{25}&=\sqrt[3]{125}\\&=5\end{aligned}$$

27.
$$\begin{aligned}(3\sqrt{2})^2&=3\sqrt{2}\cdot3\sqrt{2}\\&=3(3)\sqrt{2}\sqrt{2}\\&=9\sqrt{4}\\&=9(2)\\&=18\end{aligned}$$

29.
$$\begin{aligned}(-2\sqrt{2})^2&=-2\sqrt{2}(-2\sqrt{2})\\&=-2(-2)\sqrt{2}\sqrt{2}\\&=4\sqrt{4}\\&=4(2)\\&=8\end{aligned}$$

31.

$$\left(3\sqrt[3]{9}\right)\left(2\sqrt[3]{3}\right) = 3(2)\sqrt[3]{9}\sqrt[3]{3}$$
$$= 6\sqrt[3]{27}$$
$$= 6(3)$$
$$= 18$$

33.

$$\sqrt[3]{2} \cdot \sqrt[3]{12} = \sqrt[3]{24}$$
$$= \sqrt[3]{8}\sqrt[3]{3}$$
$$= 2\sqrt[3]{3}$$

35.

$$\sqrt{ab^3} \cdot \sqrt{ab} = \sqrt{a^2b^4}$$
$$= ab^2$$

37.

$$\sqrt{5ab}\sqrt{5a} = \sqrt{25a^2b}$$
$$= \sqrt{25a^2}\sqrt{b}$$
$$= 5a\sqrt{b}$$

39.

$$-4\sqrt[3]{5r^2s}\left(5\sqrt[3]{2r}\right) = -20\sqrt[3]{10r^3s}$$
$$= -20\sqrt[3]{r^3}\sqrt[3]{10s}$$
$$= -20r\sqrt[3]{10s}$$

41.

$$\sqrt{x(x+3)}\sqrt{x^3(x+3)} = \sqrt{x^4(x+3)^2}$$
$$= x^2(x+3)$$

43.

$$\left(\sqrt[3]{9b}\right)^3 = \sqrt[3]{9b}\sqrt[3]{9b}\sqrt[3]{9b}$$
$$= \sqrt[3]{729b^3}$$
$$= 9b$$

45.

$$\sqrt[4]{2a^3}\sqrt[4]{3a^2b} = \sqrt[4]{6a^5b}$$
$$= \sqrt[4]{a^4}\sqrt[4]{6ab}$$
$$= a\sqrt[4]{6ab}$$

47.

$$\sqrt[5]{2t}\left(\sqrt[5]{16t}\right) = \sqrt[5]{32t^2}$$
$$= \sqrt[5]{32}\sqrt[5]{t^2}$$
$$= 2\sqrt[5]{t^2}$$

49.

$$3\sqrt{5}\left(4-\sqrt{5}\right) = 3\sqrt{5}(4) - 3\sqrt{5}\left(\sqrt{5}\right)$$
$$= 3(4)\sqrt{5} - 3(5)$$
$$= 12\sqrt{5} - 15$$

51.

$$3\sqrt{2}\left(4\sqrt{6}+2\sqrt{7}\right) = 3(4)\left(\sqrt{2}\sqrt{6}\right) + 3(2)\left(\sqrt{2}\sqrt{7}\right)$$
$$= 12\sqrt{12} + 6\sqrt{14}$$
$$= 12\sqrt{4}\sqrt{3} + 6\sqrt{14}$$
$$= 12\left(2\sqrt{3}\right) + 6\sqrt{14}$$
$$= 24\sqrt{3} + 6\sqrt{14}$$

53.

$$-2\sqrt{5x}\left(4\sqrt{2x} - 3\sqrt{3}\right)$$
$$= -2(4)\sqrt{5x}\sqrt{2x} + 2(3)\sqrt{5x}\sqrt{3}$$
$$= -8\sqrt{10x^2} + 6\sqrt{15x}$$
$$= -8\sqrt{x^2}\sqrt{10} + 6\sqrt{15x}$$
$$= -8x\sqrt{10} + 6\sqrt{15x}$$

55.

$$\left(\sqrt{2}+1\right)\left(\sqrt{2}-3\right)$$
$$= \sqrt{2}\sqrt{2} + \sqrt{2}(-3) + 1\left(\sqrt{2}\right) + 1(-3)$$
$$= \sqrt{4} - 3\sqrt{2} + \sqrt{2} - 3$$
$$= 2 - 2\sqrt{2} - 3$$
$$= -1 - 2\sqrt{2}$$

57.

$$\left(\sqrt[3]{5z} + \sqrt[3]{3}\right)\left(\sqrt[3]{5z} + 2\sqrt[3]{3}\right)$$
$$= \sqrt[3]{5z}\sqrt[3]{5z} + \sqrt[3]{5z}\left(2\sqrt[3]{3}\right) + \sqrt[3]{3}\left(\sqrt[3]{5z}\right) + \sqrt[3]{3}\left(2\sqrt[3]{3}\right)$$
$$= \sqrt[3]{25z^2} + 2\sqrt[3]{15z} + \sqrt[3]{15z} + 2\sqrt[3]{9}$$
$$= \sqrt[3]{25z^2} + 3\sqrt[3]{15z} + 2\sqrt[3]{9}$$

59.

$$\left(\sqrt{3x}-\sqrt{2y}\right)\left(\sqrt{3x}+\sqrt{2y}\right)$$
$$=\sqrt{3x}\sqrt{3x}+\sqrt{3x}\sqrt{2y}-\sqrt{2y}\sqrt{3x}-\sqrt{2y}\sqrt{2y}$$
$$=\sqrt{9x^2}+\sqrt{6xy}-\sqrt{6xy}-\sqrt{4y^2}$$
$$=3x-2y$$

61.

$$\left(2\sqrt{3a}-\sqrt{b}\right)\left(\sqrt{3a}+3\sqrt{b}\right)$$
$$=2\sqrt{3a}\sqrt{3a}+2\sqrt{3a}\left(3\sqrt{b}\right)-\sqrt{b}\left(\sqrt{3a}\right)-\sqrt{b}\left(3\sqrt{b}\right)$$
$$=2\sqrt{9a^2}+6\sqrt{3ab}-\sqrt{3ab}-3\sqrt{b^2}$$
$$=2(3a)+5\sqrt{3ab}-3(b)$$
$$=6a+5\sqrt{3ab}-3b$$

63.

$$\left(3\sqrt{2r}-2\right)^2$$
$$=\left(3\sqrt{2r}-2\right)\left(3\sqrt{2r}-2\right)$$
$$=3\sqrt{2r}\left(3\sqrt{2r}\right)+3\sqrt{2r}(-2)-2\left(3\sqrt{2r}\right)-2(-2)$$
$$=9\sqrt{4r^2}-6\sqrt{2r}-6\sqrt{2r}+4$$
$$=9(2r)-12\sqrt{2r}+4$$
$$=18r-12\sqrt{2r}+4$$

65.

$$-2\left(\sqrt{3x}+\sqrt{3}\right)^2$$
$$=-2\left(\sqrt{3x}+\sqrt{3}\right)\left(\sqrt{3x}+\sqrt{3}\right)$$
$$=-2\left[\sqrt{3x}\sqrt{3x}+\sqrt{3x}\sqrt{3}+\sqrt{3}\sqrt{3x}+\sqrt{3}\sqrt{3}\right]$$
$$=-2\left(\sqrt{9x^2}+\sqrt{9x}+\sqrt{9x}+\sqrt{9}\right)$$
$$=-2\left(3x+\sqrt{9}\sqrt{x}+\sqrt{9}\sqrt{x}+3\right)$$
$$=-2\left(3x+3\sqrt{x}+3\sqrt{x}+3\right)$$
$$=-2\left(3x+6\sqrt{x}+3\right)$$
$$=-6x-12\sqrt{x}-6$$

67.

$$\left(2\sqrt[3]{4a^2}+1\right)\left(\sqrt[3]{4a^2}-3\right)$$
$$=2\sqrt[3]{16a^4}-6\sqrt[3]{4a^2}+\sqrt[3]{4a^2}-3$$
$$=2\sqrt[3]{8a^3}\sqrt[3]{2a}-5\sqrt[3]{4a^2}-3$$
$$=4a\sqrt[3]{2a}-5\sqrt[3]{4a^2}-3$$

69.

$$\sqrt{\frac{1}{7}}=\frac{\sqrt{1}}{\sqrt{7}}$$
$$=\frac{\sqrt{1}}{\sqrt{7}}\cdot\frac{\sqrt{7}}{\sqrt{7}}$$
$$=\frac{\sqrt{7}}{\sqrt{49}}$$
$$=\frac{\sqrt{7}}{7}$$

71.

$$\frac{6}{\sqrt{30}}=\frac{6}{\sqrt{30}}\cdot\frac{\sqrt{30}}{\sqrt{30}}$$
$$=\frac{6\sqrt{30}}{\sqrt{900}}$$
$$=\frac{6\sqrt{30}}{30}$$
$$=\frac{\sqrt{30}}{5}$$

73.

$$\frac{\sqrt{5}}{\sqrt{8}}=\frac{\sqrt{5}}{\sqrt{4\cdot 2}}$$
$$=\frac{\sqrt{5}}{2\sqrt{2}}$$
$$=\frac{\sqrt{5}}{2\sqrt{2}}\cdot\frac{\sqrt{2}}{\sqrt{2}}$$
$$=\frac{\sqrt{10}}{2\sqrt{4}}$$
$$=\frac{\sqrt{10}}{2(2)}$$
$$=\frac{\sqrt{10}}{4}$$

75.

$$\frac{1}{\sqrt[3]{2}}=\frac{1}{\sqrt[3]{2}}\bullet\frac{\sqrt[3]{4}}{\sqrt[3]{4}}=\frac{\sqrt[3]{4}}{\sqrt[3]{8}}=\frac{\sqrt[3]{4}}{2}$$

77.

$$\frac{3}{\sqrt[3]{9}}=\frac{3}{\sqrt[3]{9}}\bullet\frac{\sqrt[3]{3}}{\sqrt[3]{3}}=\frac{3\sqrt[3]{3}}{\sqrt[3]{27}}=\frac{3\sqrt[3]{3}}{3}=\sqrt[3]{3}$$

79.

$$\frac{\sqrt[3]{2}}{\sqrt[3]{9}}=\frac{\sqrt[3]{2}}{\sqrt[3]{9}}\bullet\frac{\sqrt[3]{3}}{\sqrt[3]{3}}=\frac{\sqrt[3]{6}}{\sqrt[3]{27}}=\frac{\sqrt[3]{6}}{3}$$

81.

$$\frac{\sqrt{8}}{\sqrt{xy}}=\frac{\sqrt{8}}{\sqrt{xy}}\bullet\frac{\sqrt{xy}}{\sqrt{xy}}=\frac{\sqrt{8xy}}{\sqrt{x^2y^2}}=\frac{\sqrt{4\bullet 2xy}}{\sqrt{x^2y^2}}=\frac{2\sqrt{2xy}}{xy}$$

83.

$$\frac{\sqrt{10xy^2}}{\sqrt{2xy^3}}=\sqrt{\frac{10xy^2}{2xy^3}}=\sqrt{\frac{5}{y}}=\frac{\sqrt{5}}{\sqrt{y}}\bullet\frac{\sqrt{y}}{\sqrt{y}}=\frac{\sqrt{5y}}{y}$$

85.

$$\frac{\sqrt[3]{4a^2}}{\sqrt[3]{2ab}}=\sqrt[3]{\frac{4a^2}{2ab}}=\sqrt[3]{\frac{2a}{b}}=\frac{\sqrt[3]{2a}}{\sqrt[3]{b}}\bullet\frac{\sqrt[3]{b^2}}{\sqrt[3]{b^2}}=\frac{\sqrt[3]{2ab^2}}{\sqrt[3]{b^3}}=\frac{\sqrt[3]{2ab^2}}{b}$$

87.

$$\frac{1}{\sqrt[4]{4}}=\frac{1}{\sqrt[4]{4}}\bullet\frac{\sqrt[4]{4}}{\sqrt[4]{4}}=\frac{\sqrt[4]{4}}{\sqrt[4]{16}}=\frac{\sqrt[4]{4}}{2}$$

89.

$$\frac{1}{\sqrt[5]{16}}=\frac{1}{\sqrt[5]{16}}\bullet\frac{\sqrt[5]{2}}{\sqrt[5]{2}}=\frac{\sqrt[5]{2}}{\sqrt[5]{32}}=\frac{\sqrt[5]{2}}{2}$$

91.

$$\frac{\sqrt[4]{s}}{\sqrt[4]{3t^2}} = \frac{\sqrt[4]{s}}{\sqrt[4]{3t^2}} \bullet \frac{\sqrt[4]{27t^2}}{\sqrt[4]{27t^2}}$$
$$= \frac{\sqrt[4]{27st^2}}{\sqrt[4]{81t^4}}$$
$$= \frac{\sqrt[4]{27st^2}}{3t}$$

93.

$$\frac{t}{\sqrt[5]{27a}} = \frac{t}{\sqrt[5]{27a}} \bullet \frac{\sqrt[5]{9a^4}}{\sqrt[5]{9a^4}}$$
$$= \frac{t\sqrt[5]{9a^4}}{\sqrt[5]{243a^5}}$$
$$= \frac{t\sqrt[5]{9a^4}}{3a}$$

95.

$$\frac{\sqrt{2}}{\sqrt{5}+3} = \frac{\sqrt{2}}{\sqrt{5}+3} \bullet \frac{\sqrt{5}-3}{\sqrt{5}-3}$$
$$= \frac{\sqrt{2}\left(\sqrt{5}-3\right)}{\left(\sqrt{5}+3\right)\left(\sqrt{5}-3\right)}$$
$$= \frac{\sqrt{10}-3\sqrt{2}}{\sqrt{25}-9}$$
$$= \frac{\sqrt{10}-3\sqrt{2}}{5-9}$$
$$= \frac{\sqrt{10}-3\sqrt{2}}{-4}$$
$$= \frac{-\sqrt{10}+3\sqrt{2}}{4}$$
$$= \frac{3\sqrt{2}-\sqrt{10}}{4}$$

97.

$$\frac{\sqrt{7}-\sqrt{2}}{\sqrt{2}+\sqrt{7}} = \frac{\sqrt{7}-\sqrt{2}}{\sqrt{2}+\sqrt{7}} \bullet \frac{\sqrt{2}-\sqrt{7}}{\sqrt{2}-\sqrt{7}}$$
$$= \frac{\left(\sqrt{7}-\sqrt{2}\right)\left(\sqrt{2}-\sqrt{7}\right)}{\left(\sqrt{2}+\sqrt{7}\right)\left(\sqrt{2}-\sqrt{7}\right)}$$
$$= \frac{\sqrt{7}\sqrt{2}+\sqrt{7}\left(-\sqrt{7}\right)-\sqrt{2}\sqrt{2}-\sqrt{2}\left(-\sqrt{7}\right)}{\sqrt{2}\sqrt{2}-\sqrt{7}\sqrt{7}}$$
$$= \frac{\sqrt{14}-\sqrt{49}-\sqrt{4}+\sqrt{14}}{\sqrt{4}-\sqrt{49}}$$
$$= \frac{\sqrt{14}-7-2+\sqrt{14}}{2-7}$$
$$= \frac{-9+2\sqrt{14}}{-5}$$
$$= \frac{-\left(9-2\sqrt{14}\right)}{-5}$$
$$= \frac{9-2\sqrt{14}}{5}$$

99.

$$\frac{3\sqrt{2}-5\sqrt{3}}{2\sqrt{3}-3\sqrt{2}}$$

$$=\frac{3\sqrt{2}-5\sqrt{3}}{2\sqrt{3}-3\sqrt{2}}\bullet\frac{2\sqrt{3}+3\sqrt{2}}{2\sqrt{3}+3\sqrt{2}}$$

$$=\frac{\left(3\sqrt{2}-5\sqrt{3}\right)\left(2\sqrt{3}+3\sqrt{2}\right)}{\left(2\sqrt{3}-3\sqrt{2}\right)\left(2\sqrt{3}+3\sqrt{2}\right)}$$

$$=\frac{3\sqrt{2}\left(2\sqrt{3}\right)+3\sqrt{2}\left(3\sqrt{2}\right)-5\sqrt{3}\left(2\sqrt{3}\right)-5\sqrt{3}\left(3\sqrt{2}\right)}{2\sqrt{3}\left(2\sqrt{3}\right)-3\sqrt{2}\left(3\sqrt{2}\right)}$$

$$=\frac{6\sqrt{6}+9\sqrt{4}-10\sqrt{9}-15\sqrt{6}}{4\sqrt{9}-9\sqrt{4}}$$

$$=\frac{6\sqrt{6}+9(2)-10(3)-15\sqrt{6}}{4(3)-9(2)}$$

$$=\frac{6\sqrt{6}+18-30-15\sqrt{6}}{12-18}$$

$$=\frac{-12-9\sqrt{6}}{-6}$$

$$=\frac{-\cancel{3}\left(4+3\sqrt{6}\right)}{-\cancel{3}(2)}$$

$$=\frac{4+3\sqrt{6}}{2}$$

$$=\frac{3\sqrt{6}+4}{2}$$

101.

$$\frac{2}{\sqrt{x}+1}=\frac{2}{\sqrt{x}+1}\bullet\frac{\sqrt{x}-1}{\sqrt{x}-1}$$

$$=\frac{2\left(\sqrt{x}-1\right)}{\left(\sqrt{x}+1\right)\sqrt{x}-1}$$

$$=\frac{2\sqrt{x}-2}{\sqrt{x^2}-1}$$

$$=\frac{2\left(\sqrt{x}-1\right)}{x-1}$$

103.

$$\frac{2z-1}{\sqrt{2z}-1}=\frac{2z-1}{\sqrt{2z}-1}\bullet\frac{\sqrt{2z}+1}{\sqrt{2z}+1}$$

$$=\frac{(2z-1)\left(\sqrt{2z}+1\right)}{\left(\sqrt{2z}-1\right)\left(\sqrt{2z}+1\right)}$$

$$=\frac{2z\sqrt{2z}+2z(1)-1\left(\sqrt{2z}\right)-1(1)}{\sqrt{2z}\sqrt{2z}-1(1)}$$

$$=\frac{2z\sqrt{2z}+2z-\sqrt{2z}-1}{\sqrt{4z^2}-1}$$

$$=\frac{2z\sqrt{2z}+2z-\sqrt{2z}-1}{2z-1}$$

$$=\frac{2z\left(\sqrt{2z}+1\right)-\left(\sqrt{2z}+1\right)}{2z-1}$$

$$=\frac{\left(\sqrt{2z}+1\right)\cancel{(2z-1)}}{\cancel{2z-1}}$$

$$=\sqrt{2z}+1$$

105.

$$\frac{\sqrt{x}-\sqrt{y}}{\sqrt{x}+\sqrt{y}}=\frac{\sqrt{x}-\sqrt{y}}{\sqrt{x}+\sqrt{y}}\bullet\frac{\sqrt{x}-\sqrt{y}}{\sqrt{x}-\sqrt{y}}$$

$$=\frac{\left(\sqrt{x}-\sqrt{y}\right)\left(\sqrt{x}-\sqrt{y}\right)}{\left(\sqrt{x}+\sqrt{y}\right)\left(\sqrt{x}-\sqrt{y}\right)}$$

$$=\frac{\sqrt{x}\sqrt{x}-\sqrt{x}\sqrt{y}-\sqrt{y}\sqrt{x}+\sqrt{y}\sqrt{y}}{\sqrt{x}\sqrt{x}-\sqrt{y}\sqrt{y}}$$

$$=\frac{\sqrt{x^2}-\sqrt{xy}-\sqrt{xy}+\sqrt{y^2}}{\sqrt{x^2}-\sqrt{y^2}}$$

$$=\frac{x-2\sqrt{xy}+y}{x-y}$$

107.

$$\frac{\sqrt{a}+3\sqrt{b}}{\sqrt{a}-3\sqrt{b}}=\frac{\sqrt{a}+3\sqrt{b}}{\sqrt{a}-3\sqrt{b}}\bullet\frac{\sqrt{a}+3\sqrt{b}}{\sqrt{a}+3\sqrt{b}}$$

$$=\frac{\left(\sqrt{a}+3\sqrt{b}\right)\left(\sqrt{a}+3\sqrt{b}\right)}{\left(\sqrt{a}-3\sqrt{b}\right)\left(\sqrt{a}+3\sqrt{b}\right)}$$

$$=\frac{\sqrt{a}\sqrt{a}+\sqrt{a}\left(3\sqrt{b}\right)+3\sqrt{b}\sqrt{a}+3\sqrt{b}\left(3\sqrt{b}\right)}{\sqrt{a}\sqrt{a}-3\sqrt{b}\left(3\sqrt{b}\right)}$$

$$=\frac{\sqrt{a^2}+3\sqrt{ab}+3\sqrt{ab}+9\sqrt{b^2}}{\sqrt{a^2}-9\sqrt{b^2}}$$

$$=\frac{a+6\sqrt{ab}+9b}{a-9b}$$

109.

$$\frac{\sqrt{x}+3}{x}=\frac{\sqrt{x}+3}{x}\cdot\frac{\sqrt{x}-3}{\sqrt{x}-3}$$
$$=\frac{\left(\sqrt{x}+3\right)\left(\sqrt{x}-3\right)}{x\left(\sqrt{x}-3\right)}$$
$$=\frac{\sqrt{x}\sqrt{x}-3(3)}{x\sqrt{x}-3x}$$
$$=\frac{\sqrt{x^2}-9}{x\sqrt{x}-3x}$$
$$=\frac{x-9}{x\sqrt{x}-3x}$$
$$=\frac{x-9}{x\left(\sqrt{x}-3\right)}$$

111.

$$\frac{\sqrt{x}+\sqrt{y}}{\sqrt{x}}=\frac{\sqrt{x}+\sqrt{y}}{\sqrt{x}}\cdot\frac{\sqrt{x}-\sqrt{y}}{\sqrt{x}-\sqrt{y}}$$
$$=\frac{\left(\sqrt{x}+\sqrt{y}\right)\left(\sqrt{x}-\sqrt{y}\right)}{\sqrt{x}\left(\sqrt{x}-\sqrt{y}\right)}$$
$$=\frac{\sqrt{x}\sqrt{x}-\sqrt{y}\sqrt{y}}{\sqrt{x}\left(\sqrt{x}-\sqrt{y}\right)}$$
$$=\frac{\sqrt{x^2}-\sqrt{y^2}}{\sqrt{x}\left(\sqrt{x}-\sqrt{y}\right)}$$
$$=\frac{x-y}{\sqrt{x}\left(\sqrt{x}-\sqrt{y}\right)}$$

APPLICATIONS

113. STATISTICS

$$\frac{1}{\sigma\sqrt{2\pi}}=\frac{1}{\sigma\sqrt{2\pi}}\cdot\frac{\sqrt{2\pi}}{\sqrt{2\pi}}$$
$$=\frac{\sqrt{2\pi}}{\sigma\sqrt{4\pi^2}}$$
$$=\frac{\sqrt{2\pi}}{\sigma(2\pi)}$$
$$=\frac{\sqrt{2\pi}}{2\pi\sigma}$$

115. TRIGONOMETRY

$$\frac{\text{length of side } AC}{\text{length of side } AB}=\frac{1}{\sqrt{2}}$$
$$=\frac{1}{\sqrt{2}}\cdot\frac{\sqrt{2}}{\sqrt{2}}$$
$$=\frac{\sqrt{2}}{\sqrt{4}}$$
$$=\frac{\sqrt{2}}{2}$$

WRITING

117. Answers will vary.

REVIEW

119.

$$\frac{8}{b-2}+\frac{3}{2-b}=-\frac{1}{b}$$
$$\frac{8}{b-2}-\frac{3}{b-2}=-\frac{1}{b}$$
$$b(b-2)\left(\frac{8}{b-2}-\frac{3}{b-2}\right)=b(b-2)\left(-\frac{1}{b}\right)$$
$$8b-3b=-(b-2)$$
$$5b=-b+2$$
$$6b=2$$
$$b=\frac{2}{6}$$
$$b=\frac{1}{3}$$

CHALLENGE PROBLEMS

121.

$$\sqrt{2}\cdot\sqrt[3]{2}=2^{1/2}\cdot2^{1/3}$$
$$=2^{(1/2)+(1/3)}$$
$$=2^{3/6+2/6}$$
$$=2^{5/6}$$
$$=\sqrt[6]{2^5}$$
$$=\sqrt[6]{32}$$

SECTION 9.5

VOCABULARY

1. Equations such as $\sqrt{x+4}-4=5$ and $\sqrt[3]{x+1}=12$ are called **radical** equations.

3. Squaring both sides of an equation can introduce **extraneous** solutions.

CONCEPTS

5. The power rule states that if *x, y,* and *n* are real numbers and $x=y$, then $x^{\boxed{n}}=y^{\boxed{n}}$.

7. a. square both sides
   b. cube both sides
   c. subtract 3 from both sides

9. a. $\left(\sqrt{x}\right)^2=x$

   b. $\left(\sqrt{x-5}\right)^2=x-5$

   c.
$$\begin{aligned}\left(4\sqrt{2x}\right)^2&=4^2\left(\sqrt{2x}\right)^2\\&=16(2x)\\&=32x\end{aligned}$$

   d. $\left(-\sqrt{x+3}\right)^2=x+3$

11. a.
$$\begin{aligned}\left(\sqrt{x}-3\right)^2&=\left(\sqrt{x}-3\right)\left(\sqrt{x}-3\right)\\&=\sqrt{x^2}-3\sqrt{x}-3\sqrt{x}+9\\&=x-6\sqrt{x}+9\end{aligned}$$

   b.
$$\begin{aligned}&\left(\sqrt{2y+1}+5\right)^2\\&=\left(\sqrt{2y+1}+5\right)\left(\sqrt{2y+1}+5\right)\\&=2y+1+5\sqrt{2y+1}+5\sqrt{2y+1}+25\\&=2y+10\sqrt{2y+1}+26\end{aligned}$$

13. The principal square root of a number, in this case $\sqrt{8x-7}$, is never negative.

NOTATION

15.
$$\begin{aligned}\sqrt{3x}-1&=5\\\sqrt{3x}&=\boxed{6}\\\left(\sqrt{3x}\right)^{\boxed{2}}&=(6)^{\boxed{2}}\\\boxed{3x}&=36\\x&=12\end{aligned}$$

$\boxed{\text{Yes}}$ it checks.

PRACTICE

17.
$$\begin{aligned}\sqrt{5x-6}&=2\\\left(\sqrt{5x-6}\right)^2&=2^2\\5x-6&=4\\5x&=10\\x&=2\end{aligned}$$

19.
$$\begin{aligned}\sqrt{6x+1}+2&=7\\\sqrt{6x+1}&=5\\\left(\sqrt{6x+1}\right)^2&=(5)^2\\6x+1&=25\\6x&=24\\x&=4\end{aligned}$$

21.
$$\begin{aligned}2\sqrt{4x+1}&=\sqrt{x+4}\\\left(2\sqrt{4x+1}\right)^2&=\left(\sqrt{x+4}\right)^2\\4(4x+1)&=x+4\\16x+4&=x+4\\15x+4&=4\\15x&=0\\x&=0\end{aligned}$$

23.

$$\sqrt[3]{7n-1}=3$$
$$\left(\sqrt[3]{7n-1}\right)^3=(3)^3$$
$$7n-1=27$$
$$7n=28$$
$$n=4$$

25.

$$\sqrt[4]{10p+1}=\sqrt[4]{11p-7}$$
$$\left(\sqrt[4]{10p+1}\right)^4=\left(\sqrt[4]{11p-7}\right)^4$$
$$10p+1=11p-7$$
$$1=p-7$$
$$8=p$$
$$p=8$$

27.

$$(6x+2)^{1/2}=(5x+3)^{1/2}$$
$$\sqrt{6x+2}=\sqrt{5x+3}$$
$$\left(\sqrt{6x+2}\right)^2=\left(\sqrt{5x+3}\right)^2$$
$$6x+2=5x+3$$
$$x+2=3$$
$$x=1$$

29.

$$(x+8)^{1/3}=-2$$
$$\sqrt[3]{x+8}=-2$$
$$\left(\sqrt[3]{x+8}\right)^3=(-2)^3$$
$$x+8=-8$$
$$x=-16$$

31.

$$\sqrt{5-x}+10=9$$
$$\sqrt{5-x}=-1$$
$$\left(\sqrt{5-x}\right)^2=(-1)^2$$
$$5-x=1$$
$$-x=-4$$
$$x=4$$

No solution

$$\sqrt{5-4}+10\overset{?}{=}9$$
$$1+10\overset{?}{=}9$$
$$11\neq 9$$

33.

$$x=\frac{\sqrt{12x-5}}{2}$$
$$2(x)=2\left(\frac{\sqrt{12x-5}}{2}\right)$$
$$2x=\sqrt{12x-5}$$
$$(2x)^2=\left(\sqrt{12x-5}\right)^2$$
$$4x^2=12x-5$$
$$4x^2-12x+5=0$$
$$(2x-5)(2x-1)=0$$
$$2x-5=0 \quad \text{or} \quad 2x-1=0$$
$$2x=5 \qquad 2x=1$$
$$x=\frac{5}{2} \qquad x=\frac{1}{2}$$

35.

$$\sqrt{x+2}-\sqrt{4-x}=0$$
$$\sqrt{x+2}=\sqrt{4-x}$$
$$\left(\sqrt{x+2}\right)^2=\left(\sqrt{4-x}\right)^2$$
$$x+2=4-x$$
$$2x+2=4$$
$$2x=2$$
$$x=1$$

37.

$$2\sqrt{x}=\sqrt{5x-16}$$
$$\left(2\sqrt{x}\right)^2=\left(\sqrt{5x-16}\right)^2$$
$$4x=5x-16$$
$$-x=-16$$
$$x=16$$

39.

$$r-9=\sqrt{2r-3}$$
$$(r-9)^2=\left(\sqrt{2r-3}\right)^2$$
$$r^2-9r-9r+81=2r-3$$
$$r^2-18r+81=2r-3$$
$$r^2-18r+81-2r+3=2r-3-2r+3$$
$$r^2-20r+84=0$$
$$(r-14)(r-6)=0$$
$$r-14=0 \quad \text{or} \quad r-6=0$$
$$r=14 \qquad r=\not{6}$$

6 is extraneous

41.

$$\left(m^4+m^2-25\right)^{1/4}=m$$
$$\sqrt[4]{m^4+m^2-25}=m$$
$$\left(\sqrt[4]{m^4+m^2-25}\right)^4=(m)^4$$
$$m^4+m^2-25=m^4$$
$$m^4+m^2-25-m^4=m^4-m^4$$
$$m^2-25=0$$
$$(m+5)(m-5)=0$$
$$m+5=0 \quad \text{or} \quad m-5=0$$
$$m=\not{-5} \qquad m=5$$

−5 is extraneous

43.

$$\sqrt{-5x+24}=6-x$$
$$\left(\sqrt{-5x+24}\right)^2=(6-x)^2$$
$$-5x+24=36-6x-6x+x^2$$
$$-5x+24=36-12x+x^2$$
$$-5x+24+5x-24=36-12x+x^2+5x-24$$
$$0=x^2-7x+12$$
$$0=(x-3)(x-4)$$
$$x-3=0 \quad \text{or} \quad x-4=0$$
$$x=3 \qquad x=4$$

45.

$$\sqrt{y+2}=4-y$$
$$\left(\sqrt{y+2}\right)^2=(4-y)^2$$
$$y+2=16-4y-4y+y^2$$
$$y+2=16-8y+y^2$$
$$y+2-y-2=16-8y+y^2-y-2$$
$$0=y^2-9y+14$$
$$0=(y-2)(y-7)$$
$$y-2=0 \quad \text{or} \quad y-7=0$$
$$y=2 \qquad y=\not{7}$$

7 is extraneous

47.

$$\sqrt[3]{x^3-7}=x-1$$
$$\left(\sqrt[3]{x^3-7}\right)^3=(x-1)^3$$
$$x^3-7=(x-1)(x-1)(x-1)$$
$$x^3-7=(x-1)(x^2-x-x+1)$$
$$x^3-7=(x-1)(x^2-2x+1)$$
$$x^3-7=x(x^2-2x+1)-1(x^2-2x+1)$$
$$x^3-7=x^3-2x^2+x-x^2+2x-1$$
$$x^3-7=x^3-3x^2+3x-1$$
$$x^3-7-x^3=x^3-3x^2+3x-1-x^3$$
$$-7=-3x^2+3x-1$$
$$3x^2-3x+1-7=0$$
$$3x^2-3x-6=0$$
$$\frac{3x^2}{3}-\frac{3x}{3}-\frac{6}{3}=\frac{0}{3}$$
$$x^3-x-2=0$$
$$(x-2)(x+1)=0$$
$$x-2=0 \quad \text{or} \quad x+1=0$$
$$x=2 \qquad\qquad x=-1$$

49.

$$\sqrt[4]{x^4+4x^2-4}=-x$$
$$\left(\sqrt[4]{x^4+4x^2-4}\right)^4=(-x)^4$$
$$x^4+4x^2-4=x^4$$
$$x^4+4x^2-4-x^4=x^4-x^4$$
$$4x^2-4=0$$
$$\frac{4x^2}{4}-\frac{4}{4}=\frac{0}{4}$$
$$x^2-1=0$$
$$(x-1)(x+1)=0$$
$$x-1=0 \quad \text{or} \quad x+1=0$$
$$x=\not{1} \qquad\qquad x=-1$$

1 is extraneous

51.

$$\sqrt[4]{12t+4}+2=0$$
$$\sqrt[4]{12t+4}=-2$$
$$\left(\sqrt[4]{12t+4}\right)^4=(-2)^4$$
$$12t+4=16$$
$$12t=12$$
$$t=\not{1}$$

No Solution

1 is extraneous

53.

$$\sqrt{2y+1}=1-2\sqrt{y}$$
$$\left(\sqrt{2y+1}\right)^2=\left(1-2\sqrt{y}\right)^2$$
$$2y+1=1-2\sqrt{y}-2\sqrt{y}+4\sqrt{y^2}$$
$$2y+1=1-4\sqrt{y}+4y$$
$$2y+1-4y-1=1-4\sqrt{y}+4y-4y-1$$
$$-2y=-4\sqrt{y}$$
$$(-2y)^2=\left(-4\sqrt{y}\right)^2$$
$$4y^2=16y$$
$$4y^2-16y=0$$
$$4y(y-4)=0$$
$$4y=0 \quad \text{or} \quad y-4=0$$
$$y=0 \qquad\qquad y=\not{4}$$

4 is extraneous

55.

$$\sqrt{n^2+6n+3}=\sqrt{n^2-6n-3}$$
$$\left(\sqrt{n^2+6n+3}\right)^2=\left(\sqrt{n^2-6n-3}\right)^2$$
$$n^2+6n+3=n^2-6n-3$$
$$n^2+6n+3-n^2=n^2-6n-3-n^2$$
$$6n+3=-6n-3$$
$$12n+3=-3$$
$$12n=-6$$
$$n=-\frac{6}{12}$$
$$n=-\frac{1}{2}$$

57.

$$\sqrt{7t^2-4}=\sqrt{7t-8t^2}$$
$$\left(\sqrt{7t^2-4}\right)^2=\left(\sqrt{7t-8t^2}\right)^2$$
$$7t^2-4=7t-8t^2$$
$$7t^2-4+8t^2-7t=7t-8t^2+8t^2-7t$$
$$15t^2-7t-4=0$$
$$(5t-4)(3t+1)=0$$
$$5t-4=0 \quad \text{or} \quad 3t+1=0$$
$$5t=4 \qquad 3t=-1$$
$$t=\frac{4}{5} \qquad t=-\frac{1}{3}$$

59.

$$\sqrt{y+7}+3=\sqrt{y+4}$$
$$\left(\sqrt{y+7}+3\right)^2=\left(\sqrt{y+4}\right)^2$$
$$y+7+3\sqrt{y+7}+3\sqrt{y+7}+9=y+4$$
$$y+16+6\sqrt{y+7}=y+4$$
$$y+16+6\sqrt{y+7}-y-16=y+4-y-16$$
$$6\sqrt{y+7}=-12$$
$$\frac{6\sqrt{y+7}}{6}=\frac{-12}{6}$$
$$\sqrt{y+7}=-2$$
$$\left(\sqrt{y+7}\right)^2=(-2)^2$$
$$y+7=4$$
$$y=-3$$

no solution

–3 is extraneous

61.

$$2+\sqrt{u}=\sqrt{2u+7}$$
$$\left(2+\sqrt{u}\right)^2=\left(\sqrt{2u+7}\right)^2$$
$$4+2\sqrt{u}+2\sqrt{u}+\sqrt{u^2}=2u+7$$
$$4+4\sqrt{u}+u=2u+7$$
$$4+4\sqrt{u}+u-4-u=2u+7-4-u$$
$$4\sqrt{u}=u+3$$
$$\left(4\sqrt{u}\right)^2=(u+3)^2$$
$$16u=u^2+3u+3u+9$$
$$16u=u^2+6u+9$$
$$16u-16u=u^2+6u+9-16u$$
$$0=u^2-10u+9$$
$$0=(u-1)(u-9)$$
$$u-1=0 \quad \text{or} \quad u-9=0$$
$$u=1 \qquad u=9$$

63.

$$\sqrt{6t+1}-3\sqrt{t}=-1$$
$$\sqrt{6t+1}=-1+3\sqrt{t}$$
$$\left(\sqrt{6t+1}\right)^2=\left(3\sqrt{t}-1\right)^2$$
$$6t+1=9\sqrt{t^2}-3\sqrt{t}-3\sqrt{t}+1$$
$$6t+1=9t-6\sqrt{t}+1$$
$$6t+1-9t-1=9t-6\sqrt{t}+1-9t-1$$
$$-3t=-6\sqrt{t}$$
$$(-3t)^2=\left(-6\sqrt{t}\right)^2$$
$$9t^2=36t$$
$$9t^2-36t=0$$
$$9t(t-4)=0$$
$$9t=0 \quad \text{or} \quad t-4=0$$
$$t=0 \qquad t=4$$

0 is extraneous

65.

$$\sqrt{2x+5}+\sqrt{x+2}=5$$
$$\sqrt{2x+5}=5-\sqrt{x+2}$$
$$\left(\sqrt{2x+5}\right)^2=\left(5-\sqrt{x+2}\right)^2$$
$$2x+5=25-5\sqrt{x+2}-5\sqrt{x+2}+x+2$$
$$2x+5=27-10\sqrt{x+2}+x$$
$$2x+5-27-x=27-10\sqrt{x+2}+x-27-x$$
$$x-22=-10\sqrt{x+2}$$
$$(x-22)^2=\left(-10\sqrt{x+2}\right)^2$$
$$x^2-22x-22x+484=100(x+2)$$
$$x^2-44x+484=100x+200$$
$$x^2-144x+284=0$$
$$(x-2)(x-142)=0$$
$$x-2=0 \quad \text{or} \quad x-142=0$$
$$x=2 \qquad x=\cancel{142}$$

142 is extraneous

67.

$$\sqrt{x-5}-\sqrt{x+3}=4$$
$$\sqrt{x-5}=4+\sqrt{x+3}$$
$$\left(\sqrt{x-5}\right)^2=\left(4+\sqrt{x+3}\right)^2$$
$$x-5=16+4\sqrt{x+3}+4\sqrt{x+3}+x+3$$
$$x-5=19+8\sqrt{x+3}+x$$
$$x-5-x-19=19+8\sqrt{x+3}+x-x-19$$
$$-24=8\sqrt{x+3}$$
$$\frac{-24}{8}=\frac{8\sqrt{x+3}}{8}$$
$$-3=\sqrt{x+3}$$
$$(-3)^2=\left(\sqrt{x+3}\right)^2$$
$$9=x+3$$
$$\cancel{6}=x$$

No Solution

6 is extraneous

69.

$$\sqrt{3x-6}=3$$
$$\left(\sqrt{3x-6}\right)^2=(3)^2$$
$$3x-6=9$$
$$3x=15$$
$$x=5$$

71.

$$\sqrt{x+8}-\sqrt{x}=2$$
$$\sqrt{x+8}=2+\sqrt{x}$$
$$\left(\sqrt{x+8}\right)^2=\left(2+\sqrt{x}\right)^2$$
$$x+8=4+2\sqrt{x}+2\sqrt{x}+\sqrt{x^2}$$
$$x+8=4+4\sqrt{x}+x$$
$$x+8-4-x=4+4\sqrt{x}+x-4-x$$
$$4=4\sqrt{x}$$
$$\frac{4}{4}=\frac{4\sqrt{x}}{4}$$
$$1=\sqrt{x}$$
$$(1)^2=\left(\sqrt{x}\right)^2$$
$$1=x$$

73.

$$v=\sqrt{2gh}$$
$$v^2=\left(\sqrt{2gh}\right)^2$$
$$v^2=2gh$$
$$\frac{v^2}{2g}=\frac{2gh}{2g}$$
$$\frac{v^2}{2g}=h$$

75.

$$T = 2\pi\sqrt{\frac{\ell}{32}}$$

$$(T)^2 = \left(2\pi\sqrt{\frac{\ell}{32}}\right)^2$$

$$T^2 = 4\pi^2 \bullet \frac{\ell}{32}$$

$$T^2 = \frac{4\pi^2\ell}{32}$$

$$T^2 = \frac{\pi^2\ell}{8}$$

$$8(T^2) = 8\left(\frac{\pi^2\ell}{8}\right)$$

$$8T^2 = \pi^2\ell$$

$$\frac{8T^2}{\pi^2} = \frac{\pi^2\ell}{\pi^2}$$

$$\frac{8T^2}{\pi^2} = \ell$$

77.

$$r = \sqrt[3]{\frac{A}{P}} - 1$$

$$r + 1 = \sqrt[3]{\frac{A}{P}}$$

$$(r+1)^3 = \left(\sqrt[3]{\frac{A}{P}}\right)^3$$

$$(r+1)^3 = \frac{A}{P}$$

$$P(r+1)^3 = P\left(\frac{A}{P}\right)$$

$$P(r+1)^3 = A$$

79.

$$L_A = L_B\sqrt{1 - \frac{v^2}{c^2}}$$

$$\frac{L_A}{L_B} = \frac{L_B\sqrt{1 - \frac{v^2}{c^2}}}{L_B}$$

$$\frac{L_A}{L_B} = \sqrt{1 - \frac{v^2}{c^2}}$$

$$\left(\frac{L_A}{L_B}\right)^2 = \left(\sqrt{1 - \frac{v^2}{c^2}}\right)^2$$

$$\frac{L_A^{\ 2}}{L_B^{\ 2}} = 1 - \frac{v^2}{c^2}$$

$$\frac{L_A^{\ 2}}{L_B^{\ 2}} - 1 = -\frac{v^2}{c^2}$$

$$-c^2\left(\frac{L_A^{\ 2}}{L_B^{\ 2}} - 1\right) = -c^2\left(-\frac{v^2}{c^2}\right)$$

$$-c^2\left(\frac{L_A^{\ 2}}{L_B^{\ 2}} - 1\right) = v^2$$

$$c^2\left(-\frac{L_A^{\ 2}}{L_B^{\ 2}} + 1\right) = v^2$$

$$c^2\left(1 - \frac{L_A^{\ 2}}{L_B^{\ 2}}\right) = v^2$$

APPLICATIONS

81. HIGHWAY DESIGN

$$s = 3\sqrt{r}$$

$$40 = 3\sqrt{r}$$

$$(40)^2 = \left(3\sqrt{r}\right)^2$$

$$1{,}600 = 9r$$

$$\frac{1{,}600}{9} = \frac{9r}{9}$$

$$178 = r$$

$$r = 178 \text{ ft}$$

83. WIND POWER

$$v=\sqrt[3]{\frac{P}{0.02}}$$
$$29=\sqrt[3]{\frac{P}{0.02}}$$
$$(29)^3=\left(\sqrt[3]{\frac{P}{0.02}}\right)^3$$
$$24{,}389=\frac{P}{0.02}$$
$$0.02(24{,}389)=0.02\left(\frac{P}{0.02}\right)$$
$$488=P$$
$$P=488 \text{ watts}$$

85. THEATER PRODUCTIONS

$$w_2=\sqrt{w_1^2+w_3^2}$$
$$12.5=\sqrt{w_1^2+(7.5)^2}$$
$$12.5=\sqrt{w_1^2+56.25}$$
$$(12.5)^2=\left(\sqrt{w_1^2+56.25}\right)^2$$
$$156.25=w_1^2+56.25$$
$$156.25-156.25=w_1^2+56.25-156.25$$
$$0=w_1^2-100$$
$$0=(w_1-10)(w_1+10)$$
$$w_1-10=0 \quad \text{or} \quad w_1+10=0$$
$$w_1=10 \qquad w_1=\cancel{-10}$$
$$w_1=10 \text{ lb.}$$

87. SUPPLY AND DEMAND

$$\sqrt{5x}=\sqrt{100-3x^2}$$
$$\left(\sqrt{5x}\right)^2=\left(\sqrt{100-3x^2}\right)^2$$
$$5x=100-3x^2$$
$$3x^2+5x-100=0$$
$$(3x+20)(x-5)=0$$
$$3x+20=0 \quad \text{or} \quad x-5=0$$
$$3x=-20 \qquad x=5$$
$$x=\cancel{-\frac{20}{3}}$$

The equilibrium price is $5.

WRITING

89. Answers will vary.

91. Isolate each radical on one side of the equation.

93. Answers will vary.

REVIEW

95. Let I = intensity and d = distance from the bulb.

$$I=\frac{k}{d^2}$$
$$40=\frac{k}{5^2}$$
$$40=\frac{k}{25}$$
$$25(40)=25\left(\frac{k}{25}\right)$$
$$1{,}000=k$$

$$I=\frac{k}{d^2}$$
$$I=\frac{1{,}000}{20^2}$$
$$=\frac{1{,}000}{400}$$
$$=2.5 \text{ foot-candles}$$

97.

$$\frac{12}{0.166044}=\frac{30}{x}$$
$$12x=30(0.166044)$$
$$12x=4.98132$$
$$x=0.41511 \text{ in.}$$

CHALLENGE PROBLEMS

99.

$$\sqrt[3]{2x} = \sqrt{x}$$
$$\left(\sqrt[3]{2x}\right)^2 = \left(\sqrt{x}\right)^2$$
$$\sqrt[3]{4x^2} = x$$
$$\left(\sqrt[3]{4x^2}\right)^3 = (x)^3$$
$$4x^2 = x^3$$
$$0 = x^3 - 4x^2$$
$$0 = x^2(x-4)$$
$$x = 0, \quad x = 0, \quad x - 4 = 0$$
$$x = 0 \qquad\qquad x = 4$$

101.

$$\sqrt{x+2} + \sqrt{2x} = \sqrt{18-x}$$
$$\left(\sqrt{x+2} + \sqrt{2x}\right)^2 = \left(\sqrt{18-x}\right)^2$$
$$x + 2 + \sqrt{x+2}\sqrt{2x} + \sqrt{x+2}\sqrt{2x} + 2x = 18 - x$$
$$3x + 2 + 2\sqrt{x+2}\sqrt{2x} = 18 - x$$
$$3x + 2 + 2\sqrt{x+2}\sqrt{2x} - 3x - 2 = 18 - x - 3x - 2$$
$$2\sqrt{x+2}\sqrt{2x} = -4x + 16$$
$$\frac{2\sqrt{x+2}\sqrt{2x}}{2} = \frac{-4x}{2} + \frac{16}{2}$$
$$\sqrt{x+2}\sqrt{2x} = -2x + 8$$
$$\left(\sqrt{x+2}\sqrt{2x}\right)^2 = (-2x+8)^2$$
$$2x(x+2) = 4x^2 - 16x - 16x + 64$$
$$2x^2 + 4x = 4x^2 - 32x + 64$$
$$2x^2 + 4x - 2x^2 - 4x = 4x^2 - 32x + 64 - 2x^2 - 4x$$
$$0 = 2x^2 - 36x + 64$$
$$\frac{0}{2} = \frac{2x^2}{2} - \frac{36x}{2} + \frac{64}{2}$$
$$0 = x^2 - 18x + 32$$
$$0 = (x-2)(x-16)$$
$$x - 2 = 0 \text{ or } x - 16 = 0$$
$$x = 2 \qquad x = \not{16}$$

SECTION 9.6

VOCABULARY

1. In a right triangle, the side opposite the 90° angle is called the **hypotenuse**.

3. The **Pythagorean** Theorem states that in any right triangle, the square of the hypotenuse is equal to the sum of the squares of the lengths of the two legs.

CONCEPTS

5. If a and b are the lengths of the legs of a right triangle and c is the length of the hypotenuse, then $a^2 + b^2 = c^2$.

7. In an isosceles right triangle, the length of the hypotenuse is the length of one leg times $\sqrt{2}$.

9. The length of the longer leg of a 30°–60°–90° triangle is the length of the shorter leg times $\sqrt{3}$.

11. In a right triangle, the shorter leg is opposite the **30°** angle, and the longer leg is opposite the **60°** angle.

13. a. $c = 8$
    b. $c = \sqrt{15}$
    c. $c = \sqrt{24} = 2\sqrt{6}$

NOTATION

15.

$$\begin{aligned}\sqrt{(-1-3)^2 + [2-(-4)]^2} &= \sqrt{(-4)^2 + [\boxed{6}]^2} \\ &= \sqrt{\boxed{52}} \\ &= \sqrt{\boxed{4} \cdot 13} \\ &= \boxed{2}\sqrt{13} \\ &\approx 7.21\end{aligned}$$

PRACTICE

17.

$$\begin{aligned}a^2 + b^2 &= c^2 \\ 6^2 + 8^2 &= c^2 \\ 36 + 64 &= c^2 \\ 100 &= c^2 \\ \sqrt{100} &= c \\ 10 &= c \\ c &= 10 \text{ ft}\end{aligned}$$

19.

$$\begin{aligned}a^2 + b^2 &= c^2 \\ a^2 + 18^2 &= 82^2 \\ a^2 + 324 &= 6,724 \\ a^2 &= 6,400 \\ a &= \sqrt{6,400} \\ a &= 80 \text{ m}\end{aligned}$$

21. In an isosceles right triangle, the length of the hypotenuse is the length of one leg times $\sqrt{2}$ and the lengths of the legs are equal.

$$x = 2$$

$$\begin{aligned}h &= x\sqrt{2} \\ &= 2\sqrt{2} \\ &\approx 2.83\end{aligned}$$

23. The shorter leg of a 30°–60°–90° triangle is half as long as the hypotenuse. The longer leg of a 30°–60°–90° triangle is the length of the shorter leg times $\sqrt{3}$.

$$x = 5\sqrt{3} \qquad h = 2(5)$$
$$x \approx 8.66 \qquad h = 10$$

25.

$$\begin{aligned}2x &= 9.37 \\ x &= 4.69\end{aligned}$$

$$\begin{aligned}y &= x\sqrt{3} \\ &= 4.69\sqrt{3} \\ &= 8.12\end{aligned}$$

27.

$$17.12 = x\sqrt{2}$$
$$\frac{17.12}{\sqrt{2}} = \frac{x\sqrt{2}}{\sqrt{2}}$$
$$12.11 = x$$

$$x = y$$
$$y = 12.11$$

29. The face of the cube is a square with sides of 7 cm. The diagonal forms a 45°–45°–90° triangle in which the diagonal is the hypotenuse.

$$h = x\sqrt{2}$$
$$= 7\sqrt{2} \text{ cm}$$

31.

$$d = \sqrt{(x_2 - x_1)^2 + (y_2 - y_1)^2}$$
$$= \sqrt{(3-0)^2 + (-4-0)^2}$$
$$= \sqrt{(3)^2 + (-4)^2}$$
$$= \sqrt{9+16}$$
$$= \sqrt{25}$$
$$= 5$$

33.

$$d = \sqrt{(x_2 - x_1)^2 + (y_2 - y_1)^2}$$
$$= \sqrt{[3-(-2)]^2 + [4-(-8)]^2}$$
$$= \sqrt{(5)^2 + (12)^2}$$
$$= \sqrt{25+144}$$
$$= \sqrt{169}$$
$$= 13$$

35.

$$d = \sqrt{(x_2 - x_1)^2 + (y_2 - y_1)^2}$$
$$= \sqrt{(12-6)^2 + (16-8)^2}$$
$$= \sqrt{(6)^2 + (8)^2}$$
$$= \sqrt{36+64}$$
$$= \sqrt{100}$$
$$= 10$$

37.

$$d = \sqrt{(x_2 - x_1)^2 + (y_2 - y_1)^2}$$
$$= \sqrt{[-5-(-3)]^2 + (-5-5)^2}$$
$$= \sqrt{(-2)^2 + (-10)^2}$$
$$= \sqrt{4+100}$$
$$= \sqrt{104}$$
$$= \sqrt{4 \cdot 26}$$
$$= 2\sqrt{26}$$

39. ISOSCELES TRIANGLES
The distance between two pairs of vertices must be equal.

Find the distance between (–2, 4) and (2, 8).

$$d = \sqrt{(x_2 - x_1)^2 + (y_2 - y_1)^2}$$
$$= \sqrt{[2-(-2)]^2 + (8-4)^2}$$
$$= \sqrt{(4)^2 + (4)^2}$$
$$= \sqrt{16+16}$$
$$= \sqrt{32}$$
$$= \sqrt{16 \cdot 2}$$
$$= 4\sqrt{2}$$

Find the distance between (2, 8) and (6, 4).

$$d = \sqrt{(x_2 - x_1)^2 + (y_2 - y_1)^2}$$
$$= \sqrt{(6-2)^2 + (4-8)^2}$$
$$= \sqrt{(4)^2 + (-4)^2}$$
$$= \sqrt{16+16}$$
$$= \sqrt{32}$$
$$= \sqrt{16 \cdot 2}$$
$$= 4\sqrt{2}$$

Since the distances are equal, the triangle must be isosceles.

APPLICATIONS

41. WASHINGTON, D.C.
The x– and y–axes divide the square into 4 isosceles triangles (45°–45°–90°). If the area of the square is 100, then the length of the sides would be 10. Thus, the hypotenuse is 10. Find the length of the legs and that would be the distance of a corner from the origin.

$$x\sqrt{2}=10$$
$$\frac{x\sqrt{2}}{\sqrt{2}}=\frac{10}{\sqrt{2}}$$
$$x=\frac{10}{\sqrt{2}}\bullet\frac{\sqrt{2}}{\sqrt{2}}$$
$$x=\frac{10\sqrt{2}}{2}$$
$$x=5\sqrt{2}$$

The coordinates would all be $5\sqrt{2}$ units from the origin in all four directions:

$$\left(5\sqrt{2},0\right)=(7.07,0)$$
$$\left(-5\sqrt{2},0\right)=(-7.07,0)$$
$$\left(0,5\sqrt{2}\right)=(0,7.07)$$
$$\left(0,-5\sqrt{2}\right)=(0,\ -7.07)$$

43. HARDWARE
If the sides of the nut are 10 mm, then half of the side is 5 mm, which would be the length of the side opposite the 30° angle. The height is one–half the side opposite the 60° angle. To find the total height, multiply 2 times the length of the side opposite the 60° angle.

$$h=2\left(x\sqrt{3}\right)$$
$$=2\left(5\sqrt{3}\right)$$
$$=10\sqrt{3}\text{ mm}$$
$$\approx 17.32\text{ mm}$$

45. BASEBALL
If the ball lands 10 feet behind 3rd base, use the following right triangle to find how far he must throw the ball.

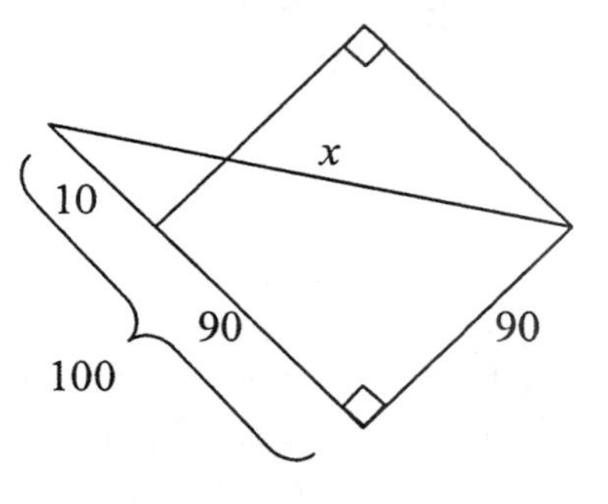

$$100^2+90^2=x^2$$
$$10{,}000+8{,}100=x^2$$
$$18{,}100=x^2$$
$$\sqrt{18{,}100}=x$$
$$\sqrt{100\bullet 181}=x$$
$$10\sqrt{181}=x$$
$$x=10\sqrt{181}\text{ ft}$$
$$x\approx 134.54\text{ ft}$$

47. CLOTHESLINES
Find one–half of the amount stretched by using the Pythagorean Theorem. The blue line (amount stretched) is the hypotenuse (c). Let a = 1, and $b=\frac{15}{2}=7.5$.

$$a^2+b^2=c^2$$
$$1^2+(7.5)^2=c^2$$
$$1+56.25=c^2$$
$$57.25=c^2$$
$$\sqrt{57.25}=c$$
$$c\approx 7.566$$
$$2c\approx 15.132$$

When stretched, the line is 15.132 ft. To find the amount the line is stretched, find the difference in the original length of the line and the length when stretched:

$$15.132-15=0.132$$
$$\approx 0.13\text{ ft.}$$

49. ART HISTORY

a. Use the ordered pairs (5, 0) and (8, 21)

$$
\begin{aligned}
d &= \sqrt{(x_2 - x_1)^2 + (y_2 - y_1)^2} \\
&= \sqrt{(8-5)^2 + (21-0)^2} \\
&= \sqrt{(3)^2 + (21)^2} \\
&= \sqrt{9 + 441} \\
&= \sqrt{450} \\
&= 21.21 \text{ units}
\end{aligned}
$$

b. Use the ordered pairs (10, 13) and (2, 11).

$$
\begin{aligned}
d &= \sqrt{(x_2 - x_1)^2 + (y_2 - y_1)^2} \\
&= \sqrt{(2-10)^2 + (11-13)^2} \\
&= \sqrt{(-8)^2 + (-2)^2} \\
&= \sqrt{64 + 4} \\
&= \sqrt{68} \\
&= 8.25 \text{ units}
\end{aligned}
$$

c. Use the ordered pairs (7, 19) and (12, 7).

$$
\begin{aligned}
d &= \sqrt{(x_2 - x_1)^2 + (y_2 - y_1)^2} \\
&= \sqrt{(12-7)^2 + (7-19)^2} \\
&= \sqrt{(5)^2 + (-12)^2} \\
&= \sqrt{25 + 144} \\
&= \sqrt{169} \\
&= 13.00 \text{ units}
\end{aligned}
$$

51. PACKAGING

Let $a = 24$, $b = 24$, and $c = 4$.

$$
\begin{aligned}
d &= \sqrt{a^2 + b^2 + c^2} \\
&= \sqrt{24^2 + 24^2 + 4^2} \\
&= \sqrt{576 + 576 + 16} \\
&= \sqrt{1{,}168} \\
&\approx 34.2 \text{ in.}
\end{aligned}
$$

Yes, the bone will fit in the box because the length of the diagonal (34.2 in.) is longer than the length of the bone (34 in.).

WRITING

53. Answers will vary.

REVIEW

55. DISCOUNT BUYING

Let c = cost of each unit and
let x = number of units purchased.
Unit cost · number = total cost

$$\text{unit cost } (c) = \frac{\text{total cost}}{\text{number of units}(x)}$$

	unit cost	number	total cost
first purchase	$c = \frac{224}{x}$	x	224
second purchase	$c - 4 = \frac{224}{x+1}$	$x + 1$	224

Solve the second unit cost for c.

$$
\begin{aligned}
c - 4 &= \frac{224}{x+1} \\
c &= \frac{224}{x+1} + 4
\end{aligned}
$$

Set the costs equal and solve the equation for x.

$$
\begin{aligned}
\frac{224}{x} &= \frac{224}{x+1} + 4 \\
x(x+1)\left(\frac{224}{x}\right) &= x(x+1)\left(\frac{224}{x+1} + 4\right) \\
224(x+1) &= 224(x) + 4x(x+1) \\
224x + 224 &= 224x + 4x^2 + 4x \\
224x + 224 - 224x &= 224x + 4x^2 + 4x - 224x \\
224 &= 4x^2 + 4x \\
224 - 224 &= 4x^2 + 4x - 224 \\
0 &= 4x^2 + 4x - 224 \\
\frac{0}{4} &= \frac{4x^2}{4} + \frac{4x}{4} - \frac{224}{4} \\
0 &= x^2 + x - 56 \\
0 &= (x-7)(x+8)
\end{aligned}
$$

$$
\begin{aligned}
x - 7 &= 0 \quad \text{or} \quad x + 8 = 0 \\
x &= 7 \qquad\qquad x = -8
\end{aligned}
$$

Since the answer cannot be negative, he originally bought 7 motors.

57.

$$\frac{16+6+10+4+5+13}{6} = \frac{54}{6}$$
$$= 9$$

CHALLENGE PROBLEMS

59. Using a 45°–45°–90° triangle on the face of the cube, you can draw a diagonal whose measure would be $a\sqrt{2}$ in. Now you have a triangle involving the following sides: base of the cubes whose measure is a, diagonal of a face whose length is $a\sqrt{2}$, and the diagonal of the cube (d). Use the Pythagorean Theorem to find the length of the diagonal (d).

$$(a)^2 + \left(a\sqrt{2}\right)^2 = d^2$$
$$a^2 + a^2(2) = d^2$$
$$a^2 + 2a^2 = d^2$$
$$3a^2 = d^2$$
$$d = \sqrt{3a^2}$$
$$d = a\sqrt{3} \text{ in.}$$

SECTION 9.7

VOCABULARY

1. The **imaginary** number i is defined as $i=\sqrt{-1}$.

3. For the complex number 2 + 5i, we call 2 the **real** part and 5 the **imaginary** part.

5. 6 + 3i and 6 – 3i are called complex **conjugates**.

CONCEPTS

7. a. $i=\boxed{\sqrt{-1}}$

   b. $i^2=\boxed{-1}$

   c. $i^3=\boxed{-i}$

   d. $i^4=\boxed{1}$

9. To add (or subtract) complex numbers, add (or subtract) their **real** parts and add (or subtract) their **imaginary** parts.

11. To divide complex numbers, multiply the numerator and denominator by the complex conjugate of the **denominator**.

13. a. $\sqrt{-16}+\sqrt{-9}=4i+3i=7i$

    b. $\sqrt{-2}\sqrt{-3}=i\sqrt{2}\cdot i\sqrt{3}$
    $=i^2\sqrt{6}$
    $=-\sqrt{6}$

15.

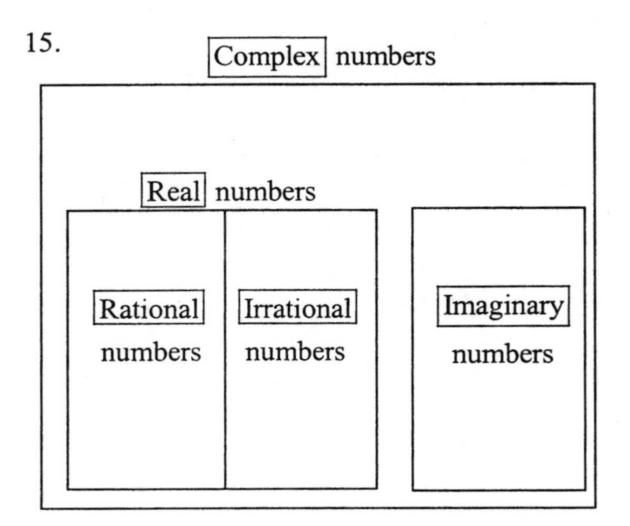

17. a. Yes

    b. Yes

NOTATION

19.

$$(3+2i)(3-i)=\boxed{9}-3i+\boxed{6i}-2i^2$$
$$=9+3i+\boxed{2}$$
$$=\boxed{11}+3i$$

21. a. true

    b. false

    c. false

    d. false

PRACTICE

23.

$$\sqrt{-9}=\sqrt{-1\cdot 9}$$
$$=3i$$

25.

$$\sqrt{-7}=\sqrt{-1\cdot 7}$$
$$=i\sqrt{7} \text{ or } \sqrt{7}i$$

27.

$$\sqrt{-24}=\sqrt{-1}\sqrt{4}\sqrt{6}$$
$$=2i\sqrt{6}$$
$$=2\sqrt{6}i$$

29.

$$-\sqrt{-24}=-\sqrt{-1}\sqrt{4}\sqrt{6}$$
$$=-2i\sqrt{6}$$
$$=-2\sqrt{6}i$$

31.

$$5\sqrt{-81}=5\sqrt{-1}\sqrt{81}$$
$$=5(9i)$$
$$=45i$$

33.

$$\sqrt{-\frac{25}{9}}=\frac{\sqrt{-1}\sqrt{25}}{\sqrt{9}}$$
$$=\frac{5i}{3}$$
$$=\frac{5}{3}i$$

35.

$$\sqrt{-1}\sqrt{-36}=(i)(6i)$$
$$=6i^2$$
$$=-6$$

37.

$$\sqrt{-2}\sqrt{-6}=\left(i\sqrt{2}\right)\left(i\sqrt{6}\right)$$
$$=i^2\sqrt{12}$$
$$=i^2\sqrt{4}\sqrt{3}$$
$$=2i^2\sqrt{3}$$
$$=-2\sqrt{3}$$

39.

$$\frac{\sqrt{-25}}{\sqrt{-64}}=\frac{5i}{8i}$$
$$=\frac{5}{8}$$

41.

$$-\frac{\sqrt{-400}}{\sqrt{-1}}=-\frac{20i}{i}$$
$$=-20$$

43.

$$(3+4i)+(5-6i)=(3+5)+(4-6)i$$
$$=8-2i$$

45.

$$(7-3i)-(4+2i)=(7-3i)+(-4-2i)$$
$$=(7-4)+(-3-2)i$$
$$=3-5i$$

47.

$$(6-i)+(9+3i)=(6+9)+(-1+3)i$$
$$=15+2i$$

49.

$$\left(8+\sqrt{-25}\right)+\left(7+\sqrt{-4}\right)=(8+5i)+(7+2i)$$
$$=(8+7)+(5+2)i$$
$$=15+7i$$

51.

$$3(2-i)=6-3i$$

53.

$$-5i(5-5i)=-25i+25i^2$$
$$=-25i-25$$
$$=-25-25i$$

55.

$$(2+i)(3-i)=6-2i+3i-i^2$$
$$=6+i-(-1)$$
$$=6+i+1$$
$$=7+i$$

57.

$$(3-2i)(2+3i)=6+9i-4i-6i^2$$
$$=6+5i-6(-1)$$
$$=6+5i+6$$
$$=12+5i$$

59.

$$(4+i)(3-i)=12-4i+3i-i^2$$
$$=12-i-(-1)$$
$$=12-i+1$$
$$=13-i$$

61.

$$\left(2-\sqrt{-16}\right)\left(3+\sqrt{-4}\right)$$
$$=(2-4i)(3+2i)$$
$$=6+4i-12i-8i^2$$
$$=6-8i-8(-1)$$
$$=6-8i+8$$
$$=14-8i$$

63.

$$\left(2+\sqrt{2}i\right)\left(3-\sqrt{2}i\right)$$
$$=6-2\sqrt{2}i+3\sqrt{2}i-2i^2$$
$$=6+\sqrt{2}i-2(-1)$$
$$=6+\sqrt{2}i+2$$
$$=8+\sqrt{2}i$$

65.

$$(2+i)^2=(2+i)(2+i)$$
$$=4+2i+2i+i^2$$
$$=4+4i-1$$
$$=3+4i$$

67.

$$(3i)^2=(3i)(3i)$$
$$=9i^2$$
$$=9(-1)$$
$$=-9$$
$$=-9+0i$$

69.

$$\left(i\sqrt{6}\right)^2=\left(i\sqrt{6}\right)\left(i\sqrt{6}\right)$$
$$=i^2\sqrt{36}$$
$$=(-1)(6)$$
$$=-6$$
$$=-6+0i$$

71.

$$\frac{1}{i}=\frac{1}{i}\bullet\frac{i}{i}$$
$$=\frac{i}{i^2}$$
$$=\frac{i}{-1}$$
$$=-i$$
$$=0-i$$

73.

$$\frac{4}{5i^3}=\frac{4}{5i^3}\bullet\frac{i}{i}$$
$$=\frac{4i}{5i^4}$$
$$=\frac{4i}{5(1)}$$
$$=\frac{4i}{5}$$
$$=0+\frac{4}{5}i$$

75.

$$\frac{3i}{8\sqrt{-9}}=\frac{\cancel{3i}}{8\left(\cancel{3i}\right)}$$
$$=\frac{1}{8}+0i$$

77.

$$\frac{-3}{5i^5}=\frac{-3}{5i^5}\bullet\frac{i}{i}$$
$$=\frac{-3i}{5i^6}$$
$$=-\frac{3i}{5(-1)}$$
$$=\frac{3i}{5}$$
$$=0+\frac{3}{5}i$$

79.

$$\frac{5}{2-i}=\frac{5}{2-i}\bullet\frac{2+i}{2+i}$$
$$=\frac{5(2+i)}{(2-i)(2+i)}$$
$$=\frac{10+5i}{4-i^2}$$
$$=\frac{10+5i}{4-(-1)}$$
$$=\frac{\cancel{5}(2+i)}{\cancel{5}}$$
$$=2+i$$

81.

$$\frac{-12}{7-\sqrt{-1}}=\frac{-12}{7-i}$$
$$=\frac{-12}{7-i}\bullet\frac{7+i}{7+i}$$
$$=\frac{-12(7+i)}{(7-i)(7+i)}$$
$$=\frac{-84-12i}{49-i^2}$$
$$=\frac{-84-12i}{49-(-1)}$$
$$=\frac{-84-12i}{50}$$
$$=-\frac{84}{50}-\frac{12}{50}i$$
$$=-\frac{42}{25}-\frac{6}{25}i$$

83.

$$\frac{5i}{6+2i}=\frac{5i}{6+2i}\bullet\frac{6-2i}{6-2i}$$
$$=\frac{5i(6-2i)}{(6+2i)(6-2i)}$$
$$=\frac{30i-10i^2}{36-4i^2}$$
$$=\frac{30i-10(-1)}{36-4(-1)}$$
$$=\frac{30i+10}{36+4}$$
$$=\frac{10+30i}{40}$$
$$=\frac{10}{40}+\frac{30}{40}i$$
$$=\frac{1}{4}+\frac{3}{4}i$$

85.

$$\frac{-2i}{3+2i}=\frac{-2i}{3+2i}\bullet\frac{3-2i}{3-2i}$$
$$=\frac{-2i(3-2i)}{(3+2i)(3-2i)}$$
$$=\frac{-6i+4i^2}{9-4i^2}$$
$$=\frac{-6i+4(-1)}{9-4(-1)}$$
$$=\frac{-6i-4}{9+4}$$
$$=\frac{-4-6i}{13}$$
$$=-\frac{4}{13}-\frac{6}{13}i$$

87.

$$\frac{3-2i}{3+2i}=\frac{3-2i}{3+2i}\bullet\frac{3-2i}{3-2i}$$
$$=\frac{(3-2i)(3-2i)}{(3+2i)(3-2i)}$$
$$=\frac{9-6i-6i+4i^2}{9-4i^2}$$
$$=\frac{9-12i+4(-1)}{9-4(-1)}$$
$$=\frac{9-12i-4}{9+4}$$
$$=\frac{5-12i}{13}$$
$$=\frac{5}{13}-\frac{12}{13}i$$

89.

$$\frac{3+2i}{3+i}=\frac{3+2i}{3+i}\bullet\frac{3-i}{3-i}$$
$$=\frac{(3+2i)(3-i)}{(3+i)(3-i)}$$
$$=\frac{9-3i+6i-2i^2}{9-i^2}$$
$$=\frac{9+3i-2(-1)}{9-(-1)}$$
$$=\frac{9+3i+2}{9+1}$$
$$=\frac{11+3i}{10}$$
$$=\frac{11}{10}+\frac{3}{10}i$$

91.

$$\frac{\sqrt{5}-\sqrt{3}i}{\sqrt{5}+\sqrt{3}i}=\frac{\sqrt{5}-\sqrt{3}i}{\sqrt{5}+\sqrt{3}i}\bullet\frac{\sqrt{5}-\sqrt{3}i}{\sqrt{5}-\sqrt{3}i}$$
$$=\frac{(\sqrt{5}-\sqrt{3}i)(\sqrt{5}-\sqrt{3}i)}{(\sqrt{5}+\sqrt{3}i)(\sqrt{5}-\sqrt{3}i)}$$
$$=\frac{5-\sqrt{15}i-\sqrt{15}i+3i^2}{5-3i^2}$$
$$=\frac{5-2\sqrt{15}i+3(-1)}{5-3(-1)}$$
$$=\frac{5-2\sqrt{15}i-3}{5+3}$$
$$=\frac{2-2\sqrt{15}i}{8}$$
$$=\frac{2}{8}-\frac{2\sqrt{15}}{8}i$$
$$=\frac{1}{4}-\frac{\sqrt{15}}{4}i$$

93.

$$i^{21}=i^{4\cdot5+1}$$
$$=(i^4)^5\bullet i^1$$
$$=(1)^5\bullet i$$
$$=1\bullet i$$
$$=i$$

95.

$$i^{27}=i^{4\cdot6+3}$$
$$=(i^4)^6\bullet i^3$$
$$=(1)^6\bullet i^3$$
$$=1\bullet i^3$$
$$=i^3$$
$$=-i$$

97.

$$i^{100}=i^{4\cdot25}$$
$$=(i^4)^{25}$$
$$=(1)^{25}$$
$$=1$$

99.

$$i^{97}=i^{4\cdot24+1}$$
$$=(i^4)^{24}\bullet i^1$$
$$=(1)^{24}\bullet i$$
$$=1\bullet i$$
$$=i$$

APPLICATIONS

101. FRACTALS

Step 1: i^2+i

Step 2 $(i^2+i)^2=(i^2+i)(i^2+i)+i$
$$=i^4+i^3+i^3+i^2+i$$
$$=i^4+2i^3+i^2+i$$
$$=1+2(-i)+(-1)+i$$
$$=1-2i-1+i$$
$$=-i$$

Step 3 $(-i)^2+i=i^2+i$
$$=-1+i$$

WRITING

103. Answers will vary.

REVIEW

105. WIND SPEEDS

Let x = the rate of the wind.

	distance	rate	time
with a tail wind	330	$200+x$	$t_1=\frac{d}{r}$ $=\frac{330}{200+x}$
against the wind	330	$200-x$	$t_1=\frac{d}{r}$ $=\frac{330}{200-x}$

Since the total time for the trip is $3\frac{1}{3}$ hours, add the times and set the sum equal to $3\frac{1}{3}$.

$$t_1+t_2=3\frac{1}{3}$$
$$\frac{330}{200+x}+\frac{330}{200-x}=\frac{10}{3}$$
$$3(200+x)(200-x)\left(\frac{330}{200+x}+\frac{330}{200-x}\right)=3(200+x)(200-x)\left(\frac{10}{3}\right)$$
$$3\cdot 330(200-x)+3\cdot 330(200+x)=10(200+x)(200-x)$$
$$990(200-x)+990(200+x)=10(40{,}000-x^2)$$
$$198{,}000-990x+198{,}000+990x=400{,}000-10x^2$$
$$396{,}000=400{,}000-10x^2$$
$$-4{,}000=-10x^2$$
$$\frac{-4{,}000}{-10}=\frac{-10x^2}{-10}$$
$$400=x^2$$
$$0=x^2-400$$
$$0=(x+20)(x-20)$$
$$x+20=0 \quad \text{or} \quad x-20=0$$
$$x=-20 \qquad x=20$$

Since the rate cannot be negative, the rate of the wind is 20 mph.

CHALLENGE PROBLEMS

107.

$$\frac{2+3i}{2-3i}=\frac{2+3i}{2-3i}\cdot\frac{2-3i}{2-3i}$$
$$=\frac{(2+3i)(2-3i)}{(2-3i)(2-3i)}$$
$$=\frac{4-9i^2}{4-6i-6i+9i^2}$$
$$=\frac{4-9(-1)}{4-12i+9(-1)}$$
$$=\frac{4+9}{4-12i-9}$$
$$=\frac{13}{-5-12i}$$

CHAPTER 9 KEY CONCEPTS

1.

$$\sqrt[3]{-54h^6} = \sqrt[3]{-27h^6}\sqrt[3]{2}$$
$$= -3h^2\sqrt[3]{2}$$

2.

$$2\sqrt[3]{64e} + 3\sqrt[3]{8e} = 2\left(4\sqrt[3]{e}\right) + 3\left(2\sqrt[3]{e}\right)$$
$$= 8\sqrt[3]{e} + 6\sqrt[3]{e}$$
$$= 14\sqrt[3]{e}$$

3.

$$\sqrt{72} - \sqrt{200} = \sqrt{36}\sqrt{2} - \sqrt{100}\sqrt{2}$$
$$= 6\sqrt{2} - 10\sqrt{2}$$
$$= -4\sqrt{2}$$

4.

$$-4\sqrt[3]{5r^2s}\left(5\sqrt[3]{2r}\right) = (-4\cdot 5)\sqrt[3]{5r^2s\cdot 2r}$$
$$= -20\sqrt[3]{10r^3s}$$
$$= -20\sqrt[3]{r^3}\sqrt[3]{10s}$$
$$= -20r\sqrt[3]{10s}$$

5.

$$\left(\sqrt{3s} - \sqrt{2t}\right)\left(\sqrt{3s} + \sqrt{2t}\right)$$
$$= 3s + \sqrt{6st} - \sqrt{6st} - 2t$$
$$= 3s - 2t$$

6.

$$-\sqrt{3}\left(\sqrt{7} - \sqrt{5}\right) = -\sqrt{21} + \sqrt{15}$$

7.

$$\left(3\sqrt{2n} - 2\right)^2 = \left(3\sqrt{2n} - 2\right)\left(3\sqrt{2n} - 2\right)$$
$$= 9\sqrt{4n^2} - 6\sqrt{2n} - 6\sqrt{2n} + 4$$
$$= 9(2n) - 12\sqrt{2n} + 4$$
$$= 18n - 12\sqrt{2n} + 4$$

8.

$$\frac{\sqrt[3]{9j}}{\sqrt[3]{3jk}} = \sqrt[3]{\frac{9j}{3jk}}$$
$$= \sqrt[3]{\frac{3}{k}}$$
$$= \sqrt[3]{\frac{3}{k}}\cdot\sqrt[3]{\frac{k^2}{k^2}}$$
$$= \frac{\sqrt[3]{3k^2}}{\sqrt[3]{k^3}}$$
$$= \frac{\sqrt[3]{3k^2}}{k}$$

9.

$$\sqrt{1-2g} = \sqrt{g+10}$$
$$\left(\sqrt{1-2g}\right)^2 = \left(\sqrt{g+10}\right)^2$$
$$1 - 2g = g + 10$$
$$1 - 3g = 10$$
$$-3g = 9$$
$$g = -3$$

10.

$$4 - \sqrt[3]{4+12x} = 0$$
$$4 = \sqrt[3]{4+12x}$$
$$(4)^3 = \left(\sqrt[3]{4+12x}\right)^3$$
$$64 = 4 + 12x$$
$$60 = 12x$$
$$5 = x$$

11.

$$\sqrt{y+2}-4=-y$$
$$\sqrt{y+2}=4-y$$
$$\left(\sqrt{y+2}\right)^2=(4-y)^2$$
$$y+2=16-4y-4y+y^2$$
$$y+2=16-8y+y^2$$
$$y+2-y-2=16-8y+y^2-y-2$$
$$0=y^2-9y+14$$
$$0=(y-2)(y-7)$$
$$y-2=0 \quad \text{or} \quad y-7=0$$
$$y=2 \qquad\qquad y=\not{7}$$

7 is extraneous

12.

$$\sqrt[4]{12t+4}+2=0$$
$$\sqrt[4]{12t+4}=-2$$
$$\left(\sqrt[4]{12t+4}\right)^4=(-2)^4$$
$$12t+4=16$$
$$12t=12$$
$$t=\not{1}$$

1 is extraneous

no solution

13.

$$\sqrt[3]{3}=3^{1/3}$$

14.

$$5a^{2/5}=5\sqrt[5]{a^2}$$

CHAPTER 9 REVIEW

SECTION 9.1
Radical Expressions and Radical Functions

1.
$\sqrt{49}=7$

2.
$-\sqrt{121}=-11$

3.
$\sqrt{\frac{225}{49}}=\frac{\sqrt{225}}{\sqrt{49}}$
$=\frac{15}{7}$

4.
$\sqrt{-4}$ is not real

5.
$\sqrt{0.01}=0.1$

6.
$\sqrt{25x^2}=|5x|$
$=5|x|$

7.
$\sqrt{x^8}=\sqrt{(x^4)^2}$
$=x^4$

8.
$\sqrt{x^2+4x+4}=\sqrt{(x+2)^2}$
$=|x+2|$

9.
$\sqrt[3]{-27}=-3$

10.
$-\sqrt[3]{216}=-6$

11.
$\sqrt[3]{64x^6y^3}=4x^2y$

12.
$\sqrt[3]{\frac{x^9}{125}}=\frac{\sqrt[3]{x^9}}{\sqrt[3]{125}}$
$=\frac{x^3}{5}$

13.
$\sqrt[4]{625}=5$

14.
$\sqrt[5]{-32}=-2$

15.
$\sqrt[4]{256x^8y^4}=|4x^2y|$
$=4x^2|y|$

16.
$\sqrt[15]{(x+1)^{15}}=x+1$

17.
$-\sqrt[4]{\frac{1}{16}}=-\frac{\sqrt[4]{1}}{\sqrt[4]{16}}$
$=-\frac{1}{2}$

18.
$\sqrt[6]{-1}$ is not real

19.
$\sqrt{0}=0$

20. $\sqrt[3]{0}=0$

21. Find the length of each side.
$s(A)=\sqrt{A}$
$s(144)=\sqrt{144}$
$=12$
Find the perimeter of a square whose side is 12 ft.
$P=4s$
$=4(12)$
$=48$ ft

22.

$$A(V) = 6\sqrt[3]{V^2}$$
$$A(8) = 6\sqrt[3]{8^2}$$
$$= 6\sqrt[3]{64}$$
$$= 6(4)$$
$$= 24 \text{ cm}^2$$

23. $f(x) = \sqrt{x+2}$

D: $[-2, \infty)$ R: $[0, \infty)$

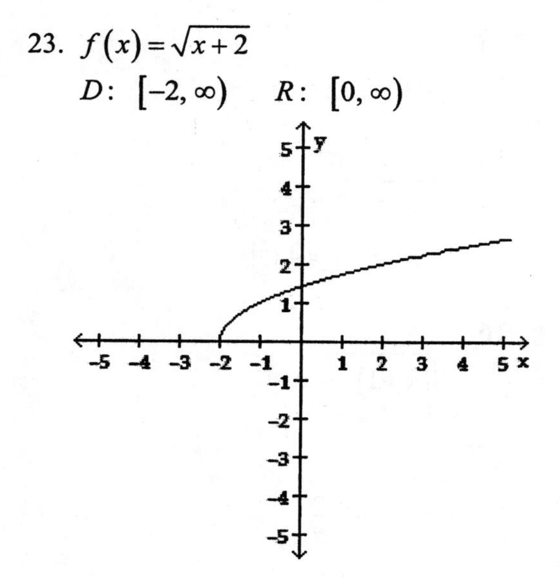

24. $f(x) = -\sqrt[3]{x} + 3$

D: $(-\infty, \infty)$ R: $(-\infty, \infty)$

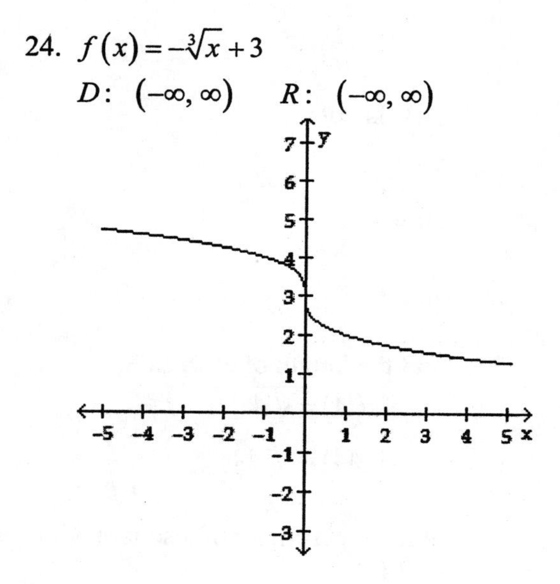

SECTION 9.2
Rational Exponents

25.

$$t^{1/2} = \sqrt{t}$$

26.

$$(5xy^3)^{1/4} = \sqrt[4]{5xy^3}$$

27.

$$25^{1/2} = \sqrt{25}$$
$$= 5$$

28.

$$-36^{1/2} = -\sqrt{36}$$
$$= -6$$

29.

$$(-36)^{1/2} = \sqrt{-36}$$

is not a real number

30.

$$1^{1/5} = \sqrt[5]{1}$$
$$= 1$$

31.

$$\left(\frac{9}{x^2}\right)^{1/2} = \sqrt{\frac{9}{x^2}}$$
$$= \frac{\sqrt{9}}{\sqrt{x^2}}$$
$$= \frac{3}{x}$$

32.

$$(-8)^{1/3} = \sqrt[3]{-8}$$
$$= -2$$

33.

$$625^{1/4} = \sqrt[4]{625}$$
$$= 5$$

34.

$$\sqrt[4]{81c^4d^4} = 3cd$$

35.

$$9^{3/2} = \left(\sqrt{9}\right)^3$$
$$= 3^3$$
$$= 27$$

36.

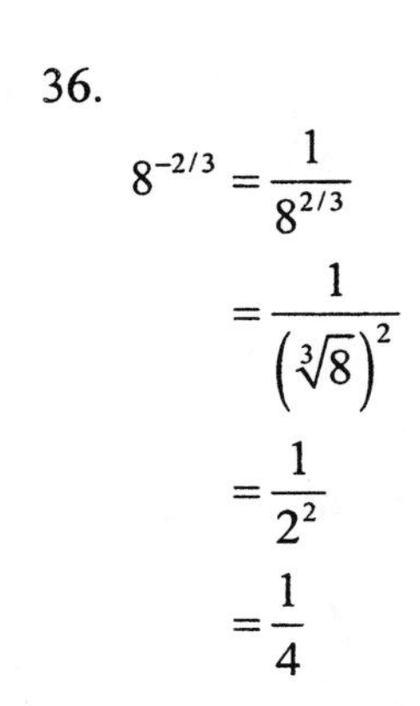

$$\begin{aligned} 8^{-2/3} &= \frac{1}{8^{2/3}} \\ &= \frac{1}{\left(\sqrt[3]{8}\right)^2} \\ &= \frac{1}{2^2} \\ &= \frac{1}{4} \end{aligned}$$

37.

$$\begin{aligned} -49^{5/2} &= -\left(\sqrt{49}\right)^5 \\ &= -(7)^5 \\ &= -16{,}807 \end{aligned}$$

38.

$$\begin{aligned} \frac{1}{100^{-1/2}} &= 100^{1/2} \\ &= \sqrt{100} \\ &= 10 \end{aligned}$$

39.

$$\begin{aligned} \left(\frac{4}{9}\right)^{-3/2} &= \left(\frac{9}{4}\right)^{3/2} \\ &= \left(\sqrt{\frac{9}{4}}\right)^3 \\ &= \left(\frac{3}{2}\right)^3 \\ &= \frac{27}{8} \end{aligned}$$

40.

$$\begin{aligned} \frac{1}{25^{5/2}} &= \frac{1}{\left(\sqrt{25}\right)^5} \\ &= \frac{1}{5^5} \\ &= \frac{1}{3{,}125} \end{aligned}$$

41.

$$\begin{aligned} \left(25x^2y^4\right)^{3/2} &= \left(\sqrt{25x^2y^4}\right)^3 \\ &= \left(5xy^2\right)^3 \\ &= 125x^3y^6 \end{aligned}$$

42.

$$\begin{aligned} \left(8u^6v^3\right)^{-2/3} &= \frac{1}{\left(8u^6v^3\right)^{2/3}} \\ &= \frac{1}{\left(\sqrt[3]{8u^6v^3}\right)^2} \\ &= \frac{1}{\left(2u^2v\right)^2} \\ &= \frac{1}{4u^4v^2} \end{aligned}$$

43.

$$\begin{aligned} 5^{1/4}5^{1/2} &= 5^{1/4+1/2} \\ &= 5^{1/4+2/4} \\ &= 5^{3/4} \end{aligned}$$

44.

$$\begin{aligned} a^{3/7}a^{-2/7} &= a^{3/7+(-2/7)} \\ &= a^{1/7} \end{aligned}$$

45.

$$\begin{aligned} \left(k^{4/5}\right)^{10} &= k^{(4/5)(10)} \\ &= k^8 \end{aligned}$$

46.

$$\begin{aligned} \frac{3^{5/6}3^{1/3}}{3^{1/2}} &= \frac{3^{5/6+1/3}}{3^{1/2}} \\ &= \frac{3^{5/6+2/6}}{3^{1/2}} \\ &= \frac{3^{7/6}}{3^{1/2}} \\ &= 3^{7/6-1/2} \\ &= 3^{7/6-3/6} \\ &= 3^{4/6} \\ &= 3^{2/3} \end{aligned}$$

47.

$$\begin{aligned} u^{1/2}\left(u^{1/2}-u^{-1/2}\right) &= u^{1/2}u^{1/2}-u^{1/2}u^{-1/2} \\ &= u^{1/2+1/2}-u^{1/2+(-1/2)} \\ &= u^{2/2}-u^{0} \\ &= u^{1}-1 \\ &= u-1 \end{aligned}$$

48.

$$\begin{aligned} v^{2/3}\left(v^{1/3}+v^{4/3}\right) &= v^{2/3}v^{1/3}+v^{2/3}v^{4/3} \\ &= v^{2/3+1/3}+v^{2/3+4/3} \\ &= v^{3/3}+v^{6/3} \\ &= v^{1}+v^{2} \\ &= v+v^{2} \end{aligned}$$

49.

$$\begin{aligned} \sqrt[4]{a^2} &= a^{2/4} \\ &= a^{1/2} \\ &= \sqrt{a} \end{aligned}$$

50.

$$\begin{aligned} \sqrt[3]{\sqrt{c}} &= \sqrt[3]{c^{1/2}} \\ &= \left(c^{1/2}\right)^{1/3} \\ &= c^{1/6} \\ &= \sqrt[6]{c} \end{aligned}$$

51.

$$\begin{aligned} d &= 1.22a^{1/2} \\ &= 1.22(22{,}500)^{1/2} \\ &= 1.22\left(\sqrt{22{,}500}\right) \\ &= 1.22(150) \\ &= 183 \text{ miles} \end{aligned}$$

52. Check (64, 64).

$$\begin{aligned} x^{2/3}+y^{2/3} &= 32 \\ (64)^{2/3}+(64)^{2/3} &\stackrel{?}{=} 32 \\ \left(\sqrt[3]{64}\right)^2+\left(\sqrt[3]{64}\right)^2 &\stackrel{?}{=} 32 \\ 4^2+4^2 &\stackrel{?}{=} 32 \\ 16+16 &\stackrel{?}{=} 32 \\ 32 &= 32 \end{aligned}$$

Check (–64, 64).

$$\begin{aligned} x^{2/3}+y^{2/3} &= 32 \\ (-64)^{2/3}+(64)^{2/3} &\stackrel{?}{=} 32 \\ \left(\sqrt[3]{-64}\right)^2+\left(\sqrt[3]{64}\right)^2 &\stackrel{?}{=} 32 \\ (-4)^2+4^2 &\stackrel{?}{=} 32 \\ 16+16 &\stackrel{?}{=} 32 \\ 32 &= 32 \end{aligned}$$

SECTION 9.3
Simplifying and Combining Radical Expressions

53.

$$\begin{aligned} \sqrt{240} &= \sqrt{16}\sqrt{15} \\ &= 4\sqrt{15} \end{aligned}$$

54.

$$\begin{aligned} \sqrt[3]{54} &= \sqrt[3]{27}\sqrt[3]{2} \\ &= 3\sqrt[3]{2} \end{aligned}$$

55.

$$\begin{aligned} \sqrt[4]{32} &= \sqrt[4]{16}\sqrt[4]{2} \\ &= 2\sqrt[4]{2} \end{aligned}$$

56.

$$\begin{aligned} -2\sqrt[5]{-96} &= -2\sqrt[5]{-32}\sqrt[5]{3} \\ &= -2(-2)\sqrt[5]{3} \\ &= 4\sqrt[5]{3} \end{aligned}$$

57.

$$\sqrt{8x^5} = \sqrt{4x^4}\sqrt{2x}$$
$$= 2x^2\sqrt{2x}$$

58.

$$\sqrt[3]{r^{17}} = \sqrt[3]{r^{15}}\sqrt[3]{r^2}$$
$$= r^5\sqrt[3]{r^2}$$

59.

$$\sqrt[3]{16x^5y^4} = \sqrt[3]{8x^3y^3}\sqrt[3]{2x^2y}$$
$$= 2xy\sqrt[3]{2x^2y}$$

60.

$$3\sqrt[3]{27j^7k} = 3\sqrt[3]{27j^6}\sqrt[3]{jk}$$
$$= 3(3j^2)\sqrt[3]{jk}$$
$$= 9j^2\sqrt[3]{jk}$$

61.

$$\frac{\sqrt{32x^3}}{\sqrt{2x}} = \sqrt{\frac{32x^3}{2x}}$$
$$= \sqrt{16x^2}$$
$$= 4x$$

62.

$$\sqrt{\frac{17xy}{64a^4}} = \frac{\sqrt{17xy}}{\sqrt{64a^4}}$$
$$= \frac{\sqrt{17xy}}{8a^2}$$

63.

$$\sqrt{2} + 2\sqrt{2} = 3\sqrt{2}$$

64.

$$6\sqrt{20} - \sqrt{5} = 6\sqrt{4}\sqrt{5} - \sqrt{5}$$
$$= 6(2)\sqrt{5} - \sqrt{5}$$
$$= 12\sqrt{5} - \sqrt{5}$$
$$= 11\sqrt{5}$$

65.

$$2\sqrt[3]{3} - \sqrt[3]{24} = 2\sqrt[3]{3} - \sqrt[3]{8}\sqrt[3]{3}$$
$$= 2\sqrt[3]{3} - 2\sqrt[3]{3}$$
$$= 0$$

66.

$$-\sqrt[4]{32a^5} - 2\sqrt[4]{162a^5}$$
$$= -\sqrt[4]{16a^4}\sqrt[4]{2a} - 2\sqrt[4]{81a^4}\sqrt[4]{2a}$$
$$= -2a\sqrt[4]{2a} - 2(3a)\sqrt[4]{2a}$$
$$= -2a\sqrt[4]{2a} - 6a\sqrt[4]{2a}$$
$$= -8a\sqrt[4]{2a}$$

67.

$$2x\sqrt{8} + 2\sqrt{200x^2} + \sqrt{50x^2}$$
$$= 2x\sqrt{4}\sqrt{2} + 2\sqrt{100x^2}\sqrt{2} + \sqrt{25x^2}\sqrt{2}$$
$$= 2x(2)\sqrt{2} + 2(10x)\sqrt{2} + 5x\sqrt{2}$$
$$= 4x\sqrt{2} + 20x\sqrt{2} + 5x\sqrt{2}$$
$$= 29x\sqrt{2}$$

68.

$$\sqrt[3]{54x^3} - 3\sqrt[3]{16x^3} + 4\sqrt[3]{128x^3}$$
$$= \sqrt[3]{27x^3}\sqrt[3]{2} - 3\sqrt[3]{8x^3}\sqrt[3]{2} + 4\sqrt[3]{64x^3}\sqrt[3]{2}$$
$$= 3x\sqrt[3]{2} - 3(2x)\sqrt[3]{2} + 4(4x)\sqrt[3]{2}$$
$$= 3x\sqrt[3]{2} - 6x\sqrt[3]{2} + 16x\sqrt[3]{2}$$
$$= 13x\sqrt[3]{2}$$

69.

a. You do not add the radicands together.
b. They are not like terms.
c. It should be $2\sqrt[3]{y^2}$.
d. You do not subtract the radicands.

70.

$$\sqrt{40} + \sqrt{32} + \sqrt{8}$$
$$= \sqrt{4}\sqrt{10} + \sqrt{16}\sqrt{2} + \sqrt{4}\sqrt{2}$$
$$= 2\sqrt{10} + 4\sqrt{2} + 2\sqrt{2}$$
$$= \left(2\sqrt{10} + 6\sqrt{2}\right) \text{ in.}$$
$$= 14.8 \text{ in.}$$

SECTION 9.4
Multiplying and Dividing Radical Expressions

71.

$$\sqrt{7}\sqrt{7} = \sqrt{49}$$
$$= 7$$

72.

$$(2\sqrt{5})(3\sqrt{2})=6\sqrt{10}$$

73.

$$\begin{aligned}(-2\sqrt{8})^2&=(-2\sqrt{8})(-2\sqrt{8})\\&=4\sqrt{64}\\&=4(8)\\&=32\end{aligned}$$

74.

$$\begin{aligned}2\sqrt{6}\sqrt{216}&=2\sqrt{1,296}\\&=2(36)\\&=72\end{aligned}$$

75.

$$\begin{aligned}\sqrt{9x}\sqrt{x}&=\sqrt{9x^2}\\&=3x\end{aligned}$$

76.

$$\left(\sqrt[3]{x+1}\right)^3=x+1$$

77.

$$\begin{aligned}-\sqrt[3]{2x^2}\sqrt[3]{4x^8}&=-\sqrt[3]{8x^{10}}\\&=-\sqrt[3]{8x^9}\sqrt[3]{x}\\&=-2x^3\sqrt[3]{x}\end{aligned}$$

78.

$$\begin{aligned}\sqrt[5]{9}\bullet\sqrt[5]{27}&=\sqrt[5]{243}\\&=3\end{aligned}$$

79.

$$\begin{aligned}&3\sqrt{7t}\left(2\sqrt{7t}+3\sqrt{3t^2}\right)\\&=6\sqrt{49t^2}+9\sqrt{21t^3}\\&=6\sqrt{49t^2}+9\sqrt{t^2}\sqrt{21t}\\&=6(7t)+9(t)\sqrt{21t}\\&=42t+9t\sqrt{21t}\end{aligned}$$

80.

$$\begin{aligned}-\sqrt[4]{256x^5y^{11}}\sqrt[4]{625x^9y^3}&=-\sqrt[4]{160,000x^{14}y^{14}}\\&=-\sqrt[4]{160,000x^{12}y^{12}}\sqrt[4]{x^2y^2}\\&=-20x^3y^3\sqrt[4]{x^2y^2}\end{aligned}$$

81.

$$\begin{aligned}\left(\sqrt{3b}+\sqrt{3}\right)^2&=\left(\sqrt{3b}+\sqrt{3}\right)\left(\sqrt{3b}+\sqrt{3}\right)\\&=3b+\sqrt{9b}+\sqrt{9b}+3\\&=3b+2\sqrt{9b}+3\\&=3b+2\sqrt{9}\sqrt{b}+3\\&=3b+2(3)\sqrt{b}+3\\&=3b+6\sqrt{b}+3\end{aligned}$$

82.

$$\begin{aligned}&\left(\sqrt[3]{3p}-2\sqrt[3]{2}\right)\left(\sqrt[3]{3p}+\sqrt[3]{2}\right)\\&=\sqrt[3]{9p^2}+\sqrt[3]{6p}-2\sqrt[3]{6p}-2\sqrt[3]{4}\\&=\sqrt[3]{9p^2}-\sqrt[3]{6p}-2\sqrt[3]{4}\end{aligned}$$

83.

$$\begin{aligned}\frac{10}{\sqrt{3}}&=\frac{10}{\sqrt{3}}\bullet\frac{\sqrt{3}}{\sqrt{3}}\\&=\frac{10\sqrt{3}}{\sqrt{9}}\\&=\frac{10\sqrt{3}}{3}\end{aligned}$$

84.

$$\begin{aligned}\sqrt{\frac{3}{5xy}}&=\frac{\sqrt{3}}{\sqrt{5xy}}\\&=\frac{\sqrt{3}}{\sqrt{5xy}}\bullet\frac{\sqrt{5xy}}{\sqrt{5xy}}\\&=\frac{\sqrt{15xy}}{\sqrt{25x^2y^2}}\\&=\frac{\sqrt{15xy}}{5xy}\end{aligned}$$

85.

$$\frac{\sqrt[3]{uv}}{\sqrt[3]{u^5v^7}}=\sqrt[3]{\frac{uv}{u^5v^7}}$$
$$=\sqrt[3]{\frac{1}{u^4v^6}}$$
$$=\sqrt[3]{\frac{1}{u^4v^6}}\bullet\sqrt[3]{\frac{u^2}{u^2}}$$
$$=\sqrt[3]{\frac{u^2}{u^6v^6}}$$
$$=\frac{\sqrt[3]{u^2}}{u^2v^2}$$

86.

$$\frac{\sqrt[4]{a}}{\sqrt[4]{3b^2}}=\frac{\sqrt[4]{a}}{\sqrt[4]{3b^2}}\bullet\frac{\sqrt[4]{27b^2}}{\sqrt[4]{27b^2}}$$
$$=\frac{\sqrt[4]{27ab^2}}{\sqrt[4]{81b^4}}$$
$$=\frac{\sqrt[4]{27ab^2}}{3b}$$

87.

$$\frac{2}{\sqrt{2}-1}=\frac{2}{\sqrt{2}-1}\bullet\frac{\sqrt{2}+1}{\sqrt{2}+1}$$
$$=\frac{2\left(\sqrt{2}+1\right)}{\left(\sqrt{2}-1\right)\left(\sqrt{2}+1\right)}$$
$$=\frac{2\sqrt{2}+2}{2-1}$$
$$=\frac{2\sqrt{2}+2}{1}$$
$$=2\sqrt{2}+2$$
$$=2\left(\sqrt{2}+1\right)$$

88.

$$\frac{4\sqrt{x}-2\sqrt{y}}{\sqrt{y}+4\sqrt{x}}=\frac{4\sqrt{x}-2\sqrt{y}}{\sqrt{y}+4\sqrt{x}}\bullet\frac{\sqrt{y}-4\sqrt{x}}{\sqrt{y}-4\sqrt{x}}$$
$$=\frac{\left(4\sqrt{x}-2\sqrt{y}\right)\left(\sqrt{y}-4\sqrt{x}\right)}{\left(\sqrt{y}+4\sqrt{x}\right)\left(\sqrt{y}-4\sqrt{x}\right)}$$
$$=\frac{40\sqrt{xy}-16\sqrt{x^2}-2\sqrt{y^2}+8\sqrt{xy}}{\sqrt{y^2}-16\sqrt{x^2}}$$
$$=\frac{-16x-2y+12\sqrt{xy}}{y-16x}$$

89.

$$\frac{\sqrt{a}-\sqrt{b}}{\sqrt{a}}=\frac{\sqrt{a}-\sqrt{b}}{\sqrt{a}}\bullet\frac{\sqrt{a}+\sqrt{b}}{\sqrt{a}+\sqrt{b}}$$
$$=\frac{\left(\sqrt{a}-\sqrt{b}\right)\left(\sqrt{a}+\sqrt{b}\right)}{\sqrt{a}\left(\sqrt{a}+\sqrt{b}\right)}$$
$$=\frac{\sqrt{a^2}-\sqrt{b^2}}{\sqrt{a^2}+\sqrt{ab}}$$
$$=\frac{a-b}{a+\sqrt{ab}}$$

90.

$$r=\sqrt[3]{\frac{3V}{4\pi}}$$
$$=\frac{\sqrt[3]{3V}}{\sqrt[3]{4\pi}}\bullet\frac{\sqrt[3]{2\pi^2}}{\sqrt[3]{2\pi^2}}$$
$$=\frac{\sqrt[3]{6\pi^2V}}{\sqrt[3]{8\pi^3}}$$
$$=\frac{\sqrt[3]{6\pi^2V}}{2\pi}$$

SECTION 9.5
Solving Radical Equations

91.

$$\sqrt{7x-10}-1=11$$
$$\sqrt{7x-10}=12$$
$$\left(\sqrt{7x-10}\right)^2=12^2$$
$$7x-10=144$$
$$7x=154$$
$$x=22$$

92.

$$u=\sqrt{25u-144}$$
$$u^2=\left(\sqrt{25u-144}\right)^2$$
$$u^2=25u-144$$
$$u^2-25u+144=0$$
$$(u-9)(u-16)=0$$
$$u-9=0 \quad \text{or} \quad u-16=0$$
$$u=9 \qquad\qquad u=16$$

93.

$$2\sqrt{y-3}=\sqrt{2y+1}$$
$$\left(2\sqrt{y-3}\right)^2=\left(\sqrt{2y+1}\right)^2$$
$$4(y-3)=2y+1$$
$$4y-12=2y+1$$
$$4y=2y+13$$
$$2y=13$$
$$y=\frac{13}{2}$$

94.

$$\sqrt{z+1}+\sqrt{z}=2$$
$$\sqrt{z+1}=2-\sqrt{z}$$
$$\left(\sqrt{z+1}\right)^2=\left(2-\sqrt{z}\right)^2$$
$$z+1=4-2\sqrt{z}-2\sqrt{z}+z$$
$$z+1=4-4\sqrt{z}+z$$
$$z+1-z-4=4-4\sqrt{z}+z-z-4$$
$$-3=-4\sqrt{z}$$
$$(-3)^2=\left(-4\sqrt{z}\right)^2$$
$$9=16z$$
$$\frac{9}{16}=z$$

95.

$$\sqrt[3]{x^3+56}-2=x$$
$$\sqrt[3]{x^3+56}=x+2$$
$$\left(\sqrt[3]{x^3+56}\right)^3=(x+2)^3$$
$$x^3+56=(x+2)(x+2)^2$$
$$x^3+56=(x+2)\left(x^2+4x+4\right)$$
$$x^3+56=x^3+4x^2+4x+2x^2+8x+8$$
$$x^3+56=x^3+6x^2+12x+8$$
$$x^3+56-x^3=x^3+6x^2+12x+8-x^3$$
$$56=6x^2+12x+8$$
$$56-56=6x^2+12x+8-56$$
$$0=6x^2+12x-48$$
$$\frac{0}{6}=\frac{6x^2}{6}+\frac{12x}{6}-\frac{48}{6}$$
$$0=x^2+2x-8$$
$$0=(x+4)(x-2)$$
$$x+4=0 \quad \text{or} \quad x-2=0$$
$$x=-4 \qquad\qquad x=2$$

96.

$$\sqrt[4]{8x-8}+2=0$$
$$\sqrt[4]{8x-8}=-2$$
$$\left(\sqrt[4]{8x-8}\right)^4=(-2)^4$$
$$8x-8=16$$
$$8x=24$$
$$x=\not{3}$$

No solution

3 is extraneous

97.

$$(x+2)^{1/2}-(4-x)^{1/2}=0$$
$$\sqrt{x+2}-\sqrt{4-x}=0$$
$$\sqrt{x+2}=\sqrt{4-x}$$
$$\left(\sqrt{x+2}\right)^2=\left(\sqrt{4-x}\right)^2$$
$$x+2=4-x$$
$$2x+2=4$$
$$2x=2$$
$$x=1$$

98.

$$\sqrt{b^2+b}=\sqrt{3-b^2}$$
$$\left(\sqrt{b^2+b}\right)^2=\left(\sqrt{3-b^2}\right)^2$$
$$b^2+b=3-b^2$$
$$b^2+b+b^2=3-b^2+b^2$$
$$2b^2+b=3$$
$$2b^2+b-3=0$$
$$(2b+3)(b-1)=0$$
$$2b+3=0 \quad \text{or} \quad b-1=0$$
$$2b=-3 \qquad b=1$$
$$b=-\frac{3}{2}$$

99.

$$\sqrt{2x^2-7x}=2$$
$$\left(\sqrt{2x^2-7x}\right)^2=2^2$$
$$2x^2-7x=4$$
$$2x^2-7x-4=0$$
$$(2x+1)(x-4)=0$$
$$2x+1=0 \quad \text{or} \quad x-4=0$$
$$2x=-1 \qquad x=4$$
$$x=-\frac{1}{2}$$

100. $x=2$

$$\sqrt{2(2)-3}\stackrel{?}{=}-2(2)+5$$
$$\sqrt{4-3}\stackrel{?}{=}-4+5$$
$$\sqrt{1}\stackrel{?}{=}1$$
$$1=1$$

101.

$$r=\sqrt{\frac{A}{P}}-1$$
$$(r+1)^2=\left(\sqrt{\frac{A}{P}}\right)^2$$
$$(r+1)^2=\frac{A}{P}$$
$$P(r+1)^2=P\left(\frac{A}{P}\right)$$
$$P(r+1)^2=A$$
$$\frac{P(r+1)^2}{(r+1)^2}=\frac{A}{(r+1)^2}$$
$$P=\frac{A}{(r+1)^2}$$

102.

$$h=\sqrt[3]{\frac{12I}{b}}$$
$$(h)^3=\left(\sqrt[3]{\frac{12I}{b}}\right)^3$$
$$h^3=\frac{12I}{b}$$
$$b\left(h^3\right)=b\left(\frac{12I}{b}\right)$$
$$h^3b=12I$$
$$\frac{h^3b}{12}=\frac{12I}{12}$$
$$\frac{h^3b}{12}=I$$

SECTION 9.6
Geometric Applications of Radicals

103. On the left right triangle, one leg is 8 and the other leg is $\frac{1}{2}(30) = 15$. The roof line, x, is the hypotenuse of the right triangle. Use the Pythagorean Theorem to find x.

$$a^2 + b^2 = c^2$$
$$8^2 + 15^2 = c^2$$
$$64 + 225 = c^2$$
$$289 = c^2$$
$$c = \sqrt{289}$$
$$c = 17$$

The roof line is 17 ft.

104. The hypotenuse is 125 and one leg is 117. Let x = the length of the other leg, which is one–half the d. Use the Pythagorean Theorem to find x.

$$a^2 + b^2 = c^2$$
$$x^2 + 117^2 = 125^2$$
$$x^2 + 13{,}689 = 15{,}625$$
$$x^2 = 1{,}936$$
$$x = \sqrt{1{,}936}$$
$$x = 44 \text{ yd}$$

$$d = 2x$$
$$= 2(44)$$
$$= 88 \text{ yd}$$

The distance the boat advances is 88 yards.

105. In an isosceles right triangle, the legs are equal. Let $a = b = 7$. Find c.

$$a^2 + b^2 = c^2$$
$$7^2 + 7^2 = c^2$$
$$49 + 49 = c^2$$
$$98 = c^2$$
$$c = \sqrt{98}$$
$$c = \sqrt{49}\sqrt{2}$$
$$c = 7\sqrt{2} \text{ meters}$$

106. The hypotenuse is $12\sqrt{3}$, which is twice the length of the shorter leg. The length of the longer leg is $\sqrt{3}$ times the length of the shorter leg. Let x = length of the shorter leg and y = length of the longer leg.

$$2x = 12\sqrt{3}$$
$$\frac{2x}{2} = \frac{12\sqrt{3}}{2}$$
$$x = 6\sqrt{3} \text{ cm}$$

$$y = 6\sqrt{3}\left(\sqrt{3}\right)$$
$$= 6(3)$$
$$= 18 \text{ cm}$$

107. The hypotenuse is $\sqrt{2}$ times the length of a leg.

$$x = 5\sqrt{2}$$
$$= 7.07 \text{ in.}$$

108. The hypotenuse is 10, which is twice the length of the shorter leg. The length of the longer leg is $\sqrt{3}$ times the length of the shorter leg. Let y = length of the shorter leg and x = length of the longer leg.

$$2y = 10$$
$$y = 5 \text{ cm}$$

$$x = 5\left(\sqrt{3}\right)$$
$$= 5\sqrt{3}$$
$$= 8.66 \text{ cm}$$

109.

$$d = \sqrt{(x_2 - x_1)^2 + (y_2 - y_1)^2}$$
$$= \sqrt{(5-0)^2 + (-12-0)^2}$$
$$= \sqrt{(5)^2 + (-12)^2}$$
$$= \sqrt{25 + 144}$$
$$= \sqrt{169}$$
$$= 13$$

110.

$$\begin{aligned} d &= \sqrt{(x_2 - x_1)^2 + (y_2 - y_1)^2} \\ &= \sqrt{[-2-(-4)]^2 + (8-6)^2} \\ &= \sqrt{(2)^2 + (2)^2} \\ &= \sqrt{4+4} \\ &= \sqrt{8} \\ &= \sqrt{4}\sqrt{2} \\ &= 2\sqrt{2} \end{aligned}$$

SECTION 9.7
Complex Numbers

111.

$$\begin{aligned} \sqrt{-25} &= \sqrt{25}\sqrt{-1} \\ &= 5i \end{aligned}$$

112.

$$\begin{aligned} 4\sqrt{-18} &= 4\sqrt{-1}\sqrt{9}\sqrt{2} \\ &= 4(3i)\sqrt{2} \\ &= 12i\sqrt{2} \end{aligned}$$

113.

$$\begin{aligned} -\sqrt{-6} &= -\sqrt{-1}\sqrt{6} \\ &= -i\sqrt{6} \end{aligned}$$

114.

$$\begin{aligned} \sqrt{-\frac{9}{64}} &= \sqrt{-1}\sqrt{\frac{9}{64}} \\ &= \frac{3}{8}i \end{aligned}$$

115. Complex Numbers

Real Numbers	Imaginary Numbers

116.
a. true
b. true
c. false
d. false

117. $3 - 6i$

118. $-1 + 7i$

119. $0 - 19i$

120. $0 + i$

121.

$$\begin{aligned} (3+4i)+(5-6i) &= (3+5)+(4-6)i \\ &= 8-2i \end{aligned}$$

122.

$$\begin{aligned} &\left(7-\sqrt{-9}\right)-\left(4+\sqrt{-4}\right) \\ &= (7-3i)-(4+2i) \\ &= (7-3i)+(-4-2i) \\ &= (7-4)+(-3-2)i \\ &= 3-5i \end{aligned}$$

123.

$$\begin{aligned} 3i(2-i) &= 6i-3i^2 \\ &= 6i-3(-1) \\ &= 6i+3 \\ &= 3+6i \end{aligned}$$

124.

$$\begin{aligned} &(2-7i)(-3+4i) \\ &= -6+8i+21i-28i^2 \\ &= -6+29i-28(-1) \\ &= -6+29i+28 \\ &= 22+29i \end{aligned}$$

125.

$$\begin{aligned} \sqrt{-3}\cdot\sqrt{-9} &= i\sqrt{3}\cdot 3i \\ &= 3i^2\sqrt{3} \\ &= 3(-1)\sqrt{3} \\ &= -3\sqrt{3} \\ &= 0-3\sqrt{3} \end{aligned}$$

126.

$$\begin{aligned} (9i)^2 &= 81i^2 \\ &= 81(-1) \\ &= -81 \\ &= -81+0i \end{aligned}$$

127.

$$\frac{3}{4i}=\frac{3}{4i}\cdot\frac{i}{i}$$
$$=\frac{3i}{4i^2}$$
$$=\frac{3i}{4(-1)}$$
$$=\frac{3i}{-4}$$
$$=0-\frac{3}{4}i$$

128.

$$\frac{2+3i}{2-3i}=\frac{2+3i}{2-3i}\cdot\frac{2+3i}{2+3i}$$
$$=\frac{(2+3i)(2+3i)}{(2-3i)(2+3i)}$$
$$=\frac{4+6i+6i+9i^2}{4-9i^2}$$
$$=\frac{4+12i+9(-1)}{4-9(-1)}$$
$$=\frac{4+12i-9}{4+9}$$
$$=\frac{-5+12i}{13}$$
$$=-\frac{5}{13}+\frac{12}{13}i$$

129.

$$i^{42}=i^{4\cdot 10+2}$$
$$=(i^4)^{10}\cdot i^2$$
$$=(1)^{10}\cdot i^2$$
$$=1\cdot i^2$$
$$=i^2$$
$$=-1$$

130.

$$i^{97}=i^{4\cdot 24+1}$$
$$=(i^4)^{24}\cdot i^1$$
$$=(1)^{24}\cdot i$$
$$=1\cdot i$$
$$=i$$

CHAPTER 9 TEST

1. $f(x)=\sqrt{x-1}$
   D: $[1, \infty)$; R: $[0, \infty)$

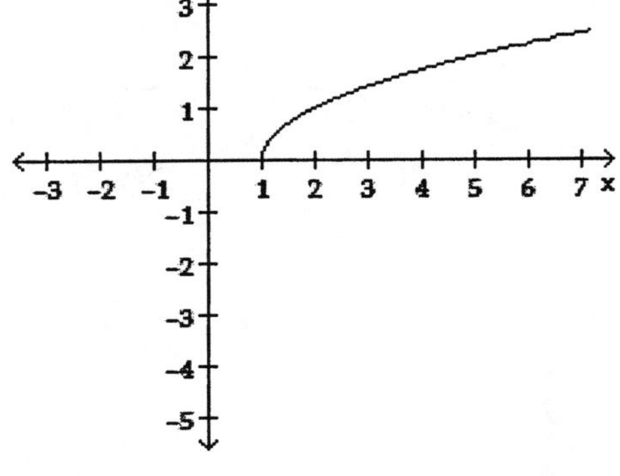

2.
$$v(d)=\sqrt{64.4d}$$
$$v(32.8)=\sqrt{64.4(32.8)}$$
$$=\sqrt{2{,}112.32}$$
$$=46 \text{ mph}$$

3. a. $f(-1)=-1$
   b. $f(8)=2$
   c. $x=1$
   d. D: $(-\infty, \infty)$, R: $(-\infty, \infty)$

4. No real number raised to the fourth power is -16.

5.
$$\left(49x^4\right)^{1/2}=\sqrt{49x^4}$$
$$=7x^2$$

6.
$$-27^{2/3}=-\left(\sqrt[3]{27}\right)^2$$
$$=-(3)^2$$
$$=-9$$

7.

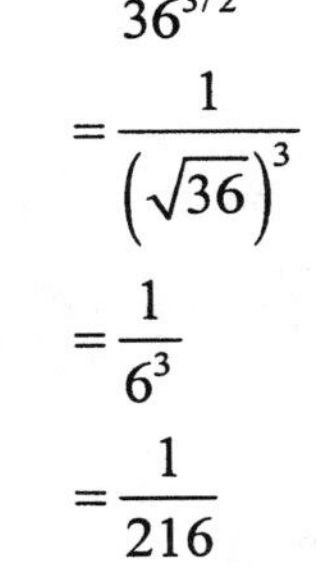

$$36^{-3/2}=\frac{1}{36^{3/2}}$$
$$=\frac{1}{\left(\sqrt{36}\right)^3}$$
$$=\frac{1}{6^3}$$
$$=\frac{1}{216}$$

8.
$$\left(-\frac{8}{125n^6}\right)^{-2/3}=\left(-\frac{125n^6}{8}\right)^{2/3}$$
$$=\left(-\sqrt[3]{\frac{125n^6}{8}}\right)^2$$
$$=\left(-\frac{5n^2}{2}\right)^2$$
$$=\frac{25n^4}{4}$$

9.
$$\frac{2^{5/3}2^{1/6}}{2^{1/2}}=\frac{2^{5/3+1/6}}{2^{1/2}}$$
$$=\frac{2^{10/6+1/6}}{2^{1/2}}$$
$$=\frac{2^{11/6}}{2^{1/2}}$$
$$=2^{11/6-1/2}$$
$$=2^{11/6-3/6}$$
$$=2^{8/6}$$
$$=2^{4/3}$$

10.
$$\left(a^{2/3}\right)^{1/6}=a^{(2/3)(1/6)}$$
$$=a^{2/18}$$
$$=a^{1/9}$$

11.

$$\sqrt{x^2} = |x|$$

12.

$$\sqrt{y^2+10y+25} = \sqrt{(y+5)^2}$$
$$= |y+5|$$

13.

$$\sqrt[3]{-64x^3y^6} = -4xy^2$$

14.

$$\sqrt{\frac{4a^2}{9}} = \frac{2}{3}a$$

15.

$$\sqrt[5]{(t+8)^5} = t+8$$

16.

$$\sqrt{250x^3y^5} = \sqrt{25x^2y^4}\sqrt{10xy}$$
$$= 5xy^2\sqrt{10xy}$$

17.

$$\frac{\sqrt[3]{24x^{15}y^4}}{\sqrt[3]{y}} = \sqrt[3]{\frac{24x^{15}y^4}{y}}$$
$$= \sqrt[3]{24x^{15}y^3}$$
$$= \sqrt[3]{8x^{15}y^3}\sqrt[3]{3}$$
$$= 2x^5y\sqrt[3]{3}$$

18.

$$\sqrt[7]{256} = \sqrt[7]{128}\sqrt[7]{2}$$
$$= 2\sqrt[7]{2}$$

19.

$$2\sqrt{48y^5} - 3y\sqrt{12y^3}$$
$$= 2\sqrt{16y^4}\sqrt{3y} - 3y\sqrt{4y^2}\sqrt{3y}$$
$$= 2\left(4y^2\right)\sqrt{3y} - 3y(2y)\sqrt{3y}$$
$$= 8y^2\sqrt{3y} - 6y^2\sqrt{3y}$$
$$= 2y^2\sqrt{3y}$$

20.

$$2\sqrt[3]{40} - \sqrt[3]{5{,}000} + 4\sqrt[3]{625}$$
$$= 2\sqrt[3]{8}\sqrt[3]{5} - \sqrt[3]{1{,}000}\sqrt[3]{5} + 4\sqrt[3]{125}\sqrt[3]{5}$$
$$= 2(2)\sqrt[3]{5} - 10\sqrt[3]{5} + 4(5)\sqrt[3]{5}$$
$$= 4\sqrt[3]{5} - 10\sqrt[3]{5} + 20\sqrt[3]{5}$$
$$= 14\sqrt[3]{5}$$

21.

$$\sqrt[4]{243z^{13}} + z\sqrt[4]{48z^9} = \sqrt[4]{81z^{12}}\sqrt[4]{3z} + z\sqrt[4]{16z^8}\sqrt[4]{3z}$$
$$= 3z^3\sqrt[4]{3z} + z\left(2z^2\right)\sqrt[4]{3z}$$
$$= 3z^3\sqrt[4]{3z} + 2z^3\sqrt[4]{3z}$$
$$= 5z^3\sqrt[4]{3z}$$

22.

$$-2\sqrt{xy}\left(3\sqrt{x} + \sqrt{xy^3}\right) = -6\sqrt{x^2y} - 2\sqrt{x^2y^4}$$
$$= -6\sqrt{x^2}\sqrt{y} - 2\sqrt{x^2y^4}$$
$$-6x\sqrt{y} - 2xy^2$$

23.

$$\left(3\sqrt{2} + \sqrt{3}\right)\left(2\sqrt{2} - 3\sqrt{3}\right)$$
$$= 6\sqrt{4} - 9\sqrt{6} + 2\sqrt{6} - 3\sqrt{9}$$
$$= 6(2) - 7\sqrt{6} - 3(3)$$
$$= 12 - 7\sqrt{6} - 9$$
$$= 3 - 7\sqrt{6}$$

24.

$$\left(\sqrt[3]{2a}+9\right)^2 = \left(\sqrt[3]{2a}+9\right)\left(\sqrt[3]{2a}+9\right)$$
$$= \sqrt[3]{4a^2} + 9\sqrt[3]{2a} + 9\sqrt[3]{2a} + 81$$
$$= \sqrt[3]{4a^2} + 18\sqrt[3]{2a} + 81$$

25.

$$\frac{8}{\sqrt{10}} = \frac{8}{\sqrt{10}} \cdot \frac{\sqrt{10}}{\sqrt{10}}$$
$$= \frac{8\sqrt{10}}{\sqrt{100}}$$
$$= \frac{8\sqrt{10}}{10}$$
$$= \frac{4\sqrt{10}}{5}$$

26.

$$\frac{3t-1}{\sqrt{3t}-1}=\frac{3t-1}{\sqrt{3t}-1}\bullet\frac{\left(\sqrt{3t}+1\right)}{\left(\sqrt{3t}+1\right)}$$
$$=\frac{(3t-1)\left(\sqrt{3t}+1\right)}{\left(\sqrt{3t}-1\right)\left(\sqrt{3t}+1\right)}$$
$$=\frac{3t\sqrt{3t}+3t-\sqrt{3t}-1}{3t-1}$$
$$=\frac{3t\left(\sqrt{3t}+1\right)-1\left(\sqrt{3t}+1\right)}{3t-1}$$
$$=\frac{(3t-1)\left(\sqrt{3t}+1\right)}{3t-1}$$
$$=\sqrt{3t}+1$$

27.

$$\sqrt[3]{\frac{9}{4a}}=\sqrt[3]{\frac{9}{4a}}\bullet\sqrt[3]{\frac{2a^2}{2a^2}}$$
$$=\frac{\sqrt[3]{18a^2}}{\sqrt[3]{8a^3}}$$
$$=\frac{\sqrt[3]{18a^2}}{2a}$$

28.

$$\frac{\sqrt{5}+3}{-4\sqrt{2}}=\frac{\sqrt{5}+3}{-4\sqrt{2}}\bullet\frac{\sqrt{5}-3}{\sqrt{5}-3}$$
$$=\frac{\left(\sqrt{5}+3\right)\left(\sqrt{5}-3\right)}{-4\sqrt{2}\left(\sqrt{5}-3\right)}$$
$$=\frac{5-9}{-4\sqrt{10}+12\sqrt{2}}$$
$$=\frac{-4}{-4\left(\sqrt{10}-3\sqrt{2}\right)}$$
$$=\frac{1}{\sqrt{10}-3\sqrt{2}}$$
$$=\frac{1}{\sqrt{2}\sqrt{5}-3\sqrt{2}}$$
$$=\frac{1}{\sqrt{2}\left(\sqrt{5}-3\right)}$$

29.

$$2\sqrt{x}=\sqrt{x+1}$$
$$\left(2\sqrt{x}\right)^2=\left(\sqrt{x+1}\right)^2$$
$$4x=x+1$$
$$3x=1$$
$$x=\frac{1}{3}$$

30.

$$\sqrt[3]{6n+4}-4=0$$
$$\sqrt[3]{6n+4}=4$$
$$\left(\sqrt[3]{6n+4}\right)^3=4^3$$
$$6n+4=64$$
$$6n=60$$
$$n=10$$

31.

$$1-\sqrt{u}=\sqrt{u-3}$$
$$\left(1-\sqrt{u}\right)^2=\left(\sqrt{u-3}\right)^2$$
$$1-\sqrt{u}-\sqrt{u}+u=u-3$$
$$1-2\sqrt{u}+u=u-3$$
$$1-2\sqrt{u}+u-1-u=u-3-1-u$$
$$-2\sqrt{u}=-4$$
$$\frac{-2\sqrt{u}}{-2}=\frac{-4}{-2}$$
$$\sqrt{u}=2$$
$$\left(\sqrt{u}\right)^2=(2)^2$$
$$u=\not{4}$$

No solution

4 is extraneous

32.

$$(2m^2-9)^{1/2}=m$$
$$\sqrt{2m^2-9}=m$$
$$\left(\sqrt{2m^2-9}\right)^2=(m)^2$$
$$2m^2-9=m^2$$
$$2m^2-9-m^2=m^2-m^2$$
$$m^2-9=0$$
$$(m+3)(m-3)=0$$
$$m+3=0 \quad \text{or} \quad m-3=0$$
$$m=-\not{3} \qquad m=3$$

−3 is extraneous

33. $\sqrt{x-8}$ is the principal square root. It cannot be negative.

34.

$$r=\sqrt[3]{\frac{GMt^2}{4\pi^2}}$$
$$(r)^3=\left(\sqrt[3]{\frac{GMt^2}{4\pi^2}}\right)^3$$
$$r^3=\frac{GMt^2}{4\pi^2}$$
$$4\pi^2(r^3)=4\pi^2\left(\frac{GMt^2}{4\pi^2}\right)$$
$$4\pi^2r^3=GMt^2$$
$$\frac{4\pi^2r^3}{Mt^2}=\frac{GMt^2}{Mt^2}$$
$$\frac{4\pi^2r^3}{Mt^2}=G$$

35. The length of the longer side is $\sqrt{3}$ times the length of the shorter side. The length of the hypotenuse is twice the length of the shorter side. First find the length of the shorter side and use that to find x, the length of the hypotenuse.

Let y = the length of the shorter side.

$$y\sqrt{3}=8$$
$$\frac{y\sqrt{3}}{\sqrt{3}}=\frac{8}{\sqrt{3}}$$
$$y=\frac{8}{\sqrt{3}}$$
$$y=\frac{8}{\sqrt{3}}\cdot\frac{\sqrt{3}}{\sqrt{3}}$$
$$y=\frac{8\sqrt{3}}{3}$$
$$=4.62\text{ cm}$$

$$x=2y$$
$$=2(4.62)$$
$$=9.24\text{ cm}$$

36. The length of the hypotenuse is $\sqrt{2}$ times the length of a leg. Let x = the length of a leg.

$$x\sqrt{2}=12.26$$
$$\frac{x\sqrt{2}}{\sqrt{2}}=\frac{12.26}{\sqrt{2}}$$
$$x=\frac{12.26}{\sqrt{2}}$$
$$x=\frac{12.26}{\sqrt{2}}\cdot\frac{\sqrt{2}}{\sqrt{2}}$$
$$x=\frac{12.26\sqrt{2}}{2}$$
$$x=6.13\sqrt{2}$$
$$x=8.67\text{ cm}$$

37.

$$d=\sqrt{(x_2-x_1)^2+(y_2-y_1)^2}$$
$$=\sqrt{[22-(-2)]^2+(12-5)^2}$$
$$=\sqrt{(24)^2+(7)^2}$$
$$=\sqrt{576+49}$$
$$=\sqrt{625}$$
$$=25$$

38. Use the Pythagorean Theorem to find h.

$$a^2+b^2=c^2$$
$$(h)^2+(45)^2=(53)^2$$
$$h^2+2{,}025=2{,}809$$
$$h^2=784$$
$$h=\sqrt{784}$$
$$h=28\text{ in.}$$

39.

$$\sqrt{-5}=i\sqrt{5}$$

40.

$$i^{22}=i^{4\cdot 5+2}$$
$$=\left(i^4\right)^5\cdot i^2$$
$$=(1)^5\cdot i^2$$
$$=1\cdot i^2$$
$$=i^2$$
$$=-1$$

41.

$$(2+4i)+(-3+7i)=(2-3)+(4+7)i$$
$$=-1+11i$$

42.

$$\left(3-\sqrt{-9}\right)-\left(-1+\sqrt{-16}\right)$$
$$=(3-3i)-(-1+4i)$$
$$=(3-3i)+(1-4i)$$
$$=(3+1)+(-3-4)i$$
$$=4-7i$$

43.

$$2i(3-4i)=6i-8i^2$$
$$=6i-8(-1)$$
$$=6i+8$$
$$=8+6i$$

44.

$$(3+2i)(-4-i)=-12-3i-8i-2i^2$$
$$=-12-11i-2(-1)$$
$$=-12-11i+2$$
$$=-10-11i$$

45.

$$\frac{1}{i\sqrt{2}}=\frac{1}{i\sqrt{2}}\cdot\frac{i\sqrt{2}}{i\sqrt{2}}$$
$$=\frac{i\sqrt{2}}{2i^2}$$
$$=\frac{i\sqrt{2}}{2(-1)}$$
$$=\frac{i\sqrt{2}}{-2}$$
$$=0-\frac{\sqrt{2}}{2}i$$

46.

$$\frac{2+i}{3-i}=\frac{2+i}{3-i}\cdot\frac{3+i}{3+i}$$
$$=\frac{(2+i)(3+i)}{(3-i)(3+i)}$$
$$=\frac{6+2i+3i+i^2}{9-i^2}$$
$$=\frac{6+5i+(-1)}{9-(-1)}$$
$$=\frac{5+5i}{10}$$
$$=\frac{5}{10}+\frac{5}{10}i$$
$$=\frac{1}{2}+\frac{1}{2}i$$

SECTION 10.1

VOCABULARY

1. An equation of the form $ax^2 + bx + c = 0$, where $a \neq 0$, is called a **quadratic** equation.

3. $x^2 + 6x + 9$ is called a **perfect** square trinomial because it factors as $(x + 3)^2$.

CONCEPTS

5. For any nonnegative number c, if $x^2 = c$, then $x = \sqrt{c}$ or $x = -\sqrt{c}$.

7. $\frac{1}{5} \pm \frac{\sqrt{2}}{5} = \frac{1 \pm \sqrt{2}}{5}$

9. Yes, it is.

$$x^2 - 18 = 0$$
$$\left(-3\sqrt{2}\right)^2 - 18 \stackrel{?}{=} 0$$
$$9(2) - 18 \stackrel{?}{=} 0$$
$$18 - 18 \stackrel{?}{=} 0$$
$$0 = 0$$

11. a. $\left(\frac{12}{2}\right)^2 = 6^2 = 36$

b. $\left(\frac{-5}{2}\right)^2 = \frac{25}{4}$

c. $\left[\frac{1}{2}\left(-\frac{1}{2}\right)\right]^2 = \left[-\frac{1}{4}\right]^2 = \frac{1}{16}$

d. $\left(\frac{1}{2} \cdot \frac{3}{4}\right)^2 = \left(\frac{3}{8}\right)^2 = \frac{9}{64}$

13. a. Subtract 35 from both sides

b. Add 36 to both sides

15.

$$x + 7 = \pm\sqrt{6}$$
$$x = -7 \pm \sqrt{6}$$

17. a. 4 is not a factor of the numerator. Only common factors of the numerator and denominator can be removed.

b. 5 is not a factor of the numerator. Only common factors of the numerator and denominator can be removed.

NOTATION

19. a. 2 solutions; $2\sqrt{5}$ and $-2\sqrt{5}$

b. ± 4.47

PRACTICE

21.

$$6x^2 + 12x = 0$$
$$6x(x+2) = 0$$
$$6x = 0 \quad \text{and} \quad x + 2 = 0$$
$$x = 0 \qquad\qquad x = -2$$

23.

$$y^2 - 25 = 0$$
$$(y-5)(y+5) = 0$$
$$y - 5 = 0 \quad \text{or} \quad y + 5 = 0$$
$$y = 5 \qquad\qquad y = -5$$

25.

$$r^2 + 6r + 8 = 0$$
$$(r+2)(r+4) = 0$$
$$r + 2 = 0 \quad \text{or} \quad r + 4 = 0$$
$$r = -2 \qquad\qquad r = -4$$

27.

$$2z^2 = -2 + 5z$$
$$2z^2 - 5z + 2 = 0$$
$$(2z-1)(z-2) = 0$$
$$2z - 1 = 0 \quad \text{or} \quad z - 2 = 0$$
$$2z = 1 \qquad\qquad z = 2$$
$$z = \frac{1}{2}$$

29.

$$x^2 = 36$$
$$x = \sqrt{36} \quad \text{or} \quad x = -\sqrt{36}$$
$$x = 6 \qquad\qquad x = -6$$

31.

$$z^2 = 5$$
$$z = \sqrt{5} \quad \text{or} \quad z = -\sqrt{5}$$

33.

$$3x^2 - 16 = 0$$
$$3x^2 = 16$$
$$x^2 = \frac{16}{3}$$
$$x = \sqrt{\frac{16}{3}} \quad \text{or} \quad x = -\sqrt{\frac{16}{3}}$$
$$x = \frac{4}{\sqrt{3}} \qquad\qquad x = -\frac{4}{\sqrt{3}}$$
$$x = \frac{4}{\sqrt{3}} \bullet \frac{\sqrt{3}}{\sqrt{3}} \qquad\qquad x = -\frac{4}{\sqrt{3}} \bullet \frac{\sqrt{3}}{\sqrt{3}}$$
$$x = \frac{4\sqrt{3}}{3} \qquad\qquad x = -\frac{4\sqrt{3}}{3}$$

35.

$$(x+1)^2 = 1$$
$$\sqrt{(x+1)^2} = \sqrt{1}$$
$$x + 1 = \pm 1$$
$$x + 1 = 1 \quad \text{or} \quad x + 1 = -1$$
$$x = 0 \qquad\qquad x = -2$$

37.

$$(s-7)^2 = 9$$
$$\sqrt{(s-7)^2} = \sqrt{9}$$
$$s - 7 = \pm 3$$
$$s - 7 = 3 \quad \text{or} \quad s - 7 = -3$$
$$s = 10 \qquad\qquad s = 4$$

39.

$$(x+5)^2 - 3 = 0$$
$$(x+5)^2 = 3$$
$$\sqrt{(x+5)^2} = \sqrt{3}$$
$$x + 5 = \sqrt{3} \quad \text{or} \quad x + 5 = -\sqrt{3}$$
$$x = -5 + \sqrt{3} \qquad\qquad x = -5 - \sqrt{3}$$
$$x = -5 \pm \sqrt{3}$$

41.

$$(a+2)^2 = 8$$
$$a + 2 = \sqrt{8} \quad \text{or} \quad a + 2 = -\sqrt{8}$$
$$a + 2 = \sqrt{4}\sqrt{2} \qquad\qquad a + 2 = -\sqrt{4}\sqrt{2}$$
$$a + 2 = 2\sqrt{2} \qquad\qquad a + 2 = -2\sqrt{2}$$
$$a = -2 + 2\sqrt{2} \qquad\qquad a = -2 - 2\sqrt{2}$$
$$a = -2 \pm 2\sqrt{2}$$

43.

$$(3x-1)^2 = 25$$
$$3x - 1 = \sqrt{25} \quad \text{or} \quad 3x - 1 = -\sqrt{25}$$
$$3x - 1 = 5 \qquad\qquad 3x - 1 = -5$$
$$3x = 6 \qquad\qquad 3x = -4$$
$$x = 2 \qquad\qquad x = -\frac{4}{3}$$

45.

$$p^2 = -16$$
$$p = \sqrt{-16} \quad \text{or} \quad p = -\sqrt{-16}$$
$$p = 4i \qquad\qquad p = -4i$$
$$p = \pm 4i$$

47.

$$4m^2 + 81 = 0$$
$$4m^2 = -81$$
$$m^2 = -\frac{81}{4}$$
$$m = \sqrt{-\frac{81}{4}} \quad \text{or} \quad m = -\sqrt{-\frac{81}{4}}$$
$$m = \frac{9}{2}i \qquad\qquad m = -\frac{9}{2}i$$
$$m = \pm\frac{9}{2}i$$

49.

$$(x-3)^2=-5$$
$$x-3=\sqrt{-5} \quad \text{or} \quad x-3=-\sqrt{-5}$$
$$x-3=i\sqrt{5} \qquad x-3=-i\sqrt{5}$$
$$x=3+i\sqrt{5} \qquad x=3-i\sqrt{5}$$
$$x=3\pm i\sqrt{5}$$

51.

$$2d^2=3h$$
$$\frac{2d^2}{2}=\frac{3h}{2}$$
$$d^2=\frac{3h}{2}$$
$$d=\sqrt{\frac{3h}{2}}$$
$$d=\frac{\sqrt{3h}}{\sqrt{2}}$$
$$d=\frac{\sqrt{3h}}{\sqrt{2}}\bullet\frac{\sqrt{2}}{\sqrt{2}}$$
$$d=\frac{\sqrt{6h}}{\sqrt{4}}$$
$$d=\frac{\sqrt{6h}}{2}$$

53.

$$E=mc^2$$
$$\frac{E}{m}=\frac{mc^2}{m}$$
$$\frac{E}{m}=c^2$$
$$c=\sqrt{\frac{E}{m}}$$
$$c=\frac{\sqrt{E}}{\sqrt{m}}\bullet\frac{\sqrt{m}}{\sqrt{m}}$$
$$c=\frac{\sqrt{Em}}{m}$$

55.

$$x^2+2x-8=0$$
$$x^2+2x=8$$
$$x^2+2x+\left(\frac{2}{2}\right)^2=8+\left(\frac{2}{2}\right)^2$$
$$x^2+2x+(1)^2=8+(1)^2$$
$$x^2+2x+1=8+1$$
$$(x+1)^2=9$$
$$x+1=\pm\sqrt{9}$$
$$x+1=\pm 3$$
$$x+1=3 \quad \text{or} \quad x+1=-3$$
$$x=2 \qquad x=-4$$

57.

$$k^2-8k+12=0$$
$$k^2-8k=-12$$
$$k^2-8k+\left(\frac{-8}{2}\right)^2=-12+\left(\frac{-8}{2}\right)^2$$
$$k^2-8k+(-4)^2=-12+(-4)^2$$
$$k^2-8k+16=-12+16$$
$$(k-4)^2=4$$
$$k-4=\pm\sqrt{4}$$
$$k-4=\pm 2$$
$$k-4=2 \quad \text{or} \quad k-4=-2$$
$$k=6 \qquad k=2$$

59.

$$g^2+5g-6=0$$
$$g^2+5g=6$$
$$g^2+5g+\left(\frac{5}{2}\right)^2=6+\left(\frac{5}{2}\right)^2$$
$$g^2+5g+\frac{25}{4}=6+\frac{25}{4}$$
$$\left(g+\frac{5}{2}\right)^2=\frac{24}{4}+\frac{25}{4}$$
$$\left(g+\frac{5}{2}\right)^2=\frac{49}{4}$$
$$g+\frac{5}{2}=\pm\sqrt{\frac{49}{4}}$$
$$g+\frac{5}{2}=\pm\frac{7}{2}$$
$$g+\frac{5}{2}=\frac{7}{2} \quad \text{or} \quad g+\frac{5}{2}=-\frac{7}{2}$$
$$g=\frac{2}{2} \qquad g=-\frac{12}{2}$$
$$g=1 \qquad g=-6$$

61.

$$x^2-3x-4=0$$
$$x^2-3x=4$$
$$x^2-3x+\left(\frac{-3}{2}\right)^2=4+\left(\frac{-3}{2}\right)^2$$
$$x^2-3x+\frac{9}{4}=4+\frac{9}{4}$$
$$x^2-3x+\frac{9}{4}=\frac{16}{4}+\frac{9}{4}$$
$$\left(x-\frac{3}{2}\right)^2=\frac{25}{4}$$
$$x-\frac{3}{2}=\pm\sqrt{\frac{25}{4}}$$
$$x-\frac{3}{2}=\pm\frac{5}{2}$$
$$x-\frac{3}{2}=\frac{5}{2} \quad \text{or} \quad x-\frac{3}{2}=-\frac{5}{2}$$
$$x=\frac{8}{2} \qquad x=\frac{-2}{2}$$
$$x=4 \qquad x=-1$$

63.

$$x^2+8x+6=0$$
$$x^2+8x=-6$$
$$x^2+8x+\left(\frac{8}{2}\right)^2=-6+\left(\frac{8}{2}\right)^2$$
$$x^2+8x+(4)^2=-6+(4)^2$$
$$x^2+8x+16=-6+16$$
$$(x+4)^2=10$$
$$x+4=\pm\sqrt{10}$$
$$x=-4\pm\sqrt{10}$$

65.

$$x^2-2x=17$$
$$x^2-2x+\left(\frac{-2}{2}\right)^2=17+\left(\frac{-2}{2}\right)^2$$
$$x^2-2x+(-1)^2=17+(-1)^2$$
$$x^2-2x+1=17+1$$
$$(x-1)^2=18$$
$$x-1=\pm\sqrt{18}$$
$$x-1=\pm\sqrt{9}\sqrt{2}$$
$$x-1=\pm3\sqrt{2}$$
$$x=1\pm3\sqrt{2}$$

67.

$$m^2 - 7m + 3 = 0$$
$$m^2 - 7m = -3$$
$$m^2 - 7m + \left(\frac{-7}{2}\right)^2 = -3 + \left(\frac{-7}{2}\right)^2$$
$$m^2 - 7m + \frac{49}{4} = -3 + \frac{49}{4}$$
$$m^2 - 7m + \frac{49}{4} = -\frac{12}{4} + \frac{49}{4}$$
$$\left(m - \frac{7}{2}\right)^2 = \frac{37}{4}$$
$$m - \frac{7}{2} = \pm\sqrt{\frac{37}{4}}$$
$$m - \frac{7}{2} = \pm\frac{\sqrt{37}}{2}$$
$$m = \frac{7}{2} \pm \frac{\sqrt{37}}{2}$$
$$m = \frac{7 \pm \sqrt{37}}{2}$$

69.

$$a^2 - a = 3$$
$$a^2 - a + \left(\frac{-1}{2}\right)^2 = 3 + \left(\frac{-1}{2}\right)^2$$
$$a^2 - a + \frac{1}{4} = 3 + \frac{1}{4}$$
$$a^2 - a + \frac{1}{4} = \frac{12}{4} + \frac{1}{4}$$
$$\left(a - \frac{1}{2}\right)^2 = \frac{13}{4}$$
$$a - \frac{1}{2} = \pm\sqrt{\frac{13}{4}}$$
$$a - \frac{1}{2} = \pm\frac{\sqrt{13}}{2}$$
$$a = \frac{1}{2} \pm \frac{\sqrt{13}}{2}$$
$$a = \frac{1 \pm \sqrt{13}}{2}$$

71.

$$2x^2 - x - 1 = 0$$
$$2x^2 - x = 1$$
$$\frac{2x^2}{2} - \frac{x}{2} = \frac{1}{2}$$
$$x^2 - \frac{1}{2}x = \frac{1}{2}$$
$$x^2 - \frac{1}{2}x + \left(\frac{1}{2}\bullet\frac{-1}{2}\right)^2 = \frac{1}{2} + \left(\frac{1}{2}\bullet\frac{-1}{2}\right)^2$$
$$x^2 - \frac{1}{2}x + \left(\frac{-1}{4}\right)^2 = \frac{1}{2} + \left(\frac{-1}{4}\right)^2$$
$$x^2 - \frac{1}{2}x + \frac{1}{16} = \frac{1}{2} + \frac{1}{16}$$
$$\left(x - \frac{1}{4}\right)^2 = \frac{8}{16} + \frac{1}{16}$$
$$\left(x - \frac{1}{4}\right)^2 = \frac{9}{16}$$
$$x - \frac{1}{4} = \pm\sqrt{\frac{9}{16}}$$
$$x - \frac{1}{4} = \pm\frac{3}{4}$$
$$x = \frac{1}{4} \pm \frac{3}{4}$$
$$x = \frac{1}{4} + \frac{3}{4} \quad \text{or} \quad x = \frac{1}{4} - \frac{3}{4}$$
$$x = \frac{4}{4} \qquad x = -\frac{2}{4}$$
$$x = 1 \qquad x = -\frac{1}{2}$$

73.

$$3x^2-6x=1$$
$$\frac{3x^2}{3}-\frac{6x}{3}=\frac{1}{3}$$
$$x^2-2x=\frac{1}{3}$$
$$x^2-2x+\left(\frac{-2}{2}\right)^2=\frac{1}{3}+\left(\frac{-2}{2}\right)^2$$
$$x^2-2x+1=\frac{1}{3}+1$$
$$x^2-2x+1=\frac{1}{3}+\frac{3}{3}$$
$$(x-1)^2=\frac{4}{3}$$
$$x-1=\pm\sqrt{\frac{4}{3}}$$
$$x-1=\pm\frac{2}{\sqrt{3}}$$
$$x-1=\pm\frac{2}{\sqrt{3}}\bullet\frac{\sqrt{3}}{\sqrt{3}}$$
$$x-1=\pm\frac{2\sqrt{3}}{3}$$
$$x=1\pm\frac{2\sqrt{3}}{3}$$
$$x=\frac{3}{3}\pm\frac{2\sqrt{3}}{3}$$
$$x=\frac{3\pm2\sqrt{3}}{3}$$

75.

$$4x^2-4x=7$$
$$\frac{4x^2}{4}-\frac{4x}{4}=\frac{7}{4}$$
$$x^2-x=\frac{7}{4}$$
$$x^2-x+\left(\frac{-1}{2}\right)^2=\frac{7}{4}+\left(\frac{-1}{2}\right)^2$$
$$x^2-x+\frac{1}{4}=\frac{7}{4}+\frac{1}{4}$$
$$\left(x-\frac{1}{2}\right)^2=2$$
$$x-\frac{1}{2}=\pm\sqrt{2}$$
$$x=\frac{1}{2}\pm\sqrt{2}$$
$$x=\frac{1}{2}\pm\frac{2\sqrt{2}}{2}$$
$$x=\frac{1\pm2\sqrt{2}}{2}$$

77.

$$2x^2+5x-2=0$$
$$\frac{2x^2}{2}+\frac{5x}{2}-\frac{2}{2}=\frac{0}{2}$$
$$x^2+\frac{5}{2}x-1=0$$
$$x^2+\frac{5}{2}x=1$$
$$x^2+\frac{5}{2}x+\left(\frac{1}{2}\bullet\frac{5}{2}\right)^2=1+\left(\frac{1}{2}\bullet\frac{5}{2}\right)^2$$
$$x^2+\frac{5}{2}x+\left(\frac{5}{4}\right)^2=1+\left(\frac{5}{4}\right)^2$$
$$x^2+\frac{5}{2}x+\frac{25}{16}=1+\frac{25}{16}$$
$$\left(x+\frac{5}{4}\right)^2=\frac{16}{16}+\frac{25}{16}$$
$$\left(x+\frac{5}{4}\right)^2=\frac{41}{16}$$
$$x+\frac{5}{4}=\pm\sqrt{\frac{41}{16}}$$
$$x+\frac{5}{4}=\pm\frac{\sqrt{41}}{4}$$
$$x=-\frac{5}{4}\pm\frac{\sqrt{41}}{4}$$
$$x=\frac{-5\pm\sqrt{41}}{4}$$

79.

$$\frac{7x+1}{5}=-x^2$$
$$\frac{7x}{5}+\frac{1}{5}=-x^2$$
$$x^2+\frac{7x}{5}=-\frac{1}{5}$$
$$x^2+\frac{7}{5}x+\left(\frac{1}{2}\bullet\frac{7}{5}\right)^2=-\frac{1}{5}+\left(\frac{1}{2}\bullet\frac{7}{5}\right)^2$$
$$x^2+\frac{7}{5}x+\left(\frac{7}{10}\right)^2=-\frac{1}{5}+\left(\frac{7}{10}\right)^2$$
$$x^2+\frac{7}{5}x+\frac{49}{100}=-\frac{1}{5}+\frac{49}{100}$$
$$\left(x+\frac{7}{10}\right)^2=-\frac{20}{100}+\frac{49}{100}$$
$$\left(x+\frac{7}{10}\right)^2=\frac{29}{100}$$
$$x+\frac{7}{10}=\pm\sqrt{\frac{29}{100}}$$
$$x+\frac{7}{10}=\pm\frac{\sqrt{29}}{10}$$
$$x=-\frac{7}{10}\pm\frac{\sqrt{29}}{10}$$
$$x=\frac{-7\pm\sqrt{29}}{10}$$

81.

$$p^2+2p+2=0$$
$$p^2+2p=-2$$
$$p^2+2p+\left(\frac{2}{2}\right)^2=-2+\left(\frac{2}{2}\right)^2$$
$$p^2+2p+(1)^2=-2+(1)^2$$
$$p^2+2p+1=-2+1$$
$$(p+1)^2=-1$$
$$p+1=\pm\sqrt{-1}$$
$$p+1=\pm i$$
$$p=-1\pm i$$

83.

$$y^2+8y+18=0$$
$$y^2+8y=-18$$
$$y^2+8y+\left(\frac{8}{2}\right)^2=-18+\left(\frac{8}{2}\right)^2$$
$$y^2+8y+(4)^2=-18+(4)^2$$
$$y^2+8y+16=-18+16$$
$$(y+4)^2=-2$$
$$y+4=\pm\sqrt{-2}$$
$$y+4=\pm i\sqrt{2}$$
$$y=-4\pm i\sqrt{2}$$

85.

$$3m^2-2m+3=0$$
$$3m^2-2m=-3$$
$$\frac{3m^2}{3}-\frac{2m}{3}=\frac{-3}{3}$$
$$m^2-\frac{2}{3}m=-1$$
$$m^2-\frac{2}{3}m+\left(\frac{1}{2}\bullet\frac{-2}{3}\right)^2=-1+\left(\frac{1}{2}\bullet\frac{-2}{3}\right)^2$$
$$m^2-\frac{2}{3}m+\left(\frac{-1}{3}\right)^2=-1+\left(\frac{-1}{3}\right)^2$$
$$m^2-\frac{2}{3}m+\frac{1}{9}=-1+\frac{1}{9}$$
$$\left(m-\frac{1}{3}\right)^2=-\frac{9}{9}+\frac{1}{9}$$
$$\left(m-\frac{1}{3}\right)^2=-\frac{8}{9}$$
$$m-\frac{1}{3}=\pm\sqrt{-\frac{8}{9}}$$
$$m-\frac{1}{3}=\pm\frac{\sqrt{-8}}{3}$$
$$m-\frac{1}{3}=\pm\frac{2\sqrt{2}}{3}i$$
$$m=\frac{1}{3}\pm\frac{2\sqrt{2}}{3}i$$

APPLICATIONS

87. FLAGS

$$\text{Area} = \text{length} \bullet \text{width}$$
$$100 = 1.9x(x)$$
$$100 = 1.9x^2$$
$$\frac{100}{1.9} = \frac{1.9x^2}{1.9}$$
$$52.63 = x^2$$
$$x = \sqrt{52.63}$$
$$x = 7.25$$
$$x = 7\frac{1}{4} \text{ ft}$$
$$1.9x = 1.9\left(7\frac{1}{4}\right)$$
$$= 13\frac{3}{4} \text{ ft}$$

89. ACCIDENTS

$$h = s - 16t^2$$
$$5 = 4(12) - 16t^2$$
$$5 = 48 - 16t^2$$
$$-43 = -16t^2$$
$$2.6875 = t^2$$
$$t = \sqrt{2.6875}$$
$$t = 1.6 \text{ sec}$$

91. AUTOMOBILE ENGINES

$$V = \pi r^2 h$$
$$47.75 = \pi r^2(5.25)$$
$$\frac{47.75}{5.25\pi} = \frac{\pi r^2(5.25)}{5.25\pi}$$
$$\frac{47.75}{16.5} = r^2$$
$$2.89 = r^2$$
$$r = \sqrt{2.89}$$
$$r = 1.70 \text{ in.}$$

93. PICTURE FRAMING
The length of the mat is $x + 5 + x = 2x + 5$.
The width of the mat is $x + 4 + x = 2x + 4$.
The total area of the mat and picture is $4(5) = 20$.

Total area – Area of picture = area of mat

$$(2x+5)(2x+4) - 20 = 20$$
$$4x^2 + 8x + 10x + 20 - 20 = 20$$
$$4x^2 + 18x = 20$$
$$\frac{4x^2}{4} + \frac{18x}{4} = \frac{20}{4}$$
$$x^2 + \frac{9}{2}x = 5$$
$$x^2 + \frac{9}{2}x + \left(\frac{1}{2} \bullet \frac{9}{2}\right)^2 = 5 + \left(\frac{1}{2} \bullet \frac{9}{2}\right)^2$$
$$x^2 + \frac{9}{2}x + \left(\frac{9}{4}\right)^2 = 5 + \left(\frac{9}{4}\right)^2$$
$$x^2 + \frac{9}{2}x + \frac{81}{16} = 5 + \frac{81}{16}$$
$$\left(x + \frac{9}{4}\right)^2 = \frac{80}{16} + \frac{81}{16}$$
$$\left(x + \frac{9}{4}\right)^2 = \frac{161}{16}$$
$$x + \frac{9}{4} = \sqrt{\frac{161}{16}}$$
$$x + \frac{9}{4} = \frac{\sqrt{161}}{4}$$
$$x = -\frac{9}{4} + \frac{\sqrt{161}}{4}$$
$$x = \frac{-9 + \sqrt{161}}{4}$$
$$x = 0.92 \text{ in.}$$

95. DIMENSIONS OF A RECTANGLE
Let the width = x, length = x + 4, and area = 20.

$$(\text{width})(\text{length}) = \text{Area}$$
$$x(x+4) = 20$$
$$x^2 + 4x = 20$$
$$x^2 + 4x + \left(\frac{4}{2}\right)^2 = 20 + \left(\frac{4}{2}\right)^2$$
$$x^2 + 4x + (2)^2 = 20 + (2)^2$$
$$x^2 + 4x + 4 = 20 + 4$$
$$(x+2)^2 = 24$$
$$x + 2 = \sqrt{24}$$
$$x + 2 = 2\sqrt{6}$$
$$x = -2 + 2\sqrt{6}$$
$$x = 2.9 \text{ ft}$$
$$x + 4 = 6.9 \text{ ft}$$

The width is 2.9 ft and the length is 6.9 ft.

WRITING

97. Answers will vary.

REVIEW

99.
$$\sqrt[3]{40a^3b^6} = \sqrt[3]{8a^3b^6}\,\sqrt[3]{5}$$
$$= 2ab^2\sqrt[3]{5}$$

101.
$$\sqrt[8]{x^{24}} = \sqrt[8]{(x^3)^8}$$
$$= x^3$$

103.
$$\sqrt{175a^2b^3} = \sqrt{25a^2b^2}\,\sqrt{7b}$$
$$= 5ab\sqrt{7b}$$

CHALLENGE PROBLEMS

105. Take one-half of the coefficient of x and square it: $\left(\frac{\sqrt{3}}{2}\right)^2 = \frac{3}{4}$

SECTION 10.2

VOCABULARY

1. An equation of the form $ax^2 + bx + c = 0$, with $a \neq 0$, is a **quadratic** equation.

CONCEPTS

3. a. $x^2 + 2x + 5 = 0$
   b. $3x^2 + 2x - 1 = 0$

5. a. true
   b. true

7. a.

$$\frac{-2 \pm \sqrt{2^2 - 4(1)(-8)}}{2(1)} = \frac{-2 \pm \sqrt{4+32}}{2} = \frac{-2 \pm \sqrt{36}}{2} = \frac{-2 \pm 6}{2}$$

$$x = \frac{-2+6}{2} \quad \text{or} \quad x = \frac{-2-6}{2}$$

$$x = \frac{4}{2} \qquad x = \frac{-8}{2}$$

$$x = 2 \qquad x = -4$$

b.

$$\frac{-(-1) \pm \sqrt{(-1)^2 - 4(2)(-4)}}{2(2)} = \frac{1 \pm \sqrt{1+32}}{4} = \frac{1 \pm \sqrt{33}}{4}$$

9. a.

$$\frac{3 \pm 6\sqrt{2}}{3} = \frac{\cancel{3}\left(1 \pm 2\sqrt{2}\right)}{\cancel{3}} = \frac{1+2\sqrt{2}}{1} = 1+2\sqrt{2}$$

b.

$$\frac{-12 \pm 4\sqrt{7}}{8} = \frac{\cancel{4}\left(-3 \pm \sqrt{7}\right)}{\cancel{4} \cdot 2} = \frac{-3 \pm \sqrt{7}}{2}$$

NOTATION

11. a. The fraction bar wasn't drawn under both parts of the numerator.
    b. A $\pm$ sign wasn't written between b and the radical.

PRACTICE

13. $x^2 + 3x + 2 = 0$

$a = 1$, $b = 3$, and $c = 2$

$$x = \frac{-b \pm \sqrt{b^2 - 4ac}}{2a} = \frac{-3 \pm \sqrt{(3)^2 - 4(1)(2)}}{2(1)} = \frac{-3 \pm \sqrt{9-8}}{2} = \frac{-3 \pm \sqrt{1}}{2} = \frac{-3 \pm 1}{2}$$

$$x = \frac{-3+1}{2} \quad \text{or} \quad x = \frac{-3-1}{2}$$

$$x = \frac{-2}{2} \qquad x = \frac{-4}{2}$$

$$x = -1 \qquad x = -2$$

15. $x^2 + 12x = -36$

$x^2 + 12x + 36 = 0$

$a = 1$, $b = 12$, and $c = 36$

$$x = \frac{-b \pm \sqrt{b^2 - 4ac}}{2a}$$

$$= \frac{-12 \pm \sqrt{(12)^2 - 4(1)(36)}}{2(1)}$$

$$= \frac{-12 \pm \sqrt{144 - 144}}{2}$$

$$= \frac{-12 \pm \sqrt{0}}{2}$$

$$= \frac{-12 \pm 0}{2}$$

$$x = \frac{-12 + 0}{2} \quad \text{or} \quad x = \frac{-12 - 0}{2}$$

$$x = \frac{-12}{2} \qquad x = \frac{-12}{2}$$

$$x = -6 \qquad x = -6$$

17. $2x^2 + 5x - 3 = 0$

$a = 2$, $b = 5$, and $c = -3$

$$x = \frac{-b \pm \sqrt{b^2 - 4ac}}{2a}$$

$$= \frac{-5 \pm \sqrt{(5)^2 - 4(2)(-3)}}{2(2)}$$

$$= \frac{-5 \pm \sqrt{25 - (-24)}}{4}$$

$$= \frac{-5 \pm \sqrt{49}}{4}$$

$$= \frac{-5 \pm 7}{4}$$

$$x = \frac{-5 + 7}{4} \quad \text{or} \quad x = \frac{-5 - 7}{4}$$

$$x = \frac{2}{4} \qquad x = \frac{-12}{4}$$

$$x = \frac{1}{2} \qquad x = -3$$

19. $5x^2 + 5x + 1 = 0$

$a = 5$, $b = 5$, and $c = 1$

$$x = \frac{-b \pm \sqrt{b^2 - 4ac}}{2a}$$

$$= \frac{-5 \pm \sqrt{(5)^2 - 4(5)(1)}}{2(5)}$$

$$= \frac{-5 \pm \sqrt{25 - 20}}{10}$$

$$x = \frac{-5 \pm \sqrt{5}}{10}$$

21. $8u = -4u^2 - 3$

$4u^2 + 8u + 3 = 0$

$a = 4$, $b = 8$, and $c = 3$

$$u = \frac{-b \pm \sqrt{b^2 - 4ac}}{2a}$$

$$= \frac{-8 \pm \sqrt{(8)^2 - 4(4)(3)}}{2(4)}$$

$$= \frac{-8 \pm \sqrt{64 - 48}}{8}$$

$$= \frac{-8 \pm \sqrt{16}}{8}$$

$$= \frac{-8 \pm 4}{8}$$

$$u = \frac{-8 + 4}{8} \quad \text{or} \quad u = \frac{-8 - 4}{8}$$

$$u = \frac{-4}{8} \qquad u = \frac{-12}{8}$$

$$u = -\frac{1}{2} \qquad u = -\frac{3}{2}$$

23. $-16y^2 - 8y + 3 = 0$

$0 = 16y^2 + 8y - 3$

$a = 16, b = 8, c = -3$

$$y = \frac{-b \pm \sqrt{b^2 - 4ac}}{2a}$$

$$= \frac{-8 \pm \sqrt{(-8)^2 - 4(16)(-3)}}{2(16)}$$

$$= \frac{-8 \pm \sqrt{64 - (-192)}}{32}$$

$$= \frac{-8 \pm \sqrt{256}}{32}$$

$$= \frac{-8 \pm 16}{32}$$

$$y = \frac{-8+16}{32} \quad \text{or} \quad y = \frac{-8-16}{32}$$

$$y = \frac{8}{32} \qquad y = \frac{-24}{32}$$

$$y = \frac{1}{4} \qquad y = -\frac{3}{4}$$

25.

$$x^2 - \frac{14}{15}x = \frac{8}{15}$$

$$15\left(x^2 - \frac{14}{15}x\right) = 15\left(\frac{8}{15}\right)$$

$$15x^2 - 14x = 8$$

$$15x^2 - 14x - 8 = 0$$

$a = 15$, $b = -14$, and $c = -8$

$$x = \frac{-b \pm \sqrt{b^2 - 4ac}}{2a}$$

$$= \frac{-(-14) \pm \sqrt{(-14)^2 - 4(15)(-8)}}{2(15)}$$

$$= \frac{14 \pm \sqrt{196 - (-480)}}{30}$$

$$= \frac{14 \pm \sqrt{676}}{30}$$

$$= \frac{14 \pm 26}{30}$$

$$x = \frac{14+26}{30} \quad \text{or} \quad x = \frac{14-26}{30}$$

$$x = \frac{40}{30} \qquad x = -\frac{12}{30}$$

$$x = \frac{4}{3} \qquad x = -\frac{2}{5}$$

27.

$$\frac{x^2}{2} + \frac{5}{2}x = -1$$

$$2\left(\frac{x^2}{2} + \frac{5}{2}x\right) = 2(-1)$$

$$x^2 + 5x = -2$$

$$x^2 + 5x + 2 = 0$$

$a = 1$, $b = 5$, and $c = 2$

$$x = \frac{-b \pm \sqrt{b^2 - 4ac}}{2a}$$

$$= \frac{-5 \pm \sqrt{(5)^2 - 4(1)(2)}}{2(1)}$$

$$= \frac{-5 \pm \sqrt{25 - 8}}{2}$$

$$x = \frac{-5 \pm \sqrt{17}}{2}$$

29. $2x^2 - 1 = 3x$

$2x^2 - 3x - 1 = 0$

$a = 2$, $b = -3$, and $c = -1$

$$x = \frac{-b \pm \sqrt{b^2 - 4ac}}{2a}$$

$$= \frac{-(-3) \pm \sqrt{(-3)^2 - 4(2)(-1)}}{2(2)}$$

$$= \frac{3 \pm \sqrt{9 - (-8)}}{4}$$

$$x = \frac{3 \pm \sqrt{17}}{4}$$

31. $-x^2 + 10x = 18$

$0 = x^2 - 10x + 18$

$a = 1$, $b = -10$, and $c = 18$

$$x = \frac{-b \pm \sqrt{b^2 - 4ac}}{2a}$$
$$= \frac{-(-10) \pm \sqrt{(-10)^2 - 4(1)(18)}}{2(1)}$$
$$= \frac{10 \pm \sqrt{100 - 72}}{2}$$
$$= \frac{10 \pm \sqrt{28}}{2}$$
$$= \frac{10 \pm \sqrt{4}\sqrt{7}}{2}$$
$$= \frac{10 \pm 2\sqrt{7}}{2}$$
$$= \frac{\not{2}\left(5 \pm \sqrt{7}\right)}{\not{2}}$$
$$x = 5 \pm \sqrt{7}$$

33. $x^2 - 6x = 391$

$x^2 - 6x - 391 = 0$

$a = 1$, $b = -6$, and $c = -391$

$$x = \frac{-b \pm \sqrt{b^2 - 4ac}}{2a}$$
$$= \frac{-(-6) \pm \sqrt{(-6)^2 - 4(1)(391)}}{2(1)}$$
$$= \frac{6 \pm \sqrt{36 - (-1{,}564)}}{2}$$
$$= \frac{6 \pm \sqrt{1{,}600}}{2}$$
$$= \frac{6 \pm 40}{2}$$
$$x = \frac{6 + 40}{2} \quad \text{or} \quad x = \frac{6 - 40}{2}$$
$$x = \frac{46}{2} \qquad x = \frac{-34}{2}$$
$$x = 23 \qquad x = -17$$

35.

$$x^2 - \frac{5}{3} = -\frac{11}{6}x$$
$$6\left(x^2 - \frac{5}{3}\right) = 6\left(-\frac{11}{6}x\right)$$
$$6x^2 - 10 = -11x$$
$$6x^2 + 11x - 10 = 0$$

$a = 6$, $b = 11$, and $c = -10$

$$x = \frac{-b \pm \sqrt{b^2 - 4ac}}{2a}$$
$$= \frac{-11 \pm \sqrt{(11)^2 - 4(6)(-10)}}{2(6)}$$
$$= \frac{-11 \pm \sqrt{121 - (-240)}}{12}$$
$$= \frac{-11 \pm \sqrt{361}}{12}$$
$$x = \frac{-11 \pm 19}{12}$$
$$x = \frac{-11 + 19}{12} \quad \text{or} \quad x = \frac{-11 - 19}{12}$$
$$x = \frac{8}{12} \qquad x = \frac{-30}{12}$$
$$x = \frac{2}{3} \qquad x = -\frac{5}{2}$$

37. $x^2 + 2x + 2 = 0$

$a = 1$, $b = 2$, and $c = 2$

$$x = \frac{-b \pm \sqrt{b^2 - 4ac}}{2a}$$
$$= \frac{-2 \pm \sqrt{(2)^2 - 4(1)(2)}}{2(1)}$$
$$= \frac{-2 \pm \sqrt{4 - 8}}{2}$$
$$= \frac{-2 \pm \sqrt{-4}}{2}$$
$$= \frac{-2 \pm 2i}{2}$$
$$= -\frac{2}{2} \pm \frac{2}{2}i$$
$$x = -1 \pm i$$

39. $2x^2 + x + 1 = 0$

$a = 2$, $b = 1$, and $c = 1$

$$x = \frac{-b \pm \sqrt{b^2 - 4ac}}{2a}$$
$$= \frac{-1 \pm \sqrt{(1)^2 - 4(2)(1)}}{2(2)}$$
$$= \frac{-1 \pm \sqrt{1-8}}{4}$$
$$= \frac{-1 \pm \sqrt{-7}}{4}$$
$$= \frac{-1 \pm i\sqrt{7}}{4}$$
$$x = -\frac{1}{4} \pm \frac{\sqrt{7}}{4}i$$

41. $3x^2 - 4x = -2$

$3x^2 - 4x + 2 = 0$

$a = 3$, $b = -4$, and $c = 2$

$$x = \frac{-b \pm \sqrt{b^2 - 4ac}}{2a}$$
$$= \frac{-(-4) \pm \sqrt{(-4)^2 - 4(3)(2)}}{2(3)}$$
$$= \frac{4 \pm \sqrt{16-24}}{6}$$
$$= \frac{4 \pm \sqrt{-8}}{6}$$
$$= \frac{4 \pm \sqrt{-4}\sqrt{2}}{6}$$
$$= \frac{4 \pm 2i\sqrt{2}}{6}$$
$$= \frac{4}{6} \pm \frac{2\sqrt{2}}{6}i$$
$$x = \frac{2}{3} \pm \frac{\sqrt{2}}{3}i$$

43. $3x^2 - 2x = -3$

$3x^2 - 2x + 3 = 0$

$a = 3$, $b = -2$, and $c = 3$

$$x = \frac{-b \pm \sqrt{b^2 - 4ac}}{2a}$$
$$= \frac{-(-2) \pm \sqrt{(-2)^2 - 4(3)(3)}}{2(3)}$$
$$= \frac{2 \pm \sqrt{4-36}}{6}$$
$$= \frac{2 \pm \sqrt{-32}}{6}$$
$$= \frac{2 \pm \sqrt{-16}\sqrt{2}}{6}$$
$$= \frac{2 \pm 4i\sqrt{2}}{6}$$
$$= \frac{2}{6} \pm \frac{4\sqrt{2}}{6}i$$
$$x = \frac{1}{3} \pm \frac{2\sqrt{2}}{3}i$$

45.

$$\frac{x^2}{8} - \frac{x}{2} + 1 = 0$$
$$8\left(\frac{x^2}{8} - \frac{x}{2} + 1\right) = 8(0)$$
$$x^2 - 4x + 8 = 0$$

$a = 1$, $b = -4$, and $c = 8$

$$x = \frac{-b \pm \sqrt{b^2 - 4ac}}{2a}$$
$$= \frac{-(-4) \pm \sqrt{(-4)^2 - 4(1)(8)}}{2(1)}$$
$$= \frac{4 \pm \sqrt{16-32}}{2}$$
$$= \frac{4 \pm \sqrt{-16}}{2}$$
$$= \frac{4 \pm 4i}{2}$$
$$= \frac{4}{2} \pm \frac{4}{2}i$$
$$x = 2 \pm 2i$$

47.

$$\frac{a^2}{10}-\frac{3a}{5}+\frac{7}{5}=0$$

$$10\left(\frac{a^2}{10}-\frac{3a}{5}+\frac{7}{5}\right)=10(0)$$

$$a^2-6a+14=0$$

$a=1$, $b=-6$, and $c=14$

$$a=\frac{-b\pm\sqrt{b^2-4ac}}{2a}$$

$$=\frac{-(-6)\pm\sqrt{(-6)^2-4(1)(14)}}{2(1)}$$

$$=\frac{6\pm\sqrt{36-56}}{2}$$

$$=\frac{6\pm\sqrt{-20}}{2}$$

$$=\frac{6\pm\sqrt{-4}\sqrt{5}}{2}$$

$$=\frac{-6\pm 2i\sqrt{5}}{2}$$

$$=-\frac{6}{2}\pm\frac{2\sqrt{5}}{2}i$$

$$a=-3\pm i\sqrt{5}$$

49.

$$50x^2+30x-10=0$$

$$\frac{50x^2}{10}+\frac{30x}{10}-\frac{10}{10}=0$$

$$5x^2+3x-1=0$$

$a=5$, $b=3$, and $c=-1$

$$x=\frac{-b\pm\sqrt{b^2-4ac}}{2a}$$

$$=\frac{-3\pm\sqrt{(3)^2-4(5)(-1)}}{2(5)}$$

$$=\frac{-3\pm\sqrt{9-(-20)}}{10}$$

$$x=\frac{-3\pm\sqrt{29}}{10}$$

51.

$$900x^2-8{,}100x=1{,}800$$

$$\frac{900x^2}{900}-\frac{8{,}100x}{900}=\frac{1{,}800}{900}$$

$$x^2-9x=2$$

$$x^2-9x-2=0$$

$a=1$, $b=-9$, and $c=-2$

$$x=\frac{-b\pm\sqrt{b^2-4ac}}{2a}$$

$$=\frac{-(-9)\pm\sqrt{(-9)^2-4(1)(-2)}}{2(1)}$$

$$=\frac{9\pm\sqrt{81-(-8)}}{2}$$

$$x=\frac{9\pm\sqrt{89}}{2}$$

53.

$$-0.6x^2-0.03=-0.4x$$

$$0=0.6x^2-0.4x+0.03$$

$$100(0)=100\left(0.6x^2-0.4x+0.03\right)$$

$$0=60x^2-40x+3$$

$a=60$, $b=-40$, and $c=3$

$$x=\frac{-b\pm\sqrt{b^2-4ac}}{2a}$$

$$=\frac{-(-40)\pm\sqrt{(-40)^2-4(60)(3)}}{2(60)}$$

$$=\frac{40\pm\sqrt{1{,}600-720}}{120}$$

$$=\frac{40\pm\sqrt{880}}{120}$$

$$=\frac{40\pm\sqrt{16}\sqrt{55}}{120}$$

$$=\frac{40\pm 4\sqrt{55}}{120}$$

$$=\frac{\cancel{4}\left(10\pm\sqrt{55}\right)}{\cancel{4}\cdot 30}$$

$$x=\frac{10\pm\sqrt{55}}{30}$$

55. $x^2 + 8x + 5 = 0$

$a = 1$, $b = 8$, and $c = 5$

$$x = \frac{-b \pm \sqrt{b^2 - 4ac}}{2a}$$
$$= \frac{-8 \pm \sqrt{(8)^2 - 4(1)(5)}}{2(1)}$$
$$= \frac{-8 \pm \sqrt{64 - 20}}{2}$$
$$= \frac{-8 \pm \sqrt{44}}{2}$$
$$= \frac{-8 \pm \sqrt{4}\sqrt{11}}{2}$$
$$= \frac{-8 \pm 2\sqrt{11}}{2}$$
$$x = \frac{-8 + 2\sqrt{11}}{2} \quad \text{or} \quad x = \frac{-8 - 2\sqrt{11}}{2}$$
$$x = -0.68 \qquad\qquad x = -7.32$$

57. $3x^2 - 2x - 2 = 0$

$a = 3$, $b = -2$, and $c = -2$

$$x = \frac{-b \pm \sqrt{b^2 - 4ac}}{2a}$$
$$= \frac{-(-2) \pm \sqrt{(-2)^2 - 4(3)(-2)}}{2(3)}$$
$$= \frac{2 \pm \sqrt{4 - (-24)}}{6}$$
$$= \frac{2 \pm \sqrt{28}}{6}$$
$$= \frac{2 \pm \sqrt{4}\sqrt{7}}{6}$$
$$= \frac{2 \pm 2\sqrt{7}}{6}$$
$$x = \frac{2 + 2\sqrt{7}}{6} \quad \text{or} \quad x = \frac{2 - 2\sqrt{7}}{6}$$
$$x = 1.22 \qquad\qquad x = -0.55$$

59.

$$0.7x^2 - 3.5x - 25 = 0$$
$$10(0.7x^2 - 3.5x - 25) = 10(0)$$
$$7x^2 - 35x - 250 = 0$$

$a = 7$, $b = -35$, and $c = -250$

$$x = \frac{-b \pm \sqrt{b^2 - 4ac}}{2a}$$
$$= \frac{-(-35) \pm \sqrt{(-35)^2 - 4(7)(-250)}}{2(7)}$$
$$= \frac{35 \pm \sqrt{1,225 - (-7,000)}}{14}$$
$$= \frac{35 \pm \sqrt{8,225}}{14}$$
$$x = \frac{35 + \sqrt{8,225}}{14} \quad \text{or} \quad x = \frac{35 - \sqrt{8,225}}{14}$$
$$x = 8.98 \qquad\qquad x = -3.98$$

APPLICATIONS

61. IMAX SCREENS

Let the width $= x$ and length $= x + 20$.

The area is 11,349.

$$(\text{width})(\text{length}) = \text{area}$$
$$x(x + 20) = 11,349$$
$$x^2 + 20x - 11,349 = 0$$

$a = 1$, $b = 20$, and $c = -11,349$

$$x = \frac{-b \pm \sqrt{b^2 - 4ac}}{2a}$$
$$= \frac{-20 \pm \sqrt{(20)^2 - 4(1)(-11,349)}}{2(1)}$$
$$= \frac{-20 \pm \sqrt{400 - (-45,396)}}{2}$$
$$= \frac{-20 \pm \sqrt{45,796}}{2}$$
$$= \frac{-20 \pm 214}{2}$$
$$x = \frac{-20 + 214}{2} \quad \text{or} \quad x = \frac{-20 - 214}{2}$$
$$x = \frac{194}{2} \qquad\qquad x = \frac{-234}{2}$$
$$x = 97 \qquad\qquad x = \cancel{-117}$$
$$x + 20 = 117$$

The dimensions are 97 ft by 117 ft.

63. PARKS

Let the width = x and length = $5x$.

Perimeter = $2x + 2(5x) = 2x + 10x = 12x$.

Area = $x(5x) = 5x^2$

Perimeter = Area + 4.75

$$12x = 5x^2 + 4.75$$
$$0 = 5x^2 - 12x + 4.75$$

$a = 5$, $b = -12$, and $c = 4.75$

$$x = \frac{-b \pm \sqrt{b^2 - 4ac}}{2a}$$
$$= \frac{-(-12) \pm \sqrt{(-12)^2 - 4(5)(4.75)}}{2(5)}$$
$$= \frac{12 \pm \sqrt{144 - 95}}{10}$$
$$= \frac{12 \pm \sqrt{49}}{10}$$
$$= \frac{12 \pm 7}{10}$$
$$x = \frac{12+7}{10} \quad \text{or} \quad x = \frac{12-7}{10}$$
$$x = \frac{19}{10} \qquad x = \frac{5}{10}$$
$$x = 1.9 \text{ (crossed out)} \qquad x = 0.5$$

$$5x = 5(0.5)$$
$$= 2.5$$

$x = 1.9$ is not possible because it states that the width is less than 1 mile. The width must be 0.5 miles and the length must be 2.5 miles.

65. BADMINTON

A right triangle is formed with the string being the hypotenuse. Let the hypotenuse be x. The height would be $(x - 4)$ and the base would be $0.5x - 1$. Use the Pythagorean Theorem.

$$a^2 + b^2 = c^2$$
$$(x-4)^2 + (0.5x-1)^2 = x^2$$
$$x^2 - 4x - 4x + 16 + 0.25x^2 - 0.5x - 0.5x + 1 = x^2$$
$$1.25x^2 - 9x + 17 = x^2$$
$$0.25x^2 - 9x + 17 = 0$$
$$100(0.25x^2 - 9x + 17) = 100(0)$$
$$25x^2 - 900x + 1{,}700 = 0$$

$a = 25$, $b = -900$, and $c = 1{,}700$

$$x = \frac{-b \pm \sqrt{b^2 - 4ac}}{2a}$$
$$= \frac{-(-900) \pm \sqrt{(-900)^2 - 4(25)(1{,}700)}}{2(25)}$$
$$= \frac{900 \pm \sqrt{810{,}000 - 170{,}000}}{50}$$
$$= \frac{900 \pm \sqrt{640{,}000}}{50}$$
$$= \frac{900 \pm 800}{50}$$
$$x = \frac{900+800}{50} \quad \text{or} \quad x = \frac{900-800}{50}$$
$$x = \frac{1{,}700}{50} \qquad x = \frac{100}{50}$$
$$x = 34 \qquad x = 2 \text{ (crossed out)}$$

$x \neq 2$ because it would make the height negative: $x - 4 = 2 - 4 = -2$.

The length of the string is 34 in.

67. DANCES

Let x = number of increases in ticket price.
New price: $4 + 0.10x$
Number tickets sold: $300 - 5x$
New price · # tickets sold = new receipts

$$(4+0.10x)(300-5x)=1{,}248$$
$$1{,}200-20x+30x-0.5x^2=1{,}248$$
$$1{,}200+10x-0.5x^2=1{,}248$$
$$0=0.5x^2-10x+48$$
$$10(0)=10(0.5x^2-10x+48)$$
$$0=5x^2-100x+480$$

$a = 5$, $b = -100$, $c = 480$

$$x=\frac{-b\pm\sqrt{b^2-4ac}}{2a}$$
$$=\frac{-(-100)\pm\sqrt{(-100)^2-4(5)(480)}}{2(5)}$$
$$=\frac{100\pm\sqrt{10{,}000-9{,}600}}{10}$$
$$=\frac{100\pm\sqrt{400}}{10}$$
$$=\frac{100\pm 20}{10}$$

$$x=\frac{100+20}{10} \quad\text{or}\quad x=\frac{100-20}{10}$$
$$x=\frac{120}{10} \qquad\qquad x=\frac{80}{10}$$
$$x=12 \qquad\qquad x=8$$

$$4+0.10x=4+0.10(12) \qquad 4+0.10x=4+0.10(8)$$
$$=4+1.20 \qquad\qquad =4+0.80$$
$$=\$5.20 \qquad\qquad =\$4.80$$

When the ticket prices are \$5.20 or \$4.80, the receipts will be \$1,248.

69. MAGAZINE SALES

Let x = number of new subscribers
New price: $20 + 0.01x$
Number subscribers: $3{,}000 + x$
New price · # subscribers = total profit

$$(20+0.01x)(3{,}000+x)=120{,}000$$
$$60{,}000+20x+30x+0.01x^2=120{,}000$$
$$60{,}000+50x+0.01x^2=120{,}000$$
$$0.01x^2+50x-60{,}000=0$$
$$100(0.01x^2+50x-60{,}000)=100(0)$$
$$x^2+5{,}000x-6{,}000{,}000=0$$

$a = 1$, $b = 5{,}000$, $c = -6{,}000{,}000$

$$x=\frac{-b\pm\sqrt{b^2-4ac}}{2a}$$
$$=\frac{-5{,}000\pm\sqrt{(5{,}000)^2-4(1)(-6{,}000{,}000)}}{2(1)}$$
$$=\frac{-5{,}000\pm\sqrt{25{,}000{,}000-(-24{,}000{,}000)}}{2}$$
$$=\frac{-5{,}000\pm\sqrt{49{,}000{,}000}}{2}$$
$$=\frac{-5{,}000\pm 7{,}000}{2}$$

$$x=\frac{-5{,}000+7{,}000}{2} \quad\text{or}\quad x=\frac{-5{,}000-7{,}000}{2}$$
$$x=\frac{2{,}000}{2} \qquad\qquad x=\frac{-12{,}000}{2}$$
$$x=1{,}000 \qquad\qquad x=\cancel{-6{,}000}$$

$$3{,}000+x=3{,}000+1{,}000$$
$$=4{,}000$$

There cannot be a negative number of subscribers.

4,000 subscribers will bring a profit of \$120,000.

71. INVESTMENT RATES

$$1{,}000(1+r)^2+2{,}000(1+r)=3{,}368.10$$
$$1{,}000(1+2r+r^2)+2{,}000(1+r)=3{,}368.10$$
$$1{,}000+2{,}000r+1{,}000r^2+2{,}000+2{,}000r=3{,}368.10$$
$$1{,}000r^2+4{,}000r+3{,}000=3{,}368.10$$
$$1{,}000r^2+4{,}000r-368.10=0$$

$a=1{,}000,\ b=4{,}000,\ c=-368.10$

$$x=\frac{-b\pm\sqrt{b^2-4ac}}{2a}$$
$$=\frac{-4{,}000\pm\sqrt{(4{,}000)^2-4(1{,}000)(-368.10)}}{2(1{,}000)}$$
$$=\frac{-4{,}000\pm\sqrt{16{,}000{,}000-(-1{,}472{,}400)}}{2{,}000}$$
$$=\frac{-4{,}000\pm\sqrt{17{,}472{,}400}}{2{,}000}$$
$$=\frac{-4{,}000\pm 4{,}180}{2{,}000}$$

$x=\frac{-4{,}000+4{,}180}{2{,}000}$ or $x=\frac{-4{,}000-4{,}180}{2{,}000}$

$x=\frac{180}{2{,}000}$ $\qquad$ $x=\frac{-8{,}180}{2{,}000}$

$x=0.09$ $\qquad$ $x=-4.09$ (crossed out)

$x=9\%$

73. RETIREMENT

Let $P=75$.

$$P=0.03x^2-1.37x+82.51$$
$$75=0.03x^2-1.37x+82.51$$
$$0=0.03x^2-1.37x+7.51$$
$$100(0)=100(0.03x^2-1.37x+7.51)$$
$$0=3x^2-137x+751$$

$a=3,\ b=-137,\ c=751$

$$x=\frac{-b\pm\sqrt{b^2-4ac}}{2a}$$
$$=\frac{-(-137)\pm\sqrt{(-137)^2-4(3)(751)}}{2(3)}$$
$$=\frac{137\pm\sqrt{18{,}769-9{,}012}}{6}$$
$$=\frac{137\pm\sqrt{9{,}757}}{6}$$
$$=\frac{137\pm 99}{6}$$

$x=\frac{137+99}{6}$ or $x=\frac{137-99}{6}$

$x=\frac{236}{6}$ $\qquad$ $x=\frac{38}{6}$

$x=39.3$ (crossed out) $\qquad$ $x=6.3$

$x=39.3$ represents the year 2009 (1970 + 39 = 2009) which hasn't occurred yet.

$x=6.3$ represents the year 1976 (1970 + 6 = 1976). Thus, the model indicates that 75% of the men ages 55–64 were part of the workforce in early 1976.

WRITING

75. Answers will vary.

REVIEW

77. $\sqrt{n}=n^{1/2}$

79. $\sqrt[4]{3b}=(3b)^{1/4}$

81. $t^{1/3} = \sqrt[3]{t}$

83. $(3t)^{1/4} = \sqrt[4]{3t}$

CHALLENGE PROBLEMS

85. $x^2 + 2\sqrt{2}x - 6 = 0$

$a = 1,\ b = 2\sqrt{2},\ c = -6$

$$x = \frac{-b \pm \sqrt{b^2 - 4ac}}{2a}$$

$$= \frac{-2\sqrt{2} \pm \sqrt{\left(2\sqrt{2}\right)^2 - 4(1)(-6)}}{2(1)}$$

$$= \frac{-2\sqrt{2} \pm \sqrt{8 - (-24)}}{2}$$

$$= \frac{-2\sqrt{2} \pm \sqrt{32}}{2}$$

$$= \frac{-2\sqrt{2} \pm \sqrt{16}\sqrt{2}}{2}$$

$$= \frac{-2\sqrt{2} \pm 4\sqrt{2}}{2}$$

$$x = \frac{-2\sqrt{2} + 4\sqrt{2}}{2} \quad \text{or} \quad x = \frac{-2\sqrt{2} - 4\sqrt{2}}{2}$$

$$x = \frac{2\sqrt{2}}{2} \qquad\qquad x = \frac{-6\sqrt{2}}{2}$$

$$x = \sqrt{2} \qquad\qquad x = -3\sqrt{2}$$

87. $x^2 - 3ix - 2 = 0$

$a = 1,\ b = -3i,\ c = -2$

$$x = \frac{-b \pm \sqrt{b^2 - 4ac}}{2a}$$

$$= \frac{-(-3i) \pm \sqrt{(-3i)^2 - 4(1)(-2)}}{2(1)}$$

$$= \frac{3i \pm \sqrt{9i^2 - (-8)}}{2}$$

$$= \frac{3i \pm \sqrt{-9 + 8}}{2}$$

$$= \frac{3i \pm \sqrt{-1}}{2}$$

$$= \frac{3i \pm i}{2}$$

$$x = \frac{3i + i}{2} \quad \text{or} \quad x = \frac{3i - i}{2}$$

$$x = \frac{4i}{2} \qquad\qquad x = \frac{2i}{2}$$

$$x = 2i \qquad\qquad x = i$$

SECTION 10.3

VOCABULARY

1. For the quadratic equation $ax^2+bx+c=0$, the **discriminant** is b^2-4ac.

CONCEPTS

3. If $b^2-4ac<0$, the solutions of the equation are nonreal complex **conjugates**.

5. If b^2-4ac is a perfect square, the solutions are **rational** numbers and **unequal**.

7. a. $y=x^2$
   b. $y=\sqrt{x}$
   c. $y=x^{1/3}$
   d. $y=\frac{1}{x}$
   e. $y=x+1$

NOTATION

9.
$$\begin{aligned}b^2-\boxed{4ac}&=\boxed{5}^2-4(1)(\boxed{6})\\&=25-\boxed{24}\\&=1\end{aligned}$$
Since *a, b,* and *c* are rational numbers and the value of the discriminant is a perfect square, the solutions are **rational** numbers and unequal.

PRACTICE

11.
$$4x^2-4x+1=0$$
$$\begin{aligned}b^2-4ac&=(-4)^2-4(4)(1)\\&=16-16\\&=0\end{aligned}$$
rational and equal

13.
$$5x^2+x+2=0$$
$$\begin{aligned}b^2-4ac&=(1)^2-4(5)(2)\\&=1-40\\&=-39\end{aligned}$$
complex conjugates

15.
$$2x^2=4x-1$$
$$2x^2-4x+1=0$$
$$\begin{aligned}b^2-4ac&=(-4)^2-4(2)(1)\\&=16-8\\&=8\end{aligned}$$
irrational and unequal

17.
$$x(2x-3)=20$$
$$2x^2-3x=20$$
$$2x^2-3x-20=0$$
$$\begin{aligned}b^2-4ac&=(-3)^2-4(2)(-20)\\&=9-(-160)\\&=169 \text{ (perfect square)}\end{aligned}$$
rational and unequal

19.
$$1{,}492x^2+1{,}776x-2{,}000=0$$
$$\begin{aligned}b^2-4ac&=(1{,}776)^2-4(1{,}492)(-2{,}000)\\&=3{,}154{,}176-(-11{,}936{,}000)\\&=15{,}090{,}176\end{aligned}$$
Yes, the solutions are real numbers.

21. Let $y=x^2$.
$$x^4-17x^2+16=0$$
$$(x^2)^2-17x^2+16=0$$
$$y^2-17y+16=0$$
$$(y-16)(y-1)=0$$
$$y-16=0 \quad \text{or} \quad y-1=0$$
$$y=16 \qquad y=1$$
$$x^2=16 \qquad x^2=1$$
$$x=\pm\sqrt{16} \qquad x=\pm\sqrt{1}$$
$$x=\pm4 \qquad x=\pm1$$
$$x=4,\ -4,\ 1,\ -1$$

23. Let $y = x^2$.

$$x^4 = 6x^2 - 5$$
$$x^4 - 6x^2 + 5 = 0$$
$$\left(x^2\right)^2 - 6x^2 + 5 = 0$$
$$y^2 - 6y + 5 = 0$$
$$(y-5)(y-1) = 0$$
$$y - 5 = 0 \quad \text{or} \quad y - 1 = 0$$
$$y = 5 \qquad y = 1$$
$$x^2 = 5 \qquad x^2 = 1$$
$$x = \pm\sqrt{5} \qquad x = \pm\sqrt{1}$$
$$x = \pm\sqrt{5} \qquad x = \pm 1$$
$$x = \sqrt{5},\ -\sqrt{5},\ 1,\ -1$$

25. Let $y = t^2$.

$$t^4 + 3t^2 = 28$$
$$t^4 + 3t^2 - 28 = 0$$
$$\left(t^2\right)^2 + 3t^2 - 28 = 0$$
$$y^2 + 3y - 28 = 0$$
$$(y-4)(y+7) = 0$$
$$y - 4 = 0 \quad \text{or} \quad y + 7 = 0$$
$$y = 4 \qquad y = -7$$
$$t^2 = 4 \qquad t^2 = -7$$
$$t = \pm\sqrt{4} \qquad t = \pm\sqrt{-7}$$
$$t = \pm 2 \qquad t = \pm i\sqrt{7}$$
$$t = 2,\ -2,\ i\sqrt{7},\ -i\sqrt{7}$$

27. Let $y = x^2$.

$$x^4 + 19x^2 + 18 = 0$$
$$\left(x^2\right)^2 + 19x^2 + 18 = 0$$
$$y^2 + 19y + 18 = 0$$
$$(y+18)(y+1) = 0$$
$$y + 18 = 0 \quad \text{or} \quad y + 1 = 0$$
$$y = -18 \qquad y = -1$$
$$x^2 = -18 \qquad x^2 = -1$$
$$x = \pm\sqrt{-18} \qquad x = \pm\sqrt{-1}$$
$$x = \pm 3i\sqrt{2} \qquad x = \pm i$$
$$x = 3i\sqrt{2},\ -3i\sqrt{2},\ i,\ -i$$

29. Let $y = \sqrt{x}$.

$$2x + \sqrt{x} - 3 = 0$$
$$2\left(\sqrt{x}\right)^2 + \sqrt{x} - 3 = 0$$
$$2y^2 + y - 3 = 0$$
$$(2y+3)(y-1) = 0$$
$$2y + 3 = 0 \quad \text{or} \quad y - 1 = 0$$
$$y = -\frac{3}{2} \qquad y = 1$$
$$\sqrt{x} = -\frac{3}{2} \qquad \sqrt{x} = 1$$
$$\left(\sqrt{x}\right)^2 = \left(-\frac{3}{2}\right)^2 \qquad \sqrt{x} = (1)^2$$
$$x = \frac{9}{4} \qquad x = 1$$

$\frac{9}{4}$ is extraneous;

$x = 1$

31. Let $y = \sqrt{x}$.

$$3x + 5\sqrt{x} + 2 = 0$$
$$3\left(\sqrt{x}\right)^2 + 5\sqrt{x} + 2 = 0$$
$$3y^2 + 5y + 2 = 0$$
$$(3y+2)(y+1) = 0$$
$$3y + 2 = 0 \quad \text{or} \quad y + 1 = 0$$
$$y = -\frac{2}{3} \qquad y = -1$$
$$\sqrt{x} = -\frac{2}{3} \qquad \sqrt{x} = -1$$
$$\left(\sqrt{x}\right)^2 = \left(-\frac{2}{3}\right)^2 \qquad \sqrt{x} = (-1)^2$$
$$x = \frac{4}{9} \qquad x = 1$$

$\frac{4}{9}$ and 1 are extraneous

No Solution

33. Let $y = x^{1/2}$.

$$x - 6x^{1/2} = -8$$
$$x - 6x^{1/2} + 8 = 0$$
$$\left(x^{1/2}\right)^2 - 6x^{1/2} + 8 = 0$$
$$y^2 - 6y + 8 = 0$$
$$(y-2)(y-4) = 0$$
$$y - 2 = 0 \quad \text{or} \quad y - 4 = 0$$
$$y = 2 \qquad y = 4$$
$$x^{1/2} = 2 \qquad x^{1/2} = 4$$
$$\left(x^{1/2}\right)^2 = (2)^2 \qquad \left(x^{1/2}\right)^2 = (4)^2$$
$$x = 4 \qquad x = 16$$

35. Let $y = \sqrt{x}$.

$$2x - \sqrt{x} = 3$$
$$2x - \sqrt{x} - 3 = 0$$
$$2\left(\sqrt{x}\right)^2 - \sqrt{x} - 3 = 0$$
$$2y^2 - y - 3 = 0$$
$$(2y-3)(y+1) = 0$$
$$2y - 3 = 0 \quad \text{or} \quad y + 1 = 0$$
$$y = \frac{3}{2} \qquad y = -1$$
$$\sqrt{x} = \frac{3}{2} \qquad \sqrt{x} = -1$$
$$\left(\sqrt{x}\right)^2 = \left(\frac{3}{2}\right)^2 \qquad \left(\sqrt{x}\right)^2 = (-1)^2$$
$$x = \frac{9}{4} \qquad x = \not{1}$$

1 is extraneous

$$x = \frac{9}{4}$$

37. Let $y = x^{1/3}$.

$$x^{2/3} + 5x^{1/3} + 6 = 0$$
$$\left(x^{1/3}\right)^2 + 5x^{1/3} + 6 = 0$$
$$y^2 + 5y + 6 = 0$$
$$(y+2)(y+3) = 0$$
$$y + 2 = 0 \quad \text{or} \quad y + 3 = 0$$
$$y = -2 \qquad y = -3$$
$$x^{1/3} = -2 \qquad x^{1/3} = -3$$
$$\left(x^{1/3}\right)^3 = (-2)^3 \qquad \left(x^{1/3}\right)^3 = (-3)^3$$
$$x = -8 \qquad x = -27$$

39. Let $y = a^{1/3}$.

$$a^{2/3} - 2a^{1/3} - 3 = 0$$
$$\left(a^{1/3}\right)^2 - 2a^{1/3} - 3 = 0$$
$$y^2 - 2y - 3 = 0$$
$$(y+1)(y-3) = 0$$
$$y + 1 = 0 \quad \text{or} \quad y - 3 = 0$$
$$y = -1 \qquad y = 3$$
$$a^{1/3} = -1 \qquad a^{1/3} = 3$$
$$\left(a^{1/3}\right)^3 = (-1)^3 \qquad \left(a^{1/3}\right)^3 = (3)^3$$
$$a = -1 \qquad a = 27$$

41. Let $y = x^{1/5}$.

$$2x^{2/5} - 5x^{1/5} = -3$$
$$2x^{2/5} - 5x^{1/5} + 3 = 0$$
$$2\left(x^{1/5}\right)^2 - 5x^{1/5} + 3 = 0$$
$$2y^2 - 5y + 3 = 0$$
$$(2y-3)(y-1) = 0$$
$$2y - 3 = 0 \quad \text{or} \quad y - 1 = 0$$
$$y = \frac{3}{2} \qquad y = 1$$
$$x^{1/5} = \frac{3}{2} \qquad x^{1/5} = 1$$
$$\left(x^{1/5}\right)^5 = \left(\frac{3}{2}\right)^5 \qquad \left(x^{1/5}\right)^5 = (1)^5$$
$$x = \frac{243}{32} \qquad x = 1$$

43. Let $y = 2x + 1$.

$$2(2x+1)^2 - 7(2x+1) + 6 = 0$$
$$2y^2 - 7y + 6 = 0$$
$$(2y-3)(y-2) = 0$$
$$2y-3=0 \quad \text{or} \quad y-2=0$$
$$y=\frac{3}{2} \qquad y=2$$
$$2x+1=\frac{3}{2} \qquad 2x+1=2$$
$$2x=\frac{1}{2} \qquad 2x=1$$
$$x=\frac{1}{4} \qquad x=\frac{1}{2}$$

45. Let $y = c + 1$.

$$(c+1)^2 - 4(c+1) + 8 = 0$$
$$y^2 - 4y + 8 = 0$$
$$y=\frac{-b\pm\sqrt{b^2-4ac}}{2a}$$
$$y=\frac{-(-4)\pm\sqrt{(-4)^2-4(1)(8)}}{2(1)}$$
$$y=\frac{4\pm\sqrt{16-32}}{2}$$
$$y=\frac{4\pm\sqrt{-16}}{2}$$
$$y=\frac{4\pm 4i}{2}$$
$$y=\frac{2(2\pm 2i)}{2}$$
$$y=2\pm 2i$$
$$c+1=2\pm 2i$$
$$c+1-1=2\pm 2i-1$$
$$c=1\pm 2i$$

47. Let $y = a^2 - 4$.

$$(a^2-4)^2 - 4(a^2-4) - 32 = 0$$
$$y^2 - 4y - 32 = 0$$
$$(y-8)(y+4) = 0$$
$$y-8=0 \quad \text{or} \quad y+4=0$$
$$y=8 \qquad y=-4$$
$$a^2-4=8 \qquad a^2-4=-4$$
$$a^2=12 \qquad a^2=0$$
$$a=\pm\sqrt{12} \qquad a=\pm\sqrt{0}$$
$$a=\pm 2\sqrt{3} \qquad a=0$$
$$a=2\sqrt{3},\ -2\sqrt{3},\ 0$$

49. Let $y = \dfrac{3m+2}{m}$.

$$9\left(\frac{3m+2}{m}\right)^2 - 30\left(\frac{3m+2}{m}\right) + 25 = 0$$
$$9y^2 - 30y + 25 = 0$$
$$(3y-5)(3y-5) = 0$$
$$3y-5=0 \quad \text{or} \quad 3y-5=0$$
$$y=\frac{5}{3} \qquad y=\frac{5}{3}$$
$$\frac{3m+2}{m}=\frac{5}{3} \qquad \frac{3m+2}{m}=\frac{5}{3}$$
$$3(3m+2)=5m \qquad 3(3m+2)=5m$$
$$9m+6=5m \qquad 9m+6=5m$$
$$6=-4m \qquad 6=-4m$$
$$-\frac{3}{2}=m \qquad -\frac{3}{2}=m$$
$$m=-\frac{3}{2},\ -\frac{3}{2}$$

51. Let $y = 8-\sqrt{a}$.

$$\left(8-\sqrt{a}\right)^2+6\left(8-\sqrt{a}\right)-7=0$$

$$y^2+6y-7=0$$

$$(y+7)(y-1)=0$$

$$y+7=0 \quad \text{or} \quad y-1=0$$

$$y=-7 \qquad y=1$$

$$8-\sqrt{a}=-7 \qquad 8-\sqrt{a}=1$$

$$-\sqrt{a}=-15 \qquad -\sqrt{a}=-7$$

$$\sqrt{a}=15 \qquad \sqrt{a}=7$$

$$\left(\sqrt{a}\right)^2=(15)^2 \quad \left(\sqrt{a}\right)^2=(7)^2$$

$$a=225 \qquad a=49$$

$$a=225,\ 49$$

53. Let $y = \dfrac{1}{x}$.

$$8x^{-2}-10x^{-1}-3=0$$

$$8\left(\frac{1}{x}\right)^2-10\left(\frac{1}{x}\right)-3=0$$

$$8y^2-10y-3=0$$

$$(2y-3)(4y+1)=0$$

$$2y-3=0 \quad \text{or} \quad 4y+1=0$$

$$y=\frac{3}{2} \qquad y=-\frac{1}{4}$$

$$\frac{1}{x}=\frac{3}{2} \qquad \frac{1}{x}=-\frac{1}{4}$$

$$2=3x \qquad 4=-x$$

$$\frac{2}{3}=x \qquad -4=x$$

$$x=\frac{2}{3},\ -4$$

55. Let $y = \dfrac{1}{t+1}$.

$$8(t+1)^{-2}-30(t+1)^{-1}+7=0$$

$$8\left(\frac{1}{t+1}\right)^2-30\left(\frac{1}{t+1}\right)+7=0$$

$$8y^2-30y+7=0$$

$$(4y-1)(2y-7)=0$$

$$4y-1=0 \quad \text{or} \quad 2y-7=0$$

$$y=\frac{1}{4} \qquad y=\frac{7}{2}$$

$$\frac{1}{t+1}=\frac{1}{4} \qquad \frac{1}{t+1}=\frac{7}{2}$$

$$4=t+1 \qquad 7(t+1)=2$$

$$3=t \qquad 7t+7=2$$

$$t=3 \qquad t=-\frac{5}{7}$$

$$t=3,\ -\frac{5}{7}$$

57. Let $y = \dfrac{1}{x^2}$.

$$x^{-4}-2x^{-2}+1=0$$

$$\frac{1}{x^4}-2\left(\frac{1}{x^2}\right)+1=0$$

$$\left(\frac{1}{x^2}\right)^2-2\left(\frac{1}{x^2}\right)+1=0$$

$$y^2-2y+1=0$$

$$(y-1)(y-1)=0$$

$$y-1=0 \quad \text{or} \quad y-1=0$$

$$y=1 \qquad y=1$$

$$\frac{1}{x^2}=\frac{1}{1} \qquad \frac{1}{x^2}=\frac{1}{1}$$

$$1=x^2 \qquad 1=x^2$$

$$x=\pm\sqrt{1} \qquad x=\pm\sqrt{1}$$

$$x=\pm 1 \qquad x=\pm 1$$

$$x=1,\ -1,\ 1,\ -1$$

59.

$$x+\frac{2}{x-2}=0$$
$$(x-2)\left(x+\frac{2}{x-2}\right)=(x-2)0$$
$$x(x-2)+2=0$$
$$x^2-2x+2=0$$
$$x=\frac{-b\pm\sqrt{b^2-4ac}}{2a}$$
$$x=\frac{-(-2)\pm\sqrt{(-2)^2-4(1)(2)}}{2(1)}$$
$$x=\frac{2\pm\sqrt{4-8}}{2}$$
$$x=\frac{2\pm\sqrt{-4}}{2}$$
$$x=\frac{2\pm2i}{2}$$
$$x=\frac{2}{2}\pm\frac{2}{2}i$$
$$x=1\pm i$$

61.

$$x+5+\frac{4}{x}=0$$
$$x\left(x+5+\frac{4}{x}\right)=x(0)$$
$$x^2+5x+4=0$$
$$(x+1)(x+4)=0$$
$$x+1=0 \quad \text{or} \quad x+4=0$$
$$x=-1 \qquad x=-4$$

63.

$$\frac{1}{x+2}+\frac{24}{x+3}=13$$
$$(x+2)(x+3)\left(\frac{1}{x+2}+\frac{24}{x+3}\right)=(x+2)(x+3)(13)$$
$$x+3+24(x+2)=13(x+2)(x+3)$$
$$x+3+24x+48=13(x^2+5x+6)$$
$$25x+51=13x^2+65x+78$$
$$0=13x^2+40x+27$$
$$0=(13x+27)(x+1)$$
$$13x+27=0 \quad \text{or} \quad x+1=0$$
$$x=-\frac{27}{13} \qquad x=-1$$

65.

$$\frac{2}{x-1}+\frac{1}{x+1}=3$$
$$(x-1)(x+1)\left(\frac{2}{x-1}+\frac{1}{x+1}\right)=3(x-1)(x+1)$$
$$2(x+1)+(x-1)=3(x^2-1)$$
$$2x+2+x-1=3x^2-3$$
$$3x+1=3x^2-3$$
$$0=3x^2-3x-4$$
$$x=\frac{-b\pm\sqrt{b^2-4ac}}{2a}$$
$$x=\frac{-(-3)\pm\sqrt{(-3)^2-4(3)(-4)}}{2(3)}$$
$$x=\frac{3\pm\sqrt{9-(-48)}}{6}$$
$$x=\frac{3\pm\sqrt{57}}{6}$$

APPLICATIONS

67. FLOWER ARRANGING

$$r=\frac{4h^2+w^2}{8h}$$
$$18=\frac{4h^2+34^2}{8h}$$
$$18=\frac{4h^2+1{,}156}{8h}$$
$$8h(18)=8h\left(\frac{4h^2+1{,}156}{8h}\right)$$
$$144h=4h^2+1{,}156$$
$$0=4h^2-144h+1{,}156$$
$$\frac{0}{4}=\frac{4h^2}{4}-\frac{144h}{4}+\frac{1{,}156}{4}$$
$$0=h^2-36h+289$$
$$h=\frac{-b\pm\sqrt{b^2-4ac}}{2a}$$
$$h=\frac{-(-36)\pm\sqrt{(-36)^2-4(1)(289)}}{2(1)}$$
$$h=\frac{36\pm\sqrt{1{,}296-1{,}156}}{2}$$
$$h=\frac{36\pm\sqrt{140}}{4}$$
$$h=\frac{36\pm2\sqrt{35}}{4}$$
$$h=\frac{18\pm\sqrt{35}}{2}$$
$$h=\frac{18+\sqrt{35}}{2} \quad\text{or}\quad h=\frac{18-\sqrt{35}}{2}$$
$$h\approx 12.0 \text{ in.} \qquad h\approx \cancel{6.0} \text{ in.}$$

The height is 12.0 inches.

69. SNOWMOBILES

Recall $t=\frac{d}{r}$.

	distance	rate	time
original rate	150	r	$t=\frac{150}{r}$
increased rate	150	$r+20$	$t-2=\frac{150}{r+20}$

Solve the second time for t.

$$t-2=\frac{150}{r+20}$$
$$t=\frac{150}{r+20}+2$$

Set the times equal and solve for r.

$$\frac{150}{r}=\frac{150}{r+20}+2$$
$$r(r+20)\left(\frac{150}{r}\right)=r(r+20)\left(\frac{150}{r+20}+2\right)$$
$$150(r+20)=150(r)+2r(r+20)$$
$$150r+3{,}000=150r+2r^2+40r$$
$$150r+3{,}000-150r=150r+2r^2+40r-150r$$
$$3{,}000=2r^2+40r$$
$$0=2r^2+40r-3{,}000$$
$$\frac{0}{2}=\frac{2r^2}{2}+\frac{40r}{2}-\frac{3{,}000}{2}$$
$$0=r^2+20h-1{,}500$$
$$0=(r+50)(r-30)$$
$$r+50=0 \quad\text{or}\quad r-30=0$$
$$r=\cancel{-50} \qquad r=30$$

Since the rate cannot be negative, her original rate, r, must be 30 mph.

71. CROWD CONTROL

Let x = the number of minutes it takes to clear the grandstand through the west exit.

West exit's work in 1 minute = $\frac{1}{x}$

East exit's work in 1 minute = $\frac{1}{x+4}$

Amount of work together in 1 minute = $\frac{1}{6}$

$$\frac{1}{x}+\frac{1}{x+4}=\frac{1}{6}$$

$$6x(x+4)\left(\frac{1}{x}+\frac{1}{x+4}\right)=6x(x+4)\left(\frac{1}{6}\right)$$

$$6(x+4)+6x=x(x+4)$$

$$6x+24+6x=x^2+4x$$

$$12x+24=x^2+4x$$

$$0=x^2-8x-24$$

$$x=\frac{-b\pm\sqrt{b^2-4ac}}{2a}$$

$$x=\frac{-(-8)\pm\sqrt{(-8)^2-4(1)(-24)}}{2(1)}$$

$$x=\frac{8\pm\sqrt{64-(-96)}}{2}$$

$$x=\frac{8\pm\sqrt{160}}{2}$$

$$x=\frac{8\pm4\sqrt{10}}{2}$$

$$x=\frac{8}{2}\pm\frac{4}{2}\sqrt{10}$$

$$x=4\pm2\sqrt{10}$$

$$x=4+2\sqrt{10} \text{ or } x=4-2\sqrt{10}$$

$$x=10.3 \qquad x=-2.3$$

$$x+4=14.3$$

It takes 14.3 minutes to clear the grandstand using only the east exit.

WRITING

73. Answers will vary.

REVIEW

75. $x=3$

77. $m=\frac{2}{3}$ and $b=(0, 0)$

$$y=mx+b$$

$$y=\frac{2}{3}x+0$$

$$y=\frac{2}{3}x$$

CHALLENGE PROBLEMS

79. Let $y = x^3$.

$$x^6+17x^3+16=0$$

$$(x^3)^2+17x^3+16=0$$

$$y^2+17y+16=0$$

$$(y+1)(y+16)=0$$

$$y+1=0 \text{ or } y+16=0$$

$$y=-1 \qquad y=-16$$

$$x^3=-1 \qquad x^3=-16$$

$$x=\sqrt[3]{-1} \qquad x=\sqrt[3]{-16}$$

$$x=1 \qquad x=\sqrt[3]{-8}\sqrt[3]{2}$$

$$x=1 \qquad x=-2\sqrt[3]{2}$$

SECTION 10.4

VOCABULARY

1. $f(x)=2x^2-4x+1$ is called a **quadratic** function. Its graph is a cup–shaped figure called a **parabola.**

3. The vertical line $x = 1$ divides the parabola into two halves. This line is called the **axis of symmetry**.

CONCEPTS

5. a. a parabola
   b. (1, 0) and (3, 0)
   c. (0, –3)
   d. (2, 1)
   e. $x = 2$

7.

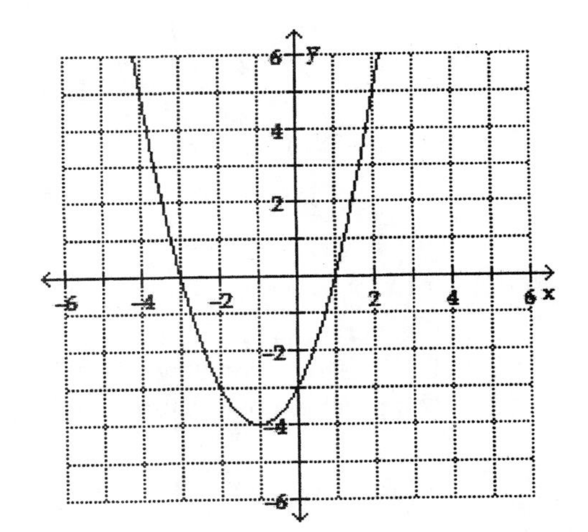

9. a. $f(x)=\boxed{2}\left(x^2+6x\right)+11$
   b. $f(x)=\boxed{2}\left(x^2+6x+\boxed{9}\right)+11-\boxed{18}$

11. The x–intercepts are (–3, 0) and (5, 0), so the solutions of the function are $x = -3$ and $x = 5$.

NOTATION

13. $h = -1$
    $f(x)=2[x-(-1)]^2+6$

PRACTICE

15.

x	$f(x) = x^2$
–3	$(-3)^2 = 9$
–2	$(-2)^2 = 4$
–1	$(-1)^2 = 1$
0	$(0)^2 = 0$
1	$(1)^2 = 1$
2	$(2)^2 = 4$
3	$(3)^2 = 9$

x	$g(x) = 2x^2$
–3	$2(-3)^2 = 18$
–2	$2(-2)^2 = 8$
–1	$2(-1)^2 = 2$
0	$2(0)^2 = 0$
1	$2(1)^2 = 2$
2	$2(2)^2 = 8$
3	$2(3)^2 = 18$

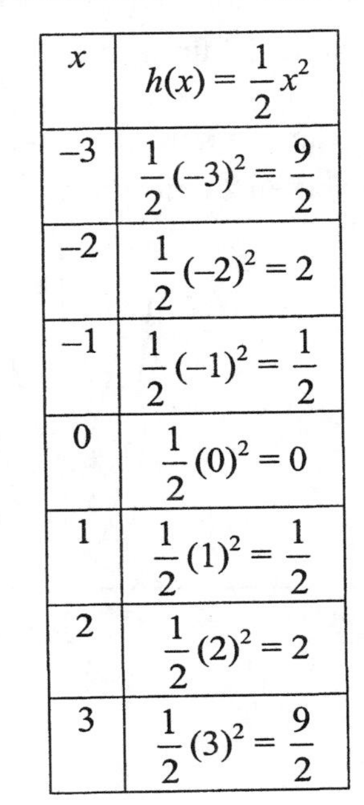

x	$h(x) = \frac{1}{2}x^2$
–3	$\frac{1}{2}(-3)^2 = \frac{9}{2}$
–2	$\frac{1}{2}(-2)^2 = 2$
–1	$\frac{1}{2}(-1)^2 = \frac{1}{2}$
0	$\frac{1}{2}(0)^2 = 0$
1	$\frac{1}{2}(1)^2 = \frac{1}{2}$
2	$\frac{1}{2}(2)^2 = 2$
3	$\frac{1}{2}(3)^2 = \frac{9}{2}$

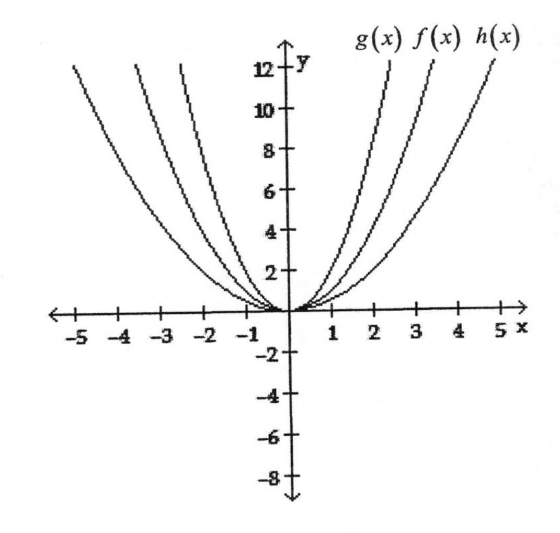

17.

x	$f(x) = 4x^2$
–3	$4(-3)^2 = 36$
–2	$4(-2)^2 = 16$
–1	$4(-1)^2 = 4$
0	$4(0)^2 = 0$
1	$4(1)^2 = 4$
2	$4(2)^2 = 16$
3	$4(3)^2 = 36$

The graph of $g(x) = 4x^2 + 3$ would be shifted 3 units up.

The graph of $h(x) = 4x^2 - 2$ would be shifted 2 units down.

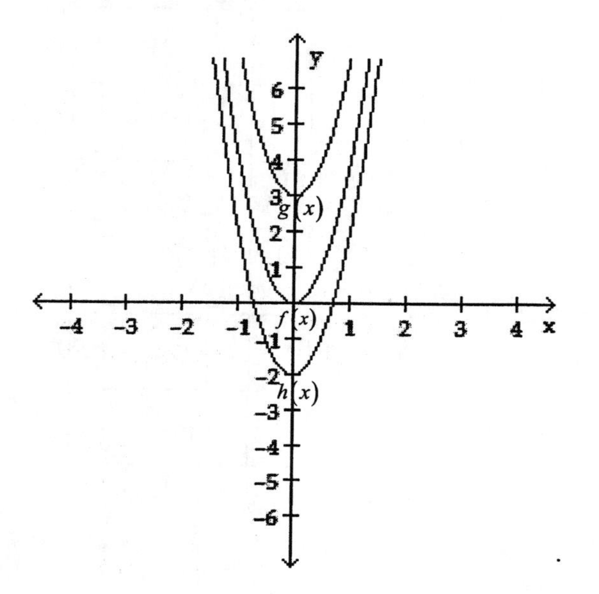

19.

x	$f(x) = -3x^2$
–3	$-3(-3)^2 = -27$
–2	$-3(-2)^2 = -12$
–1	$-3(-1)^2 = -3$
0	$-3(0)^2 = 0$
1	$-3(1)^2 = -3$
2	$-3(2)^2 = -12$
3	$-3(3)^2 = -27$

The graph of $g(x) = 3(x-2)^2 - 1$ would be shifted 2 units right and 1 unit down.

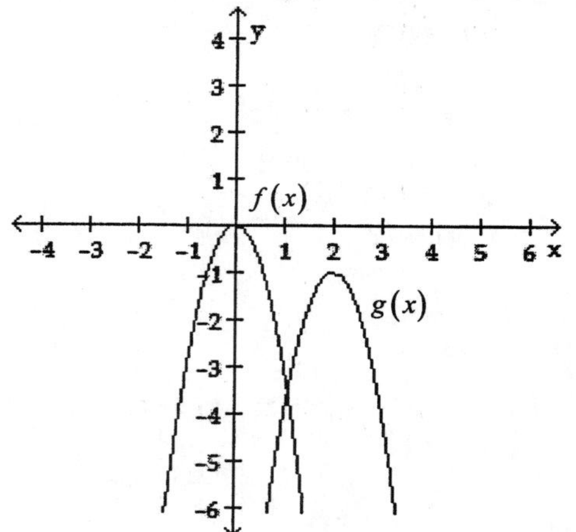

21. $f(x) = (x-1)^2 + 2$
The vertex is (1, 2).
The axis of symmetry is $x = 1$.
The parabola opens upward.

23. $f(x) = 2(x+3)^2 - 4$
The vertex is (–3, –4).
The axis of symmetry is $x = -3$.
The parabola opens upward.

25. $f(x) = 0.5(x-7.5)^2 + 8.5$
The vertex is (7.5, 8.5).
The axis of symmetry is $x = 7.5$.
The parabola opens upward.

27. $f(x) = 2x^2 - 4x$

$$f(x) = 2(x^2 - 2x)$$

$$f(x) = 2(x^2 - 2x + 1) - 2(1)$$

$$f(x) = 2(x-1)^2 - 2$$

The vertex is (1, –2).
The axis of symmetry is $x = 1$.
The parabola opens upward.

29. $f(x) = -4x^2 + 16x + 5$

$f(x) = -4(x^2 - 4x) + 5$

$f(x) = -4(x^2 - 4x + 4) + 5 - (-4)(4)$

$f(x) = -4(x-2)^2 + 5 + 16$

$f(x) = -4(x-2)^2 + 21$

The vertex is (2, 21).
The axis of symmetry is $x = 2$.
The parabola opens downward.

31. $f(x) = 3x^2 + 4x + 2$

$f(x) = 3\left(x^2 + \frac{4}{3}x\right) + 2$

$f(x) = 3\left(x^2 + \frac{4}{3}x + \frac{4}{9}\right) + 2 - 3\left(\frac{4}{9}\right)$

$f(x) = 3\left(x + \frac{2}{3}\right)^2 + 2 - \frac{4}{3}$

$f(x) = 3\left(x + \frac{2}{3}\right)^2 + \frac{2}{3}$

The vertex is $\left(-\frac{2}{3}, \frac{2}{3}\right)$.

The axis of symmetry is $x = -\frac{2}{3}$.

The parabola opens upward.

33. $f(x) = (x-3)^2 + 2$

The vertex is (3, 2).
The axis of symmetry is $x = 3$.
The parabola opens upward.

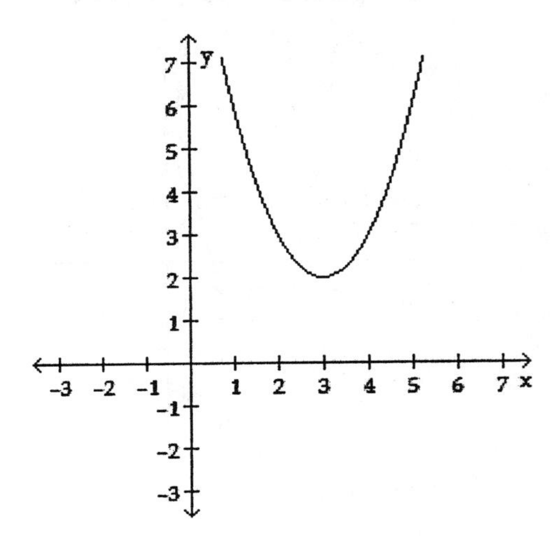

35. $f(x) = -(x-2)^2$

$f(x) = -(x-2)^2 + 0$

The vertex is (2, 0).
The axis of symmetry is $x = 2$.
The parabola opens downward.

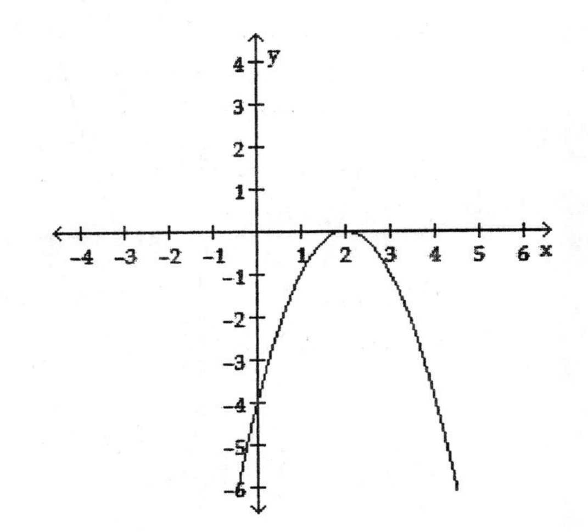

37. $f(x) = -2(x+3)^2 + 4$

The vertex is (–3, 4).
The axis of symmetry is $x = -3$.
The parabola opens downward.

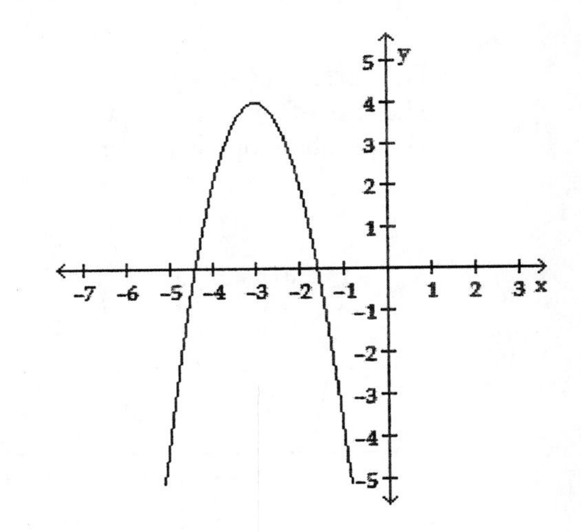

39. $f(x)=\frac{1}{2}(x+1)^2-3$

The vertex is (–1, –3).
The axis of symmetry is $x=-1$.
The parabola opens upward.

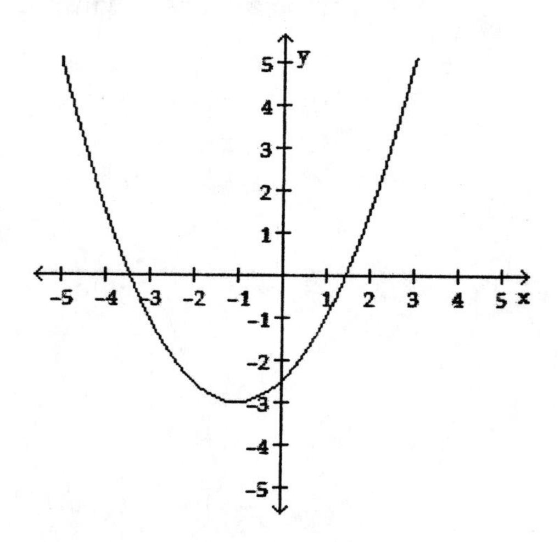

41. $f(x)=x^2+2x-3$

$f(x)=\left(x^2+2x\right)-3$

$f(x)=\left(x^2+2x+1\right)-3-1$

$f(x)=(x+1)^2-4$

The vertex is (–1, –4).
The axis of symmetry is $x=-1$.
The parabola opens upward.

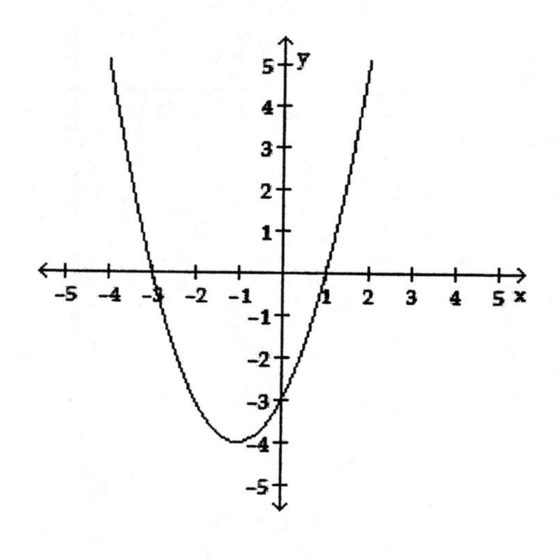

43. $f(x)=3x^2-12x+10$

$f(x)=3\left(x^2-4x\right)+10$

$f(x)=3\left(x^2-4x+4\right)+10-3(4)$

$f(x)=3(x-2)^2+10-12$

$f(x)=3(x-2)^2-2$

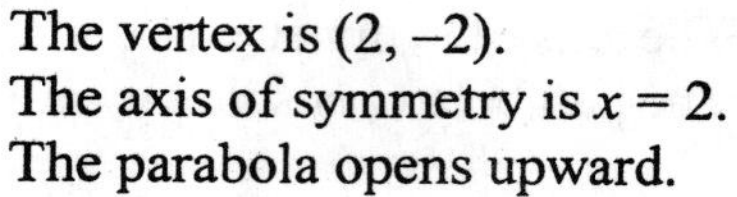

The vertex is (2, –2).
The axis of symmetry is $x=2$.
The parabola opens upward.

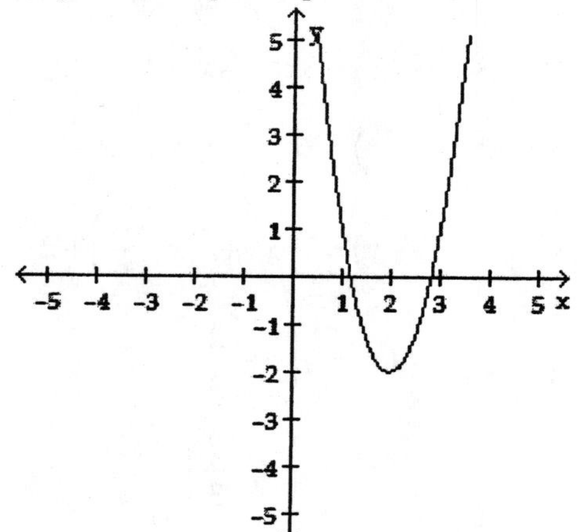

45. $f(x)=2x^2+8x+6$

$f(x)=2\left(x^2+4x\right)+6$

$f(x)=2\left(x^2+4x+4\right)+6-2(4)$

$f(x)=2(x+2)^2+6-8$

$f(x)=2(x+2)^2-2$

The vertex is (–2, –2).
The axis of symmetry is $x=-2$.
The parabola opens upward.

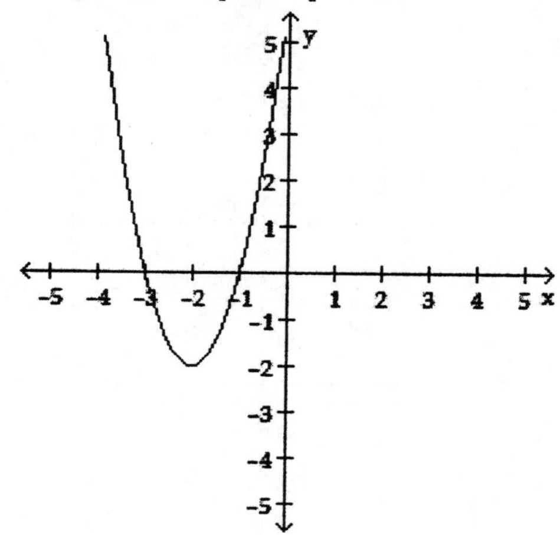

47. $f(x)=x^2+x-6$

$f(x)=(x^2+x)-6$

$f(x)=\left(x^2+x+\frac{1}{4}\right)-6-\frac{1}{4}$

$f(x)=\left(x+\frac{1}{2}\right)^2-\frac{25}{4}$

The vertex is $\left(-\frac{1}{2}, -\frac{25}{4}\right)$.

The axis of symmetry is $x = -\frac{1}{2}$.

The parabola opens upward.

49. $f(x)=-x^2-8x-17$

$f(x)=-(x^2+8x)-17$

$f(x)=-(x^2+8x+16)-17-(-1)(16)$

$f(x)=-(x+4)^2-17+16$

$f(x)=-(x+4)^2-1$

The vertex is (–4, –1).
The axis of symmetry is $x = -4$.
The parabola opens downward.

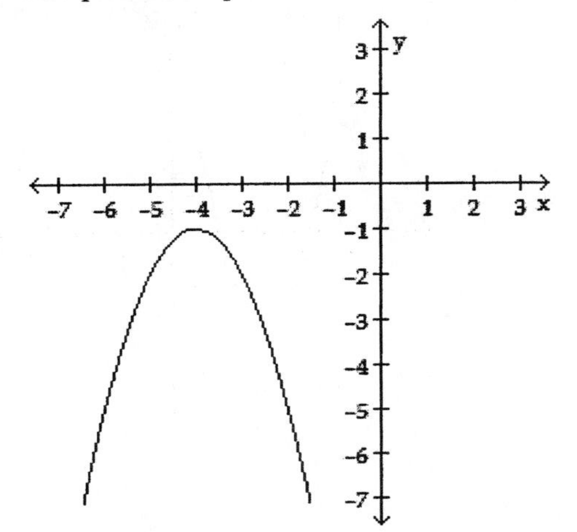

51. $f(x)=-4x^2+16x-10$

$f(x)=-4(x^2-4x)-10$

$f(x)=-4(x^2-4x+4)-10-(-4)(4)$

$f(x)=-4(x-2)^2-10+16$

$f(x)=-4(x-2)^2+6$

The vertex is (2, 6).
The axis of symmetry is $x = 2$.
The parabola opens downward.

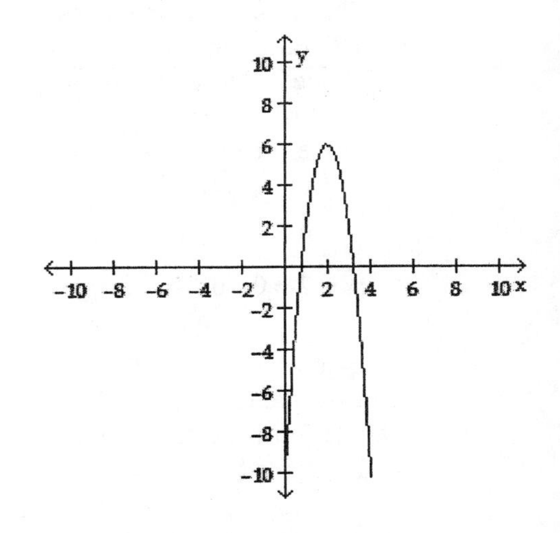

53. x–intercept: Let $f(x) = 0$.

$0=x^2-2x+1$

$0=(x-1)(x-1)$

$x-1=0$ or $x-1=0$

$x=1$ $\qquad x=1$

x–intercept is (1, 0)

y–intercept = (0, c)

$c = 1$

y–intercept is (0, 1)

55. x–intercept: Let $f(x) = 0$.

$0=-x^2-10x-21$

$x^2+10x+21=0$

$(x+7)(x+3)=0$

$x+3=0$ or $x+7=0$

$x=-3$ $\qquad x=-7$

x–intercepts are (–3, 0) and (–7, 0)

y–intercept = (0, c)

$c = -21$

y–intercept is (0, –21)

57. $f(x)=x^2+4x+4$

Step 1: Since $a=1>0$, the parabola opens upward.

Step 2: Find the vertex and axis of symmetry.

$$x=\frac{-b}{2a}=\frac{-4}{2(1)}=\frac{-4}{2}=-2$$

$$y=(-2)^2+4(-2)+4=4-8+4=0$$

vertex: (–2, 0)

The axis of symmetry is $x=-2$.

Step 3: Find the x– and y–intercepts. Since $c=4$, the y–intercept is (0, 4). To find the x–intercepts, let $y=0$ and solve the equation for x.

$$0=x^2+4x+4$$

$$0=(x+2)(x+2)$$

$$x+2=0 \quad \text{or} \quad x+2=0$$

$$x=-2 \qquad\qquad x=-2$$

x–intercept: (–2, 0)

y–intercept: (0, 4)

Step 4: Find another point(s).

Let $x=-1$.

$$y=(-1)^2+4(-1)+4=1-4+4=1$$

(–1, 1)

Use symmetry to find the points (–3, 1) and (–4, 4).

Step 5: Plot the points and draw the parabola.

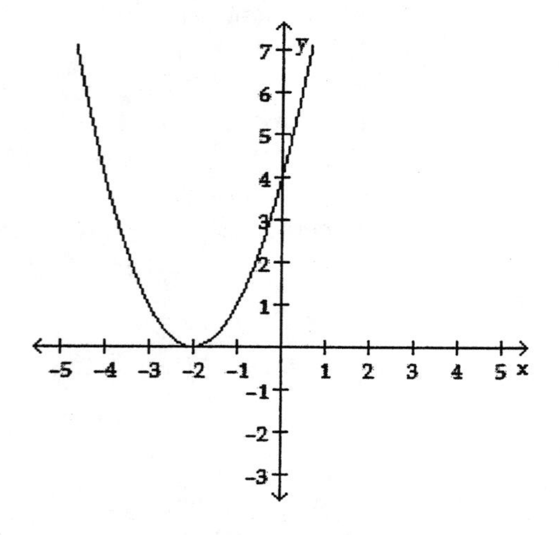

59. $f(x)=-x^2+2x-1$

Step 1: Since $a=-1<0$, the parabola opens downward.

Step 2: Find the vertex and axis of symmetry.

$$x=\frac{-b}{2a}=\frac{-2}{2(-1)}=\frac{-2}{-2}=1$$

$$y=-(1)^2+2(1)-1=-1+2-1=0$$

vertex: (1, 0)

The axis of symmetry is $x=1$.

Step 3: Find the x– and y–intercepts. Since $c=-1$, the y–intercept is (0, –1). To find the x–intercepts, let $y=0$ and solve the equation for x.

$$0=-x^2+2x-1$$

$$x^2-2x+1=0$$

$$(x-1)(x-1)=0$$

$$x-1=0 \quad \text{or} \quad x-1=0$$

$$x=1 \qquad\qquad x=1$$

x–intercept: (1, 0)

y–intercept: (0, –1)

Step 4: Find another point(s).

Let $x=-1$.

$$y=-(-1)^2+2(-1)-1=-1-2-1=-4$$

(–1, –4)

Use symmetry to find the points (2, –1) and (3, –4).

Step 5: Plot the points and draw the parabola.

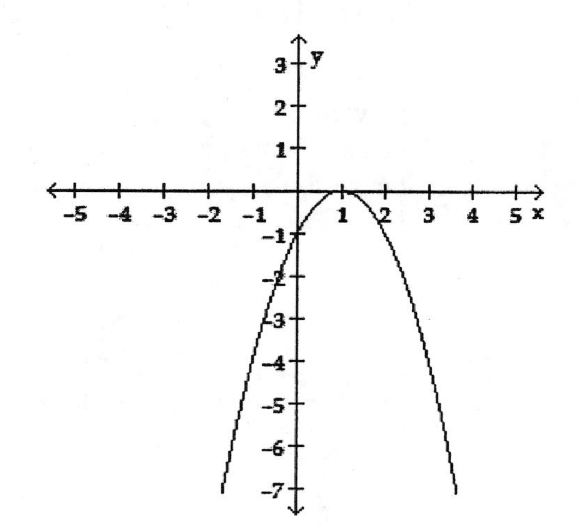

61. $f(x) = x^2 - 2x$

Step 1: Since $a = 1 > 0$, the parabola opens upward.

Step 2: Find the vertex and axis of symmetry.

$$x = \frac{-b}{2a} = \frac{-(-2)}{2(1)} = \frac{2}{2} = 1$$

$$y = (1)^2 - 2(1) = 1 - 2 = -1$$

vertex: $(1, -1)$

The axis of symmetry is $x = 1$.

Step 3: Find the x– and y–intercepts.

Since $c = 0$, the y–intercept is $(0, 0)$.

To find the x–intercepts, let $y = 0$ and solve the equation for x.

$$0 = x^2 - 2x$$

$$0 = x(x - 2)$$

$$x = 0 \quad \text{or} \quad x - 2 = 0$$

$$x = 0 \qquad\qquad x = 2$$

x–intercepts: $(0, 0)$ and $(2, 0)$

y–intercept: $(0, 0)$

Step 4: Find another point(s).

Let $x = -1$.

$$y = (-1)^2 - 2(-1) = 1 + 2 = 3$$

$(-1, 3)$

Use symmetry to find the point $(3, 3)$.

Step 5: Plot the points and draw the parabola.

63. $f(x) = 4x^2 - 12x + 9$

Step 1: Since $a = 4 > 0$, the parabola opens upward.

Step 2: Find the vertex and axis of symmetry.

$$x = \frac{-b}{2a} = \frac{-(-12)}{2(4)} = \frac{12}{8} = \frac{3}{2}$$

$$y = 4\left(\frac{3}{2}\right)^2 - 12\left(\frac{3}{2}\right) + 9$$

$$= 4\left(\frac{9}{4}\right) - 12\left(\frac{3}{2}\right) + 9$$

$$= 9 - 18 + 9$$

$$= 0$$

vertex: $\left(\frac{3}{2}, 0\right)$

The axis of symmetry is $x = \frac{3}{2}$.

Step 3: Find the x– and y–intercepts.

Since $c = 9$, the y–intercept is $(0, 9)$.

To find the x–intercepts, let $y = 0$ and solve the equation for x.

$$0 = 4x^2 - 12x + 9$$

$$0 = (2x - 3)(2x - 3)$$

$$2x - 3 = 0 \quad \text{or} \quad 2x - 3 = 0$$

$$2x = 3 \qquad\qquad 2x = 3$$

$$x = \frac{3}{2} \qquad\qquad x = \frac{3}{2}$$

x–intercept: $\left(\frac{3}{2}, 0\right)$

y–intercept: $(0, 9)$

Step 4: Find another point(s).

Let $x = 1$.

$$y = 4(1)^2 - 12(1) + 9 = 4 - 12 + 9 = 1$$

$(1, 1)$

Use symmetry to find the points $(2, 1)$ and $(3, 9)$.

Step 5: Plot the points and draw the parabola.

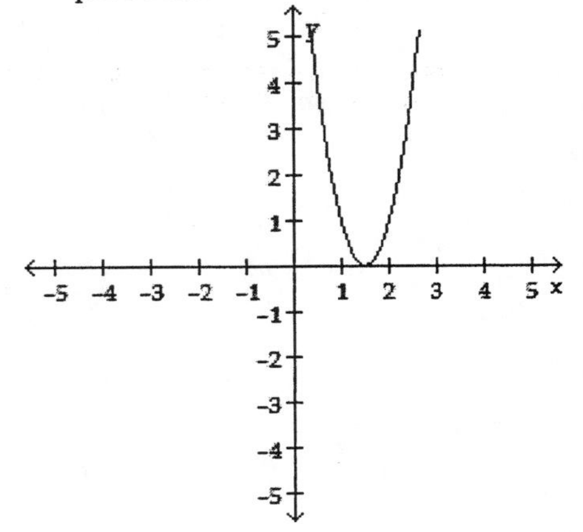

65. $f(x) = 2x^2 - 8x + 6$

Step 1: Since $a = 2 > 0$, the parabola opens upward.

Step 2: Find the vertex and axis of symmetry.

$$x = \frac{-b}{2a} = \frac{-(-8)}{2(2)} = \frac{8}{4} = 2$$

$$y = 2(2)^2 - 8(2) + 6 = 8 - 16 + 6 = -2$$

vertex: (2, –2)

The axis of symmetry is $x = 2$.

Step 3: Find the x– and y–intercepts.

Since $c = 6$, the y–intercept is (0, 6).

To find the x–intercepts, let $y = 0$ and solve the equation for x.

$$0 = 2x^2 - 8x + 6$$

$$0 = 2(x^2 - 4x + 3)$$

$$0 = (x-1)(x-3)$$

$$x - 1 = 0 \quad \text{or} \quad x - 3 = 0$$

$$x = 1 \qquad\qquad x = 3$$

x–intercepts: (1, 0) and (3, 0)

y–intercept: (0, 6)

Step 4: Use symmetry to find the point (4, 6)

Step 5: Plot the points and draw the parabola.

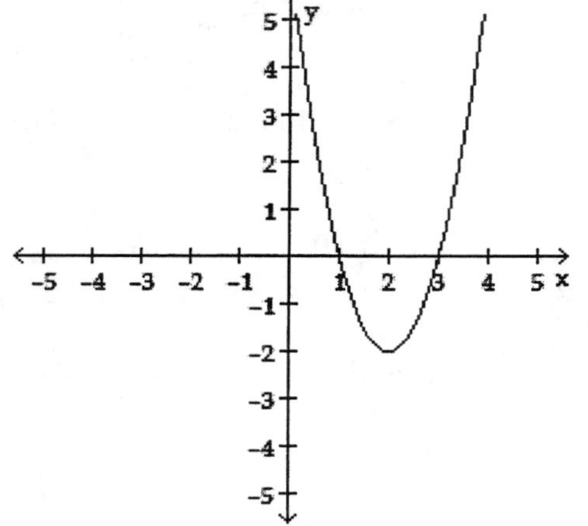

67. $f(x) = -6x^2 - 12x - 8$

Step 1: Since $a = -6 < 0$, the parabola opens downward.

Step 2: Find the vertex and axis of symmetry.

$$x = \frac{-b}{2a} = \frac{-(-12)}{2(-6)} = \frac{12}{-12} = -1$$

$$y = -6(-1)^2 - 12(-1) - 8$$
$$= -6 + 12 - 8$$
$$= -2$$

vertex: (–1, –2)

The axis of symmetry is $x = -1$.

Step 3: Find the x– and y–intercepts.

Since $c = -8$, the y–intercept is (0, –8).

Since the vertex is located below the x–axis and the parabola opens downward, there are no x–intercepts.

x–intercepts: none

y–intercept: (0, –8)

Step 4: Find another point(s).

Let $x = 1$.

$$y = -6(1)^2 - 12(1) - 8$$
$$= -6 - 12 - 8$$
$$= -26$$

(1, –26)

Use symmetry to find the points (–2, –8) and (–3, –26)

Step 5: Plot the points and draw the parabola.

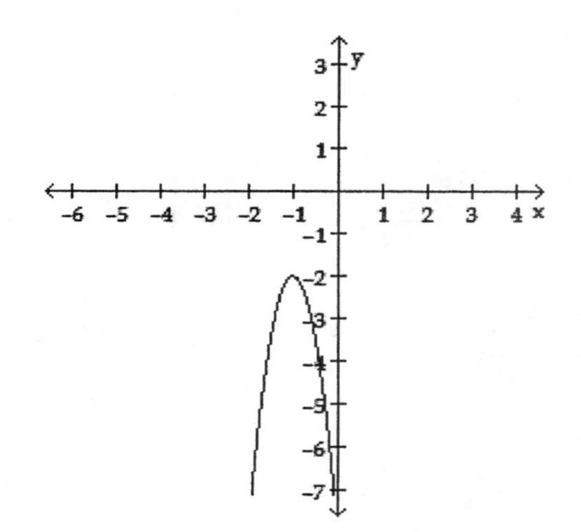

69. $f(x)=2x^2-x+1$
vertex: (0.25, 0.88)

71. $f(x)=7+x-x^2$
vertex: (0.50, 7.25)

73. $x^2+x-6=0$
The x–intercepts are (–3, 0) and (2, 0), so the solutions of the equation are $x=-3$ and $x=2$.

75. $0.5x^2-0.7x-3=0$
The x–intercepts are (–1.85, 0) and (3.25, 0), so the solutions of the equation are $x=-1.85$ and $x=3.25$.

APPLICATIONS

77. CROSSWORD PUZZLES

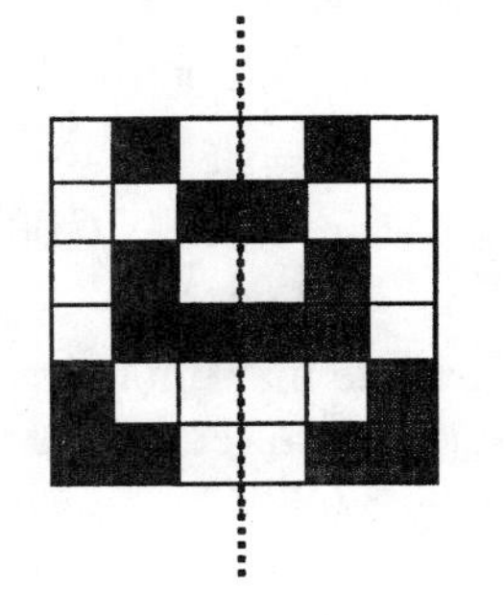

79. FIREWORKS
The vertex of the parabola would be its maximum height and time of explosion.

$$s=120t-16t^2$$
$$s=-16t^2+120t$$
$$t=\frac{-b}{2a}$$
$$=\frac{-120}{2(-16)}$$
$$=\frac{-120}{-32}$$
$$=3.75$$

$$s(3.75)=120(3.75)-16(3.75)^2$$
$$=450-16(14.0625)$$
$$=450-225$$
$$=225$$

The shell reaches its maximum height of 225 ft in 3.75 seconds.

81. FENCING A FIELD
Let the width = x and the length = y. The sum of the three sides is 1,000 ft.

$$x+y+x=1{,}000$$
$$2x+y=1{,}000$$
$$y=-2x+1{,}000$$

To find a function for the area, multiply length times width.

$$\text{Area} = \text{width} \bullet \text{length}$$
$$=x(y)$$
$$=x(-2x+1{,}000)$$
$$A(x)=-2x^2+1{,}000x$$

Use the vertex formula to find the width (x) and substitution to find the length (y).

$$A(x)=-2x^2+1{,}000x$$
$$x=\frac{-b}{2a}$$
$$=\frac{-1{,}000}{2(-2)}$$
$$=\frac{-1{,}000}{-4}$$
$$=250 \text{ ft}$$
$$y=-2x+1{,}000$$
$$=-2(250)+1{,}000$$
$$=-500+1{,}000$$
$$=500 \text{ ft}$$

The dimensions of the field are 250 ft by 500 ft.

83. OPERATING COSTS

The vertex of the parabola represented by the function would be its minimum cost.

$$C(n)=2.2n^2-66n+655$$

$$\begin{aligned} n&=\frac{-b}{2a}\\ &=\frac{-(-66)}{2(2.2)}\\ &=\frac{66}{4.4}\\ &=15 \text{ minutes} \end{aligned}$$

$$\begin{aligned} C(15)&=2.2(15)^2-66(15)+655\\ &=2.2(225)-66(15)+655\\ &=495-990+655\\ &=\$160 \end{aligned}$$

The cost of running the machine is a minimum at 15 minutes. The minimum cost is $160.

85. U.S. ARMY

The vertex of the parabola represented by the function would be its maximum strength.

$$N(x)=-0.0534x^2+0.337x+0.969$$

$$\begin{aligned} x&=\frac{-b}{2a}\\ &=\frac{-0.337}{2(-0.0534)}\\ &=\frac{-0.337}{-0.1068}\\ &=3.16 \end{aligned}$$

$$\begin{aligned} N(3.16)&=-0.0534(3.16)^2+0.337(3.16)+0.969\\ &=-0.0534(9.9856)+0.337(3.16)+0.969\\ &=-0.533+1.065+0.969\\ &=1.5 \end{aligned}$$

It maximum strength would be in the 3rd year, which represents 1968. During that year, the army's personnel strength level was 1.5 million. At this time in history, the U.S. involvement in the Vietnam War was at its peak.

87. MAXIMIZING REVENUE

The vertex of the parabola represented by the function would be its maximum revenue.

$$R(x)=-\frac{x^2}{5}+80x-1{,}000$$

$$R(x)=-\frac{1}{5}x^2+80x-1{,}000$$

$$R(x)=-0.2x^2+80x-1{,}000$$

$$\begin{aligned} x&=\frac{-b}{2a}\\ &=\frac{-80}{2(-0.2)}\\ &=\frac{-80}{-0.4}\\ &=200 \end{aligned}$$

$$\begin{aligned} R(200)&=-\frac{(200)^2}{5}+80x-1{,}000\\ &=-\frac{40{,}000}{5}+80(200)-1{,}000\\ &=-8{,}000+16{,}000-1{,}000\\ &=7{,}000 \end{aligned}$$

To obtain the maximum revenue, they must sell 200 stereos. The maximum revenue is $7,000.

WRITING

89. Answers will vary.

91. Answers will vary.

93. Answers will vary.

REVIEW

95.

$$\begin{aligned} \sqrt{8a}\sqrt{2a^3b}&=\sqrt{16a^4b}\\ &=\sqrt{16a^4}\sqrt{b}\\ &=4a^2\sqrt{b} \end{aligned}$$

97.

$$\frac{\sqrt{3}}{\sqrt{50}}=\frac{\sqrt{3}}{\sqrt{25}\sqrt{2}}$$
$$=\frac{\sqrt{3}}{5\sqrt{2}}$$
$$=\frac{\sqrt{3}}{5\sqrt{2}}\bullet\frac{\sqrt{2}}{\sqrt{2}}$$
$$=\frac{\sqrt{6}}{5\sqrt{4}}$$
$$=\frac{\sqrt{6}}{5(2)}$$
$$=\frac{\sqrt{6}}{10}$$

99.

$$3\left(\sqrt{5b}-\sqrt{3}\right)^2=3\left(\sqrt{5b}-\sqrt{3}\right)\left(\sqrt{5b}-\sqrt{3}\right)$$
$$=3\left(\sqrt{25b^2}-\sqrt{15b}-\sqrt{15b}+\sqrt{9}\right)$$
$$=3\left(5b-2\sqrt{15b}+3\right)$$
$$=15b-6\sqrt{15b}+9$$

CHALLENGE PROBLEMS

101.

$$f(x)=ax^2+bx+c$$
$$f(x)=a\left(x^2+\frac{b}{a}x\right)+c$$
$$f(x)=a\left[x^2+\frac{b}{a}x+\left(\frac{b}{2a}\right)^2\right]+c-a\left(\frac{b}{2a}\right)^2$$
$$f(x)=a\left[x^2+\frac{b}{a}x+\frac{b^2}{4a^2}\right]+c-a\left(\frac{b^2}{4a^2}\right)$$
$$f(x)=a\left(x+\frac{b}{2a}\right)^2+c-\frac{b^2}{4a}$$
$$f(x)=a\left(x+\frac{b}{2a}\right)^2+\frac{4ac}{4a}-\frac{b^2}{4a}$$
$$f(x)=a\left(x+\frac{b}{2a}\right)^2+\frac{4ac-b^2}{4a}$$
$$\text{vertex}=\left(-\frac{b}{2a},\frac{4ac-b^2}{4a}\right)$$

SECTION 10.5

VOCABULARY

1. $x^2+3x-18<0$ is an example of a **quadratic** inequality in one variable.

3. $y \le x^2-4x+3$ is an example of a nonlinear inequality in **two** variables.

CONCEPTS

5. $(-\infty, -1)$, $(1, 4)$, $(4, \infty)$

7. a. 

   b. 

9. a. $(-3, 2)$
   b. $(-\infty,-1]\cup[1,\infty)$

11. a. solid
    b. yes

$$y \le x^2+2x+1$$
$$0 \overset{?}{\le} 0^2+2(0)+1$$
$$0 \overset{?}{\le} 0+0+1$$
$$0 \le 1$$

NOTATION

13. $x^2-6x-7\ge 0$

PRACTICE

15.

$$x^2-5x+4<0$$
$$x^2-5x+4=0$$
$$(x-1)(x-4)=0$$
$$x-1=0 \quad \text{or} \quad x-4=0$$
$$x=1 \qquad\qquad x=4$$
$$(1,4)$$

17.

$$x^2-8x+15>0$$
$$x^2-8x+15=0$$
$$(x-5)(x-3)=0$$
$$x-5=0 \quad \text{or} \quad x-3=0$$
$$x=5 \qquad\qquad x=3$$
$$(-\infty,3)\cup(5,\infty)$$

19.

$$x^2+x-12\le 0$$
$$x^2+x-12=0$$
$$(x-3)(x+4)=0$$
$$x-3=0 \quad \text{or} \quad x+4=0$$
$$x=3 \qquad\qquad x=-4$$
$$[-4,3]$$

21.

$$x^2+8x<-16$$
$$x^2+8x+16<0$$
$$x^2+8x+16=0$$
$$(x+4)(x+4)=0$$
$$x+4=0 \quad \text{or} \quad x+4=0$$
$$x=-4 \qquad\qquad x=-4$$
$$\varnothing$$

No Solution

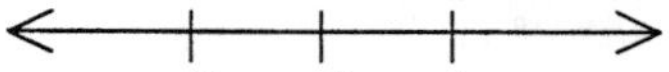

23.

$$x^2\ge 9$$
$$x^2-9\ge 0$$
$$x^2-9=0$$
$$(x+3)(x-3)=0$$
$$x+3=0 \quad \text{or} \quad x-3=0$$
$$x=-3 \qquad\qquad x=3$$
$$(-\infty,-3]\cup[3,\infty)$$

25.

$$2x^2 - 50 < 0$$
$$2(x^2 - 25) = 0$$
$$2(x+5)(x-5) = 0$$
$$x + 5 = 0 \quad \text{or} \quad x - 5 = 0$$
$$x = -5 \qquad\qquad x = 5$$
$$(-5, 5)$$

–5 5

27.

$$\frac{1}{x} < 2$$
$$\frac{1}{x} = 2$$
$$\frac{1}{x} - 2 = 0$$
$$\frac{1}{x} - \frac{2x}{x} = 0$$
$$\frac{1-2x}{x} = 0$$
$$x\left(\frac{1-2x}{x}\right) = x(0)$$
$$1 - 2x = 0$$
$$-2x = -1$$
$$\frac{1}{2} = x$$

Set the denominator $= 0$.

$$x = 0$$

Critical numbers $= 0$ and $\frac{1}{2}$

$$(-\infty, 0) \cup \left(\frac{1}{2}, \infty\right)$$

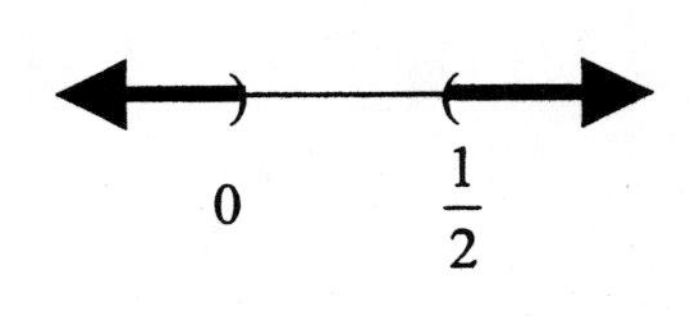

29.

$$-\frac{5}{x} < 3$$
$$-\frac{5}{x} = 3$$
$$-\frac{5}{x} - 3 = 0$$
$$-\frac{5}{x} - \frac{3x}{x} = 0$$
$$\frac{-5-3x}{x} = 0$$
$$x\left(\frac{-5-3x}{x}\right) = x(0)$$
$$-5 - 3x = 0$$
$$-3x = 5$$
$$x = -\frac{5}{3}$$

Set the denominator $= 0$.

$$x = 0$$

Critical numbers $= 0$ and $-\frac{5}{3}$

$$\left(-\infty, -\frac{5}{3}\right) \cup (0, \infty)$$

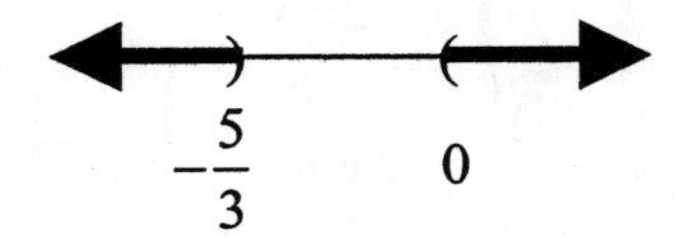

31.

$$\frac{x^2-x-12}{x-1}<0$$

$$\frac{x^2-x-12}{x-1}=0$$

$$(x-1)\left(\frac{x^2-x-12}{x-1}\right)=(x-1)(0)$$

$$x^2-x-12=0$$

$$(x-4)(x+3)=0$$

$$x-4=0 \quad \text{or} \quad x+3=0$$

$$x=4 \qquad\qquad x=-3$$

Set the denominator $=0$.

$$x-1=0$$

$$x=1$$

Critical numbers $=-3,\ 1,$ and 4

$$(-\infty,-3)\cup(1,4)$$

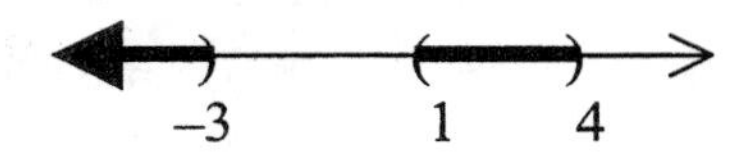

33.

$$\frac{6x^2-5x+1}{2x+1}>0$$

$$\frac{6x^2-5x+1}{2x+1}=0$$

$$(2x+1)\left(\frac{6x^2-5x+1}{2x+1}\right)=(2x+1)(0)$$

$$6x^2-5x+1=0$$

$$(2x-1)(3x-1)=0$$

$$2x-1=0 \quad \text{or} \quad 3x-1=0$$

$$x=\frac{1}{2} \qquad\qquad x=\frac{1}{3}$$

Set the denominator $=0$.

$$2x+1=0$$

$$x=-\frac{1}{2}$$

Critical numbers $=-\frac{1}{2},\ \frac{1}{3},$ and $\frac{1}{2}$

$$\left(-\frac{1}{2},\frac{1}{3}\right)\cup\left(\frac{1}{2},\infty\right)$$

$-\frac{1}{2}$ $\quad$ $\frac{1}{3}$ $\quad$ $\frac{1}{2}$

35.

$$\frac{3}{x-2}<\frac{4}{x}$$

$$\frac{3}{x-2}-\frac{4}{x}<0$$

$$\frac{3}{x-2}-\frac{4}{x}=0$$

$$\frac{3}{x-2}\cdot\frac{x}{x}-\frac{4}{x}\cdot\frac{x-2}{x-2}=0$$

$$\frac{3x-4(x-2)}{x(x-2)}=0$$

$$\frac{3x-4x+8}{x(x-2)}=0$$

$$\frac{-x+8}{x(x-2)}=0$$

$$x(x-2)\left(\frac{-x+8}{x(x-2)}\right)=x(x-2)(0)$$

$$-x+8=0$$

$$8=x$$

Set the denominator $=0$.

$$x(x-2)=0$$

$$x=0 \quad \text{or} \quad x-2=0$$

$$x=0 \qquad\qquad x=2$$

Critical numbers $=0,\ 2,$ and 8

$$(0,2)\cup(8,\infty)$$

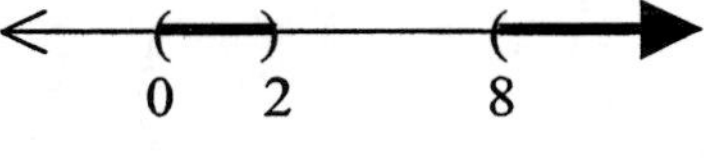

37.

$$\frac{7}{x-3} \geq \frac{2}{x+4}$$

$$\frac{7}{x-3} - \frac{2}{x+4} \geq 0$$

$$\frac{7}{x-3} - \frac{2}{x+4} = 0$$

$$\frac{7}{x-3} \bullet \frac{x+4}{x+4} - \frac{2}{x+4} \bullet \frac{x-3}{x-3} = 0$$

$$\frac{7(x+4)-2(x-3)}{(x-3)(x+4)} = 0$$

$$\frac{7x+28-2x+6}{(x-3)(x+4)} = 0$$

$$\frac{5x+34}{(x-3)(x+4)} = 0$$

$$(x-3)(x+4)\left(\frac{5x+34}{(x-3)(x+4)}\right) = (x-3)(x+4)(0)$$

$$5x+34 = 0$$

$$5x = -34$$

$$x = -\frac{34}{5}$$

Set the denominator $= 0$.

$$(x-3)(x+4) = 0$$

$$x-3=0 \quad \text{or} \quad x+4=0$$

$$x=3 \qquad\qquad x=-4$$

Critical numbers $= -\frac{34}{5}, -4,$ and 3

$$\left[-\frac{34}{5}, -4\right) \cup (3, \infty)$$

$-\frac{34}{5}$ -4 3

39.

$$\frac{x}{x+4} \leq \frac{1}{x+1}$$

$$\frac{x}{x+4} - \frac{1}{x+1} < 0$$

$$\frac{x}{x+4} - \frac{1}{x+1} = 0$$

$$\frac{x}{x+4} \bullet \frac{x+1}{x+1} - \frac{1}{x+1} \bullet \frac{x+4}{x+4} = 0$$

$$\frac{x(x+1)-(x+4)}{(x+1)(x+4)} = 0$$

$$\frac{x^2+x-x-4}{(x+1)(x+4)} = 0$$

$$\frac{x^2-4}{(x+1)(x+4)} = 0$$

$$(x+1)(x+4)\left(\frac{x^2-4}{(x+1)(x+4)}\right) = (x+1)(x+4)(0)$$

$$x^2-4=0$$

$$(x+2)(x-2)=0$$

$$x+2=0 \quad \text{or} \quad x-2=0$$

$$x=-2 \qquad\qquad x=2$$

Set the denominator $= 0$.

$$(x+1)(x+4)=0$$

$$x+1=0 \quad \text{or} \quad x+4=0$$

$$x=-1 \qquad\qquad x=-4$$

Critical numbers $= -4, -2, -1,$ and 2

$$(-4,-2] \cup (-1,2]$$

-4 -2 -1 2

41.

$$(x+2)^2 > 0$$
$$(x+2)(x+2) = 0$$
$$x+2=0 \quad \text{or} \quad x+2=0$$
$$x=-2 \qquad x=-2$$
$$(-\infty,-2)\cup(-2,\infty)$$

−2

43. $x^2 - 2x - 3 < 0$

(−1, 3)

45. $\dfrac{x+3}{x-2} > 0$

$(-\infty,-3)\cup(2,\infty)$

47. $y < x^2 + 1$

Complete a table of values and use (0, 0) as a check point.

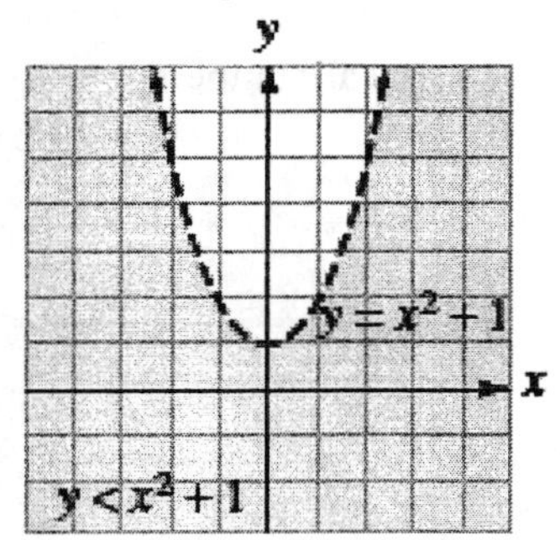

49. $y \le x^2 + 5x + 6$

Complete a table of values and use (0, 0) as a check point.

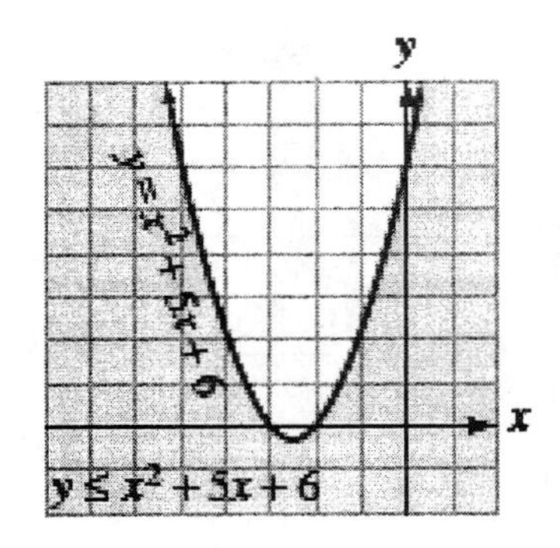

51. $y < |x+4|$

Complete a table of values and use (0, 0) as a check point.

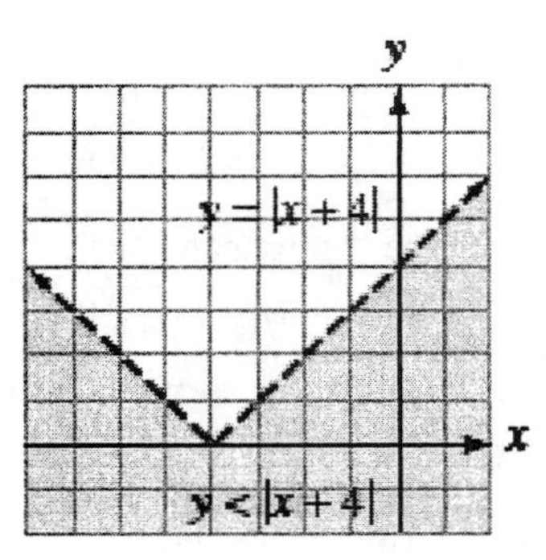

53. $y \le -|x| + 2$

Complete a table of values and use (0, 0) as a check point.

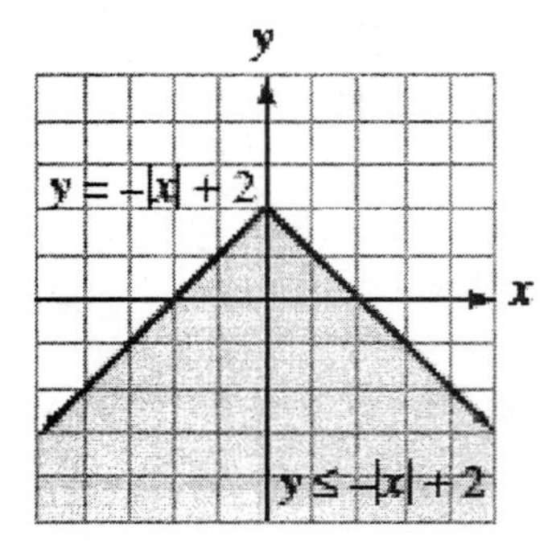

APPLICATIONS

55. BRIDGES

$$L=\frac{1}{9{,}000}x^2+5$$

$$\frac{1}{9{,}000}x^2+5>95$$

$$\frac{1}{9{,}000}x^2+5=95$$

$$\frac{1}{9{,}000}x^2-90=0$$

$$\frac{x^2}{9{,}000}-\frac{90}{1}\bullet\frac{9{,}000}{9{,}000}=0$$

$$\frac{x^2-810{,}000}{9{,}000}=0$$

$$9{,}000\left(\frac{x^2-810{,}000}{9{,}000}\right)=9{,}000(0)$$

$$x^2-810{,}000=0$$

$$(x-900)(x+900)=0$$

$$x-900=0 \quad \text{or} \quad x+900=0$$

$$x=900 \qquad x=-900$$

$$(-2{,}100,-900)\cup(900,2100)$$

WRITING

57. Answers will vary.

59. Answers will vary.

REVIEW

61. $x = ky$

63. $t = kxy$

CHALLENGE PROBLEMS

65. a.

$$x^2-x-12>0$$

$$x^2-x-12=0$$

$$(x-4)(x+3)=0$$

$$x-4=0 \quad \text{or} \quad x+3=0$$

$$x=4 \qquad x=-3$$

$$(-\infty,-3)\cup(4,\infty)$$

b. Answers will vary.

$$\frac{x-4}{x+3}>0$$

CHAPTER 10 KEY CONCEPTS

1.

$$4k^2 + 8k = 0$$
$$4k(k+2) = 0$$
$$4k = 0 \quad \text{or} \quad k+2=0$$
$$k = 0 \qquad k = -2$$

2.

$$z^2 + 8z + 15 = 0$$
$$(z+3)(z+5) = 0$$
$$z+3=0 \quad \text{or} \quad z+5=0$$
$$z=-3 \qquad z=-5$$

3.

$$2r^2 + 5r = -3$$
$$2r^2 + 5r + 3 = 0$$
$$(2r+3)(r+1) = 0$$
$$2r+3=0 \quad \text{or} \quad r+1=0$$
$$2r=-3 \qquad r=-1$$
$$r=-\frac{3}{2}$$

4.

$$u^2 = 24$$
$$u = \pm\sqrt{24}$$
$$u = \pm\sqrt{4}\sqrt{6}$$
$$u = \pm 2\sqrt{6}$$

5.

$$(s-7)^2 - 9 = 0$$
$$(s-7)^2 = 9$$
$$s-7 = \pm\sqrt{9}$$
$$s-7 = \pm 3$$
$$s-7=3 \quad \text{or} \quad s-7=-3$$
$$s=10 \qquad s=4$$

6.

$$3x^2 + 16 = 0$$
$$3x^2 = -16$$
$$x^2 = \frac{-16}{3}$$
$$x = \pm\sqrt{\frac{-16}{3}}$$
$$x = \pm\frac{\sqrt{-16}}{\sqrt{3}}$$
$$x = \pm\frac{4i}{\sqrt{3}}$$
$$x = \pm\frac{4i}{\sqrt{3}} \bullet \frac{\sqrt{3}}{\sqrt{3}}$$
$$x = \pm\frac{4i\sqrt{3}}{3}$$
$$x = 0 \pm \frac{4\sqrt{3}}{3}i$$

7.

$$x^2 + 10x - 7 = 0$$
$$x^2 + 10x = 7$$
$$x^2 + 10x + \left(\frac{10}{2}\right)^2 = 7 + \left(\frac{10}{2}\right)^2$$
$$x^2 + 10x + (5)^2 = 7 + (5)^2$$
$$x^2 + 10x + 25 = 7 + 25$$
$$(x+5)^2 = 32$$
$$x+5 = \pm\sqrt{32}$$
$$x+5 = \pm\sqrt{16}\sqrt{2}$$
$$x+5 = \pm 4\sqrt{2}$$
$$x = -5 \pm 4\sqrt{2}$$

8.

$$4x^2 - 4x - 1 = 0$$

$$\frac{4x^2}{4} - \frac{4x}{4} - \frac{1}{4} = \frac{0}{4}$$

$$x^2 - x - \frac{1}{4} = 0$$

$$x^2 - x = \frac{1}{4}$$

$$x^2 - x + \left(\frac{-1}{2}\right)^2 = \frac{1}{4} + \left(\frac{-1}{2}\right)^2$$

$$x^2 - x + \frac{1}{4} = \frac{1}{4} + \frac{1}{4}$$

$$\left(x - \frac{1}{2}\right)^2 = \frac{1}{2}$$

$$x - \frac{1}{2} = \pm\sqrt{\frac{1}{2}}$$

$$x - \frac{1}{2} = \pm\frac{\sqrt{1}}{\sqrt{2}}$$

$$x - \frac{1}{2} = \pm\frac{\sqrt{1}}{\sqrt{2}} \bullet \frac{\sqrt{2}}{\sqrt{2}}$$

$$x - \frac{1}{2} = \pm\frac{\sqrt{2}}{2}$$

$$x = \frac{1}{2} \pm \frac{\sqrt{2}}{2}$$

$$x = \frac{1 \pm \sqrt{2}}{2}$$

9.

$$x^2 + 2x + 2 = 0$$

$$x^2 + 2x = -2$$

$$x^2 + 2x + \left(\frac{2}{2}\right)^2 = -2 + \left(\frac{2}{2}\right)^2$$

$$x^2 + 2x + (1)^2 = -2 + (1)^2$$

$$x^2 + 2x + 1 = -2 + 1$$

$$(x+1)^2 = -1$$

$$x + 1 = \pm\sqrt{-1}$$

$$x + 1 = \pm i$$

$$x = -1 \pm i$$

10.

$$2x^2 - 1 = 3x$$

$$2x^2 - 3x - 1 = 0$$

$$x = \frac{-b \pm \sqrt{b^2 - 4ac}}{2a}$$

$$= \frac{-(-3) \pm \sqrt{(-3)^2 - 4(2)(-1)}}{2(2)}$$

$$= \frac{3 \pm \sqrt{9 - (-8)}}{4}$$

$$= \frac{3 \pm \sqrt{17}}{4}$$

11.

$$x^2 - 6x - 391 = 0$$

$$x = \frac{-b \pm \sqrt{b^2 - 4ac}}{2a}$$

$$= \frac{-(-6) \pm \sqrt{(-6)^2 - 4(1)(-391)}}{2(1)}$$

$$= \frac{6 \pm \sqrt{36 - (-1{,}564)}}{2}$$

$$= \frac{6 \pm \sqrt{1{,}600}}{2}$$

$$= \frac{6 \pm 40}{2}$$

$$x = \frac{6+40}{2} \quad \text{or} \quad x = \frac{6-40}{2}$$

$$x = \frac{46}{2} \qquad x = \frac{-34}{2}$$

$$x = 23 \qquad x = -17$$

12.

$3x^2 + 2x + 1 = 0$

$$x = \frac{-b \pm \sqrt{b^2 - 4ac}}{2a}$$

$$= \frac{-2 \pm \sqrt{(2)^2 - 4(3)(1)}}{2(3)}$$

$$= \frac{-2 \pm \sqrt{4 - 12}}{6}$$

$$= \frac{-2 \pm \sqrt{-8}}{6}$$

$$= \frac{-2 \pm \sqrt{-4}\sqrt{2}}{6}$$

$$x = \frac{-2 \pm 2i\sqrt{2}}{6}$$

$$x = -\frac{2}{6} \pm \frac{2\sqrt{2}}{6}i$$

$$x = -\frac{1}{3} \pm \frac{\sqrt{2}}{3}i$$

13. The x–intercepts of the graph are (–2, 0) and (1, 0), so the solutions to the equation are $x = -2$ and $x = 1$.

CHAPTER 10 REVIEW

SECTION 10.1
The Square Root Property and Completing the Square

1.

$$x^2+9x+20=0$$
$$(x+4)(x+5)=0$$
$$x+4=0 \quad \text{or} \quad x+5=0$$
$$x=-4 \qquad x=-5$$

2.

$$6x^2+17x+5=0$$
$$(2x+5)(3x+1)=0$$
$$2x+5=0 \quad \text{or} \quad 3x+1=0$$
$$2x=-5 \qquad 3x=-1$$
$$x=-\frac{5}{2} \qquad x=-\frac{1}{3}$$

3.

$$x^2=28$$
$$x=\pm\sqrt{28}$$
$$x=\pm\sqrt{4}\sqrt{7}$$
$$x=\pm2\sqrt{7}$$

4.

$$(t+2)^2=36$$
$$t+2=\pm\sqrt{36}$$
$$t+2=\pm6$$
$$t+2=6 \quad \text{or} \quad t+2=-6$$
$$t=4 \qquad t=-8$$

5.

$$5a^2+11a=0$$
$$a(5a+11)=0$$
$$a=0 \quad \text{or} \quad 5a+11=0$$
$$a=0 \qquad 5a=-11$$
$$a=0 \qquad a=-\frac{11}{5}$$

6.

$$5x^2-49=0$$
$$5x^2=49$$
$$x^2=\frac{49}{5}$$
$$x=\pm\sqrt{\frac{49}{5}}$$
$$x=\pm\frac{7}{\sqrt{5}}$$
$$x=\pm\frac{7}{\sqrt{5}}\cdot\frac{\sqrt{5}}{\sqrt{5}}$$
$$x=\pm\frac{7\sqrt{5}}{5}$$

7.

$$a^2+25=0$$
$$a^2=-25$$
$$a=\pm\sqrt{-25}$$
$$a=\pm5i$$

8. $\frac{1}{4}$ must be added to both sides because

$$\left(\frac{-1}{2}\right)^2=\frac{1}{4}$$

9.

$$x^2+6x+8=0$$
$$x^2+6x=-8$$
$$x^2+6x+\left(\frac{6}{2}\right)^2=-8+\left(\frac{6}{2}\right)^2$$
$$x^2+6x+(3)^2=-8+(3)^2$$
$$x^2+6x+9=-8+9$$
$$(x+3)^2=1$$
$$x+3=\pm\sqrt{1}$$
$$x+3=\pm1$$
$$x+3=1 \quad \text{or} \quad x+3=-1$$
$$x=-2 \qquad x=-4$$

10.

$$2x^2-6x+3=0$$
$$2x^2-6x=-3$$
$$\frac{2x^2}{2}-\frac{6x}{2}=\frac{-3}{2}$$
$$x^2-3x=-\frac{3}{2}$$
$$x^2-3x+\left(\frac{-3}{2}\right)^2=-\frac{3}{2}+\left(\frac{-3}{2}\right)^2$$
$$x^2-3x+\frac{9}{4}=-\frac{3}{2}+\frac{9}{4}$$
$$\left(x-\frac{3}{2}\right)^2=-\frac{6}{4}+\frac{9}{4}$$
$$\left(x-\frac{3}{2}\right)^2=\frac{3}{4}$$
$$x-\frac{3}{2}=\pm\sqrt{\frac{3}{4}}$$
$$x-\frac{3}{2}=\pm\frac{\sqrt{3}}{\sqrt{4}}$$
$$x-\frac{3}{2}=\pm\frac{\sqrt{3}}{2}$$
$$x=\frac{3}{2}\pm\frac{\sqrt{3}}{2}$$
$$x=\frac{3\pm\sqrt{3}}{2}$$

11.

$$x^2-2x+13=0$$
$$x^2-2x=-13$$
$$x^2-2x+\left(\frac{-2}{2}\right)^2=-13+\left(\frac{-2}{2}\right)^2$$
$$x^2-2x+(-1)^2=-13+(-1)^2$$
$$x^2-2x+1=-13+1$$
$$(x-1)^2=-12$$
$$x-1=\pm\sqrt{-12}$$
$$x-1=\pm\sqrt{-4}\sqrt{3}$$
$$x-1=\pm 2i\sqrt{3}$$
$$x=1\pm 2i\sqrt{3}$$

12.

$$A=\pi r^2$$
$$\frac{A}{\pi}=\frac{\pi r^2}{\pi}$$
$$\frac{A}{\pi}=r^2$$
$$r^2=\frac{A}{\pi}$$
$$r=\sqrt{\frac{A}{\pi}}$$
$$r=\frac{\sqrt{A}}{\sqrt{\pi}}$$
$$r=\frac{\sqrt{A}}{\sqrt{\pi}}\cdot\frac{\sqrt{\pi}}{\sqrt{\pi}}$$
$$r=\frac{\sqrt{A\pi}}{\pi}$$

13. 2 is not a factor of the numerator. Only common factors of the numerator and denominator can be removed.

14.

$$d=16t^2$$
$$605=16t^2$$
$$\frac{605}{16}=\frac{16t^2}{16}$$
$$37.8125=t^2$$
$$t=\pm\sqrt{37.8125}$$
$$t=\pm 6$$

Since the time cannot be negative, the ball should be dropped 6 seconds before midnight.

SECTION 10.2
The Quadratic Formula

15.

$$x^2 - 10x = 0$$

$$x = \frac{-b \pm \sqrt{b^2 - 4ac}}{2a}$$

$$= \frac{-(-10) \pm \sqrt{(-10)^2 - 4(1)(0)}}{2(1)}$$

$$= \frac{10 \pm \sqrt{100 - 0}}{2}$$

$$= \frac{10 \pm \sqrt{100}}{2}$$

$$= \frac{10 \pm 10}{2}$$

$$x = \frac{10 + 10}{2} \quad \text{or} \quad x = \frac{10 - 10}{2}$$

$$x = \frac{20}{2} \qquad x = \frac{0}{2}$$

$$x = 10 \qquad x = 0$$

16.

$$-x^2 + 10x - 18 = 0$$

$$0 = x^2 - 10x + 18$$

$$x = \frac{-b \pm \sqrt{b^2 - 4ac}}{2a}$$

$$= \frac{-(-10) \pm \sqrt{(-10)^2 - 4(1)(18)}}{2(1)}$$

$$= \frac{10 \pm \sqrt{100 - 72}}{2}$$

$$= \frac{10 \pm \sqrt{28}}{2}$$

$$= \frac{10 \pm \sqrt{4}\sqrt{7}}{2}$$

$$= \frac{10 \pm 2\sqrt{7}}{2}$$

$$= \frac{10}{2} \pm \frac{2\sqrt{7}}{2}$$

$$x = 5 \pm \sqrt{7}$$

17.

$$2x^2 + 13x = 7$$

$$2x^2 + 13x - 7 = 0$$

$$x = \frac{-b \pm \sqrt{b^2 - 4ac}}{2a}$$

$$= \frac{-13 \pm \sqrt{(13)^2 - 4(2)(-7)}}{2(2)}$$

$$= \frac{-13 \pm \sqrt{169 - (-56)}}{4}$$

$$= \frac{-13 \pm \sqrt{225}}{4}$$

$$= \frac{-13 \pm 15}{4}$$

$$x = \frac{-13 + 15}{4} \quad \text{or} \quad x = \frac{-13 - 15}{4}$$

$$x = \frac{2}{4} \qquad x = \frac{-28}{4}$$

$$x = \frac{1}{2} \qquad x = -7$$

18.

$$26y - 3y^2 = 2$$

$$0 = 3y^2 - 26y + 2$$

$$y = \frac{-b \pm \sqrt{b^2 - 4ac}}{2a}$$

$$= \frac{-(-26) \pm \sqrt{(-26)^2 - 4(3)(2)}}{2(3)}$$

$$= \frac{26 \pm \sqrt{676 - 24}}{6}$$

$$= \frac{26 \pm \sqrt{652}}{6}$$

$$= \frac{26 \pm \sqrt{4}\sqrt{163}}{6}$$

$$= \frac{26 \pm 2\sqrt{163}}{6}$$

$$= \frac{\cancel{2}(13 \pm \sqrt{163})}{\cancel{2} \cdot 3}$$

$$y = \frac{13 \pm \sqrt{163}}{3}$$

19.

$$\frac{p^2}{3}+\frac{p}{2}+\frac{1}{2}=0$$

$$6\left(\frac{p^2}{3}+\frac{p}{2}+\frac{1}{2}\right)=6(0)$$

$$2p^2+3p+3=0$$

$$p=\frac{-b\pm\sqrt{b^2-4ac}}{2a}$$

$$=\frac{-3\pm\sqrt{(3)^2-4(2)(3)}}{2(2)}$$

$$=\frac{-3\pm\sqrt{9-24}}{4}$$

$$=\frac{-3\pm\sqrt{-15}}{4}$$

$$=\frac{-3\pm i\sqrt{15}}{4}$$

$$p=-\frac{3}{4}\pm\frac{\sqrt{15}}{4}i$$

20.

$$3{,}000t^2-4{,}000t=-2{,}000$$

$$\frac{3{,}000t^2}{1{,}000}-\frac{4{,}000t}{1{,}000}=\frac{-2{,}000}{1{,}000}$$

$$3t^2-4t=-2$$

$$3t^2-4t+2=0$$

$$t=\frac{-b\pm\sqrt{b^2-4ac}}{2a}$$

$$=\frac{-(-4)\pm\sqrt{(-4)^2-4(3)(2)}}{2(3)}$$

$$=\frac{4\pm\sqrt{16-24}}{6}$$

$$=\frac{4\pm\sqrt{-8}}{6}$$

$$=\frac{4\pm\sqrt{-4}\sqrt{2}}{6}$$

$$=\frac{4\pm 2i\sqrt{2}}{6}$$

$$=\frac{4}{6}\pm\frac{2\sqrt{2}}{6}i$$

$$t=\frac{2}{3}\pm\frac{\sqrt{2}}{3}i$$

21.

$$0.5x^2+0.3x-0.1=0$$

$$10(0.5x^2+0.3x-0.1)=10(0)$$

$$5x^2+3x-1=0$$

$$x=\frac{-b\pm\sqrt{b^2-4ac}}{2a}$$

$$=\frac{-3\pm\sqrt{(3)^2-4(5)(-1)}}{2(5)}$$

$$=\frac{-3\pm\sqrt{9-(-20)}}{10}$$

$$=\frac{-3\pm\sqrt{29}}{10}$$

22. The # of students would be 300 – 5x.
The price per student would be 20 + 0.50x.
(# of students)(price per student) = revenue

$$(300-5x)(20+0.5x)=6{,}240$$

$$6{,}000+150x-100x-2.5x^2=6{,}240$$

$$6{,}000+50x-2.5x^2=6{,}240$$

$$0=2.5x^2-50x+240$$

$$\frac{0}{2.5}=\frac{2.5x^2}{2.5}-\frac{50x}{2.5}+\frac{240}{2.5}$$

$$0=x^2-20x+96$$

$$x=\frac{-b\pm\sqrt{b^2-4ac}}{2a}$$

$$=\frac{-(-20)\pm\sqrt{(-20)^2-4(1)(96)}}{2(1)}$$

$$=\frac{20\pm\sqrt{400-384}}{2}$$

$$=\frac{20\pm\sqrt{16}}{2}$$

$$=\frac{20\pm4}{2}$$

$$x=\frac{20+4}{2} \quad \text{or} \quad x=\frac{20-4}{2}$$

$$x=\frac{24}{2} \qquad x=\frac{16}{2}$$

$$x=12 \qquad x=8$$

$$\text{price}=20+0.5x \qquad \text{price}=20+0.5x$$

$$=20+0.5(12) \qquad =20+0.5(8)$$

$$=20+6 \qquad =20+4$$

$$=\$26 \qquad =\$24$$

23. Let the border on top/bottom be x. Then the borders on the side would be $\frac{1}{2}x$. The length of the picture is $35 - 2(x)$ and the width of the picture is $23 - 2(\frac{1}{2}x)$.

$$(\text{length})(\text{width}) = \text{Area}$$
$$(35-2x)\left(23-2\cdot\frac{1}{2}x\right)=615$$
$$(35-2x)(23-x)=615$$
$$805-35x-46x+2x^2=615$$
$$2x^2-81x+805=615$$
$$2x^2-81x+190=0$$
$$x=\frac{-b\pm\sqrt{b^2-4ac}}{2a}$$
$$=\frac{-(-81)\pm\sqrt{(-81)^2-4(2)(190)}}{2(2)}$$
$$=\frac{81\pm\sqrt{6,561-1,520}}{4}$$
$$=\frac{81\pm\sqrt{5,041}}{4}$$
$$=\frac{81\pm 71}{4}$$
$$x=\frac{81+71}{4} \quad \text{or} \quad x=\frac{81-71}{4}$$
$$x=\frac{152}{4} \qquad x=\frac{10}{4}$$
$$x=38 \qquad x=2.5$$

Since the total length is only 35 inches, the border cannot be 38 inches. So, the border on top and bottom is 2.5 inches and the border on the sides is ½ of 2.5, which is 1.25 inches.

24.
$$d=-16t^2+40t+5$$
$$25=-16t^2+40t+5$$
$$16t^2-40t+20=0$$
$$t=\frac{-b\pm\sqrt{b^2-4ac}}{2a}$$
$$=\frac{-(-40)\pm\sqrt{(-40)^2-4(16)(20)}}{2(16)}$$
$$=\frac{40\pm\sqrt{1,600-1,280}}{32}$$
$$=\frac{40\pm\sqrt{320}}{32}$$
$$=\frac{40\pm\sqrt{64}\sqrt{5}}{32}$$
$$=\frac{40\pm 8\sqrt{5}}{32}$$
$$t=\frac{40+8\sqrt{5}}{32} \quad \text{or} \quad t=\frac{40-8\sqrt{5}}{32}$$
$$t=1.8 \qquad t=0.7$$

He will be able to grab it in 0.7 seconds and 1.8 seconds.

SECTION 10.3

The Discriminant and Equations That Can be Written in Quadratic Form

25. irrational and unequal
$$b^2-4ac=(4)^2-4(3)(-3)$$
$$=16-(-36)$$
$$=52$$

26. complex conjugates
$$b^2-4ac=(-5)^2-4(4)(7)$$
$$=25-112$$
$$=-87$$

27. equal rational numbers
$$b^2-4ac=(-12)^2-4(9)(4)$$
$$=144-144$$
$$=0$$

28. rational and unequal

$$m(2m-3)=20$$
$$2m^2-3m=20$$
$$2m^2-3m-20=0$$
$$b^2-4ac=(-3)^2-4(2)(-20)$$
$$=9-(-160)$$
$$=169$$

29. Let $y=\sqrt{x}$.

$$x-13\sqrt{x}+12=0$$
$$\left(\sqrt{x}\right)^2-13\sqrt{x}+12=0$$
$$y^2-13y+12=0$$
$$(y-12)(y-1)=0$$
$$y-12=0 \quad \text{and} \quad y-1=0$$
$$y=12 \qquad y=1$$
$$\sqrt{x}=12 \qquad \sqrt{x}=1$$
$$\left(\sqrt{x}\right)^2=(12)^2 \quad \left(\sqrt{x}\right)^2=(1)^2$$
$$x=144 \qquad x=1$$

30. Let $y=a^{1/3}$.

$$a^{2/3}+a^{1/3}-6=0$$
$$\left(a^{1/3}\right)^2+a^{1/3}-6=0$$
$$y^2+y-6=0$$
$$(y+3)(y-2)=0$$
$$y+3=0 \quad \text{or} \quad y-2=0$$
$$y=-3 \qquad y=2$$
$$a^{1/3}=-3 \qquad a^{1/3}=2$$
$$\left(a^{1/3}\right)^3=(-3)^3 \quad \left(a^{1/3}\right)^3=(2)^3$$
$$a=-27 \qquad a=8$$

31. Let $y=x^2$.

$$3x^4+x^2-2=0$$
$$3\left(x^2\right)^2+x^2-2=0$$
$$3y^2+y-2=0$$
$$(3y-2)(y+1)=0$$
$$3y-2=0 \quad \text{or} \quad y+1=0$$
$$3y=2$$
$$y=\frac{2}{3} \qquad y=-1$$
$$x^2=\frac{2}{3} \qquad x^2=-1$$
$$x=\pm\sqrt{\frac{2}{3}} \qquad x=\pm\sqrt{-1}$$
$$x=\pm\sqrt{\frac{2}{3}\bullet\frac{3}{3}} \qquad x=\pm i$$
$$x=\pm\frac{\sqrt{6}}{3}$$
$$x=\frac{\sqrt{6}}{3},\ -\frac{\sqrt{6}}{3},\ i,\ -i$$

32.

$$\frac{6}{x+2}+\frac{6}{x+1}=5$$
$$(x+2)(x+1)\left(\frac{6}{x+2}+\frac{6}{x+1}\right)=5(x+2)(x+1)$$
$$6(x+1)+6(x+2)=5\left(x^2+2x+x+2\right)$$
$$6x+6+6x+12=5\left(x^2+3x+2\right)$$
$$12x+18=5x^2+15x+10$$
$$0=5x^2+3x-8$$
$$0=(5x+8)(x-1)$$
$$5x+8=0 \quad \text{or} \quad x-1=0$$
$$5x=-8 \qquad x=1$$
$$x=-\frac{8}{5}$$

33. Let $y = (x - 7)$.

$$(x-7)^2 + 6(x-7) + 10 = 0$$
$$y^2 + 6y + 10 = 0$$
$$y = \frac{-b \pm \sqrt{b^2 - 4ac}}{2a}$$
$$y = \frac{-6 \pm \sqrt{(6)^2 - 4(1)(10)}}{2(1)}$$
$$y = \frac{-6 \pm \sqrt{36-40}}{2}$$
$$y = \frac{-6 \pm \sqrt{-4}}{2}$$
$$y = \frac{-6 \pm 2i}{2}$$
$$y = \frac{-6}{2} \pm \frac{2}{2}i$$
$$y = -3 \pm i$$
$$x - 7 = -3 \pm i$$
$$x = -3 \pm i + 7$$
$$x = 4 \pm i$$

34. Let $y = \dfrac{1}{m^2}$.

$$m^{-4} - 2m^{-2} + 1 = 0$$
$$\frac{1}{m^4} - 2\left(\frac{1}{m^2}\right) + 1 = 0$$
$$\left(\frac{1}{m^2}\right)^2 - 2\left(\frac{1}{m^2}\right) + 1 = 0$$
$$y^2 - 2y + 1 = 0$$
$$(y-1)(y-1) = 0$$
$$y - 1 = 0 \quad \text{or} \quad y - 1 = 0$$
$$y = 1 \qquad y = 1$$
$$\frac{1}{m^2} = \frac{1}{1} \qquad \frac{1}{m^2} = \frac{1}{1}$$
$$m^2 = 1 \qquad m^2 = 1$$
$$m = \pm\sqrt{1} \qquad m = \pm\sqrt{1}$$
$$m = \pm 1 \qquad m = \pm 1$$
$$m = 1,\ -1,\ 1,\ -1$$

35. Let $y = \dfrac{x+1}{x}$.

$$4\left(\frac{x+1}{x}\right)^2 + 12\left(\frac{x+1}{x}\right) + 9 = 0$$
$$4y^2 + 12y + 9 = 0$$
$$(2y+3)(2y+3) = 0$$
$$2y + 3 = 0 \quad \text{or} \quad 2y + 3 = 0$$
$$2y = -3 \qquad 2y = -3$$
$$y = -\frac{3}{2} \qquad y = -\frac{3}{2}$$
$$\frac{x+1}{x} = -\frac{3}{2} \qquad \frac{x+1}{x} = -\frac{3}{2}$$
$$2(x+1) = -3x \qquad 2(x+1) = -3x$$
$$2x + 2 = -3x \qquad 2x + 2 = -3x$$
$$2 = -5x \qquad 2 = -5x$$
$$-\frac{2}{5} = x \qquad -\frac{2}{5} = x$$

36. Let x = the number of minutes it takes the younger girl to do the yard work.

Younger girl's work in 1 minute = $\frac{1}{x}$

Older girl's work in 1 minute = $\frac{1}{x-20}$

Work together in 1 minute = $\frac{1}{45}$

Younger girl + older girl = work together

$$\frac{1}{x}+\frac{1}{x-20}=\frac{1}{45}$$

$$45x(x-20)\left(\frac{1}{x}+\frac{1}{x-20}\right)=45x(x-20)\left(\frac{1}{45}\right)$$

$$45(x-20)+45x=x(x-20)$$

$$45x-900+45x=x^2-20x$$

$$90x-900=x^2-20x$$

$$0=x^2-110x+900$$

$$x=\frac{-b\pm\sqrt{b^2-4ac}}{2a}$$

$$x=\frac{-(-110)\pm\sqrt{(-110)^2-4(1)(900)}}{2(1)}$$

$$x=\frac{110\pm\sqrt{12{,}100-3{,}600}}{2}$$

$$x=\frac{110\pm\sqrt{8{,}500}}{2}$$

$$x=\frac{110\pm92.2}{2}$$

$$x=\frac{110+92.2}{2} \quad \text{or} \quad x=\frac{110-92.2}{2}$$

$$x=\frac{202}{2} \qquad x=\frac{17}{2}$$

$$x=101 \qquad x=8$$

$$x-20=101-20 \qquad x-20=8-20$$

$$=81 \qquad =-12$$

Since the older sister's time cannot be negative, her time must be about 81 min.

SECTION 10.4
Quadratic Functions and Their Graphs

37. Since 1997 – 1993 = 4, let $x = 4$.

$$S(x)=2.2x^2-7.7x+39.9$$

$$S(4)=2.2(4)^2-7.7(4)+39.9$$

$$=2.2(16)-7.7(4)+39.9$$

$$=35.2-30.8+39.9$$

$$=44.3$$

The annual sale for 1997 was $44.3 billion.

38. The graph of $g(x)$ is shifted down 3 units.

x	$f(x) = 2x^2$
-2	$2(-2)^2 = 8$
-1	$2(-1)^2 = 2$
0	$2(0)^2 = 0$
1	$2(1)^2 = 2$
2	$2(2)^2 = 8$

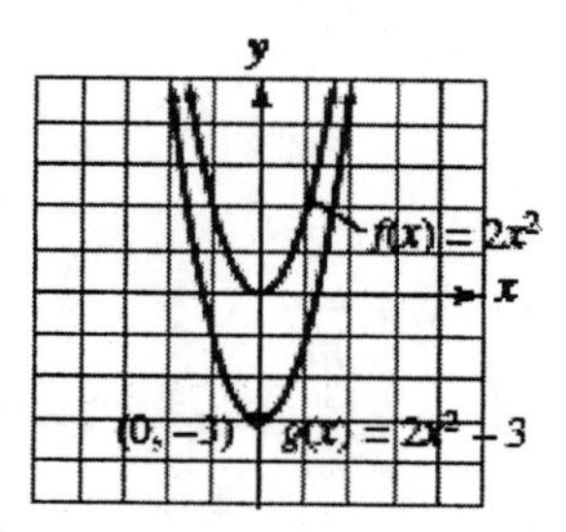

39. The graph of $g(x)$ is shifted right 2 units and up 1 unit.

x	$f(x) = -4x^2$
-2	$-4(-2)^2 = -16$
-1	$-4(-1)^2 = -4$
0	$-4(0)^2 = 0$
1	$-4(1)^2 = -4$
2	$-4(2)^2 = -16$

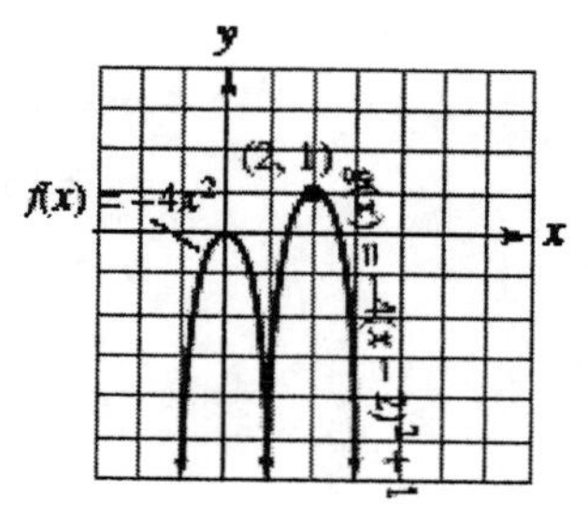

40. $f(x) = -2(x-1)^2 + 4$

Vertex: (1, 4)

Axis of symmetry: $x = 1$

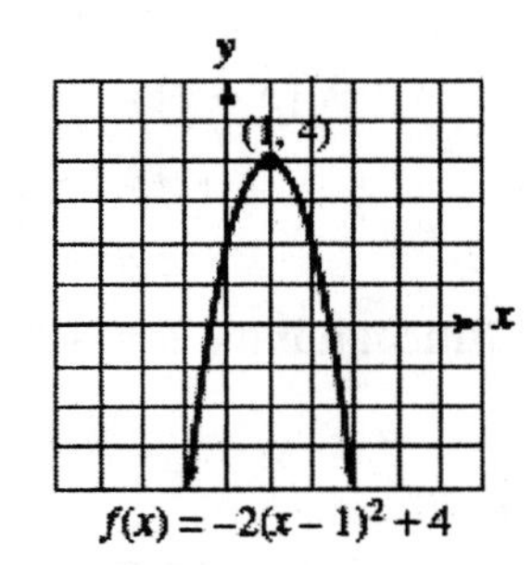

41. $f(x) = 4x^2 + 16x + 9$

$f(x) = 4(x^2 + 4x) + 9$

$f(x) = 4(x^2 + 4x + 4) + 9 - 4(4)$

$f(x) = 4(x+2)^2 + 9 - 16$

$f(x) = 4(x+2)^2 - 7$

Vertex: (–2, 7)

Axis of symmetry: $x = -2$

42. $f(x) = x^2 + x - 2$

Step 1: Since $a = 1 > 0$, the parabola opens upward.

Step 2: Find the vertex and axis of symmetry.

$$x = \frac{-b}{2a} = \frac{-1}{2(1)} = -\frac{1}{2}$$

$$y = \left(-\frac{1}{2}\right)^2 + \left(-\frac{1}{2}\right) - 2$$

$$= \frac{1}{4} - \frac{1}{2} - 2$$

$$= -\frac{9}{4}$$

vertex: $\left(-\frac{1}{2}, -\frac{9}{4}\right)$

The axis of symmetry is $x = -\frac{1}{2}$.

Step 3: Find the x– and y–intercepts. Since $c = -2$, the y–intercept is (0, –2). To find the x–intercepts, let $y = 0$ and solve the equation for x.

$$0 = x^2 + x - 2$$

$$0 = (x+2)(x-1)$$

$$x + 2 = 0 \quad \text{or} \quad x - 1 = 0$$

$$x = -2 \qquad\qquad x = 1$$

x–intercepts: (–2, 0) and (1, 0)

y–intercept: (0, –2)

Step 4: Use symmetry to find the point (–1, –2).

Step 5: Plots the points and draw the parabola.

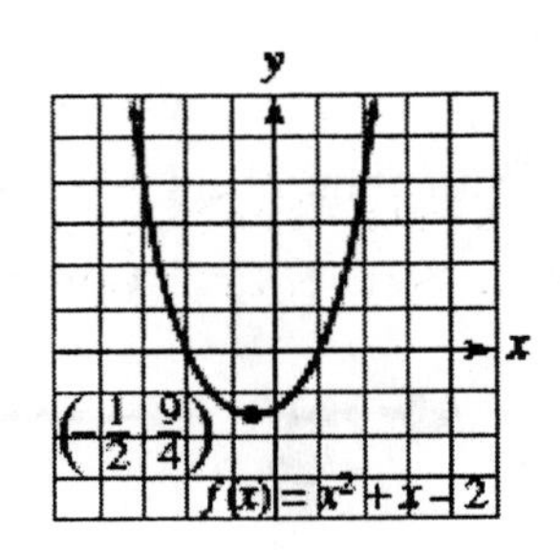

43. The vertex would represent the maximum of the function. Use the vertex formula to find the value of x and use substitution to find the number of farms.

$$\begin{aligned} x &= \frac{-b}{2a} \\ &= \frac{-155{,}652}{2(-1{,}526)} \\ &= \frac{-155{,}652}{-3{,}052} \\ &= 51 \end{aligned}$$

$$1870 + 51 = 1921$$

$$\begin{aligned} N(51) &= -1{,}526(51)^2 + 155{,}652(51) + 2{,}500{,}200 \\ &= -1{,}526(2{,}601) + 155{,}652(51) + 2{,}500{,}200 \\ &= -3{,}969{,}126 + 7{,}938{,}252 + 2{,}500{,}200 \\ &= 6{,}469{,}326 \text{ farms} \end{aligned}$$

44. The x–intercepts are $(-2, 0)$ and $\left(\frac{1}{3}, 0\right)$, so the solutions to the equation are $x = -2$ and $x = \frac{1}{3}$.

SECTION 10.5
Quadratic and Other Nonlinear Inequalities

45.

$$\begin{aligned} x^2 + 2x - 35 &> 0 \\ x^2 + 2x - 35 &= 0 \\ (x+7)(x-5) &= 0 \end{aligned}$$

$$x + 7 = 0 \quad \text{or} \quad x - 5 = 0$$

$$x = -7 \qquad x = 5$$

Critical numbers = -7 and 5.

$$(-\infty, -7) \cup (5, \infty)$$

46.

$$\begin{aligned} x^2 &\le 81 \\ x^2 - 81 &\le 0 \\ x^2 - 81 &= 0 \\ (x+9)(x-9) &= 0 \end{aligned}$$

$$x + 9 = 0 \quad \text{or} \quad x - 9 = 0$$

$$x = -9 \qquad x = 9$$

Critical numbers = -9 and 9

$$[-9, 9]$$

47.

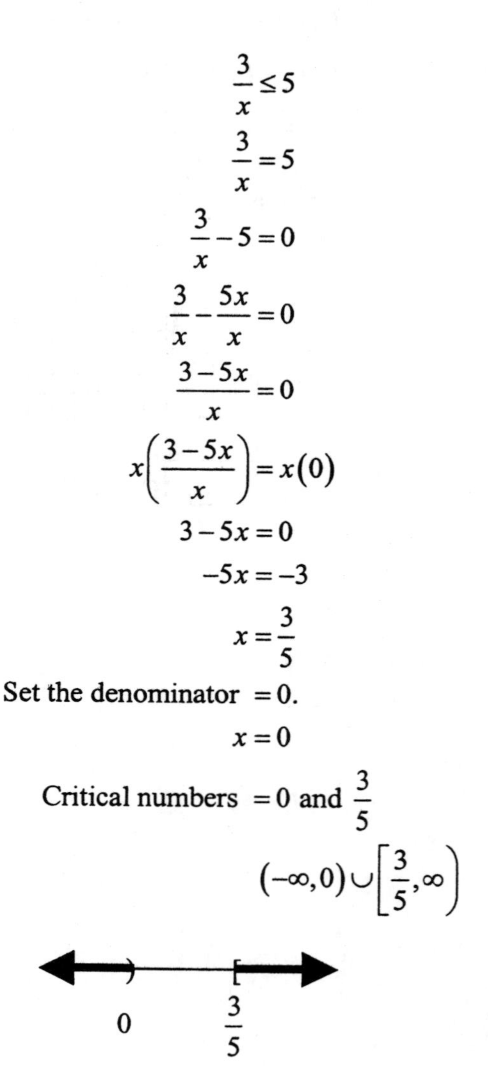

$$\begin{aligned} \frac{3}{x} &\le 5 \\ \frac{3}{x} &= 5 \\ \frac{3}{x} - 5 &= 0 \\ \frac{3}{x} - \frac{5x}{x} &= 0 \\ \frac{3-5x}{x} &= 0 \\ x\left(\frac{3-5x}{x}\right) &= x(0) \\ 3 - 5x &= 0 \\ -5x &= -3 \\ x &= \frac{3}{5} \end{aligned}$$

Set the denominator $= 0$.

$$x = 0$$

Critical numbers $= 0$ and $\frac{3}{5}$

$$(-\infty, 0) \cup \left[\frac{3}{5}, \infty\right)$$

48.

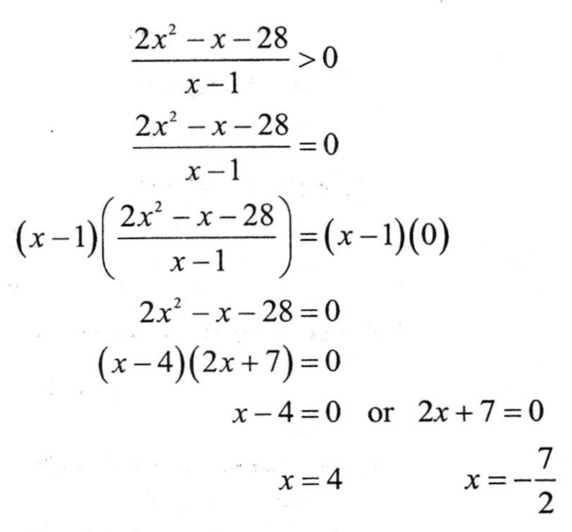

$$\frac{2x^2 - x - 28}{x - 1} > 0$$

$$\frac{2x^2 - x - 28}{x - 1} = 0$$

$$(x-1)\left(\frac{2x^2 - x - 28}{x - 1}\right) = (x-1)(0)$$

$$2x^2 - x - 28 = 0$$

$$(x-4)(2x+7) = 0$$

$$x - 4 = 0 \quad \text{or} \quad 2x + 7 = 0$$

$$x = 4 \qquad\qquad x = -\frac{7}{2}$$

Set the denominator $= 0$.

$$x - 1 = 0$$

$$x = 1$$

Critical numbers $= -\frac{7}{2}$, 1, and 4

$$\left(-\frac{7}{2}, 1\right) \cup (4, \infty)$$

$-\frac{7}{2}$ 1 4

49. $\left[-4, \frac{2}{3}\right]$

50. $(-\infty, 0) \cup (1, \infty)$

51. $y < \frac{1}{2}x^2 - 1$

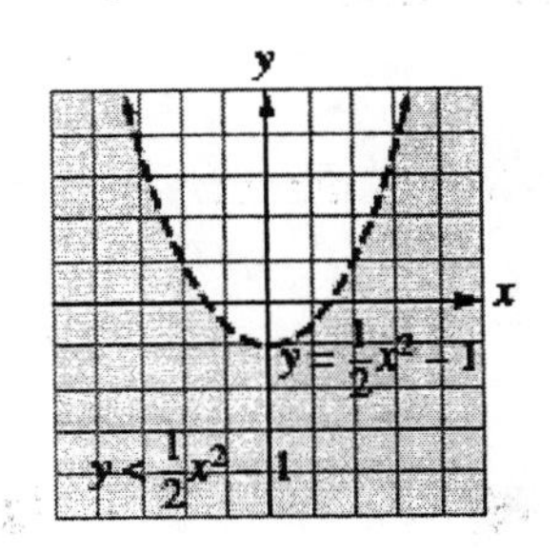

52. $y \geq -|x|$

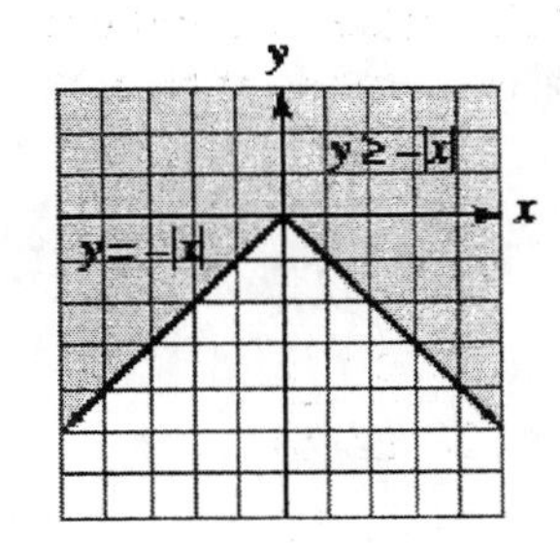

CHAPTER 10 TEST

1.

$$3x^2 + 18x = 0$$
$$3x(x+6) = 0$$
$$3x = 0 \quad \text{or} \quad x + 6 = 0$$
$$x = 0 \qquad\qquad x = -6$$

2.

$$m^2 + 4 = 0$$
$$m^2 = -4$$
$$m = \pm\sqrt{-4}$$
$$m = \pm 2i$$

3.

$$(a+7)^2 = 50$$
$$a + 7 = \pm\sqrt{50}$$
$$a + 7 = \pm\sqrt{25}\sqrt{2}$$
$$a + 7 = \pm 5\sqrt{2}$$
$$a = -7 \pm 5\sqrt{2}$$

4.

$$x(6x+19) = -15$$
$$6x^2 + 19x = -15$$
$$6x^2 + 19x + 15 = 0$$
$$(2x+3)(3x+5) = 0$$
$$2x + 3 = 0 \quad \text{or} \quad 3x + 5 = 0$$
$$2x = -3 \qquad\qquad 3x = -5$$
$$x = -\frac{3}{2} \qquad\qquad x = -\frac{5}{3}$$

5. 144

$$\left(\frac{24}{2}\right)^2 = (12)^2 = 144$$

6.

$$3x^2 + x - 24 = 0$$
$$3x^2 + x = 24$$
$$\frac{3x^2}{3} + \frac{x}{3} = \frac{24}{3}$$
$$x^2 + \frac{1}{3}x = 8$$
$$x^2 + \frac{1}{3}x + \left(\frac{1}{2} \bullet \frac{1}{3}\right)^2 = 8 + \left(\frac{1}{2} \bullet \frac{1}{3}\right)^2$$
$$x^2 + \frac{1}{3}x + \left(\frac{1}{6}\right)^2 = 8 + \left(\frac{1}{6}\right)^2$$
$$x^2 + \frac{1}{3}x + \frac{1}{36} = 8 + \frac{1}{36}$$
$$\left(x + \frac{1}{6}\right)^2 = \frac{288}{36} + \frac{1}{36}$$
$$\left(x + \frac{1}{6}\right)^2 = \frac{289}{36}$$
$$x + \frac{1}{6} = \sqrt{\frac{289}{36}}$$
$$x + \frac{1}{6} = \pm\frac{17}{6}$$
$$x = -\frac{1}{6} \pm \frac{17}{6}$$
$$x = -\frac{1}{6} + \frac{17}{6} \quad \text{or} \quad x = -\frac{1}{6} - \frac{17}{6}$$
$$x = \frac{16}{6} \qquad\qquad x = -\frac{18}{6}$$
$$x = \frac{8}{3} \qquad\qquad x = -3$$

7.

$$2x^2 - 8x + 5 = 0$$

$$x = \frac{-b \pm \sqrt{b^2 - 4ac}}{2a}$$

$$= \frac{-(-8) \pm \sqrt{(-8)^2 - 4(2)(5)}}{2(2)}$$

$$= \frac{8 \pm \sqrt{64 - 40}}{4}$$

$$= \frac{8 \pm \sqrt{24}}{4}$$

$$= \frac{8 \pm \sqrt{4}\sqrt{6}}{4}$$

$$= \frac{8 \pm 2\sqrt{6}}{4}$$

$$= \frac{\cancel{2}\left(4 \pm \sqrt{6}\right)}{\cancel{2} \cdot 2}$$

$$x = \frac{4 \pm \sqrt{6}}{2}$$

8.

$$\frac{t^2}{8} - \frac{t}{4} = \frac{1}{2}$$

$$8\left(\frac{t^2}{8} - \frac{t}{4}\right) = 8\left(\frac{1}{2}\right)$$

$$t^2 - 2t = 4$$

$$t^2 - 2t - 4 = 0$$

$$t = \frac{-b \pm \sqrt{b^2 - 4ac}}{2a}$$

$$= \frac{-(-2) \pm \sqrt{(-2)^2 - 4(1)(-4)}}{2(1)}$$

$$= \frac{2 \pm \sqrt{4 - (-16)}}{2}$$

$$= \frac{2 \pm \sqrt{20}}{2}$$

$$= \frac{2 \pm \sqrt{4}\sqrt{5}}{2}$$

$$= \frac{2 \pm 2\sqrt{5}}{2}$$

$$= \frac{\cancel{2}\left(1 \pm \sqrt{5}\right)}{\cancel{2}}$$

$$t = 1 \pm \sqrt{5}$$

9.

$$-t^2 + 4t - 13 = 0$$

$$0 = t^2 - 4t + 13$$

$$t = \frac{-b \pm \sqrt{b^2 - 4ac}}{2a}$$

$$= \frac{-(-4) \pm \sqrt{(-4)^2 - 4(1)(13)}}{2(1)}$$

$$= \frac{4 \pm \sqrt{16 - 52}}{2}$$

$$= \frac{4 \pm \sqrt{-36}}{2}$$

$$= \frac{4 \pm 6i}{2}$$

$$= \frac{4}{2} \pm \frac{6}{2}i$$

$$t = 2 \pm 3i$$

10.

$$0.01x^2 = -0.08x - 0.15$$

$$100\left(0.01x^2\right) = 100(-0.08x - 0.15)$$

$$x^2 = -8x - 15$$

$$x^2 + 8x + 15 = 0$$

$$x = \frac{-b \pm \sqrt{b^2 - 4ac}}{2a}$$

$$= \frac{-8 \pm \sqrt{(8)^2 - 4(1)(15)}}{2(1)}$$

$$= \frac{-8 \pm \sqrt{64 - 60}}{2}$$

$$= \frac{-8 \pm \sqrt{4}}{2}$$

$$= \frac{-8 \pm 2}{2}$$

$$x = \frac{-8 + 2}{2} \quad \text{or} \quad x = \frac{-8 - 2}{2}$$

$$x = \frac{-6}{2} \qquad x = \frac{-10}{2}$$

$$x = -3 \qquad x = -5$$

11. Let $x=\sqrt{y}$

$$2y-3\sqrt{y}+1=0$$
$$2\left(\sqrt{y}\right)^2-3\sqrt{y}+1=0$$
$$2x^2-3x+1=0$$
$$(2x-1)(x-1)=0$$
$$2x-1=0 \quad \text{or} \quad x-1=0$$
$$2x=1 \qquad x-1+1=0+1$$
$$x=\frac{1}{2} \qquad x=1$$
$$\sqrt{y}=\frac{1}{2} \qquad \sqrt{y}=1$$
$$\left(\sqrt{y}\right)^2=\left(\frac{1}{2}\right)^2 \quad \left(\sqrt{y}\right)^2=(1)^2$$
$$y=\frac{1}{4} \qquad y=1$$

12. Let $y=\frac{1}{m}$.

$$m^{-2}+m^{-1}=-1$$
$$\frac{1}{m^2}+\frac{1}{m}=-1$$
$$\left(\frac{1}{m}\right)^2+\frac{1}{m}+1=0$$
$$y^2+y+1=0$$
$$y=\frac{-b\pm\sqrt{b^2-4ac}}{2a}$$
$$y=\frac{-1\pm\sqrt{1^2-4(1)(1)}}{2(1)}$$
$$y=\frac{-1\pm\sqrt{1-4}}{2}$$
$$y=\frac{-1\pm\sqrt{-3}}{2}$$
$$y=\frac{-1\pm i\sqrt{3}}{2}$$

$$\frac{1}{m}=\frac{-1\pm i\sqrt{3}}{2}$$
$$2=m\left(-1\pm i\sqrt{3}\right)$$
$$\frac{2}{-1\pm i\sqrt{3}}=\frac{m\left(-1\pm i\sqrt{3}\right)}{-1\pm i\sqrt{3}}$$
$$\frac{2}{-1\pm i\sqrt{3}}=m$$
$$m=\frac{2}{-1\pm i\sqrt{3}}\bullet\frac{-1\pm i\sqrt{3}}{-1\pm i\sqrt{3}}$$
$$m=\frac{2\left(-1-i\sqrt{3}\right)}{1\pm 3i^2}$$
$$m=\frac{2\left(-1\pm i\sqrt{3}\right)}{1+3}$$
$$m=\frac{2\left(-1\pm i\sqrt{3}\right)}{4}$$
$$m=\frac{-1\pm i\sqrt{3}}{2}$$

13. Let $y = x^2$

$$x^4 - x^2 - 12 = 0$$
$$(x^2)^2 - x^2 - 12 = 0$$
$$y^2 - y - 12 = 0$$
$$(y-4)(y+3) = 0$$
$$y - 4 = 0 \quad \text{or} \quad y + 3 = 0$$
$$y = 4 \qquad y = -3$$
$$x^2 = 4 \qquad x^2 = -3$$
$$x = \pm\sqrt{4} \qquad x = \pm\sqrt{-3}$$
$$x = \pm 2 \qquad x = \pm i\sqrt{3}$$
$$x = 2,\ -2,\ i\sqrt{3},\ -i\sqrt{3}$$

14. Let $y = \dfrac{x+2}{3x}$.

$$4\left(\frac{x+2}{3x}\right)^2 - 4\left(\frac{x+2}{3x}\right) - 3 = 0$$
$$4y^2 - 4y - 3 = 0$$
$$(2y+1)(2y-3) = 0$$
$$2y + 1 = 0 \quad \text{or} \quad 2y - 3 = 0$$
$$2y = -1 \qquad 2y = 3$$
$$y = -\frac{1}{2} \qquad y = \frac{3}{2}$$
$$\frac{x+2}{3x} = -\frac{1}{2} \qquad \frac{x+2}{3x} = \frac{3}{2}$$
$$2(x+2) = -1(3x) \qquad 2(x+2) = 3(3x)$$
$$2x + 4 = -3x \qquad 2x + 4 = 9x$$
$$4 = -5x \qquad 4 = 7x$$
$$-\frac{4}{5} = x \qquad \frac{4}{7} = x$$

15.

$$E = mc^2$$
$$\frac{E}{m} = \frac{mc^2}{m}$$
$$\frac{E}{m} = c^2$$
$$c = \sqrt{\frac{E}{m}}$$
$$c = \frac{\sqrt{E}}{\sqrt{m}} \cdot \frac{\sqrt{m}}{\sqrt{m}}$$
$$c = \frac{\sqrt{Em}}{m}$$

16. a. complex conjugates

$$b^2 - 4ac = (5)^2 - 4(3)(17)$$
$$= 25 - 204$$
$$= -179$$

b. rational and equal

$$b^2 - 4ac = (-12)^2 - 4(9)(4)$$
$$= 144 - 144$$
$$= 0$$

17. Let the width $= x$.
Then length = $332x + 8$.

$$(\text{width})(\text{length}) = \text{Area}$$
$$x(332x + 8) = 6{,}759$$
$$332x^2 + 8x - 6{,}759 = 0$$
$$x = \frac{-b \pm \sqrt{b^2 - 4ac}}{2a}$$
$$= \frac{-8 \pm \sqrt{(8)^2 - 4(332)(-6{,}759)}}{2(332)}$$
$$= \frac{-8 \pm \sqrt{64 - (-8{,}975{,}952)}}{664}$$
$$= \frac{-8 \pm \sqrt{8{,}976{,}016}}{664}$$
$$= \frac{-8 \pm 2{,}996}{664}$$
$$x = \frac{-8 + 2{,}996}{664} \quad \text{or} \quad x = \frac{-8 - 2{,}996}{664}$$
$$x = \frac{2{,}988}{664} \qquad x = \frac{-3{,}004}{664}$$
$$x = 4.5\ \text{ft} \qquad x = -4.52$$
$$332x + 8 = 332(4.5) + 8$$
$$= 1{,}502\ \text{ft}$$

The dimensions are 4.5 ft by 1,502 ft.

18. Let x = the number of minutes it takes the assistant to make the pastry dessert.

Assistant's work in 1 minute = $\dfrac{1}{x}$

Chef's work in 1 minute = $\dfrac{1}{x-8}$

Work together in 1 minute = $\dfrac{1}{25}$

Assistant + chef = work together

$$\frac{1}{x}+\frac{1}{x-8}=\frac{1}{25}$$

$$25x(x-8)\left(\frac{1}{x}+\frac{1}{x-8}\right)=25x(x-8)\left(\frac{1}{25}\right)$$

$$25(x-8)+25x=x(x-8)$$

$$25x-200+25x=x^2-8x$$

$$50x-200=x^2-8x$$

$$0=x^2-58x+200$$

$$x=\frac{-b\pm\sqrt{b^2-4ac}}{2a}$$

$$x=\frac{-(-58)\pm\sqrt{(-58)^2-4(1)(200)}}{2(1)}$$

$$x=\frac{58\pm\sqrt{3{,}364-800}}{2}$$

$$x=\frac{58\pm\sqrt{2{,}564}}{2}$$

$$x=\frac{58\pm 50.6}{2}$$

$$x=\frac{58+50.6}{2} \quad \text{or} \quad x=\frac{58-50.6}{2}$$

$$x=\frac{108.6}{2} \qquad x=\frac{8.6}{2}$$

$$x=54.3 \qquad x=4.3$$

$$x-8=54.3-8 \qquad x-8=4.3-8$$

$$=46.3 \qquad = \cancel{-3.7}$$

It takes the assistant about 54 minutes and the chef about 46 minutes to make the pastry dessert.

19. Let x = shorter side of a triangle.
Then $x + 14$ = longer leg of a triangle.
Use Pythagorean Theorem to find x.

$$x^2+(x+14)^2=26^2$$

$$x^2+x^2+28x+196=676$$

$$2x^2+28x-480=0$$

$$\frac{2x^2}{2}+\frac{28x}{2}-\frac{480}{2}=\frac{0}{2}$$

$$x^2+14x-240=0$$

$$(x-10)(x+24)=0$$

$$x-10=0 \quad \text{or} \quad x+24=0$$

$$x=10 \qquad x=-24$$

The segment extending from ground to the top of the building is the sum of 2 shorter legs. Since the length of one shorter leg is 10 in., the length of the segment is 20 in.

20. The vertex is (0, 6) and the parabola is going down. The mathematical model that has those characteristics is **iii.**

21. $f(x)=-3(x-1)^2-2$

Vertex: (1, –2); Axis of Symmetry: $x = 1$

22. $f(x)=5x^2+10x-1$

$$f(x)=5(x^2+2x)-1$$

$$f(x)=5(x^2+2x+1)-1-5(1)$$

$$f(x)=5(x+1)^2-1-5$$

$$f(x)=5(x+1)^2-6$$

Vertex: (–1, –6)

Axis of Symmetry: $x = -1$

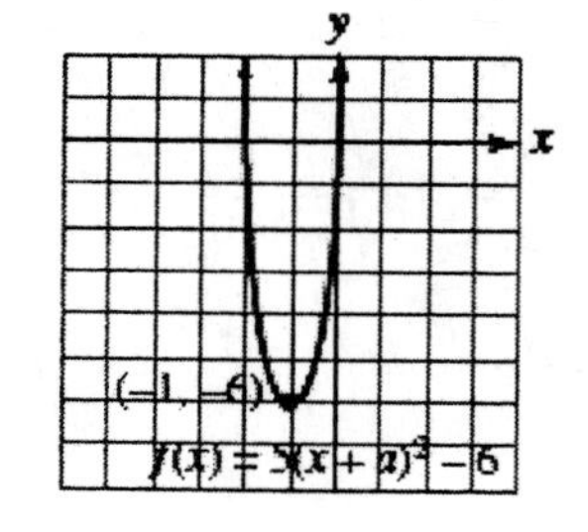

23. $f(x) = 2x^2 + x - 1$

Step 1: Since $a = 2 > 0$, the parabola opens upward.

Step 2: Find the vertex and axis of symmetry.

$$x = \frac{-b}{2a} = \frac{-1}{2(2)} = -\frac{1}{4}$$

$$y = 2\left(-\frac{1}{4}\right)^2 + \left(-\frac{1}{4}\right) - 1$$

$$= 2\left(\frac{1}{16}\right) - \frac{1}{4} - 1$$

$$= \frac{1}{8} - \frac{2}{8} - \frac{8}{8}$$

$$= -\frac{9}{8}$$

vertex: $\left(-\frac{1}{4}, -\frac{9}{8}\right)$

The axis of symmetry is $x = -\frac{1}{4}$.

Step 3: Find the x– and y–intercepts. Since $c = -1$, the y–intercept is (0, –1). To find the x–intercepts, let $y = 0$ and solve the equation for x.

$$0 = 2x^2 + x - 1$$

$$0 = (2x - 1)(x + 1)$$

$$2x - 1 = 0 \quad \text{or} \quad x + 1 = 0$$

$$x = \frac{1}{2} \qquad\qquad x = -1$$

x–intercepts: $\left(\frac{1}{2}, 0\right)$ and (–1, 0)

y–intercept: (0, –1)

Step 4: Let $x = 1$ to find another point.

$$y = 2(1)^2 + (1) - 1 = 2 + 1 - 1 = 2$$

(1, 2)

Use symmetry to find the points (–1, –1) and (–2, 2).

Step 5: Plot the points and draw the parabola.

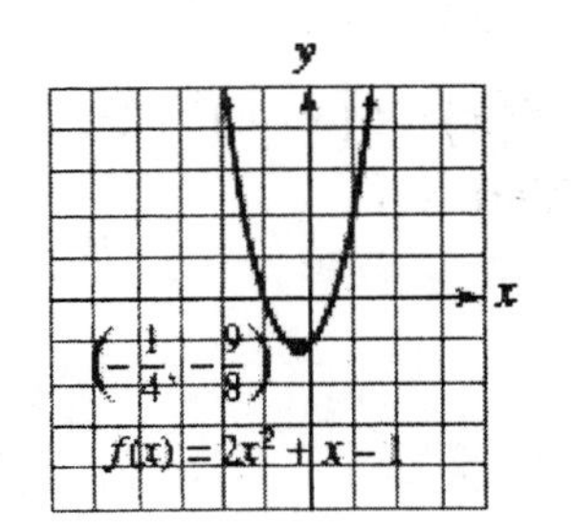

24. The vertex of the parabola represented by the equation would be the highest point. Use the vertex formula to find the vertex.

$$h = -16t^2 + 112t + 15$$

$$x = \frac{-b}{2a} = \frac{-112}{2(-16)} = \frac{-112}{-32} = 3.5$$

$$y = -16(3.5)^2 + 112(3.5) + 15$$

$$= -16(12.25) + 112(3.5) + 15$$

$$= -196 + 392 + 15$$

$$= 211 \text{ ft}$$

The flare explodes at 211 ft.

25.

$$x^2 - 2x > 8$$

$$x^2 - 2x - 8 > 0$$

$$x^2 - 2x - 8 = 0$$

$$(x - 4)(x + 2) = 0$$

$$x - 4 = 0 \quad \text{or} \quad x + 2 = 0$$

$$x = 4 \qquad\qquad x = -2$$

critical numbers = 4 and -2

$$(-\infty, -2) \cup (4, \infty)$$

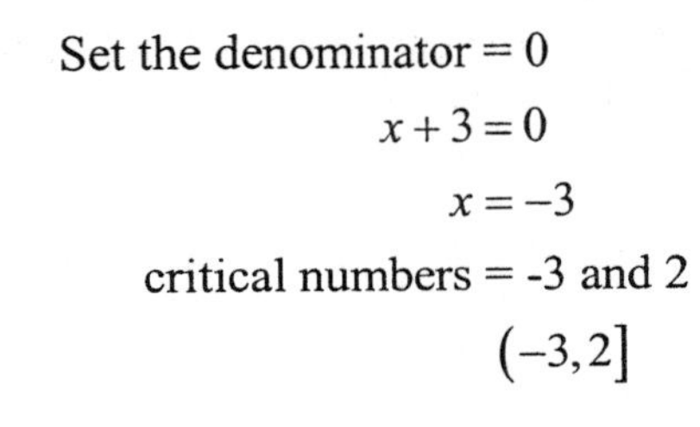

26.

$$\frac{x-2}{x+3} \le 0$$

$$\frac{x-2}{x+3} = 0$$

$$(x+3)\left(\frac{x-2}{x+3}\right) = (x+3)(0)$$

$$x - 2 = 0$$

$$x = 2$$

Set the denominator = 0

$$x + 3 = 0$$

$$x = -3$$

critical numbers = -3 and 2

$$(-3, 2]$$

27. We do not know whether x is positive or negative. When we multiply both sides by x we don't know whether or not to reverse the inequality symbol.

28. $y \le -x^2 + 3$

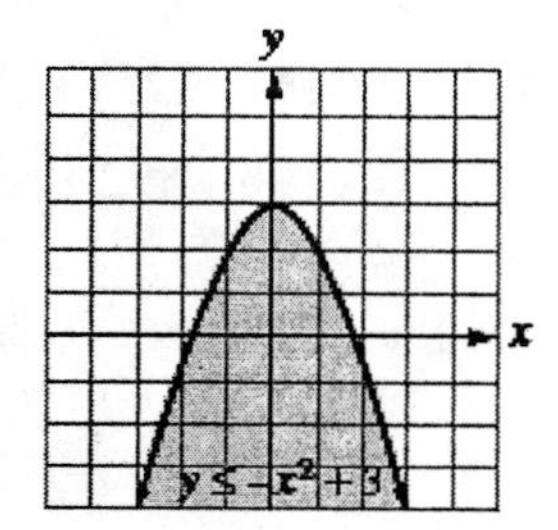

29. The x–intercepts are (–3, 0) and (2, 0), so the solutions to the equation are $x = -3$ and $x = 2$.

30. [–3, 2]

CHAPTER 10 CUMULATIVE REVIEW

1. Use point–slope formula.

$$\begin{aligned} y-y_1 &= m(x-x_1) \\ y-(-4) &= 3[x-(-2)] \\ y+4 &= 3(x+2) \\ y+4 &= 3x+6 \\ y &= 3x+2 \end{aligned}$$

2. Parallel lines have the same slope, so find the slope of the given equation by solving the equation for y.

$$\begin{aligned} 2x+3y &= 6 \\ 3y &= -2x+6 \\ y &= -\frac{2}{3}x+2 \\ m &= -\frac{2}{3} \end{aligned}$$

Use the point–slope formula.

$$\begin{aligned} y-y_1 &= m(x-x_1) \\ y-(-2) &= -\frac{2}{3}(x-0) \\ y+2 &= -\frac{2}{3}x-0 \\ y &= -\frac{2}{3}x-2 \end{aligned}$$

3. Find the slope using the ordered pairs (2000, 800) and (2015, 1,000).

$$\begin{aligned} m &= \frac{y_2-y_1}{x_2-x_1} \\ &= \frac{1{,}000-800}{2015-2000} \\ &= \frac{200}{15} \\ &= 13.333 \end{aligned}$$

The projected average rate is an increase of about 13,333 a year.

4.

$$\begin{cases} y = -\frac{5}{2}x + \frac{1}{2} \\ 2x - \frac{3}{2}y = 5 \end{cases}$$

The lines intersect at (1, –2).

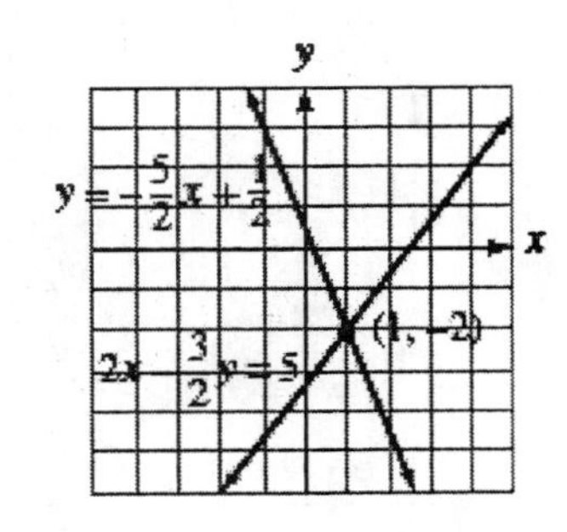

5.

$$x = \frac{D_x}{D}$$

$$= \frac{\begin{vmatrix} 4 & -1 & 1 \\ -1 & 2 & -1 \\ -2 & 1 & -3 \end{vmatrix}}{\begin{vmatrix} 1 & -1 & 1 \\ 1 & 2 & -1 \\ 1 & 1 & -3 \end{vmatrix}}$$

$$= \frac{4\begin{vmatrix} 2 & -1 \\ 1 & -3 \end{vmatrix} + 1\begin{vmatrix} -1 & -1 \\ -2 & -3 \end{vmatrix} + 1\begin{vmatrix} -1 & 2 \\ -2 & 1 \end{vmatrix}}{1\begin{vmatrix} 2 & -1 \\ 1 & -3 \end{vmatrix} + 1\begin{vmatrix} 1 & -1 \\ 1 & -3 \end{vmatrix} + 1\begin{vmatrix} 1 & 2 \\ 1 & 1 \end{vmatrix}}$$

$$= \frac{4(-6+1)+1(3-2)+1(-1+4)}{1(-6+1)+1(-3+1)+1(1-2)}$$

$$= \frac{4(-5)+1(1)+1(3)}{1(-5)+1(-2)+1(-1)}$$

$$= \frac{-20+1+3}{-5-2-1}$$

$$= \frac{-16}{-8}$$

$$= 2$$

$$y = \frac{D_y}{D}$$

$$= \frac{\begin{vmatrix} 1 & 4 & 1 \\ 1 & -1 & -1 \\ 1 & -2 & -3 \end{vmatrix}}{\begin{vmatrix} 1 & -1 & 1 \\ 1 & 2 & -1 \\ 1 & 1 & -3 \end{vmatrix}}$$

$$= \frac{1\begin{vmatrix} -1 & -1 \\ -2 & -3 \end{vmatrix} - 4\begin{vmatrix} 1 & -1 \\ 1 & -3 \end{vmatrix} + 1\begin{vmatrix} 1 & -1 \\ 1 & -2 \end{vmatrix}}{1\begin{vmatrix} 2 & -1 \\ 1 & -3 \end{vmatrix} + 1\begin{vmatrix} 1 & -1 \\ 1 & -3 \end{vmatrix} + 1\begin{vmatrix} 1 & 2 \\ 1 & 1 \end{vmatrix}}$$

$$= \frac{1(3-2)-4(-3+1)+1(-2+1)}{1(-6+1)+1(-3+1)+1(1-2)}$$

$$= \frac{1(1)-4(-2)+1(-1)}{1(-5)+1(-2)+1(-1)}$$

$$= \frac{1+8-1}{-5-2-1}$$

$$= \frac{8}{-8}$$

$$= -1$$

$$z = \frac{D_z}{D}$$

$$= \frac{\begin{vmatrix} 1 & -1 & 4 \\ 1 & 2 & -1 \\ 1 & 1 & -2 \end{vmatrix}}{\begin{vmatrix} 1 & -1 & 1 \\ 1 & 2 & -1 \\ 1 & 1 & -3 \end{vmatrix}}$$

$$= \frac{1\begin{vmatrix} 2 & -1 \\ 1 & -2 \end{vmatrix} + 1\begin{vmatrix} 1 & -1 \\ 1 & -2 \end{vmatrix} + 4\begin{vmatrix} 1 & 2 \\ 1 & 1 \end{vmatrix}}{1\begin{vmatrix} 2 & -1 \\ 1 & -3 \end{vmatrix} + 1\begin{vmatrix} 1 & -1 \\ 1 & -3 \end{vmatrix} + 1\begin{vmatrix} 1 & 2 \\ 1 & 1 \end{vmatrix}}$$

$$= \frac{1(-4+1)+1(-2+1)+4(1-2)}{1(-6+1)+1(-3+1)+1(1-2)}$$

$$= \frac{1(-3)+1(-1)+4(-1)}{1(-5)+1(-2)+1(-1)}$$

$$= \frac{-3-1-4}{-5-2-1}$$

$$= \frac{-8}{-8}$$

$$= 1$$

The solution is (2, –1, 1).

6.

$$\begin{cases} 3x + 2y > 6 \\ x + 3y \le 2 \end{cases}$$

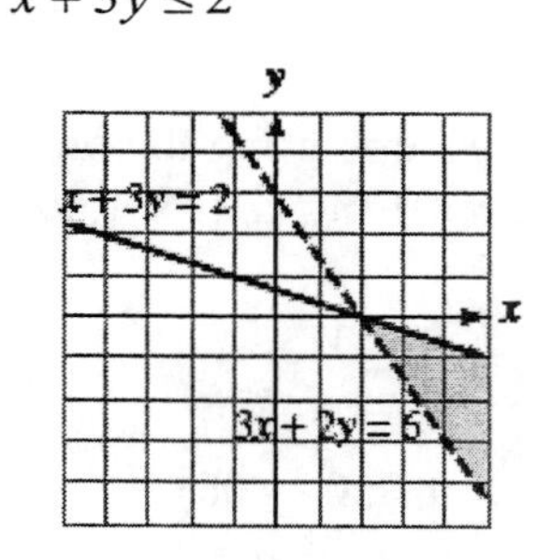

7.

$$5(-2x+2) > 20 - x$$
$$-10x + 10 > 20 - x$$
$$-9x + 10 > 20$$
$$-9x > 10$$
$$x < -\frac{10}{9}$$
$$\left(-\infty, -\frac{10}{9}\right)$$

$-\frac{10}{9}$

8.

$$|2x-5| \geq 25$$
$$2x - 5 \geq 25 \quad \text{or} \quad 2x - 5 \leq -25$$
$$2x \geq 30 \qquad 2x \leq -20$$
$$x \geq 15 \qquad x \leq -10$$
$$(-\infty, -10] \cup [15, \infty)$$

-10 15

9.

$$\left(\frac{-3x^4y^2}{-9x^5y^{-2}}\right)^2 = \left(\frac{1}{3}x^{4-5}y^{2-(-2)}\right)^2$$
$$= \left(\frac{1}{3}x^{-1}y^4\right)^2$$
$$= \left(\frac{y^4}{3x}\right)^2$$
$$= \frac{y^{4 \cdot 2}}{3^2x^2}$$
$$= \frac{y^8}{9x^2}$$

10. a. all real numbers from 0 through 24
b. $f(3) = 0.5$
c. $f(6) = 1.5$
d. $f(15) =$ about -1.4
e. The low tide mark was -2.5 m.
f. 0, 2, 9, 17

11.

$$(a+2)(3a^2 + 4a - 2)$$
$$= a(3a^2 + 4a - 2) + 2(3a^2 + 4a - 2)$$
$$= 3a^3 + 4a^2 - 2a + 6a^2 + 8a - 4$$
$$= 3a^3 + 10a^2 + 6a - 4$$

12.

$$\frac{x^3 + y^3}{x^3 - y^3} \div \frac{x^2 - xy + y^2}{x^2 + xy + y^2}$$
$$= \frac{x^3 + y^3}{x^3 - y^3} \bullet \frac{x^2 + xy + y^2}{x^2 - xy + y^2}$$
$$= \frac{(x+y)\cancel{(x^2 - xy + y^2)}\cancel{(x^2 + xy + y^2)}}{(x-y)\cancel{(x^2 + xy + y^2)}\cancel{(x^2 - xy + y^2)}}$$
$$= \frac{x+y}{x-y}$$

13.

$$\frac{1}{x+y} - \frac{1}{x-y} - \frac{2y}{y^2 - x^2}$$
$$= \frac{1}{x+y} - \frac{1}{x-y} - \frac{2y}{(y-x)(y+x)}$$
$$= \frac{1}{x+y} - \frac{1}{x-y} - \frac{-2y}{(x-y)(x+y)}$$
$$= \frac{1}{x+y}\left(\frac{x-y}{x-y}\right) - \frac{1}{x-y}\left(\frac{x+y}{x+y}\right) - \frac{-2y}{(x-y)(x+y)}$$
$$= \frac{x-y}{(x+y)(x-y)} - \frac{x+y}{(x+y)(x-y)} - \frac{-2y}{(x+y)(x-y)}$$
$$= \frac{x-y-(x+y)-(-2y)}{(x+y)(x-y)}$$
$$= \frac{x-y-x-y+2y}{(x+y)(x-y)}$$
$$= \frac{0}{(x+y)(x-y)}$$
$$= 0$$

14.

$$\frac{\frac{1}{r^2+4r+4}}{\frac{r}{r+2}+\frac{r}{r+2}}=\frac{\frac{1}{(r+2)(r+2)}}{\frac{r}{r+2}+\frac{r}{r+2}}$$

$$=\frac{\frac{1}{(r+2)(r+2)}}{\frac{r}{r+2}+\frac{r}{r+2}}\bullet\frac{(r+2)(r+2)}{(r+2)(r+2)}$$

$$=\frac{\frac{(r+2)(r+2)}{(r+2)(r+2)}}{\frac{r(r+2)(r+2)}{r+2}+\frac{r(r+2)(r+2)}{r+2}}$$

$$=\frac{1}{r(r+2)+r(r+2)}$$

$$=\frac{1}{r^2+2r+r^2+2r}$$

$$=\frac{1}{2r^2+4r}$$

$$=\frac{1}{2r(r+2)}$$

15.

$$x^4-16y^4=(x^2+4y^2)(x^2-4y^2)$$
$$=(x^2+4y^2)(x+2y)(x-2y)$$

16.

$$30a^4-4a^3-16a^2=2a^2(15a^2-2a-8)$$
$$=2a^2(3a+2)(5a-4)$$

17.

$$x^2+4y-xy-4x=x^2-xy-4x+4y$$
$$=x(x-y)-4(x-y)$$
$$=(x-y)(x-4)$$

18.

$$8x^6+125y^3$$
$$=(2x^2+5y)\left[(2x^2)^2-(2x^2)(5y)+(5y)^2\right]$$
$$=(2x^2+5y)(4x^4-10x^2y+25y^2)$$

19.

$$x^2+10x+25-y^8=(x^2+10x+25)-y^8$$
$$=(x+5)^2-y^8$$
$$=(x+5-y^4)(x+5+y^4)$$

20.

$$49s^6-84s^3n^2+36n^4=(7s^3-6n^2)^2$$

21.

$$(m+4)(2m+3)-22=10m$$
$$2m^2+3m+8m+12-22=10m$$
$$2m^2+11m-10=10m$$
$$2m^2+m-10=0$$
$$(2m+5)(m-2)=0$$
$$2m+5=0 \quad \text{or} \quad m-2=0$$
$$m=-\frac{5}{2} \qquad m=2$$

22.

$$6a^3-2a=a^2$$
$$6a^3-a^2-2a=0$$
$$a(6a^2-a-2)=0$$
$$a(2a+1)(3a-2)=0$$
$$a=0 \quad 2a+1=0 \quad \text{or} \quad 3a-2=0$$
$$a=0 \qquad a=-\frac{1}{2} \qquad a=\frac{2}{3}$$

23.

$$\frac{x-4}{x-3}+\frac{x-2}{x-3}=x-3$$
$$x-3\left(\frac{x-4}{x-3}+\frac{x-2}{x-3}\right)=(x-3)(x-3)$$
$$x-4+x-2=x^2-3x-3x+9$$
$$2x-6=x^2-6x+9$$
$$0=x^2-8x+15$$
$$0=(x-3)(x-5)$$
$$x-3=0 \quad \text{or} \quad x-5=0$$
$$x=\cancel{3} \qquad x=5$$

3 is extraneous

24.

$$P+\frac{a}{V^2}=\frac{RT}{V-b}$$
$$V^2(V-b)\left(P+\frac{a}{V^2}\right)=V^2(V-b)\left(\frac{RT}{V-b}\right)$$
$$PV^2(V-b)+a(V-b)=RTV^2$$
$$PV^3-PV^2b+aV-ab=RTV^2$$
$$-PV^2b-ab=RTV^2-PV^3-aV$$
$$b(-PV^2-a)=RTV^2-PV^3-aV$$
$$\frac{b(-PV^2-a)}{-PV^2-a}=\frac{RTV^2-PV^3-aV}{-PV^2-a}$$
$$b=-\frac{RTV^2-PV^3-aV}{PV^2+a}$$
$$b=\frac{-RTV^2+PV^3+aV}{PV^2+a}$$

25. $f(x)=x^3+x^2-6x$

$D:(-\infty,\infty), R:(-\infty,\infty)$

26. $f(x)=\frac{4}{x}$

$D:(0,\infty), R:(0,\infty)$

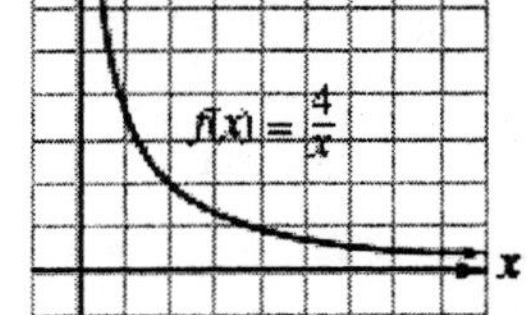

27. Let I = intensity and d = distance from the source.

$$I=kd^2$$
$$4=k(2)^2$$
$$4=k(4)$$
$$4=4k$$
$$1=k$$

Find I for $k=1$ and $d=3$.

$$I=kd^2$$
$$I=(1)(3)^2$$
$$I=(1)(9)$$
$$I=9$$

28. $f(x)=\sqrt{x-2}$

$D:[2,\infty), R:[0,\infty)$

29.

$$\sqrt[3]{-27x^3}=-3x$$

30.

$$\sqrt{48t^3}=\sqrt{16t^2}\sqrt{3t}$$
$$=4t\sqrt{3t}$$

31.

$$64^{-2/3}=\frac{1}{64^{2/3}}$$
$$=\frac{1}{\left(\sqrt[3]{64}\right)^2}$$
$$=\frac{1}{(4)^2}$$
$$=\frac{1}{16}$$

32.

$$\frac{x^{5/3}x^{1/2}}{x^{3/4}} = \frac{x^{5/3+1/2}}{x^{3/4}}$$
$$= \frac{x^{10/6+3/6}}{x^{3/4}}$$
$$= \frac{x^{13/6}}{x^{3/4}}$$
$$= x^{13/6-3/4}$$
$$= x^{26/12-9/12}$$
$$= x^{17/12}$$

33.

$$-3\sqrt[4]{32} - 2\sqrt[4]{162} + 5\sqrt[4]{48}$$
$$= -3\sqrt[4]{16}\sqrt[4]{2} - 2\sqrt[4]{81}\sqrt[4]{2} + 5\sqrt[4]{16}\sqrt[4]{3}$$
$$= -3(2)\sqrt[4]{2} - 2(3)\sqrt[4]{2} + 5(2)\sqrt[4]{3}$$
$$= -6\sqrt[4]{2} - 6\sqrt[4]{2} + 10\sqrt[4]{3}$$
$$= -12\sqrt[4]{2} + 10\sqrt[4]{3}$$

34.

$$3\sqrt{2}\left(2\sqrt{3} - 4\sqrt{12}\right) = 6\sqrt{6} - 12\sqrt{24}$$
$$= 6\sqrt{6} - 12\sqrt{4}\sqrt{6}$$
$$= 6\sqrt{6} - 12(2)\sqrt{6}$$
$$= 6\sqrt{6} - 24\sqrt{6}$$
$$= -18\sqrt{6}$$

35.

$$\frac{\sqrt{x}+2}{\sqrt{x}-1} = \frac{\sqrt{x}+2}{\sqrt{x}-1} \bullet \frac{\sqrt{x}+1}{\sqrt{x}+1}$$
$$= \frac{\left(\sqrt{x}+2\right)\left(\sqrt{x}+1\right)}{\left(\sqrt{x}-1\right)\left(\sqrt{x}+1\right)}$$
$$= \frac{\sqrt{x^2} + \sqrt{x} + 2\sqrt{x} + 2}{\sqrt{x^2} - 1}$$
$$= \frac{x + 3\sqrt{x} + 2}{x - 1}$$

36.

$$\frac{5}{\sqrt[3]{x}} = \frac{5}{\sqrt[3]{x}} \bullet \frac{\sqrt[3]{x^2}}{\sqrt[3]{x^2}}$$
$$= \frac{5\sqrt[3]{x^2}}{\sqrt[3]{x^3}}$$
$$= \frac{5\sqrt[3]{x^2}}{x}$$

37.

$$5\sqrt{x+2} = x+8$$
$$\left(5\sqrt{x+2}\right)^2 = (x+8)^2$$
$$25(x+2) = x^2 + 8x + 8x + 64$$
$$25x + 50 = x^2 + 16x + 64$$
$$0 = x^2 - 9x + 14$$
$$0 = (x-2)(x-7)$$
$$x - 2 = 0 \quad \text{or} \quad x - 7 = 0$$
$$x = 2 \qquad\qquad x = 7$$

38.

$$\sqrt{x} + \sqrt{x+2} = 2$$
$$\sqrt{x+2} = 2 - \sqrt{x}$$
$$\left(\sqrt{x+2}\right)^2 = \left(2 - \sqrt{x}\right)^2$$
$$x + 2 = 4 - 2\sqrt{x} - 2\sqrt{x} + \sqrt{x^2}$$
$$x + 2 = 4 - 4\sqrt{x} + x$$
$$x + 2 - 4 - x = 4 - 4\sqrt{x} + x - 4 - x$$
$$-2 = -4\sqrt{x}$$
$$(-2)^2 = \left(-4\sqrt{x}\right)^2$$
$$4 = 16x$$
$$\frac{4}{16} = x$$
$$\frac{1}{4} = x$$

39. The length of the hypotenuse in an isosceles right triangles is the length of the leg times $\sqrt{2}$.
The length of the hypotenuse is $3\sqrt{2}$ in.

40. The length of the longer leg is $\sqrt{3}$ times the length of the shorter leg. The length of the hypotenuse is 2 times the length of the shorter side.

Let x = length of the shorter leg.

longer leg:

$$x\sqrt{3}=3$$

$$\frac{x\sqrt{3}}{\sqrt{3}}=\frac{3}{\sqrt{3}}$$

$$x=\frac{3}{\sqrt{3}}\bullet\frac{\sqrt{3}}{\sqrt{3}}$$

$$x=\frac{3\sqrt{3}}{3}$$

$$x=\sqrt{3}\text{ in.}$$

hypotenuse:

$$2x=2\sqrt{3}\text{ in.}$$

41.

$$d=\sqrt{(x_1-x_2)^2+(y_1-y_2)^2}$$

$$=\sqrt{[4-(-2)]^2+(14-6)^2}$$

$$=\sqrt{(6)^2+(8)^2}$$

$$=\sqrt{36+64}$$

$$=\sqrt{100}$$

$$=10$$

42.

$$i^{43}=i^{40}i^2i$$

$$=1(-1)i$$

$$=-i$$

43.

$$\left(-7+\sqrt{-81}\right)-\left(-2-\sqrt{-64}\right)$$

$$=(-7+9i)-(-2-8i)$$

$$=-7+9i+2+8i$$

$$=-5+17i$$

44.

$$\frac{5}{3-i}=\frac{5}{3-i}\bullet\frac{3+i}{3+i}$$

$$=\frac{5(3+i)}{(3-i)(3+i)}$$

$$=\frac{15+5i}{9-i^2}$$

$$=\frac{15+5i}{9-(-1)}$$

$$=\frac{15+5i}{10}$$

$$=\frac{15}{10}+\frac{5}{10}i$$

$$=\frac{3}{2}+\frac{1}{2}i$$

45.

$$(2+i)^2=(2+i)(2+i)$$

$$=4+2i+2i+i^2$$

$$=4+4i+(-1)$$

$$=3+4i$$

46.

$$\frac{-4}{6i^7}=\frac{-4}{6i^7}\bullet\frac{i}{i}$$

$$=\frac{-4i}{6i^8}$$

$$=\frac{-4i}{6(1)}$$

$$=-\frac{4}{6}i$$

$$=-\frac{2}{3}i$$

$$=0-\frac{2}{3}i$$

47. You must add 9 to make a perfect square trinomial.

$$\left(\frac{6}{2}\right)^2=(3)^2=9$$

48.

$$2x^2+x-3=0$$
$$2x^2+x=3$$
$$\frac{2x^2}{2}+\frac{x}{2}=\frac{3}{2}$$
$$x^2+\frac{1}{2}x=\frac{3}{2}$$
$$x^2+\frac{1}{2}x+\left(\frac{1}{2}\bullet\frac{1}{2}\right)^2=\frac{3}{2}+\left(\frac{1}{2}\bullet\frac{1}{2}\right)^2$$
$$x^2+\frac{1}{2}x+\left(\frac{1}{4}\right)^2=\frac{3}{2}+\left(\frac{1}{4}\right)^2$$
$$x^2+\frac{1}{2}x+\frac{1}{16}=\frac{3}{2}+\frac{1}{16}$$
$$\left(x+\frac{1}{4}\right)^2=\frac{24}{16}+\frac{1}{16}$$
$$\left(x+\frac{1}{4}\right)^2=\frac{25}{16}$$
$$x+\frac{1}{4}=\pm\sqrt{\frac{25}{16}}$$
$$x+\frac{1}{4}=\pm\frac{5}{4}$$
$$x=-\frac{1}{4}\pm\frac{5}{4}$$
$$x=-\frac{1}{4}+\frac{5}{4} \quad \text{or} \quad x=-\frac{1}{4}-\frac{5}{4}$$
$$x=\frac{4}{4} \qquad x=-\frac{6}{4}$$
$$x=1 \qquad x=-\frac{3}{2}$$

49.

$$\frac{a^2}{8}-\frac{a}{2}=-1$$
$$8\left(\frac{a^2}{8}-\frac{a}{2}\right)=8(-1)$$
$$a^2-4a=-8$$
$$a^2-4a+8=0$$
$$a=\frac{-(-4)\pm\sqrt{(-4)^2-4(1)(8)}}{2(1)}$$
$$=\frac{4\pm\sqrt{16-32}}{2}$$
$$=\frac{4\pm\sqrt{-16}}{2}$$
$$=\frac{4\pm 4i}{2}$$
$$=\frac{4}{2}\pm\frac{4}{2}i$$
$$=2\pm 2i$$

50. Let x = width of the uniform sidewalk.
To find the length or width, you must take the total length or width and subtract $2x$ since there is a sidewalk of width x on both ends of the garden.
The length is $(24 - 2x)$ and the width is $(16 - 2x)$. Solve the inequality to find x.

$$(\text{length})(\text{width}) \le 180$$
$$(24-2x)(16-2x) \le 180$$
$$384-48x-32x+4x^2 \le 180$$
$$4x^2-80x+384 \le 180$$
$$4x^2-80x+204 \le 0$$
$$4x^2-80x+204 = 0$$
$$\frac{4x^2}{4}-\frac{80x}{4}+\frac{204}{4}=\frac{0}{4}$$
$$x^2-20x+51=0$$
$$(x-3)(x-17)=0$$
$$x-3=0 \quad \text{or} \quad x-17=0$$
$$x=3 \qquad\qquad x=17$$

17 is not possible: it would make the length and width negative.

Length:
$$24-2(3)=24-6$$
$$=18 \text{ ft}$$

Width:
$$16-2(3)=16-6$$
$$=10 \text{ ft}$$

The dimension of the largest possible garden is 10 ft by 18 ft.

51. Let x = length of shorter leg.
Then $170 - x$ = length of longer leg.
Use the Pythagorean Theorem to find x.

$$x^2+(170-x)^2=130^2$$
$$x^2+28{,}900-170x-170x+x^2=16{,}900$$
$$2x^2-340x+28{,}900=16{,}900$$
$$2x^2-340x+12{,}000=0$$
$$\frac{2x^2}{2}-\frac{340x}{2}+\frac{12{,}000}{2}=0$$
$$x^2-170x+6{,}000=0$$
$$x=\frac{-(-170)\pm\sqrt{(-170)^2-4(1)(6{,}000)}}{2(1)}$$
$$x=\frac{170\pm\sqrt{28{,}900-24{,}000}}{2}$$
$$x=\frac{170\pm\sqrt{4{,}900}}{2}$$
$$x=\frac{170\pm 70}{2}$$
$$x=\frac{170+70}{2} \quad \text{or} \quad x=\frac{170-70}{2}$$
$$x=\frac{240}{2} \qquad\qquad x=\frac{100}{2}$$
$$x=120 \qquad\qquad x=50$$

The two segments are 50 m and 120 m.

52. $f(x) = -x^2 - 4x$

Step 1: Since $a = -1 < 0$, the parabola opens downward.

Step 2: Find the vertex and axis of symmetry.

$$x = \frac{-b}{2a} = \frac{-(-4)}{2(-1)} = \frac{4}{-2} = -2$$

$$y = -(-2)^2 - 4(-2) = -4 + 8 = 4$$

vertex: (–2, 4)

The axis of symmetry is $x = -2$.

Step 3: Find the x– and y–intercepts.

Since $c = 0$, the y–intercept is (0, 0).

To find the x–intercept, let $y = 0$.

$$0 = -x^2 - 4x$$
$$0 = -x(x+4)$$
$$-x = 0 \quad \text{or} \quad x + 4 = 0$$
$$x = 0 \qquad\qquad x = -4$$

x–intercepts: (0, 0) and (–4, 0)

y–intercepts: (0, 0)

Step 4: Find another point(s).

Let $x = -1$.

$$y = -(-1)^2 - 4(-1) = -1 + 4 = 3$$

(–1, 3)

Use symmetry to find the point (–3, 3)

Step 5: Plots the points and draw the parabola.

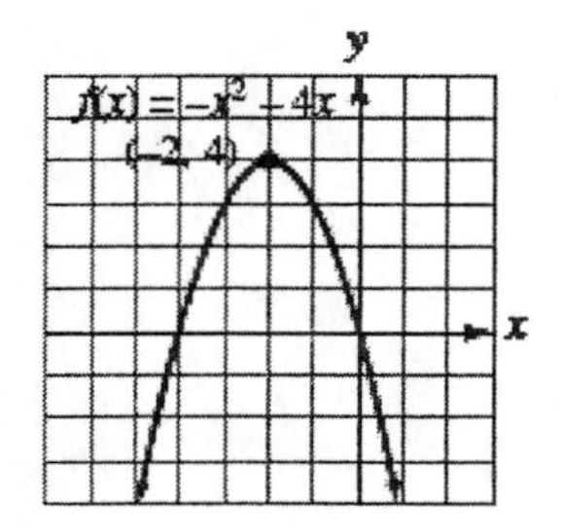

53. Let $y = a^{1/2}$.

$$a - 7a^{1/2} + 12 = 0$$
$$\left(a^{1/2}\right)^2 - 7a^{1/2} + 12 = 0$$
$$y^2 - 7y + 12 = 0$$
$$(y-3)(y-4) = 0$$
$$y - 3 = 0 \quad \text{or} \quad y - 4 = 0$$
$$y = 3 \qquad\qquad y = 4$$
$$a^{1/2} = 3 \qquad\qquad a^{1/2} = 4$$
$$\left(a^{1/2}\right)^2 = (3)^2 \qquad \left(a^{1/2}\right)^2 = (4)^2$$
$$a = 9 \qquad\qquad a = 16$$

54. Let $y = \frac{1}{x^2}$.

$$x^{-4} - 2x^{-2} + 1 = 0$$
$$\frac{1}{x^4} - 2\left(\frac{1}{x^2}\right) + 1 = 0$$
$$\left(\frac{1}{x^2}\right)^2 - 2\left(\frac{1}{x^2}\right) + 1 = 0$$
$$y^2 - 2y + 1 = 0$$
$$(y-1)(y-1) = 0$$
$$y - 1 = 0 \quad \text{or} \quad y - 1 = 0$$
$$y = 1 \qquad\qquad y = 1$$
$$\frac{1}{x^2} = 1 \qquad\qquad \frac{1}{x^2} = 1$$
$$x^2 = 1 \qquad\qquad x^2 = 1$$
$$x = \pm\sqrt{1} \qquad\qquad x = \pm\sqrt{1}$$
$$x = \pm 1 \qquad\qquad x = \pm 1$$
$$x = 1,\ -1,\ 1,\ -1$$

55. The x–intercept is $\left(-\frac{3}{4}, 0\right)$, so the solution to the equation is $x = -\frac{3}{4}$.

56. Use $x = -\frac{3}{4}$ as the critical number. The answer would be **<u>no solution</u>**.

SECTION 11.1

VOCABULARY

1. The **sum** of f and g, denoted as $f + g$, is defined by $(f+g)(x) = \boxed{f(x)+g(x)}$ and the **difference** of f and g, denoted as $f - g$, is defined by $(f-g)(x) = \boxed{f(x)-g(x)}$.

3. The **domain** of the function $f + g$ is the set of real numbers x that are in the domain of both f and g.

5. Under the **identity** function, the value that is assigned to any real number x is x itself.

CONCEPTS

7. a. $(f \circ g)(3) = f\left(\boxed{g(3)}\right)$

   b. To find $f(g(3))$, we first find $\boxed{g(3)}$ and then substitute that value for x in $f(x)$.

9.

$$\begin{aligned} f(-2) &= 2-3(-2)^2 \\ &= 2-3(4) \\ &= 2-12 \\ &= -10 \\ g(-10) &= -10+10 \\ &= 0 \end{aligned}$$

11. a.

$$\begin{aligned} (f+g)(1) &= f(1)+g(1) \\ &= 3+4 \\ &= 7 \end{aligned}$$

b.

$$\begin{aligned} (f-g)(5) &= f(5)-g(5) \\ &= 8-0 \\ &= 8 \end{aligned}$$

c.

$$\begin{aligned} (f \bullet g)(1) &= f(1)g(1) \\ &= 3(4) \\ &= 12 \end{aligned}$$

d.

$$\begin{aligned} (g/f)(5) &= \frac{g(5)}{f(5)} \\ &= \frac{0}{8} \\ &= 0 \end{aligned}$$

NOTATION

13.

$$\begin{aligned} (f \bullet g)(x) &= f(x) \bullet \boxed{g(x)} \\ &= \boxed{(3x-1)}(2x+3) \\ &= 6x^2 + \boxed{9x} - \boxed{2x} - 3 \\ (f \bullet g)(x) &= 6x^2 + 7x - 3 \end{aligned}$$

PRACTICE

15. $f + g$

$$\begin{aligned} (f+g)(x) &= f(x)+g(x) \\ &= 3x+4x \\ &= 7x \\ D&:(-\infty,\infty) \end{aligned}$$

17. $f \cdot g$

$$\begin{aligned} (f \bullet g)(x) &= f(x) \bullet g(x) \\ &= (3x)(4x) \\ &= 12x^2 \\ D&:(-\infty,\infty) \end{aligned}$$

19. $g-f$

$$\begin{aligned}(g-f)(x)&=g(x)-f(x)\\&=4x-3x\\&=x\\D&:(-\infty,\infty)\end{aligned}$$

21. g/f

$$\begin{aligned}(g/f)(x)&=\frac{g(x)}{f(x)}\\&=\frac{4x}{3x}\\&=\frac{4}{3}\\D&:(-\infty,0)\cup(0,\infty)\end{aligned}$$

23. $f+g$

$$\begin{aligned}(f+g)(x)&=f(x)+g(x)\\&=2x+1+x-3\\&=3x-2\\D&:(-\infty,\infty)\end{aligned}$$

25. $f\cdot g$

$$\begin{aligned}(f\bullet g)(x)&=f(x)\bullet g(x)\\&=(2x+1)(x-3)\\&=2x^2-6x+x-3\\&=2x^2-5x-3\\D&:(-\infty,\infty)\end{aligned}$$

27. $g-f$

$$\begin{aligned}(g-f)(x)&=g(x)-f(x)\\&=(x-3)-(2x+1)\\&=x-3-2x-1\\&=-x-4\\D&:(-\infty,\infty)\end{aligned}$$

29. g/f

$$\begin{aligned}(g/f)(x)&=\frac{g(x)}{f(x)}\\&=\frac{x-3}{2x+1}\\D&:\left(-\infty,-\frac{1}{2}\right)\cup\left(-\frac{1}{2},\infty\right)\end{aligned}$$

31. $f-g$

$$\begin{aligned}(f-g)(x)&=f(x)-g(x)\\&=(3x-2)-\left(2x^2+1\right)\\&=3x-2-2x^2-1\\&=-2x^2+3x-3\\D&:(-\infty,\infty)\end{aligned}$$

33. f/g

$$\begin{aligned}(f/g)(x)&=\frac{f(x)}{g(x)}\\&=\frac{3x-2}{2x^2+1}\\D&:(-\infty,\infty)\end{aligned}$$

35. $f-g$

$$\begin{aligned}(f-g)(x)&=f(x)-g(x)\\&=\left(x^2-1\right)-\left(x^2-4\right)\\&=x^2-1-x^2+4\\&=3\\D&:(-\infty,\infty)\end{aligned}$$

37. g/f

$$\begin{aligned}(g/f)(x)&=\frac{g(x)}{f(x)}\\&=\frac{x^2-4}{x^2-1}\\D&:(-\infty,-1)\cup(-1,1)\cup(1,\infty)\end{aligned}$$

39. $(f\circ g)(2)$

$$\begin{aligned}(f\circ g)(2)&=f(g(2))\\&=f\left(2^2-1\right)\\&=f(4-1)\\&=f(3)\\&=2(3)+1\\&=6+1\\&=7\end{aligned}$$

41. $(g \circ f)(-3)$

$$\begin{aligned}(g \circ f)(-3) &= g(f(-3)) \\ &= g(2(-3)+1) \\ &= g(-6+1) \\ &= g(-5) \\ &= (-5)^2 - 1 \\ &= 25 - 1 \\ &= 24\end{aligned}$$

43. $(f \circ g)(0)$

$$\begin{aligned}(f \circ g)(0) &= f(g(0)) \\ &= f(0^2 - 1) \\ &= f(0-1) \\ &= f(-1) \\ &= 2(-1)+1 \\ &= -2+1 \\ &= -1\end{aligned}$$

45. $(f \circ g)\left(\frac{1}{2}\right)$

$$\begin{aligned}(f \circ g)\left(\frac{1}{2}\right) &= f\left(g\left(\frac{1}{2}\right)\right) \\ &= f\left(\left(\frac{1}{2}\right)^2 - 1\right) \\ &= f\left(\frac{1}{4} - 1\right) \\ &= f\left(-\frac{3}{4}\right) \\ &= 2\left(-\frac{3}{4}\right) + 1 \\ &= \frac{-3}{2} + 1 \\ &= -\frac{1}{2}\end{aligned}$$

47. $(f \circ g)(x)$

$$\begin{aligned}(f \circ g)(x) &= f(g(x)) \\ &= f(x^2 - 1) \\ &= 2(x^2 - 1) + 1 \\ &= 2x^2 - 2 + 1 \\ &= 2x^2 - 1\end{aligned}$$

49. $(g \circ f)(2x)$

$$\begin{aligned}(g \circ f)(2x) &= g(f(2x)) \\ &= g(2(2x)+1) \\ &= g(4x+1) \\ &= (4x+1)^2 - 1 \\ &= 16x^2 + 8x + 1 - 1 \\ &= 16x^2 + 8x\end{aligned}$$

51. $(f \circ g)(4)$

$$\begin{aligned}(f \circ g)(4) &= f(g(4)) \\ &= f((4)^2 + 4) \\ &= f(16+4) \\ &= f(20) \\ &= 3(20) - 2 \\ &= 60 - 2 \\ &= 58\end{aligned}$$

53. $(g \circ f)(-3)$

$$\begin{aligned}(g \circ f)(-3) &= g(f(-3)) \\ &= g(3(-3)-2) \\ &= g(-9-2) \\ &= g(-11) \\ &= (-11)^2 - 11 \\ &= 121 - 11 \\ &= 110\end{aligned}$$

55. $(g\circ f)(0)$

$$\begin{aligned}(g\circ f)(0)&=g(f(0))\\&=g(3(0)-2)\\&=g(0-2)\\&=g(-2)\\&=(-2)^2-2\\&=4-2\\&=2\end{aligned}$$

57. $(g\circ f)(x)$

$$\begin{aligned}(g\circ f)(x)&=g(f(x))\\&=g(3x-2)\\&=(3x-2)^2+3x-2\\&=9x^2-12x+4+3x-2\\&=9x^2-9x+2\end{aligned}$$

59. $(f\circ g)(4)$

$$\begin{aligned}(f\circ g)(4)&=f(g(4))\\&=f\left(\frac{1}{4^2}\right)\\&=f\left(\frac{1}{16}\right)\\&=\frac{1}{\frac{1}{16}}\\&=\frac{1}{1}\div\frac{1}{16}\\&=\frac{1}{1}\bullet\frac{16}{1}\\&=16\end{aligned}$$

61. $(g\circ f)\left(\frac{1}{3}\right)$

$$\begin{aligned}(g\circ f)\left(\frac{1}{3}\right)&=g\left(f\left(\frac{1}{3}\right)\right)\\&=g\left(\frac{1}{\frac{1}{3}}\right)\\&=g\left(1\div\frac{1}{3}\right)\\&=g\left(1\bullet\frac{3}{1}\right)\\&=g(3)\\&=\frac{1}{(3)^2}\\&=\frac{1}{9}\end{aligned}$$

63. $(g\circ f)(8x)$

$$\begin{aligned}(g\circ f)(8x)&=g(f(8x))\\&=g\left(\frac{1}{8x}\right)\\&=\frac{1}{\left(\frac{1}{8x}\right)^2}\\&=\frac{1}{\frac{1}{64x^2}}\\&=1\div\frac{1}{64x^2}\\&=1\bullet\frac{64x^2}{1}\\&=64x^2\end{aligned}$$

65.

$$
\begin{aligned}
(f\circ g)(x) &= f(g(x)) & (g\circ f)(x) &= g(f(x)) \\
&= f(2x-5) & &= g(x+1) \\
&= (2x-5)+1 & &= 2(x+1)-5 \\
&= 2x-4 & &= 2x+2-5 \\
& & &= 2x-3
\end{aligned}
$$

So, $(f\circ g)(x) \neq (g\circ f)(x)$.

APPLICATIONS

67. $f(3) = 500$ and $g(3) = 503$

$$
\begin{aligned}
(f+g)(3) &= f(3)+g(3) \\
&= 500+503 \\
&= 1003
\end{aligned}
$$

69. $f(10) = 505$ and $g(10) = 514$

$$
\begin{aligned}
(f+g)(10) &= f(10)+g(10) \\
&= 505+514 \\
&= 1{,}019
\end{aligned}
$$

In 2000, the average combined score on the SAT was 1,019.

71. METALLURGY

$$F(t) = -200t + 2{,}700$$

$$C(F) = \frac{5}{9}(F-32)$$

$$
\begin{aligned}
(C\circ F)(t) &= C(F(t)) \\
&= C(-200t+2{,}700) \\
&= \frac{5}{9}\left[(-200t+2{,}700)-32\right] \\
&= \frac{5}{9}(-200t+2{,}668)
\end{aligned}
$$

$$C(t) = \frac{5}{9}(2{,}668-200t)$$

73. VACATION MILEAGE COSTS

a. On the first graph, 500 miles corresponds with 25 gallons consumed. On the second graph, 25 gallons consumed corresponds with about \$37.50.

b. Find $(C\circ G)(m)$.

$$
\begin{aligned}
(C\circ G)(m) &= C(G(m)) \\
&= C\left(\frac{m}{20}\right) \\
&= 1.50\left(\frac{m}{20}\right)
\end{aligned}
$$

$$C(m) = \frac{1.50m}{20}$$

$$C(m) = 0.075m$$

WRITING

75. Answers will vary.

77. Answers will vary.

REVIEW

79.

$$
\begin{aligned}
\frac{\frac{ac-ad-c+d}{a^3-1}}{\frac{c^2-2cd+d^2}{a^2+a+1}} &= \frac{ac-ad-c+d}{a^3-1} \div \frac{c^2-2cd+d^2}{a^2+a+1} \\
&= \frac{ac-ad-c+d}{a^3-1} \bullet \frac{a^2+a+1}{c^2-2cd+d^2} \\
&= \frac{a(c-d)-1(c-d)}{(a-1)(a^2+a+1)} \bullet \frac{a^2+a+1}{(c-d)(c-d)} \\
&= \frac{\cancel{(c-d)}\cancel{(a-1)}}{\cancel{(a-1)}\cancel{(a^2+a+1)}} \bullet \frac{\cancel{a^2+a+1}}{\cancel{(c-d)}(c-d)} \\
&= \frac{1}{c-d}
\end{aligned}
$$

CHALLENGE PROBLEMS

81. If $f(x) = x^2$ and $g(x) = \boxed{2x+5}$, then $(f\circ g)(x) = 4x^2 + 20x + 25$.

SECTION 11.2

VOCABULARY

1. For a **one–to–one** function, each input is assigned exactly one output, and each output corresponds to exactly one input.

3. The functions f and f^{-1} are **inverses**.

5. The graphs of a function and its inverse are mirror **images** of each other with respect to $y = x$. We also say that their graphs are **symmetric** with respect to the line $y = x$.

CONCEPTS

7. a. If every horizontal line that intersects the graph of a function does so only **once**, the function is one–to–one.
   b. If any horizontal line that intersects the graph of a function does so more than once, the function is not **one–to–one**.

9. The graphs of a function and its inverse are symmetrical about the line $y = \boxed{x}$.

11. To find the inverse of the function $f(x) = 2x - 3$, we begin by replacing $f(x)$ with **y** and then we **interchange** x and y.

13. It is a function because every x value only has one y value. It is not one–to–one because both 3 and 7 in the domain correspond to 4 in the range.

15. Yes.

17. $f^{-1}(6) = 2$

19. The graphs are not symmetric about the line $y = x$.

21.

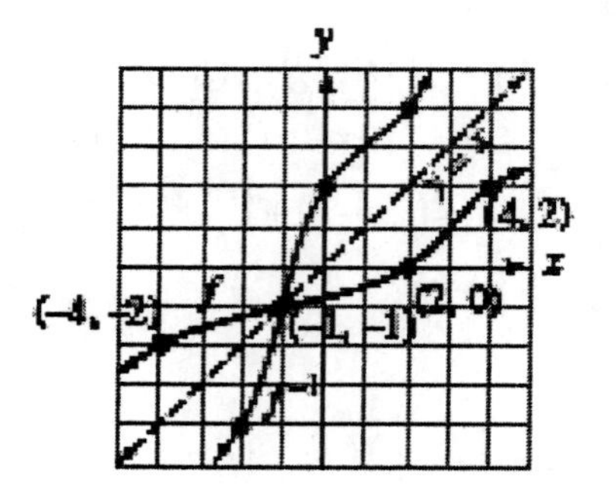

NOTATION

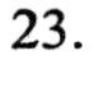

23.

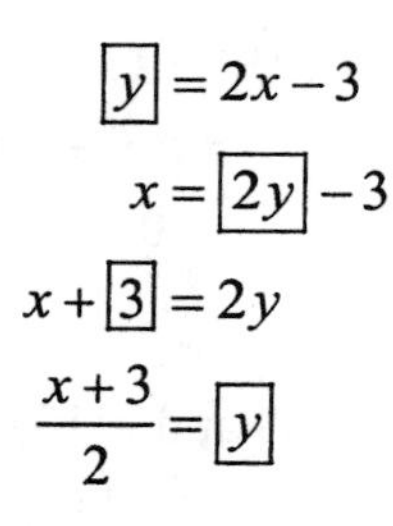

$$\boxed{y} = 2x - 3$$
$$x = \boxed{2y} - 3$$
$$x + \boxed{3} = 2y$$
$$\frac{x+3}{2} = \boxed{y}$$

The inverse of $f(x) = 2x - 3$ is $f^{-1}(x) = \dfrac{x+3}{2}$.

25. The symbol f^{-1} is read as "the **inverse of** f" or "f **inverse**."

PRACTICE

27. Yes, each output corresponds to exactly one input, so the function is one–to–one.

29. No. $2^4 = (-2)^4 = 16$ so the output 16 corresponds to two different inputs, 2 and –2.

31. No. $-(3)^2 + 3(3) = -(0) + 3(0) = 0$ so the output 0 corresponds to two different inputs, 0 and –3.

33. No. The output 1 corresponds to more than one input, 1, 2, 3 and 4.

35. one–to–one

37. not one–to–one

39. not one–to–one

41.

$$f(x) = 2x + 4$$
$$y = 2x + 4$$
$$x = 2y + 4$$
$$x - 4 = 2y$$
$$\frac{x-4}{2} = y$$
$$y = \frac{x-4}{2}$$
$$f^{-1}(x) = \frac{x-4}{2}$$

43.

$$f(x) = \frac{x}{5} + \frac{4}{5}$$
$$y = \frac{x}{5} + \frac{4}{5}$$
$$x = \frac{y}{5} + \frac{4}{5}$$
$$5(x) = 5\left(\frac{y}{5} + \frac{4}{5}\right)$$
$$5x = y + 4$$
$$5x - 4 = y$$
$$y = 5x - 4$$
$$f^{-1}(x) = 5x - 4$$

45.

$$f(x) = \frac{x-4}{5}$$
$$y = \frac{x-4}{5}$$
$$x = \frac{y-4}{5}$$
$$5(x) = 5\left(\frac{y-4}{5}\right)$$
$$5x = y - 4$$
$$5x + 4 = y$$
$$y = 5x + 4$$
$$f^{-1}(x) = 5x + 4$$

47.

$$f(x) = \frac{2}{x-3}$$
$$y = \frac{2}{x-3}$$
$$x = \frac{2}{y-3}$$
$$(y-3)(x) = (y-3)\left(\frac{2}{y-3}\right)$$
$$xy - 3x = 2$$
$$xy = 2 + 3x$$
$$y = \frac{2+3x}{x}$$
$$y = \frac{2}{x} + \frac{3x}{x}$$
$$y = \frac{2}{x} + 3$$
$$f^{-1}(x) = \frac{2}{x} + 3$$

49.

$$f(x) = \frac{4}{x}$$
$$y = \frac{4}{x}$$
$$x = \frac{4}{y}$$
$$y(x) = y\left(\frac{4}{y}\right)$$
$$xy = 4$$
$$y = \frac{4}{x}$$
$$f^{-1}(x) = \frac{4}{x}$$

51.

$$\begin{aligned} f(x) &= x^3+8 \\ y &= x^3+8 \\ x &= y^3+8 \\ x-8 &= y^3 \\ y^3 &= x-8 \\ \sqrt[3]{y^3} &= \sqrt[3]{x-8} \\ y &= \sqrt[3]{x-8} \\ f^{-1}(x) &= \sqrt[3]{x-8} \end{aligned}$$

53.

$$\begin{aligned} f(x) &= \sqrt[3]{x} \\ y &= \sqrt[3]{x} \\ x &= \sqrt[3]{y} \\ (x)^3 &= \left(\sqrt[3]{y}\right)^3 \\ x^3 &= y \\ y &= x^3 \\ f^{-1}(x) &= x^3 \end{aligned}$$

55.

$$\begin{aligned} f(x) &= (x+10)^3 \\ y &= (x+10)^3 \\ x &= (y+10)^3 \\ \sqrt[3]{x} &= \sqrt[3]{(y+10)^3} \\ \sqrt[3]{x} &= y+10 \\ \sqrt[3]{x}-10 &= y \\ y &= \sqrt[3]{x}-10 \\ f^{-1}(x) &= \sqrt[3]{x}-10 \end{aligned}$$

57.

$$\begin{aligned} f(x) &= 2x^3-3 \\ y &= 2x^3-3 \\ x &= 2y^3-3 \\ x+3 &= 2y^3 \\ \frac{x+3}{2} &= y^3 \\ \sqrt[3]{\frac{x+3}{2}} &= \sqrt[3]{y^3} \\ \sqrt[3]{\frac{x+3}{2}} &= y \\ y &= \sqrt[3]{\frac{x+3}{2}} \\ f^{-1}(x) &= \sqrt[3]{\frac{x+3}{2}} \end{aligned}$$

59.

$$\begin{aligned} (f \circ f^{-1})(x) &= f\left(f^{-1}(x)\right) \\ &= f\left(\frac{x-9}{2}\right) \\ &= 2\left(\frac{x-9}{2}\right)+9 \\ &= x-9+9 \\ &= x \end{aligned}$$

$$\begin{aligned} (f^{-1} \circ f)(x) &= f^{-1}\left(f(x)\right) \\ &= f^{-1}(2x+9) \\ &= \frac{2x+9-9}{2} \\ &= \frac{2x}{2} \\ &= x \end{aligned}$$

61.

$$
\begin{aligned}
(f \circ f^{-1})(x) &= f\left(f^{-1}(x)\right) \\
&= f\left(\frac{2}{x}+3\right) \\
&= \frac{2}{\left(\frac{2}{x}+3\right)-3} \\
&= \frac{2}{\frac{2}{x}} \\
&= \frac{2}{\frac{2}{x}} \cdot \frac{x}{x} \\
&= \frac{2x}{2} \\
&= x
\end{aligned}
$$

$$
\begin{aligned}
(f^{-1} \circ f)(x) &= f^{-1}\left(f(x)\right) \\
&= f^{-1}\left(\frac{2}{x-3}\right) \\
&= \frac{2}{\frac{2}{x-3}}+3 \\
&= \frac{2}{1} \cdot \frac{x-3}{2}+3 \\
&= \frac{2(x-3)}{2}+3 \\
&= x-3+3 \\
&= x
\end{aligned}
$$

63.

$$
\begin{aligned}
f(x) &= 2x \\
y &= 2x \\
x &= 2y \\
\frac{x}{2} &= y \\
y &= \frac{x}{2} \\
f^{-1}(x) &= \frac{x}{2}
\end{aligned}
$$

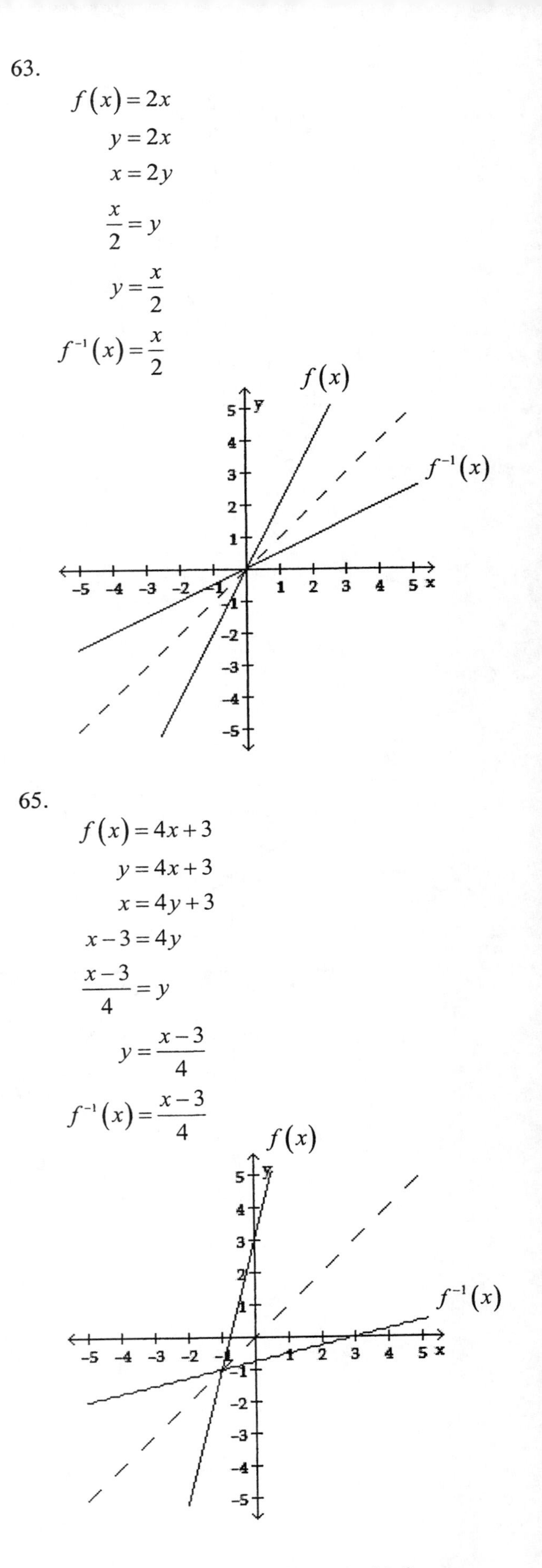

65.

$$
\begin{aligned}
f(x) &= 4x+3 \\
y &= 4x+3 \\
x &= 4y+3 \\
x-3 &= 4y \\
\frac{x-3}{4} &= y \\
y &= \frac{x-3}{4} \\
f^{-1}(x) &= \frac{x-3}{4}
\end{aligned}
$$

67.

$$f(x)=-\frac{2}{3}x+3$$
$$y=-\frac{2}{3}x+3$$
$$x=-\frac{2}{3}y+3$$
$$3(x)=3\left(-\frac{2}{3}y+3\right)$$
$$3x=-2y+9$$
$$3x-9=-2y$$
$$\frac{3x-9}{-2}=y$$
$$-\frac{3}{2}x+\frac{9}{2}=y$$
$$y=-\frac{3}{2}x+\frac{9}{2}$$
$$f^{-1}(x)=-\frac{3}{2}x+\frac{9}{2}$$

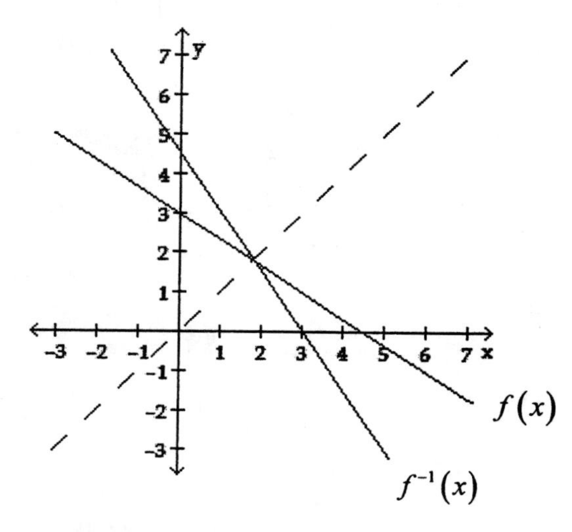

69.

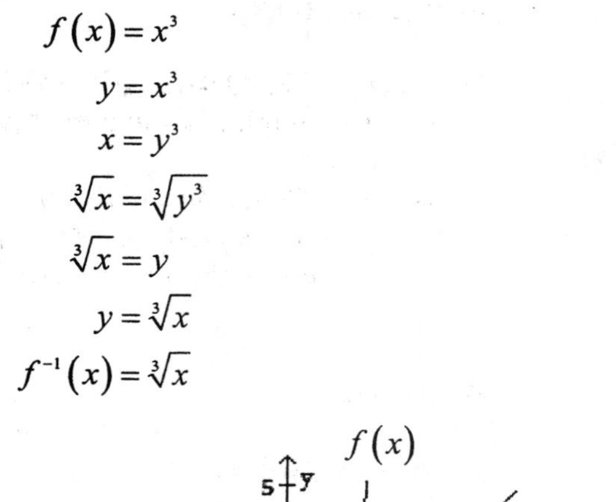

$$f(x)=x^3$$
$$y=x^3$$
$$x=y^3$$
$$\sqrt[3]{x}=\sqrt[3]{y^3}$$
$$\sqrt[3]{x}=y$$
$$y=\sqrt[3]{x}$$
$$f^{-1}(x)=\sqrt[3]{x}$$

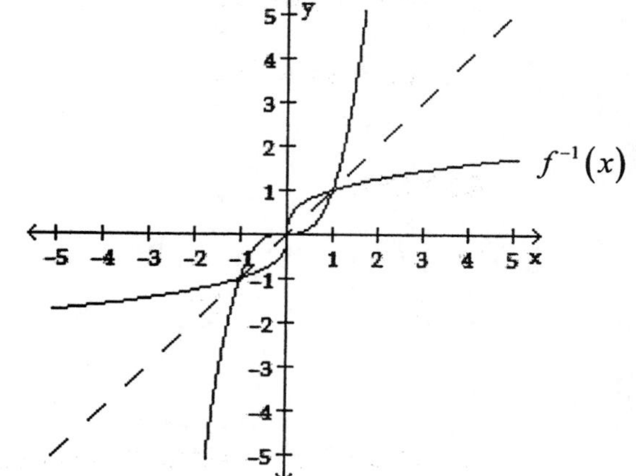

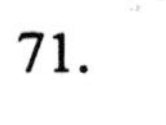

71.

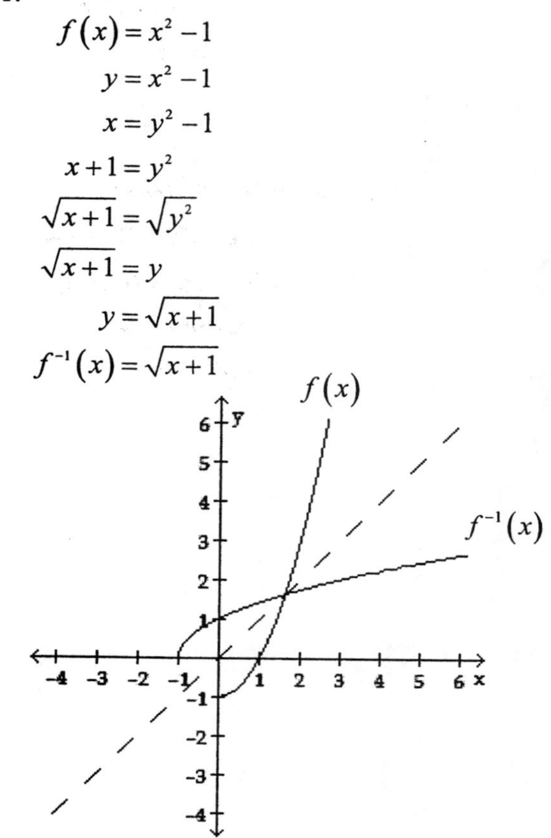

$$f(x)=x^2-1$$
$$y=x^2-1$$
$$x=y^2-1$$
$$x+1=y^2$$
$$\sqrt{x+1}=\sqrt{y^2}$$
$$\sqrt{x+1}=y$$
$$y=\sqrt{x+1}$$
$$f^{-1}(x)=\sqrt{x+1}$$

APPLICATIONS

73. INTERPERSONAL RELATIONSHIPS
    a. The graph is a function, but its inverse is not.
    b. No. Twice during this period, the person's anxiety level was at the maximum threshold value.

WRITING

75. Answers will vary.

77. Answers will vary.

REVIEW

79.

$$3-\sqrt{-64}=3-8i$$

81.

$$\begin{aligned}(3+4i)(2-3i)&=6-9i+8i-12i^2\\&=6-i-12(-1)\\&=6-i+12\\&=18-i\end{aligned}$$

83.

$$\begin{aligned}(6-8i)^2&=(6-8i)(6-8i)\\&=36-48i-48i+64i^2\\&=36-96i+64(-1)\\&=36-96i-64\\&=-28-96i\end{aligned}$$

CHALLENGE PROBLEMS

85.

$$\begin{aligned}f(x)&=\frac{x+1}{x-1}\\y&=\frac{x+1}{x-1}\\x&=\frac{y+1}{y-1}\\(y-1)(x)&=(y-1)\left(\frac{y+1}{y-1}\right)\\xy-x&=y+1\\xy&=y+1+x\\xy-y&=x+1\\y(x-1)&=x+1\\y&=\frac{x+1}{x-1}\\f^{-1}(x)&=\frac{x+1}{x-1}\end{aligned}$$

SECTION 11.3

VOCABULARY

1. exponential

3. (0, ∞)

5. yes

7. increasing

CONCEPTS

9. $f(x)=x^2$ and $g(x)=2^x$

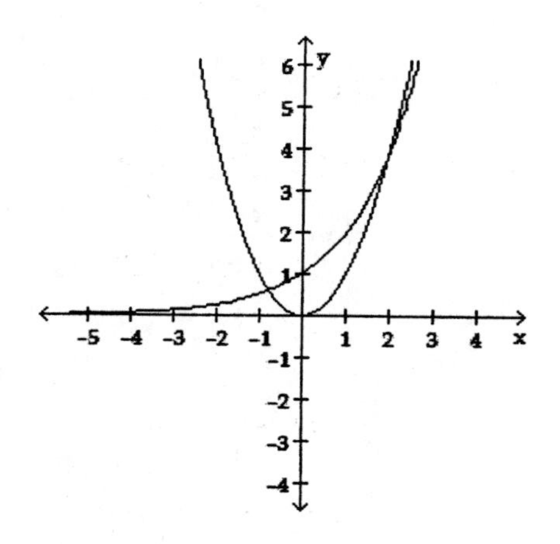

11. $A=P\left(1+\frac{r}{k}\right)^{kt}$ and $FV=PV(1+i)^n$

13. $g(x)=2^x+3$ and $h(x)=2^x-2$

NOTATION

15. The base is $\left(1+\frac{r}{k}\right)$ and the exponent is *kt*.

PRACTICE

17. $2^{\sqrt{2}}=2.6651$

19. $5^{\sqrt{5}}=36.5548$

21.
$$\begin{aligned}\left(2^{\sqrt{3}}\right)^{\sqrt{3}}&=2^{\sqrt{3}\cdot\sqrt{3}}\\&=2^{\sqrt{9}}\\&=2^3\\&=8\end{aligned}$$

23.
$$\begin{aligned}7^{\sqrt{3}}7^{\sqrt{12}}&=7^{\sqrt{3}+\sqrt{12}}\\&=7^{\sqrt{3}+2\sqrt{3}}\\&=7^{3\sqrt{3}}\end{aligned}$$

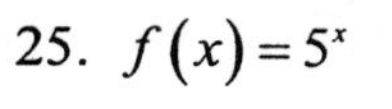

25. $f(x)=5^x$

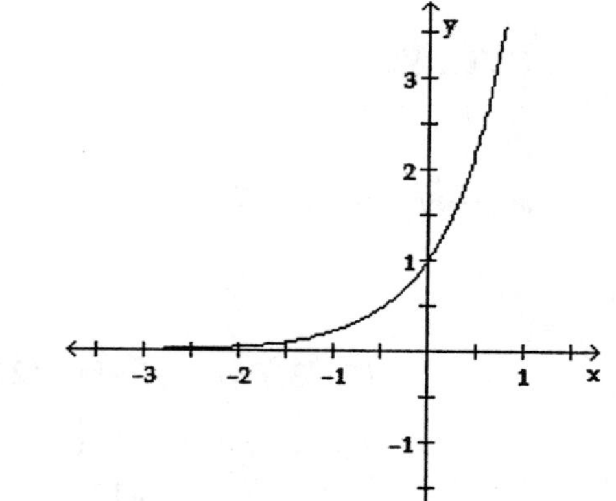

27. $y=\left(\frac{1}{4}\right)^x$

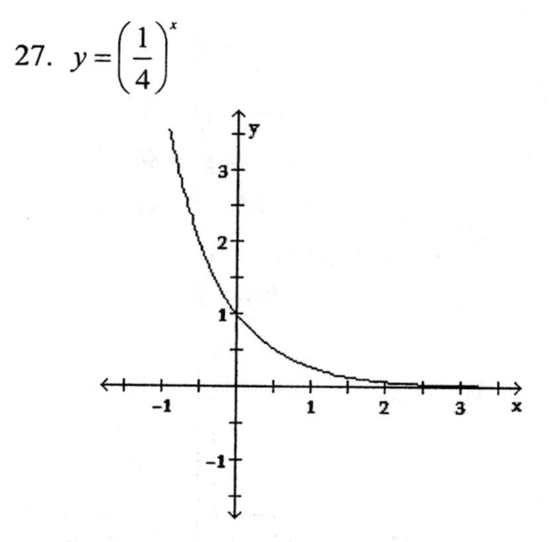

29. $f(x) = 3^x - 2$

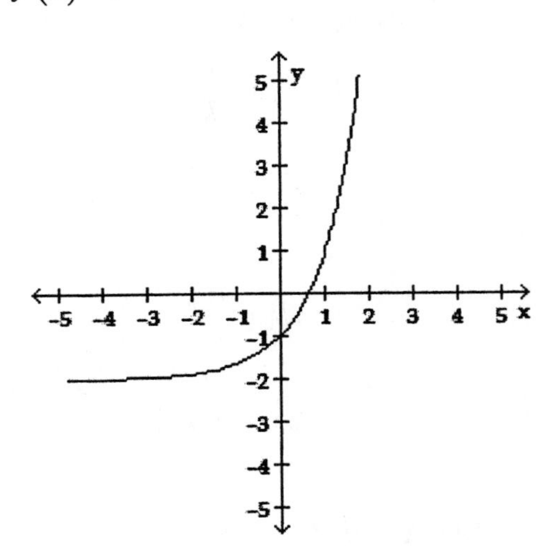

31. $f(x) = 3^{x-1}$

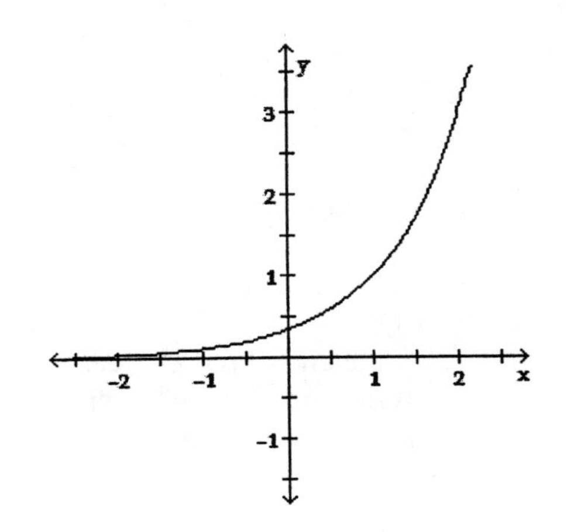

33. $f(x) = \frac{1}{2}(3^{x/2})$

The function is increasing.

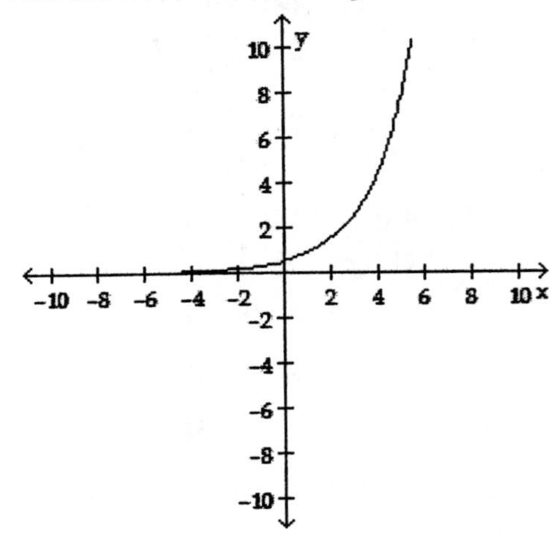

35. $y = 2(3^{-x/2})$

The function is decreasing.

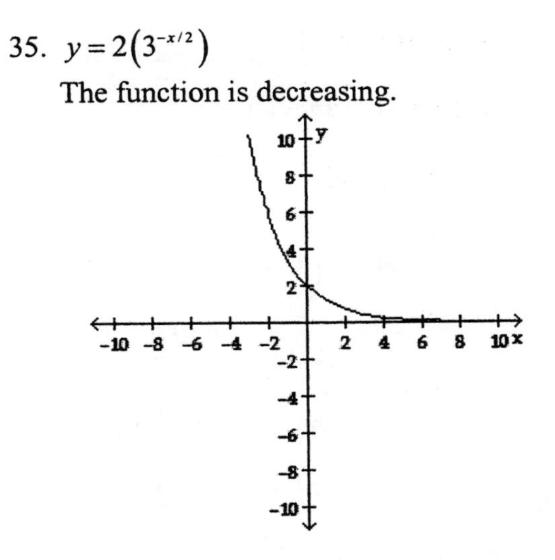

APPLICATIONS

37. COMPOUND INTEREST

$P = 10{,}000$, $r = 0.08$, $k = 4$, $t = 10$

$$
\begin{aligned}
A &= P\left(1 + \frac{r}{k}\right)^{kt} \\
&= 10{,}000\left(1 + \frac{0.08}{4}\right)^{4(10)} \\
&= 10{,}000(1 + 0.02)^{40} \\
&= 10{,}000(1.02)^{40} \\
&= 10{,}000(2.20804) \\
&= \$22{,}080.40
\end{aligned}
$$

39. COMPARING INTEREST RATES

Find the amount earned at $5\frac{1}{2}\%$.

$P = 1{,}000$, $r = 0.055$, $k = 4$, $t = 5$

$$\begin{aligned}A &= P\left(1+\frac{r}{k}\right)^{kt}\\&= 1{,}000\left(1+\frac{0.055}{4}\right)^{4(5)}\\&= 1{,}000(1+0.01375)^{20}\\&= 1{,}000(1.01375)^{20}\\&= 1{,}000(1.314066)\\&= \$1{,}314.07\end{aligned}$$

Find the amount earned at 5%.
$P = 1{,}000$, $r = 0.05$, $k = 4$, $t = 5$

$$\begin{aligned}A &= P\left(1+\frac{r}{k}\right)^{kt}\\&= 1{,}000\left(1+\frac{0.05}{4}\right)^{4(5)}\\&= 1{,}000(1+0.0125)^{20}\\&= 1{,}000(1.0125)^{20}\\&= 1{,}000(1.282037)\\&= \$1{,}282.04\end{aligned}$$

Find the difference in the two amounts.

$$\$1{,}314.07 - 1{,}282.04 = \$32.03$$

41. COMPOUND INTEREST

$P = 1$, $r = 0.05$, $k = 1$, $t = 300$

$$\begin{aligned}A &= P\left(1+\frac{r}{k}\right)^{kt}\\&= 1\left(1+\frac{0.05}{1}\right)^{1(300)}\\&= 1(1+0.05)^{300}\\&= 1(1.05)^{300}\\&= 1(2{,}273{,}996.129)\\&= \$2{,}273{,}996.13\end{aligned}$$

43. WORLD POPULATION

a. The population reached $\frac{1}{2}$ billion about 1500 and it reach 1 billion about 1825.

b. The population was about 6.5 billion in 2000.

c. exponential

45. VALUE OF A CAR

a. at the end of the 2nd year
b. at the end of the 4th year
c. during the 7th year

47. BACTERIAL CULTURES

Let $t = 4$.

$$\begin{aligned}P &= (6\times10^6)(2.3)^t\\&= (6\times10^6)(2.3)^4\\&= (6\times10^6)(27.9841)\\&= 1.679046\times10^8\end{aligned}$$

49. DISCHARGING A BATTERY

Let $t = 5$.

$$\begin{aligned}C &= (3\times10^{-4})(0.7)^t\\&= (3\times10^{-4})(0.7)^5\\&= (3\times10^{-4})(0.16807)\\&= 5.0421\times10^{-5} \text{ coulombs}\end{aligned}$$

51. SALVAGE VALUES
If the computer is worth only 75% of its preceding year's value, then it is losing 25% of its value each year.
Let $n = 5$, $i = -0.25$, and $PV = 4{,}700$

$$\begin{aligned} FV &= PV(1+i)^n \\ &= 4{,}700(1-0.25)^5 \\ &= 4{,}700(0.75)^5 \\ &= 4{,}700(0.2373047) \\ &= \$1{,}115.33 \end{aligned}$$

WRITING

53. Answers will vary.

55. Answers will vary.

REVIEW

57. The sum of same–side interior angles is 180°.

$$\begin{aligned} 3x + (2x - 20) &= 180 \\ 5x - 20 &= 180 \\ 5x &= 200 \\ x &= 40^\circ \end{aligned}$$

59. Angle 2 is equal to $3x$ because they are alternate interior angles.

$$\begin{aligned} 3x &= 3(40) \\ &= 120^\circ \end{aligned}$$

CHALLENGE PROBLEMS

61. Answers will vary.

SECTION 11.4

VOCABULARY

1. the natural exponential function

3. (0, ∞)

5. yes

7. increasing

CONCEPTS

9. In **continuous** compound interest, the number of compoundings is infinitely large.

11. To two decimal places, the value of e is **2.72**.

13. $\sqrt{2} \approx 1.41$; $e \approx 2.718$; $\pi \approx 3.14$

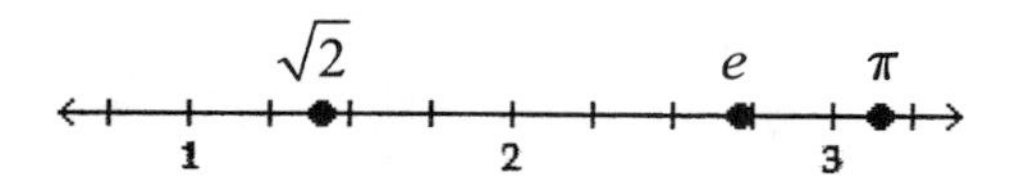

15. an exponential function

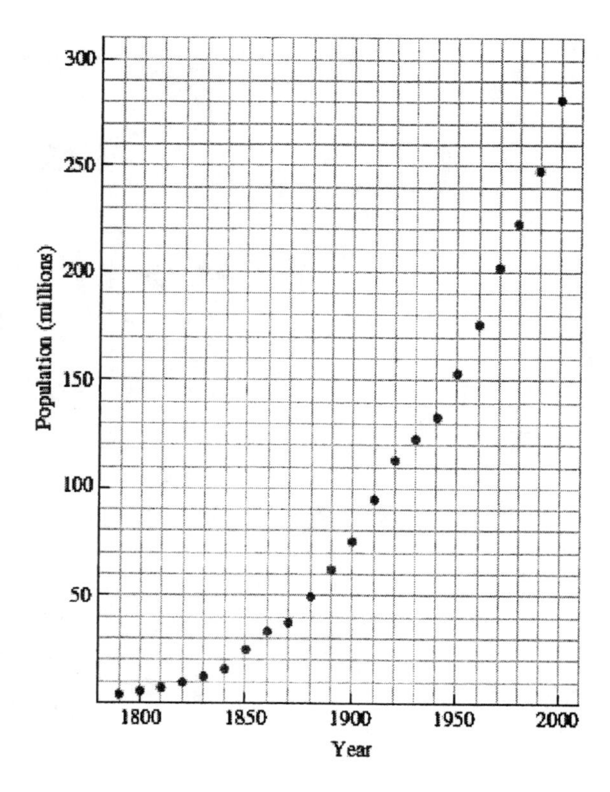

17. The y–coordinate of the point on the graph having an x–coordinate of 1 is 2.7182818… The name of that number is e.

NOTATION

19. $P = 1{,}000$, $r = 0.09$, and $t = 10$

$$A = \boxed{1{,}000}e^{(0.09)(\boxed{10})}$$
$$= 1{,}000e^{\boxed{0.9}}$$
$$\approx \boxed{1{,}000}(2.459603111)$$
$$\approx 2{,}459.603111$$

PRACTICE

21. $f(x) = e^x + 1$

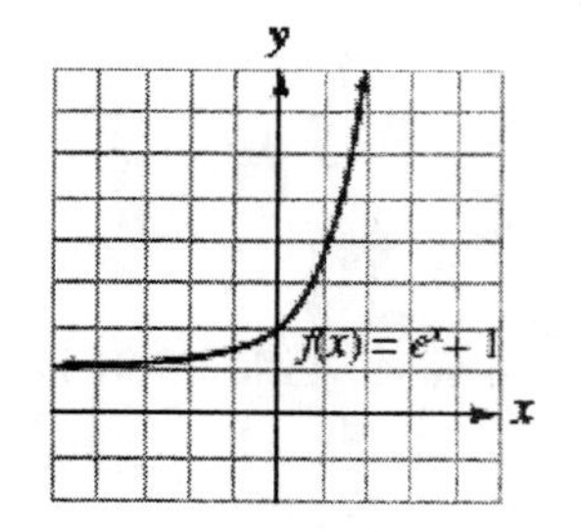

23. 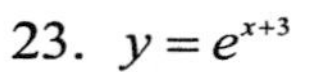

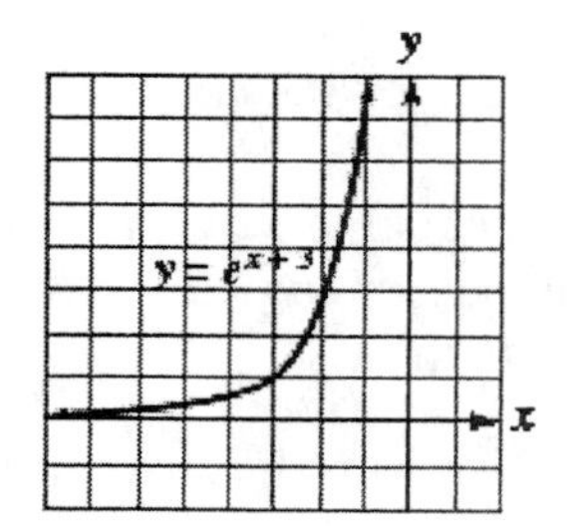

25. 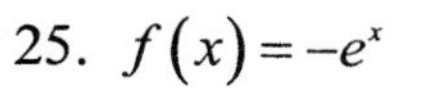

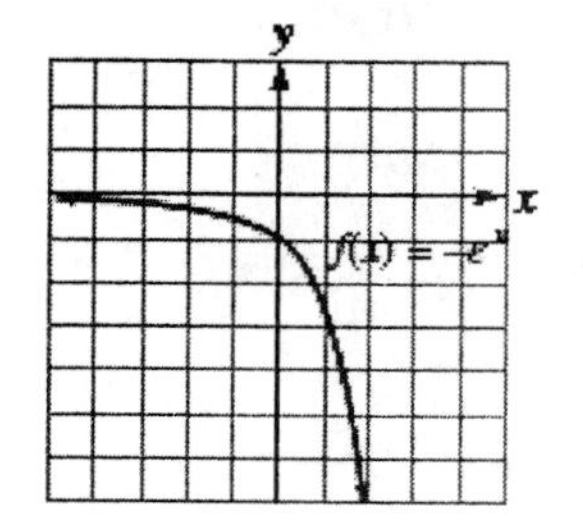

27. $f(x) = 2e^x$

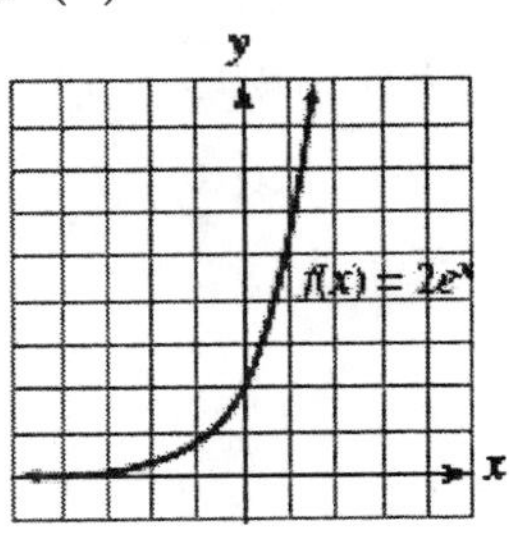

APPLICATIONS

29. CONTINUOUS COMPOUND INTEREST

$P = 5{,}000$, $r = 0.082$, and $t = 12$

$$A = Pe^{rt}$$
$$= 5{,}000e^{(0.082)(12)}$$
$$= 5{,}000e^{0.984}$$
$$\approx 5{,}000(2.675135411)$$
$$\approx \$13{,}375.68$$

31. COMPARISON OF COMPOUNDING METHODS

Annual Compounding:

$PV = 5{,}000$, $i = 0.085$, and $n = 5$

$$FV = PV(1+i)^n$$
$$= 5{,}000(1+0.085)^5$$
$$= 5{,}000(1.085)^5$$
$$= 5{,}000(1.5037)$$
$$= \$7{,}518.28$$

Continuous Compounding:

$P = 5{,}000$, $r = 0.085$, and $t = 5$

$$A = Pe^{rt}$$
$$= 5{,}000e^{(0.085)(5)}$$
$$= 5{,}000e^{0.425}$$
$$\approx 5{,}000(1.52959042)$$
$$\approx \$7{,}647.95$$

33. DETERMINING THE INITIAL DEPOSIT

$A = 11{,}180$, $r = 0.07$, and $t = 7$

$$A = Pe^{rt}$$
$$11{,}180 = Pe^{(0.07)(7)}$$
$$11{,}180 = Pe^{0.49}$$
$$\frac{11{,}180}{e^{0.49}} = \frac{Pe^{0.49}}{e^{0.49}}$$
$$\frac{11{,}180}{e^{0.49}} = P$$
$$P \approx \frac{11{,}180}{1.63231622}$$
$$P \approx \$6{,}849.16$$

35. WORLD POPULATION GROWTH

$P = 6.1$, $r = 0.014$, and $t = 30$

$$A = Pe^{rt}$$
$$= 6.1e^{(0.014)(30)}$$
$$= 6.1e^{0.42}$$
$$\approx 6.1(1.521961556)$$
$$\approx 9.3 \text{ billion}$$

37. POPULATION GROWTH

$$P = 173e^{0.03t}$$
$$= 173e^{0.03(20)}$$
$$= 173e^{0.6}$$
$$\approx 173(1.8221188)$$
$$\approx 315$$

39. EPIDEMICS

$$P = P_0e^{0.27t}$$
$$= 2e^{0.27(12)}$$
$$= 2e^{3.24}$$
$$\approx 2(25.53372175)$$
$$\approx 51$$

41. HALF–LIFE OF A DRUG

According to the graph, it take 12 hours for the initial dosage to be 50%.

43. SKYDIVING

$$\begin{aligned} v &= 50\left(1-e^{-0.2t}\right) \\ &= 50\left(1-e^{-0.2(20)}\right) \\ &= 50\left(1-e^{-4}\right) \\ &\approx 50(1-0.018315639) \\ &\approx 50(0.981684361) \\ &\approx 49 \text{ mps} \end{aligned}$$

45. THE MALTHUSIAN MODEL

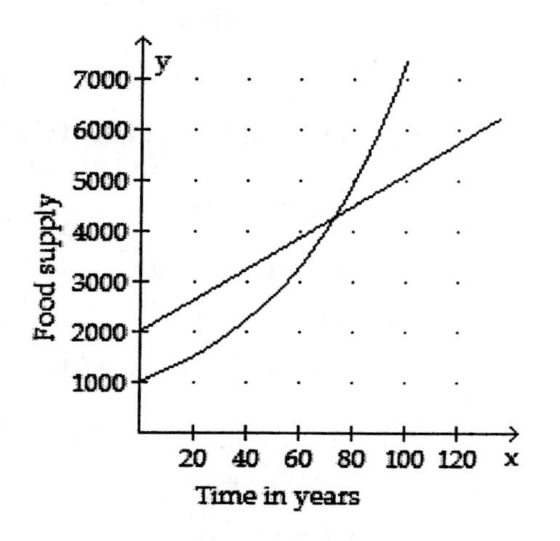

The two graphs intersect at (72, 4,232) so the food supply would be adequate for about 72 years.

WRITING

47. Answers will vary.

49. Answers will vary.

REVIEW

51.

$$\begin{aligned} \sqrt{240x^5} &= \sqrt{16x^4}\sqrt{15x} \\ &= 4x^2\sqrt{15x} \end{aligned}$$

53.

$$\begin{aligned} 4\sqrt{48y^3} - 3y\sqrt{12y} &= 4\sqrt{16y^2}\sqrt{3y} - 3y\sqrt{4}\sqrt{3y} \\ &= 4(4y)\sqrt{3y} - 3y(2)\sqrt{3y} \\ &= 16y\sqrt{3y} - 6y\sqrt{3y} \\ &= 10y\sqrt{3y} \end{aligned}$$

CHALLENGE PROBLEMS

55. False.

$$e^e \overset{?}{>} e^3$$

$$e^{2.71829} \overset{?}{>} e^3$$

$$15.15 \not> 20.09$$

57.

$$\begin{aligned} e^{t+5} &= ke^t \\ \frac{e^{t+5}}{e^t} &= \frac{ke^t}{e^t} \\ e^{t+5-t} &= k \\ e^5 &= k \\ k &= e^5 \end{aligned}$$

SECTION 11.5

VOCABULARY

1. logarithmic

3. $(-\infty, \infty)$

5. yes

7. increasing

CONCEPTS

9. The equation $y = \log_b x$ is equivalent to the exponential equation $\underline{x = b^y}$.

11. The functions $f(x) = \log_{10} x$ and $f(x) = 10^x$ are **inverse** functions.

13. $y = \log x$

x	y
$\frac{1}{100}$	-2
$\frac{1}{10}$	-1
1	0
10	1
100	2

15. $f(x) = \log_6 x$

input	output
-6	none
0	none
$\frac{1}{216}$	-3
$\sqrt{6}$	$\frac{1}{2}$
6^8	8

17. $f(x) = \log x$

x	$f(x)$
0.5	-0.30
1	0
2	0.30
4	0.60
6	0.78
8	0.90
10	1

19. They decrease.

NOTATION

21. a. $\log x = \log_{\boxed{10}} x$

b. $\log_{10} 10^x = \boxed{x}$

PRACTICE

23.
$$\log_3 81 = 4$$
$$3^4 = 81$$

25.
$$\log 10 = 1$$
$$10^1 = 10$$

27.
$$\log_4 \frac{1}{64} = -3$$
$$4^{-3} = \frac{1}{64}$$

29.
$$\log_5 \sqrt{5} = \frac{1}{2}$$
$$5^{1/2} = \sqrt{5}$$

31.
$$8^2 = 64$$
$$\log_8 64 = 2$$

33.

$$4^{-2} = \frac{1}{16}$$

$$\log_4 \frac{1}{16} = -2$$

35.

$$\left(\frac{1}{2}\right)^{-5} = 32$$

$$\log_{1/2} 32 = -5$$

37.

$$x^y = z$$

$$\log_x z = y$$

39.

$$\log_2 8 = x$$

$$2^x = 8$$

$$2^x = 2^3$$

$$x = 3$$

41.

$$\log_4 16 = x$$

$$4^x = 16$$

$$4^x = 4^2$$

$$x = 2$$

43.

$$\log_{1/2} \frac{1}{32} = x$$

$$\left(\frac{1}{2}\right)^x = \frac{1}{32}$$

$$2^{-x} = 2^{-5}$$

$$-x = -5$$

$$x = 5$$

45.

$$\log_9 3 = x$$

$$9^x = 3$$

$$3^{2x} = 3^1$$

$$2x = 1$$

$$x = \frac{1}{2}$$

47.

$$\log \frac{1}{10} = x$$

$$10^x = \frac{1}{10}$$

$$10^x = 10^{-1}$$

$$x = -1$$

49.

$$\log 1{,}000{,}000 = x$$

$$10^x = 1{,}000{,}000$$

$$10^x = 10^6$$

$$x = 6$$

51.

$$\log_8 x = 2$$

$$8^2 = x$$

$$64 = x$$

53.

$$\log_{25} x = \frac{1}{2}$$

$$25^{1/2} = x$$

$$\sqrt{25} = x$$

$$5 = x$$

55.

$$\log_5 x = -2$$

$$5^{-2} = x$$

$$\frac{1}{5^2} = x$$

$$\frac{1}{25} = x$$

57.

$$\log_{36} x = -\frac{1}{2}$$

$$36^{-1/2} = x$$

$$\frac{1}{36^{1/2}} = x$$

$$\frac{1}{\sqrt{36}} = x$$

$$\frac{1}{6} = x$$

59.

$$\begin{aligned} \log_x 0.01 &= -2 \\ x^{-2} &= 0.01 \\ \left(x^{-2}\right)^{-1/2} &= (0.01)^{-1/2} \\ x &= \left(\frac{1}{100}\right)^{-1/2} \\ x &= (100)^{1/2} \\ x &= \sqrt{100} \\ x &= 10 \end{aligned}$$

61.

$$\begin{aligned} \log_{27} 9 &= x \\ 27^x &= 9 \\ 3^{3x} &= 3^2 \\ 3x &= 2 \\ x &= \frac{2}{3} \end{aligned}$$

63.

$$\begin{aligned} \log_x 5^3 &= 3 \\ x^3 &= 5^3 \\ x^3 &= 125 \\ \sqrt[3]{x^3} &= \sqrt[3]{125} \\ x &= 5 \end{aligned}$$

65.

$$\begin{aligned} \log_x \frac{\sqrt{3}}{3} &= \frac{1}{2} \\ x^{1/2} &= \frac{\sqrt{3}}{3} \\ \left(x^{1/2}\right)^2 &= \left(\frac{\sqrt{3}}{3}\right)^2 \\ x &= \frac{\sqrt{9}}{9} \\ x &= \frac{3}{9} \\ x &= \frac{1}{3} \end{aligned}$$

67.

$$\begin{aligned} \log_{100} x &= \frac{3}{2} \\ 100^{3/2} &= x \\ \left(\sqrt{100}\right)^3 &= x \\ (10)^3 &= x \\ 1{,}000 &= x \end{aligned}$$

69.

$$\begin{aligned} \log_x \frac{1}{64} &= -3 \\ x^{-3} &= \frac{1}{64} \\ \left(x^{-3}\right)^{-1/3} &= \left(\frac{1}{64}\right)^{-1/3} \\ x &= (64)^{1/3} \\ x &= \sqrt[3]{64} \\ x &= 4 \end{aligned}$$

71.

$$\begin{aligned} \log_8 x &= 0 \\ 8^0 &= x \\ 1 &= x \end{aligned}$$

73.

$\log 3.25 = 0.5119$

75.

$\log 0.00467 = -2.3307$

77.

$\log x = 1.4023$
$x = 25.25$

79.

$\log x = -1.71$
$x = 0.02$

81. $f(x) = \log_3 x$
increasing

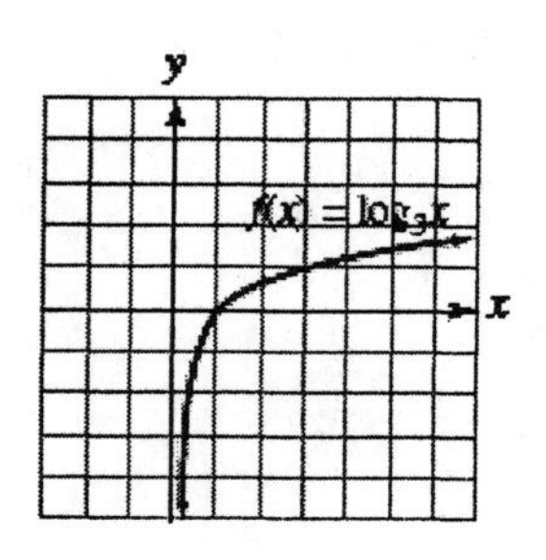

83. $y = \log_{1/2} x$
decreasing

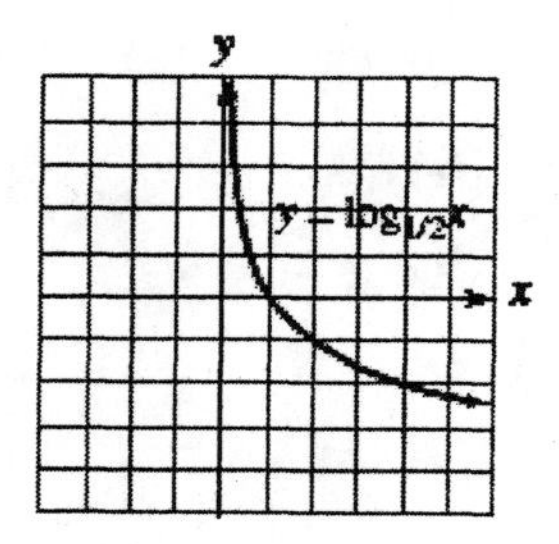

85. $f(x) = 3 + \log_3 x$

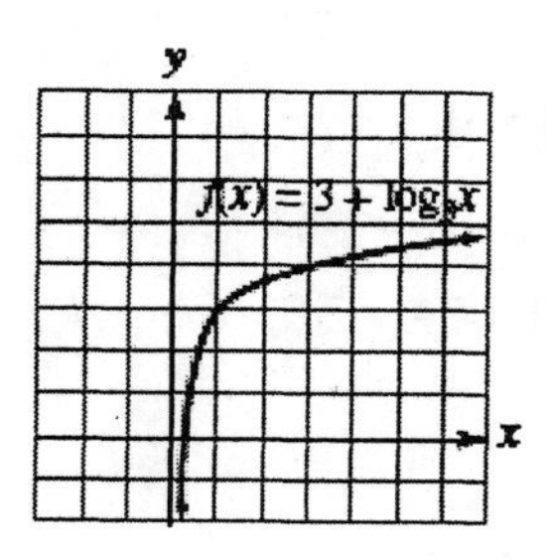

87. $y = \log_{1/2}(x - 2)$

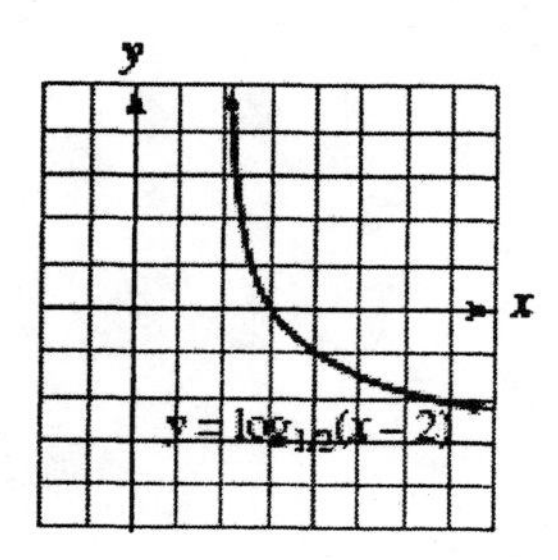

89.
$f(x) = 6^x$
$f^{-1}(x) = \log_6 x$

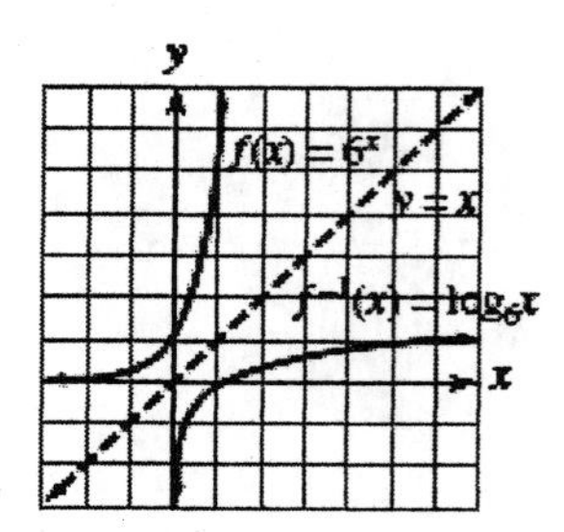

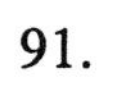
91.
$f(x) = 5^x$
$f^{-1}(x) = \log_5 x$

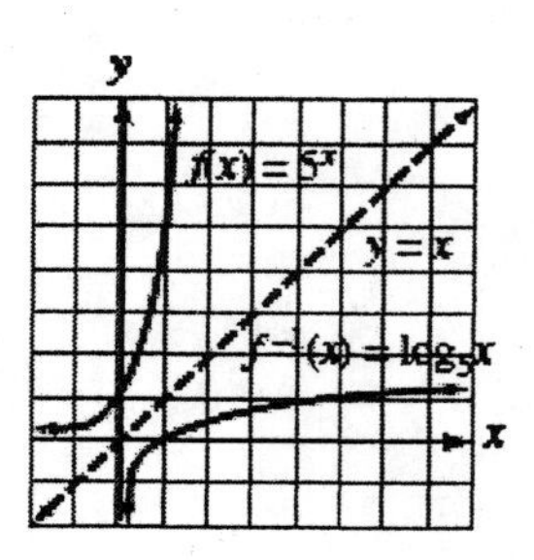

APPLICATIONS

93. INPUT VOLTAGE
$E_O = 20$ and $E_I = 0.71$

$$\text{db gain} = 20\left(\log\frac{E_O}{E_I}\right)$$
$$= 20\left(\log\frac{20}{0.71}\right)$$
$$= 20(\log 28.169)$$
$$= 20(1.45)$$
$$= 29 \text{ db}$$

95. db GAIN

$E_O = 30$ and $E_I = 0.1$

$$\begin{aligned}\text{db gain} &= 20\left(\log\frac{E_o}{E_I}\right)\\ &= 20\left(\log\frac{30}{0.1}\right)\\ &= 20(\log 300)\\ &= 20(2.477)\\ &= 49.5 \text{ db}\end{aligned}$$

97. THE RICHTER SCALE

$A = 5{,}000$ and $P = 0.2$

$$\begin{aligned}R &= \log\frac{A}{P}\\ &= \log\frac{5{,}000}{0.2}\\ &= \log 25{,}000\\ &\approx 4.4\end{aligned}$$

99. EARTHQUAKES

$R = 4$ and $P = \frac{1}{4} = 0.25$

$$\begin{aligned}R &= \log\frac{A}{P}\\ 4 &= \log\frac{x}{0.25}\\ \log\frac{x}{0.25} &= 4\\ 10^4 &= \frac{x}{0.25}\\ 10{,}000 &= \frac{x}{0.25}\\ 0.25(10{,}000) &= 0.25\left(\frac{x}{0.25}\right)\\ 2{,}500 &= x\\ x &= 2{,}500 \text{ micrometers}\end{aligned}$$

101. DEPRECIATION

$V = 8{,}000$, $C = 37{,}000$, and $N = 5$

$$\begin{aligned}n &= \frac{\log V - \log C}{\log\left(1 - \frac{2}{N}\right)}\\ &= \frac{\log 8{,}000 - \log 37{,}000}{\log\left(1 - \frac{2}{5}\right)}\\ &\approx \frac{3.90 - 4.57}{\log\left(\frac{3}{5}\right)}\\ &\approx \frac{-0.67}{-0.22}\\ &\approx 3 \text{ years old}\end{aligned}$$

103. INVESTING

$P = 1{,}000$, $A = 20{,}000$, and $r = 0.12$

$$\begin{aligned}n &= \frac{\log\left(\frac{Ar}{P} + 1\right)}{\log(1 + r)}\\ &= \frac{\log\left(\frac{20{,}000 \cdot 0.12}{1{,}000} + 1\right)}{\log(1 + 0.12)}\\ &= \frac{\log(2.4 + 1)}{\log(1 + 0.12)}\\ &= \frac{\log 3.4}{\log 1.12}\\ &\approx \frac{0.5315}{0.0492}\\ &\approx 10.8 \text{ years}\end{aligned}$$

WRITING

105. Answers will vary.

REVIEW

107.

$$\begin{aligned}\sqrt[3]{6x + 4} &= 4\\ \left(\sqrt[3]{6x + 4}\right)^3 &= (4)^3\\ 6x + 4 &= 64\\ 6x &= 60\\ x &= 10\end{aligned}$$

109.

$$\sqrt{a+1}-1=3a$$
$$\sqrt{a+1}=3a+1$$
$$\left(\sqrt{a+1}\right)^2=(3a+1)^2$$
$$a+1=9a^2+3a+3a+1$$
$$a+1=9a^2+6a+1$$
$$0=9a^2+5a$$
$$0=a(9a+5)$$
$$a=0 \quad \text{or} \quad 9a+5=0$$
$$a=0 \qquad\qquad a=-\not{5}/\not{9}$$

CHALLENGE PROBLEMS

111. The domain would be all real number except 1, since $(1^2 - 1) = 0$.
$(-\infty,1)\cup(1,\infty)$

SECTION 11.6

VOCABULARY

1. $f(x) = \ln x$ is called the **natural** logarithmic function. The base is $\boxed{e}$.

CONCEPTS

3. $f(x) = \ln x$

x	$f(x) = \ln x$
0.5	–0.69
1	0
2	0.69
4	1.39
6	1.79
8	2.08
10	2.30

5. The graph of $f(x) = \ln x$ has the ***y*–axis** as an asymptote.

7. The range of the function $f(x) = \ln x$ is the interval $(-\infty, \infty)$.

9. The graph of $f(x) = \ln x$ has the *x*–intercept $(\boxed{1}, 0)$. The *y*–axis is an **asymptote** of the graph.

11. The logarithm of a negative number or zero is not defined.

13. The *x*–coordinate of the point having a *y*–coordinate of 1 is 2.7182818... This number is *e*.

NOTATION

15. We read ln *x* letter–by–letter as "$\boxed{\ell} - \boxed{n}$ of *x*."

17. If a population grows exponentially at a rate *r*, the time it will take the population to double is given by the formula $t = \boxed{\dfrac{\ln 2}{r}}$.

19. $\ln e = 1$

21. $\ln e^6 = 6$

23. $\ln \dfrac{1}{e} = \ln e^{-1} = -1$

25. $\ln \sqrt[4]{e} = \ln e^{1/4} = \dfrac{1}{4}$

PRACTICE

27. ln 35.15 = 3.5596

29. ln 0.00465 = –5.3709

31. ln 1.72 = 0.5423

33. ln (–0.1) = undefined

35.
$$\ln x = 1.4023$$
$$e^{1.4023} = x$$
$$4.0645 = x$$

37.
$$\ln x = 4.24$$
$$e^{4.24} = x$$
$$69.4079 = x$$

39.
$$\ln x = -3.71$$
$$e^{-3.71} = x$$
$$0.0245 = x$$

41.
$$1.001 = \ln x$$
$$x = e^{1.001}$$
$$x = 2.7210$$

43. $y = \ln\left(\frac{1}{2}x\right)$

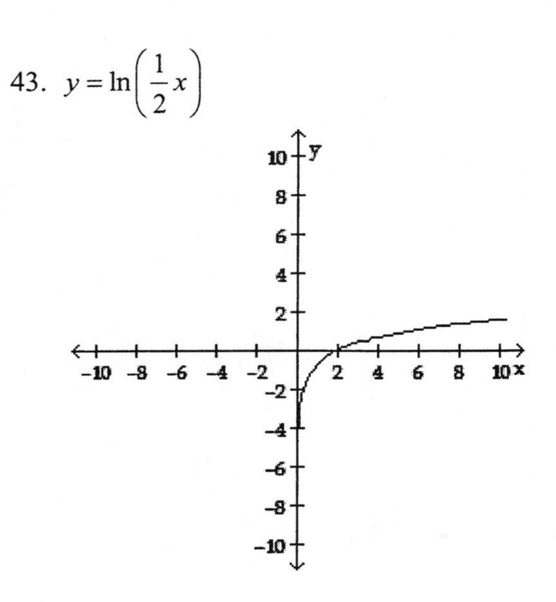

45. $f(x) = \ln(-x)$

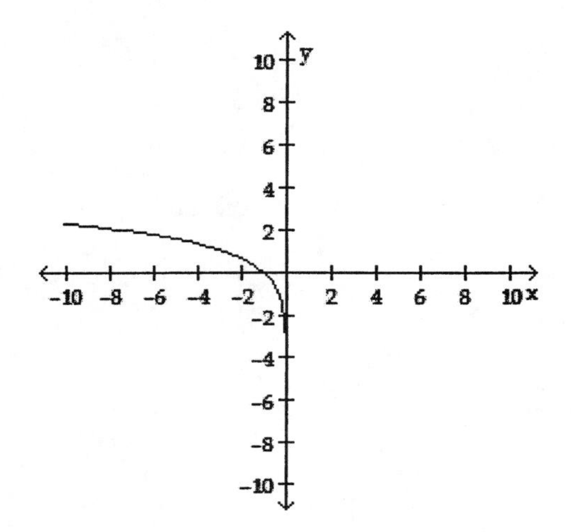

APPLICATIONS

47. POPULATION GROWTH
Let $r = 0.12$.

$$t = \frac{\ln 2}{r}$$
$$= \frac{\ln 2}{0.12}$$
$$\approx \frac{0.6931}{0.12}$$
$$\approx 5.8 \text{ years}$$

49. POPULATION GROWTH
Let $r = 0.12$.

$$t = \frac{\ln 3}{r}$$
$$= \frac{\ln 3}{0.12}$$
$$\approx \frac{1.0986}{0.12}$$
$$\approx 9.2 \text{ years}$$

51. FORENSIC MEDICINE
Let $T_s = 70$.

$$t = \frac{1}{0.25}\ln\frac{98.6 - T_s}{82 - T_s}$$
$$= \frac{1}{0.25}\ln\frac{98.6 - 70}{82 - 70}$$
$$= \frac{1}{0.25}\ln\frac{28.6}{12}$$
$$= \frac{1}{0.25}\ln 2.383$$
$$\approx \frac{1}{0.25}(0.8684)$$
$$\approx 3.5 \text{ hours}$$

WRITING

53. Answers will vary.

55. Answers will vary.

REVIEW

57. The slope of the line is 5 and the point it passes through is (0, 0). Use point slope–formula.

$$y - y_1 = m(x - x_1)$$
$$y - 0 = 5(x - 0)$$
$$y = 5x - 0$$
$$y = 5x$$

59. The slope of the line is $-\frac{3}{2}$ and the point it passes through is (3, 2). Use point slope–formula.

$$y - y_1 = m(x - x_1)$$
$$y - 2 = -\frac{3}{2}(x - 3)$$
$$y - 2 = -\frac{3}{2}x + \frac{9}{2}$$
$$y - 2 + 2 = -\frac{3}{2}x + \frac{9}{2} + 2$$
$$y = -\frac{3}{2}x + \frac{13}{2}$$

61. $x = 2$

CHALLENGE PROBLEMS

63.

$$P = P_0 e^{rt}$$
$$P = P_0 e^{r\left(\frac{\ln 2}{r}\right)}$$
$$P = P_0 e^{\ln 2}$$
$$P = P_0(2)$$
$$P = 2P_0$$

65. $t = \frac{\ln 4}{r}$

SECTION 11.7

VOCABULARY

1. The expression $\log_3 (4x)$ is the logarithm of a **product**.

3. The expression $\log 4^x$ is the logarithm of a **power**.

CONCEPTS

5. $\log_b 1 = \boxed{0}$

7. $\log_b MN = \log_b \boxed{M} + \log_b \boxed{N}$

9. $\log_b \frac{M}{N} = \log_b M - \log_b N$

11. $\log_b b^x = \boxed{x}$

13. $\log_b \frac{M}{N} \boxed{\neq} \frac{\log_b M}{\log_b N}$

15. $\log_b x = \frac{\log_a x}{\boxed{\log_a b}}$

17. $10^0 = 1$; $10^1 = 10$; $10^2 = 100$

19. $\log_4 1 = 0$

21. $\log_4 4^7 = 7$

23. $5^{\log_5 10} = 10$

25. $\log_5 5^2 = 2$

27. $\ln e = 1$

29. $\log_3 3^7 = 7$

NOTATION

31.
$$\begin{aligned}\log_b rst &= \log_b \left(\boxed{rs}\right)t \\ &= \log_b (rs) + \log_b \boxed{t} \\ &= \log_b \boxed{r} + \log_b \boxed{s} + \log_b t\end{aligned}$$

PRACTICE

33.
$$\begin{aligned}\log\left[(2.5)(3.7)\right] &= \log 2.5 + \log 3.7 \\ \log(9.25) &= 0.3979 + 0.5682 \\ 0.9661 &= 0.9661\end{aligned}$$

35.
$$\begin{aligned}\ln(2.25)^4 &= 4\ln 2.25 \\ \ln(25.6289) &= 4(0.8109) \\ 3.2437 &= 3.2437\end{aligned}$$

37.
$$\begin{aligned}\log\sqrt{24.3} &= \frac{1}{2}\log 24.3 \\ \log(4.9295) &= \frac{1}{2}(1.3856) \\ 0.6928 &= 0.6928\end{aligned}$$

39.
$$\begin{aligned}\log_2 (4 \cdot 5) &= \log_2 4 + \log_2 5 \\ &= \log_2 2^2 + \log_2 5 \\ &= 2 + \log_2 5\end{aligned}$$

41.
$$\begin{aligned}\log_6 \frac{x}{36} &= \log_6 x - \log_6 36 \\ &= \log_6 x - \log_6 6^2 \\ &= \log_6 x - 2\end{aligned}$$

43.
$$\ln y^4 = 4\ln y$$

45.
$$\begin{aligned}\log\sqrt{5} &= \log 5^{1/2} \\ &= \frac{1}{2}\log 5\end{aligned}$$

47.
$$\log xyz = \log(xy) + \log z$$
$$= \log x + \log y + \log z$$

49.
$$\log_2 \frac{2x}{y} = \log_2(2x) - \log_2 y$$
$$= \log_2 2 + \log_2 x - \log_2 y$$
$$= 1 + \log_2 x - \log_2 y$$

51.
$$\log x^3 y^2 = \log x^3 + \log y^2$$
$$= 3\log x + 2\log y$$

53.
$$\log_b (xy)^{1/2} = \frac{1}{2}\log_b(xy)$$
$$= \frac{1}{2}(\log_b x + \log_b y)$$

55.
$$\log_a \frac{\sqrt[3]{x}}{\sqrt[4]{yz}} = \log_a \frac{x^{1/3}}{(yz)^{1/4}}$$
$$= \log_a x^{1/3} - \log_a (yz)^{1/4}$$
$$= \frac{1}{3}\log_a x - \frac{1}{4}\log_a(yz)$$
$$= \frac{1}{3}\log_a x - \frac{1}{4}(\log_a y + \log_a z)$$
$$= \frac{1}{3}\log_a x - \frac{1}{4}\log_a y - \frac{1}{4}\log_a z$$

57.
$$\ln x\sqrt{z} = \ln x + \ln\sqrt{z}$$
$$= \ln x + \ln z^{1/2}$$
$$= \ln x + \frac{1}{2}\ln z$$

59.
$$\log_2(x+1) - \log_2 x = \log_2 \frac{x+1}{x}$$

61.
$$2\log x + \frac{1}{2}\log y = \log x^2 + \log y^{1/2}$$
$$= \log x^2 y^{1/2}$$

63.
$$-3\log_b x - 2\log_b y + \frac{1}{2}\log_b z$$
$$= \frac{1}{2}\log_b z - 3\log_b x - 2\log_b y$$
$$= \log_b z^{1/2} - \log_b x^3 - \log_b y^2$$
$$= \log_b z^{1/2} - \left(\log_b x^3 + \log_b y^2\right)$$
$$= \log_b z^{1/2} - \left(\log_b x^3 y^2\right)$$
$$= \log_b \frac{z^{1/2}}{x^3 y^2}$$

65.
$$\ln\left(\frac{x}{z} + x\right) - \ln\left(\frac{y}{z} + y\right) = \ln\left(\frac{\frac{x}{z} + x}{\frac{y}{z} + y}\right)$$
$$= \ln\left(\frac{\frac{x}{z} + x}{\frac{y}{z} + y} \cdot \frac{z}{z}\right)$$
$$= \ln\left(\frac{x + xz}{y + yz}\right)$$
$$= \ln\left(\frac{x\cancel{(1+z)}}{y\cancel{(1+z)}}\right)$$
$$= \ln\frac{x}{y}$$

67. False
$$\log xy = \log x + \log y$$

69. False
$$\log_b \frac{A}{B} = \log_b A - \log_b B$$

71. True

73.
$$\log_b 28 = \log_b(4 \cdot 7)$$
$$= \log_b 4 + \log_b 7$$
$$= 0.6021 + 0.8451$$
$$= 1.4472$$

75.

$$\log_b \frac{4}{63} = \log_b 4 - \log_b 63$$
$$= \log_b 4 - \log_b (7 \cdot 9)$$
$$= \log_b 4 - (\log_b 7 + \log_b 9)$$
$$= \log_b 4 - \log_b 7 - \log_b 9$$
$$= 0.6021 - 0.8451 - 0.9542$$
$$= -1.1972$$

77.

$$\log_b \frac{63}{4} = \log_b 63 - \log_b 4$$
$$= \log_b (7 \cdot 9) - \log_b 4$$
$$= \log_b 7 + \log_b 9 - \log_b 4$$
$$= 0.8451 + 0.9542 - 0.6021$$
$$= 1.1972$$

79.

$$\log_b 64 = \log_b 4^3$$
$$= 3\log_b 4$$
$$= 3(0.6021)$$
$$= 1.8063$$

81.

$$\log_3 7 = \frac{\log 7}{\log 3}$$
$$\approx \frac{0.84509}{0.47712}$$
$$\approx 1.7712$$

83.

$$\log_{1/3} 3 = \frac{\log 3}{\log \frac{1}{3}}$$
$$\approx \frac{0.47712}{-0.47712}$$
$$\approx -1.0000$$

85.

$$\log_3 8 = \frac{\log 8}{\log 3}$$
$$\approx \frac{0.90309}{0.47712}$$
$$\approx 1.8928$$

87.

$$\log_{\sqrt{2}} \sqrt{5} = \frac{\log \sqrt{5}}{\log \sqrt{2}}$$
$$\approx \frac{0.349485}{0.150514}$$
$$\approx 2.3219$$

APPLICATIONS

89. pH OF A SOLUTION

$$\text{pH} = -\log[\text{H}^+]$$
$$= -\log[1.7 \times 10^{-5}]$$
$$= -(\log 1.7 + \log 10^{-5})$$
$$= -(\log 1.7 - 5\log 10)$$
$$= -\log 1.7 + 5\log 10$$
$$= -0.2304 + 5 \cdot 1$$
$$= -0.2304 + 5$$
$$= 4.8$$

91. AQUARIUMS

For pH of 6.8:

$$\text{pH} = -\log[\text{H}^+]$$
$$6.8 = -\log[\text{H}^+]$$
$$-6.8 = \log[\text{H}^+]$$
$$[\text{H}^+] = 10^{-6.8}$$
$$[\text{H}^+] \approx 1.6 \times 10^{-7}$$

For pH of 7.6:

$$\text{pH} = -\log[\text{H}^+]$$
$$7.6 = -\log[\text{H}^+]$$
$$-7.6 = \log[\text{H}^+]$$
$$[\text{H}^+] = 10^{-7.6}$$
$$[\text{H}^+] \approx 2.5 \times 10^{-8}$$

The corresponding range in the hydrogen ion concentration is from 2.5×10^{-8} to 1.6×10^{-7}.

WRITING

93. Answers will vary.

REVIEW

95.

$$\begin{aligned} m &= \frac{y_2 - y_1}{x_2 - x_1} \\ &= \frac{-4-3}{4-(-2)} \\ &= \frac{-7}{6} \\ &= -\frac{7}{6} \end{aligned}$$

97.

$$\begin{aligned} \left(\frac{x_1+x_2}{2}, \frac{y_1+y_2}{2}\right) &= \left(\frac{-2+4}{2}, \frac{3+(-4)}{2}\right) \\ &= \left(\frac{2}{2}, \frac{-1}{2}\right) \\ &= \left(1, -\frac{1}{2}\right) \end{aligned}$$

CHALLENGE PROBLEMS

99. Answers will vary.

101.

$$\log_{b^2} x \stackrel{?}{=} \frac{1}{2}\log_b x$$

$$\begin{aligned} \log_{b^2} x &= \frac{\log_b x}{\log_b b^2} \\ &= \frac{\log_b x}{2\log_b b} \\ &= \frac{\log_b x}{2(1)} \\ &= \frac{\log_b x}{2} \\ &= \frac{1}{2}\log_b x \end{aligned}$$

SECTION 11.8

VOCABULARY

1. An equation with a variable in its exponent, such as $3^{2x}=8$, is called an **exponential** equation.

CONCEPTS

3. Yes, it is a solution.

$$5^{2x+3}=\frac{1}{5}$$
$$5^{2(-2)+3}\stackrel{?}{=}\frac{1}{5}$$
$$5^{-4+3}\stackrel{?}{=}\frac{1}{5}$$
$$5^{-1}\stackrel{?}{=}\frac{1}{5}$$
$$\frac{1}{5}=\frac{1}{5}$$

5. Yes, it is a solution.

$$2^{2x+1}=70$$
$$2^{2(2.5646)+1}\stackrel{?}{=}70$$
$$2^{5.1292+1}\stackrel{?}{=}70$$
$$2^{6.1292}\stackrel{?}{=}70$$
$$69.9969\approx 70$$

7. Both sides of the exponential equation $5^{x-3}=125$ can be written as a power of $\boxed{5}$.

9. a. $\log(x+1)=2$
   $10^2=x+1$
   b. $\ln(x+1)=2$
   $e^2=x+1$

11. If $2^x=9$, then $\log 2^x=\boxed{\log 9}$.

13. $x\log 7=12$

15. a. $\dfrac{\log 8}{\log 5}\approx 1.2920$
   b. $\dfrac{2\ln 12}{\ln 9}\approx 2.2619$

17. a. $A=\boxed{A_0 2^{-t/h}}$
   b. $P=\boxed{P_0 e^{kt}}$

NOTATION

19.

$$2^x=7$$
$$\boxed{\log}2^x=\log 7$$
$$x\boxed{\log 2}=\log 7$$
$$x=\frac{\log 7}{\log 2}$$

PRACTICE

21.

$$2^{x-2}=64$$
$$2^{x-2}=2^6$$
$$x-2=6$$
$$x=8$$

23.

$$5^{4x}=\frac{1}{125}$$
$$5^{4x}=\frac{1}{5^3}$$
$$5^{4x}=5^{-3}$$
$$4x=-3$$
$$x=-\frac{3}{4}$$

25.

$$2^{x^2-2x}=8$$
$$2^{x^2-2x}=2^3$$
$$x^2-2x=3$$
$$x^2-2x-3=0$$
$$(x-3)(x+1)=0$$
$$x-3=0 \quad\text{or}\quad x+1=0$$
$$x=3 \qquad\qquad x=-1$$

27.

$$3^{x^2+4x} = \frac{1}{81}$$
$$3^{x^2+4x} = \frac{1}{3^4}$$
$$3^{x^2+4x} = 3^{-4}$$
$$x^2 + 4x = -4$$
$$x^2 + 4x + 4 = 0$$
$$(x+2)(x+2) = 0$$
$$x+2=0 \quad \text{or} \quad x+2=0$$
$$x=-2 \qquad\qquad x=-2$$

29.

$$4^x = 5$$
$$\log 4^x = \log 5$$
$$x\log 4 = \log 5$$
$$\frac{x\log 4}{\log 4} = \frac{\log 5}{\log 4}$$
$$x \approx 1.1610$$

31.

$$13^{x-1} = 2$$
$$\log 13^{x-1} = \log 2$$
$$(x-1)\log 13 = \log 2$$
$$x\log 13 - \log 13 = \log 2$$
$$x\log 13 = \log 2 + \log 13$$
$$\frac{x\log 13}{\log 13} = \frac{\log 2 + \log 13}{\log 13}$$
$$x \approx 1.2702$$

33.

$$2^{x+1} = 3^x$$
$$\log 2^{x+1} = \log 3^x$$
$$(x+1)\log 2 = x\log 3$$
$$x\log 2 + \log 2 = x\log 3$$
$$\log 2 = x\log 3 - x\log 2$$
$$\log 2 = x(\log 3 - \log 2)$$
$$\frac{\log 2}{\log 3 - \log 2} = \frac{x(\log 3 - \log 2)}{\log 3 - \log 2}$$
$$x = \frac{\log 2}{\log 3 - \log 2}$$
$$x \approx 1.7095$$

35.

$$2^x = 3^x$$
$$\log 2^x = \log 3^x$$
$$x\log 2 = x\log 3$$
$$x\log 2 - x\log 3 = x\log 3 - x\log 3$$
$$x\log 2 - x\log 3 = 0$$
$$x(\log 2 - \log 3) = 0$$
$$\frac{x(\log 2 - \log 3)}{\log 2 - \log 3} = \frac{0}{\log 2 - \log 3}$$
$$x = \frac{0}{\log 2 - \log 3}$$
$$x = 0$$

37.

$$7^{x^2} = 10$$
$$\log 7^{x^2} = \log 10$$
$$x^2 \log 7 = \log 10$$
$$x^2 = \frac{\log 10}{\log 7}$$
$$x = \pm\sqrt{\frac{\log 10}{\log 7}}$$
$$x \approx \pm 1.0878$$

39.

$$8^{x^2} = 9^x$$
$$\log 8^{x^2} = \log 9^x$$
$$x^2 \log 8 = x\log 9$$
$$x^2 \log 8 - x\log 9 = 0$$
$$x(x\log 8 - \log 9) = 0$$
$$x = 0 \quad \text{or} \quad x\log 8 - \log 9 = 0$$
$$x = 0 \qquad x\log 8 = \log 9$$
$$x = 0 \qquad x = \frac{\log 9}{\log 8}$$
$$x = 0 \qquad x \approx 1.0566$$

41.

$$\begin{aligned} e^{3x} &= 9 \\ \ln e^{3x} &= \ln 9 \\ 3x \ln e &= \ln 9 \\ 3x \cdot 1 &= \ln 9 \\ 3x &= \ln 9 \\ x &= \frac{\ln 9}{3} \\ x &\approx 0.7324 \end{aligned}$$

43.

$$\begin{aligned} e^{-0.2t} &= 14.2 \\ \ln e^{-0.2t} &= \ln 14.2 \\ -0.2t \ln e &= \ln 14.2 \\ -0.2t \cdot 1 &= \ln 14.2 \\ -0.2t &= \ln 14.2 \\ t &= \frac{\ln 14.2}{-0.2} \\ t &\approx -13.2662 \end{aligned}$$

45.

$$\begin{aligned} 2^{x+1} &= 7 \\ 2^{x+1} - 7 &= 0 \\ y &= 2^{x+1} - 7 \end{aligned}$$

The x–intercept is (1.8, 0) so the solution is $x \approx 1.8$.

47.

$$\begin{aligned} 4\left(2^{x^2}\right) &= 8^{3x} \\ 4\left(2^{x^2}\right) - 8^{3x} &= 0 \\ y &= 4\left(2^{x^2}\right) - 8^{3x} \end{aligned}$$

The x–intercepts are (8.8, 0) and (0.2, 0) so the solutions are $x \approx 8.8$ and $x \approx 0.2$.

49.

$$\begin{aligned} \log(x+2) &= 4 \\ 10^4 &= x+2 \\ 10{,}000 &= x+2 \\ 9{,}998 &= x \end{aligned}$$

51.

$$\begin{aligned} \log(7-x) &= 2 \\ 10^2 &= 7-x \\ 100 &= 7-x \\ 93 &= -x \\ -93 &= x \end{aligned}$$

53.

$$\begin{aligned} \ln x &= 1 \\ e^1 &= x \\ x &= e \\ x &\approx 2.7183 \end{aligned}$$

55.

$$\begin{aligned} \ln(x+1) &= 3 \\ e^3 &= x+1 \\ x+1 &= e^3 \\ x &= e^3 - 1 \\ x &\approx 20.0855 - 1 \\ x &\approx 19.0855 \end{aligned}$$

57.

$$\begin{aligned} \log 2x &= \log 4 \\ 2x &= 4 \\ x &= 2 \end{aligned}$$

59.

$$\begin{aligned} \ln(3x+1) &= \ln(x+7) \\ 3x+1 &= x+7 \\ 2x+1 &= 7 \\ 2x &= 6 \\ x &= 3 \end{aligned}$$

61.

$$\begin{aligned} \log(3-2x) - \log(x+24) &= 0 \\ \log(3-2x) &= \log(x+24) \\ 3-2x &= x+24 \\ 3-3x &= 24 \\ -3x &= 21 \\ x &= -7 \end{aligned}$$

63.

$$\log\frac{4x+1}{2x+9}=0$$
$$10^0=\frac{4x+1}{2x+9}$$
$$1=\frac{4x+1}{2x+9}$$
$$2x+9=4x+1$$
$$9=2x+1$$
$$8=2x$$
$$4=x$$

65.

$$\log x^2=2$$
$$10^2=x^2$$
$$100=x^2$$
$$x^2=100$$
$$x=\pm\sqrt{100}$$
$$x=\pm 10$$
$$x=10,-10$$

67.

$$\log x+\log(x-48)=2$$
$$\log_{10}\left[x(x-48)\right]=2$$
$$10^2=x(x-48)$$
$$100=x^2-48x$$
$$0=x^2-48x-100$$
$$0=(x-50)(x+2)$$
$$x-50=0 \quad \text{or} \quad x+2=0$$
$$x=50 \qquad x=\cancel{-2}$$

A negative number does not have a logarithm, so the only possible answer is 50.

69.

$$\log x+\log(x-15)=2$$
$$\log_{10}\left[x(x-15)\right]=2$$
$$10^2=x(x-15)$$
$$100=x^2-15x$$
$$0=x^2-15x-100$$
$$0=(x-20)(x+5)$$
$$x-20=0 \quad \text{or} \quad x+5=0$$
$$x=20 \qquad x=\cancel{-5}$$

A negative number does not have a logarithm, so the only possible answer is 20.

71.

$$\log(x+90)=3-\log x$$
$$\log(x+90)+\log x=3$$
$$\log_{10}\left[x(x+90)\right]=3$$
$$10^3=x(x+90)$$
$$1{,}000=x^2+90x$$
$$0=x^2+90x-1{,}000$$
$$0=(x+100)(x-10)$$
$$x+100=0 \quad \text{or} \quad x-10=0$$
$$x=\cancel{-100} \qquad x=10$$

A negative number does not have a logarithm, so the only possible answer is 10.

73.

$$\log(x-6)-\log(x-2)=\log\frac{5}{x}$$
$$\log\left(\frac{x-6}{x-2}\right)=\log\frac{5}{x}$$
$$\frac{x-6}{x-2}=\frac{5}{x}$$
$$x(x-2)\left(\frac{x-6}{x-2}\right)=x(x-2)\left(\frac{5}{x}\right)$$
$$x(x-6)=5(x-2)$$
$$x^2-6x=5x-10$$
$$x^2-6x-5x+10=0$$
$$x^2-11x+10=0$$
$$(x-1)(x-10)=0$$
$$x-1=0 \quad \text{or} \quad x-10=0$$
$$x=\not{1} \qquad x=10$$

When $x = 1$, the two numbers on the left side would be negative, which is not possible. The only possible answer for x is 10.

75.

$$\frac{\log(3x-4)}{\log x}=2$$
$$(\log x)\left(\frac{\log(3x-4)}{\log x}\right)=(\log x)(2)$$
$$\log(3x-4)=2\log x$$
$$\log(3x-4)=\log x^2$$
$$3x-4=x^2$$
$$0=x^2-3x+4$$

No Solution

The discriminant indicates that the answers are not real. There are no solutions.

77.

$$\frac{\log(5x+6)}{2}=\log x$$
$$2\left(\frac{\log(5x+6)}{2}\right)=2(\log x)$$
$$\log(5x+6)=\log x^2$$
$$5x+6=x^2$$
$$0=x^2-5x-6$$
$$0=(x-6)(x+1)$$
$$x-6=0 \quad \text{or} \quad x+1=0$$
$$x=6 \qquad x=\not{-1}$$

A negative number does not have a logarithm, so the only answer is $x = 6$.

79.

$$\log_3 x=\log_3\left(\frac{1}{x}\right)+4$$
$$\log_3 x-\log_3\left(\frac{1}{x}\right)=4$$
$$\log_3\frac{x}{\frac{1}{x}}=4$$
$$\log_3 x^2=4$$
$$3^4=x^2$$
$$81=x^2$$
$$x^2=81$$
$$x=\pm\sqrt{81}$$
$$x=\pm 9$$
$$x=9,\not{-9}$$

A negative number does not have a logarithm, so the only answer is $x = 9$.

81.

$$2\log_2 x = 3 + \log_2(x-2)$$
$$2\log_2 x - \log_2(x-2) = 3$$
$$\log_2 x^2 - \log_2(x-2) = 3$$
$$\log_2 \frac{x^2}{x-2} = 3$$
$$2^3 = \frac{x^2}{x-2}$$
$$8 = \frac{x^2}{x-2}$$
$$(x-2)(8) = (x-2)\left(\frac{x^2}{x-2}\right)$$
$$8x - 16 = x^2$$
$$0 = x^2 - 8x + 16$$
$$0 = (x-4)(x-4)$$
$$x - 4 = 0 \quad \text{or} \quad x - 4 = 0$$
$$x = 4 \qquad\qquad x = 4$$

83.

$$\log(7y+1) = 2\log(y+3) - \log 2$$
$$\log(7y+1) = \log(y+3)^2 - \log 2$$
$$\log(7y+1) = \log\frac{(y+3)^2}{2}$$
$$7y+1 = \frac{(y+3)^2}{2}$$
$$2(7y+1) = 2\left(\frac{(y+3)^2}{2}\right)$$
$$14y + 2 = (y+3)^2$$
$$14y + 2 = y^2 + 6y + 9$$
$$0 = y^2 + 6y + 9 - 14y - 2$$
$$0 = y^2 - 8y + 7$$
$$0 = (y-1)(y-7)$$
$$y - 1 = 0 \quad \text{or} \quad y - 7 = 0$$
$$y = 1 \qquad\qquad y = 7$$

85.

$$\log x + \log(x-15) = 2$$
$$\log x + \log(x-15) - 2 = 0$$
$$y = \log x + \log(x-15) - 2$$

The graph of the function has an x–intercept of (20, 0), so the answer is $x = 20$.

87.

$$\ln(2x+5) - \ln 3 = \ln(x-1)$$
$$\ln(2x+5) - \ln 3 - \ln(x-1) = 0$$
$$y = \ln(2x+5) - \ln 3 - \ln(x-1)$$

The graph of the function has an x–intercept of (8, 0), so the answer is $x = 8$.

APPLICATIONS

89. TRITIUM DECAY

If 25% of the tritium has decayed, then 75%, or 0.75, of the tritium remains.
(100% – 25% = 75%)

Let $h = 12.4$ and $A = 0.75A_0$.

$$A = A_0 2^{-t/h}$$
$$0.75A_0 = A_0 2^{-t/12.4}$$
$$\frac{0.75A_0}{A_0} = \frac{A_0 2^{-t/12.4}}{A_0}$$
$$0.75 = 2^{-t/12.4}$$
$$\log 0.75 = \log 2^{-t/12.4}$$
$$\log 0.75 = -\frac{t}{12.4}\log 2$$
$$-12.4(\log 0.75) = -12.4\left(-\frac{t}{12.4}\log 2\right)$$
$$-12.4(\log 0.75) = t\log 2$$
$$\frac{-12.4(\log 0.75)}{\log 2} = \frac{t\log 2}{\log 2}$$
$$\frac{-12.4(\log 0.75)}{\log 2} = t$$
$$t = \frac{-12.4(\log 0.75)}{\log 2}$$
$$t \approx 5.1 \text{ years}$$

91. THORIUM DECAY
If 80% of the thorium has decayed, then 20%, or 0.20, of the thorium remains.
(100% – 80% = 20%)

Let $h = 18.4$ and $A = 0.20A_0$.

$$A = A_0 2^{-t/h}$$
$$0.20A_0 = A_0 2^{-t/18.4}$$
$$\frac{0.20A_0}{A_0} = \frac{A_0 2^{-t/18.4}}{A_0}$$
$$0.20 = 2^{-t/18.4}$$
$$\log 0.20 = \log 2^{-t/18.4}$$
$$\log 0.20 = -\frac{t}{18.4}\log 2$$
$$-18.4(\log 0.20) = -18.4\left(-\frac{t}{18.4}\log 2\right)$$
$$-18.4(\log 0.20) = t\log 2$$
$$\frac{-18.4(\log 0.20)}{\log 2} = \frac{t\log 2}{\log 2}$$
$$\frac{-18.4(\log 0.20)}{\log 2} = t$$
$$t = \frac{-18.4(\log 0.20)}{\log 2}$$
$$t \approx 42.7 \text{ days}$$

93. CARBON–14 DATING
60% of the carbon–14 remains and the half–life is 5,700 years (from Example 10).

Let $h = 5{,}700$ and $A = 0.60A_0$.

$$A = A_0 2^{-t/h}$$
$$0.60A_0 = A_0 2^{-t/5{,}700}$$
$$\frac{0.60A_0}{A_0} = \frac{A_0 2^{-t/5{,}700}}{A_0}$$
$$0.60 = 2^{-t/5{,}700}$$
$$\log 0.60 = \log 2^{-t/5{,}700}$$
$$\log 0.60 = -\frac{t}{5{,}700}\log 2$$
$$-5{,}700(\log 0.60) = -5{,}700\left(-\frac{t}{5{,}700}\log 2\right)$$
$$-5{,}700(\log 0.60) = t\log 2$$
$$\frac{-5{,}700(\log 0.60)}{\log 2} = \frac{t\log 2}{\log 2}$$
$$\frac{-5{,}700(\log 0.60)}{\log 2} = t$$
$$t = \frac{-5{,}700(\log 0.60)}{\log 2}$$
$$t \approx 4{,}200 \text{ years}$$

95. COMPOUND INTEREST

Let $FV = 800$, $PV = 500$, and $i = 0.085$.

$$FV = PV(1+i)^n$$
$$800 = 500(1+0.085)^n$$
$$800 = 500(1.085)^n$$
$$\frac{800}{500} = \frac{500(1.085)^n}{500}$$
$$1.6 = 1.085^n$$
$$\log 1.6 = \log 1.085^n$$
$$\log 1.6 = n\log 1.085$$
$$\frac{\log 1.6}{\log 1.085} = n$$
$$n \approx 5.7 \text{ years}$$

97. COMPOUND INTEREST

Let $P = 2{,}100$, $P_0 = 1{,}300$, $r = 0.09$, and $k = 4$.

$$P = P_0\left(1+\frac{r}{k}\right)^{kt}$$
$$2{,}100 = 1{,}300\left(1+\frac{0.09}{4}\right)^{4t}$$
$$2{,}100 = 1{,}300(1+0.0225)^{4t}$$
$$2{,}100 = 1{,}300(1.0225)^{4t}$$
$$\frac{2{,}100}{1{,}300} = \frac{1{,}300(1.0225)^{4t}}{1{,}300}$$
$$1.6154 = (1.0225)^{4t}$$
$$\log 1.6154 = \log(1.0225)^{4t}$$
$$\log 1.6154 = 4t\log 1.0225$$
$$\frac{\log 1.6154}{\log 1.0225} = 4t$$
$$4t = \frac{\log 1.6154}{\log 1.0225}$$
$$t = \frac{\log 1.6154}{4\log 1.0225}$$
$$t \approx 5.4 \text{ years}$$

99. RULE OF SEVENTY

This formula works because $\ln 2 \approx 0.7$.

101. RODENT CONTROL

Let P_0=30,000, $P = 2(30{,}000) = 60{,}000$, and $t = 5$.

$$P = P_0e^{kt}$$
$$60{,}000 = 30{,}000e^{k5}$$
$$60{,}000 = 30{,}000e^{5k}$$
$$\frac{60{,}000}{30{,}000} = \frac{30{,}000e^{5k}}{30{,}000}$$
$$2 = e^{5k}$$
$$5k = \ln 2$$
$$k = \frac{\ln 2}{5}$$

Let $P_0 = 30{,}000$, $P = 1{,}000{,}000$, and $k = \frac{\ln 2}{5}$.

$$P = P_0e^{kt}$$
$$1{,}000{,}000 = 30{,}000e^{\left(\frac{\ln 2}{5}\right)t}$$
$$\frac{1{,}000{,}000}{30{,}000} = \frac{30{,}000e^{\left(\frac{\ln 2}{5}\right)t}}{30{,}000}$$
$$\frac{100}{3} = e^{\left(\frac{\ln 2}{5}\right)t}$$
$$\left(\frac{\ln 2}{5}\right)t = \ln\frac{100}{3}$$
$$\left(\frac{5}{\ln 2}\right)\left(\frac{\ln 2}{5}\right)t = \left(\frac{5}{\ln 2}\right)\left(\ln\frac{100}{3}\right)$$
$$t \approx 25.3 \text{ years}$$

103. BACTERIA CULTURE

Let $P_0 = 1$, $P = 2(1) = 2$, and $t = 24$.

$$P = P_0 e^{kt}$$
$$2 = 1e^{k24}$$
$$2 = e^{24k}$$
$$\ln 2 = \ln e^{24k}$$
$$\ln 2 = 24k$$
$$24k = \ln 2$$
$$k = \frac{\ln 2}{24}$$

Let $P_0 = 1$, $t = 36$, and $k = \dfrac{\ln 2}{24}$.

$$P = P_0 e^{kt}$$
$$P = 1e^{\left(\frac{\ln 2}{24}\right)(36)}$$
$$P = e^{\frac{3\ln 2}{2}}$$
$$P \approx 2.828 \text{ times larger}$$

105. MEDICINE

Use the formula derived in Example 12.

Let $B = 5\times10^6$ and $b = 500$.

$$n = \frac{1}{\ln 2}\left(\ln\frac{B}{b}\right)$$
$$n = \frac{1}{\ln 2}\left(\ln\frac{5\times10^6}{500}\right)$$
$$n \approx 13.3 \text{ generations}$$

107. NEWTON'S LAW OF COOLING

Let $t = 3$ and $T = 90$.

$$T = 60 + 40e^{kt}$$
$$90 = 60 + 40e^{3k}$$
$$90 - 60 = 60 + 40e^{3k} - 60$$
$$30 = 40e^{3k}$$
$$\frac{30}{40} = \frac{40e^{3k}}{40}$$
$$0.75 = e^{3k}$$
$$\ln 0.75 = \ln e^{3k}$$
$$\ln 0.75 = 3k$$
$$\frac{\ln 0.75}{3} = \frac{3k}{3}$$
$$\frac{\ln 0.75}{3} = k$$
$$k = \frac{\ln 0.75}{3}$$
$$k = \frac{1}{3}\log 0.75$$

WRITING

109. Answers will vary.

111. Answers will vary.

REVIEW

113. Use the Pythagorean Theorem.

$$a^2 + b^2 = c^2$$
$$\left(\sqrt{7}\right)^2 + b^2 = (12)^2$$
$$7 + b^2 = 144$$
$$b^2 = 137$$
$$b = \sqrt{137} \text{ in.}$$

CHALLENGE PROBLEMS

115. $x \le 3$; The number 3 would make the first expression the logarithm of 0, which is not possible, and any numbers less than 3 would make the first expression a logarithm of a negative number, which is not possible.

CHAPTER 11 KEY CONCEPTS

1. No. The output of 4 corresponds with two inputs, 2 and –2.

2. No.

3. Yes.

4.

$$f(x) = -2x - 1$$
$$y = -2x - 1$$
$$x = -2y - 1$$
$$x + 1 = -2y$$
$$\frac{x+1}{-2} = y$$
$$-\frac{x+1}{2} = y$$
$$y = -\frac{x+1}{2}$$
$$f^{-1}(x) = -\frac{x+1}{2}$$

5.

x	4	–2	–6
$f^{-1}(x)$	–2	1	3

6. $\log 1{,}000 = 3$

7. $2^{-3} = \frac{1}{8}$

8.

$$\log_4 x = \frac{1}{2}$$
$$4^{1/2} = x$$
$$\sqrt{4} = x$$
$$2 = x$$
$$x = 2$$

9.

$$\log_x \frac{9}{4} = 2$$
$$x^2 = \frac{9}{4}$$
$$\sqrt{x^2} = \sqrt{\frac{9}{4}}$$
$$x = \frac{3}{2}$$

10. 2.71828

11. e

12.

$$\ln x = -0.28$$
$$e^{\ln x} = e^{-0.28}$$
$$x = 0.7558$$

13.

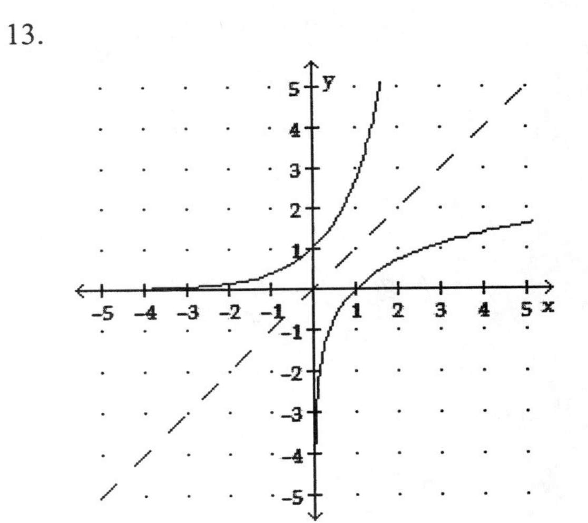

CHAPTER 11 REVIEW

SECTION 11.1
Algebra and Composition of Functions

1. $f+g$

$$\begin{aligned}(f+g)(x)&=f(x)+g(x)\\&=(2x)+(x+1)\\&=3x+1\end{aligned}$$

$D:(-\infty,\infty)$

2. $f-g$

$$\begin{aligned}(f-g)(x)&=f(x)-g(x)\\&=(2x)-(x+1)\\&=2x-x-1\\&=x-1\end{aligned}$$

$D:(-\infty,\infty)$

3. $f\cdot g$

$$\begin{aligned}(f\cdot g)(x)&=f(x)\cdot g(x)\\&=(2x)(x+1)\\&=2x^2+2x\end{aligned}$$

$D:(-\infty,\infty)$

4. f/g

$$\begin{aligned}(f/g)(x)&=\frac{f(x)}{g(x)}\\&=\frac{2x}{x+1}\end{aligned}$$

$D:(-\infty,-1)\cup(-1,\infty)$

5.

$$\begin{aligned}(f\circ g)(-1)&=f(g(-1))\\&=f(2(-1)+1)\\&=f(-2+1)\\&=f(-1)\\&=(-1)^2+2\\&=1+2\\&=3\end{aligned}$$

6.

$$\begin{aligned}(g\circ f)(0)&=g(f(0))\\&=g(0^2+2)\\&=g(0+2)\\&=g(2)\\&=2(2)+1\\&=4+1\\&=5\end{aligned}$$

7.

$$\begin{aligned}(f\circ g)(x)&=f(g(x))\\&=f(2x+1)\\&=(2x+1)^2+2\\&=4x^2+2x+2x+1+2\\&=4x^2+4x+3\end{aligned}$$

8.

$$\begin{aligned}(g\circ f)(x)&=g(f(x))\\&=g(x^2+2)\\&=2(x^2+2)+1\\&=2x^2+4+1\\&=2x^2+5\end{aligned}$$

9. a.

$$\begin{aligned}(f+g)(2)&=f(2)+g(2)\\&=3+9\\&=12\end{aligned}$$

b.

$$\begin{aligned}(f\cdot g)(2)&=f(2)\cdot g(2)\\&=3(9)\\&=27\end{aligned}$$

c.

$$\begin{aligned}(f\circ g)(2)&=f(g(2))\\&=f(9)\\&=7\end{aligned}$$

d.

$$\begin{aligned}(g\circ f)(2)&=g(f(2))\\&=g(3)\\&=0\end{aligned}$$

10.

$$\begin{aligned}(C \circ f)(m) &= C(f(m)) \\ &= C\left(\frac{m}{8}\right) \\ &= 1.85\left(\frac{m}{8}\right) \\ C(m) &= \frac{1.85m}{8}\end{aligned}$$

SECTION 11.2
Inverse Functions

11. No. The output 4 corresponds with more than one input, 1 and –1.

12. Yes.

13. Yes.

14. No. The output –5 corresponds with more than one input, 0 and 4.

15. Yes.

16. No. It does not pass the horizontal line test.

17.

x	$f^{-1}(x)$
–6	–6
–3	–1
12	7
3	20

18.

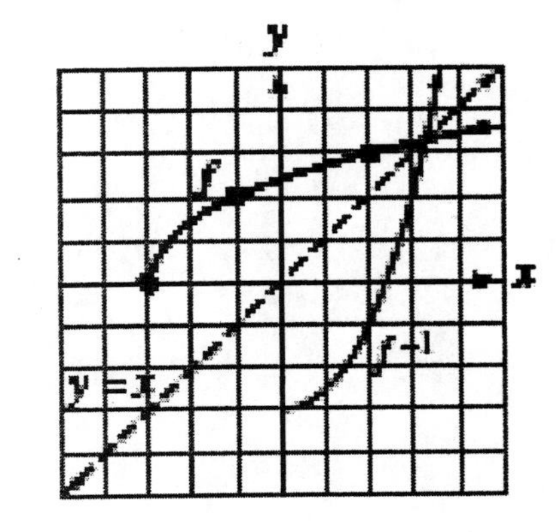

19.

$$\begin{aligned}f(x) &= 6x - 3 \\ y &= 6x - 3 \\ x &= 6y - 3 \\ x + 3 &= 6y \\ \frac{x+3}{6} &= y \\ y &= \frac{x+3}{6} \\ f^{-1}(x) &= \frac{x+3}{6}\end{aligned}$$

20.

$$\begin{aligned}f(x) &= \frac{4}{x-1} \\ y &= \frac{4}{x-1} \\ x &= \frac{4}{y-1} \\ (y-1)(x) &= (y-1)\left(\frac{4}{y-1}\right) \\ xy - x &= 4 \\ xy &= 4 + x \\ y &= \frac{4+x}{x} \\ y &= \frac{4}{x} + \frac{x}{x} \\ y &= \frac{4}{x} + 1 \\ f^{-1}(x) &= \frac{4}{x} + 1\end{aligned}$$

21.

$$\begin{aligned}f(x) &= (x+2)^3 \\ y &= (x+2)^3 \\ x &= (y+2)^3 \\ \sqrt[3]{x} &= \sqrt[3]{(y+2)^3} \\ \sqrt[3]{x} &= y + 2 \\ \sqrt[3]{x} - 2 &= y \\ y &= \sqrt[3]{x} - 2 \\ f^{-1}(x) &= \sqrt[3]{x} - 2\end{aligned}$$

22.

$$f(x)=\frac{x}{6}-\frac{1}{6}$$
$$y=\frac{x}{6}-\frac{1}{6}$$
$$x=\frac{y}{6}-\frac{1}{6}$$
$$x+\frac{1}{6}=\frac{y}{6}$$
$$6\left(x+\frac{1}{6}\right)=6\left(\frac{y}{6}\right)$$
$$6x+1=y$$
$$y=6x+1$$
$$f^{-1}(x)=6x+1$$

23.

$$f(x)=\sqrt[3]{x-1}$$
$$y=\sqrt[3]{x-1}$$
$$x=\sqrt[3]{y-1}$$
$$(x)^3=\left(\sqrt[3]{y-1}\right)^3$$
$$x^3=y-1$$
$$x^3+1=y$$
$$y=x^3+1$$
$$f^{-1}(x)=x^3+1$$

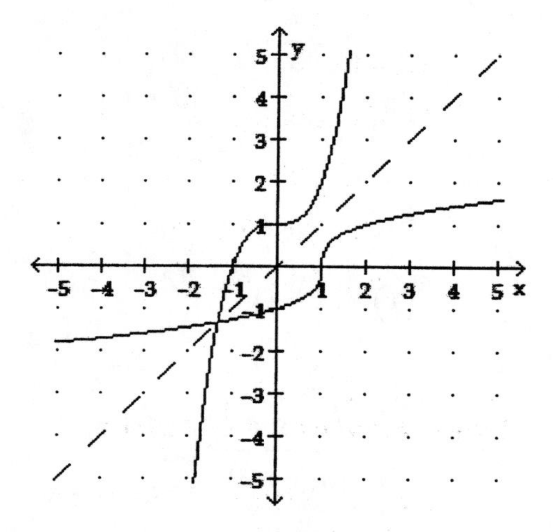

24.

$$(f\circ f^{-1})(x)=f(f^{-1}(x))$$
$$=f\left(-\frac{x-5}{4}\right)$$
$$=5-4\left(-\frac{x-5}{4}\right)$$
$$=5+x-5$$
$$=x$$

$$(f^{-1}\circ f)(x)=f^{-1}(f(x))$$
$$=f^{-1}(5-4x)$$
$$=-\frac{(5-4x)-5}{4}$$
$$=-\frac{-4x}{4}$$
$$=\frac{4x}{4}$$
$$=x$$

SECTION 11.3
Exponential Functions

25.

$$5^{\sqrt{6}}\bullet 5^{3\sqrt{6}}=5^{\sqrt{6}+3\sqrt{6}}$$
$$=5^{4\sqrt{6}}$$

26.

$$\left(2^{\sqrt{14}}\right)^{\sqrt{2}}=2^{\sqrt{14}\bullet\sqrt{2}}$$
$$=2^{\sqrt{28}}$$
$$=2^{\sqrt{4}\sqrt{7}}$$
$$=2^{2\sqrt{7}}$$

27. $f(x) = 3^x$

D: $(-\infty, \infty)$, R: $(0, \infty)$

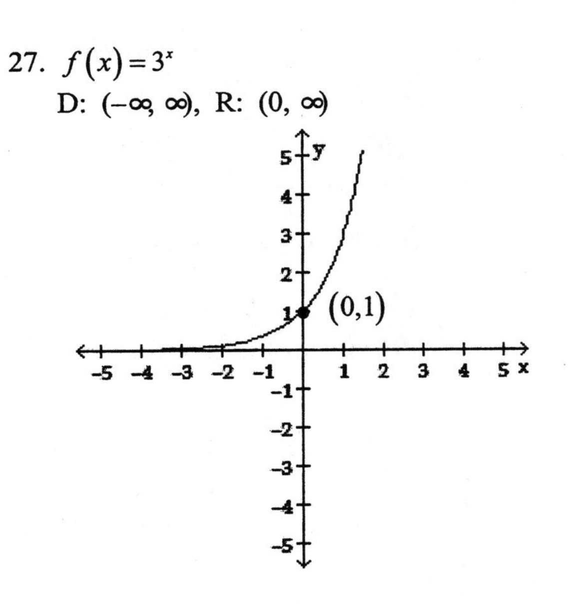

28. $f(x) = \left(\frac{1}{3}\right)^x$

D: $(-\infty, \infty)$, R: $(0, \infty)$

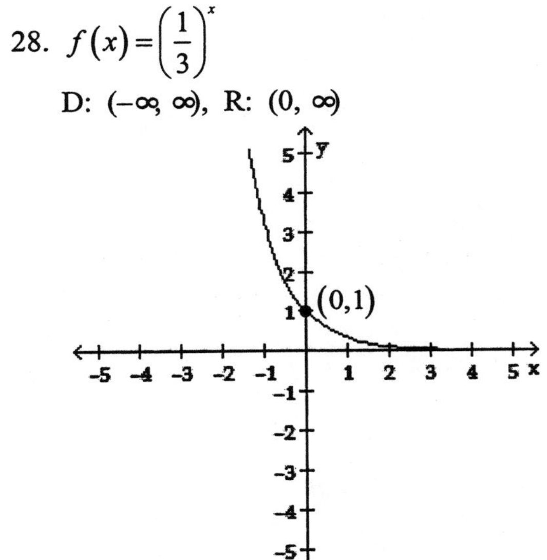

29. $f(x) = \left(\frac{1}{2}\right)^x - 2$

D: $(-\infty, \infty)$, R: $(-2, \infty)$

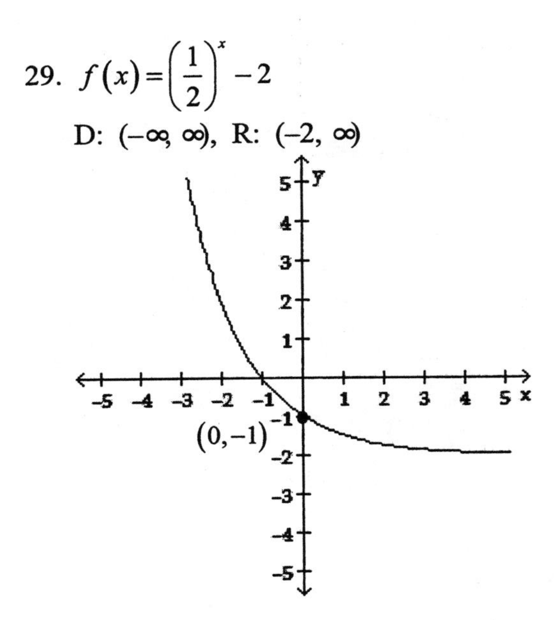

30. $f(x) = 3^{x-1}$

D: $(-\infty, \infty)$, R: $(0, \infty)$

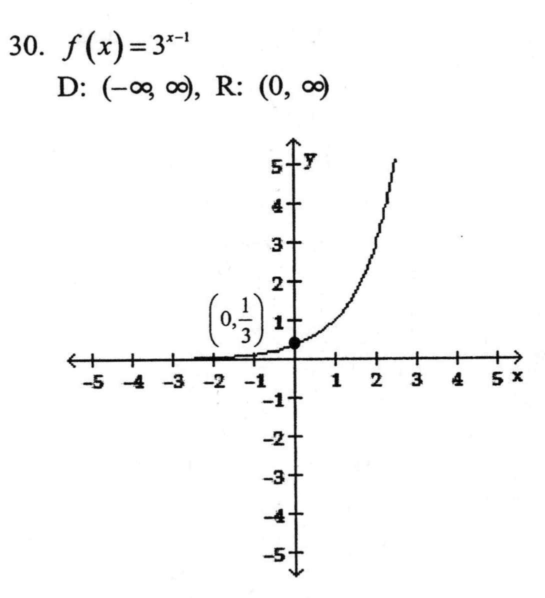

31. the x–axis ($y = 0$)

32. an exponential function

33. Let $P = 10{,}500$, $r = 0.09$, $k = 4$, and $t = 6$

$$A = 10{,}500\left(1 + \frac{0.09}{4}\right)^{4(60)}$$
$$= 10{,}500(1 + 0.0225)^{240}$$
$$= 10{,}500(1.0225)^{240}$$
$$= 10{,}500(208.543186)$$
$$= \$2{,}189{,}703.45$$

34. Let $t = 5$.

$$V(t) = 12{,}000\left(10^{-0.155t}\right)$$
$$V(5) = 12{,}000\left(10^{-0.155(5)}\right)$$
$$= 12{,}000\left(10^{-0.775}\right)$$
$$= 12{,}000(0.16788)$$
$$\approx \$2{,}014.56$$

SECTION 11.4
Base–e Exponential Functions

35. Let $t = 15$. $(1995 - 1980 = 15)$

$$r(t) = 13.9e^{-0.035t}$$
$$r(15) = 13.9e^{-0.035(15)}$$
$$= 13.9e^{-0.525}$$
$$\approx 13.9(0.59155)$$
$$\approx 8.22\%$$

36. Let $P = 10{,}500$, $r = 0.09$, and $t = 60$.

$$A = Pe^{rt}$$
$$= 10{,}500e^{0.09(60)}$$
$$= 10{,}500e^{5.4}$$
$$\approx 10{,}500(221.4064)$$
$$\approx \$2{,}324{,}767.37$$

37. $f(x) = e^x + 1$

D: $(-\infty, \infty)$, R: $(1, \infty)$

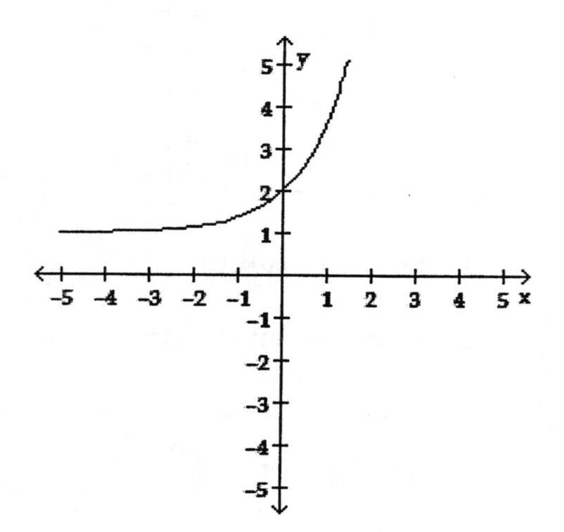

38. $f(x) = e^{x-3}$

D: $(-\infty, \infty)$, R: $(0, \infty)$

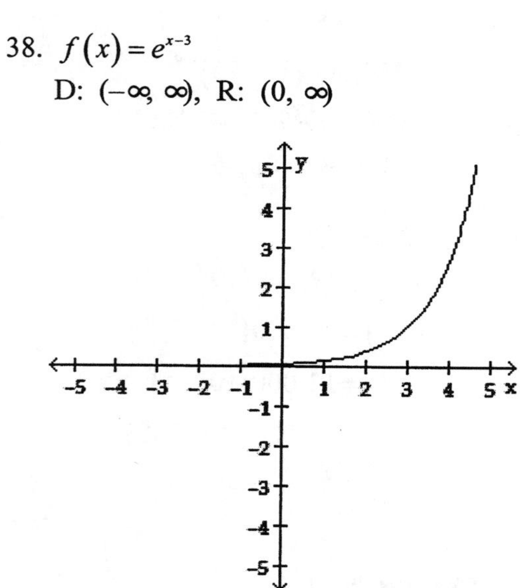

39. Let $P_0 = 290{,}340{,}000$, $k = 0.0092$, and $t = 50$.

$$P = P_0e^{kt}$$
$$= 290{,}340{,}000e^{0.0092(50)}$$
$$= 290{,}340{,}000e^{0.46}$$
$$\approx 290{,}340{,}000(1.584074)$$
$$\approx 459{,}920{,}041$$

40. The exponent on the base e is negative.

SECTION 11.5
Logarithmic Functions

41. D: $(0, \infty)$, R: $(-\infty, \infty)$

42. Since there is no real number such that $10^? = 0$, $\log 0$ is undefined.

43. $4^3 = 64$

44. $\log_7 \frac{1}{7} = -1$

45.

$$\log_3 9 = x$$
$$3^x = 9$$
$$3^x = 3^2$$
$$x = 2$$

46.

$$\log_9 \frac{1}{81} = x$$
$$9^x = \frac{1}{81}$$
$$9^x = \frac{1}{9^2}$$
$$9^x = 9^{-2}$$
$$x = -2$$

47.

$$\log_{1/2} 1 = x$$
$$\left(\frac{1}{2}\right)^x = 1$$
$$x = 0$$

48.

$$\log_5(-25) = x$$
$$5^x = -25$$

not possible

49.

$$\log_6 \sqrt{6} = x$$
$$6^x = \sqrt{6}$$
$$6^x = 6^{1/2}$$
$$x = \frac{1}{2}$$

50.

$$\log 1,000 = x$$
$$\log_{10} 1,000 = x$$
$$10^x = 1,000$$
$$10^x = 10^3$$
$$x = 3$$

51.

$$\log_2 x = 5$$
$$2^5 = x$$
$$32 = x$$

52.

$$\log_3 x = -4$$
$$3^{-4} = x$$
$$\frac{1}{3^4} = x$$
$$\frac{1}{81} = x$$

53.

$$\log_x 16 = 2$$
$$x^2 = 16$$
$$\sqrt{x^2} = \sqrt{16}$$
$$x = 4$$

54.

$$\log_x \frac{1}{100} = -2$$
$$x^{-2} = \frac{1}{100}$$
$$\left(x^{-2}\right)^{-1/2} = \left(\frac{1}{100}\right)^{-1/2}$$
$$x = (100)^{1/2}$$
$$x = \sqrt{100}$$
$$x = 10$$

55.

$$\log_9 3 = x$$
$$9^x = 3$$
$$\left(3^2\right)^x = 3^1$$
$$3^{2x} = 3^1$$
$$2x = 1$$
$$x = \frac{1}{2}$$

56.

$$\log_{27} x = \frac{2}{3}$$
$$27^{2/3} = x$$
$$\left(\sqrt[3]{27}\right)^2 = x$$
$$(3)^2 = x$$
$$9 = x$$

57.

$$\log 4.51 = x$$
$$0.6542 = x$$

58.

$$\log x = 1.43$$
$$10^{1.43} = x$$
$$x = 26.9153$$

59. $f(x) = \log_4 x$ and $g(x) = 4^x$

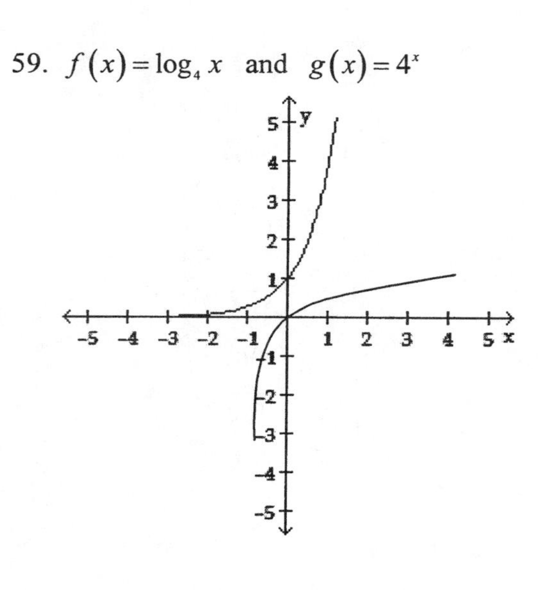

60. $f(x) = \log_{1/3} x$ and $g(x) = \left(\frac{1}{3}\right)^x$

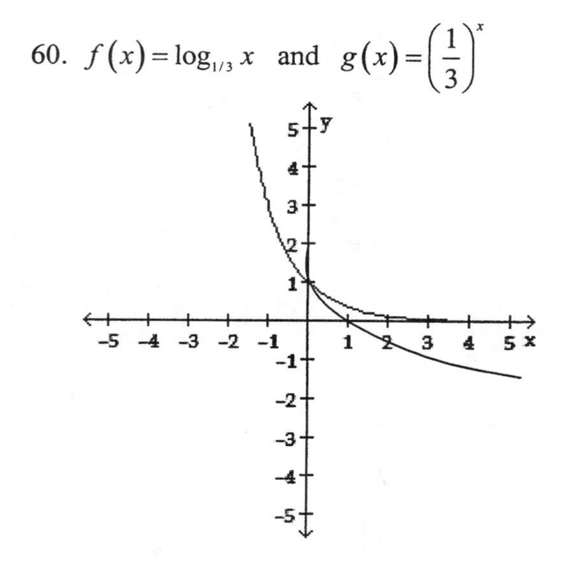

61. $f(x) = \log(x-2)$

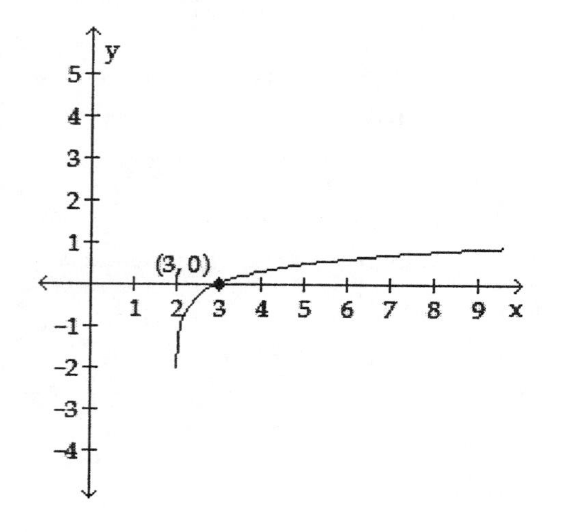

62. $f(x) = 3 + \log x$

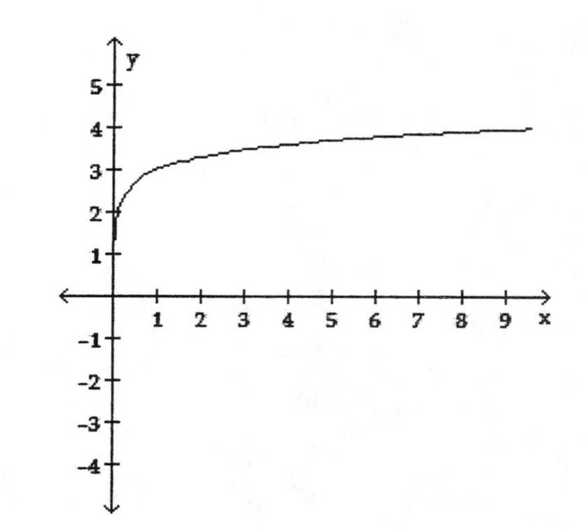

63. Let $E_O = 18$ and $E_I = 0.04$.

$$\begin{aligned} \text{db gain} &= 20\log\frac{E_O}{E_I} \\ &= 20\log\frac{18}{0.04} \\ &= 20\log 450 \\ &\approx 20(2.65321) \\ &\approx 53 \end{aligned}$$

64. Let $P = 0.3$ and $A = 7{,}500$.

$$\begin{aligned} R &= \log\frac{A}{P} \\ &= \log\frac{7{,}500}{0.3} \\ &= \log 25{,}000 \\ &\approx 4.4 \end{aligned}$$

SECTION 11.6
Base–e Logarithms

65.

$\ln e = 1$

66.

$\ln e^2 = 2$

67.

$$\begin{aligned} \ln\frac{1}{e^5} &= \ln e^{-5} \\ &= -5 \end{aligned}$$

68.

$$\ln\sqrt{e} = \ln e^{1/2} = \frac{1}{2}$$

69.

$\ln(-e) = \text{undefined}$

70.

$\ln 0 = \text{undefined}$

71.

$\ln 1 = 0$

72.

$\ln e^{-7} = -7$

73.

$\ln 452 = 6.1137$

74.

$\ln 0.85 = -0.1625$

75.

$$\ln x = 2.336$$
$$e^{\ln x} = e^{2.336}$$
$$x = 10.3398$$

76.

$$\ln x = -8.8$$
$$e^{\ln x} = e^{-8.8}$$
$$x = 0.0002$$

77. $\log x = \log_{10} x$ and $\ln x = \log_e x$

78. $f(x) = e^x$

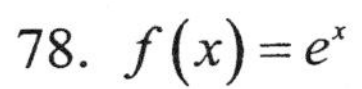

79. $f(x) = 1 + \ln x$

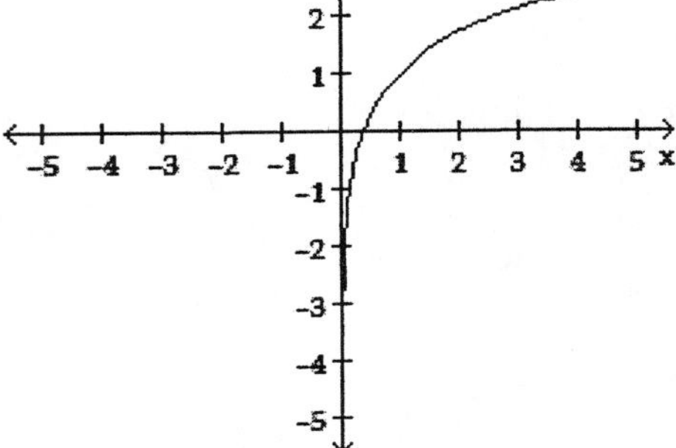

80. $f(x) = \ln(x+1)$

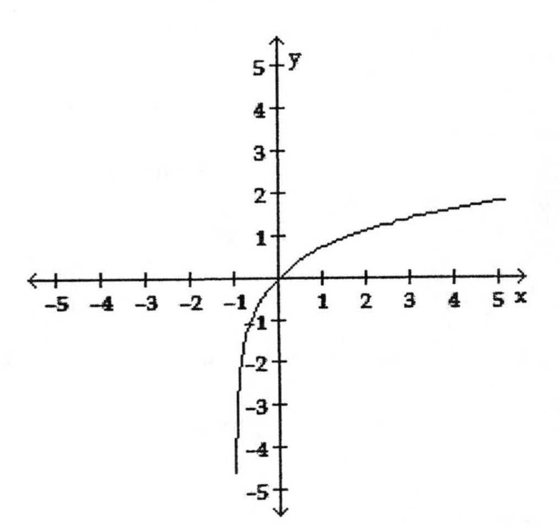

81. Let $r = 0.0143$.

$$t = \frac{\ln 2}{0.0143} \approx 48.5 \text{ years}$$

82. Let $a = 19$.

$$H(a) = 13 + 20.03\ln a$$
$$H(19) = 13 + 20.03\ln 19$$
$$\approx 13 + 20.03(2.94444)$$
$$\approx 72 \text{ in. or } 6 \text{ ft.}$$

SECTION 11.7
Properties of Logarithms

83.
$$\log_2 1 = 0$$

84.
$$\log_9 9 = 1$$

85.
$$\log 10^3 = 3$$

86.
$$7^{\log_7 4} = 4$$

87.
$$\log_3 27x = \log_3 27 + \log_3 x$$
$$= 3 + \log_3 x$$

88.
$$\log \frac{100}{x} = \log 100 - \log x$$
$$= 2 - \log x$$

89.
$$\log_5 \sqrt{27} = \log_5 27^{1/2}$$
$$= \frac{1}{2} \log_5 27$$

90.
$$\log_b 10ab = \log_b 10 + \log_b a + \log_b b$$
$$= \log_b 10 + \log_b a + 1$$

91.
$$\log_b \frac{x^2 y^3}{z} = \log_b x^2 + \log_b y^3 - \log_b z$$
$$= 2\log_b x + 3\log_b y - \log_b z$$

92.
$$\ln \sqrt{\frac{x}{yz^2}} = \ln\left(\frac{x}{yz^2}\right)^{1/2}$$
$$= \frac{1}{2} \ln \frac{x}{yz^2}$$
$$= \frac{1}{2}\left(\ln x - \ln y - \ln z^2\right)$$
$$= \frac{1}{2}\left(\ln x - \ln y - 2\ln z\right)$$

93.
$$3\log_2 x - 5\log_2 y + 7\log_2 z$$
$$= \log_2 x^3 - \log_2 y^5 + \log_2 z^7$$
$$= \log_2 \frac{x^3 z^7}{y^5}$$

94.
$$-3\log_b y - 7\log_b z + \frac{1}{2}\log_b (x+2)$$
$$= -\log_b y^3 - \log_b z^7 + \log_b (x+2)^{1/2}$$
$$= -\log_b y^3 - \log_b z^7 + \log_b \sqrt{x+2}$$
$$= \log_b \frac{\sqrt{x+2}}{y^3 z^7}$$

95.
$$\log_b 40 = \log_b (5 \bullet 8)$$
$$= \log_b 5 + \log_b 8$$
$$= 1.1609 + 1.5000$$
$$= 2.6609$$

96.
$$\log_b 64 = \log_b (8 \bullet 8)$$
$$= \log_b 8 + \log_b 8$$
$$= 1.5000 + 1.5000$$
$$= 3.0000$$

97.
$$\log_5 17 = \frac{\log 17}{\log 5}$$
$$\approx \frac{1.23045}{0.69897}$$
$$\approx 1.7604$$

98.
$$\text{pH} = -\log\left[\text{H}^+\right]$$
$$3.1 = \log\left[\text{H}^+\right]^{-1}$$
$$\log\left[\text{H}^+\right]^{-1} = 3.1$$
$$\left(10^{3.1}\right)^{-1} = \left(\left[\text{H}^+\right]^{-1}\right)^{-1}$$
$$10^{-3.1} = \left[\text{H}^+\right]$$
$$\left[\text{H}^+\right] = 0.00079$$
$$\left[\text{H}^+\right] = 7.9 \times 10^{-4} \text{ grams-ions/liter}$$

SECTION 11.8
Exponential and Logarithmic Equations

99.

$$5^{x+2} = 625$$
$$5^{x+2} = 5^4$$
$$x+2 = 4$$
$$x = 2$$

100.

$$2^{x^2+4x} = \frac{1}{8}$$
$$2^{x^2+4x} = 2^{-3}$$
$$x^2 + 4x = -3$$
$$x^2 + 4x + 3 = 0$$
$$(x+1)(x+3) = 0$$
$$x+1 = 0 \quad \text{or} \quad x+3 = 0$$
$$x = -1 \qquad x = -3$$

101.

$$3^x = 7$$
$$\log 3^x = \log 7$$
$$x \log 3 = \log 7$$
$$x = \frac{\log 7}{\log 3}$$
$$x \approx 1.7712$$

102.

$$2^x = 3^{x-1}$$
$$\log 2^x = \log 3^{x-1}$$
$$x \log 2 = (x-1)\log 3$$
$$x \log 2 = x \log 3 - \log 3$$
$$x \log 2 - x \log 3 = -\log 3$$
$$x(\log 2 - \log 3) = -\log 3$$
$$x = \frac{-\log 3}{\log 2 - \log 3}$$
$$x \approx 2.7095$$

103.

$$e^x = 7$$
$$\ln e^x = \ln 7$$
$$x = \ln 7$$
$$x = 1.9459$$

104.

$$e^{-0.4t} = 25$$
$$\ln e^{-0.4t} = \ln 25$$
$$-0.4t = \ln 25$$
$$t = \frac{\ln 25}{-0.4}$$
$$t = -8.0472$$

105.

$$\log(x-4) = 2$$
$$10^2 = x-4$$
$$100 = x-4$$
$$104 = x$$

106.

$$\ln(2x-3) = \ln 15$$
$$e^{\ln(2x-3)} = e^{\ln 15}$$
$$2x-3 = 15$$
$$2x = 18$$
$$x = 9$$

107.

$$\log x + \log(29-x) = 2$$
$$\log[x(29-x)] = 2$$
$$10^2 = x(29-x)$$
$$100 = 29x - x^2$$
$$x^2 - 29x + 100 = 0$$
$$(x-25)(x-4) = 0$$
$$x-25 = 0 \quad \text{or} \quad x-4 = 0$$
$$x = 25 \qquad x = 4$$

108.

$$\log_2 x + \log_2 (x-2) = 3$$
$$\log_2 x(x-2) = 3$$
$$2^3 = x(x-2)$$
$$8 = x^2 - 2x$$
$$0 = x^2 - 2x - 8$$
$$0 = (x-4)(x+2)$$
$$x-4 = 0 \quad \text{or} \quad x+2 = 0$$
$$x = 4 \qquad x = \not{-2}$$

109.

$$\frac{\log(7x-12)}{\log x}=2$$
$$\log x\left(\frac{\log(7x-12)}{\log x}\right)=2\log x$$
$$\log(7x-12)=\log x^2$$
$$7x-12=x^2$$
$$0=x^2-7x+12$$
$$0=(x-3)(x-4)$$
$$x-3=0 \quad \text{or} \quad x-4=0$$
$$x=3 \qquad\qquad x=4$$

110.

$$\log_2(x+2)+\log_2(x-1)=2$$
$$\log_2[(x+2)(x-1)]=2$$
$$2^2=(x+2)(x-1)$$
$$4=x^2-x+2x-2$$
$$4=x^2+x-2$$
$$0=x^2+x-6$$
$$0=(x+3)(x-2)$$
$$x+3=0 \quad \text{or} \quad x-2=0$$
$$x=\cancel{-3} \qquad\qquad x=2$$

111.

$$\log x+\log(x-5)=\log 6$$
$$\log x(x-5)=\log 6$$
$$x(x-5)=6$$
$$x^2-5x=6$$
$$x^2-5x-6=0$$
$$(x-6)(x+1)=0$$
$$x-6=0 \quad \text{or} \quad x+1=0$$
$$x=6 \qquad\qquad x=\cancel{-1}$$

112.

$$\log 3-\log(x-1)=-1$$
$$\log\frac{3}{x-1}=-1$$
$$10^{-1}=\frac{3}{x-1}$$
$$\frac{1}{10}=\frac{3}{x-1}$$
$$10(x-1)\left(\frac{1}{10}\right)=10(x-1)\left(\frac{3}{x-1}\right)$$
$$x-1=30$$
$$x=31$$

113.

$$\frac{\log 8}{\log 15}\neq\log 8-\log 15$$
$$0.7679\neq-0.2730$$

114. Let $h=5{,}700$ and $A=\frac{2}{3}A_0$.

$$A=A_0 2^{-t/h}$$
$$\frac{2}{3}A_0=A_0 2^{-t/5{,}700}$$
$$\frac{2A_0}{3A_0}=\frac{A_0 2^{-t/5{,}700}}{A_0}$$
$$\frac{2}{3}=2^{-t/5{,}700}$$
$$\log\frac{2}{3}=\log 2^{-t/5{,}700}$$
$$\log\frac{2}{3}=-\frac{t}{5{,}700}\log 2$$
$$-5{,}700\left(\log\frac{2}{3}\right)=-5{,}700\left(-\frac{t}{5{,}700}\log 2\right)$$
$$-5{,}700\left(\log\frac{2}{3}\right)=t\log 2$$
$$\frac{-5{,}700\left(\log\frac{2}{3}\right)}{\log 2}=\frac{t\log 2}{\log 2}$$
$$\frac{-5{,}700\left(\log\frac{2}{3}\right)}{\log 2}=t$$
$$t=\frac{-5{,}700\left(\log\frac{2}{3}\right)}{\log 2}$$
$$t\approx 3{,}300 \text{ years}$$

115. Let $P_0 = 800$, $P = 3(800) = 2{,}400$, and $t = 14$.

$$P = P_0 e^{kt}$$
$$2{,}400 = 800e^{k(14)}$$
$$3 = e^{14k}$$
$$\ln 3 = \ln e^{14k}$$
$$\ln 3 = 14k$$
$$14k = \ln 3$$
$$k = \frac{\ln 3}{14}$$

Let $P_0 = 800$, $P = 1{,}000{,}000$, and $k = \frac{\ln 3}{14}$.

$$P = P_0 e^{kt}$$
$$1{,}000{,}000 = 800e^{\left(\frac{\ln 3}{14}\right)t}$$
$$1{,}250 = e^{\left(\frac{\ln 3}{14}\right)t}$$
$$\ln 1{,}250 = \ln e^{\left(\frac{\ln 3}{14}\right)t}$$
$$\ln 1{,}250 = \left(\frac{\ln 3}{14}\right)t$$
$$\left(\frac{14}{\ln 3}\right)(\ln 1{,}250) = \left(\frac{14}{\ln 3}\right)\left(\frac{\ln 3}{14}\right)t$$
$$\frac{14\ln 1{,}250}{\ln 3} = t$$
$$t = \frac{14\ln 1{,}250}{\ln 3}$$
$$t \approx 91 \text{ days}$$

116. $x = 2$ and $x = 5$

For $x = 2$:

$$\log x = 1 - \log(7 - x)$$
$$\log 2 \stackrel{?}{=} 1 - \log(7 - 2)$$
$$\log 2 \stackrel{?}{=} 1 - \log 5$$
$$0.3010 = 0.3010$$

For $x = 5$:

$$\log x = 1 - \log(7 - x)$$
$$\log 5 \stackrel{?}{=} 1 - \log(7 - 5)$$
$$\log 5 \stackrel{?}{=} 1 - \log 2$$
$$0.6990 = 0.6990$$

CHAPTER 11 TEST

1. $f+g$

$$\begin{aligned}(f+g)(x)&=f(x)+g(x)\\&=x+9+4x^2-3x+2\\&=4x^2-2x+11\end{aligned}$$

$D:(-\infty,\infty)$

2. g/f

$$\begin{aligned}(g/f)(x)&=\frac{g(x)}{f(x)}\\&=\frac{4x^2-3x+2}{x+9}\end{aligned}$$

$D:(-\infty,-9)\cup(-9,\infty)$

3.

$$\begin{aligned}(g\circ f)(-3)&=g(f(-3))\\&=g\left(2(-3)^2+3\right)\\&=g(2(9)+3)\\&=g(18+3)\\&=g(21)\\&=4(21)-8\\&=84-8\\&=76\end{aligned}$$

4.

$$\begin{aligned}(f\circ g)(x)&=f(g(x))\\&=f(4x-8)\\&=2(4x-8)^2+3\\&=2\left(16x^2-32x-32x+64\right)+3\\&=2\left(16x^2-64x+64\right)+3\\&=32x^2-128x+128+3\\&=32x^2-128x+131\end{aligned}$$

5.

$$\begin{aligned}(f\cdot g)(9)&=f(9)\cdot g(9)\\&=-1\cdot 16\\&=-16\end{aligned}$$

6.

$$\begin{aligned}(f\circ g)(-3)&=f(g(-3))\\&=f(10)\\&=17\end{aligned}$$

7. Yes, it is one–to–one.

8. No, it doesn't pass the horizontal line test.

9.

$$\begin{aligned}f(x)&=-\frac{1}{3}x\\y&=-\frac{1}{3}x\\x&=-\frac{1}{3}y\\3(x)&=3\left(-\frac{1}{3}y\right)\\3x&=-y\\-3x&=y\\y&=-3x\\f^{-1}(x)&=-3x\end{aligned}$$

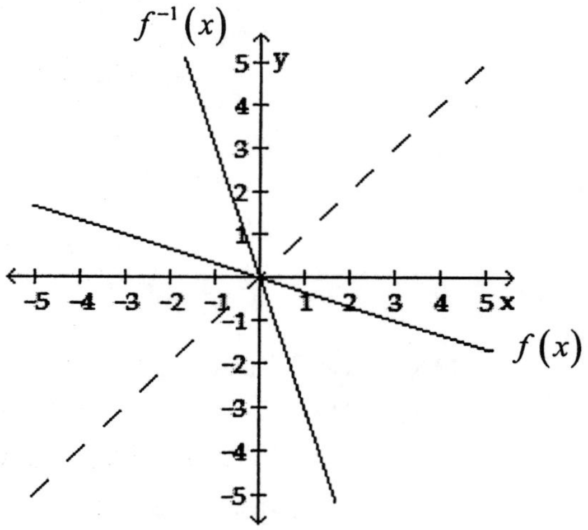

10.

$$f(x)=(x-15)^3$$
$$y=(x-15)^3$$
$$x=(y-15)^3$$
$$\sqrt[3]{x}=\sqrt[3]{(y-15)^3}$$
$$\sqrt[3]{x}=y-15$$
$$\sqrt[3]{x}+15=y$$
$$y=\sqrt[3]{x}+15$$
$$f^{-1}(x)=\sqrt[3]{x}+15$$

11.

$$\begin{aligned}(f\circ f^{-1})(x)&=f(f^{-1}(x))\\&=f\left(\frac{x-4}{4}\right)\\&=4\left(\frac{x-4}{4}\right)+4\\&=x-4+4\\&=x\end{aligned}$$

$$\begin{aligned}(f^{-1}\circ f)(x)&=f^{-1}(f(x))\\&=f^{-1}(4x+4)\\&=\frac{4x+4-4}{4}\\&=\frac{4x}{4}\\&=x\end{aligned}$$

12. a. Yes.

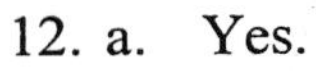

b. Yes.

c. $f^{-1}(260)=80$; when the temperature of the tire tread is 260°, the vehicle is traveling 80 mph.

13. $f(x)=2^x+1$

D: $(-\infty, \infty)$, R: $(1, \infty)$

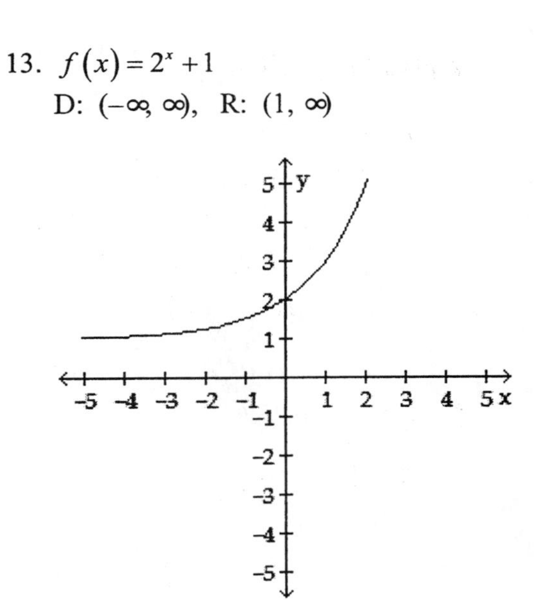

14. $f(x)=3^{-x}$

D: $(-\infty, \infty)$, R: $(0, \infty)$

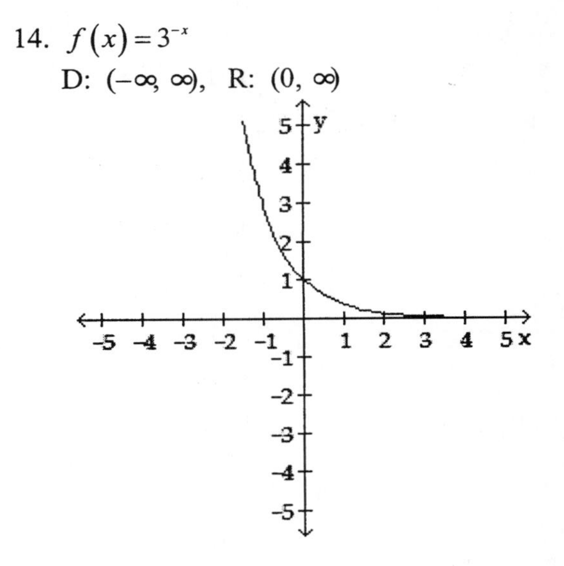

15. Let $t=6$ and $A_0=3$.

$$\begin{aligned}A&=A_0(2)^{-t}\\&=3(2)^{-6}\\&=3(0.015625)\\&=0.046875\text{ g}\end{aligned}$$

16. Let $P = 1{,}000$, $r = 0.06$, $k = 2$, and $t = 1$.

$$A = P\left(1+\frac{r}{k}\right)^{kt}$$
$$= 1{,}000\left(1+\frac{0.06}{2}\right)^{2(1)}$$
$$= 1{,}000(1+0.03)^2$$
$$= 1{,}000(1.03)^2$$
$$= 1{,}000(1.0609)$$
$$= \$1{,}060.90$$

17. $f(x) = e^x$

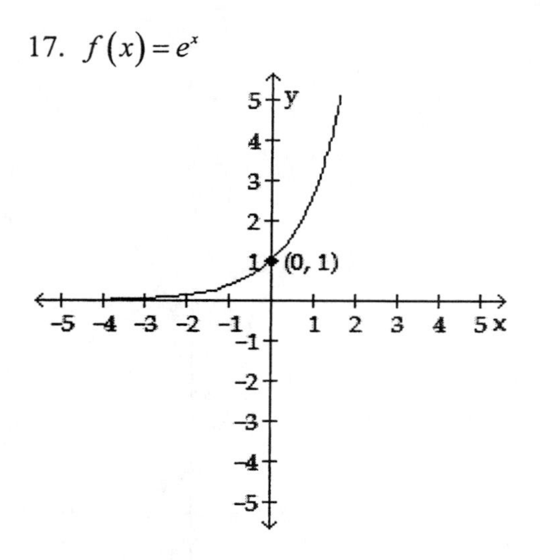

18. Let $P_0 = 1{,}050{,}000{,}000$, $r = 0.0147$, and $t = 30$.

$$P = P_0 e^{kt}$$
$$= (1{,}050{,}000{,}000)e^{0.0147(30)}$$
$$= (1{,}050{,}000{,}000)e^{0.441}$$
$$\approx (1{,}050{,}000{,}000)(1.55426)$$
$$\approx 1{,}631{,}973{,}737$$

19.

$$\log_6 \frac{1}{36} = -2$$
$$6^{-2} = \frac{1}{36}$$

20. a. D: $(0, \infty)$, R: $(-\infty, \infty)$

b. $f^{-1}(x) = 10^x$

21. $\log_5 25 = 2$ since $5^2 = 25$

22. $\log_9 \frac{1}{81} = -2$ since $9^{-2} = \frac{1}{81}$

23. $\log(-100)$ is undefined since $10^? \neq -100$

24.

$$\ln\frac{1}{e^6} = \ln e^{-6}$$
$$= -6$$

25. $\log_4 2 = \frac{1}{2}$ since $4^{1/2} = 2$

26. $\log_{1/3} 1 = 0$ since $\left(\frac{1}{3}\right)^0 = 1$

27.

$$\log_x 32 = 5$$
$$x^5 = 32$$
$$\sqrt[5]{x^5} = \sqrt[5]{32}$$
$$x = 2$$

28.

$$\log_8 x = \frac{4}{3}$$
$$8^{4/3} = x$$
$$\left(\sqrt[3]{8}\right)^4 = x$$
$$(2)^4 = x$$
$$16 = x$$

29.

$$\log_3 x = -3$$
$$3^{-3} = x$$
$$\frac{1}{3^3} = x$$
$$\frac{1}{27} = x$$

30.

$$\ln x = 1$$
$$e^{\ln x} = e^1$$
$$x = e$$

31. $f(x) = -\log_3 x$

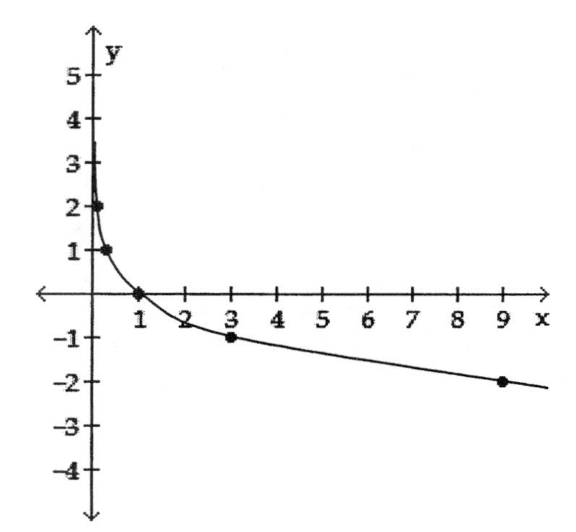

32. $f(x) = \ln x$

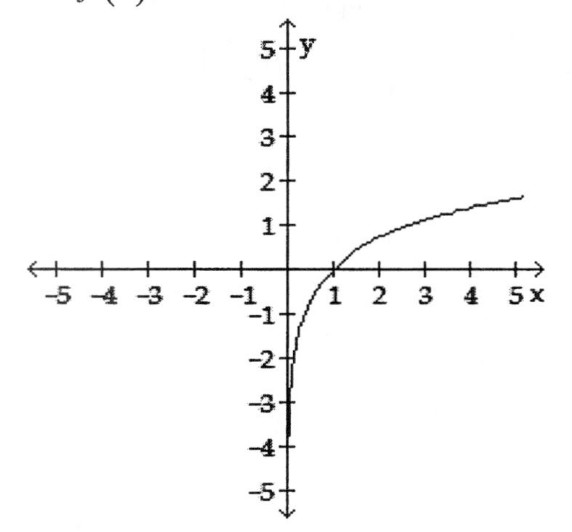

33. Let $H^+ = 3.7 \times 10^{-7}$.

$$\begin{aligned} \text{pH} &= -\log\left[\text{H}^+\right] \\ &= -\log\left(3.7 \times 10^{-7}\right) \\ &\approx -(-6.4) \\ &\approx 6.4 \end{aligned}$$

34. Let $E_O = 60$ and $E_I = 0.3$.

$$\begin{aligned} \text{db gain} &= 20\log\left(\frac{E_O}{E_I}\right) \\ &= 20\log\left(\frac{60}{0.3}\right) \\ &= 20\log 200 \\ &\approx 20(2.3) \\ &\approx 46 \end{aligned}$$

35.

$$\log x = -1.06$$
$$10^{-1.06} = x$$
$$x = 10^{-1.06}$$
$$x \approx 0.0871$$

36.

$$\log_7 3 = \frac{\log 3}{\log 7} \approx 0.5646$$

37.

$$\begin{aligned} \log_b a^2bc^3 &= \log_b a^2 + \log_b b + \log_b c^3 \\ &= 2\log_b a + 1 + 3\log_b c \end{aligned}$$

38.

$$\begin{aligned} &\frac{1}{2}\ln(a+2) + \ln b - 3\ln c \\ &= \ln(a+2)^{1/2} + \ln b - \ln c^3 \\ &= \ln\sqrt{a+2} + \ln b - \ln c^3 \\ &= \ln b\sqrt{a+2} - \ln c^3 \\ &= \ln\frac{b\sqrt{a+2}}{c^3} \end{aligned}$$

39.

$$5^x = 3$$
$$\log 5^x = \log 3$$
$$x\log 5 = \log 3$$
$$x = \frac{\log 3}{\log 5}$$
$$x \approx 0.6826$$

40.

$$3^{x-1}=27$$
$$3^{x-1}=3^3$$
$$x-1=3$$
$$x=4$$

41.

$$\ln(5x+2)=\ln(2x+5)$$
$$5x+2=2x+5$$
$$3x+2=5$$
$$3x=3$$
$$x=1$$

42.

$$\log x+\log(x-9)=1$$
$$\log x(x-9)=1$$
$$10^1=x(x-9)$$
$$10=x^2-9x$$
$$0=x^2-9x-10$$
$$0=(x-10)(x+1)$$
$$x-10=0 \quad \text{or} \quad x+1=0$$
$$x=10 \qquad x=\not{-1}$$

-1 is extraneous

43.

$$x\approx 5$$
$$\frac{1}{2}\ln(x-1)=\ln 2$$
$$\frac{1}{2}\ln(5-1)\stackrel{?}{=}\ln 2$$
$$\frac{1}{2}\ln 4\stackrel{?}{=}\ln 2$$
$$\frac{1}{2}(1.3863)\stackrel{?}{=}0.6931$$
$$0.6931=0.6931$$

44. Let $P_0=5$, $P=4(5)=20$, and $t=6$.

$$P=P_0e^{kt}$$
$$20=5e^{k(6)}$$
$$4=e^{6k}$$
$$\ln 4=\ln e^{6k}$$
$$\ln 4=6k$$
$$6k=\ln 4$$
$$k=\frac{\ln 4}{6}$$

Let $P_0=5$, $P=500$, and $k=\dfrac{\ln 4}{6}$.

$$P=P_0e^{kt}$$
$$500=5e^{\left(\frac{\ln 4}{6}\right)t}$$
$$100=e^{\left(\frac{\ln 4}{6}\right)t}$$
$$\ln 100=\ln e^{\left(\frac{\ln 4}{6}\right)t}$$
$$\ln 100=\left(\frac{\ln 4}{6}\right)t$$
$$\left(\frac{6}{\ln 4}\right)(\ln 100)=\left(\frac{6}{\ln 4}\right)\left(\frac{\ln 4}{6}\right)t$$
$$\frac{6\ln 100}{\ln 4}=t$$
$$t=\frac{6\ln 100}{\ln 4}$$
$$t\approx 20 \text{ minutes}$$

SECTION 12.1

VOCABULARY

1. The pair of equations $\begin{cases} x - y = -1 \\ 2x - y = 1 \end{cases}$ is called a **system** of equations.

3. When the graphs of the equations of a system are identical lines, the equations are called dependent and the system has **infinitely** many solutions.

CONCEPTS

5. a. true
   b. false
   c. true
   d. true

7. a. 3, –4 (answers will vary)
   b. 2; –3 (answers will vary)

NOTATION

9. Solve $\begin{cases} y = 3x - 7 \\ x + y = 5 \end{cases}$.

$$x + \left(\boxed{3x-7}\right) = 5$$
$$x + 3x - 7 = 5$$
$$\boxed{4}x - 7 = 5$$
$$4x = \boxed{12}$$
$$x = 3$$

$$y = 3x - 7$$
$$y = 3\left(\boxed{3}\right) - 7$$
$$y = \boxed{2}$$

The solution is (3, 2).

PRACTICE

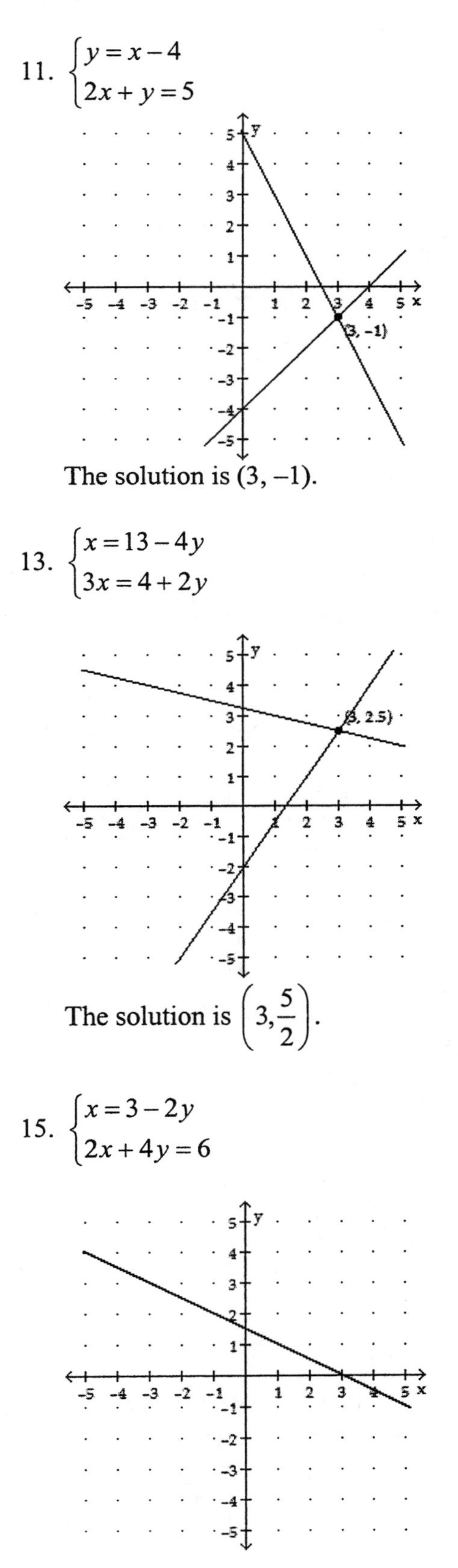

11. $\begin{cases} y = x - 4 \\ 2x + y = 5 \end{cases}$

The solution is (3, –1).

13. $\begin{cases} x = 13 - 4y \\ 3x = 4 + 2y \end{cases}$

The solution is $\left(3, \frac{5}{2}\right)$.

15. $\begin{cases} x = 3 - 2y \\ 2x + 4y = 6 \end{cases}$

There are infinitely many solutions; the equations are dependent.

17. $\begin{cases} y = 3 \\ x = 2 \end{cases}$

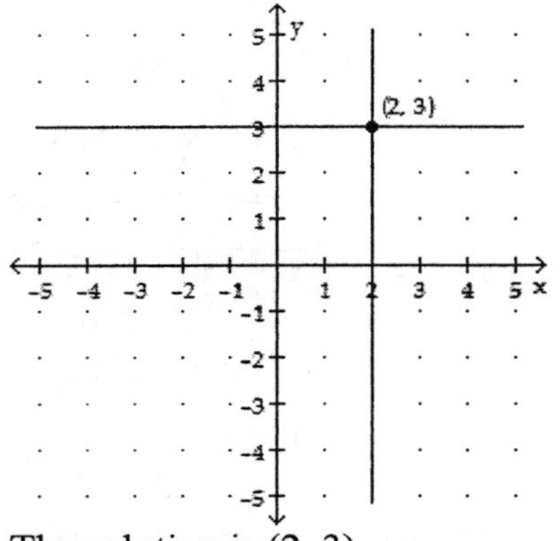

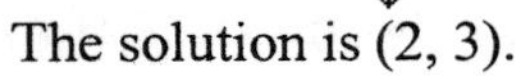
The solution is (2, 3).

19. $\begin{cases} x = \dfrac{11-2y}{3} \\ y = \dfrac{11-6x}{4} \end{cases}$

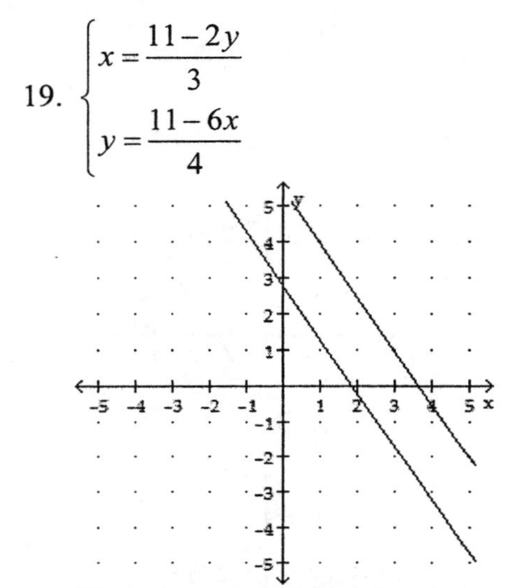

There is no solution; it is an inconsistent system.

21. $\begin{cases} y = -\dfrac{5}{2}x + \dfrac{1}{2} \\ 2x - \dfrac{3}{2}y = 5 \end{cases}$

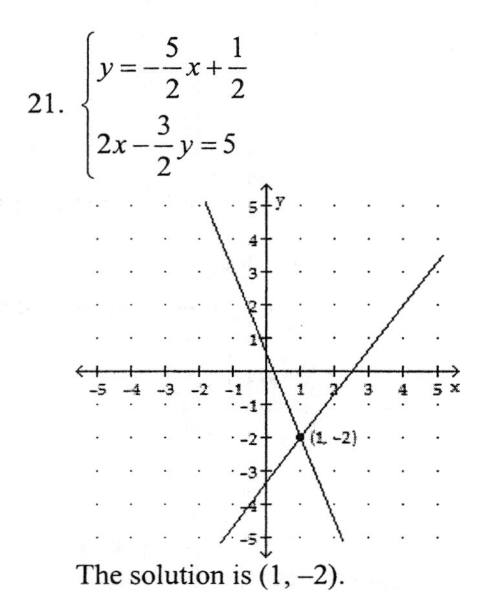

The solution is (1, –2).

23. $\begin{cases} y = 3.2x - 1.5 \\ y = -2.7x - 3.7 \end{cases}$

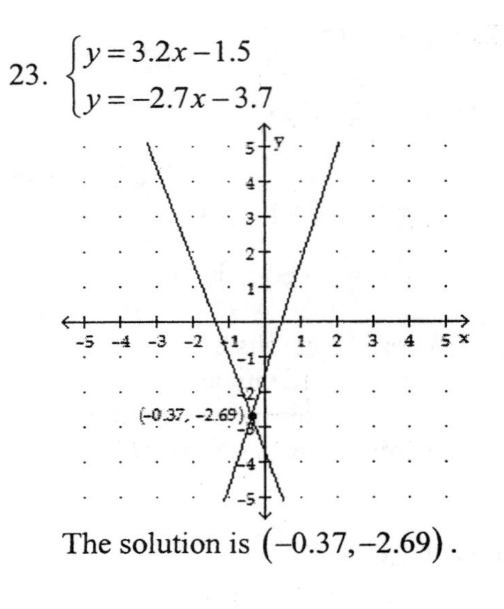

The solution is $(-0.37, -2.69)$.

25. $\begin{cases} 1.7x + 2.3y = 3.2 \\ y = 0.25x + 8.95 \end{cases}$

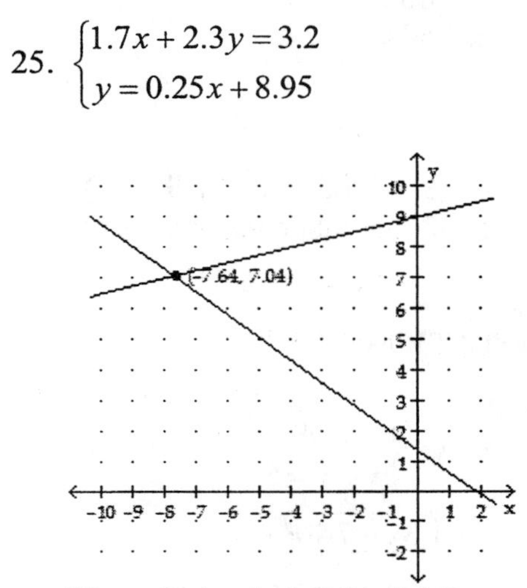

The solution is (–7.64, 7.04).

27.

$\begin{cases} y = x \\ x + y = 4 \end{cases}$

$x + y = 4$ The 2nd equation

$x + (x) = 4$

$2x = 4$

$\boxed{x = 2}$

$y = x$ The 1st equation

$\boxed{y = 2}$

The solution is (2, 2).

29.

$$\begin{cases} x = 2 + y \\ 2x + y = 13 \end{cases}$$

$2x + y = 13$ The 2nd equation

$2(2 + y) + y = 13$

$4 + 2y + y = 13$

$4 + 3y = 13$

$3y = 9$

$\boxed{y = 3}$

$x = 2 + y$ The 1st equation

$x = 2 + 3$

$\boxed{x = 5}$

The solution is $(5, 3)$.

31.

$$\begin{cases} x + 2y = 6 \\ 3x - y = -10 \end{cases}$$

Solve the first equation for x.

$$\begin{cases} x = -2y + 6 \\ 3x - y = -10 \end{cases}$$

$3x - y = -10$ The 2nd equation

$3(-2y + 6) - y = -10$

$-6y + 18 - y = -10$

$-7y + 18 = -10$

$-7y = -28$

$\boxed{y = 4}$

$x = -2y + 6$ The 1st equation

$x = -2(4) + 6$

$x = -8 + 6$

$\boxed{x = -2}$

The solution is $(-2, 4)$.

33.

$$\begin{cases} y - 3x = -5 \\ 21x = 7y + 35 \end{cases}$$

Solve the first equation for y.

$$\begin{cases} y = -5 + 3x \\ 21x = 7y + 35 \end{cases}$$

$21x = 7y + 35$ The 2nd equation

$21x = 7(-5 + 3x) + 35$

$21x = -35 + 21x + 35$

$21x = 21x$

$21x - 21x = 21x - 21x$

$0 = 0$

true

Infinitely many solutions

Dependent Equations

35.

$$\begin{cases} 3x - 4y = 9 \\ x + 2y = 8 \end{cases}$$

Solve the second equation for x.

$$\begin{cases} 3x - 4y = 9 \\ x = -2y + 8 \end{cases}$$

Substitute $x = -2y + 8$ into the first equation and solve for y.

$3x - 4y = 9$ the 1st equation

$3(-2y + 8) - 4y = 9$

$-6y + 24 - 4y = 9$

$-10y + 24 = 9$

$-10y = -15$

$y = \frac{15}{10}$

$y = \frac{3}{2}$

Substitute $y = \frac{3}{2}$ into the second equation and solve for x.

$x + 2y = 8$ the 2nd equation

$x + 2\left(\frac{3}{2}\right) = 8$

$x + 3 = 8$

$x = 5$

The solution is $\left(5, \frac{3}{2}\right)$.

37.

$$\begin{cases} x - \dfrac{4y}{5} = 4 \\ \dfrac{y}{3} = \dfrac{x}{2} - \dfrac{5}{2} \end{cases}$$

clear both equations of fractions

$$\begin{cases} 5(x) - 5\left(\dfrac{4y}{5}\right) = 5(4) \\ 6\left(\dfrac{y}{3}\right) = 6\left(\dfrac{x}{2}\right) - 6\left(\dfrac{5}{2}\right) \end{cases}$$

$$\begin{cases} 5x - 4y = 20 \\ 2y = 3x - 15 \end{cases}$$

solve for y in 2nd equation

$$\begin{cases} 5x - 4y = 20 \\ y = \dfrac{3}{2}x - \dfrac{15}{2} \end{cases}$$

$5x - 4y = 20$ The 1st equation

$$5x - 4\left(\frac{3}{2}x - \frac{15}{2}\right) = 20$$
$$5x - 6x + 30 = 20$$
$$-x + 30 = 20$$
$$-x + 30 - 30 = 20 - 30$$
$$-x = -10$$
$$\frac{-x}{-1} = \frac{-10}{-1}$$
$$\boxed{x = 10}$$

$y = \dfrac{3}{2}x - \dfrac{15}{2}$ The 2nd equation

$$y = \frac{3}{2}(10) - \frac{15}{2}$$
$$y = \frac{30}{2} - \frac{15}{2}$$
$$\boxed{y = \frac{15}{2}}$$

The solution is $\left(10, \dfrac{15}{2}\right)$.

39.

$$\begin{cases} x - y = 3 \\ x + y = 7 \end{cases}$$

Add the equations to eliminate y.

$$x - y = 3$$
$$\underline{x + y = 7}$$
$$2x = 10$$
$$x = 5$$

Substitute $x = 5$ into the second equation and solve for y.

$x + y = 7$ 2nd equation

$$5 + y = 7$$
$$5 + y - 5 = 7 - 5$$
$$y = 2$$

The solution is (5, 2).

41.

$$\begin{cases} 2x + y = -10 \\ 2x - y = -6 \end{cases}$$

Add the equations to eliminate y.

$$2x + y = -10$$
$$\underline{2x - y = -6}$$
$$4x = -16$$
$$x = -4$$

Substitute $x = -4$ into the first equation and solve for y.

$2x + y = -10$ 1st equation

$$2(-4) + y = -10$$
$$-8 + y = -10$$
$$-8 + y + 8 = -10 + 8$$
$$y = -2$$

The solution is $(-4, -2)$.

43.

$$\begin{cases} 2x+3y=8 \\ 3x-2y=-1 \end{cases}$$

Multiply the 1st equation by 2 and multiply the 2nd equation by 3.

$$\begin{cases} 4x+6y=16 \\ 9x-6y=-3 \end{cases}$$

Add the equations to eliminate y.

$$\begin{aligned} 4x+6y&=16 \\ \underline{9x-6y}&\underline{=-3} \\ 13x&=13 \\ x&=1 \end{aligned}$$

Substitute $x=1$ into the first equation and solve for y.

$$\begin{aligned} 2x+3y&=8 \quad \text{1st equation} \\ 2(1)+3y&=8 \\ 2+3y&=8 \\ 2+3y-2&=8-2 \\ 3y&=6 \\ y&=2 \end{aligned}$$

The solution is (1, 2).

45.

$$\begin{cases} 4(x-2)=-9y \\ 2(x-3y)=-3 \end{cases}$$

$$\begin{cases} 4x-8=-9y \\ 2x-6y=-3 \end{cases}$$

Write the 1st equation in standard form.

$$\begin{cases} 4x+9y=8 \\ 2x-6y=-3 \end{cases}$$

Multiply the 2nd equation by -2.

$$\begin{cases} 4x+9y=8 \\ -4x+12y=6 \end{cases}$$

Add the equations to eliminate x.

$$\begin{aligned} 4x+9y&=8 \\ \underline{-4x+12y}&\underline{=6} \\ 21y&=14 \\ y&=\frac{14}{21} \\ y&=\frac{2}{3} \end{aligned}$$

Substitute $y=\frac{2}{3}$ into the first equation and solve for x.

$$\begin{aligned} 4x+9y&=8 \quad \text{1st equation} \\ 4x+9\left(\frac{2}{3}\right)&=8 \\ 4x+6&=8 \\ 4x+6-6&=8-6 \\ 4x&=2 \\ x&=\frac{2}{4} \\ x&=\frac{1}{2} \end{aligned}$$

The solution is $\left(\frac{1}{2},\frac{2}{3}\right)$.

47.

$$\begin{cases} 8x-4y=16 \\ 2x-4=y \end{cases}$$

Write the second equation in standard form.

$$\begin{cases} 8x-4y=16 \\ 2x-y=4 \end{cases}$$

Multiply the second equation by -4.

$$\begin{array}{r} 8x-4y=16 \\ \underline{-8x+4y=-16} \\ 0=0 \end{array}$$

true

Infinitely many solutions

Dependent equations

49.

$$\begin{cases} x=\frac{3}{2}y+5 \\ 2x-3y=8 \end{cases}$$

Multiply the first equation by 2 to eliminate fractions and write in standard form.

$$\begin{cases} 2x-3y=10 \\ 2x-3y=8 \end{cases}$$

Multiply the second equation by -1 and add the equations to eliminate x.

$$\begin{array}{r} 2x-3y=10 \\ \underline{-2x+3y=-8} \\ 0\neq 2 \end{array}$$

no solution

inconsistent system

51.

$$\begin{cases} \frac{x}{2}+\frac{y}{2}=6 \\ \frac{x}{2}-\frac{y}{2}=-2 \end{cases}$$

Multiply both equations by 2 to eliminate fractions.

$$\begin{cases} x+y=12 \\ x-y=-4 \end{cases}$$

Add the equations to eliminate y.

$$\begin{array}{r} x+y=12 \\ \underline{x-y=-4} \\ 2x=8 \\ x=4 \end{array}$$

Substitute $x=4$ into the 1st equation and solve for y.

$$\begin{array}{r} x+y=12 \\ 4+y=12 \\ y=8 \end{array}$$

The solution is $(4,8)$.

53.

$$\begin{cases} \frac{2}{3}x - \frac{1}{4}y = -8 \\ 0.5x - 0.375y = -9 \end{cases}$$

Multiply the 1st equation by 12 to eliminate fractions and the 2nd equation by 1,000 to eliminate decimals.

$$\begin{cases} 8x - 3y = -96 \\ 500x - 375y = -9{,}000 \end{cases}$$

Multiply the 1st equation by -125 and add the equations to eliminate y.

$$\begin{aligned} -1{,}000x + 375y &= 12{,}000 \\ 500x - 375y &= -9{,}000 \\ \hline -500x &= 3{,}000 \\ x &= -6 \end{aligned}$$

Substitute $x = -6$ into the 1st equation and solve for y.

$$\begin{aligned} 8x - 3y &= -96 \\ 8(-6) - 3y &= -96 \\ -48 - 3y &= -96 \\ -3y &= -48 \\ y &= 16 \end{aligned}$$

The solution is $(-6, 16)$.

55.

$$\begin{cases} \frac{3}{2}p + \frac{1}{3}q = 2 \\ \frac{2}{3}p + \frac{1}{9}q = 1 \end{cases}$$

Multiply the 1st equation by 6 and the 2nd equation by 9 to eliminate fractions.

$$\begin{cases} 9p + 2q = 12 \\ 6p + q = 9 \end{cases}$$

Multiply the 2nd equation by -2 and add the equations to eliminate q.

$$\begin{aligned} 9p + 2q &= 12 \\ -12p - 2q &= -18 \\ \hline -3p &= -6 \\ p &= 2 \end{aligned}$$

Substitute $p = 2$ into the 2nd equation and solve for q.

$$\begin{aligned} 6p + q &= 9 \\ 6(2) + q &= 9 \\ 12 + q &= 9 \\ q &= -3 \end{aligned}$$

The solution is $(2, -3)$.

57.

$$\begin{cases} \dfrac{m-n}{5}+\dfrac{m+n}{2}=6 \\ \dfrac{m-n}{2}-\dfrac{m+n}{4}=3 \end{cases}$$

Multiply the 1st equation by 10 and the 2nd equation by 4 to eliminate fractions.

$$\begin{cases} 2m-2n+5m+5n=60 \\ 2m-2n-m-n=12 \end{cases}$$

Combine like terms and write in general form.

$$\begin{cases} 7m+3n=60 \\ m-3n=12 \end{cases}$$

Add the equations to eliminate n.

$$7m+3n=60$$
$$\underline{m-3n=12}$$
$$8m=72$$
$$m=9$$

Substitute $m=9$ into the 2nd equation and solve for n.

$$m-3n=12$$
$$9-3n=12$$
$$-3n=3$$
$$n=-1$$

The solution is $(9,-1)$.

59.

$$\begin{cases} \dfrac{1}{x}+\dfrac{1}{y}=\dfrac{5}{6} \\ \dfrac{1}{x}-\dfrac{1}{y}=\dfrac{1}{6} \end{cases}$$

Let $a=\dfrac{1}{x}$ and $b=\dfrac{1}{y}$.

$$\begin{cases} a+b=\dfrac{5}{6} \\ a-b=\dfrac{1}{6} \end{cases}$$

Add equations to eliminate b.

$$a+b=\frac{5}{6}$$
$$\underline{a-b=\frac{1}{6}}$$
$$2a=\frac{6}{6}$$
$$2a=1$$
$$a=\frac{1}{2}$$

Substitute $a=\dfrac{1}{2}$ in the 1st equation and solve for b.

$$a+b=\frac{5}{6}$$
$$\frac{1}{2}+b=\frac{5}{6}$$
$$\frac{1}{2}+b-\frac{1}{2}=\frac{5}{6}-\frac{1}{2}$$
$$b=\frac{1}{3}$$
$$a=\frac{1}{2}=\frac{1}{x} \text{ so } x=2$$
$$b=\frac{1}{3}=\frac{1}{y} \text{ so } y=3$$

The solution is $(2, 3)$.

61.
$$\begin{cases} \frac{1}{x}+\frac{2}{y}=-1 \\ \frac{2}{x}-\frac{1}{y}=-7 \end{cases}$$

Let $a=\frac{1}{x}$ and $b=\frac{1}{y}$.

$$\begin{cases} a+2b=-1 \\ 2a-b=-7 \end{cases}$$

Multiply the 2nd equation by 2 and add equations to eliminate b.

$$a+2b=-1$$
$$\underline{4a-2b=-14}$$
$$5a=-15$$
$$a=-3$$

Substitute $a=-3$ in the 1st equation and solve for b.

$$a+2b=-1$$
$$(-3)+2b=-1$$
$$2b=2$$
$$b=1$$

$$a=\frac{-3}{1}=\frac{1}{x} \text{ so } x=-\frac{1}{3}$$
$$b=\frac{1}{1}=\frac{1}{y} \text{ so } y=1$$

The solution is $\left(-\frac{1}{3},1\right)$.

63. $\begin{cases} y<3x+2 \\ y<-2x+3 \end{cases}$

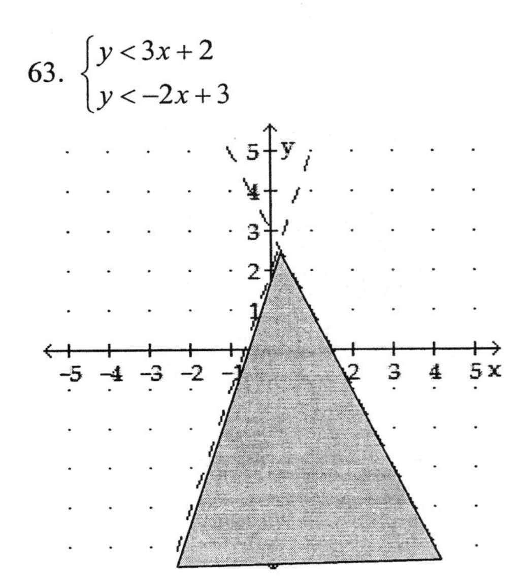

65. $\begin{cases} 3x+2y>6 \\ x+3y\le 2 \end{cases}$

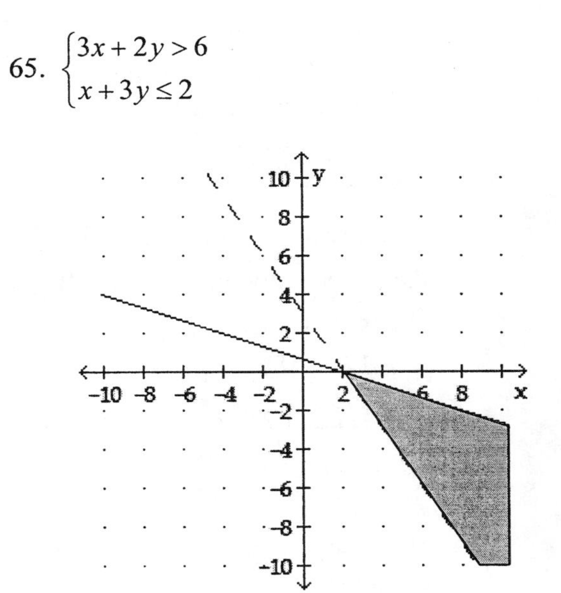

APPLICATIONS

67. HEARING TESTS
(2,000, 50)

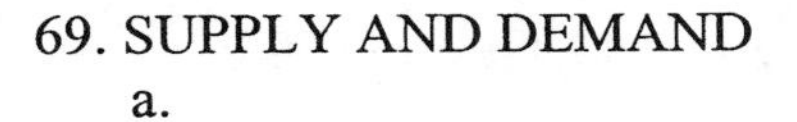
69. SUPPLY AND DEMAND

a.

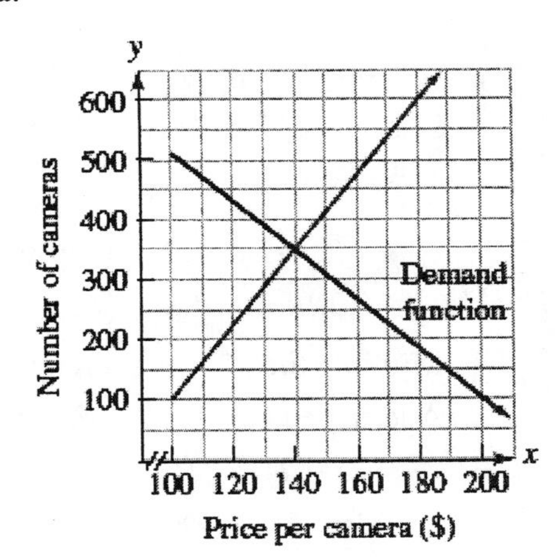

b. \$140 – the x–value at the point of intersection

c. As supply increases, demand decreases.

71. BUSINESS
The breakeven point is (250, 2,500). If the company makes 250 widgets, the cost to make them and the revenue obtained from their sale will be equal: \$2,500.

73. ADVERTISING

Let x = 30 second spot and y = 15 second spot.

$$\begin{cases} 4x+6y=6{,}050 \\ 3x+5y=4{,}775 \end{cases}$$

Multiply the 1st equation by 3 and the 2nd equation by -4. Add the equations to eliminate x.

$$\begin{aligned} 12x+18y&=18{,}150 \\ -12x-20y&=-19{,}100 \\ \hline -2y&=-950 \\ y&=475 \end{aligned}$$

Substitute $y=475$ in the 1st equation and solve for x.

$$\begin{aligned} 4x+6y&=6{,}050 \\ 4x+6(475)&=6{,}050 \\ 4x+2{,}850&=6{,}050 \\ 4x&=3{,}200 \\ x&=800 \end{aligned}$$

The 30-second spot costs \$475 and the 15-second spot costs \$800.

75. FENCING A FIELD

Let x = width of field and y = length of the field.

$$\begin{cases} 2x+2y=72 & \text{(perimeter)} \\ 3x+2y=88 & \text{(with partition)} \end{cases}$$

Multiply the 2nd equation by -1. Add the equations to eliminate y.

$$\begin{aligned} 2x+2y&=72 \\ -3x-2y&=-88 \\ \hline -x&=-16 \\ x&=16 \end{aligned}$$

Substitute $x=16$ in the 1st equation and solve for y.

$$\begin{aligned} 2x+2y&=72 \\ 2(16)+2y&=72 \\ 32+2y&=72 \\ 2y&=40 \\ y&=20 \end{aligned}$$

The width is 16 m and the length is 20 m.

77. TRAFFIC SIGNALS

Angles x and y make a straight line, so their sum is 180°. Brace A and Brace B are perpendicular so their sum is 90°.

$$\begin{cases} x+y=180 \\ \frac{2}{5}x+y=90 \end{cases}$$

Multiply the 1st equation by -1.

$$\begin{aligned} -x-y&=-180 \\ \frac{2}{5}x+y&=90 \\ \hline -\frac{3}{5}x&=-90 \\ -\frac{5}{3}\left(-\frac{3}{5}x\right)&=-\frac{5}{3}(-90) \\ x&=150 \end{aligned}$$

Substitute $x=150$ in the 1st equation and solve for y.

$$\begin{aligned} x+y&=180 \\ 150+y&=180 \\ y&=30 \end{aligned}$$

x = 150° and y = 30°.

79. RETIREMENT INCOME

Let x = amount at 6% and y = amount at 7.5%.

$$\begin{cases} x+y=12{,}000 \\ 0.06x+0.075y=810 \end{cases}$$

Multiply the 1st equation by -6 and multiply the 2nd equation by 1,000.

$$\begin{aligned} -60x-60y&=-720{,}000 \\ 60x+75y&=810{,}000 \\ \hline 15y&=90{,}000 \\ y&=6{,}000 \end{aligned}$$

Substitute $y=6{,}000$ in the 1st equation and solve for x.

$$\begin{aligned} x+y&=12{,}000 \\ x+6{,}000&=12{,}000 \\ x&=6{,}000 \end{aligned}$$

\$6,000 was invested at 6% and \$6,000 was invested at 7.5%.

81. DELIVERY SERVICES

Let x = rate of the delivery truck.
The $x + 143$ = rate of the plane.
Let the traveling time = y.

	distance	rate	time
truck	50	x	y
plane	180	$x + 143$	y

$$\begin{cases} xy = 50 \\ (x+143)y = 180 \end{cases}$$

Solve the equations for y.

$$\begin{cases} y = \dfrac{50}{x} \\ y = \dfrac{180}{x+143} \end{cases}$$

Use substitution and cross-multiply.

$$\frac{50}{x} = \frac{180}{x+143}$$
$$50(x+143) = 180(x)$$
$$50x + 7{,}150 = 180x$$
$$7{,}150 = 130x$$
$$55 = x$$

The truck was going 55 mph.

83. FARMING

Let x = grams in Mix A and y = grams in Mix B.

	Mix A	Mix B	mixture
Protein	$0.12x$	$0.15y$	15
Carb.	$0.09x$	$0.05y$	7.5

$$\begin{cases} 0.12x + 0.15y = 15 \\ 0.09x + 0.05y = 7.5 \end{cases}$$

Multiply both equations by 100 to eliminate decimals.

$$\begin{cases} 12x + 15y = 1{,}500 \\ 9x + 5y = 750 \end{cases}$$

Multiply the 2nd equation by -3.
Add equations to eliminate y.

$$12x + 15y = 1{,}500$$
$$\underline{-27x - 15y = -2{,}250}$$
$$-15x = -750$$
$$x = 50$$

Substitute $x = 50$ in the 1st equation and solve for y.

$$12x + 15y = 1{,}500$$
$$12(50) + 15y = 1{,}500$$
$$600 + 15y = 1{,}500$$
$$15x = 900$$
$$y = 60$$

50 grams of Mix A and
60 grams of Mix B

85. COSMETOLOGY

Let x = number of permanents.

$$\begin{cases} C_1 = 23.60x + 2{,}101.20 \\ C_2 = 44x \end{cases}$$

Breakeven point:

$$C_1 = C_2$$
$$23.60x + 2{,}101.20 = 44x$$
$$2{,}101.20 = 20.4x$$
$$103 = x$$

The breakeven point is 103 permanents.

87. DERMATOLOGY

Let $x =$ # of grams of 0.2% cream
and $y =$ # of grams of 0.7% cream

	0.2% solution	0.7% solution	0.3% mixture
# of grams	x	y	185
% Triclosan	$0.002x$	$0.007y$	0.003(185)

$$\begin{cases} x+y=185 \\ 0.002x+0.007y=0.003(185) \end{cases}$$

Multiply the 2nd equation by 1,000 to eliminate decimals.

$$\begin{cases} x+y=185 \\ 2x+7y=555 \end{cases}$$

Multiply the 1st equation by -2.
Add equations to eliminate x.

$$\begin{aligned} -2x-2y&=-370 \\ 2x+7y&=555 \\ \hline 5y&=185 \\ y&=37 \end{aligned}$$

Substitute $y=37$ in the 1st equation.

$$\begin{aligned} x+y&=185 \\ x+37&=185 \\ x&=148 \end{aligned}$$

148 grams of 0.2% and 37 grams of 0.7%

WRITING

89. Answers will vary.

91. Answers will vary.

REVIEW

93.

$$\begin{aligned} A&=p+prt \\ A-p&=prt \\ \frac{A-p}{pt}&=\frac{prt}{pt} \\ \frac{A-p}{pt}&=r \\ r&=\frac{A-p}{pt} \end{aligned}$$

95.

$$\begin{aligned} \frac{V_2}{V_1}&=\frac{P_1}{P_2} \\ P_2V_1\left(\frac{V_2}{V_1}\right)&=P_2V_1\left(\frac{P_1}{P_2}\right) \\ P_2V_2&=P_1V_1 \\ \frac{P_2V_2}{V_1}&=\frac{P_1V_1}{V_1} \\ \frac{P_2V_2}{V_1}&=P_1 \\ P_1&=\frac{P_2V_2}{V_1} \end{aligned}$$

CHALLENGE PROBLEMS

97.

$$\begin{cases} Ax+By=-2 \\ Bx-Ay=-26 \end{cases}$$

Let $x=-3$ and $y=5$.

$$\begin{cases} A(-3)+B(5)=-2 \\ B(-3)-A(5)=-26 \end{cases}$$

$$\begin{cases} -3A+5B=-2 \\ -5A-3B=-26 \end{cases}$$

Multiply the first equation by 3 and the second by 5.

$$\begin{aligned} -9A+15B&=-6 \\ -25A-15B&=-130 \\ \hline -34A&=-136 \\ A&=4 \end{aligned}$$

Substitute $A=4$ into the first equation and solve for B.

$$\begin{aligned} -3A+5B&=-2 \\ -3(4)+5B&=-2 \\ -12+5B&=-2 \\ 5B&=10 \\ B&=2 \end{aligned}$$

$A=4$ and $B=2$.

SECTION 12.2

VOCABULARY

1. $\begin{cases} 2x+y-3z=0 \\ 3x-y+4z=5 \\ 4x+2y-6z=0 \end{cases}$ is called a **system** of three linear equations.

3. The equation $2x + 3y + 4z = 5$ is a linear equation with **three** variables.

5. When three planes coincide, the equations of the system are **dependent**, and there are infinitely many solutions.

CONCEPTS

7. a. no solution
   b. no solution

NOTATION

9. $x + 2y - 3z = -6$

PRACTICE

11. Yes.

$$\begin{aligned} x-y+z&=2 \\ 2-1+1&\stackrel{?}{=}2 \\ 2&=2 \end{aligned}$$

$$\begin{aligned} 2x+y-z&=4 \\ 2(2)+1-1&\stackrel{?}{=}4 \\ 4+1-1&\stackrel{?}{=}4 \\ 4&=4 \end{aligned}$$

$$\begin{aligned} 2x-3y+z&=2 \\ 2(2)-3(1)+1&\stackrel{?}{=}2 \\ 4-3+1&\stackrel{?}{=}2 \\ 2&=2 \end{aligned}$$

13.

$$\begin{cases} x+y+z=4 & (1) \\ 2x+y-z=1 & (2) \\ 2x-3y+z=1 & (3) \end{cases}$$

Add Equations 1 and 2 to eliminate z.

$$\begin{aligned} x+y+z&=4 \quad (1) \\ 2x+y-z&=1 \quad (2) \\ \hline 3x+2y\quad&=5 \quad (4) \end{aligned}$$

Add Equations 3 and 2 to eliminate z.

$$\begin{aligned} 2x-3y+z&=1 \quad (3) \\ 2x+y-z&=1 \quad (2) \\ \hline 4x-2y\quad&=2 \quad (5) \end{aligned}$$

Add Equations 4 and 5 to eliminate y.

$$\begin{aligned} 3x+2y&=5 \quad (4) \\ 4x-2y&=2 \quad (5) \\ \hline 7x\quad&=7 \\ x&=1 \end{aligned}$$

Substitute $x=1$ into Equation 4 to find y.

$$\begin{aligned} 3x+2y&=5 \quad (4) \\ 3(1)+2y&=5 \\ 3+2y&=5 \\ 2y&=2 \\ y&=1 \end{aligned}$$

Substitute $x=1$ and $y=1$ into Equation 1 to find z.

$$\begin{aligned} x+y+z&=4 \quad (1) \\ 1+1+z&=4 \\ 2+z&=4 \\ z&=2 \end{aligned}$$

The solution is (1, 1, 2).

15.

$$\begin{cases} 2x+2y+3z=10 & (1) \\ 3x+y-z=0 & (2) \\ x+y+2z=6 & (3) \end{cases}$$

Multiply Equation 2 by 3 and add Equations 1 and 2 to eliminate z.

$$\begin{array}{rl} 2x+2y+3z=10 & (1) \\ \underline{9x+3y-3z=0} & (2) \\ 11x+5y \quad\quad =10 & (4) \end{array}$$

Multiply Equation 2 by 2 and add Equations 2 and 3 to eliminate z.

$$\begin{array}{rl} 6x+2y-2z=0 & (2) \\ \underline{x+\ y+2z=6} & (3) \\ 7x+3y \quad\quad =6 & (5) \end{array}$$

Write equations 4 and 5 as a system.

$$\begin{cases} 11x+5y=10 & (4) \\ 7x+3y=6 & (5) \end{cases}$$

Multiply Equation 4 by -3, multiply Equation 5 by 5, and add to eliminate y.

$$\begin{array}{r} -33x-15y=-30 \\ \underline{35x+15y=30} \\ 2x \quad\quad =0 \\ x=0 \end{array}$$

Substitute $x=0$ into Equation 4 to find y.

$$\begin{array}{rl} 11x+5y=10 & (4) \\ 11(0)+5y=10 & \\ 5y=10 & \\ y=2 & \end{array}$$

Substitute $x=0$ and $y=2$ into Equation 1 to find z.

$$\begin{array}{rl} 2x+2y+3z=10 & (1) \\ 2(0)+2(2)+3z=10 & \\ 0+4+3z=10 & \\ 4+3z=10 & \\ 3z=6 & \\ z=2 & \end{array}$$

The solution is (0, 2, 2).

17.

$$\begin{cases} b+2c=7-a \\ a+c=8-2b \\ 2a+b+c=9 \end{cases}$$

Write all equations in general form.

$$\begin{cases} a+b+2c=7 & (1) \\ a+2b+c=8 & (2) \\ 2a+b+c=9 & (3) \end{cases}$$

Multiply Equation 2 by -2 and add Equations 1 and 2 to eliminate c.

$$\begin{array}{rl} a+\ \ b+2c=7 & (1) \\ \underline{-2a-4b-2c=-16} & (2) \\ -a-3b \quad\quad =-9 & (4) \end{array}$$

Multiply Equation 2 by -1 and add Equations 2 and 3 to eliminate c.

$$\begin{array}{rl} -a-2b-c=-8 & (2) \\ \underline{2a+\ b+\ c=9} & (3) \\ a-\ b \quad\quad =1 & (5) \end{array}$$

Add Equations 4 and 5 to eliminate a.

$$\begin{array}{rl} -a-3b=-9 & (4) \\ \underline{a-\ b=1} & (5) \\ -4b=-8 & \\ b=2 & \end{array}$$

Substitute $b=2$ into Equation 5 and solve for a.

$$\begin{array}{rl} a-b=1 & (5) \\ a-2=1 & \\ a=3 & \end{array}$$

Substitute $a=3$ and $b=2$ into Equation 1 and solve for c.

$$\begin{array}{rl} a+b+2c=7 & (1) \\ 3+2+2c=7 & \\ 5+2c=7 & \\ 2c=2 & \\ c=1 & \end{array}$$

The solution is $(3, 2, 1)$.

19.

$$\begin{cases} 2x+y-z=1 & (1) \\ x+2y+2z=2 & (2) \\ 4x+5y+3z=3 & (3) \end{cases}$$

Multiply Equation 1 by 2 and add Equations 1 and 2 to eliminate z.

$$\begin{array}{rl} 4x+2y-2z=2 & (1) \\ \underline{x+2y+2z=2} & (2) \\ 5x+4y=4 & (4) \end{array}$$

Multiply Equation 1 by 3 and add Equations 1 and 3 to eliminate z.

$$\begin{array}{rl} 6x+3y-3z=3 & (1) \\ \underline{4x+5y+3z=3} & (3) \\ 10x+8y \quad\quad =6 & (5) \end{array}$$

Multiply Equation 4 by -2 and add Equations 4 and 5 to eliminate x.

$$\begin{array}{r} -10x-8y=-8 \\ \underline{10x+8y=6} \\ 0\neq-2 \end{array}$$

Since $0\neq-2$, the system is inconsistent and has no solution.

21.

$$\begin{cases} a+b+c=180 & (1) \\ \frac{a}{4}+\frac{b}{2}+\frac{c}{3}=60 & (2) \\ 2b+3c-330=0 & (3) \end{cases}$$

Multiply Equation 2 by 12 to eliminate fractions and write Equation 3 in standard form.

$$\begin{cases} a+b+c=180 & (1) \\ 3a+6b+4c=720 & (2) \\ 2b+3c=330 & (3) \end{cases}$$

Multiply Equation 1 by -3 and add Equations 1 and 2 to eliminate a.

$$\begin{array}{rl} -3a-3b-3c=-540 & (1) \\ \underline{3a+6b+4c=720} & (2) \\ 3b+\ c=180 & (4) \end{array}$$

Multiply Equation 4 by -3 and add Equations 4 and 3 to eliminate c.

$$\begin{array}{rl} 2b+3c=330 & (3) \\ \underline{-9b-3c=-540} & (4) \\ -7b=-210 & \\ b=30 & \end{array}$$

Substitute $b=30$ into Equation 3 and solve for c.

$$\begin{array}{rl} 2b+3c=330 & (3) \\ 2(30)+3c=330 & \\ 60+3c=330 & \\ 3c=270 & \\ c=90 & \end{array}$$

Substitute $c=90$ and $b=30$ into Equation 1 and solve for a.

$$\begin{array}{rl} a+b+c=180 & (1) \\ a+30+90=180 & \\ a+120=180 & \\ a=60 & \end{array}$$

The solution is $(60,30,90)$.

23.

$$\begin{cases} 0.5a + 0.3b = 2.2 & (1) \\ 1.2c - 8.5b = -24.4 & (2) \\ 3.3c + 1.3a = 29 & (3) \end{cases}$$

Multiply each equation by 10 to eliminate decimials and write in standard form.

$$\begin{cases} 5a + 3b = 22 & (1) \\ -85b + 12c = -244 & (2) \\ 13a + 33c = 290 & (3) \end{cases}$$

Multiply Equation 1 by 85 and Equation 2 by 3. Add the equations.

$$\begin{array}{rr} 425a + 255b \qquad = 1{,}870 & (1) \\ \underline{-255b + 36c = -732} & (2) \\ 425a + 36c = 1{,}138 & (4) \end{array}$$

Multiply Equation 3 by -36 and Equation 4 by 33. Add the equations.

$$\begin{array}{rr} -468a - 1{,}188c = -10{,}440 & (3) \\ \underline{14{,}025a + 1{,}188c = 37{,}554} & (4) \\ 13{,}557a = 27{,}114 & \\ a = 2 & \end{array}$$

Substitute $a = 2$ into Equation 1 and solve for b.

$$\begin{aligned} 5a + 3b &= 22 \quad (1) \\ 5(2) + 3b &= 22 \\ 10 + 3b &= 22 \\ 3b &= 12 \\ b &= 4 \end{aligned}$$

Substitute $a = 2$ into Equation 3 and solve for c.

$$\begin{aligned} 13a + 33c &= 290 \quad (3) \\ 13(2) + 33c &= 290 \\ 26 + 33c &= 290 \\ 33c &= 264 \\ c &= 8 \end{aligned}$$

The solution is $(2, 4, 8)$.

25.

$$\begin{cases} 2x + 3y + 4z = 6 & (1) \\ 2x - 3y - 4z = -4 & (2) \\ 4x + 6y + 8z = 12 & (3) \end{cases}$$

Multiply Equation 1 by -2 and add Equations 1 and 3.

$$\begin{array}{rr} -4x - 6y - 8z = -12 & (1) \\ \underline{4x + 6y + 8z = 12} & (3) \\ 0 = 0 & \end{array}$$

The equations of the system are dependent, and there are infinitely many solutions.

27.

$$\begin{cases} x+\frac{1}{3}y+z=13 \\ \frac{1}{2}x-y+\frac{1}{3}z=-2 \\ x+\frac{1}{2}y-\frac{1}{3}z=2 \end{cases}$$

Write each equation in standard form by muliplying by a LCD to eliminate fractions.

$$\begin{cases} 3x+y+3z=39 & (1) \\ 3x-6y+2z=-12 & (2) \\ 6x+3y-2z=12 & (3) \end{cases}$$

Multiply Equation 2 by -1 and add Equations 1 and 2.

$$\begin{array}{rl} -3x+6y-2z=12 & (2) \\ \underline{3x+\ \ y+3z=39} & (1) \\ 7y+z=51 & (4) \end{array}$$

Multiply Equation 1 by -2 and add Equations 1 and 3.

$$\begin{array}{rl} -6x-2y-6z=-78 & (1) \\ \underline{6x+3y-2z=12} & (3) \\ y-8z=-66 & (5) \end{array}$$

Multiply Equation 4 by 8 and add Equations 4 and 5.

$$\begin{array}{rl} 56y+8z=408 & (4) \\ \underline{y-8z=-66} & (5) \\ 57y=342 & \\ y=6 & \end{array}$$

Substitute $y=6$ into Equation 4.

$$\begin{array}{rl} 7y+z=51 & (4) \\ 7(6)+z=51 & \\ 42+z=51 & \\ z=9 & \end{array}$$

Substitute $y=6$ and $z=9$ into Equation 1.

$$\begin{array}{rl} 3x+y+3z=39 & (1) \\ 3x+6+3(9)=39 & \\ 3x+33=39 & \\ 3x=6 & \\ x=2 & \end{array}$$

The solution is $(2,\ 6,\ 9)$.

APPLICATIONS

29. MAKING STATUES

Let x = expensive type, y = middle–priced type, and z = inexpensive type.

$$\begin{cases} x+y+z=180 & (1) \\ 5x+4y+3z=650 & (2) \\ 20x+12y+9z=2{,}100 & (3) \end{cases}$$

Multiply Equation 1 by -5 and add Equations 1 and 2.

$$\begin{array}{rl} -5x-5y-5z=-900 & (1) \\ \underline{5x+4y+3z=650} & (2) \\ -y-2z=-250 & (4) \end{array}$$

Multiply Equation 1 by -20 and add Equations 1 and 3.

$$\begin{array}{rl} -20x-20y-20z=-3{,}600 & (1) \\ \underline{20x+12y+\ 9z=2{,}100} & (3) \\ -8y-11z=-1{,}500 & (5) \end{array}$$

Multiply Equation 4 by -8 and add Equations 4 and 5.

$$\begin{array}{rl} 8y+16z=2{,}000 & (4) \\ \underline{-8y-11z=-1{,}500} & (5) \\ 5z=500 & \\ z=100 & \end{array}$$

Substitute $z=100$ into Equation 4.

$$\begin{array}{rl} -y-2z=-250 & (4) \\ -y-2(100)=-250 & \\ -y-200=-250 & \\ -y=-50 & \\ y=50 & \end{array}$$

Substitute $z=100$ and $y=50$ into Equation 4.

$$\begin{array}{rl} x+y+z=180 & (1) \\ x+50+100=180 & \\ x+150=180 & \\ x=30 & \end{array}$$

He must sell 30 expensive types, 50 middle–priced types, and 100 inexpensive types.

31. NUTRITION

Let x = # oz of Food A, y = # oz of Food B, and z = # oz of Food C.

$$\begin{cases} 2x+3y+z=14 & (1) \\ x+2y+z=9 & (2) \\ 2x+y+2z=9 & (3) \end{cases}$$

Multiply Equation 3 by -1 and add Equations 1 and 3.

$$\begin{aligned} 2x+3y+z&=14 && (1) \\ -2x-y-2z&=-9 && (3) \\ \hline 2y-z&=5 && (4) \end{aligned}$$

Multiply Equation 2 by -2 and add Equations 1 and 2.

$$\begin{aligned} 2x+3y+\ z&=14 && (1) \\ -2x-4y-2z&=-18 && (2) \\ \hline -y-z&=-4 && (5) \end{aligned}$$

Multiply Equation 5 by -1 and add Equations 4 and 5.

$$\begin{aligned} 2y-z&=5 && (4) \\ y+z&=4 && (5) \\ \hline 3y&=9 \\ y&=3 \end{aligned}$$

Substitute $y=3$ into Equation 4.

$$\begin{aligned} 2y-z&=5 && (4) \\ 2(3)-z&=5 \\ 6-z&=5 \\ -z&=-1 \\ z&=1 \end{aligned}$$

Substitute $y=3$ and $z=1$ into Equation 2.

$$\begin{aligned} x+2y+z&=9 && (2) \\ x+2(3)+1&=9 \\ x+6+1&=9 \\ x+7&=9 \\ x&=2 \end{aligned}$$

She needs 2 oz of Food A, 3 oz of Food B, and 1 oz of Food C.

33. CHAINSAW SCULPTING

Let x = # of totem poles, y = # of bears, and z = # of deer.

$$\begin{cases} 2x+2y+z=14 & (1) \\ x+2y+2z=15 & (2) \\ 3x+2y+2z=21 & (3) \end{cases}$$

Multiply Equation 2 by -1 and add Equations 1 and 2.

$$\begin{aligned} 2x+2y+z&=14 && (1) \\ -x-2y-2z&=-15 && (2) \\ \hline x-z&=-1 && (4) \end{aligned}$$

Multiply Equation 2 by -1 and add Equations 3 and 2.

$$\begin{aligned} 3x+2y+2z&=21 && (3) \\ -x-2y-2z&=-15 && (2) \\ \hline 2x&=6 \\ x&=3 \end{aligned}$$

Substitute $x=3$ into Equation 4.

$$\begin{aligned} x-z&=-1 && (4) \\ 3-z&=-1 \\ -z&=-4 \\ z&=4 \end{aligned}$$

Substitute $x=3$ and $z=4$ into Equation 1.

$$\begin{aligned} 2x+2y+z&=14 && (1) \\ 2(3)+2y+4&=14 \\ 6+2y+4&=14 \\ 2y+10&=14 \\ 2y&=4 \\ y&=2 \end{aligned}$$

3 totem poles, 2 bears, and 4 deer should be produced.

35. EARTH'S ATMOSPHERE

Let x = % nitrogen, y = % oxygen, and z = % other gases.

$$\begin{cases} x = 12 + 3(y+z) & (1) \\ z = y - 20 & (2) \\ x + y + z = 100 & (3) \end{cases}$$

Write each equation in standard form.

$$\begin{cases} x - 3y - 3z = 12 & (1) \\ y - z = 20 & (2) \\ x + y + z = 100 & (3) \end{cases}$$

Multiply Equation 1 by -1 and add Equations 1 and 3.

$$\begin{array}{rl} -x + 3y + 3z = -12 & (1) \\ \underline{x + \ y + \ z = 100} & (3) \\ 4y + 4z = 88 & (4) \end{array}$$

Multiply Equation 2 by 4 and add Equations 2 and 4.

$$\begin{array}{rl} 4y + 4z = 88 & (4) \\ \underline{4y - 4z = 80} & (2) \\ 8y = 168 & \\ y = 21 & \end{array}$$

Substitute $y = 21$ into Equation 2.

$$\begin{array}{rl} y - z = 20 & (2) \\ 21 - z = 20 & \\ -z = -1 & \\ z = 1 & \end{array}$$

Substitute $z = 1$ and $y = 21$ into Equation 3.

$$\begin{array}{rl} x + y + z = 100 & (3) \\ x + 21 + 1 = 100 & \\ x + 22 = 100 & \\ x = 78 & \end{array}$$

The Earth's atmosphere is 78% nitrogen, 21 % oxygen, and 1% other gases.

37. GRAPHS OF SYSTEMS

a. infinitely many solutions, all lying on the line running down the binding
b. 3 parallel planes (shelves); no solution
c. each pair of planes (cards) intersect; no solution
d. 3 planes (faces of die) intersect at a corner; 1 solution

39. ASTRONOMY

Substitute the coordinates of (–2, 5), (2, –3), and (4, –1) into $y = ax^2 + bx + c$.

$$\begin{cases} a(-2)^2 + b(-2) + c = 5 \\ a(2)^2 + b(2) + c = -3 \\ a(4)^2 + b(4) + c = -1 \end{cases}$$

Simplify.

$$\begin{cases} 4a - 2b + c = 5 & (1) \\ 4a + 2b + c = -3 & (2) \\ 16a + 4b + c = -1 & (3) \end{cases}$$

Add Equations 1 and 2.

$$\begin{array}{rl} 4a - 2b + c = 5 & (1) \\ \underline{4a + 2b + c = -3} & (2) \\ 8a + 2c = 2 & (4) \end{array}$$

Multiply Equation 1 by 2 and add Equations 1 and 3.

$$\begin{array}{rl} 8a - 4b + 2c = 10 & (1) \\ \underline{16a + 4b + c = -1} & (3) \\ 24a + 3c = 9 & (5) \end{array}$$

Multiply Equation 4 by -3 and add Equations 4 and 5.

$$\begin{array}{rl} -24a - 6c = -6 & (4) \\ \underline{24a + 3c = 9} & (5) \\ -3c = 3 & \\ c = -1 & \end{array}$$

Substitute $c = -1$ into Equation 4.

$$\begin{array}{rl} 8a + 2c = 2 & (4) \\ 8a + 2(-1) = 2 & \\ 8a - 2 = 2 & \\ 8a = 4 & \\ a = \frac{1}{2} & \end{array}$$

Substitute $c = -1$ and $a = \frac{1}{2}$ into Equation 2.

$$\begin{array}{rl} 4a + 2b + c = -3 & (2) \\ 4\left(\frac{1}{2}\right) + 2b + (-1) = -3 & \\ 2 + 2b - 1 = -3 & \\ 2b + 1 = -3 & \\ 2b = -4 & \\ b = -2 & \end{array}$$

The equation is $y = \frac{1}{2}x^2 - 2x - 1$.

41. WALKWAYS

Substitute the coordinates of (1, 3), (3, 1), and (1, –1) into $x^2 + y^2 + Cx + Dy + E = 0$.

$$\begin{cases}(1)^2+(3)^2+C(1)+D(3)+E=0\\(3)^2+(1)^2+C(3)+D(1)+E=0\\(1)^2+(-1)^2+C(1)+D(-1)+E=0\end{cases}$$

Simplify.

$$\begin{cases}1+9+C+3D+E=0 & (1)\\9+1+3C+D+E=0 & (2)\\1+1+C-D+E=0 & (3)\end{cases}$$

$$\begin{cases}C+3D+E=-10 & (1)\\3C+D+E=-10 & (2)\\C-D+E=-2 & (3)\end{cases}$$

Multiply Equation 2 by -1 and add Equations 1 and 2.

$$\begin{aligned}C+3D+E&=-10 && (1)\\-3C-D-E&=10 && (2)\\\hline -2C+2D&=0 && (4)\end{aligned}$$

Multiply Equation 2 by -1 and add Equations 3 and 2.

$$\begin{aligned}C-D+E&=-2 && (3)\\-3C-D-E&=10 && (2)\\\hline -2C-2D&=8 && (5)\end{aligned}$$

Add Equations 4 and 5.

$$\begin{aligned}-2C+2D&=0 && (4)\\-2C-2D&=8 && (5)\\\hline -4C&=8\\C&=-2\end{aligned}$$

Substitute $C=-2$ into Equation 4.

$$\begin{aligned}-2C+2D&=0 && (4)\\-2(-2)+2D&=0\\4+2D&=0\\2D&=-4\\D&=-2\end{aligned}$$

Substitute $C=-2$ and $D=-2$ into Equation 3.

$$\begin{aligned}C-D+E&=-2 && (3)\\-2-(-2)+E&=-2\\-2+2+E&=-2\\E&=-2\end{aligned}$$

The equation is $x^2+y^2-2x-2y-2=0$.

43. TRIANGLES

$$\begin{cases}A+B+C=180\\A=(B+C)-100\\C=2B-40\end{cases}$$

Simplify.

$$\begin{cases}A+B+C=180 & (1)\\A-B-C=-100 & (2)\\-2B+C=-40 & (3)\end{cases}$$

Add Equations 1 and 2.

$$\begin{aligned}A+B+C&=180 && (1)\\A-B-C&=-100 && (2)\\\hline 2A&=80\\A&=40\end{aligned}$$

Substitute $A=40$ into Equation 2.

$$\begin{aligned}A-B-C&=-100 && (2)\\40-B-C&=-100\\-B-C&=-140 && (4)\end{aligned}$$

Add Equations 3 and 4.

$$\begin{aligned}-2B+C&=-40 && (3)\\-B-C&=-140 && (4)\\\hline -3B&=-180\\B&=60\end{aligned}$$

Substitute $B=60$ into Equation 3.

$$\begin{aligned}-2B+C&=-40 && (3)\\-2(60)+C&=-40\\-120+C&=-40\\C&=80\end{aligned}$$

The measure of the angles are $\angle A=40^\circ, \angle B=60^\circ$, and $\angle C=80^\circ$.

45. INTEGER PROBLEM

Let x = 1st integer, y = 2nd integer, and z = 3rd integer.

$$\begin{cases} x+y+z=48 & (1) \\ 2x+y+z=60 & (2) \\ x+2y+z=63 & (3) \end{cases}$$

Multiply Equation 1 by -1 and add Equations 1 and 2.

$$\begin{aligned} -x-y-z&=-48 \quad (1) \\ \underline{2x+y+z} &\underline{=60} \quad (2) \\ x&=12 \end{aligned}$$

Multiply Equation 1 by -1 and add Equations 1 and 3.

$$\begin{aligned} -x-y-z&=-48 \quad (1) \\ \underline{x+2y+z} &\underline{=63} \quad (3) \\ y&=15 \end{aligned}$$

Substitute $x=12$ and $y=15$ into Equation 1.

$$\begin{aligned} x+y+z&=48 \quad (1) \\ 12+15+z&=48 \\ 27+z&=48 \\ z&=21 \end{aligned}$$

The integers are 12, 15, and 21.

WRITING

47. Answers will vary.

REVIEW

49. $f(x)=|x|$

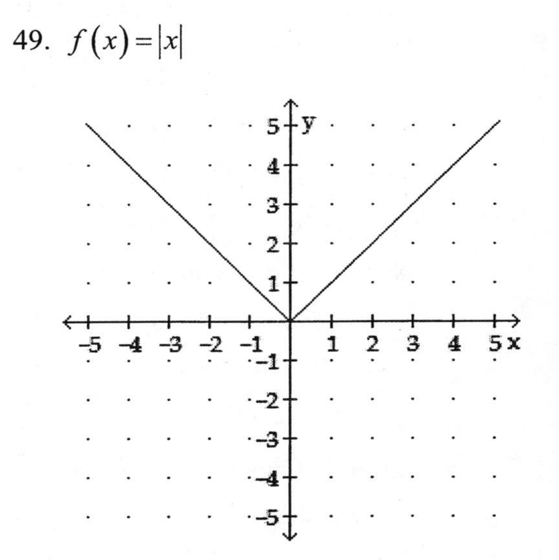

51. $h(x)=x^3$

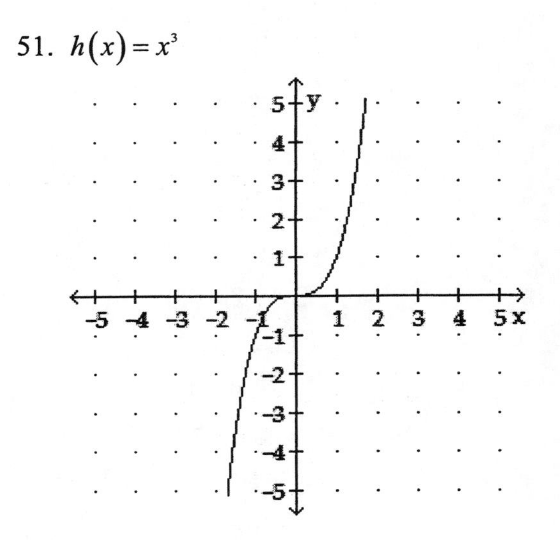

CHALLENGE PROBLEMS

53.

$$\begin{cases} w+x+y+z=3 & (1) \\ w-x+y+z=1 & (2) \\ w+x-y+z=1 & (3) \\ w+x+y-z=3 & (4) \end{cases}$$

Add Equations 3 and 4.

$$\begin{array}{rl} w+x-y+z=1 & (3) \\ \underline{w+x+y-z=3} & (4) \\ 2w+2x=4 & (5) \end{array}$$

Multiply Equation 1 by -1 and add Equations 1 and 2.

$$\begin{array}{rl} -w-x-y-z=-3 & (1) \\ \underline{w-x+y+z=1} & (2) \\ -2x=-2 & \\ x=1 & \end{array}$$

Substitute $x=1$ into Equation 5.

$$\begin{aligned} 2w+2x&=4 \quad (5) \\ 2w+2(1)&=4 \\ 2w+2&=4 \\ 2w&=2 \\ w&=1 \end{aligned}$$

Add Equations 2 and 3.

$$\begin{array}{rl} w-x+y+z=1 & (2) \\ \underline{w+x-y+z=1} & (3) \\ 2w+2z=2 & (6) \end{array}$$

Substitute $w=1$ into Equation 6.

$$\begin{aligned} 2w+2z&=2 \quad (6) \\ 2(1)+2z&=2 \\ 2+2z&=2 \\ 2z&=0 \\ z&=0 \end{aligned}$$

53. (cont.)

Substitute $x=1$, $w=1$, and $z=0$ into Equation 1.

$$\begin{aligned} w+x+y+z&=3 \quad (1) \\ 1+1+y+0&=3 \\ y+2&=3 \\ y&=1 \end{aligned}$$

The solution is $(1,\ 1,\ 1,\ 0)$.

SECTION 12.3

VOCABULARY

1. A **matrix** is a rectangular array of numbers.

3. A 3×4 matrix has 3 **rows** and 4 **columns**.

5. A matrix that represents the equations of a system is called an **augmented** matrix.

CONCEPTS

7. a. 2×3
   b. 3×4

9. $\begin{cases} x-y=-10 \\ y=6 \end{cases}$

$$y=6$$
$$x-y=-10$$
$$x-6=-10$$
$$x=-4$$
$$(-4,6)$$

11. It has no solution. The system is inconsistent.

NOTATION

13. a. Multiply row 1 by $\frac{1}{3}$.

$$\left[\begin{array}{ccc|c} 1 & 2 & -3 & 0 \\ 1 & 5 & -2 & 1 \\ -2 & 2 & -2 & 5 \end{array}\right]$$

b. To row 2, add -1 times row 1

$$\left[\begin{array}{ccc|c} 1 & 2 & -3 & 0 \\ 0 & 3 & 1 & 1 \\ -2 & 2 & -2 & 5 \end{array}\right]$$

15.

$$\left[\begin{array}{cc|c} 4 & \boxed{-1} & 14 \\ 1 & 1 & 6 \end{array}\right]$$

$$\left[\begin{array}{cc|c} \boxed{1} & 1 & 6 \\ 4 & -1 & 14 \end{array}\right]$$

$$\left[\begin{array}{cc|c} 1 & 1 & 6 \\ 0 & \boxed{-5} & -10 \end{array}\right]$$

$$\left[\begin{array}{cc|c} 1 & 1 & 6 \\ 0 & 1 & \boxed{2} \end{array}\right]$$

This matrix represents the system

$$\begin{cases} x+y=6 \\ \boxed{y}=2 \end{cases}$$

The solution is $(\boxed{4},2)$.

PRACTICE

17.

$$\left[\begin{array}{cc|c} 1 & 1 & 2 \\ 1 & -1 & 0 \end{array}\right]$$

$-R_1+R_2$

$$\left[\begin{array}{cc|c} 1 & 1 & 2 \\ 0 & -2 & -2 \end{array}\right]$$

$-\frac{1}{2}R_2$

$$\left[\begin{array}{cc|c} 1 & 1 & 2 \\ 0 & 1 & 1 \end{array}\right]$$

This matrix represents the system

$$\begin{cases} x+y=2 \\ y=1 \end{cases}$$

$$x+1=2$$
$$x=1$$

The solution is $(1, 1)$.

19.

$$\left[\begin{array}{cc|c} 2 & 1 & 1 \\ 1 & 2 & -4 \end{array}\right]$$

$R_1 \leftrightarrow R_2$

$$\left[\begin{array}{cc|c} 1 & 2 & -4 \\ 2 & 1 & 1 \end{array}\right]$$

$-2R_1 + R_2$

$$\left[\begin{array}{cc|c} 1 & 2 & -4 \\ 0 & -3 & 9 \end{array}\right]$$

$-\frac{1}{3}R_2$

$$\left[\begin{array}{cc|c} 1 & 2 & -4 \\ 0 & 1 & -3 \end{array}\right]$$

This matrix represents the system

$$\begin{cases} x + 2y = -4 \\ y = -3 \end{cases}$$

$x - 6 = -4$

$x = 2$

The solution is $(2, -3)$.

21.

$$\left[\begin{array}{cc|c} 2 & -1 & -1 \\ 1 & -2 & 1 \end{array}\right]$$

$R_1 \leftrightarrow R_2$

$$\left[\begin{array}{cc|c} 1 & -2 & 1 \\ 2 & -1 & -1 \end{array}\right]$$

$-2R_1 + R_2$

$$\left[\begin{array}{cc|c} 1 & -2 & 1 \\ 0 & 3 & -3 \end{array}\right]$$

$\frac{1}{3}R_2$

$$\left[\begin{array}{cc|c} 1 & -2 & 1 \\ 0 & 1 & -1 \end{array}\right]$$

This matrix represents the system

$$\begin{cases} x - 2y = 1 \\ y = -1 \end{cases}$$

$x + 2 = 1$

$x = -1$

The solution is $(-1, -1)$.

23.

$$\left[\begin{array}{cc|c} 3 & 4 & -12 \\ 9 & -2 & 6 \end{array}\right]$$

$\frac{1}{3}R_1$

$$\left[\begin{array}{cc|c} 1 & \frac{4}{3} & -4 \\ 9 & -2 & 6 \end{array}\right]$$

$-9R_1 + R_2$

$$\left[\begin{array}{cc|c} 1 & \frac{4}{3} & -4 \\ 0 & -14 & 42 \end{array}\right]$$

$-\frac{1}{14}R_2$

$$\left[\begin{array}{cc|c} 1 & \frac{4}{3} & -4 \\ 0 & 1 & -3 \end{array}\right]$$

This matrix represents the system

$$\begin{cases} x + \frac{4}{3}y = -4 \\ y = -3 \end{cases}$$

$x - 4 = -4$

$x = 0$

The solution is $(0, -3)$.

25.

$$\left[\begin{array}{ccc|c} 1 & 1 & 1 & 6 \\ 1 & 2 & 1 & 8 \\ 1 & 1 & 2 & 9 \end{array}\right]$$

$-1R_1 + R_2$

$$\left[\begin{array}{ccc|c} 1 & 1 & 1 & 6 \\ 0 & 1 & 0 & 2 \\ 1 & 1 & 2 & 9 \end{array}\right]$$

$-1R_1 + R_3$

$$\left[\begin{array}{ccc|c} 1 & 1 & 1 & 6 \\ 0 & 1 & 0 & 2 \\ 0 & 0 & 1 & 3 \end{array}\right]$$

This matrix represents the system

$$\begin{cases} x + y + z = 6 \\ y = 2 \\ z = 3 \end{cases}$$

$x + 2 + 3 = 6$

$x + 5 = 6$

$x = 1$

The solution is $(1, 2, 3)$.

27.

$$\left[\begin{array}{ccc|c} 3 & 1 & -3 & 5 \\ 1 & -2 & 4 & 10 \\ 1 & 1 & 1 & 13 \end{array}\right]$$

$R_1 \leftrightarrow R_3$

$$\left[\begin{array}{ccc|c} 1 & 1 & 1 & 13 \\ 1 & -2 & 4 & 10 \\ 3 & 1 & -3 & 5 \end{array}\right]$$

$-1R_1 + R_2$

$$\left[\begin{array}{ccc|c} 1 & 1 & 1 & 13 \\ 0 & -3 & 3 & -3 \\ 3 & 1 & -3 & 5 \end{array}\right]$$

$-\frac{1}{3}R_2$

$$\left[\begin{array}{ccc|c} 1 & 1 & 1 & 13 \\ 0 & 1 & -1 & 1 \\ 3 & 1 & -3 & 5 \end{array}\right]$$

$-3R_1 + R_3$

$$\left[\begin{array}{ccc|c} 1 & 1 & 1 & 13 \\ 0 & 1 & -1 & 1 \\ 0 & -2 & -6 & -34 \end{array}\right]$$

$2R_2 + R_3$

$$\left[\begin{array}{ccc|c} 1 & 1 & 1 & 13 \\ 0 & 1 & -1 & 1 \\ 0 & 0 & -8 & -32 \end{array}\right]$$

$-\frac{1}{8}R_3$

$$\left[\begin{array}{ccc|c} 1 & 1 & 1 & 13 \\ 0 & 1 & -1 & 1 \\ 0 & 0 & 1 & 4 \end{array}\right]$$

This matrix represents the system

$$\begin{cases} x+y+z=13 \\ y-z=1 \\ z=4 \end{cases}$$

$$y-4=1$$
$$y=5$$
$$x+5+4=13$$
$$x+9=13$$
$$x=4$$

The solution is $(4, 5, 4)$.

29.

$$\left[\begin{array}{ccc|c} 3 & -2 & 4 & 4 \\ 1 & 1 & 1 & 3 \\ 6 & -2 & -3 & 10 \end{array}\right]$$

$R_1 \leftrightarrow R_2$

$$\left[\begin{array}{ccc|c} 1 & 1 & 1 & 3 \\ 3 & -2 & 4 & 4 \\ 6 & -2 & -3 & 10 \end{array}\right]$$

$-3R_1 + R_2$

$$\left[\begin{array}{ccc|c} 1 & 1 & 1 & 3 \\ 0 & -5 & 1 & -5 \\ 6 & -2 & -3 & 10 \end{array}\right]$$

$-6R_1 + R_3$

$$\left[\begin{array}{ccc|c} 1 & 1 & 1 & 3 \\ 0 & -5 & 1 & -5 \\ 0 & -8 & -9 & -8 \end{array}\right]$$

$-\frac{1}{5}R_2$

$$\left[\begin{array}{ccc|c} 1 & 1 & 1 & 3 \\ 0 & 1 & -\frac{1}{5} & 1 \\ 0 & -8 & -9 & -8 \end{array}\right]$$

$8R_2 + R_3$

$$\left[\begin{array}{ccc|c} 1 & 1 & 1 & 3 \\ 0 & 1 & -\frac{1}{5} & 1 \\ 0 & 0 & -\frac{53}{5} & 0 \end{array}\right]$$

$-\frac{5}{53}R_3$

$$\left[\begin{array}{ccc|c} 1 & 1 & 1 & 3 \\ 0 & 1 & -\frac{1}{5} & 1 \\ 0 & 0 & 1 & 0 \end{array}\right]$$

This matrix represents the system

$$\begin{cases} x+y+z=3 \\ y-\frac{1}{5}z=1 \\ z=0 \end{cases}$$

$$y-0=1$$
$$y=1$$
$$x+1+0=3$$
$$x+1=3$$
$$x=2$$

The solution is $(2, 1, 0)$.

31.

$$\left[\begin{array}{ccc|c} 2 & 1 & 3 & 3 \\ -2 & -1 & 1 & 5 \\ 4 & -2 & 2 & 2 \end{array}\right]$$

$R_1 + R_2$

$$\left[\begin{array}{ccc|c} 0 & 0 & 4 & 8 \\ -2 & -1 & 1 & 5 \\ 4 & -2 & 2 & 2 \end{array}\right]$$

$R_1 \leftrightarrow R_3$

$$\left[\begin{array}{ccc|c} 4 & -2 & 2 & 2 \\ -2 & -1 & 1 & 5 \\ 0 & 0 & 4 & 8 \end{array}\right]$$

$2R_2 + R_1$

$$\left[\begin{array}{ccc|c} 4 & -2 & 2 & 2 \\ 0 & -4 & 4 & 12 \\ 0 & 0 & 4 & 8 \end{array}\right]$$

$\frac{1}{4}R_1$

$$\left[\begin{array}{ccc|c} 1 & -\frac{1}{2} & \frac{1}{2} & \frac{1}{2} \\ 0 & -4 & 4 & 12 \\ 0 & 0 & 4 & 8 \end{array}\right]$$

$-\frac{1}{4}R_2$

$$\left[\begin{array}{ccc|c} 1 & -\frac{1}{2} & \frac{1}{2} & \frac{1}{2} \\ 0 & 1 & -1 & -3 \\ 0 & 0 & 4 & 8 \end{array}\right]$$

$\frac{1}{4}R_3$

$$\left[\begin{array}{ccc|c} 1 & -\frac{1}{2} & \frac{1}{2} & \frac{1}{2} \\ 0 & 1 & -1 & -3 \\ 0 & 0 & 1 & 2 \end{array}\right]$$

This matrix represents the system

$$\begin{cases} x - \frac{1}{2}y + \frac{1}{2}z = \frac{1}{2} \\ y - z = -3 \\ z = 2 \end{cases}$$

$$y - 2 = -3$$
$$y = -1$$
$$x + \frac{1}{2} + 1 = \frac{1}{2}$$
$$x + \frac{3}{2} = \frac{1}{2}$$
$$x = -1$$

The solution is $(-1, -1, 2)$.

33.

$$\left[\begin{array}{ccc|c} 2 & 1 & -3 & -7 \\ 3 & -1 & 2 & -9 \\ -2 & -1 & -1 & 3 \end{array}\right]$$

$R_1 + R_3$

$$\left[\begin{array}{ccc|c} 2 & 1 & -3 & -7 \\ 3 & -1 & 2 & -9 \\ 0 & 0 & -4 & -4 \end{array}\right]$$

$\frac{1}{2}R_1$

$$\left[\begin{array}{ccc|c} 1 & \frac{1}{2} & -\frac{3}{2} & -\frac{7}{2} \\ 3 & -1 & 2 & -9 \\ 0 & 0 & -4 & -4 \end{array}\right]$$

$-3R_1 + R_2$

$$\left[\begin{array}{ccc|c} 1 & \frac{1}{2} & -\frac{3}{2} & -\frac{7}{2} \\ 0 & -\frac{5}{2} & \frac{13}{2} & \frac{3}{2} \\ 0 & 0 & -4 & -4 \end{array}\right]$$

$-\frac{2}{5}R_2$

$$\left[\begin{array}{ccc|c} 1 & \frac{1}{2} & -\frac{3}{2} & -\frac{7}{2} \\ 0 & 1 & -\frac{13}{5} & -\frac{3}{5} \\ 0 & 0 & -4 & -4 \end{array}\right]$$

$-\frac{1}{4}R_3$

$$\left[\begin{array}{ccc|c} 1 & \frac{1}{2} & -\frac{3}{2} & -\frac{7}{2} \\ 0 & 1 & -\frac{13}{5} & -\frac{3}{5} \\ 0 & 0 & 1 & 1 \end{array}\right]$$

This matrix represents the system

$$\begin{cases} x + \frac{1}{2}y - \frac{3}{2}z = -\frac{7}{2} \\ y - \frac{13}{5}z = -\frac{3}{5} \\ z = 1 \end{cases}$$

$$y - \frac{13}{5} = -\frac{3}{5}$$
$$y = 2$$
$$x + 1 - \frac{3}{2} = -\frac{7}{2}$$
$$x - \frac{1}{2} = -\frac{7}{2}$$
$$x = -3$$

The solution is $(-3, 2, 1)$.

35.

$$\left[\begin{array}{ccc|c} 2 & 1 & -2 & 6 \\ 4 & -1 & 1 & -1 \\ 6 & -2 & 3 & -5 \end{array}\right]$$

$\frac{1}{2}R_1$

$$\left[\begin{array}{ccc|c} 1 & \frac{1}{2} & -1 & 3 \\ 4 & -1 & 1 & -1 \\ 6 & -2 & 3 & -5 \end{array}\right]$$

$-4R_1 + R_2$

$$\left[\begin{array}{ccc|c} 1 & \frac{1}{2} & -1 & 3 \\ 0 & -3 & 5 & -13 \\ 6 & -2 & 3 & -5 \end{array}\right]$$

$-6R_1 + R_3$

$$\left[\begin{array}{ccc|c} 1 & \frac{1}{2} & -1 & 3 \\ 0 & -3 & 5 & -13 \\ 0 & -5 & 9 & -23 \end{array}\right]$$

$-\frac{1}{3}R_2$

$$\left[\begin{array}{ccc|c} 1 & \frac{1}{2} & -1 & 3 \\ 0 & 1 & -\frac{5}{3} & \frac{13}{3} \\ 0 & -5 & 9 & -23 \end{array}\right]$$

$5R_2 + R_3$

$$\left[\begin{array}{ccc|c} 1 & \frac{1}{2} & -1 & 3 \\ 0 & 1 & -\frac{5}{3} & \frac{13}{3} \\ 0 & 0 & \frac{2}{3} & -\frac{4}{3} \end{array}\right]$$

$\frac{3}{2}R_3$

$$\left[\begin{array}{ccc|c} 1 & \frac{1}{2} & -1 & 3 \\ 0 & 1 & -\frac{5}{3} & \frac{13}{3} \\ 0 & 0 & 1 & -2 \end{array}\right]$$

This matrix represents the system

$$\begin{cases} x + \frac{1}{2}y - z = 3 \\ y - \frac{5}{3}z = \frac{13}{3} \\ z = -2 \end{cases}$$

$$y + \frac{10}{3} = \frac{13}{3} \qquad x + \frac{1}{2} + 2 = 3$$

$$y = 1 \qquad x + \frac{5}{2} = 3$$

$$x = \frac{1}{2}$$

The solution is $\left(\frac{1}{2}, 1, -2\right)$.

37.

$$\left[\begin{array}{cc|c} 1 & -3 & 9 \\ -2 & 6 & 18 \end{array}\right]$$

$2R_1 + R_2$

$$\left[\begin{array}{cc|c} 1 & -3 & 9 \\ 0 & 0 & 36 \end{array}\right]$$

This matrix represents the system

$$\begin{cases} x - 3y = 9 \\ 0 + 0 = 36 \end{cases}$$

No solution. The system is inconsistent.

39.

$$\left[\begin{array}{cc|c} 4 & 4 & 12 \\ -1 & -1 & -3 \end{array}\right]$$

$\frac{1}{4}R_1$

$$\left[\begin{array}{cc|c} 1 & 1 & 3 \\ -1 & -1 & -3 \end{array}\right]$$

$R_1 + R_2$

$$\left[\begin{array}{cc|c} 1 & 1 & 3 \\ 0 & 0 & 0 \end{array}\right]$$

This matrix represents the system

$$\begin{cases} x + y = 3 \\ 0 + 0 = 0 \end{cases}$$

Infinitely many solutions.

The equations are dependent.

41.

$$\left[\begin{array}{ccc|c} 6 & 1 & -1 & -2 \\ 1 & 2 & 1 & 5 \\ 0 & 5 & -1 & 2 \end{array}\right]$$

$R_1 \leftrightarrow R_2$

$$\left[\begin{array}{ccc|c} 1 & 2 & 1 & 5 \\ 6 & 1 & -1 & -2 \\ 0 & 5 & -1 & 2 \end{array}\right]$$

$-6R_1 + R_2$

$$\left[\begin{array}{ccc|c} 1 & 2 & 1 & 5 \\ 0 & -11 & -7 & -32 \\ 0 & 5 & -1 & 2 \end{array}\right]$$

$-\frac{1}{11}R_2$

$$\left[\begin{array}{ccc|c} 1 & 2 & 1 & 5 \\ 0 & 1 & \frac{7}{11} & \frac{32}{11} \\ 0 & 5 & -1 & 2 \end{array}\right]$$

$-5R_2 + R_3$

$$\left[\begin{array}{ccc|c} 1 & 2 & 1 & 5 \\ 0 & 1 & \frac{7}{11} & \frac{32}{11} \\ 0 & 0 & -\frac{46}{11} & -\frac{138}{11} \end{array}\right]$$

$-\frac{11}{46}R_3$

$$\left[\begin{array}{ccc|c} 1 & 2 & 1 & 5 \\ 0 & 1 & \frac{7}{11} & \frac{32}{11} \\ 0 & 0 & 1 & 3 \end{array}\right]$$

This matrix represents the system

$$\begin{cases} x + 2y + z = 5 \\ y + \frac{7}{11}z = \frac{32}{11} \\ z = 3 \end{cases}$$

$$y + \frac{21}{11} = \frac{32}{11}$$

$$y = 1$$

$$x + 2 + 3 = 5$$

$$x + 5 = 5$$

$$x = 0$$

The solution is $(0, 1, 3)$.

43.

$$\left[\begin{array}{ccc|c} 2 & 1 & -1 & 1 \\ 1 & 2 & 2 & 2 \\ 4 & 5 & 3 & 3 \end{array}\right]$$

$R_1 \leftrightarrow R_2$

$$\left[\begin{array}{ccc|c} 1 & 2 & 2 & 2 \\ 2 & 1 & -1 & 1 \\ 4 & 5 & 3 & 3 \end{array}\right]$$

$-2R_1 + R_2$

$$\left[\begin{array}{ccc|c} 1 & 2 & 2 & 2 \\ 0 & -3 & -5 & -3 \\ 4 & 5 & 3 & 3 \end{array}\right]$$

$-\frac{1}{3}R_2$

$$\left[\begin{array}{ccc|c} 1 & 2 & 2 & 2 \\ 0 & 1 & \frac{5}{3} & 1 \\ 4 & 5 & 3 & 3 \end{array}\right]$$

$-4R_1 + R_3$

$$\left[\begin{array}{ccc|c} 1 & 2 & 2 & 2 \\ 0 & 1 & \frac{5}{3} & 1 \\ 0 & -3 & -5 & -5 \end{array}\right]$$

$3R_2 + R_3$

$$\left[\begin{array}{ccc|c} 1 & 2 & 2 & 2 \\ 0 & 1 & \frac{5}{3} & 1 \\ 0 & 0 & 0 & -2 \end{array}\right]$$

This matrix represents the system

$$\begin{cases} x + 2y + 2z = 2 \\ y + \frac{5}{3}z = 1 \\ 0 + 0 + 0 = -2 \end{cases}$$

This system is inconsistent.

No solution.

45.

$$\left[\begin{array}{ccc|c} 5 & 3 & 0 & 4 \\ 0 & 3 & -4 & 4 \\ 1 & 0 & 1 & 1 \end{array}\right]$$

$\frac{1}{5}R_1$

$$\left[\begin{array}{ccc|c} 1 & \frac{3}{5} & 0 & \frac{4}{5} \\ 0 & 3 & -4 & 4 \\ 1 & 0 & 1 & 1 \end{array}\right]$$

$-R_1 + R_3$

$$\left[\begin{array}{ccc|c} 1 & \frac{3}{5} & 0 & \frac{4}{5} \\ 0 & 3 & -4 & 4 \\ 0 & -\frac{3}{5} & 1 & \frac{1}{5} \end{array}\right]$$

$\frac{1}{3}R_2$

$$\left[\begin{array}{ccc|c} 1 & \frac{3}{5} & 0 & \frac{4}{5} \\ 0 & 1 & -\frac{4}{3} & \frac{4}{3} \\ 0 & -\frac{3}{5} & 1 & \frac{1}{5} \end{array}\right]$$

$\frac{3}{5}R_2 + R_3$

$$\left[\begin{array}{ccc|c} 1 & \frac{3}{5} & 0 & \frac{4}{5} \\ 0 & 1 & -\frac{4}{3} & \frac{4}{3} \\ 0 & 0 & \frac{1}{5} & 1 \end{array}\right]$$

$5R_3$

$$\left[\begin{array}{ccc|c} 1 & \frac{3}{5} & 0 & \frac{4}{5} \\ 0 & 1 & -\frac{4}{3} & \frac{4}{3} \\ 0 & 0 & 1 & 5 \end{array}\right]$$

This matrix represents the system

$$\begin{cases} x + \frac{3}{5}y = \frac{4}{5} \\ y - \frac{4}{3}z = \frac{4}{3} \\ z = 5 \end{cases}$$

$$y - \frac{20}{3} = \frac{4}{3}$$
$$y = 8$$
$$x + \frac{24}{5} = \frac{4}{5}$$
$$x = -4$$

The solution is $(-4, 8, 5)$.

47.

$$\left[\begin{array}{ccc|c} 1 & -1 & 0 & 1 \\ 2 & 0 & -1 & 0 \\ 0 & 2 & -1 & -2 \end{array}\right]$$

$-2R_1 + R_2$

$$\left[\begin{array}{ccc|c} 1 & -1 & 0 & 1 \\ 0 & 2 & -1 & -2 \\ 0 & 2 & -1 & -2 \end{array}\right]$$

$\frac{1}{2}R_2$

$$\left[\begin{array}{ccc|c} 1 & -1 & 0 & 1 \\ 0 & 1 & -\frac{1}{2} & -1 \\ 0 & 2 & -1 & -2 \end{array}\right]$$

$\frac{1}{2}R_3$

$$\left[\begin{array}{ccc|c} 1 & -1 & 0 & 1 \\ 0 & 1 & -\frac{1}{2} & -1 \\ 0 & 1 & -\frac{1}{2} & -1 \end{array}\right]$$

$-R_2 + R_3$

$$\left[\begin{array}{ccc|c} 1 & -1 & 0 & 1 \\ 0 & 1 & -\frac{1}{2} & -1 \\ 0 & 0 & 0 & 0 \end{array}\right]$$

This matrix represents the system

$$\begin{cases} x - y = 1 \\ y - \frac{1}{2}z = -1 \\ 0 + 0 + 0 = 0 \end{cases}$$

Infinitely many solutions.

Dependent equations.

49. Let the angles be x and y.

$$\begin{cases} x+y=90 \\ x=46+y \end{cases}$$

Write the equations in standard form.

$$\begin{cases} x+y=90 \\ x-y=46 \end{cases}$$

$$\left[\begin{array}{cc|c} 1 & 1 & 90 \\ 1 & -1 & 46 \end{array}\right]$$

$-R_1+R_2$

$$\left[\begin{array}{cc|c} 1 & 1 & 90 \\ 0 & -2 & -44 \end{array}\right]$$

$-\frac{1}{2}R_2$

$$\left[\begin{array}{cc|c} 1 & 1 & 90 \\ 0 & 1 & 22 \end{array}\right]$$

$$\begin{cases} x+y=90 \\ y=22 \end{cases}$$

$$x+22=90$$
$$x=68$$

The measures of the angles are 22° and 68°.

51.

$$\begin{cases} A+B+C=180 \\ B=25+A \\ C=2A-5 \end{cases}$$

Write the equations in standard form.

$$\begin{cases} A+B+C=180 \\ A-B=-25 \\ 2A-C=5 \end{cases}$$

$$\left[\begin{array}{ccc|c} 1 & 1 & 1 & 180 \\ 1 & -1 & 0 & -25 \\ 2 & 0 & -1 & 5 \end{array}\right]$$

$-R_1+R_2$

$$\left[\begin{array}{ccc|c} 1 & 1 & 1 & 180 \\ 0 & -2 & -1 & -205 \\ 2 & 0 & -1 & 5 \end{array}\right]$$

$-\frac{1}{2}R_2$

$$\left[\begin{array}{ccc|c} 1 & 1 & 1 & 180 \\ 0 & 1 & \frac{1}{2} & 102.5 \\ 2 & 0 & -1 & 5 \end{array}\right]$$

$-2R_1+R_3$

$$\left[\begin{array}{ccc|c} 1 & 1 & 1 & 180 \\ 0 & 1 & \frac{1}{2} & 102.5 \\ 0 & -2 & -3 & -355 \end{array}\right]$$

$2R_2+R_3$

$$\left[\begin{array}{ccc|c} 1 & 1 & 1 & 180 \\ 0 & 1 & \frac{1}{2} & 102.5 \\ 0 & 0 & -2 & -150 \end{array}\right]$$

$-\frac{1}{2}R_3$

$$\left[\begin{array}{ccc|c} 1 & 1 & 1 & 180 \\ 0 & 1 & \frac{1}{2} & 102.5 \\ 0 & 0 & 1 & 75 \end{array}\right]$$

$$\begin{cases} A+B+C=180 \\ B+\frac{1}{2}C=102.5 \\ C=75 \end{cases}$$

$$B+\frac{75}{2}=102.5$$
$$B=65$$
$$A+65+75=180$$
$$A+140=180$$
$$A=40$$

The measures of the angles are 40°, 65°, and 75°.

APPLICATIONS

53. DIGITAL PHOTOGRAPHY
$512(512) = 262{,}144$

55. PHYSICAL THERAPY
Let the measure of angle 1 $= x$ and the measure of angle 2 $= y$.

$$\begin{cases} x + y = 180 \\ x = y - 28 \end{cases}$$

Write the equations in standard form.

$$\begin{cases} x + y = 180 \\ x - y = -28 \end{cases}$$

$$\left[\begin{array}{cc|c} 1 & 1 & 180 \\ 1 & -1 & -28 \end{array}\right]$$

$-R_1 + R_2$

$$\left[\begin{array}{cc|c} 1 & 1 & 180 \\ 0 & -2 & -208 \end{array}\right]$$

$-\frac{1}{2}R_2$

$$\left[\begin{array}{cc|c} 1 & 1 & 180 \\ 0 & 1 & 104 \end{array}\right]$$

$$\begin{cases} x + y = 180 \\ y = 104 \end{cases}$$

$$x + 104 = 180$$
$$x = 76$$

The measure of angle 1 is 76° and the measure of angle 2 is 104°.

57. THEATER SEATING
Let x = number of seats in the founder's circle, y = number of box seats, and z = promenade seats.

$$\begin{cases} x + y + z = 800 \\ 30x + 20y + 10z = 13{,}000 \\ 40x + 30y + 25z = 23{,}000 \end{cases}$$

$$\left[\begin{array}{ccc|c} 1 & 1 & 1 & 800 \\ 30 & 20 & 10 & 13{,}000 \\ 40 & 30 & 25 & 23{,}000 \end{array}\right]$$

$-30R_1 + R_2$

$$\left[\begin{array}{ccc|c} 1 & 1 & 1 & 800 \\ 0 & -10 & -20 & -11{,}000 \\ 40 & 30 & 25 & 23{,}000 \end{array}\right]$$

57. (cont.)

$-\frac{1}{10}R_2$

$$\left[\begin{array}{ccc|c} 1 & 1 & 1 & 800 \\ 0 & 1 & 2 & 1{,}100 \\ 40 & 30 & 25 & 23{,}000 \end{array}\right]$$

$-40R_1 + R_3$

$$\left[\begin{array}{ccc|c} 1 & 1 & 1 & 800 \\ 0 & 1 & 2 & 1{,}100 \\ 0 & -10 & -15 & -9{,}000 \end{array}\right]$$

$10R_2 + R_3$

$$\left[\begin{array}{ccc|c} 1 & 1 & 1 & 800 \\ 0 & 1 & 2 & 1{,}100 \\ 0 & 0 & 5 & 2{,}000 \end{array}\right]$$

$\frac{1}{5}R_3$

$$\left[\begin{array}{ccc|c} 1 & 1 & 1 & 800 \\ 0 & 1 & 2 & 1{,}100 \\ 0 & 0 & 1 & 400 \end{array}\right]$$

This matrix represents the system

$$\begin{cases} x + y + z = 800 \\ y + 2z = 1{,}100 \\ z = 400 \end{cases}$$

$$y + 2(400) = 1{,}100$$
$$y + 800 = 1{,}100$$
$$y = 300$$
$$x + 300 + 400 = 800$$
$$x + 700 = 800$$
$$x = 100$$

There are 100 Founder's circle seats, 300 box seats, and 400 promenade seats.

WRITING

59. Answers will vary.

REVIEW

61. $m = \dfrac{y_2 - y_1}{x_2 - x_1}(x_2 \neq x_1)$

63. $y - y_1 = m(x - x_1)$

CHALLENGE PROBLEMS

65. $k \neq 0$

SECTION 12.4

VOCABULARY

1. A determinant is a **number** that is associated with a square matrix.

3. The **minor** of b_1 in $\begin{vmatrix} a_1 & b_1 & c_1 \\ a_2 & b_2 & c_2 \\ a_3 & b_3 & c_3 \end{vmatrix}$ is

$\begin{vmatrix} a_2 & c_2 \\ a_3 & c_3 \end{vmatrix}$.

5. A 3 × 3 determinant has 3 **rows** and 3 **columns**.

CONCEPTS

7. If the denominator determinant D for a system of equations is zero, the equations of the system are **dependent** or the system is **inconsistent**.

9.
$$\begin{vmatrix} a & b \\ c & d \end{vmatrix} = \boxed{ad} - \boxed{bc}$$

11. $\begin{vmatrix} 3 & 4 \\ 2 & -3 \end{vmatrix}$

13. $\left(\frac{7}{11}, -\frac{5}{11}\right)$

NOTATION

15.
$$\begin{vmatrix} 5 & -2 \\ -2 & 6 \end{vmatrix} = 5(\boxed{6}) - (-2)(-2)$$
$$= \boxed{30} - 4$$
$$= 26$$

PRACTICE

17.
$$\begin{vmatrix} 2 & 3 \\ -2 & 1 \end{vmatrix} = 2(1) - (3)(-2)$$
$$= 2 + 6$$
$$= 8$$

19.
$$\begin{vmatrix} -1 & 2 \\ 3 & -4 \end{vmatrix} = (-1)(-4) - (3)(2)$$
$$= 4 - 6$$
$$= -2$$

21.
$$\begin{vmatrix} 10 & 0 \\ 1 & 20 \end{vmatrix} = 10(20) - 0(1)$$
$$= 200 - 0$$
$$= 200$$

23.
$$\begin{vmatrix} -6 & -2 \\ 15 & 4 \end{vmatrix} = -6(4) - (-2)(15)$$
$$= -24 - (-30)$$
$$= 6$$

25.
$$\begin{vmatrix} 1 & 2 & 0 \\ 0 & 1 & 2 \\ 0 & 0 & 1 \end{vmatrix} = 1\begin{vmatrix} 1 & 2 \\ 0 & 1 \end{vmatrix} - 2\begin{vmatrix} 0 & 2 \\ 0 & 1 \end{vmatrix} + 0\begin{vmatrix} 0 & 1 \\ 0 & 0 \end{vmatrix}$$
$$= 1(1-0) - 2(0-0) + 0(0-0)$$
$$= 1(1) - 2(0) + 0(0)$$
$$= 1 - 0 + 0$$
$$= 1$$

27.
$$\begin{vmatrix} 1 & -2 & 3 \\ -2 & 1 & 1 \\ -3 & -2 & 1 \end{vmatrix} = 1\begin{vmatrix} 1 & 1 \\ -2 & 1 \end{vmatrix} - (-2)\begin{vmatrix} -2 & 1 \\ -3 & 1 \end{vmatrix} + 3\begin{vmatrix} -2 & 1 \\ -3 & -2 \end{vmatrix}$$
$$= 1(1+2) + 2(-2+3) + 3(4+3)$$
$$= 1(3) + 2(1) + 3(7)$$
$$= 3 + 2 + 21$$
$$= 26$$

29.

$$\begin{vmatrix} 1 & 0 & 1 \\ 0 & 1 & 0 \\ 1 & 1 & 1 \end{vmatrix} = 1\begin{vmatrix} 1 & 0 \\ 1 & 1 \end{vmatrix} - 0\begin{vmatrix} 0 & 0 \\ 1 & 1 \end{vmatrix} + 1\begin{vmatrix} 0 & 1 \\ 1 & 1 \end{vmatrix}$$
$$= 1(1-0) - 0(0-0) + 1(0-1)$$
$$= 1(1) - 0(0) + 1(-1)$$
$$= 1 - 0 - 1$$
$$= 0$$

31.

$$\begin{vmatrix} 1 & 2 & 1 \\ -3 & 7 & 3 \\ -4 & 3 & -5 \end{vmatrix} = 1\begin{vmatrix} 7 & 3 \\ 3 & -5 \end{vmatrix} - 2\begin{vmatrix} -3 & 3 \\ -4 & -5 \end{vmatrix} + 1\begin{vmatrix} -3 & 7 \\ -4 & 3 \end{vmatrix}$$
$$= 1(-35-9) - 2(15+12) + 1(-9+28)$$
$$= 1(-44) - 2(27) + 1(19)$$
$$= -44 - 54 + 19$$
$$= -79$$

33.

$$x = \frac{D_x}{D} = \frac{\begin{vmatrix} 6 & 1 \\ 2 & -1 \end{vmatrix}}{\begin{vmatrix} 1 & 1 \\ 1 & -1 \end{vmatrix}} = \frac{6(-1)-1(2)}{1(-1)-1(1)} = \frac{-6-2}{-1-1} = \frac{-8}{-2} = 4$$

$$y = \frac{D_y}{D} = \frac{\begin{vmatrix} 1 & 6 \\ 1 & 2 \end{vmatrix}}{\begin{vmatrix} 1 & 1 \\ 1 & -1 \end{vmatrix}} = \frac{1(2)-6(1)}{1(-1)-1(1)} = \frac{2-6}{-1-1} = \frac{-4}{-2} = 2$$

The solution is (4, 2).

35.

$$x = \frac{D_x}{D} = \frac{\begin{vmatrix} 0 & 3 \\ -4 & -6 \end{vmatrix}}{\begin{vmatrix} 2 & 3 \\ 4 & -6 \end{vmatrix}} = \frac{0(-6)-3(-4)}{2(-6)-3(4)} = \frac{0+12}{-12-12} = \frac{12}{-24} = -\frac{1}{2}$$

$$y = \frac{D_y}{D} = \frac{\begin{vmatrix} 2 & 0 \\ 4 & -4 \end{vmatrix}}{\begin{vmatrix} 2 & 3 \\ 4 & -6 \end{vmatrix}} = \frac{2(-4)-0(4)}{2(-6)-3(4)} = \frac{-8-0}{-12-12} = \frac{-8}{-24} = \frac{1}{3}$$

The solution is $\left(-\frac{1}{2}, \frac{1}{3}\right)$.

37.

$$x = \frac{D_x}{D} = \frac{\begin{vmatrix} 11 & 2 \\ 11 & 4 \end{vmatrix}}{\begin{vmatrix} 3 & 2 \\ 6 & 4 \end{vmatrix}} = \frac{11(4)-2(11)}{3(4)-2(6)} = \frac{44-22}{12-12} = \frac{22}{0}$$
undefined

$$y = \frac{D_y}{D} = \frac{\begin{vmatrix} 3 & 11 \\ 6 & 11 \end{vmatrix}}{\begin{vmatrix} 3 & 2 \\ 6 & 4 \end{vmatrix}} = \frac{3(11)-11(6)}{3(4)-2(6)} = \frac{33-66}{12-12} = \frac{-33}{0}$$
undefined

The system has no solution and is an inconsistent system.

39. Write the first equation in standard form.

$$y=\frac{-2x+1}{3}$$
$$3y=-2x+1$$
$$2x+3y=1$$

$$\begin{cases}2x+3y=1\\3x-2y=8\end{cases}$$

$$x=\frac{D_x}{D}=\frac{\begin{vmatrix}1 & 3\\8 & -2\end{vmatrix}}{\begin{vmatrix}2 & 3\\3 & -2\end{vmatrix}}=\frac{1(-2)-3(8)}{2(-2)-3(3)}=\frac{-2-24}{-4-9}=\frac{-26}{-13}=2$$

$$y=\frac{D_y}{D}=\frac{\begin{vmatrix}2 & 1\\3 & 8\end{vmatrix}}{\begin{vmatrix}2 & 3\\3 & -2\end{vmatrix}}=\frac{2(8)-1(3)}{2(-2)-3(3)}=\frac{16-3}{-4-9}=\frac{13}{-13}=-1$$

The solution is (2, –1).

41.

$$x=\frac{D_x}{D}=\frac{\begin{vmatrix}4 & 1 & 1\\0 & 1 & -1\\2 & -1 & 1\end{vmatrix}}{\begin{vmatrix}1 & 1 & 1\\1 & 1 & -1\\1 & -1 & 1\end{vmatrix}}$$

$$=\frac{4\begin{vmatrix}1 & -1\\-1 & 1\end{vmatrix}-1\begin{vmatrix}0 & -1\\2 & 1\end{vmatrix}+1\begin{vmatrix}0 & 1\\2 & -1\end{vmatrix}}{1\begin{vmatrix}1 & -1\\-1 & 1\end{vmatrix}-1\begin{vmatrix}1 & -1\\1 & 1\end{vmatrix}+1\begin{vmatrix}1 & 1\\1 & -1\end{vmatrix}}$$

$$=\frac{4(1-1)-1(0+2)+1(0-2)}{1(1-1)-1(1+1)+1(-1-1)}=\frac{4(0)-1(2)+1(-2)}{1(0)-1(2)+1(-2)}=\frac{0-2-2}{0-2-2}=\frac{-4}{-4}=1$$

41. (cont.)

$$y=\frac{D_y}{D}=\frac{\begin{vmatrix}1 & 4 & 1\\1 & 0 & -1\\1 & 2 & 1\end{vmatrix}}{\begin{vmatrix}1 & 1 & 1\\1 & 1 & -1\\1 & -1 & 1\end{vmatrix}}$$

$$=\frac{1\begin{vmatrix}0 & -1\\2 & 1\end{vmatrix}-4\begin{vmatrix}1 & -1\\1 & 1\end{vmatrix}+1\begin{vmatrix}1 & 0\\1 & 2\end{vmatrix}}{1\begin{vmatrix}1 & -1\\-1 & 1\end{vmatrix}-1\begin{vmatrix}1 & -1\\1 & 1\end{vmatrix}+1\begin{vmatrix}1 & 1\\1 & -1\end{vmatrix}}$$

$$=\frac{1(0+2)-4(1+1)+1(2-0)}{1(1-1)-1(1+1)+1(-1-1)}=\frac{1(2)-4(2)+1(2)}{1(0)-1(2)+1(-2)}=\frac{2-8+2}{0-2-2}=\frac{-4}{-4}=1$$

$$z=\frac{D_z}{D}=\frac{\begin{vmatrix}1 & 1 & 4\\1 & 1 & 0\\1 & -1 & 2\end{vmatrix}}{\begin{vmatrix}1 & 1 & 1\\1 & 1 & -1\\1 & -1 & 1\end{vmatrix}}$$

$$=\frac{1\begin{vmatrix}1 & 0\\-1 & 2\end{vmatrix}-1\begin{vmatrix}1 & 0\\1 & 2\end{vmatrix}+4\begin{vmatrix}1 & 1\\1 & -1\end{vmatrix}}{1\begin{vmatrix}1 & -1\\-1 & 1\end{vmatrix}-1\begin{vmatrix}1 & -1\\1 & 1\end{vmatrix}+1\begin{vmatrix}1 & 1\\1 & -1\end{vmatrix}}$$

$$=\frac{1(2-0)-1(2-0)+4(-1-1)}{1(1-1)-1(1+1)+1(-1-1)}=\frac{1(2)-1(2)+4(-2)}{1(0)-1(2)+1(-2)}=\frac{2-2-8}{0-2-2}=\frac{-8}{-4}=2$$

The solution is (1, 1, 2).

43.

$$x=\frac{D_x}{D}$$

$$=\frac{\begin{vmatrix}7&1&2\\8&2&1\\9&1&1\end{vmatrix}}{\begin{vmatrix}1&1&2\\1&2&1\\2&1&1\end{vmatrix}}$$

$$=\frac{7\begin{vmatrix}2&1\\1&1\end{vmatrix}-1\begin{vmatrix}8&1\\9&1\end{vmatrix}+2\begin{vmatrix}8&2\\9&1\end{vmatrix}}{1\begin{vmatrix}2&1\\1&1\end{vmatrix}-1\begin{vmatrix}1&1\\2&1\end{vmatrix}+2\begin{vmatrix}1&2\\2&1\end{vmatrix}}$$

$$=\frac{7(2-1)-1(8-9)+2(8-18)}{1(2-1)-1(1-2)+2(1-4)}$$

$$=\frac{7(1)-1(-1)+2(-10)}{1(1)-1(-1)+2(-3)}$$

$$=\frac{7+1-20}{1+1-6}$$

$$=\frac{-12}{-4}$$

$$=3$$

$$y=\frac{D_y}{D}$$

$$=\frac{\begin{vmatrix}1&7&2\\1&8&1\\2&9&1\end{vmatrix}}{\begin{vmatrix}1&1&2\\1&2&1\\2&1&1\end{vmatrix}}$$

$$=\frac{1\begin{vmatrix}8&1\\9&1\end{vmatrix}-7\begin{vmatrix}1&1\\2&1\end{vmatrix}+2\begin{vmatrix}1&8\\2&9\end{vmatrix}}{1\begin{vmatrix}2&1\\1&1\end{vmatrix}-1\begin{vmatrix}1&1\\2&1\end{vmatrix}+2\begin{vmatrix}1&2\\2&1\end{vmatrix}}$$

$$=\frac{1(8-9)-7(1-2)+2(9-16)}{1(2-1)-1(1-2)+2(1-4)}$$

$$=\frac{1(-1)-7(-1)+2(-7)}{1(1)-1(-1)+2(-3)}$$

$$=\frac{-1+7-14}{1+1-6}$$

$$=\frac{-8}{-4}$$

$$=2$$

43. (cont.)

$$z=\frac{D_z}{D}$$

$$=\frac{\begin{vmatrix}1&1&7\\1&2&8\\2&1&9\end{vmatrix}}{\begin{vmatrix}1&1&2\\1&2&1\\2&1&1\end{vmatrix}}$$

$$=\frac{1\begin{vmatrix}2&8\\1&9\end{vmatrix}-1\begin{vmatrix}1&8\\2&9\end{vmatrix}+7\begin{vmatrix}1&2\\2&1\end{vmatrix}}{1\begin{vmatrix}2&1\\1&1\end{vmatrix}-1\begin{vmatrix}1&1\\2&1\end{vmatrix}+2\begin{vmatrix}1&2\\2&1\end{vmatrix}}$$

$$=\frac{1(18-8)-1(9-16)+7(1-4)}{1(2-1)-1(1-2)+2(1-4)}$$

$$=\frac{1(10)-1(-7)+7(-3)}{1(1)-1(-1)+2(-3)}$$

$$=\frac{10+7-21}{1+1-6}$$

$$=\frac{-4}{-4}$$

$$=1$$

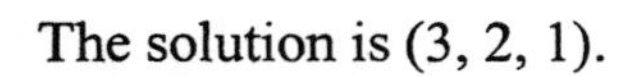
The solution is (3, 2, 1).

45.

$$x=\frac{D_x}{D}$$

$$=\frac{\begin{vmatrix}5 & 1 & 1\\ 10 & -2 & 3\\ -3 & 1 & -4\end{vmatrix}}{\begin{vmatrix}2 & 1 & 1\\ 1 & -2 & 3\\ 1 & 1 & -4\end{vmatrix}}$$

$$=\frac{5\begin{vmatrix}-2 & 3\\ 1 & -4\end{vmatrix}-1\begin{vmatrix}10 & 3\\ -3 & -4\end{vmatrix}+1\begin{vmatrix}10 & -2\\ -3 & 1\end{vmatrix}}{2\begin{vmatrix}-2 & 3\\ 1 & -4\end{vmatrix}-1\begin{vmatrix}1 & 3\\ 1 & -4\end{vmatrix}+1\begin{vmatrix}1 & -2\\ 1 & 1\end{vmatrix}}$$

$$=\frac{5(8-3)-1(-40+9)+1(10-6)}{2(8-3)-1(-4-3)+1(1+2)}$$

$$=\frac{5(5)-1(-31)+1(4)}{2(5)-1(-7)+1(3)}$$

$$=\frac{25+31+4}{10+7+3}$$

$$=\frac{60}{20}$$

$$=3$$

$$y=\frac{D_y}{D}$$

$$=\frac{\begin{vmatrix}2 & 5 & 1\\ 1 & 10 & 3\\ 1 & -3 & -4\end{vmatrix}}{\begin{vmatrix}2 & 1 & 1\\ 1 & -2 & 3\\ 1 & 1 & -4\end{vmatrix}}$$

$$=\frac{2\begin{vmatrix}10 & 3\\ -3 & -4\end{vmatrix}-5\begin{vmatrix}1 & 3\\ 1 & -4\end{vmatrix}+1\begin{vmatrix}1 & 10\\ 1 & -3\end{vmatrix}}{2\begin{vmatrix}-2 & 3\\ 1 & -4\end{vmatrix}-1\begin{vmatrix}1 & 3\\ 1 & -4\end{vmatrix}+1\begin{vmatrix}1 & -2\\ 1 & 1\end{vmatrix}}$$

$$=\frac{2(-40+9)-5(-4-3)+1(-3-10)}{2(8-3)-1(-4-3)+1(1+2)}$$

$$=\frac{2(-31)-5(-7)+1(-13)}{2(5)-1(-7)+1(3)}$$

$$=\frac{-62+35-13}{10+7+3}$$

$$=\frac{-40}{20}$$

$$=-2$$

45. (cont.)

$$z=\frac{D_z}{D}$$

$$=\frac{\begin{vmatrix}2 & 1 & 5\\ 1 & -2 & 10\\ 1 & 1 & -3\end{vmatrix}}{\begin{vmatrix}2 & 1 & 1\\ 1 & -2 & 3\\ 1 & 1 & -4\end{vmatrix}}$$

$$=\frac{2\begin{vmatrix}-2 & 10\\ 1 & -3\end{vmatrix}-1\begin{vmatrix}1 & 10\\ 1 & -3\end{vmatrix}+5\begin{vmatrix}1 & -2\\ 1 & 1\end{vmatrix}}{2\begin{vmatrix}-2 & 3\\ 1 & -4\end{vmatrix}-1\begin{vmatrix}1 & 3\\ 1 & -4\end{vmatrix}+1\begin{vmatrix}1 & -2\\ 1 & 1\end{vmatrix}}$$

$$=\frac{2(6-10)-1(-3-10)+5(1+2)}{2(8-3)-1(-4-3)+1(1+2)}$$

$$=\frac{2(-4)-1(-13)+5(3)}{2(5)-1(-7)+1(3)}$$

$$=\frac{-8+13+15}{10+7+3}$$

$$=\frac{20}{20}$$

$$=1$$

The solution is (3, –2, 1).

47.

$$x=\frac{D_x}{D}$$

$$=\frac{\begin{vmatrix}1 & -3 & 0\\ 1 & 0 & -8\\ 0 & 2 & -4\end{vmatrix}}{\begin{vmatrix}4 & -3 & 0\\ 6 & 0 & -8\\ 0 & 2 & -4\end{vmatrix}}$$

$$=\frac{1\begin{vmatrix}0 & -8\\ 2 & -4\end{vmatrix}-(-3)\begin{vmatrix}1 & -8\\ 0 & -4\end{vmatrix}-0\begin{vmatrix}1 & 0\\ 0 & 2\end{vmatrix}}{4\begin{vmatrix}0 & -8\\ 2 & -4\end{vmatrix}-(-3)\begin{vmatrix}6 & -8\\ 0 & -4\end{vmatrix}-0\begin{vmatrix}6 & 0\\ 0 & 2\end{vmatrix}}$$

$$=\frac{1(0+16)+3(-4-0)-0(2-0)}{4(0+16)+3(-24-0)-0(12-0)}$$

$$=\frac{1(16)+3(-4)-0(2)}{4(16)+3(-24)-0(12)}$$

$$=\frac{16-12-0}{64-72-0}$$

$$=\frac{4}{-8}$$

$$=-\frac{1}{2}$$

$$y=\frac{D_y}{D}$$

$$=\frac{\begin{vmatrix}4 & 1 & 0\\ 6 & 1 & -8\\ 0 & 0 & -4\end{vmatrix}}{\begin{vmatrix}4 & -3 & 0\\ 6 & 0 & -8\\ 0 & 2 & -4\end{vmatrix}}$$

$$=\frac{4\begin{vmatrix}1 & -8\\ 0 & -4\end{vmatrix}-1\begin{vmatrix}6 & -8\\ 0 & -4\end{vmatrix}-0\begin{vmatrix}6 & 1\\ 0 & 0\end{vmatrix}}{4\begin{vmatrix}0 & -8\\ 2 & -4\end{vmatrix}-(-3)\begin{vmatrix}6 & -8\\ 0 & -4\end{vmatrix}-0\begin{vmatrix}6 & 0\\ 0 & 2\end{vmatrix}}$$

$$=\frac{4(-4-0)-1(-24-0)-0(0-0)}{4(0+16)+3(-24-0)-0(12-0)}$$

$$=\frac{4(-4)-1(-24)-0(0)}{4(16)+3(-24)-0(12)}$$

$$=\frac{-16+24-0}{64-72-0}$$

$$=\frac{8}{-8}$$

$$=-1$$

47. (cont.)

$$z=\frac{D_z}{D}$$

$$=\frac{\begin{vmatrix}4 & -3 & 1\\ 6 & 0 & 1\\ 0 & 2 & 0\end{vmatrix}}{\begin{vmatrix}4 & -3 & 0\\ 6 & 0 & -8\\ 0 & 2 & -4\end{vmatrix}}$$

$$=\frac{4\begin{vmatrix}0 & 1\\ 2 & 0\end{vmatrix}-(-3)\begin{vmatrix}6 & 1\\ 0 & 0\end{vmatrix}+1\begin{vmatrix}6 & 0\\ 0 & 2\end{vmatrix}}{4\begin{vmatrix}0 & -8\\ 2 & -4\end{vmatrix}-(-3)\begin{vmatrix}6 & -8\\ 0 & -4\end{vmatrix}-0\begin{vmatrix}6 & 0\\ 0 & 2\end{vmatrix}}$$

$$=\frac{4(0-2)+3(0-0)+1(12-0)}{4(0+16)+3(-24-0)-0(12-0)}$$

$$=\frac{4(-2)+3(0)+1(12)}{4(16)+3(-24)-0(12)}$$

$$=\frac{-8+0+12}{64-72-0}$$

$$=\frac{4}{-8}$$

$$=-\frac{1}{2}$$

The solution is $\left(-\frac{1}{2},-1,-\frac{1}{2}\right)$.

49.

$$x=\frac{D_x}{D}$$

$$=\frac{\begin{vmatrix}6 & 3 & 4\\ -4 & -3 & -4\\ 12 & 6 & 8\end{vmatrix}}{\begin{vmatrix}2 & 3 & 4\\ 2 & -3 & -4\\ 4 & 6 & 8\end{vmatrix}}$$

$$=\frac{6\begin{vmatrix}-3 & -4\\ 6 & 8\end{vmatrix}-3\begin{vmatrix}-4 & -4\\ 12 & 8\end{vmatrix}+4\begin{vmatrix}-4 & -3\\ 12 & 6\end{vmatrix}}{2\begin{vmatrix}-3 & -4\\ 6 & 8\end{vmatrix}-3\begin{vmatrix}2 & -4\\ 4 & 8\end{vmatrix}+4\begin{vmatrix}2 & -3\\ 4 & 6\end{vmatrix}}$$

$$=\frac{6(-24+24)-3(-32+48)+4(-24+36)}{2(-24+24)-3(16+16)+4(12+12)}$$

$$=\frac{6(0)-3(16)+4(12)}{2(0)-3(32)+4(24)}$$

$$=\frac{0-48+48}{0-96+96}$$

$$=\frac{0}{0}$$

Since the denominator determinant D is 0 and the numerator determinant D_x is 0, the system is consistent, but the equations are dependent. There are infinitely many solutions.

51. Write the equations in standard form.

$$\begin{cases}2x+y-z=1\\ x+2y+2z=2\\ 4x+5y+3z=3\end{cases}$$

$$x=\frac{D_x}{D}$$

$$=\frac{\begin{vmatrix}1 & 1 & -1\\ 2 & 2 & 2\\ 3 & 5 & 3\end{vmatrix}}{\begin{vmatrix}2 & 1 & -1\\ 1 & 2 & 2\\ 4 & 5 & 3\end{vmatrix}}$$

$$=\frac{1\begin{vmatrix}2 & 2\\ 5 & 3\end{vmatrix}-1\begin{vmatrix}2 & 2\\ 3 & 3\end{vmatrix}-1\begin{vmatrix}2 & 2\\ 3 & 5\end{vmatrix}}{2\begin{vmatrix}2 & 2\\ 5 & 3\end{vmatrix}-1\begin{vmatrix}1 & 2\\ 4 & 3\end{vmatrix}-1\begin{vmatrix}1 & 2\\ 4 & 5\end{vmatrix}}$$

$$=\frac{1(6-10)-1(6-6)-1(10-6)}{2(6-10)-1(3-8)-1(5-8)}$$

$$=\frac{1(-4)-1(0)-1(4)}{2(-4)-1(-5)-1(-3)}$$

$$=\frac{-4-0-4}{-8+5+3}$$

$$=\frac{-8}{0}\text{ is undefined}$$

Since the denominator determinant D is 0 and the numerator determinant D_x is not 0, the system is inconsistent and has no solutions.

53. Write the equations in standard form. Multiply the second equation by 2 to eliminate fractions.

$$\begin{cases} x+y=1 \\ y+2z=5 \\ x-z=-3 \end{cases}$$

$$x=\frac{D_x}{D}$$

$$=\frac{\begin{vmatrix} 1 & 1 & 0 \\ 5 & 1 & 2 \\ -3 & 0 & -1 \end{vmatrix}}{\begin{vmatrix} 1 & 1 & 0 \\ 0 & 1 & 2 \\ 1 & 0 & -1 \end{vmatrix}}$$

$$=\frac{1\begin{vmatrix} 1 & 2 \\ 0 & -1 \end{vmatrix}-1\begin{vmatrix} 5 & 2 \\ -3 & -1 \end{vmatrix}+0\begin{vmatrix} 5 & 1 \\ -3 & 0 \end{vmatrix}}{1\begin{vmatrix} 1 & 2 \\ 0 & -1 \end{vmatrix}-1\begin{vmatrix} 0 & 2 \\ 1 & -1 \end{vmatrix}+0\begin{vmatrix} 0 & 1 \\ 1 & 0 \end{vmatrix}}$$

$$=\frac{1(-1-0)-1(-5+6)+0(0+3)}{1(-1-0)-1(0-2)+0(0-1)}$$

$$=\frac{1(-1)-1(1)+0(3)}{1(-1)-1(-2)+0(-1)}$$

$$=\frac{-1-1+0}{-1+2+0}$$

$$=\frac{-2}{1}$$

$$=-2$$

53. (cont.)

$$y=\frac{D_y}{D}$$

$$=\frac{\begin{vmatrix} 1 & 1 & 0 \\ 0 & 5 & 2 \\ 1 & -3 & -1 \end{vmatrix}}{\begin{vmatrix} 1 & 1 & 0 \\ 0 & 1 & 2 \\ 1 & 0 & -1 \end{vmatrix}}$$

$$=\frac{1\begin{vmatrix} 5 & 2 \\ -3 & -1 \end{vmatrix}-1\begin{vmatrix} 0 & 2 \\ 1 & -1 \end{vmatrix}+0\begin{vmatrix} 0 & 5 \\ 1 & -3 \end{vmatrix}}{1\begin{vmatrix} 1 & 2 \\ 0 & -1 \end{vmatrix}-1\begin{vmatrix} 0 & 2 \\ 1 & -1 \end{vmatrix}+0\begin{vmatrix} 0 & 1 \\ 1 & 0 \end{vmatrix}}$$

$$=\frac{1(-5+6)-1(0-2)+0(0-5)}{1(-1-0)-1(0-2)+0(0-1)}$$

$$=\frac{1(1)-1(-2)+0(-5)}{1(-1)-1(-2)+0(-1)}$$

$$=\frac{1+2+0}{-1+2+0}$$

$$=\frac{3}{1}$$

$$=3$$

$$z=\frac{D_z}{D}$$

$$=\frac{\begin{vmatrix} 1 & 1 & 1 \\ 0 & 1 & 5 \\ 1 & 0 & -3 \end{vmatrix}}{\begin{vmatrix} 1 & 1 & 0 \\ 0 & 1 & 2 \\ 1 & 0 & -1 \end{vmatrix}}$$

$$=\frac{1\begin{vmatrix} 1 & 5 \\ 0 & -3 \end{vmatrix}-1\begin{vmatrix} 0 & 5 \\ 1 & -3 \end{vmatrix}+1\begin{vmatrix} 0 & 1 \\ 1 & 0 \end{vmatrix}}{1\begin{vmatrix} 1 & 2 \\ 0 & -1 \end{vmatrix}-1\begin{vmatrix} 0 & 2 \\ 1 & -1 \end{vmatrix}+0\begin{vmatrix} 0 & 1 \\ 1 & 0 \end{vmatrix}}$$

$$=\frac{1(-3-0)-1(0-5)+1(0-1)}{1(-1-0)-1(0-2)+0(0-1)}$$

$$=\frac{1(-3)-1(-5)+1(-1)}{1(-1)-1(-2)+0(-1)}$$

$$=\frac{-3+5-1}{-1+2+0}$$

$$=\frac{1}{1}$$

$$=1$$

The solution is (–2, 3, 1).

55. INVENTORIES

Let x = the number of phones that sold for \$67 and y = the number of phones that sold for \$100.

$$\begin{cases} x+y=360 \\ 67x+100y=29,400 \end{cases}$$

$$x=\frac{D_x}{D}$$
$$=\frac{\begin{vmatrix} 360 & 1 \\ 29,400 & 100 \end{vmatrix}}{\begin{vmatrix} 1 & 1 \\ 67 & 100 \end{vmatrix}}$$
$$=\frac{360(100)-1(29,400)}{1(100)-1(67)}$$
$$=\frac{36,000-29,400}{100-67}$$
$$=\frac{6,600}{33}$$
$$=200$$

$$y=\frac{D_y}{D}$$
$$=\frac{\begin{vmatrix} 1 & 360 \\ 67 & 29,400 \end{vmatrix}}{\begin{vmatrix} 1 & 1 \\ 67 & 100 \end{vmatrix}}$$
$$=\frac{1(29,400)-360(67)}{1(100)-1(67)}$$
$$=\frac{29,400-24,120}{100-67}$$
$$=\frac{5,280}{33}$$
$$=160$$

There were 200 of the \$67 phones and 160 of the \$100 phones.

57. INVESTING

Let x = amount in HiTech, y = amount in SaveTel, and z = amount in OilCo.

6.6% of $20,000 = $1,320

$$\begin{cases} x+y+z=20{,}000 \\ y+z=3x \\ 0.10x+0.05y+0.06z=1{,}320 \end{cases}$$

Write the second equation in standard form and multiply the third equation by 100 to eliminate decimals.

$$\begin{cases} x+y+z=20{,}000 \\ -3x+y+z=0 \\ 10x+5y+6z=132{,}000 \end{cases}$$

$$x=\frac{D_x}{D}=\frac{\begin{vmatrix} 20{,}000 & 1 & 1 \\ 0 & 1 & 1 \\ 132{,}000 & 5 & 6 \end{vmatrix}}{\begin{vmatrix} 1 & 1 & 1 \\ -3 & 1 & 1 \\ 10 & 5 & 6 \end{vmatrix}}$$

$$=\frac{20{,}000\begin{vmatrix} 1 & 1 \\ 5 & 6 \end{vmatrix}-1\begin{vmatrix} 0 & 1 \\ 132{,}000 & 6 \end{vmatrix}+1\begin{vmatrix} 0 & 1 \\ 132{,}000 & 5 \end{vmatrix}}{1\begin{vmatrix} 1 & 1 \\ 5 & 6 \end{vmatrix}-1\begin{vmatrix} -3 & 1 \\ 10 & 6 \end{vmatrix}+1\begin{vmatrix} -3 & 1 \\ 10 & 5 \end{vmatrix}}$$

$$=\frac{20{,}000(6-5)-1(0-132{,}000)+1(0-132{,}000)}{1(6-5)-1(-18-10)+1(-15-10)}$$

$$=\frac{20{,}000(1)-1(-132{,}000)+1(-132{,}000)}{1(1)-1(-28)+1(-25)}$$

$$=\frac{20{,}000+132{,}000-132{,}000}{1+28-25}$$

$$=\frac{20{,}000}{4}$$

$$=5{,}000$$

57. (cont.)

$$y=\frac{D_y}{D}=\frac{\begin{vmatrix} 1 & 20{,}000 & 1 \\ -3 & 0 & 1 \\ 10 & 132{,}000 & 6 \end{vmatrix}}{\begin{vmatrix} 1 & 1 & 1 \\ -3 & 1 & 1 \\ 10 & 5 & 6 \end{vmatrix}}$$

$$=\frac{1\begin{vmatrix} 0 & 1 \\ 132{,}000 & 6 \end{vmatrix}-20{,}000\begin{vmatrix} -3 & 1 \\ 10 & 6 \end{vmatrix}+1\begin{vmatrix} -3 & 0 \\ 10 & 132{,}000 \end{vmatrix}}{1\begin{vmatrix} 1 & 1 \\ 5 & 6 \end{vmatrix}-1\begin{vmatrix} -3 & 1 \\ 10 & 6 \end{vmatrix}+1\begin{vmatrix} -3 & 1 \\ 10 & 5 \end{vmatrix}}$$

$$=\frac{1(0-132{,}000)-20{,}000(-18-10)+1(-396{,}000-0)}{1(6-5)-1(-18-10)+1(-15-10)}$$

$$=\frac{1(-132{,}000)-20{,}000(-28)+1(-396{,}000)}{1(1)-1(-28)+1(-25)}$$

$$=\frac{-132{,}000+560{,}000-396{,}000}{1+28-25}$$

$$=\frac{32{,}000}{4}$$

$$=8{,}000$$

$$z=\frac{D_z}{D}=\frac{\begin{vmatrix} 1 & 1 & 20{,}000 \\ -3 & 1 & 0 \\ 10 & 5 & 132{,}000 \end{vmatrix}}{\begin{vmatrix} 1 & 1 & 1 \\ -3 & 1 & 1 \\ 10 & 5 & 6 \end{vmatrix}}$$

$$=\frac{1\begin{vmatrix} 1 & 0 \\ 5 & 132{,}000 \end{vmatrix}-1\begin{vmatrix} -3 & 0 \\ 10 & 132{,}000 \end{vmatrix}+20{,}000\begin{vmatrix} -3 & 1 \\ 10 & 5 \end{vmatrix}}{1\begin{vmatrix} 1 & 1 \\ 5 & 6 \end{vmatrix}-1\begin{vmatrix} -3 & 1 \\ 10 & 6 \end{vmatrix}+1\begin{vmatrix} -3 & 1 \\ 10 & 5 \end{vmatrix}}$$

$$=\frac{1(132{,}000-0)-1(-396{,}000-0)+20{,}000(-15-10)}{1(6-5)-1(-18-10)+1(-15-10)}$$

$$=\frac{1(132{,}000)-1(-396{,}000)+20{,}000(-25)}{1(1)-1(-28)+1(-25)}$$

$$=\frac{132{,}000+396{,}000-500{,}000}{1+28-25}$$

$$=\frac{28{,}000}{4}$$

$$=7{,}000$$

He should invest $5,000 in HiTech, $8,000 in SaveTel, and $7,000 in OilCo.

59. Use a calculator with matrix capabilities.

$$\begin{vmatrix} 2 & -3 & 4 \\ -1 & 2 & 4 \\ 3 & -3 & 1 \end{vmatrix} = -23$$

61. Use a calculator with matrix capabilities.

$$\begin{vmatrix} 2 & 1 & -3 \\ -2 & 2 & 4 \\ 1 & -2 & 2 \end{vmatrix} = 26$$

WRITING

63. Answers will vary.

65. Answers will vary.

REVIEW

67. Find the slope of each using $y = mx + b$.

$$y = 2x - 7 \qquad m = 2$$

$$\begin{aligned} x - 2y &= 7 \\ -2y &= -x + 7 \\ y &= \frac{1}{2}x - \frac{7}{2} \\ m &= \frac{1}{2} \end{aligned}$$

No. Since the slopes are not opposite reciprocals, the lines are NOT perpendicular.

69. The graph of g is 2 units below the graph of f. $g(x)$ is shifted down 2 units.

71. The y–intercept since the x–coordinate is 0.

73. The independent variable is x and the dependent variable is y.

CHALLENGE PROBLEMS

75.

$$\begin{vmatrix} x & y & 1 \\ -2 & 3 & 1 \\ 3 & 5 & 1 \end{vmatrix} = 0$$

$$x\begin{vmatrix} 3 & 1 \\ 5 & 1 \end{vmatrix} - y\begin{vmatrix} -2 & 1 \\ 3 & 1 \end{vmatrix} + 1\begin{vmatrix} -2 & 3 \\ 3 & 5 \end{vmatrix} = 0$$

$$\begin{aligned} x(3-5) - y(-2-3) + 1(-10-9) &= 0 \\ x(-2) - y(-5) + 1(-19) &= 0 \\ -2x + 5y - 19 &= 0 \\ 2x - 5y + 19 &= 0 \\ 2x - 5y &= -19 \end{aligned}$$

For (–2, 3):

$$\begin{aligned} 2(-2) - 5(3) &\stackrel{?}{=} -19 \\ -4 - 15 &\stackrel{?}{=} -19 \\ -19 &= -19 \end{aligned}$$

For (3, 5):

$$\begin{aligned} 2(3) - 5(5) &\stackrel{?}{=} -19 \\ 6 - 25 &\stackrel{?}{=} -19 \\ -19 &= -19 \end{aligned}$$

CHAPTER 12 KEY CONCEPTS

1. Yes.

$$2x - y = 1$$
$$2\left(\frac{1}{4}\right) - \left(-\frac{1}{2}\right) \stackrel{?}{=} 1$$
$$\frac{1}{2} + \frac{1}{2} \stackrel{?}{=} 1$$
$$1 = 1$$

$$4x + 2y = 0$$
$$4\left(\frac{1}{4}\right) + 2\left(-\frac{1}{2}\right) \stackrel{?}{=} 0$$
$$1 - 1 \stackrel{?}{=} 0$$
$$0 = 0$$

2. No.

$$2(4) - 0 + 1 \stackrel{?}{=} 9$$
$$8 - 0 + 1 \stackrel{?}{=} 9$$
$$9 = 9$$

$$3(4) + 0 - 4(1) \stackrel{?}{=} 8$$
$$12 + 0 - 4 \stackrel{?}{=} 8$$
$$8 = 8$$

$$2(4) - 7(0) \stackrel{?}{=} -1$$
$$8 - 0 \stackrel{?}{=} -1$$
$$8 \neq -1$$

3. $\begin{cases} 2x + 5y = 8 \\ y = 3x + 5 \end{cases}$

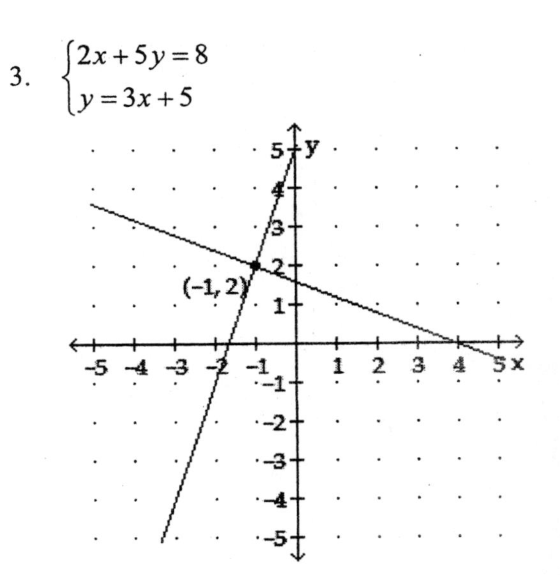

The solution is (−1, 2).

4.

$$\begin{cases} 9x - 8y = 1 & (1) \\ 6x + 12y = 5 & (2) \end{cases}$$

Multiply Equation 1 by 2 and Equation 2 by -3 to create opposites.

$$\begin{aligned} 18x - 16y &= 2 \\ -18x - 36y &= -15 \\ \hline -52y &= -13 \\ y &= \frac{1}{4} \end{aligned}$$

Substitute $y = \frac{1}{4}$ into Equation 1 and solve the equation for x.

$$9x - 8y = 1 \qquad (1)$$
$$9x - 8\left(\frac{1}{4}\right) = 1$$
$$9x - 2 = 1$$
$$9x = 3$$
$$x = \frac{1}{3}$$

The solution is $\left(\frac{1}{3}, \frac{1}{4}\right)$.

5.

$$\begin{cases} 4x - y - 10 = 0 & (1) \\ 3x + 5y = 19 & (2) \end{cases}$$

Solve Equation 1 for y.

$$4x - y - 10 = 0 \quad (1)$$
$$4x - y - 10 + y = 0 + y$$
$$4x - 10 = y$$
$$y = 4x - 10$$

Substitute $y = 4x - 10$ into Equation 2 and solve the equation for x.

$$3x + 5y = 19 \quad (2)$$
$$3x + 5(4x - 10) = 19$$
$$3x + 20x - 50 = 19$$
$$23x - 50 = 19$$
$$23x = 69$$
$$x = 3$$

Subsitute $x = 3$ into Equation 2 and solve the equation for y.

$$3x + 5y = 19 \quad (2)$$
$$3(3) + 5y = 19$$
$$9 + 5y = 19$$
$$5y = 10$$
$$y = 2$$

The solution is $(3, 2)$.

6.

$$\begin{cases} -x + 3y + 2z = 5 & (1) \\ 3x + 2y + z = -1 & (2) \\ 2x - y + 3z = 4 & (3) \end{cases}$$

Multiply Equation 1 by 3 and add Equations 1 and 2 to eliminate x.

$$\begin{array}{rl} -3x + 9y + 6z = 15 & (1) \\ \underline{3x + 2y + z = -1} & (2) \\ 11y + 7z = 14 & (4) \end{array}$$

Multiply Equation 1 by 2 and add Equations 1 and 3 to eliminate x.

$$\begin{array}{rl} -2x + 6y + 4z = 10 & (1) \\ \underline{2x - y + 3z = 4} & (3) \\ 5y + 7z = 14 & (5) \end{array}$$

6. (cont.)

Multiply Equation 4 by -1 and add Equations 4 and 5 to eliminate y.

$$\begin{array}{rl} -11y - 7z = -14 & (4) \\ \underline{5y + 7z = 14} & (5) \\ -6y = 0 & \\ y = 0 & \end{array}$$

Substitute $y = 0$ into Equation 5 and solve for z.

$$5y + 7z = 14 \quad (5)$$
$$5(0) + 7z = 14$$
$$0 + 7z = 14$$
$$7z = 14$$
$$z = 2$$

Substitute $y = 0$ and $z = 2$ into Equation 1 and solve for z.

$$-x + 3y + 2z = 5 \quad (1)$$
$$-x + 3(0) + 2(2) = 5$$
$$-x + 0 + 4 = 5$$
$$-x + 4 = 5$$
$$-x = 1$$
$$x = -1$$

The solution is $(-1, 0, 2)$.

7.

$$\left[\begin{array}{cc|c} 1 & -6 & 3 \\ 1 & 3 & 21 \end{array}\right]$$

$-1R_1 + R_2$

$$\left[\begin{array}{cc|c} 1 & -6 & 3 \\ 0 & 9 & 18 \end{array}\right]$$

$\frac{1}{9}R_2$

$$\left[\begin{array}{cc|c} 1 & -6 & 3 \\ 0 & 1 & 2 \end{array}\right]$$

This matrix represents the system

$$\begin{cases} x - 6y = 3 \\ y = 2 \end{cases}$$

$$x - 6y = 3$$
$$x - 6(2) = 3$$
$$x - 12 = 3$$
$$x = 15$$

The solution is $(15, 2)$.

8.

$$x=\frac{D_x}{D}=\frac{\begin{vmatrix}7 & 0 & 2\\ 9 & -1 & 3\\ 1 & 1 & -1\end{vmatrix}}{\begin{vmatrix}1 & 0 & 2\\ 2 & -1 & 3\\ 0 & 1 & -1\end{vmatrix}}$$

$$=\frac{7\begin{vmatrix}-1 & 3\\ 1 & -1\end{vmatrix}-0\begin{vmatrix}9 & 3\\ 1 & -1\end{vmatrix}+2\begin{vmatrix}9 & -1\\ 1 & 1\end{vmatrix}}{1\begin{vmatrix}-1 & 3\\ 1 & -1\end{vmatrix}-0\begin{vmatrix}2 & 3\\ 0 & -1\end{vmatrix}+2\begin{vmatrix}2 & -1\\ 0 & 1\end{vmatrix}}$$

$$=\frac{7(1-3)-0(-9-3)+2(9+1)}{1(1-3)-0(-2-0)+2(2-0)}$$

$$=\frac{7(-2)-0(-12)+2(10)}{1(-2)-0(-2)+2(2)}$$

$$=\frac{-14-0+20}{-2-0+4}$$

$$=\frac{6}{2}$$

$$=3$$

$$y=\frac{D_y}{D}=\frac{\begin{vmatrix}1 & 7 & 2\\ 2 & 9 & 3\\ 0 & 1 & -1\end{vmatrix}}{\begin{vmatrix}1 & 0 & 2\\ 2 & -1 & 3\\ 0 & 1 & -1\end{vmatrix}}$$

$$=\frac{1\begin{vmatrix}9 & 3\\ 1 & -1\end{vmatrix}-7\begin{vmatrix}2 & 3\\ 0 & -1\end{vmatrix}+2\begin{vmatrix}2 & 9\\ 0 & 1\end{vmatrix}}{1\begin{vmatrix}-1 & 3\\ 1 & -1\end{vmatrix}-0\begin{vmatrix}2 & 3\\ 0 & -1\end{vmatrix}+2\begin{vmatrix}2 & -1\\ 0 & 1\end{vmatrix}}$$

$$=\frac{1(-9-3)-7(-2-0)+2(2-0)}{1(1-3)-0(-2-0)+2(2-0)}$$

$$=\frac{1(-12)-7(-2)+2(2)}{1(-2)-0(-2)+2(2)}$$

$$=\frac{-12+14+4}{-2-0+4}$$

$$=\frac{6}{2}$$

$$=3$$

8. (cont.)

$$z=\frac{D_z}{D}=\frac{\begin{vmatrix}1 & 0 & 7\\ 2 & -1 & 9\\ 0 & 1 & 1\end{vmatrix}}{\begin{vmatrix}1 & 0 & 2\\ 2 & -1 & 3\\ 0 & 1 & -1\end{vmatrix}}$$

$$=\frac{1\begin{vmatrix}-1 & 9\\ 1 & 1\end{vmatrix}-0\begin{vmatrix}2 & 9\\ 0 & 1\end{vmatrix}+7\begin{vmatrix}2 & -1\\ 0 & 1\end{vmatrix}}{1\begin{vmatrix}-1 & 3\\ 1 & -1\end{vmatrix}-0\begin{vmatrix}2 & 3\\ 0 & -1\end{vmatrix}+2\begin{vmatrix}2 & -1\\ 0 & 1\end{vmatrix}}$$

$$=\frac{1(-1-9)-0(2-0)+7(2-0)}{1(1-3)-0(-2-0)+2(2-0)}$$

$$=\frac{1(-10)-0(2)+7(2)}{1(-2)-0(-2)+2(2)}$$

$$=\frac{-10-0+14}{-2-0+4}$$

$$=\frac{4}{2}$$

$$=2$$

The solution is (3, 3, 2).

9. The equations of the system are dependent. There are infinitely many solutions.

10. The system is inconsistent. There are no solutions.

CHAPTER 12 REVIEW

SECTION 12.1
Systems with Two Variables

1. a. Answers will vary.
   (1, 3), (2, 1), and (4, –3)

   b. Answers will vary.
   (0, –4), (2, –2), and (4, 0)

   c. The lines intersect at the point (3, –1).

2. President Clinton's job approval and disapproval ratings were the same; approximately 47% in 5/94 and approximately 48% in 5/95.

3. $\begin{cases} 2x + y = 11 \\ -x + 2y = 7 \end{cases}$

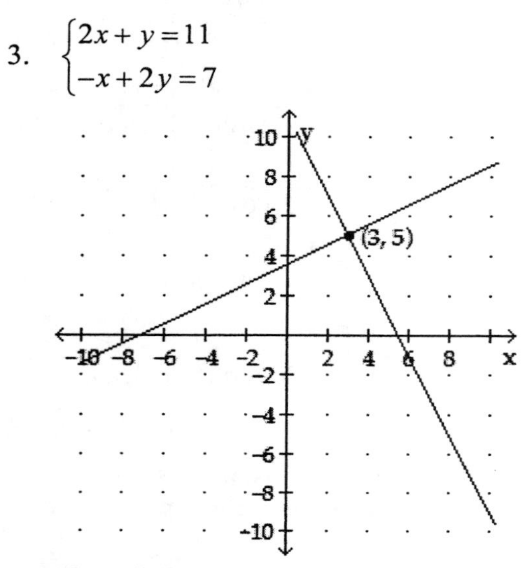

The solution is (3, 5).

4. $\begin{cases} y = -\dfrac{3}{2}x \\ 2x - 3y + 13 = 0 \end{cases}$

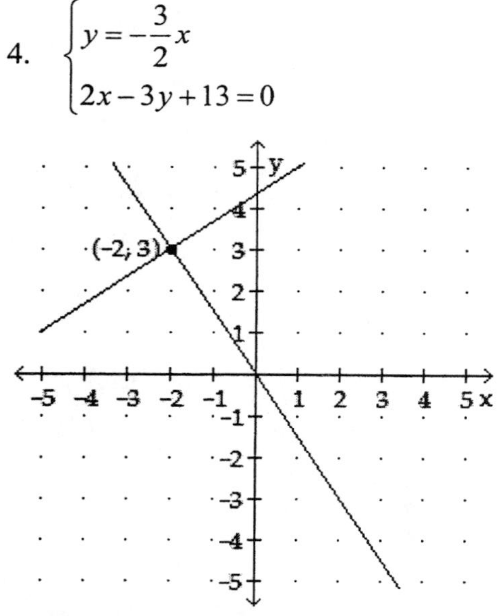

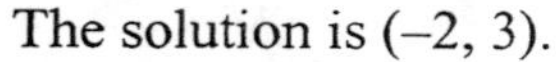
The solution is (–2, 3).

5. $\begin{cases} \dfrac{1}{2}x + \dfrac{1}{3}y = 2 \\ y = 6 - \dfrac{3}{2}x \end{cases}$

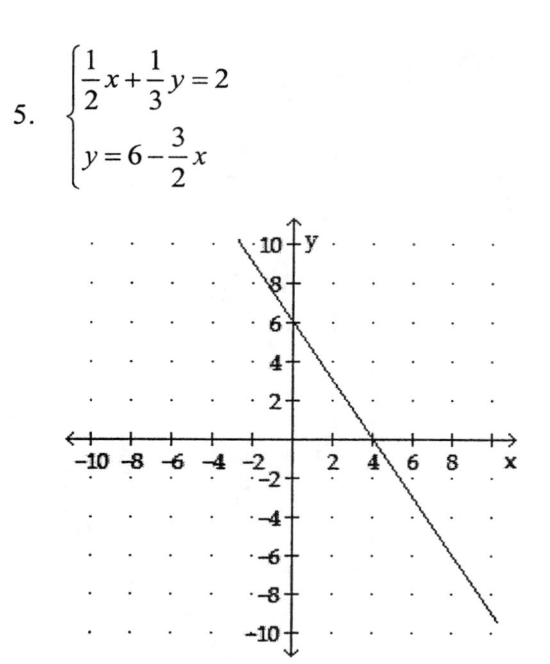

There are infinitely many solutions; the equations are dependent.

6. $\begin{cases} \dfrac{x}{3} - \dfrac{y}{2} = 1 \\ 6x - 9y = 3 \end{cases}$

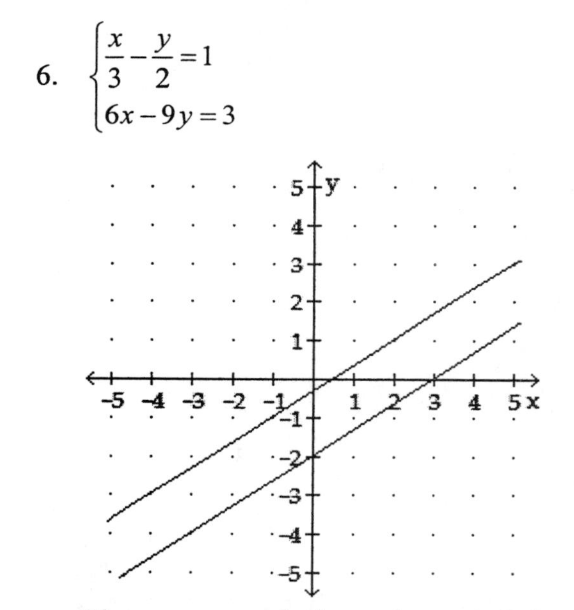

There are no solutions; the system is inconsistent.

7. The lines intersect at $x = 2$.

8. The lines intersect at $x = -1$.

SECTION 12.2
Systems with Three Variables

9. No.

$$\begin{aligned} x-y+z &= 4 \\ 2-(-1)+1 &\stackrel{?}{=} 4 \\ 2+1+1 &\stackrel{?}{=} 4 \\ 4 &= 4 \end{aligned}$$

$$\begin{aligned} x+2y-z &= -1 \\ 2+2(-1)-1 &\stackrel{?}{=} -1 \\ 2-2-1 &\stackrel{?}{=} -1 \\ -1 &= -1 \end{aligned}$$

$$\begin{aligned} x+y-3z &= -1 \\ 2+(-1)-3(1) &\stackrel{?}{=} -1 \\ 2-1-3 &\stackrel{?}{=} -1 \\ -2 &\neq -1 \end{aligned}$$

10.

$$\begin{cases} x+y+z=6 & (1) \\ x-y-z=-4 & (2) \\ -x+y-z=-2 & (3) \end{cases}$$

Multiply Equation 1 by -1 and add Equations 1 and 2 to eliminate x.

$$\begin{aligned} -x-y-z&=-6 \quad (1) \\ x-y-z&=-4 \quad (2) \\ \hline -2y-2z&=-10 \quad (4) \end{aligned}$$

Add Equations 2 and 3 to eliminate x and y.

$$\begin{aligned} x-y-z&=-4 \quad (2) \\ -x+y-z&=-2 \quad (3) \\ \hline -2z&=-6 \\ z&=3 \end{aligned}$$

Substitute $z=3$ into Equation 4 and solve for y.

$$\begin{aligned} -2y-2z&=-10 \quad (4) \\ -2y-2(3)&=-10 \\ -2y-6&=-10 \\ -2y&=-4 \\ y&=2 \end{aligned}$$

Substitute $z=3$ and $y=2$ into Equation 1 and solve for x.

$$\begin{aligned} x+y+z&=6 \quad (1) \\ x+2+3&=6 \\ x+5&=6 \\ x&=1 \end{aligned}$$

The solution is $(1, 2, 3)$.

11.

$$\begin{cases} 2x+3y+z=-5 & (1) \\ -x+2y-z=-6 & (2) \\ 3x+y+2z=4 & (3) \end{cases}$$

Multiply Equation 2 by 2 and add Equations 1 and 2 to eliminate x.

$$\begin{array}{rl} 2x+3y+z=-5 & (1) \\ \underline{-2x+4y-2z=-12} & (2) \\ 7y-z=-17 & (4) \end{array}$$

Multiply Equation 2 by 3 and add Equations 2 and 3 to eliminate x.

$$\begin{array}{rl} -3x+6y-3z=-18 & (2) \\ \underline{3x+y+2z=4} & (3) \\ 7y-z=-14 & (5) \end{array}$$

Multiply Equation 4 by -1 and add Equations 4 and 5 to eliminate y and z.

$$\begin{array}{rl} -7y+z=17 & (4) \\ \underline{7y-z=-14} & (5) \\ 0\neq 3 & \end{array}$$

No solution

Inconsistent system

12.

$$\begin{cases} x+y-z=-3 & (1) \\ x+z=2 & (2) \\ 2x-y+2z=3 & (3) \end{cases}$$

Add Equations 1 and 3 to eliminate y.

$$\begin{array}{rl} x+y-z=-3 & (1) \\ \underline{2x-y+2z=3} & (3) \\ 3x+z=0 & (4) \end{array}$$

Multiply Equation 2 by -1 and add Equations 2 and 4 to eliminate z.

$$\begin{array}{rl} -x-z=-2 & (2) \\ \underline{3x+z=0} & (4) \\ 2x=-2 & \\ x=-1 & \end{array}$$

Substitute $x=-1$ into Equation 4 and solve for z.

$$\begin{array}{rl} 3x+z=0 & (4) \\ 3(-1)+z=0 & \\ -3+z=0 & \\ z=3 & \end{array}$$

Substitute $z=3$ and $x=-1$ into Equation 1 and solve for y.

$$\begin{array}{rl} x+y-z=-3 & (1) \\ -1+y-3=-3 & \\ y-4=-3 & \\ y=1 & \end{array}$$

The solution is $(-1, 1, 3)$.

13.

$$\begin{cases} 3x+3y+6z=-6 & (1) \\ -x-y-2z=2 & (2) \\ 2x+2y+4z=-4 & (3) \end{cases}$$

Multiply Equation 2 by 3 and add Equations 1 and 2 to eliminate x.

$$\begin{array}{rl} 3x+3y+6z=-6 & (1) \\ \underline{-3x-3y-6z=6} & (2) \\ 0=0 & \end{array}$$

Infinitely many solutions; Dependent equations

14. yes; infinitely many solutions

15. Let x = number of pounds of peanuts, y = number of pounds of cashews, and z = number of pounds of Brazil nuts.

$$\begin{cases} x+y+z=50 & (1) \\ 3x+9y+9z=6(50) & (2) \\ y=x-15 & (3) \end{cases}$$

Write each equation in standard form.

$$\begin{cases} x+y+z=50 & (1) \\ 3x+9y+9z=300 & (2) \\ -x+y=-15 & (3) \end{cases}$$

Add Equations 1 and 3 to eliminate x.

$$\begin{array}{rl} x+y+z=50 & (1) \\ \underline{-x+y\quad\ =-15} & (3) \\ 2y+z=35 & (4) \end{array}$$

Multiply Equation 1 by -3 and add Equations 1 and 2 to eliminate x.

$$\begin{array}{rl} -3x-3y-3z=-150 & (1) \\ \underline{3x+9y+9z=300} & (2) \\ 6y+6z=150 & (5) \end{array}$$

Multiply Equation 4 by -6 and add Equations 4 and 5 to eliminate z.

$$\begin{array}{rl} -12y-6z=-210 & (4) \\ \underline{6y+6z=150} & (5) \\ -6y=-60 & \\ y=10 & \end{array}$$

Substitute $y=10$ into Equation 3 and solve for x.

$$\begin{array}{rl} y=x-15 & (3) \\ 10=x-15 & \\ 25=x & \end{array}$$

Substitute $y=10$ and $x=25$ into Equation 1 and solve for x.

$$\begin{array}{rl} x+y+z=50 & (1) \\ 25+10+z=50 & \\ 35+z=50 & \\ z=15 & \end{array}$$

She used 25 lbs of peanuts, 10 lbs of cashews, and 15 lbs of Brazil nuts.

SECTION 12.3
Solving Systems Using Matrices

16. $\left[\begin{array}{cc|c} 5 & 4 & 3 \\ 1 & -1 & -3 \end{array}\right]$

17. $\left[\begin{array}{ccc|c} 1 & 2 & 3 & 6 \\ 1 & -3 & -1 & 4 \\ 6 & 1 & -2 & -1 \end{array}\right]$

18.

$$\left[\begin{array}{cc|c} 1 & -1 & 4 \\ 3 & 7 & -18 \end{array}\right]$$

$-3R_1+R_2$

$$\left[\begin{array}{cc|c} 1 & -1 & 4 \\ 0 & 10 & -30 \end{array}\right]$$

$\frac{1}{10}R_2$

$$\left[\begin{array}{cc|c} 1 & -1 & 4 \\ 0 & 1 & -3 \end{array}\right]$$

This matrix represents the system

$$\begin{cases} x-y=4 \\ y=-3 \end{cases}$$

$$\begin{aligned} x-y&=4 \\ x-(-3)&=4 \\ x+3&=4 \\ x&=1 \end{aligned}$$

The solution is $(1,-3)$.

19.

$$\left[\begin{array}{ccc|c} 1 & 2 & -3 & 5 \\ 1 & 1 & 1 & 0 \\ 3 & 4 & 2 & -1 \end{array}\right]$$

$R_1 - R_2$

$$\left[\begin{array}{ccc|c} 1 & 2 & -3 & 5 \\ 0 & 1 & -4 & 5 \\ 3 & 4 & 2 & -1 \end{array}\right]$$

$-3R_1 + R_3$

$$\left[\begin{array}{ccc|c} 1 & 2 & -3 & 5 \\ 0 & 1 & -4 & 5 \\ 0 & -2 & 11 & -16 \end{array}\right]$$

$2R_2 + R_3$

$$\left[\begin{array}{ccc|c} 1 & 2 & -3 & 5 \\ 0 & 1 & -4 & 5 \\ 0 & 0 & 3 & -6 \end{array}\right]$$

$\frac{1}{3}R_3$

$$\left[\begin{array}{ccc|c} 1 & 2 & -3 & 5 \\ 0 & 1 & -4 & 5 \\ 0 & 0 & 1 & -2 \end{array}\right]$$

This matrix represents the system

$$\begin{cases} x + 2y - 3z = 5 \\ y - 4z = 5 \\ z = -2 \end{cases}$$

$$y - 4(-2) = 5$$
$$y + 8 = 5$$
$$y = -3$$
$$x + 2(-3) - 3(-2) = 5$$
$$x - 6 + 6 = 5$$
$$x = 5$$

The solution is $(5,\ -3,\ -2)$.

20.

$$\left[\begin{array}{cc|c} 16 & -8 & 32 \\ -2 & 1 & -4 \end{array}\right]$$

$\frac{1}{16}R_1$

$$\left[\begin{array}{cc|c} 1 & -\frac{1}{2} & 2 \\ -2 & 1 & -4 \end{array}\right]$$

$2R_1 + R_2$

$$\left[\begin{array}{cc|c} 1 & -\frac{1}{2} & 2 \\ 0 & 0 & 0 \end{array}\right]$$

This matrix represents the system

$$\begin{cases} x - \frac{1}{2}y = 2 \\ 0 + 0 = 0 \end{cases}$$

Infinitely many solutions
Dependent equations

21.

$$\left[\begin{array}{ccc|c} 1 & 2 & 2 & 2 \\ 4 & 5 & 3 & 3 \\ 2 & 1 & -1 & 1 \end{array}\right]$$

$-4R_1 + R_2$

$$\left[\begin{array}{ccc|c} 1 & 2 & 2 & 2 \\ 0 & -3 & -5 & -5 \\ 2 & 1 & -1 & 1 \end{array}\right]$$

$-2R_1 + R_3$

$$\left[\begin{array}{ccc|c} 1 & 2 & 2 & 2 \\ 0 & -3 & -5 & -5 \\ 0 & -3 & -5 & -3 \end{array}\right]$$

$-R_2 + R_3$

$$\left[\begin{array}{ccc|c} 1 & 2 & 2 & 2 \\ 0 & -3 & -5 & -5 \\ 0 & 0 & 0 & 2 \end{array}\right]$$

This matrix represents the system

$$\begin{cases} x + 2y + 2z = 2 \\ -3y - 5z = -5 \\ 0 + 0 + 0 = 2 \end{cases}$$

No solution
Inconsistent system

22. Let x = amount invested at 6% and y = amount invested at 12%.

$$\begin{cases} x + y = 10{,}000 \\ 0.06x + 0.12y = 960 \end{cases}$$

Multiply the second equation by 100 to eliminate decimals.

$$\begin{cases} x + y = 10{,}000 \\ 6x + 12y = 96{,}000 \end{cases}$$

$$\left[\begin{array}{cc|c} 1 & 1 & 10{,}000 \\ 6 & 12 & 96{,}000 \end{array}\right]$$

$-6R_1 + R_2$

$$\left[\begin{array}{cc|c} 1 & 1 & 10{,}000 \\ 0 & 6 & 36{,}000 \end{array}\right]$$

$\frac{1}{6}R_2$

$$\left[\begin{array}{cc|c} 1 & 1 & 10{,}000 \\ 0 & 1 & 6{,}000 \end{array}\right]$$

This matrix represents the system

$$\begin{cases} x + y = 10{,}000 \\ y = 6{,}000 \end{cases}$$

$$x + 6{,}000 = 10{,}000$$
$$x = 4{,}000$$

$4,000 was invested at 6% and $6,000 was invested at 12%.

SECTION 12.4
Solving Systems Using Determinants

23.

$$2(3) - 3(-4) = 6 + 12$$
$$= 18$$

24.

$$-3(-6) - (-4)(5) = 18 + 20$$
$$= 38$$

25.

$$-1\begin{vmatrix} -1 & 3 \\ -2 & 2 \end{vmatrix} - 2\begin{vmatrix} 2 & 3 \\ 1 & 2 \end{vmatrix} - 1\begin{vmatrix} 2 & -1 \\ 1 & -2 \end{vmatrix}$$
$$= -1(-2+6) - 2(4-3) - 1(-4+1)$$
$$= -1(4) - 2(1) - 1(-3)$$
$$= -4 - 2 + 3$$
$$= -3$$

26.

$$3\begin{vmatrix} -2 & -2 \\ 1 & -1 \end{vmatrix} + 2\begin{vmatrix} 1 & -2 \\ 2 & -1 \end{vmatrix} + 2\begin{vmatrix} 1 & -2 \\ 2 & 1 \end{vmatrix}$$
$$= 3(2+2) + 2(-1+4) + 2(1+4)$$
$$= 3(4) + 2(3) + 2(5)$$
$$= 12 + 6 + 10$$
$$= 28$$

27.

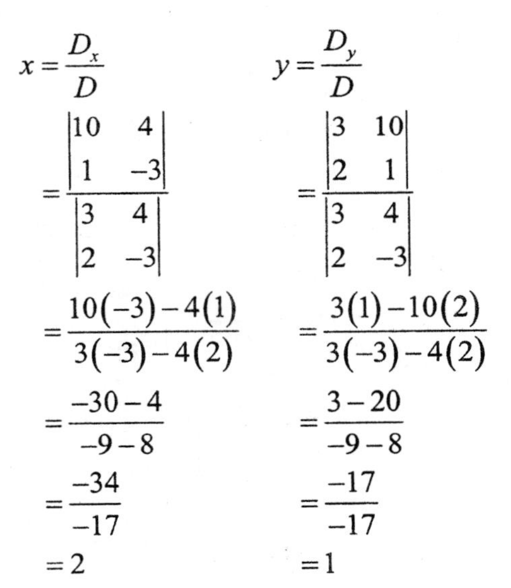

$$x = \frac{D_x}{D} = \frac{\begin{vmatrix} 10 & 4 \\ 1 & -3 \end{vmatrix}}{\begin{vmatrix} 3 & 4 \\ 2 & -3 \end{vmatrix}} = \frac{10(-3) - 4(1)}{3(-3) - 4(2)} = \frac{-30 - 4}{-9 - 8} = \frac{-34}{-17} = 2$$

$$y = \frac{D_y}{D} = \frac{\begin{vmatrix} 3 & 10 \\ 2 & 1 \end{vmatrix}}{\begin{vmatrix} 3 & 4 \\ 2 & -3 \end{vmatrix}} = \frac{3(1) - 10(2)}{3(-3) - 4(2)} = \frac{3 - 20}{-9 - 8} = \frac{-17}{-17} = 1$$

The solution is (2, 1).

28.

$$x=\frac{D_x}{D}=\frac{\begin{vmatrix}-6 & -4\\ 5 & 2\end{vmatrix}}{\begin{vmatrix}-6 & -4\\ 3 & 2\end{vmatrix}}=\frac{-6(2)-(-4)(5)}{-6(2)-(-4)(3)}=\frac{-12+20}{-12+12}=\frac{8}{0}$$

No solution; inconsistent system

29.

$$x=\frac{D_x}{D}=\frac{\begin{vmatrix}0 & 2 & 1\\ 3 & 1 & 1\\ 5 & 1 & 2\end{vmatrix}}{\begin{vmatrix}1 & 2 & 1\\ 2 & 1 & 1\\ 1 & 1 & 2\end{vmatrix}}=\frac{0\begin{vmatrix}1 & 1\\ 1 & 2\end{vmatrix}-2\begin{vmatrix}3 & 1\\ 5 & 2\end{vmatrix}+1\begin{vmatrix}3 & 1\\ 5 & 1\end{vmatrix}}{1\begin{vmatrix}1 & 1\\ 1 & 2\end{vmatrix}-2\begin{vmatrix}2 & 1\\ 1 & 2\end{vmatrix}+1\begin{vmatrix}2 & 1\\ 1 & 1\end{vmatrix}}$$

$$=\frac{0(2-1)-2(6-5)+1(3-5)}{1(2-1)-2(4-1)+1(2-1)}=\frac{0(1)-2(1)+1(-2)}{1(1)-2(3)+1(1)}=\frac{0-2-2}{1-6+1}=\frac{-4}{-4}=1$$

29. (cont.)

$$y=\frac{D_y}{D}=\frac{\begin{vmatrix}1 & 0 & 1\\ 2 & 3 & 1\\ 1 & 5 & 2\end{vmatrix}}{\begin{vmatrix}1 & 2 & 1\\ 2 & 1 & 1\\ 1 & 1 & 2\end{vmatrix}}=\frac{1\begin{vmatrix}3 & 1\\ 5 & 2\end{vmatrix}-0\begin{vmatrix}2 & 1\\ 1 & 2\end{vmatrix}+1\begin{vmatrix}2 & 3\\ 1 & 5\end{vmatrix}}{1\begin{vmatrix}1 & 1\\ 1 & 2\end{vmatrix}-2\begin{vmatrix}2 & 1\\ 1 & 2\end{vmatrix}+1\begin{vmatrix}2 & 1\\ 1 & 1\end{vmatrix}}$$

$$=\frac{1(6-5)-0(4-1)+1(10-3)}{1(2-1)-2(4-1)+1(2-1)}=\frac{1(1)-0(3)+1(7)}{1(1)-2(3)+1(1)}=\frac{1-0+7}{1-6+1}=\frac{8}{-4}=-2$$

$$z=\frac{D_z}{D}=\frac{\begin{vmatrix}1 & 2 & 0\\ 2 & 1 & 3\\ 1 & 1 & 5\end{vmatrix}}{\begin{vmatrix}1 & 2 & 1\\ 2 & 1 & 1\\ 1 & 1 & 2\end{vmatrix}}=\frac{1\begin{vmatrix}1 & 3\\ 1 & 5\end{vmatrix}-2\begin{vmatrix}2 & 3\\ 1 & 5\end{vmatrix}+0\begin{vmatrix}2 & 1\\ 1 & 1\end{vmatrix}}{1\begin{vmatrix}1 & 1\\ 1 & 2\end{vmatrix}-2\begin{vmatrix}2 & 1\\ 1 & 2\end{vmatrix}+1\begin{vmatrix}2 & 1\\ 1 & 1\end{vmatrix}}$$

$$=\frac{1(5-3)-2(10-3)+0(2-1)}{1(2-1)-2(4-1)+1(2-1)}=\frac{1(2)-2(7)+0(1)}{1(1)-2(3)+1(1)}=\frac{2-14+0}{1-6+1}=\frac{-12}{-4}=3$$

The solution is (1, –2, 3).

30.

$$x=\frac{D_x}{D}$$

$$=\frac{\begin{vmatrix}2&3&1\\7&3&2\\-7&-1&-1\end{vmatrix}}{\begin{vmatrix}2&3&1\\1&3&2\\1&-1&-1\end{vmatrix}}$$

$$=\frac{2\begin{vmatrix}3&2\\-1&-1\end{vmatrix}-3\begin{vmatrix}7&2\\-7&-1\end{vmatrix}+1\begin{vmatrix}7&3\\-7&-1\end{vmatrix}}{2\begin{vmatrix}3&2\\-1&-1\end{vmatrix}-3\begin{vmatrix}1&2\\1&-1\end{vmatrix}+1\begin{vmatrix}1&3\\1&-1\end{vmatrix}}$$

$$=\frac{2(-3+2)-3(-7+14)+1(-7+21)}{2(-3+2)-3(-1-2)+1(-1-3)}$$

$$=\frac{2(-1)-3(7)+1(14)}{2(-1)-3(-3)+1(-4)}$$

$$=\frac{-2-21+14}{-2+9-4}$$

$$=\frac{-9}{3}$$

$$=-3$$

$$y=\frac{D_y}{D}$$

$$=\frac{\begin{vmatrix}2&2&1\\1&7&2\\1&-7&-1\end{vmatrix}}{\begin{vmatrix}2&3&1\\1&3&2\\1&-1&-1\end{vmatrix}}$$

$$=\frac{2\begin{vmatrix}7&2\\-7&-1\end{vmatrix}-2\begin{vmatrix}1&2\\1&-1\end{vmatrix}+1\begin{vmatrix}1&7\\1&-7\end{vmatrix}}{2\begin{vmatrix}3&2\\-1&-1\end{vmatrix}-3\begin{vmatrix}1&2\\1&-1\end{vmatrix}+1\begin{vmatrix}1&3\\1&-1\end{vmatrix}}$$

$$=\frac{2(-7+14)-2(-1-2)+1(-7-7)}{2(-3+2)-3(-1-2)+1(-1-3)}$$

$$=\frac{2(7)-2(-3)+1(-14)}{2(-1)-3(-3)+1(-4)}$$

$$=\frac{14+6-14}{-2+9-4}$$

$$=\frac{6}{3}$$

$$=2$$

30. (cont.)

$$z=\frac{D_z}{D}$$

$$=\frac{\begin{vmatrix}2&3&2\\1&3&7\\1&-1&-7\end{vmatrix}}{\begin{vmatrix}2&3&1\\1&3&2\\1&-1&-1\end{vmatrix}}$$

$$=\frac{2\begin{vmatrix}3&7\\-1&-7\end{vmatrix}-3\begin{vmatrix}1&7\\1&-7\end{vmatrix}+2\begin{vmatrix}1&3\\1&-1\end{vmatrix}}{2\begin{vmatrix}3&2\\-1&-1\end{vmatrix}-3\begin{vmatrix}1&2\\1&-1\end{vmatrix}+1\begin{vmatrix}1&3\\1&-1\end{vmatrix}}$$

$$=\frac{2(-21+7)-3(-7-7)+2(-1-3)}{2(-3+2)-3(-1-2)+1(-1-3)}$$

$$=\frac{2(-14)-3(-14)+2(-4)}{2(-1)-3(-3)+1(-4)}$$

$$=\frac{-28+42-8}{-2+9-4}$$

$$=\frac{6}{3}$$

$$=2$$

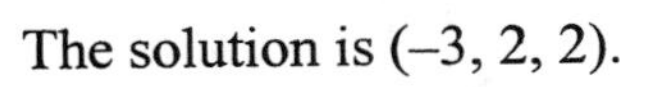
The solution is (–3, 2, 2).

31. Let x = cups of mix A, y = cups of mix B, and z = cups of mix C.

$$\begin{cases} 5x+6y+8z=24 \\ 2x+3y+3z=10 \\ x+2y+z=5 \end{cases}$$

$$x=\frac{D_x}{D}$$
$$=\frac{\begin{vmatrix} 24 & 6 & 8 \\ 10 & 3 & 3 \\ 5 & 2 & 1 \end{vmatrix}}{\begin{vmatrix} 5 & 6 & 8 \\ 2 & 3 & 3 \\ 1 & 2 & 1 \end{vmatrix}}$$
$$=\frac{24\begin{vmatrix} 3 & 3 \\ 2 & 1 \end{vmatrix}-6\begin{vmatrix} 10 & 3 \\ 5 & 1 \end{vmatrix}+8\begin{vmatrix} 10 & 3 \\ 5 & 2 \end{vmatrix}}{5\begin{vmatrix} 3 & 3 \\ 2 & 1 \end{vmatrix}-6\begin{vmatrix} 2 & 3 \\ 1 & 1 \end{vmatrix}+8\begin{vmatrix} 2 & 3 \\ 1 & 2 \end{vmatrix}}$$
$$=\frac{24(3-6)-6(10-15)+8(20-15)}{5(3-6)-6(2-3)+8(4-3)}$$
$$=\frac{24(-3)-6(-5)+8(5)}{5(-3)-6(-1)+8(1)}$$
$$=\frac{-72+30+40}{-15+6+8}$$
$$=\frac{-2}{-1}$$
$$=2$$

31. (cont.)

$$y=\frac{D_y}{D}$$
$$=\frac{\begin{vmatrix} 5 & 24 & 8 \\ 2 & 10 & 3 \\ 1 & 5 & 1 \end{vmatrix}}{\begin{vmatrix} 5 & 6 & 8 \\ 2 & 3 & 3 \\ 1 & 2 & 1 \end{vmatrix}}$$
$$=\frac{5\begin{vmatrix} 10 & 3 \\ 5 & 1 \end{vmatrix}-24\begin{vmatrix} 2 & 3 \\ 1 & 1 \end{vmatrix}+8\begin{vmatrix} 2 & 10 \\ 1 & 5 \end{vmatrix}}{5\begin{vmatrix} 3 & 3 \\ 2 & 1 \end{vmatrix}-6\begin{vmatrix} 2 & 3 \\ 1 & 1 \end{vmatrix}+8\begin{vmatrix} 2 & 3 \\ 1 & 2 \end{vmatrix}}$$
$$=\frac{5(10-15)-24(2-3)+8(10-10)}{5(3-6)-6(2-3)+8(4-3)}$$
$$=\frac{5(-5)-24(-1)+8(0)}{5(-3)-6(-1)+8(1)}$$
$$=\frac{-25+24+0}{-15+6+8}$$
$$=\frac{-1}{-1}$$
$$=1$$

$$z=\frac{D_z}{D}$$
$$=\frac{\begin{vmatrix} 5 & 6 & 24 \\ 2 & 3 & 10 \\ 1 & 2 & 5 \end{vmatrix}}{\begin{vmatrix} 5 & 6 & 8 \\ 2 & 3 & 3 \\ 1 & 2 & 1 \end{vmatrix}}$$
$$=\frac{5\begin{vmatrix} 3 & 10 \\ 2 & 5 \end{vmatrix}-6\begin{vmatrix} 2 & 10 \\ 1 & 5 \end{vmatrix}+24\begin{vmatrix} 2 & 3 \\ 1 & 2 \end{vmatrix}}{5\begin{vmatrix} 3 & 3 \\ 2 & 1 \end{vmatrix}-6\begin{vmatrix} 2 & 3 \\ 1 & 1 \end{vmatrix}+8\begin{vmatrix} 2 & 3 \\ 1 & 2 \end{vmatrix}}$$
$$=\frac{5(15-20)-6(10-10)+24(4-3)}{5(3-6)-6(2-3)+8(4-3)}$$
$$=\frac{5(-5)-6(0)+24(1)}{5(-3)-6(-1)+8(1)}$$
$$=\frac{-25-0+24}{-15+6+8}$$
$$=\frac{-1}{-1}$$
$$=1$$

2 cups of mix A, 1 cup of mix B, and 1 cup of mix C are needed.

CHAPTER 12 TEST

1. $\begin{cases} 2x + y = 5 \\ y = 2x - 3 \end{cases}$

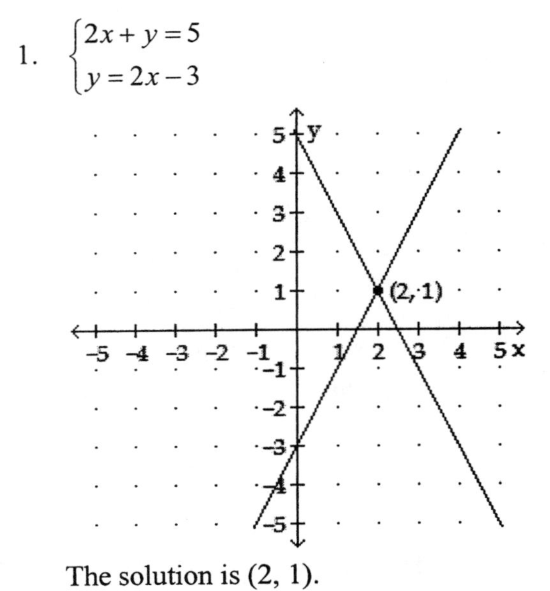

The solution is (2, 1).

2. The two lines intersect at the point (3, 1), so $x = 3$.

3.
$\begin{cases} 2x - 4y = 14 \\ x + 2y = 7 \end{cases}$

Solve the second equation for x.

$\begin{cases} 2x - 4y = 14 \\ x = -2y + 7 \end{cases}$

Substitute $x = -2y + 7$ into the first equation and solve for y.

$$2x - 4y = 14$$
$$2(-2y + 7) - 4y = 14$$
$$-4y + 14 - 4y = 14$$
$$-8y + 14 = 14$$
$$-8y = 0$$
$$y = 0$$

Substitute $y = 0$ into the second equation and solve for x.

$$x = -2y + 7$$
$$x = -2(0) + 7$$
$$x = 0 + 7$$
$$x = 7$$

The solution is $(7, 0)$.

4.
$\begin{cases} 2x + 3y = -5 \\ 3x - 2y = 12 \end{cases}$

Multiply the first equation by -3 and the second equation by 2 to eliminate x.

$$-6x - 9y = 15$$
$$\underline{6x - 4y = 24}$$
$$-13y = 39$$
$$y = -3$$

Substitute $y = -3$ into the first equation and solve for x.

$$2x + 3(-3) = -5$$
$$2x - 9 = -5$$
$$2x = 4$$
$$x = 2$$

The solution is $(2, -3)$.

5.
$\begin{cases} 3(x + y) = x - 3 \\ -y = \dfrac{2x + 3}{3} \end{cases}$

Write the equations in standard form.

$\begin{cases} 3x + 3y = x - 3 \\ -3y = 2x + 3 \end{cases}$

$\begin{cases} 2x + 3y = -3 \\ 2x + 3y = -3 \end{cases}$

The equations are the same so they are dependent.

6. No.

$$x - 2y + z = 5 \qquad\qquad 2x + 4y = -4$$
$$-1 - 2\left(-\frac{1}{2}\right) + 5 \stackrel{?}{=} 5 \qquad 2(-1) + 4\left(-\frac{1}{2}\right) \stackrel{?}{=} -4$$
$$-1 + 1 + 5 \stackrel{?}{=} 5 \qquad\qquad -2 + (-2) \stackrel{?}{=} -4$$
$$5 = 5 \qquad\qquad -4 = -4$$

$$-6y + 4z = 22$$
$$-6\left(-\frac{1}{2}\right) + 4(5) \stackrel{?}{=} 22$$
$$3 + 20 \stackrel{?}{=} 22$$
$$23 \neq 22$$

7.

$$\begin{cases} x+y+z=4 & (1) \\ x+y-z=6 & (2) \\ 2x-3y+z=-1 & (3) \end{cases}$$

Multiply Equation 1 by -1 and add Equations 1 and 2.

$$\begin{array}{rl} -x-y-z=-4 & (1) \\ \underline{x+y-z=6} & (2) \\ -2z=2 & \\ z=-1 & \end{array}$$

Multiply Equation 1 by -2 and add Equations 1 and 3 to eliminate x.

$$\begin{array}{rl} -2x-2y-2z=-8 & (1) \\ \underline{2x-3y+z=-1} & (3) \\ -5y-z=-9 & (4) \end{array}$$

Substitute $z=-1$ into Equation 4 and solve for y.

$$\begin{array}{rl} -5y-z=-9 & (4) \\ -5y-(-1)=-9 & \\ -5y+1=-9 & \\ -5y=-10 & \\ y=2 & \end{array}$$

Substitute $z=-1$ and $y=2$ into Equation 1 and solve for x.

$$\begin{array}{rl} x+y+z=4 & (1) \\ x+2-1=4 & \\ x+1=4 & \\ x=3 & \end{array}$$

The solution is $(3, 2, -1)$.

8. The sum of the angles of a triangle is 180°.

$$\begin{cases} 2x+y=180 \\ y=x+15 \end{cases}$$

Substitute $y=x+15$ into the first equation and solve for x.

$$\begin{aligned} 2x+(x+15)&=180 \\ 3x+15&=180 \\ 3x&=165 \\ x&=55 \end{aligned}$$

Substitute $x=55$ into the second equation and solve for y.

$$\begin{aligned} y&=55+15 \\ y&=70 \end{aligned}$$

The measure of the angles are 55° and 70°.

9. Let x = gallons of 40% solution and y = gallons of 80% solution.

$$\begin{cases} x+y=20 \\ 0.40x+0.80y=0.50(20) \end{cases}$$

Multiply the second equation by 100 to eliminate decimals.

$$\begin{cases} x+y=20 \\ 40x+80y=50(20) \end{cases}$$

$$\begin{cases} x+y=20 \\ 40x+80y=1,000 \end{cases}$$

Multiply the first equation by -40 and add the equations to eliminate x.

$$\begin{array}{r} -40x-40y=-800 \\ \underline{40x+80y=1,000} \\ 40y=200 \\ y=5 \end{array}$$

Substitute $y=5$ into the first equation and solve for x.

$$\begin{aligned} x+5&=20 \\ x&=15 \end{aligned}$$

15 gallons of 40% solution and 5 gallons of 80% solution are needed.

10. Let x = number of impressions.

Total Cost = setup + cost per impression

$$\begin{cases} C_1 = 1{,}775 + 5.75x \\ C_2 = 3{,}975 + 4.15x \end{cases}$$

Breakpoint:

$$\begin{aligned} C_1 &= C_2 \\ 1{,}775 + 5.75x &= 3{,}975 + 4.15x \\ 1{,}775 + 5.75x - 4.15x &= 3{,}975 + 4.15x - 4.15x \\ 1{,}775 + 1.6x &= 3{,}975 \\ 1{,}775 + 1.6x - 1{,}775 &= 3{,}975 - 1{,}775 \\ 1.6x &= 2{,}200 \\ x &= 1{,}375 \text{ impressions} \end{aligned}$$

11.

$$\left[\begin{array}{cc|c} 1 & 1 & 4 \\ 2 & -1 & 2 \end{array}\right]$$

$-2R_1 + R_2$

$$\left[\begin{array}{cc|c} 1 & 1 & 4 \\ 0 & -3 & -6 \end{array}\right]$$

$-\frac{1}{3}R_2$

$$\left[\begin{array}{cc|c} 1 & 1 & 4 \\ 0 & 1 & 2 \end{array}\right]$$

This matrix represents the system

$$\begin{cases} x + y = 4 \\ y = 2 \end{cases}$$

$$\begin{aligned} x + 2 &= 4 \\ x &= 2 \end{aligned}$$

The solution is $(2, 2)$.

12.

$$\left[\begin{array}{ccc|c} 2 & 1 & -1 & 1 \\ 1 & 2 & 2 & 2 \\ 4 & 5 & 3 & 3 \end{array}\right]$$

$R_1 \leftrightarrow R_2$

$$\left[\begin{array}{ccc|c} 1 & 2 & 2 & 2 \\ 2 & 1 & -1 & 1 \\ 4 & 5 & 3 & 3 \end{array}\right]$$

$-2R_1 + R_2$

$$\left[\begin{array}{ccc|c} 1 & 2 & 2 & 2 \\ 0 & -3 & -5 & -3 \\ 4 & 5 & 3 & 3 \end{array}\right]$$

$-4R_1 + R_3$

$$\left[\begin{array}{ccc|c} 1 & 2 & 2 & 2 \\ 0 & -3 & -5 & -3 \\ 0 & -3 & -5 & -5 \end{array}\right]$$

$-R_2 + R_3$

$$\left[\begin{array}{ccc|c} 1 & 2 & 2 & 2 \\ 0 & -3 & -5 & -3 \\ 0 & 0 & 0 & -2 \end{array}\right]$$

This matrix represents the system

$$\begin{cases} x + 2y + 2z = 2 \\ -3y - 5z = -3 \\ 0 + 0 + 0 = -2 \end{cases}$$

No solution

Inconsistent system

13.

$$\begin{aligned} 2(5) - (-3)(4) &= 10 + 12 \\ &= 22 \end{aligned}$$

14.

$$\begin{aligned} &1\begin{vmatrix} 0 & 3 \\ -2 & 2 \end{vmatrix} - 2\begin{vmatrix} 2 & 3 \\ 1 & 2 \end{vmatrix} + 0\begin{vmatrix} 2 & 0 \\ 1 & -2 \end{vmatrix} \\ &= 1(0+6) - 2(4-3) + 0(-4-0) \\ &= 1(6) - 2(1) + 0(-4) \\ &= 6 - 2 + 0 \\ &= 4 \end{aligned}$$

15. a. $\begin{vmatrix} -6 & -1 \\ -6 & 1 \end{vmatrix}$

b. $\begin{vmatrix} 1 & -1 \\ 3 & 1 \end{vmatrix}$

16.

$$x = \frac{D_x}{D}$$
$$= \frac{\begin{vmatrix} -6 & -1 \\ -6 & 1 \end{vmatrix}}{\begin{vmatrix} 1 & -1 \\ 3 & 1 \end{vmatrix}}$$
$$= \frac{-6(1) - (-1)(-6)}{1(1) - (-1)(3)}$$
$$= \frac{-6 - 6}{1 + 3}$$
$$= \frac{-12}{4}$$
$$= -3$$

17.

$$y = \frac{D_y}{D}$$
$$= \frac{\begin{vmatrix} 1 & -6 \\ 3 & -6 \end{vmatrix}}{\begin{vmatrix} 1 & -1 \\ 3 & 1 \end{vmatrix}}$$
$$= \frac{1(-6) - (-6)(3)}{1(1) - (-1)(3)}$$
$$= \frac{-6 + 18}{1 + 3}$$
$$= \frac{12}{4}$$
$$= 3$$

18.

$$z = \frac{D_z}{D}$$
$$= \frac{\begin{vmatrix} 1 & 1 & 4 \\ 1 & 1 & 6 \\ 2 & -3 & -1 \end{vmatrix}}{\begin{vmatrix} 1 & 1 & 1 \\ 1 & 1 & -1 \\ 2 & -3 & 1 \end{vmatrix}}$$
$$= \frac{1\begin{vmatrix} 1 & 6 \\ -3 & -1 \end{vmatrix} - 1\begin{vmatrix} 1 & 6 \\ 2 & -1 \end{vmatrix} + 4\begin{vmatrix} 1 & 1 \\ 2 & -3 \end{vmatrix}}{1\begin{vmatrix} 1 & -1 \\ -3 & 1 \end{vmatrix} - 1\begin{vmatrix} 1 & -1 \\ 2 & 1 \end{vmatrix} + 1\begin{vmatrix} 1 & 1 \\ 2 & -3 \end{vmatrix}}$$
$$= \frac{1(-1+18) - 1(-1-12) + 4(-3-2)}{1(1-3) - 1(1+2) + 1(-3-2)}$$
$$= \frac{1(17) - 1(-13) + 4(-5)}{1(-2) - 1(3) + 1(-5)}$$
$$= \frac{17 + 13 - 20}{-2 - 3 - 5}$$
$$= \frac{10}{-10}$$
$$= -1$$

19. Let x = children's tickets sold, y = general admission tickets, and z = seniors tickets.

$$\begin{cases} x+y+z=100 \\ 3x+6y+5z=410 \\ x=2y \end{cases}$$

Write the third equation in standard form.

$$\begin{cases} x+y+z=100 & (1) \\ 3x+6y+5z=410 & (2) \\ x-2y=0 & (3) \end{cases}$$

Multiply Equation 1 by -5 to eliminate z and add Equations 1 and 2.

$$\begin{aligned} -5x-5y-5z&=-500 && (1) \\ 3x+6y+5z&=410 && (2) \\ \hline -2x+y&=-90 && (4) \end{aligned}$$

Multiply Equation 3 by 2 and add Equations 3 and 4 to eliminate x.

$$\begin{aligned} 2x-4y&=0 && (3) \\ -2x+y&=-90 && (4) \\ \hline -3y&=-90 \\ y&=30 \end{aligned}$$

Substitute $y=30$ into Equation 3 and solve for x.

$$\begin{aligned} x-2y&=0 && (3) \\ x-2(30)&=0 \\ x-60&=0 \\ x&=60 \end{aligned}$$

Substitute $y=30$ and $x=60$ into Equation 1 and solve for z.

$$\begin{aligned} x+y+z&=100 && (1) \\ 60+30+z&=100 \\ 90+z&=100 \\ z&=10 \end{aligned}$$

They sold 60 children's tickets, 30 general admission tickets, and 10 seniors tickets.

20. Answers will vary.

21. The system has no solution.

22. Answers will vary.
(2035, 25); in the year 2035, the percent of the U.S. population that is children under age 18 and the percent of the U.S. population that is adults 65 and older will be the same, about 25%.

CHAPTER 12 CUMULATIVE REVIEW

1. a. true
   b. true
   c. true

2.
$$-4+2[-7-3(-9)]=-4+2[-7+27]$$
$$=-4+2[20]$$
$$=-4+40$$
$$=36$$

3. Let x = # of employees in early 2000.
5,200 is 5.3% of # of employees
$$5{,}200=0.053 \cdot x$$
$$\frac{5{,}200}{0.053}=\frac{0.053 \cdot x}{0.053}$$
$$98{,}113 \approx x$$
Xerox had approximately 98,113 employees.

4.
$$\frac{5}{6}k=10$$
$$\frac{6}{5}\left(\frac{5}{6}k\right)=\frac{6}{5}(10)$$
$$k=12$$

5.
$$-(3a+1)+a=2$$
$$-3a-1+a=2$$
$$-2a-1=2$$
$$-2a=3$$
$$a=-\frac{3}{2}$$

6.
$$\frac{2z+3}{3}+\frac{3z-4}{6}=\frac{z-2}{2}$$
$$6\left(\frac{2z+3}{3}+\frac{3z-4}{6}\right)=6\left(\frac{z-2}{2}\right)$$
$$2(2z+3)+(3z-4)=3(z-2)$$
$$4z+6+3z-4=3z-6$$
$$7z+2=3z-6$$
$$4z+2=-6$$
$$4z=-8$$
$$z=-2$$

7.
$$5x+7<2x+1$$
$$3x+7<1$$
$$3x<-6$$
$$x<-2$$
$$(-\infty,-2)$$

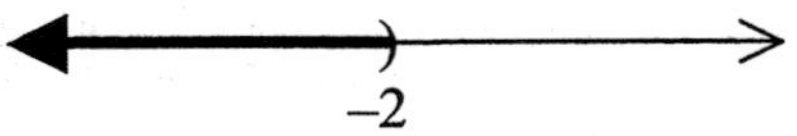

−2

8.

	Principal	· rate	· time	= Interest
7%	x	0.07	1	$0.07x$
10%	$28{,}000-x$	0.10	1	$0.10(28{,}000-x)$
combined interest				2,560

7% interest + 10% interest = combined interest
$$0.07x+0.10(28{,}000-x)=2{,}560$$
$$0.07x+2{,}800-0.10x=2{,}560$$
$$-0.03x+2{,}800=2{,}560$$
$$-0.03x=-240$$
$$x=8{,}000$$

$$28{,}000-x=28{,}000-8{,}000$$
$$=20{,}000$$
They invested \$8,000 at 7% and \$20,000 at 10%.

9. $y=-3$

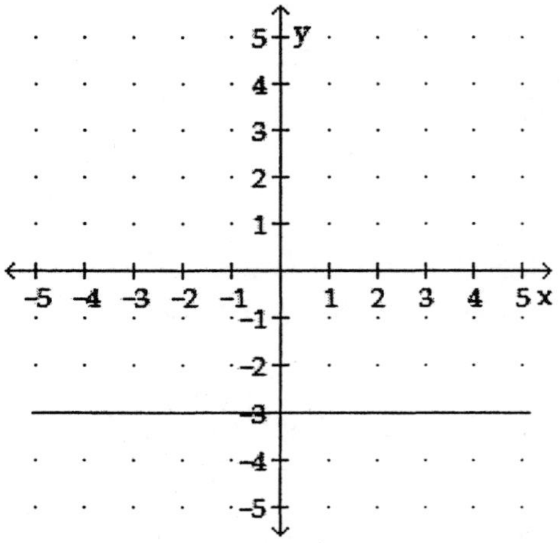

y
5 4 3 2 1 −1 −2 −3 −4 −5
−5 −4 −3 −2 −1 1 2 3 4 5 x

10. $(-2, -1)$ and $m = \frac{4}{3}$

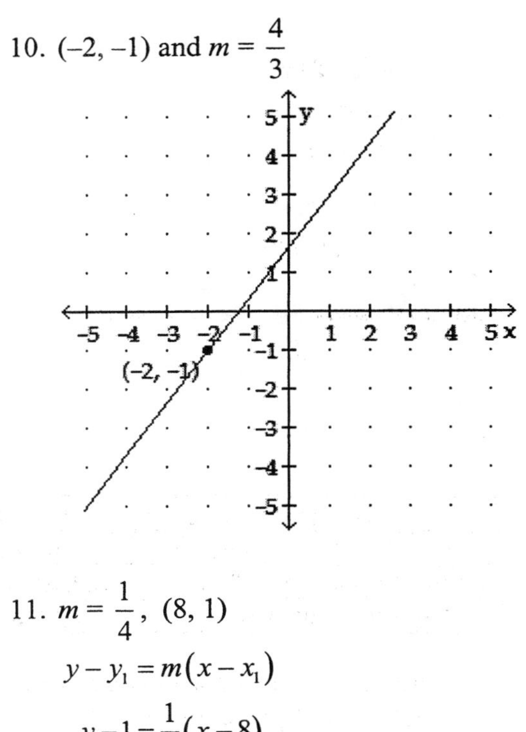

11. $m = \frac{1}{4}$, $(8, 1)$

$$y - y_1 = m(x - x_1)$$
$$y - 1 = \frac{1}{4}(x - 8)$$
$$y - 1 = \frac{1}{4}x - 2$$
$$y = \frac{1}{4}x - 1$$

12. Find the rate of change (slope) using the ordered pairs (1995, 418) and (2002, 530).

$$m = \frac{y_2 - y_1}{x_2 - x_1}$$
$$= \frac{530 - 418}{2002 - 1995}$$
$$= \frac{112}{7}$$
$$= 16$$

It was an increase of \$16 per year.

13.

$$y^3\left(y^2y^4\right)^3 = y^3\left(y^{2+4}\right)^3$$
$$= y^3\left(y^6\right)^3$$
$$= y^3\left(y^{6\cdot 3}\right)$$
$$= y^3\left(y^{18}\right)$$
$$= y^{3+18}$$
$$= y^{21}$$

14.

$$\left(\frac{21x^{-2}y^2z^{-2}}{7x^3y^{-1}}\right)^{-2} = \left(\frac{7x^3y^{-1}}{21x^{-2}y^2z^{-2}}\right)^2$$
$$= \left(\frac{7}{21}x^{3-(-2)}y^{-1-2}z^{0-(-2)}\right)^2$$
$$= \left(\frac{1}{3}x^5y^{-3}z^2\right)^2$$
$$= \left(\frac{x^5z^2}{3y^3}\right)^2$$
$$= \frac{x^{5\cdot 2}z^{2\cdot 2}}{3^2y^{3\cdot 2}}$$
$$= \frac{x^{10}z^4}{9y^6}$$

15. 2,600,000 to 1

16. 7.3×10^{-4}

17.

$$4\left(4x^3 + 2x^2 - 3x - 8\right) - 5\left(2x^3 - 3x + 8\right)$$
$$= 16x^3 + 8x^2 - 12x - 32 - 10x^3 + 15x - 40$$
$$= 6x^3 + 8x^2 + 3x - 72$$

18.

$$\left(-2a^3\right)\left(3a^2\right) = -6a^{3+2}$$
$$= -6a^5$$

19.

$$(2b - 1)(3b + 4) = 6b^2 + 8b - 3b - 4$$
$$= 6b^2 + 5b - 4$$

20.

$$(2x + 5y)^2 = (2x + 5y)(2x + 5y)$$
$$= 4x^2 + 10xy + 10xy + 25y^2$$
$$= 4x^2 + 20xy + 25y^2$$

21.

$$(3x+y)(2x^2-3xy+y^2)$$
$$=3x(2x^2-3xy+y^2)+y(2x^2-3xy+y^2)$$
$$=6x^3-9x^2y+3xy^2+2x^2y-3xy^2+y^3$$
$$=6x^3-7x^2y+y^3$$

22. Write the problem in descending order before dividing.

$$\begin{array}{r} 2x+1 \\ x-3\overline{)2x^2-5x-3} \\ \underline{2x^2-6x} \\ x-3 \\ \underline{x-3} \\ 0 \end{array}$$

The answer is $2x + 1$.

23.

$$6a^2-12a^3b+36ab=6a(a-2a^2b+6b)$$

24.

$$2x+2y+ax+ay=(2x+2y)+(ax+ay)$$
$$=2(x+y)+a(x+y)$$
$$=(x+y)(2+a)$$

25.

$$b^3+125=(b)^3+(5)^3$$
$$=(b+5)(b^2-5b+25)$$

26.

$$t^4-16=(t^2+4)(t^2-4)$$
$$=(t^2+4)(t+2)(t-2)$$

27.

$$3x^2+8x=0$$
$$x(3x+8)=0$$
$$x=0 \quad \text{or} \quad 3x+8=0$$
$$3x=-8$$
$$x=-\frac{8}{3}$$

28.

$$15x^2-2=7x$$
$$15x^2-7x-2=0$$
$$(3x-2)(5x+1)=0$$
$$3x-2=0 \quad \text{or} \quad 5x+1=0$$
$$3x=2 \qquad 5x=-1$$
$$x=\frac{2}{3} \qquad x=-\frac{1}{5}$$

29.

$$P=2L+2W$$
$$=2(x^3+3x)+2(2x^3-x)$$
$$=2x^3+6x+4x^3-2x$$
$$=6x^3+4x$$

30.

$$A=\frac{1}{2}bh$$
$$22.5=\frac{1}{2}(x+4)(x)$$
$$2(22.5)=2\left(\frac{1}{2}\right)(x+4)(x)$$
$$45=x(x+4)$$
$$45=x^2+4x$$
$$0=x^2+4x-45$$
$$0=(x+9)(x-5)$$
$$x+9=0 \quad \text{or} \quad x-5=0$$
$$x=\cancel{-9} \qquad x=5$$

Since the height cannot be negative, the height of the triangle is 5 inches.

31.

$$\frac{x^2-x-6}{2x^2+9x+10}\div\frac{x^2-25}{2x^2+15x+25}$$
$$=\frac{x^2-x-6}{2x^2+9x+10}\bullet\frac{2x^2+15x+25}{x^2-25}$$
$$=\frac{(x-3)\cancel{(x+2)}}{\cancel{(2x+5)}\cancel{(x+2)}}\bullet\frac{\cancel{(2x+5)}\cancel{(x+5)}}{\cancel{(x+5)}(x-5)}$$
$$=\frac{x-3}{x-5}$$

32.

$$\frac{x+5}{xy}-\frac{x-1}{x^2y}=\frac{x+5}{xy}\left(\frac{x}{x}\right)-\frac{x-1}{x^2y}$$
$$=\frac{x^2+5x}{x^2y}-\frac{x-1}{x^2y}$$
$$=\frac{x^2+5x-(x-1)}{x^2y}$$
$$=\frac{x^2+5x-x+1}{x^2y}$$
$$=\frac{x^2+4x+1}{x^2y}$$

33.

$$\frac{3x^2-27}{x^2+3x-18}=\frac{3(x^2-9)}{x^2+3x-18}$$
$$=\frac{3(x+3)\cancel{(x-3)}}{(x+6)\cancel{(x-3)}}$$
$$=\frac{3(x+3)}{x+6}$$

34.

$$\frac{\frac{5}{y}+\frac{4}{y+1}}{\frac{4}{y}-\frac{5}{y+1}}=\frac{\frac{5}{y}+\frac{4}{y+1}}{\frac{4}{y}-\frac{5}{y+1}}\bullet\frac{y(y+1)}{y(y+1)}$$
$$=\frac{\frac{5y(y+1)}{y}+\frac{4y(y+1)}{y+1}}{\frac{4y(y+1)}{y}-\frac{5y(y+1)}{y+1}}$$
$$=\frac{5(y+1)+4y}{4(y+1)-5y}$$
$$=\frac{5y+5+4y}{4y+4-5y}$$
$$=\frac{9y+5}{-y+4}$$
$$=\frac{9y+5}{4-y}$$

35.

$$\frac{7}{q^2-q-2}+\frac{1}{q+1}=\frac{3}{q-2}$$
$$\frac{7}{(q+1)(q-2)}+\frac{1}{q+1}=\frac{3}{q-2}$$
$$(q+1)(q-2)\left(\frac{7}{(q+1)(q-2)}+\frac{1}{q+1}\right)=(q+1)(q-2)\left(\frac{3}{q-2}\right)$$
$$7+(q-2)=3(q+1)$$
$$5+q=3q+3$$
$$5-2q=3$$
$$-2q=-2$$
$$q=1$$

36. Let x = # days it will take them to roof the house together.

Homeowner's work in 1 day = $\frac{1}{7}$

Roofer's work in 1 day = $\frac{1}{4}$

Work together in 1 day = $\frac{1}{x}$

Owner's work + roofer's work = work together

$$\frac{1}{7}+\frac{1}{4}=\frac{1}{x}$$
$$28x\left(\frac{1}{7}+\frac{1}{4}\right)=28x\left(\frac{1}{x}\right)$$
$$4x+7x=28$$
$$11x=28$$
$$x=\frac{28}{11}$$
$$x=2\frac{6}{11}$$

It will take $2\frac{6}{11}$ days to roof the house working together.

37. Let x = days to lose 25 pounds.

$$\frac{10 \text{ pounds}}{350 \text{ days}}=\frac{25 \text{ pounds}}{x \text{ days}}$$
$$10(x)=25(350)$$
$$10x=8,750$$
$$x=875$$

38.

$$\frac{6}{4}=\frac{x}{26}$$
$$6(26)=x(4)$$
$$156=4x$$
$$39=x$$

39. $\begin{cases} x+y=1 \\ y=x+5 \end{cases}$

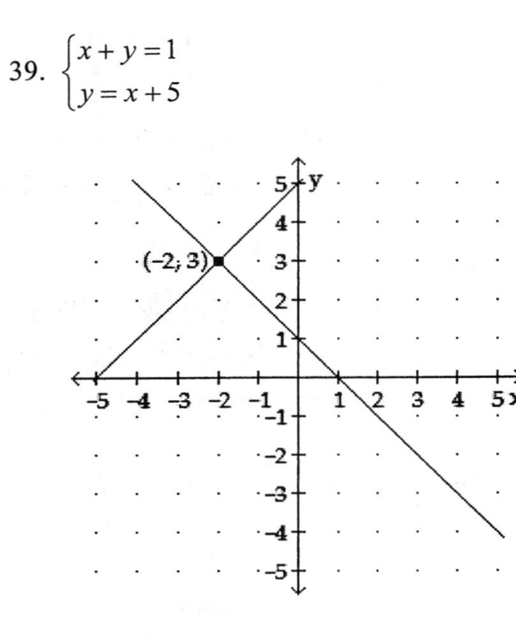

The solution is (–2, 3).

40.

$$\begin{cases} y=2x+5 \\ x+2y=-5 \end{cases}$$

Substitute $y=2x+5$ into the second equation and solve for x.

$$x+2y=-5$$
$$x+2(2x+5)=-5$$
$$x+4x+10=-5$$
$$5x+10=-5$$
$$5x=-15$$
$$x=-3$$

Substitute $x=-3$ into the first equation and solve for y.

$$y=2x+5$$
$$y=2(-3)+5$$
$$y=-6+5$$
$$y=-1$$

The solution is (–3, –1).

41.

$$\begin{cases} \frac{3}{5}s+\frac{4}{5}t=1 \\ -\frac{1}{4}s+\frac{3}{8}t=1 \end{cases}$$

Multiply each equation by the LCD to eliminate fractions.

$$\begin{cases} 5\left(\frac{3}{5}s+\frac{4}{5}t\right)=5(1) \\ 8\left(-\frac{1}{4}s+\frac{3}{8}t\right)=8(1) \end{cases}$$

$$\begin{cases} 3s+4t=5 \\ -2s+3t=8 \end{cases}$$

Multiply the first equation by 2 and the second equation by 3 to create opposites, and then add the two equations.

$$6s+8t=10$$
$$\underline{-6s+9t=24}$$
$$17t=34$$
$$t=2$$

Substitute $t=2$ into the first equation and solve for s.

$$3s+4t=5$$
$$3s+4(2)=5$$
$$3s+8=5$$
$$3s=-3$$
$$s=-1$$

The solution is (–1, 2).

42.

	Sweet-n-Sour	Bit-O-Honey	Mixture
# of pounds	x	y	48
Value	$3x$	$4y$	4(48) = 192

$$\begin{cases} x+y=48 \\ 3x+6y=192 \end{cases}$$

Multiply the first equation by -3, and add the equations to eliminate x.

$$-3x-3y=-144$$
$$\underline{3x+6y=192}$$
$$3y=48$$
$$y=16$$

Substitute $y=16$ into the first equation and solve for x.

$$x+y=48$$
$$x+16=48$$
$$x=32$$

32 pounds of Sweet-n-Sour candy and 16 pounds of Bit-O-Honey candy are needed.

43.

$$\left|\frac{x-2}{3}\right|-4\le 0$$
$$\left|\frac{x-2}{3}\right|\le 4$$
$$-4\le\frac{x-2}{3}\le 4$$
$$3(-4)\le 3\left(\frac{x-2}{3}\right)\le 3(4)$$
$$-12\le x-2\le 12$$
$$-12+2\le x-2+2\le 12+2$$
$$-10\le x\le 14$$
$$[-10, 14]$$

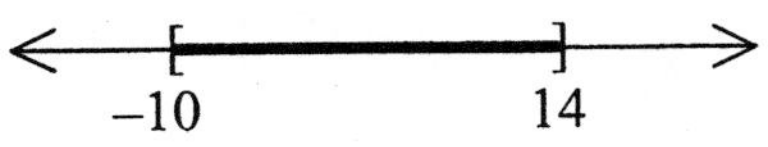

44.

$$3x+2<8 \quad\text{or}\quad 2x-3>11$$
$$3x<6 \qquad 2x>14$$
$$x<2 \qquad x>7$$
$$(-\infty,\ 2)\cup(7,\ \infty)$$

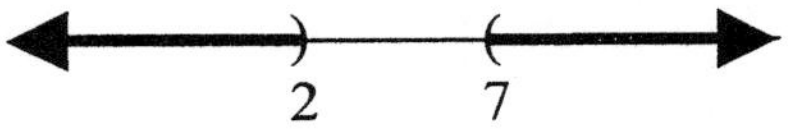

45.

$$x^2+4x+4-y^2=(x+2)^2-y^2$$
$$=(x+2-y)(x+2+y)$$

46.

$$b^4-17b^2+16=(b^2-1)(b^2-16)$$
$$=(b+1)(b-1)(b+4)(b-4)$$

47. $\begin{cases} 3x+4y\ge -7 \\ 2x-3y\ge 1 \end{cases}$

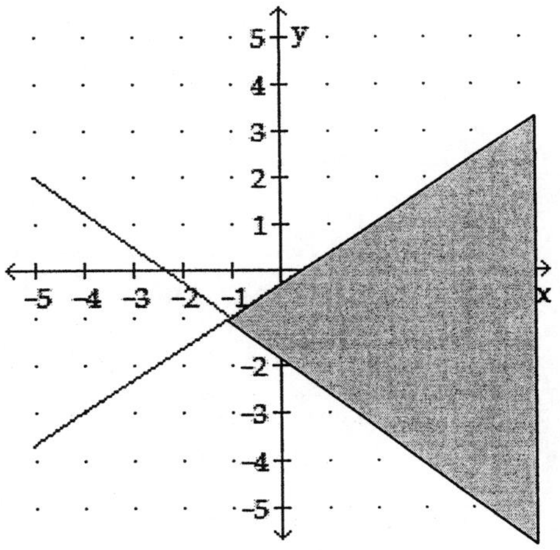

48. $f(x)=3x^2+3x-8$

$$f(-1)=3(-1)^2+3(-1)-8$$
$$=3(1)+3(-1)-8$$
$$=3-3-8$$
$$=-8$$

49. a. $f(-3)=4$

b. $x=-1$ and $x=-5$

50. Yes, it passes the vertical line test.

51. Let s = speed of the gears and t = # of teeth.
Find k when $s = 3$ and $t = 10$

$$s = \frac{k}{t}$$
$$3 = \frac{k}{10}$$
$$10(3) = 10\left(\frac{k}{10}\right)$$
$$30 = k$$

Find s when $k = 30$ and $t = 25$.

$$s = \frac{k}{t}$$
$$s = \frac{30}{25}$$
$$s = 1.2$$

52. $f(x) = (x-1)^3$

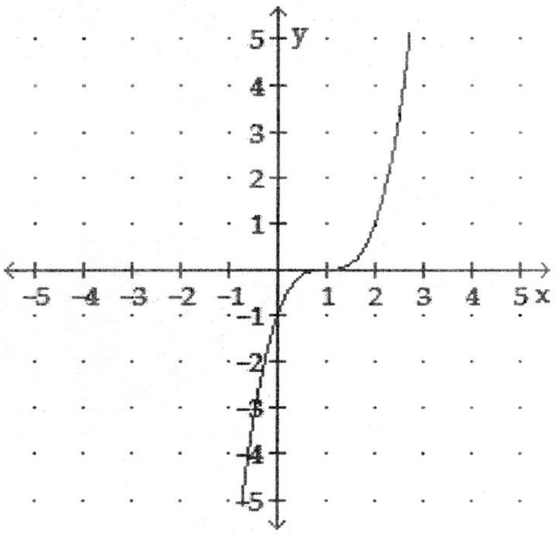

$D:(-\infty,\infty);\ R:(-\infty,\infty)$

53.

$$\begin{array}{r|rrrr} 3 & 3 & -10 & 5 & -6 \\ & & 9 & -3 & 6 \\ \hline & 3 & -1 & 2 & 0 \end{array}$$

The solution is $3x^2 - x + 2$.

54.

$$\sqrt{50x^2} = \sqrt{25x^2}\sqrt{2} = 5x\sqrt{2}$$

55.

$$\sqrt{100a^6b^4} = 10a^3b^2$$

56.

$$\begin{aligned} 3\sqrt{24} + \sqrt{54} &= 3\sqrt{4}\sqrt{6} + \sqrt{9}\sqrt{6} \\ &= 3(2)\sqrt{6} + 3\sqrt{6} \\ &= 6\sqrt{6} + 3\sqrt{6} \\ &= 9\sqrt{6} \end{aligned}$$

57.

$$\begin{aligned} \sqrt{\frac{72x^3}{y^2}} &= \frac{\sqrt{72x^3}}{\sqrt{y^2}} \\ &= \frac{\sqrt{36x^2}\sqrt{2x}}{\sqrt{y^2}} \\ &= \frac{6x\sqrt{2x}}{y} \end{aligned}$$

58.

$$\begin{aligned} &\sqrt[5]{x^6y^2} + \sqrt[5]{32x^6y^2} + \sqrt[5]{x^6y^2} \\ &= \sqrt[5]{x^5}\sqrt[5]{xy^2} + \sqrt[5]{32x^5}\sqrt[5]{xy^2} + \sqrt[5]{x^5}\sqrt[5]{xy^2} \\ &= x\sqrt[5]{xy^2} + 2x\sqrt[5]{xy^2} + x\sqrt[5]{xy^2} \\ &= 4x\sqrt[5]{xy^2} \end{aligned}$$

59.

$$\begin{aligned} \sqrt[3]{\frac{27m^3}{8n^6}} &= \frac{\sqrt[3]{27m^3}}{\sqrt[3]{8n^6}} \\ &= \frac{3m}{2n^2} \end{aligned}$$

60.

$$(-8)^{-4/3} = \left(-\frac{1}{8}\right)^{4/3}$$
$$= \left(\sqrt[3]{-\frac{1}{8}}\right)^4$$
$$= \left(-\frac{1}{2}\right)^4$$
$$= \frac{1}{16}$$

61.

$$\frac{2}{\sqrt[3]{a}} = \frac{2}{\sqrt[3]{a}} \bullet \frac{\sqrt[3]{a^2}}{\sqrt[3]{a^2}}$$
$$= \frac{2\sqrt[3]{a^2}}{\sqrt[3]{a^3}}$$
$$= \frac{2\sqrt[3]{a^2}}{a}$$

62.

$$\frac{\sqrt{x}-\sqrt{y}}{\sqrt{x}+\sqrt{y}} = \frac{\sqrt{x}-\sqrt{y}}{\sqrt{x}+\sqrt{y}}\left(\frac{\sqrt{x}-\sqrt{y}}{\sqrt{x}-\sqrt{y}}\right)$$
$$= \frac{\sqrt{x^2}-\sqrt{xy}-\sqrt{xy}+\sqrt{y^2}}{\sqrt{x^2}-\sqrt{y^2}}$$
$$= \frac{x-2\sqrt{xy}+y}{x-y}$$

63.

$$2+\sqrt{u} = \sqrt{2u+7}$$
$$\left(2+\sqrt{u}\right)^2 = \left(\sqrt{2u+7}\right)^2$$
$$4+2\sqrt{u}+2\sqrt{u}+\sqrt{u^2} = 2u+7$$
$$4+4\sqrt{u}+u = 2u+7$$
$$4+4\sqrt{u}+u-4-u = 2u+7-4-u$$
$$4\sqrt{u} = u+3$$
$$\left(4\sqrt{u}\right)^2 = (u+3)^2$$
$$16u = u^2+6u+9$$
$$0 = u^2-10u+9$$
$$0 = (u-1)(u-9)$$
$$u-1=0 \quad \text{or} \quad u-9=0$$
$$u=1 \qquad u=9$$

64.

$$x^2+8x+12=0$$
$$x^2+8x=-12$$
$$x^2+8x+16=-12+16$$
$$(x+4)^2=4$$
$$\sqrt{(x+4)^2}=\pm\sqrt{4}$$
$$x+4=2 \quad \text{or} \quad x+4=-2$$
$$x=-2 \qquad x=-6$$

65. Use the Pythagorean Theorem to find x.

$$x^2+x^2=15^2$$
$$2x^2=225$$
$$x^2=112.5$$
$$x=\sqrt{112.5}$$
$$x\approx 10.6$$

The height of the entire storage arrangement is $x + x = 2x$.

$$2x \approx 2(10.6)$$
$$\approx 21.2$$

The height of the entire storage arrangement is about 21.2 inches.

66. $4x^2-x-2=0$

$a=4$, $b=-1$, $c=-2$

$$x=\frac{-b\pm\sqrt{b^2-4ac}}{2a}$$
$$=\frac{-(-1)\pm\sqrt{(-1)^2-4(4)(-2)}}{2(4)}$$
$$=\frac{1\pm\sqrt{1+32}}{8}$$
$$=\frac{1\pm\sqrt{33}}{8}$$
$$x=\frac{1+\sqrt{33}}{8} \quad \text{or} \quad x=\frac{1-\sqrt{33}}{8}$$
$$x\approx 0.84 \qquad x\approx -0.59$$

67.

$$\sqrt{-49}=7i$$

68.

$$\sqrt{-54} = \sqrt{-9}\sqrt{6}$$
$$= 3i\sqrt{6}$$

69.

$$(2+3i)-(1-2i) = 2+3i-1+2i$$
$$= 1+5i$$

70.

$$(7-4i)+(9+2i) = 16-2i$$

71.

$$(3-2i)(4-3i) = 12-9i-8i+6i^2$$
$$= 12-17i+6(-1)$$
$$= 12-17i-6$$
$$= 6-17i$$

72.

$$\frac{3-i}{2+i} = \frac{3-i}{2+i} \bullet \frac{2-i}{2-i}$$
$$= \frac{6-3i-2i+i^2}{4-i^2}$$
$$= \frac{6-5i+(-1)}{4-(-1)}$$
$$= \frac{5-5i}{5}$$
$$= \frac{5}{5} - \frac{5}{5}i$$
$$= 1-i$$

73.

$$x^2 + 16 = 0$$
$$x^2 = -16$$
$$x = \pm\sqrt{-16}$$
$$x = \pm 4i$$
$$x = 0 \pm 4i$$

74.

$$x^2 - 4x = -5$$
$$x^2 - 4x + 5 = 0$$
$$a = 1, b = -4, \ c = 5$$
$$x = \frac{-b \pm \sqrt{b^2 - 4ac}}{2a}$$
$$= \frac{-(-4) \pm \sqrt{(-4)^2 - 4(1)(5)}}{2(1)}$$
$$= \frac{4 \pm \sqrt{16-20}}{2}$$
$$= \frac{4 \pm \sqrt{-4}}{2}$$
$$= \frac{4 \pm 2i}{2}$$
$$= \frac{4}{2} \pm \frac{2}{2}i$$
$$x = 2 \pm i$$

75. $y = 2x^2 + 8x + 6$

Step 1: Since $a = 2 > 0$, the parabola opens upward.

Step 2: Find the vertex and axis of symmetry.

$$x = \frac{-b}{2a} = \frac{-8}{2(2)} = -\frac{8}{4} = -2$$

$$y = 2(-2)^2 + 8(-2) + 6$$
$$= 8 - 16 + 6$$
$$= -2$$

vertex: $(-2,\ -2)$

The axis of symmetry is $x = -2$.

Step 3: Find the x– and y–intercepts. Since $c = 6$, the y–intercept is (0, 6). To find the x–intercepts, let $y = 0$ and solve the equation for x.

$$0 = 2x^2 + 8x + 6$$
$$0 = 2(x^2 + 4x + 3)$$
$$0 = (x+3)(x+1)$$
$$x + 3 = 0 \quad \text{or} \quad x + 1 = 0$$
$$x = -3 \qquad\qquad x = -1$$

x–intercepts: (–3, 0) and (–1, 0)
y–intercept: (0, 6)

Step 4: Use symmetry to find the point (–4, 6).

Step 5: Plot the points and draw the parabola.

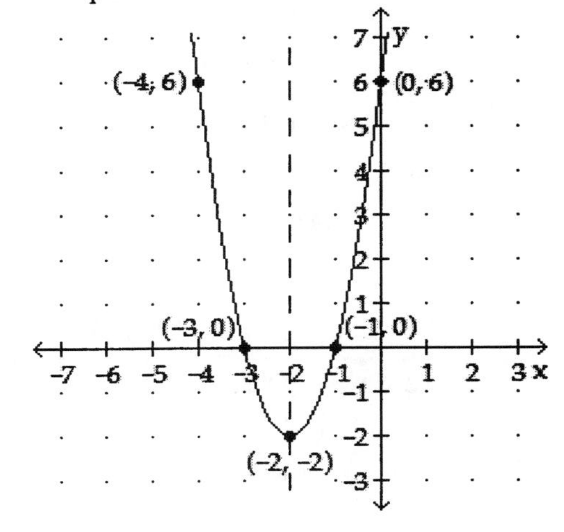

76. Let $x = a^{1/3}$.

$$a^{2/3} + a^{1/3} - 6 = 0$$
$$\left(a^{1/3}\right)^2 + a^{1/3} - 6 = 0$$
$$x^2 + x - 6 = 0$$
$$(x+3)(x-2) = 0$$
$$x + 3 = 0 \quad \text{or} \quad x - 2 = 0$$
$$x = -3 \qquad x = 2$$
$$a^{1/3} = -3 \qquad a^{1/3} = 2$$
$$\left(a^{1/3}\right)^3 = (-3)^3 \qquad \left(a^{1/3}\right)^3 = (2)^3$$
$$a = -27 \qquad a = 8$$

77.

$$f(x) = -\frac{3}{2}x + 3$$
$$y = -\frac{3}{2}x + 3$$
$$x = -\frac{3}{2}y + 3$$
$$x - 3 = -\frac{3}{2}y$$
$$-\frac{2}{3}(x-3) = -\frac{2}{3}\left(-\frac{3}{2}y\right)$$
$$-\frac{2}{3}x + 2 = y$$
$$y = -\frac{2}{3}x + 2$$
$$f^{-1}(x) = -\frac{2}{3}x + 2$$

78.

$$(f \circ g)(-3) = f(g(-3))$$
$$= f\left((-3)^2 + (-3)\right)$$
$$= f(9-3)$$
$$= f(6)$$
$$= 3(6) - 2$$
$$= 18 - 2$$
$$= 16$$

79. $f(x) = 5^x$

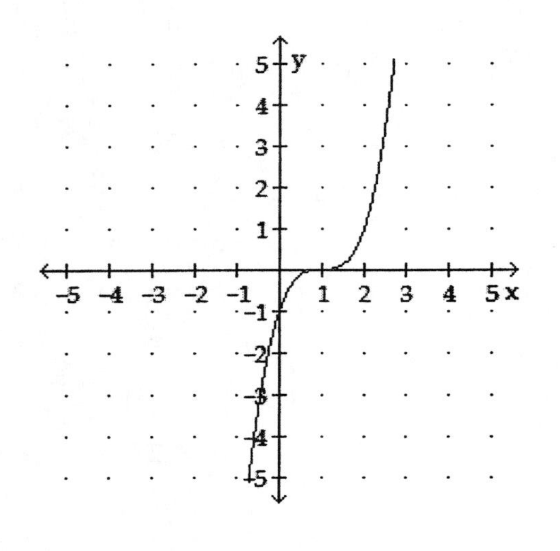

$D:(-\infty, \infty);\ R:(0, \infty)$

80. $f(x) = \ln x$

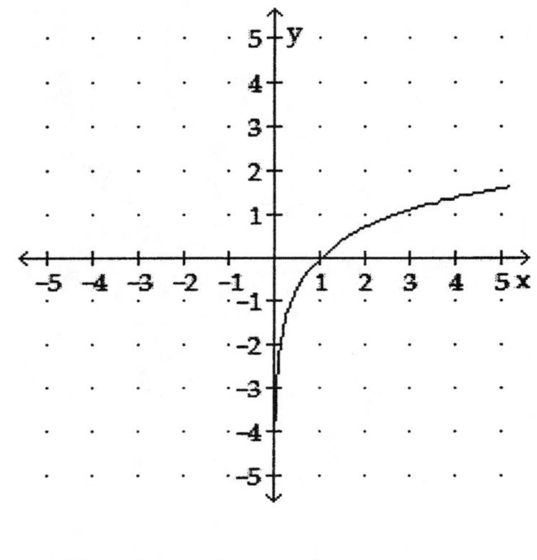

$D:(0, \infty);\ R:(-\infty, \infty)$

81.

$$\log_x 5 = 1$$
$$x^1 = 5$$
$$x = 5$$

82.

$$\log_8 x = 2$$
$$8^2 = x$$
$$64 = x$$

83.

$$\log_9 \frac{1}{81} = x$$
$$9^x = \frac{1}{81}$$
$$9^x = 9^{-2}$$
$$x = -2$$

84.

$$\ln \frac{y^3\sqrt{x}}{z} = \ln \frac{y^3 x^{1/2}}{z}$$
$$= \ln\left(y^3 x^{1/2}\right) - \ln z$$
$$= \ln y^3 + \ln x^{1/2} - \ln z$$
$$= 3\ln y + \frac{1}{2}\ln x - \ln z$$

85.

$$5^{x-3} = 3^{2x}$$
$$\log 5^{x-3} = \log 3^{2x}$$
$$(x-3)\log 5 = (2x)\log 3$$
$$x\log 5 - 3\log 5 = 2x\log 3$$
$$-3\log 5 = 2x\log 3 - x\log 5$$
$$-3\log 5 = x(2\log 3 - \log 5)$$
$$\frac{-3\log 5}{2\log 3 - \log 5} = \frac{x(2\log 3 - \log 5)}{2\log 3 - \log 5}$$
$$\frac{-3\log 5}{2\log 3 - \log 5} = x$$
$$x = \frac{-3\log 5}{2\log 3 - \log 5}$$
$$x \approx -8.2144$$

86.

$$\log(x+90) = 3 - \log x$$
$$\log(x+90) + \log x = 3$$
$$\log x(x+90) = 3$$
$$\log_{10}(x^2 + 90x) = 3$$
$$10^3 = x^2 + 90x$$
$$1{,}000 = x^2 + 90x$$
$$0 = x^2 + 90x - 1{,}000$$
$$0 = (x+100)(x-10)$$
$$x + 100 = 0 \quad \text{or} \quad x - 10 = 0$$
$$x = \cancel{-100} \qquad x = 10$$

Since the log of a negative number is undefined, $x \neq -100$. So, $x = 10$.

87.

$$\begin{cases} 5x - 4y = 10 \\ x - 7y = 2 \end{cases}$$

$$\left[\begin{array}{cc|c} 5 & -4 & 10 \\ 1 & -7 & 2 \end{array}\right]$$

$R_1 \leftrightarrow R_2$

$$\left[\begin{array}{cc|c} 1 & -7 & 2 \\ 5 & -4 & 10 \end{array}\right]$$

$-5R_1 + R_2$

$$\left[\begin{array}{cc|c} 1 & -7 & 2 \\ 0 & 31 & 0 \end{array}\right]$$

$\frac{1}{31}R_2$

$$\left[\begin{array}{cc|c} 1 & -7 & 2 \\ 0 & 1 & 0 \end{array}\right]$$

This matrix represents the system

$$\begin{cases} x - 7y = 2 \\ y = 0 \end{cases}$$

Use substitution to find x.

$$x - 7(0) = 2$$
$$x = 2$$

The solution is $(2, 0)$.

88. $\begin{cases} 3x+2y-z=-8 \\ 2x-y+7z=10 \\ 2x+2y-3z=-10 \end{cases}$

$$\frac{D_x}{D}=\frac{\begin{vmatrix} -8 & 2 & -1 \\ 10 & -1 & 7 \\ -10 & 2 & -3 \end{vmatrix}}{\begin{vmatrix} 3 & 2 & -1 \\ 2 & -1 & 7 \\ 2 & 2 & -3 \end{vmatrix}}$$

$$=\frac{-8\begin{vmatrix} -1 & 7 \\ 2 & -3 \end{vmatrix}-2\begin{vmatrix} 10 & 7 \\ -10 & -3 \end{vmatrix}-1\begin{vmatrix} 10 & -1 \\ -10 & 2 \end{vmatrix}}{3\begin{vmatrix} -1 & 7 \\ 2 & -3 \end{vmatrix}-2\begin{vmatrix} 2 & 7 \\ 2 & -3 \end{vmatrix}-1\begin{vmatrix} 2 & -1 \\ 2 & 2 \end{vmatrix}}$$

$$=\frac{-8(3-14)-2(-30+70)-1(20-10)}{3(3-14)-2(-6-14)-1(4+2)}$$

$$=\frac{-8(-11)-2(40)-1(10)}{3(-11)-2(-20)-1(6)}$$

$$=\frac{88-80-10}{-33+40-6}$$

$$=\frac{-2}{1}$$

$$=-2$$

88. (cont.)

$$\frac{D_y}{D}=\frac{\begin{vmatrix} 3 & -8 & -1 \\ 2 & 10 & 7 \\ 2 & -10 & -3 \end{vmatrix}}{\begin{vmatrix} 3 & 2 & -1 \\ 2 & -1 & 7 \\ 2 & 2 & -3 \end{vmatrix}}$$

$$=\frac{3\begin{vmatrix} 10 & 7 \\ -10 & -3 \end{vmatrix}+8\begin{vmatrix} 2 & 7 \\ 2 & -3 \end{vmatrix}-1\begin{vmatrix} 2 & 10 \\ 2 & -10 \end{vmatrix}}{3\begin{vmatrix} -1 & 7 \\ 2 & -3 \end{vmatrix}-2\begin{vmatrix} 2 & 7 \\ 2 & -3 \end{vmatrix}-1\begin{vmatrix} 2 & -1 \\ 2 & 2 \end{vmatrix}}$$

$$=\frac{3(-30+70)+8(-6-14)-1(-20-20)}{3(3-14)-2(-6-14)-1(4+2)}$$

$$=\frac{3(40)+8(-20)-1(-40)}{3(-11)-2(-20)-1(6)}$$

$$=\frac{120-160+40}{-33+40-6}$$

$$=\frac{0}{1}$$

$$=0$$

$$\frac{D_z}{D}=\frac{\begin{vmatrix} 3 & 2 & -8 \\ 2 & -1 & 10 \\ 2 & 2 & -10 \end{vmatrix}}{\begin{vmatrix} 3 & 2 & -1 \\ 2 & -1 & 7 \\ 2 & 2 & -3 \end{vmatrix}}$$

$$=\frac{3\begin{vmatrix} -1 & 10 \\ 2 & -10 \end{vmatrix}-2\begin{vmatrix} 2 & 10 \\ 2 & -10 \end{vmatrix}-8\begin{vmatrix} 2 & -1 \\ 2 & 2 \end{vmatrix}}{3\begin{vmatrix} -1 & 7 \\ 2 & -3 \end{vmatrix}-2\begin{vmatrix} 2 & 7 \\ 2 & -3 \end{vmatrix}-1\begin{vmatrix} 2 & -1 \\ 2 & 2 \end{vmatrix}}$$

$$=\frac{3(10-20)-2(-20-20)-8(4+2)}{3(3-14)-2(-6-14)-1(4+2)}$$

$$=\frac{3(-10)-2(-40)-8(6)}{3(-11)-2(-20)-1(6)}$$

$$=\frac{-30+80-48}{-33+40-6}$$

$$=\frac{2}{1}$$

$$=2$$

The solution is (–2, 0, 2).

SECTION 13.1

VOCABULARY

1. The curves formed by the intersection of a plane with an infinite right–circular cone are called **conic sections**.

3. A **circle** is the set of all points in a plane that are a fixed distance from a fixed point called its center. The fixed distance is called the **radius**.

CONCEPTS

5. a. $(x-h)^2+(y-k)^2=r^2$
   b. $x^2+y^2=r^2$

7. a. $(2,-1)$; $r=4$
   b. $(x-2)^2+(y-(-1))^2=4^2$
   $(x-2)^2+(y+1)^2=16$

9. a. $y=a(x-h)^2+k$
   b. $x=a(y-k)^2+h$

11. a. circle
    b. parabola
    c. parabola
    d. circle

NOTATION

13. $h=6$, $k=-2$, and $r=3$

PRACTICE

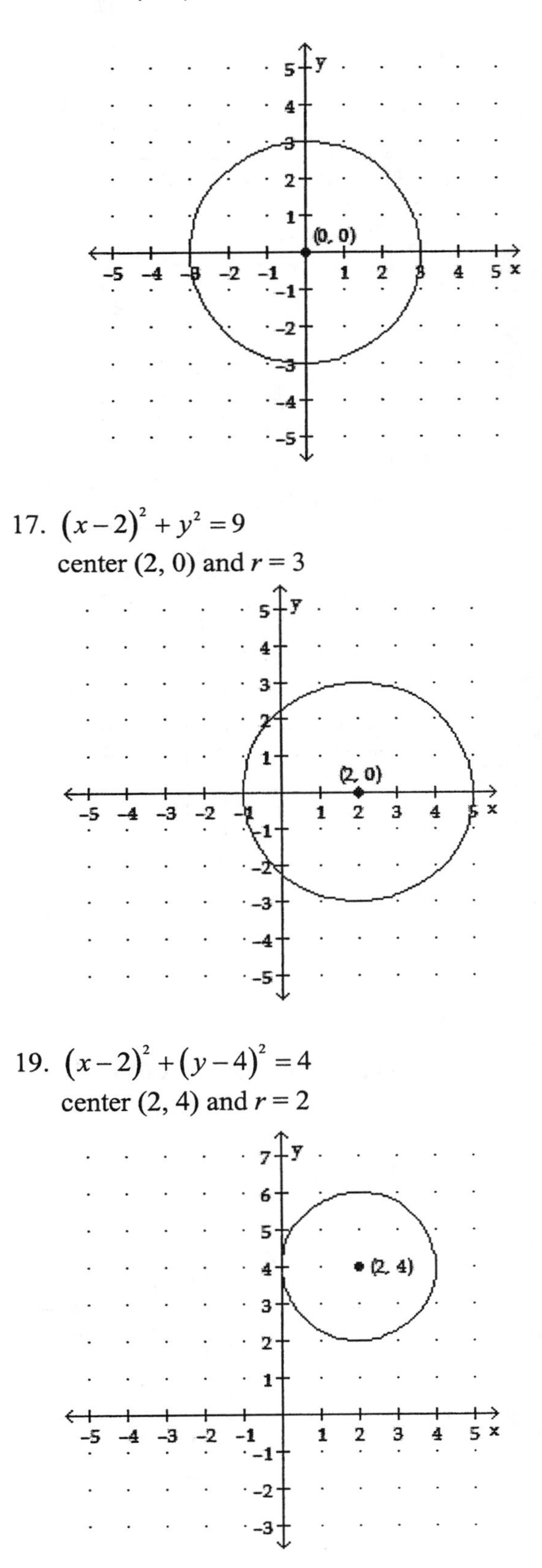

21. $(x+3)^2+(y-1)^2=16$

center (–3, 1) and $r = 4$

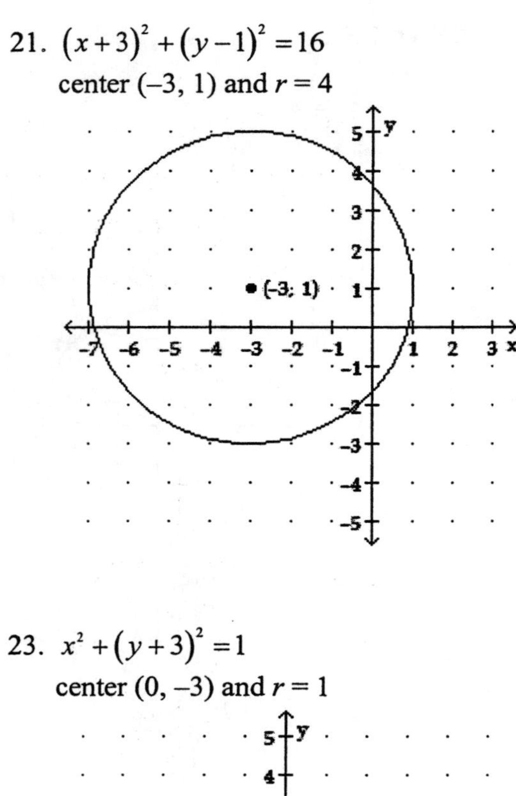

23. $x^2+(y+3)^2=1$

center (0, –3) and $r = 1$

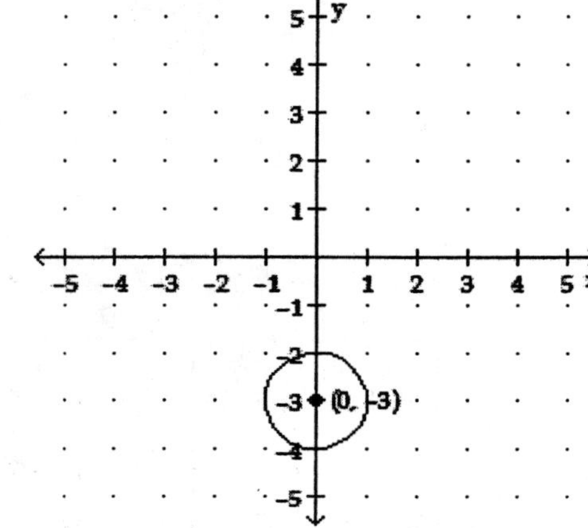

25. $x^2+y^2=6$

center (0, 0) and $r = \sqrt{6} \approx 2.4$

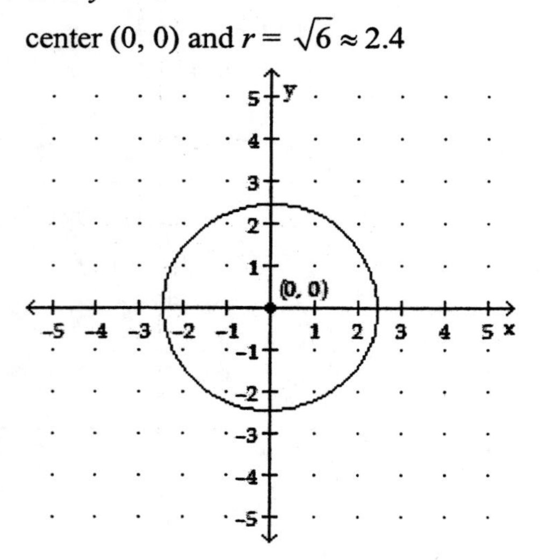

27. $(x-1)^2+(y-3)^2=15$

center (1, 3) and $r = \sqrt{15} \approx 3.9$

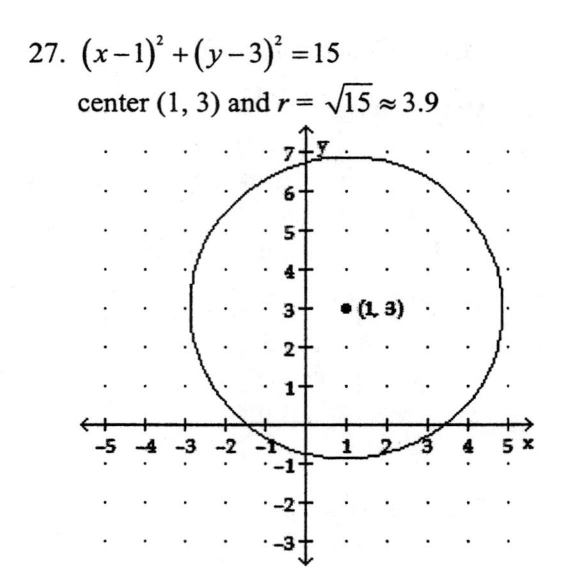

29. $x^2+y^2=7$

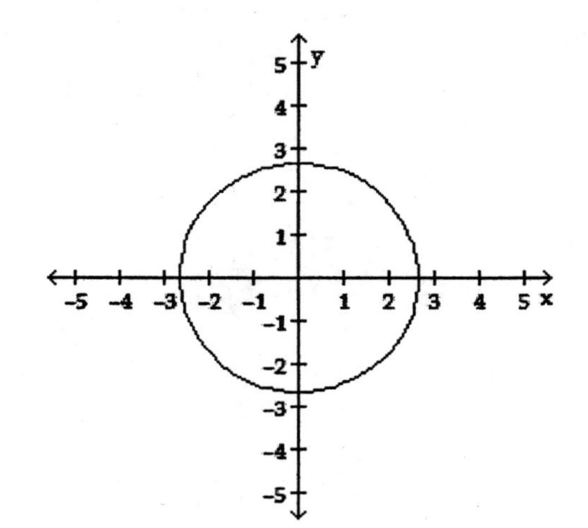

31. $(x+1)^2+y^2=16$

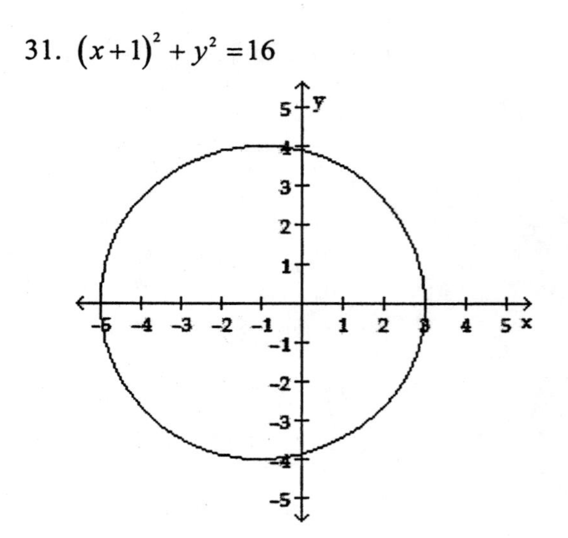

33. $h = 0, k = 0, r = 1$

$$(x-h)^2 + (y-k)^2 = r^2$$
$$(x-0)^2 + (y-0)^2 = 1^2$$
$$x^2 + y^2 = 1$$

35. $h = 6, k = 8, r = 5$

$$(x-h)^2 + (y-k)^2 = r^2$$
$$(x-6)^2 + (y-8)^2 = 5^2$$
$$(x-6)^2 + (y-8)^2 = 25$$

37. $h = -2, k = 6, r = 12$

$$(x-h)^2 + (y-k)^2 = r^2$$
$$(x-(-2))^2 + (y-6)^2 = 12^2$$
$$(x+2)^2 + (y-6)^2 = 144$$

39. $h = 0, k = 0, r = \dfrac{1}{4}$

$$(x-h)^2 + (y-k)^2 = r^2$$
$$(x-0)^2 + (y-0)^2 = \left(\frac{1}{4}\right)^2.$$
$$x^2 + y^2 = \frac{1}{16}$$

41. $h = \dfrac{2}{3}, k = -\dfrac{7}{8}, r = \sqrt{2}$

$$(x-h)^2 + (y-k)^2 = r^2$$
$$\left(x-\frac{2}{3}\right)^2 + \left(y-\left(-\frac{7}{8}\right)\right)^2 = \left(\sqrt{2}\right)^2.$$
$$\left(x-\frac{2}{3}\right)^2 + \left(y+\frac{7}{8}\right)^2 = 2$$

43. $h=0,\ k=0,\ d=4\sqrt{2}$ so $r=\dfrac{4\sqrt{2}}{2}=2\sqrt{2}$

$$(x-h)^2 + (y-k)^2 = r^2$$
$$(x-0)^2 + (y-0)^2 = \left(2\sqrt{2}\right)^2$$
$$x^2 + y^2 = 4(2)$$
$$x^2 + y^2 = 8$$

45.

$$x^2 + y^2 - 2x + 4y = -1$$
$$\left(x^2 - 2x\right) + \left(y^2 + 4y\right) = -1$$
$$\left(x^2 - 2x + 1\right) + \left(y^2 + 4y + 4\right) = -1 + 1 + 4$$
$$(x-1)^2 + (y+2)^2 = 4$$

Center at $(1,\ -2)$; radius = 2

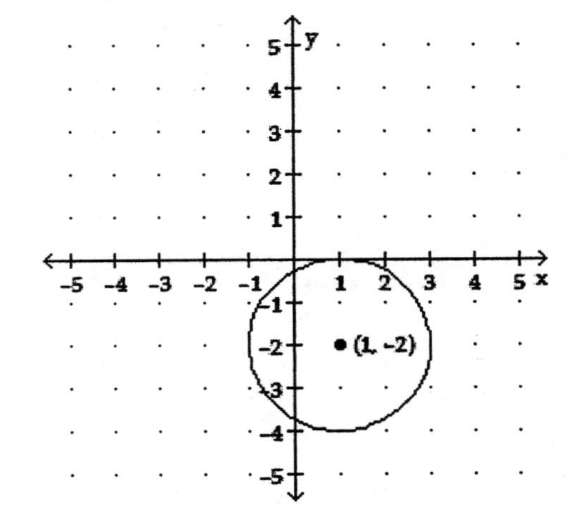

47.

$$x^2 + y^2 + 4x + 2y = 4$$
$$\left(x^2 + 4x\right) + \left(y^2 + 2y\right) = 4$$
$$\left(x^2 + 4x + 4\right) + \left(y^2 + 2y + 1\right) = 4 + 4 + 1$$
$$(x+2)^2 + (y+1)^2 = 9$$

Center at $(-2,\ -1)$; radius = 3

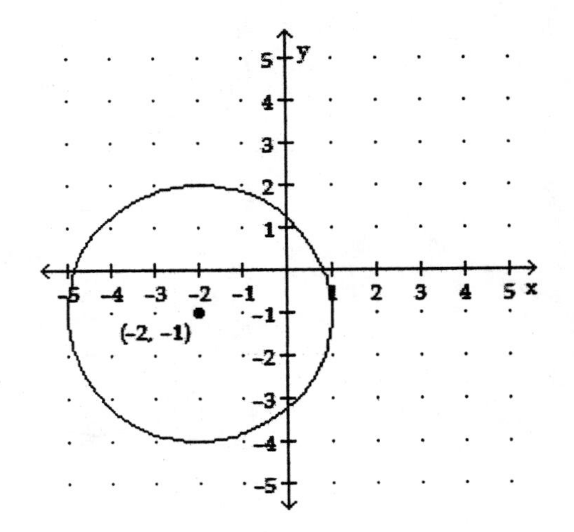

49.

$$x^2 + y^2 + 2x - 8 = 0$$
$$\left(x^2 + 2x\right) + y^2 = 0 + 8$$
$$\left(x^2 + 2x + 1\right) + y^2 = 8 + 1$$
$$\left(x + 1\right)^2 + y^2 = 9$$

Center at $(-1, 0)$; radius = 3

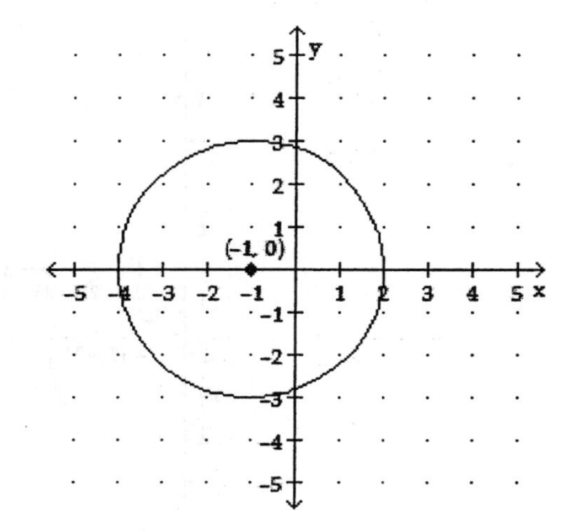

51.

$$x^2 + y^2 - 6x + 8y + 18 = 0$$
$$\left(x^2 - 6x\right) + \left(y^2 + 8y\right) = -18$$
$$\left(x^2 - 6x + 9\right) + \left(y^2 + 8y + 16\right) = -18 + 9 + 16$$
$$\left(x - 3\right)^2 + \left(y + 4\right)^2 = 7$$

Center at $(3, -4)$; radius = $\sqrt{7}$

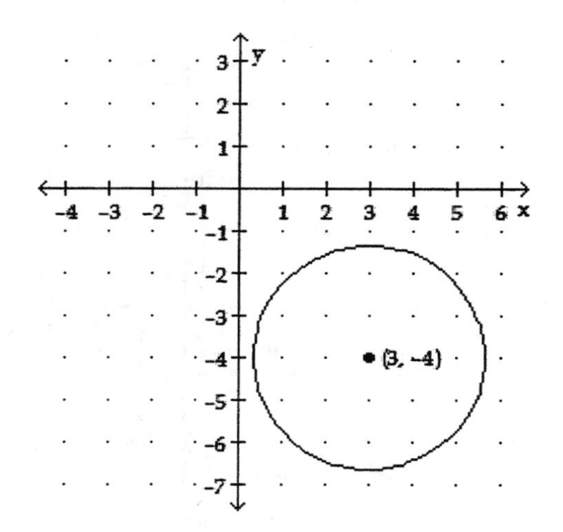

53.

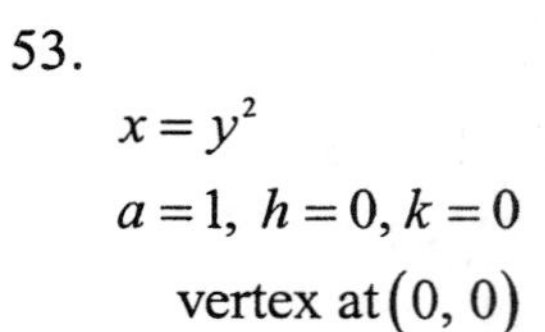

$x = y^2$

$a = 1,\ h = 0, k = 0$

vertex at $(0, 0)$

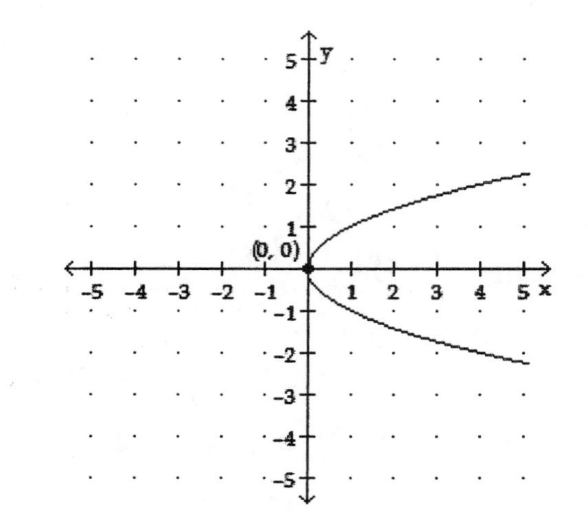

55.

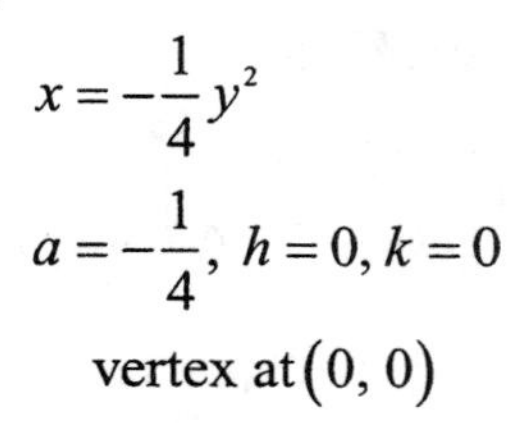

$x = -\dfrac{1}{4}y^2$

$a = -\dfrac{1}{4},\ h = 0, k = 0$

vertex at $(0, 0)$

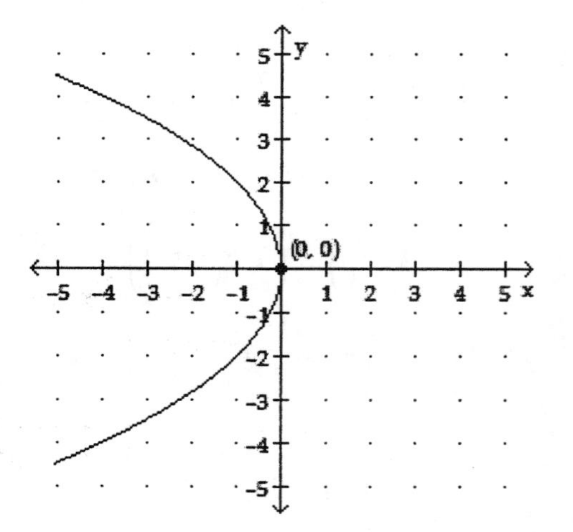

57.

$$x = 2(y+1)^2 + 3$$
$$a = 2,\ h = 3, k = -1$$
vertex at $(3,\ -1)$

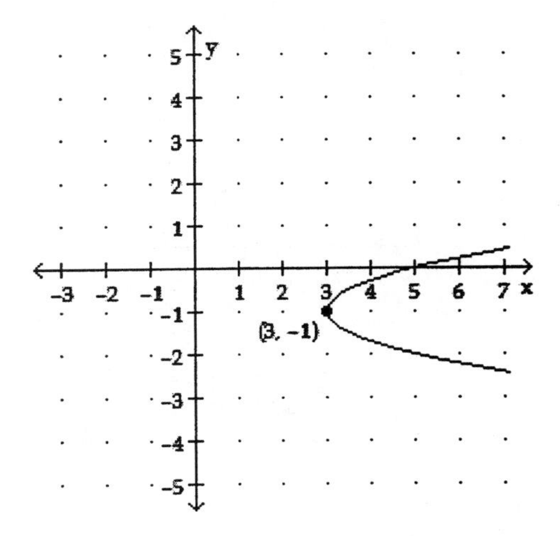

59.

$$x = -3y^2 + 18y - 25$$
$$x = -3(y^2 - 6y) - 25$$
$$x = -3(y^2 - 6y + 9) - 25 - (-3)(9)$$
$$x = -3(y-3)^2 - 25 + 27$$
$$x = -3(y-3)^2 + 2$$
$$a = -3,\ h = 2, k = 3$$
vertex at $(2,\ 3)$

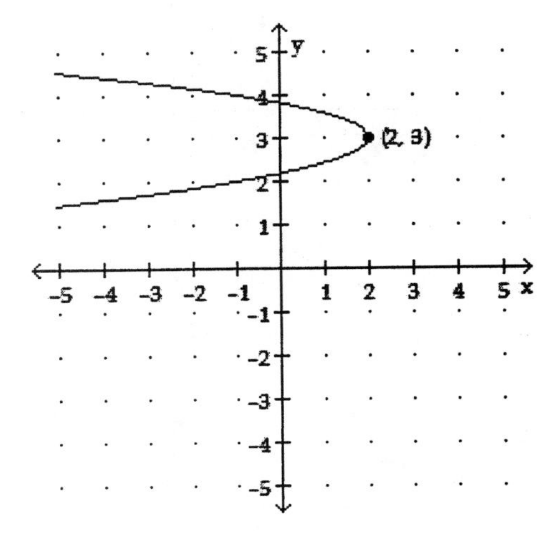

61.

$$x = \frac{1}{2}y^2 + 2y$$
$$x = \frac{1}{2}(y^2 + 4y)$$
$$x = \frac{1}{2}(y^2 + 4y + 4) - \frac{1}{2}(4)$$
$$x = \frac{1}{2}(y+2)^2 - 2$$
$$a = \frac{1}{2},\ h = -2, k = -2$$
vertex at $(-2, -2)$

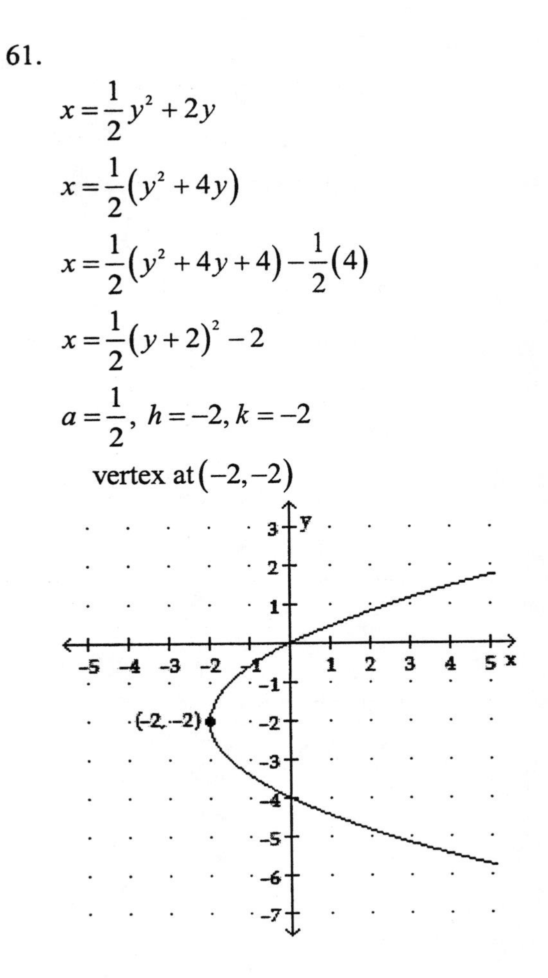

63.

$$y = 2x^2 - 4x + 5$$
$$y = 2(x^2 - 2x) + 5$$
$$y = 2(x^2 - 2x + 1) + 5 - 2(1)$$
$$y = 2(x-1)^2 + 5 - 2$$
$$y = 2(x-1)^2 + 3$$
$$a = 2,\ h = 1, k = 3$$
vertex at $(1,\ 3)$

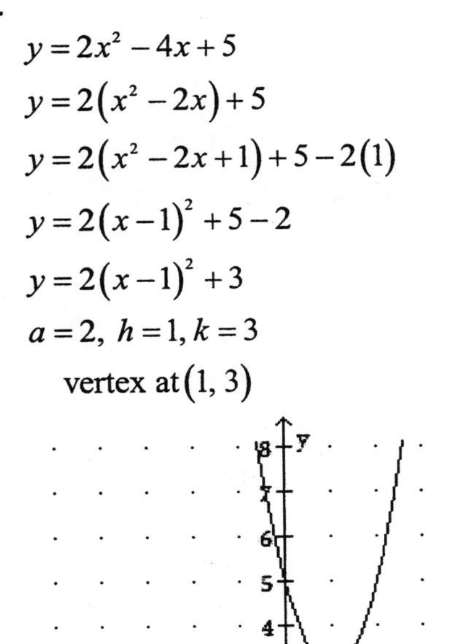

65.

$$y = -x^2 - 2x + 3$$
$$y = -(x^2 + 2x) + 3$$
$$y = -(x^2 + 2x + 1) + 3 - (-1)(1)$$
$$y = -(x+1)^2 + 3 + 1$$
$$y = -(x+1)^2 + 4$$
$$a = -1,\ h = -1, k = 4$$

vertex at $(-1, 4)$

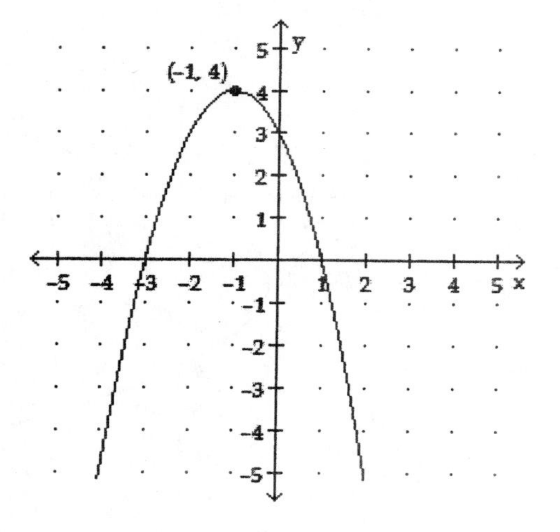

67.

$$y^2 + 4x - 6y = -1$$
$$4x = -y^2 + 6y - 1$$
$$x = -\frac{1}{4}y^2 + \frac{3}{2}y - \frac{1}{4}$$
$$x = -\frac{1}{4}(y^2 - 6y) - \frac{1}{4}$$
$$x = -\frac{1}{4}(y^2 - 6y + 9) - \frac{1}{4} - \left(-\frac{1}{4}\right)(9)$$
$$x = -\frac{1}{4}(y-3)^2 - \frac{1}{4} + \frac{9}{4}$$
$$x = -\frac{1}{4}(y-3)^2 + 2$$

$$a = -\frac{1}{4},\ h = 2, k = 3$$

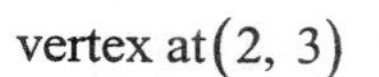
vertex at $(2, 3)$

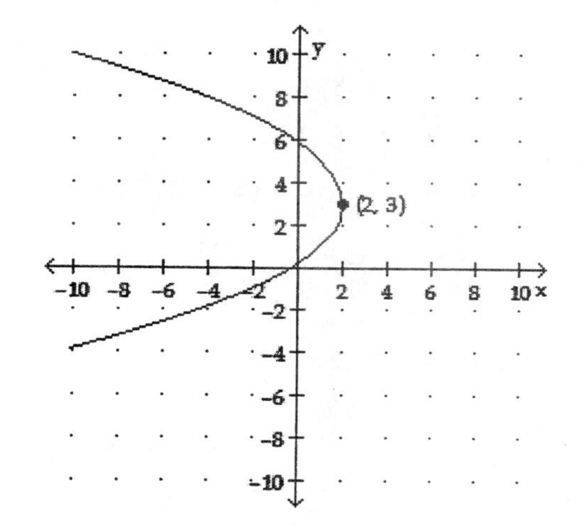

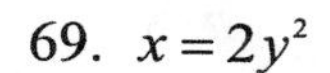
69. $x = 2y^2$

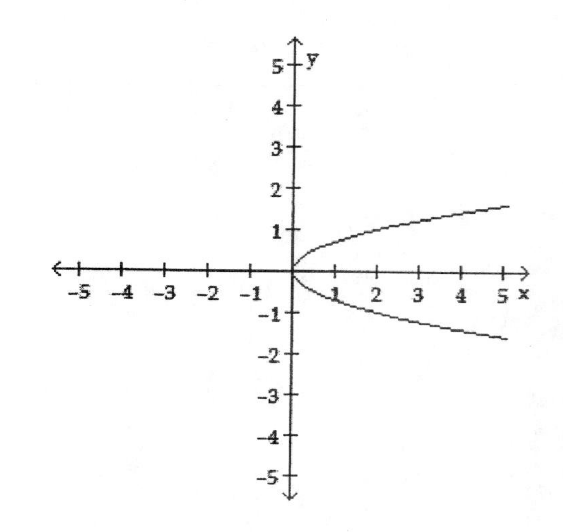

71. $x^2 - 2x + y = 6$

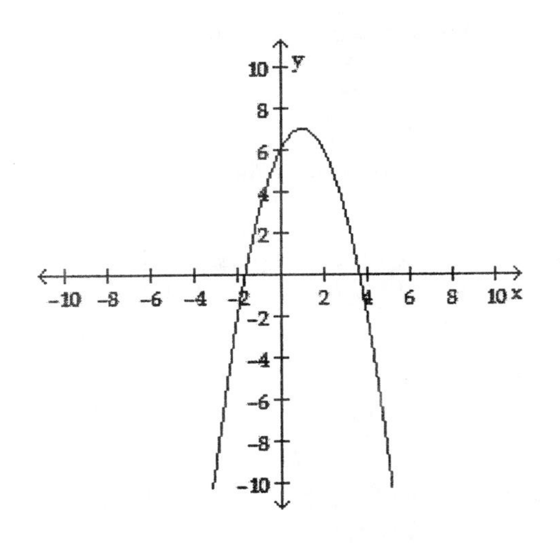

APPLICATIONS

73. MESHING GEARS

The radius of the larger gear is 4, so the two gears touch at the point (4, 0). The center of the smaller gear is (7, 0). So the radius of the smaller gear is 7 – 4 = 3. The equation of the smaller gear is $(x-7)^2 + y^2 = 3^2$ or $(x-7)^2 + y^2 = 9$.

$$d = \sqrt{(x_2 - x_1)^2 + (y_2 - y_1)^2}$$
$$= \sqrt{(-50 - (-20))^2 + (0 - 0)^2}$$
$$= \sqrt{(-30)^2 + (0 - 0)^2}$$
$$= \sqrt{900}$$
$$= 30 \text{ ft.}$$

75. BROADCAST RANGES

Find the standard equation for each coverage and graph to see if they overlap.

$$x^2 + y^2 - 8x - 20y + 16 = 0$$
$$(x^2 - 8x) + (y^2 - 20y) = -16$$
$$(x^2 - 8x + 16) + (y^2 - 20y + 100) = -16 + 16 + 100$$
$$(x-4)^2 + (y-10)^2 = 100$$

center at $(4, 10)$; radius = 10

$$x^2 + y^2 + 2x + 4y - 11 = 0$$
$$(x^2 + 2x) + (y^2 + 4y) = 11$$
$$(x^2 + 2x + 1) + (y^2 + 4y + 4) = 11 + 1 + 4$$
$$(x+1)^2 + (y+2)^2 = 16$$

center at $(-1, -2)$; radius = 4

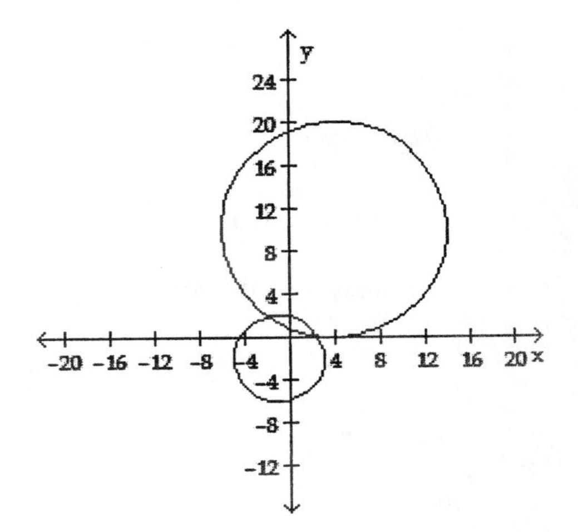

Since the coverage areas overlap, they cannot be licensed for the same frequency.

77. PROJECTILES

Its landing position is an x–intercept, so let $y = 0$, and solve the equation for x.

$$y = 30x - x^2$$
$$0 = 30x - x^2$$
$$0 = x(30 - x)$$
$$x = 0 \quad \text{or} \quad 30 - x = 0$$
$$x = 0 \qquad\qquad 30 = x$$

If it lands 30 feet away from its starting point, it is 35 – 30 = 5 ft from the castle.

79. COMETS

Find the vertex of the comet's orbit.

$$2y^2 - 9x = 18$$
$$-9x = -2y^2 + 18$$
$$x = \frac{2}{9}y^2 - 2$$
$$k = 0 \text{ and } h = -2$$
$$\text{vertex} = (-2, 0)$$

Find the distance between the sun (0, 0) and the comet (–2, 0) using the distance formula.

$$d = \sqrt{(x_2 - x_1)^2 + (y_2 - y_1)^2}$$
$$= \sqrt{(-2-0)^2 + (0-0)^2}$$
$$= \sqrt{4}$$
$$= 2 \text{ AU}$$

WRITING

81. Answers will vary.

83. Answers will vary.

REVIEW

85.

$$|3x - 4| = 11$$
$$3x - 4 = 11 \quad \text{or} \quad 3x - 4 = -11$$
$$3x = 15 \qquad 3x = -7$$
$$x = 5 \qquad x = -\frac{7}{3}$$

87.

$$|3x + 4| = |5x - 2|$$
$$3x + 4 = 5x - 2 \quad \text{or} \quad 3x + 4 = -(5x - 2)$$
$$3x + 4 = 5x - 2 \qquad 3x + 4 = -5x + 2$$
$$-2x + 4 = -2 \qquad 8x + 4 = 2$$
$$-2x = -6 \qquad 8x = -2$$
$$x = 3 \qquad x = -\frac{1}{4}$$

CHALLENGE PROBLEMS

89. Answers will vary.

91. To find the center of the circle, find the midpoint of the points (–2, –6) and (8, 10).

$$\left(\frac{x_1 + x_2}{2}, \frac{y_1 + y_2}{2}\right) = \left(\frac{-2+8}{2}, \frac{-6+10}{2}\right)$$
$$= \left(\frac{6}{2}, \frac{4}{2}\right)$$
$$= (3, 2)$$

To find the radius, use the distance formula to find the distance from (8, 10) and the center (3, 2).

$$d = \sqrt{(x_2 - x_1)^2 + (y_2 - y_1)^2}$$
$$= \sqrt{(3-8)^2 + (2-10)^2}$$
$$= \sqrt{(-5)^2 + (-8)^2}$$
$$= \sqrt{25 + 64}$$
$$= \sqrt{89}$$

Write the equation of the circle with center (3, 2) and radius $\sqrt{89}$.

$$(x-3)^2 + (y-2)^2 = \left(\sqrt{89}\right)^2$$
$$(x-3)^2 + (y-2)^2 = 89$$

SECTION 13.2

VOCABULARY

1. The curve graphed below is an **ellipse**.

3. In the graph above, F_1 and F_2 are the **foci** of the ellipse. Each one is called a **focus** of the ellipse.

5. The line segment joining the vertices of an ellipse is called the **major axis** of the ellipse.

CONCEPTS

7. $$\frac{x^2}{a^2}+\frac{y^2}{b^2}=1$$

9. x–intercepts: $(a, 0)$ and $(-a, 0)$
   y–intercepts: $(0, b)$ and $(0, -b)$

11. a. $(-2, 1)$; $a = 2$ and $b = 5$
    b. vertical
    c. $$\frac{(x+2)^2}{4}+\frac{(y-1)^2}{25}=1$$

13.
$$4(x-1)^2+64(y+5)^2=64$$
$$\frac{4(x-1)^2}{64}+\frac{64(y+5)^2}{64}=\frac{64}{64}$$
$$\frac{(x-1)^2}{16}+\frac{(y+5)^2}{1}=1$$

NOTATION

15. $h=-8$, $k=6$, $a=\sqrt{100}=10$,
    $b=\sqrt{144}=12$

PRACTICE

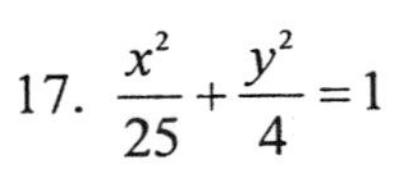

17. $$\frac{x^2}{25}+\frac{y^2}{4}=1$$

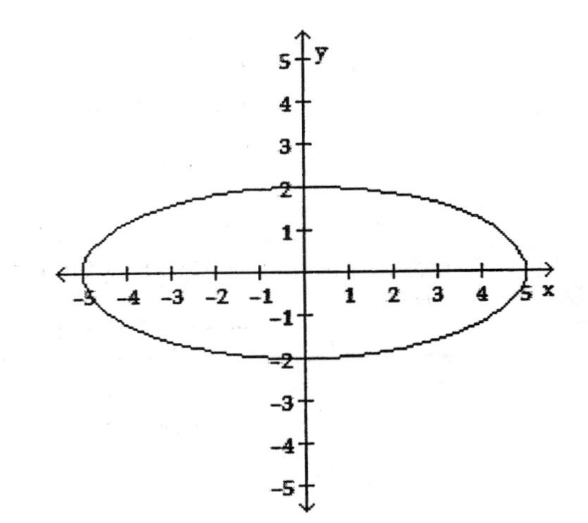

19. $$\frac{x^2}{4}+\frac{y^2}{9}=1$$

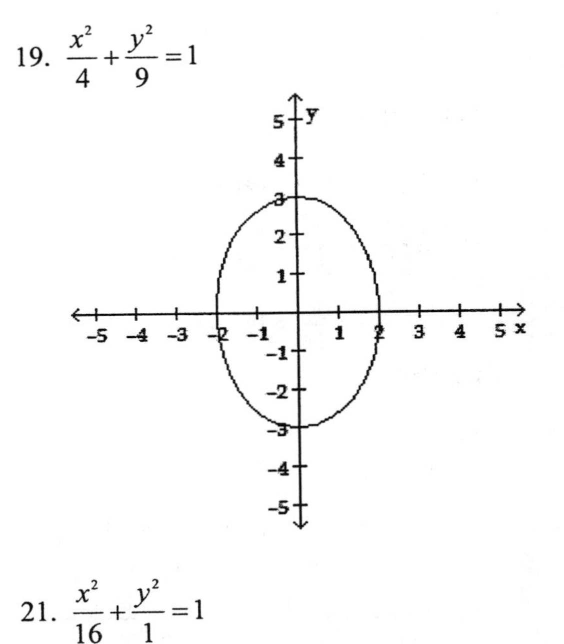

21. $$\frac{x^2}{16}+\frac{y^2}{1}=1$$

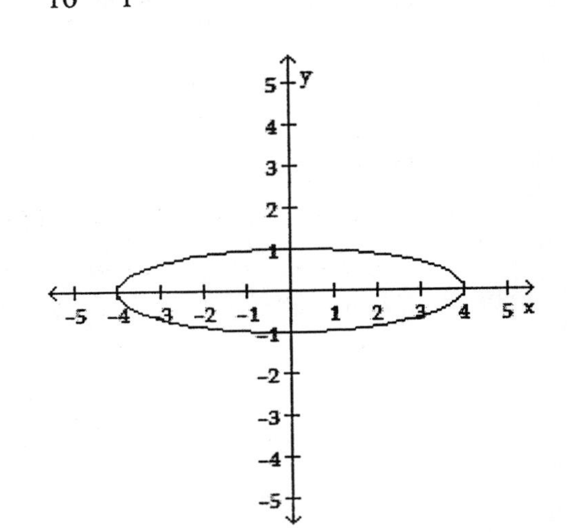

23.

$$x^2 + 9y^2 = 9$$

$$\frac{x^2}{9} + \frac{9y^2}{9} = \frac{9}{9}$$

$$\frac{x^2}{9} + \frac{y^2}{1} = 1$$

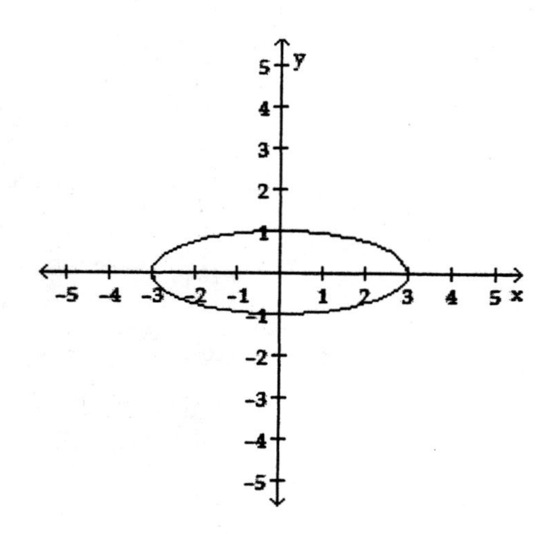

25.

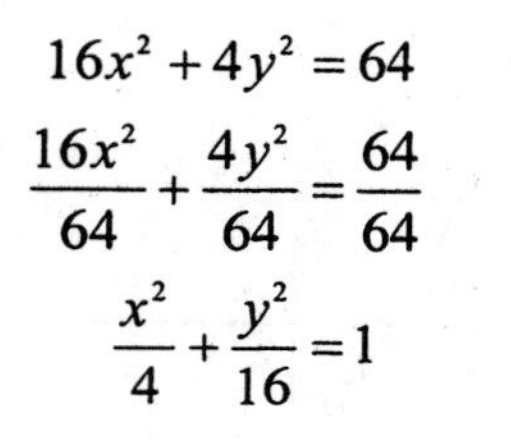

$$16x^2 + 4y^2 = 64$$

$$\frac{16x^2}{64} + \frac{4y^2}{64} = \frac{64}{64}$$

$$\frac{x^2}{4} + \frac{y^2}{16} = 1$$

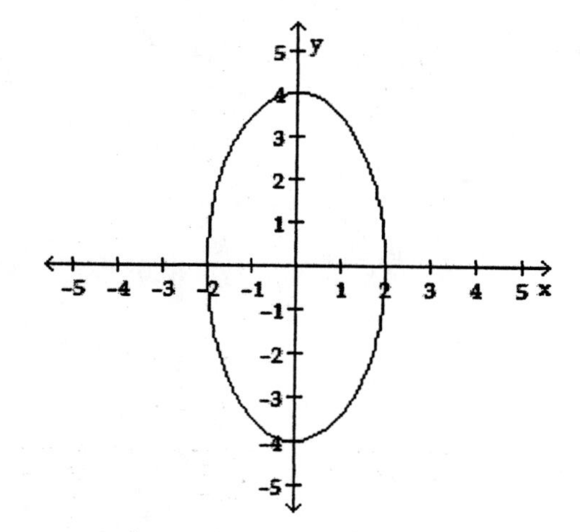

27.

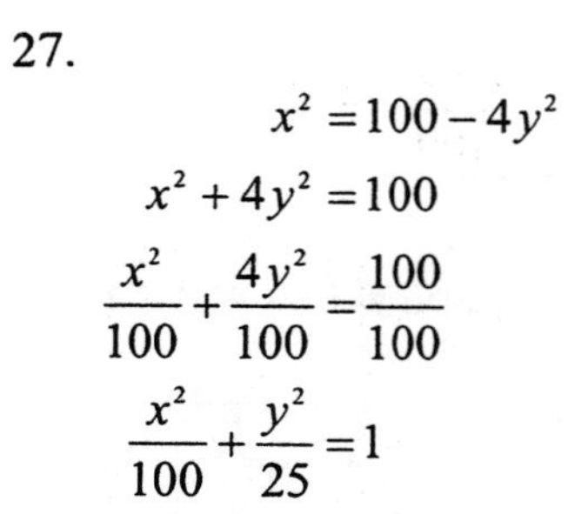

$$x^2 = 100 - 4y^2$$

$$x^2 + 4y^2 = 100$$

$$\frac{x^2}{100} + \frac{4y^2}{100} = \frac{100}{100}$$

$$\frac{x^2}{100} + \frac{y^2}{25} = 1$$

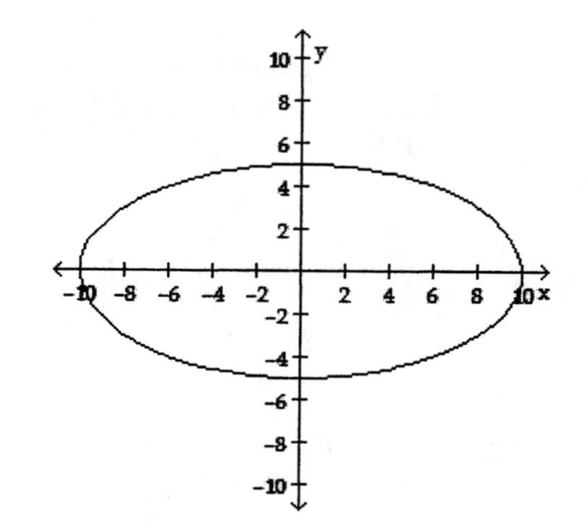

29. 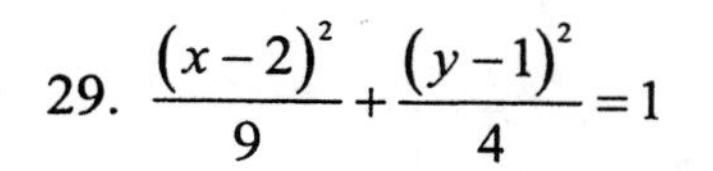

$$\frac{(x-2)^2}{9} + \frac{(y-1)^2}{4} = 1$$

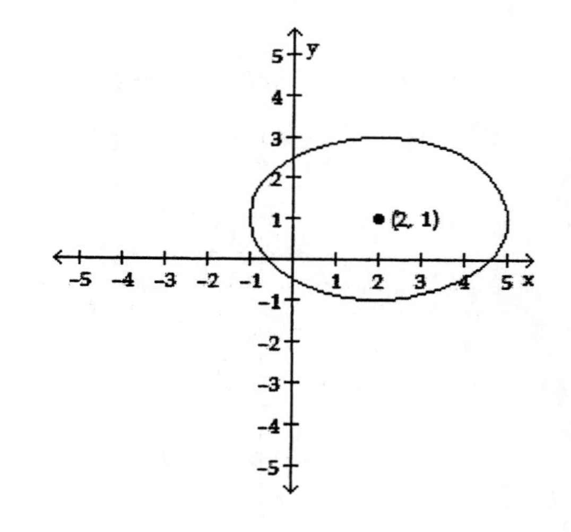

31. $\dfrac{(x+2)^2}{64}+\dfrac{(y-2)^2}{100}=1$

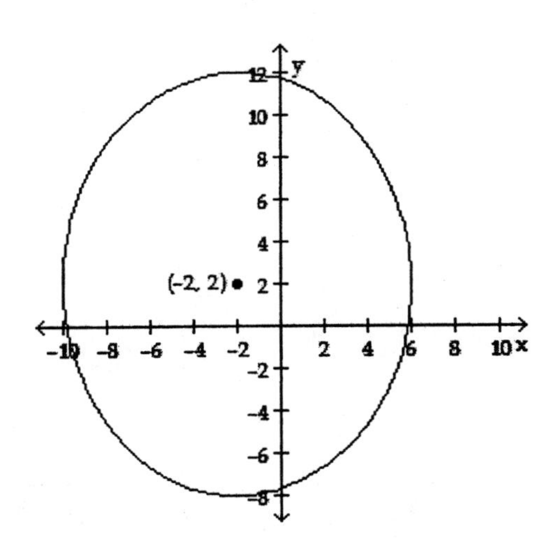

33.

$$(x+1)^2+4(y+2)^2=4$$

$$\frac{(x+1)^2}{4}+\frac{4(y+2)^2}{4}=\frac{4}{4}$$

$$\frac{(x+1)^2}{4}+\frac{(y+2)^2}{1}=1$$

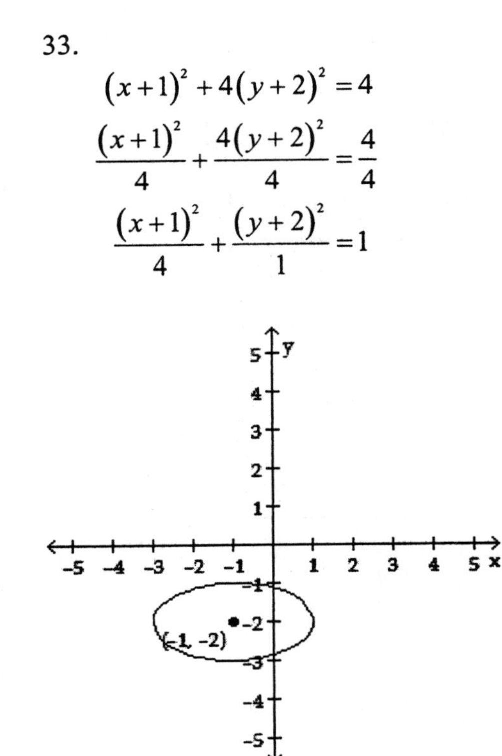

35.

$$(x-2)^2+4(y+1)^2=4$$

$$\frac{(x-2)^2}{4}+\frac{4(y+1)^2}{4}=\frac{4}{4}$$

$$\frac{(x-2)^2}{4}+\frac{(y+1)^2}{1}=1$$

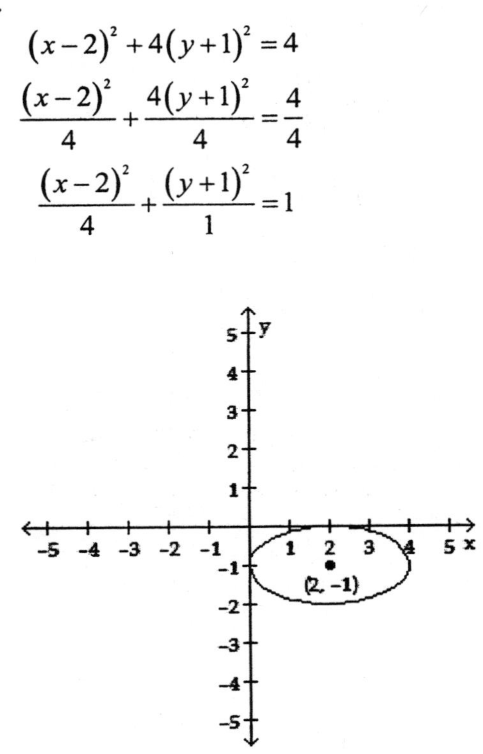

37.

$$9(x-1)^2=36-4(y+2)^2$$

$$9(x-1)^2+4(y+2)^2=36$$

$$\frac{9(x-1)^2}{36}+\frac{4(y+2)^2}{36}=\frac{36}{36}$$

$$\frac{(x-1)^2}{4}+\frac{(y+2)^2}{9}=1$$

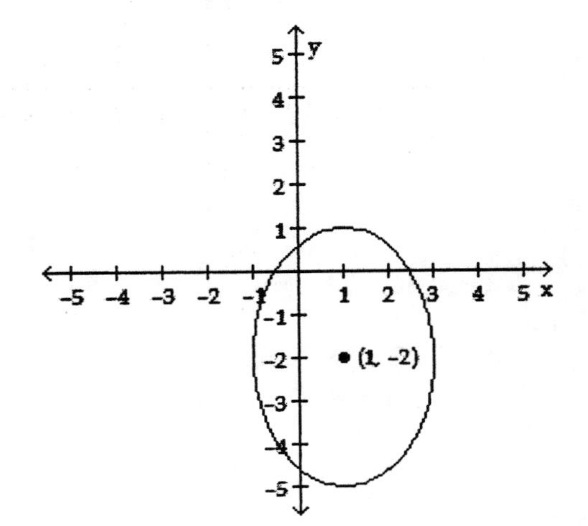

39.

$$\frac{x^2}{9}+\frac{y^2}{4}=1$$

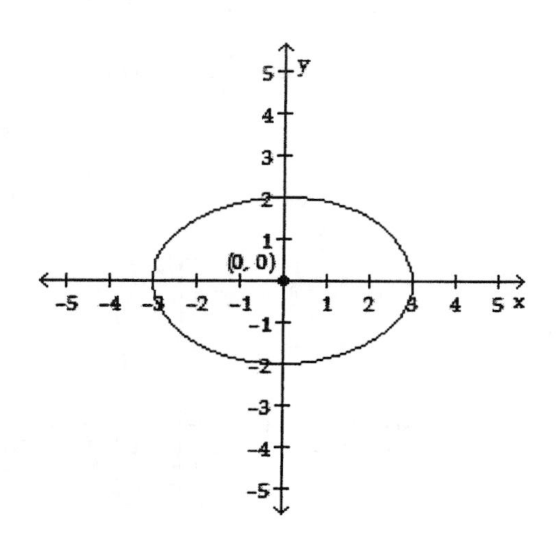

41.

$$\frac{x^2}{4}+\frac{(y-1)^2}{9}=1$$

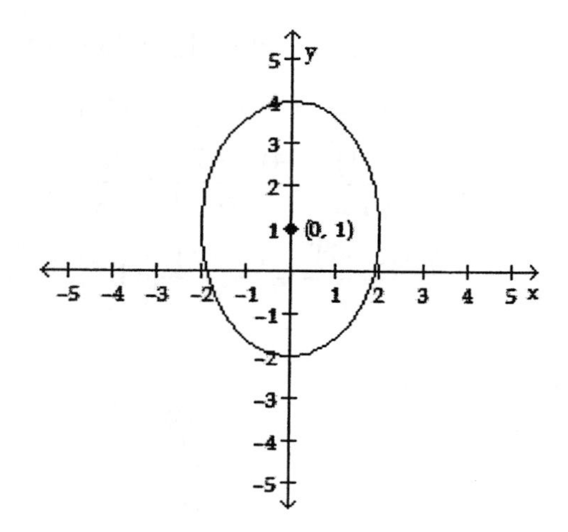

APPLICATIONS

43. FITNESS EQUIPMENT

$a=\pm 12,\ b=\pm 5$

$$\frac{x^2}{144}+\frac{y^2}{25}=1$$

45. CALCULATING CLEARANCE

Let $x=10$.

$$\frac{x^2}{400}+\frac{y^2}{100}=1$$
$$\frac{10^2}{400}+\frac{y^2}{100}=1$$
$$\frac{100}{400}+\frac{y^2}{100}=1$$
$$\frac{1}{4}+\frac{y^2}{100}=1$$
$$100\left(\frac{1}{4}+\frac{y^2}{100}\right)=100(1)$$
$$25+y^2=100$$
$$y^2=75$$
$$y=\sqrt{75}$$
$$y=\sqrt{25}\sqrt{3}$$
$$y=5\sqrt{3}\text{ ft}$$

47. AREA OF AN ELLIPSE

$$9x^2+16y^2=144$$
$$\frac{9x^2}{144}+\frac{16y^2}{144}=\frac{144}{144}$$
$$\frac{x^2}{16}+\frac{y^2}{9}=1$$

$a=4$ and $b=3$

$$A=\pi ab$$
$$=\pi(4)(3)$$
$$=12\pi\text{ sq. units}$$
$$\approx 37.7\text{ sq. units}$$

WRITING

49. Answers will vary.

51. Answers will vary.

REVIEW

53.

$$3x^{-2}y^2\left(4x^2+3y^{-2}\right)=12x^0y^2+9x^{-2}y^0$$
$$=12(1)y^2+9x^{-2}(1)$$
$$=12y^2+\frac{9}{x^2}$$

55.

$$\frac{x^{-2}+y^{-2}}{x^{-2}-y^{-2}}=\frac{\frac{1}{x^2}+\frac{1}{y^2}}{\frac{1}{x^2}-\frac{1}{y^2}}$$

$$=\frac{\frac{1}{x^2}+\frac{1}{y^2}}{\frac{1}{x^2}-\frac{1}{y^2}}\bullet\frac{x^2y^2}{x^2y^2}$$

$$=\frac{\frac{x^2y^2}{x^2}+\frac{x^2y^2}{y^2}}{\frac{x^2y^2}{x^2}-\frac{x^2y^2}{y^2}}$$

$$=\frac{y^2+x^2}{y^2-x^2}$$

$$=\frac{y^2+x^2}{(y+x)(y-x)}$$

CHALLENGE PROBLEMS

57. It forms a circle.

59.

$$9x^2+4y^2-18x+16y=11$$

$$(9x^2-18x)+(4y^2+16y)=11$$

$$9(x^2-2x)+4(y^2+4y)=11$$

$$9(x^2-2x+1)+4(y^2+4y+4)=11+9(1)+4(4)$$

$$9(x-1)^2+4(y+2)^2=36$$

$$\frac{9(x-1)^2}{36}+\frac{4(y+2)^2}{36}=\frac{36}{36}$$

$$\frac{(x-1)^2}{4}+\frac{(y+2)^2}{9}=1$$

SECTION 13.3

VOCABULARY

1. The two–branch curve graphed below is a **hyperbola**.

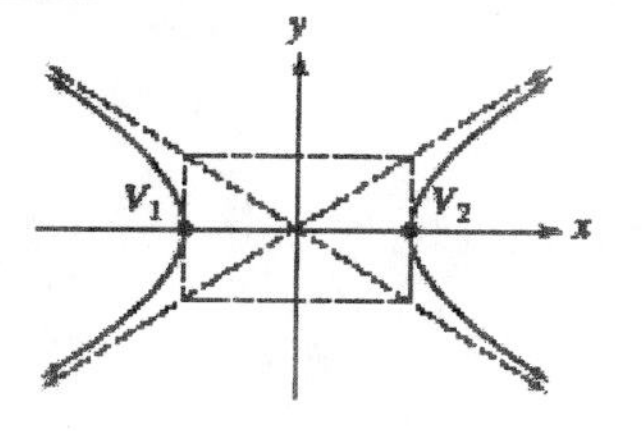

3. In the graph above, V_1 and V_2 are called the **vertices** of the hyperbola.

5. The extended **diagonals** of the central rectangle are asymptotes of the hyperbola.

CONCEPTS

7. $\frac{x^2}{a^2}-\frac{y^2}{b^2}=1$

9. $\frac{(x-h)^2}{a^2}-\frac{(y-k)^2}{b^2}=1$

11. a. $(-1,-2);\ a=3,\ b=1$

    b. $\frac{(y+2)^2}{9}-\frac{(x+1)^2}{1}=1$

13.

$$100(x+1)^2-25(y-5)^2=100$$

$$\frac{100(x+1)^2}{100}-\frac{25(y-5)^2}{100}=\frac{100}{100}$$

$$\frac{(x+1)^2}{1}-\frac{(y-5)^2}{4}=1$$

NOTATION

15. $h=5,\ k=-11,\ a=5,\ b=6$

PRACTICE

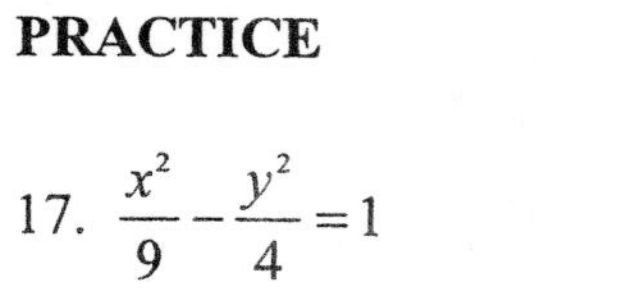

17. $\frac{x^2}{9}-\frac{y^2}{4}=1$

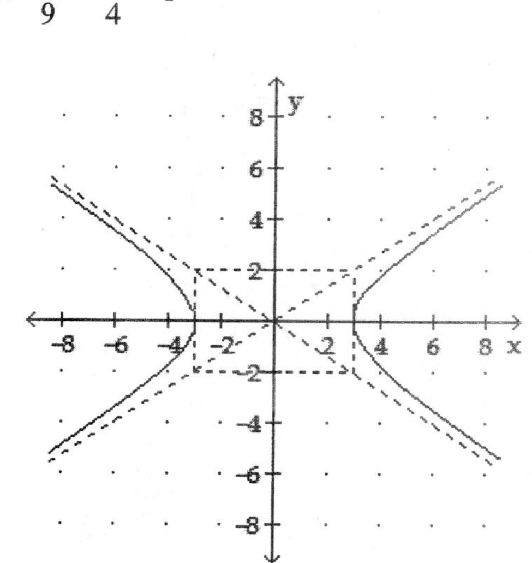

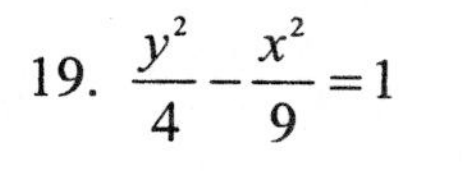

19. $\frac{y^2}{4}-\frac{x^2}{9}=1$

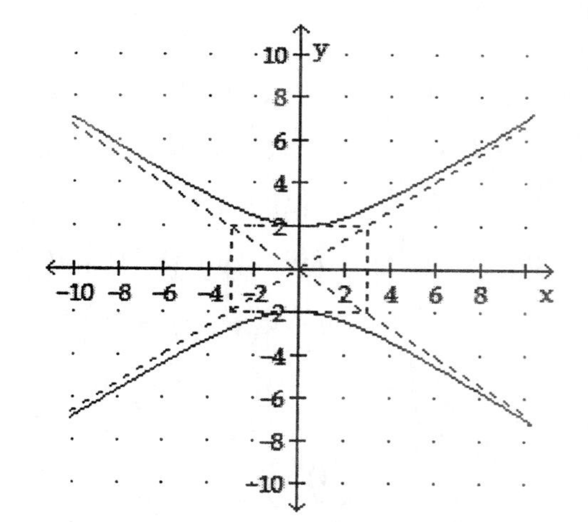

21.

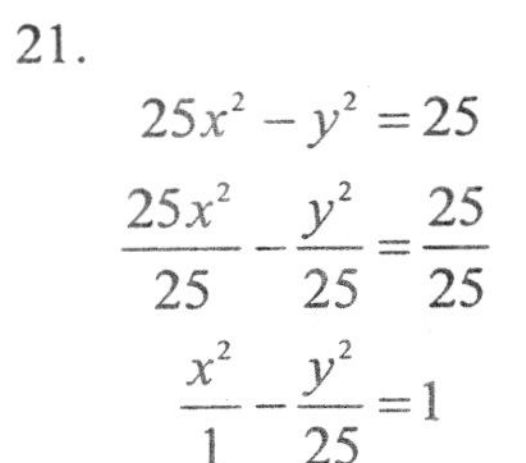

$$25x^2 - y^2 = 25$$

$$\frac{25x^2}{25} - \frac{y^2}{25} = \frac{25}{25}$$

$$\frac{x^2}{1} - \frac{y^2}{25} = 1$$

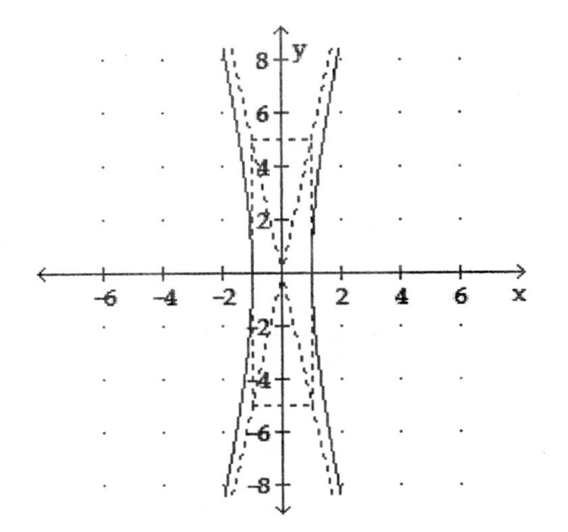

23. $\dfrac{(x-2)^2}{9} - \dfrac{y^2}{16} = 1$

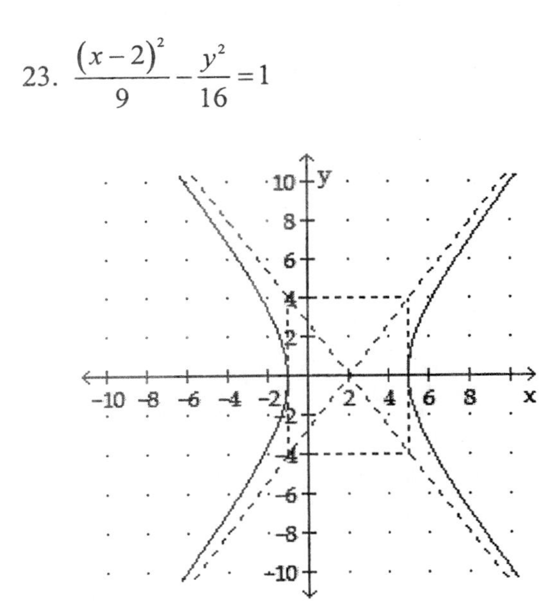

25. $\dfrac{(y+1)^2}{1} - \dfrac{(x-2)^2}{4} = 1$

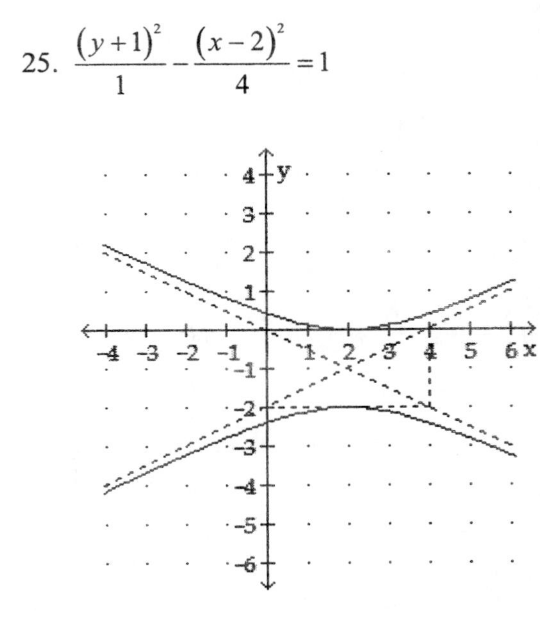

27.

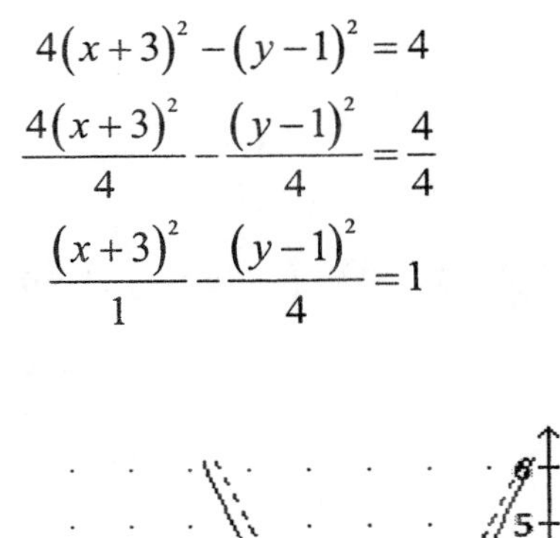

$$4(x+3)^2 - (y-1)^2 = 4$$

$$\frac{4(x+3)^2}{4} - \frac{(y-1)^2}{4} = \frac{4}{4}$$

$$\frac{(x+3)^2}{1} - \frac{(y-1)^2}{4} = 1$$

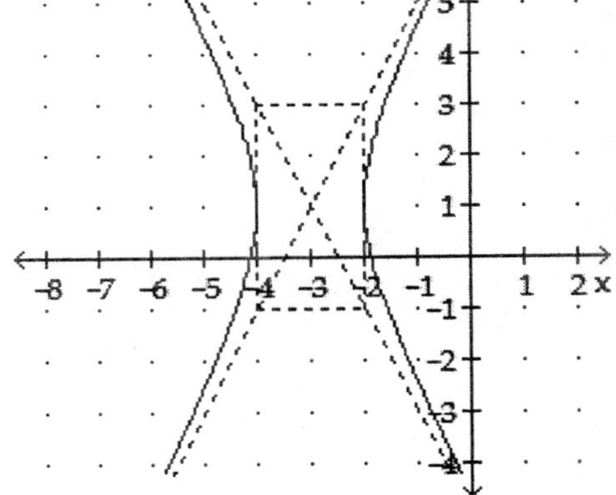

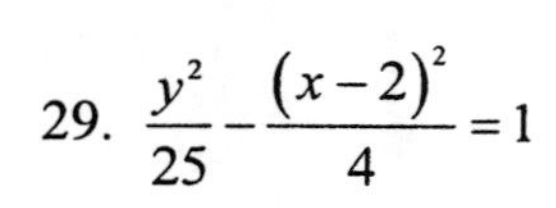

29. $\dfrac{y^2}{25}-\dfrac{(x-2)^2}{4}=1$

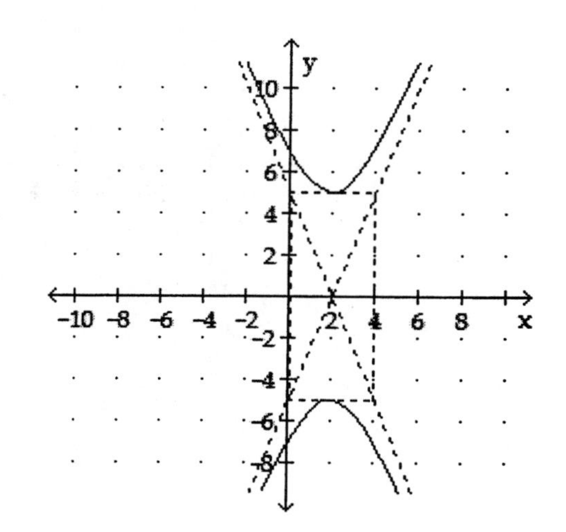

31.

$$x^2+2x-y^2-2y=9$$
$$(x^2+2x)-(y^2+2y)=9$$
$$(x^2+2x+1)-(y^2+2y+1)=9+1-1$$
$$(x+1)^2-(y+1)^2=9$$
$$\frac{(x+1)^2}{9}-\frac{(y+1)^2}{9}=1$$

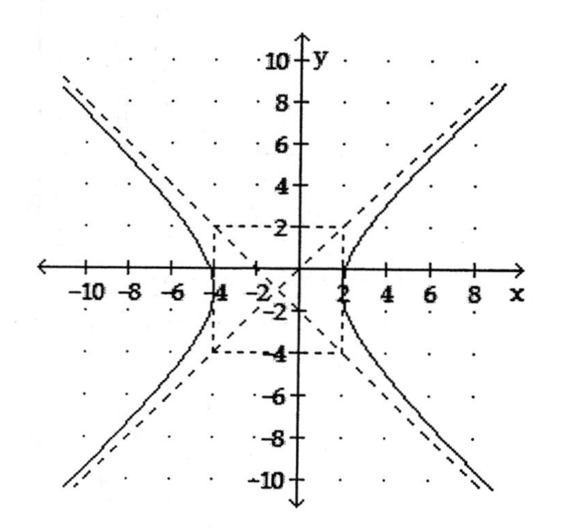

33.

$$x^2-y^2-6y=34$$
$$x^2-(y^2+6y)=34$$
$$x^2-(y^2+6y+9)=34-9$$
$$x^2-(y+3)^2=25$$
$$\frac{x^2}{25}-\frac{(y+3)^2}{25}=1$$

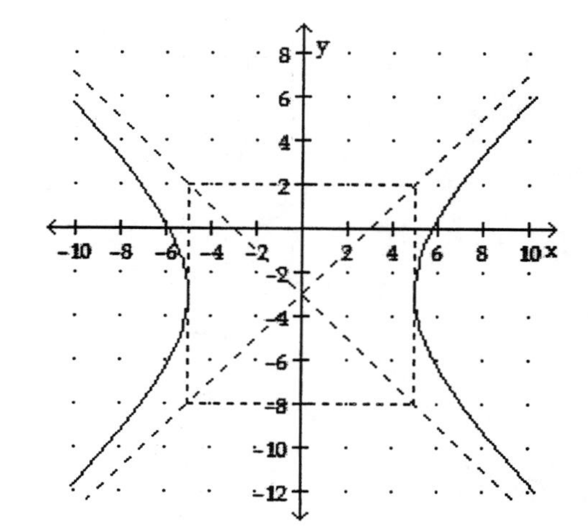

35.

$$xy = 8$$

$$y = \frac{8}{x}$$

x	$y = \frac{8}{x}$
1	$\frac{8}{1} = 8$
2	$\frac{8}{2} = 4$
4	$\frac{8}{4} = 2$
–4	$\frac{8}{-4} = -2$
–2	$\frac{8}{-2} = -4$
–1	$\frac{8}{-1} = -8$

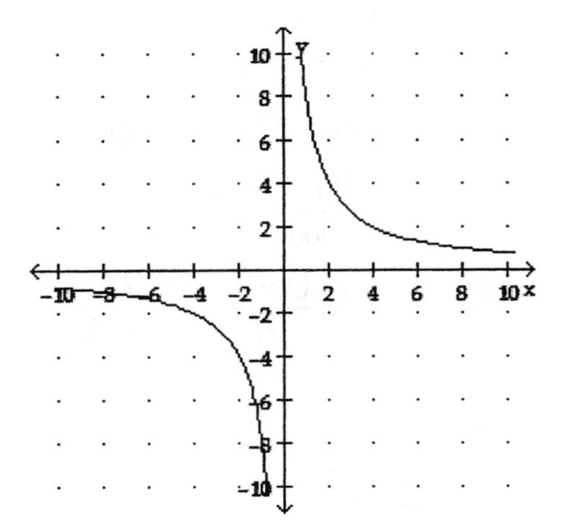

37.

$$\frac{x^2}{9} - \frac{y^2}{4} = 1$$

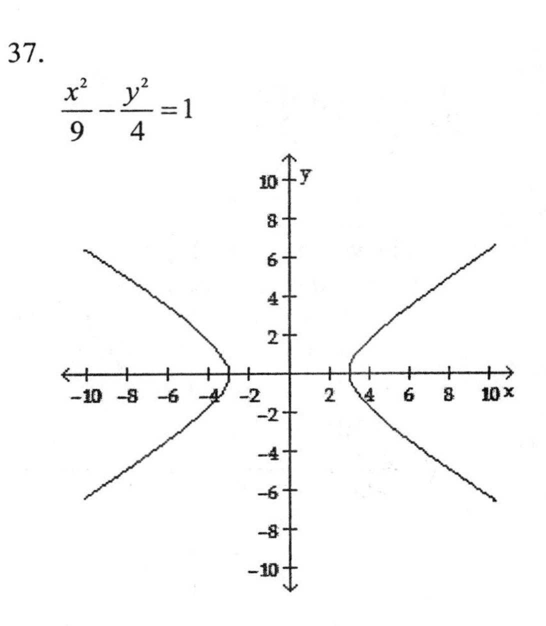

39.

$$\frac{x^2}{4} - \frac{(y-1)^2}{9} = 1$$

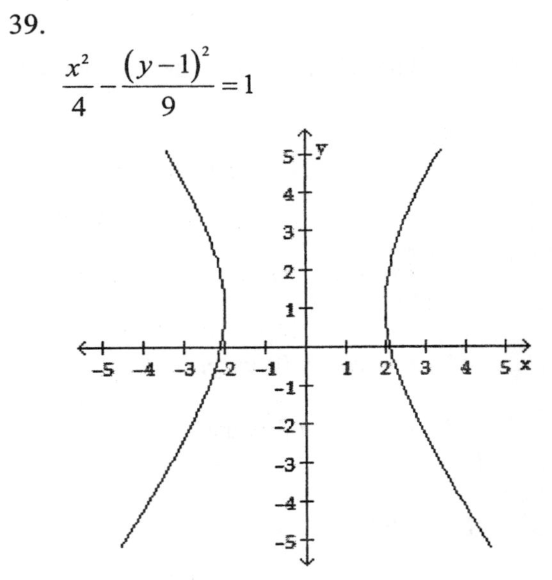

APPLICATIONS

41. ALPHA PARTICLES

Find the coordinates of the vertex.

$$9y^2 - x^2 = 81$$

$$\frac{9y^2}{81} - \frac{x^2}{81} = \frac{81}{81}$$

$$\frac{y^2}{9} - \frac{x^2}{81} = 1$$

$$a = 3 \text{ and } b = 9$$

The vertex is (3, 0), so the particle comes within 3 units of the nucleus.

43. SONIC BOOM

$$y^2 - x^2 = 25$$

$$\frac{y^2}{25} - \frac{x^2}{25} = 1$$

The vertex is (0, 5). The y-value five miles from the vertex is 5 + 5 = 10.
Let $y = 10$ and solve for x to find half of the width of the hyperbola.

$$y^2 - x^2 = 25$$
$$10^2 - x^2 = 25$$
$$100 - x^2 = 25$$
$$-x^2 = -75$$
$$x^2 = 75$$
$$x = \sqrt{75}$$
$$x = \pm 5\sqrt{3}$$

The width of the hyperbola would be $2x$.

$$2x = 2\left(5\sqrt{3}\right)$$
$$= 10\sqrt{3} \text{ miles}$$

WRITING

45. Answers will vary.

47. Answers will vary.

REVIEW

49.

$$\log_8 x = 2$$
$$8^2 = x$$
$$64 = x$$

51.

$$\log_{1/2} \frac{1}{8} = x$$
$$\left(\frac{1}{2}\right)^x = \frac{1}{8}$$
$$2^{-x} = 2^{-3}$$
$$-x = -3$$
$$x = 3$$

53.

$$\log_x \frac{9}{4} = 2$$
$$x^2 = \frac{9}{4}$$
$$x = \sqrt{\frac{9}{4}}$$
$$x = \frac{3}{2}$$

CHALLENGE PROBLEMS

55.

$$36x^2 - 25y^2 - 72x - 100y = 964$$
$$\left(36x^2 - 72x\right) - \left(25y^2 + 100y\right) = 964$$
$$36\left(x^2 - 2x\right) - 25\left(y^2 + 4y\right) = 964$$
$$36\left(x^2 - 2x + 1\right) - 25\left(y^2 + 4y + 4\right) = 964 + 36(1) - 25(4)$$
$$36(x-1)^2 - 25(y+2)^2 = 900$$
$$\frac{36(x-1)^2}{900} - \frac{25(y+2)^2}{900} = \frac{900}{900}$$
$$\frac{(x-1)^2}{25} - \frac{(y+2)^2}{36} = 1$$

57.

$$\frac{x^2}{1} - \frac{y^2}{25} = 1$$

SECTION 13.4

VOCABULARY

1. $\begin{cases} 4x^2 + 6y^2 = 24 \\ 9x^2 - y^2 = 9 \end{cases}$ is a **system** of two nonlinear equations.

3. When solving a system by **graphing**, it is often difficult to determine the coordinates of the intersection points.

5. A **secant** is a line that intersects a circle at two points.

CONCEPTS

7. a. At most, a line can intersect an ellipse at **two** points.
   b. At most, a line can intersect a parabola at **four** points.
   c. At most, an ellipse can intersect a circle at **four** points.
   d. At most, a hyperbola can intersect a circle at **four** points.

9. They intersect at (–3, 2), (3, 2), (–3, –2), and (3, –2).

11. a. –4
    b. –2

NOTATION

13.

$$x^2 + y^2 = 5$$
$$x^2 + \left(\boxed{2x}\right)^2 = 5$$
$$x^2 + 4x^2 = \boxed{5}$$
$$\boxed{5}x^2 = 5$$
$$x^2 = \boxed{1}$$
$$x = 1 \quad \text{or} \quad x = -1$$

If $x = 1$, then

$$y = 2\left(\boxed{1}\right) = 2$$

If $x = -1$, then

$$y = 2\left(\boxed{-1}\right) = -2$$

The solutions are $(1, 2)$ and $\left(-1, \boxed{-2}\right)$.

PRACTICE

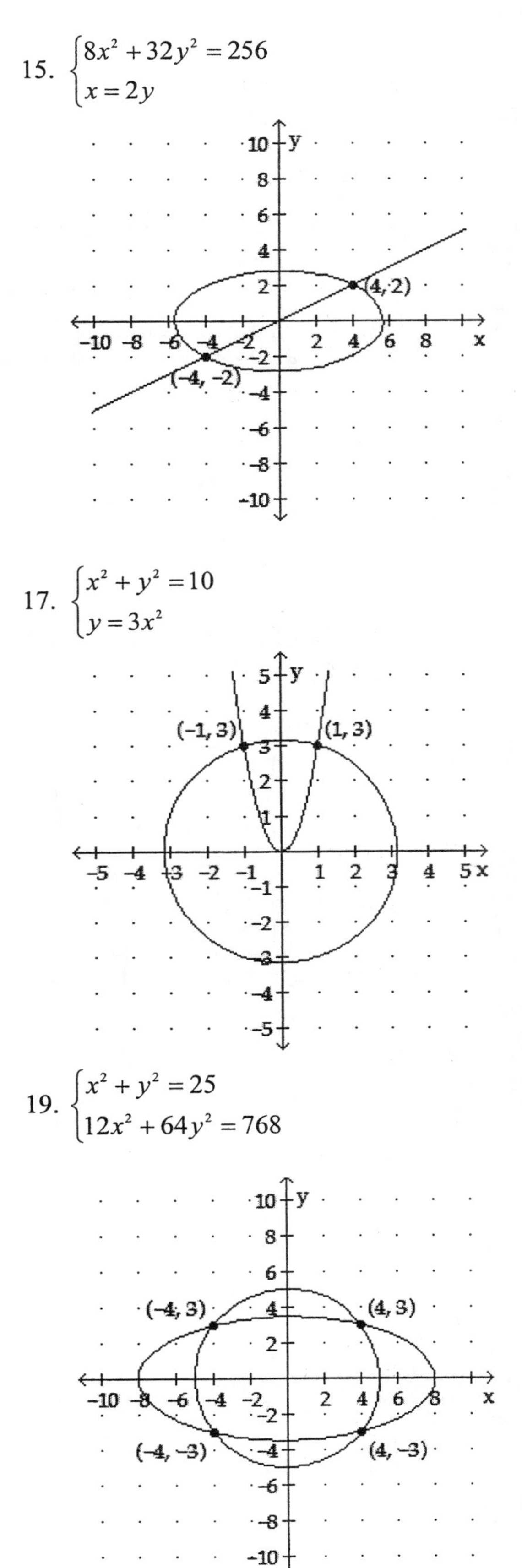

15. $\begin{cases} 8x^2 + 32y^2 = 256 \\ x = 2y \end{cases}$

17. $\begin{cases} x^2 + y^2 = 10 \\ y = 3x^2 \end{cases}$

19. $\begin{cases} x^2 + y^2 = 25 \\ 12x^2 + 64y^2 = 768 \end{cases}$

21. $\begin{cases} x^2 - 13 = -y^2 \\ y = \dfrac{2}{3}x \end{cases}$

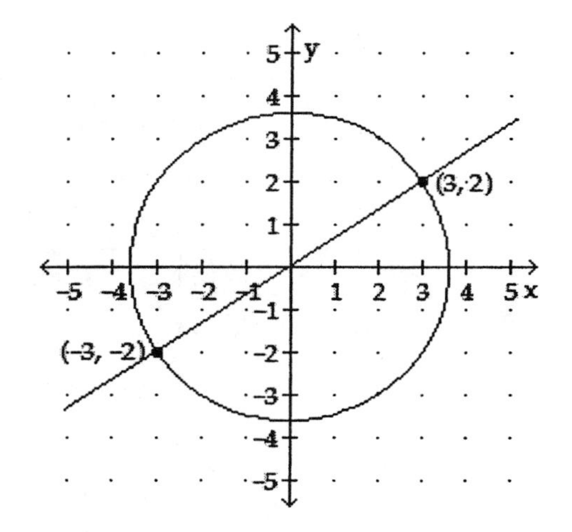

23. $\begin{cases} x^2 - 6x - y = -5 \\ x^2 - 6x + y = -5 \end{cases}$

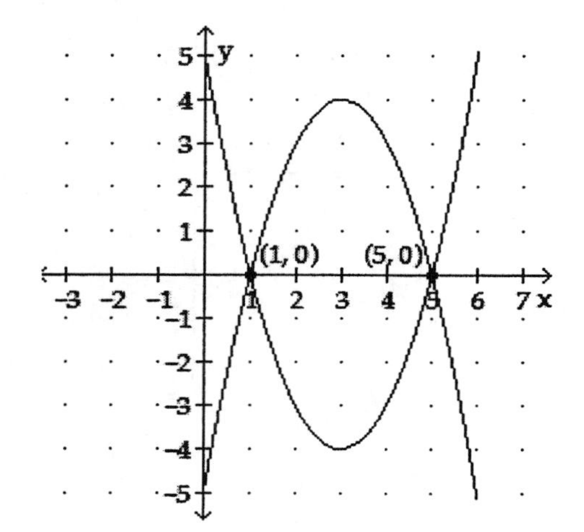

25.

$\begin{cases} 25x^2 + 9y^2 = 225 \\ 5x + 3y = 15 \end{cases}$

Solve the second equation for y.

$$5x + 3y = 15$$
$$3y = -5x + 15$$
$$y = -\frac{5}{3}x + 5$$

Substitute $y = -\frac{5}{3}x + 5$ into the first equation and solve for x.

$$25x^2 + 9y^2 = 225$$
$$25x^2 + 9\left(-\frac{5}{3}x + 5\right)^2 = 225$$
$$25x^2 + 9\left(\frac{25}{9}x^2 - \frac{25}{3}x - \frac{25}{3}x + 25\right) = 225$$
$$25x^2 + 9\left(\frac{25}{9}x^2 - \frac{50}{3}x + 25\right) = 225$$
$$25x^2 + 25x^2 - 150x + 225 = 225$$
$$50x^2 - 150x = 0$$
$$50x(x - 3) = 0$$
$$50x = 0 \quad \text{or} \quad x - 3 = 0$$
$$x = 0 \qquad\qquad x = 3$$

Substitute $x = 0$ and $x = 3$ into $5x + 3y = 15$ and solve for y.

$$5(0) + 3y = 15 \qquad 5(3) + 3y = 15$$
$$3y = 15 \qquad 15 + 3y = 15$$
$$y = 5 \qquad 3y = 0$$
$$y = 0$$

The solutions are $(0, 5)$ and $(3, 0)$.

27.

$$\begin{cases} x^2 + y^2 = 2 \\ x + y = 2 \end{cases}$$

Solve the second equation for y.

$x + y = 2$

$y = -x + 2$

Substitute $y = -x + 2$ into the first equation and solve for x.

$$\begin{aligned} x^2 + y^2 &= 2 \\ x^2 + (-x+2)^2 &= 2 \\ x^2 + (x^2 - 2x - 2x + 4) &= 2 \\ x^2 + (x^2 - 4x + 4) &= 2 \\ 2x^2 - 4x + 4 &= 2 \\ 2x^2 - 4x + 2 &= 0 \\ 2(x^2 - 2x + 1) &= 0 \\ 2(x-1)(x-1) &= 0 \end{aligned}$$

$x - 1 = 0$ or $x - 1 = 0$

$x = 1$ $\qquad$ $x = 1$

Substitute $x = 1$ into $x + y = 2$ and solve for y.

$(1) + y = 2$ $\qquad$ $(1) + y = 2$

$1 + y = 2$ $\qquad$ $1 + y = 2$

$y = 1$ $\qquad$ $y = 1$

The solution is $(1, 1)$.

29.

$$\begin{cases} x^2 + y^2 = 5 \\ x + y = 3 \end{cases}$$

Solve the second equation for y.

$x + y = 3$

$y = -x + 3$

Substitute $y = -x + 3$ into the first equation and solve for x.

$$\begin{aligned} x^2 + y^2 &= 5 \\ x^2 + (-x+3)^2 &= 5 \\ x^2 + (x^2 - 3x - 3x + 9) &= 5 \\ x^2 + (x^2 - 6x + 9) &= 5 \\ 2x^2 - 6x + 9 &= 5 \\ 2x^2 - 6x + 4 &= 0 \\ 2(x^2 - 3x + 2) &= 0 \\ 2(x-1)(x-2) &= 0 \end{aligned}$$

$x - 1 = 0$ or $x - 2 = 0$

$x = 1$ $\qquad$ $x = 2$

Substitute $x = 1$ and $x = 2$ into $x + y = 2$ and solve for y.

$(1) + y = 3$ $\qquad$ $(2) + y = 3$

$y = 2$ $\qquad$ $y = 1$

The solutions are $(1, 2)$ and $(2, 1)$.

31.

$$\begin{cases} x^2 + y^2 = 13 \\ y = x^2 - 1 \end{cases}$$

Substitute $y = x^2 - 1$ into the first equation and solve for x.

$$x^2 + y^2 = 13$$
$$x^2 + \left(x^2 - 1\right)^2 = 13$$
$$x^2 + \left(x^4 - x^2 - x^2 + 1\right) = 13$$
$$x^2 + \left(x^4 - 2x^2 + 1\right) = 13$$
$$x^4 - x^2 + 1 = 13$$
$$x^4 - x^2 - 12 = 0$$
$$\left(x^2 - 4\right)\left(x^2 + 3\right) = 0$$

$x^2 - 4 = 0$ or $x^2 + 3 = 0$

$x^2 = 4$	$x^2 = -3$
$x = \sqrt{4}$	$x = \sqrt{-3}$
$x = \pm 2$	not real

Substitute $x = 2$ and $x = -2$ into $y = x^2 - 1$ and solve for y.

$y = (2)^2 - 1$	$y = (-2)^2 - 1$
$y = 4 - 1$	$y = 4 - 1$
$y = 3$	$y = 3$

The solutions are $(2, 3)$ and $(-2, 3)$.

33.

$$\begin{cases} x^2 + y^2 = 30 \\ y = x^2 \end{cases}$$

Write the second equation in standard form and use elimination.

$$\begin{array}{r} x^2 + y^2 = 30 \\ -x^2 + y = 0 \\ \hline y^2 + y = 30 \end{array}$$
$$y^2 + y - 30 = 0$$
$$(y + 6)(y - 5) = 0$$

$y + 6 = 0$ or $y - 5 = 0$

$y = -6$ $\quad$ $y = 5$

Substitute $y = -6$ and $y = 5$ into $y = x^2$ and solve for x.

$-6 = x^2$	$5 = x^2$
$x^2 = -6$	$x^2 = 5$
$x = \sqrt{-6}$	$x = \sqrt{5}$
not real	$x = \pm\sqrt{5}$

The solutions are $\left(\sqrt{5}, 5\right)$ and $\left(-\sqrt{5}, 5\right)$.

35.

$$\begin{cases} x^2 + y^2 = 13 \\ x^2 - y^2 = 5 \end{cases}$$

Add the equations and use elimination to solve for x.

$$\begin{array}{r} x^2 + y^2 = 13 \\ x^2 - y^2 = 5 \\ \hline 2x^2 = 18 \end{array}$$
$$x^2 = 9$$
$$x = \sqrt{9}$$
$$x = \pm 3$$

Substitute $x = 3$ and $x = -3$ into $x^2 + y^2 = 13$ and solve for y.

$x^2 + y^2 = 13$	$x^2 + y^2 = 13$
$(3)^2 + y^2 = 13$	$(-3)^2 + y^2 = 13$
$9 + y^2 = 13$	$9 + y^2 = 13$
$y^2 = 4$	$y^2 = 4$
$y = \sqrt{4}$	$y = \sqrt{4}$
$y = \pm 2$	$y = \pm 2$

The solutions are $(3, 2)$, $(3, -2)$, $(-3, 2)$, and $(-3, -2)$.

37.

$$\begin{cases} x^2 + y^2 = 20 \\ x^2 - y^2 = -12 \end{cases}$$

Add the equations and use elimination to solve for x.

$$\begin{aligned} x^2 + y^2 &= 20 \\ x^2 - y^2 &= -12 \\ \hline 2x^2 &= 8 \\ x^2 &= 4 \\ x &= \sqrt{4} \\ x &= \pm 2 \end{aligned}$$

Substitute $x = 2$ and $x = -2$ into $x^2 + y^2 = 20$ and solve for y.

$$\begin{aligned} x^2 + y^2 &= 20 \\ (2)^2 + y^2 &= 20 \\ 4 + y^2 &= 20 \\ y^2 &= 16 \\ y &= \sqrt{16} \\ y &= \pm 4 \end{aligned} \qquad \begin{aligned} x^2 + y^2 &= 20 \\ (-2)^2 + y^2 &= 20 \\ 4 + y^2 &= 20 \\ y^2 &= 16 \\ y &= \sqrt{16} \\ y &= \pm 4 \end{aligned}$$

The solutions are $(2,4)$, $(2,-4)$, $(-2,4)$ and $(-2,-4)$.

39.

$$\begin{cases} y^2 = 40 - x^2 \\ y = x^2 - 10 \end{cases}$$

Write both equations in standard form and use elimination.

$$\begin{aligned} x^2 + y^2 &= 40 \\ -x^2 + y &= -10 \\ \hline y^2 + y &= 30 \\ y^2 + y - 30 &= 0 \\ (y+6)(y-5) &= 0 \end{aligned}$$

$$y + 6 = 0 \quad \text{or} \quad y - 5 = 0$$

$$y = -6 \qquad\qquad y = 5$$

Substitute $y = -6$ and $y = 5$ into $y = x^2 - 10$ and solve for x.

$$\begin{aligned} -6 &= x^2 - 10 \\ x^2 - 10 &= -6 \\ x^2 &= 4 \\ x &= \sqrt{4} \\ x &= \pm 2 \end{aligned} \qquad \begin{aligned} 5 &= x^2 - 10 \\ x^2 - 10 &= 5 \\ x^2 &= 15 \\ x &= \sqrt{15} \\ x &= \pm\sqrt{15} \end{aligned}$$

The solutions are $(2,-6)$, $(-2,-6)$, $(\sqrt{15},5)$, and $(-\sqrt{15},5)$.

41.

$$\begin{cases} y = x^2 - 4 \\ x^2 - y^2 = -16 \end{cases}$$

Write the first equation in standard form and use elimination.

$$\begin{aligned} -x^2 + y &= -4 \\ x^2 - y^2 &= -16 \\ \hline -y^2 + y &= -20 \\ y^2 - y - 20 &= 0 \\ (y+4)(y-5) &= 0 \end{aligned}$$

$y + 4 = 0$ or $y - 5 = 0$

$y = -4$ $\quad y = 5$

Substitute $y = -4$ and $y = 5$ into $y = x^2 - 4$ and solve for x.

$$\begin{aligned} -4 &= x^2 - 4 & \quad 5 &= x^2 - 4 \\ x^2 - 4 &= -4 & x^2 - 4 &= 5 \\ x^2 &= 0 & x^2 &= 9 \\ x &= \sqrt{0} & x &= \sqrt{9} \\ x &= 0 & x &= \pm 3 \end{aligned}$$

The solutions are $(0, -4), (3, 5)$, and $(-3, 5)$.

43.

$$\begin{cases} x^2 - y^2 = -5 \\ 3x^2 + 2y^2 = 30 \end{cases}$$

Multiply the first equation by 2 and use elimination to solve for x.

$$\begin{aligned} 2x^2 - 2y^2 &= -10 \\ 3x^2 + 2y^2 &= 30 \\ \hline 5x^2 &= 20 \\ x^2 &= 4 \\ x &= \sqrt{4} \\ x &= \pm 2 \end{aligned}$$

Substitute $x = 2$ and $x = -2$ into $x^2 - y^2 = -5$ and solve for y.

$$\begin{aligned} x^2 - y^2 &= -5 & \quad x^2 - y^2 &= -5 \\ (2)^2 - y^2 &= -5 & (-2)^2 - y^2 &= -5 \\ 4 - y^2 &= -5 & 4 - y^2 &= -5 \\ -y^2 &= -9 & -y^2 &= -9 \\ y^2 &= 9 & y^2 &= 9 \\ y &= \pm\sqrt{9} & y &= \pm\sqrt{9} \\ y &= \pm 3 & y &= \pm 3 \end{aligned}$$

The solutions are $(2, 3)$, $(2, -3)$, $(-2, 3)$, and $(-2, -3)$.

45.

$$\begin{cases} \frac{1}{x}+\frac{2}{y}=1 \\ \frac{2}{x}-\frac{1}{y}=\frac{1}{3} \end{cases}$$

Multiply the second equation by 2 and add the two equations to eliminate y.

$$\frac{1}{x}+\frac{2}{y}=1$$
$$\frac{4}{x}-\frac{2}{y}=\frac{2}{3}$$
$$\frac{5}{x}=\frac{5}{3}$$
$$3x\left(\frac{5}{x}\right)=3x\left(\frac{5}{3}\right)$$
$$15=5x$$
$$3=x$$

Substitute $x=3$ into the first equation and solve for y.

$$\frac{1}{3}+\frac{2}{y}=1$$
$$\frac{2}{y}=\frac{2}{3}$$
$$3y\left(\frac{2}{y}\right)=3y\left(\frac{2}{3}\right)$$
$$6=2y$$
$$3=y$$

The solution is $(3, 3)$.

47.

$$\begin{cases} 3y^2=xy \\ 2x^2+xy-84=0 \end{cases}$$

Solve the first equation for x.

$$3y^2=xy$$
$$\frac{3y^2}{y}=x$$
$$3y=x$$

Substitute $3y=x$ into the second equation and solve for y.

$$2(3y)^2+(3y)y-84=0$$
$$2(9y^2)+3y^2-84=0$$
$$18y^2+3y^2-84=0$$
$$21y^2-84=0$$
$$21y^2=84$$
$$y^2=4$$
$$y=\sqrt{4}$$
$$y=\pm 2$$

Substitute $y=2$ and $y=-2$ into the first equation to solve for x.

$$3y^2=xy \qquad 3y^2=xy$$
$$3(2)^2=x(2) \qquad 3(-2)^2=x(-2)$$
$$12=2x \qquad 12=-2x$$
$$6=x \qquad -6=x$$

Set the denominator of $\frac{3y^2}{y}=0$ and you also get $y=0$. Substitute $y=0$ into the second equation and solve for x.

$$2x^2+xy-84=0$$
$$2x^2+x(0)-84=0$$
$$2x^2-84=0$$
$$2(x^2-42)=0$$
$$x^2=42$$
$$x=\pm\sqrt{42}$$

The solutions are $(6,2),(-6,-2),(\sqrt{42},0)$ and $(-\sqrt{42},0)$.

49.

$$\begin{cases} xy = \frac{1}{6} \\ y + x = 5xy \end{cases}$$

Solve the first equation for y.

$$xy = \frac{1}{6}$$
$$\frac{1}{x}(xy) = \frac{1}{x}\left(\frac{1}{6}\right)$$
$$y = \frac{1}{6x}$$

Substitute $y = \frac{1}{6x}$ into the second equation and solve for x.

$$y + x = 5xy$$
$$\frac{1}{6x} + x = 5x\left(\frac{1}{6x}\right)$$
$$\frac{1}{6x} + x = \frac{5}{6}$$
$$6x\left(\frac{1}{6x} + x\right) = 6x\left(\frac{5}{6}\right)$$
$$1 + 6x^2 = 5x$$
$$6x^2 - 5x + 1 = 0$$
$$(2x - 1)(3x - 1) = 0$$
$$2x - 1 = 0 \quad \text{or} \quad 3x - 1 = 0$$
$$x = \frac{1}{2} \qquad\qquad x = \frac{1}{3}$$

Substitute $x = \frac{1}{2}$ and $x = \frac{1}{3}$ into $xy = \frac{1}{6}$ and solve for y.

$$xy = \frac{1}{6} \qquad\qquad xy = \frac{1}{6}$$
$$\frac{1}{2}y = \frac{1}{6} \qquad\qquad \frac{1}{3}y = \frac{1}{6}$$
$$6\left(\frac{1}{2}y\right) = 6\left(\frac{1}{6}\right) \qquad\qquad 6\left(\frac{1}{3}y\right) = 6\left(\frac{1}{6}\right)$$
$$3y = 1 \qquad\qquad 2y = 1$$
$$y = \frac{1}{3} \qquad\qquad y = \frac{1}{2}$$

The solutions are $\left(\frac{1}{2}, \frac{1}{3}\right)$ and $\left(\frac{1}{3}, \frac{1}{2}\right)$.

51. INTEGER PROBLEM

Let the integers be x and y.

$$\begin{cases} xy = 32 \\ x + y = 12 \end{cases}$$

Solve the first equation for y.

$$xy = 32$$
$$y = \frac{32}{x}$$

Substitute $y = \frac{32}{x}$ into $x + y = 12$ and solve for x.

$$x + y = 12$$
$$x + \frac{32}{x} = 12$$
$$x\left(x + \frac{32}{x}\right) = x(12)$$
$$x^2 + 32 = 12x$$
$$x^2 - 12x + 32 = 0$$
$$(x - 4)(x - 8) = 0$$
$$x - 4 = 0 \quad \text{or} \quad x - 8 = 0$$
$$x = 4 \qquad\qquad x = 8$$

Substitute $x = 4$ and $x = 8$ into the first equation to solve for y.

$$xy = 32 \qquad\qquad xy = 32$$
$$4y = 32 \qquad\qquad 8y = 32$$
$$y = 8 \qquad\qquad y = 4$$

The two integers are 4 and 8.

APPLICATIONS

53. GEOMETRY

Let x = length and y = width.
Area = Length · Width
Perimeter = 2(Length) + 2(Width)

$$\begin{cases} xy = 63 \\ 2x + 2y = 32 \end{cases}$$

Solve the first equation for y.

$$xy = 63$$
$$y = \frac{63}{x}$$

Substitute $y = \frac{63}{x}$ into the second equation and solve for x.

$$2x + 2y = 32$$
$$2x + 2\left(\frac{63}{x}\right) = 32$$
$$2x + \frac{126}{x} = 32$$
$$x\left(2x + \frac{126}{x}\right) = x(32)$$
$$2x^2 + 126 = 32x$$
$$2x^2 - 32x + 126 = 0$$
$$2(x^2 - 16x + 63) = 0$$
$$2(x-7)(x-9) = 0$$
$$x - 7 = 0 \quad \text{or} \quad x - 9 = 0$$
$$x = 7 \qquad\qquad x = 9$$

Substitute $x = 7$ and $x = 9$ into the first equation and solve for y.

$$xy = 63 \qquad xy = 63$$
$$7y = 63 \qquad 9y = 63$$
$$y = 9 \qquad y = 7$$

The dimensions are 7 cm by 9 cm.

55. INVESTING

Let x = the amount invested and y = rate of investment.
Carol's investment:
$xy = 67.50$
John's investment:
$(x + 150)(y + 0.015) = 94.50$

$$\begin{cases} xy = 67.50 \\ (x+150)(y+0.015) = 94.50 \end{cases}$$

Solve the first equation for y and simplify the second equation

$$\begin{cases} y = \frac{67.50}{x} \\ xy + 0.015x + 150y + 2.25 = 94.50 \end{cases}$$

$$\begin{cases} y = \frac{67.5}{x} \\ xy - 0.015x + 150y = 92.25 \end{cases}$$

$$xy + 0.015x + 150y = 92.25$$
$$x\left(\frac{67.5}{x}\right) + 0.015x + 150\left(\frac{67.5}{x}\right) = 92.25$$
$$67.5 + 0.015x + \frac{10{,}125}{x} = 92.25$$
$$0.015x + \frac{10{,}125}{x} = 24.75$$
$$x\left(0.015x + \frac{10{,}125}{x}\right) = x(24.75)$$
$$0.015x^2 + 10{,}125 = 24.75x$$
$$0.015x^2 - 24.75x + 10{,}125 = 0$$
$$1{,}000(0.015x^2 - 24.75x + 10{,}125) = 1{,}000(0)$$
$$15x^2 - 24{,}750x + 10{,}125{,}000 = 0$$
$$x^2 - 1{,}650x + 675{,}000 = 0$$
$$(x - 900)(x - 750) = 0$$
$$x - 900 = 0 \quad \text{or} \quad x - 750 = 0$$
$$x = 900 \qquad\qquad x = 750$$

$$y = \frac{67.5}{x} \qquad y = \frac{67.5}{x}$$
$$y = \frac{67.5}{900} \qquad y = \frac{67.5}{750}$$
$$y = 0.075 \qquad y = 0.09$$
$$y = 7.5\% \qquad y = 9\%$$

She invested \$900 at 7.5% or \$750 at 9%.

57. DRIVING RATES

Let r = the rate and t = time.

Distance = rate · time

Jim's trip:

$$rt = 306$$

His brother's trip:

$$(r - 17)(t + 1.5) = 306$$

$$\begin{cases} rt = 306 \\ (r-17)(t+1.5) = 306 \end{cases}$$

Solve the first equation for t and simplify the second equation

$$\begin{cases} t = \dfrac{306}{r} \\ rt + 1.5r - 17t - 25.5 = 306 \end{cases}$$

$$\begin{cases} t = \dfrac{306}{r} \\ rt + 1.5r - 17t = 331.5 \end{cases}$$

$$rt + 1.5r - 17t = 331.5$$

$$r\left(\frac{306}{r}\right) + 1.5r - 17\left(\frac{306}{r}\right) = 331.5$$

$$306 + 1.5r - \frac{5,202}{r} = 331.5$$

$$1.5r - \frac{5,202}{r} = 25.5$$

$$r\left(1.5r - \frac{5,202}{r}\right) = r(25.5)$$

$$1.5r^2 - 5,202 = 25.5r$$

$$1.5r^2 - 25.5r - 5,202 = 0$$

$$r^2 - 17r - 3,468 = 0$$

$$(r-68)(r+51) = 0$$

$$r - 68 = 0 \quad \text{or} \quad r + 51 = 0$$

$$r = 68 \qquad r = \cancel{-51}$$

$$t = \frac{306}{68}$$

$$t = 4.5$$

Jim's rate was 68 mph and his driving time was 4.5 hours.

WRITING

59. Answers will vary.

REVIEW

61.

$$\log 5x = 4$$

$$10^4 = 5x$$

$$10,000 = 5x$$

$$2,000 = x$$

63.

$$\frac{\log(8x-7)}{\log x} = 2$$

$$\log(8x-7) = 2\log x$$

$$\log(8x-7) = \log x^2$$

$$8x - 7 = x^2$$

$$0 = x^2 - 8x + 7$$

$$0 = (x-7)(x-1)$$

$$x - 7 = 0 \quad \text{or} \quad x - 1 = 0$$

$$x = 7 \qquad x = \cancel{1}$$

$x \neq 1$ because it make the denominator undefined

CHALLENGE PROBLEMS

65. a. The system may have 0, 1, 2, 3, or 4 solutions.
    b. The system may have 0, 1, 2, 3, or 4 solutions.

CHAPTER 13 KEY CONCEPTS

1. ellipse

2. circle

3. parabola

4. hyperbola

5. hyperbola

6. ellipse

7. ellipse

8. parabola

9. circle

10. a. $(-1, 2)$
    b. $\sqrt{16} = 4$

11. a. $(-2, 1)$
    b. right

12. a. $(0, 0)$
    b. vertical
    c. $(0, 4)$ and $(0, -4)$

13. a. $(-2, 1)$
    b. left and right
    c. 6 units horizontally and 4 units vertically

14.

$(x+1)^2 + (y-2)^2 = 16$

center at $(-1,2)$ and radius = 4

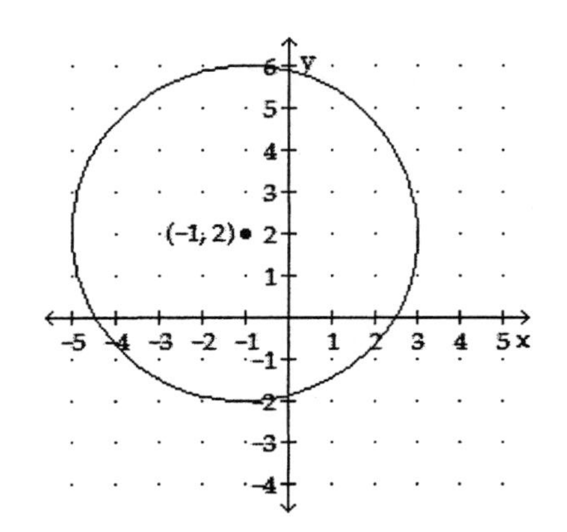

15.

$x = \frac{1}{2}(y-1)^2 - 2$

vertex at $(-2,1)$

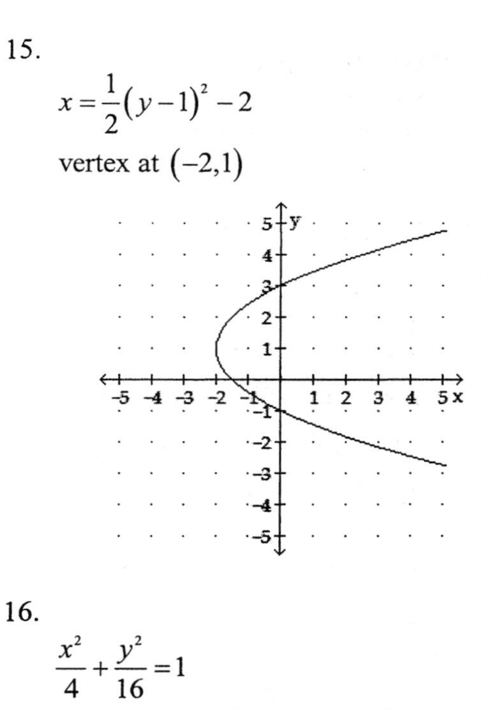

16.

$\frac{x^2}{4} + \frac{y^2}{16} = 1$

vertices at $(0,4)$ and $(0,-4)$

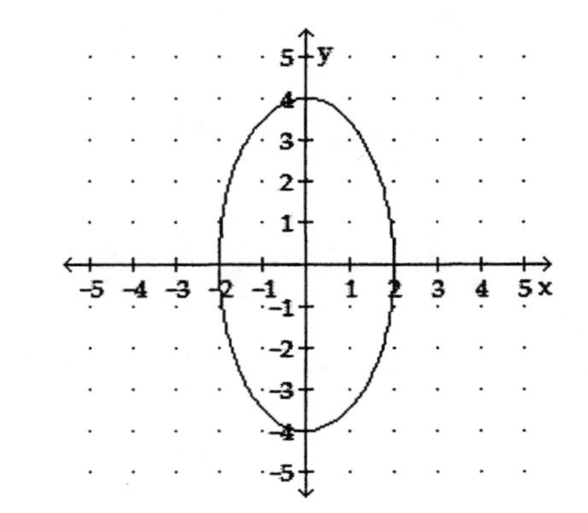

17.

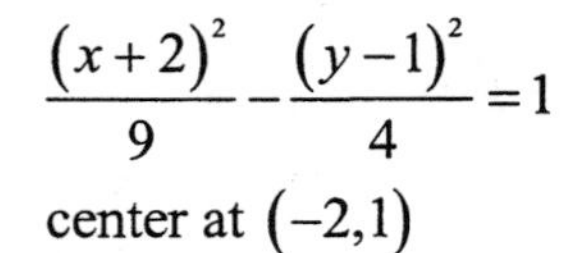

$\frac{(x+2)^2}{9} - \frac{(y-1)^2}{4} = 1$

center at $(-2,1)$

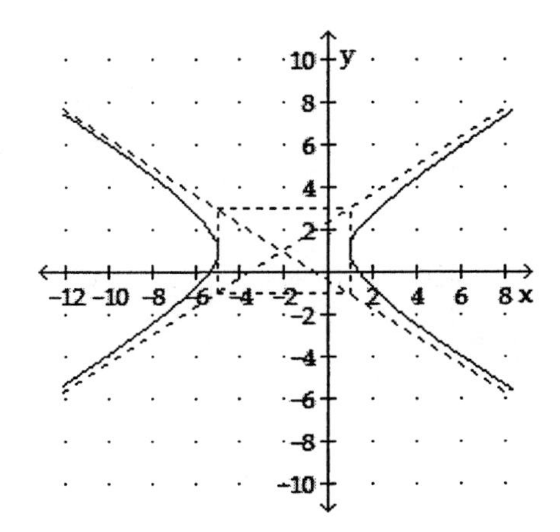

CHAPTER 13 REVIEW

SECTION 13.1
The Circle and the Parabola

1. $x^2 + y^2 = 16$

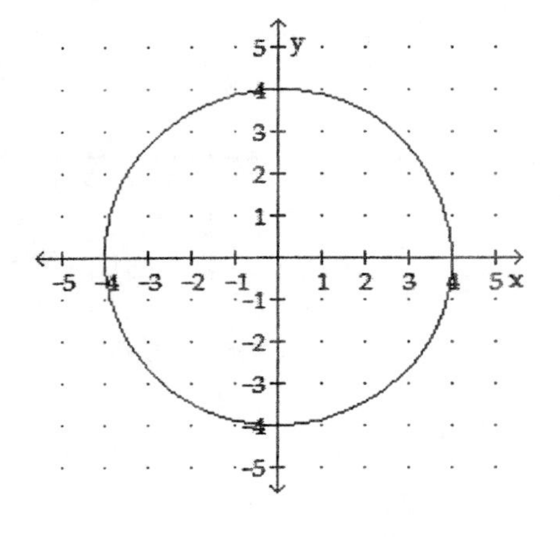

2. $(x-1)^2 + (y+2)^2 = 4$

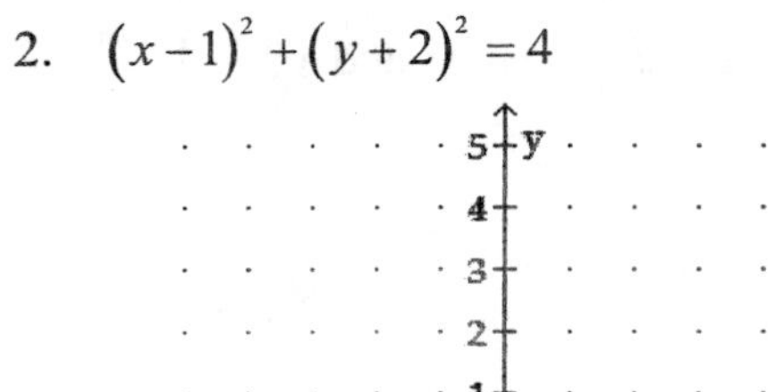

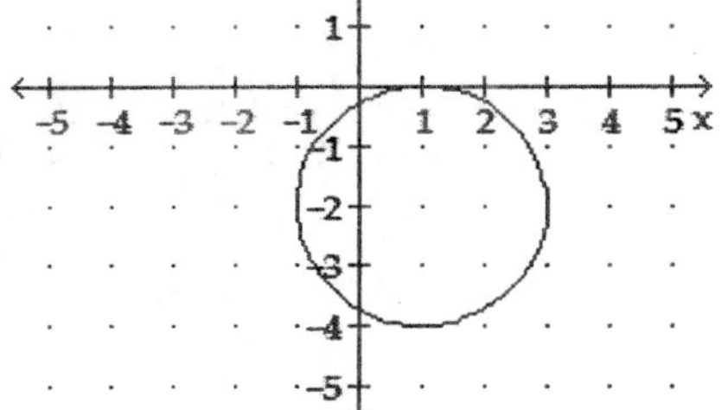

3.

$$x^2 + y^2 + 4x - 2y = 4$$
$$(x^2 + 4x) + (y^2 - 2y) = 4$$
$$(x^2 + 4x + 4) + (y^2 - 2y + 1) = 4 + 4 + 1$$
$$(x+2)^2 + (y-1)^2 = 9$$

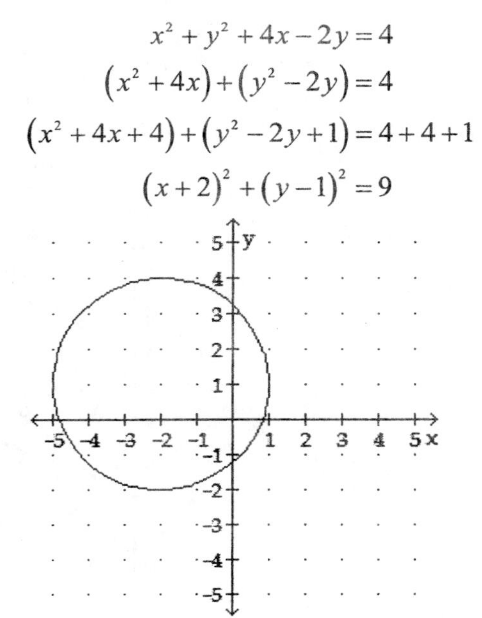

4.

$$\left(x - \frac{17}{2}\right)^2 + \left(y - \frac{17}{2}\right)^2 = \left(\frac{17}{2}\right)^2$$
$$(x-8.5)^2 + (y-8.5)^2 = (8.5)^2$$

5. center: $(-6, 0)$
   radius: $\sqrt{24} = 2\sqrt{6}$

6. A circle is the set of all points in a plane that are a fixed distance from a point called its **center**. The fixed distance is called the **radius** of the circle.

7. $x = y^2$
   vertex: $(0, 0)$

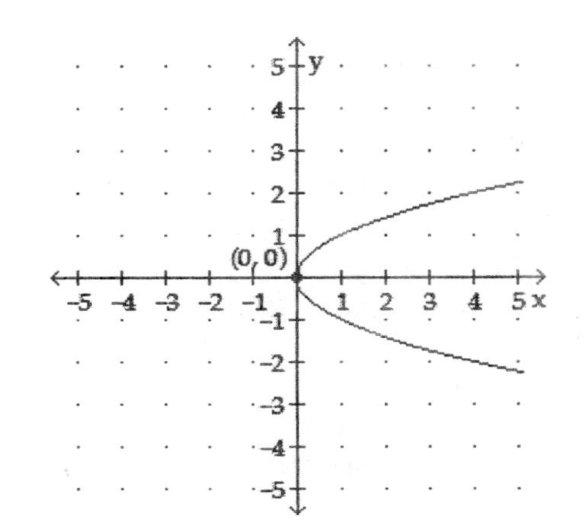

8. $x = 2(y+1)^2 - 2$
   vertex: $(-2, -1)$

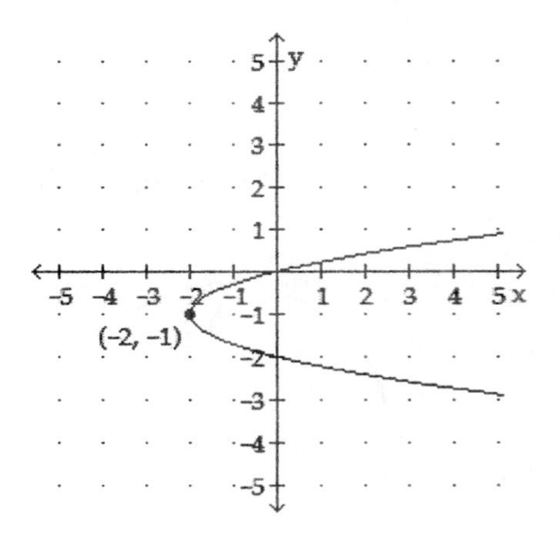

9.

$$x = -3y^2 + 12y - 7$$
$$x = -3(y^2 - 4y) - 7$$
$$x = -3(y^2 - 4y + 4) - 7 - (-3)(4)$$
$$x = -3(y-2)^2 + 5$$

vertex: $(5, 2)$

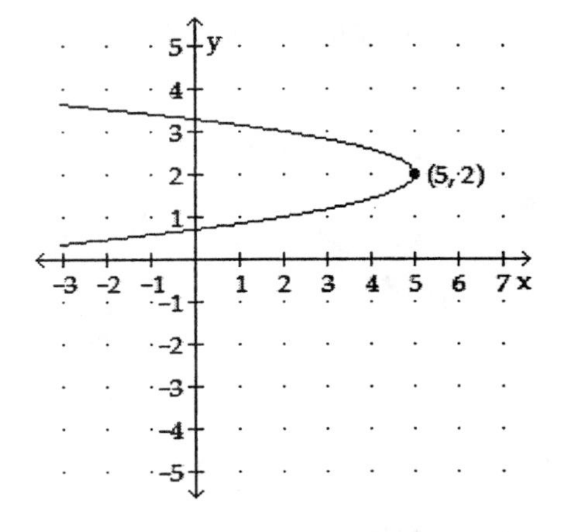

10. a. vertex: $(1, 4)$
axis of symmetry: $y = 4$

b. $y = 2x^2 - 4x + 5$
$$y = 2(x^2 - 2x) + 5$$
$$y = 2(x^2 - 2x + 1) + 5 - 2(1)$$
$$y = 2(x-1)^2 + 3$$

vertex: $(1, 3)$

axis of symmetry: $x = 1$

11. Use symmetry about the axis of symmetry to find the points $(2, -3)$ and $(6, -4)$.

12. When $x = 22$, $y = 0$.
$$y = -\frac{5}{121}(x-11)^2 + 5$$
$$0 \stackrel{?}{=} -\frac{5}{121}(22-11)^2 + 5$$
$$0 \stackrel{?}{=} -\frac{5}{121}(11)^2 + 5$$
$$0 \stackrel{?}{=} -\frac{5}{121}(121) + 5$$
$$0 \stackrel{?}{=} -5 + 5$$
$$0 = 0$$

SECTION 13.2
The Ellipse

13.
$$9x^2 + 16y^2 = 144$$
$$\frac{9x^2}{144} + \frac{16y^2}{144} = \frac{144}{144}$$
$$\frac{x^2}{16} + \frac{y^2}{9} = 1$$

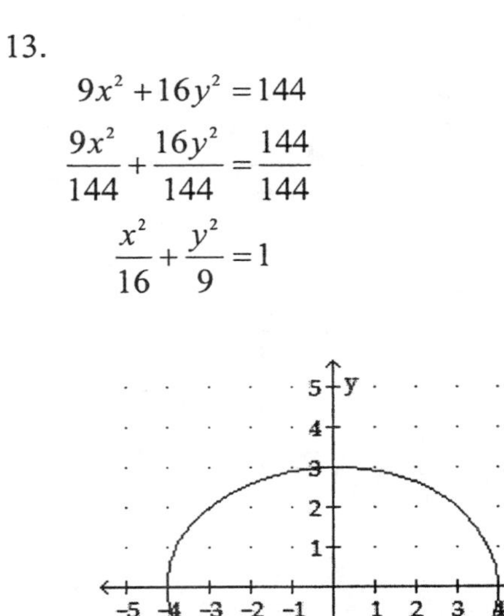

14.
$$\frac{(x-2)^2}{4} + \frac{(y-1)^2}{25} = 1$$

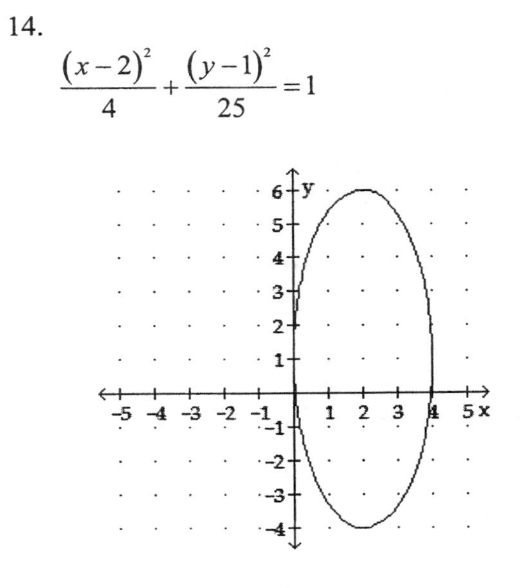

15.

$$4(x+1)^2+9(y-1)^2=36$$
$$\frac{4(x+1)^2}{36}+\frac{9(y-1)^2}{36}=\frac{36}{36}$$
$$\frac{(x+1)^2}{9}+\frac{(y-1)^2}{4}=1$$

16.

$$\frac{x^2}{144}+y^2=1$$
$$\frac{x^2}{12^2}+\frac{y^2}{1^2}=1$$

17.

$$\frac{x^2}{9}+\frac{y^2}{4}=1$$
$$\frac{2^2}{9}+\frac{y^2}{4}=1$$
$$\frac{4}{9}+\frac{y^2}{4}=1$$
$$\frac{y^2}{4}=\frac{5}{9}$$
$$9y^2=20$$
$$y^2=\frac{20}{9}$$
$$y=\pm\sqrt{\frac{20}{9}}$$
$$y=\pm\frac{\sqrt{20}}{\sqrt{9}}$$
$$y=\pm\frac{2\sqrt{5}}{3}$$

The ordered pairs are $\left(2,\frac{2\sqrt{5}}{3}\right)=(2,1.5)$ and $\left(2,-\frac{2\sqrt{5}}{3}\right)=(2,-1.5)$.

18.

$$\frac{x^2}{5^2}+\frac{y^2}{3^2}=1$$
$$\frac{x^2}{25}+\frac{y^2}{9}=1$$

19. An **ellipse** is the set of all points in a plane for which the sum of the distances from two fixed points is a constant. Each of the fixed points is called a **focus**.

20.

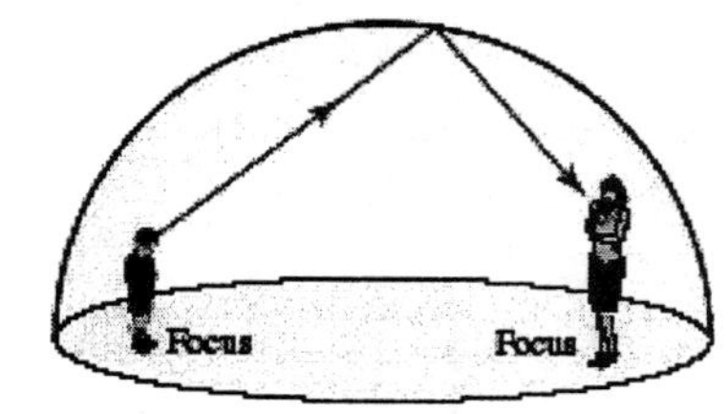

SECTION 13.3
The Hyperbola

21.

$$\frac{y^2}{9}-\frac{x^2}{1}=1$$

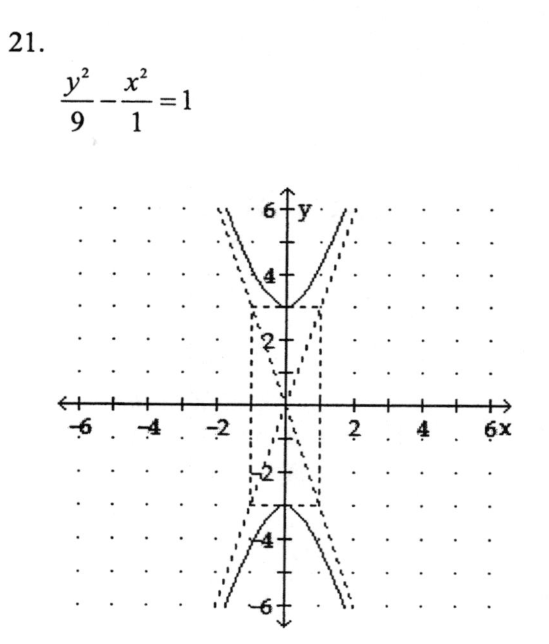

22.

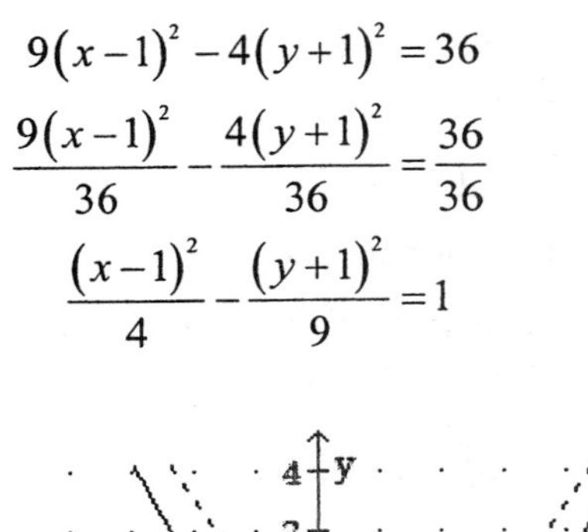

$$9(x-1)^2 - 4(y+1)^2 = 36$$
$$\frac{9(x-1)^2}{36} - \frac{4(y+1)^2}{36} = \frac{36}{36}$$
$$\frac{(x-1)^2}{4} - \frac{(y+1)^2}{9} = 1$$

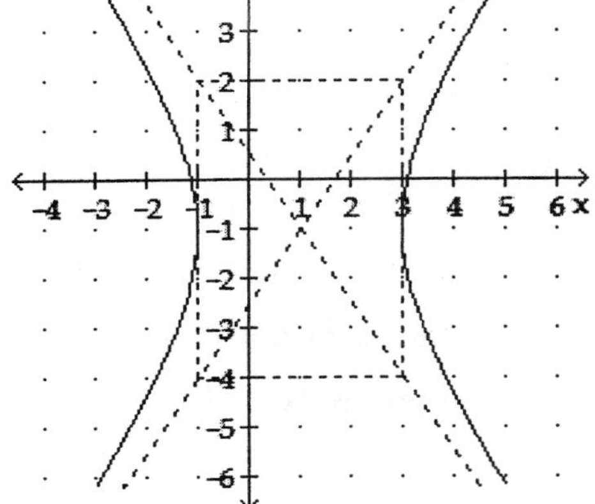

23. $xy = 9$

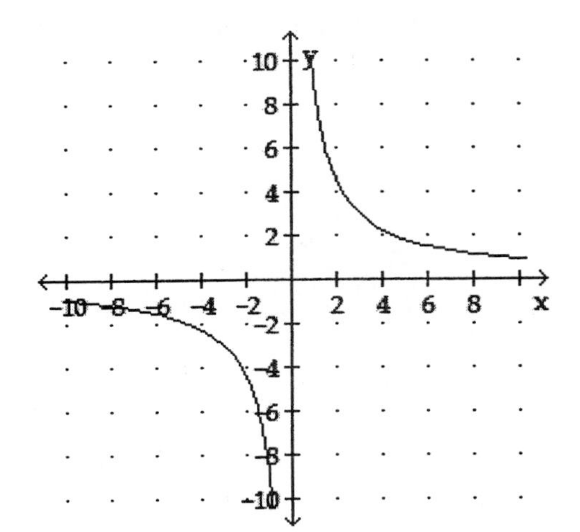

24.

$$y^2 - 4y - x^2 - 2x - 22 = 0$$
$$(y^2 - 4y) - (x^2 + 2x) = 22$$
$$(y^2 - 4y + 4) - (x^2 + 2x + 1) = 22 + 4 - 1$$
$$(y-2)^2 - (x+1)^2 = 25$$
$$\frac{(y-2)^2}{25} - \frac{(x+1)^2}{25} = 1$$

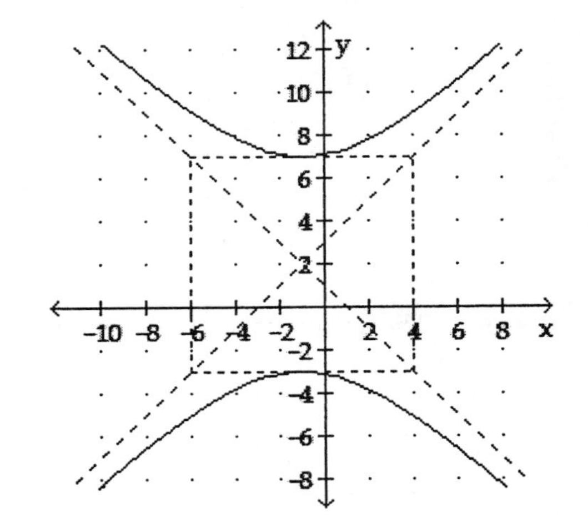

25.

$$x^2 - 4y^2 = 4$$
$$\frac{x^2}{4} - \frac{4y^2}{4} = \frac{4}{4}$$
$$\frac{x^2}{4} - \frac{y^2}{1} = 1$$

$a = 2$, so they were $2^2 = 4$ units apart.

26. a. ellipse
b. hyperbola
c. parabola
d. circle

SECTION 13.4
Solving Systems of Nonlinear Equations

27. Yes, it is a solution.

$$x^2 + y^2 = 20 \qquad\qquad x^2 - y^2 = 2$$
$$\left(-\sqrt{11}\right)^2 + (-3)^2 \stackrel{?}{=} 20 \qquad \left(-\sqrt{11}\right)^2 - (-3)^2 \stackrel{?}{=} 2$$
$$11 + 9 \stackrel{?}{=} 20 \qquad\qquad 11 - 9 \stackrel{?}{=} 2$$
$$20 = 20 \qquad\qquad 2 = 2$$

28. The graphs intersect at (0, 3) and (0, –3).

29. a. 2
b. 4
c. 4
d. 4

30. Let $x = 0$ in both equations to find the y–coordinate.

$$x^2 + y^2 = 1 \qquad 4y^2 - x^2 = 4$$
$$0^2 + y^2 = 1 \qquad 4y^2 - 0^2 = 4$$
$$y^2 = 1 \qquad 4y^2 = 4$$
$$y = \sqrt{1} \qquad y^2 = 1$$
$$y = \pm 1 \qquad y = \sqrt{1}$$
$$y = \pm 1$$

The solutions are (0, 1) and (0, –1).

31.
$$\begin{cases} y^2 - x^2 = 16 \\ y + 4 = x^2 \end{cases}$$

Substitute $y + 4 = x^2$ into the first equation and solve for y.

$$y^2 - x^2 = 16$$
$$y^2 - (y + 4) = 16$$
$$y^2 - y - 4 = 16$$
$$y^2 - y - 20 = 0$$
$$(y - 5)(y + 4) = 0$$
$$y - 5 = 0 \quad \text{or} \quad y + 4 = 0$$
$$y = 5 \qquad y = -4$$

Substitute $y = 5$ and $y = -4$ into the second equation and solve for x.

$$y + 4 = x^2 \qquad y + 4 = x^2$$
$$5 + 4 = x^2 \qquad -4 + 4 = x^2$$
$$9 = x^2 \qquad 0 = x^2$$
$$x = \sqrt{9} \qquad x = \sqrt{0}$$
$$x = \pm 3 \qquad x = 0$$

The solutions are $(3,\ 5), (-3,\ 5)$, and $(0, -4)$.

32.
$$\begin{cases} y = -x^2 + 2 \\ x^2 - y - 2 = 0 \end{cases}$$

Substitute $y = -x^2 + 2$ into the second equation and solve for x.

$$x^2 - y - 2 = 0$$
$$x^2 - (-x^2 + 2) - 2 = 0$$
$$x^2 + x^2 - 2 - 2 = 0$$
$$2x^2 - 4 = 0$$
$$2(x^2 - 2) = 0$$
$$x^2 - 2 = 0$$
$$x^2 = 2$$
$$x = \pm\sqrt{2}$$

Substitute $x = \sqrt{2}$ and $x = -\sqrt{2}$ into the first equation and solve for y.

$$y = -x^2 + 2 \qquad y = -x^2 + 2$$
$$y = -\left(\sqrt{2}\right)^2 + 2 \qquad y = -\left(-\sqrt{2}\right)^2 + 2$$
$$y = -2 + 2 \qquad y = -2 + 2$$
$$y = 0 \qquad y = 0$$

The solutions are $\left(\sqrt{2}, 0\right)$ and $\left(-\sqrt{2}, 0\right)$.

33.

$$\begin{cases} x^2+2y^2=12 \\ 2x-y=2 \end{cases}$$

Solve the second equation for y.

$$\begin{aligned} 2x-y&=2 \\ -y&=-2x+2 \\ y&=2x-2 \end{aligned}$$

Substitute $y=2x-2$ into the first equation and solve for x.

$$\begin{aligned} x^2+2y^2&=12 \\ x^2+2(2x-2)^2&=12 \\ x^2+2(4x^2-4x-4x+4)&=12 \\ x^2+2(4x^2-8x+4)&=12 \\ x^2+8x^2-16x+8&=12 \\ 9x^2-16x-4&=0 \\ (9x+2)(x-2)&=12 \end{aligned}$$

$$\begin{aligned} 9x+2&=0 \quad \text{or} \quad x-2=0 \\ 9x&=-2 \qquad\qquad x=2 \\ x&=-\frac{2}{9} \end{aligned}$$

Substitute $x=-\frac{2}{9}$ and $x=2$ into the second equation and solve for y.

$$\begin{aligned} y&=2x-2 & y&=2x-2 \\ y&=2\left(-\frac{2}{9}\right)-2 & y&=2(2)-2 \\ y&=-\frac{4}{9}-2 & y&=4-2 \\ y&=-\frac{22}{9} & y&=2 \end{aligned}$$

The solutions are $\left(-\frac{2}{9},-\frac{22}{9}\right)$ and $(2,2)$.

34.

$$\begin{cases} 3x^2+y^2=52 \\ x^2-y^2=12 \end{cases}$$

Add the equations and solve for x.

$$\begin{aligned} 3x^2+y^2&=52 \\ \underline{x^2-y^2}&\underline{=12} \\ 4x^2&=64 \\ x^2&=16 \\ x&=\sqrt{16} \\ x&=\pm 4 \end{aligned}$$

Substitute $x=4$ and $x=-4$ into the second equation and solve for y.

$$\begin{aligned} x^2-y^2&=12 & x^2-y^2&=12 \\ (4)^2-y^2&=12 & (-4)^2-y^2&=12 \\ 16-y^2&=12 & 16-y^2&=12 \\ -y^2&=-4 & -y^2&=-4 \\ y^2&=4 & y^2&=4 \\ y&=\sqrt{4} & y&=\sqrt{4} \\ y&=\pm 2 & y&=\pm 2 \end{aligned}$$

The solutions are $(4,2),(4,-2)$, $(-4,2)$, and $(-4,-2)$

35.

$$\begin{cases} \dfrac{x^2}{16}+\dfrac{y^2}{12}=1 \\ x^2-\dfrac{y^2}{3}=1 \end{cases}$$

Multiply the first equation by 48 and the second equation by 3 to eliminate fractions.

$$\begin{cases} 3x^2+4y^2=48 \\ 3x^2-y^2=3 \end{cases}$$

Multiply the second equation by 4.
Add the equations and solve for x.

$$\begin{aligned} 3x^2+4y^2&=48 \\ \underline{12x^2-4y^2}&\underline{=12} \\ 15x^2&=60 \\ x^2&=4 \\ x&=\sqrt{4} \\ x&=\pm 2 \end{aligned}$$

Substitute $x=2$ and $x=-2$ into the second equation and solve for y.

$$\begin{aligned} 3x^2-y^2&=3 \\ 3(2)^2-y^2&=3 \\ 3(4)-y^2&=3 \\ 12-y^2&=3 \\ -y^2&=-9 \\ y^2&=9 \\ y&=\sqrt{9} \\ y&=\pm 3 \end{aligned} \qquad \begin{aligned} 3x^2-y^2&=3 \\ 3(-2)^2-y^2&=3 \\ 3(4)-y^2&=3 \\ 12-y^2&=3 \\ -y^2&=-9 \\ y^2&=9 \\ y&=\sqrt{9} \\ y&=\pm 3 \end{aligned}$$

The solutions are $(2, 3), (2, -3)$, $(-2, 3)$, and $(-2, -3)$

36.

$$\begin{cases} xy=4 \\ x^2+\dfrac{y^2}{2}=9 \end{cases}$$

Solve the first equation for y and multiply the second equation by 2 to eliminate fractions.

$$\begin{cases} y=\dfrac{4}{x} \\ 2x^2+y^2=18 \end{cases}$$

Substitute $y=\dfrac{4}{x}$ into the second equation and solve for x.

$$\begin{aligned} 2x^2+y^2&=18 \\ 2x^2+\left(\frac{4}{x}\right)^2&=18 \\ 2x^2+\frac{16}{x^2}&=18 \\ x^2\left(2x^2+\frac{16}{x^2}\right)&=x^2(18) \\ 2x^4+16&=18x^2 \\ 2x^4-18x^2+16&=0 \\ 2\left(x^4-9x^2+8\right)&=0 \\ 2\left(x^2-1\right)\left(x^2-8\right)&=0 \end{aligned}$$

$$\begin{aligned} x^2-1&=0 \\ x^2&=1 \\ x&=\sqrt{1} \\ x&=\pm 1 \end{aligned} \quad \text{or} \quad \begin{aligned} x^2-8&=0 \\ x^2&=8 \\ x&=\sqrt{8} \\ x&=\pm 2\sqrt{2} \end{aligned}$$

Substitute all four values into the first equation and solve for y.

$$\begin{aligned} xy&=4 \\ 1y&=4 \\ y&=4 \end{aligned} \qquad \begin{aligned} xy&=4 \\ -1y&=4 \\ y&=-4 \end{aligned}$$

$$\begin{aligned} xy&=4 \\ 2\sqrt{2}\cdot y&=4 \\ y&=\frac{4}{2\sqrt{2}} \\ y&=\frac{2}{\sqrt{2}} \\ y&=\frac{2\sqrt{2}}{2} \\ y&=\sqrt{2} \end{aligned} \qquad \begin{aligned} xy&=4 \\ -2\sqrt{2}\cdot y&=4 \\ y&=\frac{4}{-2\sqrt{2}} \\ y&=-\frac{2}{\sqrt{2}} \\ y&=-\frac{2\sqrt{2}}{2} \\ y&=-\sqrt{2} \end{aligned}$$

The solutions are $(1, 4), (-1, -4)$, $(2\sqrt{2}, \sqrt{2})$, and $(-2\sqrt{2}, -\sqrt{2})$

CHAPTER 13 TEST

1. A circle is the set of all points in a plane that are a fixed distance from a point called its **center**. The fixed distance is called the **radius** of the circle.

2. center: (0, 0); radius: $\sqrt{100}=10$

3.

$$x^2+y^2+4x-6y=5$$
$$\left(x^2+4x\right)+\left(y^2-6y\right)=5$$
$$\left(x^2+4x+4\right)+\left(y^2-6y+9\right)=5+4+9$$
$$\left(x+2\right)^2+\left(y-3\right)^2=18$$

center: $(-2,3)$

radius: $\sqrt{18}=3\sqrt{2}$

4. The center is (4, 3) and the radius is 3. The equation of the circle is

$$\left(x-4\right)^2+\left(y-3\right)^2=3^2$$
$$\left(x-4\right)^2+\left(y-3\right)^2=9$$

5. $\left(x+2\right)^2+\left(y-1\right)^2=9$

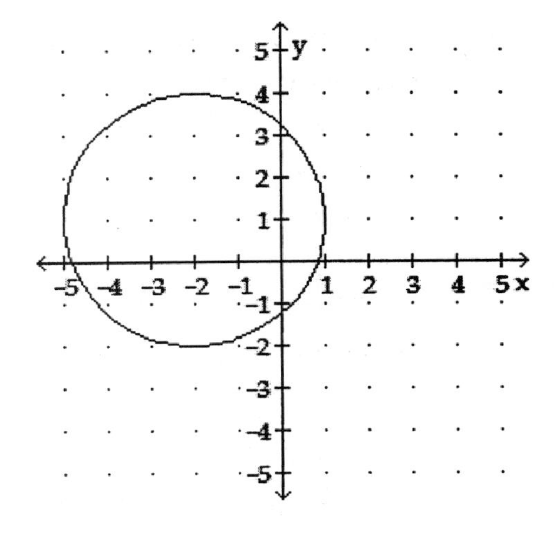

6.

$$x=y^2-2y+3$$
$$x=\left(y^2-2y\right)+3$$
$$x=\left(y^2-2y+1\right)+3-1$$
$$x=\left(y-1\right)^2+2$$

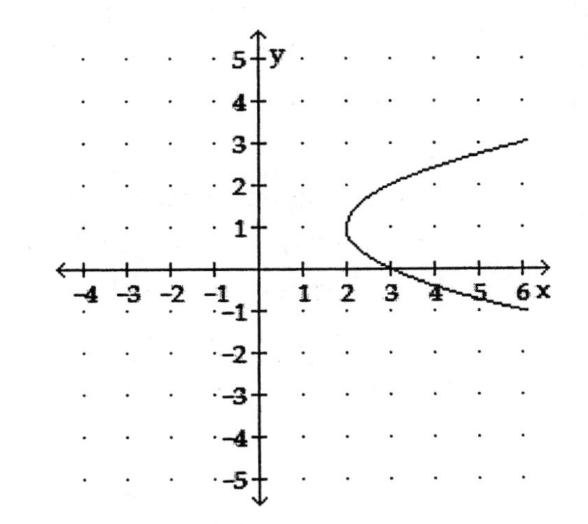

7.

$$y=-2x^2-4x+5$$
$$y=-2\left(x^2+2x\right)+5$$
$$y=-2\left(x^2+2x+1\right)+5-(-2)(1)$$
$$y=-2\left(x+1\right)^2+5+2$$
$$y=-2\left(x+1\right)^2+7$$

vertex: $(-1,7)$

axis of symmetry: $x=-1$

8.

$x = -\frac{1}{10}y^2$	y
$-\frac{1}{10}(4)^2 = -1.6$	4
$-\frac{1}{10}(3)^2 = -0.9$	3
$-\frac{1}{10}(2)^2 = -0.4$	2
$-\frac{1}{10}(1)^2 = -0.1$	1
$-\frac{1}{10}(0)^2 = 0$	0
$-\frac{1}{10}(-1)^2 = -0.1$	-1
$-\frac{1}{10}(-2)^2 = -0.4$	-2
$-\frac{1}{10}(-3)^2 = -0.9$	-3
$-\frac{1}{10}(-4)^2 = -1.6$	-4

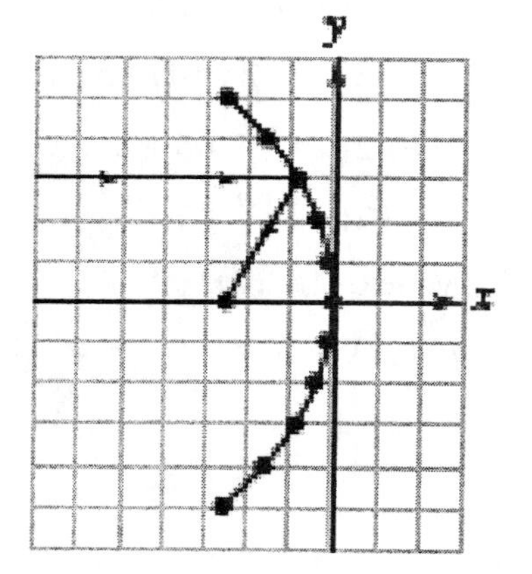

9.

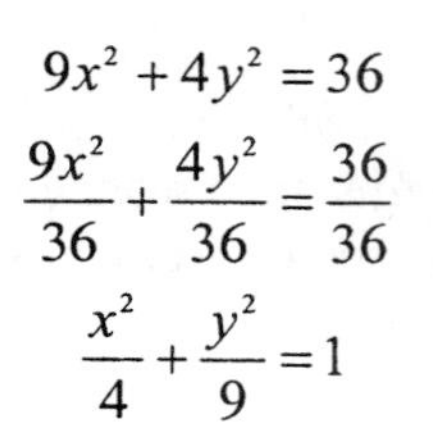

$$9x^2 + 4y^2 = 36$$

$$\frac{9x^2}{36} + \frac{4y^2}{36} = \frac{36}{36}$$

$$\frac{x^2}{4} + \frac{y^2}{9} = 1$$

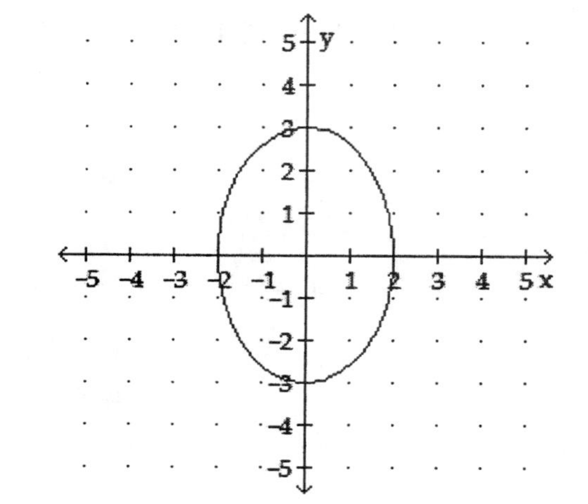

10.

$$\frac{(x-2)^2}{9} - \frac{y^2}{1} = 1$$

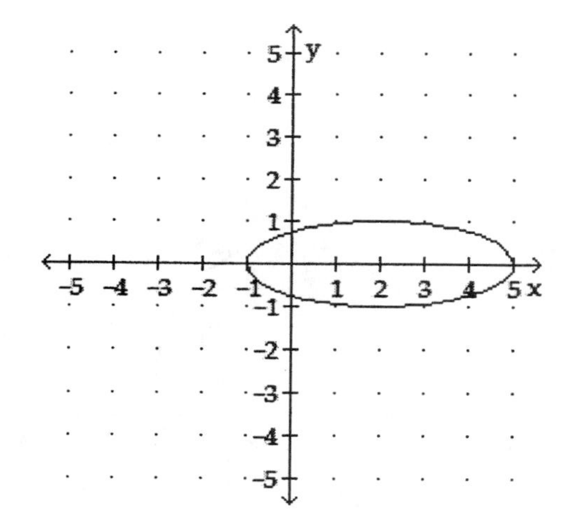

11. The center is $(1, -2)$, $a = 4$, and $b = 3$. The equation of the ellipse is

$$\frac{(x-1)^2}{16} + \frac{(y+2)^2}{9} = 1.$$

12.

$$25(x+8)^2+36(y-10)^2=900$$
$$\frac{25(x+8)^2}{900}+\frac{36(y-10)^2}{900}=\frac{900}{900}$$
$$\frac{(x+8)^2}{36}+\frac{(y-10)^2}{25}=1$$

The center is (–8, 10) and the vertices are (–2, 10) and (–14, 10)

13. Let $x=-2$.

$$\frac{x^2}{36}+\frac{y^2}{9}=1$$
$$\frac{(-2)^2}{36}+\frac{y^2}{9}=1$$
$$\frac{4}{36}+\frac{y^2}{9}=1$$
$$\frac{1}{9}+\frac{y^2}{9}=1$$
$$\frac{y^2}{9}=\frac{8}{9}$$
$$9y^2=72$$
$$y^2=8$$
$$y=\sqrt{8}$$
$$y=\pm2\sqrt{2}$$

14. Answers will vary.

15.

$$x^2+2x-y^2+2y-4=0$$
$$(x^2+2x)-(y^2-2y)=4$$
$$(x^2+2x+1)-(y^2-2y+1)=4+1-1$$
$$(x+1)^2-(y-1)^2=4$$

center: $(-1,\ 1)$

horizonal dimensions: 4 units

vertical dimensions: 4 units

16. $xy=-4$

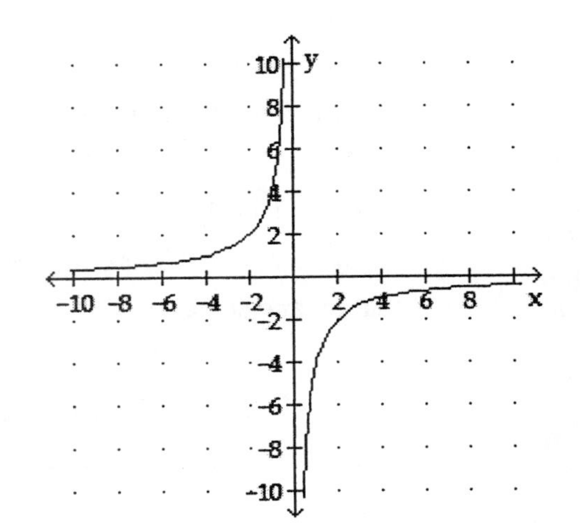

17. center: (0, 0); $a=6$, $b=4$

$$\frac{y^2}{4^2}-\frac{x^2}{6^2}=1$$
$$\frac{y^2}{16}-\frac{x^2}{36}=1$$

18. a. ellipse
b. hyperbola
c. circle
d. parabola

19. $\begin{cases}x^2+y^2=25\\ y-x=1\end{cases}$

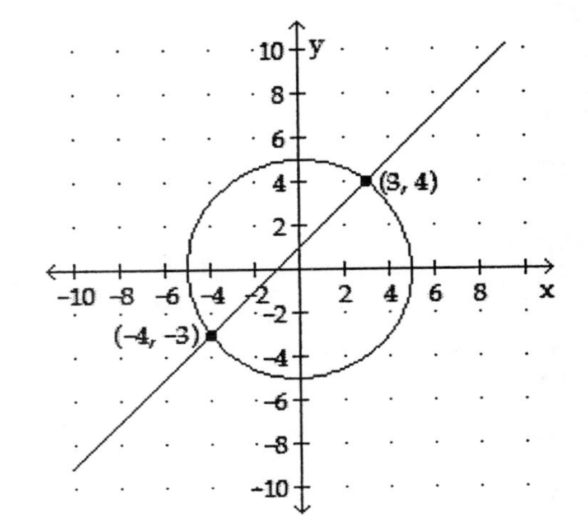

20.

$$\begin{cases} 2x - y = -2 \\ x^2 + y^2 = 16 + 4y \end{cases}$$

Solve the first equation for y.

$$\begin{cases} y = 2x + 2 \\ x^2 + y^2 = 16 + 4y \end{cases}$$

Substitute $y = 2x + 2$ into the second equation and solve for x.

$$x^2 + y^2 = 16 + 4y$$
$$x^2 + (2x+2)^2 = 16 + 4(2x+2)$$
$$x^2 + (4x^2 + 4x + 4x + 4) = 16 + 8x + 8$$
$$5x^2 + 8x + 4 = 8x + 24$$
$$5x^2 - 20 = 0$$
$$5(x^2 - 4) = 0$$
$$5(x-2)(x+2) = 0$$
$$x - 2 = 0 \quad \text{or} \quad x + 2 = 0$$
$$x = 2 \qquad\qquad x = -2$$

Substitute $x = 2$ and $x = -2$ into the first equation and solve for y.

$$y = 2x + 2 \qquad y = 2x + 2$$
$$y = 2(2) + 2 \qquad y = 2(-2) + 2$$
$$y = 4 + 2 \qquad y = -4 + 2$$
$$y = 6 \qquad y = -2$$

The solutions are $(2,6)$ and $(-2,-2)$.

21.

$$\begin{cases} 5x^2 - y^2 - 3 = 0 \\ x^2 + 2y^2 = 5 \end{cases}$$

Write the first equation in standard form

$$\begin{cases} 5x^2 - y^2 = 3 \\ x^2 + 2y^2 = 5 \end{cases}$$

Multiply the second equation by -5 and add the equations to solve for y.

$$5x^2 - y^2 = 3$$
$$\underline{-5x^2 - 10y^2 = -25}$$
$$-11y^2 = -22$$
$$y^2 = 2$$
$$y = \pm\sqrt{2}$$

Substitute $y = \sqrt{2}$ and $y = -\sqrt{2}$ into the second equation and solve for x.

$$x^2 + 2y^2 = 5 \qquad x^2 + 2y^2 = 5$$
$$x^2 + 2\left(\sqrt{2}\right)^2 = 5 \qquad x^2 + 2\left(-\sqrt{2}\right)^2 = 5$$
$$x^2 + 2(2) = 5 \qquad x^2 + 2(2) = 5$$
$$x^2 + 4 = 5 \qquad x^2 + 4 = 5$$
$$x^2 = 1 \qquad x^2 = 1$$
$$x = \sqrt{1} \qquad x = \sqrt{1}$$
$$x = \pm 1 \qquad x = \pm 1$$

The solutions are $\left(1,\sqrt{2}\right), \left(-1,\sqrt{2}\right)$, $\left(1,-\sqrt{2}\right)$, and $\left(-1,-\sqrt{2}\right)$.

22.

$$\begin{cases} xy = -\dfrac{9}{2} \\ 3x + 2y = 6 \end{cases}$$

Solve the first equation for y.

$$\begin{cases} y = -\dfrac{9}{2x} \\ 3x + 2y = 6 \end{cases}$$

Substitute $y = -\dfrac{9}{2x}$ into the second equation and solve for x.

$$3x + 2y = 6$$
$$3x + 2\left(-\frac{9}{2x}\right) = 6$$
$$3x - \frac{9}{x} = 6$$
$$x\left(3x - \frac{9}{x}\right) = x(6)$$
$$3x^2 - 9 = 6x$$
$$3x^2 - 6x - 9 = 0$$
$$3(x^2 - 2x - 3) = 0$$
$$3(x-3)(x+1) = 0$$
$$x - 3 = 0 \quad \text{or} \quad x + 1 = 0$$
$$x = 3 \qquad\qquad x = -1$$

Substitute $x = 3$ and $x = -1$ into the first equation and solve for y.

$$xy = -\frac{9}{2} \qquad\qquad xy = -\frac{9}{2}$$
$$3y = -\frac{9}{2} \qquad\qquad -1y = -\frac{9}{2}$$
$$\frac{1}{3}(3y) = \frac{1}{3}\left(-\frac{9}{2}\right) \qquad -1(-1y) = -1\left(-\frac{9}{2}\right)$$
$$y = -\frac{3}{2} \qquad\qquad y = \frac{9}{2}$$

The solutions are $\left(3, -\dfrac{3}{2}\right)$ and $\left(-1, \dfrac{9}{2}\right)$.

SECTION 14.1

VOCABULARY

1. The two-term polynomial expression $a + b$ is called a **binomial**.

3. We can use the **binomial** theorem to raise binomials to positive-integer powers without doing the actual multiplication.

5. $n!$ (read as "n **factorial**") is the product of consecutively **decreasing** natural numbers from n to 1.

CONCEPTS

7. Every binomial expansion has **one** more term than the power of the binomial.

9. The first term of the expansion of $(r+s)^{20}$ is $\boxed{r^{20}}$ and the last term is $\boxed{s^{20}}$.

11. The coefficients of the terms of the expansion of $(c + d)^{20}$ begin with $\boxed{1}$, increase through some values, and then decrease through those same values, back to $\boxed{1}$.

13. $n\cdot\boxed{(n-1)!} = n!$

15. $0! = \boxed{1}$

17. The coefficient of the fourth term of the expansion of $(a + b)^9$ is 9! divided by **3!(9 – 3)!**.

19. The exponent on a in the fifth term of the expansion of $(a + b)^6$ is $\boxed{2}$ and the exponent on b is $\boxed{4}$.

21. $(x + y)^3$

$$= x^{\boxed{3}} + \frac{\boxed{3!}}{1!(3-1)!}x^2\boxed{y} + \frac{\boxed{3!}}{\boxed{2}!(3-2)!}xy^{\boxed{2}} + y^{\boxed{3}}$$

NOTATION

23. $n! = n\cdot\left(\boxed{n-1}\right)(n-2)\ldots3\cdot2\cdot1$

PRACTICE

25.
$$\begin{aligned} 3! &= 3\cdot2\cdot1 \\ &= 6 \end{aligned}$$

27.
$$\begin{aligned} 5! &= 5\cdot4\cdot3\cdot2\cdot1 \\ &= 120 \end{aligned}$$

29.
$$\begin{aligned} 3!+4! &= 3\cdot2\cdot1 + 4\cdot3\cdot2\cdot1 \\ &= 6+24 \\ &= 30 \end{aligned}$$

31.
$$\begin{aligned} 3!(4!) &= (3\cdot2\cdot1)(4\cdot3\cdot2\cdot1) \\ &= (6)(24) \\ &= 144 \end{aligned}$$

33.
$$\begin{aligned} 8(7!) &= 8(7\cdot6\cdot5\cdot4\cdot3\cdot2\cdot1) \\ &= 8(5,040) \\ &= 40,320 \end{aligned}$$

35.
$$\begin{aligned} \frac{9!}{11!} &= \frac{\cancel{9}\cdot\cancel{8}\cdot\cancel{7}\cdot\cancel{6}\cdot\cancel{5}\cdot\cancel{4}\cdot\cancel{3}\cdot\cancel{2}\cdot\cancel{1}}{11\cdot10\cdot\cancel{9}\cdot\cancel{8}\cdot\cancel{7}\cdot\cancel{6}\cdot\cancel{5}\cdot\cancel{4}\cdot\cancel{3}\cdot\cancel{2}\cdot\cancel{1}} \\ &= \frac{1}{11\cdot10} \\ &= \frac{1}{110} \end{aligned}$$

37.
$$\begin{aligned} \frac{49!}{47!} &= \frac{49\cdot48\cdot\cancel{47!}}{\cancel{47!}} \\ &= \frac{49\cdot48}{1} \\ &= 2,352 \end{aligned}$$

39.

$$\frac{9!}{7!0!} = \frac{9 \cdot 8 \cdot \cancel{7!}}{\cancel{7!}0!} = \frac{9 \cdot 8}{1} = 72$$

41.

$$\frac{5!}{1!(5-1)!} = \frac{5!}{1!4!} = \frac{5 \cdot \cancel{4!}}{1 \cdot \cancel{4!}} = 5$$

43.

$$\frac{5!}{3!(5-3)!} = \frac{5!}{3!2!} = \frac{5 \cdot 4 \cdot \cancel{3!}}{\cancel{3!}(2 \cdot 1)} = \frac{20}{2} = 10$$

45.

$$\frac{7!}{5!(7-5)!} = \frac{7!}{5!2!} = \frac{7 \cdot 6 \cdot \cancel{5!}}{\cancel{5!}(2 \cdot 1)} = \frac{42}{2} = 21$$

47.

$$\frac{5!(8-5)!}{4!7!} = \frac{\cancel{5!}\,\cancel{3!}}{(4 \cdot \cancel{3!})(7 \cdot 6 \cdot \cancel{5!})} = \frac{1}{4 \cdot 7 \cdot 6} = \frac{1}{168}$$

49.

$11! = 39{,}916{,}800$

51.

$20! = 2.432902008 \times 10^{18}$

53.

$$\begin{aligned}
&(x+y)^4 \\
&= x^4 + \frac{4!}{1!(4-1)!}x^3y + \frac{4!}{2!(4-2)!}x^2y^2 \\
&\quad + \frac{4!}{3!(4-3)!}xy^3 + y^4 \\
&= x^4 + \frac{4 \cdot \cancel{3!}}{1!\cancel{3!}}x^3y + \frac{4 \cdot 3 \cdot \cancel{2!}}{2!\cancel{2!}}x^2y^2 \\
&\quad + \frac{4 \cdot \cancel{3!}}{\cancel{3!}1!}xy^3 + y^4 \\
&= x^4 + \frac{4}{1}x^3y + \frac{12}{2}x^2y^2 + \frac{4}{1}xy^3 + y^4 \\
&= x^4 + 4x^3y + 6x^2y^2 + 4xy^3 + y^4
\end{aligned}$$

55.

$$\begin{aligned}
&(c-d)^5 \\
&= (c+(-d))^5 \\
&= c^5 + \frac{5!}{1!(5-1)!}c^4(-d) + \frac{5!}{2!(5-2)!}c^3(-d)^2 \\
&\quad + \frac{5!}{3!(5-3)!}c^2(-d)^3 + \frac{5!}{4!(5-4)!}c(-d)^4 + (-d)^5 \\
&= c^5 + \frac{5 \cdot \cancel{4!}}{1!\cancel{4!}}c^4(-d) + \frac{5 \cdot 4 \cdot \cancel{3!}}{2!\cancel{3!}}c^3(-d)^2 \\
&\quad + \frac{5 \cdot 4 \cdot \cancel{3!}}{\cancel{3!}2!}c^2(-d)^3 + \frac{5 \cdot \cancel{4!}}{\cancel{4!}1!}c(-d)^4 + (-d)^5 \\
&= c^5 - \frac{5}{1}c^4d + \frac{20}{2}c^3d^2 - \frac{20}{2}c^2d^3 + \frac{5}{1}cd^4 - d^5 \\
&= c^5 - 5c^4d + 10c^3d^2 - 10c^2d^3 + 5cd^4 - d^5
\end{aligned}$$

57.

$$(s+t)^6$$

$$=s^6+\frac{6!}{1!(6-1)!}s^5t+\frac{6!}{2!(6-2)!}s^4t^2$$

$$+\frac{6!}{3!(6-3)!}s^3t^3+\frac{6!}{4!(6-4)!}s^2t^4$$

$$+\frac{6!}{5!(6-5)!}st^5+t^6$$

$$=s^6+\frac{6!}{1!5!}s^5t+\frac{6!}{2!4!}s^4t^2$$

$$+\frac{6!}{3!3!}s^3t^3+\frac{6!}{4!2!}s^2t^4$$

$$+\frac{6!}{5!1!}st^5+t^6$$

$$=s^6+\frac{6}{1}s^5t+\frac{30}{2}s^4t^2+\frac{120}{6}s^3t^3+\frac{30}{2}s^2t^4$$

$$+\frac{6}{1}st^5+t^6$$

$$=s^6+6s^5t+15s^4t^2+20s^3t^3+15s^2t^4+6st^5+t^6$$

59.

$$(a-b)^9=(a+(-b))^9$$

$$=a^9+\frac{9!}{1!(9-1)!}a^8(-b)+\frac{9!}{2!(9-2)!}a^7(-b)^2$$

$$+\frac{9!}{3!(9-3)!}a^6(-b)^3+\frac{9!}{4!(9-4)!}a^5(-b)^4$$

$$+\frac{9!}{5!(9-5)!}a^4(-b)^5+\frac{9!}{6!(9-6)!}a^3(-b)^6$$

$$+\frac{9!}{7!(9-7)!}a^2(-b)^7+\frac{9!}{8!(9-8)!}a(-b)^8+(-b)^9$$

$$=a^9+\frac{9!}{1!8!}a^8(-b)+\frac{9!}{2!7!}a^7(-b)^2+\frac{9!}{3!6!}a^6(-b)^3$$

$$+\frac{9!}{4!5!}a^5(-b)^4+\frac{9!}{5!4!}a^4(-b)^5+\frac{9!}{6!3!}a^3(-b)^6$$

$$+\frac{9!}{7!2!}a^2(-b)^7+\frac{9!}{8!1!}a(-b)^8+(-b)^9$$

$$=a^9-\frac{9}{1}a^8b+\frac{72}{2}a^7b^2-\frac{504}{6}a^6b^3+\frac{3{,}024}{24}a^5b^4$$

$$-\frac{3{,}024}{24}a^4b^5+\frac{504}{6}a^3b^6-\frac{72}{2}a^2b^7+\frac{9}{1}ab^8-b^9$$

$$=a^9-9a^8b+36a^7b^2-84a^6b^3+126a^5b^4-126a^4b^5$$

$$+84a^3b^6-36a^2b^7+9ab^8-b^9$$

61.

$$(2x+y)^3$$

$$=(2x)^3+\frac{3!}{1!(3-1)!}(2x)^2y+\frac{3!}{2!(3-2)!}(2x)y^2+y^3$$

$$=8x^3+\frac{3!}{1!2!}(4x^2)y+\frac{3!}{2!1!}(2x)y^2+y^3$$

$$=8x^3+\frac{3}{1}(4x^2)y+\frac{3}{1}(2x)y^2+y^3$$

$$=8x^3+12x^2y+6xy^2+y^3$$

63.

$$(2t-3)^5$$

$$=(2t)^5+\frac{5!}{1!(5-1)!}(2t)^4(-3)+\frac{5!}{2!(5-2)!}(2t)^3(-3)^2$$

$$+\frac{5!}{3!(5-3)!}(2t)^2(-3)^3+\frac{5!}{4!(5-4)!}(2t)(-3)^4+(-3)^5$$

$$=(2t)^5+\frac{5\cdot\cancel{4!}}{1!\cancel{4!}}(2t)^4(-3)+\frac{5\cdot4\cdot\cancel{3!}}{2!\cancel{3!}}(2t)^3(-3)^2$$

$$+\frac{5\cdot4\cdot\cancel{3!}}{\cancel{3!}2!}(2t)^2(-3)^3+\frac{5\cdot\cancel{4!}}{\cancel{4!}1!}(2t)(-3)^4+(-3)^5$$

$$=32t^5+\frac{5}{1}(16t^4)(-3)+\frac{20}{2}(8t^3)(9)+\frac{20}{2}(4t^2)(-27)$$

$$+\frac{5}{1}(2t)(81)-243$$

$$=32t^5-240t^4+720t^3-1{,}080t^2+810t-243$$

65.

$$(5m-2n)^4$$

$$=(5m)^4+\frac{4!}{1!(4-1)!}(5m)^3(-2n)$$

$$+\frac{4!}{2!(4-2)!}(5m)^2(-2n)^2+\frac{4!}{3!(4-3)!}(5m)(-2n)^3$$

$$+(-2n)^4$$

$$=625m^4+\frac{4\cdot\cancel{3!}}{1!\cancel{3!}}(125m^3)(-2n)$$

$$+\frac{4\cdot3\cdot\cancel{2!}}{2!\cancel{2!}}(25m^2)(4n^2)+\frac{4\cdot\cancel{3!}}{\cancel{3!}1!}(5m)(-8n^3)+16n^4$$

$$=625m^4+\frac{4}{1}(-250m^3n)+\frac{12}{2}(100m^2n^2)$$

$$+\frac{4}{1}(-40mn^3)+16n^4$$

$$=625m^4-1{,}000m^3n+600m^2n^2-160mn^3+16n^4$$

67.

$$\left(\frac{x}{3}+\frac{y}{2}\right)^3$$

$$=\left(\frac{x}{3}\right)^3+\frac{3!}{1!(3-1)!}\left(\frac{x}{3}\right)^2\left(\frac{y}{2}\right)+\frac{3!}{2!(3-2)!}\left(\frac{x}{3}\right)\left(\frac{y}{2}\right)^2+\left(\frac{y}{2}\right)^3$$

$$=\frac{x^3}{27}+\frac{3!}{1!2!}\left(\frac{x^2}{9}\right)\left(\frac{y}{2}\right)+\frac{3!}{2!1!}\left(\frac{x}{3}\right)\left(\frac{y^2}{4}\right)+\frac{y^3}{8}$$

$$=\frac{x^3}{27}+\frac{3}{1}\left(\frac{x^2y}{18}\right)+\frac{3}{1}\left(\frac{xy^2}{12}\right)+\frac{y^3}{8}$$

$$=\frac{x^3}{27}+\frac{x^2y}{6}+\frac{xy^2}{4}+\frac{y^3}{8}$$

69.

$$\left(\frac{x}{3}-\frac{y}{2}\right)^4$$

$$=\left(\frac{x}{3}\right)^4+\frac{4!}{1!(4-1)!}\left(\frac{x}{3}\right)^3\left(-\frac{y}{2}\right)+\frac{4!}{2!(4-2)!}\left(\frac{x}{3}\right)^2\left(-\frac{y}{2}\right)^2+\frac{4!}{3!(4-3)!}\left(\frac{x}{3}\right)\left(-\frac{y}{2}\right)^3+\left(-\frac{y}{2}\right)^4$$

$$=\frac{x^4}{81}+\frac{4\cdot\not{3!}}{1!\not{3!}}\left(\frac{x^3}{27}\right)\left(-\frac{y}{2}\right)+\frac{4\cdot3\cdot\not{2!}}{2!\not{2!}}\left(\frac{x^2}{9}\right)\left(\frac{y^2}{4}\right)+\frac{4\cdot\not{3!}}{\not{3!}1!}\left(\frac{x}{3}\right)\left(-\frac{y^3}{8}\right)+\frac{y^4}{16}$$

$$=\frac{x^4}{81}+\frac{4}{1}\left(-\frac{x^3y}{54}\right)+\frac{12}{2}\left(\frac{x^2y^2}{36}\right)+\frac{4}{1}\left(-\frac{xy^3}{24}\right)+\frac{y^4}{16}$$

$$=\frac{x^4}{81}-\frac{2x^3y}{27}+\frac{x^2y^2}{6}-\frac{xy^3}{6}+\frac{y^4}{16}$$

71.

$$\left(c^2-d^2\right)^5$$

$$=\left(c^2\right)^5+\frac{5!}{1!(5-1)!}\left(c^2\right)^4\left(-d^2\right)+\frac{5!}{2!(5-2)!}\left(c^2\right)^3\left(-d^2\right)^2+\frac{5!}{3!(5-3)!}\left(c^2\right)^2\left(-d^2\right)^3+\frac{5!}{4!(5-4)!}\left(c^2\right)\left(-d^2\right)^4+\left(-d^2\right)^5$$

$$=c^{10}+\frac{5\cdot\not{4!}}{1!\not{4!}}\left(c^8\right)\left(-d^2\right)+\frac{5\cdot4\cdot\not{3!}}{2!\not{3!}}\left(c^6\right)\left(d^4\right)+\frac{5\cdot4\cdot\not{3!}}{\not{3!}2!}\left(c^4\right)\left(-d^6\right)+\frac{5\cdot\not{4!}}{\not{4!}1!}\left(c^2\right)\left(d^8\right)-d^{10}$$

$$=c^{10}+\frac{5}{1}\left(-c^8d^2\right)+\frac{20}{2}\left(c^6d^4\right)+\frac{20}{2}\left(-c^4d^6\right)+\frac{5}{1}\left(c^2d^8\right)-d^{10}$$

$$=c^{10}-5c^8d^2+10c^6d^4-10c^4d^6+5c^2d^8-d^{10}$$

73. The 4th term of $(x-y)^4$

$$\frac{4!}{3!(4-3)!}x(-y)^3=\frac{4\cdot\not{3!}}{\not{3!}1!}x(-y)^3=-\frac{4}{1}xy^3=-4xy^3$$

75. The 5th term of $(r+s)^6$

$$\frac{6!}{4!(6-4)!}r^2s^4=\frac{6\cdot5\cdot\not{4!}}{\not{4!}2!}r^2s^4=\frac{30}{2}r^2s^4=15r^2s^4$$

77. The 3rd term of $(x-y)^8$

$$\frac{8!}{2!(8-2)!}x^6(-y)^2=\frac{8\cdot7\cdot\not{6!}}{2!\not{6!}}x^6y^2=\frac{56}{2}x^6y^2=28x^6y^2$$

79. The 2nd term of $(x - 3y)^4$

$$\frac{4!}{1!(4-1)!}x^3(-3y) = \frac{4 \cdot 3!}{1!3!}(-3x^3y)$$
$$= \frac{4}{1}(-3x^3y)$$
$$= -12x^3y$$

81. The 4th term of $(2t - 5)^7$

$$\frac{7!}{3!(7-3)!}(2t)^4(-5)^3 = \frac{7 \cdot 6 \cdot 5 \cdot 4!}{3!4!}(16t^4)(-125)$$
$$= \frac{210}{6}(-2{,}000t^4)$$
$$= 35(-2{,}000t^4)$$
$$= -70{,}000t^4$$

83. The 5th term of $(2x - 3y)^5$

$$\frac{5!}{4!(5-4)!}(2x)(-3y)^4 = \frac{5 \cdot 4!}{4!1!}(2x)(81y^4)$$
$$= \frac{5}{1}(162xy^4)$$
$$= 810xy^4$$

85. The 2nd term of $\left(\frac{c}{2} - \frac{d}{3}\right)^4$

$$\frac{4!}{1!(4-1)!}\left(\frac{c}{2}\right)^3\left(-\frac{d}{3}\right) = \frac{4 \cdot 3!}{1!3!}\left(\frac{c^3}{8}\right)\left(-\frac{d}{3}\right)$$
$$= \frac{4}{1}\left(-\frac{c^3d}{24}\right)$$
$$= -\frac{c^3d}{6}$$
$$= -\frac{1}{6}c^3d$$

87. The 2nd term of $(a^2 - b^2)^6$

$$\frac{6!}{1!(6-1)!}(a^2)^5(-b^2) = \frac{6 \cdot 5!}{1!5!}(a^{10})(-b^2)$$
$$= \frac{6}{1}(-a^{10}b^2)$$
$$= -6a^{10}b^2$$

WRITING

89. Answers will vary.

91. Answers will vary.

REVIEW

93.

$$2\log x + \frac{1}{2}\log y = \log x^2 + \log y^{1/2}$$
$$= \log x^2 y^{1/2}$$

95.

$$\ln(xy + y^2) - \ln(xz + yz) + \ln z$$
$$= \ln z(xy + y^2) - \ln(xz + yz)$$
$$= \ln\frac{z(xy + y^2)}{(xz + yz)}$$
$$= \ln\frac{zy(x+y)}{z(x+y)}$$
$$= \ln y$$

CHALLENGE PROBLEMS

97. The constant term is the coefficient of $x^5x^{-5} = x^0 = 1$. It is the 6th term in the expansion.

$$\frac{10!}{5!(10-5)!}(x)^5\left(\frac{1}{x}\right)^5 = \frac{10 \cdot 9 \cdot 8 \cdot 7 \cdot 6 \cdot 5!}{5!5!}(x)^5(x^{-1})^5$$
$$= \frac{30{,}240}{120}(x^5)(x^{-5})$$
$$= 252x^{5+(-5)}$$
$$= 252x^0$$
$$= 252(1)$$
$$= 252$$

99. a. Let $n = 1$.

$$\frac{n!}{0!(n-0)!} = \frac{1!}{0!(1-0)!}$$
$$= \frac{\cancel{1!}}{0!\cancel{1!}}$$
$$= \frac{1}{0!}$$
$$= \frac{1}{1}$$
$$= 1$$

b.

$$\frac{n!}{n!(n-n)!} = \frac{n!}{n!0!}$$
$$= \frac{\cancel{n!}}{\cancel{n!}0!}$$
$$= \frac{1}{0!}$$
$$= \frac{1}{1}$$
$$= 1$$

SECTION 14.2

VOCABULARY

1. A **sequence** is a function whose domain is the set of natural numbers.

3. Each term of an **arithmetic** sequence is found by adding the same number to the previous term.

5. If a single number is inserted between a and b to form an arithmetic sequence, the number is called the arithmetic **mean** between a and b.

CONCEPTS

7. 1, 7, 13

9. a. $a_n = a_1 + (n-1)d$

   b. $S_n = \dfrac{n(a_1 + a_n)d}{2}$

NOTATION

11. The notation a_n represents the ***n*th** term of a sequence.

13. The symbol Σ is the Greek letter **sigma**.

15. We read $\sum_{k=1}^{10} 3k$ as "the **summation** of $3k$ as k **runs** from 1 to 10."

PRACTICE

17. $a_n = 4n - 1$
$a_1 = 4(1) - 1 = 3$
$a_2 = 4(2) - 1 = 7$
$a_3 = 4(3) - 1 = 11$
$a_4 = 4(4) - 1 = 15$
$a_5 = 4(5) - 1 = 19$
The first five terms are 3, 7, 11, 15, and 19.

19. $a_n = -3n + 1$
$a_1 = -3(1) + 1 = -2$
$a_2 = -3(2) + 1 = -5$
$a_3 = -3(3) + 1 = -8$
$a_4 = -3(4) + 1 = -11$
$a_5 = -3(5) + 1 = -14$
The first five terms are –2, –5, –8, –11 and –14.

21. $a_n = a_1 + (n-1)d$
$a_n = 3 + 2(n-1)$
$a_1 = 3 + 2(1-1) = 3 + 0 = 3$
$a_2 = 3 + 2(2-1) = 3 + 2 = 5$
$a_3 = 3 + 2(3-1) = 3 + 4 = 7$
$a_4 = 3 + 2(4-1) = 3 + 6 = 9$
$a_5 = 3 + 2(5-1) = 3 + 8 = 11$
The first five terms are 3, 5, 7, 9, and 11.

23. $a_n = a_1 + (n-1)d$
$a_n = -5 - 3(n-1)$
$a_1 = -5 - 3(1-1) = -5 + 0 = -5$
$a_2 = -5 - 3(2-1) = -5 - 3 = -8$
$a_3 = -5 - 3(3-1) = -5 - 6 = -11$
$a_4 = -5 - 3(4-1) = -5 - 9 = -14$
$a_5 = -5 - 3(5-1) = -5 - 12 = -17$
The first five terms are –5, –8, –11, –14, and –17.

25. $a_n = a_1 + (n-1)d$
Find d.
$29 = 5 + d(5-1)$
$29 = 5 + d(4)$
$24 = 4d$
$6 = d$
$a_1 = 5 + 6(1-1) = 5 + 0 = 5$
$a_2 = 5 + 6(2-1) = 5 + 6 = 11$
$a_3 = 5 + 6(3-1) = 5 + 12 = 17$
$a_4 = 5 + 6(4-1) = 5 + 18 = 23$
$a_5 = 5 + 6(5-1) = 5 + 24 = 29$
The first five terms are 5, 11, 17, 23, and 29.

27. $a_n = a_1 + (n-1)d$

Find d.

$$-39 = -4 + d(6-1)$$
$$-39 = -4 + d(5)$$
$$-35 = 5d$$
$$-7 = d$$
$$a_1 = -4 - 7(1-1) = -4 - 0 = -4$$
$$a_2 = -4 - 7(2-1) = -4 - 7 = -11$$
$$a_3 = -4 - 7(3-1) = -4 - 14 = -18$$
$$a_4 = -4 - 7(4-1) = -4 - 21 = -25$$
$$a_5 = -4 - 7(5-1) = -4 - 28 = -32$$

The first five terms are –4, –11, –18, –25, and –32.

29. $a_n = a_1 + (n-1)d$

Find a_1.

$$-83 = a_1 + 7(6-1)$$
$$-83 = a_1 + 7(5)$$
$$-83 = a_1 + 35$$
$$-118 = a_1$$
$$a_1 = -118 + 7(1-1) = -118 + 0 = -118$$
$$a_2 = -118 + 7(2-1) = -118 + 7 = -111$$
$$a_3 = -118 + 7(3-1) = -118 + 14 = -104$$
$$a_4 = -118 + 7(4-1) = -118 + 21 = -97$$
$$a_5 = -118 + 7(5-1) = -118 + 28 = -90$$

The first five terms are –118, –111, –104, –97, and –90.

31. $a_n = a_1 + (n-1)d$

Find a_1.

$$16 = a_1 - 3(7-1)$$
$$16 = a_1 - 3(6)$$
$$16 = a_1 - 18$$
$$34 = a_1$$
$$a_1 = 34 - 3(1-1) = 34 - 0 = 34$$
$$a_2 = 34 - 3(2-1) = 34 - 3 = 31$$
$$a_3 = 34 - 3(3-1) = 34 - 6 = 28$$
$$a_4 = 34 - 3(4-1) = 34 - 9 = 25$$
$$a_5 = 34 - 3(5-1) = 34 - 12 = 22$$

The first five terms are 34, 31, 28, 25, and 22.

33. Find d.

$138 - 131 = 7$, so $d = 7$.

Find a_1 if $a_n = 138$ and $n = 20$.

$$138 = a_1 + 7(20-1)$$
$$138 = a_1 + 7(19)$$
$$138 = a_1 + 133$$
$$5 = a_1$$
$$a_1 = 5 + 7(1-1) = 5 + 0 = 5$$
$$a_2 = 5 + 7(2-1) = 5 + 7 = 12$$
$$a_3 = 5 + 7(3-1) = 5 + 14 = 19$$
$$a_4 = 5 + 7(4-1) = 5 + 21 = 26$$
$$a_5 = 5 + 7(5-1) = 5 + 28 = 33$$

The first five terms are 5, 12, 19, 26, and 33.

35.

$$a_{30} = 7 + 12(30-1)$$
$$= 7 + 12(29)$$
$$= 7 + 348$$
$$= 355$$

37. Find d.

$-9 - (-4)$, so $d = -5$.

Find a_1 if $a_n = -4$ and $n = 2$.

$$-4 = a_1 - 5(2-1)$$
$$-4 = a_1 - 5(1)$$
$$-4 = a_1 - 5$$
$$1 = a_1$$

Find a_{37}.

$$a_{37} = 1 - 5(37-1)$$
$$= 1 - 5(36)$$
$$= 1 - 180$$
$$= -179$$

39. Let $d = 11$, $n = 27$ and $a_n = 263$.

$$263 = a_1 + 11(27-1)$$
$$263 = a_1 + 11(26)$$
$$263 = a_1 + 286$$
$$-23 = a_1$$

41. Let $n = 44$, $a_n = 556$, and $a_1 = 40$.

$$556 = 40 + d(44-1)$$
$$556 = 40 + d(43)$$
$$516 = 43d$$
$$12 = d$$

43. Let $a_1 = 2$ and $a_5 = 11$.

$$a_2 = 2 + d, a_3 = 2 + 2d, a_4 = 2 + 3d$$
$$a_5 = a_1 + (5-1)d$$
$$11 = 2 + 4d$$
$$9 = 4d$$
$$\frac{9}{4} = d$$

$$a_2 = 2 + d = 2 + \frac{9}{4} = \frac{17}{4}$$
$$a_3 = 2 + 2d = 2 + 2\left(\frac{9}{4}\right) = \frac{13}{2}$$
$$a_4 = 2 + 3d = 2 + 3\left(\frac{9}{4}\right) = \frac{35}{4}$$

45. Let $a_1 = 10$ and $a_6 = 20$.

$$a_2 = 10 + d, a_3 = 10 + 2d, a_4 = 10 + 3d, a_5 = 10 + 4d$$
$$a_6 = a_1 + (6-1)d$$
$$20 = 10 + 5d$$
$$10 = 5d$$
$$2 = d$$

$$a_2 = 10 + d = 10 + 2 = 12$$
$$a_3 = 10 + 2d = 10 + 2(2) = 14$$
$$a_4 = 10 + 3d = 10 + 3(2) = 16$$
$$a_5 = 10 + 4d = 10 + 4(2) = 18$$

47. Let $a_1 = 10$ and $a_3 = 19$.

$$a_2 = 10 + d$$
$$a_3 = a_1 + (3-1)d$$
$$19 = 10 + 2d$$
$$9 = 2d$$
$$\frac{9}{2} = d$$

$$a_2 = 10 + d = 10 + \frac{9}{2} = \frac{29}{2}$$

49.

$$\sum_{k=1}^{4}(3k) = 3(1) + 3(2) + 3(3) + 3(4)$$
$$= 3 + 6 + 9 + 12$$

51.

$$\sum_{k=2}^{4} k^2 = 2^2 + 3^2 + 4^2$$
$$= 4 + 9 + 16$$

53.

$$1 + 4 + 9 + 16 + 25 = 1^2 + 2^2 + 3^2 + 4^2 + 5^2$$
$$= \sum_{k=1}^{5} k^2$$

55.

$$3 + 4 + 5 + 6 = \sum_{k=3}^{6} k$$

57.

$$a_1 = 1, \quad d = 4 - 1 = 3, \quad n = 30$$
$$a_n = 1 + 3(30-1)$$
$$= 1 + 3(29)$$
$$= 88$$

$$S_n = \frac{n(a_1 + a_n)}{2}$$
$$S_{30} = \frac{30(1+88)}{2}$$
$$= \frac{30(89)}{2}$$
$$= 1{,}335$$

59.

$$a_1 = -5, \quad d = -1 - (-5) = 4, \quad n = 17$$
$$a_n = -5 + 4(17-1)$$
$$= -5 + 4(16)$$
$$= 59$$

$$S_n = \frac{n(a_1 + a_n)}{2}$$
$$S_{17} = \frac{17(-5+59)}{2}$$
$$= \frac{17(54)}{2}$$
$$= 459$$

61.

$$a_2 = 7, \quad a_3 = 12, \quad d = 12 - 7 = 5, \quad n = 12$$

$$\begin{aligned} 7 &= a_1 + 5(2-1) \\ 7 &= a_1 + 5(1) \\ 7 &= a_1 + 5 \\ 2 &= a_1 \end{aligned}$$

$$\begin{aligned} a_n &= 2 + 5(12-1) \\ &= 2 + 5(11) \\ &= 57 \end{aligned}$$

$$\begin{aligned} S_n &= \frac{n(a_1 + a_n)}{2} \\ S_{12} &= \frac{12(2+57)}{2} \\ &= \frac{12(59)}{2} \\ &= 354 \end{aligned}$$

63.

$$\begin{aligned} a_n &= 2n + 1 \\ 31 &= 2n + 1 \\ 30 &= 2n \\ 15 &= n \end{aligned}$$

$$a_1 = 2(1) + 1 = 3$$

$$\begin{aligned} S_n &= \frac{n(a_1 + a_n)}{2} \\ &= \frac{15(3+31)}{2} \\ &= 255 \end{aligned}$$

65.

$$a_1 = 1, \quad a_{50} = 50, \quad n = 50$$

$$\begin{aligned} S_n &= \frac{n(a_1 + a_n)}{2} \\ &= \frac{50(1+50)}{2} \\ &= \frac{50(51)}{2} \\ &= 1{,}275 \end{aligned}$$

67.

$$a_1 = 1, \quad a_n = 99, \quad n = 50$$

$$\begin{aligned} S_n &= \frac{n(a_1 + a_n)}{2} \\ &= \frac{50(1+99)}{2} \\ &= \frac{50(100)}{2} \\ &= 2{,}500 \end{aligned}$$

69.

$$\begin{aligned} \sum_{k=1}^{4}(6k) &= 6(1) + 6(2) + 6(3) + 6(4) \\ &= 6 + 12 + 18 + 24 \\ &= 60 \end{aligned}$$

71.

$$\begin{aligned} \sum_{k=3}^{4} k^3 &= 3^3 + 4^3 \\ &= 27 + 64 \\ &= 91 \end{aligned}$$

73.

$$\begin{aligned} \sum_{k=3}^{4}(k^2 + 3) &= (3^2 + 3) + (4^2 + 3) \\ &= (9+3) + (16+3) \\ &= (12) + (19) \\ &= 31 \end{aligned}$$

75.

$$\begin{aligned} \sum_{k=4}^{4}(2k + 4) &= 2(4) + 4 \\ &= 8 + 4 \\ &= 12 \end{aligned}$$

APPLICATIONS

77. SAVING MONEY
Let $a_1 = 60$ and $d = 50$.
Let $n =$ the month. Find a_n to find the amount in the account each month with $a_1 = 60$ and $d = 50$.

$a_n = a_1 + d(n-1)$
$a_1 = 60$
$a_2 = 60 + 50(2-1) = 60 + 50(1) = 110$
$a_3 = 60 + 50(3-1) = 60 + 50(2) = 160$
$a_4 = 60 + 50(4-1) = 60 + 50(3) = 210$
$a_5 = 60 + 50(5-1) = 60 + 50(4) = 260$
$a_6 = 60 + 50(6-1) = 60 + 50(5) = 310$

To find the amount after 10 years, or $10(12) = 120$ months, plus the first deposit, let $n = 120 + 1 = 121$ and find a_{121}.

$a_{121} = 60 + 50(121-1) = 60 + 50(120) = 6{,}060$

She will have a savings of $6,060 after 10 years.

79. DESIGNING PATIOS
$a_1 = 1, \quad a_n = 150, \quad n = 150$

$$S_n = \frac{n(a_1 + a_n)}{2} = \frac{150(1+150)}{2} = \frac{150(151)}{2} = 11{,}325$$

11,325 bricks are needed.

81. FALLING OBJECTS
To find the distance the object fell during the 12th second, find the difference in the distance at 11 seconds and 12 seconds. Find the distance it had fallen after 11 seconds by letting $t = 11$.

$S = 16t^2 = 16(11)^2 = 16(121) = 1{,}936$ ft.

Find the distance it had fallen after 12 seconds by letting $t = 12$.

$S = 16t^2 = 16(12)^2 = 16(144) = 2{,}304$ ft.

Distance it fell during the 12th second:

$2{,}304 - 1{,}936 = 368$ ft.

The object fell 368 feet during the 12th second.

WRITING

83. Answers will vary.

85. Answers will vary.

REVIEW

87.
$$\log_2 \frac{2x}{y} = \log_2 2 + \log_2 x - \log_2 y = 1 + \log_2 x - \log_2 y$$

89.
$$\log x^3y^2 = \log x^3 + \log y^2 = 3\log x + 2\log y$$

CHALLENGE PROBLEMS

91.
$$\sum_{k=1}^{5} 5k = 5(1) + 5(2) + 5(3) + 5(4) + 5(5) = 5(1+2+3+4+5) = 5\sum_{k=1}^{5} k$$

93.
$$\sum_{k=1}^{n} 3 = \sum_{k=1}^{n} 3k^0 = 3(1)^0 + 3(2)^0 + 3(3)^0 + ... + 3(n)^0 = 3(1) + 3(1) + 3(1) + ... + 3(1) = 3(1+1+1+...+1) = 3n$$

SECTION 14.3

VOCABULARY

1. Each term of a **geometric** sequence is found by multiplying the previous term by the same number.

3. If a single number is inserted between a and b to form a geometric sequence, the number is called the geometric **mean** between a and b.

CONCEPTS

5.
$$a_1 = 16$$
$$a_2 = 16\left(\frac{1}{4}\right) = 4$$
$$a_3 = 16\left(\frac{1}{4}\right)^2 = 1$$
16, 4, 1

7. $a_n = a_1 r^{n-1}$

9. a. yes
   b. no, $3 \not\leq 1$
   c. no, $6 \not\leq 1$
   d. yes

NOTATION

11. An infinite geometric sequence is of the form $a_1, a_1 r, \boxed{a_1 r^2}, a_1 r^3, \boxed{a_1 r^4}, \ldots$

13. To find the common ration of a geometric sequence, we use the formula $r = \frac{a_{\boxed{n+1}}}{a_{\boxed{n}}}$.

PRACTICE

15. $a_1 = 3, r = 2$
$$a_1 = 3$$
$$a_2 = a_1 r = 3(2) = 6$$
$$a_3 = a_1 r^2 = 3(2)^2 = 12$$
$$a_4 = a_1 r^3 = 3(2)^3 = 24$$
$$a_5 = a_1 r^4 = 3(2)^4 = 48$$
3, 6, 12, 24, 48

17. $a_1 = -5, r = \frac{1}{5}$
$$a_1 = -5$$
$$a_2 = a_1 r = -5\left(\frac{1}{5}\right) = -1$$
$$a_3 = a_1 r^2 = -5\left(\frac{1}{5}\right)^2 = -\frac{1}{5}$$
$$a_4 = a_1 r^3 = -5\left(\frac{1}{5}\right)^3 = -\frac{1}{25}$$
$$a_5 = a_1 r^4 = -5\left(\frac{1}{5}\right)^4 = -\frac{1}{125}$$
$$-5, -1, -\frac{1}{5}, -\frac{1}{25}, -\frac{1}{125}$$

19. $a_1 = 2, r > 0$, third term is 32

Use $a_n = ar^{n-1}$ to find r.
$$a = 2,\ a_3 = 32 \text{ and } n = 3$$
$$32 = 2r^{3-1}$$
$$32 = 2r^2$$
$$16 = r^2$$
$$4 = r$$
$$a_1 = 2$$
$$a_2 = a_1 r = 2(4) = 8$$
$$a_3 = a_1 r^2 = 2(4)^2 = 32$$
$$a_4 = a_1 r^3 = 2(4)^3 = 128$$
$$a_5 = a_1 r^4 = 2(4)^4 = 512$$
2, 8, 32, 128, 512

21. $a_1 = -3$, fourth term is -192

Use $a_n = ar^{n-1}$ to find r.

$$a = -3,\ a_4 = -192 \text{ and } n = 4$$
$$-192 = -3r^{4-1}$$
$$-192 = -3r^3$$
$$64 = r^3$$
$$4 = r$$
$$a_1 = -3$$
$$a_2 = a_1r = -3(4) = -12$$
$$a_3 = a_1r^2 = -3(4)^2 = -48$$
$$a_4 = a_1r^3 = -3(4)^3 = -192$$
$$a_5 = a_1r^4 = -3(4)^4 = -768$$
$$-3, -12, -48, -192, -768$$

23. $a_1 = -64$, $r < 0$, fifth term is -4

Use $a_n = ar^{n-1}$ to find r.

$$a = -64,\ a_5 = -4 \text{ and } n = 5$$
$$-4 = -64r^{5-1}$$
$$-4 = -64r^4$$
$$\frac{1}{16} = r^4$$
$$-\frac{1}{2} = r \text{ since } r < 0$$
$$a_1 = -64$$
$$a_2 = a_1r = -64\left(-\frac{1}{2}\right) = 32$$
$$a_3 = a_1r^2 = -64\left(-\frac{1}{2}\right)^2 = -16$$
$$a_4 = a_1r^3 = -64\left(-\frac{1}{2}\right)^3 = 8$$
$$a_5 = a_1r^4 = -64\left(-\frac{1}{2}\right)^4 = -4$$
$$-64,\ 32, -16,\ 8, -4$$

25. $a_1 = -64$, sixth term is -2

Use $a_n = ar^{n-1}$ to find r.

$$a = -64,\ a_6 = -2 \text{ and } n = 6$$
$$-2 = -64r^{6-1}$$
$$-2 = -64r^5$$
$$\frac{1}{32} = r^5$$
$$\frac{1}{2} = r$$
$$a_1 = -64$$
$$a_2 = a_1r = -64\left(\frac{1}{2}\right) = -32$$
$$a_3 = a_1r^2 = -64\left(\frac{1}{2}\right)^2 = -16$$
$$a_4 = a_1r^3 = -64\left(\frac{1}{2}\right)^3 = -8$$
$$a_5 = a_1r^4 = -64\left(\frac{1}{2}\right)^4 = -4$$
$$-64, -32, -16, -8, -4$$

27. second term is 10, third term is 50

Use $r = \frac{a_{n+1}}{a_n}$ to find r.

$$a_3 = 50 \text{ and } a_2 = 10$$
$$r = \frac{50}{10}$$
$$r = 5$$

Use $a_n = ar^{n-1}$ to find a.

$$r = 5,\ a_2 = 10 \text{ and } n = 2$$
$$10 = a(5)^{2-1}$$
$$10 = a(5)^1$$
$$10 = 5a$$
$$2 = a$$

$$a_1 = 2$$
$$a_2 = a_1r = 2(5) = 10$$
$$a_3 = a_1r^2 = 2(5)^2 = 50$$
$$a_4 = a_1r^3 = 2(5)^3 = 250$$
$$a_5 = a_1r^4 = 2(5)^4 = 1{,}250$$
$$2,\ 10,\ 50,\ 250,\ 1{,}250$$

29. $a_1 = 7, r = 2$

Use $a_n = ar^{n-1}$ to find a_{10}.

$r = 2,\ a_1 = 7$ and $n = 10$

$a_{10} = 7(2)^{10-1}$

$a_{10} = 7(2)^9$

$a_{10} = 7(512)$

$a_{10} = 3{,}584$

31. $r = -3, a_8 = -81$

Use $a_n = ar^{n-1}$ to find a_1.

$r = -3,\ a_8 = -81$ and $n = 8$

$-81 = a_1(-3)^{8-1}$

$-81 = a_1(-3)^7$

$-81 = a_1(-2{,}187)$

$\frac{-81}{-2{,}187} = a_1$

$\frac{1}{27} = a_1$

33. $a_1 = -8$, $a_6 = -1{,}944$

Use $a_n = ar^{n-1}$ to find a_1.

$a_1 = -8,\ a_6 = -1{,}944$ and $n = 6$

$-1{,}944 = -8(r)^{6-1}$

$-1{,}944 = -8r^5$

$243 = r^5$

$3 = r$

35. $a_1 = 2, a_5 = 162$ for three geometric means between 2 and 162

Use $a_n = ar^{n-1}$ to find r.

$a_1 = 2,\ a_5 = 162$ and $n = 5$

$162 = 2(r)^{5-1}$

$162 = 2r^4$

$81 = r^4$

$3 = r$

$a_1 = 2$

$a_2 = 2(3) = 6$

$a_3 = 2(3)^2 = 18$

$a_4 = 2(3)^3 = 54$

$a_5 = 2(3)^4 = 162$

6, 18, 54

37. $a_1 = -4, a_6 = -12{,}500$ for four geometric means between -4 and $-12{,}500$

Use $a_n = ar^{n-1}$ to find r.

$a_1 = -4,\ a_6 = -12{,}500$ and $n = 6$

$-12{,}500 = -4(r)^{6-1}$

$-12{,}500 = -4r^5$

$3{,}125 = r^5$

$5 = r$

$a_1 = -4$

$a_2 = -4(5) = -20$

$a_3 = -4(5)^2 = -100$

$a_4 = -4(5)^3 = -500$

$a_5 = -4(5)^4 = -2{,}500$

$a_6 = -4(5)^5 = -12{,}500$

$-20, -100, -500, -2{,}500$

39. $a_1 = 2, a_3 = 128$ for the negative geometric mean between 2 and 128

Use $a_n = ar^{n-1}$ to find r.

$a_1 = 2,\ a_3 = 128$ and $n = 3$

$128 = 2(r)^{3-1}$

$128 = 2r^2$

$64 = r^2$

$\pm 8 = r$

$-8 = r$ for $r < 0$ or the negative geometric mean

$a_1 = 2$

$a_2 = 2(-8) = -16$

The geometric mean is -16.

41. $a_1 = 10$, $a_3 = 20$ for the positive geometric mean between 10 and 20

Use $a_n = ar^{n-1}$ to find r.

$$a_1 = 10,\ a_3 = 20 \text{ and } n = 3$$
$$20 = 10(r)^{3-1}$$
$$20 = 10r^2$$
$$2 = r^2$$
$$\pm\sqrt{2} = r$$
$$\sqrt{2} = r \text{ for } r > 0 \text{ or}$$

the positive geometric mean

$$a_1 = 10$$
$$a_2 = 10\left(\sqrt{2}\right) = 10\sqrt{2}$$

The geometric mean is $10\sqrt{2}$.

43. $a_1 = -50$, $a_3 = 10$ for the geometric mean between -50 and 10

Use $a_n = ar^{n-1}$ to find r.

$$a_1 = -50,\ a_3 = 10 \text{ and } n = 3$$
$$10 = -50(r)^{3-1}$$
$$10 = -50r^2$$
$$-\frac{1}{5} = r^2$$

There is no geometric mean because no real number squared is $-\frac{1}{5}$. When you square a real number, it is always positive.

45. Find r.

$$r = \frac{a_2}{a_1} = \frac{6}{2} = 3$$

Find the sum with $a_1 = 2$, $r = 3$, and $n = 6$.

$$S_n = \frac{a_1(1-r^n)}{1-r}$$
$$= \frac{2(1-3^6)}{1-3}$$
$$= \frac{2(1-729)}{-2}$$
$$= \frac{2(-728)}{-2}$$
$$= 728$$

47. Find r.

$$r = \frac{a_2}{a_1} = \frac{-6}{2} = -3$$

Find the sum with $a_1 = 2$, $r = -3$, and $n = 5$.

$$S_n = \frac{a_1(1-r^n)}{1-r}$$
$$= \frac{2(1-(-3)^5)}{1-(-3)}$$
$$= \frac{2(1+243)}{4}$$
$$= \frac{2(244)}{4}$$
$$= 122$$

49. Find r.

$$r = \frac{a_2}{a_1} = \frac{-6}{3} = -2$$

Find the sum with $a_1 = 3$, $r = -2$, and $n = 8$.

$$S_n = \frac{a_1(1-r^n)}{1-r}$$
$$= \frac{3(1-(-2)^8)}{1-(-2)}$$
$$= \frac{3(1-256)}{3}$$
$$= \frac{3(-255)}{3}$$
$$= -255$$

51. Find r.

$$r = \frac{a_2}{a_1} = \frac{6}{3} = 2$$

Find the sum with $a_1 = 3$, $r = 2$, and $n = 7$.

$$S_n = \frac{a_1(1-r^n)}{1-r}$$
$$= \frac{3(1-2^7)}{1-2}$$
$$= \frac{3(1-128)}{-1}$$
$$= \frac{3(-127)}{-1}$$
$$= 381$$

53. Find r.

$$r = \frac{a_3}{a_2} = \frac{\frac{1}{5}}{1} = \frac{1}{5}$$

Use $a_n = ar^{n-1}$ to find a_1.

$$r = \frac{1}{5},\ a_3 = \frac{1}{5} \text{ and } n = 3$$

$$\frac{1}{5} = a_1\left(\frac{1}{5}\right)^{3-1}$$

$$\frac{1}{5} = a_1\left(\frac{1}{5}\right)^{2}$$

$$\frac{1}{5} = \frac{1}{25}a_1$$

$$25 \bullet \frac{1}{5} = 25 \bullet \frac{1}{25}a_1$$

$$5 = a_1$$

Find the sum with $a_1 = 5$, $r = \frac{1}{5}$, and $n = 4$.

$$S_n = \frac{a_1(1-r^n)}{1-r}$$

$$= \frac{5\left(1-\frac{1}{5}^4\right)}{1-\frac{1}{5}}$$

$$= \frac{5\left(1-\frac{1}{625}\right)}{\frac{4}{5}}$$

$$= \frac{5\left(\frac{624}{625}\right)}{\frac{4}{5}}$$

$$= \frac{\frac{624}{125}}{\frac{4}{5}}$$

$$= \frac{624}{125} \bullet \frac{5}{4}$$

$$= \frac{156}{25}$$

55. Find r.

$$r = \frac{a_4}{a_3} = \frac{1}{-2} = -\frac{1}{2}$$

Use $a_n = ar^{n-1}$ to find a_1.

$$r = -\frac{1}{2},\ a_3 = -2 \text{ and } n = 3$$

$$-2 = a_1\left(-\frac{1}{2}\right)^{3-1}$$

$$-2 = a_1\left(-\frac{1}{2}\right)^{2}$$

$$-2 = \frac{1}{4}a_1$$

$$4 \bullet -2 = 4 \bullet \frac{1}{4}a_1$$

$$-8 = a_1$$

Find the sum with $a_1 = -8$, $r = -\frac{1}{2}$, and $n = 6$.

$$S_n = \frac{a_1(1-r^n)}{1-r}$$

$$= \frac{-8\left(1-\left(-\frac{1}{2}\right)^6\right)}{1-\left(-\frac{1}{2}\right)}$$

$$= \frac{-8\left(1-\frac{1}{64}\right)}{1+\frac{1}{2}}$$

$$= \frac{-8\left(\frac{63}{64}\right)}{\frac{3}{2}}$$

$$= \frac{-\frac{63}{8}}{\frac{3}{2}}$$

$$= -\frac{63}{8} \bullet \frac{2}{3}$$

$$= -\frac{21}{4}$$

57. Find r.

$$r=\frac{a_2}{a_1}=\frac{4}{8}=\frac{1}{2}$$

$$|r|=\left|\frac{1}{2}\right|=\frac{1}{2}<1 \text{ so the sum does exist}$$

Find the sum with $a_1 = 8$ and $r=\frac{1}{2}$.

$$\begin{aligned} S&=\frac{a_1}{1-r}\\ &=\frac{8}{1-\frac{1}{2}}\\ &=\frac{8}{\frac{1}{2}}\\ &=8\bullet\frac{2}{1}\\ &=16 \end{aligned}$$

59. Find r.

$$r=\frac{a_2}{a_1}=\frac{18}{54}=\frac{1}{3}$$

$$|r|=\left|\frac{1}{3}\right|=\frac{1}{3}<1 \text{ so the sum does exist}$$

Find the sum with $a_1 = 54$ and $r=\frac{1}{3}$.

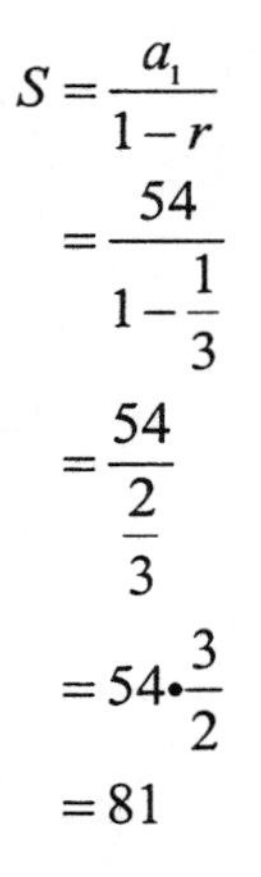

$$\begin{aligned} S&=\frac{a_1}{1-r}\\ &=\frac{54}{1-\frac{1}{3}}\\ &=\frac{54}{\frac{2}{3}}\\ &=54\bullet\frac{3}{2}\\ &=81 \end{aligned}$$

61. Find r.

$$r=\frac{a_2}{a_1}=\frac{-6}{12}=-\frac{1}{2}$$

$$|r|=\left|-\frac{1}{2}\right|=\frac{1}{2}<1 \text{ so the sum does exist}$$

Find the sum with $a_1 = 12$ and $r=-\frac{1}{2}$.

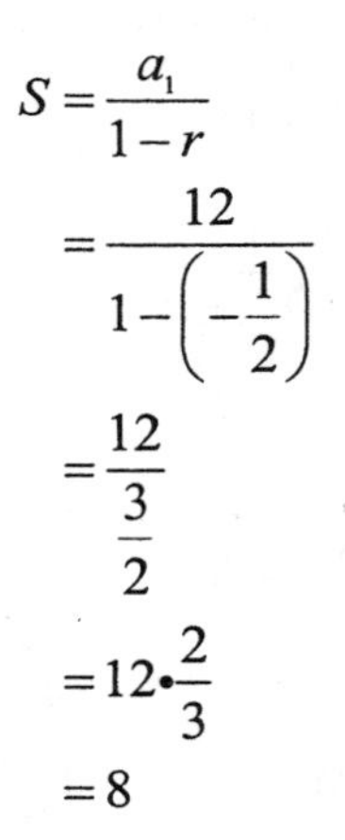

$$\begin{aligned} S&=\frac{a_1}{1-r}\\ &=\frac{12}{1-\left(-\frac{1}{2}\right)}\\ &=\frac{12}{\frac{3}{2}}\\ &=12\bullet\frac{2}{3}\\ &=8 \end{aligned}$$

63. Find r.

$$r=\frac{a_2}{a_1}=\frac{15}{-45}=-\frac{1}{3}$$

$$|r|=\left|-\frac{1}{3}\right|=\frac{1}{3}<1 \text{ so the sum does exist}$$

Find the sum with $a_1 = -45$ and $r=-\frac{1}{3}$.

$$\begin{aligned} S&=\frac{a_1}{1-r}\\ &=\frac{-45}{1-\left(-\frac{1}{3}\right)}\\ &=\frac{-45}{\frac{4}{3}}\\ &=-45\bullet\frac{3}{4}\\ &=-\frac{135}{4} \end{aligned}$$

65. Find r.

$$r=\frac{a_2}{a_1}=\frac{6}{\frac{9}{2}}=\frac{4}{3}$$

Since $|r|=\left|\frac{4}{3}\right|=\frac{4}{3}\geq 1$, the sum of the terms of the sequence, S, does not exist.

67. Find r.

$$r=\frac{a_2}{a_1}=\frac{-9}{-\frac{27}{2}}=\frac{2}{3}$$

$$|r|=\left|\frac{2}{3}\right|=\frac{2}{3}<1 \text{ so the sum does exist}$$

Find the sum with $a_1=-\frac{27}{2}$ and $r=\frac{2}{3}$.

$$\begin{aligned}S&=\frac{a_1}{1-r}\\&=\frac{-\frac{27}{2}}{1-\frac{2}{3}}\\&=\frac{-\frac{27}{2}}{\frac{1}{3}}\\&=-\frac{27}{2}\bullet 3\\&=-\frac{81}{2}\end{aligned}$$

69.

$$0.\bar{1}=0.111...=\frac{1}{10}+\frac{1}{100}+\frac{1}{1{,}000}+...$$

where $a_1=\frac{1}{10}$ and $r=\frac{1}{10}$.

$|r|=\left|\frac{1}{10}\right|=\frac{1}{10}<1$ so the sum does exist.

Find the sum with $a_1=\frac{1}{10}$ and $r=\frac{1}{10}$.

$$\begin{aligned}S&=\frac{a_1}{1-r}\\&=\frac{\frac{1}{10}}{1-\frac{1}{10}}\\&=\frac{\frac{1}{10}}{\frac{9}{10}}\\&=\frac{1}{10}\bullet\frac{10}{9}\\&=\frac{1}{9}\end{aligned}$$

$$0.\bar{1}=\frac{1}{9}$$

71.

$$0.\bar{3} = 0.333... = \frac{3}{10} + \frac{3}{100} + \frac{3}{1,000} + ...$$

where $a_1 = \frac{3}{10}$ and $r = \frac{1}{10}$.

$|r| = \left|\frac{1}{10}\right| = \frac{1}{10} < 1$ so the sum does exist.

Find the sum with $a_1 = \frac{3}{10}$ and $r = \frac{1}{10}$.

$$\begin{aligned} S &= \frac{a_1}{1-r} \\ &= \frac{\frac{3}{10}}{1-\frac{1}{10}} \\ &= \frac{\frac{3}{10}}{\frac{9}{10}} \\ &= \frac{3}{10} \bullet \frac{10}{9} \\ &= \frac{1}{3} \\ 0.\bar{3} &= \frac{1}{3} \end{aligned}$$

73.

$$0.\overline{12} = 0.121212... = \frac{12}{100} + \frac{12}{10,000} + \frac{12}{1,000,000} + ...$$

where $a_1 = \frac{12}{100}$ and $r = \frac{1}{100}$.

$|r| = \left|\frac{1}{100}\right| = \frac{1}{100} < 1$ so the sum does exist.

Find the sum with $a_1 = \frac{12}{100}$ and $r = \frac{1}{100}$.

$$\begin{aligned} S &= \frac{a_1}{1-r} \\ &= \frac{\frac{12}{100}}{1-\frac{1}{100}} \\ &= \frac{\frac{12}{100}}{\frac{99}{100}} \\ &= \frac{12}{100} \bullet \frac{100}{99} \\ &= \frac{12}{99} \\ &= \frac{4}{33} \\ 0.\bar{1} &= \frac{4}{33} \end{aligned}$$

75.

$$0.\overline{75} = 0.757575... = \frac{75}{100} + \frac{75}{10,000} + \frac{75}{1,000,000} + ...$$

where $a_1 = \frac{75}{100}$ and $r = \frac{1}{100}$.

$|r| = \left|\frac{1}{100}\right| = \frac{1}{100} < 1$ so the sum does exist.

Find the sum with $a_1 = \frac{75}{100}$ and $r = \frac{1}{100}$.

$$\begin{aligned} S &= \frac{a_1}{1-r} \\ &= \frac{\frac{75}{100}}{1-\frac{1}{100}} \\ &= \frac{\frac{75}{100}}{\frac{99}{100}} \\ &= \frac{75}{100} \cdot \frac{100}{99} \\ &= \frac{75}{99} \\ &= \frac{25}{33} \\ 0.\overline{75} &= \frac{25}{33} \end{aligned}$$

APPLICATIONS

77. DECLINING SAVINGS

If he spends 12% of the funds each year, then 88% of the funds remain. The amount of money in the savings box after 15 years is represented by the 16th term of a geometric series, where $a_1 = 10,000$, $n = 16$, and $r = 0.88$.

$$\begin{aligned} a_n &= a_1 r^{n-1} \\ a_{16} &= (10,000)(0.88)^{16-1} \\ a_{16} &= (10,000)(0.88)^{15} \\ a_{16} &= (10,000)(0.1469738) \\ a_{16} &= \$1,469.74 \end{aligned}$$

He will have \$1,469.74 after 15 years.

79. HOUSE APPRECIATION

If the house appreciates 6% each year, then it will be worth 106% of the preceding years' value (100% + 6% interest). The value of the house after 12 years is represented by the 13th term of a geometric series, where $a_1 = 70,000$, $n = 13$, and $r = 1.06$.

$$\begin{aligned} a_n &= a_1 r^{n-1} \\ a_{13} &= (70,000)(1.06)^{13-1} \\ a_{13} &= (70,000)(1.06)^{12} \\ a_{13} &= (70,000)(2.01219647) \\ a_{13} &= \$140,853.75 \end{aligned}$$

In 12 years, the house will be worth \$140,853.75.

81. INSCRIBED SQUARES

The area of each new is ½ the area of the previous square. The area of the 12th square is represented by the 12th term of a geometric series, where $a_1 = 1$, $r = ½$, and $n = 12$.

$$\begin{aligned} a_n &= a_1 r^{n-1} \\ a_{12} &= (1)\left(\frac{1}{2}\right)^{12-1} \\ a_{12} &= (1)\left(\frac{1}{2}\right)^{11} \\ a_{12} &= (1)(0.00048828) \\ a_{12} &\approx 0.0005 \end{aligned}$$

83. BOUNCING BALLS

The total distance the ball travels is the sum of two motions, falling and rebounding. The distance the ball falls is given by the sum $10+\frac{1}{2}\bullet 10+\frac{1}{2}\left(\frac{1}{2}\bullet 10\right)+...$ or $10+5+\frac{5}{2}+...$. The distance the ball rebounds begins with the first bounce up which is 5 m, and then is one-half the distance afterwards or $5+\frac{5}{2}+\frac{5}{4}+...$.

Since these are infinite geometric series, use the formula $S=\frac{a_1}{1-r}$ to find the sum.

Falling:

$a_1 = 10$ and $r = \frac{1}{2}$

$$\begin{aligned} S &= \frac{a_1}{1-r} \\ &= \frac{10}{1-\frac{1}{2}} \\ &= \frac{10}{\frac{1}{2}} \\ &= 10\bullet\frac{2}{1} \\ &= 20 \end{aligned}$$

Rebounding:

$a_1 = 5$ and $r = \frac{1}{2}$

$$\begin{aligned} S &= \frac{a_1}{1-r} \\ &= \frac{5}{1-\frac{1}{2}} \\ &= \frac{5}{\frac{1}{2}} \\ &= 5\bullet\frac{2}{1} \\ &= 10 \end{aligned}$$

The distance the ball travels is the sum of the distances falling and rebounding:

20 + 10 = 30 m.

The ball travels a total of 30 m.

85. PEST CONTROL

To find the sum of the given infinite geometric series use the formula $S=\frac{a_1}{1-r}$ with $a_1 = 1{,}000$ and $r = 0.8$.

$$\begin{aligned} S &= \frac{a_1}{1-r} \\ &= \frac{1{,}000}{1-0.8} \\ &= \frac{1{,}000}{0.2} \\ &= 5{,}000 \end{aligned}$$

The long-term population is 5,000.

WRITING

87. Answers will vary.

89. Answers will vary.

REVIEW

91.

$$\begin{aligned} x^2-5x-6 &\le 0 \\ (x-6)(x+1) &= 0 \end{aligned}$$

$$x-6=0 \quad\text{or}\quad x+1=0$$
$$x=6 \qquad\qquad x=-1$$

The critical numbers are 6 and –1. Pick numbers in all three parts and substitute to determine which areas to shade.

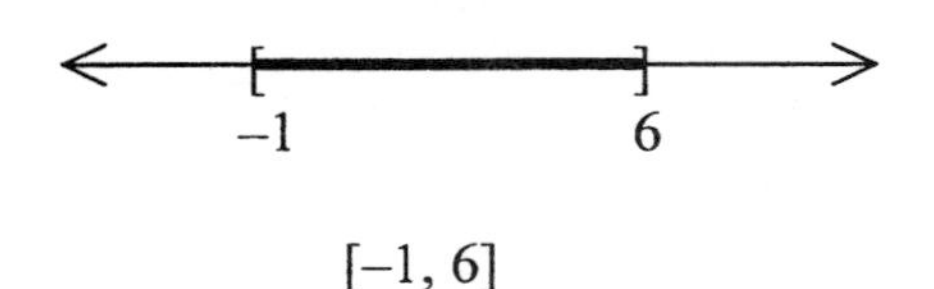

[–1, 6]

93.

$$\frac{x-4}{x+3}>0$$

$$\frac{x-4}{x+3}=0$$

$$(x+3)\left(\frac{x-4}{x+3}\right)=(x+3)(0)$$

$$x-4=0$$

$$x=4$$

Set the denominator $=0$.

$$x+3=0$$

$$x=-3$$

Critical numbers $=-3$ and 4

$$(-\infty,-3)\cup(4,\infty)$$

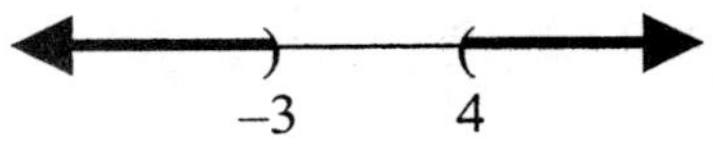

CHALLENGE PROBLEMS

95.

$$f(x)=1+x+x^2+x^3+x^4+\ldots$$

$$f\left(\frac{1}{2}\right)=1+\frac{1}{2}+\left(\frac{1}{2}\right)^2+\left(\frac{1}{2}\right)^3+\left(\frac{1}{2}\right)^4+\ldots$$

$$f\left(\frac{1}{2}\right)=1+\text{ infinite geometric series }\left(a=\frac{1}{2},r=\frac{1}{2}\right)$$

$$f\left(\frac{1}{2}\right)=1+\frac{\frac{1}{2}}{1-\frac{1}{2}}$$

$$f\left(\frac{1}{2}\right)=1+\frac{\frac{1}{2}}{\frac{1}{2}}$$

$$f\left(\frac{1}{2}\right)=1+1$$

$$f\left(\frac{1}{2}\right)=2$$

$$f\left(-\frac{1}{2}\right)=1+\left(-\frac{1}{2}\right)+\left(-\frac{1}{2}\right)^2+\left(-\frac{1}{2}\right)^3+\left(-\frac{1}{2}\right)^4+\ldots$$

$$f\left(-\frac{1}{2}\right)=1+\text{ infinite geometric series }\left(a=-\frac{1}{2},r=-\frac{1}{2}\right)$$

$$f\left(-\frac{1}{2}\right)=1+\frac{-\frac{1}{2}}{1-\left(-\frac{1}{2}\right)}$$

$$f\left(-\frac{1}{2}\right)=1+\frac{-\frac{1}{2}}{\frac{3}{2}}$$

$$f\left(-\frac{1}{2}\right)=1-\frac{1}{3}$$

$$f\left(-\frac{1}{2}\right)=\frac{2}{3}$$

97. the arithmetic mean

SECTION 14.4

VOCABULARY

1. A **tree** diagram like that shown below can be used to count the number of possible outcomes.

3. A **permutation** is an arrangement of objects.

CONCEPTS

5. If an event E_1 can be done in p ways and (after it occurs) a second event E_2 can be done in q ways, the event E_1 followed by E_2 can be done in $\underline{p \cdot q}$ ways.

7. The formula for the number of permutations of n things taken r at a time is $\underline{P(n,r)=\frac{n!}{(n-r)!}}$.

9. They symbol $C(n, r)$ or $\left(\boxed{\frac{n}{r}}\right)$ means the number of **combinations** of n things taken r at a time.

11. $0! = \boxed{1}$

NOTATION

13. $P(6, 2)$

$$P(6,2)=\frac{\boxed{6!}}{(6-2)!}$$
$$=\frac{6\cdot 5\cdot 4!}{\boxed{4!}}$$
$$=6\cdot\boxed{5}$$
$$=30$$

PRACTICE

15. $P(3, 3)$

$$P(3,3)=\frac{3!}{(3-3)!}$$
$$=\frac{3\cdot 2\cdot 1}{0!}$$
$$=\frac{6}{1}$$
$$=6$$

17. $P(5, 3)$

$$P(5,3)=\frac{5!}{(5-3)!}$$
$$=\frac{5\cdot 4\cdot 3\cdot 2!}{2!}$$
$$=\frac{60}{1}$$
$$=60$$

19. $P(2, 2)\cdot P(3, 3)$

$$P(2,2)\cdot P(3,3)=\frac{2!}{(2-2)!}\cdot\frac{3!}{(3-3)!}$$
$$=\frac{2\cdot 1}{0!}\cdot\frac{3\cdot 2\cdot 1}{0!}$$
$$=\frac{2}{1}\cdot\frac{6}{1}$$
$$=12$$

21. $\frac{P(5,3)}{P(4,2)}$

$$\frac{P(5,3)}{P(4,2)}=\frac{\frac{5!}{(5-3)!}}{\frac{4!}{(4-2)!}}$$
$$=\frac{\frac{5\cdot 4\cdot 3\cdot 2!}{2!}}{\frac{4\cdot 3\cdot 2!}{2!}}$$
$$=\frac{\frac{60}{1}}{\frac{12}{1}}$$
$$=\frac{60}{12}$$
$$=5$$

23. $\dfrac{P(6,2)\cdot P(7,3)}{P(5,1)}$

$$\frac{P(6,2)\cdot P(7,3)}{P(5,1)}=\frac{\frac{6!}{(6-2)!}\cdot\frac{7!}{(7-3)!}}{\frac{5!}{(5-1)!}}$$
$$=\frac{\frac{6\cdot5\cdot4!}{4!}\cdot\frac{7\cdot6\cdot5\cdot4!}{4!}}{\frac{5\cdot4!}{4!}}$$
$$=\frac{\frac{30}{1}\cdot\frac{210}{1}}{\frac{5}{1}}$$
$$=\frac{6,300}{5}$$
$$=1,260$$

25. $C(5, 3)$

$$C(5,3)=\frac{5!}{3!(5-3)!}$$
$$=\frac{5\cdot4\cdot3!}{3!\cdot2!}$$
$$=\frac{20}{2}$$
$$=10$$

27. $\dbinom{6}{3}$

$$\binom{6}{3}=\frac{6!}{3!(6-3)!}$$
$$=\frac{6\cdot5\cdot4\cdot3!}{3!\cdot3!}$$
$$=\frac{120}{6}$$
$$=20$$

29. $\dbinom{5}{4}\dbinom{5}{3}$

$$\binom{5}{4}\binom{5}{3}=\frac{5!}{4!(5-4)!}\cdot\frac{5!}{3!(5-3)!}$$
$$=\frac{5\cdot4!}{4!\cdot1!}\cdot\frac{5\cdot4\cdot3!}{3!\cdot2!}$$
$$=\frac{5}{1}\cdot\frac{20}{2}$$
$$=50$$

31. $\dfrac{C(38,37)}{C(19,18)}$

$$\frac{C(38,37)}{C(19,18)}=\frac{\frac{38!}{37!(38-37)!}}{\frac{19!}{18!(19-18)!}}$$
$$=\frac{\frac{38\cdot37!}{37!\cdot1!}}{\frac{19\cdot18!}{18!\cdot1!}}$$
$$=\frac{\frac{38}{1}}{\frac{19}{1}}$$
$$=\frac{38}{19}$$
$$=2$$

33. $C(12, 0)\cdot C(12, 12)$

$$C(12,0)\cdot C(12,12)=\frac{12!}{0!(12-0)!}\cdot\frac{12!}{12!(12-12)!}$$
$$=\frac{12!}{0!12!}\cdot\frac{12!}{12!0!}$$
$$=\frac{1}{1}\cdot\frac{1}{1}$$
$$=1$$

35. $C(n, 2)$

$$C(n,2)=\frac{n!}{2!(n-2)!}$$
$$=\frac{n!}{2(n-2)!}$$

37. $(x + y)^4$

$$= \binom{4}{0}x^4 + \binom{4}{1}x^3y + \binom{4}{2}x^2y^2 + \binom{4}{3}xy^3 + \binom{4}{4}y^4$$
$$= x^4 + 4x^3y + 6x^2y^2 + 4xy^3 + y^4$$

39. $(2x + y)^3$

$$= \binom{3}{0}(2x)^3 + \binom{3}{1}(2x)^2 y + \binom{3}{2}(2x)y^2 + \binom{3}{3}y^3$$
$$= 8x^3 + 3(4x^2)y + 3(2x)y^2 + y^3$$
$$= 8x^3 + 12x^2y + 6xy^2 + y^3$$

41. $(3x - 2)^4$

$$= \binom{4}{0}(3x)^4 + \binom{4}{1}(3x)^3(-2) + \binom{4}{2}(3x)^2(-2)^2 + \binom{4}{3}(3x)(-2)^3 + \binom{4}{4}(-2)^4$$
$$= 81x^4 + 4(27x^3)(-2) + 6(9x^2)(4) + 4(3x)(-8) + 16$$
$$= 81x^4 - 216x^3 + 216x^2 - 96x + 16$$

43.

$$\binom{5}{3}x^2(-5y)^3 = 10x^2(-125y^3)$$
$$= -1,250x^2y^3$$

45.

$$\binom{4}{1}(x^2)^3(-y^3) = 4x^6(-y^3)$$
$$= -4x^6y^3$$

APPLICATIONS

47. PLANNING AN EVENING
Let E_1 be the event "seeing a movie" and E_2 be the event "eating dinner." Because there are 5 ways to accomplish E_1 and 7 ways to accomplish E_2, the number of choices that Kristy has is $5 \cdot 7 = 35$.

49. LICENSE PLATES
If there are 10 choices for each digit and 6 digits, then the number of license plates is $10 \cdot 10 \cdot 10 \cdot 10 \cdot 10 \cdot 10 = 1,000,000$.

51. LICENSE PLATES
If the first digit cannot be zero, then there are only 9 choices for the first digit, 9 choices for the 2nd digit, 8 choices for the 3rd digit, 7 choices for the 4th digit, 6 choices for the 5th digit, and 5 choices for the 6th digit.

$$9 \cdot 9 \cdot 8 \cdot 7 \cdot 6 \cdot 5 = 136,080$$

There can be 136,080 license plates.

53. PHONE NUMBERS
There are 8 options for the first digit, and 10 options for each of the other 6 digits:

$$8 \cdot 10 \cdot 10 \cdot 10 \cdot 10 \cdot 10 \cdot 10 = 8,000,000$$

55. LINING
The 1st person can be any one of the 6, the 2nd person is one of the 5 left, the 3rd is one of the 4 left, the 4th person is one of the 3 left, the 2nd person can be one of the 2 left, and the last person is the only one left:

$$6 \cdot 5 \cdot 4 \cdot 3 \cdot 2 \cdot 1 = 720 \text{ ways}$$

57. ARRANGING BOOKS
The novels can be arranged in $4 \cdot 3 \cdot 2 \cdot 1 = 24$ ways, and the biographies can be arranged in $5 \cdot 4 \cdot 3 \cdot 2 \cdot 1 = 120$ ways. So the total number of ways to arrange both would be $24 \cdot 120 = 2,880$ ways.

59. LOCKS

Find $P(25, 3)$

$$P(25,3)=\frac{25!}{(25-3)!}$$
$$=\frac{25 \cdot 24 \cdot 23 \cdot 22!}{22!}$$
$$=\frac{13,800}{1}$$
$$=13,800$$

There are 13,800 different combinations.

61. ARRANGING APPOINTMENTS

Find $P(10, 3)$

$$P(10,3)=\frac{10!}{(10-3)!}$$
$$=\frac{10 \cdot 9 \cdot 8 \cdot 7!}{7!}$$
$$=\frac{720}{1}$$
$$=720$$

There are 720 ways to fill the appointments.

63. PALINDROMES

There are 9 choices for the 1st position (since you cannot use 0 for the 1st position), 10 choices for the 2nd position, and 10 choices for the 3rd position. There is only 1 choice for the 4th position since it must be the same as the second, and only 1 choice for the 5th position since it must be the same as the first:

$$9 \cdot 10 \cdot 10 \cdot 1 \cdot 1 = 900$$

There are 900 5 digit palindromes.

65. PICNICS

Find $C(14, 3)$

$$C(14,3)=\frac{14!}{3!(14-3)!}$$
$$=\frac{14 \cdot 13 \cdot 12 \cdot 11!}{3! \cdot 11!}$$
$$=\frac{2,184}{6}$$
$$=364$$

There are 364 possible committees.

67. COMMITTEES

Let n = # of people in the group. Since they are forming a 3-person committee, n must be larger than 3. Find a combination using trial-and-error.

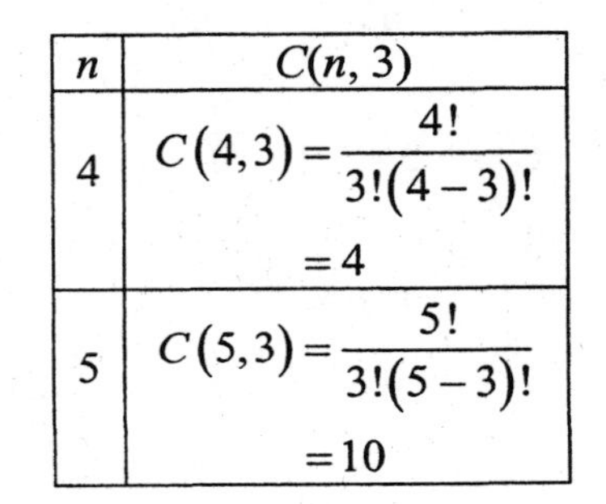

n	$C(n, 3)$
4	$C(4,3)=\frac{4!}{3!(4-3)!}$ $=4$
5	$C(5,3)=\frac{5!}{3!(5-3)!}$ $=10$

There are 5 people in the group, since

69. LOTTERIES

There are 100 digits from 0 to 99. Find $C(100,6)$ to find the number of possible choices.

$$C(100,6)=\frac{100!}{6!(100-6)!}$$
$$=\frac{100 \cdot 99 \cdot 98 \cdot 97 \cdot 96 \cdot 95 \cdot \cancel{94!}}{6 \cdot 5 \cdot 4 \cdot 3 \cdot 2 \cdot 1 \cdot \cancel{94!}}$$
$$=\frac{8.58277728 \times 10^{11}}{720}$$
$$=1,192,052,400$$

There are 1,192,052,400 possible choices.

71. COMMITTEES
Find the product of $C(3, 2)$ for the selection of the men and $C(4, 2)$ for the selection of the women.

$$C(3,2)\cdot C(4,2)=\frac{3!}{2!(3-2)!}\cdot\frac{4!}{2!(4-2)!}$$
$$=\frac{3\cdot \cancel{2!}}{\cancel{2!}1!}\cdot\frac{4\cdot 3\cdot \cancel{2!}}{\cancel{2!}2\cdot 1}$$
$$=\frac{3}{1}\cdot\frac{12}{2}$$
$$=18$$

There are 18 ways to select the committee.

73. CHOOSING CLOTHES
Find the product of $C(12, 2)$ for the selection of the shirts and $C(10, 3)$ for the selection of the neckties.

$$C(12,2)\cdot C(10,3)=\frac{12!}{2!(12-2)!}\cdot\frac{10!}{3!(10-3)!}$$
$$=\frac{12\cdot 11\cdot \cancel{10!}}{2\cdot 1\cdot \cancel{10!}}\cdot\frac{10\cdot 9\cdot 8\cdot \cancel{7!}}{3\cdot 2\cdot 1\cdot \cancel{7!}}$$
$$=\frac{132}{2}\cdot\frac{720}{6}$$
$$=7{,}920$$

There are 7,920 ways to select the clothes.

WRITING

75. Answers will vary.

REVIEW

77.

$$2^{x+1}=3^x$$
$$\log 2^{x+1}=\log 3^x$$
$$(x+1)\log 2=x\log 3$$
$$x\log 2+\log 2=x\log 3$$
$$\log 2=x\log 3-x\log 2$$
$$\log 2=x(\log 3-\log 2)$$
$$\frac{\log 2}{\log 3-\log 2}=\frac{x(\log 3-\log 2)}{\log 3-\log 2}$$
$$\frac{\log 2}{\log 3-\log 2}=x$$
$$x=\frac{\log 2}{\log 3-\log 2}$$
$$x\approx 1.7095$$

79.

$$e^{3x}=9$$
$$\ln e^{3x}=\ln 9$$
$$3x=\ln 9$$
$$x=\frac{\ln 9}{3}$$
$$x\approx 0.7324$$

CHALLENGE PROBLEMS

81. Let A and B be the people who must stand together and C, D, and E be the other people in the group. The sample space would be (A or B) in one position and C, D, and E in one of the other 3 positions or $4\cdot 3\cdot 2\cdot 1$. Since there are two ways to position A and B (AB or BA), then the possible combinations are

$$2\cdot(4\cdot 3\cdot 2\cdot 1)=48.$$

SECTION 14.5

VOCABULARY

1. An **experiment** is any activity for which the outcome is uncertain.

CONCEPTS

3. The probability of an event E is defined as $P(E) = \dfrac{s}{n}$.

5. If an event cannot happen, its probability is **0**.

NOTATION

7. a. The number of black face cards is $\boxed{6}$.
   b. The number of cards in the deck is $\boxed{52}$.
   c. The probability is $\boxed{\dfrac{6}{52}}$ or $\boxed{\dfrac{3}{26}}$.

PRACTICE

9. {(1, H), (2, H), (3, H), (4, H), (5, H), (6, H), (1, T), (2, T), (3, T), (4, T), (5, T), (6, T)}

11. {a, b, c, d, e, f, g, h, i, j, k, l, m, n, o, p, q, r, s, t, u, v, w, x, y, z}

13. $s = 1$ and $n = 6$

$$P(2) = \frac{s}{n} = \frac{1}{6}$$

15. You could roll a 2, 3, 4, or 5, so $s = 4$ and $n = 6$.

$$P(>1 \text{ or } <6) = \frac{s}{n} = \frac{4}{6} = \frac{2}{3}$$

17. There are 19 numbers less than 20, so $s = 19$ and $n = 42$.

$$P(<20) = \frac{s}{n} = \frac{19}{42}$$

19. The prime numbers less than 42 are 1, 3, 5, 7, 11, 13, 17, 19, 23, 29, 31, 37, and 41. So, $s = 13$ and $n = 42$.

$$P(\text{prime numbers}) = \frac{s}{n} = \frac{13}{42}$$

21. There are 3 red sections and 8 total sections, so $s = 3$ and $n = 8$.

$$P(\text{red section}) = \frac{s}{n} = \frac{3}{8}$$

23. There are 0 brown sections and 8 total sections, so $s = 0$ and $n = 8$.

$$P(\text{brown section}) = \frac{s}{n} = \frac{0}{8} = 0$$

25. You could roll (1, 3), (2, 2), or (3, 1) to get a sum of 4. From Example 1, we see that there are 36 possible outcomes when rolling two dice. So, $s = 3$ and $n = 36$.

$$P(\text{sum of } 4) = \frac{s}{n} = \frac{3}{36} = \frac{1}{12}$$

27. There are 5 red eggs in a basket of 12 eggs. So, $s = 5$ and $n = 12$.

$$P(\text{red egg}) = \frac{s}{n} = \frac{5}{12}$$

29. The number of ways we can draw 6 diamonds from the 13 diamonds in the deck is $C(13, 6)$, and the number of ways we can draw 6 cards from the 52 cards in the deck is $C(52, 6)$. The probability of drawing 6 diamonds is the ratio of the number of favorable outcomes to the number of possible outcomes.

$$P(6\text{ diamonds}) = \frac{s}{n}$$
$$= \frac{C(13,6)}{C(52,6)}$$
$$= \frac{\frac{13!}{6!7!}}{\frac{52!}{6!46!}}$$
$$= \frac{13!}{6!7!} \cdot \frac{6!46!}{52!}$$
$$= \frac{13 \cdot 12 \cdot 11 \cdot 10 \cdot 9 \cdot 8 \cdot \cancel{7!}}{\cancel{6!} \cdot \cancel{7!}} \cdot \frac{\cancel{6!}\,\cancel{46!}}{52 \cdot 51 \cdot 50 \cdot 49 \cdot 48 \cdot 47 \cdot \cancel{46!}}$$
$$= \frac{13 \cdot 12 \cdot 11 \cdot 10 \cdot 9 \cdot 8}{52 \cdot 51 \cdot 50 \cdot 49 \cdot 48 \cdot 47}$$
$$= \frac{33}{391{,}510}$$

31. The number of ways we can draw 5 clubs from the 13 clubs in the deck is $C(13, 5)$, and the number of ways we can draw 5 cards from the 26 black cards in the deck is $C(26, 5)$. The probability of drawing 5 clubs is the ratio of the number of favorable outcomes to the number of possible outcomes.

$$P(5\text{ clubs}) = \frac{s}{n}$$
$$= \frac{C(13,5)}{C(26,5)}$$
$$= \frac{\frac{13!}{5!8!}}{\frac{26!}{5!21!}}$$
$$= \frac{13!}{5!8!} \cdot \frac{5!21!}{26!}$$
$$= \frac{13 \cdot 12 \cdot 11 \cdot 10 \cdot 9 \cdot \cancel{8!}}{\cancel{5!} \cdot \cancel{8!}} \cdot \frac{\cancel{5!}\,\cancel{21!}}{26 \cdot 25 \cdot 24 \cdot 23 \cdot 22 \cdot \cancel{21!}}$$
$$= \frac{13 \cdot 12 \cdot 11 \cdot 10 \cdot 9}{26 \cdot 25 \cdot 24 \cdot 23 \cdot 22}$$
$$= \frac{9}{460}$$

33. Let f represent failure of an engine and s represent survival of an engine. The sample space would be:
{ (f, f, f, f), (f, f, f, s), (f, f, s, f), (f, s, f, f), (s, f, f, f), (f, f, s, s), (f, s, f, s), (s, f, f, s), (f, s, s, f), (s, f, s, f), (s, s, f, f), (f, s, s, s), (s, f, s, s), (s, s, f, s), (s, s, s, f), (s, s, s, s)}

35. Using the sample space from Exercise 33, there are 4 out of 16 chances that exactly one engine will survive. So, let $s = 4$, and $n = 16$.

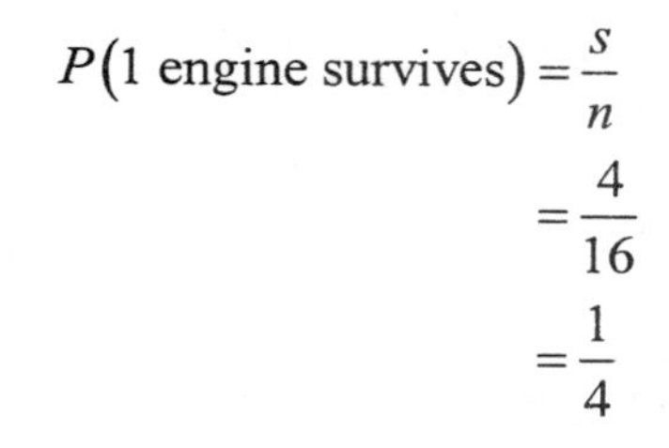

$$P(1\text{ engine survives}) = \frac{s}{n} = \frac{4}{16} = \frac{1}{4}$$

37. Using the sample space from Exercise 33, there are 4 out of 16 chances that exactly 3 engines will survive. So, let $s = 4$, and $n = 16$.

$$P(3\text{ engines survive}) = \frac{s}{n} = \frac{4}{16} = \frac{1}{4}$$

39.

$$\frac{1}{16} + \frac{1}{4} + \frac{3}{8} + \frac{1}{4} + \frac{1}{16} = \frac{1}{16} + \frac{4}{16} + \frac{6}{16} + \frac{4}{16} + \frac{1}{16}$$
$$= \frac{16}{16}$$
$$= 1$$

41. Let $s = 32$ (the number of doctors who oppose the legislation) and $n = 119$ (the total number of doctors surveyed).

$$P(\text{doctors who oppose}) = \frac{s}{n} = \frac{32}{119}$$

APPLICATIONS

43. QUALITY CONTROL
If 2 tires are defective, then 10 – 2 = 8 tires are good. The number of ways we can choose 4 good tires from the 8 good tires available is $C(8,4)$, and the number of ways we can choose 4 good tires from the 10 tires available is $C(10, 4)$. The probability of drawing 4 good tires is the ratio of the number of favorable outcomes to the number of possible outcomes.

$$\begin{aligned} P(4 \text{ good tires}) &= \frac{s}{n} \\ &= \frac{C(8,4)}{C(10,4)} \\ &= \frac{\frac{8!}{4!4!}}{\frac{10!}{4!6!}} \\ &= \frac{8!}{4!4!} \cdot \frac{4!6!}{10!} \\ &= \frac{\not{8} \cdot \not{7} \cdot 6 \cdot 5 \cdot \not{4}!}{\not{4}! \cdot \not{4}!} \cdot \frac{\not{4}!\,\not{6}!}{10 \cdot 9 \cdot \not{8} \cdot \not{7} \cdot \not{6}!} \\ &= \frac{30}{90} \\ &= \frac{1}{3} \end{aligned}$$

WRITING

45. Answers will vary.

REVIEW

47.

$$\begin{aligned} 5^{4x} &= \frac{1}{125} \\ 5^{4x} &= 5^{-3} \\ 4x &= -3 \\ x &= -\frac{3}{4} \end{aligned}$$

49.

$$\begin{aligned} 2^{x^2-2x} &= 8 \\ 2^{x^2-2x} &= 2^3 \\ x^2 - 2x &= 3 \\ x^2 - 2x - 3 &= 0 \\ (x-3)(x+1) &= 0 \end{aligned}$$

$$x - 3 = 0 \quad \text{or} \quad x + 1 = 0$$

$$x = 3 \qquad\qquad x = -1$$

51.

$$\begin{aligned} 3^{x^2+4x} &= \frac{1}{81} \\ 3^{x^2+4x} &= 3^{-4} \\ x^2 + 4x &= -4 \\ x^2 + 4x + 4 &= 0 \\ (x+2)(x+2) &= 0 \end{aligned}$$

$$x + 2 = 0 \quad \text{or} \quad x + 2 = 0$$

$$x = -2 \qquad\qquad x = -2$$

CHALLENGE PROBLEMS

53. Let $P(A)$ = probability students are gifted in math, $P(A + B)$ = probability students are gifted in math and art, and $P(B \mid A)$ = probability students who are gifted in math are also gifted in art.

$P(A) = 30\% = 0.30$

$P(A + B) = 10\% = 0.10$

$$\begin{aligned} P(A \text{ and } B) &= P(A) \cdot P(B|A) \\ 0.10 &= 0.30 \cdot P(B|A) \\ \frac{0.10}{0.30} &= \frac{0.30 \cdot P(B|A)}{0.30} \\ \frac{1}{3} &= P(B|A) \\ P(B|A) &= \frac{1}{3} \end{aligned}$$

The probability that students who are gifted in math are also gifted in art is $\frac{1}{3}$.

CHAPTER 14 KEY CONCEPTS

1. g
2. o
3. i
4. l
5. u
6. y
7. p
8. x
9. m
10. d
11. f
12. c
13. w
14. b
15. j
16. s
17. e
18. z
19. n
20. k
21. h
22. v
23. q
24. a
25. r
26. t

CHAPTER 14 REVIEW

SECTION 14.1
The Binomial Theorem

1. The coefficients for the expansion of $(a + b)^5$ are 1, 5, 10, 10, 5, 1.

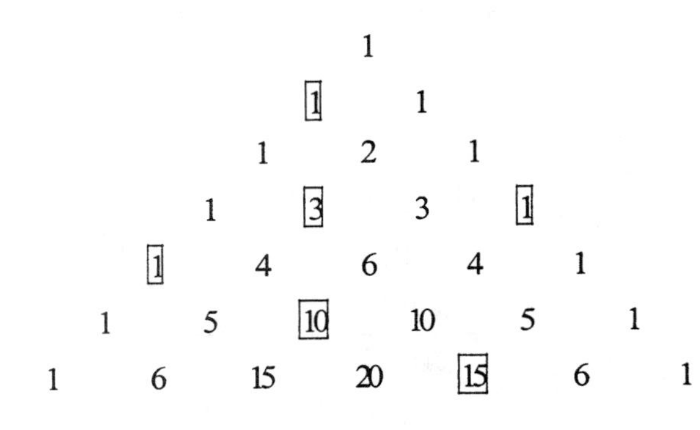

2. a. 13 terms
   b. 12
   c. The first term is a^{12} and the last term is b^{12}.
   d. The exponents of a decrease and the exponents of b increase.

3.
$$\begin{aligned}(4!)(3!) &= (4\cdot3\cdot2\cdot1)(3\cdot2\cdot1)\\ &= (24)(6)\\ &= 144\end{aligned}$$

4.
$$\begin{aligned}\frac{5!}{3!} &= \frac{5\cdot4\cdot\cancel{3!}}{\cancel{3!}}\\ &= 20\end{aligned}$$

5.
$$\begin{aligned}\frac{6!}{2!(6-2)!} &= \frac{6!}{2!4!}\\ &= \frac{6\cdot5\cdot\cancel{4!}}{2\cdot1\cdot\cancel{4!}}\\ &= \frac{30}{2}\\ &= 15\end{aligned}$$

6.
$$\begin{aligned}\frac{12!}{3!(12-3)!} &= \frac{12!}{3!9!}\\ &= \frac{12\cdot11\cdot10\cdot\cancel{9!}}{3\cdot2\cdot1\cdot\cancel{9!}}\\ &= \frac{1{,}320}{6}\\ &= 220\end{aligned}$$

7.
$$\begin{aligned}(n-n)! &= 0!\\ &= 1\end{aligned}$$

8.
$$\begin{aligned}\frac{8!}{7!} &= \frac{8\cdot\cancel{7!}}{\cancel{7!}}\\ &= 8\end{aligned}$$

9.
$$\begin{aligned}&(x+y)^5\\ &= x^5 + \frac{5!}{1!(5-1)!}x^4y + \frac{5!}{2!(5-2)!}x^3y^2\\ &\quad + \frac{5!}{3!(5-3)!}x^2y^3 + \frac{5!}{4!(5-4)!}xy^4 + y^5\\ &= x^5 + \frac{5\cdot\cancel{4!}}{1!\cancel{4!}}x^4y + \frac{5\cdot4\cdot\cancel{3!}}{2!\cancel{3!}}x^3y^2\\ &\quad + \frac{5\cdot4\cdot\cancel{3!}}{\cancel{3!}2!}x^2y^3 + \frac{5\cdot\cancel{4!}}{\cancel{4!}1!}xy^4 + y^5\\ &= x^5 + \frac{5}{1}x^4y + \frac{20}{2}x^3y^2 + \frac{20}{2}x^2y^3 + \frac{5}{1}xy^4 + y^5\\ &= x^5 + 5x^4y + 10x^3y^2 + 10x^2y^3 + 5xy^4 + y^5\end{aligned}$$

10.

$$
\begin{aligned}
(x-y)^9 &= (x+(-y))^9 \\
&= x^9 + \frac{9!}{1!(9-1)!}x^8(-y) + \frac{9!}{2!(9-2)!}x^7(-y)^2 \\
&\quad + \frac{9!}{3!(9-3)!}x^6(-y)^3 + \frac{9!}{4!(9-4)!}x^5(-y)^4 \\
&\quad + \frac{9!}{5!(9-5)!}x^4(-y)^5 + \frac{9!}{6!(9-6)!}x^3(-y)^6 \\
&\quad + \frac{9!}{7!(9-7)!}x^2(-y)^7 + \frac{9!}{8!(9-8)!}x(-y)^8 + (-y)^9 \\
&= x^9 + \frac{9!}{1!8!}x^8(-y) + \frac{9!}{2!7!}x^7(-y)^2 + \frac{9!}{3!6!}x^6(-y)^3 \\
&\quad + \frac{9!}{4!5!}x^5(-y)^4 + \frac{9!}{5!4!}x^4(-y)^5 + \frac{9!}{6!3!}x^3(-y)^6 \\
&\quad + \frac{9!}{7!2!}x^2(-y)^7 + \frac{9!}{8!1!}x(-y)^8 + (-y)^9 \\
&= x^9 - \frac{9}{1}x^8y + \frac{72}{2}x^7y^2 - \frac{504}{6}x^6y^3 + \frac{3{,}024}{24}x^5y^4 \\
&\quad - \frac{3{,}024}{24}x^4y^5 + \frac{504}{6}x^3y^6 - \frac{72}{2}x^2y^7 + \frac{9}{1}xy^8 - y^9 \\
&= x^9 - 9x^8y + 36x^7y^2 - 84x^6y^3 + 126x^5y^4 - 126x^4y^5 \\
&\quad + 84x^3y^6 - 36x^2y^7 + 9xy^8 - y^9
\end{aligned}
$$

11.

$$
\begin{aligned}
&(4x-y)^3 \\
&= (4x)^3 + \frac{3!}{1!(3-1)!}(4x)^2(-y) + \frac{3!}{2!(3-2)!}(4x)(-y)^2 + (-y)^3 \\
&= 64x^3 + \frac{3!}{1!2!}(16x^2)(-y) + \frac{3!}{2!1!}(4x)(y^2) - y^3 \\
&= 64x^3 - \frac{3}{1}(16x^2)y + \frac{3}{1}(4x)y^2 - y^3 \\
&= 64x^3 - 48x^2y + 12xy^2 - y^3
\end{aligned}
$$

12.

$$
\begin{aligned}
&\left(\frac{c}{2}+\frac{d}{3}\right)^4 \\
&= \left(\frac{c}{2}\right)^4 + \frac{4!}{1!(4-1)!}\left(\frac{c}{2}\right)^3\left(\frac{d}{3}\right) \\
&\quad + \frac{4!}{2!(4-2)!}\left(\frac{c}{2}\right)^2\left(\frac{d}{3}\right)^2 + \frac{4!}{3!(4-3)!}\left(\frac{c}{2}\right)\left(\frac{d}{3}\right)^3 \\
&\quad + \left(\frac{d}{3}\right)^4 \\
&= \frac{c^4}{16} + \frac{4\cdot\cancel{3!}}{1!\cancel{3!}}\left(\frac{c^3}{8}\right)\left(\frac{d}{3}\right) + \frac{4\cdot3\cdot\cancel{2!}}{2!\cancel{2!}}\left(\frac{c^2}{4}\right)\left(\frac{d^2}{9}\right) \\
&\quad + \frac{4\cdot\cancel{3!}}{\cancel{3!}1!}\left(\frac{c}{2}\right)\left(\frac{d^3}{27}\right) + \frac{d^4}{80} \\
&= \frac{c^4}{16} + \frac{4}{1}\left(\frac{c^3d}{24}\right) + \frac{12}{2}\left(\frac{c^2d^2}{36}\right) \\
&\quad + \frac{4}{1}\left(\frac{cd^3}{54}\right) + \frac{d^4}{81} \\
&= \frac{c^4}{16} + \frac{c^3d}{6} + \frac{c^2d^2}{6} + \frac{2cd^3}{27} + \frac{d^4}{81}
\end{aligned}
$$

13. The 3rd term of $(x + y)^4$

$$
\begin{aligned}
\frac{4!}{2!(4-2)!}x^2y^2 &= \frac{4\cdot3\cdot\cancel{2!}}{\cancel{2!}2!}x^2y^2 \\
&= \frac{12}{2}x^2y^2 \\
&= 6x^2y^2
\end{aligned}
$$

14. The 4th term of $(x - y)^6$

$$
\begin{aligned}
\frac{6!}{3!(6-3)!}x^3(-y)^3 &= \frac{6\cdot5\cdot4\cdot\cancel{3!}}{\cancel{3!}3\cdot2\cdot1}x^3(-y)^3 \\
&= -\frac{120}{6}x^3y^3 \\
&= -20x^3y^3
\end{aligned}
$$

15. The 2nd term of $(3x - 4y)^3$

$$
\begin{aligned}
\frac{3!}{1!(3-1)!}(3x)^2(-4y) &= \frac{3\cdot\cancel{2!}}{1!\cancel{2!}}(9x^2)(-4y) \\
&= \frac{3}{1}(-36x^2y) \\
&= -108x^2y
\end{aligned}
$$

16. The 5th term of $(u^2 - v^2)^5$

$$\frac{5!}{4!(5-4)!}(u^2)(-v^3)^4 = \frac{5 \cdot \cancel{4!}}{\cancel{4!}1!}(u^2)(v^{12})$$
$$= \frac{5}{1}(u^2 v^{12})$$
$$= 5u^2 v^{12}$$

SECTION 14.2
Arithmetic Sequences and Series

17. $a_n = 2n - 4$

$a_1 = 2(1) - 4 = -2$
$a_2 = 2(2) - 4 = 0$
$a_3 = 2(3) - 4 = 2$
$a_4 = 2(4) - 4 = 4$

The first four terms are –2, 0, 2, and 4.

18. Find a_8 if $a_1 = 7$ and $d = 5$.

$a_8 = a_1 + (n-1)(d)$
$a_8 = 7 + (8-1)(5)$
$a_8 = 7 + (7)(5)$
$a_8 = 7 + 35$
$a_8 = 42$

19. Find d.

$242 - 212 = 30$, so $2d = 30$ and $d = 15$.

Find a_1 if $a_n = 212$ and $n = 7$.

$212 = a_1 + 15(7-1)$
$212 = a_1 + 15(6)$
$212 = a_1 + 90$
$122 = a_1$
$a_1 = 122 + 15(1-1) = 122 + 0 = 122$
$a_2 = 122 + 15(2-1) = 122 + 15 = 137$
$a_3 = 122 + 15(3-1) = 122 + 30 = 152$
$a_4 = 122 + 15(4-1) = 122 + 45 = 167$
$a_5 = 122 + 15(5-1) = 122 + 60 = 182$

20. Find d.

$-6 - 6 = -12$, so $d = -12$.

Find a_{101} if $a_1 = 6$, $d = -12$ and $n = 101$.

$a_n = a_1 + d(n-1)$
$a_{101} = 6 - 12(101-1)$
$= 6 - 12(100)$
$= 6 - 1{,}200$
$= -1{,}194$

21. Let $n = 23$ and $a_n = -625$, $a_1 = -515$.

$a_n = a_1 + d(n-1)$
$-625 = -515 + d(23-1)$
$-625 = -515 + d(22)$
$-110 = 22d$
$-5 = d$

22. Let $a_1 = 8$ and $a_4 = 25$.

$a_2 = 2 + d, a_3 = 2 + 2d$
$a_n = a_1 + d(n-1)$
$a_4 = a_1 + (4-1)d$
$25 = 8 + 3d$
$17 = 3d$
$\frac{17}{3} = d$

$a_2 = 8 + d = 8 + \frac{17}{3} = \frac{41}{3}$
$a_3 = 8 + 2d = 8 + 2\left(\frac{17}{3}\right) = \frac{58}{3}$

23.

$$a_1 = 9,\ \ d = 6\frac{1}{2} - 9 = -2\frac{1}{2} = -\frac{5}{2},\ \ n = 10$$

$$a_{10} = 9 + \left(-\frac{5}{2}\right)(10-1)$$
$$= 9 + \left(-\frac{5}{2}\right)(9)$$
$$= 9 - \frac{45}{2}$$
$$= -13.5$$

$$S_n = \frac{n(a_1 + a_n)}{2}$$
$$S_{10} = \frac{10(9-13.5)}{2}$$
$$= \frac{10(-4.5)}{2}$$
$$= -22.5$$
$$= -\frac{45}{2}$$

24. Find d.

$22 - 6 = 16$, so $4d = 16$ and $d = 4$.

Find a_1 if $a_n = 6$ and $n = 2$.

$$6 = a_1 + 4(2-1)$$
$$6 = a_1 + 4(1)$$
$$6 = a_1 + 4$$
$$2 = a_1$$

Find a_{28}.

$$a_1 = 2,\ \ d = 4,\ \ n = 28$$
$$a_{28} = 2 + 4(28-1)$$
$$= 2 + 4(27)$$
$$= 110$$

$$S_n = \frac{n(a_1 + a_n)}{2}$$
$$S_{28} = \frac{28(2+110)}{2}$$
$$= \frac{28(112)}{2}$$
$$= 1{,}568$$

25.

$$\sum_{k=4}^{6} \frac{1}{2}k = \frac{1}{2}(4) + \frac{1}{2}(5) + \frac{1}{2}(6)$$
$$= 2 + \frac{5}{2} + 3$$
$$= \frac{15}{2}$$

26.

$$\sum_{k=2}^{5} 7k^2 = 7(2^2) + 7(3^2) + 7(4^2) + 7(5^2)$$
$$= 7(4) + 7(9) + 7(16) + 7(25)$$
$$= 28 + 63 + 112 + 175$$
$$= 378$$

27.

$$\sum_{k=1}^{4}(3k-4) = (3\cdot1-4) + (3\cdot2-4) + (3\cdot3-4) + (3\cdot4-4)$$
$$= -1 + 2 + 5 + 8$$
$$= 14$$

28.

$$\sum_{k=10}^{10} 36k = 36(10)$$
$$= 360$$

29.

$$a_1 = 1,\ \ a_n = 100,\ \ n = 100,$$
$$S_n = \frac{n(a_1 + a_n)}{2}$$
$$S_{100} = \frac{100(1+100)}{2}$$
$$= \frac{100(101)}{2}$$
$$= 5{,}050$$

30. $a_1 = 10$, $a_2 = 12$, so $d = 2$.
Let $n = 30$ and find a_{30} to find the number of seats on the last row.

$$\begin{aligned} a_{30} &= 10 + 2(30-1) \\ &= 10 + 2(29) \\ &= 68 \end{aligned}$$

Find the sum of all of the seats.

$$S_n = \frac{n(a_1 + a_n)}{2}$$

$$\begin{aligned} S_{30} &= \frac{30(10+68)}{2} \\ &= \frac{30(78)}{2} \\ &= 1{,}170 \end{aligned}$$

SECTION 14.3
Geometric Sequences and Series

31. $a_1 = \frac{1}{8}, r = 2$

$$\begin{aligned} a_n &= a_1 r^{n-1} \\ a_6 &= a_1 r^{6-1} \\ &= \frac{1}{8}(2)^{6-1} \\ &= \frac{1}{8}(2)^5 \\ &= \frac{1}{8}(32) \\ &= 4 \end{aligned}$$

32. 4th term is 3, 5th term is $\frac{3}{2}$

Use $r = \frac{a_{n+1}}{a_n}$ to find r.

$a_4 = 3$ and $a_5 = \frac{3}{2}$

$$\begin{aligned} r &= \frac{\frac{3}{2}}{3} \\ r &= \frac{3}{2} \bullet \frac{1}{3} \\ r &= \frac{1}{2} \end{aligned}$$

Use $a_n = a_1 r^{n-1}$ to find a_1.

$r = \frac{1}{2}$, $a_4 = 3$ and $n = 4$

$$\begin{aligned} 3 &= a_1 \left(\frac{1}{2}\right)^{4-1} \\ 3 &= a_1 \left(\frac{1}{2}\right)^3 \\ 3 &= \frac{1}{8} a_1 \\ 24 &= a_1 \end{aligned}$$

$$\begin{aligned} a_1 &= 24 \\ a_2 &= a_1 r = 24\left(\frac{1}{2}\right) = 12 \\ a_3 &= a_1 r^2 = 24\left(\frac{1}{2}\right)^2 = 6 \\ a_4 &= a_1 r^3 = 24\left(\frac{1}{2}\right)^3 = 3 \\ a_5 &= a_1 r^4 = 24\left(\frac{1}{2}\right)^4 = \frac{3}{2} \end{aligned}$$

$24, 12, 6, 3, \frac{3}{2}$

33. $r = -3$, $a_9 = 243$

Use $a_n = ar^{n-1}$ to find a_1.

$$r = -3,\ a_9 = 243 \text{ and } n = 9$$

$$243 = a_1(-3)^{9-1}$$

$$243 = a_1(-3)^8$$

$$243 = a_1(6{,}561)$$

$$\frac{243}{6{,}561} = a_1$$

$$\frac{1}{27} = a_1$$

34. $a_1 = -6$, $a_4 = 384$ for two geometric means between -6 and 384

Use $a_n = ar^{n-1}$ to find r.

$$a_1 = -6,\ a_4 = 384 \text{ and } n = 4$$

$$384 = -6(r)^{4-1}$$

$$384 = -6r^3$$

$$-64 = r^3$$

$$-4 = r$$

$$a_1 = -6$$

$$a_2 = -6(-4) = 24$$

$$a_3 = -6(-4)^2 = -96$$

$$a_4 = -6(-4)^3 = 384$$

The geometric means are 24 and -96.

35. Find r.

$$r = \frac{a_2}{a_1} = \frac{54}{162} = \frac{1}{3}$$

Find the sum of $a_1 = 162$, $r = \frac{1}{3}$, and $n = 7$.

$$S_n = \frac{a_1(1-r^n)}{1-r}$$

$$= \frac{162\left(1-\left(\frac{1}{3}\right)^7\right)}{1-\frac{1}{3}}$$

$$= \frac{162\left(1-\frac{1}{2{,}187}\right)}{\frac{2}{3}}$$

$$= \frac{162\left(\frac{2{,}186}{2{,}187}\right)}{\frac{2}{3}}$$

$$= \frac{\frac{4{,}372}{27}}{\frac{2}{3}}$$

$$= \frac{4{,}372}{27} \cdot \frac{3}{2}$$

$$= \frac{2{,}186}{9}$$

36. Find r.

$$r=\frac{a_2}{a_1}=\frac{-\frac{1}{4}}{\frac{1}{8}}=-2$$

Find the sum of $a_1=\frac{1}{8}$, $r=2$, and $n=8$.

$$S_n=\frac{a_1(1-r^n)}{1-r}$$
$$=\frac{\frac{1}{8}\left(1-(-2)^8\right)}{1-(-2)}$$
$$=\frac{\frac{1}{8}(1-256)}{3}$$
$$=\frac{\frac{1}{8}(-255)}{3}$$
$$=\frac{-255}{8}\bullet\frac{1}{3}$$
$$=\frac{-255}{24}$$
$$=-\frac{85}{8}$$

37. If he uses 25% of the birdseed each month, then 75% of the birdseed remains. The amount in the bag after 12 months is represented by the 13th term of a geometric series, where $a_1=50$, $n=13$, and $r=0.75$.

$$a_n=a_1r^{n-1}$$
$$a_{13}=(50)(0.75)^{13-1}$$
$$a_{13}=(50)(0.75)^{12}$$
$$a_{13}=(50)(0.031676352)$$
$$a_{13}\approx 1.6\text{ lb}$$

He will have about 1.6 lbs of birdseed left.

38. Find r.

$$r=\frac{a_2}{a_1}=\frac{20}{25}=\frac{4}{5}$$

Find the sum of the infinite geometric sequence with $a_1=25$ and $r=\frac{4}{5}$.

$$S=\frac{a_1}{1-r}$$
$$=\frac{25}{1-\frac{4}{5}}$$
$$=\frac{25}{\frac{1}{5}}$$
$$=25\bullet\frac{5}{1}$$
$$=125$$

39.

$$0.\overline{05}=0.050505\ldots=\frac{5}{100}+\frac{5}{10{,}000}+\frac{5}{1{,}000{,}000}+\ldots$$

where $a_1=\frac{5}{100}$ and $r=\frac{1}{100}$.

Find the sum with $a_1=\frac{5}{100}$ and $r=\frac{1}{100}$.

$$S=\frac{a_1}{1-r}$$
$$=\frac{\frac{5}{100}}{1-\frac{1}{100}}$$
$$=\frac{\frac{5}{100}}{\frac{99}{100}}$$
$$=\frac{5}{100}\bullet\frac{100}{99}$$
$$=\frac{5}{99}$$
$$0.\overline{05}=\frac{5}{99}$$

40. The total distance the ball travels is the sum of two motions, falling and rebounding. The distance the ball falls is given by the sum

$$10+\frac{9}{10}\bullet 10+\frac{9}{10}\left(\frac{9}{10}\bullet 10\right)+\ldots \text{ or}$$

$10+9+\frac{81}{10}+\ldots.$ The distance the ball rebounds begins with the first bounce up which is 9 ft, and then is nine-tenths the distance afterwards or $9+\frac{81}{10}+\frac{729}{100}+\ldots.$ Since these are infinite geometric series, use the formula $S=\frac{a_1}{1-r}$ to find the sum.

Falling:

$a_1 = 10$ and $r = \frac{9}{10}$

$$S=\frac{a_1}{1-r}=\frac{10}{1-\frac{9}{10}}=\frac{10}{\frac{1}{10}}=10\bullet\frac{10}{1}=100$$

Rebounding:

$a_1 = 9$ and $r = \frac{9}{10}$

$$S=\frac{a_1}{1-r}=\frac{9}{1-\frac{9}{10}}=\frac{9}{\frac{1}{10}}=9\bullet\frac{10}{1}=90$$

The distance the ball travels is the sum of the distances falling and rebounding:

100 + 90 = 190 ft.

The ball travels a total of 190 ft.

SECTION 14.4
Permutations and Combinations

41. $17 \cdot 8 = 136$

42. There are 26 letters and 10 digits possible. So the number of combinations are

$$26 \cdot 26 \cdot 26 \cdot 10 \cdot 10 \cdot 10 = 17{,}576{,}000$$

43. $P(7, 7)$

$$P(7,7)=\frac{7!}{(7-7)!}=\frac{7\bullet6\bullet5\bullet4\bullet3\bullet2\bullet1}{0!}=\frac{7\bullet6\bullet5\bullet4\bullet3\bullet2\bullet1}{1}=5{,}040$$

44. $P(7, 0)$

$$P(7,0)=\frac{7!}{(7-0)!}=\frac{7!}{7!}=1$$

45. $P(8, 6)$

$$P(8,6)=\frac{8!}{(8-6)!}=\frac{8!}{2!}=\frac{8\bullet7\bullet6\bullet5\bullet4\bullet3\bullet\cancel{2!}}{\cancel{2!}}=\frac{8\bullet7\bullet6\bullet5\bullet4\bullet3}{1}=20{,}160$$

46. $\dfrac{P(9,6)}{P(10,7)}$

$$\frac{P(9,6)}{P(10,7)}=\frac{\dfrac{9!}{(9-6)!}}{\dfrac{10!}{(10-7)!}}$$
$$=\frac{\dfrac{9!}{3!}}{\dfrac{10!}{3!}}$$
$$=\frac{9!}{3!}\cdot\frac{3!}{10!}$$
$$=\frac{\cancel{9!}}{\cancel{3!}}\cdot\frac{\cancel{3!}}{10\cdot\cancel{9!}}$$
$$=\frac{1}{10}$$

47. $C(7, 7)$

$$C(7,7)=\frac{7!}{7!(7-7)!}$$
$$=\frac{\cancel{7!}}{\cancel{7!}\cdot 0!}$$
$$=\frac{1}{1}$$
$$=1$$

48. $C(7, 0)$

$$C(7,0)=\frac{7!}{0!(7-0)!}$$
$$=\frac{\cancel{7!}}{0!\cancel{7!}}$$
$$=\frac{1}{1}$$
$$=1$$

49. $\dbinom{8}{6}$

$$\binom{8}{6}=\frac{8!}{6!(8-6)!}$$
$$=\frac{8\cdot 7\cdot\cancel{6!}}{\cancel{6!}\cdot 2!}$$
$$=\frac{56}{2}$$
$$=28$$

50. $C(6, 3)\cdot C(7, 3)$

$$C(6,3)\cdot C(7,3)=\frac{6!}{3!(6-3)!}\cdot\frac{7!}{3!(7-3)!}$$
$$=\frac{6!}{3!3!}\cdot\frac{7!}{3!4!}$$
$$=\frac{6\cdot 5\cdot 4\cdot\cancel{3!}}{3\cdot 2\cdot 1\cdot\cancel{3!}}\cdot\frac{7\cdot 6\cdot 5\cdot\cancel{4!}}{3\cdot 2\cdot 1\cdot\cancel{4!}}$$
$$=\frac{120}{6}\cdot\frac{210}{6}$$
$$=700$$

51. $\dfrac{C(7,3)}{C(6,3)}$

$$\frac{C(7,3)}{C(6,3)}=\frac{\dfrac{7!}{3!(7-3)!}}{\dfrac{6!}{3!(6-3)!}}$$
$$=\frac{\dfrac{7!}{3!4!}}{\dfrac{6!}{3!3!}}$$
$$=\frac{\dfrac{7\cdot 6\cdot 5\cdot\cancel{4!}}{3\cdot 2\cdot 1\cdot\cancel{4!}}}{\dfrac{6\cdot 5\cdot 4\cdot\cancel{3!}}{3\cdot 2\cdot 1\cdot\cancel{3!}}}$$
$$=\frac{\dfrac{210}{6}}{\dfrac{120}{6}}$$
$$=\frac{35}{20}$$
$$=\frac{7}{4}$$

52. $(3y-2z)^4$

$$=\binom{4}{0}(3y)^4+\binom{4}{1}(3y)^3(-2z)$$
$$+\binom{4}{2}(3y)^2(-2z)^2+\binom{4}{3}(3y)(-2z)^3$$
$$+\binom{4}{4}(-2z)^4$$
$$=81y^4+4(27y^3)(-2z)+6(9y^2)(4z^2)$$
$$+4(3y)(-8z^3)+16z^4$$
$$=81y^4-216y^3z+216y^2z^2-96yz^3+16z^4$$

53. $5! = 5\cdot4\cdot3\cdot2\cdot1$
$= 120$

54. $5!\cdot3! = 5\cdot4\cdot3\cdot2\cdot1\cdot3\cdot2\cdot1$
$= 120\cdot6$
$= 720$

55. $C(10, 3)$

$$C(10,3)=\frac{10!}{3!(10-3)!}=\frac{10\cdot9\cdot8\cdot\cancel{7!}}{3\cdot2\cdot1\cdot\cancel{7!}}=\frac{720}{6}=120$$

56. $\binom{5}{2}\binom{6}{2}$

$$\binom{5}{2}\binom{6}{2}=\frac{5!}{2!(5-2)!}\cdot\frac{6!}{2!(6-2)!}=\frac{5\cdot4\cdot\cancel{3!}}{2!\cdot\cancel{3!}}\cdot\frac{6\cdot5\cdot\cancel{4!}}{2!\cdot\cancel{4!}}=\frac{20}{2}\cdot\frac{30}{2}=150$$

SECTION 14.5
Probability

57. There are 6 blue areas and 16 total areas, so $s = 6$ and $n = 16$.

$$P(\text{blue area})=\frac{s}{n}=\frac{6}{16}=\frac{3}{8}$$

58. There are 8 even-numbered areas and 16 total areas, so $s = 8$ and $n = 16$.

$$P(\text{even number})=\frac{s}{n}=\frac{8}{16}=\frac{1}{2}$$

59. There are 14 areas whose number is greater than 2 and 16 total areas, so $s = 14$ and $n = 16$.

$$P(>2)=\frac{s}{n}=\frac{14}{16}=\frac{7}{8}$$

60. There are 2 ways to get 11 from rolling 2 dice: $\{(5, 6) \text{ and } (6, 5)\}$. There are 36 possible combinations when rolling 2 dice, so $s = 2$ and $n = 36$.

$$P(11)=\frac{s}{n}=\frac{2}{36}=\frac{1}{18}$$

61. The probability of living forever is 0.

62. There are 4 tens in a standard deck of cards and 52 cards in the deck, so $s = 4$ and $n = 52$.

$$
\begin{aligned}
P(10) &= \frac{s}{n} \\
&= \frac{4}{52} \\
&= \frac{1}{13}
\end{aligned}
$$

63. The number of ways we can draw 3 aces from the 4 aces in the deck is $C(4, 3)$, and the number of ways we can draw 5 cards from the 52 cards in the deck is $C(52, 5)$. The probability of drawing 3 aces is the ratio of the number of favorable outcomes to the number of possible outcomes.

$$
\begin{aligned}
P(5 \text{ aces}) &= \frac{s}{n} \\
&= \frac{C(4,3)}{C(52,5)} \\
&= \frac{\frac{4!}{3!1!}}{\frac{52!}{5!47!}} \\
&= \frac{4!}{3!1!} \cdot \frac{5!47!}{52!} \\
&= \frac{4 \cdot \cancel{3!}}{\cancel{3!} \cdot 1} \cdot \frac{5 \cdot 4 \cdot 3 \cdot 2 \cdot 1 \cdot \cancel{47!}}{52 \cdot 51 \cdot 50 \cdot 49 \cdot 48 \cdot \cancel{47!}} \\
&= \frac{4}{1} \cdot \frac{120}{311{,}875{,}200} \\
&= \frac{1}{649{,}740}
\end{aligned}
$$

64. The number of ways we can draw 5 spades from the 13 spades in the deck is $C(13, 5)$, and the number of ways we can draw 5 cards from the 52 cards in the deck is $C(52, 5)$. The probability of drawing 5 spades is the ratio of the number of favorable outcomes to the number of possible outcomes.

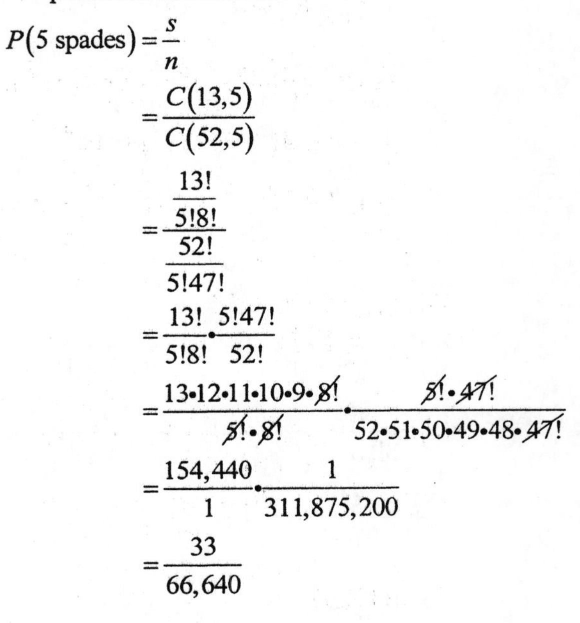

$$
\begin{aligned}
P(5 \text{ spades}) &= \frac{s}{n} \\
&= \frac{C(13,5)}{C(52,5)} \\
&= \frac{\frac{13!}{5!8!}}{\frac{52!}{5!47!}} \\
&= \frac{13!}{5!8!} \cdot \frac{5!47!}{52!} \\
&= \frac{13 \cdot 12 \cdot 11 \cdot 10 \cdot 9 \cdot \cancel{8!}}{\cancel{5!} \cdot \cancel{8!}} \cdot \frac{\cancel{5!} \cdot \cancel{47!}}{52 \cdot 51 \cdot 50 \cdot 49 \cdot 48 \cdot \cancel{47!}} \\
&= \frac{154{,}440}{1} \cdot \frac{1}{311{,}875{,}200} \\
&= \frac{33}{66{,}640}
\end{aligned}
$$

CHAPTER 14 TEST

1. $a_n = -6n + 8$

   $a_1 = -6(1) + 8 = 2$

   $a_2 = -6(2) + 8 = -4$

   $a_3 = -6(3) + 8 = -10$

   $a_4 = -6(4) + 8 = -16$

   The first four terms are 2, –4, –10, and –16.

2.

$$(a-b)^6 = (a+(-b))^6$$
$$= a^6 + \frac{6!}{1!(6-1)!}a^5(-b) + \frac{6!}{2!(6-2)!}a^4(-b)^2$$
$$+ \frac{6!}{3!(6-3)!}a^3(-b)^3 + \frac{6!}{4!(6-4)!}a^2(-b)^4$$
$$+ \frac{6!}{5!(6-5)!}a(-b)^5 + (-b)^6$$
$$= a^6 + \frac{6!}{1!5!}a^5(-b) + \frac{6!}{2!4!}a^4(-b)^2$$
$$+ \frac{6!}{3!3!}a^3(-b)^3 + \frac{6!}{4!2!}a^2(-b)^4$$
$$+ \frac{6!}{5!1!}a(-b)^5 + (-b)^6$$
$$= a^6 - \frac{6}{1}a^5b + \frac{30}{2}a^4b^2 - \frac{120}{6}a^3b^3 + \frac{30}{2}a^2b^4$$
$$- \frac{6}{1}ab^5 + b^6$$
$$= a^6 - 6a^5b + 15a^4b^2 - 20a^3b^3 + 15a^2b^4 - 6ab^5 + b^6$$

3. The 3rd term of $(x^2 + 2y)$

$$\frac{4!}{2!(4-2)!}(x^2)^2(2y)^2 = \frac{4 \cdot 3 \cdot \cancel{2!}}{2!\cancel{2!}}(x^4)(4y^2)$$
$$= \frac{12}{2}(4x^4y^2)$$
$$= 24x^4y^2$$

4. Find d.

   $10 - 3 = 7$, so $d = 7$.

   Find a_{10} if $a_1 = 3$, $d = 7$ and $n = 10$.

$$a_n = a_1 + d(n-1)$$
$$a_{10} = 3 + 7(10-1)$$
$$= 3 + 7(9)$$
$$= 3 + 63$$
$$= 66$$

5.

$$a_1 = -2, \quad d = 3 - (-2) = 5, \quad n = 12$$
$$a_{12} = -2 + 5(12-1)$$
$$= -2 + 5(11)$$
$$= -2 + 55$$
$$= 53$$

$$S_n = \frac{n(a_1 + a_n)}{2}$$
$$S_{12} = \frac{12(-2+53)}{2}$$
$$= \frac{12(51)}{2}$$
$$= 306$$

6. Let $a_1 = 2$ and $a_4 = 98$.

$$a_4 = a_1 + (4-1)d$$
$$98 = 2 + 3d$$
$$96 = 3d$$
$$32 = d$$

$$a_n = a_1 + (n-1)d$$
$$a_2 = 2 + (2-1)(32) = 2 + 1(32) = 34$$
$$a_3 = 2 + (3-1)(32) = 2 + 2(32) = 66$$

   The arithmetic means are 34 and 66.

7.

$$\left(5-\frac{5}{4}\right)=(17-2)d$$

$$\frac{15}{4}=15d$$

$$\frac{1}{15}\bullet\frac{15}{4}=\frac{1}{15}\bullet 15d$$

$$\frac{1}{4}=d$$

8. Find d.

$$(-75-(-11))=(20-4)d$$

$$-64=16d$$

$$-4=d$$

Find a_1 if $a_n=-11$, $d=-4$ and $n=4$.

$$-11=a_1-4(4-1)$$

$$-11=a_1-4(3)$$

$$-11=a_1-12$$

$$1=a_1$$

Find a_{27}.

$$a_1=1,\quad d=4,\quad n=27$$

$$a_{27}=1-4(27-1)$$

$$=1-4(26)$$

$$=-103$$

$$S_n=\frac{n(a_1+a_n)}{2}$$

$$S_{27}=\frac{27(1-103)}{2}$$

$$=\frac{27(-102)}{2}$$

$$=-1{,}377$$

9. The sum of all 25 rows minus the sum of the first 15 rows equals the number of pipe left in the stack after removing the first 15 rows. For S_{25}, let $a_1 = 1$, $a_{25} = 25$, and $n = 25$. For S_{15}, let $a_1 = 1$, $a_{15} = 15$, and $n = 15$.

$$S_n=\frac{n(a_1+a_n)}{2}$$

$$S_{25}-S_{15}=\frac{25(1+25)}{2}-\frac{15(1+15)}{2}$$

$$=\frac{25(26)}{2}-\frac{15(16)}{2}$$

$$=325-120$$

$$=205$$

There are 205 pipes remaining in the stack.

10.

$$\sum_{k=1}^{3}(2k-3)=(2\cdot1-3)+(2\cdot2-3)+(2\cdot3-3)$$

$$=-1+1+3$$

$$=3$$

11.

Use $r=\frac{a_{n+1}}{a_n}$ to find r.

$$a_2=-\frac{1}{3}\text{ and }a_1=-\frac{1}{9}$$

$$r=\frac{-\frac{1}{3}}{-\frac{1}{9}}$$

$$r=-\frac{1}{3}\bullet-\frac{9}{1}$$

$$r=3$$

Use $a_n=a_1r^{n-1}$ to find a_7.

$$r=3,\ a_1=-\frac{1}{9}\text{ and }n=7$$

$$a_7=-\frac{1}{9}(3)^{7-1}$$

$$a_7=-\frac{1}{9}(3)^6$$

$$a_7=-\frac{1}{9}(729)$$

$$a_7=-81$$

12. Find r.

$$r = \frac{a_2}{a_1} = \frac{\frac{1}{9}}{\frac{1}{27}} = \frac{1}{9} \bullet \frac{27}{1} = 3$$

Find the sum of $a_1 = \frac{1}{27}$, $r = 3$, and $n = 6$.

$$S_n = \frac{a_1\left(1-r^n\right)}{1-r}$$
$$= \frac{\frac{1}{27}\left(1-3^6\right)}{1-3}$$
$$= \frac{\frac{1}{27}(1-729)}{-2}$$
$$= \frac{\frac{1}{27}(-728)}{-2}$$
$$= \frac{-728}{27} \bullet \frac{1}{-2}$$
$$= \frac{364}{27}$$

13. Let $r = -\frac{2}{3}$ and $a_4 = -\frac{16}{9}$.

$$a_n = a_1 r^{n-1}$$
$$a_4 = a_1 r^{4-1}$$
$$a_4 = a_1 r^3$$
$$-\frac{16}{9} = a_1\left(-\frac{2}{3}\right)^3$$
$$-\frac{16}{9} = -\frac{8}{27} a_1$$
$$-\frac{27}{8} \bullet -\frac{16}{9} = -\frac{27}{8} \bullet -\frac{8}{27} a_1$$
$$6 = a_1$$

14. $a_1 = 3$, $a_4 = 648$ for two geometric means between 3 and 648

Use $a_n = ar^{n-1}$ to find r.

$a_1 = 3$, $a_4 = 648$ and $n = 4$

$$648 = 3(r)^{4-1}$$
$$648 = 3r^3$$
$$216 = r^3$$
$$\sqrt[3]{216} = \sqrt[3]{r^3}$$
$$6 = r$$
$$a_1 = 3$$
$$a_2 = 3(6) = 18$$
$$a_3 = 3(6)^2 = 108$$
$$a_4 = 3(6)^3 = 648$$

The geometric means are 18 and 108.

15. Find r.

$$r = \frac{a_2}{a_1} = \frac{3}{9} = \frac{1}{3}$$

Find the sum of the infinite geometric series with $a_1 = 9$ and $r = \frac{1}{3}$.

$$S = \frac{a_1}{1-r}$$
$$= \frac{9}{1-\frac{1}{3}}$$
$$= \frac{9}{\frac{2}{3}}$$
$$= 9 \bullet \frac{3}{2}$$
$$= \frac{27}{2}$$

16. Find d if $a_1 = 16$, $a_2 = 48$, and $n = 2$.

$$a_n = a_1 + d(n-1)$$
$$48 = 16 + d(2-1)$$
$$48 = 16 + d$$
$$32 = d$$

Find a_{10} to find out how far the object fell during the 10th second.

$$a_1 = 16, \quad d = 32, \quad n = 10$$
$$a_{10} = 16 + 32(10-1)$$
$$= 16 + 32(9)$$
$$= 304$$

Find the sum of the distances fell during the first 10 seconds.

$$S_n = \frac{n(a_1 + a_n)}{2}$$
$$S_{10} = \frac{10(16+304)}{2}$$
$$= \frac{10(320)}{2}$$
$$= 1{,}600 \text{ ft}$$

17.

$$\frac{7!}{4!} = \frac{7 \cdot 6 \cdot 5 \cdot \cancel{4!}}{\cancel{4!}}$$
$$= \frac{7 \cdot 6 \cdot 5}{1}$$
$$= 210$$

18.

$$0! = 1$$

19.

$$P(5,4) = \frac{5!}{(5-4)!}$$
$$= \frac{5!}{1!}$$
$$= \frac{5 \cdot 4 \cdot 3 \cdot 2 \cdot 1}{1}$$
$$= 120$$

20.

$$C(6,4) = \frac{6!}{4!(6-4)!}$$
$$= \frac{6 \cdot 5 \cdot \cancel{4!}}{\cancel{4!}2!}$$
$$= \frac{30}{2}$$
$$= 15$$

21.

$$\binom{8}{3} = \frac{8!}{3!(8-3)!}$$
$$= \frac{8 \cdot 7 \cdot 6 \cdot \cancel{5!}}{3 \cdot 2 \cdot 1 \cdot \cancel{5!}}$$
$$= \frac{336}{6}$$
$$= 56$$

22.

$$C(6,0) \cdot P(3,3) = \frac{6!}{0!(6-0)!} \cdot \frac{3!}{(3-3)!}$$
$$= \frac{\cancel{6!}}{1 \cdot \cancel{6!}} \cdot \frac{3!}{0!}$$
$$= \frac{1}{1} \cdot \frac{6}{1}$$
$$= 6$$

23. Since the first digit cannot be 0, 1, or, 2, there are only 7 choices for the first digit. There are 10 choices for each of the other 6 digits. The number of possible choices is $7 \cdot 10 \cdot 10 \cdot 10 \cdot 10 \cdot 10 \cdot 10 = 7{,}000{,}000$.

24. The number of possible seating arrangements is $9! = 362{,}880$.

25.

$$C(7,3)=\frac{7!}{3!(7-3)!}$$
$$=\frac{7\cdot 6\cdot 5\cdot \cancel{4!}}{3!\cdot \cancel{4!}}$$
$$=\frac{210}{6}$$
$$=35$$

26.

$$C(5,1)\cdot C(4,2)=\frac{5!}{1!(5-1)!}\cdot\frac{4!}{2!(4-2)!}$$
$$=\frac{5!}{1\cdot\cancel{4!}}\cdot\frac{\cancel{4!}}{2!2!}$$
$$=\frac{120}{1}\cdot\frac{1}{4}$$
$$=30$$

27. Let $s = 1$ and $n = 6$.

$$P(5)=\frac{s}{n}$$
$$=\frac{1}{6}$$

28. There are 4 jacks and 4 queens and 52 total cards in a standard deck. Let $s = 8$ and $n = 52$.

$$P(\text{jack or queen})=\frac{s}{n}$$
$$=\frac{8}{52}$$
$$=\frac{2}{13}$$

29. The number of ways we can draw 5 hearts from the 13 hearts in the deck is $C(13, 5)$, and the number of ways we can draw 5 cards from the 52 cards in the deck is $C(52, 5)$. The probability of drawing 5 hearts is the ratio of the number of favorable outcomes to the number of possible outcomes.

$$P(5\text{ hearts})=\frac{s}{n}$$
$$=\frac{C(13,5)}{C(52,5)}$$
$$=\frac{\frac{13!}{5!8!}}{\frac{52!}{5!47!}}$$
$$=\frac{13!}{5!8!}\cdot\frac{5!47!}{52!}$$
$$=\frac{13\cdot 12\cdot 11\cdot 10\cdot 9\cdot\cancel{8!}}{\cancel{5!}\cdot\cancel{8!}}\cdot\frac{\cancel{5!}\cdot\cancel{47!}}{52\cdot 51\cdot 50\cdot 49\cdot 48\cdot\cancel{47!}}$$
$$=\frac{154{,}440}{1}\cdot\frac{1}{311{,}875{,}200}$$
$$=\frac{33}{66{,}640}$$

30. The number of ways we can get 2 heads from the 5 possible heads is $C(13, 5)$, and the number of choices is 2 each toss or 2^5.

$$P(\text{heads})=\frac{s}{n}$$
$$=\frac{C(5,2)}{2^5}$$
$$=\frac{\frac{5!}{2!3!}}{32}$$
$$=\frac{10}{32}$$
$$=\frac{5}{16}$$

31. You can shade any 7 sections to end up with 9 that are not shaded.

32. a. The probability of an event that cannot happen is **0**.
    b. The probability of an event that is guaranteed to happen is **1**.

CHAPTER 14 CUMULATIVE REVIEW

1. 0

2. $-\frac{4}{3}, 5, 6, 0, -23$

3. $\pi, \sqrt{2}, e$

4. $-\frac{4}{3}, \pi, 5, 6, \sqrt{2}, 0, -23, e$

5. Let x = the amount of money she has to invest. Then $x + 3{,}000$ = the needed amount of money to earn the greater interest.

	principal	rate	Interest
original	x	0.075	$0.075x$
larger amount	$x + 3{,}000$	0.11	$0.11(x + 3{,}000)$

$$11\% \text{ investment} = 2(7.5\% \text{ investment})$$
$$0.11(x+3{,}000) = 2(0.075x)$$
$$0.11x + 330 = 0.15x$$
$$330 = 0.04x$$
$$8{,}250 = x$$

She had \$8,250 to invest originally.

6. Use the points (1,600, 68) and (2,800, 78).

$$\text{rate of change} = \frac{y_2 - y_1}{x_2 - x_1} = \frac{78-68}{2{,}800-1{,}600} = \frac{10}{1{,}200} = \frac{1}{120} \text{ db/rpm}$$

7. Find the slope of each line.

$$3x - 4y = 12$$
$$-4y = -3x + 12$$
$$y = \frac{3}{4}x - 3$$
$$m = \frac{3}{4}$$

$$y = \frac{3}{4}x - 5$$
$$m = \frac{3}{4}$$

Since the slopes for both lines are the same, the lines must be parallel.

8. Find the slope of each line.

$$y = 3x + 4$$
$$m = 3$$

$$x = -3y + 4$$
$$x - 4 = -3y$$
$$y = -\frac{1}{3}x + \frac{4}{3}$$
$$m = -\frac{1}{3}$$

Since the slopes for both lines are the opposite reciprocals, the lines must be perpendicular.

9. $y = mx + b$; $m = -2$ and $b = 5$
$y = -2x + 5$

10. Find m.

$$m = \frac{y_2 - y_1}{x_2 - x_1} = \frac{4-(-5)}{-5-8} = \frac{9}{-13} = -\frac{9}{13}$$

Use $m = -\frac{9}{13}$ and $(8, -5)$.

$$y - y_1 = m(x - x_1)$$
$$y - (-5) = -\frac{9}{13}(x-8)$$
$$y + 5 = -\frac{9}{13}x + \frac{72}{13}$$
$$y = -\frac{9}{13}x + \frac{7}{13}$$

11.

$$\begin{cases} 2x - y = -21 \\ 4x + 5y = 7 \end{cases}$$

Solve the first equation for y.

$$2x - y = -21$$
$$-y = -2x - 21$$
$$y = 2x + 21$$

Substitute $y = 2x + 21$ into the second equation and solve for x.

$$4x + 5y = 7$$
$$4x + 5(2x + 21) = 7$$
$$4x + 10x + 105 = 7$$
$$14x = -98$$
$$x = -7$$
$$y = 2x + 21$$
$$y = 2(-7) + 21$$
$$y = -14 + 21$$
$$y = 7$$

The solution is $(-7, 7)$.

12.

$$\begin{cases} 4y + 5x - 7 = 0 \\ \frac{10}{7}x - \frac{4}{9}y = \frac{17}{21} \end{cases}$$

Write the equations in standard form. Multiply the second equation by 63 to eliminate fractions.

$$\begin{cases} 5x + 4y = 7 \\ 90x - 28y = 51 \end{cases}$$

Multiply the first equation by 7 and add the equations to solve for x.

$$35x + 28y = 49$$
$$\underline{90x - 28y = 51}$$
$$125x = 100$$
$$x = \frac{4}{5}$$

Substitute $x = \frac{4}{5}$ into the first equation and solve for y.

$$5x + 4y = 7$$
$$5\left(\frac{4}{5}\right) + 4y = 7$$
$$4 + 4y = 7$$
$$4y = 3$$
$$y = \frac{3}{4}$$

The solution is $\left(\frac{4}{5}, \frac{3}{4}\right)$.

13.

Write the first equation in standard form.

$$\begin{cases} 2x+2y=-1 \\ 3x+4y=0 \end{cases}$$

$$x=\frac{D_x}{D} = \frac{\begin{vmatrix} -1 & 2 \\ 0 & 4 \end{vmatrix}}{\begin{vmatrix} 2 & 2 \\ 3 & 4 \end{vmatrix}} = \frac{-1(4)-2(0)}{2(4)-2(3)} = \frac{-4-0}{8-6} = \frac{-4}{2} = -2$$

$$y=\frac{D_y}{D} = \frac{\begin{vmatrix} 2 & -1 \\ 3 & 0 \end{vmatrix}}{\begin{vmatrix} 2 & 2 \\ 3 & 4 \end{vmatrix}} = \frac{2(0)-(-1)(3)}{2(4)-2(3)} = \frac{0+3}{8-6} = \frac{3}{2}$$

The solution is $\left(-2, \frac{3}{2}\right)$.

14.

$$\begin{cases} b+2c=7-a \\ a+c=8-2b \\ 2a+b+c=9 \end{cases}$$

Write the equations in standard form.

$$\begin{cases} a+b+2c=7 & (1) \\ a+2b+c=8 & (2) \\ 2a+b+c=9 & (3) \end{cases}$$

Multiply Equation 2 by -2 and add Equations 1 and 2.

$$\begin{aligned} a+b+2c&=7 && (1) \\ \underline{-2a-4b-2c}&\underline{=-16} && (2) \\ -a-3b&=-9 && (4) \end{aligned}$$

Multiply Equation 2 by -1 and add Equations 2 and 3.

$$\begin{aligned} -a-2b-c&=-8 && (2) \\ \underline{2a+b+c}&\underline{=9} && (3) \\ a-b&=1 && (5) \end{aligned}$$

Add Equations 4 and 5.

$$\begin{aligned} -a-3b&=-9 && (4) \\ \underline{a-b}&\underline{=1} && (5) \\ -4b&=-8 \\ b&=2 \end{aligned}$$

Substitute $b=2$ into Equation 5.

$$\begin{aligned} a-b&=1 && (5) \\ a-2&=1 \\ a&=3 \end{aligned}$$

Substitute $b=2$ and $a=3$ into Equation 1.

$$\begin{aligned} a+b+2c&=7 && (1) \\ 3+2+2c&=7 \\ 5+2c&=7 \\ 2c&=2 \\ c&=1 \end{aligned}$$

The solution is $(3,2,1)$.

15. The solution would be the x-coordinate of the point of intersection. From the graph, they intersect at $(3, -13)$, so the solution to the equation is $x = 3$.

16. Let x = measure of angle C.
Then $\angle B = 5 + 5x$ and
$\angle A = 5 + (5+5x) = 10 + 5x$. The sum of the angles of the triangle is 180°.

$$\angle A + \angle B + \angle C = 180$$
$$(10+5x)+(5+5x)+(x) = 180$$
$$11x + 15 = 180$$
$$11x = 165$$
$$x = 15$$
$$\angle C = 15^\circ$$
$$\angle B = 5 + 5(15) = 80^\circ$$
$$\angle A = 10 + 5(15) = 85^\circ$$

17. It doesn't pass the vertical line test.

18.

$$f(x) = 3x^5 - 2x^2 + 1$$
$$f(-1) = 3(-1)^5 - 2(-1)^2 + 1$$
$$f(-1) = 3(-1) - 2(1) + 1$$
$$f(-1) = -3 - 2 + 1$$
$$f(-1) = -4$$

$$f(a) = 3a^5 - 2a^2 + 1$$

19. a. $h(-3) = 4$
b. $h(4) = 3$
c. $x = 0$ and $x = 2$
d. $x = 1$

20. 1.73×10^{14}; 4.6×10^{-8}

21.

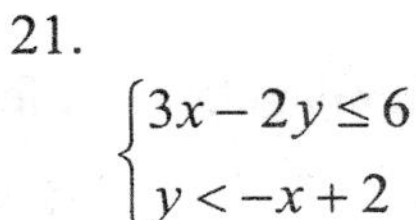

$$\begin{cases} 3x - 2y \le 6 \\ y < -x + 2 \end{cases}$$

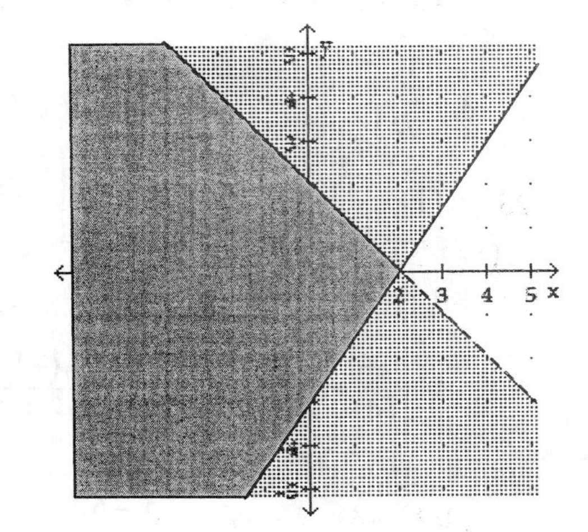

22.

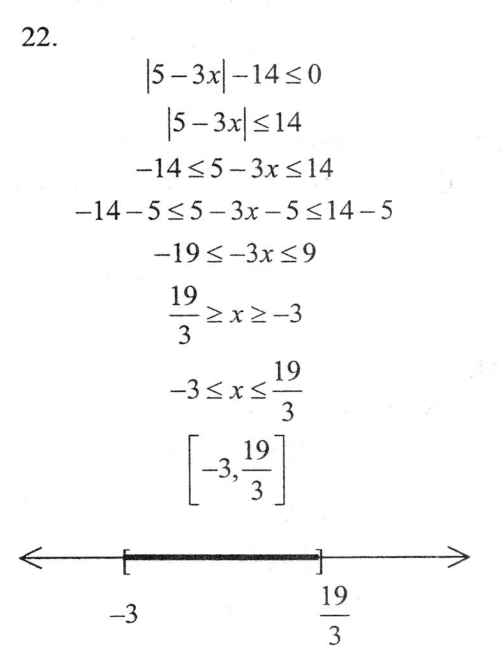

$$|5 - 3x| - 14 \le 0$$
$$|5 - 3x| \le 14$$
$$-14 \le 5 - 3x \le 14$$
$$-14 - 5 \le 5 - 3x - 5 \le 14 - 5$$
$$-19 \le -3x \le 9$$
$$\frac{19}{3} \ge x \ge -3$$
$$-3 \le x \le \frac{19}{3}$$
$$\left[-3, \frac{19}{3}\right]$$

23.

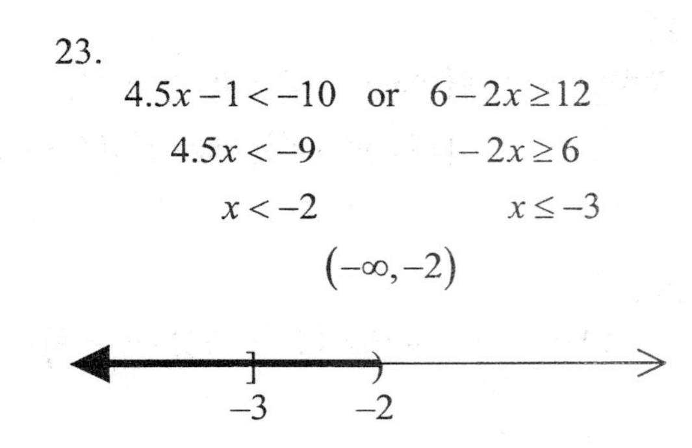

$$4.5x - 1 < -10 \quad \text{or} \quad 6 - 2x \ge 12$$
$$4.5x < -9 \qquad -2x \ge 6$$
$$x < -2 \qquad x \le -3$$
$$(-\infty, -2)$$

24.

$$
\begin{aligned}
&(x-3y)(x^2+3xy+9y^2)\\
&=x(x^2+3xy+9y^2)-3y(x^2+3xy+9y^2)\\
&=x^3+3x^2y+9xy^2-3x^2y-9xy^2-27y^3\\
&=x^3-27y^3
\end{aligned}
$$

25.

$$
\begin{aligned}
&(-2x^2y^3+6xy+5y^2)-(-4x^2y^3-7xy+2y^2)\\
&=-2x^2y^3+6xy+5y^2+4x^2y^3+7xy-2y^2\\
&=2x^2y^3+13xy+3y^2
\end{aligned}
$$

26.

$$
\begin{aligned}
(9ab^2-4)^2&=(9ab^2-4)(9ab^2-4)\\
&=81a^2b^4-36ab^2-36ab^2+16\\
&=81a^2b^4-72ab^2+16
\end{aligned}
$$

27.

$$
\begin{aligned}
ab^{-2}c^{-3}(a^{-4}bc^3+a^{-3}b^4c^3)&=a^{-3}b^{-1}c^0+a^{-2}b^2c^0\\
&=\frac{1}{a^3b}+\frac{b^2}{a^2}
\end{aligned}
$$

28.

$$
\begin{aligned}
&3x^3y-4x^2y^2-6x^2y+8xy^2\\
&=xy(3x^2-4xy-6x+8y)\\
&=xy[(3x^2-4xy)-(6x-8y)]\\
&=xy[x(3x-4y)-2(3x-4y)]\\
&=xy(3x-4y)(x-2)
\end{aligned}
$$

29.

$$
\begin{aligned}
256x^4y^4-z^8&=(16x^2y^2+z^4)(16x^2y^2-z^4)\\
&=(16x^2y^2+z^4)(4xy+z^2)(4xy-z^2)
\end{aligned}
$$

30.

$$
12y^6+23y^3+10=(3y^3+2)(4y^3+5)
$$

31.

$$
\begin{aligned}
\frac{A\lambda}{2}+1&=2d+3\lambda\\
\frac{A\lambda}{2}-3\lambda&=2d-1\\
2\left(\frac{A\lambda}{2}-3\lambda\right)&=2(2d-1)\\
A\lambda-6\lambda&=4d-2\\
\lambda(A-6)&=4d-2\\
\frac{\lambda(A-6)}{A-6}&=\frac{4d-2}{A-6}\\
\lambda&=\frac{4d-2}{A-6}
\end{aligned}
$$

32. Let $u=(x+7)$.

$$
\begin{aligned}
(x+7)^2&=-2(x+7)-1\\
(x+7)^2+2(x+7)+1&=0\\
u^2+2u+1&=0\\
(u+1)(u+1)&=0\\
u+1=0 \quad &\text{or} \quad u+1=0\\
u=-1 \quad &\quad\quad u=-1\\
x+7=-1 \quad &\quad\quad x+7=-1\\
x=-8 \quad &\quad\quad x=-8
\end{aligned}
$$

33.

$$
\begin{aligned}
x^3+x^2&=0\\
x^2(x+1)&=0\\
x^2=0 \quad &\text{or} \quad x+1=0\\
x=0,\ x=0,\ &x=-1
\end{aligned}
$$

34.

x	$f(x) = -x^3 - x^2 + 6x$
-4	$-(-4)^3 - (-4)^2 + 6(-4) = 64 - 16 - 24$ $= 24$
-3	$-(-3)^3 - (-3)^2 + 6(-3) = 27 - 9 - 18$ $= 0$
-2	$-(-2)^3 - (-2)^2 + 6(-2) = 8 - 4 - 12$ $= -8$
-1	$-(-1)^3 - (-1)^2 + 6(-1) = 1 - 1 - 6$ $= -6$
0	$-(0)^3 - (0)^2 + 6(0) = 0 - 0 + 0$ $= 0$
1	$-(1)^3 - (1)^2 + 6(1) = -1 - 1 + 6$ $= 4$
2	$-(2)^3 - (2)^2 + 6(2) = -8 - 4 + 12$ $= 0$
3	$-(3)^3 - (3)^2 + 6(3) = -27 - 9 + 18$ $= -18$

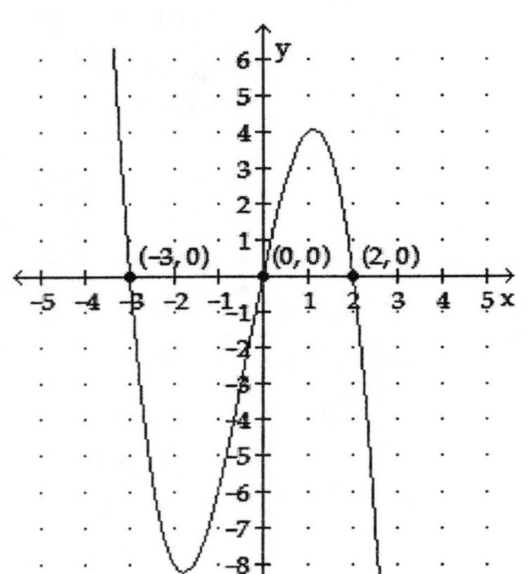

35.

$$\left(\frac{3x^5y^2}{6x^5y^{-2}}\right)^{-4} = \left(\frac{1}{2}x^{5-5}y^{2-(-2)}\right)^{-4}$$
$$= \left(\frac{1}{2}x^0y^4\right)^{-4}$$
$$= \left(\frac{y^4}{2}\right)^{-4}$$
$$= \left(\frac{2}{y^4}\right)^4$$
$$= \frac{16}{y^{16}}$$

36.

$$\frac{6x^2 + 13x + 6}{6 - 5x - 6x^2} = \frac{(2x+3)(3x+2)}{(3+2x)(2-3x)}$$
$$= \frac{3x+2}{-3x+2}$$
$$= -\frac{3x+2}{3x-2}$$

37.

$$\frac{p^3 - q^3}{q^2 - p^2} \cdot \frac{q^2 + pq}{p^3 + p^2q + pq^2}$$
$$= \frac{(p-q)(p^2+pq+q^2)}{(q-p)(q+p)} \cdot \frac{q(q+p)}{p(p^2+pq+q^2)}$$
$$= \frac{(p-q)(p+q)(p+q)}{-(p-q)(q+p)} \cdot \frac{q(q+p)}{p(p+q)(p+q)}$$
$$= -\frac{q}{p}$$

38.

$$\frac{2}{a-2}+\frac{3}{a+2}-\frac{a-1}{a^2-4}$$
$$=\frac{2}{a-2}+\frac{3}{a+2}-\frac{a-1}{(a+2)(a-2)}$$
$$=\frac{2}{a-2}\left(\frac{a+2}{a+2}\right)+\frac{3}{a+2}\left(\frac{a-2}{a-2}\right)-\frac{a-1}{(a+2)(a-2)}$$
$$=\frac{2a+4}{(a+2)(a-2)}+\frac{3a-6}{(a+2)(a-2)}-\frac{a-1}{(a+2)(a-2)}$$
$$=\frac{2a+4+3a-6-(a-1)}{(a+2)(a-2)}$$
$$=\frac{2a+4+3a-6-a+1}{(a+2)(a-2)}$$
$$=\frac{4a-1}{(a+2)(a-2)}$$

39.

$$\frac{x-4}{x-3}+\frac{x-2}{x-3}=x-3$$
$$(x-3)\left(\frac{x-4}{x-3}+\frac{x-2}{x-3}\right)=(x-3)(x-3)$$
$$x-4+x-2=x^2-3x-3x+9$$
$$2x-6=x^2-6x+9$$
$$0=x^2-8x+15$$
$$0=(x-3)(x-5)$$
$$x-3=0 \quad \text{or} \quad x-5=0$$
$$x=\not{3} \qquad x=5$$

3 is extraneous

40.

$$\frac{1}{R}=\frac{1}{R_1}+\frac{1}{R_2}+\frac{1}{R_3}$$
$$(RR_1R_2R_3)\left(\frac{1}{R}\right)=(RR_1R_2R_3)\left(\frac{1}{R_1}+\frac{1}{R_2}+\frac{1}{R_3}\right)$$
$$R_1R_2R_3=RR_2R_3+RR_1R_3+RR_1R_2$$
$$R_1R_2R_3=R(R_2R_3+R_1R_3+R_1R_2)$$
$$\frac{R_1R_2R_3}{R_2R_3+R_1R_3+R_1R_2}=\frac{R(R_2R_3+R_1R_3+R_1R_2)}{R_2R_3+R_1R_3+R_1R_2}$$
$$\frac{R_1R_2R_3}{R_2R_3+R_1R_3+R_1R_2}=R$$
$$R=\frac{R_1R_2R_3}{R_2R_3+R_1R_3+R_1R_2}$$

41. a. a quadratic function
b. at about 85% and 120% of the suggested inflation

42. Use the Pythagorean Theorem. The legs of the triangle formed are both 16 (sides of the square) and the waist size is the hypotenuse, c.

$$a^2+b^2=c^2$$
$$16^2+16^2=c^2$$
$$256+256=c^2$$
$$512=c^2$$
$$\sqrt{512}=\sqrt{c^2}$$
$$22.6=c$$

Subtract 1 inch from 22.6 to allow for the pin, and the largest waist size the diaper can wrap around is about 21.6 inches.

43.

$$\begin{array}{r} -x^2+x+5+\frac{8}{x-1} \\ x-1\overline{)-x^3+2x^2+4x+3} \\ \underline{-x^3+x^2} \\ x^2+4x \\ \underline{x^2-x} \\ 5x+3 \\ \underline{5x-5} \\ 8 \end{array}$$

44. Let I = intensity of the light and d = distance from the source.

$$I = \frac{k}{d^2}$$
$$18 = \frac{k}{4^2}$$
$$18 = \frac{k}{16}$$
$$16(18) = 16\left(\frac{k}{16}\right)$$
$$288 = k$$

Find I if $d = 12$ and $k = 288$.

$$I = \frac{k}{d^2}$$
$$I = \frac{288}{12^2}$$
$$I = \frac{288}{144}$$
$$I = 2 \text{ lumens}$$

45. $f(x) = \sqrt{x} + 2$

$D:[0,\infty);\ R:[2,\infty)$

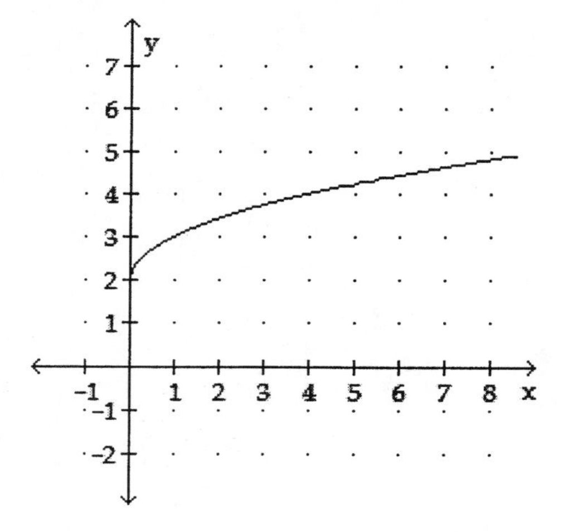

46.

$$\sqrt{98} + \sqrt{8} - \sqrt{32} = \sqrt{49}\sqrt{2} + \sqrt{4}\sqrt{2} - \sqrt{16}\sqrt{2}$$
$$= 7\sqrt{2} + 2\sqrt{2} - 4\sqrt{2}$$
$$= 5\sqrt{2}$$

47.

$$3\left(\sqrt{5x} - \sqrt{3}\right)^2 = 3\left(\sqrt{5x} - \sqrt{3}\right)\left(\sqrt{5x} - \sqrt{3}\right)$$
$$= 3\left(\sqrt{25x^2} - \sqrt{15x} - \sqrt{15x} + \sqrt{9}\right)$$
$$= 3\left(5x - 2\sqrt{15x} + 3\right)$$
$$= 15x - 6\sqrt{15x} + 9$$

48.

$$12\sqrt[3]{648x^4} + 3\sqrt[3]{81x^4} = 12\sqrt[3]{216x^3}\sqrt[3]{3x} + 3\sqrt[3]{27x^3}\sqrt[3]{3x}$$
$$= 12(6x)\left(\sqrt[3]{3x}\right) + 3(3x)\left(\sqrt[3]{3x}\right)$$
$$= 72x\sqrt[3]{3x} + 9x\sqrt[3]{3x}$$
$$= 81x\sqrt[3]{3x}$$

49.

$$\left(\frac{25}{49}\right)^{-3/2} = \left(\frac{49}{25}\right)^{3/2}$$
$$= \left(\sqrt{\frac{49}{25}}\right)^3$$
$$= \left(\frac{7}{5}\right)^3$$
$$= \frac{343}{125}$$

50.

$$\frac{\sqrt[3]{4a^2}}{\sqrt[3]{2ab}} = \sqrt[3]{\frac{4a^2}{2ab}}$$
$$= \sqrt[3]{\frac{2a}{b}}$$
$$= \sqrt[3]{\frac{2a}{b} \cdot \frac{b^2}{b^2}}$$
$$= \sqrt[3]{\frac{2ab^2}{b^3}}$$
$$= \frac{\sqrt[3]{2ab^2}}{b}$$

51.

$$\frac{3t-1}{\sqrt{3t}+1}=\frac{3t-1}{\sqrt{3t}+1}\bullet\frac{\sqrt{3t}-1}{\sqrt{3t}-1}$$
$$=\frac{3t\sqrt{3t}-3t-\sqrt{3t}+1}{\sqrt{9t^2}-1}$$
$$=\frac{3t\left(\sqrt{3t}-1\right)-\left(\sqrt{3t}-1\right)}{3t-1}$$
$$=\frac{\left(\sqrt{3t}-1\right)\cancel{(3t-1)}}{\cancel{3t-1}}$$
$$=\sqrt{3t}-1$$

52.

$$\left(-7+\sqrt{-81}\right)-\left(-2-\sqrt{-64}\right)$$
$$=(-7+9i)-(-2-8i)$$
$$=-7+9i+2+8i$$
$$=-5+17i$$

53.

$$\frac{2-5i}{2+5i}=\frac{2-5i}{2+5i}\bullet\frac{2-5i}{2-5i}$$
$$=\frac{4-10i-10i+25i^2}{4-25i^2}$$
$$=\frac{4-20i+25(-1)}{4-25(-1)}$$
$$=\frac{4-20i-25}{4+25}$$
$$=\frac{-21-20i}{29}$$
$$=-\frac{21}{29}-\frac{20}{29}i$$

54.

$$i^{42}=i^{40}\bullet i^2$$
$$=\left(i^4\right)^{10}\bullet i^2$$
$$=(1)^{10}\bullet(-1)$$
$$=1\bullet(-1)$$
$$=-1$$

55.

$$\sqrt{3a+1}=a-1$$
$$\left(\sqrt{3a+1}\right)^2=(a-1)^2$$
$$3a+1=a^2-2a+1$$
$$0=a^2-5a$$
$$0=a(a-5)$$
$$a=0 \quad \text{or} \quad a-5=0$$
$$a=\cancel{0} \qquad a=5$$

0 is extraneous

56.

$$\sqrt{x+3}-\sqrt{3}=\sqrt{x}$$
$$\sqrt{x+3}=\sqrt{x}+\sqrt{3}$$
$$\left(\sqrt{x+3}\right)^2=\left(\sqrt{x}+\sqrt{3}\right)^2$$
$$x+3=\sqrt{x^2}+\sqrt{3x}+\sqrt{3x}+\sqrt{9}$$
$$x+3=x+2\sqrt{3x}+3$$
$$0=2\sqrt{3x}$$
$$0=\sqrt{3x}$$
$$0^2=\left(\sqrt{3x}\right)^2$$
$$0=3x$$
$$0=x$$

57.

$$x^4+19x^2+18=0$$
$$\left(x^2+1\right)\left(x^2+18\right)=0$$
$$x^2+1=0 \quad \text{or} \quad x^2+18=0$$
$$x^2=-1 \qquad x^2=-18$$
$$x=\sqrt{-1} \qquad x=\sqrt{-18}$$
$$x=\pm i \qquad x=\pm 3i\sqrt{2}$$
$$x=-i,i,3i\sqrt{2},-3i\sqrt{2}$$

58.

$$4w^2+6w+1=0$$

$$w=\frac{-b\pm\sqrt{b^2-4ac}}{2a}$$

$$w=\frac{-6\pm\sqrt{6^2-4(4)(1)}}{2(4)}$$

$$w=\frac{-6\pm\sqrt{36-16}}{8}$$

$$w=\frac{-6\pm\sqrt{20}}{8}$$

$$w=\frac{-6\pm2\sqrt{5}}{8}$$

$$w=-\frac{6}{8}\pm\frac{2\sqrt{5}}{8}$$

$$w=-\frac{3}{4}\pm\frac{\sqrt{5}}{4}$$

$$w=\frac{-3\pm\sqrt{5}}{4}$$

59. Let $u=2x+1$.

$$2(2x+1)^2-7(2x+1)+6=0$$

$$2u^2-7u+6=0$$

$$(2u-3)(u-2)=0$$

$$2u-3=0 \quad\text{or}\quad u-2=0$$

$$u=\frac{3}{2} \qquad u=2$$

$$2x+1=\frac{3}{2} \qquad 2x+1=2$$

$$2x=\frac{1}{2} \qquad 2x=1$$

$$x=\frac{1}{4} \qquad x=\frac{1}{2}$$

60.

$$3x^2-4x=-2$$

$$3x^2-4x+2=0$$

$$x=\frac{-b\pm\sqrt{b^2-4ac}}{2a}$$

$$x=\frac{-(-4)\pm\sqrt{(-4)^2-4(3)(2)}}{2(3)}$$

$$x=\frac{4\pm\sqrt{16-24}}{6}$$

$$x=\frac{4\pm\sqrt{-8}}{6}$$

$$x=\frac{4\pm2i\sqrt{2}}{6}$$

$$x=\frac{4}{6}\pm\frac{2\sqrt{2}}{6}i$$

$$x=\frac{2}{3}\pm\frac{\sqrt{2}}{3}i$$

61. $f(x)=-6x^2-12x-8$

Step 1: Since $a=-6<0$, the parabola opens downward.

Step 2: Find the vertex and axis of symmetry.

$$x=\frac{-b}{2a}=\frac{-(-12)}{2(-6)}=\frac{12}{-12}=-1$$

$$\begin{aligned}f(-1)&=-6(-1)^2-12(-1)-8\\&=-6+12-8\\&=-2\end{aligned}$$

vertex: $(-1,-2)$

The axis of symmetry is $x=-1$.

Step 3: Find the x- and y-intercepts.

Since $c=-8$, the y-intercept is $(0,-8)$. The vertex is located below the x-axis and it opens downward, so there are no x-intercepts.

Step 4: Use symmetry to find the point $(-2,-8)$.

Step 5: Plot the points and draw the parabola.

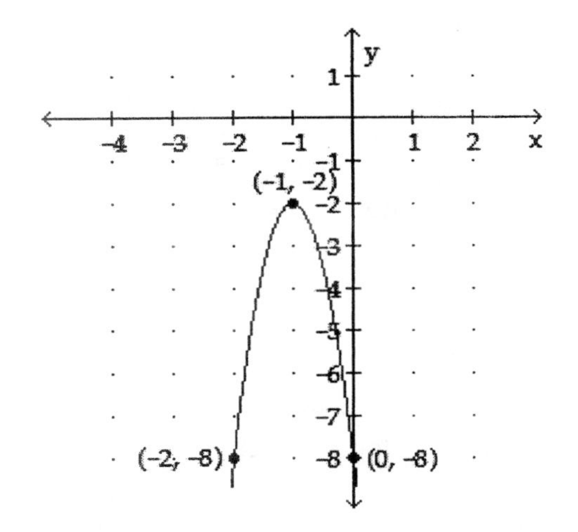

62.

$$\begin{aligned}(f\circ g)(x)&=f(g(x))\\&=f(2x+1)\\&=(2x+1)^2-2\\&=4x^2+4x+1-2\\&=4x^2+4x-1\end{aligned}$$

63.

$$\begin{aligned}f(x)&=2x^3-1\\y&=2x^3-1\\x&=2y^3-1\\x+1&=2y^3\\\frac{x+1}{2}&=y^3\\\sqrt[3]{\frac{x+1}{2}}&=y\\y&=\sqrt[3]{\frac{x+1}{2}}\\f^{-1}(x)&=\sqrt[3]{\frac{x+1}{2}}\end{aligned}$$

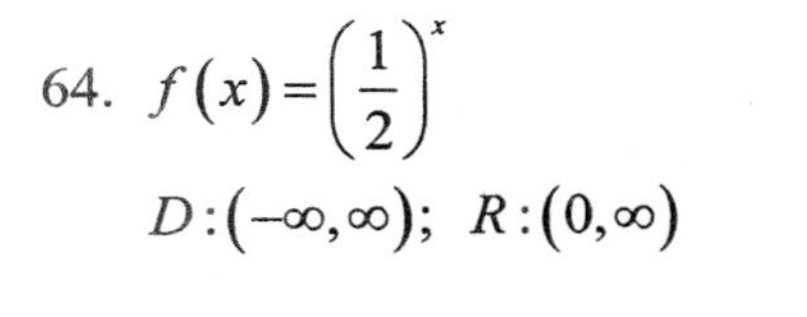

64. $f(x)=\left(\frac{1}{2}\right)^x$

$D:(-\infty,\infty);\ R:(0,\infty)$

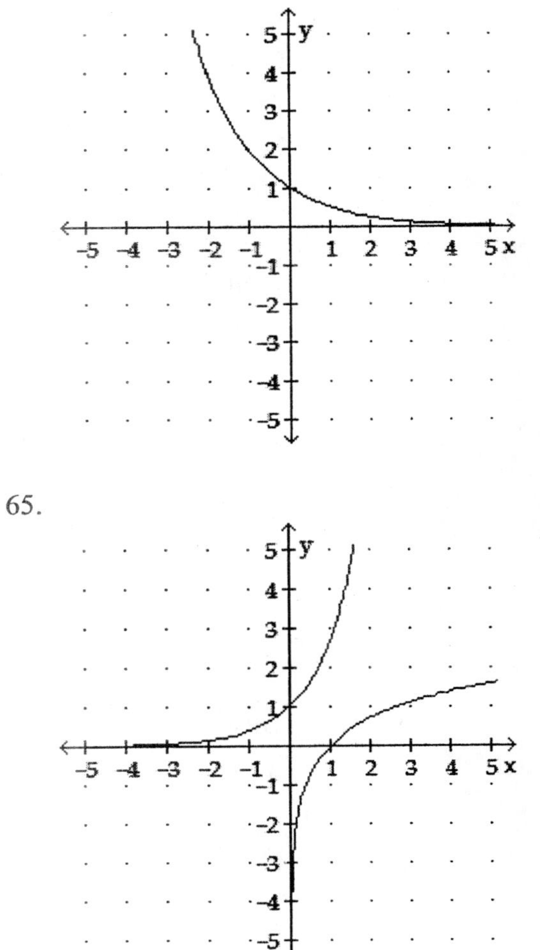

65.

66.

$$\log_6 \frac{36}{x^3} = \log_6 36 - \log_6 x^3$$
$$= \log_6 6^2 - \log_6 x^3$$
$$= 2\log_6 6 - 3\log_6 x$$
$$= 2(1) - 3\log_6 x$$
$$= 2 - 3\log_6 x$$

67.

$$\frac{1}{2}\ln x + \ln y - \ln z = \ln x^{1/2} + \ln y - \ln z$$
$$= \ln \frac{x^{1/2}y}{z}$$
$$= \ln \frac{\sqrt{x} \cdot y}{z}$$
$$= \ln \frac{y\sqrt{x}}{z}$$

68. Let $P = 119$ million, $r = 0.0143$, and $t = 25$.

$$A = Pe^{rt}$$
$$= 117e^{0.0143(25)}$$
$$= 117e^{0.3575}$$
$$\approx 117(1.429750566)$$
$$\approx 167.28$$
$$\approx 170 \text{ million}$$

69.

$$\log_x 25 = 2$$
$$x^2 = 25$$
$$x = 5$$

70.

$$\log 1{,}000 = x$$
$$10^x = 1{,}000$$
$$10^x = 10^3$$
$$x = 3$$

71.

$$\log_3 x = -3$$
$$3^{-3} = x$$
$$\frac{1}{27} = x$$

72.

$$\ln e = x$$
$$1 = x$$

73.

$$\log 98 = \log(7 \cdot 14)$$
$$= \log 7 + \log 14$$
$$= 0.8451 + 1.1461$$
$$= 1.9912$$

74.

$\log 0$ is undefined

75.

$$2^{x+2} = 3^x$$
$$\log 2^{x+2} = \log 3^x$$
$$(x+2)\log 2 = x\log 3$$
$$x\log 2 + 2\log 2 = x\log 3$$
$$2\log 2 = x\log 3 - x\log 2$$
$$2\log 2 = x(\log 3 - \log 2)$$
$$\frac{2\log 2}{\log 3 - \log 2} = x$$
$$x = \frac{2\log 2}{\log 3 - \log 2}$$
$$x \approx 3.4190$$

76.

$$\log x + \log(x+9) = 1$$
$$\log(x^2 + 9x) = 1$$
$$10^1 = x^2 + 9x$$
$$0 = x^2 + 9x - 10$$
$$0 = (x+10)(x-1)$$
$$x + 10 = 0 \quad \text{or} \quad x - 1 = 0$$
$$x = \cancel{-10} \qquad x = 1$$

77.

$$5^{4x} = \frac{1}{125}$$
$$5^{4x} = 5^{-3}$$
$$4x = -3$$
$$x = -\frac{3}{4}$$

78.

$$\log_3 x = \log_3 \left(\frac{1}{x}\right) + 4$$
$$\log_3 x - \log_3 \left(\frac{1}{x}\right) = 4$$
$$\log_3 \frac{x}{\frac{1}{x}} = 4$$
$$\log_3 x^2 = 4$$
$$3^4 = x^2$$
$$81 = x^2$$
$$9 = x$$

79. Use the distance formula to find the radius.

$$D = \sqrt{(x_2 - x_1)^2 + (y_2 - y_1)^2}$$
$$= \sqrt{(-1-3)^2 + (-2-1)^2}$$
$$= \sqrt{(-4)^2 + (-3)^2}$$
$$= \sqrt{16+9}$$
$$= \sqrt{25}$$
$$= 5$$

Use the equation of a circle with center (1, 3) and $r = 5$.

$$(x-h)^2 + (y-k)^2 = r^2$$
$$(x-1)^2 + (y-3)^2 = 5^2$$
$$x^2 - 2x + 1 + y^2 - 6y + 9 = 25$$
$$x^2 + y^2 - 2x - 6y - 15 = 0$$

80. $\dfrac{(x-2)^2}{9} - \dfrac{y}{1} = 1$

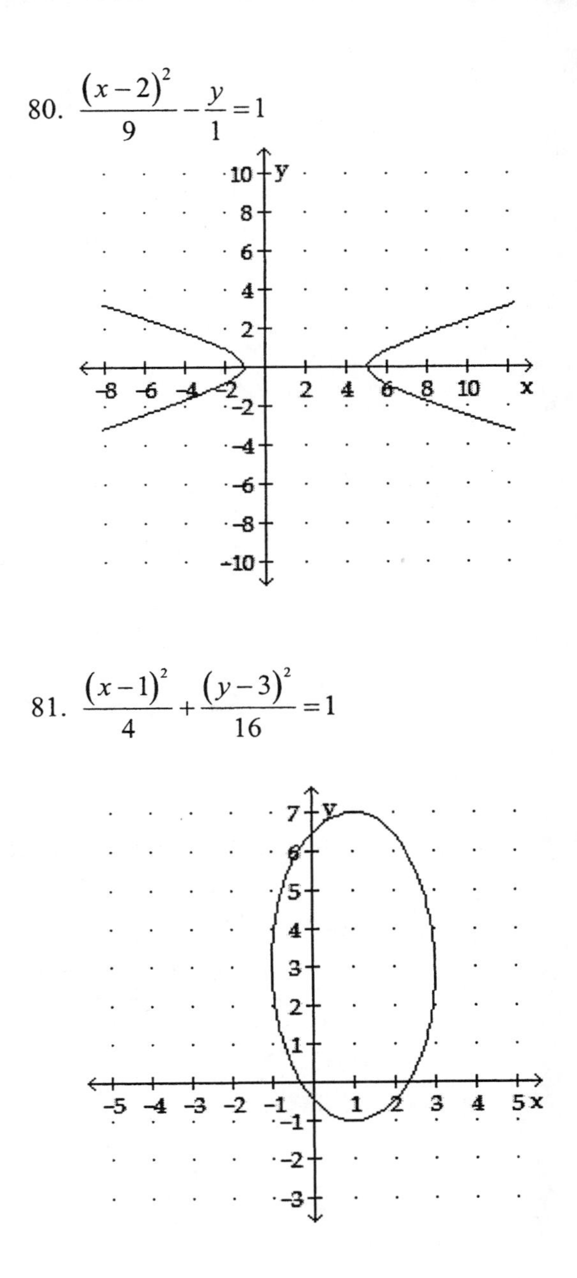

81. $\dfrac{(x-1)^2}{4} + \dfrac{(y-3)^2}{16} = 1$

82.

$$y^2+4x-6y=-1$$
$$4x=-y^2+6y-1$$
$$x=-\frac{1}{4}y^2+\frac{3}{2}y-\frac{1}{4}$$
$$x=-\frac{1}{4}(y^2-6y)-\frac{1}{4}$$
$$x=-\frac{1}{4}(y^2-6y+9)-\frac{1}{4}+\frac{1}{4}(9)$$
$$x=-\frac{1}{4}(y-3)^2+2$$

vertex: $(2,3)$

axis of symmetry: $y=3$

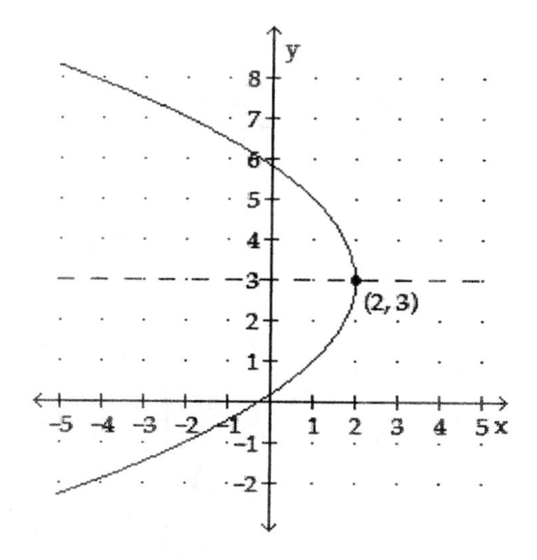

83.

$$(3a-b)^4$$
$$=(3a)^4+\frac{4!}{1!(4-1)!}(3a)^3(-b)$$
$$+\frac{4!}{2!(4-2)!}(3a)^2(-b)^2+\frac{4!}{3!(4-3)!}(3a)(-b)^3$$
$$+(-b)^4$$
$$=81a^4+\frac{4\cdot\cancel{3!}}{1!\cancel{3!}}(27a^3)(-b)+\frac{4\cdot 3\cdot\cancel{2!}}{2!\cancel{2!}}(9a^2)(b^2)$$
$$+\frac{4\cdot\cancel{3!}}{\cancel{3!}1!}(3a)(-b^3)+b^4$$
$$=81a^4-108a^3b+54a^2b^2-12ab^3+b^4$$

84. The 7th term of $(2x-y)^8$

$$\frac{8!}{6!(8-6)!}(2x)^2(-y)^6=\frac{8\cdot 7\cdot\cancel{6!}}{\cancel{6!}2!}(4x^2)(y^6)$$
$$=\frac{56}{2}(4x^2y^6)$$
$$=28(4x^2y^6)$$
$$=112x^2y^6$$

85. Let $n=20$, $a_1=-11$, and $d=6$.

$$a_n=a_1+d(n-1)$$
$$a_{20}=-11+6(20-1)$$
$$a_{20}=-11+6(19)$$
$$a_{20}=-11+114$$
$$a_{20}=103$$

86. Let $n=20$, $a_1=6$, and $d=3$.

$$a_n=a_1+d(n-1)$$
$$a_{20}=6+3(20-1)$$
$$a_{20}=6+3(19)$$
$$a_{20}=6+57$$
$$a_{20}=63$$

Find S if $a_{20}=63$ and $a_1=6$.

$$S_n=\frac{n(a_1+a_n)}{2}$$
$$S_{20}=\frac{20(6+63)}{2}$$
$$=\frac{20(69)}{2}$$
$$=690$$

87.

$$\sum_{k=3}^{5}(2k+1)=(2\cdot 3+1)+(2\cdot 4+1)+(2\cdot 5+1)$$
$$=7+9+11$$
$$=27$$

88. Let $a_1 = \frac{1}{27}$ and $r = 3$.

$$a_n = a_1 r^{n-1}$$
$$a_7 = \left(\frac{1}{27}\right)(3)^{7-1}$$
$$a_7 = \left(\frac{1}{27}\right)(3)^6$$
$$a_7 = \left(\frac{1}{27}\right)(729)$$
$$a_7 = 27$$

89. If the boat depreciates 12% of its value each year, then 88% of its value remains. The value of the boat in 9 years is represented by the 10th term of a geometric series, where $a_1 = 9{,}000$, $n = 10$, and $r = 0.88$.

$$a_n = a_1 r^{n-1}$$
$$a_{10} = (9{,}000)(0.88)^{10-1}$$
$$a_{10} = (9{,}000)(0.88)^9$$
$$a_{10} = (9{,}000)(0.316478)$$
$$a_{10} = \$2{,}848.31$$

The boat is worth $2,848.31 after 9 years.

90. Find r.

$$r = \frac{a_2}{a_1} = \frac{\frac{1}{32}}{\frac{1}{64}} = \frac{1}{32} \bullet \frac{64}{1} = 2$$

Find the sum with $a_1 = \frac{1}{64}$, $n = 10$, and $r = 2$.

$$S_n = \frac{a_1(1 - r^n)}{1 - r}$$
$$= \frac{\frac{1}{64}(1 - 2^{10})}{1 - 2}$$
$$= \frac{\frac{1}{64}(1 - 1{,}024)}{-1}$$
$$= \frac{\frac{1}{64}(-1{,}023)}{-1}$$
$$= \frac{-\frac{1{,}023}{64}}{-1}$$
$$= \frac{1{,}023}{64}$$

91. Find r.

$$r = \frac{a_2}{a_1} = \frac{3}{9} = \frac{1}{3}$$

Find the sum with $a_1 = 9$ and $r = \frac{1}{3}$.

$$S = \frac{a_1}{1 - r}$$
$$= \frac{9}{1 - \frac{1}{3}}$$
$$= \frac{9}{\frac{2}{3}}$$
$$= \frac{9}{1} \bullet \frac{3}{2}$$
$$= \frac{27}{2}$$

92. $7! = 7 \cdot 6 \cdot 5 \cdot 4 \cdot 3 \cdot 2 \cdot 1$
$= 5{,}040$

93. $C(9, 3)$

$$\frac{9!}{3!(9-3)!} = \frac{9!}{3!6!}$$
$$= \frac{9 \cdot 8 \cdot 7 \cdot \cancel{6!}}{3!\cancel{6!}}$$
$$= \frac{504}{6}$$
$$= 84$$

94. There are 12 face cards and 52 total cards in a standard deck. Let $s = 12$ and $n = 52$.

$$P(\text{face card}) = \frac{s}{n}$$
$$= \frac{12}{52}$$
$$= \frac{3}{13}$$

APPENDIX II: Synthetic Division

VOCABULARY

1. The method of dividing $x^2 + 2x - 9$ by $x - 4$ shown below is called **synthetic** division.

3. In Exercise 1, the synthetic **divisor** is 4.

5. The factor **theorem** tells us how to find one factor of a polynomial if the remainder of a certain division is 0.

CONCEPTS

7. a. $(5x^3 + x - 3) \div (x + 2)$

   b. $5x^2 - 10x + 21 - \dfrac{45}{x+2}$

9. Rather than substituting 8 for x in $P(x) = 6x^3 - x^2 - 17x + 9$, we can divide the polynomial **$6x^3 - x^2 - 17x + 9$** by **$x - 8$** to find $P(8)$.

NOTATION

11. 

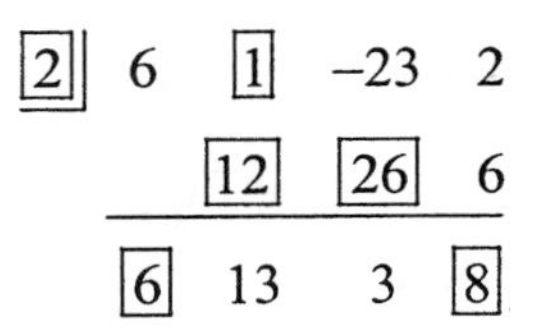

PRACTICE

13.

$$\begin{array}{r|rrr} 1 & 1 & 1 & -2 \\ & & 1 & 2 \\ \hline & 1 & 2 & 0 \end{array}$$

$$\frac{x^2 + x - 2}{x - 1} = x + 2$$

15.

$$\begin{array}{r|rrr} 4 & 1 & -7 & 12 \\ & & 4 & -12 \\ \hline & 1 & -3 & 0 \end{array}$$

$$\frac{x^2 - 7x + 12}{x - 4} = x - 3$$

17. Write the numerator in descending order.

$$\begin{array}{r|rrr} -4 & 1 & 6 & 8 \\ & & -4 & -8 \\ \hline & 1 & 2 & 0 \end{array}$$

$$\frac{x^2 + 6x + 8}{x + 4} = x + 2$$

19.

$$\begin{array}{r|rrr} -2 & 1 & -5 & 14 \\ & & -2 & 14 \\ \hline & 1 & -7 & 28 \end{array}$$

$$\frac{x^2 - 5x + 14}{x + 2} = x - 7 + \frac{28}{x + 2}$$

21.

$$\begin{array}{r|rrrr} 3 & 3 & -10 & 5 & -6 \\ & & 9 & -3 & 6 \\ \hline & 3 & -1 & 2 & 0 \end{array}$$

$$\frac{3x^3 - 10x^2 + 5x - 6}{x - 3} = 3x^2 - x + 2$$

23. Write the numerator in descending order and insert 0 for any missing terms.

$$\begin{array}{r|rrrr} 2 & 2 & 0 & -5 & -6 \\ & & 4 & 8 & 6 \\ \hline & 2 & 4 & 3 & 0 \end{array}$$

$$\frac{2x^3 - 5x - 6}{x - 2} = 2x^2 + 4x + 3$$

25. Write the numerator in descending order and insert 0 for any missing terms.

$$\begin{array}{r|rrrr} -1 & 6 & 5 & 0 & 4 \\ & & -6 & 1 & -1 \\ \hline & 6 & -1 & 1 & 3 \end{array}$$

$$\frac{6x^3+5x^2+4}{x+1}=6x^2-x+1+\frac{3}{x+1}$$

27.

$$\begin{array}{r|rrrr} -1 & 1 & 1 & 1 & 2 \\ & & -1 & 0 & -1 \\ \hline & 1 & 0 & 1 & 1 \end{array}$$

$$\frac{t^3+t^2+t+2}{t+1}=t^2+1+\frac{1}{t+1}$$

29.

$$\begin{array}{r|rrrrrr} 1 & 1 & 0 & 0 & 0 & 0 & -1 \\ & & 1 & 1 & 1 & 1 & 1 \\ \hline & 1 & 1 & 1 & 1 & 1 & 0 \end{array}$$

$$\frac{a^5-1}{a-1}=a^4+a^3+a^2+a+1$$

31.

$$\begin{array}{r|rrrrrr} 3 & -5 & 4 & 30 & 2 & 20 & 3 \\ & & -15 & -33 & -9 & -21 & -3 \\ \hline & -5 & -11 & -3 & -7 & -1 & 0 \end{array}$$

$$\frac{-5x^5+4x^4+30x^3+2x^2+20x+3}{x-3}$$
$$=-5x^4-11x^3-3x^2-7x-1$$

33.

$$\begin{array}{r|rrrr} \frac{1}{2} & 8 & -4 & 2 & -1 \\ & & 4 & 0 & 1 \\ \hline & 8 & 0 & 2 & 0 \end{array}$$

$$\frac{8t^3-4t^2+2t-1}{t-\frac{1}{2}}=8t^2+2$$

35.

$$\begin{array}{r|rrrrr} 8 & 1 & -1 & -56 & -2 & 16 \\ & & 8 & 56 & 0 & -16 \\ \hline & 1 & 7 & 0 & -2 & 0 \end{array}$$

$$\frac{x^4-x^3-56x^2-2x+16}{x-8}=x^3+7x^2-2$$

37.

$$\begin{array}{r|rrr} 0.2 & 7.2 & -2.1 & 0.5 \\ & & 1.44 & -0.132 \\ \hline & 7.2 & -0.66 & 0.368 \end{array}$$

$$\frac{7.2x^2-2.1x+0.5}{x-0.2}=7.2x-0.66+\frac{0.368}{x-0.2}$$

39.

$$\begin{array}{r|rrrr} -57 & 9 & 0 & 0 & -25 \\ & & -513 & 29{,}241 & -1{,}666{,}737 \\ \hline & 9 & -513 & 29{,}241 & -1{,}666{,}762 \end{array}$$

$$\frac{9x^3-25}{x+57}=9x^2-513x+29{,}241-\frac{1{,}666{,}762}{x+57}$$

41.

$$P(x)=2x^3-4x^2+2x-1$$
$$P(1)=2(1)^3-4(1)^2+2(1)-1$$
$$=2(1)-4(1)+2(1)-1$$
$$=2-4+2-1$$
$$=-1$$

$$\begin{array}{r|rrrr} 1 & 2 & -4 & 2 & -1 \\ & & 2 & -2 & 0 \\ \hline & 2 & -2 & 0 & -1 \end{array}$$

The remainder, -1, equals $P(1)$.

43.

$$P(x) = 2x^3 - 4x^2 + 2x - 1$$
$$P(-2) = 2(-2)^3 - 4(-2)^2 + 2(-2) - 1$$
$$= 2(-8) - 4(4) + 2(-2) - 1$$
$$= -16 - 16 - 4 - 1$$
$$= -37$$

$$\begin{array}{r|rrrr} -2 & 2 & -4 & 2 & -1 \\ & & -4 & 16 & -36 \\ \hline & 2 & -8 & 18 & -37 \end{array}$$

The remainder, -37, equals $P(-2)$.

45.

$$P(x) = 2x^3 - 4x^2 + 2x - 1$$
$$P(3) = 2(3)^3 - 4(3)^2 + 2(3) - 1$$
$$= 2(27) - 4(9) + 2(3) - 1$$
$$= 54 - 36 + 6 - 1$$
$$= 23$$

$$\begin{array}{r|rrrr} 3 & 2 & -4 & 2 & -1 \\ & & 6 & 6 & 24 \\ \hline & 2 & 2 & 8 & 23 \end{array}$$

The remainder, 23, equals $P(3)$.

47.

$$P(x) = 2x^3 - 4x^2 + 2x - 1$$
$$P(0) = 2(0)^3 - 4(0)^2 + 2(0) - 1$$
$$= 2(0) - 4(0) + 2(0) - 1$$
$$= 0 - 0 + 0 - 1$$
$$= -1$$

$$\begin{array}{r|rrrr} 0 & 2 & -4 & 2 & -1 \\ & & 0 & 0 & 0 \\ \hline & 2 & -4 & 2 & -1 \end{array}$$

The remainder, -1, equals $P(0)$.

49.

$$Q(x) = x^4 - 3x^3 + 2x^2 + x - 3$$
$$Q(-1) = (-1)^4 - 3(-1)^3 + 2(-1)^2 + (-1) - 3$$
$$= 1 - 3(-1) + 2(1) - 1 - 3$$
$$= 1 + 3 + 2 - 1 - 3$$
$$= 2$$

$$\begin{array}{r|rrrrr} -1 & 1 & -3 & 2 & 1 & -3 \\ & & -1 & 4 & -6 & 5 \\ \hline & 1 & -4 & 6 & -5 & 2 \end{array}$$

The remainder, 2, equals $Q(-1)$.

51.

$$Q(x) = x^4 - 3x^3 + 2x^2 + x - 3$$
$$Q(2) = (2)^4 - 3(2)^3 + 2(2)^2 + (2) - 3$$
$$= 16 - 3(8) + 2(4) + 2 - 3$$
$$= 16 - 24 + 8 + 2 - 3$$
$$= -1$$

$$\begin{array}{r|rrrrr} 2 & 1 & -3 & 2 & 1 & -3 \\ & & 2 & -2 & 0 & 2 \\ \hline & 1 & -1 & 0 & 1 & -1 \end{array}$$

The remainder, -1, equals $Q(2)$.

53.

$$Q(x) = x^4 - 3x^3 + 2x^2 + x - 3$$
$$Q(3) = (3)^4 - 3(3)^3 + 2(3)^2 + (3) - 3$$
$$= 81 - 3(27) + 2(9) + 3 - 3$$
$$= 81 - 81 + 18 + 3 - 3$$
$$= 18$$

$$\begin{array}{r|rrrrr} 3 & 1 & -3 & 2 & 1 & -3 \\ & & 3 & 0 & 6 & 21 \\ \hline & 1 & 0 & 2 & 7 & 18 \end{array}$$

The remainder, 18, equals $Q(3)$.

55.

$$Q(x)=x^4-3x^3+2x^2+x-3$$

$$\begin{aligned}Q(-3)&=(-3)^4-3(-3)^3+2(-3)^2+(-3)-3\\&=81-3(-27)+2(9)-3-3\\&=81+81+18-3-3\\&=174\end{aligned}$$

$$\begin{array}{r|rrrrr} -3 & 1 & -3 & 2 & 1 & -3 \\ & & -3 & 18 & -60 & 177 \\ \hline & 1 & -6 & 20 & -59 & 174 \end{array}$$

The remainder, 174, equals $Q(-3)$.

57.

$$\begin{array}{r|rrrr} 2 & 1 & -4 & 1 & -2 \\ & & 2 & -4 & -6 \\ \hline & 1 & -2 & -3 & -8 \end{array}$$

The remainder is -8 so $P(2)=-8$.

59.

$$\begin{array}{r|rrrr} 3 & 2 & 0 & 1 & 2 \\ & & 6 & 18 & 57 \\ \hline & 2 & 6 & 19 & 59 \end{array}$$

The remainder is 59 so $P(3)=59$.

61.

$$\begin{array}{r|rrrrr} -2 & 1 & -2 & 1 & -3 & 2 \\ & & -2 & 8 & -18 & 42 \\ \hline & 1 & -4 & 9 & -21 & 44 \end{array}$$

The remainder is 44 so $P(-2)=44$.

63.

$$\begin{array}{r|rrrrrr} -\frac{1}{2} & 3 & 0 & 0 & 0 & 0 & 1 \\ & & -\frac{3}{2} & \frac{3}{4} & -\frac{3}{8} & \frac{3}{16} & -\frac{3}{32} \\ \hline & 3 & -\frac{3}{2} & \frac{3}{4} & -\frac{3}{8} & \frac{3}{16} & \frac{29}{32} \end{array}$$

The remainder is $\frac{29}{32}$ so $P\left(-\frac{1}{2}\right)=\frac{29}{32}$.

65.

$$\begin{array}{r|rrrr} 3 & 1 & -3 & 5 & -15 \\ & & 3 & 0 & 15 \\ \hline & 1 & 0 & 5 & 0 \end{array}$$

The remainder is 0, so $x-3$ is a factor.

67.

$$\begin{array}{r|rrr} -2 & 3 & -7 & 4 \\ & & -6 & 26 \\ \hline & 3 & -13 & 30 \end{array}$$

The remainder is 30,
so $x+2$ is NOT a factor.

WRITING

69. Answers will vary.

71. Answers will vary.

REVIEW

73.

$$\begin{aligned}x^2z\left(y^3-z\right)&=(-3)^2(0)\left((-5)^3-0\right)\\&=9(0)(-125-0)\\&=9(0)(-125)\\&=0\end{aligned}$$

75.

$$\begin{aligned}\frac{x-y^2}{2y-1+x}&=\frac{-3-(-5)^2}{2(-5)-1+(-3)}\\&=\frac{-3-25}{-10-1-3}\\&=\frac{-28}{-14}\\&=2\end{aligned}$$

CHALLENGE PROBLEMS

77. During the synthetic division, every column would add to be zero except for the last, which would stay 1. Thus, the remainder would be 1.